HANDBOOK OF Water Resources and Pollution Control

HANDBOOK OF
Water Resources and Pollution Control

EDITED BY
HARRY W. GEHM and
JACOB I. BREGMAN

GENE V. BEELAND
TECHNICAL EDITOR

VNR VAN NOSTRAND REINHOLD COMPANY
NEW YORK CINCINNATI TORONTO LONDON MELBOURNE

Van Nostrand Reinhold Company Regional Offices:
New York Cincinnati Atlanta Dallas San Francisco

Van Nostrand Reinhold Company International Offices:
London Toronto Melbourne

Library of Congress Catalog Card Number: 75-29050
ISBN: 0-442-21041-8

Manufactured in the United States of America

Published by Van Nostrand Reinhold Company
450 West 33rd Street, New York, N.Y. 10001

Published simultaneously in Canada by Van Nostrand Reinhold Ltd.

15 14 13 12 11 10 9 8 7 6 5 4 3 2 1

Library of Congress Cataloging in Publication Data

Main entry under title:

Handbook of water resources and pollution control.

Includes bibliographical references and index.
1. Water-supply engineering. 2. Sewage disposal.
3. Water–Pollution. I. Gehm, Harry W.
II. Bregman, J. I.
TD345.H26 363.6'1 75-29050
ISBN 0-442-21041-8

Preface

Technology as it relates to water is in rapid evolution. This is due primarily to new national policies and goals toward improved water quality which take expression as federal and state water pollution control regulations. The ability to meet their requirements is, in many instances, dependent upon the development of new technologies or refinement of those which have been practiced in the water field for many years. This situation has sparked intensified research and development programs which are in turn contributing to the technological evolutionary process.

Another result of expanding federal and state control of water quality is that many public officials as well as private citizens, untrained in the field, have been thrust into positions of water management responsibility. Others have acquired new responsibilities under increasingly stringent regulations which expand their scope of operation. Personnel at all levels of government are charged with solving water-related problems which are entirely new to their experience. Elected officials, planners, and those appointed to decision-making boards or commissions at the local level, too, are confronted, many for the first time, with the necessity to make difficult choices among pollution control alternatives.

It was in this climate that this book on water was written. It responds both to changing technology and the needs of those who will employ it. The book embraces the constant principles of the scientific and technical disciplines involved in water pollution control and within their framework, changing control technologies. The sciences covered which either affect or are affected by water pollution are global hydrology, water chemistry, microbiology, aquatic biology, and the waste assimilative capacity of streams.

Advanced techniques for treatment of both municipal and industrial wastes, many of which are still in experimental stages, are presented. These techniques, for the most part, are designed not only to reduce pollution loads, but to increase available water supplies through the reuse of reclaimed wastewater.

Advanced techniques of cost allocation among users of joint municipal/industrial waste treatment facilities are also discussed along with alternative methods of recovering costs. Joint treatment is frequently undertaken to obtain economies of scale, and federal law now requires industry to pay its full share of treatment in plants constructed with federal funding.

The chapter on acceptable techniques for disposing of sludge resulting from waste treatment is of increasing importance as populations grow, treatment requirements become more stringent, and more solids are removed. This is one of the most difficult and expensive facets of dealing with municipal and industrial wastewaters.

The book, however, is not confined to wastewater, its treatment, and reuse. It addresses water supply and use as well, both for human and industrial needs. These chapters are especially appropriate in an era when water supply concerns have often been overshadowed by preoccupation with wastewater problems.

As a guide to potential water demand, data are presented on actual water use by nearly 200 public water supplies in the U.S., information which is not tabulated in one source elsewhere. The treatment and distribution of potable water are covered in two chapters. The former presents the most modern practice in the preparation of potable water and provides a basis for meeting future requirements. The chapter on distribution will be especially useful to those responsible for extending water systems to serve burgeoning metropolitan areas. This is also true of the chapter "Wastewater in the City," which is concerned with projecting anticipated quantities of sewage, how to confine it, and how to remove it from its source.

The potential for increasing the world's usable water supply through desalination of seawater is explored. This chapter discussed theoretical, practical, and economic considerations of the more promising processes of desalination.

Innovative techniques for beneficial use of thermal discharges from power plants are described. These include increased-productivity greenhouses for vegetables as well as flowers, improved livestock production, and space heating.

The book closes with a section-by-section analysis of the 1972 Amendments to the Federal Water Pollution Control Act which establish, for the foreseeable future, the goals anticipated to be achieved in the nation's drive toward clean waterways. It is imperative that engineers understand the law and its requirements in order to gauge the magnitude of the effort and the technology it will require.

Two important areas are not covered individually by chapters in this volume, although they are indirectly addressed in other chapters. One is the federal drinking water standards which are presently undergoing a thorough revision within the Environmental Protection Agency. The other is federal water quality criteria for various uses of water which are similarly being revised. Sufficient change is anticipated that discussion of the existing standards would, in both cases, serve no useful purpose in this reference volume.

Recognized experts were selected to author the chapters in their specialty areas. We extend our deep appreciation for their contributions in the full knowledge that this "extracurricular" assignment placed a heavy burden on already busy schedules.

Our especial thanks go to Mrs. Gene Beeland of WAPORA, Inc., technical editor of the *Handbook of Water Resources and Pollution Control*, who for over three years has borne the day-to-day responsibility for its production. She has dealt with over 3000 pages of manuscript, over 200 permissions to publish copyrighted materials, author schedules, and the myriad other details attendant to a project of this scope. The book would not be a reality today without her assistance.

H. W. GEHM
J. I. BREGMAN
Editors

Contents

1

Hydrology

Raymond L. Nace

Research Hydrologist
U.S. Geological Survey
Raleigh, N. C.

THE WATER CYCLE

Since very ancient times it has been empirically known that there is a water cycle—a series of recurring events that can be depended on to repeat themselves. The water cycle (hydrological cycle) is the central concept of hydrology, the science of water. Water circulates naturally through five principal realms—oceans, atmosphere, lakes and rivers, icecaps and glaciers, and underground. Hydrology concerns water and its behavior in all these realms.

Air and water are the most mobile physical elements of the earth system, and this mobility permits the water cycle to operate. Constant circulation of water from ocean to atmosphere (evaporation), from the atmosphere to land areas or directly back to the ocean (precipitation), and from the land back to the ocean and atmosphere (runoff and evaporation) may be called the planetary water cycle. Within it are many regional and local subcycles. For practical reasons, most resource-oriented studies of water have concerned regional and local phenomena. However, as water supply and water quality problems proliferate, they must be viewed in a larger context—national, continental, and even global.

Large-scale phenomena of the hydrological cycle may seem to be far removed from the purview of sanitary engineers, most of whom are concerned with specific local or regional problems. However, large-scale phenomena are critical, whether we deal with water as a supply resource or as a vehicle for the transportation of wastes. Worldwide contamination of oceans and the atmosphere is a good example. Where water and air are concerned, events in any part of the world are consequences of events in all other parts of the world. Because the water cycle is a global phenomenon, it will be considered from that standpoint.

A serious handicap to all who work with water is the inadequacy of meteorological and hydrological forecasting for periods of more than a few hours. Forecasting can be improved through better means for interpreting meteorological data, but more and better data from large areas also are needed. This was one of the reasons for establishing the IHD (International Hydrological Decade, 1965-1974), a joint effort of more than one hundred nations to improve hydrological knowledge. The IHD is symbolic. It is a special decade during which attention turned from man's frail waterwheels to the great waterwheel of nature—the global hydrological cycle.

Oceans and Seas. Hydrologists generally treat the world ocean as though it were only of peripheral interest to hydrology. For local studies this is generally true, except in coastal areas where salt water invades river estuaries and may also invade coastal aquifers. The ocean, however, is tremendously important in the hydrological cycle.

The world ocean contains about 317×10^6 mi^3 of water (Table 1-1). A more significant characteristic of the ocean, however, is its vast expanse of 139.5×10^6 mi^2. The ocean averages 2.27 mi in depth but the widths of the main oceans are 900 to 2500 times that amount. Because of this geometry and the great heat-storage capacity of the ocean, the sea-atmosphere interface is the most important of all interfaces involved in the hydrological cycle. The system ocean-atmosphere-continents is a great heat engine driven by solar energy. Solar energy and the oceans form a giant natural water-distillation plant and the atmosphere provides the water-distribution system. Owing to its immense reserve of thermal energy, circulation of oceanic water, coupled with atmospheric circulation, provides the climate-control system of the world.

Thermal energy transactions, including evaporation, occur throughout the world ocean. Therefore, the entire ocean area is a source of atmospheric vapor and hence of precipitation on land. The total volume of oceanic water is about 3000 times the annual volume of oceanic evaporation. Hence, the average residence time of water molecules in the sea is about 3000 years. Residence in abyssal zones is much longer and in shallow zones much less.

Heat Fluxes. Heat is the driving force of the water cycle, and evaporation is the first stage of the cycle. The solar constant may be defined for present purposes as solar radiation per unit area reaching the top of the atmosphere on a surface perpendicular to the solar beam. The "top" of the atmosphere cannot be defined precisely in terms of altitude because it varies with latitude and air movements. In general, it is the level above which no significant absorption or backscattering of radiation occurs. Solar energy which reaches the earth eventually

TABLE 1-1. Estimated Long-Term Average Value for Elements and Parameters of the Water Cycle[a]

Parameter	Volume (mi^3)	Equivalent Depth[b]	Average Residence Time
Atmospheric water	3.1×10^3	1.0 in.	10 days
World ocean	$317{,}000 \times 10^3$	1.6 mi	>3,000 yrs
Freshwater lakes	30×10^3	9.6 in.	Days to 100s yrs
Saline lakes and inland seas	25×10^3	8.0 in.	Days to 100s yrs
Swamps and bogs	8.6×10^3	2.8 in.	Days to yrs
River channels	0.41×10^3	0.13 in.	~2 weeks
Soil water; vadose water	16×10^3	5.1 in.	Days to yrs
Groundwater	$1{,}000 \times 10^3$	26 ft	Days to 100,000s yrs
Icecaps and glaciers	$7{,}300 \times 10^3$	196 ft	Days to 100,000s yrs
Permafrost	Unknown	Unknown	Unknown
Biological water	0.17×10^3[c]	0.04 in.	Days to weeks

[a]Adapted from Nace.[1]
[b]Computed as though it were spread uniformly over the entire earth.
[c]Some estimates are as high as 1.2×10^3 mi^3.

is reradiated into space as heat. Meanwhile, circulation of oceans and the atmosphere distributes the heat around the earth.

Differential heating is the key to the entire cycle. Heating is different at different latitudes, at different seasons, and at different times of day. Heating of land areas differs from that at sea. Within land areas it differs depending on soil color, type of vegetation or lack of it, and so on *ad infinitum*. Heating of the ocean also is differential. The results of these and other differentials are imbalances of thermal energy, which cause imbalances of density or pressure energy, and these in turn contribute to air flow and ocean currents. Friction, the earth's rotation, and gravity also have profound effects. The Coriolis force, for example, operating because of the earth's rotation, sets up the principal wind patterns in the lower atmosphere.

Solar radiation is generally treated as the sole source of heat that drives the water cycle because heat conducted to the earth's surface from its interior is negligible by comparison. A commonly used value for the solar constant (S) is 2.0 ± 0.04 ly (langley; 1 ly = 1 gm-cal/cm^2). Only part of the atmosphere receives radiation at any one time, and little of its outer surface is perpendicular to the incident radiation. Using a solar constant value of 2.0 ly/min, total energy intercepted by the earth in unit time is 2.55×10^{18} cal/min.[2] Total energy intercepted by the earth is equal to $\pi r^2 S$, where r is the mean radius of the earth (6.371×10^8 cm) and S is the solar constant of 2 ly:

$$\pi r^2 S = 2.55 \times 10^{18} \text{ cal/min} \qquad (1)$$

About 30 percent of the incident radiation is reflected and scattered in space by clouds and atmospheric constituents; 17 percent is absorbed by the same media; 22 percent reaches the earth's surface as diffuse sky radiation; and 31 percent reaches the surface as direct-beam radiation. Thus, about 53 percent of intercepted solar radiation actually reaches the earth's surface.

If the intercepted energy were spread uniformly over the full spheroidal surface of the earth, the average intensity of radiation per unit area and time ($\overline{E}_s$) at the top of the atmosphere would be only one-fourth of the solar constant:

$$\overline{E}_s = \frac{\pi r^2 S}{4 \pi r^2} = \frac{S}{4} = 0.5 \text{ ly/min}$$
$$= 263 \times 10^3 \text{ ly/yr} \qquad (2)$$

Owing to absorption, backscattering, and other losses, the effective radiation for producing evaporation is 0.078 ly over the oceans and 0.098 ly over the earth's surface as a whole.[3]

The mean global values are significant for the global hydrological cycle. The distribution of energy, however, is not uniform and the average values cannot be applied in local or regional studies. At the equator, for example, the annual value of $\overline{E}_s$ is about 2.4 times that near the poles. A good summary of the energy cycle of the earth is contained in Ref. 4.

Vapor Fluxes. The flow of vapor in air is called a vapor flux. According to a recent estimate,[5] the earth's mean upward flux of water (evaporation, which includes transpiration) is about 1000 mm/yr (about 1250 mm/yr from the sea and 410 mm/yr from land). Many other estimates are extant and the differences among them are significant, but no means is available to judge which estimate is "best." For working purposes, the following values probably are acceptable:

World average
1016 mm = 39.8 in. (125,000 mi^3)
Oceanic area
1230 mm = 48.4 in. (107,800 mi^3)
Land
480 mm = 18.9 in. (17,200 mi^3)

The implication of three to four significant figures is, of course, erroneous. They are retained only for bookkeeping purposes. According to this estimate only about 14 percent of world evaporation is from land areas. The methods and problems of measuring and estimating oceanic evaporation and precipitation and the current state of knowledge have recently been analyzed and described.[6]

Because of the accessibility of inhabited land areas, the state of knowledge there is somewhat better than for the oceans but still is unsatisfactory. Estimation of terrestrial evaporation is a formidable problem because no generally acceptable or usable method is available for direct measurement of evaporation for areas larger than a few hundred square feet. A commonly accepted value for evaporation, averaged for the conterminous United States, is 21 in.

Advective atmospheric vapor fluxes have attracted considerable attention in recent years, but they need much more study. Analysis of vapor flux divergences is a relatively new method for indirect determination of net precipitation and evaporation, but stations for determination of vertically integrated vapor fluxes are far too few for hydrological purposes.

A study of vapor transport over the North American air space[5] indicated that vapor inflow exceeded outflow by about 210×10^9 gm/sec in winter, this being the excess of winter precipitation over evaporation. Summer outflow exceeded inflow by about 80×10^9 gm/sec, representing excess of summer evaporation over precipitation.

Table 1-2 is derived from an analysis[7] which shows the total annual flux of water vapor into and out of the conterminous United States during each of 2 years. The differences between income and outgo were 2420 and 2240×10^{12} kg/yr in the 2 years, respectively. This is equivalent to 11.8 in. or less of yearly precipitation over the United States. Actual precipitation of water derived from oceanic sources cannot be computed in that way from such generalized data, but the computation is a way of giving meaning to the amounts of water involved.

TABLE 1-2. Vapor Flux over the United States[a]

Source or Destination	Flux (10^{12} kg/yr) 1961–62[b]	1962–63[b]
Pacific Ocean to West Coast	4,320	4,480
Gulf of California to Southwest	670	730
Gulf of Mexico to midcontinent	4,580	3,790
Totals, income	9,570	9,000
East Coast to Atlantic Ocean	6,480	6,260
Midcontinent to Canada	670	500
Totals, outgo	7,150	6,760
From West Coast across Rocky Mountains (and their southern extension) into midcontinent and Canada	4,420	4,800

[a]Compiled from a figure by Rasmusson.[7]
[b]Each year is from May through April.

Evaporation of a single molecule of water is a simple process in which the molecule is torn free from others after absorbing radiant energy. Gross evaporation from ocean surfaces or land surfaces, however, is a result of complex reactions among water, soils, vegetation, air, and radiation. It occurs wherever water is available and the air adjacent to it is not already saturated. Evaporation occurs even at freezing temperatures and the great icecaps, glaciers, sea ice, and perennial snow fields of the world are not immune. By the process of sublimation, ice and snow may vaporize without passing through a liquid phase. Sublimation accounts for about 10 percent of evaporation from snow and ice fields, but it is probably largely offset by hoar frost deposited by the reverse process (W. J. Campbell, personal communication, June 14, 1970).

Evaporation from free water surfaces such as ponds and lakes cannot be measured directly, even by calculation from stage-capacity tables and from measurement of surface inflow and outflow. Measurement of each factor has inherent errors. In addition, a water body may lose by seepage into the ground or gain by effluent seepage of groundwater. Wind has a great effect on evaporation from water surfaces because it removes humid air and commonly introduces drier air. Floating pans containing water do not give a true measure of lake evaporation because the protruding rim of the pan changes the wind pattern and hence the eddy diffusion above the pan.

Similarly, land pans do not give a true measure of lake evaporation because air near the surface over land generally is less humid than that over a lake. Experimental evidence indicates that under many conditions lake evaporation is about 70 percent of evaporation from the U. S. National Weather Service Class A land pan. The highest reported pan evaporation rate in the United States is about 155 in. at a location in Death Valley, California, in the year ending on May 1, 1959. Weather Service data indicate that average annual Class A pan evaporation ranges from 140 in. in parts of the arid Southwest to only 25 in. in the far Northwest and Northeast. Recommended pan coefficients range from 60 to 80 percent depending on location.[8] Evaporation is measured regularly at only about 540 stations in the United States, and records for periods longer than 40 years are available for only about 54 stations.

Vegetation has an important role in the water cycle. Plants send their roots into the soil, withdraw water, transpire it, and thus put the root zone of the earth in communication with the atmosphere. No satisfactory method is available for differentiating direct evaporation from plant transpiration, so the term evapotranspiration has been coined to designate the combined processes. Transpiration is evaporation, and so long as the two types of evaporation cannot be estimated or measured separately, evapotranspiration is a useless term.

Measurement of evaporation from lysimeters is a common and useful practice to obtain data for estimating local evaporation from vegetated areas. For large areas, however, evaporation must be computed on the basis of theory, using observations of insolation, temperature, duration of daylight, wind, and other factors. These calculations generally yield a value that expresses potential evaporation, the amount of evaporation that would occur if water were plentiful, as in a pond or water-saturated land area. Soil, however, is seldom fully saturated. Actual evaporation in most land areas is less than the potential, commonly much less. From dry soil it approaches nil.

The percent saturation of the air with water vapor is called the relative humidity. Humidity varies with time, place, and altitude. At a given time and place, the amount of liquid water equivalent in the overlying column of air is called the precipitable water content.

Air masses of varying humidity may move upward and downward as well as laterally. Vertical humidity profiles can be obtained by various rocket-borne or balloon-borne sensors. Knowing the direction and speed of air movement, vertically integrated horizontal vapor fluxes may be determined. With a sufficient number of determinations around an area, the net vapor balance may be determined.

Thus far, such determinations have been made in research and experiments, but not routinely. The results indicate that the method is reliable, within acceptable limits, only for areas of about 200,000 mi^2 or more. Little success has been achieved with smaller areas. The method does not define either precipitation or evaporation. Rather, it discloses the net vapor balance. That is, if a given amount of vapor moves into an area, and a different amount moves out, it can be determined whether evaporation exceeded precipitation, or the reverse.

Precipitation. Precipitation is the beginning of the land phase of the water cycle. Disposition of the precipitated water depends on many factors such as rate of rainfall or snowfall, rate of snow melt, topography, vegetation, nature and condition of the soil or rock at the land surface, and relative humidity.

Worldwide, the estimated number of rain gauges of all kinds in operation is somewhat more than 100,000, and the number of snow gauges is about 25,000. This is an average of 1 gauge per 460 mi^2 of land area. Understandably, therefore, it is difficult to estimate total precipitation on land areas, much less for the vast area of the world ocean. None but crude approximations can be achieved. A commonly accepted value for average annual precipitation averaged for the conterminous United States is 30 in.

For many purposes in meteorology and hydrology, standard elevated rain gauges are sufficiently accurate. They are not adequate for careful quantitative hydrology. The hydrologist needs to know how much precipitation actually reaches the ground, not how much is caught in a vessel some distance above ground.

Pit gauges, with the aperture at ground level, have been widely used in the United Kingdom, the Soviet Union, and elsewhere. They are considered to be superior to elevated gauges for measuring rain, but they have little value for measuring snowfall.

Contrary to the common working assumption, rainfall measured at individual gauges is not truly representative of the area between gauges. Even where gauge networks are relatively dense (on the order of 1 per 40 mi^2, as in Japan and Western Europe), measured rainfall is about 2×10^{-10} of the total areal amount. In most areas it is 10^{-11} to 10^{-13} or even smaller. Many thunderstorm cells are only 2 to 4 mi in horizontal diameter, so many rain showers cannot be measured accurately by an ordinary station network.

For the earth as a whole, averaged for long periods of time, the amount of atmospheric moisture is constant. Average long-term precipitation, therefore, must equal evaporation, though the distribution of precipitation differs from that of evaporation. World average precipitation probably is about as shown below:

World total
1016 mm = 39.8 in. (125,000 mi^3)
Ocean area
1138 mm = 44.8 in. (99,800 mi^3)
Land area
724 mm = 28.5 in. (25,200 mi^3)

According to this estimate, only 20 percent of all precipitation falls on land areas.

Precipitation has been observed systematically more widely and during longer periods of time than most other meteorological parameters. Much of the early work was done by individuals to satisfy personal curiosity. For some parts of the eastern United States, a few data are available even for pre-Revolutionary periods. However, very few systematic records are available for periods longer than 100 years. Coverage is sparse in many parts of the conterminous United States, especially in wilderness and high-altitude areas. No data are available for vast areas of Alaska. Puerto Rico, on the other hand, has the most intensive network of observations of any area of comparable size in the Western Hemisphere.

The most common rain gauge in the United States during many years has been a cylindrical tube about 2 ft high, with a horizontal circular catchment orifice 8 in. in diameter. The depth of rainfall is measured with a calibrated dip stick. This gauge is still the standard for routine measurement of daily precipitation in regional observation networks. More complicated equipment is available to measure and record rates of precipitation and other factors in experimental areas.

Observational coverage in the United States and elsewhere is being improved constantly, but it is unlikely that coverage ever will be adequate for precise estimation of precipitation. The U. S. National Weather Service and other agencies maintain only about 10,500 permanent precipitation gauges, of which only 5500 have records longer than 40 years.[9] Coverage by 1 gauge per square mile would require more than 3×10^6 gauges for the conterminous states. This still would sample only about 0.3×10^{-6} percent of the total area.

Gauge records are generally biased toward the low side.[10] Experience with standard rain gauges[11] indicates that they are generally negatively biased about 1.5 percent owing to evaporation, adhesion to the instrument, and several other factors. In addition, gauges in exposed positions may be in error by 5 to 80 percent, depending on wind speed. No absolute measurement of precipitation is possible at present, and field measurements are only comparative.

The physics and hydrodynamics of precipitation are beyond the scope of this discussion. It is necessary, however, to reconcile the small amount of atmospheric moisture with the comparatively large volume of precipitation.

Firstly, atmospheric moisture is being constantly renewed by evaporation and advection. Secondly, for most of the world, precipitation is a rare event in terms of hours per year during which it occurs. At Paris, France, a fairly typical temperate zone location, average annual duration of precipitation is 577 hr, or about 7 percent of the time.[12] Thirdly, the air mass in a storm contains considerably more than the global average of moisture. Finally, the atmosphere is on the move and moisture is renewed in the area over which rain-producing conditions persist.

Intensity of precipitation is an important factor, because the disposition of precipitation among evaporation, infiltration, and runoff is strongly affected by the rate of precipitation. Precipitation of low intensity and long duration has the advantage that there is ample opportunity for infiltration. Total precipitation of an inch or so during a few hours may produce little or no direct runoff. The same amount of rain within a few minutes may produce heavy runoff.

Interception. Much precipitation is intercepted by vegetation and evaporates without ever reaching the ground. This is called interception loss. Many studies of the phenomenon have been made but no wholly satisfactory means have been developed for estimation of regional interception. Rain gauges generally are purposely sited in the open, where they theoretically measure total precipitation. Only for experimental purposes are companion gauges placed under tree canopies.

Although it is considered that open-space rain gauge records are generally biased on the low side, the bias may be accentuated in forested areas in terms of precipitation that actually reaches the ground. The extent of permanent interception depends strongly on the intensity and duration of precipitation. Heavy precipitation quickly saturates foliage and, if continued, full precipitation after saturation reaches the ground. This is called throughfall. Light precipitation of short duration may only wet the vegetation and no moisture will reach the ground except in open areas. Light precipitation during an extended period, however, may largely reach the ground.

Some good data on interception are available for experimental areas, but only estimates are available for large areas. Owing to the variables noted above, it is readily apparent that direct measurement of large-area interception is impossible.

The amount of interception varies with the nature of the foliage and bark of the vegetation. Some intercepted precipitation may be absorbed by foliage and bark, and a certain amount is required to wet leaves and bark before throughfall and stemflow occur. Wind, by shaking leaves, may cause greater throughfall than would occur without wind. Specialists in this field commonly consider that 0.02 to 0.10 in. of rain is necessary for throughfall to occur, depending on the type and height of vegetation. It may be assumed that a nearly equal amount evaporates from foliage after the storm. In addition, some water may evaporate from foliage even while rain is falling.

Most interception is satisfied by rain falling in the early part of a storm. However, many storms yield only small amounts of precipitation and, where vegetation is dense, interception may amount to 25 to 50 percent.[13]

Some observers believe that interception of snow by conifer forests is highly significant. A sizable fraction of snowfall may be intercepted and evaporate without reaching the ground. Various studies of conifer forest areas report annual interception rates of 4 to 15 in. of precipitation and amounts of up to 38 percent of total precipitation.[14] Hardwood forests intercept little snow.

In one experiment, snowpack accumulation under trees was 30 percent less than in the open. In other experiments, rainfall interception ranged from 100 percent from a rain shower of 0.01 in. to 16 percent for showers of 0.5 to 1.0 in.[15]

On the other hand, it has been noted[16] that calculation of an energy budget for snow in the trees shows that little energy is available for sublimation. Also, the moisture gradient is small and very little snow actually evaporates. The whole topic of interception, especially of snow, needs much more study.

Infiltration. Infiltration, the movement of water downward through the soil surface, is an elusive parameter in the hydrological cycle and it can be measured directly only in small experimental plots. Various kinds of soil moisture probes are available with various degrees of accuracy. Some of them, by their very presence, disturb the soil so that they do not correctly define infiltration under natural conditions.

Infiltrometers have been also widely used to determine rates and amounts of infiltration. Again, the infiltrometer itself disturbs natural conditions and the results may be erroneous.

Infiltrometers in some areas show no deep

infiltration where other lines of evidence indicate that infiltration must occur.

Infiltration is one of the more critical phenomena of the hydrological cycle for several reasons. Where the rate is slow, the ratio of runoff to infiltration is high and restoration of soil moisture is low. Where the surface soil is highly permeable, infiltration may be rapid and runoff from a light but steady rain may be negligible. No recharge of groundwater by deep percolation can occur until the moisture demand of the soil mantle has been satisfied. Thereafter, further infiltration leads to deep percolation and groundwater recharge.

Infiltration rates in very tight soil range from 0.0 to 0.5 in. water/hr. In very permeable soils, rates may range up to several inches per hour.

Deep Percolation. Deep percolation refers to the process whereby water percolates downward below the soil or root zone and may reach the water table. In order for percolating water to reach the water table, the supply of water must exceed the field capacity of the soil and subsoil. In this zone, capillary forces dominate when the earth material is not saturated. At the saturation and near-saturation levels, however, gravitational forces dominate and water can move downward to recharge groundwater in the zone of saturation.

The hydrodynamics of deep percolation is not well understood, but the fact of deep percolation is well recognized. Direct recharge by vertical percolation downward through permeable earth is widespread where the supply of water from precipitation is adequate. Where the soil and subsoil are poorly permeable, recharge may be small even in areas of high precipitation.

Arid areas and semi-arid areas contain some very large aquifers, but some of them receive very little direct annual recharge. An example is the Ogallala Formation of the Southern High Plains of Texas and New Mexico. Estimates of the average annual rate of recharge are in the range of a few tenths of an inch. The aquifer is geologically isolated and receives no replenishment of water by underflow from surrounding sources. The huge amount of water in this aquifer is an inheritance from the past, when a wetter climate prevailed.

Another great aquifer, the basalt aquifer of the Snake River group in southern Idaho, probably receives only a few tenths of an inch annually by direct infiltration of precipitation. However, the aquifer has copious sources of replenishment: several rivers which debouch onto the plain from surrounding mountains sink into the basalt and disappear. Mass underflow of groundwater from surrounding mountains into the plain must be significant in amount. Some reaches of the main stream of the area, the Snake River, are perched above the water table, to which the river loses water by seepage. Some reservoirs and lakes also are leaky. Finally, excess water applied to irrigated areas finds its way to the water table.

Gravel-filled valleys in arid areas are another special case. Rain and melting snow in adjacent mountains feed streams which are perennial in the hard-rock mountains but disappear when they reach their permeable alluvial cones or fans. Some of the water may be disposed of by deep percolation, but in some cases the water is sufficient merely to a wet shallow layer of alluvium from which the water evaporates.

No direct measurement of deep percolation is possible, but several methods are available for estimation. One of the most common is to observe fluctuations of the water table, represented by the water levels in observation wells. Following a rainstorm or melting of snow, water levels will rise after a lag period that depends on the permeability of the soil and the depth to water. If the porosity of the aquifer is known or estimated, the rise in water table, suitably adjusted for previous moisture content, can be converted to equivalent depth or volume of water added.

Relatively little is known about rates of deep percolation except in local areas. A study on Long Island, New York showed that beneath one recharge basin in coarse sand and gravel the downward rate of movement averaged 5 ft/hr.[17] In a recharge basin with a few feet of head of water in the basin, the rate of infiltration and hence of downward percolation would be higher than through undisturbed land surfaces in many areas.

No infallible method is available for accurate estimation in large areas. Under natural conditions during long periods of time, natural discharge of groundwater equals natural recharge. Natural discharge is through seeps and springs, by evaporation where the water table is at shallow depth, and by transpiration where plant roots reach the water table or the capillary fringe above it. It would not be possible to measure all these parameters. Several indirect methods have been used to estimate recharge in the Southern High Plains of New Mexico.[18] Results from each method differed, but they seemed to converge on a value of about 0.5 in./yr. In arid areas, deep percolation may be nil in most years. In humid areas it may amount to several inches or more per year.

Groundwater Underflow. Underflow is the lateral movement of groundwater, whether in confined or unconfined aquifers. Rates of underflow cover a wide range, depending on the permeability of the aquifer, the rate of recharge, and the hydraulic gradient. In coarse sand and gravel, rates of 50 to 60 ft/day are considered to be in the upper range. Rates in cavernous limestone and basalt may be higher, and even under low gradients rates may average 20 to 40 ft/day. In less permeable materials such as sand, silt, and clay, rates are much less, ranging down to a fraction of a foot per day or nil.

Flow rate is not a direct function of porosity. Clay, for example, is among the most porous of earth materials, with porosities up to 60 percent. However, the pores are so small that capillary and molecular forces dominate and may prevent gravity flow.

Rates of flow have only been sparsely studied in artesian aquifers. The water from artesian wells is derived by elastic compression of the aquifer. In a non-artesian well, a dye or other tracer may be introduced in one well and its time of arrival observed in another well down gradient. In the artesian situation, the tracer would have to be introduced under pressure. This would disrupt the pressure gradients in the aquifer and the time of appearance of the tracer in another well would not be indicative of natural rates of movement.

Application of specialized techniques permits estimation of movement rates and residence time of water in some aquifers. Analysis of ^{14}C in dissolved carbon compounds in groundwater is an example. Results for water from various deep aquifers in North Africa and Arabia indicate residence times in the range of 20,000 to 30,000 years. These indicate movement rates on the order of a few hundred feet per year from distant recharge areas in the plateaus and mountains. These rates are consistent with the known properties of the aquifer such as permeability and pressure gradients.

Rate of movement of water in an aquifer should not be confused with rate of transfer of changes in hydraulic head. Transfer of head is analogous to what occurs when there is a sudden influx of water at the head of a reservoir. This influx creates a surge or bore whose celerity depends on its height and the depth of water below and ahead of it. Under some circumstances celerities of many miles per hour may be achieved. Flow over the spillway of a dam may increase within a few minutes after the influx of water while the new water itself is still at the head of the reservoir.

Analogous phenomena occur in unconfined aquifers. An influx of recharge water in one area may create a so-called recharge wave which is propagated outward from the area. The aquifer material, however, has a damping effect on the wave, and the lower the permeability the greater the damping. Even in highly permeable material, the celerity may be no more than a few hundred feet per day through short distances. Recharge waves generally are completely damped out within a few miles, often much less. Meanwhile, the recharge water remains at or near the place where it entered the aquifer. The recharge wave is merely a transfer of head.

In terms of rate of circulation, subsurface water-bearing materials may be grouped in several zones; as described below.

Zone of Rapid Circulation. The zone of rapid circulation extends from the land surface downward a few tens or hundreds of feet. It includes the soil zone, vadose zone and zone of relatively shallow unconfined groundwater. Water of meteoric origin in this zone has a residence time of a few hours or days to a few years. Natural discharge from this zone is a principal source of water that sustains the flow of streams during rainless periods (base flow).

Zone of Delayed Circulation. In the zone of delayed circulation the water is generally fresh and of meteoric origin; it may be unconfined but commonly is confined under artesian pressure. While the water circulates rather freely, circulation is delayed and residence time of water in the aquifer may range from a few decades to a few centuries. The depth to or through this zone may range from a few hundred to a few thousand feet.

Zone of Lethargic Circulation. The zone of lethargic circulation contains water which is largely or wholly of meteoric origin but which has been out of the normal hydrological cycle during many hundreds or thousands of years. The water commonly is highly saline. Its movement is generally considered to be hydrodynamic, but geochemical osmosis and electrochemical forces also may operate. Such waters commonly are at depths exceeding 5000 ft, but they may occur within the upper few hundred feet. For practical purposes, most of these waters may be considered to be nonparticipants in the modern hydrological cycle.

Other Zones. Other zones (stagnant and dry) exist, but they are of no concern in the water cycle. They may be of interest as zones for the storage of radioactive and industrial liquid or gaseous wastes.

Exceptions occur in all of the depth and circulation zones noted above. Saline water, for example, occurs at shallow depth in large areas and may feed saline springs, as in the Permian Basin in the southwestern United States.

Groundwater Outflow. Under natural conditions, most unconfined aquifers are full to overflowing. Overflow occurs through springs and seeps and by diffused percolation into stream channels. Were this not so, many streams would be dry during rainless periods. At high-flow and flood stages of rivers, some surface water seeps into river banks and becomes groundwater. When river stages fall, the water returns slowly to the river. Bank storage and return flow were first recognized and described, though not named, 300 years ago,[19] but the phenomena received relatively little study until modern times.

The important phenomenon of base flow has been studied chiefly in connection with flow-duration analyses of streams. It has been largely neglected as a means of estimating the sustained annual yield of aquifers, although it could be helpful in that respect. Conjunctive use of groundwater and surface water inevitably will become widespread. This will require much more attention to groundwater flow, including contributions to base flow in streams.

In some places the discharge of groundwater through springs is spectacular. An example is the springs along the valley walls of a 40-mi reach of the Snake River in southern Idaho. The aggregate average discharge of these springs is about 6000 ft^3/sec. One group, the Thousand Springs, alone discharges about 750 ft^3/sec.

The existence of oceanic shoreline and submarine springs has been known since antiquity. For the most part, these have been treated as curiosities. However, it is now known that they have considerable importance in some localities. An outstanding example is Hawaii, where an aerial survey by infrared imagery disclosed the existance of at least 219 springs with discharges sufficiently large to affect noticeably the temperature of the receiving seawater. Discharge from individual longshore springs of Hawaii range up to 238×10^6 gpd (2260 $ft^3/sec.$)[20]

Longshore and submarine springs have been found at many other places: along the Atlantic and Gulf coasts of Florida, many Mediterranean countries, in the Red Sea, along the shore of Normandy in the English Channel, in the West Indies, near the Sandwich Islands, and elsewhere. It has been estimated[21] that about 2×10^9 m^3 (70.6×10^9 ft^3) of water discharges annually through coastal and submarine springs in the karstic areas of Greece; this is about 22 percent of estimated annual groundwater recharge (9×10^9 m^3 = 318×10^9 ft^3).

The karstic area is about 45,000 km^2 (17,380 mi^2), island karst included. Karst is a distinctive type of landform in regions of readily-soluble limestone bedrock. Many depressions and sinks form and surface streams are few or lacking. Large springs may occur in deep valleys and in submarine areas nearby.

Large oceanic springs generally issue from hard-rock aquifers such as creviced basalt and cavernous limestone. Groundwater also reaches the sea, however, by mass seepage through granular aquifers. This has been demonstrated for Long Island and New York and has been asserted for part of the coast of southern California.

Artesian aquifers that extend beneath the floors of continental shelves also contain fresh water offshore at some places. Ship-based test drilling disclosed artesian fresh water in the Floridan aquifer at least 75 mi offshore.[22] The water came from about 425 ft below the sea floor and rose in the casing to nearly 30 ft above sea level.

Many other submarine areas of groundwater discharge have been described. It is not implied that this is a major phenomenon, except locally, but it does indicate that this mode of discharge is not negligible. The total, world-wide or continent-wide, probably is equivalent to only a few percent of surface runoff, and hence is well within the margin of error of river discharge measurements.

Lakes and Swamps. Serious modern study of lakes begin with the Swiss scientist Forel, who coined the term limnology. This he defined as the oceanography of lakes. By limnology, many current workers mean fresh water biology.

In the Northern Hemisphere, most lakes occur north of the 40th parallel. The Southern Hemisphere contains relatively few lakes except in Africa, which has some of the world's larger ones. Swamps and wetlands are more widely distributed than lakes, but they have received comparatively little study because they are commonly regarded as nuisances rather than assets. Only in relatively recent years has the ecological importance of swamps and wetlands begun to receive adequate recognition.

Lakes and swamps, wherever they are abundant or large, play a major role in local water balances. Large lakes, such as the Great Lakes of North America, are sufficiently extensive to affect regional climate and hence the hydrological cycle. The Great Lakes are an immense heat reservoir, helping to cool the surrounding area in summer and to warm it in winter. In doing so they contribute to high humidity in summer and to fog in winter.

Many lakes are merely wide places in rivers, but that characterization is too general for all. Many, if not most, of the hundreds of thousands of small lakes in northern North America and Scandinavia owe their origin to continental glaciation during the Ice Age. Glacial action left an irregular topography containing many depressions that are now occupied by lakes. The Great Lakes basin is, at least in part, an isostatically low area that was depressed by the weight of the Pleistocene icecap. Lakes Pend Oreille and Coeur d'Alene in Idaho occupy valleys which have been dammed by glacial moraines. Many other physical classes of lakes occur—too many to describe here.

The important thing about lakes and swamps is that they do occur and each must be studied in relation to local geology, climate, and ecology in order to be understood. In the hydrological cycle, they function as reservoirs that regulate runoff, as infiltration basins that recharge groundwater, and as evaporators.

Many small lakes may be considered as outcrops of the water table, being fed by groundwater in the same way that groundwater contributes to the base flow of streams. Some lakes, on the other hand, are perched above the water table and may contribute to groundwater recharge by percolation losses and deep seepage.

Lakes of the world cover about 330,000 mi^2 of area and contain about 30,000 mi^3 of water. Most of this is in large lakes, large being defined as a volume of 5 mi^3 or more. The hundreds of thousands of smaller lakes, although locally important, add but little to the total. Large lakes of North America, Africa, and Asia contain nearly three-fourths of all the fresh liquid surface water on the continents.

A reservoir, once impounded, functions much like a natural lake. Apparently no data on the world total volume and area of man-made lakes are available. These, however, are obviously only a small percentage of the values for natural lakes.

Saline lakes and inland seas are nearly equal in area and volume to freshwater lakes (270,000 mi^2 and 25,000 mi^3). However, about 65 percent of the area and about 95 percent of the volume of inland saline surface water is in the Caspian Sea.

Swamps have received little attention from hydrologists. Swamps are areas where the water table is at or near the land surface and open-water areas may be considered as outcrops of the water table. It has been estimated that swamp areas of the world, including the marshy areas of tundras, contain about 4300 mi^3 of water, while peat bogs contain an equivalent amount. Swamps and marshes have great importance to ecologists, and their significance in sanitary engineering is growing. Inland water areas of the world probably approach 400,000 mi^2.

Rivers: Their Entrained and Dissolved Solids. It is said that some Chinese people call rivers "threads of the sea." Blaise Pascal called them "roads that move." Both characterizations are apt. The water in every river (except in endorheic basins) is on its way to the sea and each is a conveyor belt for the transport of suspended and dissolved solids to the sea.

In the conterminous United States, surface runoff is measured regularly by state, federal, and local agencies at about 24,500 stations on principal rivers and many of their tributaries. However, only about 7000 of these stations have more than 15 years of record, and only about 3600 have more than 40 years of record.

Water, being a universal solvent, dissolves mineral matter wherever the two come in contact. The nature and solubility of minerals in the watershed, therefore, and the duration of contact determine the chemical characteristics acquired by the water. Rainwater contains a small amount of dissolved atmospheric carbon dioxide, which forms carbonic acid, and even this weak acid solution is effective in dissolving carbonate minerals such as limestone.

Rivers differ through an extreme range in their length, channel characteristics, velocity of flow, discharge rate, variability of discharge rate, susceptibility to flooding, and dissolved and entrained load of solids.

The largest river in the world in terms of discharge is the Amazon, with an average annual discharge of about 6.2×10^6 ft^3/sec. The estimated discharge of all rivers of the world into the sea is on the order of 32.6×10^6 ft^3/sec. This estimate is based on classical studies, modified by data from many sources. The Amazon discharges about 19 percent of the total.

The number of rivers whose discharge to the sea exceeds 1000 m^3/sec (about 35,300 ft^3/sec) is about 75, and their aggregate discharge is 630,500 m^3/sec, or about 68 percent of the estimated total of all rivers. These 75 rivers drain about 52 percent of the land areas that drain to the sea. The remaining land obviously has a lesser average unit-area yield of runoff than the area of the 75 great rivers. The unit-area yield of the great rivers averages about 12 liters/sec/km^2. For a sample of 200 lesser rivers, the unit-area yield is somewhat greater.

Calculations of channel storage in certain reaches of some rivers are common for water management purposes. Total storage in all river channels of the world would be impossible to calculate and probably would have little practical value. For hydrological perspective, however, it is useful to make an order-of-magnitude estimate. One crude estimation[23] is about 1700 km^3 (about 410 mi^3). This is equivalent to the average sea-mouth discharge of all rivers during about 2 weeks.

Travel time of water in river channels is significant in this connection. A common assumption is that current velocity decreases downstream in a river even though discharge may be augmented by tributary inflow. Discharge is the product of channel cross-sectional area times water velocity, and area is the product of width times depth. Increased discharge may be accommodated by an increase in any of these three factors or by some combination of them. Actually, width tends to increase downstream faster than depth and depth faster than velocity. In general, velocity actually tends to remain constant or to increase downstream.[24]

The thread of maximum velocity in a river has no fixed position but it is near the center in straight channels and near the concave side of curves or meanders. Also, the maximum velocity is a short distance below the surface, commonly 20 to 25 percent of the depth. For this reason, if a mass of pollutant or other soluble substance enters a stream, it does not travel downstream as a uniform mass but becomes mixed and diluted, chiefly by hydraulic dispersion. Owing to cross-sectional differences in velocity, an attenuated solution of the mass will travel fastest in the high velocity thread of the stream. The maximum concentration will move more slowly, and a dilute solution will follow even more slowly. A study on the Mississippi River, when the discharge rate was 240,000 ft^3/sec, showed that the leading edge of a dye tracer traveled 120 mi in about 81 hr, the peak concentration traveled the same distance in a little over 90 hr, and the trailing edge required about 109 hr. The travel time and longitudinal dispersion would, of course, differ at different discharge rates. Assuming uniform rate of flow throughout the Mississippi's 1800-mi length at that velocity, an average molecule of water would require only 56 days to travel the entire length.

The short residence times of river water and atmospheric vapor lend emphasis to the precarious nature of the land phase of the water cycle and to the need to understand it better. They also point out the importance of the base-flow component of river flow.

Data are scant on the chemical load of rivers where they discharge to the sea. In some, the water may carry only a few milligrams per liter of dissolved solids (DS). Loads of 100 up to 300 mg/l are common. At the other extreme, a few streams carry several thousand milligrams per liter of DS. The rate of chemical denudation of the continents is very difficult to estimate but this has been done by various authors. A recent estimate is summarized in Table 1-3.[25] The estimated mean chemical composition of rivers of the world is 120 mg/l of DS.

Data on suspended solids in rivers that enter the sea are even more scanty then those for dissolved solids. Many estimates of total discharge have been made, based on limited data and varying assumptions. A recent estimate[26] indicates an average annual measured discharge of somewhat more than 8×10^9 tons from a drainage area of 15.4×10^6 mi^2, and an extrapolated total of 20×10^9 tons from the world total land area. Assuming that a cubic foot of dried river sediment weighs 90 lb, the dry volume of the total annual transport is about 444×10^9 ft^3.

TABLE 1-3. Estimated Annual Rate of Chemical Denudation of Continents[a]

Continent	Area (10^3 mi^2)	Rate (pkg/mi^2)	Total (10^6 pkg)
North America	8,172	85	695
Europe	4,211	110	463
Asia	17,985	83	1,493
Africa	11,500	63	724
Australia	2,970	6	178
South America	7,551	73	551
World total (rounded)			4,000

[a]Based on Livingstone.[25] Denudation rates in many areas have been accelerated by human activity in recent time. The rates shown are not indicative of those that prevailed during recent geological millenia.

Sediment transport to the sea has great practical importance. However, most eroded material is trapped in reservoirs, lakes, and floodplains and does not reach the sea or even tidewater areas. Hence, the effects of human activity on transport to the sea and to tidewater can be estimated only approximately.

Sediment includes nutrients (in addition to dissolved nutrients) which enter the food cycle of aquatic organisms. Not all sediment that reaches tidewater continues to the open sea, because much of it is deposited in deltas, bays, and estuaries. Thus, bays and estuaries tend to be filled in and deltas to be built up. The coarse fraction of sediments that reaches the sea tends to be deposited near shore and to be transported by long-shore currents which build beaches and spits. Where extensive dam construction has provided effective sediment traps and has reduced transport of solids to the sea, unanticipated side effects have occurred. Long-shore currents, denied a supply of river sediments to use their transport energy, have begun to erode beaches, as in southern California.

The fine-grained component of suspended solids that reaches tidewater is apt to be carried far out to sea. Photographs from orbiting vehicles show that off the mouths of the Mississippi and Orinoco Rivers and off the coast of southwestern Africa, turbidity from suspended sediment is evident 20 to 90 mi offshore.

Within land areas, the suspended load ranges widely from stream to stream and from time to time in a given stream. Peak concentrations of suspended load in some streams may range as high as 646,000 ppm.[27] That is, more than half of the weight of the water-sediment mixture may be sediment. If the sediment particles have an average specific gravity of 2.6, then about 41 percent of the volume of the mixture is sediment. Under such circumstances, measurement of stream discharge would be seriously in error if all the volume were treated as water yield.

Streams in the humid part of the United States commonly contain 10 to 50 mg/l of suspended sediment but rarely as much as 100mg/l. Midwestern subhumid streams seldom contain more than 1000 mg/l. Southwestern streams commonly contain 1000 to 10,000 mg/l, and waters in a flash flood may carry up to 35,000 mg/l.

It has been demonstrated[28] that total sediment yield is rather closely related to precipitation on watersheds. Yield is at a maximum where precipitation averages 10 to 14 in. yearly. Where precipitation is low, runoff is too small to yield much sediment. Where it is high, the density of vegetation effectively holds the soil. However, geological, topographic, and other conditions affect sediment yield. Basins that are otherwise comparable but differ in geology and topography vary about tenfold among themselves in the rate of sediment yield.

Human activity has radically altered erosion rates and transport of sediment to the sea in many parts of the world. In many areas, alteration has been progressively increasing erosion and transportation–so much so that historical records are no longer a good index of what to expect in the future. Much has been written on this topic, but a sample study of a small area is indicative. The area is the Scott Run basin (4.5 mi^2) in Fairfax County, Virginia. During 1961-1964, 11 percent of the area was the scene of highway construction. Studies and measurements showed that 88 storm events accounted for 37 percent of runoff and 99 percent of sediment movement during 3 percent of the time. Construction areas, varying from 1 to 10 percent of the basin at given times, contributed 85 percent of the sediment. Precipitation during the study was 12 percent below normal. With normal precipitation, erosion from construction areas would have been 20 percent greater, yielding an average of 151 tons per acre per year. This is 10 times the amount normally expected from cultivated land, 200 times that expected from grassland, and 2000 times that expected from forest land.[29] Phenomena of this sort would have little effect on total sediment yield of a basin 100 mi^2 or more in area, and they would persist during only a few years.

It has been found[30] that in urban and suburban areas of Maryland, from areas cleared for construction, the equivalent of many decades of natural erosion may occur within a single year. The average natural sediment yield is 200 to 500 tons/mi^2/yr, and some predominantly wooded areas yield even less. Farmed watersheds may yield up to 1000 tons/mi^2/yr. Sediment from construction areas ranges from 3000 to 150,000 mg/l (several thousand to 140,000 tons/mi^2/yr).

Estuaries. Estuaries have been called places where rivers meet the sea. A typical estuary is the drowned mouth and lower channel reach of a river that empties to the sea. The channel is cut below sea level and is invaded by the sea. Chesapeake Bay, in the eastern United States, is one of the world's greatest estuaries. During the Ice Age, so much water was stored as ice on the continents that sea level at times was several hundred feet lower than at present. During low stands of the sea, the Susquehanna, one of the world's great rivers, carved a valley well below present sea level. The Potomac, James, and other rivers that enter Chesapeake Bay were tributaries of the Susquehanna.

Both tidal action and gravitational forces send seawater up estuaries, where it mixes with fresh water. The mixing pattern may be complex or simple, depending on tide range, depth of the estuary, configuration of the estuary, and rate of discharge of the river. Each estuary requires individual study.

Estuaries in general are being filled by alluvial sediment and organic accumulations. Filling has been in progress during many thousands of years and in parts of Chesapeake Bay accumulations of sediment are as much as 100 ft thick. Near Annapolis, the old thalweg of the Susquehanna River beneath the sediments is about 200 ft below present sea level.[31]

Not all coastal embayments are estuaries. The Carolina sounds such as Pamlico are enclosed by offshore bars formed by marine wave and current action. Puget Sound is a fjord gouged out by glacial action. Some embayments have been formed by downwarping of the earth's crust. Nearly all embayments, however, have been affected by the rise of sea level during the last 15,000 years, and rivers that enter sounds in general have true estuaries.

The ecology of estuaries is extremely complex and their biotas are very varied and versatile. They are highly susceptible to change in stresses and have prominent roles in the water cycle and in the disposition of the dissolved and entrained solids in the rivers that enter them.

Chesapeake Bay, being the largest estuary in the United States, is worthy of further note, and is described by the following:[32]

Length	195 mi
Width	up to 30 mi
Surface area	4,300 mi^2
Length of tidal shoreline	4,600 mi
Maximum depth	175 ft
Average depth	28 ft
Water content	18.52×10^{12} gal
Tributary drainage area	74,000 mi^2

The bay is fed by four major rivers, fifty large tributaries, and many smaller streams. The yearly sediment contribution from the two principal rivers is 0.6×10^6 tons from the Susquehanna and 2.5×10^6 tons from the Potomac. However, a dam on the Susquehanna River traps much sediment that otherwise would reach the bay.

Average annual evaporation from the Chesapeake Bay is about 45 in. annually, so the total volume is about 782×10^6 gal. Although this is more than offset by rain on the bay, both the rain and the evaporation are a water loss so far as water supply is concerned.

Snow and Ice. Snow and ice accumulations include ephemeral winter accumulations of snow; alpine, piedmont, and valley glaciers; the great icecaps of Greenland and Antarctica; and sea ice. The importance of winter snow cover has long been recognized, and extensive effort has been expended to extend and improve snow surveys as an aid to water supply forecasting.

The flow of many streams in many areas of the world consists chiefly of high-altitude snowmelt. Until the 1950s, however, what little was known about polar regions and the great icecaps had been learned largely as a result of scientific curiosity and from expeditions that had international prestige value. It is now well recognized that the polar and tropical zones are the weather factories of the earth system and that, therefore, events in these regions are the dominant controls of weather, and hence of water supply throughout the world.

Antarctica contains the world's largest single mass of ice. The size of this mass is difficult to conceive. Less than half of the world land area is in the Southern Hemisphere. Antarctica contains nearly 30 percent of all the land area in the Southern Hemisphere. Almost all of it is covered with ice. Various authors attribute various areas to Antarctica, probably because some include ice-shelf areas and others do not. For present purposes, it is sufficiently accurate to assume an area of 5.5×10^6 mi^2. The main ice mass is equivalent to about 5.6×10^6 mi^3 of water. Perennial shelf ice amounts to about 0.18×10^6 mi^3.

The next larger icecap is that of Greenland, whose main icecap has an area of 0.67×10^6 mi^2 and a water-equivalent volume of about 0.635×10^6 mi^3. Various lesser icecaps, such as those of Iceland, Ellesmere Land, and various Arctic and Antarctic islands cover about 0.04×10^6 mi^2 and contain the equivalent of about 0.02×10^6 mi^3 of water. Oceanic pack ice varies widely in extent from season to season and less so from year to year. The area of sea ice is one of the most variable physical features of the earth. In the Antarctic, for example, the winter area of sea ice is almost as great as the area of Antarctica itself. In the North Atlantic area, the pack ice boundary migrates back and forth through a distance of about 200 mi. The total average area of pack ice is on the order of 11×10^6 mi^2.

Evidently, about 17.5×10^6 mi^2 of the earth's surface (about 9 percent) is covered by continuous masses of ice. This, however, is only part of the refrigeration system.

Alpine, valley, and piedmont glaciers are widespread in North America, Europe, Asia, and South America. These glaciers cover about 0.152×10^6 mi^2 and the contained water volume is about 0.035×10^6 mi^3.

Permafrost (permanently frozen ground) is another major feature of the continents because it occurs in at least 15 percent of the world's land areas (some estimates are as high as 24 percent). The total area is about 8.5×10^6 mi^2. This does not include the ice-covered areas of Greenland and Antarctica.

The world occurrence of permafrost areas has been summarized as follows:[33] 85 percent of Alaska, 40-50 percent of Canada, 47 percent of the Soviet Union and large adjacent segments of China and Mongolia, and the Arctic archipelagos. Other occurrences are in ice-free sections of Antarctica, Greenland, and scattered areas in Iceland, Scandinavia, high mountains of central Asia, the North American Cordillera, the Alps, and the Chilean Andes.

Biological Waters. Water stored in organisms, chiefly plants, is estimated to be about 170 mi^3, an amount so small as to be negligible in water balances of regional or larger scale. The flux of water through vegetation, however, is large among parameters of the land phase of the water cycle. A quite ordinary tree may extract from the soil and transpire 50 gal of water per day.

The albedos of various kinds of vegetation vary widely and hence their rates of transpiration, which depend on use of radiant energy, vary. Also, deciduous vegetation uses practically no water when dormant, while evergreens use water the year round. It would be extremely laborious, even if sufficient data were available, to calculate average worldwide transpiration. A controlled guess puts the process in perspective, however. Evaporation from land areas is about 25,200 mi^3 annually. If only a third of this is transpired, the volume is about 840 mi^3. This is equal to about 12 percent of the sea discharge of all rivers of the world. This estimate probably is conservatively low, but it helps to illustrate a topic discussed later—per capita water demand for human purposes.

Conclusions. About 15×10^6 mi^2 (26 percent) of the world's land area is under permanent refrigeration. An equivalent amount is arid to extremely arid. In addition, vast high-mountain areas are not suitable for habitation for many people. Therefore, considerably less than half of the world's land area is hospitable for normal human occupation. These vast areas, however, are major controlling factors in world weather and climate. Therefore, they must receive increasing attention.

WATER BALANCES

Study of the origin and interrelationships of waters in the hydrological cycle has deep historical roots. Anaxagoras of Clazomene (500-428 B.C.), a highly original thinker, formed a basically correct idea of the hydrological cycle: heat from the sun raises water into the atmosphere, from which it falls as rain. Rainwater collects in subterranean reservoirs which feed rivers. The reservoirs fill during wet periods. Perennial streams flow from large reservoirs and ephemeral streams from small ones.

Plato (c. 428-348 B.C.) disagreed. He supposed that a series of interconnecting subterranean channels communicate with their underground source, the vast reservoir of mythical Tartarus. Perpetual surging to and fro in the subterranean reservoirs causes springs and rivers to flow. All waters of rivers and seas return finally to Tartarus.

Aristotle (384-322 B.C.) disagreed with Plato, his mentor. According to him, some springs are fed by meteoric water, but the principal flow of rivers originates in great underground caverns where coldness transforms air into water.

Living in a semi-arid region, Aristotle could not conceive that rain could be more than a minor source of water.

Aristotle's ideas, with fanciful variations, persisted for nearly 2000 years. In the sixteenth century, Jacques Besson and Bernard Palissy both asserted the pluvial origin of groundwater, but they made no measurements and their ideas seem to have had no practical or scientific impact at that time. In 1674, Perrault[19] made the first quantitative approach to the hydrological cycle, based on the water balance concept. He showed that annual precipitation on the upper basin of the Seine River in France was at least 6 or 7 times the volume discharge of the upper river. He correctly established the concept that precipitation is disposed of by evaporation, transpiration, runoff, and groundwater recharge. His concept was independently confirmed within a few years. Perrault had some erroneous notions about groundwater recharge, but these were logical, based on observation. Recharge could not be well understood until stratigraphic and structural geology achieved a firm footing early in the nineteenth century.

A significant contribution to the water balance concept was made by John Dalton in 1802, who studied the quantitative relations between precipitation, evaporation, and runoff and the origin of springs. He was the first person to try to determine the water balance of England and the first to develop a formula for calculation of evaporation from free water surfaces.

During the remainder of the nineteenth century, the problem of water balances attracted much attention, chiefly in Europe. In America, F. H. Newell in 1894 demonstrated the direct relation between runoff and precipitation and prepared isopleth maps for their mean values in the United States. During the twentieth century, much work has been done in many countries on water balances and the measurement and calculation of water balance parameters.

The principal parameters of the hydrological cycle are listed in Table 1-4. This table and the discussion that follows are based largely on a nontechnical summary by T. G. Chapman (see Ref. 1) of water balance methodology. Table 1-5 shows the status of networks for collection of some of the principal data that are needed for water balance studies.

TABLE 1-4. Principal Parameters of the Hydrological Cycle[a]

Storages	Fluxes
Atmospheric water (W)	Evaporation (E)
Oceans and seas (O)	From bare and vegetated land
Lakes and reservoirs (L)	From oceans and other water bodies
River channels (V)	From snow and ice
Swamps (S)	Horizontal vapor flux ($\vec{Q}$)
Biological water (B)	Precipitation (P)
Moisture in soil and unsaturated zone (M)	Runoff (R)
Groundwater (G)	Infiltration; deep percolation
Frozen water	Recharge, underflow, and discharge (G_d)
ice caps	Rates of movement
glaciers	
pack ice	
ice shelves	
permafrost	

[a]Other fluxes affect the hydrological cycle, such as ocean currents and heat fluxes. Though not included in the discussion that follows, they would not be overlooked in water balance studies. Heat fluxes, for example, affect evaporation, but we deal here only with evaporation, once it has been determined or estimated; similarly for other factors.

Study of water balances is based on the principle of conservation of mass. For any given three-dimensional region, total input equals total output or, if they are not equal, the difference will be balanced by a change of water storage within the region. In general, therefore, study of water balances implies measurements both of storages and of fluxes (rates of flow), as listed in Table 1-4.

Symbols in the table will be used in balance equations on following pages. In these equations, the symbols given with the list of parameters represent average values per unit area of the region considered. The symbol Δ denotes change in storage during the period considered, and the symbol $\oint$ indicates integration around a boundary of length λ enclosing an area A. For large areas and long periods of time, it is usually possible to eliminate the need for some measurements. Taking all applications into consideration, the number of possible combinations of parameters, area sizes, and time scales

TABLE 1-5. Hydrological and Related Data Collection Networks in the United States[a]

Parameter	Number of Stations[b] Total	>15 Yrs of Record	>40 Yrs of Record
Air temperature	6,481	5,000	4,000
Relative humidity	360	250	180
Barometric	620	550	200
Windspeed	1,190	300	180
Sunshine duration	178	174	150
Solar radiation	85	75	8
Precipitation	10,535	9,500	5,500
Evaporation	540	323	54
Snow cover	9,785[c]	8,800	4,800
Glacier variations	239	222	142
Lysimeters	–[d]	–[d]	–[d]
Surface runoff	24,500	7,080	3,600
Groundwater levels	30,670[e]	–[f]	–[f]
Dissolved solids			
Precipitation	80	0	0
Surface water	4,500	109	–[f]
Groundwater	3,000	–[f]	–[f]
Suspended solids (surface water)	1,400	42	–[f]
Radiometric data (all realms)	2,600	–[f]	–[f]

[a]From Nace.[9]
[b]Includes only stations in current operation in permanent networks. Excludes stations for special purposes such as research, temporary stations, and stations having only local significance.
[c]Does not include 1,200 snow courses in western states.
[d]Information not available.
[e]About 2170 continuous or daily measurements and 27,500 less frequent measurements.
[f]Information not available in readily retrievable form.

is very large. Examples will be given here only for large-scale balances.

Atmosphere. Among many contemporary estimates of the long-term average amount of water stored in the atmosphere, differences are only a few percent. The value of 3100 mi^3 is widely accepted, equivalent to 1 in. of depth if condensed as a uniform layer over the entire earth. Because world average annual precipitation and evaporation each is equal to about 40 times the average storage value, the mean residence time of an average molecule of water in the atmosphere is about 9 days. Most authorities postulate 8 to 12 days, so 10 days seems to be an acceptable value.

The gross values for storage and residence time vary seasonally somewhat, and local values vary widely from time to time and from place to place. Because of these variations, observation and analysis of atmospheric moisture fluxes is an important part of the aerological approach to computations of regional water balances. The following equation represents the atmospheric water balance:

$$P + \frac{\lambda}{A} \oint \vec{Q} = E + \Delta W \qquad (3)$$

Because the residence time of atmospheric water is only a few days, the last term is negligible for long periods. Thus, for periods of one month or more, the equation becomes

$$E - P = \frac{\lambda}{A} \oint \vec{Q} \qquad (4)$$

This is the basis of the aerological method for determining the difference between evaporation and precipitation within a region. The right-hand side of Eq. 4 can be substituted for $(E - P)$ in any of the following equations relating to extended periods of time, and it can be used as a general test of the accuracy of conventional measurements of precipitation and evaporation. As a working method in water balance computations, it has most promise for estimating $(E - P)$ over the oceans. For these areas, the prospects of radical improvements in conventional observations seem less promising than improvements in the network of coastal aerological measurement stations.

Oceans. The storage volume and mean residence time of oceanic water are important because the oceans store and distribute tremendous amounts of energy. The equation for the oceanic water balance is simple:

$$P + R_o + \Delta O + G_o = E \qquad (5)$$

in which R_o and G_o are runoff and groundwater discharge to the oceans, respectively.

The amount of water stored in land areas varies seasonally. Seasonal storage changes in the Southern Hemisphere do not offset those in the north because the north has more land area. Hence, there is a calculable seasonal change in sea level. However, ΔO is so small that it can be ignored unless there is definite evidence of a general change of sea level. Discharge of groundwater to the sea is diffuse and relatively small along most of the coastal areas of the world. Most of this kind of discharge occurs in special areas like the volcanic islands of Hawaii and from limestone aquifers like those around parts of the Mediterranean basin. Some of these aquifers have relatively long residence times. In general, it is safe to assume that G_o is constant from year to year.

Continents and Large Islands. For continents and large islands the parameters for water balances are numerous:

$$P = R_o + E + \Delta L + \Delta V + \Delta S + \Delta B + \Delta M + \Delta G + G_d \quad (6)$$

For any period, changes in storage of biological water will be negligible compared with the other terms, and ΔB may be omitted from the equation.

For periods of one year or more, regional changes in all storages above groundwater will be small compared with the other terms. Therefore, for annual and long-term balances:

$$P = R_o + E + \Delta G + G_d \quad (7)$$

If there is evidence that major groundwater storages are in long-term equilibrium, with inputs equal to outputs, then ΔG may also be omitted from Eq. 7. Where groundwater is being heavily mined or built up by recharge, ΔG must be retained.

Large River Basins. Water balance parameters are numerous also for large river basins, but under some circumstances the formula for calculation can be greatly simplified. The basic formula is:

$$P = R + E + \Delta L + \Delta V + \Delta S + \Delta B + \Delta M + \Delta G + G_n \quad (8)$$

where G_n is the net discharge of groundwater from the basin. As before, ΔB may be neglected for all cases, and changes in river channel storage, (ΔV) also may be neglected for periods of one year or more. The annual balance may then be expressed as

$$P = R + E + \Delta L + \Delta S + \Delta M + \Delta G + G_n \quad (9)$$

For the special case where the boundaries of an aquifer coincide with those of the river basin, and where all groundwater discharge appears as base flow, $G_n = 0$. In this situation, assuming no trend in potentiometric levels and no trend in storage in lakes, reservoirs, and swamps, the equation for the long-term balance is

$$P = R + E \quad (10)$$

Lakes, Swamps, and Reservoirs. Lakes and reservoirs have a relatively simple balance equation:

$$P + R_n = E + G_n + \Delta L \quad (11)$$

where R_n is the difference between surface inflow and outflow, and G_n is the difference between subsurface outflow and inflow. It will not usually be possible to neglect ΔL for periods of one year or less, but the long-term balance will be:

$$P + R_n = E + G_n \quad (12)$$

When a lake is ephemeral, the short-term situation is much more complex, because the area of water surface varies through a wide range. For that reason, it is necessary to account for variations in soil moisture and vadose water near the edges of an ephemeral lake.

Equations for swamps are similar to those for lakes, but details of computation with the formulas may be different.

Frozen Water. Glaciologists have developed special methods for computing mass balances of glaciers and icecaps. The methods are described in guidance materials prepared by the International Commission on Snow and Ice[34] of the International Association of Hydrological Sciences.

AMOUNT AND AVAILABILITY OF WATER SOURCES

From data given in the previous sections on continents, large islands, and large river basins, it is evident that the water cycle is largely an oceanic and atmospheric affair. From a global standpoint, water supplies in the land areas are incidental effects of the water cycle.

The amount of water can be defined more readily than its availability, because not all water is available for direct use. Use of water may be direct or by withdrawal from its source. Navigation and hydropower generation are examples of direct use. Diversion of river water and pumping of groundwater are examples of withdrawal use. The amount of water that is available for use is only a small fraction of the total amount of water in existence.

Oceanic and Estuarine Waters. It is necessary to mention only in passing the extensive direct use of oceanic and estuarine water for navigation, recreation, and culture of aquatic organisms. Oceans and estuaries are used also as sinks for wastes.

Among withdrawal uses, the cooling of machinery, condensers, and other equipment, is a long-standing practice, and many other uses may be found. These waters have been used extensively for centuries also as diluents or degradants of wastes, notably sewage near shore, and oily wastes, trash, and garbage both near shore and on the high seas.

The coastline of the conterminous United States includes nearly 900 estuarine systems, ranging from minor to major in size and importance. Bays, inlets, and tidal marshes are affected by much more than what goes on in their immediate vicinity. About 50 million people live near these areas, but they are affected by activities of people who live hundreds to thousands of miles away. The land furnishes nutrients to estuaries, but it also furnishes freshwater, and this is vital to salinity regimes of the coastal areas. Oysters and some clams, for example, live only in brackish water. Young shrimp spend several months in coastal marshes before they migrate to the open sea. Many species of fish spend part of their life cycles in estuaries. Maintenance of the supply of fresh water to coastal areas thus is vital to estuarine ecology.

Seas have been widely used as repositories for noxious and toxic chemicals and solid materials on the assumption that ocean water will render them innocuous by chemical reaction, by dilution, or otherwise. No one knows how many millions of tons of such materials have been dumped at sea or what their ultimate impact on the marine environment may be.

Evidence is ample that the capacity of even the mighty world ocean to dilute, degrade, or destroy wastes has a finite limit much lower than had been taken for granted until recent years. This problem has been adequately treated in conservationist literature and need not be belabored here. It is a problem, however, in which sanitary engineers inevitably will become deeply involved.

Oceans and estuaries are available as sources of water for desalination (see Chapter 16). The problem here is not supply, which is limitless in a practical sense, but technology, economics, and ecology. The technology and economics are beyond the scope of hydrology, but hydrology is deeply involved in ecology. Many elements of the biotas of bays and estuaries are highly sensitive even to small variations in salinity and temperature of water. The discharge of heated concentrated brine from desalination plants could have profound effects on estuarine and bay ecology. Availability of oceanic water for desalination may hinge on the ability of the environment to withstand the impact of waste discharges and thermal loading. Bays, estuaries, and sounds could become dead seas. On the other hand, thermal loading in some circumstances has beneficial effects, as by increasing ice-free areas in winter and accelerating biological productivity at temperatures higher than those that would prevail at lower temperatures.

Atmospheric Water and Precipitation. The previous section referred to the average liquid equivalent of atmospheric water vapor (about 3100 mi^3 of water). Annual worldwide precipitation is about 125,000 mi^3, or about 40 times the instantaneous amount of vapor. The short average residence time of water vapor molecules in the atmosphere (about 10 days) illustrates the crucial importance of continual evaporation from the seas and advective transfer of vapor over the continents. Equally important from the standpoint of continental precipitation is the occurrence of conditions that produce precipitation. These conditions obviously are rare in dry regions. It has been calculated, for example, that during an average week in July about 15×10^6 acre-ft of water, as vapor, is transported across the dry state of Arizona. This exceeds the average annual flow of the Colorado River at Lee's Ferry.

According to a statistical analysis,[35] weighted average annual rainfall in Arizona during 1925 to 1958 ranged from a low of 6.44 in. in 1956 to a high of 18.34 in. in 1931. The computed average for the 34-year period was 13.03 in. per year. Using an average value of 13 in. and an area of 113,909 mi^2, total annual precipitation on the state is a little less than 79×10^6 acre-ft of water, or only about 5 times the value of atmospheric vapor transport during one week. Precipitation during July averages about 2 in. (12×10^6 acre-ft), or about 20 percent of vapor transport during that month.

For the conterminous United States as a whole, it has been estimated that about 20×10^9 tons (4.8×10^{12} gal) of water flows daily across the country as vapor in the prevailing winds. Only about 10 percent of this is precipitated.

The gross values for atmospheric water are useful in the present context chiefly to provide perspective. Only a small amount of atmospheric water is available for precipitation at any one time and place. A given area may receive inches or even feet of precipitation from a given storm within a few hours or days. This is because, as already noted, vapor-laden air masses move and renew the supply of vapor over the area in which precipitation-producing conditions prevail.

Average annual rainfall over the conterminous United States differs widely from region to region. The highest average is 140 in. in the Olympic Mountains of Washington, and the lowest is about 5 in. in the desert areas of parts of California, Arizona, and Nevada. The 20-in. isohyet of the mid-continent area is generally accepted as the boundary between the subhumid eastern part of the Great Plains and the western semi-arid zone. Although this isohyet is an irregular curve, it follows approximately along the 100th meridian. Here, as with all averages, however, caution is necessary because of annual and periodic variations in weather. For example, the average 20-in. isohyet for Kansas extends through the western part of the state. During a wet period, however, it may shift westward into Colorado, and during a drought, it may shift to the eastern part of Kansas.

Based on records of the National Weather Service and other data, precipitation, averaged for the entire conterminous United States, is about 30 in. per year, equivalent to about 1430 mi^3 of water. Owing to the negative bias of conventional rain gauges and to the paucity of snow data for high-altitude areas, actual precipitation may be greater.

Since about 1937, meteorologists have commonly assumed that no important percentage of continental precipitation is directly recycled water evaporated from land areas. This concept, however, has been challenged and persuasive evidence presented that rainfall at a given location may be increased substantially by moisture evaporated from nearby land areas.[36] It is becoming rather widely accepted that for large areas a significant percentage of precipitation has been evaporated from the land within or near the area.

Atmospheric vapor becomes available only when condensed and precipitated. Owing to water deficiencies at some times and places, the intensity of interest in weather modification and rainmaking is understandable. Even a moderate deficiency in precipitation represents a tremendous amount of water. A 10 percent deficiency for the conterminous United States would be 483×10^6 acre-ft or 157×10^{12} gal. This is equivalent to more than half of actual artificial consumptive use. At current water prices, the lacking water would be worth billions of dollars for some uses. No weather experts believe, however, that rainfall could be increased by 10 percent over the United States, or even over a large area.

Evaporation and Soil Moisture. Worldwide evaporation has already been mentioned. Only about 14 percent of it is derived from land and inland water areas (some estimates range from 10 to 16 percent). The fact that land areas comprise less than a third of the earth's surface is only one of the reasons for this. Evaporation rates depend partly on availability of water for evaporation; hence, they vary much more widely than precipitation. The type of vegetative cover or lack of it is also important. Moreover, evaporation from soil, even when wet, is less than that from free water surfaces because water molecules have to overcome greater resistance to escape from soils.

The commonly-used computed value of 21 in. for actual annual evaporation, averaged for the entire conterminous United States, is equivalent to a volume of 1000 mi^3. The rate is considerably less than potential evaporation (indicated by landpan data on an earlier page), but it can be considered only as an approximation. This value has no significance for local or regional purposes because the rate varies widely from area to area and within these areas from year to year. Reports on regional and local studies must be consulted to obtain values for engineering purposes at specific sites.

The high percentage of annual water income that is consumed by evaporation lends emphasis to the importance of soil moisture conservation, because most evaporated water becomes soil moisture before it evaporates.

Estimation of total soil moisture is very difficult, because only point data can be obtained. Intensive field sampling and measuring give a good measure of the amount of soil moisture in field plots, but for large areas only rough estimates can be made. The estimate in Table 1-1 of total world soil moisture plus vadose water is only a crude guess. Summaries of modern methods of soil moisture measurement[37] and direct methods of soil moisture estimation[38] are available.

Soil is a vast water storage reservoir. An average silt-loam soil to a depth of 5 ft will hold about 20 in. of water. Half of this will be held so tightly that plants cannot extract it and it does not evaporate readily.

Soil moisture is available for use only where it occurs and only for vegetative growth. Soil moisture conservation, therefore, is a means of water use. The aim is to promote trapping of water as soil moisture and to assure maximum use by plants and minimal dissipation by direct evaporation. Irrigation, it should be recalled, is merely a method of supplementing the natural supply of soil moisture available to vegetation.

High evaporation rates also account for the interest of water managers in efforts to suppress evaporation. In some eastern states, precipitation exceeds evaporation from free water surfaces. But in much of the West, evaporation exceeds precipitation on water surfaces. With an evaporation rate of 48 in./yr, each square mile of a reservoir or lake would lose about 830×10^6 gal of water per year. The current state of knowledge about evaporation and of methods for its measurement and estimation are discussed in the literature.[39,40]

Attempts to suppress evaporation from water surfaces have their counterparts in attempts to conserve soil moisture. Mulching of soil probably is a practice nearly as old as irrigation. Other practices are the reverse of attempts to build up soil moisture by inhibiting infiltration. In dry areas, where natural rainfall and soil moisture are insufficient to grow profitable or useful vegetation, where no supplemental water is available for irrigation, or where the soil or the terrain are unsuitable for cultivation, certain practices have been developed to suppress infiltration and promote runoff. Soil surfaces can be chemically treated to make them virtually impermeable and yield surface water. At the extreme, surfaces may be paved over or covered with plastic films.

The construction of a city produces a virtual tin roof effect. One inch of rain on one acre (43,560 ft^2) of area amounts to 27,154 gal of water. An inch on a city the size of Chicago (141,888 acres) would amount to about 3.86×10^9 gal. Owing to the tin roof effect of urbanized areas, the percentage of runoff can be very high. In this case, increase of runoff is inevitable but inadvertent. The principal aim of street and highway engineers is to get rid of water, rather than to conserve it. This reduces soil moisture and groundwater recharge. Moreover, it increases runoff, variability of streamflow, and flood incidents.[41] Urbanization and highway construction generally cause wastage, as excess runoff, of water that would otherwise have contributed to recharge of soil moisture and groundwater.

Like any such generalization, the statement above is over-simplified. Under special circumstances, cities may increase water storage. In arid areas, little or no rainfall reaches the water table. Infiltration occurs largely along stream channels and an increase of water in channels may increase infiltration.

Lakes. Some of the data for about 250 principal freshwater lakes in the United States, in-

TABLE 1-6. Estimated Area and Water Volume in Selected Freshwater Lakes in North America

Name and Location	Area (mi^2)	Mean Depth (ft)	Volume (mi^3)
Chelan (Washington)	55	535(?)	6(?)
Crater (Oregon)	20	1,608	6
Erie (USA, Canada)	9,940	80	150
Huron (USA, Canada)	23,010	250	1,090
Michigan (USA)	22,400	325	1,380
Ontario (USA, Canada)	7,540	330	470
Pend Oreille (Idaho)	148	400(?)	~12
Superior (USA, Canada)	31,820	487	2,940
Tahoe (California)	193	627	30
Totals	95,116		~6,084

cluding Alaska, which have surface areas of 10 mi^2 or more, have been summarized.[42] Tens of thousands of lesser lakes occur. Though no complete census is available, storage in natural lakes far exceeds that in man-made reservoirs. Table 1-6 shows the areas and storage volumes of some principal natural lakes whose storage exceeds 5 mi^3.

The average lake, aside from the natural water cycle functions noted previously, traditionally has been valued chiefly for recreational and aesthetic purposes. The water generally has not been available for withdrawal use because this would conflict with other traditional uses.

It is obvious from Table 1-6 that nearly all the natural fresh lake water in the United States is in the Great Lakes. If we add to the total shown, the storage in other exclusively Canadian large lakes, the sum is about 7800 mi^3.

Probably a few other American lakes exceed 5 mi^3 in volume, but their number cannot be large. The nation is remarkably deficient in basic data on its lake resources, probably because they have less recognized economic value than some other water resources. The Great Lakes are an exception, having tremendous economic importance in addition to their value for recreation and sources of water for municipal and industrial uses. A few other lakes serve multiple purposes also. Pend Oreille in Idaho has both sport and commercial fishing. The basin in which the lake occurs has a large, unfilled capacity. By means of a dam on the Pend Oreille River, the lake level can be raised and lowered for flow regulation and hydropower generation.

According to census data, the total inland water area of the United States, exclusive of the Great Lakes, is about 68,975 mi^2. This includes permanent water areas of lakes, reservoirs, and ponds more than 40 acres in extent, plus streams, sloughs, and canals more than one-eighth of a mile in width.

It is impossible from agglomerated data like these to estimate natural lake area, but it cannot be more than a few tens of thousands of square miles. Most individual lakes occupy only a few acres or tens of acres and contain only a few tens or hundreds of acre-feet of water. The aggregate water volume in the many thousands of lesser lakes is quite small.

Saline lakes in the United States contain relatively insignificant amounts of water. The largest is Great Salt Lake, Utah, with an average area of 1750 mi^2 and a volume of only about 6.6 mi^3. Pyramid Lake, Nevada is much smaller in area but contains about as much water. The water in saline lakes is, for the most part, not available except for recreation or, in a few cases, as a source of extractable minerals.

Reservoirs. Construction and use of dams and reservoirs has a long and generally successful history. Early dams and reservoirs were small, and the Egyptian pharaohs were probably the first to impound a large reservoir. During the XII Dynasty, Pharaoh Menes used the Fayum depression as an offstream reservoir into which excess Nile water was diverted, forming Lake Moeris in the desert 50 mi southwest of Cairo. During years of deficient flood water, stored water was led back to the valley.

Middle Eastern peoples experimented with on-stream dams without notable success. In humid climates, until rather recently, the principal use of dams was to create small hydraulic heads for turning machinery, principally with waterwheels. Invention of the hydraulic turbine and electrical power generator to replace steam engines spurred the development of dams with higher heads and reservoirs with larger capacities. In the United States, the main impetus for large dams and reservoirs early in the twentieth century was to develop water for irrigation.

Large dams and reservoirs are now a worldwide phenomenon. A dam in Kariba Gorge on the Zambesi River between Southern and Northern Rhodesia impounds the largest reservoir in the world. Kariba Dam, 420 ft high and 80 ft thick, impounds a lake having a capacity of 140×10^6 acre-ft of water. This is about 6 times the capacity of Lake Mead in the United States, the second largest man-made lake in the world.

Dams and reservoirs have many purposes: to provide water for agriculture, industry, and municipalities; navigation; flood mitigation; hydroelectric power generation; aquatic food production; recreation; and others. Reservoirs range in size from small stock ponds a few tens of feet in diameter to the giants like Kariba.

Most larger dams now have multiple purposes, whereas earlier dams commonly had a single purpose, such as power generation or water supply. As a general rule, however, it may be said that the purpose of dams and reservoirs is to regulate the flow of streams in order to equalize the supply of water or have more available when it is wanted. There are, however, definite limits to the degree of regulation that can be achieved.

As has been pointed out,[43] change in storage is a measure of the work done by a reservoir. The degree of regulation of flow by a reservoir depends on the ratio of reservoir capacity to the volume of river discharge. Langbein showed that, for any given river, water control by storage follows a law of diminishing returns:

> "Each successive increment of control requires a larger amount of storage space than the preceding increment . . . Storage capacity equal to a year's flow can regulate almost 40 percent of the flow, but doubling the capacity increases regulation only by about one-third."

Similarly, additional increments of storage will increase regulation by successively smaller amounts, and new dams become uneconomical because of the small increase of regulation.

In eastern states, this problem is not so obvious because of the relatively high cost of water—generally about \$0.40 to \$0.50/1000 gal. In the dry western states, the problem is critical because of the generally lower price of water—commonly less than \$0.01/1000 gal. As urbanization and industrialization increase in the West, this cost disparity tends to disappear, and the process of evaporation becomes critical.

In most eastern areas, precipitation exceeds evaporation, so reservoir surfaces do not introduce serious losses of water. In dry climates, however, evaporation imposes a ceiling on river regulation.

In the process of adding new reservoirs, a stage will be reached where gain in regulation is offset by increased evaporation. Beyond that point, new reservoirs will actually lead to a net loss of water. For example, Pickwick Lake, a reservoir on the Tennessee River, has an average surface area of 46,800 acres and its estimated evaporation loss is 0.12×10^6 acre-ft/yr. This is equivalent to about 31 in. Precipitation on the reservoir is in the neighborhood of 50 in., so precipitation more than offsets evaporation. In contrast, evaporation from Lake Mead on the Colorado River averages about 85 in., while precipitation is only about a tenth of that amount. With a surface area of 157,740 acres, the annual net loss of water is about 1.1×10^6 acre-ft.

Data has been compiled for principal reservoirs in the United States and Puerto Rico, including all reservoirs and controlled natural lakes with a capacity of 5000 acre-ft or more as of January, 1963.[44] The survey showed that the fifty states and the Commonwealth of Puerto Rico contained 1562 lakes and reservoirs above the minimum size; their usable storage space was 359,360,000 acre-ft, and the corresponding surface area was 14,831,000 acres. A significant number of these reservoirs are natural lakes whose storage capacity has been increased by artificial structures.

Small reservoirs are unimportant, except locally, in terms of storage capacity. From 1948 to 1964, about 3100 small reservoirs were constructed for watershed protection and flood control. Their average capacity was about 1000 acre-ft.

In 1963, 31 reservoirs in the conterminous United States (mostly in western states) had ca-

TABLE 1-7. Area and Capacity of Selected Large Reservoirs in the United States

Name	Location	Surface Area (acres)	Total Capacity (10^3 acre-ft)
Lake Mead (Hoover Dam)	Arizona–Nevada	157,736	29,827
Lake Powell (Glen Canyon)	Arizona	161,390	27,000
Oahe	South Dakota	376,000	26,630
Garrison	North Dakota	390,000	24,800
Fort Peck	Montana	245,000	19,410
F. D. Roosevelt (Grand Coulee)	Washington	79,400	9,562
Fort Randall	South Dakota	102,600	6,093
Lake Cumberland	Kentucky	63,530	6,089
Kentucky Lake	Kentucky	261,000	6,003
Lake Texoma (Denison)	Oklahoma–Texas	144,100	5,530
Bull Shoals	Arkansas	71,240	5,408
Shasta Lake	California	6,665	4,498
Sam Rayburn	Texas	114,500	4,479

pacities of 2×10^6 acre-ft or more. The largest is Lake Mead, with a total capacity of 27.2×10^6 acre-ft (about 8 mi^3). The largest man-made reservoir in the United States, therefore, is in the lower range of size of large natural lakes. None even approaches the earth's really large natural lakes. Table 1-7 lists all United States reservoirs having individual total storage capacities of 4×10^6 acre-ft or more. The total capacity of most reservoirs is somewhat larger than their usable capacity.

In an estimate of actual evaporation from water surfaces in the seventeen states that lie wholly or partly west of the 100th meridian, the water surfaces were classified as shown in Table 1-8.[45] The estimate includes natural and artificial lakes, reservoirs and ponds, surfaces of streams and canals, and "coastal waters" which include enclosed bays, lagoons, and estuaries more than 1000 acres in area.

Reservoirs and regulated lakes in these states number about 700, with a mean water surface area of about 3,446,000 acres, or about one-half of all the total freshwater surfaces. The estimate of gross evaporation from these western states is 23,641,000 acre-ft, of which 12,299,000 acre-ft is from the 51 principal reservoirs. Annual evaporation from reservoirs was about 6.4 percent of their usable storage capacity. Principal rivers of the western area evaporate nearly 45,000,000 acre-ft annually, or about 19 percent of total evaporation from the seventeen western states. The percentage of total evaporation from rivers tends to increase downstream, especially if large reservoirs are present. From the Missouri River in its long course across the Great Plains and through five large reservoirs, evaporation amounts to about 20 percent of streamflow.[45]

Reservoirs have been dealt with at some length in this section in order to illustrate that they have drawbacks as well as advantages so far as water supply is concerned. It will be noted that the first three items in Table 1-8

TABLE 1-8. Evaporation from Seventeen Western States in the United States (Data from Meyers, 1962[45])

Water Surface	Area (10^3 acres)	Annual Evaporation Rate (ft)	Annual Evaporation Amount (10^3 acre-ft)
Principal reservoirs and regulated lakes	3,446	3.5	12,299
Other lakes >500 acres	653	3.5	1,987
Small ponds and reservoirs	977	3.5	3,369
Principal rivers	1,209	3.5	4,421
Small streams and canals	457	3.5	1,565
Subtotal	6,742		23,641
Four principal interior saline lakes	1,328		5,220
Coastal bays and sounds	4,193		12,345
Subtotal	5,521		17,565
Total	12,263		41,206

amount to 17,755,000 acre-ft of water annually. This is equivalent to a steady year-round flow of somewhat more than 24,000 ft^3 sec, which would be equivalent to a moderately large river.

Runoff and Groundwater Recharge. Because evaporation, as described, includes water evaporated directly from the soil and open water surfaces along with water evaporated by vegetative transpiration, the balance of precipitated water is available for deep percolation to recharge groundwater and for surface runoff in streams. The balance for the conterminous United States is 9 in., equivalent to about 430 mi^3 of water at an average flow rate of 1,880,000 ft^3/sec.

An independent estimate[46] of river discharge to the sea is as follows:

	Annual Total	Average Rate
To Atlantic Ocean	77 mi^3	499,200 ft^3/sec
To Gulf of Mexico	190	887,400
To Pacific Ocean	107	359,400
Total	374 mi^3	1,746,000 ft^3/sec

The sum is considerably less than total runoff computed above as a residual, but the tabulation does not include discharge northward into Canada, into the Great Lakes, or directly into the St. Lawrence River. The estimate also does not include discharge of groundwater directly to the sea, but this probably is negligible compared to runoff, except in specialized local areas. The value of 430 mi^3 of runoff and groundwater discharge directly to the sea seems to be acceptable.

In addition, Alaskan streams discharge about 1,510,000 ft^3/sec, nearly as much as from the conterminous United States. Data are not available for Hawaii, where much of the discharge is by groundwater underflow.

The total length of river channels in the conterminous United States, mappable on a scale of 1:62,500, is about 3×10^6 mi.[24]

Flow data for all principal rivers of the United States and many smaller ones are readily available in reports of the U.S. Geological Survey. Several other federal agencies such as the Corps of Engineers, Forest Service, Soil Conservation Service, and Agricultural Research Service collect flow data for special purposes but do not regularly publish them. Some state agencies collect and regularly publish flow data.

As was noted earlier, information is relatively poor about the dissolved and suspended load discharged by rivers into the sea. Interest of water users, and therefore of hydrologists, has centered largely on water quality within the continent. Even there, data coverage is spotty and few records of long continuous duration are available (see Table 1-8). Many data are stored in the files of federal, state, and local agencies, but it is likely that most data have not been published.

Examples of the average dissolved solids in principal U.S. rivers are as follows:[25]

	mg/l
St. Lawrence River at Sorel	161
Hudson River at Hudson	116
Tombigbee River at Epes	155
Mississippi River at New Orleans	223
Rio Grande at Laredo	881
Colorado River, Yuma main canal	711
Sacramento River at Sacramento	152
Columbia River at Cascade Locks	125

Values in the range indicated are common among large rivers, but by no means do they cover the range of composition that may be found locally. Some streams are nearly as pure as rainwater, while others contain several thousand miligrams per liter of dissolved solids. This is equally true of springs.

The sources, quality, temperature, treatment, and cost of public water supplies in the United States and Puerto Rico are discussed extensively in the literature.[47,48,49,50,51]

The above estimate of continental discharge of surface water indicates that about 50 percent goes into the Gulf of Mexico, chiefly through the Mississippi River. Average annual discharge of the Mississippi, ninth largest river in the world, is about 140 mi^3. It ranks as eighth in terms of annual load of sediment delivered to the sea, delivering about 344×10^6 short tons. In contrast, the Amazon, largest river in the world, with a discharge nearly 10 times larger, delivers only about 400×10^6 tons, being a relatively clear river.

An estimate was mentioned previously of channel storage in rivers of the world. While the estimate has no validity as scientific data, it illustrates the importance of the hydrological cycle. Channel storage is about one-fifteenth of the estimated amount of water stored as soil moisture, and probably is equivalent to the total flow of American rivers during about 2 or 3 weeks.

Evidently, channel storage must be replaced continually. This cannot be accomplished directly by rain alone because, in terms of percentage of total time available, precipitation is a rare event. Rivers flow during rainless periods because of natural discharge of groundwater through springs, by mass seepage into river channels, and discharge from upstream areas. This contribution is called base flow. Lakes and swamps also contribute to base flow, but for the nation as a whole, groundwater provides the principal component. On the average, perhaps 50 to 60 percent of river water that reaches the sea has passed through at least one aquifer.

Storage is the principal means for time and space equilization of water supply, and traditional reliance has been on man-made lakes. However, as already noted, the law of diminishing returns sets in with the multiplication of dams on stream systems. Moreover, the capacities of surface reservoirs are small compared to those of natural underground reservoirs—aquifers.

The capacities of aquifers have received relatively little study, but their aggregate is immense. A striking example is the basalt aquifer of the Snake River in southern Idaho east of King Hill. The main aquifer occupies an area of about 12,000 mi^2, is about 200 mi in length (east-west), and exceeds 70 mi in width at some places. The conservatively estimated volume of water stored in the aquifer is about 320 mi^3. About 80 mi^3 of this water is potentially extractable and hence is analogous to so-called live storage in a surface reservoir. The Snake River above King Hill, Idaho, which is at the western edge of the aquifer, is one of the most completely regulated sizable rivers in North America. Yet all the surface reservoirs on the upper Snake River have an aggregate storage capacity of only about 1 mi^3. The sustained yield of the basalt aquifer is on the order of 6000 ft^3/sec.

Another impressive aquifer is the Ogallala Formation and associated alluvial and eolian sediments of the Southern High Plains of Texas and New Mexico, occupying an area of about 32,000 mi^2 south of the Canadian River. The estimated volume of stored water in the aquifer in 1938, before the onset of heavy pumping, was about 150 mi^3. There is a very low rate of recharge to this aquifer from precipitation as mentioned previously. The aquifer is a large but isolated blanket of sediment and receives no replenishment by underflow from adjacent areas. Hence, virtually all water withdrawn and consumed amounts to permanent depletion of the resource. The aquifer has been heavily depleted since 1938.

It has been estimated[52] that by 1967 the formation contained about 184×10^6 acre-ft of storage available for pumping. This value was based on an estimated specific yield of 15 percent for the formation as a whole. Using a total porosity value of 30 percent, the volume of the contained water amounts to 368×10^6 acre-ft, or somewhat more than 100 mi^3, and about 50 mi^3 is recoverable at 15 percent specific yield.

Another area containing vast reservoirs of groundwater is the Mississippi Embayment, a crudely triangular area whose apex is in southern Illinois and whose base is approximately along the 32nd parallel of latitude. The embayment includes about 100,000 mi^2 in the states of Alabama, Arkansas, Illinois, Kentucky, Louisiana, Mississippi, Missouri, Tennessee, and Texas. In geologically ancient times, the area was depressed periodically and occupied by a sea in which a great thickness of terrigenous sediments accumulated. Although the pores of these sediments originally were filled with seawater, much of this has been flushed out and replaced by a large reserve of fresh groundwater. Considering water to be fresh if it contains less than 1000 mg/l of DS, it is estimated[53] that the eighteen regional and several local aquifers in the embayment have a potential sustained yield of 30×10^9 gpd, while current use is only about a tenth of that amount.

In the embayment, potential sustained yield is about 10 mi^3/yr, meaning that if that amount were withdrawn it would be restored by natural recharge. The total amount of water in storage, of course is much greater. The aquifers contain freshwater at some places to depths as great as 2000 ft. Taking a conservative average thickness of 1000 ft, the volume of saturated earth in the embayment is nearly 19,000 mi^3. At 30 percent porosity, the contained volume of water is about 5700 mi^3. For comparison, Lake Superior contains about 2900 mi^3 of water.

One of the larger groundwater supplies in the United States is the aggregate of Tertiary limestones in the southeastern states of Florida and parts of Alabama, Georgia, and North and South Carolina. The principal aquifers there are limestones, which carry water under artesian pressure. These aquifers discharge water naturally through seeps and large springs, many of which are submarine (offshore). The largest onshore springs, Silver and Rainbow Springs in Marion County, Florida, have a combined flow of 2310 ft^3/sec. About 1.5×10^9 gpd of water is used in the area, chiefly from the limestone aquifer, but a much larger amount discharges naturally through springs. No tenable estimate is available of total storage in the aquifer. The principal aquifer, however, ranges up to 2000 ft in thickness and occurs in an area exceeding 60,000 mi^2. With an average saturated thickness of 1000 ft and porosity of 20 percent, the contained amount of water would be nearly 2300 mi^3.

These few illustrations admittedly are some of the larger aquifers in the United States. However, many other copious aquifers exist such as those in the Central Valley of California, in the Great Plains and High Plains, in the Atlantic Coastal Plain, and elsewhere. The point should be clear, however, that even small aquifers have large storage capacity.

Snow and Ice. The amount of water (as liquid) in deep freeze in icecaps and glaciers of the world is about 7×10^6 mi^3, or enough water to feed all the rivers of the world at their present discharge rates for about 900 years. Glaciers are not widespread within the conterminous United States, but they are much more numerous and significant than is commonly realized.

Alpine, piedmont, and valley glaciers occur in eight western states: Washington, Oregon, California, Nevada, Idaho, Montana, Wyoming, and Colorado. The total number of these glaciers is 1117 and they occur an aggregate area of about 204 mi^2. The chief concentration of these glaciers is in the northern Cascade Range of Washington.[54]

Glaciers are, in effect, nonstructural water storage reservoirs. The water yield from melting of seasonal snow is about the same as the water-equivalent of the snowpack. Glacial meltwater, however, can be much more or much less than the water-equivalent of winter snow, depending on the area and depth of the snowpack and on the area and rate of glacial melting. Glaciers of British Columbia contribute a considerable part of the flow of the Columbia River, and shrinkage of these glaciers has been supplying an extra increment of water during recent decades. Regrettably, no accurate data are available on the mass balances of these glaciers, so the magnitude of the runoff bonus is not known (M. F. Meier, personal communication, 1965).

The great importance of glaciers, like that of structural reservoirs, is regulation of streamflow. They grow and store water in winter when it is less needed for use, and release it during summer when it is most needed. An excellent example is South Cascade Glacier. Runoff from the main meltwater stream peaks in July and August, discharging 50 percent of total annual yield during those two months.

Glaciers are highly sensitive to climate. Thus, during hot, dry terms of years they melt more rapidly and ameliorate the effects of drought on water supply. During cold, wet periods the glaciers tend to grow and store water when it is less needed for use. Precipitation that is higher than normal on glaciers produces less than normal runoff because of the high albedo of snow (about 0.75) compared to that of ice (about 0.35). The net effect of these and other factors is that runoff from glaciers is regulated naturally in ways that tend to counterbalance extremes in precipitation and temperature from year to year.

Glacier effects occasionally are nonbeneficial.

Internal meltwater channels in a glacier sometimes become stopped up and water is stored in the glacier. In another case, the glacier in a tributary valley may spread across a main valley and a dam a stream. The stored or dammed water thereafter may break out suddenly, causing a flood. One flood from Nisqually Glacier at Mt. Ranier discharged about 70,000 ft^3/sec at its peak. Fortunately, outburst floods are rare.

Thousands of glaciers occur in Alaska, but relatively few of these have been studied individually and no complete census is available. The aggregate amount of water stored as ice, however, is large.

Vast amounts of permafrost create serious water supply and sanitation problems, especially where the ground is frozen to considerable depth. The greatest known thickness of permafrost is 2850 ft in eastern Siberia. Depths as great as 1000 ft are not rare. Development and use of water is very difficult in permafrost areas. Rivers are apt to freeze solid in winter and discharge liquid water during only a few weeks or months of the year. Where groundwater of suitable quality is available, it is generally the preferred source of supply in permafrost areas.

Owing to the presence of vast mineral resources in the permafrost areas of the world, human pressure on those areas will increase rapidly in the near future. These areas are peculiarly sensitive to environmental disturbances, and attempts to apply methods of development that are suitable in temperate climates may have disastrous consequences.

Concerning snow, accumulations in the high country are the greatest natural surface water reservoir in the western states. All the great river systems–Columbia, Missouri, Colorado, Rio Grande, and Sacramento–originate in high mountains. Hence, snow surveys are a major element in water supply forecasting.

Total Available Water. Within any area, total water and total available water may be two wholly different quantities. In general, the first quantity is much larger, but locally the reverse may be true, owing to reuse of water. "Available," on the other hand, is a flexible term. Owing to sparsity of population in the dry Southwest and to its density in the moist Northeast, the available per capita supply of water is larger in the Southwest.

We have noted that soil moisture is available for use only as soil moisture. Use can be varied by varying the kind of vegetation; use can be increased by increasing the soil moisture (irrigation, land-management practices); or use can be decreased by preventing or suppressing soil moisture recharge. Flexibility in the use of soil moisture, however, is very limited, though use of soil moisture is the largest single use of water.

Soil moisutre is not generally listed as a water-use category but, as will be discussed later, it is the most important single use. Soil water has the great disadvantage in that it can be used only once.

The residual of precipitated water that does not evaporate directly or through vegetation is what is generally called the water resource of an area. In a rather far-fetched sense, all surface water is available for use of some kind. A lake or river, even though utilized only for recreation, sports, and scenic attraction, is being used. In a practical sense, however, water use means multiple uses, some of which relegate recreation and sport to the background, while others make these possible where they did not exist before.

The share of the resource that is available for use depends on time, place, circumstances, economics, technology, and social and political considerations.

When a high flood is in progress in a river, the prime concern is to get the water off the land and out to sea, even though upstream source areas of the flood waters may later need the water that had to be let go.

The place where water is available may differ from the place where it is wanted. This is the reason for the well-known California Water Plan and many others.

Circumstances also may make water unavailable. In a coastal area, low streamflow during drought may allow upstream incursion of seawater much farther than usual. If an industry or city has its water intake in the zone of incursion, it may be faced with a serious problem even though the upstream supply of water is adequate. Precisely this situation threatened to develop in Philadelphia late in the drought of the early 1960s. The city made preparations to install temporary upstream intakes for fresh water. Fortunately, a plan was developed whereby streamflow was regulated so as to keep the salt water tongue below the regular intakes.

A common circumstance that interposes a wide margin between total water and available water is lack of water facilities. Many smaller communities in the Northeast had serious water supply problems even though water was obtainable but had not been tapped and therefore was not available. In 1965, somewhat more than 100 communities in the Northeast were critically short of water. Nevertheless, it has been authoritatively said,[55] "There is not and never has been an overall shortage of water for public supply in the region. Where water shortages have occurred the cause has been a shortage of adequate facilities for the collection, storage, treatment, and delivery of water." The city of Camden, New Jersey, for example, had a water shortage problem even though suitable groundwater was present but unavailable because it was undeveloped.

The matter of water unavailability for economic reasons hardly needs discussion. The cost of developing and delivering a water supply may simply exceed the value of the water for the purpose for which it would be used. On the other hand, as the economy of an area develops, it may become feasible to proceed with a formerly uneconomic development.

Technology commonly is tied closely to economics. The centrifugal pump, for example, is relatively inefficient for pumping groundwater from anything but shallow depths. Development of the efficient modern high-speed turbine pump, however, permits pumping of groundwater from depths of several thousand feet. Widespread networks of three-phase electrical power have encouraged the use of electrical pump motors instead of the less efficient steam, gasoline, and diesel engines that once were so common, especially in rural areas.

Social and political limitations on the availability of water are many and varied. The drought in the Northeast has been called a political drought because it was largely a problem of water distribution and allocation within certain political constraints. When members of the Delaware River Basin Commission and governors of the states involved reached compromise agreements on water allocations, water management, and facilities for water use, the problem was alleviated for the time being. The problem itself was soon eliminated by the ending of the drought.

The New York City water shortage during the drought was a result of a combination of factors. The city water deliveries are unmetered so the public had no incentive to conserve water. Much of the system is old and leaky and a large amount of water is wasted. The Hudson River, adjacent to the city, is badly polluted and would require more treatment than water from the Delaware River watershed. Moreover, there is public opposition to use of Hudson River water because of its polluted condition.

In general, it is an unacceptable practice to dry up a river completely, although this has been done with a few streams. In irrigated areas, much of the water applied to land is consumed by evaporation and incorporation in the vegetal produce. However, sufficient water must be applied to flush the soil and maintain a favorable salt balance. The excess water reaches the water table and in many areas reappears as base flow of streams where it is reusable, at least for some purposes. An entire river may be diverted and used in this way, but the return flow is diminished in quantity and impaired in quality.

In contrast, a hydroelectric plant may use the entire flow of a river with virtually no consumptive use. Where a regulating reservoir is involved, however, evaporation losses occur and the quality of the water may be altered by limnological processes.

Industrial and municipal water uses commonly involve little consumptive use. Water is diverted, used, and returned to rivers time after time so that, in effect, the available supply may be several to many times the total flow of the streams. In principle, therefore, all water in a lake or stream is available for use and reuse. As the water deteriorates in quality, however, it becomes a technological and economic problem whether it can be reused downstream from users who pollute it, change its chemical character, or raise its temperature. Nondiverting uses also may be in conflict. For example, it may be necessary to restrict upstream use of water in order to maintain downstream pool levels for navigation.

The subject of groundwater availability requires special mention. Not all groundwater is actually available, because not all of it can be extracted from the aquifer by pumping. Assuming a porosity of 25 percent, a cubic mile of aquifer would contain about 275×10^9 gal of water. If the specific yield of the aquifer were 10 percent, recoverable water would be 27.5×10^9 gal. Some aquifers, however, contain huge amounts of water, and even if only 10 percent can be extracted, the available amount is still very large.

Most of the large economically feasible surface reservoir sites in the United States have already been developed. In the future, therefore, natural underground reservoirs will receive increased attention. They may be used in several ways. Water may be pumped out at rates such that, over the long-term, average natural recharge equals pumpage. During droughts, aquifers may be depleted by pumping and subsequently allowed to recover during years of abundant precipitation and recharge.

The available share of water may be increased also by artificially-induced groundwater recharge. That is, excess surface water may be stored underground at times when it is not needed and pumped out at times when surface supplies are short. Alternatively, the stored groundwater may simply be allowed, by natural processes, to augment the base flow of streams. Extensive experience with artificial recharge has already been accumulated on Long Island, New York, in alluvial-fan areas of southern California, in parts of the Midwest, in the rice-growing area of Arkansas, and at many overseas locations.

Some aquifers, even though not depleted, have large unused storage capacities. Depth to water in the Snake River Plain is 100 to 500 ft beneath large areas. Each unfilled layer in this aquifer 10 ft thick has a live storage capacity (effective porosity) of at least 1 mi^3 of water, assuming a specific yield of about 5 percent.

The basalt aquifer of the Snake River Plain is physically unique, but it is not unique in being capacious. Many other capacious aquifer systems occur in the United States. Even the smaller ones, in the aggregate, provide huge storage capacity. A very crude order-of-magnitude estimate of the amount of groundwater stored in the upper half mile of the earth's crust beneath the conterminous United States is about 53,000 mi^3. This is not an accurate value, but it helps to give perspective. For comparison, large lakes in the North American continent contain about 8000 mi^3 of water, and this is about one-fourth of all the fresh surface water on the globe.

Another factor that affects the availability of water is its chemical quality. The U.S. Public Health Service set the rather low limit of 500 mg/l of total dissolved solids for desirable potable water. The U.S. Geological Survey classifies water as saline if it contains more than 100 mg/l of DS. While these definitions have a rationale, they are unrealistic for many areas. Evaluation of water supplies under those standards would lead to underestimation of water resources.[56]

Millions of people throughout the world regularly drink water containing much more than 500 mg/l of dissolved solids (DS). Some animals, such as sheep, are tolerant of salts in excess of 2000 mg/l. More than a hundred public water supplies in the United States contain 2000 mg/l or more of DS. In many places, both public and rural water supplies contain as much as 3000 mg/l of DS. In Tunisia, irrigation has been successful in suitable well-drained soils with water containing up to 6000 mg/l of DS.

Evidently, the amount of water that is available for use depends partly on the nature of the use, the circumstances of use, and the user technology. Traditionally, in much of the United States, water containing more than 1000 mg/l of DS has been regarded as a nuisance rather than an asset.

In some regions, fresh water ($<$1000 mg/l of DS) occurs at depths down to several thousand feet. In others, however, saline water is present at shallow depth or even near the land surface.

Of the total groundwater in the United States, a huge amount is saline by definition. This is misleading, however, in an assessment of usable water. Saline water, without treatment, is useful for some industrial purposes such as cooling. With treatment it can be converted to potable water. The amount of usable water, therefore, changes as technology for use advances and as the value of water increases.

The principal long-term problem with saline waters is their disposition after use. If discharged to streams, they degrade the quality of surface waters. If used for cooling and then discharged, they also impose a heat load on streams. In all desalination processes, on the other hand, the end products are (1) a stream of fresh water and (2) a stream of concentrated residual brine. Disposition of the latter is a major problems.

Possibilities for Augmenting Water Supplies. Several possibilities are available, and some are already in actual use, for increasing the absolute amount of water available in land areas. Also, what may be called the effective supply may be increased by various conservation measures. The total supply in the United States may be increased by rainmaking, desalination of salt water, and importation of water from Canada. The effective supply of water already available may be increased by evaporation suppression, control of water-loving vegetation, reclamation of wastewater, and by less profligate use of water for all purposes. None of these, singly or in conjunction with others, will be a panacea for water problems.

Rainmaking. Inasmuch as nearly three-fourths of the earth's area is ocean, the widespread occurrence of deserts is surprising unless we understand the relations among air, water vapor, land, and oceans. Molecular diffusion of water vapor in the atmosphere is negligible. Dispersal of water vapor therefore depends largely upon eddy diffusion and the circulation of the atmosphere.

Air tends to flow from cold regions to warm, but moisture tends to evaporate from warm regions and to condense in cold ones. Air movement tends to prevail, however, because of higher partial pressures. Water vapor tends also to move down thermal gradients, which explains polar ice and the tendency of rainy periods to be in the winter season in many parts of the world. Since the distribution of moisture depends on atmospheric circulation, distribution is sporadic and seems capricious, but there is no real caprice in nature. Atmospheric circulation is a complex result of the interplay of many factors: the earth's rotation, boundaries between land and water, land topography, ocean currents, local and general solar radiation, and others.

Artificial control of atmospheric circulation would require application of energy in amounts several orders of magnitude greater than any now available. So-called trigger mechanisms, such as cloud seeding, have been widely used. But since air does not contain much moisture, induced precipitation cannot be large in amount unless air currents are present or can be induced to renew the supply of moisture over the land area.

It is well established that under suitable conditions precipitation may be artificially induced within small areas. However, the specious prophecies of a decade ago about weather modification and changing of climates have been discredited.

It is not within the scope of this chapter to discuss this topic in detail. It is necessary only to note that rainmaking may confer local benefits. Inducing high-mountain precipitation may help to build up winter snowpacks and add to the sustained flow of some streams. The percentage increase of total water supply, however, would be minute. Furthermore, it would be unreasonable to expect that tampering with the weather will appreciably increase water supply for the nation as a whole or for any major river basin within the foreseeable future.

Data for an experimental cloud-seeding area of about 100,000 mi^2 in the midcontinent region indicated that during the experimental 5-year period, precipitation actually decreased by about 20 percent.[57] Others, however, cling to the hope that improved techniques of seeding will yield positive results.

Seemingly, no analysis has yet shown satisfactorily whether cloud seeding has actually caused a net increase in precipitation or only a redistribution.[59] Thus, weather modification and rainmaking remain controversial topics. Aside from technical and scientific problems, legal and political problems arise. Some states have banned cloud seeding and other tampering with the weather.

Desalination. Desalination technology is discussed in Chapter 16. Desalination as a source of augmented water supply, however, must be kept in perspective. Worldwide daily production of desalted water in 1970 was only about 300×10^6 gal. By the end of 1968, desalting plants in operation or under construction in the United States numbered 288, with an aggregate freshwater production capacity of 39.6×10^6 gpd. This was equivalent to 0.0147 percent of the national fresh water use in 1965. An increase of nearly 1000-fold would be necessary to add 10 percent to the total supply at a total cost of some billions of dollars. Actual production of desalted water (degrees of salinity vary widely) in the United States in 1969 averaged only 7.5×10^6 gpd at an average cost of \$0.80/1000 gal. This would generally be considered as a high cost compared to the average cost of water from municipal systems. However, most cost comparisons are highly oversimplified, and only by very involved computations could a realistic direct comparison be made.

The charge to the customer in some metered municipal systems is not the true cost of the water. Bond issues are generally voted to finance a water works system. Householders commonly have to pay a fee to connect into the system. In at least some cases, the charge for water is not sufficient to cover the cost of production and delivery plus amortization of plant and facilities. In some cases water supply is subsidized by taxes which are not identified with water costs.

Considerable confusion prevails about the difference between cost and price paid for water production and delivery. For example, commonly quoted figures for the average cost of municipal system water in the United States are \$0.30 to \$0.40/1000 gal. This is misleading, however. Domestic water use accounts for only about 1 percent of all water use in the United States. The principal two users of water are industry and irrigation agriculture. Most industrial use of water is for cooling, and the prices paid for this use range from about \$0.05 to \$0.11/1000 gal. Higher prices may be paid (\$0.12 to \$0.28/1000 gal) where the volume required is small to moderate. The prices paid for irrigation water are in the range of less than \$0.01 to several tens of cents per 1000 gal.

For these and other reasons, it is a gross error to compare the cost of desalinized water with \$0.30 or \$0.40 municipal water. According to one estimate,[60] on a national basis, any competitive means for increasing water supply must do so at a cost of \$0.10 to \$0.20/1000 gal. In fact, however, there is no rationale for comparisons on a national basis. Comparisons should be made on the basis of costs and prices at the place where water is wanted. The world has about 20,000 mi of arid coastline. Coastal areas are likely candidates for desalination of seawater, especially for municipal supply.

It is also pointless to argue whether water can be desalted at costs competitive with those for fresh water. This has been done at some places for many years. Desalted water is competitive where it is cheaper to produce than to obtain fresh water by other means.

Annual streamflow in the United States averages about 1.2×10^{12} gpd. Fresh water withdrawal use amounts to about 25 percent of the total. But consumptive use is only 78×10^9 gpd, or about 10 percent of the total supply. Although many areas are short of water, it is obvious that the nation as a whole is well-endowed. Ninety percent of its water supply serves chiefly to carry wastes out to the sea. It is equally obvious that even a small increase in the efficiency of water use could increase the effective supply immediately in an amount far greater than the addition by desalination over a long period of time.

This discussion, though peripheral to the main topic of hydrology, seems warranted because the nation has vast reserves of saline waters. Formerly considered as a nuisance (as they always will be at some places), they are now recognized as a resource. As previously noted, waters with a salinity of 1000 mg/l or more underlie fresh water throughout much of the United States. The depth to saline water is less than 500 ft below the land surface in somewhat more than one-third of the area of the conterminous United States. Saline water as a resource has been thoroughly discussed in a symposium volume[61] which describes a variety of ways in which saline water and saline aquifers may be used directly and indirectly.

Currently, interest is considerable in underground storage or disposal of industrial and other wastes. Some writers and investigators tend to consider that any aquifer containing saline water is potentially available as a waste sink. The symposium volume demonstrates that this use of saline aquifers must be weighed relative to many other possible uses.

Economics of management of the residual or waste liquid from desalination is apt to be the controlling factor in use of saline water resources. To pump fresh water long distances inland from coastal desalinization plants will be prohibitively costly, especially because the water must be lifted to elevations of hundreds of feet or more. On the other hand, if inland sources of saline water are used, the cost of pipelines to dispose of the saline residue at sea may be excessive.

Importation of Water. During most of human history, when water supplies became unsatisfactory or inadequate, people migrated to new locations. During the twentieth century, advancing engineering technology has made it possible to avoid exporting people, but rather to export the problem. New York City exported its water problem to the Delaware River basin. Southern California exported its problems to the Colorado River basin, and more lately to northern California.

For years some planners have dreamed of tapping the Columbia River system to export water to the Southwest. Still others have viewed the vast water resources of Canada and part of Alaska, proposing ideas to divert water as far east as the Great Lakes region and as far south as the Sonoran Desert of Mexico.

These ideas (they have not advanced to the stage where they could be called plans) are too complex to be discussed here, but some general implications merit attention.

During some 5000 years of history, engineers have been able to accomplish most of the tasks set before them. It seems logical to assume, therefore, that they could deliver water to the Southwest and to the Sonora Desert from sources as far away as the Yukon and Fraser Rivers of Alaska and Canada. Whether this would be economically justifiable is a wholly different question which would require years of study. The social desirability or acceptability of the action is a still different and probably more difficult problem. International political problems might be insurmountable. The most thorny problem, however, would be the ecological effects. These might be disastrous. The delicate heat balance regime of the Northern Hemisphere, especially in the arctic and subarctic zones, would be altered. This, coupled with the activation of similar grandiose ideas that have been entertained in the Soviet Union, might alter the climate of at least the Northern Hemisphere. Knowledge is not sufficient to predict the climate change, but it might be for the worse. Ecological effects would be almost equally unpredictable.

Redistribution of water on the scale already practiced has and will produce unforseen and unwanted side effects. Redistribution on ever larger scales may lead to tragedy or it may lead to benefits. Our lack of knowledge of large-scale phenomena and effects is appalling.

Evaporation Suppression. Suppression of evaporation was mentioned briefly in a previous section. Evaporation from tanks and small ponds or lakes may be suppressed by roofing them over with floating rafts or with plastic sheets. For larger water bodies, chemical suppressants seem to hold the most promise.

Although high suppression rates have been reported by some laboratory workers, field results are less encouraging. A 20 percent reduction in evaporation under field conditions from a large water body would be phenomenal.

Various chemical substances, chiefly in the cetyl alcohol group, have been used to form monomolecular films on water surfaces. Chemical films actually suppress evaporation markedly, but the chief problem is to maintain the film. Wind piles it up against the lee shore and waves break it up. On reservoirs open to recreational boating, boats constantly break up and scatter the film. Plankton organisms also thrive on the substance, which requires regular replacement. Reduction of evaporation by about 9 percent has been reported for Lake Hefner (2587 acres) in Oklahoma. However, the cost of maintaining the suppressant film on large lakes has been high. Some observers believe that improved technology within a few years may reduce costs to about $10/acre-ft of water saved. Evaporation suppression raises water temperature and in some cases the rise may be detrimental.

Evaporation suppression does not offer much hope of materially augmenting water supplies unless or until methods are found that are effective on large lakes and reservoirs since smaller ones form only a small component of total water supplies.

Control of Vegetation. In general, use of water by vegetation is beneficial except where water is scarce and the vegetation itself is undesirable. The latter two conditions are widespread in the seventeen states. Hydrophytes (plants which live in marshes or open water) and phreatophytes (which live where their roots can reach the water table), may be grouped together as water-loving vegetation. Hundreds of species belong in these groups, but about a score of species are the worst offenders.

Water-loving vegetation of nonbeneficial species occupies more than 10 million acres in the western states and is said to consume about 25 million acre-ft of water annually. This estimate may be somewhat high. These plants occur along river channels and canals, around lakes and reservoirs, and on deltas and flood plains. It has been estimated that in the Copiapó Valley of Chile, about one-third of the water supply is consumed by nonbeneficial vegetation.

The literature on water-loving vegetation is extensive and much experimental work has been done on its control and eradication. Efforts have also been made to substitute beneficial vegetation. Eradication and control have met with some measure of success, but in general the problem of phreatophytes is largely unsolved. One of the worst offenders is a tree, salt cedar (Tamarisk), which was introduced to North America from abroad some time before the middle of the nineteenth century. About forty species occur in the United States, of which only two or three species are really significant as phreatophytes. They have spread throughout the West and Southwest, chiefly south of the 37th parallel and below an altitude of 5000 ft.

A field study[62] near Yuma, Arizona showed that at that latitude and temperature regime, with a long growing season, a phreatophyte such as arrowweed (*Pluchea sericea*) may use up to 96 in. of water annually with the water table at a depth of 5.5 ft. An intensive recent study of water use by phreatophytes and results of their control has been made in an area of 6000 acres along a 15-mi reach of the Gila River in Arizona. Final results of the study are not available. It seems likely that groundwater management, by denial of water, may be the most effective low-cost means for phreatophyte control.

Water-loving vegetation is abundant in eastern states also, but the problem there is less acute because of the greater abundance of water. In the East, one of the principal problems is hydrophytes such as milfoil and water hyacinth, which grow directly in the water. Milfoil is especially widespread in shallow water where, at some times and places, it is a serious obstruction to navigation and recreational use of brackish water.

Reclamation of Wastewater. The Colorado River is the only large river in the United States whose flow has been seriously depleted. This arises from high evaporation rates from large reservoirs, high consumptive use, low return flow from irrigation, and diversions to southern California. On the whole, however, consumptive depletion of rivers is small. Many so-called water shortages are shortages of water of suitable quality. The principal solution to this is more effective management of water—reduction of pollution and improved technology for water purification. The ever-increasing variety of exotic toxic and polluting wastes in water will make heavy demands on reclamation technology.

Less Profligate Use of Water. With growth of population, agriculture, and industry, use of water is bound to increase. As will be seen in a later section, however, not only has the absolute amount of water use increased, but the *per capita* use has increased. This cannot go on indefinitely because the supply of water is finite and most of the cheaper sources of water have been developed already. New supplies will be more expensive.

New household appliances such as dishwashers, clothes washers, and others have greatly increased household use of water. Economies could be effected there. However, domestic use of water is not a major share of total water use. Agriculture and industry are the principal water users and they will have to make the main economies.

Innovations in Water Management. Various practices in water management, while not all new, would be innovative if they became widespread. They would not necessarily add to the total supply of water, but they would add to the effective supply by lowering demand.

One large use of water is for cooling, and the potential demand for the thermal-electrical power industry alone is staggering. This could be alleviated by more effective cooling towers and cooling ponds. These involve more loss of water by evaporation than once-through use of water, but they reduce thermal loading of water supplies. Air cooling rather than water cooling also can reduce water demand.

Many industries can reduce water demand by purification and recycling of water, by change in process operations, or simply by elimination of wasteful practices.

Many examples are available of water saving, along with dollar saving, through more efficient use of water. According to the October 5, 1959 issue of "Water Newsletter," General Motors Corporation launched a program several years earlier to conserve and control its use of water. The United States and Canadian plants of General Motors used more than 70 billion gal/yr. The conservation program consisted of four phases: (1) inspection of plant utilities, (2) instruction of personnel on the need and benefits of water conservation, (3) installation of safeguards against leaks and faulty controls, and (4) a study of methods for designing future facilities with an eye to water conservation.

According to the news article, the company had been saving $5000 a month in water bills through a simple change in the nozzle size at one plant. Water requirements were reduced by 1 million gpd in a supply obtained from an overtaxed municipal system. This was accomplished by replacing manual controls with automatic water controls at one point. At another plant, a change in process operations produced a significant reduction in discharge of wastewater.

According to news reports, a steel mill in California developed a modern recirculation system that reduced its water requirements for processing a ton of steel from 40,000 to 1600 gal. A soap manufacturing plant reduced its daily water needs from 6,000,000 to about 400,000 gpd.

In most systems, the charge per thousand gallons of water decreases for successive increments of use. The effect is to encourage profligate use of water. Water use could be reduced by revision of rate structures, possibly under the control of an agency analogous to those that control rates for power and communication.

It seems obvious that no panacea for water problems is in sight and that none will arise. The nearest thing to a panacea will be a radical change in human attitudes and ways of managing water. Water can no longer be treated as a free good. Rainmaking may add a few percent to total water supply in some areas, but a significant increase in the total national water supply would be impossible without changing the climate of the continent and hence of the world.

Desalination must be kept in perspective. The major limitation on desalination other than costs will not be limitation on the available supply of saline water, but limitation of energy supply and disposal of waste heat and residual brine. Salt water differs from fresh water thermodynamically by the free energy of the dissolved salt. For average seawater, the vapor pressure of the brine is about 2 percent less than that of pure water at 25°C. The free energy difference is only about 13 cal/mole. However, this small difference is very significant when considered as the decrease in free energy where fresh water enters the ocean. It is equivalent to descent over a waterfall about 1000 ft high. Reversal of the process (desalination) requires application of at least the same amount of energy. In actual conversion processes this amount is exceeded considerably because of thermodynamic inefficiency.

For most conversion processes, the theoretical maximum efficiency of converting energy to useful work is in the neighborhood of 50 percent. With that degree of efficiency, many processes would be economical in terms of energy consumption. About 25 percent efficiency, however, seems to be the practical limit. The reason for this is that every gain in efficiency entails an increase in the cost of equipment. Above about 25 percent, the cost of equipment more than offsets the reduction of energy cost.

The United States may be on the verge of an energy crisis but this may be averted or alleviated for the time being by a major overhaul of management and planning in the power-producing industry and by more conservative use of power.

Waste heat also will be a major problem. All processes of conversion, whether by distillation, chemical processes, ion exchange, or others, require an input of energy and this energy ultimately becomes waste heat. The ultimate sink for the heat is the atmosphere, whether the heat is put there directly by radiation or air cooling or by dumping warmed water, from which the heat enters the atmosphere. The world-wide rise in ambient temperature from all heat sources from any conceivable population probably would be insignificant.[63] Local thermal effects would be a much greater problem. These effects are already serious problems in some areas and some industries.

The foregoing remarks lead up to two principal points: (1) technology will not add a significant amount, percentage-wise, to the total water supply of the nation; (2) therefore, future water demand must be met by supplies already at hand. This can be accomplished only by more rational and efficient use of water. For that purpose, it is important to look not at the sky and the sea, but at the ground beneath our feet.

Aquifers as Reservoirs. The great water-holding capacity of aquifers has already been illustrated. Recognition of the importance of aquifers is becoming widespread. Traditionally, groundwater and surface water were developed independently of each other and they were regarded as two separate resources. Now that the importance of groundwater has received wide recognition, there is a tendency to regard it as a new and added source. This is not true. We have a single resource, some of which happens to be underground in natural reservoirs. The concept of underground rivers is erroneous except in special geological situations such as cavernous limestone and Hawaiian-type basalt. Most groundwater occurs in the pore spaces of ordinary rocks and sediments. A water-saturated box of sand is a good illustration of a common type of aquifer.

As was noted earlier, on the regional and continental scale most of the water yield of a continent or region reaches the sea in surface streams. Much of this water has passed *through* aquifers, which serve the same storage and regulatory functions as lakes and surface reservoirs. Thus, groundwater is not a new or additional source of water. Aquifers, as reservoirs, merely increase the possibilities for water storage and multiply the alternative choices in water management.

The previous estimate that the amount of groundwater stored in the upper half mile of the earth's crust beneath the United States is about 53,000 mi^3 has no value or significance except to illustrate the tremendous storage reserve that exists and the rich possibilities that are available for better water management. For example, aquifers provide more choices to the water manager on when and where to store or use water.

The sustained yield of an aquifer is the amount of water it can yield, on the long-term average, that will be replaced by recharge. The sustained yield can be increased by artificial (man-induced) recharge.

Pumpage of groundwater in the United States nearly doubled between 1950 and 1965. In 1970, the latest year for which reliable estimates are available, aquifers supplied about 69×10^9 gpd, about 17 percent of all withdrawals in the United States. Much of this water will not be replaced by natural recharge in the arid and semi-arid western states.

Use of aquifers as reservoirs, including artificial recharge, implies (1) a source of water available for recharge and (2) ability to put the water underground. Both are knotty problems. Diversion has been proposed of water from the Mississippi River to recharge the Ogallala aquifer in the Southern High Plains, but this raises difficult political, social, and economic problems. Moreover, experimental recharge studies in the High Plains have been only mildly encouraging. In much of the area, the subsoil does not readily transmit water downward to the water table. Recharge through wells is difficult because of clogging by sediment and bacterial action.

Nevertheless, the need for artificial recharge is urgent at many places and development of effective methods under varied conditions is essential. The concept of using surface reservoirs to stabilize flow must be extended to subsurface reservoirs. All but the most highly regulated streams flood at times and for many streams the discharge during a flood is a considerable percentage of total annual yield. Flood waters are an important potential source of water for artificial recharge.

Some of the more intensive studies of induced recharge in the United States which have been made on Long Island, in Arkansas, the southern High Plains, and southern California are identified in a comprehensive annotated bibliography on artificial recharge.[64] This topic has special importance for sanitary engineers because artificial recharge with reclaimed sanitary wastewater is successfully practiced now at a few places and this practice is bound to grow.

Use of geothermal power along with conversion of saline water is an inadequately studied possibility for helping to meet both the energy and water supply demands. A nontechnical summary of the status and prospects for use of geothermal energy appears in Ref. 65.

For most parts of the world, the increase of temperature with depth in the earth is about 2°F per 100 ft. In areas of recent volcanic activity, or where highly heated rocks have been brought to relatively shallow depth by tectonic movements, the geothermal gradient is much steeper. Where the surficial crust is porous, contained water may have temperatures in the range of several hundred to 800°F. An example of such an area is the Salton Trough, part of which is occupied by the Salton Sea in southern California. The trough is 20 to 90 mi wide and 150 mi long—actually a filled-in northward extension of the Gulf of California. In the trough, porous sedimentary rocks extend to depths as great as 20,000 ft.

Water remains liquid even at elevated temperatures because of the pressure of the overburden. The amount of saline water in the Salton Trough has been variously estimated at 3 to 10×10^9 acre-ft. Holes drilled into the salaquifer provide pressure relief and an avenue of escape so that a drill hole becomes a source of mixed steam and water under high pressure. The water and steam can provide three resources: steam to drive electrical turbines; heat to convert saline water to fresh; and residual minerals, some of which might have commercial value. Power generation is of special interest because of growing power demand and because generation would be accomplished without burning fuels that pollute the atmosphere. Desalination is of special interest because of the source of free heat. The value of mineral byproducts is not evident. It has been estimated that the supply of hot water in the U.S. part of the trough would supply annually about 20,000 Mw of electricity and 5 to 7×10^6 acre-ft of water during at least several decades and perhaps for a much longer period. These estimates have yet to be verified.

Development of geothermal resources in the Salton Sea area has been chiefly in the Buttes Thermal area, southeast of the Salton Sea, involving fifteen geothermal wells. Development has been hindered by lack of sites for disposal of brine, bittern wastes, and accumulated salt deposits.

Geothermal areas occur extensively in other parts of the western states—perhaps several million acres. The Salton Trough area extends southward into Mexico where active development is already in progress.

Use of geothermal energy is not new. It has been practiced during many years in Iceland, Italy, New Zealand, and other countries. Reportedly, the geothermal energy potential of the Soviet Union exceeds all its other energy sources combined, but this potential may be exaggerated. The United States has not been aggressive in using this resource, although it has been developed at a few places such as the Geysers, 90 miles north of San Francisco, and a few other localities. In the Salton Trough, the potential for development varies from place to place. In Mexico, on the flank of the trough, salinity is sufficiently low that corrosion can be controlled. In the California area, salinities are commonly 10 or more times that of seawater, and the water is so corrosive that drilling tools and casing simply disintegrate. Here again, we find that the idea of a panacea is illusory.

FUTURE WATER NEEDS

Rise of living standards and increase of technology for the control and use of water combine to increase the per capita use of water. In terms of weight, water is by far the largest resource used by man. In industrial nations, per capita tonnage use exceeds by a large multiple that of all other resources combined.

Where water is abundant, it is economically sound to use it copiously in productive processes. Estimates of future water needs have become very fashionable during recent decades. Many of these estimates are quite naive. They consist essentially of plotting historical data on population growth and rising per capita use of water. The curve derived is then projected into the future. Such estimates take so little account of other pertinent factors that they are meaningless. Meaningful estimates must take account of sociologic, economic, technological, and other factors.[66]

It has been said that the conterminous U.S. land area, if populated at the same density as Belgium, would support a population of 2×10^9 people but that the water supply would support only 1×10^9 people. Such a statement is utterly pointless. Withdrawal uses of fresh water in the United States amount to about 1800 gpd/person. At the same rate, 1×10^9 people would require withdrawal of 1.8×10^{12} gpd. Streamflow averages only 1.2×10^{12} gpd. Even allowing for repeated use of much water, supply would not be adequate for reasons that seem obvious. On the other hand, that level of water demand will never be reached. Long before the population reached 1×10^9 (1 billion), the mineral and other non-water resources would be so depleted that the national economy could not support the necessary water developments.

How much water is needed to maintain one human life? It is much more than his household use because it includes water for crops, domestic animals, industry, and all other adjuncts. Direct withdrawal use of liquid water is only a small part of the total. Evaporation from the whole land area actually is part of water use because this evaporation is from precipitation that is necessary to maintain the land as we know it and the products we reap from it. For example, the amount of water required to raise a steer is not just the water he drinks. It includes the vastly greater amount of water used to grow his feed.

The 21 in. of rainfall on the conterminous United States that evaporates amounts to an average of about 2.87×10^{12} gpd. Divided by population (200×10^6), this gives a daily per capita use of somewhat more than 14,000 gal. The 1800 gpd/person withdrawal use, therefore, is only a small fraction of the water actually used *for*, if not by, each individual. The question, how much water is needed to maintain one life, therefore, is highly pertinent but to answer it is extremely difficult. What kind of a life? At what standard of living? With what amenities?

Estimates of future water needs for the most part have in back of them the unspoken assumption that water is the item that will place the ultimate limit on population growth. This assumption probably is erroneous. Depletion of other essential resources probably will reverse the upward spiral of living standards and hence the upward trend of per capita water use will halt and perhaps decline. Other factors probably will limit development long before water shortage. In the opionion of some analysts, the United States is already overpopulated in relation to its standard of living and total non-renewable natural resources.

Estimated use of fresh water in the fifty states and Puerto Rico in 1970 (the latest year for which reliable estimates have been published) was 320×10^9 gpd. This was withdrawal use and does not include water used to generate hydroelectric power. In addition, about 54×10^9 gpd of saline water was used, chiefly for industrial cooling. Total water use in 1970 was about 19 percent greater than in 1966. Neither the annual percentage increase in total use nor per capita increase is constant. (For detailed analyses and projections of water-use data, see Refs. 67 and 68.)

During the past decade many attempts have been made to estimate water demand by 1980, 2000, 2020, and even further in the future. These have been based on population trends, water-use trends, trends in the national economy, expected developments in technology, and other factors. Some estimates of water needs for supply and waste-dilution by the year 2000 exceed 900×10^9 gpd. Such projections (they cannot qualify as predictions) are highly conjectural. Moreover, reduction of pollution by dilution may be controlled by legislation. However, barring some catastrophic halting of population growth or stalling of the economy, it is certain that the water problem will be very serious. The demand estimated for the year 2000 under present practices would require active storage capacity of about 5 to 10 times the present surface storage capacity, which is about 400×10^6 acre ft (130×10^{12} gal). It is unreasonable to expect that surface storage of that magnitude can be developed.

While total stream discharge to the sea might be regarded as a measure of water available for use beyond present consumptive use, the situation is not that simple. Almost two-thirds of streamflow is discharged by streams at flood stage during relatively short periods of time. Only about a third of the total is actually available during most of the year.

Increased future water use will be heavily dependent upon storage reservoirs. In a 1966 analysis of water shortage requirements,[69] it was calculated that the maximum net supply that could be made available for 98 percent of years is 965×10^9 gpd, for which a storage capacity of 3.6×10^9 acre-ft would be required. It was estimated that an assured net supply of 922×10^9 gpd could be made available for 95 percent of years with 2×10^9 acre-ft of storage. These estimates were based solely on surface reservoir storage, although the authors called attention to the alternative possibility of underground storage at some places.

Probably at least a third of the estimated amount of water would have to be used for navigation, salinity control, and disposal of wastes, despite control of waste discharge. This would leave a balance of 615 to 643×10^9 gpd for consumptive use, compared to estimated current consumptive use of 310×10^9 gpd.

Reservoirs do not increase total water supply. They actually deplete it to some extent by increasing evaporation. Reservoirs can increase only the usable supply of water. Considering the practical limits of streamflow regulation by surface reservoirs, it has been concluded that the limit of continuous regulated yield is about equal to median natural flow, regionwide.[68] Nationwide, somewhat more than half of natural streamflow will be perennially usable. Evidently there is divergence of opinion about the amount of water that can be made perennially available.

Large-scale (regional or continental) water phenomena almost defy imagination. One cannot visualize 7×10^{15} kg (7.5×10^{12} tons) of water leaving the United States in vapor form each year, but the numbers afford a means for thinking about it. Owing to equality or excess of precipitation over evaporation in the eastern United States, evaporation from eastern reservoirs traditionally has been considered to be relatively unimportant in water use studies. Nevertheless, evaporation is a water loss in any climate.

One aim in site selection for reservoirs is to have a low ratio of area to depth of water, thus minimizing the percentage of evaporation. However, in a flat prairie area such as Illinois, the average depth of water supply reservoirs is a mere 10 ft, and in most of these half of the reservoir capacity is in the top 3 ft.

Average annual evaporation in Illinois is nearly 3 ft, and most reservoirs would fail in times of drought if the factor of evaporation were not seriously considered in the plans. During the drought of the early 1950s, thirty-nine Illinois communities had water shortages because evaporation far exceeded runoff.[70] Many cities that curtailed water use found that daily evaporation from their reservoirs exceeded by 2 to 3 times the amount of water normally used by their communities.

The obvious alternatives to surface storage are massive reclamation and reuse of wastewater and extensive use of underground storage. The technology of waste treatment will be discussed in later chapters of this book. Technology is or will be equal to the task of purifying almost any kind of waste. In some cases this can actually be done at a profit by recovering useful materials. It must be remembered that many, if not most, waste materials are wastes only because our society is profligate and disdains to recover them until obliged to do so.

Use of aquifers—natural underground reservoirs—seems to hold great promise. Their tremendous storage capacity has already been noted. Much capacity beyond that occupied naturally by groundwater is available in addition. This possibility has received little study and no capacity data are available for the nation as a whole.

Rational answers to local water problems are possible only within a broad framework of water facts covering wider areas. For many areas, development has already outreached information. Consuming and polluting upstream uses in many regions deny good water to downstream areas. Ultimately, it may be necessary to resort to broad-scale imaginative zoning.[68] For example, uses requiring high purity of water could be given priority over other uses, or nonconsuming and nonpolluting uses could be given priority in sequence or in places of use.

Conclusion. Attempts to forecast future water needs generally are unrealistic because they are usually based on traditional modes of water use and traditional methods of planning. Little has been done except in isolated cases, for example, to improve the efficiency of water use and reuse. A new approach toward water and planning is necessary or the situation will indeed be serious by or before the end of this century.

The river-basin as a unit for planning is a rather deeply entrenched concept. A river basin is a natural hydrological unit, but the river basin approach presupposes that water is the governing element in development and hence in planning. This clearly is not true. Growth always has been concentrated where there is an abundant and diverse endowment of natural resources, a supply of skilled labor and back-up facilities, and a favorable social and political climate.

Water is not the governing element in development because few problems are solved by water alone. Water-rich Appalachia is a dedepressed area. Flourishing southern California is water-deficient. One of the more significant properties of water is its mobility. Since prehistoric time this property has permitted environmental manipulations on ever-increasing scales. This has culminated in the bad environment that is characteristic of many inhabited areas of the world today. The great problems of the future will be restoration and preservation of acceptable environmental quality. Water will be prominent but not the dominant factor.

Finally, planning choices should no longer be made solely on a traditional benefit-cost ratio. Ecological impacts must also be considered. First, however, agreement must be reached on what is ecologically sound. The "return to nature" cry of some environmentalists is wholly impractical and further environmental modifications are inevitable. In the long run, ecology and economics must be brought into harmony. What is ecologically unsound inevitably will be economically unsound, and *vice versa*. Both economic and ecological systems are changeable and eventually the two must be congruent.

REFERENCES

1. Nace, R. L., "Scientific Framework of World Water Balance," UNESCO Technical Papers in Hydrology, No. 7 (1971).
2. Sellers, W. D., *Physical Climatology*, University of Chicago Press, pp. 11-39 (1965).
3. Nimmo, W. H. R., "The World's Water Supply and Australia's Portion of It," *Journal Institute of Engineers Australia* **21**, 40 (1949).
4. Oort, A. H., "The Energy Cycle of the Earth," *Scientific American* **223**, 54-63 (1970).
5. Miller, D. H., "The Heat and Water Budget of the Earth's Surface," *Advances in Geophysics*, Vol. II, Academic Press, Inc., New York, p. 212 (1965).
6. Laevastu, T., Clarke, L., and Wolff, P. M., "Oceanic Part of the Hydrological Cycle," World Meteorological Organization Reports on WMO/IHD Projects, Report No. 11 (1969).

7. Rasmusson, E. M., "Atmospheric Water Vapor Transport Over North America. Part 1. Characteristics of the Water Vapor Flux Field," *U.S. Dept. of Commerce Monthly Weather Review* **95** (7), 403–406 (1967).
8. Kohler, M. A., Nordensen, T. J., and Baker, D. R., "Evaporation Maps for the United States," U.S. Weather Bureau Technical Paper No. 37 (1959).
9. Nace, R. L., "Hydrological and Related Data Programs in the United States of America," *Proceedings, CENTO Seminar on Evaluation of Water Resources with Scarce Data, Tehran*, pp. 87–101 (1970).
10. Weiss, L. L., and Wilson, W. T., "Precipitation Gauge Shields," *Transactions of the International Association of Scientific Hydrology, Toronto* **1**, 462–484 (1958).
11. Corbett, E. S., "Measurement and Estimation of Precipitation on Experimental Watersheds," *Forest Hydrology*, Pergamon Press, New York, pp. 107–127 (1967).
12. Péguy, C. P., *Précis de Climatologie*, Masson et Cie., Paris, p. 189 (1961).
13. Lull, H. W., "Ecological and Silvicultural Aspects," *Handbook of Applied Hydrology*, McGraw-Hill, New York, pp. 6–8 (1964).
14. Storey, H. C., "Increasing Water Yields by Forest Management," *Science* **62**, 109 (1960).
15. Zinke, P. J., "Forest Interception Studies in the United States," *Forest Hydrology*, Pergamon Press, New York, pp. 137–161 (1967).
16. Miller, D. H., "The Energy and Mass Budget—A Means for Analyzing Changes Caused by Technology in the Surface of the Earth," Moscow Institute of Geography, U.S.S.R. Academy of Sciences, p. 4 (1969) (xerox print). Published in Russian in *Izvestiya Akad. Nauk U.S.S.R., Seriya Geographicheskaya* No. 2, pp. 118–133 (1971).
17. Seaburn, G. E., "Preliminary Analysis of Rate of Movement of Storm Runoff Through the Zone of Aeration Beneath a Recharge Basin on Long Island, New York," USGS Professional Paper 700-B, pp. 196–198 (1970).
18. Theis, C. V., "Amount of Groundwater Recharge in the Southern High Plains," *Transactions of the American Geophysical Union, 18th Annual Meeting*, pp. 564–568 (1937).
19. Perrault, Pierre, *De l 'Origine des Fontaines,* Pierre le Petit, Paris (1674).
20. Fischer, W. A., Davis, D. A., and Sousa, T. A., "Freshwater from Springs of Hawaii from Infrared Images," USGS Hydrologic Investigations Atlas HA-218 (1966).
21. Papakis, N., "Karst Groundwater Investigations in Greece," International Conference on Water for Peace, Washington, D.C., Abstract No. A/64 (1967).
22. Manheim, F. T., "Evidence for Submarine Discharge of Water on the Continental Slope of the Southern United States, and Suggestions for Further Research," *Transactions New York Academy of Sciences, Series II* **29** (7), 839–853 (1967).
23. Nace, R. L., "World Water Inventory and Control," *Water, Earth and Man*, Methuen and Co., Ltd., London, pp. 31–42 (1969).
24. Leopold, L. B., "Rivers," *American Scientist* **4,** 518–519 (1962).
25. Livingstone, D. A., "Chemical Composition of Rivers and Lakes," USGS Professional Paper 440-G (1963).
26. Holeman, J. N., "The Sediment Yield of Major Rivers of the World," *Water Resources Research* **4** (4), 737–747 (1969).
27. Beverage, J. P., and Culbertson, S. K., "Hyperconcentrations of Suspended Sediment," *Proceedings of the American Society of Civil Engineers, Journal Hydraulics Division* **HY6**, 117–128 (1964); USGS Bulletin 1181-A (1963).
28. Langbein, W. B., and Schum, S. A., "Yield of Sediment in Relation to Mean Annual Precipitation," *Transactions of the American Geophysical Union* **39**, 1076–1084 (1958).
29. Vice, R. B., Guy, H. P., and Ferguson, G. E., "Sediment Movement in an Area of Suburban Highway Construction, Scott Run Basin, Fairfax County, Virginia, 1961–64," USGS Water Supply Paper 1591-E (1969).
30. Wolman, M. G., and Schnick, A. P., "Effects of Construction on Fluvial Sediments, Urban and Suburban Areas of Maryland," *Water Resources Research* **3** (2), 451–465 (1967).
31. Hack, J. T., "Submerged River System of Chesapeake Bay," *Geological Society of America Bulletin* **68**, 817–830 (1957).
32. Cronin, L. E., "The Condition of the Chesapeake Bay," *Transcript 32nd North American Wildlife and Natural Resources Conference*, pp. 137–150 (1967).
33. Williams, J. R., "Groundwater in Permafrost Regions—An Annotated Bibliography," USGS Water Supply Paper 1972 (1965).
34. International Commission on Snow and Ice, "Combined Ice, Heat and Water Balances at Selected Glacier Basins," UNESCO Technical Papers in Hydrology, No. 5, Paris (1969).
35. Green, C. R., "Arizona Statewide Runoff," University of Arizona Institute of Atmospheric Physics, Technical Reports on the Meteorology and Climatology of Arid Regions, No. 7 (1959).
36. Stidd, C. K., "Local Moisture and Precipitation," University of Nevada Desert Research Institute, Preprint Series No. 45 (1967).
37. Wilson, R. G., "Methods of Measuring Soil Moisture," National Research Council of Canada, International Field Year for the Great Lakes, Technical Manual Series No. 1 (1971).
38. Kutilek, M., "Direct Methods of Soil Moisture Estimation for Water Balance Purposes," World Meteorological Organization Reports on WMO/IHD Projects, Report No. 14, Geneva (1971).
39. Veihmeyer, F. I., "Evapotranspiration," *Hand-*

book of Applied Hydrology, McGraw-Hill, New York, pp. 11-1 to 11-38 (1964).
40. Hounam, C. E., "Problems of Evaporation Assessment in the Water Balance," World Meteorological Organization Reports on WMO/IHD Projects, Report No. 13, Geneva (1971).
41. Rantz, S. E., "Urban Sprawl and Flooding in Southern California," USGS Circular 601-B (1970).
42. Bue, C. D., "Principal Lakes of the United States," USGS Circular 476 (1963).
43. Langbein, W. B., "Water Yield and Reservoir Storage in the United States," USGS Circular 409 (1959).
44. Martin, R. O. R., and Hanson, R. L., "Reservoirs in the United States," USGS Water Supply Paper 1838 (1966).
45. Meyers, J. S., "Evaporation from the 17 Western States," USGS Professional Paper 272-D, pp. 71–100 (1962).
46. Wilson, A., and Iseri, K. T., "River Discharge to the Sea from the Shores of the Conterminous United States, Alaska, and Puerto Rico," USGS Hydrologic Investigations Atlas HA-282 (1969).
47. Durfor, C. N., and Becker, E., "Chemical Quality of Public Water Supplies of the United States and Puerto Rico, 1962," USGS Hydrologic Investigations Atlas HA-200 (1964).
48. Rainwater, F. H., "Stream Composition of the Conterminous United States," USGS Hydrologic Investigations Atlas HA-61 (1962).
49. Blakey, J. F., "Temperature of Surface Waters in the Conterminous United States," USGS Hydrologic Investigations Atlas HA-235 (1966).
50. Schneider, W. J., "Water Data for Metropolitan Areas," USGS Water Supply Paper 1871 (1968).
51. Durfor, C. N., and Becker, E., "Public Water Supplies of the 100 Largest Cities in the United States, 1962," USGS Water Supply Paper 1812 (1964).
52. Cronin, J. G., "Groundwater in the Ogallala Formation in the Southern High Plains of Texas and New Mexico," USGS Hydrologic Investigations Atlas HA-330 (1969).
53. Cushing, E. M. *et al.*, "Availability of Water in the Mississippi Embayment," USGS Professional Paper 448-A (1970).
54. Meier, M. F., "Glaciers and Water Supply," *Journal AWWA* **61** (1), 8–12 (1969).
55. Barksdale, H. C., O'Bryan, D., and Schneider, W. J., "Effect of Drought on Water Resources in the Northeast," USGS Hydrologic Investigations Atlas HA-243 (1966).
56. Feth, J. H. *et al.*, "Preliminary Map of the Conterminous United States Showing Depth to and Quality of Shallowest Groundwater Containing more than 1000 Part per Million Dissolved Solids," USGS Hydrologic Investigations Atlas HA-199 (1965).
57. Neyman, J., Scott, E., and Smith, J. A., "Areal Spread of the Effect of Cloud Seeding at the Whitetop Experiment," *Science* **163** (3874), 1445–1449 (1969).
58. Malone, T. F., "Weather Modification, Present and Future," *Journal AWWA* **60** (10), 1143–1148 (1968).
59. Woodley, W. L., "Rainfall Enhancement by Dynamic Cloud Modification," *Science* **170** (3954), 127–132 (1970).
60. Michel, C., "Evaporation Control: A New Tool for Water Management," *Journal of Soil and Water Conservation* **18** (4), 223–229 (1963).
61. Kohout, F. A. (ed.), "Saline Water, A Valuable Resource," *Water Resources Research* **6** (5), 1441–1531 (1970).
62. McDonald, C. C., and Hughes, G. H., "Studies of Consumptive Use of Water by Phreatophytes Near Yuma, Arizona," USGS Professional Paper 486-F (1968).
63. Hammond, R. P., "Energy: The Ultimate Raw Material," *Science* **167** (3924), 1439 (1970).
64. Signor, D. C., Growitz, D. J., and Kam, W., Annotated Bibliography on Artificial Recharge of Groundwater, 1955–67," USGS Water Supply Paper 1990 (1970).
65. Lear, J., "Clean Power from Inside the Earth," *Saturday Review*, pp. 53–61, Dec. 5, (1970).
66. Sewell, W. R. D. and Bower, B. T., "Forecasting the Demands for Water," Policy and Planning Branch, Dept. of Energy, Mines, and Resources, Ottawa, Canada (1968).
67. Murray, C. R., and Reeves, E. B., "Estimated Use of Water in the United States in 1970," USGS Circular 676 (1972).
68. Piper, A. M., "Has the United States Enough Water?," USGS Water Supply Paper 1797 (1965).
69. Löf, G.O.G., and Hardison, C. H., "Storage Requirements for Water in the United States," *Water Resources Research* **2** (3), 333 (1966).
70. Roberts, W. J., "Significance of Evaporation in Hydrologic Education in the Progress of Hydrology," *Proceedings, 1st International Seminar for Hydrology Professors*, Urbana, Illinois, pp. 666–693 (1969).

2

Water Quantity Requirements for Public Supplies and Other Uses

John C. Kammerer

Water Resources Division
U. S. Geological Survey
Washington, D. C.

PRINCIPAL CATEGORIES AND TYPES OF WATER USE

A common way of classifying water uses is based on the major purposes for which the water is used, such as for public supply, for industrial use by facilities having their own water systems, for irrigation of agriculture, and for rural use by households and livestock. These are all termed *withdrawal* (or "off-channel") uses because water is withdrawn or diverted from a source, such as a stream or water well, in order to reach its point of usage. *Nonwithdrawal* (or "on-site") uses are those where the water is used without removal from where it occurs; such uses include navigation, recreation, wildlife habitat, and waste dilution and transport. The use of water to run the turbines in run-of-the-river hydroelectric power plants is sometimes considered a nonwithdrawal use because the water continues on its way downstream virtually undiminished in either quantity or quality.

This chapter is concerned entirely with the quantities of water required for withdrawal uses, especially the use of water for public supplies. Table 2-1 lists these major uses and also shows the average quantities withdrawn for each use in the United States in 1970. Although these data do reflect the relative magnitudes of water demands, the data may be misleading in making comparisons of supply with demand unless three facts are clearly understood and evaluated:

1. These are *average* rates of use and therefore by themselves give no indication of the substantial variations that occur from place to place as well as from day to day, month to month, and season to season.
2. A significant but unknown portion of the water included in the quantities shown in Table 2-1 has been used two or more

TABLE 2-1. Major Types of Water Supplies and Average Withdrawals for These Supplies in the United States[a1]

Note: all water data are rounded to two significant figures; some totals as shown below do not equal the sum of their parts as a result of rounding of more detailed data

	AVERAGE WITHDRAWALS (INCLUDING SALINE WATER)			
Type of Water Supply	Million gallons per day	Cubic feet per second	Million cubic meters per day	Thousand acre-feet per year
Public or municipal system (serving 165 million people)	27,000	42,000	100	30,000
Residential or domestic (sometimes subdivided into household uses and lawn-sprinkling); System losses and public uses (fire-fighting, street-cleaning, public parks and pools, etc.)	19,000	29,000	72	21,000
Commercial; Industrial (served by municipal systems)	8,800	14,000	33	9,900
Rural use (self-supplied; excluding irrigation)	4,500	7,000	17	5,000
Domestic	2,600	4,000	10	2,900
Livestock	1,900	2,900	7.2	2,100
Irrigated agriculture[b]	130,000	200,000	490	140,000
Delivered at farm	110,000	170,000	420	120,000
Conveyance loss (between source and delivery at farm)	22,000	34,000	83	25,000
Industry (self-supplied)[c,d]	210,000	330,000	790	240,000
Thermoelectric power generation (utilities)[c]	170,000	260,000	640	190,000
Other industrial use[d]	47,000	73,000	180	53,000
Total (total population, 206 million people)	370,000	570,000	1,400	410,000

[a]Including Puerto Rico.
[b]Includes 370 mgd of reclaimed sewage.
[c]Includes 46,000 mgd of saline water.
[d]Includes 8100 mgd of saline water and 150 mgd of reclaimed sewage.

times. This is especially true along major rivers in metropolitan and industrialized areas where some of the same water is removed from and returned to the same stream repeatedly as the water flows downstream, and the amount used is measured (or estimated) and counted each time in order to compute the total amount withdrawn.

3. Also not shown in the table is the amount of water *consumed*, and therefore unavailable for further use of any kind. The water may be consumed by evaporation and transpiration back to the atmosphere such as by crops, or may be consumed by people and animals, or may be incorporated into a manufactured product. Irrigated agriculture and rural uses consume much of the water withdrawn.

The terms *consumed* and *consumptive use* both refer to water that is no longer available, and care should be taken that the terms are not confused with water *consumption*, the word often used to describe the amount of water delivered to customers (or withdrawn from the supply source) by a municipal water system.

The quantitative units that are used to express totals or rates of water use are sometimes unclear or unwieldy for various comparative purposes because they lack uniformity and are in units other than those used to express quantities of water available, such as in streams. Some of the varieties of units are shown in the

TABLE 2-2. Hydraulic Equivalents
(Equivalent values, to three significant figures, are on the same horizontal line)

Million gallons per day (mgd)	Billion gallons per day (bgd)	Thousand acre-feet per year	Thousand cubic feet per second	Thousand gallons per minute	Million cubic meters per day	Billion gallons per year
1.0	0.001	1.12	0.00155	0.694	0.00379	0.365
1,000	1.0	1,120	1.55	694	3.79	365
0.893	0.000893	1.0	0.00138	0.620	0.00338	0.326
646	0.646	724	1.0	449	2.45	236
1.44	0.00144	1.61	0.00223	1.0	0.00545	0.526
264	0.264	296	0.409	184	1.0	96.4
2.74	0.00274	3.07	0.00424	1.90	0.0104	1.0

Example of how this table may be used: to convert from data given in 1000 ft^3/sec to data in mgd, multiply by the 646 shown (under mgd) on the horizontal line where 1000 ft^3/sec = 1.0; thus, 76,000 ft^3/sec (roughly the average flow of the Missouri River at its mouth) equals about 49,000 mgd (646 × 76).

column headings of Table 2-1. The hydraulic equivalents of these units are given in Table 2-2. Undoubtedly there will be greater uniformity once the metric system has been universally adopted.

Most of the data in this chapter are expressed in million gallons per day (mgd), gallons per day (gpd), or gallons per day per capita (gpd/person). Inasmuch as there are 3.785 liters (l) (rounded to four significant figures) in 1 U.S. gallon, the liter equivalent of any gallonage data in this chapter may be roughly approximated by multiplying by 4. A few of the most relevant approximations that may be helpful when comparing water supply with water demand are as follows:

$$1000 \text{ acre-ft/yr} \times 0.893 = 1 \text{ mgd}$$
$$1 \text{ ft}^3\text{/sec} \times 0.646 = 1 \text{ mgd}$$
$$21 \text{ in./yr (such as precipitation)} = 1 \text{ mgd/mi}^2$$
(prior to evapotranspiration)

Availability and Sources of Water Use Data. There is a great profusion of published information on water use, but the data are not only widely dispersed in the technical and scientific literature but also widely variable in their accuracy. When data of this kind are cited occasionally in the news media, the information often lacks adequate definition and sometimes even appears to be contradictory. The most common example—the average use of water per person in the United States—as variously cited in the press, commonly ranges from 50 or 60 gpd/person (a not unreasonable figure for domestic uses alone) to as much as 1800 gpd/person, an average daily amount that includes all water for municipal, rural, irrigation, and industrial purposes (other than for hydroelectric power).

The paragraphs that follow point out some of the larger compilations of data on water use that are available in published form. Complete citations of all these reports, listed by name of author, are given at the end of this chapter.

The best documented information on water use is data on municipal water supplies. A large proportion of these supplies is metered. Also, the majority of them are documented by annual reports to city or town governments or to investors in privately-owned water utilities. For many years these data have been compiled by state departments of health or similar agencies, and in turn the information has been published periodically in federal reports. However, the most recent such compilation is data as of January 1, 1964, by the U. S. Public Health Service (USPHS) for communities of populations of 25,000 and over;[2] the more comprehensive, nine-volume report of the USPHS contains essentially complete nationwide coverage as of January 1, 1963.[3] A ten-volume statistical report privately published in 1968 lists total water use and population served by each of some 23,000 water supply systems in the United States.[4] The year represented by the data is given for each system; most of the data are for either 1967, 1965, or 1962.

At 5-year intervals from 1945 to 1970, the American Water Works Association (AWWA) carried out detailed statistical surveys of water works data throughout the country and published the compiled information. The survey for 1960 contained the results of responses from 1256 water utilities.[5,6] In 1973, AWWA published the results of its 1965 and 1970 surveys of 861 and 768 water utilities, respectively. Data from the AWWA surveys and from the USPHS report for January 1, 1963 (and comparable USPHS reports of prior years) include information on the domestic use segment of municipal water supplies. Another excellent reference on this phase of water use is a 1967 report[7] prepared for the Federal Housing Administration (FHA) at Johns Hopkins University.

Quantitative information on industrial use of water is less complete and less consolidated in published form than comparable data on municipal water use. Furthermore, the use of water varies widely from industry to industry and even from one establishment to another within the same industry. Extrapolation from a relatively few comprehensive water-use studies at individual plants of a particular industry in order to estimate total water use (or average water use per unit of product) for that industry may lead to inaccurate and difficult-to-evaluate answers. The compilations that best present a major share of industrial water-use information on a total annual basis, as distinct from the per-unit-of-product basis, are the "Census of Manufactures" water-use reports prepared by the U.S. Bureau of the Census.[8] The most recent report (published in 1971) contains statistics on water use in 1968 by group and type of industry and by state and region as well as nationwide. Data include amounts of total and fresh water intake, amounts by purpose of intake, amounts by type of source, and amounts discharged.

Published compilations of water-use statistics expressed in gallons per unit of product which are documented with respect to the sources of information are relatively few in number. Three of the best are: (1) the 1969 manual prepared by the American Society for Testing and Materials (ASTM; pages 320-322, 329);[9] (2) the 1965 report on water in industry by the National Association of Manufacturers and the Chamber of Commerce of the United States;[10] and (3) the 1958 report on water for industrial use prepared for the United Nations Department of Economic and Social Affairs.[11]

Between 1956 and 1967, the U.S. Geological Survey (USGS) published a series of detailed reports on water requirements of selected industries, including summaries of industrial processes involving water use, the amounts of use per unit of product, and estimates of the total water usage by each group of industries. The authors and subjects of these reports are as follows:

Author	Date	Industry
Conklin[12]	1956	Aluminum
Conklin[13]	1956	Carbon black
Durfor[14]	1963	Styrene, butadiene, and synthetic rubber
Mussey[15]	1957	Rayon and acetate fiber
Mussey[16]	1961	Copper
Mussey[17]	1961	Pulp and paper
Otts[18]	1963	Petroleum refining
Walling and Otts[19]	1967	Iron and steel

Some information on industrial water use may be gained from reports that are primarily concerned with quantities of wastes discharged and the means of treating or reducing those wastes. One broad-scope compilation of this kind is a ten-part series that comprises Volume 3 of the report, *Cost of Clean Water*, published in 1967 by the Federal Water Pollution Control Administration (FWPCA) of the U.S. Department of the Interior.[20] One of the earliest and best examples of this kind of comprehensive report is contained in the third of three volumes on Ohio River pollution control prepared by USPHS and published in 1944 as House Document No. 266 (78th Congress, 1st Session).[21]

Every fifth year beginning in 1950, the USGS has compiled nationwide estimates of water use by source of water (surface water or groundwater) and principal types of use. The most recent report[1] contains data by states and by the 21 water resources regions designated by the federal Water Resources Council (WRC).

NATIONAL AND REGIONAL SUMMARIES OF WATER USE

In the mid-1960s the Water Resources Council delineated the boundaries of these major water resources regions for purposes of river basin planning and analysis. Each of the regions either is dominated by a single major source of water or encompasses several sources that are similar in magnitude and variability. The boundaries of the regions are shown in Fig. 2-1.

The report "Estimated Use of Water in the United States in 1970"[1] shows that nearly two-thirds of the water withdrawn during that year was in 6 of the 21 regions–4 in the East and 2 in the West:

Middle Atlantic	Columbia-North Pacific
South Atlantic-Gulf	California-South Pacific
Great Lakes	
Ohio	

Table 2-3 from that report shows the sources and quantities of all water withdrawn, by regions, as well as the regional population and per capita use. Tables 2-4 and 2-5 contain similar information for public supplies only, public supply data by states being shown in Table 2-5. The regional data reflect in a broad manner some of the basic water use characteristics of the various areas and the primary factors that result in these characteristics, such as climate, population density, and type of economy (agricultural, industrial, etc.). Part of the contrast between the East and the West is evident in the per capita data of Table 2-6, where the information is aggregated into the 9 eastern WRC regions and the 9 western WRC regions. The population is higher in the East whereas the water use per capita is greater in the West, and especially so with respect to water consumed. This situation is caused mainly by the more arid conditions in the West (except in the Pacific Northwest) and the extensive development of irrigated agriculture in the West. The greater per capita use of public supplies in the West results at least partly from the substantial

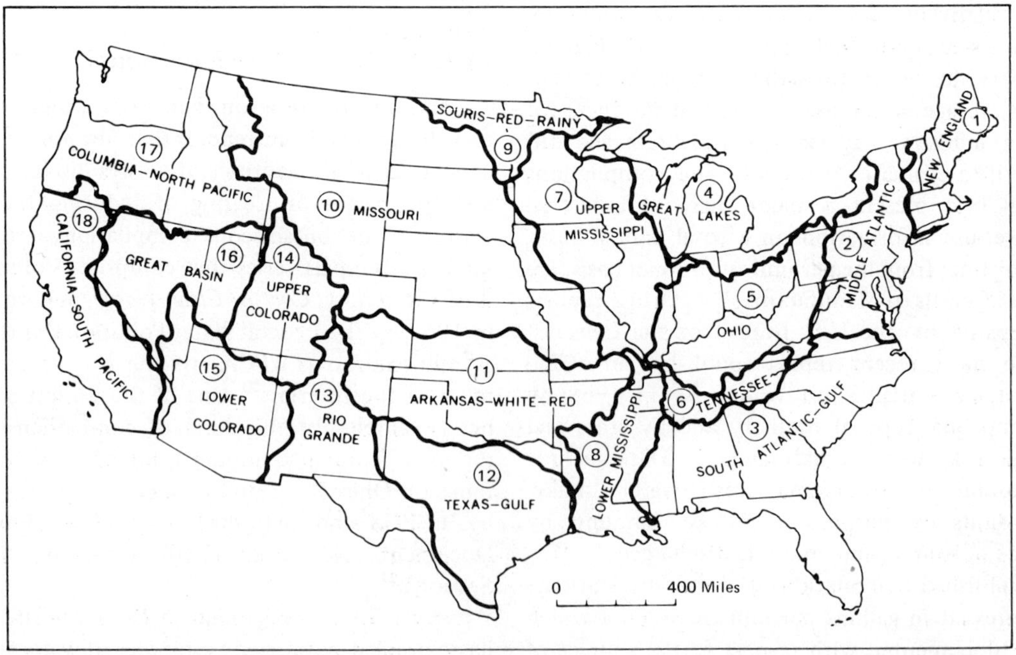

Fig. 2-1 Map of the conterminous United States showing 18 of the 21 water resources regions designated by the (U.S.) Water Resources Council. Regions 19, 20, and 21 are Alaska, Hawaii, and Puerto Rico, respectively.

TABLE 2-3. Summary of Water Withdrawn, except for Hydroelectric Power, in Million Gallons per Day, by Regions, 1970[1] (Partial figures may not add to totals because of independent rounding)

				WATER WITHDRAWN INCLUDING IRRIGATION CONVEYANCE LOSSES											
				GROUNDWATER			SURFACE WATER				ALL SOURCES				
Region Number	Water Resources Council Region	Population (thousands)	Per Capita Use (gpd)	Fresh	Saline	Fresh and Saline	Fresh	Saline	Fresh and Saline	Reclaimed Sewage	Fresh	Saline	Fresh and Saline	Convey-ance Losses	Fresh Water Consumed[a]
1	New England	11,520	840	630	1.4	640	4,100	4,900	9,000	0	4,700	4,900	9,700	0	410
2	Middle Atlantic	38,401	1,200	2,600	1.7	2,600	25,000	17,000	42,000	130	28,000	17,000	45,000	0	1,400
3	South Atlantic-Gulf	23,445	1,500	4,700	92	4,800	20,000	11,000	30,000	0	24,000	11,000	35,000	150	3,300
4	Great Lakes	29,104	1,400	1,400	400	1,800	38,000	1.0	38,000	0	39,000	400	39,000	0.1	1,200
5	Ohio	20,009	1,800	1,700	42	1,800	34,000	16	34,000	0	35,000	58	36,000	0	920
6	Tennessee	3,234	2,400	170	0	170	7,700	0	7,700	0	7,900	0	7,900	0	240
7	Upper Mississippi	12,733	1,300	2,200	19	2,300	14,000	0	14,000	0	16,000	19	16,000	.6	760
8	Lower Mississippi	6,265	2,100	3,600	41	3,600	8,500	880	9,400	0	12,000	920	13,000	550	3,600
9	Souris-Red-Rainy	808	390	68	4.8	73	250	0	250	0	310	4.8	320	1.0	71
10	Missouri	8,489	2,900	5,900	190	6,100	18,000	6.4	18,000	88	24,000	200	24,000	4,800	12,000
11	Arkansas-White-Red	6,670	1,800	6,600	38	6,700	5,200	18	5,200	11	12,000	55	12,000	310	6,800
12	Texas-Gulf	9,499	2,200	6,200	0	6,200	7,400	7,200	15,000	15	14,000	7,200	21,000	180	6,200
13	Rio Grande	1,617	3,900	2,400	0	2,400	3,800	150	3,900	17	6,200	150	6,300	590	3,300
14	Upper Colorado	404	20,000	110	4.4	110	8,000	0	8,000	0	8,100	4.4	8,100	1,100	4,100
15	Lower Colorado	2,223	3,300	4,500	0	4,500	2,800	0	2,800	5.1	7,200	0	7,200	300	5,000
16	Great Basin	1,213	5,500	1,100	9.7	1,100	5,500	5.5	5,500	53	6,700	15	6,700	2,000	3,200
17	Columbia-North Pacific	6,468	4,700	4,300	0	4,300	26,000	37	26,000	5.6	30,000	37	30,000	7,200	11,000
18	California-South Pacific	20,009	2,400	18,000	180	18,000	21,000	8,700	30,000	120	39,000	8,800	48,000	5,100	22,000
19	Alaska	302	830	39	0	39	210	1.0	210	0	250	1.0	250	0	16
20	Hawaii	770	3,500	910	13	920	850	860	1,700	66	1,800	870	2,700	220	810
21	Puerto Rico	2,712	1,100	140	1.5	140	430	2,400	2,900	0	580	2,400	3,000	42	170
	United States[b]	205,897	1,800	68,000	1,000	69,000	250,000	53,000	300,000	520	320,000	54,000	370,000	22,000	87,000

[a]Excluding irrigation conveyance losses by evaporation.
[b]Including Puerto Rico.

TABLE 2-4. Water Used for Public Supplies, by Regions, 1970[1]
(Partial figures may not add to totals because of independent rounding)

		POPULATION SERVED			WATER WITHDRAWN				WATER DELIVERED		WATER CONSUMED
Region Number	Water Resources Council Region	Ground-water (thousands)	Surface Water (thousands)	All Water (thousands)	Groundwater (mgd)	Surface Water (mgd)	All Water (mgd)	Per Capita (gpd)	Industrial and Commercial Uses (mgd)	Domestic Use and Losses[a] (mgd)	(mgd)
1	New England	2,720	7,360	10,100	330	1,100	1,400	139	610	790	190
2	Middle Atlantic	8,670	24,900	33,600	1,100	4,100	5,200	154	1,600	3,600	750
3	South Atlantic–Gulf	7,960	6,560	14,500	1,300	1,400	2,700	187	980	1,700	590
4	Great Lakes	4,000	19,400	23,400	700	3,700	4,400	188	1,700	2,700	500
5	Ohio	4,710	10,300	15,000	620	1,500	2,100	139	710	1,400	270
6	Tennessee	532	1,550	2,080	64	240	300	146	120	180	36
7	Upper Mississippi	6,350	4,530	10,900	870	690	1,600	143	560	990	190
8	Lower Mississippi	3,170	1,260	4,430	390	220	610	137	150	460	240
9	Souris–Red–Rainy	192	209	402	20	25	48	118	6.2	41	19
10	Missouri	2,850	3,860	6,710	430	590	1,000	152	300	720	250
11	Arkansas–White–Red	1,780	3,160	4,940	250	490	730	149	210	520	250
12	Texas–Gulf	3,970	3,950	7,920	590	550	1,100	144	370	780	380
13	Rio Grande	876	499	1,370	180	130	310	228	110	200	150
14	Upper Colorado	80	116	196	28	30	58	294	8.1	49	19
15	Lower Colorado	1,220	576	1,790	250	140	390	219	66	330	190
16	Great Basin	546	558	1,100	160	160	320	289	45	270	140
17	Columbia–North Pacific	1,840	2,800	4,630	460	830	1,300	278	580	710	260
18	California–South Pacific	8,030	10,700	18,700	1,600	1,800	3,400	181	620	2,800	1,400
19	Alaska	62	64	126	24	35	60	473	7.0	53	11
20	Hawaii	662	32	694	120	12	140	197	26	110	46
21	Puerto Rico	396	1,940	2,330	34	170	200	86	77	130	43
	United States[b]	60,600	104,000	165,000	9,400	18,000	27,000	166	8,800	19,000	5,900

[a]Includes public use.
[b]Including Puerto Rico.

TABLE 2-5. Water Used for Public Supplies, by States, 1970[1]
(Partial figures may not add to totals because of independent rounding)

	POPULATION SERVED			WATER WITHDRAWN				WATER DELIVERED		WATER CONSUMED
State	Groundwater (thousands)	Surface Water (thousands)	All Water (thousands)	Groundwater (mgd)	Surface Water (mgd)	All Water (mgd)	Per Capita (gpd)	Industrial and Commercial Uses (mgd)	Domestic Use and Losses[a] (mgd)	(mgd)
Alabama	765	1,420	2,190	100	360	470	213	260	210	36
Alaska	62	64	126	24	35	60	473	7.0	53	11
Arizona	989	509	1,500	190	120	310	208	47	260	160
Arkansas	605	637	1,240	71	95	170	133	63	100	36
California	8,000	10,700	18,700	1,600	1,800	3,400	181	620	2,800	1,400
Colorado	306	1,670	1,980	79	310	390	197	91	300	97
Connecticut	590	1,900	2,490	86	270	360	143	170	190	120
Delaware	217	193	410	30	46	76	185	32	44	20
Florida	4,820	592	5,410	760	120	880	163	170	720	230
Georgia	825	1,350	2,170	190	350	540	250	310	240	130
Hawaii	662	32	694	120	12	140	197	26	110	46
Idaho	407	63	470	96	15	110	237	5.8	110	29
Illinois	3,880	6,790	10,700	720	1,500	2,200	204	570	1,600	210
Indiana	1,500	1,990	3,480	210	280	490	141	140	350	120
Iowa	1,520	505	2,030	180	73	250	123	60	190	37
Kansas	817	823	1,640	130	120	250	155	61	190	74
Kentucky	304	1,860	2,160	24	160	180	83	78	100	26
Louisiana	1,290	1,370	2,670	140	240	380	144	58	330	200
Maine	153	599	752	20	89	110	146	38	72	22
Maryland	368	2,750	3,120	42	380	420	136	85	340	23
Massachusetts	1,460	3,940	5,400	170	590	750	140	320	440	38
Michigan	1,360	5,500	6,870	230	920	1,200	168	640	520	95
Minnesota	1,500	1,250	2,750	160	180	340	125	140	210	34
Mississippi	1,180	213	1,390	160	30	190	134	63	120	70
Missouri	942	3,020	3,970	92	420	510	128	260	250	95
Montana	152	354	507	26	85	110	219	32	79	55
Nebraska	908	211	1,120	150	34	190	168	46	140	38
Nevada	265	176	441	81	54	130	305	39	95	46

TABLE 2.5 Continued

State	POPULATION SERVED			WATER WITHDRAWN				WATER DELIVERED		WATER CONSUMED
	Groundwater (thousands)	Surface Water (thousands)	All Water (thousands)	Groundwater (mgd)	Surface Water (mgd)	All Water (mgd)	Per Capita (gpd)	Industrial and Commercial Uses (mgd)	Domestic Use and Losses[a] (mgd)	(mgd)
New Hampshire	261	283	544	32	38	70	128	22	48	3.6
New Jersey	3,030	3,390	6,420	340	560	900	139	110	780	180
New Mexico	645	67	712	130	16	150	204	41	100	66
New York	4,150	12,300	16,500	460	2,200	2,600	161	790	1,900	450
North Carolina	660	2,020	2,680	80	380	460	170	120	340	91
North Dakota	189	206	395	24	26	50	126	1.6	48	28
Ohio	2,480	5,980	8,460	320	1,000	1,300	157	340	990	170
Oklahoma	551	1,450	2,000	72	190	260	130	86	170	100
Oregon	355	842	1,200	67	160	230	188	90	140	48
Pennsylvania	1,350	8,300	9,650	250	1,500	1,800	181	880	870	180
Rhode Island	213	634	847	18	85	100	122	62	41	5.2
South Carolina	221	1,010	1,230	55	240	300	242	120	180	45
South Dakota	284	126	410	42	18	60	145	19	41	16
Tennessee	1,190	1,930	3,120	160	240	400	129	140	260	43
Texas	4,580	4,660	9,250	690	740	1,400	155	470	970	510
Utah	502	444	945	150	130	280	294	25	250	130
Vermont	100	189	289	14	29	42	146	13	30	1.7
Virginia	477	3,040	3,520	74	320	390	111	160	230	42
Washington	1,070	1,780	2,860	290	610	910	317	470	440	160
West Virginia	380	794	1,170	36	140	180	150	71	110	0.2
Wisconsin	1,540	1,570	3,110	220	270	480	155	230	250	48
Wyoming	130	120	250	24	25	49	197	13	37	12
District of Columbia	0	757	757	0	160	160	211	50	110	16
Puerto Rico	396	1,940	2,330	34	170	200	86	77	130	43
United States[b]	60,600	104,000	165,000	9,400	18,000	27,000	166	8,800	19,000	5,900

[a]Includes public use.
[b]Including Puerto Rico.

TABLE 2-6. Comparative per Capita Water Withdrawals and Water Consumed for Eastern and Western Water Resources Council Regions and Total United States, in Gallons per Day, 1970[1]
(All per capita data in this table have been rounded to two significant figures)

	PUBLIC SUPPLIES ONLY				ALL WITHDRAWAL USES		
	Population Served (millions)	All Uses	Domestic and Public Uses Only[a]	Total Population 1970 (millions)	Excluding Water Power	Including Water Power	Fresh Water Consumed, All Off-Channel Uses
9 Eastern WRC regions	114.4	160	100	145.5	1,400	10,000	81
9 Western WRC regions	47.4	180	130	56.6	2,900	30,000	1,300
50 States and District of Columbia	162.6	170	110	203.2	1,800	16,000	430

[a]Includes water losses in systems.

quantities of water required to keep lawns green. For this particular comparison of eastern and western regions, the 9 WRC western regions include the Missouri, Arkansas-White-Red, Texas-Gulf, and the other 6 regions to the west of these regions.

The trend in water usage nationwide, 1950 to 1970, is shown in Fig. 2-2; also included is the source of water—whether surface water (from streams and lakes) or groundwater (from wells and springs). The largest rates of increase each 5 years have been in water for self-supplied in-

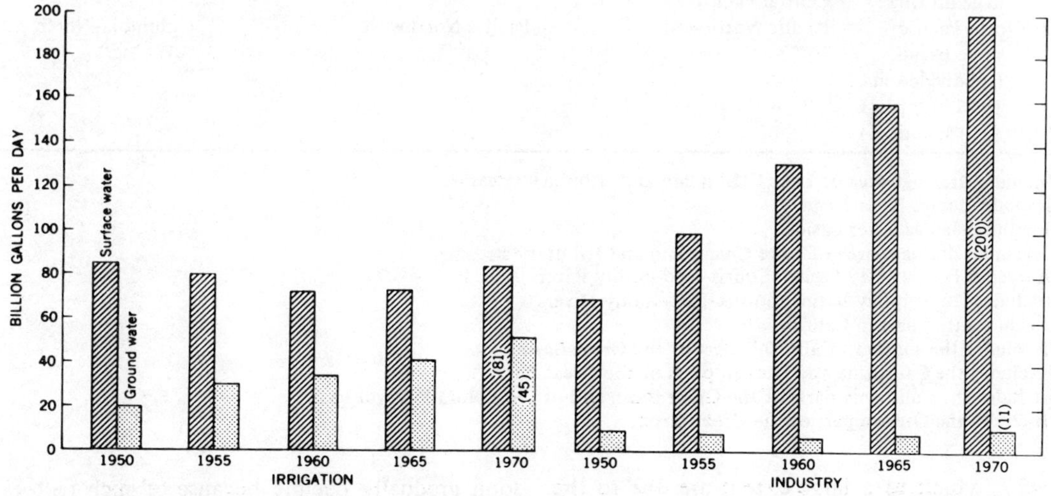

Fig. 2-2 Graphs showing trends in use of water for public supplies, rural supplies, irrigation, and self-supplied industry (1950-1970).[1] *(Prepared by the U.S. Geological Survey.)*

TABLE 2-7. Approximate Correlation of Water Resource Regions Described in Selected Reports on the Nation's Water Resources[22]
(Footnotes indicate principal differences in areas represented by regional names)

	REGIONAL NAMES USED IN THE FOLLOWING REPORTS (IDENTIFIED BY AUTHOR AND DATE OF PUBLICATION)		
U.S. Geological Survey, 1910 (WSP 241–252) to Date	Wollman and Bonem, 1971	A.M. Piper, 1965 (data to 1960)	(U.S.) Water Resources Council, 1968 (data to 1965)
1. North Atlantic slope basins (St. John River to York River)	New England Delaware–Hudson Chesapeake Bay[b]	New England Delaware–Hudson Chesapeake	North Atlantic[a,b]
2. South Atlantic slope and eastern Gulf of Mexico basins	Southeast[c]	South Atlantic–Eastern Gulf	South Atlantic–Gulf[c]
3. Ohio River basin	Ohio Cumberland Tennessee	Ohio Cumberland–Tennessee	Ohio(–Cumberland) Tennessee
4. St. Lawrence River basin	Western Great Lakes Eastern Great Lakes	Western Great Lakes Eastern Great Lakes	Great Lakes[d]
5. Hudson Bay and upper Mississippi River basins	Upper Mississippi[e]	Upper Mississippi[e]	Souris–Red–Rainy Upper Mississippi
6. Missouri River basin	Upper Missouri[f] Lower Missouri	Upper Missouri–Hudson Bay[f] Lower Missouri	Missouri
7. Lower Mississippi River basin	Upper Arkansas–White–Red Lower Arkansas–White–Red Lower Mississippi	Upper Arkansas–Red Lower Arkansas–Red–White Lower Mississippi	Arkansas–White–Red Lower Mississippi
8. Western Gulf of Mexico basins	Western Gulf Upper Rio Grande–Pecos	Western Gulf–Rio Grande–Pecos	Texas–Gulf Rio Grande
9. Colorado River basin	Colorado	Colorado	Upper Colorado Lower Colorado[g]
10. The Great Basin	Great Basin[h]	Great Basin	Great Basin[i]
11. Pacific slope basin in California	South Pacific Central Pacific	Central and South Pacific	California[j]
12. North Pacific slope basins (subdivided in 1935 into parts 12, 13, and 14)	Pacific Northwest	Pacific Northwest	Columbia–North Pacific[k]

[a]Includes drainage area of Lake Champlain and tributary streams.
[b]Includes James River basin.
[c]Excludes James River basin.
[d]Excludes drainage area of Lake Champlain and tributary streams.
[e]Excludes Hudson Bay basins (Souris–Red–Rainy Rivers).
[f]Includes Hudson Bay basins (Souris–Red–Rainy Rivers).
[g]Excludes that part in California.
[h]Excludes the southern California part of the Great Basin.
[i]Excludes the California and Oregon parts of the Great Basin.
[j]Includes the California parts of the Great Basin and of the Colorado River basin.
[k]Includes the Oregon part of the Great Basin.

dustry, which to a large extent are due to the prodigious use of cooling water in electric utility generation of thermoelectric power. Probably the rate of increase of water use will soon gradually decline because of such factors as reduced rates of population growth, increased cost of energy generation, and increased costs of developing and treating water supplies.

As the cost of water increases, the consumer is likely to seek ways of eliminating unnecessary and inefficient uses of water.

There are four principal detailed reports that consider the analyses and forecasts of the adequacies and deficiencies of present and future water supplies which have been published within the past 10 years:

1. Piper, A. M., "Has the United States Enough Water?" U. S. Geological Survey Water Supply Paper 1797, 27 pages (1965).
2. Wollman, N., and Bonem, G. W., *The Outlook for Water–Quality, Quantity, and National Growth*, Johns Hopkins Press, Baltimore (published for Resources for the Future, Inc.) 286 pages (1971).
3. (U. S.) National Water Commision, *Water Policies for the Future*, National Water Commission (final report to the President and to the Congress) 578 pages (1973).
4. (U. S.) Water Resources Council, *The Nation's Water Resources–the First National Assessment*, Water Resources Council, 414 pages (1968).

Much of the content of the reports published in 1965, 1968, and 1971 is specifically organized along regional lines, somewhat comparable to the WRC regions described earlier. The information on availability of water supplies is based largely on streamflow data contained in annual basic-data reports of the U. S. Geological Survey. The approximate correlation of regions in the USGS basic-data reports to those used in the other reports is shown in Table 2-7. (The WRC North Atlantic region has subsequently been divided into the New England and Middle Atlantic regions, and its California region has become the California-South Pacific region and no longer includes small parts of the Colorado and Great Basin regions.) The regions covered by separate parts of the series of basic-data reports on streamflow are shown in Figure 2-3.

An excellent five-volume report containing economic projections useful in forecasting water demands was published in 1972: *OBERS Projections–Economic Activity in the United States by Economic Area, Water Resources Region and Subarea, and State; Historical and Pro-*

Fig. 2-3 Map showing areas covered by the parts of the periodic basic-data reports on surface water supply (mainly stream-flow) of the United States. (Prepared by the U.S. Geological Survey.)

jected, 1929–2020. The report was prepared for the Water Resources Council by the U. S. Department of Commerce, Bureau of Economic Analysis (formerly Office of Business Economics–OBE), and the U. S. Department of Agriculture, Economic Research Service (ERS). Volume 1 describes concepts, methodology, and summary of data. Volume 2, the most useful one for metropolitan area planning, contains historical data on population, employment, personal income, and earnings by industry for each of the 173 BEA economic areas for 1950, 1959, 1962, 1968, and 1969, and projections for 1980, 1990, 2000, 2010, and 2020; income and earnings are expressed in terms of 1967-year dollars. The data in Volumes 3 and 4 are by Water Resources Council regions and Volume 5 is by states.

WATER REQUIREMENTS FOR PUBLIC SUPPLIES

The term "public supplies" applies to all municipal water supplies no matter whether they are publicly or privately owned. This water use includes residential, commercial, industrial, and public uses, as well as water losses in the system from leaks and breaks in underground pipes. The total does not include self-supplied industrial water–water that industries obtain from their own wells or by pumping or diversion from a stream, lake, reservoir, or other surface source.

The statistical information in the next several tables includes those per capita and unit-use factors most commonly dealt with when public supply water use is considered in terms of averages. Variations in use are discussed in a later section.

The average person in the United States may consume no more than a gallon a day in liquid and solid foods, including a quart or two in the form of water, milk, and other beverages. However, a household of four persons uses nearly 300 gpd from public supplies because of the many other residential uses to which the water is put, including cooking, dishwashing, laundering, flushing toilets, bathing, washing cars, and lawn and garden watering. Per capita use for residential (household) uses ranges from less than 10 gpd (annual average) in some rural homes without running water (and thus not served by a public supply) to more than 200 gpd in some areas where there is extensive lawn watering.

Regarding residential water uses, Appendix G (prepared by AVCO Corp. subsidiary, Economic Systems Corp.) of a report by the Urban Hydrology Research Council of ASCE contained a hypothetical example of ultimate water use by an "average family of the future: four persons, two automobiles, two bathrooms, one garbage disposal unit, one dishwasher, and one automatic laundry."[2,3] This example (prepared in 1965 by G. W. Reid) of use by a family of four, with weekly, monthly, and seasonal variations averaged and prorated to a daily basis, is shown in Table 2-8; the rate of use for various purposes is representative of present water-use practices and technology. Note that lawn watering constitutes a substantial part of the total use. The daily indoor uses average 59 gpd/person.

Estimates of average per capita use of public supply water for various purposes in 1970 are as follows:

	gpd
Residential use	75
Industrial use	30
Commercial use	25
Public use	20
Water system losses	15
Total	165

Public uses include water for fire fighting, street sprinkling, public swimming pools and fountains, lawn watering in public parks, and drinking, sanitation, and other uses in public buildings and facilities.

Data from various sources, including a compilation from plumbing manufacturers and contractors, indicate that indoor water use may be roughly divided percentagewise as follows:[24]

Drinking and cooking	4-5%
Bathing and personal	20-34
Laundry and dishes	14-23
Toilets	41-56
	100%

TABLE 2-8. Estimated per Capita Daily Consumption of Water for Various Purposes in a Family of Four

Family Use per Day	DAILY (AVERAGED FOR ENTIRE YEAR) (gpd/family)	(gpd/person)
Drinking and water used in kitchen:		
2 gal/person × 4	8	2
Dishwasher:		
5 gal/load × 3 loads/day	15	3.75
Toilet:		
(2) 4 flushes/person × 6 gal/flush × 4	96	24
Bathing:		
20 gal/bath × 4	80	20
Laundering:		
$\frac{6\ \text{loads/week} \times 40\ \text{gal/wash}}{7}$	34	8.5
Automobile washing:		
$\frac{2\ \text{cars} \times 1\ \text{wash/car/month} \times 150\ \text{gal/wash}}{30}$	10	2.5
Lawn watering:		
2 hr/day × 90 days/yr × 200 gal/hr	100	25
Garbage disposal unit:		
+ 1 percent × 343 gal/day	3	.75
Total	346	86.5

Source: Economic Systems Corporation, Appendix G in *Urban Water Resources Research*, American Society of Civil Engineers, Urban Hydrology Research Council, New York City, p. G 34 (1968).

A slightly more detailed breakdown of water use some years ago in an average home in Akron, Ohio, contains comparable information, but also includes outdoor uses:

Drinking	5%
Cooking and dishwashing	6
Tub and shower bathing	37
Toilet flushing	41
Clothes washing	4
Household cleaning	3
Car washing	1
Lawn and garden watering	3
	100%

Two tables prepared by the U.S. Public Health Service (1962)[25] list substantial data on unit uses and flows of water—statistics that are helpful in fully understanding and estimating water uses. These data are given in Tables 2-9 and 2-10.

Several additional unit use factors are given below:[24,26,27,28]

Indoor Use	Gallons
Each toilet flushing	3–8 (most commonly 5-6)
Each tub bath	20–40
Each shower bath	20–30
Each manual dishwashing	5-10
Each automatic dishwashing	5-12
Each washing-machine load	20-60
Food waste disposer	1-5/min of operation

Yard watering of 8000 sq ft

Humid areas	6 in./yr =	30,000 gal/yr
Semi-arid areas	20 in./yr =	100,000 gal/yr
Desert areas	36 in./yr =	180,000 gal/yr

Many economic factors influence the amount of water used in a city or in a community within a city. Eleven such considerations are:[29]

1. Size and type of community. The larger the community is, the narrower the range

TABLE 2-9. Planning Guide for Water Use[25]

Types of Establishments	gpd
Airports (per passenger)	3-5
Apartments, multiple family (per resident)	60
Bath houses (per bather)	10
Camps:	
Construction, semipermanent (per worker)	50
Day with no meals served (per camper)	15
Luxury (per camper)	100-150
Resorts, day and night, with limited plumbing (per camper)	50
Tourist with central bath and toilet facilities (per person)	35
Cottages with seasonal occupancy (per resident)	50
Courts, tourist with individual bath units (per person)	50
Clubs:	
Country (per resident member)	100
Country (per nonresident member present)	25
Dwellings:	
Boarding houses (per boarder)	50
Additional kitchen requirements for nonresident boarders	10
Luxury (per person)	100-150
Multiple family apartments (per resident)	40
Rooming houses (per resident)	60
Single family (per resident)	50-75
Estates (per resident)	100-150
Factories (gal per person per shift)	15-35
Hotels with private baths (two persons per room)	60
Hotels without private baths (per person)	50
Institutions other than hospitals (per person)	75-125
Hospitals (per bed)	250-400
Laundries, self-serviced (gal per washing, i.e., per customer)	50
Motels with bath, toilet, and kitchen facilities (per bed space)	50
With bed and toilet (per bed space)	40
Parks:	
Overnight with flush toilets (per camper)	25
Trailers with individual bath units (per camper)	50
Picnic areas:	
With bath houses, showers, and flush toilets (per picnicker)	20
With toilet facilities only (gal per picnicker)	10
Restaurants with toilet facilities (per patron)	7-10
Without toilet facilities (per patron)	$2\frac{1}{2}$-3
With bars and cocktail lounge (additional quantity per patron)	2
Schools:	
Boarding (per pupil)	75-100
Day with cafeteria, gymnasiums, and showers (per pupil)	25
Day with cafeteria but no gymnasiums or showers (per pupil)	20
Day without cafeteria, gymnasiums, or showers (per pupil)	15
Service stations (per vehicle)	10
Stores (per toilet room)	400
Swimming pools (per swimmer)	10
Theaters:	
Drive-in (per car space)	5
Movie (per auditorium seat)	5
Workers:	
Construction (per person per shift)	50
Day (school or offices per person per shift)	15

TABLE 2-10. Rates of Flow for Certain Plumbing, Household, and Farm Fixtures[25]

Location	Flow Pressure[a] (psi)	Flow Rate (gpm)
Ordinary basin faucet	8	2.0
Self-closing basin faucet	8	2.5
Sink faucet, $\frac{3}{8}$ in.	8	4.5
Sink faucet, $\frac{1}{2}$ in.	8	4.5
Bathtub faucet	8	6.0
Laundry tub faucet, $\frac{1}{2}$ in.	8	5.0
Shower	8	5.0
Ball-cock for closet	8	3.0
Flush valve for closet	15	15–40[b]
Flushometer valve for urinal	15	15.0
Garden hose (50 ft, $\frac{3}{4}$-in. sill cock)	30	5.0
Garden hose (50 ft, $\frac{5}{8}$-in. outlet)	15	3.33
Drinking fountains	15	0.75
Fire hose $1\frac{1}{2}$ in., $\frac{1}{2}$-in. nozzle	30	40.0

[a]Flow pressure is the pressure in the supply near the faucet or water outlet while the faucet or water outlet is wide open and flowing.
[b]Wide range due to variation in design and type of closet flush valves.

of fluctuation in demand. The fluctuation in use in an industrial community is normally much smaller than in a residential community.

2. Standard of living. The higher the standard of living is, the higher the demand and the greater the variation in demand.
3. Location of community. Extremes of heat and cold cause variations in demand.
4. Quality. Poor-quality water results in a reduction in use.
5. Pressure. High pressure results in increased use; low pressure in decreased use.
6. Sewers. Use of water increases when sewers are installed in an area previously without them.
7. Metering. Use of water decreases when meters are installed in a community previously without them.
8. Age of community. Older, more stable communities use less water than rapidly developing communities where new homes are being constructed and owners are planting new lawns.
9. Lawn sprinkling. Enforcement of lawn sprinkling regulations can reduce peak demands significantly.
10. Cost. An increase in water rates will reduce consumption in metered areas.
11. Air conditioning. The passage and enforcement of regulations that will prevent excessive use of water for air conditioning is probably desirable for most water supply systems.

Water Use Statistics on Public Supply Systems of Selected Cities. Data on actual water use by individual public water supplies are helpful guides as to averages and ranges of water use in various parts of the United States. Table 2-11 contains such information for some 200 public supply systems, including at least 2 for each state. Most of the data are for 1963, but more recent statistics are also shown for more than half the systems. The majority of the systems listed are the largest ones in each state. Most of the data (except in Column 7) represent total withdrawals by the public supply systems for all purposes; however, the data for 1970 (Column 13) for some cities do not include water losses in the system nor water for public uses.

TABLE 2-11. Public Supply Data for Selected Cities in the United States[a]

Note: Locations of regions (last column) are shown in Fig. 2-1; region listed is for point of use and not necessarily the point of withdrawal.

				WITHDRAWALS IN 1963						WITHDRAWALS IN 1969 OR LATER			
				AVERAGE DAILY USE			MAXIMUM DAILY USE						
State Column (1)	Name of City or Water System (2)	City Population in 1970–U.S. Census (1000 people) (3)	Population Served (1000 people) (4)	(mgd) (5)	Per Capita (gpd/person) (6)	Residential Use Only–per Capita (gpd/person) (7)	(mgd) (8)	Ratio to Average Use (9)	Maximum Hourly Use (mgd) (10)	Year (11)	Population Served (1000 people) (12)	Average Use (mgd) (13)	Region Number (Water Resources Council Regions) (14)
Alabama	Birmingham	301	600	59.0	98	29	89.3	1.5/1	127	1969	–	74.0	03
	Huntsville	138	94	12.5	133	64	14.2	1.1/1	19	1970	136	16.0	06
	Mobile	190	205	25.0	122	–	28.0	1.1/1	–	1970	–	25.0	03
	Montgomery	133	145	20.0	138	79	24.0	1.2/1	72	–	–	–	03
Alaska	Anchorage	48	33	17.7	536	–	20.0	1.1/1	28	–	–	–	19
	Fairbanks	15	–	–	–	–	–	–	–	1970	20	4.0	19
	Juneau	6	7	2.5 (1962)	357	–	–	–	–	–	–	–	19
Arizona	Mesa	63	46	10.2	222	151	20.2	2.0/1	32	1969	70	14.0	15
	Phoenix	582	555	107.0	193	–	–	–	–	1969	585	126.0	15
	Tucson	263	150	40.7	271	179	85.7	2.1/1	123	1971	–	54.0	15
	Yuma	29	30	7.0	233	126	15.0	2.1/1	22	1970	–	8.9	15
Arkansas	Fayetteville	31	23	2.5 (1962)	109	–	–	–	–	1970	–	4.3	11
	Fort Smith	63	75	13.0	173	–	–	–	–	–	–	–	11
	Hot Springs	36	42	4.8	114	–	–	–	–	1971	40	7.3	08
	Little Rock	132	200	24.0	120	–	–	–	–	1969	–	32.0	11
	Pine Bluff	57	50	5.5	110	–	–	–	–	–	–	–	08
California	Anaheim	167	101	19.0	188	–	–	–	–	1970	175	35.0	18
	Bakersfield	70	119	39.4	331	–	81.3	2.1/1	122	–	–	–	18
	East Bay Municipal Utility District (Oakland, etc.)	362	1,000	143.4	143	–	232.1	1.6/1	242	1970 –	1,160	220.0	18
	Fresno	166	153	53.4	349	275	117.1	2.2/1	161	–	–	–	18
	Long Beach	359	350	56.0	160	–	–	–	–	1971	361	58.0	18
	Los Angeles	2,816	–	–	–	–	–	–	–	1970	2,828	523.0	18
	Riverside	140	102	–	–	–	–	–	–	1971	–	33.0	18

	Sacramento	254	190	48.8	257	–	–	–	–	–	–	–	18
	San Bernardino	104	96	10.0	104	–	–	–	–	–	–	–	18
	San Diego	697	623	60.8	98	70	108.0	1.8/1	–	1971	723	116.0	18
	San Francisco	716	1,600	177.8	111	–	246.0	1.4/1	–	1971	1,840	249.0	18
	San Jose	446	362	77.0	213	100	89.0	1.2/1	142	1971	540	97.0	18
	Santa Ana	157	100	15.0	150	–	–	–	–	1971	–	24.0	18
	Santa Barbara	70	42	–	–	181	17.6	–	66	–	–	–	18
	Vallejo	67	72	14.4	200	–	19.9	1.4/1	26	1970	85	12.0	18
	Ventura	56	45	7.1	158	–	–	–	–	–	–	–	18
Colorado	Boulder	67	46	–	–	–	–	–	–	1970	–	11.0	10
	Colorado Springs	135	93	15.0	161	95	34.2	2.3/1	38	1970	–	21.0	11
	Denver	515	645	140.6	218	129	327.0	2.3/1	1,210	1971	750	178.0	10
	Pueblo	97	115	20.0	174	–	47.0	2.4/1	–	1970	100	21.0	11
Connecticut	Bridgeport	157	305	54.5	179	83	89.3	1.6/1	122	1971	350	68.0	01
	Hartford County Metropolitan District (Hartford, etc.)	158	360	44.8	124	57	66.0	1.5/1	–	1970	393	57.0	01
	New Haven	138	325	47.6	146	54	67.9	1.4/1	106	1971	400	57.0	01
Delaware	Artesian Water Co.	–	135	4.8	36	31	8.5	1.8/1	10	1972	–	10.0	02
	Dover	17	14	2.0 (1962)	143	–	–	–	–	–	–	– –	02
	Newark	21	12	2.0 (1962)	167	–	–	–	–	–	–	–	02
	Wilmington	80	156	25.0	160	–	–	–	–	–	–	–	02
D.C.	Washington	757	1,100	166.5	151	–	239.7	1.4/1	304	1973	–	214.0	02
Florida	Fort Lauderdale	140	165	19.0	115	75	38.7	2.0/1	58	1971	180	42.0	03
	Jacksonville	529	250	36.7	147	–	61.7	1.7/1	363	–	–	–	03
	Miami	335	624	104.8	168	36	156.5	1.5/1	159	1970	790	137.0	03
	Orlando	99	135	22.2	164	97	41.7	1.9/1	62	1970	177	32.0	03
	Saint Petersburg	216	245	19.0	78	–	23.0	1.2/1	26	1971	250	29.0	03
	Tampa	278	300	30.5	102	–	49.4	1.6/1	70	–	–	–	03
Georgia	Atlanta	497	610	68.2	112	53	112.0	1.6/1	–	–	–	–	03
	Columbus	154	150	18.5	123	97	29.1	1.6/1	42	1969	150	24.0	03
	Macon	122	100	20.0	200	60	42.0	2.1/1	–	–	–	–	03
	Savannah	118	150	–	–	107	45.0	–	–	–	–	–	03
Hawaii	Hilo	26	25	3.5	140	–	5.0	1.4/1	–	–	–	–	20
	Honolulu	325	338	50.2	149	–	68.7	1.4/1	100	1973	606	126.0	20

[a]Source of data for 1962 and 1963 was U.S. Public Health Service;[3,2] for 1970 was partly AWWA Committee on Water Use, 1973.[32] Remaining data for 1969–73 were from water utility reports.

TABLE 2-11. Continued[a]

Note: Locations of regions (last column) are shown in Fig. 2-1; region listed is for point of use and not necessarily the point of withdrawal.

			WITHDRAWALS IN 1963										
				AVERAGE DAILY USE			MAXIMUM DAILY USE			WITHDRAWALS IN 1969 OR LATER			
State Column (1)	Name of City or Water System (2)	City Population in 1970– U.S. Census (1000 people) (3)	Population Served (1000 people) (4)	(mgd) (5)	Per Capita (gpd/person) (6)	Residential Use Only– per Capita (gpd/person) (7)	(mgd) (8)	Ratio to Average Use (9)	Maximum Hourly Use (mgd) (10)	Year (11)	Population Served (1000 people) (12)	Average Use (mgd) (13)	Region Number (Water Resources Council Regions) (14)
Idaho	Boise	75	66	10.0	152	111	24.4	2.4/1	46	–	–	–	17
	Idaho Falls	36	36	21.4	594	–	25.0	1.2/1	–	1971	36	22.0	17
	Pocatello	40	41	8.2	200	163	20.2	2.5/1	34	–	–	–	17
Illinois	Chicago	3,367	4,501	1,044.0	232	–	1,513.0	1.4/1	–	1970	4,506	1,035.0	04
	Peoria	127	168	18.7	111	35	34.0	1.8/1	38	1971	–	25.0	07
	Rockford	147	135	24.5	181	58	46.2	1.9/1	74	1971	145	37.0	07
	Springfield	92	125	16.0	128	–	26.4	1.6/1	31	–	–	–	07
Indiana	Bloomington	43	45	5.8	129	–	8.4	1.4/1	–	1970	–	7.6	05
	Evansville	139	163	22.0	135	–	31.3	1.4/1	38	–	–	–	05
	Fort Wayne	178	169	23.6	140	50	37.8	1.6/1	54	–	–	–	04
	Gary	175	250	25.2	101	35	49.4	2.0/1	56	–	–	–	04
	Indianapolis	745	616	75.5	123	47	119.3	1.6/1	157	1971	–	93.0	05
Iowa	Cedar Rapids	111	94	11.8	126	45	22.3	1.9/1	23	1970	111	17.0	07
	Davenport	98	104	11.8	113	46	21.4	1.8/1	26	–	–	–	07
	Des Moines	201	243	28.0	115	–	48.0	1.7/1	70	1971	247	34.0	07
Kansas	Kansas City	168	200	32.0	160	26	45.8	1.4/1	53	1971	–	32.0	10
	Lawrence	46	35	4.8	137	–	8.2	1.7/1	–	–	–	–	10
	Topeka	125	135	16.0	119	–	30.4	1.9/1	41	1971	140	18.0	10
	Wichita	277	255	29.0	114	51	61.0	2.1/1	100	1970	–	38.0	11
Kentucky	Covington	53	90	7.9	88	–	11.2	1.4/1	12	–	–	–	05
	Lexington	108	135	14.8	110	34	22.0	1.5/1	58	–	–	–	05
	Louisville	361	550	86.0	156	38	126.0	1.5/1	178	1971	711	93.0	05
	Owensboro	50	60	5.5	92	33	8.3	1.5/1	10	–	–	–	05
Louisiana	Baton Rouge	166	190	22.3	117	–	30.0	1.3/1	–	–	–	–	08
	New Orleans	593	656	122.0	186	–	252.7	2.1/1	284	1970	–	116.0	08
	Shreveport	182	190	25.0	132	65	40.3	1.6/1	65	1971	–	29.0	11

Maine	Bangor	33	40	3.8	95	48	5.6	1.5/1	–	1971	–	4.0	01
	Biddeford and Saco Water Co. (Biddeford, etc.)	20	38	3.7	97	53	8.7	2.4/1	10	1970	–	4.4	01
	Lewiston	42	48	6.0	125	67	7.2	1.2/1	7.9	–	–	–	01
	Portland Water District (Portland, etc.)	65	145	20.2	139	60	30.3	1.5/1	46	1971	–	23.0	01
Maryland	Baltimore	906	1,407	221.8	158	–	316.2	1.4/1	343	1970	1,500	240.0	02
	Hagerstown	36	57	6.7	118	–	10.1	1.5/1	–	1969	70	9.4	02
	Washington Suburban Sanitary Commission	–	722	72.3	100	–	115.9	1.6/1	–	1971	1,100	124.0	02
Massachusetts	Metropolitan District Commission (Boston, etc.)	641	1,706	214.7	126	–	–	–	–	–	–	–	01
	Springfield	164	235	32.0	136	–	47.9	1.5/1	–	–	–	–	01
	Worcester	177	200	23.5	118	–	35.9	1.5/1	–	1970	–	26.0	01
Michigan	Detroit	1,511	3,800	479.3	126	49	835.8	1.7/1	1,253	1971	3,854	670.0	04
	Flint	193	200	32.9	164	–	53.8	1.6/1	122	–	–	–	04
	Grand Rapids	198	202	34.0	168	–	67.7	2.0/1	110	–	–	–	04
Minnesota	Duluth	101	114	15.7	138	–	22.7	1.4/1	33	1970	–	18.0	07
	Minneapolis	434	595	58.7	99	76	147.0	2.5/1	204	1970	618	77.0	07
	St. Paul	310	350	42.8	122	–	88.6	2.1/1	–	1970	390	54.0	07
Mississippi	Biloxi	48	49	5.5	112	84	7.2	1.3/1	11	–	–	–	03
	Greenville	40	52	4.6	88	63	7.1	1.5/1	–	–	–	–	08
	Jackson	154	150	18.3	122	71	32.0	1.7/1	53	–	–	–	03
	Meridian	45	50	5.5	110	–	8.2	1.5/1	–	–	–	–	03
Missouri	Kansas City	507	670	88.0	131	32	154.0	1.8/1	216	–	–	–	10
	St. Joseph	73	92	15.0	163	30	21.3	1.4/1	23	–	–	–	10
	St. Louis (city system)	622	870	168.7	194	–	–	–	353	–	–	–	10
	St. Louis County	–	690	67.0	97	–	122.0	1.8/1	150	1971	–	109.0	07
	Springfield	120	102	8.9	87	34	14.0	1.6/1	19	1971	–	11.0	10
Montana	Billings	62	62	12.9	208	119	31.1	2.4/1	62	1970	–	12.0	10
	Butte	23	45	13.5	300	167	24.5	1.8/1	43	1971	–	13.0	17

[a]Source of data for 1962 and 1963 was U.S. Public Health Service;[3,2] for 1970 was partly AWWA Committee on Water Use, 1973.[32] Remaining data for 1969–73 were from water utility reports.

TABLE 2-11. Continued[a]

Note: Locations of regions (last column) are shown in Fig. 2-1; region listed is for point of use and not necessarily the point of withdrawal.

			WITHDRAWALS IN 1963							WITHDRAWALS IN 1969 OR LATER			
				AVERAGE DAILY USE			MAXIMUM DAILY USE						
State Column (1)	Name of City or Water System (2)	City Population in 1970–U.S. Census (1000 people) (3)	Population Served (1000 people) (4)	(mgd) (5)	Per Capita (gpd/person) (6)	Residential Use Only–per Capita (gpd/person) (7)	(mgd) (8)	Ratio to Average Use (9)	Maximum Hourly Use (mgd) (10)	Year (11)	Population Served (1000 people) (12)	Average Use (mgd) (13)	Region Number (Water Resources Council Regions) (14)
	Great Falls	60	57	13.0	228	137	38.0	2.9/1	53	1971	–	14.0	10
	Missoula	29	42	18.0	429	214	31.7	1.8/1	42	–	–	–	17
Nebraska	Grand Island	31	26	7.0	269	–	18.5	2.6/1	–	1971	–	22.0	10
	Lincoln	150	148	21.6	146	72	58.9	2.7/1	93	1971	157	31.0	10
	Omaha	347	366	61.0	167	56	120.0	2.0/1	180	1970	–	70.0	10
Nevada	Las Vegas	126	120	34.8	290	–	–	–	–	1971	–	55.0	15
	North Las Vegas	36	30	6.0	200	–	11.0	1.8/1	–	–	–	–	15
	Sierra Pacific Power Co. (Reno, etc.)	73	98	25.0	255	–	58.0	2.3/1	115	1971	–	34.0	16
New Hampshire	Concord	30	29	3.9	134	–	6.1	1.6/1	10	–	–	–	01
	Manchester	88	96	11.9	124	60	24.0	2.0/1	36	1970	104	14.0	01
	Nashua	56	38	6.1	160	–	13.0	2.1/1	15	1971	55	9.6	01
	Portsmouth	26	34	3.6	106	–	4.0	1.1/1	5	1970	–	3.7	01
New Jersey	Jersey City	261	364	60.0	165	79	84.0	1.4/1	194	–	–	–	02
	Newark (Pequannock River source)	382	351	58.0	165	40	68.7	1.2/1	–	1971	–	65.0	02
	Passaic Valley Water Commission (Paterson, etc.)	145	–	–	–	–	–	–	–	1970	500	93.0	02
New Mexico	Albuquerque	244	241	45.0	186	100	92.5	2.1/1	161	1971	280	58.0	13
	Las Cruces	38	34	6.8	200	–	14.6	2.1/1	34	–	–	–	13
	Roswell	34	48	12.2	254	152	–	–	38	–	–	–	13
	Santa Fe	41	36	4.5	125	67	9.7	2.2/1	15	–	–	–	13
New York	Albany	116	130	26.0	200	–	–	–	–	–	–	–	02
	Buffalo	463	535	125.0	234	116	192.0	1.5/1	230	–	–	–	04
	Erie County Water Authority	–	275	33.0	120	–	–	–	–	–		–	04

	Monroe County Water Authority	–	240	30.0	125	–	–	–	–	–	–	–	04
	New York	7,895	7,750	1,151.0	149	–	1,470.0	1.3/1	–	–	–	–	02
	Rochester	296	258	56.0	217	–	–	–	–	–	–	–	04
	Syracuse	197	255	41.6	163	–	–	–	–	1970	225	49.0	04
	Yonkers	204	129	27.0	209	–	–	–	–	1971	–	29.0	02
North Carolina	Charlotte	241	215	23.4	109	71	37.0	1.6/1	59	–	–	–	03
	Greensboro	144	125	14.3	114	60	21.7	1.5/1	20	–	–	–	03
	Raleigh	122	100	10.9	109	75	15.6	1.4/1	20	–	–	–	03
	Winston-Salem	133	139	17.9	129	39	26.6	1.5/1	30	–	–	–	03
North Dakota	Fargo	53	49	4.6	94	–	12.4	2.7/1	18	–	–	–	09
	Grand Forks	39	47	3.0	64	45	6.2	2.1/1	–	–	–	–	09
	Minot	32	32	3.6	112	–	–	–	–	1971	–	3.9	09
Ohio	Akron	275	340	44.2	130	–	67.0	1.5/1	72	1970	–	50.0	04
	Cincinnati	453	800	100.0	125	76	153.0	1.5/1	234	1969	900	120.0	05
	Cleveland	751	1,822	318.0	175	–	492.0	1.5/1	983	1970	–	350.0	04
	Columbus	540	670	76.5	114	–	114.1	1.5/1	156	–	–	–	05
	Dayton	244	320	50.4	158	47	87.6	1.7/1	110	1970	380	70.0	05
	Mahoning Valley Sanitary District (Youngstown, etc.)	140	250	29.3	117	–	49.9	1.7/1	–	–	–	–	05
	Middletown	49	51	7.2	141	86	10.2	1.4/1	–	–	–	–	05
	Toledo	384	415	68.0	164	–	–	–	–	1970	–	79.0	04
Oklahoma	Lawton	74	90	11.8	131	–	23.6	2.0/1	–	–	–	–	11
	Oklahoma City	366	324	36.5	113	70	70.0	1.9/1	103	–	–	–	11
	Tulsa	332	310	50.9	164	–	86.3	1.7/1	151	–	–	–	11
Oregon	Eugene	76	72	13.1	182	86	38.0	2.9/1	55	–	–	–	17
	Portland	383	578	82.0	142	59	223.0	2.7/1	528	1971	648	107.0	17
	Salem	68	69	13.2	191	58	31.3	2.4/1	47	–	–	–	17
Pennsylvania	Erie	129	160	38.0	238	105	65.6	1.7/1	–	1971	185	46.0	04
	Philadelphia	1,949	2,005	373.0	186	–	–	–	–	1973	–	373.0	02
	Philadelphia Suburban Water Co.	–	700	46.4	66	55	72.0	1.6/1	120	1971	–	64.0	02
	Pittsburgh	520	650	95.0	146	–	130.5	1.4/1	–	1970	–	79.0	05
Rhode Island	East Providence	48	43	4.5	105	–	8.2	1.8/1	–	1971	48	5.7	01
	Pawtucket	77	112	14.4	129	–	26.0	1.8/1	–	–	–	–	01
	Providence	179	387	47.9	124	35	87.2	1.8/1	131	–	–	–	01
	Woonsocket	47	52	5.0	96	–	10.4	2.1/1	12	–	–	–	01

[a]Source of data for 1962 and 1963 was U.S. Public Health Service;[3,2] for 1970 was partly AWWA Committee on Water Use, 1973.[32] Remaining data for 1969–73 were from water utility reports.

TABLE 2-11. Continued[a]

Note: Locations of regions (last column) are shown in Fig. 2-1; region listed is for point of use and not necessarily the point of withdrawal.

State Column (1)	Name of City or Water System (2)	City Population in 1970– U.S. Census (1000 people) (3)	Withdrawals in 1963: Population Served (1000 people) (4)	Average Daily Use (mgd) (5)	Average Daily Use Per Capita (gpd/person) (6)	Average Daily Use Residential Use Only–per Capita (gpd/person) (7)	Maximum Daily Use (mgd) (8)	Maximum Daily Use Ratio to Average Use (9)	Maximum Hourly Use (mgd) (10)	Withdrawals in 1969 or Later: Year (11)	Population Served (1000 people) (12)	Average Use (mgd) (13)	Region Number (Water Resources Council Regions) (14)
South Carolina	Charleston	67	160	20.0	125	42	24.3	1.2/1	31	1971	225	28.0	03
	Columbia	114	150	16.2	108	–	30.3	1.9/1	48	1971	210	35.0	03
	Greenville	61	175	23.0	131	42	28.1	1.2/1	36	1970	–	28.0	03
South Dakota	Aberdeen	26	25	2.5	100	80	6.5	2.6/1	–	–	–	–	10
	Rapid City	44	30	6.4	213	147	20.0	3.1/1	–	–	–	–	10
	Sioux Falls	72	69	9.3	135	47	20.5	2.2/1	29	1970	–	11.0	10
Tennessee	Chattanooga	119	225	40.3	179	–	51.2	1.3/1	–	1971	–	51.0	06
	Knoxville	175	175	20.0	114	–	29.0	1.4/1	33	1971	–	24.0	06
	Memphis	624	525	68.8	131	63	82.6	1.2/1	132	–	–	–	08
	Nashville	448	315	43.4	138	34	57.0	1.3/1	–	1971	406	58.0	05
Texas	Amarillo	127	155	25.5	165	94	60.4	2.4/1	106	1971	–	25.0	11
	Austin	252	221	37.9	171	–	84.1	2.2/1	139	–	–	–	12
	Corpus Christi	205	230	51.1	266	98	76.0	1.5/1	96	1970	271	67.0	12
	Dallas	844	980	114.9	117	–	216.5	1.9/1	–	1971	1,154	170.0	12
	El Paso	322	300	51.0	170	–	104.5	2.0/1	–	1970	322	63.0	13
	Fort Worth	393	384	53.4	139	–	114.3	2.1/1	140	1970	–	69.0	12
	Houston	1,233	1,050	130.1	124	–	192.9	1.5/1	207	–	–	–	12
	Lubbock	149	150	23.0	153	93	55.6	2.4/1	86	–	–	–	12
	San Antonio	654	519	89.2	172	108	166.2	1.9/1	259	–	–	–	12
Utah	Ogden	69	60	22.6	377	–	38.2	1.7/1	–	1971	–	17.0	16
	Provo	53	40	11.8	295	–	19.2	1.6/1	–	–	–	–	16
	Salt Lake City	176	284	69.1	243	125	140.7	2.0/1	–	1972	311	82.0	16
	Salt Lake County Water Conservancy District (Midvale, etc.)	8	85	–	–	–	–	–	–	1971	–	15.0	16

Vermont	Brattleboro	12	11	1.7 (1962)	155	–	–	–	–	–	–	–	01
	Burlington	39	50	4.3	86	41	6.5	1.5/1	–	1970	65	6.0	02
	Rutland	19	18	3.0 (1962)	167	–	–	–	–	–	–	–	02
Virginia	Alexandria	111	300	24.5	82	–	37.0	1.5/1	65	–	–	–	02
	Norfolk	308	295	48.6	165	–	62.9	1.3/1	86	–	–	–	02
	Richmond	250	305	35.9	118	29	55.9	1.6/1	66	1971	425	45.0	02
	Roanoke	92	116	12.9	111	76	18.1	1.4/1	27	–	–	–	03
Washington	Seattle	531	750	–	–	–	–	–	–	1971	927	127.0	17
	Spokane	171	183	59.0	322	–	145.0	2.5/1	276	–	–	–	17
	Tacoma	155	160	58.3	364	61	87.1	1.5/1	–	1971	175	73.0	17
West Virginia	Charleston	72	141	18.5	131	50	20.5	1.1/1	22	–	–	–	05
	Huntington	74	109	12.5	115	32	18.6	1.5/1	22	–	–	–	05
	Parkersburg	44	60	5.2	87	28	6.1	1.2/1	–	–	–	–	05
	Wheeling	48	70	8.0	114	30	11.4	1.4/1	18	1971	103	7.7	05
Wisconsin	Green Bay	88	68	10.6	156	40	17.2	1.6/1	–	1971	90	16.0	04
	Madison	173	155	21.5	139	–	38.4	1.8/1	55	1971	185	30.0	07
	Milwaukee	717	903	152.0	168	47	278.0	1.8/1	405	1971	936	171.0	04
	Racine	95	100	16.1	161	45	28.6	1.8/1	48	–	–	–	04
Wyoming	Casper	39	40	6.3	158	–	17.0	2.7/1	30	–	–	–	10
	Cheyenne	41	47	10.5	223	115	22.0	2.1/1	30	–	–	–	10
	Laramie	23	21	6.0 (1962)	286	–	–	–	–	–	–	–	10
Puerto Rico	Ponce	128	113	14.5	128	–	18.0	1.2/1	–	–	–	–	21
	San Juan	453	729	66.1	91	–	73.7	1.1/1	–	–	–	–	21

[a]Source of data for 1962 and 1963 was U.S. Public Health Service;[3,2] for 1970 was partly AWWA Committee on Water Use, 1973.[32] Remaining data for 1969–73 were from water utility reports.

In 1963, the total population served by the 195 systems listed in Table 2-11 was about 66 million people and total water use was 10,393,000 gpd, an average per capita use of 157 gpd. Estimates of nationwide water use from all public supplies indicated a per capita use of 151 gpd in 1960, 155 in 1965, and 166 in 1970.[1,30,31]

The last column in Table 2-11 identifies the code number of the water use region where each city is located (see Fig. 2-1). Totals and per capita use for all public supplies in each of these regions are given in Table 2-4.

Variations in Water Demands. Flow rates of water use—in this case, the demands on the water system—vary not only from year to year and from season to season, but more importantly, from day to day and hour to hour. Some examples of short-term variations are given in Figs. 2-4, 2-5, and 2-6. Figure 2-4 contains data for Creekside Acres, East Bay Municipal Utility District, California. The upper graph shows the seasonal trend in daily use for 15 months and the lower graph shows hourly variations during 1 week in winter and 1 week in summer. Lawn watering accounts for the bulk of the much greater water use in summer. Figure 2-5 shows the components, hour by hour, of public supply use at Kansas City, Missouri, on the day of maximum use in 1954. Figure 2-6 shows hourly variations during a maximum-use day for two smaller water systems—Strathmore, N.J. (serving about 8000 people in 1966 and operated by the Matawan Township Municipal Utilities Authority) and Levittown, Pennsylvania (about 4 times the size of the Strathmore system).

Two of the more important facts to determine with respect to variations in water use are maximum daily use and maximum hourly use. These peak demands are often expressed as ratios to the average rate of water use. Table 2-11 contains this information for many of the large water systems in the United States. Table 2-12 lists average daily use and maximum daily use for these supplies combined so as to show regional characteristics in water use,

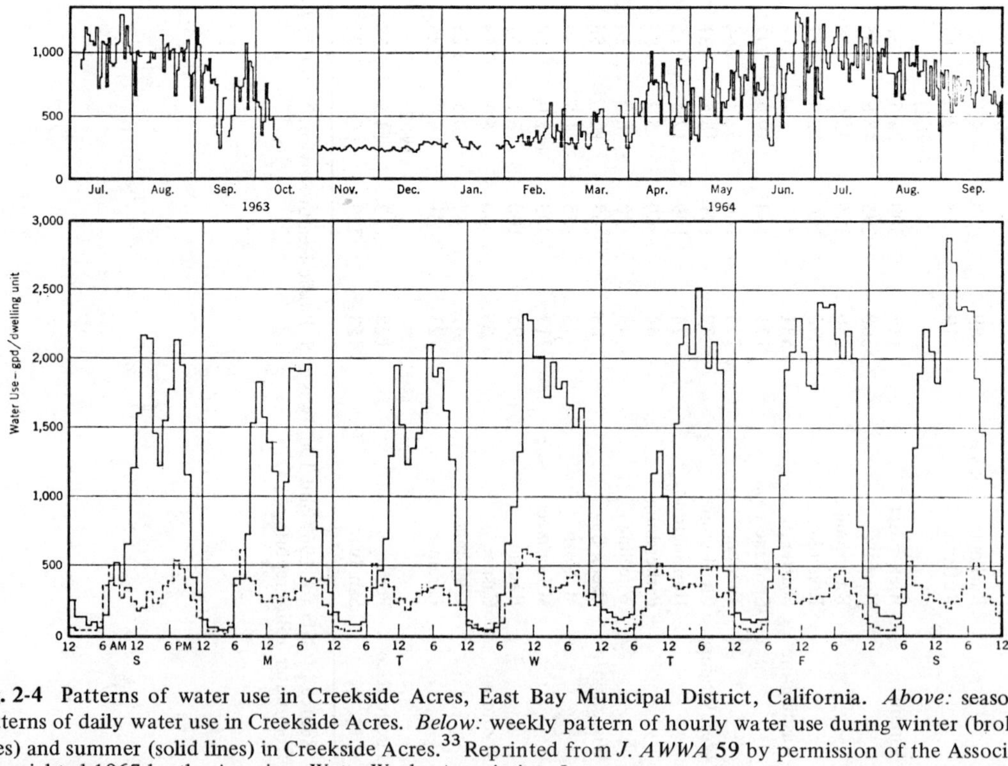

Fig. 2-4 Patterns of water use in Creekside Acres, East Bay Municipal District, California. *Above:* seasonal patterns of daily water use in Creekside Acres. *Below:* weekly pattern of hourly water use during winter (broken lines) and summer (solid lines) in Creekside Acres.[33] Reprinted from *J. AWWA* 59 by permission of the Association. Copyrighted 1967 by the American Water Works Association, Inc.

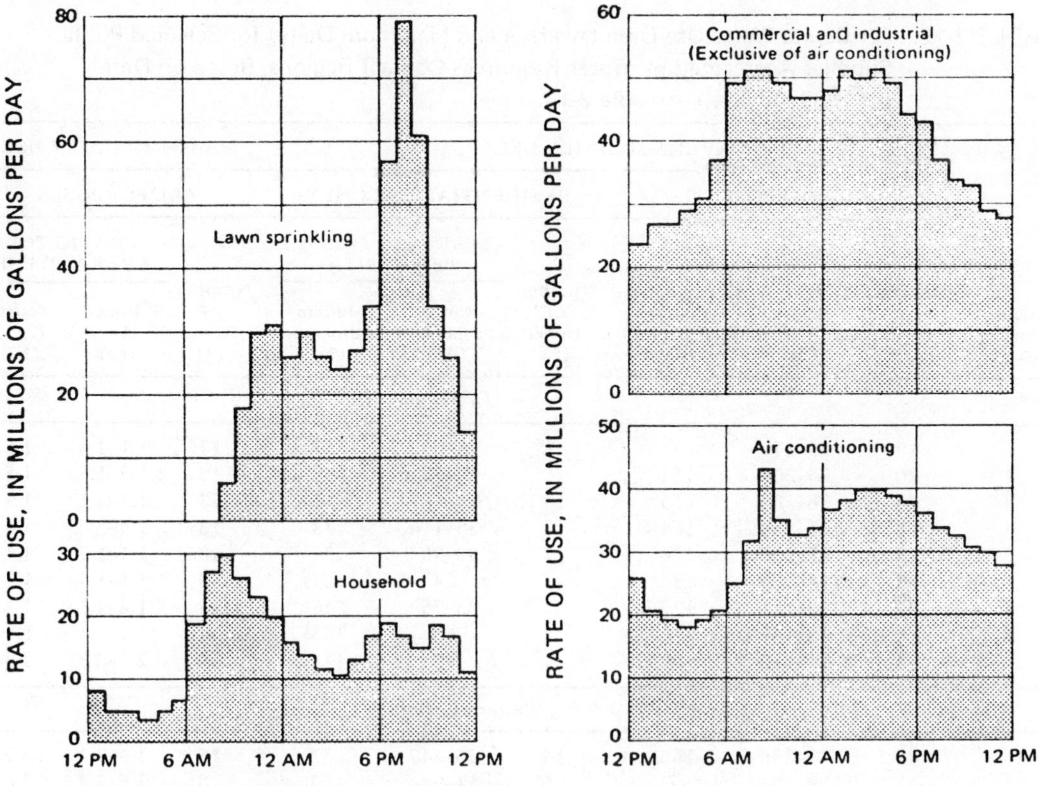

Fig. 2-5 Hourly trends in water use on maximum day of use (July 12) in 1954 of public water supply of Kansas City.[34] (*Based on data by M. P. Hatcher.*)

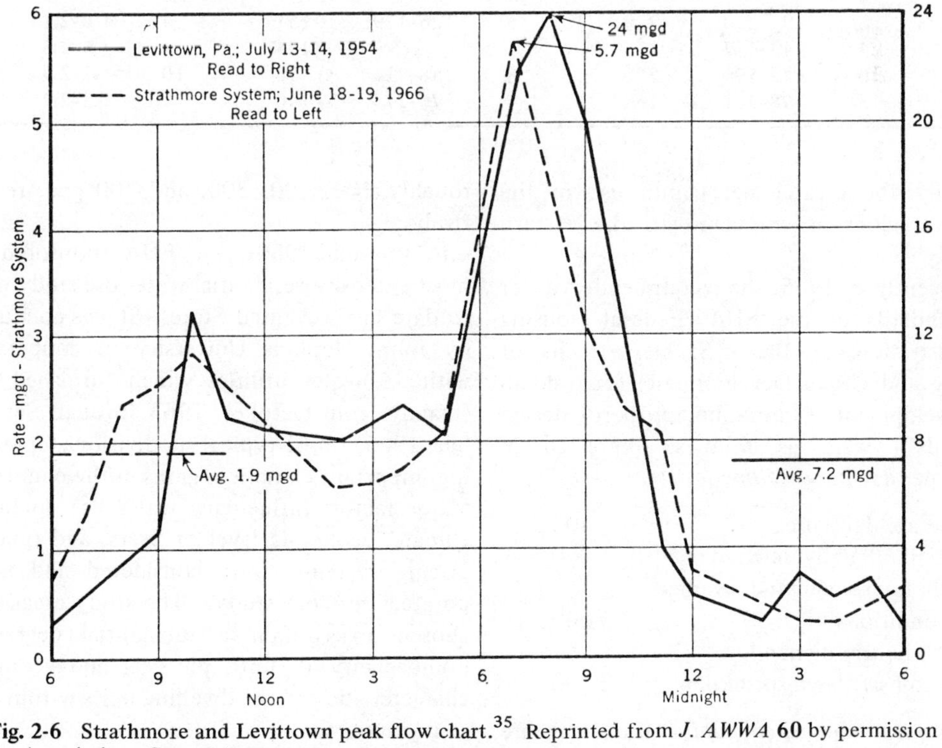

Fig. 2-6 Strathmore and Levittown peak flow chart.[35] Reprinted from *J. AWWA* **60** by permission of the Association. Copyrighted 1968 by the American Water Works Association, Inc.

TABLE 2-12. Per Capita Water Use Data (Average and Maximum Daily) for Selected Public Supplies Aggregated by Water Resources Council Regions, Based on Data for 1962-63 given in Table 2-11.

	AVERAGE DAILY USE						MAXIMUM DAILY USE		
	ALL PURPOSES			RESIDENTIAL USE ONLY			ALL PURPOSES		
		PER CAPITA			PER CAPITA			RATIO TO AVERAGE USE	
Region Number Col. (1)	Number of Items (2)	Range (gpd/person) (3)	Median (gpd/person) (4)	Number of Items (5)	Range (gpd/person) (6)	Median (gpd/person) (7)	Number of Items (8)	Range (/1- /1) (9)	Median (/1) (10)
				Eastern Regions					
1	19	95-179	125	9	35-83	57	17	1.1-2.4	1.6
2	20	36-209	154	6	29-79	40	13	1.2-1.8	1.5
3	23	78-200	115	18	29-107	71	23	1.1-2.1	1.5
4	18	101-238	164	8	35-116	48	13	1.4-2.0	1.6
5	18	87-158	124	12	28-86	36	18	1.1-1.7	1.4
6	3	114-179	133	1	64	(64)	3	1.1-1.4	1.3
7	11	99-181	122	5	35-76	46	11	1.4-2.5	1.8
8	6	88-186	116	2	63	(63)	4	1.2-2.1	1.4
9	3	64-112	94	1	45	(45)	2	2.1-2.7	2.4
				Western Regions					
10	19	87-286	163	13	26-147	72	17	1.4-3.1	2.2
11	11	109-174	132	5	51-95	70	8	1.6-2.4	2.0
12	7	117-266	153	3	93-108	98	7	1.5-2.4	1.9
13	4	125-254	193	3	67-152	100	3	2.1-2.2	2.1
14	0	–	–	0	–	–	0	–	–
15	6	193-290	228	3	126-179	151	4	1.8-2.1	2.0
16	4	243-377	275	1	125	(125)	4	1.6-2.3	1.8
17	10	142-594	250	8	58-214	98	10	1.2-2.9	2.4
18	13	98-331	160	4	70-275	140	7	1.2-2.2	1.6

especially the greater per capita use in the western regions as compared to the eastern regions.

As recently as 1965, the recommended water use standards of the FHA (Federal Housing Administration of the U.S. Department of Housing and Urban Development) for residential developments ("minimum property design standards") were as follows, expressed in gallons per day *per dwelling unit*:

Average daily use	400
Maximum daily demand	800
Peak hourly demand, average conditions	2,000
Peak hourly demand, extensive lawn sprinkling	2,800

The corresponding *per capita* use rates are roughly 100, 200, 500, and 700 gpd, respectively.

In the mid-1960s, the FHA sponsored the most extensive residential water use study made to date in the United States. It was conducted by Johns Hopkins University in cooperation with 16 water utilities which furnished data (mainly from October, 1963 through September, 1965) on 41 typical subdivisions representing important climatic regions in the country.[33] Major factors influencing water use, including climate, economic level of users, and types of pricing systems, were considered and incorporated into the study. The study areas were chosen to contain a substantial degree of homogeneity in both physical and economic characteristics of the dwelling units within each respective area.

According to the authors, the results of this research "show that residential water demands in any given area are influenced by the total number of consumers, their economic level based on the market value of their homes, the average irrigable area of their lawns, the rate of moisture loss through evapotranspiration, and the quantity of effective precipitation."

The revised FHA standards of 1968 and 1973 for residential and related water demands no longer contain the specific water use rates (gallons per day per dwelling unit) cited above, but

TABLE 2-13. List of Study Areas Included in FHA–Johns Hopkins Project on Residential Water Use[7]

Type of Study Area	Metropolitan Area	Participating Utility	Study Area
Metered public water and public sewers (western U.S.)	Oakland, Cal.	East Bay Municipal Utility District	San Lorenzo
			Creekside Acres
			Burton Valley
			Chabot Park Estates
	San Diego, Cal.	City of San Diego	Rancho Hills
			Ruffin Road residential
			Muirlands
		Helix Irrigation District at La Mesa	El Cajon
			Lemon Grove
		California Water and Telephone Co. at Chula Vista	Minot Avenue
Metered public water and public sewers (eastern U.S.)	Des Moines, Iowa	Des Moines Water Works	Patricia Park
	Forth Worth, Tex.	City of Fort Worth	Sunset Heights
			Edgecliff Village
	Little Rock, Ark.	Little Rock Municipal Water Works	Richland Subdivision
	Washington, D.C.	Washington Suburban Sanitary District	Palmer Park
			Glenmont
			English Manor
			Northwest Branch Estates
	Baltimore, Md.	Baltimore City and Baltimore County	Country Club Park
			Pine Valley
			Campus Hills
	Philadelphia, Pa.	City of Philadelphia	Normandy
			Benton Street area
Metered public water and septic tanks (eastern U.S.)	Des Moines, Iowa	Des Moines Water Works	Clive
	Baltimore, Md.	Baltimore City and Baltimore County	Hampton
	Philadelphia, Pa.	Philadelphia Suburban Water Co.	Downeast
			St. Albans
			Dogwood Lane
Flat-rate public water and public sewers (western U.S.)	Sacramento, Cal.	City of Sacramento	Golf Course Terrace
	Great Falls, Mont.	City of Great Falls	South Side
			Riverview
			Country Club Addition
	Denver, Colo.	Denver Board of Water Commissioners	57th and Tennyson
			47th and Milwaukee
			11th and Glencoe
			3rd and Jasmine
Apartment areas	San Diego, Cal.	City of San Diego	Apartment area
	Denver, Colo.	Denver Board of Water Commissioners	Apartment area
	Washington, D.C.	District of Columbia, Department of Sanitary Engineering	Apartment area
		Washington Suburban Sanitary District	Cheverly Apartments
			Blair House Apartments

instead state that if reliable local records of water use by existing water supply systems are not available, the criteria developed in the Johns Hopkins study may be used.

Although the procedures recommended by the study are difficult to apply because of the costliness and uncertain reliability of some of the data required for the calculations (inter-area housing values; actual evaporation; potential evapotranspiration; effective precipitation), the methods have been commended as substantial efforts in a rational attempt to determine water use on a theoretical basis. Furthermore, the factual water use information on which the study is based is unusually detailed and well documented. Some of the data are summarized in the next three tables.

The 41 subdivisions (study areas) analyzed are listed in the far right column of Table 2-13. The far left column shows the five groups of water-sewer categories into which the 41 study areas were classified. Economic and climatic characteristics of each category group are presented in Table 2-14 and per capita use data for each group are shown in Table 2-15. Note that a major feature of this study was differentiating quantitatively between residential uses that are predominantly indoors (domestic) and that which is often a very large use outdoors—lawn sprinkling.

One result of this study was a set of curves for predicting the expected maximum daily demand for a development having a specific number of dwelling units and a given housing density expressed in dwellings per acre (Fig. 2-7). An underlying assumption was that the development's water supply would be metered and that there would be public sewers. The formula used in preparing the curves of expected maximum daily demand and a sample calculation are as follows:

$$\overline{Q}_{(MXDY)}/a = \overline{Q}_d/a + 0.6c\,\overline{L}_s\overline{E}_{pot} \qquad (1)$$

TABLE 2-14. Characteristics of Study Areas[7]

Column	Type of Study Area (1)	Number of Dwelling Units (2)	Average Market Value (dollars) (3)	Population Density (persons per dwelling) (4)	Housing Density (dwellings per acre) (5)	Irrigable Area (acres per dwelling) (6)	Annual Precipitation (inches) (7)	Annual Temperature (°F) (8)
Metered public water and public sewers								
	West (10 areas)							
	Minimum	63	13,200	2.7	1.28	0.079	7.09	63.3
	Mean	158	24,400	3.8	3.61	0.205	14.9	67.4
	Maximum	295	42,400	4.5	6.14	0.582	27.2	72.4
	East (13 areas)							
	Minimum	124	9,500	3.4	1.32	0.046	29.2	72.2
	Mean	235	18,500	4.1	4.16	0.183	39.0	75.3
	Maximum	410	35,300	4.9	9.35	0.595	48.7	84.7
Metered public water and septic tanks (5 areas)								
	Minimum	44	19,700	3.1	0.94	0.243	30.4	70.8
	Mean	178	27,600	4.1	1.66	0.516	43.3	72.5
	Maximum	307	36,000	4.9	2.73	0.868	47.5	75.0
Flat-rate public water and public sewers (8 areas)								
	Minimum	91	11,100	2.7	2.75	0.094	14.1	65.4
	Mean	184	19,100	3.7	3.99	0.146	14.7	68.8
	Maximum	326	53,500	4.7	4.98	0.191	16.3	73.3
Apartment areas (5 areas)								
	Minimum	34	6,500	1.8	23.4	–	10.4	69.4
	Mean	789	9,200	2.6	36.2	–	28.5	72.6
	Maximum	2,373	15,200	3.0	50.0	–	43.5	76.3
All 41 study areas								
	Minimum	34	6,500	1.8	0.94	0.046	7.09	63.3
	Mean	267	20,000	3.8	7.6	0.227	27.6	71.4
	Maximum	2,373	53,500	4.9	50.0	0.868	48.7	84.7

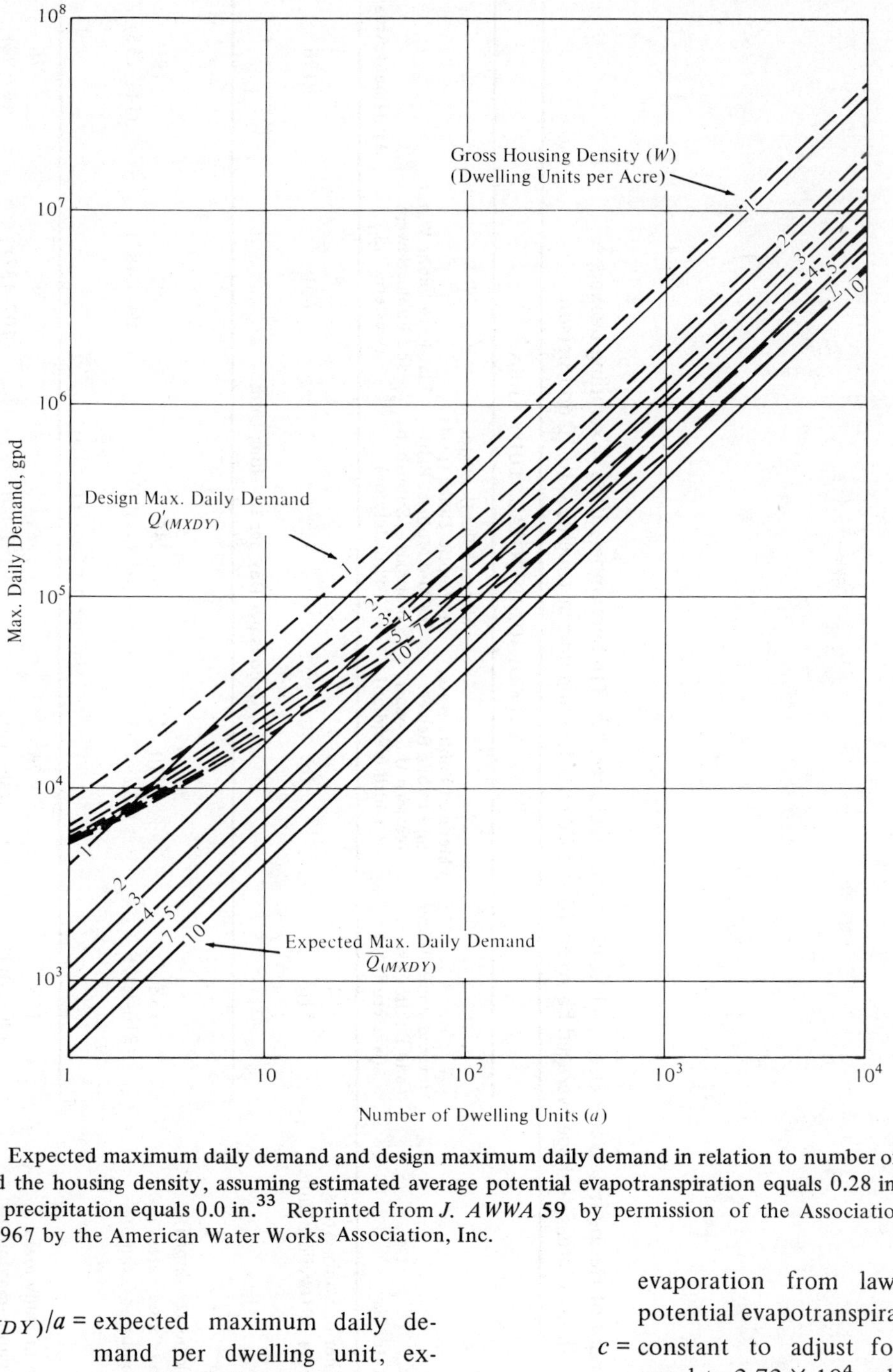

Fig. 2-7 Expected maximum daily demand and design maximum daily demand in relation to number of dwelling units and the housing density, assuming estimated average potential evapotranspiration equals 0.28 in./day and effective precipitation equals 0.0 in.[33] Reprinted from *J. AWWA* **59** by permission of the Association. Copyrighted 1967 by the American Water Works Association, Inc.

where

$\overline{Q}_{(MXDY)}/a$ = expected maximum daily demand per dwelling unit, expressed in gallons per day

a = number of dwelling units

$\overline{Q}_d$ = expected average domestic (household) use, in gallons per day

0.6 = a coefficient to adjust for the difference between actual evaporation from lawns and potential evapotranspiration

c = constant to adjust for units, equal to 2.72×10^4 gallons per acre-inch of water

$\overline{L}_s$ = average irrigable area, in acres per dwelling unit

$\overline{E}_{pot}$ = estimated average potential evapotranspiration for the period of demand in question, in

TABLE 2-15. Summary of Residential Water Use in 41 Study Areas, October, 1963 to September, 1965, Excluding Leakage[7]
Note: in most areas, leakage averaged between 10 and 30 gpd/dwelling unit or roughly 3 to 10 gpd/person.

	TYPE OF RESIDENTIAL STUDY AREA				
Type of Water Use	Metered Public Water and Public Sewers (western U.S.)	Metered Public Water and Public Sewers (eastern U.S.–east of 100th meridian)	Metered Public Water and Septic Tanks (parts of Des Moines, Baltimore area, Suburban Philadelphia)	Flat-Rate Public Water and Public Sewers (western U.S.)	Apartment Areas
Number of Study Areas	10	13	5	8	5
Number of Dwelling Units	516	769	529	601	3,196
	Gallons per Day per Dwelling Unit				
Residential use (domestic and sprinkling combined)					
Average annual: mean of study areas	458	310	245	692	191
(Minimum–maximum)	(316–679)	(212–421)	(189–309)	(504–1,487)	(141–255)
	Gallons per Capita per Day				
Average annual: mean of study areas	123	76	61	193	76
(Minimum–maximum)	(90–201)	(50–113)	(51–80)	(107-437)	(47–92)
Domestic use only					
Average annual: mean	67	51	47	66	62
(Minimum–maximum)	(41–127)	(39–66)	(44–51)	(49–111)	(45–73)

Sprinkling use only					
Average annual: mean	66	25	14	127	14
(Minimum–maximum)	(29–97)	(11–52)	(7–29)	(58–326)	(2–26)
	Ratios of Maximum Day or Peak Hour to Average Annual				
Domestic use only					
Ratio of maximum day to average annual:					
mean	1.84	1.31	1.29	1.80	1.42
(Minimum–maximum)	(1.4–2.4)	(1.1–1.5)	(1.2–1.4)	(1.2–2.2)	(1.2–2.1)
Ratio of peak hour to average annual:					
mean	4.94	2.60	2.76	4.14	4.53
(Minimum–maximum)	(3.3–6.3)	(2.1–3.8)	(2.2–3.1)	(1.8–6.2)	(1.9–13.6)
Sprinkling use only					
Ratio of maximum day to average annual:					
mean	3.85	8.59	13.2	5.37	16.1
(Minimum–maximum)	(3.2–5.2)	(5.0–13.9)	(7.4–17.0)	(3.4–8.5)	(6.2–31.0)
Ratio of peak hour to average annual:					
mean	11.8	24.2	46.0	13.3	90.6
(Minimum–maximum)	(9.1–16.4)	(11.8–43.6)	(22.2–76.4)	(7.0–22.3)	(15.8–362)
Residential use					
(domestic and sprinkling combined)					
Ratio of maximum day to average annual:					
mean	2.12	2.52	2.93	3.45	1.96
(Minimum–maximum)	(1.7–2.4)	(1.6–3.7)	(2.0–5.4)	(2.4–5.1)	(1.7–2.3)
Ratio of peak hour to average annual:					
mean	5.42	5.86	7.52	7.93	5.78
(Minimum–maximum)	(4.4–6.4)	(2.9–8.6)	(5.6–13.0)	(5.1–12.8)	(2.5–16.5)

inches of water per day (assumed for most purposes to be about 0.28 in. on the maximum day of use)

Assuming as an example, an area with a

- gross housing density of 3 dwellings per acre $(=W)$; (see third item below)
- regionally adjusted average market value of \$25,000 per dwelling unit $(=V)$; used in computing next item
- expected average domestic use of 244 gallons per day per dwelling unit, derived from the expression $(157 + 3.46\ V)$, where V equals the regionally adjusted market value per home in thousands of dollars
- irrigable area per dwelling unit $(=\overline{L}_s)$ of 0.20 acre, derived from the expression $(0.803\ W^{-1.26})$, where W equals the gross housing density in dwellings per acre
- estimated potential evapotranspiration $(=\overline{E}_{pot})$ of 0.28 inch on maximum day of use

the expected maximum daily demand is:

$$244 + (0.6)(2.72) \times 10^4)(0.20)(0.28)$$

$$= 1160 \text{ gpd/dwelling unit}$$

Some of the more recent papers on methods and criteria for estimating and forecasting peak water demands take into consideration the effects on water use of changes in the price structure of water rates.[36,37,38]

An excellent report on public supplies was prepared by the state of California in 1968.[39] The report is not only well documented in terms of definitions and methods, but also contains substantial information on monthly variations in water use.

Fire Control. Water needs for fire control are normally added to the rate of use of water on the maximum day. The required fire flow is expressed in gpm (gallons per minute) and in duration of time of continuous flow, ranging from 2 to 10 hr. In the past, recommendations for municipal fire control promulgated by the National Board of Fire Underwriters have included the following data, relating fire-flow requirements to population:

Population	Fire flow, in gpm
6,000	2,500
10,000	3,000
17,000	4,000
40,000	6,000
75,000	8,000
95,000	9,000
120,000	10,000
150,000	11,000
200,000	12,000

The Insurance Services Office, an insurance-rating organization that now performs the functions formerly carried out by the National Board, uses the following formula as a preliminary guideline in estimating required fire flows for individual facilities or structures, based on the type of construction and the total floor area:

$$F = 18\ C(A)^{0.5} \qquad (2)$$

where

F = the required flow in gpm
C = coefficient related to the type of construction
A = total floor area (including all stories but excluding basements)
(For fire-resistive buildings, consider only the 6 largest successive floor areas if the vertical openings are unprotected; if properly protected, consider only the 3 largest successive floor areas.)

Many other factors are considered before a final estimate is made, including whether a particular structure has a low-hazard or high-hazard fireloading, the separation between sturctures, and whether or not there is a sprinkler system. Some types of facilities by their very nature require greater than normal fire flows, such as lumber yards, petroleum storage tanks, refineries, grain elevators, and large chemical plants.

A "Guide for Determination of Required Flow" was published by the Insurance Services Office on June 1, 1972, upon which the graph in Fig. 2-8 is based. It indicates the range in values of required fire flows.[40]

TABLE 2-16. Commercial Demands on a Typical Summer Day at 5 Sites in Baltimore, Md.[41]

Establishment	Average Demand (gpd)	RATE OF MAXIMUM HOURLY DEMAND (gph)	(gpd)	Hour of Peak Occurrence	Ratio of Maximum Hourly Commercial Demand to R-40 Demand[a]
Gasoline service station	636	522	12,500	7–9 PM[b]	2.3/1
Washmobile[c]	9.370	3,140	75,500	11–12 AM	14/1
Laundromat[d]	3,670	522	12,500	12–1 PM	2.3/1
Commercial laundry[e]	3,570	825	19,800	10–12 AM	3.7/1
Motel (110 units)	5,500	762	18,250	8–9 AM	3.4/1
Apartment (22 units)	4,300	486	11,650	5–6 PM	2.2/1

[a]Lot type R-40 (home on 1-acre lot) peak hourly demand for single service is 5400 gpd.
[b]Erratic, with lower peaks occurring 1–2 PM and 5–7 PM.
[c]Cars washed at rate of 24 per hour. All facilities—steam cleaning of tires, automatic rinsing—available.
[d]Ten washers. Open 8 AM–8 PM. [e]Facilities equivalent to ten washers. Open 8 AM–8 PM.

Tables 2-16 and 2-17 reprinted from *J. AWWA* **53** by permission of the Association. Copyrighted 1961 by the American Water Works Association, Inc.

Public Supplies for Commercial Uses and Institutions. Most businesses and institutions in a city depend on the public system for their water supply. Guides for estimating water use for these purposes, in gallons per day, were included in Table 2-9 under the headings:

Airports	Laundries
Bath houses	Motels
Clubs, country	Restaurants
Courts, tourist	Schools
Factories	Service stations
Hotels	Stores
Institutions other than hospitals	Swimming pools
Institutions, hospitals	Theaters

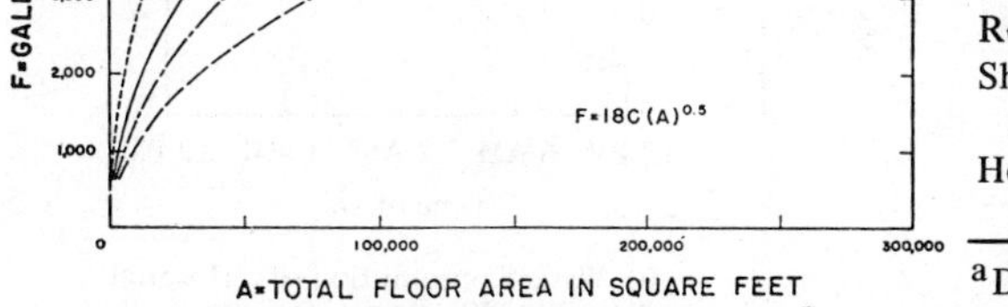

Fig. 2-8 Representative fire flows required for various types of construction in terms of total floor area. From the fire flow guide prepared by Insurance Services Office.[40] Reprinted from *J. AWWA* **65** by permission of the Association. Copyrighted 1973 by the American Water Works Association, Inc.

More detailed information on several types of commercial uses in the Baltimore area is contained in Table 2-16.[41] Water use data developed during February and March, 1960, at several kinds of institutions in the Gainesville, Florida area—schools, motels, shopping centers, hospitals, and restaurants—are summarized in Table 2-17.[42] Two of the reasons for choosing February and March were that showers in schools were heavily used because sports activi-

TABLE 2-17. Summary of Water Consumption Data, Gainesville, Florida[42]

Institution	Suggested 24-hr Average Flow (gpd[a])	FLOW FACTORS[b] 12-hr Maximum	4-hr Maximum	1-hr Maximum
Schools				
Elementary (9)	10	2.0	4.0	5.0
Junior High (2)	12	2.0	4.0	5.0
Senior High (2)	20	1.75	2.5	3.25
Combined (1)	14	1.75	2.5	3.25
Motels (6)	70	1.5	2.5	4.5
Restaurants (3)	9	1.4	1.7	2.0
Shopping centers[c] (1)	250	1.5	1.9	2.1
Hospitals[d] (1)	275	1.25	1.75	1.9

[a]Denominator for schools is student; for motels, guest; for restaurants, meal; for shopping centers, 1,000 sq ft; for hospitals, bed.
[b]Factors by which 24-hr average flows are multiplied to obtain various maximum values.
[c]Based on a shopping center with floor area of 133,458 sq. ft.
[d]Based on a 120-bed hospital.

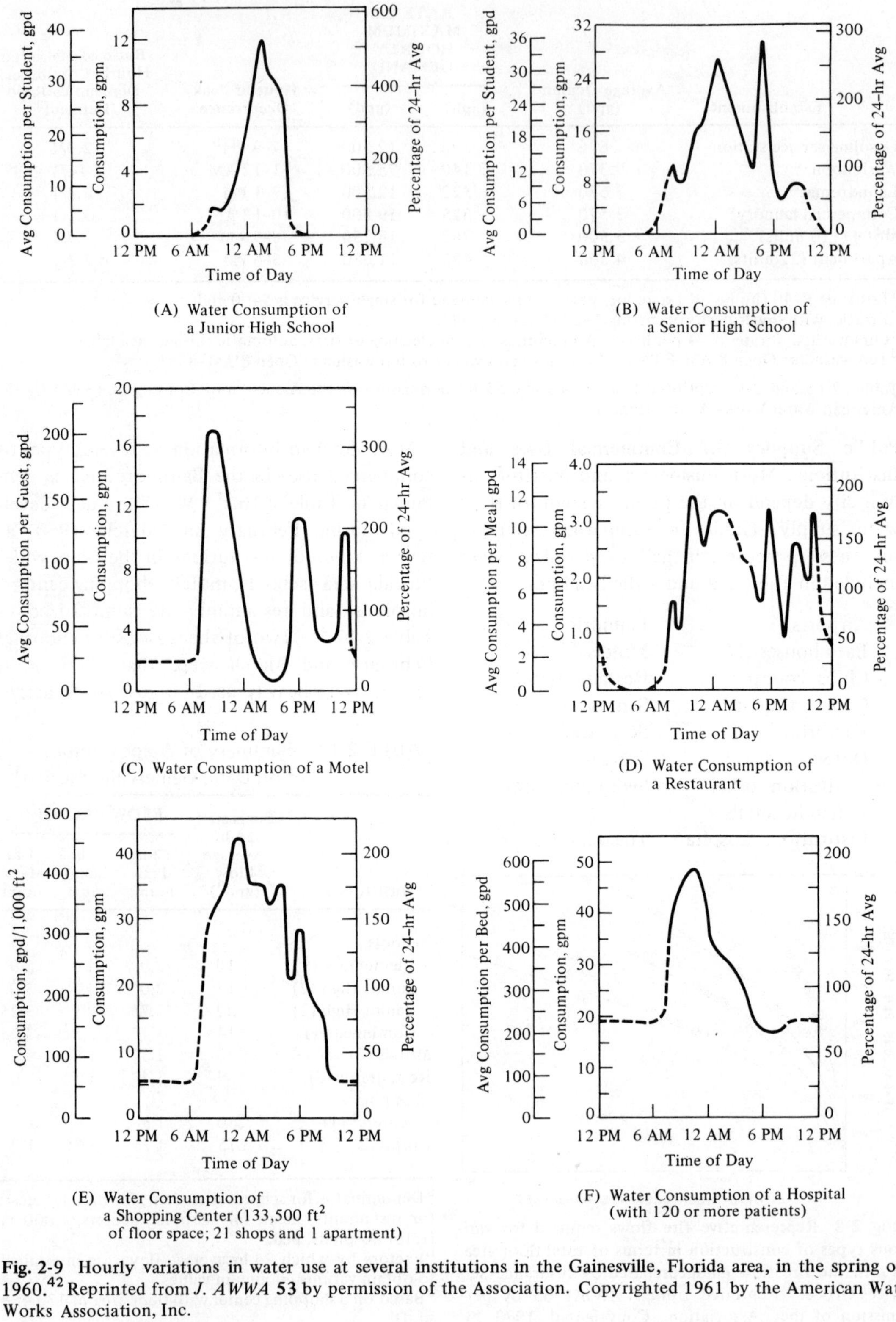

Fig. 2-9 Hourly variations in water use at several institutions in the Gainesville, Florida area, in the spring of 1960.[42] Reprinted from *J. AWWA* **53** by permission of the Association. Copyrighted 1961 by the American Water Works Association, Inc.

ties were at their peak and that motels and restaurants were serving a maximum number of guests because the tourist season was at its height. Figure 2-9 shows some of the hourly variations at several of these institutions.

WATER FOR LIVESTOCK

The water needs for livestock listed below, expressed in gallons per day per animal, are largely adapted from USPHS data:[25]

	gpd/animal
Beef cattle and steers	12
Dairy cows	
ordinary dairy herd (dry and producing cows)	15
high-producing cows (12 gal of milk or more per day	35
Goats	2
Hogs	4
Horses and mules	12
Sheep	2
Chickens	0.05–0.10
Ducks	0.06–0.18
Turkeys	0.10–0.18

Per capita rates of water use by livestock are shown by individual states in Table 2-18.[31] They were suggested in 1965 by district offices of the U.S. Geological Survey in those states.

MAJOR WATER-USING INDUSTRIES

Most industries that use large quantities of water, such as a million or more gallons a day per plant site, have their own water systems and sources of water and thus are self-supplied. Some of these same industries may, however, also be connected to a public water supply system and use that water for drinking and possibly for sanitary services.

The types of industries that use the largest quantities of water in the United States are thermoelectric utilities (mainly steam electric power plants), iron and steel, chemicals, pulp and paper, petroleum refining, and food. These industries account for more than 60 percent of the water used by all manufacturing industries. Water uses for industry are for air-conditioning, for other cooling and condensing purposes, for processing, for boiler feed, for sanitary services, and for drinking. Cooling water (for uses other than air-conditioning) accounts for about 65 percent of all industrial use for manufacturing.

The principal source of information on industrial water use for manufacturing is the U.S. Census Bureau's Census of Manufactures conducted every 5 years.[8] Most of the data compiled and analyzed are for industrial plants that have an intake (withdrawal) of at least 20×10^6 gal of water a year, equivalent to 0.0548×10^6 gal a day; such plants account for more than 95 percent of all water withdrawn

TABLE 2-18. Per Capita Rates of Water Use by Livestock, in Gallons per Day per Animal, 1965[31]

	Milk Cows	Beef Cattle	Horses, Mules	Hogs	Sheep	Goats	Chickens	Turkeys
Arizona	30	–	–	–	–	–	–	–
Arkansas	30	15	15	2	–	–	0.05	0.08
California	25	15	15	–	–	–	–	–
Colorado	15	11	–	1.3	–	–	0.09	0.18
Connecticut	–	–	–	–	–	–	0.05	0.08
Kentucky	40	8	12	–	–	–	–	0.10
Maryland	35	–	12	4	–	–	–	–
Missouri	30	20	–	–	–	–	–	0.08
Nevada	–	6	–	–	–	–	–	–
New Mexico	25	–	–	2	–	–	0.05	–
North Carolina	60	30	30	6	–	–	0.12	0.30
South Dakota	35	15	12	5	–	–	0.10	0.15
All other states	20	10	10	3	2	2	0.04	0.06[a]

[a]Also used for ducks.

TABLE 2-19. Water Intake, Gross Water Used, and Water Discharged by Industry Group and Selected Industries, 1968 and 1964[8]

		1968				1964
SIC Code Number	Industry Group / Industry / Industry Subclassification	Total Water Intake, Fresh and Brackish (mgd)	Fresh Water Intake (mgd)	Gross Water Used Including Water Recirculated or Reused (mgd)	Total Water Discharged (mgd)	Total Water Intake (mgd)
20	Food and kindred products	2,222	2,027	3,688	2,063	2,082
201	Meat products	288	288	466	274	288
202	Dairy products	153	151	214	145	170
203	Canned, cured, and frozen foods	353	345	619	329	252
204	Grain mill products	205	197	334	184	159
2046	Wet corn milling	170	162	293	153	132
206	Sugar	666	521	981	638	663
2061	Raw cane sugar	312	263	452	307	337
2062	Cane sugar refining	230	134	274	222	178
2063	Beet sugar	123	123	258	110	148
208	Beverages	307	293	578	266	296
2082	Malt liquors	181	170	386	159	162
2085	Distilled liquor, except brandy	60	60	79	49	82
21	Tobacco manufactures	16	16	211	14	8
22	Textile mill products	422	416	899	373	405
24	Lumber and wood products	323	277	562	255	414
25	Furniture and fixtures	11	11	16	11	8
26	Paper and allied products	6,170	5,707	17,868	5,693	5,655
2611	Pulpmills	742	707	2,225	726	764
2621	Papermills, except building paper	3,271	3,126	8,548	2,949	3,099
2631	Paperboard mills	1,981	1,699	6,657	1,882	1,581
28	Chemicals and allied products	12,263	8,082	25,797	11,438	10,682
281	Industrial chemicals	9,227	5,589	18,238	8,677	8,301
2812	Alkalies and chlorine	1,058	805	1,573	797	1,447
2815	Cyclic intermediates and crudes	381	285	1,003	353	304
2818	Industrial organic chemicals, not elsewhere classified	5,863	2,786	10,863	5,570	4,679
2819	Industrial inorganic chemicals, not elsewhere classified	1,411	1,271	3,923	1,282	1,386
282	Plastics materials and synthetics	1,740	1,460	4,452	1,595	1,452
2821	Plastics materials and resins	633	521	1,373	545	597
2822	Synthetic rubber	170	170	926	159	167
2823	Cellulosic man-made fibers	499	499	1,079	482	416
2824	Organic fibers, noncellulosic	438	271	1,077	408	268
29	Petroleum and coal products	3,931	2,247	19,972	3,334	3,830
2911	Petroleum refining	3,910	2,225	19,942	3,315	3,803
295	Paving and roofing materials	22	19	27	16	27
30	Rubber and plastics products, not elsewhere classified	370	359	737	351	340
3011	Tires and inner tubes	170	170	392	162	164
31	Leather and leather products	44	44	55	41	38
32	Stone, clay, and glass products	688	586	1,131	597	682
3241	Cement, hydraulic	323	285	430	271	312
33	Primary metal industries	13,712	12,290	21,315	12,866	12,603
331	Blast furnace and basic steel products	12,033	11,041	17,819	11,312	11,088
3312	Blast furnaces and steel mills	11,153	10,170	16,860	10,441	10,452
332	Iron and steel foundries	134	104	225	107	107
333	Primary nonferrous metals	1,211	816	2,512	1,129	1,140
3331	Primary copper	159	104	460	101	170

TABLE 2-19. cont.

		1968				1964
SIC Code Number	Industry Group Industry Industry Subclassification	Total Water Intake, Fresh and Brackish (mgd)	Fresh Water Intake (mgd)	Gross Water Used Including Water Recirculated or Reused (mgd)	Total Water Discharged (mgd)	Total Water Intake (mgd)
3334	Primary aluminum	562	559	1,405	548	540
34	Fabricated metal products	186	170	463	178	142
35	Machinery, except electrical	518	449	926	496	433
36	Electrical equipment and supplies	348	342	1,014	323	279
37	Transportation equipment	858	808	2,496	803	663
371	Motor vehicles and equipment	392	378	1,674	362	334
372	Aircraft and parts	392	392	721	370	263
38	Instruments and related products	104	101	323	99	79
39	Miscellaneous manufacturing industries	38	30	66	36	36
	United States total, all manufacturing industries	42,375	34,120	97,810	39,112	38,375

Note: data above, expressed in million gallons per day (mgd), have been converted from data given to nearest 1 billion gallons per year.

for manufacturing. A summary of the data for 1968 are given in Table 2-19, with emphasis on those types of industry using the most water. The industries are listed in order by SIC (Standard Industrial Classification) code numbers. A complete list and identification of such numbers is contained in a manual prepared by the U.S. Office of Management and Budget (1972).[43] The major industrial groups are indicated by two-digit codes; the total water use by an industrial group includes the water use by all three-digit industries that have the same first two digits as the industrial group two-digit code. However, in this table only the largest water users among the "three-digit" and subclass "four-digit" industries are shown.

Estimates of water use by industry for planning and forecasting purposes are usually made on the basis of one or more of the following three types of factors:

1. gal per unit (such as a ton) of product
2. gal per employee
3. gal per dollar value added in manufacture or per dollar value of production

All three methods have some limitations and difficulties of application because of wide variations in water use practices (and industrial operations and processes); even within a single industry there are continuing changes in water use technology, and changing economic impacts such as price structures and environmental protection constraints. For example, 1960 data for steam-electric power plants, which use water almost entirely for cooling purposes, showed that their water intake (withdrawal) ranged regionally from 0.3 to 67 gal/kw-hr produced, depending primarily upon the extent to which the water was recirculated.[44] This example is extracted from the most comprehensive study to date that presents national and regional forecasts of water use, including industrial use; data are given for each of 22 regions in the conterminous United States.

One of the best concise tabulations showing variations in water use in individual types of industries is presented in Table 2-20.[45]

REFERENCES

1. Murray, C. R., and Reeves, E. B., "Estimated Use of Water in the United States in 1970," USGS Circular 676 (1972).
2. U.S. Public Health Service, "Municipal Water Facilities, Communities of 25,000 Population and Over, as of January 1, 1964," USPHS Pub. 661 (1965).
3. U.S. Public Health Service, "Municipal Water Facilities Inventory as of January 1, 1963," USPHS Pub. 775 (revised) (1964).

TABLE 2-20. Variance in Unit Industrial Water Requirements[45]

	WITHDRAWAL, GALLONS PER UNIT			
Product or User	Maximum	Average	Minimum	Conservation Technique
Steam power generation, kw-hr	170[a]	80[b]	0.45 (cooling)[c]	Recycling of cooling water through natural draft cooling towers
			1.20 (theoretical)[d]	Recycling through cooling towers
			1.32 (actual)[e]	Recycling through cooling towers
Petroleum refining, gal crude oil	44.5[f]	18.3[g]	0.8 (actual)[h]	Information insufficient
			1.0 (actual)[h]	Information insufficient
			1.73 (actual)[i]	Complete recirculation through both cooling towers and cooling ponds
Finished steel, tons	65,000[j]	40,000[k]	1,400[l]	Complete recycling through cooling towers, reuse of all flows
			4,000[m]	Recycling through cooling pond
Soaps, edible oils, lb	7.5[o]		1.57[n]	Recycling through cooling towers and colloidair separation
Carbon black, lb	14.[o]	4[o]	0.25[o]	Maximum use of air cooling devices
Natural rubber (processing), lb	6[h]		2.54[h]	Insufficient information
Butadiene, lb	305[p]	160[g]	13[q]	Theoretical estimate based on complete recycling through cooling towers
Glass containers, tons	667[r]		118[r]	Cooling water recirculated
	507[r]		192[r]	Cooling water recirculated
Automobiles (passenger), units	16,000[s]		12,000[s]	Sprays substituted for immersion rinses
Trucks, buses, units	20,000[s]		15,000[s]	Sprays substituted for immersion rinses

Source: Data assembled in cooperation with the Department of Geography, University of Illinois.

[a]NAM, *Water in Industry* (1950). Maximum of firms reporting.
[b]Ohio Water Resources Board.
[c]Farmer, A. E., "Why and How of England's Big New Cooling Towers," *Power* (July, 1952) (computed).
[d]Powell, S. T., and H. E. Bacon, "Magnitude of Industrial Demand for Process Water," *J. Am. Water Works Assoc.* (August, 1950) (computed).
[e]Salzman, M. G., and L. Elliot, "Type of Water Supply Influences Location and Layout of Texas Steam-electric Plants," *Civil Eng.* (May, 1951) (computed).
[f]Simonsen, R. N., "How Four Oil Refineries Use Water," *Sewage and Ind. Wastes* (November, 1952) (computed for Toledo plant).
[g]Jordan, H. E., "Industrial Requirements for Water," *J. Am. Water Works Assoc.* (January, 1946).
[h]Partin, J. L., "Water Conservation–A By-product of Industrial Water Control," *Sewage and Ind. Wastes*, (September, 1953).
[i]Simonsen, R. N., *op. cit.* (Latonia, Kentucky plant).
[j]Ohio Water Resources Board.
[k]Iron and Steel Institute.
[l]Fontana Division of Kaiser Steel, California (2.8% make-up).
[m]Computed for Provo, Utah steel plant (6.9% make-up).
[n]Partin, J. L., *op. cit.*, Lever Bros. Two plants, only one conserving water.
[o]Barnes, J. R., Water for industry, Rept. 9, *Resources of Freedom*, Vol. 5, Government Printing Office, Washington, 1952. Borger, Texas plant of Phillips Petroleum used for minimum.
[p]Unpublished communication from E. D. Kelly, director, Office of Synthetic Rubber, Reconstruction Finance Corporation, November 9, 1953. Figure is based on theoretical once-through use.
[q]Kelly, E. D., *op. cit.* Based on theoretical maximum conservation.
[r]Partin, J. L., *op. cit.* Two firms before and after conservation.
[s]Unpublished communication from L. A. Danse, General Motors Corporation, October 23, 1953. Experience in General Motors before and after conservation.

4. Fenerty, M. R. (ed.), *American Directory of Water Utilities 1968-69*, Public Service Research, Inc., Plainfield, N.J. (1968).
5. American Water Works Association, "A Survey of Operating Data for Water Works in 1960," AWWA Staff Report (1964).
6. American Water Works Association, "Supplement to a Survey of Operating Data for Water Works in 1960," AWWA Staff Report (1964).
7. Linaweaver, F. P., Jr. *et al.*, "A Study of Residential Water Use," U.S. Dept. of Housing and Urban Development, HUD TS-12 (Feb. 1967).

8. U.S. Bureau of the Census, "Water Use in Manufacturing," *Census of Manufactures, 1967*, MC67(1)-7 (1971).
9. American Society for Testing Materials, *Manual on Water*, 3rd ed., ASTM Special Technical Pub. 442 (1969).
10. National Association of Manufacturers, *Water in Industry* (1965).
11. United Nations Department of Economic and Social Affiars, "Water for Industrial Use," United Nations Pub. E/3058 ST/ECA/50 (1958).
12. Conklin, H. L., "Water Requirements of the Aluminum Industry," USGS Water Supply Paper 1330-C (1956).
13. Conklin, H. L., "Water Requirements of the Carbon-Black Industry," USGS Water Supply Paper 1330-B (1956).
14. Durfor, C. N., "Water Requirements of the Styrene, Butadiene, and Synthetic Rubber Industries," USGS Water Supply Paper 1330-F (1963).
15. Mussey, O. D., "Water Requirements of the Rayon and Acetate Fiber Industry," USGS Water Supply Paper 1330-D (1957).
16. Mussey, O. D., "Water Requirements of the Copper Industry," USGS Water Supply Paper 1330-E (1961).
17. Mussey, O. D., "Water Requirements of the Pulp and Paper Industry," USGS Water Supply Paper 1330-A (1961).
18. Otts, L. E., Jr., "Water Requirements of the Petroleum Refining Industry," USGS Water Supply Paper 1330-G (1963).
19. Walling, F. B., and Otts, L. E., Jr., "Water Requirements of the Iron and Steel Industry," USGS Water Supply Paper 1330-H (1967).
20. Federal Water Population Control Administration, *The Cost of Clean Water*, Vol. 3–Industrial Waste Profiles, FWPCA, U.S. Dept. of the Interior (1967).
21. U.S. Public Health Service, Ohio River Pollution Control, 78th Congress, 1st Session, House Document 266, pp. 847–1368 (1944).
22. Kammerer, J. C., "Delineation of Major Water Regions in the Conterminous United States, 1908–68," *Water Resources Review* (Jan. 1969).
23. American Society of Civil Engineers, "Urban Water Resources Research," U.S. Dept. of the Interior, Office of Water Resources Research, Contract no. 14-01-0001-1585, Project no. C-1123 (Sept., 1968).
24. Howe, C. W., and Vaughan, W. J., "In-house Water Savings," *Journal AWWA* **64**, 2 (1972).
25. U.S. Public Health Service, "Manual of Individual Water Supply Systems," USPHS Pub. 24 (revised) (1962).
26. Feth, J. H., "Water Facts and Figures for Planners and Managers," USGS Circular 601-I (1973).
27. Leopold, L. B., and Langbein, W. B., "A Primer on Water," USGS Special Pub. (1960).
28. Watson, K. S., "Water Requirements of Dishwashers and Food Waste Disposers," *Journal AWWA* **55**, 5 (1963).
29. Bogue, S. H., "Trends in Water Use," *Journal AWWA* **55**, 5 (1963).
30. MacKichan, K. A., and Kammerer, J. C., "Estimated Use of Water in the United States, 1960," USGS Circular 456 (1961).
31. Murray, C. R., "Estimated Use of Water in the United States, 1965," USGS Circular 556 (1968).
32. American Water Works Association Committee on Water Use, "Water Use (Committee Report)," *Journal AWWA* **65**, 5 (1973).
33. Linaweaver, F. P., Jr., Geyer, J. C., and Wolff, J. B., "Summary Report on the Residential Water Use Research Project," *Journal AWWA* **59**, 3 (1967).
34. Savini, J., and Kammerer, J. C., "Urban Growth and the Water Regimen," USGS Water Supply Paper 1591-A (1961).
35. Allen, H. S., "Meeting the Peak Hour Demand in a Small Water System," *Journal AWWA* **60**, 12 (1968).
36. Howe, C. W., and Linaweaver, F. P., Jr., "The Impact of Price on Residential Water Demand and Its Relation to System Design and Price Structure," *Water Resources Research* **3**, 1 (1967).
37. Hanke, S. H., "Demand for Water Under Dynamic Conditions," *Water Resources Research* **6**, 5 (1970).
38. Hanke, S. H., and Boland, J. J., "Water Requirements or Water Demands," *Journal AWWA* **63**, 11 (1971).
39. California Dept. of Water Resources, "Municipal and Industrial Water Use," Bulletin No. 166-1 (1968).
40. Carl, K. J., Young, R. A., and Anderson, G. C., "Guidelines for Determining Fire-Flow Requirements," *Journal AWWA* **65**, 5 (1973).
41. Wolff, J. B., "Peak Demands in Residential Areas," *Journal AWWA* **53**, 10 (1961).
42. Searcy, P. E., and Furman, T. deS., "Water Consumption by Institutions," *Journal AWWA* **53**, 9 (1961).
43. *Standard Industrial Classification Manual*, U.S. Office of Management and Budget (1972).
44. Wollman, N., and Bonem, G. W., *The Outlook for Water: Quality, Quantity, and National Growth*, Johns Hopkins Press (published for Resources for the Future, Inc.), Baltimore, Md. (1971).
45. Hudson, H. E., Jr., and Abu-Lughod, J., "Water Requirements," *Water for Industry*, American Association for the Advancement of Science, pp. 19–21, Washington, D.C. (1956).

3

Water Chemistry

Samuel D. Faust
Professor, Department of Environment Sciences
Rutgers, The State University
New Brunswick, New Jersey

James G. McWhorter
Senior Geologist
Dames and Moore
Cranford, New Jersey

Water chemistry may be considered as the definition of chemical systems and chemical reactions which occur in aquatic environments. In turn, an aquatic environment may be natural—i.e., a lake, stream, pond, river, etc.—or may be artificially created—i.e., a water or wastewater treatment plant. These aquatic environments may be composed of one or more solid phases in contact with the water phase which may or may not be open to the atmosphere. On the other hand, many chemical reactions may occur in the aqueous phase where solid and gaseous phases have little or no influence. Such chemistry may be encountered in water and wastewater treatment plants.

For the most part, this chapter is concerned with the chemistry which affects water quality and various treatment aspects of water and wastewater. It emphasizes the applied and interpretative aspects of water chemistry and no attempt is made to include all phases of water chemistry. More comprehensive and detailed information will be found in Refs. 1 and 2.

MODEL REACTION

The Equilibrium Theory. Many water chemistry reactions may be fitted into a generalized model:

$$aA + bB = cC + dD$$

This reaction reads, from left to right, that *a* moles of A react with *b* moles of B to yield *c* moles of C and *d* moles of D. In turn, A may represent some impurity or pollutant that must be removed from the water phase with constituent B. C represents the reaction product of A which, in many cases, is a solid phase that is removed from the water phase by sedimentation or filtration. D represents, usually, a by-product constituent of the reaction. Another interpretation of this generalized model may be that A represents an ionic constituent of water and B is some sort of a solid phase. In this case, A and B react, perhaps by ion exchange, to yield another ionic constituent C and another solid phase D, whereupon the original water quality characteristics are altered.

Another important aspect may be derived from the generalized reaction model. This is the use of the symbol = which denotes a condition or state of equilibrium. A frequent interpretation of equilibrium is one in which the *forward* reaction (left to right, A + B) and the *backward* reaction (right to left, C + D) are occurring simultaneously. That is, there is a dynamic balance between these opposing reactions. If a reaction reaches a state of equilibrium, then a mathematical expression may be formulated:

$$K_{eq} = \frac{[C]^c\,[D]^d}{[A]^a\,[B]^b} \qquad (1)$$

Thus, if one multiplies the molar concentrations of the products (raised first to their molar coefficients) and then divides by the product of the molar concentrations of the reactants (raised first to their molar coefficients), a numerical value results which is called the equilibrium constant. That is, K_{eq} values are unique and constant for the many reactions that may be fitted to A + B = C + D. As seen below, the predictive use of K_{eq} is a most important application in water chemistry.

The Kinetic Approach. There is another approach to equilibrium in which there may be more interest and which is more pertinent to water chemistry. This is the kinetic approach. Oftentimes, water chemistry reactions are brought from a nonequilibrium condition to an equilibrium condition. That is, B is added to water to react with A (in a treatment plant, perhaps) to produce C and D. We are concerned now with the velocity of the forward reaction which may be formulated:

$$V_f = K_f[A]^a\,[B]^b \qquad (2)$$

where V_f is the velocity of the forward reaction and K_f is the proportionality constant. The product, $[A]^a\,[B]^b$, now acquires a different connotation. It is the driving force for the forward reaction. As [A] and [B] decrease with time, the V_f decreases and [C] and [D] increase. Thus, the driving force for the backward reaction increases with time and may be formulated:

$$V_b = K_b\,[C]^c\,[D]^d \qquad (3)$$

where V_b is the velocity of the backward reaction and K_b is the proportionality constant. Very rarely are we able to measure exclusively V_f and V_b, but rather a V_{net} value is determined:

$$V_{net} = V_f - V_b \qquad (4)$$

The graphical relationship of V_{net}, V_f, and V_b is seen in Fig. 3-1. Eventually, $V_{net} = 0$ which

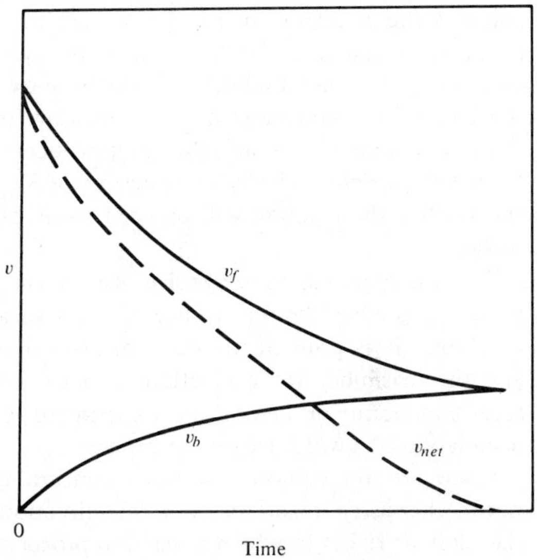

Fig. 3-1 Kinetic approach to chemical equilibrium.

means that $V_f = V_b$ and a condition of equilibrium is reached from:

$$K_{eq} = \frac{K_f}{K_b} = \frac{[C]^c\,[D]^d}{[A]^a\,[B]^b} \qquad (5)$$

The equilibrium constant is now the ratio of the forward and backward proportionality constants. Under these conditions, the concentrations of A, B, C, and D do not change analytically with time. The keys to water chemistry are the conditions of chemical equilibrium and the equilibrium constant.

CHEMICAL EQUILIBRIUM

Criteria for Chemical Equilibrium. The thermodynamic concepts underlying the condition of chemical equilibrium are utilized in this chapter mainly to denote the feasibility for a reaction to occur and to describe the water quality characteristics of ground and surface waters. A full discussion of chemical thermodynamics may be found in Refs. 2, 3, and 4.

Several criteria may be cited for a system in a condition of equilibrium: (1) the mechanical properties must be uniform and constant; that is, there must be no unbalanced forces acting on or within the system, (2) the chemical composition of a system at equilibrium must be uniform and there must be no net change in the concentrations of the reactants and products, and (3) the temperature of the system must be uniform and must be the same as the temperature of the surroundings. As noted above, the equilibrium constant, K_{eq}, is a measure of the completeness to which the reaction occurs from *left* to *right*. The larger the value of K_{eq}, the further the reaction will proceed from left to right.

This chapter makes extensive use of K_{eq} values to predict the occurrence of a chemical reaction. It is prudent, on occasion, to calculate the feasibility for a reaction to occur before conducting a laboratory experiment or installation of a water treatment process.

There are three basic questions concerning water chemistry reactions: (1) the direction, (i.e., left or right) in which a reaction proceeds under the applicable environmental conditions, (2) the extent to which the reaction proceeds, and (3) the velocity with which the reaction occurs. Chemical thermodynamics answers the first two questions whereas chemical kinetics answers the third.

State Variables. In order to describe the conditions of chemical equlibrium, one must know the state variables of the system. The mathematical relationship among the state variables is the well known equation of state:

$$PV = nRT \tag{6}$$

where P = pressure, V = volume, n = number of moles, R = gas constant, and T = temperature. As concentrations are used in water chemistry, Eq. 6 may be rearranged:

$$P = \frac{n}{V} RT = CRT \tag{7}$$

where C = concentration or, in this case, the number of moles, n, per unit of volume V. It is common practice to consider chemical reactions in aqueous systems to occur at constant P and T with the concentrations of the reactants and products as the variables. If P and T are not accepted as remaining constant in the real world, then at least they vary very little so that the feasibility or the extent of the reaction is not affected appreciably.

Reaction Feasibility. In water chemistry, the concern is with the assemblage of various reactants from a nonequilibrium state to a state of equilibrium. First, however, one should determine the feasibility (most thermodynamic texts use the term spontaneity) of the reaction. That is, can the left-to-right reaction between A and B occur? The feasibility of a proposed chemical reaction may be calculated from:

$$\Delta G = \Delta H - T\Delta S \tag{8}$$

where, for a given system, ΔG is the change in the Gibbs' free energy, ΔH is the enthalpy change, and ΔS is the entropy change. This expression comes out of the second law of thermodynamics and is applicable for systems under conditions of constant temperature and pressure and where only pressure-volume work is involved. The values of ΔG actually represent the difference between the two driving forces of a chemical reaction: ΔH and ΔS. For finite changes of ΔG, three cases may be distinguished:

1. when equilibrium exists:

$$\Delta G = 0 \tag{9}$$

2. for a feasible reaction:

$$\Delta G < 0 \tag{10}$$

3. for a nonfeasible reaction:

$$\Delta G > 0 \tag{11}$$

By convention, we say when $\Delta G = 0$, or when $\Delta H = T\Delta S$, there is no change in free energy of the reaction and a condition of equilibrium exists. By convention, we say that when $\Delta G < 0$, a negative free energy change occurs and the left-to-right reaction is feasible and should occur. By convention, we say that when $\Delta G > 0$, a positive free energy change occurs and the left-to-right reaction is not feasible and should not occur. A $+\Delta G$ value does not forbid a reaction to occur, but does indicate that the left-to-right reaction might proceed only to a very small extent. There are four possible combinations of the ΔH and $T\Delta S$ terms leading to ΔG values:

ΔH Sign	ΔS Sign	ΔG Sign	Interpretation[a]
-	+	-	Feasible
-	-	+ or -	Depends on relative magnitude of the two terms
+	+	+ or -	Depends on relative magnitude of the two terms
+	-	+	Not feasible

[a]At constant T and P.

In order to calculate free energy changes of a reaction on a uniform basis, arbitrary conditions of pressure and temperature for chemical systems are assigned. These conditions, called the standard state, are a pressure of 1 atm and a temperature of 298.15°K (25°C). In addition, every chemical element is assigned a free energy value of 0 in its standard state. In other words, elements are considered to be in their most stable physical form at 1 atm and 298.15°K. For compounds, *standard molar free energies of formation* are employed. That is, whenever a compound is formed from an element, the free energy change accompanying the formation of *1 mole* of compound is the standard molar free energy of formation. Therefore, every compound will have a ΔG° value for 1 mole, 1 atm, and 298.15°K. Now if we wish to react compounds, the free energy change of the reaction may be computed from the conventional expression:

$$\Delta G^\circ_{\text{Rex}} = \Sigma\Delta G^\circ_{\text{Products}} - \Sigma\Delta G^\circ_{\text{Reactants}} \quad (12)$$

The sign of $\Delta G^\circ_{\text{Rex}}$ now indicates whether or not the process

$$\text{Reactants (1 atm, 1 M, 298.15°K)} \longrightarrow \text{Products (1 atm, 1 M, 298.15°K)}$$

is feasible. If $\Delta G^\circ_{\text{Rex}}$ is negative, then reactants in their standard state should be converted to products in their standard state. If $\Delta G^\circ_{\text{Rex}}$ is positive, then reactants in their standard state should not be converted to products in their standard state. However, some products may be formed but not in a standard state condition. If $\Delta G^\circ_{\text{Rex}}$ is 0, then the reactants and products are already in equilibrium in their standard states.

Free Energy and the Equilibrium Constant. The relationship between $\Delta G^\circ_{\text{Rex}}$ and K_{eq} may be shown from:

$$\Delta G^\circ = -RT \ln K_{eq} = -1.364 \log K_{eq} \quad (13)$$

This expression is valid for the standard state conditions cited above. The full mathematical derivation may be found in Refs. 3 and 4. The interpretation of Eq. 13 is: (1) if ΔG° is negative, then $K_{eq} > 1$ and the left-to-right reaction is feasible. It follows that the larger the K_{eq} value, the more negative is the ΔG° value. (2) if ΔG° is positive, then $K_{eq} < 1$ and the left-to-right reaction is not feasible. Again, this does not forbid products to form from the reactants, but rather the left-to-right reaction may occur only to a small extent. Another interpretation is that the right-to-left reaction is preferred and is feasible. It follows that the smaller the K_{eq} value, the more positive is the ΔG° value. (3) if ΔG° is 0, then $K_{eq} = 1$ and the reactants and products are in equilibrium under the standard state conditions. Consequently, if the ΔG° values are known for the reactants and products, it becomes a simple mathematical exercise to calculate the K_{eq} value.

Effect of Temperature. Water chemistry reactions are rarely, if ever, conducted at 25°C. In computation of the feasibility of a reaction, it is necessary to use equilibrium constants and ΔG° values at nonstandard state temperatures. Temperature, therefore, is a factor that affects K_{eq} and ΔG° values and its effect may be easily computed. The van't Hoff relation equates temperature and the equilibrium constant. If one combines Eqs. 8 and 13 under standard state conditions, the following expression is obtained:

$$\ln K_{eq} = \frac{-\Delta H^\circ}{RT} + \frac{\Delta S^\circ}{R} \qquad (14)$$

In application of Eq. 14, it is assumed that ΔH° and ΔS° values remain constant and are independent of temperature over the range of interest. It follows, then, that a plot of $\ln K_{eq}$ versus $1/T$ should be a line whose slope is ΔH°. If the left-hand side of Eq. 14 is integrated between K_1 and K_2 and the right-hand side is integrated between T_1 and T_2, one obtains:

$$\log \frac{K_{eq2}}{K_{eq1}} = \frac{-\Delta H^\circ}{2.303R}\left(\frac{T_1 - T_2}{T_1 T_2}\right) \qquad (15)$$

Therefore, if we wish to calculate, for example, K_{eq2} at T_2, it is a simple matter provided, of course, that $\Delta H^\circ_{\text{Rex}}$ is known. It follows also that $\Delta G^\circ_{\text{Rex}}$ at T_2 may be calculated from Eq. 13. A more frequent use of Eq. 15 is to calculate $\Delta H^\circ_{\text{Rex}}$ after measurement of K_{eq1} and K_{eq2}.

Temperature is a factor that also affects the value of ΔG. The Gibbs-Helmholtz equation permits calculation of ΔG at any temperature provided, of course, that it is known at one specific temperature. From Eq. 8, one may derive the expression (at constant pressure):

$$\Delta G = \Delta H + T\left(\frac{\partial \Delta G}{\partial T}\right)_P \qquad (16)$$

(Ref. 4 gives a more thorough discussion of the Gibbs-Helmholtz equation.) A more convenient form of Eq. 16 may be obtained by rearrangement and dividing through by T^2:

$$\left(\frac{\partial\left(\frac{\Delta G}{T}\right)}{\partial T}\right)_P = -\frac{\Delta H}{T^2} \qquad (17)$$

From Eq. 17, it follows that:

$$\int d\left(\frac{\Delta G}{T}\right)_P = -\int \frac{\Delta H}{T^2}\, dt \qquad (18)$$

If ΔH is assumed constant over the temperature range of interest, one obtains upon integration:

$$\Delta G \simeq \Delta H + CT \qquad (19)$$

where C is the integration constant. Therefore, if ΔG is known at any one temperature, C may be evaluated so that ΔG at any other temperature may be calculated. Equation 19 must be regarded as an approximation and the temperature range over which it holds is limited where ΔH and C do not change appreciably. If one substitutes Eq. 13 into Eq. 19, it follows that:

$$\ln K_{eq} = \frac{-\Delta H}{RT} + \frac{C}{R} \qquad (20)$$

Note the similarity between Eqs. 14 and 20. An approximation of C may be obtained from ΔS.

Thermodynamic Calculations. Some thermodynamic values of H°, S°, and G° are seen in Table 3-1 for compounds and ions of interest in water chemistry. These data are available in most thermodynamic textbooks and in Refs. 2, 3, and 4. A source of special note is U. S. Geological Survey Bulletin 1259 which contains the thermodynamic properties of minerals and related substances.[5] Other sources may be found in Refs. 6 and 26.

Since the above thermodynamic equations are rather easy to apply, only two examples are cited here. Also, this chapter makes extensive use of these equations. For Eq. 12 we may use:

$$Ca^{2+}(aq) + CO_3{}^{2-}(aq) = CaCO_3(s)$$

ΔG° values: $-132.18 \qquad -126.22 \qquad -269.78$

$$\Sigma\Delta G^\circ_{\text{Rex}} = -269.78 - (-132.18 + -126.22)$$
$$= -11.38 \text{ kcal} \qquad (21)$$

Since the $\Delta G^\circ_{\text{Rex}}$ is negative, we can state that the left-to-right reaction is feasible. That is, Ca^{2+} and $CO_3{}^{2-}$ ions in aqueous solutions combine to form $CaCO_3$ in a solid phase. Another interpretation of this reaction would be that the right-to-left reaction would yield a positive ΔG° value and that this reaction would not be feasible. However, the right-to-left reaction occurs to some extent as witnessed by the slight

TABLE 3-1. Values of $\Delta H°$, $\Delta G°$, and $\Delta S°$ of Some Substances of Interest in Water Chemistry[a]

Substance	$\Delta H°$ (kcal/mole)	$\Delta G°$ (kcal/mole)	$\Delta S°$ (cal/mole-deg)
$Br^-(aq)$	- 28.90	- 24.57	19.29
Br_2 (l)	0.00	0.00	36.4
CH_4 (g)	- 17.89	- 12.14	44.50
CO_2 (g)	- 94.05	- 94.26	51.06
$CO_3{}^{2-}$(aq)	-161.63	-126.22	-12.7
H_2CO_3 (aq)	-167.0	-149.0	45.7
$HCO_3{}^-$(aq)	-165.18	-140.31	22.7
HCOOH (aq)	- 98.0	- 85.1	39.1
$HCOO^-$(aq)	- 98.0	- 80.0	21.9
Ca^{2+}(aq)	-129.77	-132.18	-13.2
$CaCO_3$ (calcite)	-288.45	-269.78	22.2
CaO (s)	-151.9	-144.4	9.5
$Ca(OH)_2$ (s)	-235.80	-214.33	18.2
Cl^-(aq)	- 40.02	- 31.35	13.17
Cl_2 (g)	0.00	0.00	53.29
Cu^{2+}(aq)	15.39	15.53	-23.6
Fe^{2+}(aq)	- 21.0	- 20.30	-27.1
Fe^{3+}(aq)	- 11.4	- 2.52	-70.1
H_2O (g)	- 57.80	- 54.64	45.11
H_2O (l)	- 68.32	- 56.69	16.72
HS^-(aq)	- 4.22	3.01	14.6
S^{2-}(aq)	10.0	20.0	5.3
I^-(aq)	- 13.37	- 12.35	26.14
I_2 (g)	14.88	4.63	62.28
K^+(aq)	- 60.04	- 67.47	24.5
Mg^{2+}(aq)	-110.41	-108.99	-28.2
Mn^{2+}(aq)	- 52.3	- 53.4	-20.0
$MnO_4{}^-$(aq)	-129.7	-107.4	45.4
$NH_4{}^+$(aq)	- 31.4	- 19.00	26.97
Na^+(aq)	- 57.28	- 62.59	14.4
OH^-(aq)	- 54.96	- 37.60	- 2.52
$SO_4{}^{2-}$(aq)	-216.90	-177.34	4.1

[a]Ref. 3, 25°C.

solubility of $CaCO_3$ in water. For Eq. 13 we may cite:

$$HOCl + H_2O = H_3O^+ + OCl^-$$

$$\Delta G° \text{ values: } -19.110 \quad -56.69 \quad -56.69 \quad -8.9$$

$$\Sigma \Delta G°_{Rex} = (-56.69 + -8.9) - (-19.110 - 56.69) = +10.20$$

$$\log K_{eq} = \frac{10.20}{-1.364} = -7.48; \quad K_{eq} = 10^{-7.48} \quad (22)$$

Thus, the protolysis constant for hypochlorous acid has been calculated from the free energy values of the individual constituents in Eq. 22. Since K_{eq} is less than 1, the protolysis of HOCl is not feasible and occurs only to a small extent.

ACID-BASE EQUILIBRIA

Brönsted Acids and Bases. Acidity and alkalinity are among the most common analyses applied to water and wastewaters. At the same time, these analyses are the most misunderstood and most erroneously used. In the Brönsted concept, an aqueous acid is a proton donor and an aqueous base is a proton acceptor. Furthermore, a proton must be transferred from one Brönsted base to another:

$$\underset{\text{Acid}_1}{HA} + \underset{\text{Base}_2}{B^-} = \underset{\text{Acid}_2}{HB} + \underset{\text{Base}_1}{A^-} \quad (23)$$

Thus, acid-base reactions become proton transfers from one Brönsted base to another. If one places a proton donor into water, the proton transfer reaction is:

$$HA + H_2O = H_3O^+ + A^- \quad (24)$$

In this case, the water molecule acts as the proton acceptor and competes with the base A^-, for the proton. These proton transfers to the water molecule are called protolysis or protolytic reactions. In this case, the Brönsted acid on the right-hand side of Eq. 24 is the hydronium ion, H_3O^+.

The Protolysis Constant. There have been attempts to label the strengths of acids or bases as "weak" or "strong." In the Brönsted system, it is the tendency to donate or to accept a proton that denotes the strength of an acid or base. "Weak" acids have a weak proton-donating tendency whereas a strong base would be one that has a strong tendency to accept a proton.

Perhaps a better way to order the strengths of acids and bases is to do so by the change in free energy involved in the proton transfer reaction. From Eq. 24, one may obtain a protolysis constant:

$$K_a = \frac{[H_3O^+][A^-]}{[HA][H_2O]} \quad (25)$$

In this case, K_a is the equilibrium constant K_{eq} from which the free energy change, $\Delta G°$, may be calculated from Eq. 13, $\Delta G^\circ = -1.364 \log K_a$. If the acid is hypochlorous acid, HOCl, the free energy change is $-1.364 \times -7.48 = +10.20$ kcal/mole. On the other hand, if the acid is acetic acid, HAc, the free energy change is $-1.364 \times -4.7 = +6.41$ kcal/mole. Thus, one may say that acetic acid is a stronger acid than hypochlorous acid because less energy is required to transfer a mole of protons from Ac^- to H_2O than from OCl^- to H_2O. Conversely, one can say that OCl^- is a "stronger" base than Ac^-. (It might be noted here that since the ΔG° values are positive for HOCl and HAc, the left-to-right reaction is not feasible but does occur to some extent.)

Proton Energy Level Diagrams. One may now order the strengths of acids and bases according to the free energy changes involved in the proton transfer. This order sometimes may be illustrated in a proton energy level diagram as seen in Fig. 3-2.

The scale on the left gives the energy required for the transfer of 1 mole of protons from a Brönsted acid to H_2O. As one moves from top to bottom, more energy is required and the acids become "weaker." Conversely, the right-hand scale gives the energy required for the transfer for 1 mole of protons from water to a Brönsted base. As one moves upward on this scale, the bases become "weaker." In order for a Brönsted acid to be "neutralized" by a Brönsted base, one must use a base whose proton transfer level lies below that of the acid.

Another feature of Fig. 3-2 is that pH may be a measure of the proton free energy level. For example, when $[HA] = [A^-]$, the pH is given by the pK of HA. Thus, this pH value gives the average proton free energy level which depends, of course, on the acid-base pair.

Another interpretation of Fig. 3-2 is that the energy levels on the left are occupied with protons. If, for example, an aqueous system contains several proton acceptors, e.g., CO_3^{2-}, OH^-, OCl^-, HCO_3^-, etc., the lower or unoccupied energy levels will be filled first upon the addition of protons. The order of occupation will be OH^-, CO_3^{2-}, OCl^-, and HCO_3^-. By the same reasoning, depopulation of proton levels

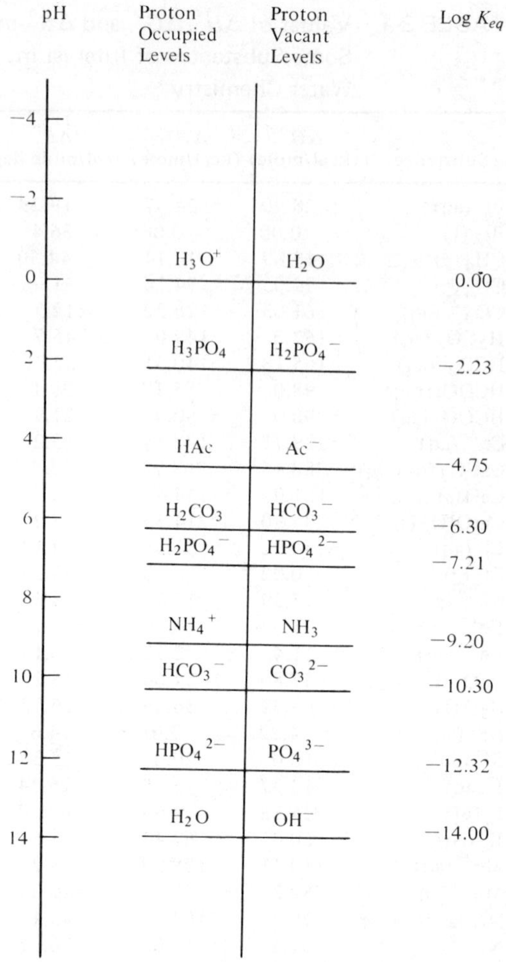

Fig. 3-2 Proton energy level diagram.

will occur in the same order from high to low proton free energy levels.

Table 3-2 gives the K_a and K_b values of some acids and bases frequently encountered in water and wastewater chemistry. The Brönsted acid-base equilibrium of water should be noted here:

$$H_2O + H_2O = H_3O^+ + OH^- \quad (26)$$

This equation is the so-called autoprotolysis of water. The K_{eq} or K_w is given by:

$$K_w = \frac{[H_3O^+][OH^-]}{[H_2O]^2} \quad (27)$$

Because the concentration of water is essentially constant in dilute aqueous solution, the $[H_2O]^2$ term in equation is arbitrarily given the value of 1. Thus, the K_w of water is = $[H_3O^+]$

TABLE 3-2. Acidity and Basicity Constants (25°C) of Interest to Water and Wastewater Chemistry[a]

Acids	pK_a	Bases	pK_b
HCl	~ −3	Cl^-	17
H_2SO_4	~ −3	HSO_4^-	17
HNO_3	−1	NO_3^-	15
H_3O^+	0	H_2O	14
H_3PO_4	2.1	$H_2PO_4^-$	11.9
CH_3COOH	4.7	CH_3COO^-	9.3
H_2CO_3	6.3	HCO_3^-	7.7
H_2S	7.1	HS^-	6.9
$H_2PO_4^-$	7.2	HPO_4^{2-}	6.8
HOCl	7.6	OCl^-	6.4
NH_4^+	9.3	NH_3	4.7
H_4SiO_4	9.5	$H_3SiO_4^-$	4.5
HCO_3^-	10.3	CO_3^{2-}	3.7
$H_3SiO_4^-$	12.6	$H_2SiO_4^{2-}$	1.4
H_2O	14	OH^-	0

[a]Ref. 2.

$\cdot$ [OH⁻]. Likewise, we consider the hydrated proton equilibria to be:

$$H_2O + H^+ = H_3O^+ \tag{28}$$

The K_{eq} for this equation is placed by convention to be 1.0 so that $\Delta G° = 0$. Consequently, Eq. 27 becomes:

$$K_w = [H^+][OH^-] \tag{29}$$

At 25°C, $K_w = 1.008 \times 10^{-14}$. For K_w values at other temperatures, see p. 76 of Ref. 2. Furthermore, it can be shown that:

$$K_w = K_a \cdot K_b \tag{30}$$

from which K_a or K_b may be calculated provided that one or the other value is given.

Distribution Diagrams—pH as the Master Variable.

The Cl_2-H_2O System. In accord with Eq. 24, a proton is transferred from the conjugate base A^- to water. The extent of this transfer is regulated by the K_a expression as noted in Eq. 25. Two observations may be made from Eq. 24: (1) the conjugate base, A^-, is distributed between HA and A^-, and (2) the H^+ ion concentration (symbol H^+ used for convenience) affects the extent of this distribution. If one were to prepare a 10^{-4} M (5.25 mg/l) aqueous solution of hypochlorous acid (HOCl), a mass balance on OCl^- would be:

$$C_t = [HOCl] + [OCl^-] \tag{31}$$

where C_t is the total concentration, moles/liter, of HOCl and OCl^-. The extent of the distribution of OCl^- between HOCl and OCl^- is regulated by:

$$K_a = \frac{[H^+][OCl^-]}{[HOCl]} = 10^{-7.48}\ (25°C) \tag{32}$$

Consequently, the fraction present of HOCl at a given $[H^+]$ may be computed from:

$$\alpha_{HOCl} = \frac{[HOCl]}{C_t} \tag{33}$$

Likewise, the fraction present of OCl^- would be:

$$\alpha_{OCl^-} = \frac{[OCl^-]}{C_t} \tag{34}$$

It must be noted, at a given $[H^+]$, that:

$$\alpha_{HOCl} + \alpha_{OCl^-} = 1.0 \tag{35}$$

By combination of Eqs. 31 and 32, one obtains:

$$\alpha_{HOCl} = \frac{[H^+]}{K_a + [H^+]} \tag{36}$$

and

$$\alpha_{OCl^-} = \frac{K_a}{K_a + [H^+]} \tag{37}$$

Thus, these fractions may be calculated at any $[H^+]$. For example, at a pH value of 7.0:

$$\alpha_{HOCl} = \frac{[H^+]}{K_a + [H^+]} = \frac{10^{-7.0}}{3.31 \times 10^{-8} + 10^{-7.0}}$$

$$= \frac{10^{-7.0}}{1.331 \times 10^{-7}} = 0.75$$

$$\alpha_{OCl^-} = 1.000 - \alpha_{HOCl} = 1.000 - 0.75 = 0.25$$

Therefore, 75 percent of the solution is HOCl and 25 percent is OCl^-. If C_t is 10^{-4} M, then:

$$HOCl = \alpha_{HOCl} C_t = 0.75 \times 10^{-4}\ M$$

and

$$OCl^- = \alpha_{OCl^-} C_t = 0.25 \times 10^{-4}\ M$$

Sometimes it is convenient to show α_{HOCl} and α_{OCl^-} as a function of pH (the master variable) in a distribution diagram that is illustrated in Fig. 3-3.

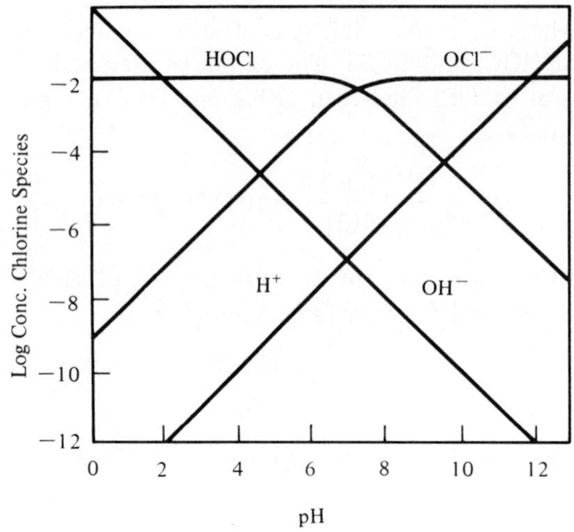

Fig. 3-3 Concentration distribution diagram for hypochlorous acid.

An excellent and very practical application of the α-value concept may be made with HOCl, OCl^-, and the disinfection of water. HOCl and OCl^- differ in their ability to kill microorganisms. Morris[7] has provided an example which is modified here. If an additive efficiency of disinfection is assumed for HOCl and OCl^-, one may produce a theoretical curve for the total amount of chlorine required to give a percentage kill of *E. coli;* for example, for a specified contact time at various pH values. Let R = the total concentration of chlorine required for the stipulated kill, A = the concentration of HOCl necessary to produce this kill (this must be evaluated experimentally at a pH value where $\alpha_{HOCl} \simeq 1.0$), and B = an efficiency ratio of $[HOCl]/[OCl^-]$. (This $[OCl^-]$ would be the concentration necessary to produce the same percentage kill that must be evaluated experimentally at a pH value where $\alpha_{OCl^-} \simeq 1.0$). Then:

$$R = [HOCl] + [OCl^-] = C_t(\alpha_{HOCl} + \alpha_{OCl^-}) \tag{38}$$

$$A = [HOCl] + B[OCl^-] = C_t(\alpha_{HOCl} + B\alpha_{OCl^-}) \tag{39}$$

$$C_t = A/(\alpha_{HOCl} + B\alpha_{OCl^-}) \tag{40}$$

Combining Eqs. 38 and 40 leads to:

$$R = \frac{A}{(\alpha_{HOCl} + B\alpha_{OCl^-})} \tag{41}$$

For a kill of 99 percent of *E. coli* in 30 min at 2-5°C, at a pH of 8.0, $[HOCl] = 0.005$ mg/l, $[OCl^-] = 0.042$ mg/l, $B = 0.12$. At pH 8.0, $\alpha_{HOCl} = .286$ and $\alpha_{OCl^-} = 0.714$. Whereupon:

$$R = \frac{0.005}{(0.286 + 0.12 \times 0.714)} = \frac{0.005}{0.372}$$

$$= 0.013 \text{ mg/l}$$

Thus, a total concentration of 0.013 mg/l Cl_2 is required for the stipulated kill. One may extend this to a plot of total chlorine versus pH as seen in Fig. 3-4. This curve is applicable,

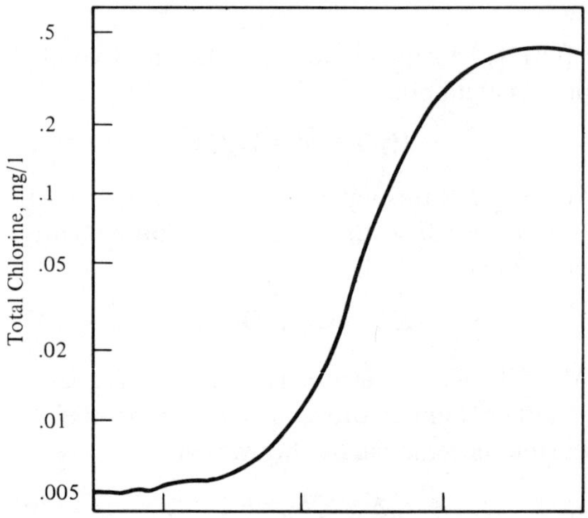

Fig. 3-4 Concentration of free available chlorine required for 99 percent kill of *E. coli* in 30 min at 2 to 5°C. Reproduced from Ref. 7 by courtesy of John Wiley and Sons, Inc., New York.

of course, only to the conditions cited. One may construct a curve for other temperatures, contact times, and organisms.

The CO_2-H_2O System. Carbonic acid, H_2CO_3, is a weak, diprotic acid of some importance in water chemistry. As seen below, total acidity and total alkalinity of aqueous systems are defined around this acid. H_2CO_3 may be formed from:

$$CO_2(g) = CO_2(aq) \quad (42)$$

and

$$CO_2(aq) + H_2O = H_2CO_3 \quad (43)$$

In turn, H_2CO_3 enters into two Brönsted acid-base equilibria (water of hydration omitted):

$$H_2CO_3 = H^+ + HCO_3^- \quad K_1 = \frac{[H^+][HCO_3^-]}{[H_2CO_3]} \quad (44)$$

$$HCO_3^- = H^+ + CO_3^{2-} \quad K_2 = \frac{[H^+][CO_3^{2-}]}{[HCO_3^-]} \quad (45)$$

The equilibrium constants for Eqs. 44 and 45 are given in Table 3-3 for several temperatures.[8,9] A mass balance on carbonate species is:

TABLE 3-3. Acidity Constants for H_2CO_3 and HCO_3^- at Several Temperatures[a]

Temperature °C	Eq. 44 $-\log K_1$	Eq. 45 $-\log K_2$
0	6.579	10.625
5	6.517	10.557
10	6.464	10.490
15	6.419	10.430
20	6.381	10.377
25	6.352	10.329
30	6.327	10.290
35	6.309	10.250
40	6.298	10.220
50	6.285	10.172

[a]Ref. 8 and 9.

$$C_t = [H_2CO_3] + [HCO_3^-] + [CO_3^{2-}] \quad (46)$$

whereupon the distribution of species as influenced by the $[H^+]$ would be:

$$\alpha_{H_2CO_3} = \frac{[H_2CO_3]}{C_t} \quad (47)$$

$$\alpha_{HCO_3^-} = \frac{[HCO_3^-]}{C_t} \quad (48)$$

$$\alpha_{CO_3^{2-}} = \frac{[CO_3^{2-}]}{C_t} \quad (49)$$

These alpha values may be computed from various pH values by combining Eqs. 44, 45, 46, 47, 48, and 49. The resultant equations are:

$$\alpha_{H_2CO_3} = \frac{[H^+]^2}{[H^+]^2 + K_1[H^+] + K_1K_2} \quad (50)$$

$$\alpha_{HCO_3^-} = \frac{K_1[H^+]}{[H^+]^2 + K_1[H^+] + K_1K_2} \quad (51)$$

$$\alpha_{CO_3^{2-}} = \frac{K_1K_2}{[H^+]^2 + K_1[H^+] + K_1K_2} \quad (52)$$

Table 3-4 gives the α values for H_2CO_3, HCO_3^-, and CO_3^{2-} that were computed from Eqs. 50, 51, and 52 using the 25°C K_1 and K_2 values from Table 3-3. Note that $\alpha_{H_2CO_3} + \alpha_{HCO_3^-} + \alpha_{CO_3^{2-}} = 1.0$.

Some useful qualitative interpretations of the carbonate α values may be obtained through the use of a distribution diagram. In this case, the log α value is plotted against pH as the master variable. In order to construct this plot, it is convenient to divide the pH abscissa into three regions: 0.00-6.35 (I), 6.35-10.33 (II), and 10.33-14.00 (III). The boundaries between regions I and II, and II and III are established by the pK_1 and pK_2 values of H_2CO_3. In region I, the approximation that $[H^+]^2 >> K_1\,[H^+] >> K_1K_2$ in the denominators of Eqs. 50, 51, and 52 is made, whereupon:

$$\alpha_{H_2CO_3} \simeq \frac{[H^+]^2}{[H^+]^2} \simeq 1.0 \quad (53)$$

$$\alpha_{HCO_3^-} \simeq \frac{K_1}{[H^+]} \quad (54)$$

$$\alpha_{CO_3^{2-}} \simeq \frac{K_1K_2}{[H^+]^2} \quad (55)$$

In region II, the approximation that $K_1[H^+] >> [H^+]^2 >> K_1K_2$ is made, whereupon:

$$\alpha_{H_2CO_3} \simeq \frac{[H^+]}{K_1} \quad (56)$$

TABLE 3-4.

ALPHA VALUES FOR WATER-CARBON DIOXIDE SYSTEM PAGE 1

TEMPERATURE 25 DEGREES CENTIGRADE

PH	ALPHA H2CO3	ALPHA HCO3	ALPHA CO3
0.00	1.000000	0.000000	0.000000
0.05	1.000000	0.000001	0.000000
0.10	0.999999	0.000001	0.000000
0.15	0.999999	0.000001	0.000000
0.20	0.999999	0.000001	0.000000
0.25	0.999999	0.000001	0.000000
0.30	0.999999	0.000001	0.000000
0.35	0.999999	0.000001	0.000000
0.40	0.999999	0.000001	0.000000
0.45	0.999999	0.000001	0.000000
0.50	0.999999	0.000001	0.000000
0.55	0.999998	0.000002	0.000000
0.60	0.999998	0.000002	0.000000
0.65	0.999998	0.000002	0.000000
0.70	0.999998	0.000002	0.000000
0.75	0.999997	0.000003	0.000000
0.80	0.999997	0.000003	0.000000
0.85	0.999997	0.000003	0.000000
0.90	0.999996	0.000004	0.000000
0.95	0.999996	0.000004	0.000000
1.00	0.999996	0.000004	0.000000
1.05	0.999995	0.000005	0.000000
1.10	0.999994	0.000006	0.000000
1.15	0.999994	0.000006	0.000000
1.20	0.999993	0.000007	0.000000
1.25	0.999992	0.000008	0.000000
1.30	0.999991	0.000009	0.000000
1.35	0.999990	0.000010	0.000000
1.40	0.999989	0.000011	0.000000
1.45	0.999987	0.000013	0.000000
1.50	0.999986	0.000014	0.000000
1.55	0.999984	0.000016	0.000000
1.60	0.999982	0.000018	0.000000
1.65	0.999980	0.000020	0.000000
1.70	0.999978	0.000022	0.000000
1.75	0.999975	0.000025	0.000000
1.80	0.999972	0.000028	0.000000
1.85	0.999968	0.000032	0.000000
1.90	0.999965	0.000035	0.000000
1.95	0.999960	0.000040	0.000000
2.00	0.999955	0.000045	0.000000
2.05	0.999950	0.000050	0.000000
2.10	0.999944	0.000056	0.000000
2.15	0.999937	0.000063	0.000000
2.20	0.999929	0.000071	0.000000
2.25	0.999921	0.000079	0.000000
2.30	0.999911	0.000089	0.000000
2.35	0.999900	0.000100	0.000000
2.40	0.999888	0.000112	0.000000
2.45	0.999874	0.000126	0.000000

TABLE 3-4. cont.

ALPHA VALUES FOR WATER-CARBON DIOXIDE SYSTEM PAGE 2

TEMPERATURE 25 DEGREES CENTIGRADE

PH	ALPHA H2CO3	ALPHA HCO3	ALPHA CO3
2.50	0.999859	0.000141	0.000000
2.55	0.999842	0.000158	0.000000
2.60	0.999823	0.000177	0.000000
2.65	0.999801	0.000199	0.000000
2.70	0.999777	0.000223	0.000000
2.75	0.999749	0.000251	0.000000
2.80	0.999719	0.000281	0.000000
2.85	0.999685	0.000315	0.000000
2.90	0.999646	0.000354	0.000000
2.95	0.999603	0.000397	0.000000
3.00	0.999555	0.000445	0.000000
3.05	0.999500	0.000500	0.000000
3.10	0.999439	0.000561	0.000000
3.15	0.999371	0.000629	0.000000
3.20	0.999294	0.000706	0.000000
3.25	0.999208	0.000792	0.000000
3.30	0.999112	0.000888	0.000000
3.35	0.999003	0.000997	0.000000
3.40	0.998882	0.001118	0.000000
3.45	0.998746	0.001254	0.000000
3.50	0.998593	0.001407	0.000000
3.55	0.998421	0.001579	0.000000
3.60	0.998229	0.001771	0.000000
3.65	0.998013	0.001987	0.000000
3.70	0.997771	0.002229	0.000000
3.75	0.997500	0.002500	0.000000
3.80	0.997196	0.002804	0.000000
3.85	0.996855	0.003145	0.000000
3.90	0.996473	0.003527	0.000000
3.95	0.996044	0.003956	0.000000
4.00	0.995563	0.004437	0.000000
4.05	0.995025	0.004975	0.000000
4.10	0.994421	0.005579	0.000000
4.15	0.993744	0.006256	0.000000
4.20	0.992986	0.007014	0.000000
4.25	0.992137	0.007863	0.000000
4.30	0.991186	0.008814	0.000000
4.35	0.990122	0.009878	0.000000
4.40	0.988930	0.011070	0.000000
4.45	0.987595	0.012404	0.000000
4.50	0.986103	0.013897	0.000000
4.55	0.984434	0.015566	0.000000
4.60	0.982567	0.017433	0.000000
4.65	0.980482	0.019518	0.000000
4.70	0.978152	0.021848	0.000000
4.75	0.975552	0.024448	0.000000
4.80	0.972650	0.027350	0.000000
4.85	0.969415	0.030585	0.000000
4.90	0.965810	0.034189	0.000000
4.95	0.961798	0.038202	0.000000

TABLE 3-4. cont.

ALPHA VALUES FOR WATER-CARBON DIOXIDE SYSTEM PAGE 3

TEMPERATURE 25 DEGREES CENTIGRADE

PH	ALPHA H2CO3	ALPHA HCO3	ALPHA CO3
5.00	0.957336	0.042664	0.000000
5.05	0.952378	0.047622	0.000000
5.10	0.946875	0.053124	0.000000
5.15	0.940777	0.059222	0.000000
5.20	0.934028	0.065972	0.000000
5.25	0.926569	0.073431	0.000001
5.30	0.918340	0.081659	0.000001
5.35	0.909280	0.090719	0.000001
5.40	0.899325	0.100674	0.000001
5.45	0.888411	0.111587	0.000001
5.50	0.876477	0.123521	0.000002
5.55	0.863463	0.136535	0.000002
5.60	0.849313	0.150684	0.000003
5.65	0.833979	0.166018	0.000003
5.70	0.817419	0.182576	0.000004
5.75	0.799605	0.200390	0.000005
5.80	0.780519	0.219474	0.000006
5.85	0.760161	0.239831	0.000008
5.90	0.738547	0.261444	0.000010
5.95	0.715713	0.284275	0.000012
6.00	0.691717	0.308268	0.000014
6.05	0.666640	0.333343	0.000018
6.10	0.640582	0.359397	0.000021
6.15	0.613667	0.386307	0.000026
6.20	0.586039	0.413930	0.000031
6.25	0.557859	0.442104	0.000037
6.30	0.529301	0.470655	0.000044
6.35	0.500549	0.499398	0.000052
6.40	0.471794	0.528144	0.000062
6.45	0.443224	0.556703	0.000074
6.50	0.415024	0.584889	0.000087
6.55	0.387370	0.612528	0.000102
6.60	0.360423	0.639458	0.000119
6.65	0.334326	0.665534	0.000139
6.70	0.309205	0.690632	0.000162
6.75	0.285163	0.714649	0.000188
6.80	0.262279	0.737503	0.000218
6.85	0.240613	0.759135	0.000252
6.90	0.220202	0.779508	0.000290
6.95	0.201063	0.798604	0.000334
7.00	0.183195	0.816422	0.000383
7.05	0.166584	0.832978	0.000438
7.10	0.151199	0.848300	0.000501
7.15	0.137001	0.862428	0.000571
7.20	0.123940	0.875410	0.000650
7.25	0.111962	0.887299	0.000740
7.30	0.101007	0.898153	0.000840
7.35	0.091013	0.908034	0.000953
7.40	0.081916	0.917004	0.001080
7.45	0.073655	0.925123	0.001222

TABLE 3-4. cont.

ALPHA VALUES FOR WATER-CARBON DIOXIDE SYSTEM PAGE 4

TEMPERATURE 25 DEGREES CENTIGRADE

PH	ALPHA H2CO3	ALPHA HCO3	ALPHA CO3
7.50	0.066165	0.932453	0.001382
7.55	0.059387	0.939051	0.001562
7.60	0.053262	0.944974	0.001764
7.65	0.047736	0.950274	0.001990
7.70	0.042757	0.955000	0.002244
7.75	0.038274	0.959197	0.002529
7.80	0.034244	0.962906	0.002848
7.85	0.030623	0.966170	0.003207
7.90	0.027374	0.969018	0.003609
7.95	0.024459	0.971482	0.004059
8.00	0.021846	0.973589	0.004564
8.05	0.019506	0.975363	0.005131
8.10	0.017411	0.976824	0.005765
8.15	0.015536	0.977988	0.006476
8.20	0.013859	0.978868	0.007273
8.25	0.012359	0.979475	0.008166
8.30	0.011019	0.979816	0.009165
8.35	0.009822	0.979894	0.010284
8.40	0.008752	0.979711	0.011537
8.45	0.007797	0.979265	0.012939
8.50	0.006944	0.978549	0.014507
8.55	0.006182	0.977557	0.016261
8.60	0.005503	0.976276	0.018221
8.65	0.004896	0.974693	0.020411
8.70	0.004355	0.972788	0.022857
8.75	0.003873	0.970541	0.025587
8.80	0.003442	0.967926	0.028631
8.85	0.003058	0.964917	0.032025
8.90	0.002716	0.961479	0.035805
8.95	0.002411	0.957579	0.040011
9.00	0.002139	0.953175	0.044686
9.05	0.001896	0.948225	0.049878
9.10	0.001680	0.942683	0.055637
9.15	0.001488	0.936496	0.062016
9.20	0.001316	0.929612	0.069072
9.25	0.001163	0.921973	0.076863
9.30	0.001027	0.913521	0.085451
9.35	0.000906	0.904195	0.094899
9.40	0.000799	0.893932	0.105270
9.45	0.000703	0.882670	0.116627
9.50	0.000618	0.870351	0.129031
9.55	0.000542	0.856917	0.142541
9.60	0.000475	0.842317	0.157208
9.65	0.000415	0.826505	0.173080
9.70	0.000362	0.809447	0.190190
9.75	0.000316	0.791119	0.208565
9.80	0.000274	0.771512	0.228214
9.85	0.000238	0.750632	0.249130
9.90	0.000206	0.728505	0.271289
9.95	0.000178	0.705178	0.294645

TABLE 3-4. cont.

ALPHA VALUES FOR WATER-CARBON DIOXIDE SYSTEM PAGE 5

TEMPERATURE 25 DEGREES CENTIGRADE

PH	ALPHA H2CO3	ALPHA HCO_3	ALPHA CO_3
10.00	0.000153	0.680718	0.319130
10.05	0.000131	0.655215	0.344654
10.10	0.000112	0.628781	0.371107
10.15	0.000096	0.601549	0.398356
10.20	0.000081	0.573671	0.426248
10.25	0.000069	0.545314	0.454618
10.30	0.000058	0.516658	0.483284
10.35	0.000049	0.487890	0.512061
10.40	0.000041	0.459201	0.540758
10.45	0.000034	0.430779	0.569187
10.50	0.000029	0.402805	0.597166
10.55	0.000024	0.375449	0.624527
10.60	0.000020	0.348865	0.651115
10.65	0.000016	0.323189	0.676795
10.70	0.000013	0.298536	0.701450
10.75	0.000011	0.274999	0.724990
10.80	0.000009	0.242650	0.747341
10.85	0.000007	0.231536	0.768456
10.90	0.000006	0.211687	0.788307
10.95	0.000005	0.193112	0.806883
11.00	0.000004	0.175804	0.824192
11.05	0.000003	0.159739	0.840257
11.10	0.000003	0.144885	0.855112
11.15	0.000002	0.131196	0.868802
11.20	0.000002	0.118621	0.881377
11.25	0.000001	0.107103	0.892896
11.30	0.000001	0.096580	0.903419
11.35	0.000001	0.086991	0.913008
11.40	0.000001	0.078271	0.921728
11.45	0.000001	0.070358	0.929641
11.50	0.000000	0.063190	0.936809
11.55	0.000000	0.056708	0.943291
11.60	0.000000	0.050855	0.949145
11.65	0.000000	0.045577	0.954423
11.70	0.000000	0.040822	0.959177
11.75	0.000000	0.036545	0.963455
11.80	0.000000	0.032701	0.967299
11.85	0.000000	0.029249	0.970751
11.90	0.000000	0.026151	0.973849
11.95	0.000000	0.023374	0.976626
12.00	0.000000	0.020885	0.979115
12.05	0.000000	0.018656	0.981344
12.10	0.000000	0.016661	0.983339
12.15	0.000000	0.014876	0.985124
12.20	0.000000	0.013280	0.986720
12.25	0.000000	0.011853	0.986147
12.30	0.000000	0.010577	0.989423
12.35	0.000000	0.009438	0.990562
12.40	0.000000	0.008420	0.991580
12.45	0.000000	0.007511	0.992489

TABLE 3-4. cont.

ALPHA VALUES FOR WATER-CARBON DIOXIDE SYSTEM PAGE 6

TEMPERATURE 25 DEGREES CENTIGRADE

PH	ALPHA H2CO3	ALPHA HCO_3	ALPHA CO_3
12.50	0.000000	0.006700	0.993300
12.55	0.000000	0.005976	0.994024
12.60	0.000000	0.005329	0.994671
12.65	0.000000	0.004753	0.885247
12.70	0.000000	0.004238	0.995762
12.75	0.000000	0.003779	0.996221
12.80	0.000000	0.003369	0.996631
12.85	0.000000	0.003004	0.996996
12.90	0.000000	0.002678	0.997322
12.95	0.000000	0.002388	0.997612
13.00	0.000000	0.002129	0.997871
13.05	0.000000	0.001897	0.998103
13.10	0.000000	0.001691	0.998309
13.15	0.000000	0.001508	0.998492
13.20	0.000000	0.001344	0.998656
13.25	0.000000	0.001198	0.998802
13.30	0.000000	0.001068	0.998932
13.35	0.000000	0.000952	0.999048
13.40	0.000000	0.000848	0.999152
13.45	0.000000	0.000756	0.999244
13.50	0.000000	0.000674	0.999326
13.55	0.000000	0.000601	0.999399
13.60	0.000000	0.000536	0.999464
13.65	0.000000	0.000477	0.999523
13.70	0.000000	0.000425	0.999575
13.75	0.000000	0.000379	0.999621
13.80	0.000000	0.000338	0.999662
13.85	0.000000	0.000301	0.999699
13.90	0.000000	0.000268	0.999732
13.95	0.000000	0.000239	0.999761
14.00	0.000000	0.000213	0.999787

$$\alpha_{HCO_3^-} \simeq \frac{K_1[H^+]}{K_1[H^+]} \simeq 1.0 \qquad (57)$$

$$\alpha_{CO_3^{2-}} \simeq \frac{K_2}{[H^+]} \qquad (58)$$

In region III, the approximation that $K_1K_2 >> K_1[H^+] >> [H^+]^2$ is made, whereupon:

$$\alpha_{H_2CO_3} \simeq \frac{[H^+]^2}{K_1K_2} \qquad (59)$$

$$\alpha_{HCO_3^-} \simeq \frac{[H^+]}{K_2} \qquad (60)$$

$$\alpha_{CO_3^{2-}} \simeq \frac{K_1K_2}{K_1K_2} \simeq 1.0 \qquad (61)$$

When Eqs. 50 through 61 are placed in a logarithmic form, the plots, as seen in Fig. 3-5, result. It should be noted that the above approximations are not valid as the equations approach the pH values equal to pK_1 and pK_2. In this case, Eqs. 50, 51, and 52 must be solved in full with the result that a curvature exists as the respective lines pass thru the pH = pK_1, pK_2 boundaries. It should be obvious that H_2CO_3 is the predominant species in region I, HCO_3^- in region II, and CO_3^{2-} in region III. Furthermore, the respective curves for H_2CO_3, HCO_3^-, and CO_3^{2-} approach α = 1.0 asymptotically in regions I, II, and III. More comment and application of the α values for the carbonate system are made in the next section under total acidity and alkalinity of natural waters.

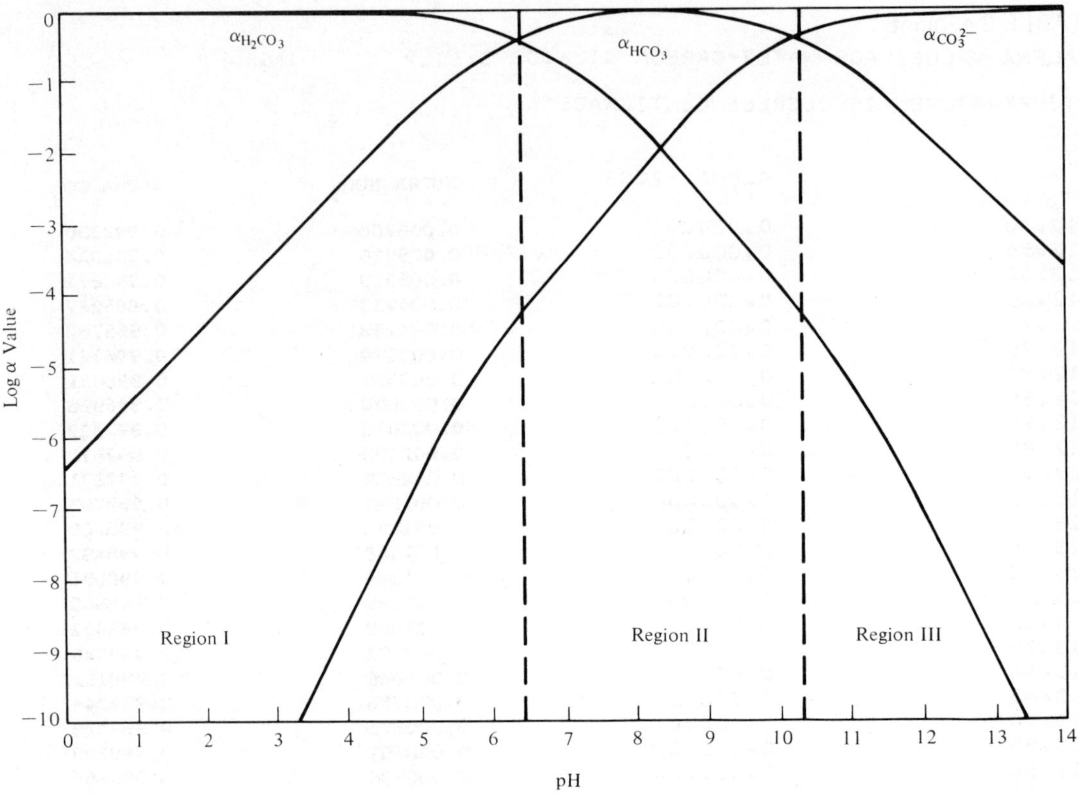

Fig. 3-5 Alpha-value distribution diagram for carbonic acid.

The H_3PO_4 System. Phosphoric acid, H_3PO_4, is a triprotic acid of some importance to water and wastewater chemistry because of the possible role of phosphorus in accelerating eutrophic processes. The appropriate acid-base equilibria are:

$$H_3PO_4 = H^+ + H_2PO_4^-; \; K_1 = \frac{[H^+][H_2PO_4^-]}{[H_3PO_4]} \quad (62)$$

$$H_2PO_4^- = H^+ + HPO_4^{2-}; \; K_2 = \frac{[H^+][HPO_4^{2-}]}{[H_2PO_4^-]} \quad (63)$$

$$HPO_4^{2-} = H^+ + PO_4^{3-}; \; K_3 = \frac{[H^+][PO_4^{3-}]}{[HPO_4^{2-}]} \quad (64)$$

Whereupon the mass balance on phosphate is:

$$C_t = [H_3PO_4] + [H_2PO_4^-] + [HPO_4^{2-}] + [PO_4^{3-}] \quad (65)$$

Again, the α values for the various phosphate species are:

$$\alpha_{H_3PO_4} = \frac{[H_3PO_4]}{C_t} \quad (66)$$

$$\alpha_{H_2PO_4^-} = \frac{[H_2PO_4^-]}{C_t} \quad (67)$$

$$\alpha_{HPO_4^{2-}} = \frac{[HPO_4^{2-}]}{C_t} \quad (68)$$

$$\alpha_{PO_4^{3-}} = \frac{[PO_4^{3-}]}{C_t} \quad (69)$$

Whereupon a combination of Eqs. 62 through 69 should yield:

$$\alpha_{H_3PO_4} = \frac{[H^+]^3}{[H^+]^3 + K_1[H^+]^2 + K_1K_2[H^+] + K_1K_2K_3} \quad (70)$$

$$\alpha_{H_2PO_4^-} = \frac{K_1 [H^+]^2}{[H^+]^3 + K_1 [H^+]^2 + K_1K_2 [H^+] + K_1K_2K_3} \quad (71)$$

$$\alpha_{HPO_4^{2-}} = \frac{K_1K_2 [H^+]}{[H^+]^3 + K_1 [H^+]^2 + K_1K_2 [H^+] + K_1K_2K_3} \quad (72)$$

$$\alpha_{PO_4^{3-}} = \frac{K_1K_2K_3}{[H^+]^3 + K_1 [H^+]^2 + K_1K_2 [H^+] + K_1K_2K_3} \quad (73)$$

Note that

$$\alpha_{H_3PO_4} + \alpha_{H_2PO_4^-} + \alpha_{HPO_4^{2-}} + \alpha_{PO_4^{3-}} = 1.0$$

at any one pH value. The appropriate solution of Eqs. 70 through 73 will yield the $-\log \alpha$ versus pH plot as seen in Fig. 3-6. In this case of a triprotic acid, four regions, I, II, III, and IV may be devised by the boundaries of where the pH value is equal to pK_1, pK_2, and pK_3 of phosphoric acid. It follows that H_3PO_4 predominates in region I, $H_2PO_4^-$ in region II, HPO_4^{2-} in region III, and PO_4^{3-} in region IV. This distribution of phosphate species with respect to pH becomes important when PO_4^{3-}, for example, enters into precipitation reactions.

ACIDITY AND ALKALINITY IN NATURAL WATERS

Conceptual Definitions. Classically, acidity and alkalinity in water and wastewater are determined by titration with standardized solutions of NaOH and H_2SO_4. The endpoints of these titrations are usually detected by the chlorimetric indicators methyl orange and phenolphthalein. There are, however, some conceptual and operational errors in these titrations.

Conceptually, acidity is defined as the capacity to neutralize a given quantity of a base and

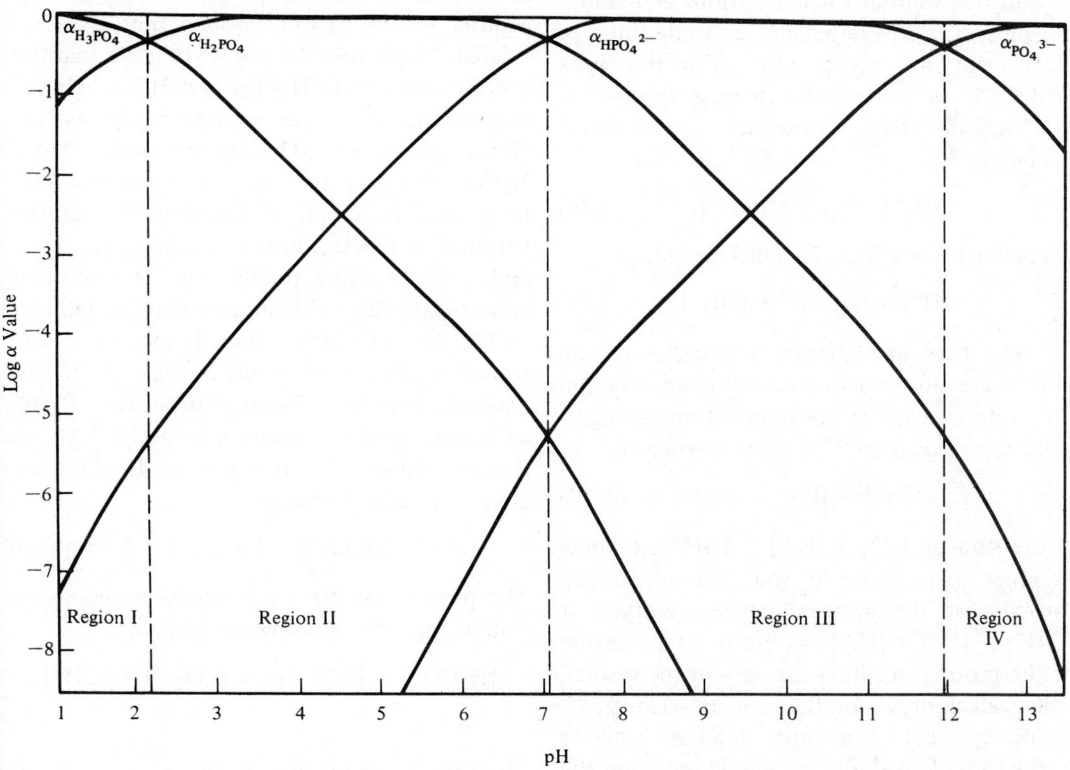

Fig. 3-6 Alpha-value distribution diagram for phosphoric acid.

alkalinity is the capacity to neutralize a given quantity of an acid. In the Brönsted system, acidity is the sum of the concentrations of the proton donors whereas alkalinity is the sum of the concentrations of proton acceptors. From Eq. 24, the proton donors in the monoprotic $HA\text{-}H_2O$ system are HA and H_3O^+, whereas the proton acceptors are A^- and H_2O. The addition of a strong Brönsted base such as NaOH to Eq. 24 yields:

$$HA + NaOH = NaA + H_2O \quad (74)$$

If the quantity of NaOH is equivalent to the quantity of HA, then a measure of the proton-donating capacity is obtained which is said to be the acidity of the system.

Another approach to acidity and alkalinity is through the concept of the proton condition of a system containing proton acceptors and proton donors. We have seen from Eq. 31, for example, that the mass balance for a weak acid may be expressed as:

$$C_t = HA + A^- \quad (75)$$

In aqueous solutions where cations and anions are present, a charge balance or a condition of electroneutrality must exist. For the weak acid, HA, in water, the charged species are H^+, A^-, and OH^-. Consequently, the charge balance is:

$$[H^+] = [A^-] + [OH^-] \quad (76)$$

A combination of Eqs. 75 and 76 yields:

$$[H^+] + HA = C_t + [OH^-] \quad (77)$$

We now have an equation where the proton donors are summed on the left-hand side and the proton acceptors are summed on the right-hand side. Equation 77 may be rearranged:

$$C_t = [H^+] + [HA] - [OH^-] \quad (78)$$

If the sum of $[H^+] + [HA] > [OH^-]$, C_t now becomes a measure of the proton-donating capacity of the system, hence acidity. If $[OH^-] > [H^+] + [HA]$, C_t becomes a measure of the proton-accepting capacity of the system, hence alkalinity. In the case of acidity, C_t would represent the quantity of a base, whereas in the case of alkalinity, C_t would represent the quantity of an acid.

The proton conditions at the beginning and end of the titration of HA are important also to the concepts of acidity and alkalinity. Equation 76 represents the proton condition at the beginning of the titration where the proton donors are balanced by the proton acceptors. Under these conditions, the acid-neutralizing capacity, hence alkalinity, of the system is zero. At the end of the titration of HA, the proton condition is obtained by substitution of the equivalent quantity of $[Na^+]$ (from NaOH) equal to C_t into Eq. 78 which, when combined with Eq. 75, yields:

$$[HA] + [H^+] = [OH^-] \quad (79)$$

In this case, the proton donors are balanced by the proton acceptor and the base-neutralizing capacity, or acidity, of the system is zero. It follows that the quantity of $[Na^+]$ added to reach the conditions expressed by Eq. 79 is a measure of the acidity.

Acidity and Alkalinity in the CO_2-H_2O System. In the measurement of acidity and alkalinity in natural waters, the proton donors and acceptors in the H_2CO_3 system must be considered. Equations 44 and 45 indicate that the proton donors are H_2CO_3 and HCO_3^- (H_3O^+ understood) and the proton acceptors are HCO_3^- and CO_3^{2-} (OH^- understood). Since H_2CO_3 is a diprotic acid, three proton conditions may be cited: at the beginning of the titration of H_2CO_3, at the midpoint, and at the end. These three points may be illustrated conveniently by a distribution diagram (similar to Fig. 3-5) in which carbonate species concentration is plotted versus pH in Fig. 3-7. In this diagram, the mass balance from Eq. 46 on carbonate species is assumed to be 10^{-3} M. At the beginning of a titration of H_2CO_3, the charge balance would be:

$$[H^+] = [HCO_3^-] + 2[CO_3^{2-}] + [OH^-] \quad (80)$$

The proton condition is obtained by combination of Eqs. 80 and 46 which yields:

$$[H_2CO_3] + [H^+] = C_t + [CO_3^{2-}] + [OH^-] \quad (81)$$

Titration is begun now with a C_t solution of NaOH. At the midpoint, one obtains a 10^{-3} M

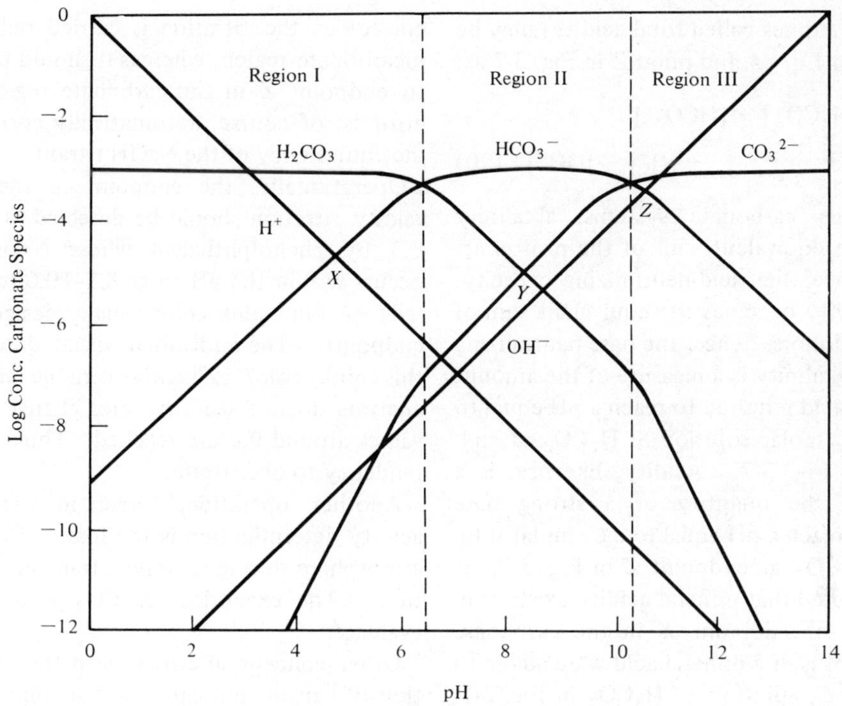

Fig. 3-7 Concentration distribution diagram for carbonic acid.

solution of $NaHCO_3$ for which the charge balance is:

$$[Na^+] + [H^+] = [HCO_3^-] + 2[CO_3^{2-}] + [OH^-] \quad (82)$$

Since $C_t = [Na^+]$, the proton condition becomes:

$$[H_2CO_3] + [H^+] = [CO_3^{2-}] + [OH^-] \quad (83)$$

Continuation of the titration to the end with NaOH yields a 10^{-3} M solution of Na_2CO_3 for which the charge balance is the same as Eq. 82. In this case, however, $2C_t = [Na^+]$, which leads to the proton condition:

$$2[H_2CO_3] + [HCO_3^-] + [H^+] = [OH^-] \quad (84)$$

Conceptual definitions for three types of alkalinity may be cited for carbonate system. Caustic alkalinity may be defined from Eq. 84 and point Z in Fig. 3-7 as:

$$[OH^-\text{-Alk}] = [OH^-] - [H^+] - [HCO_3^-] - 2[H_2CO_3] \quad (85)$$

Carbonate alkalinity may be defined from Eq. 83 and point Y in Fig. 3-7 as:

$$[CO_3^{2-}\text{-Alk}] = [CO_3^{2-}] + [OH^-] - [H^+] - [H_2CO_3] \quad (86)$$

Alkalinity (sometimes called total alkalinity) may be defined from Eq. 80 and point X in Fig. 3-7 as:

$$[\text{Alk}] = [HCO_3^-] + 2[CO_3^{2-}] + [OH^-] - [H^+] \quad (87)$$

Likewise, conceptual definitions for three types of acidity may be cited for the carbonate system. Mineral acidity may be defined from Eq. 80 and point X in Fig. 3-7 as:

$$[\text{Min-Acy}] = [H^+] - [HCO_3^-] - 2[CO_3^{2-}] - [OH^-] \quad (88)$$

Carbon dioxide acidity may be defined from Eq. 83 and point Y in Fig. 3-7 as:

$$[CO_2\text{-Acy}] = [H_2CO_3] + [H^+] - [CO_3^{2-}] - [OH^-] \quad (89)$$

Acidity (sometimes called total acidity) may be defined from Eq. 84 and point Z in Fig. 3-7 as:

$$[\text{Acy}] = 2[H_2CO_3] + [HCO_3^-] + [H^+] - [OH^-] \quad (90)$$

In aqueous carbonate systems, alkalinity becomes the equivalent sum of the proton acceptors, hence, the acid-neutralizing capacity, whereas acidity becomes the equivalent sum of the proton donors, hence, the base-neutralizing capacity. Alkalinity is a measure of the amount of a strong acid required to reach a pH equal to that of a C_t molar solution of H_2CO_3 at endpoint X in Fig. 3-7. Acidity, likewise, is a measure of the quantity of a strong base required to reach a pH equal to a C_t molar solution of Na_2CO_3 at endpoint Z in Fig. 3-7. It should be noted that mineral acidity exists only to the left of endpoint X in the carbonate system. That is, if a mineral acid were added to the 10^{-3} M C_t solution of H_2CO_3 in Fig. 3-7, the resultant pH values would lie between 0 and 4.7. Likewise, caustic alkalinity exists only to the right of endpoint Z in the carbonate system, which would be 10.7 in Fig. 3-7.

Conceptual and Operational Errors. There are several conceptual and operational errors in the analytical determination of acidity and alkalinity in the carbonate system of natural waters. In the alkalinity titration, the colorimetric indicator methyl orange is used frequently to detect the endpoint X in Fig. 3-7. The color change of methyl orange occurs over the pH range of 3.1 to 4.4 with the colors proceeding in the order red-pink-orange-yellow. The orange color is generally chosen as the endpoint.

Figure 3-7 shows that the pH of a C_t solution of H_2CO_3 is concentration dependent. For example, a 10^{-1} M solution of H_2CO_3 would have a pH value of 3.67 (25°C), whereas a 10^{-5} M solution would have a pH of 5.67 (25°C). Consequently, there would be a tendency to undertitrate waters with low alkalinities.

In the acidity titration, the colorimetric indicator, phenolphthalein, is used to detect the endpoint Y in Fig. 3-7. The determination of H_2CO_3 acidity, therefore, is conceptually incorrect as the titration is carried only to the bicarbonate region, whereas it should be carried to endpoint Z in the carbonate region. This error is, of course, automatically corrected by the equivalency of the NaOH titrant.

Operationally, the endpoint in the H_2CO_3 acidity titration should be detected at a pH of 8.3 by phenolphthalein whose color change occurs within the pH range 8.2–10.0 (colorless–red). A faint pink color usually designates the endpoint. The individual visual detection of this pink color is highly variable and many analysts do not see any color change until pH values around 9.0 are reached. Thus, there is a tendency to overtitrate.

Another operational error in the H_2CO_3 acidity determination is the loss of CO_2 to the atmosphere during sampling, transfer, and titration. The exact loss of CO_2 is difficult to evaluate.

Other conceptual errors lie in the determination of caustic and carbonate alkalinities in the carbonate system. For example, caustic alkalinity is carried to the phenolphthalein endpoint Y and corrections are made for the presence of carbonate alkalinity.[10] It should be obvious from Fig. 3-7 that caustic alkalinity should be titrated to endpoint Z, whereupon carbonate alkalinity then would lie between endpoints Y and Z. Such a determination of caustic alkalinity could be easily accomplished since the indicator, alizarin yellow R, operates in the pH range 10–12.

A frequent conceptual error occurs where it is stated that one-half of the carbonate alkalinity is titrated at the phenolphthalein endpoint.[10] Figure 3-7 shows that all of the carbonate alkalinity is titrated at endpoint Y, whereupon the bicarbonate alkalinity is titrated to endpoint X.

Calculation of Acidity and Alkalinity. The conceptual definitions of acidity and alkalinity expressed in Eqs. 85, 86, 87, 88, 89, and 90 may be combined with Eqs. 47, 48, and 49 to yield:

$$[OH^-\text{-Alk}] = [OH^-] - [H^+] - C_t(\alpha_{HCO_3^-} - 2\alpha_{H_2CO_3}) \quad (91)$$

$$[CO_3^{2-}\text{-Alk}] = [OH^-] - [H^+] + C_t(\alpha_{CO_3^{2-}} - \alpha_{H_2CO_3}) \quad (92)$$

$$[\text{Alk}] = [OH^-] - [H^+] + C_t(\alpha_{HCO_3^-} + 2\alpha_{CO_3^{2-}}) \quad (93)$$

$$[\text{Min-Acy}] = [H^+] - [OH^-] - C_t(\alpha_{HCO_3^-} - 2\alpha_{CO_3^{2-}}) \quad (94)$$

$$[CO_2\text{-Acy}] = [H^+] - [OH^-] + C_t(\alpha_{H_2CO_3} - \alpha_{CO_3^{2-}}) \quad (95)$$

$$[\text{Acy}] = [H^+] - [OH^-] + C_t(2\alpha_{H_2CO_3} + \alpha_{HCO_3^-}) \quad (96)$$

These equations apply to aqueous systems closed to the atmosphere and H_2CO_3 is considered to be a nonvolatile acid. There may be occasions, however, when an open system to the atmosphere must be considered. In this case, the equilibria expressed in Eqs. 42 and 43 are included in the acidity and alkalinity equations. The equilibrium expression for Eq. 42 is:

$$K_{97} = \frac{[H_2CO_3]}{[CO_2(g)]} \quad (97)$$

Dalton's law of partial pressure is used for the $[CO_2(g)]$ term:

$$[CO_2(g)] = \frac{P_{CO_2}}{RT} \quad (98)$$

where $[CO_2(g)]$ is the concentration of CO_2 in the gas phase, P_{CO_2} is the partial pressure of CO_2, R is the ideal gas law constant, and T is the temperature. Combination of Eqs. 97 and 98 yields:

$$[H_2CO_3] = K_H P_{CO_2} \quad (99)$$

where $K_H = K_{97}/RT$. The dimensions of K_H are mole/liter/atm, for which values at several temperatures appear in Table 3-5. In the α-value system, Eq. 99 may be rewritten:

$$C_t \alpha_{H_2CO_3} = K_H P_{CO_2} \quad (100)$$

The various forms of acidity and alkalinity may be calculated from Eqs. 91 through 96 from the water quality analyses of pH, bicarbonate, and temperature. For example, spring water near Jefferson City, Tennessee contains 213 mg/l HCO_3^-, has a pH value of 7.4, and a temperature of 58°F (14.4°C).[11] (See Table 3A-1, Appendix.) Table 3-6 shows the calculations of the mass balance on the carbonate species, the concentrations of H_2CO_3, CO_3^{2-}, alkalinity, acidity, and CO_2-acidity. Also, the partial pressure of CO_2 required to maintain the $[H_2CO_3]$ is computed. This is a typical example of how one may obtain the carbonate chemistry, acidity, and alkalinity for a natural water from only three water quality analyses. It should be noted that this water does not contain any mineral acidity or caustic and carbonate alkalinities. This may be deduced from the three water quality analyses and the carbonate species distribution diagram. Also, the approximation may be made for this water that the $[HCO_3^-] \simeq [\text{Alk}]$.

TABLE 3-5. Henry Law Constants for CO_2 (g) Solubility in Water[a]

Temperature (°C)	K_H (moles/liter/atm)
0	7.69×10^{-2}
5	6.40×10^{-2}
10	5.36×10^{-2}
15	4.55×10^{-2}
20	3.92×10^{-2}
25	3.45×10^{-2}
30	3.02×10^{-2}
40	2.29×10^{-2}
50	1.91×10^{-2}

[a]Ref. 2.

DISSOLUTION AND PRECIPITATION EQUILIBRIA

Dissolution and precipitation equilibria are encountered quite frequently in water and wastewater chemistry. The natural chemical constituents of water are the dissolution products of rocks. A variety of chemical weathering reactions are responsible for the dissolution. Precipitation reactions are employed in the softening of hard water, in the chemical coagulation of water and wastewaters, and in the control of corrosion of metallic distribution systems for water. All of these dissolution and precipitation reactions involve the principles of chemical equilibria.

TABLE 3-6. Calculations in the Carbonate System

1. Water quality analyses of a spring water sample: pH = 7.4, T = 58°F (14.4°C), $[HCO_3^-] = 213$ mg/l.
2. At pH 7.4, $\alpha_{H_2CO_3} = 0.0947$, $\alpha_{HCO_3^-} = 0.9045$, and $\alpha_{CO_3^{2-}} = 0.0008$, (15°C values).
3. From Eq. 48:

$$C_t = \frac{[HCO_3^-]}{\alpha_{HCO_3^-}} = \frac{3.49 \times 10^{-3}\ M}{0.9045} = 3.86 \times 10^{-3}\ M$$

4. From Eq. 47:

$$[H_2CO_3] = C_t \alpha_{H_2CO_3} = 3.86 \times 10^{-3} \times 0.0947 = 3.65 \times 10^{-4}\ M$$

5. From Eq. 49:

$$[CO_3^{2-}] = C_t \alpha_{CO_3^{2-}} = 3.86 \times 10^{-3} \times 0.0008 = 3 \times 10^{-6}\ M$$

6. From Eq. 93 ($[H^+]$ and $[OH^-]$ neglected):

$$[Alk] = C_t(\alpha_{HCO_3^-} + 2\alpha_{CO_3^{2-}}) = 3.86 \times 10^{-3}\ (0.9045 + 2 \times 0.0008)$$

$$[Alk] = 3.50 \times 10^{-3}\ M$$

7. From Eq. 95 ($[H^+]$ and $[OH^-]$ neglected):

$$[CO_2\text{-}Acy] = C_t(\alpha_{H_2CO_3} - \alpha_{CO_3^{2-}}) = 3.86 \times 10^{-3}\ (0.0947 - 0.0008)$$

$$[CO_2\text{-}Acy] = 3.62 \times 10^{-4}\ M$$

8. From Eq. 96 ($[H^+]$ and $[OH^-]$ neglected):

$$[Acy] = C_t(2\alpha_{H_2CO_3} + \alpha_{HCO_3^-}) = 3.86 \times 10^{-3}\ (2 \times 0.0947 + 0.9045)$$

$$[Acy] = 4.22 \times 10^{-3}\ M$$

9. From Eq. 100:

$$P_{CO_2} = \frac{[H_2CO_3]}{K_H} = \frac{3.65 \times 10^{-4}}{4.55 \times 10^{-2}} = 8.03 \times 10^{-3}\ \text{atm}$$

$CaCO_3(s)$ Solubility. As cited in Eq. 21, the reaction of Ca^{2+}(aq) and CO_3^{2-}(aq) leads to the precipitation of $CaCO_3(s)$. Furthermore, the net negative free energy of the reaction indicates that the left-to-right reaction is feasible. The equilibrium constant for Eq. 21 may be computed from Eq. 13 and is $10^{+8.34}$ at 25°C. Equation 21 is written as a precipitation reaction, whereas the dissolution reaction would be:

$$CaCO_3(s) = Ca^{2+}(aq) + CO_3^{2-}(aq) \quad (101)$$

The free energy value for this reaction is +11.38 kcal and the equilibrium constant is $10^{-8.34}$ This indicates that the left-to-right reaction is not feasible and occurs only to a very small extent. For precipitation and dissolution reactions, these equilibrium constants are frequently called solubility product constants and, for Eq. 101, may be expressed:

$$K_{sCaCO_3} = [Ca^{2+}][CO_3^{2-}] \quad (102)$$

This expression suggests that the equilibrium concentrations of Ca^{2+} and CO_3^{2-} are equal and are computed to be $10^{-4.17}$ M. Furthermore, if the product of the molar concentrations of Ca^{2+} and CO_3^{2-} exceeds $10^{-8.34}$ ($10^{-6.00}$, for example), then precipitation of $CaCO_3$ occurs via Eq. 21. On the other hand, if the product of the molar concentrations of Ca^{2+} and CO_3^{2-} is less than $10^{-8.34}$ ($10^{-10.00}$, for example), then dissolution of $CaCO_3(s)$ occurs via Eq. 101. The key to dissolution and precipitation reactions, therefore, is the multiplication of the molar concentrations of the cationic and anionic constituents in the aqueous phase. The resultant product will indicate the tendency for dissolution of the solid phase, for precipitation of the solid phase, or for equilibrium with the solid phase. Frequently there are more than two ionic constituents. Dolomite, for example, is a calcium–magnesium carbonate whose dissolution reaction is:

$$CaMg(CO_3)_2\,(s) = Ca^{2+} + Mg^{2+} + 2CO_3^{2-} \quad (103)$$

whereupon the K_s equation would be:

$$K_s = [Ca^{2+}]\,[Mg^{2+}]\,[CO_3^{2-}]^2 \quad (104)$$

Another example would be the dissolution of hydroxylopatite:

$$Ca_5OH(PO_4)_3\,(s) = 5Ca^{2+} + 3PO_4^{3-} + OH^- \quad (105)$$

whereupon the K_s equation would be:

$$K_s = [Ca^{2+}]^5\,[PO_4^{3-}]^3\,[OH]^- \quad (106)$$

It should be noted that the molar coefficients in the dissolution reactions become the exponents in the solubility product equations.

Another very important point from Eq. 102 is that the actual $[CO_3^{2-}]$ must be used in the solubility product expression for $CaCO_3(s)$. It was noted from the carbonate species distribution diagram (Fig. 3-5) that $[CO_3^{2-}]$ decreases as $[H^+]$ increases for a given C_t value. Therefore, Eq. 102 must be modified in order to obtain the equilibrium of $Ca^{2+}(aq)$ for $CaCO_3(s)$ at all pH values. Substitution of Eq. 49 into Eq. 102 yields:

$$K_s = [Ca^{2+}]\,C_t\,\alpha_{CO_3^{2-}} \quad (107)$$

In this equation, $[Ca^{2+}] = C_t$ (sum of concentrations of carbonate species), whereupon:

$$K_s = [Ca^{2+}]^2\,\alpha_{CO_3^{2-}} \quad (108)$$

or:

$$[Ca^{2+}] = \left(\frac{K_s}{\alpha_{CO_3^{2-}}}\right)^{0.5} \quad (109)$$

Equation 109 now gives the equilibrium concentration of $Ca^{2+}(aq)$ for $\alpha_{CO_3^{2-}}$ values. It follows that, as $[H^+]$ increases, $\alpha_{CO_3^{2-}}$ values decrease and the equilibrium concentration of $Ca^{2+}(aq)$ increases so that the product of $[Ca^{2+}] \times [CO_3^{2-}]$ remains equal to $K_{s_{CaCO_3}}$.

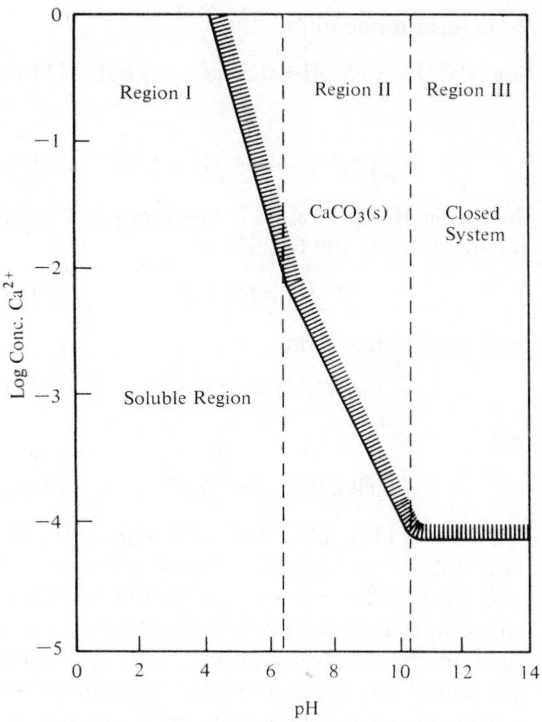

Fig. 3-8 Solubility diagram for calcium carbonate in a closed system.

The effect of $[H^+]$ on $CaCO_3(s)$ may be illustrated through the use of a solubility diagram. In Fig. 3-8, the $-\log\,[Ca^{2+}]$ is plotted versus pH at 25°C for $CaCO_3(s)$ in a closed system. This diagram was constructed in the following manner. In region I (pH values $<pK_1$ of H_2CO_3), $\alpha_{CO_3^{2-}} \simeq K_1K_2/[H^+]^2$, whereupon Eq. 109 becomes:

$$[Ca^{2+}] = [H^+]\left(\frac{K_s}{K_1K_2}\right)^{0.5} \quad (110)$$

or in logarithmic form:

$$\log\,[Ca^{2+}] = -pH + 0.5\,(pK_1 + pK_2 - pK_s) \quad (111)$$

or:

$$\log\,[Ca^{2+}] = -pH + 4.17 \quad (112)$$

In region II (pH values between pK_1 and pK_2), $\alpha_{CO_3^{2-}} \simeq K_2/[H^+]$, whereupon Eq. 109 becomes:

$$[Ca^{2+}] = \left(\frac{K_s[H^+]}{K_2}\right)^{0.5} \quad (113)$$

or in logarithmic form:

$$\log [Ca^{2+}] = -0.5 \text{ pH} + 0.5 (pK_2 - pK_s) \quad (114)$$

or:

$$\log [Ca^{2+}] = -0.5 \text{ pH} + 0.99 \quad (115)$$

In region III (pH values $> pK_2$), $\alpha_{CO_3^{2-}} \simeq 1.0$, whereupon Eq. 109 becomes:

$$[Ca^{2+}] = (K_s)^{0.5} \quad (116)$$

or in logarithmic form:

$$\log [Ca^{2+}] = -0.5 \text{ p}K_s \quad (117)$$

or:

$$\log [Ca^{2+}] = -4.17 \quad (118)$$

Equations 112, 115, and 118 represent the equilibrium boundary lines between the soluble and solid phase regions. (It should be noted that a more accurate diagram would have curved lines in the areas around pK_1 and $pK_2 \pm 0.5$ pH units.) In regions I and II, the solubility of $CaCO_3$(s) is dependent upon $[H^+]$, whereas in region III it is independent of $[H^+]$. Temperature, of course, affects the solubility of $CaCO_3$(s) as seen by the solubility constants in Table 3-7.

TABLE 3-7. Solubility Constants for $CaCO_3$ (s)[a]

Temperature (°C)	$-\log K_s$[b]
0	8.023
5	8.087
10	8.150
15	8.215
20	8.280
25	8.342
30	8.395
40	8.515
50	8.625
80	8.975

[a]Refs. 2 and 12.
[b]Infinite dilution.

The interpretation and use of Fig. 3-8 are relatively easy. Several examples follow: water A in Appendix Table 3A-1 has a pH value of 7.0, a calcium content of 46 mg/l ($10^{-2.94}$ M), and a temperature of 16°C. This is a region II water, whereupon (using 15°C constants) $\log [Ca^{2+}] = -0.5 \times 7.0 + 1.105 = -2.393$. Water A, therefore, would be undersaturated with respect to $CaCO_3$(s). Also, this water should not have any $CaCO_3$(s) scale-forming tendencies. Another example would be water B from Appendix Table 3A-1, where pH = 7.8 and $[Ca^{2+}] = 96$ mg/l ($10^{-2.62}$ M). This is also a region II water, whereupon at 15°C the equilibrium concentration of Ca^{2+} is calculated from Eq. 114 to be $10^{-2.79}$ M. Since the actual $[Ca^{2+}]$ is $10^{-2.62}$ M, this water is oversaturated with respect to $CaCO_3$(s) and would have a tendency to form scales. Another use of the solubility diagram lies in the prediction of a water to be a $CaCO_3$(s) scale-former at the higher temperatures employed in domestic hot water heaters. At 50°C (122°F), water A would have an equilibrium calcium content of $10^{-2.73}$ M, whereupon the actual concentration is $10^{-2.94}$ M. This water, therefore, should not form $CaCO_3$(s) scales in a hot water heater. At 50°C, water B would have an equilibrium calcium content of $10^{-3.126}$ M, whereas the actual concentration is $10^{-2.62}$ M. This water, therefore, should form $CaCO_3$(s) scales in a hot water heater.

$CaCO_3$(s) Saturation Indices. Figure 3-8 may apply also to the saturation or stability index of Langelier.[13] This index attempts to predict the $CaCO_3$ scale-forming or scale-dissolving tendencies of water in distribution systems. It is used in connection with corrosion control of cast iron pipes. This index is based upon this reaction:

$$CaCO_3(s) + H^+ = Ca^{2+} + HCO_3^- \quad (119)$$

from which may be derived:

$$pH_s = pCa^{2+} + pAlk - pK_{119} \quad (120)$$

where pH_s is the pH value at which the calcium and the alkalinity (assumed here to be equal to $[HCO_3^-]$) are in equilibrium with $CaCO_3$(s), and $pK_{119} = pK_{s_{CaCO_3}} - pK_{2H_2CO_3}$. The saturation index is calculated from:

$$S.I. = pH_{actual} - pH_s \quad (121)$$

where pH_{actual} is the measured pH of the water. If the pH_{actual} is greater than pH_s, then the water will have a tendency to form $CaCO_3(s)$; i.e., the right-to-left direction of Eq. 119 would be favored. If the actual pH is less than pH_s, then the water will have a tendency to dissolve $CaCO_3(s)$; i.e., the left-to-right direction of Eq. 119 would be favored. If the $pH_{actual} = pH_s$, then the water is stable with respect to $CaCO_3(s)$, or Eq. 119 is in equilibrium. Some empirical information about the saturation index is provided by Flentje.[14] As seen in Fig. 3-9, A is the zone of $CaCO_3(s)$ deposition, zone

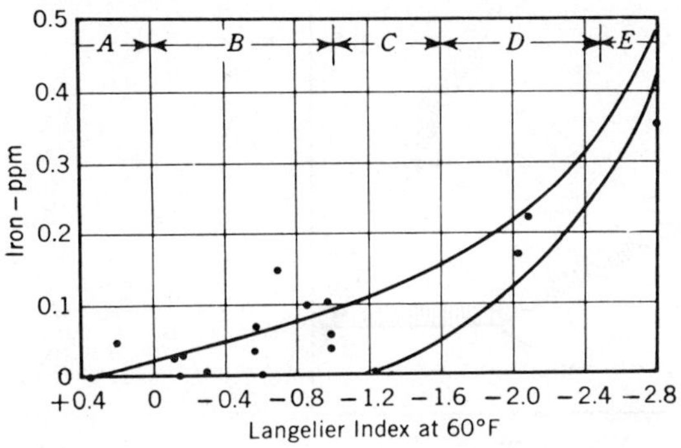

Fig. 3-9 Saturation index and iron pickup in water distribution systems. Reproduced from Ref. 14 by courtesy of the American Water Works Association, New York. *A* is zone of $CaCO_3$ deposition; *B*, of little pickup; *C*, of red and dirty water; *D*, of red water unless polyphosphates are used; *E*, of red and dirty water even if polyphosphates are used.

B shows very little iron pickup, zone C shows "red and dirty" water, zone D will show "red" water unless polyphosphates are used, and zone E will show "red" water even if polyphosphates are used. It appears, therefore, that S.I. values in the positive, 0, and slightly negative region correlate with reasonably good corrosion control and very little iron pickup in the distribution system.

Langelier[13] presents some rather cumbersome diagrams for the computation of S.I. values. Figure 3-10 shows a $CaCO_3(s)$ solubility diagram at 15°C from which S.I. values may be calculated more easily. For example, the Delaware River at Trenton has a pH value of 7.5, a calcium content of 15 mg/l as Ca^{2+}, and an alkalinity of 42.0 mg/l. The pH_{actual} versus 15 mg/l Ca^{2+} plots in the soluble region of the diagram. The pH_s is obtained by moving hori-

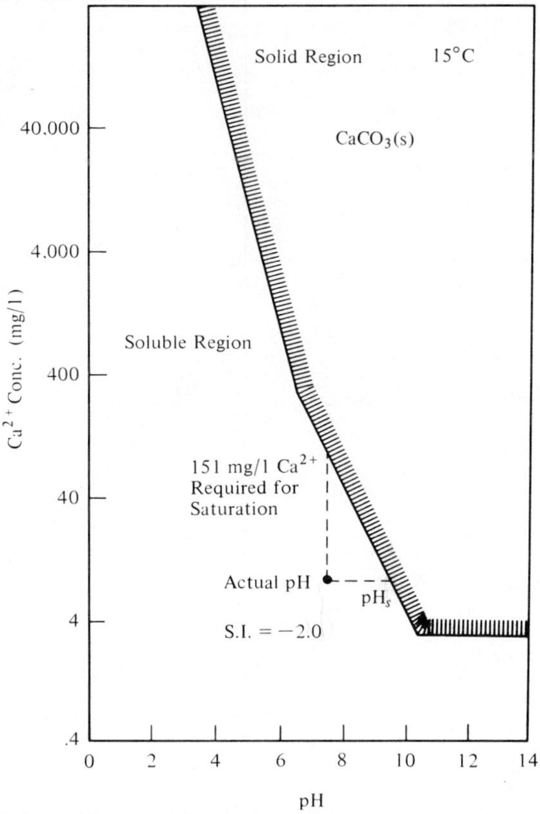

Fig. 3-10 Simplified saturation index diagram for calcium carbonate.

zontally along the pH ordinate until the solid phase boundary line is reached. For this water, pH_s is 9.5, whereupon the S.I. is −2.0 as computed from Eq. 121. Consequently, the Delaware River water would be corrosive to iron pipes as indicated by the S.I. value.

It is also possible to calculate the equilibrium concentration of calcium for the pH_{actual} from Fig. 3-10. This is accomplished by moving vertically along the Ca^{2+} ordinate until the solid phase boundary line is reached. For this water, the $[Ca^{2+}]$ required for equilibrium is 151 mg/l. Consequently, 136 mg/l of Ca^{2+} would have to be added to this water to reach a S.I. value of 0.0 at pH 7.5. Similar diagrams may be drawn for other temperatures.

Water Softening. Calcium and magnesium ions, the traditional constituents of hard water, are reduced in concentration frequently through precipitation reactions. Calcium is precipitated

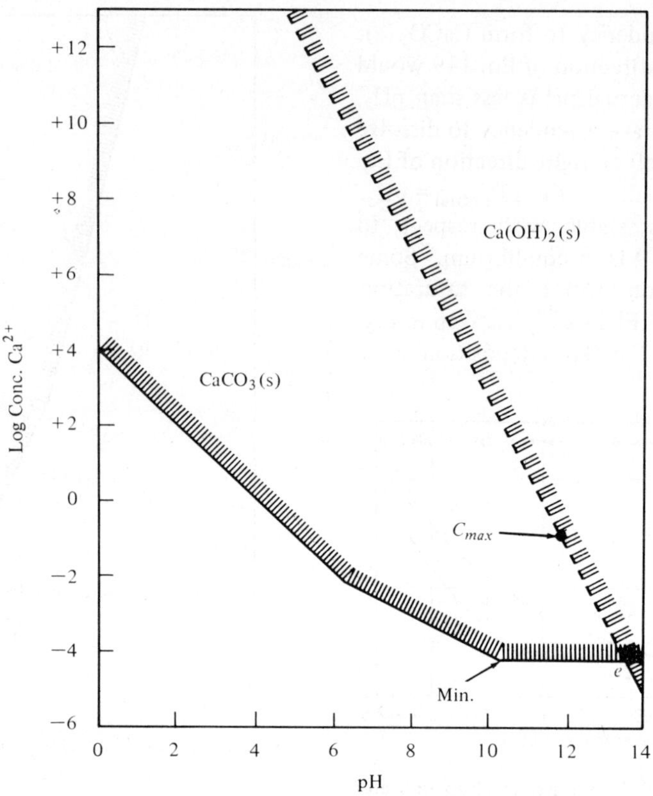

Fig. 3-11 Solubility diagrams for calcium carbonate and calcium hydroxide.

as its carbonate, whereas magnesium is removed as its hydroxide. Both precipitation reactions require adjustment of the pH value into the alkaline range. Lime ($Ca(OH)_2$) and soda ash (Na_2CO_3) are the chemicals most commonly used. As seen in Fig. 3-11, solubility of $CaCO_3(s)$ is independent of pH in the range 10.3–14.0. Also, it may be observed that calcium precipitates as the carbonate and not as the hydroxide until, at least, a pH of 13.6 is reached. On the other hand, Fig. 3-12 shows that Mg^{2+} is removed as the hydroxide at pH values greater than 9.1 rather than as the carbonate. The chemical reactions for these precipitations are:

$$H_2CO_3 + Ca(OH)_2 = CaCO_3(s) + 2H_2O \quad (122)$$

$$Ca^{2+} + 2HCO_3^- + Ca(OH)_2 = 2CaCO_3(s) + 2H_2O \quad (123)$$

$$Ca^{2+} + Na_2CO_3 = CaCO_3(s) + 2Na^+ \quad (124)$$

$$Mg^{2+} + 2HCO_3^- + 2Ca(OH)_2 = 2CaCO_3(s) + Mg(OH)_2(s) + 2H_2O \quad (125)$$

$$Mg^{2+} + Ca(OH)_2 + Na_2CO_3 = CaCO_3(s) + Mg(OH)_2(s) + 2Na^+ \quad (126)$$

Carbonic acid is included in Eq. 122 because it reacts with lime and must be included in the calculation of the chemical dosages required for softening. Equation 123 represents the so-called carbonate hardness assigned to the $Ca^{2+}(aq)$, where 1 equivalent of lime is required for precipitation. Equation 124 represents the so-called noncarbonate hardness (usually SO_4^{2-} or Cl^-) assigned to the $Ca^{2+}(aq)$, where 1 equivalent of soda ash is required for precipitation. Equation 125 represents the carbonate hardness assigned to $Mg^{2+}(aq)$, whereupon 2 equivalents of lime are required for precipitation. Equation 126 represents the noncarbonate hardness assigned to $Mg^{2+}(aq)$, where 1 equivalent of lime

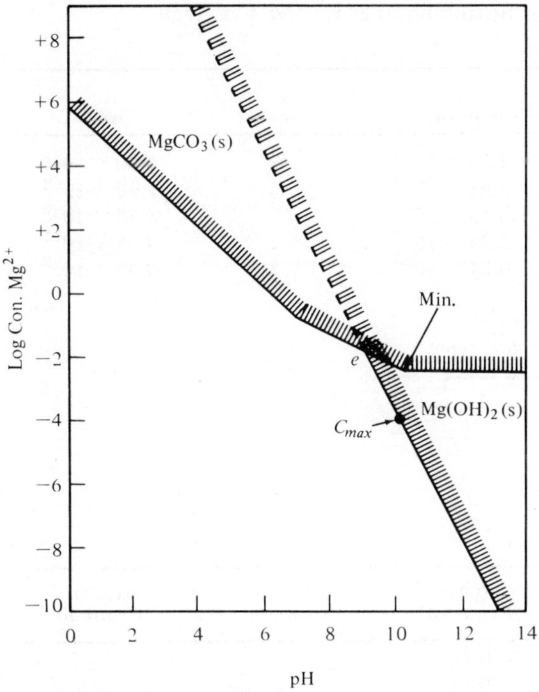

Fig. 3-12 Solubility diagrams for magnesium carbonate and magnesium hydroxide.

and 1 equivalent of soda ash are required for precipitation.

In the removal of Ca^{2+} and Mg^{2+} from water by precipitation, pH adjustment into the alkaline range is accomplished with the lime, whereas the soda ash supplies the CO_3^{2-} ions. The following condition must be met after the addition of the lime and soda ash:

$$Ca^{2+}_{initial} + Ca^{2+}_{added} = C_{t_{initial}} + CO^{2-}_{3\,added}. \tag{127}$$

from which it may be seen that $[Ca^{2+}]_{final} = C_{t_{final}}$; i.e., the $[Ca^{2+}]_{final}$ must equal the $[CO_3^{2-}]_{final}$. In calculation of the necessary lime and soda ash dosages, an equivalency expression is employed:

$$2[Ca^{2+}] + 2[Mg^{2+}] + [H^+] = [HCO_3^-] + 2[CO_3^{2-}] + [Cl^-] + 2[SO_4^{2-}] + [OH^-] \tag{128}$$

From this expression, one may easily calculate and tabulate the equivalents of lime and soda ash required.

Table 3-8 gives an example calculation for a groundwater sample in New Jersey. First, the equivalents per liter (eq/l) are calculated for each of the ionic constituents. Next, an equivalence scale is drawn (1 unit = 1 eq in this example). Then the cationic scale is drawn with the equivalents of Ca^{2+} placed first (reading left to right). The equivalents of Mg^{2+} are then placed after the Ca^{2+}. The anionic scale is placed beneath the cationic scale with the equivalents of bicarbonate drawn in first. Then the equivalents of SO_4^{2-} are placed after the HCO_3^-. If Cl^- were present, these equivalents would then be drawn after the SO_4^{2-}.

From Eq. 123, it was calculated that 3.15×10^{-3} eq/l $Ca(OH)_2$ are required for precipitation of the Ca^{2+}-carbonate hardness. This leaves 2.15×10^{-3} eq/l of Ca^{2+}-noncarbonate hardness for precipitation by Na_2CO_3 in accord with Eq. 124. The Mg^{2+}-noncarbonate hardness requires 1.90×10^{-3} eq/l of $Ca(OH)_2$ and 1.90×10^{-3} eq/l of Na_2CO_3 for precipitation in accord with Eq. 126. The nonionic H_2CO_3 is not included, of course, in the equivalency diagram. In this case, the H_2CO_3 was computed from the appropriate C_t and α-value equations as seen in the table. The quantity of lime required for neutralization of the H_2CO_3 is then calculated from Eq. 122. The total quantities of $Ca(OH)_2$ and Na_2CO_3 are indicated in the table.

The residual concentrations of Ca^{2+}, Mg^{2+}, CO_3^{2-}, and OH^- are controlled by the appropriate solid phase in Figs. 3-11 and 3-12. It is seldom, however, that the theoretical equilibrium concentrations are reached because insufficient detention time is designed into the softening units. Sometimes a 5–10 percent excess of the stoichiometric lime and soda ash requirements is added to accelerate the precipitation reactions. Consequently, this excess should be added to the quantities seen in the sample calculation in Table 3-8. Design details, sludge handling problems, recarbonation techniques, etc. in connection with the lime-soda softening processes are described in Refs. 15 and 16.

There are occasions in water chemistry where an open $CaCO_3(s)$ system should be considered. Such an open system would indicate the effect

TABLE 3-8. Lime and Soda Ash Requirements for Softening of Elmwood Ave. Well, Flemington, New Jersey

Ion	Concentration (mg/l)	MW	Moles/liter	# eq	eq/l
Ca^{2+}	106	40	2.65×10^{-3}	2	5.30×10^{-3}
Mg^{2+}	23	24.3	9.48×10^{-4}	2	1.90×10^{-3}
HCO_3^-	192	61	3.15×10^{-3}	1	3.15×10^{-3}
SO_4^{2-}	225	96	2.34×10^{-3}	2	4.68×10^{-3}
H_2CO_3	29.8	62	0.24×10^{-3}	2	0.48×10^{-3}

Equivalents scale: $\times 10^{-3}$, 1 unit = 1 eq

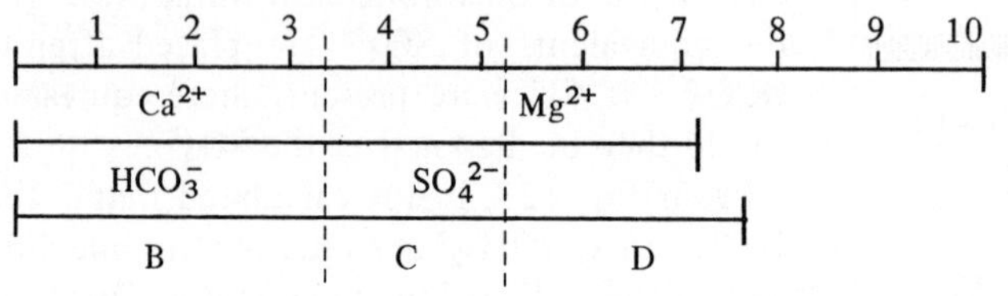

Block	Equation	To Be Removed (eq/l × 10^{-3})	$Ca(OH)_2$ Required (eq/l × 10^{-3})	To Be Removed (eq/l × 10^{-3})	Na_2CO_3 Required (eq/l × 10^{-3})
A*	122	0.48	0.48	–	–
B	123	3.15	3.15	–	–
C	124	–	–	2.15	2.15
D	126	1.90	1.90	1.90	1.90
			5.53		4.05
			or 2.765×10^{-3} M or 204.6 mg/l		or 2.025×10^{-3} M or 214.6 mg/l

*For Block A, $[H_2CO_3]$ was calculated from:

$$C_t = \frac{[HCO_3^-]}{\alpha_{HC^-}} = \frac{3.15 \times 10^{-3}}{0.93} = 3.39 \times 10^{-3}\ M$$

Neglecting $[CO_3^{2-}]$ ∴ $[H_2CO_3] = C_t - [HCO_3^-] = 3.39 \times 10^{-3} - 3.15 \times 10^{-3} = 0.24 \times 10^{-3}$ M

pH = 7.5, $\alpha_{HCO_3^-} = 0.93$

of the partial pressure of $CO_2(g)$ on the solubility of $CaCO_3(s)$. A combination of Eqs. 100 and 107 yields:

$$[Ca^{2+}] = \frac{K_s \alpha_{H_2CO_3}}{K_H P_{CO_2} \alpha_{CO_3^{2-}}} = \frac{K_s \alpha_{H_2CO_3}}{[H_2CO_3]\, \alpha_{CO_3^{2-}}} \quad (129)$$

where $CaCO_3(s)$ solubility is now a function of the partial pressure of $CO_2(g)$. This is shown in Fig. 3-13 where the solubility contours are seen for carbonic acid concentrations of 10^{-1}, 10^{-3}, and 10^{-5} M. The solubility contour for the closed system is included for comparison. The intersection of the 10^{-5} M H_2CO_3 contour with the closed system (point *A*) represents $CaCO_3(s)$ in an open system which is in equilibrium with the partial pressure of $CO_2(g)$ in the atmosphere (2.9×10^{-4} atm, 25°C). The pH value is 8.3 and $[Ca^{2+}] = 10^{-3.1}$ M. Point *B* represents $CaCO_3(s)$ in an open system which is in equilibrium with a partial pressure of 2.9×10^{-2} atm of $CO_2(g)$ (10^{-3} M H_2CO_3). The pH value is 6.8 and $[Ca^{2+}] = 10^{-2.4}$ M. Point *C* represents $CaCO_3(s)$ in an open system which is in equilibrium with a partial pressure of 2.9 atm of $CO_2(g)$ (10^{-1} M H_2CO_3). The pH value is 5.1 and $[Ca^{2+}] = 10^{-0.9}$ M. It may be readily seen that as the partial pressure of $CO_2(g)$ is increased, the pH value decreases and $[Ca^{2+}]$ is

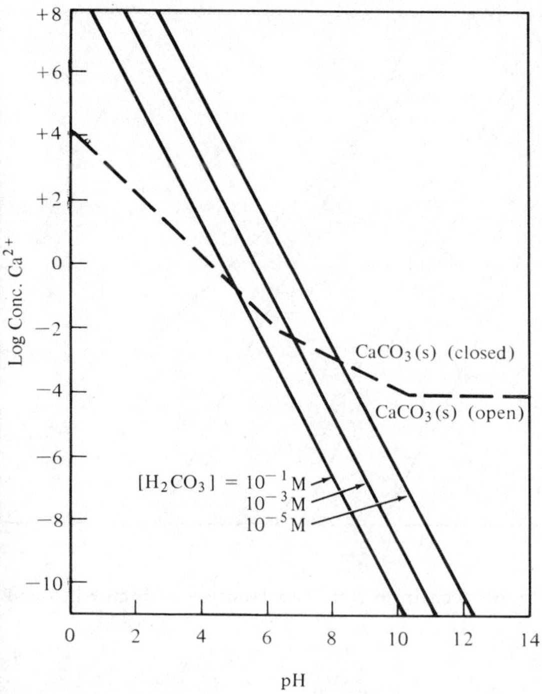

Fig. 3-13 Solubility diagram for calcium carbonate in an open system.

increased. This kind of interaction among pH, $P_{CO_2(g)}$, and $[Ca^{2+}]$ is often encountered in a natural lake during extensive photosynthetic activities of algae:

$$nCO_2 + nH_2O \underset{\text{respiration}}{\overset{\text{photosynthesis}}{\rightleftharpoons}} (CH_2O)n + nO_2 \quad (130)$$

During the daylight hours, CO_2 is removed from the water and Eq. 130 shows a shift to the right. This results in an increase in the pH value and, if sufficient Ca^{2+}(aq) is present to exceed the solubility product, $CaCO_3$(s) should precipitate. During the nighttime hours, respiration occurs and Eq. 130 shows a shift to the left. This increases $[H_2CO_3]$ with a concomitant decrease in pH value and increase in $[Ca^{2+}]$.

Solubility of Metallic Oxides and Hydroxides. Metallic oxides and hydroxides are encountered quite frequently in water and wastewater chemistry. For example, the hydrous oxides of aluminum and iron are employed in chemical coagulation of water, wastewaters, and sludges. The oxides of iron and manganese are precipitated from oxidation reactions employed for their removal from water. The metallic contaminants of copper and zinc form hydrous oxides that are relatively insoluble in water.

A generalized equation may be written for the dissolution of metallic hydroxides:

$$Me(OH)_x(s) = Me^{x+} + xOH^- \quad (131)$$

whereupon:

$$K_s = [Me^{x+}][OH^-]^x \quad (132)$$

or:

$$K_s = \frac{[Me^{x+}]K_w^x}{[H^+]^x} \quad (133)$$

rearranging:

$$\log [Me^{x+}] = \log K_s + xpK_w - xpH \quad (134)$$

This general equation shows that the solubility of metallic hydroxides is dependent upon $[H^+]$. There is an increase of $[Me^{x+}]$ as the $[H^+]$ is increased. This relationship is seen in Fig. 3-14 for some of the more commonly encountered aqueous metals. Solubility product constants are seen in Table 3-9.

The chemical coagulation of particulate and colloidal matter involves the injection of alum ($Al_2(SO_4)_3$) into water with the subsequent precipitation of a solid phase of aluminum hydroxide. The empirical formula, $Al(OH)_3$(s), is used in this chapter. Many investigators, however, believe that a polymeric form of $Al(OH)_3$(s) is actually formed. The various arguments surrounding this scientific controversy are discussed in Refs. 1, 2, and 17.

In any event, a solid phase form of $Al(OH)_3$(s) is precipitated which, in turn, reacts with the colloidal matter by a variety of mechanisms.[1,2] This results in a destabilized particle which is then easily separated from the water phase by sedimentation and/or filtration.

The solubility product constant of $Al(OH)_3$(s) (microcrystalline gibbsite) has been reported to be $10^{-32.65}$.[17]

It follows that OH^- ions are required for the precipitation of the Al^{3+} ion. Consequently, the

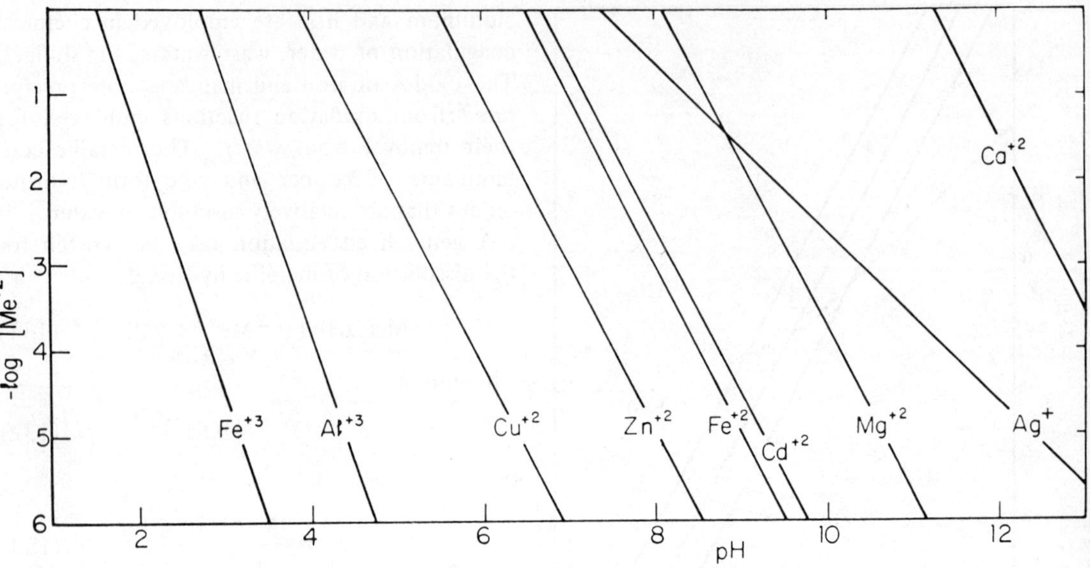

Fig. 3-14 Solubility of metal oxides and hydroxides. Reproduced from Ref. 2 by courtesy of John Wiley and Sons, Inc., New York.

solubility of $Al(OH)_3$ is dependent upon the $[H^+]$ in accord with:

$$\log [Al^{3+}] = 9.35 - 3pH \qquad (143)$$

Equation 143 is plotted as line *a* in Fig. 3-15. However, $Al(OH)_3(s)$ dissolves amphoterically under alkaline conditions to yield the $Al(OH)_4^-$ anion:

$$Al(OH)_3(s) + H_2O = Al(OH)_4^- + H^+ \qquad (144)$$

Equation 144 has an equilibrium constant of $10^{-12.71}$ [17] from which the following equation may be derived:

$$\log [Al(OH)_4^-] = pH - 12.71 \qquad (145)$$

This is plotted as line *b* in Fig. 3-15 which intersects with line *a* at a pH value of 5.5 and at a log [Al species] of $10^{-7.25}$ M. Lines *a* and *b* now define the solubility boundaries for $Al(OH)_3(s)$ with respect to pH and their intersection represents the point of minimum solubility (sometimes called the isoelectric point).

The Al^{3+} cation, however, may act as a Brönsted acid because of its ability to become hydrated and coordinate 6 molecules of water:[1,2]

$$Al(H_2O)_6^{3+} + H_2O = H_3^+O + Al(OH)(H_2O)_5^{2+} \qquad (146)$$

TABLE 3-9. Solubility Product Constants of Aqueous Metallic Hydroxides

Equation No.	Reaction	log K_S (25°C)	Reference
(135)	$Fe(OH)_3(s)$ (am) = $Fe^{3+} + 3OH^-$	-38.7^a	2
(136)	$Fe(OH)_2$ (active) = $Fe^{2+} + 2OH^-$	−14.0	2
(137)	$Fe(OH)_2$ (inactive) = $Fe^{2+} + 2OH^-$	−14.5(−15.1)	2
(138)	$Al(OH)_3(s)$ (gibbsite) = $Al^{3+} + 3OH^-$	−32.65	17
(139)	$Cu(OH)_2(s) = Cu^{2+} + 2OH^-$	−18.1	Computed
(140)	$Zn(OH)_2(s) = Zn^{2+} + 2OH^-$	−17.16	18
(141)	$Mn(OH)_2(s) = Mn^{2+} + 2OH^-$	−12.8	2
(142)	$Pb(OH)_2(s) = Pb^{2+} + 2OH^-$	−14.4	18

[a]In 3.0 M $NaClO_4$.

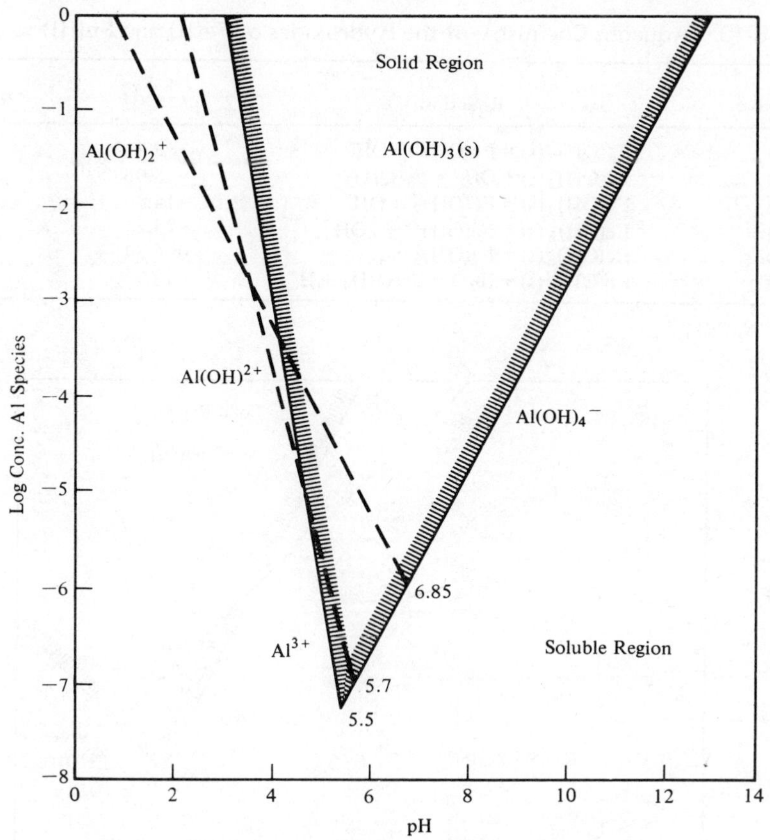

Fig. 3-15 Solubility diagram for aluminum hydroxide.

It is usual to omit the water of hydration when reporting such reactions. When Eq. 146 ($K_{eq} = 10^{-5.02}$) is combined with Eq. 143:

$$\log [Al(OH^{2+})] = 4.33 - 2pH \quad (147)$$

If this equation is plotted as line *c* in Fig. 3-15, an intersection with line *b* occurs at a pH value of 5.7 and a log [Al species] of $10^{-7.0}$ M. Thus, the point of minimum solubility is shifted slightly. Another Brönsted acid reaction of Al^{3+} is:

$$Al^{3+} + 2H_2O = Al(OH^+)_2 + 2H^+ \quad (148)$$

which, when combined with Eq. 143, yields:

$$\log [Al(OH^+)_2] = 0.79 - pH \quad (149)$$

When Eq. 149 is plotted as line *d* in Fig. 3-15, an intersection occurs at a pH value of 6.85 and a log [Al species] of $10^{-5.9}$ M. Thus, the point of minimum solubility is shifted quite sharply with respect to pH and concentration of soluble species. Equation 148 has the effect of reducing the solid phase region under acid pH values. Therefore, these complex cationic species of Al^{3+} play a significant role in the formation of $Al(OH)_3$ (s). Some investigators believe in such gigantic cationic species as $Al_7(OH)_{17}^{4+}$, etc.[1,2,17] Their effect on the formation of $Al(OH)_3$ (s) is, however, questionable.

Solubilities of the hydroxides of Fe(II) and Fe(III) are important also in water and wastewater chemistry. For example, $Fe(OH)_3$ (s) is occasionally used as a chemical coagulant in wastewater treatment and is the oxidative product of Fe(II) in iron removal processes. $Fe(OH)_2$ (s) may precipitate in waters where the oxidation of Fe(II) to Fe(III) is not feasible. The aqueous chemistries of the hydroxides of Fe(II) and Fe(III) are similar to $Al(OH)_3$ (s). These reactions are summarized in Table 3-10.

As seen in Fig. 3-16, $Fe(OH)_2$ (s) is somewhat soluble in water with, for example, 1 mole of

TABLE 3-10. Aqueous Chemistry of the Hydroxides of Fe(II) and Fe(III)

Equation No.	Reaction	log K_{eq} (25°C)	Reference
(150)	$Fe(OH)_2(s) = Fe(OH)^+ + OH^-$	−9.4	19
(151)	$Fe(OH)_2(s) + OH^- = Fe(OH)_3^-$	−5.08	19
(152)	$Fe(OH)_3(s) = Fe(OH)_2^+ + OH^-$	−14.8	19
(153)	$Fe(OH)_3(s) = Fe(OH)^{2+} + 2OH^-$	−24.2	19
(154)	$Fe(OH)_3(s) = Fe(OH)_3(aq)$	−6.54	19
(155)	$Fe(OH)_3(s) + H_2O = Fe(OH)_4^- + H^+$	−18.7[a]	2

[a]In 3.0 M $NaClO_4$.

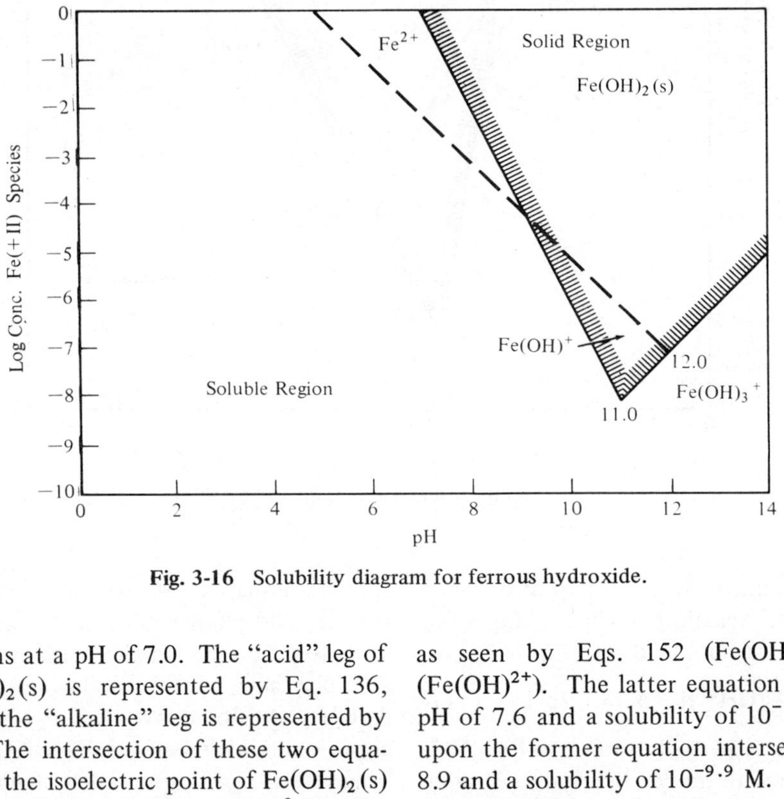

Fig. 3-16 Solubility diagram for ferrous hydroxide.

Fe^{2+}(aq) ions at a pH of 7.0. The "acid" leg of the $Fe(OH)_2$(s) is represented by Eq. 136, whereupon the "alkaline" leg is represented by Eq. 151. The intersection of these two equations yields the isoelectric point of $Fe(OH)_2$(s) at a pH of 11.0 and a solubility of 10^{-8} M. The hydrolysis of Fe^{2+} yields $Fe(OH)^+$ via Eq. 150, which may be significant over a pH range of 9.1 to 12.0. This reaction would shift the isoelectric point to a pH of 12.0 and a solubility of $10^{-7.1}$ M.

$Fe(OH)_3$(s), on the other hand, is somewhat more insoluble than $Fe(OH)_2$(s). Equation 135 forms the "acid" leg of the solubility diagram (Fig. 3-17), whereupon Eq. 155 forms the alkaline leg. These equations intersect at a pH of 5.6 and a solubility of $10^{-13.4}$ M to form the isoelectric point of $Fe(OH)_3$(s).

The hydrolysis of Fe^{3+}(aq) is quite significant as seen by Eqs. 152 ($Fe(OH)_2^+$) and 153 ($Fe(OH)^{2+}$). The latter equation intersects at a pH of 7.6 and a solubility of $10^{-11.3}$ M, whereupon the former equation intersects at a pH of 8.9 and a solubility of $10^{-9.9}$ M.

The equilibrium between the dissolved molecular species of $Fe(OH)_3$(aq) and $Fe(OH)_3$(s) is given by Eq. 154. If this reaction is operative, it has the effect of increasing the solubility of the solid phase to $10^{-6.54}$ M.

Solubility of the Phosphates. The occurrence of phosphates in surface and groundwaters has caused some concern in water pollution circles. Most of this concern is centered upon the role of phosphorus in the biological process of eutrophication. In the presence of Ca^{2+}, Al^{3+}, and Fe^{3+}, phosphate has a tendency to form solid phases. The significance of phosphate precipi-

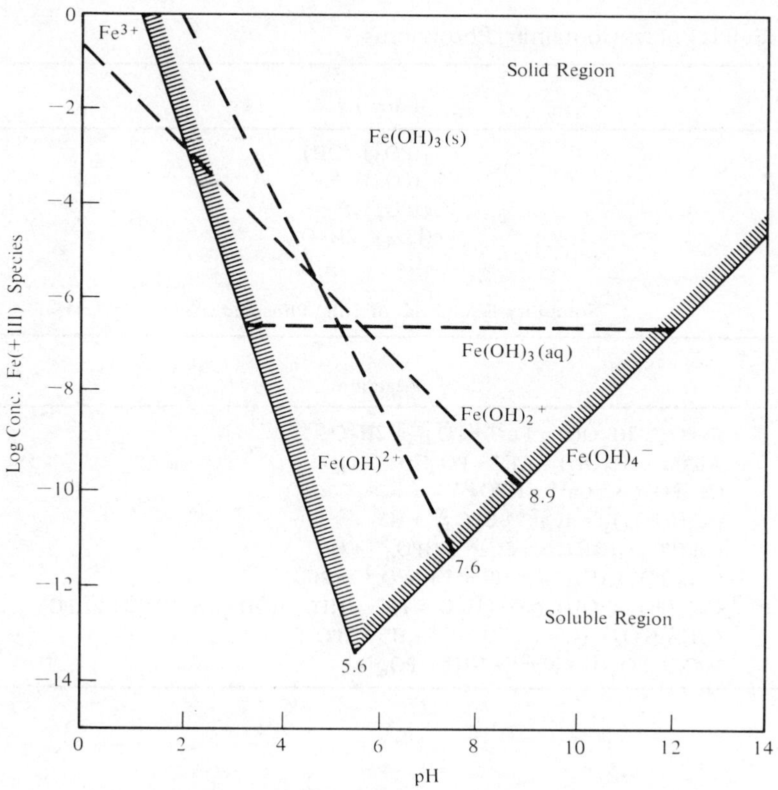

Fig. 3-17 Solubility diagram for ferric hydroxide.

tation in surface waters lies, perhaps, not in the removal of this element from the aqueous phase that would lower the soluble concentration, but that there would be a reservoir of phosphorus in the bottom sediments. As soluble phosphorus is removed from the aqueous phase by photosynthetic activities, there would be a tendency to dissolve previously precipitated phosphates. This would serve as a continuous source of this nutrient to eutrophic processes.[20]

Table 3-11 gives some of the forms of phosphorus that may have some significance in natural waters and their solubility equilibria. There is considerable scientific controversy surrounding the precipitation reactions of phosphorus in natural and artificial aqueous environments. Therefore, the solubility equilibria in Table 3-11 should be interpreted as a qualitative indication of what occurs in nature. There are, however, some well-defined phosphate solid phases: $Ca_{10}(PO_4)_6(OH)_2$ (s) (hydroxyapatite), $AlPO_4 \cdot 2H_2O$ (cariscite), and $FePO_4 \cdot 2H_2O$ (strengite). Figure 3-18 shows the solubility diagrams for six phosphate solid phases as a function of pH. It would appear that Fe^{3+} and Al^{3+} would tend to control the aqueous phos-

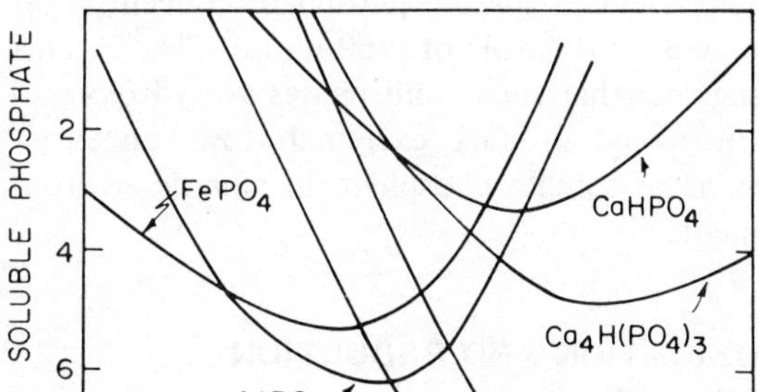

Fig. 3-18 Solubility diagrams for solid phosphate phases. Reproduced from Ref. 20 by courtesy of the University of Texas Press, Austin, Texas.

TABLE 3-11. Solids Phases Containing Phosphorus

Solid Phase	Formula	$\Delta G°$ (kcal/mole)[a]
Hydroxyapatite	$Ca_5(PO_4)_3(OH)$	−3123.5
Whitlockite	$Ca_3(PO_4)_2$	− 932.8
Fluorapatite	$Ca_5(PO_4)_3F$	—
Strengite	$Fe(PO_4) \cdot 2H_2O$	− 397.7
Berlinite	$AlPO_4$	− 388.0

Solubility Equilibria for Solid Phosphates[b]

Equation No.	Reaction	log K_{eq} (25°C)
(156)	$FePO_4 \cdot 2H_2O(s) = Fe^{3+} + PO_4^{3-} + 2H_2O$	− 23
(157)	$AlPO_4 \cdot 2H_2O(s) = Al^{3+} + PO_4^{3-} + 2H_2O$	− 21
(158)	$CaHPO_4(s) = Ca^{2+} + HPO_4^{2-}$	− 6.6
(159)	$Ca_4H(PO_4)_3 = 4Ca^{2+} + 3PO_4^{3-} + H^+$	− 46.9
(160)	$Ca_5(PO_4)_3(OH)(s) = 5Ca^{2+} + 3PO_4^{3-} + OH^-$	− 55.6
(161)	$Ca_{10}(PO_4)_6(F)_2(s) = 10Ca^{2+} + 6PO_4^{3-} + 2F$	− 118
(162)	$Ca_{10}(PO_4)_6(OH)_2(s) + 6H_2O = 4[Ca_2(HPO_4)(OH)_2] + 2Ca^{2+} + 2HPO_4^-$	− 17
(163)	$CaHAl(PO_4)_2(s) = Ca^{2+} + Al^{3+} + H^+ + 2PO_4^{3-}$	− 39
(164)	$MgNH_4PO_4(s) + Mg^{2+} + NH_4^+ + PO_4^{3-}$	− 12.6

[a] After Ref. 5.
[b] After Ref. 20.

phate concentration under acid conditions, whereas the Ca^{2+} compounds would be the controlling phases under alkaline conditions. Equation 160 may give some qualitative indication of the phosphate concentration at a pH of 7.0 and a calcium content of 40 mg/l (10^{-3} M). This yields a soluble phosphate concentration of 6.3×10^{-12} M, or 0.60 ng/l as PO_4^{3-}. This suggests that such solid phases as hydroxyapatite would produce extremely low concentrations of soluble phosphate in aquatic environments.

OXIDATION AND REDUCTION EQUILIBRIA

Conceptual Definitions. Many reactions in water and wastewater chemistry involve the exchange or transfer of electrons from one element to another or from one compound to another. Usually, this transfer of electrons is employed to change the oxidation state of a pollutant in order to facilitate its removal from the aqueous phase. For example, ferrous iron (Fe(II)) is oxidized to ferric iron (Fe(III)), whereupon $Fe(OH)_3(s)$ is formed which is removed easily from solution by sedimentation or filtration.

Conceptually, oxidation–reduction equilibria may be approached the same as acid–base equilibria. In the case at hand, an electron donor is labelled the reductant whereas an electron acceptor is labelled the oxidant. An excellent example would be:

$$Fe^{3+} + e^- = Fe^{2+} \tag{165}$$

in which the ferric iron, an oxidant, can accept an electron to produce ferrous iron, a reductant. (Since this is a half-cell reaction, another half-cell, or reductant, is needed to donate the electron as seen below.) Please note the convention employed where the electron appears on the left-hand side of Eq. 165. This indicates a reduction reaction.[21] It follows that, if the electron appears on the right-hand side, the reaction is considered as an oxidation.

Equation 165 may be rearranged to denote a condition of equilibrium:

$$K_{red} = \frac{[Fe^{2+}]}{[Fe^{3+}]\,[e^-]} \tag{166}$$

The equilibrium constant in this case is designated as K_{red} for reduction. An additional rearrangement and placement in log form leads to:

$$-\log [e^-] = \log K_{red} + \log \frac{[Fe^{3+}]}{[Fe^{2+}]} \quad (167)$$

where $[e^-]$ denotes the electron concentration. (A more precise expression would be electron activity.) We may now say that:

$$pE = -\log [e^-] \quad (168)$$

which is analogous to the more commonly known and used equation:

$$pH = -\log [H_3O^+] \quad (169)$$

Furthermore, when a condition of equilibrium exists, the $[Fe^{3+}] = [Fe^{2+}]$. Equation 167 becomes:

$$pE^\circ = \log K_{red} \quad (170)$$

where pE° denotes standard state conditions.

Since oxidation-reduction is a process of electron transfer, then the reduction in Eq. 165 must be combined with an oxidation equation which may be, for example:

$$\tfrac{1}{2}H_2 = H^+ + e^- \quad (171)$$

By combination, the complete redox reaction would be:

$$Fe^{3+} + \tfrac{1}{2}H_2 = Fe^{2+} + H^+ \quad (172)$$

By convention, the ΔG° of Eq. 171 is arbitrarily assigned a value of 0.00 and K_{eq} value of 1.0. Therefore, the equilibrium constant for Eq. 172 would be numerically the same as for Eq. 165. It should be indicated that Eq. 172 transfers only 1 mole of electrons under standard state conditions. If more than 1 mole is transferred, then Eq. 170 would read:

$$npE^\circ = \log K_{red} \quad (173)$$

A combination of Eqs. 13 and 173 leads to:

$$npE^\circ = -\frac{\Delta G^\circ}{2.3\,RT} \quad (174)$$

which now relates the free energy change, hence feasibility, of an oxidation-reduction reaction to the pE° value. Since a negative ΔG° value indicates that a left-to-right reaction is feasible, it follows, from Eq. 174, that a positive pE° value denotes the same feasibility.

Sign Conventions. Table 3-12 shows some reduction reactions commonly encountered in water and wastewater chemistry. The sign conventions follow the International Union of Pure and Applied Chemistry (IUPAC) suggestions. The symbol V is used to denote the electrode potential. References 2 and 21 provide a fuller explanation of the confusion surrounding electrode sign conventions. The IUPAC convention calls for all half-cell reactions to be written as reduction with the pE° and V values being sign invariant. All such reduction reactions are referred to or combined with the H_2 reduction reaction (Eq. 171). If one wishes to combine two reactions in order to determine the feasibility of the resultant redox reaction, the IUPAC convention utilizes the arbitrary equation:

$$pE^\circ_{rex} = pE^\circ_{red} - pE^\circ_{ox} \quad (181)$$

This equation states that the pE° value of the final reaction is the algebraic difference between the pE° value of the reduction reaction and the pE° value of the oxidation reaction. Please note that the sign of the pE° value of the oxidation reaction does not change from its sign as a reduction reaction. An example illustrates this point:

$$Cl_2(g) + 2e^- = 2Cl^- \qquad pE^\circ_{red} = +23.0$$
$$2Fe^{2+} = 2Fe^{3+} + 2e^- \qquad pE^\circ_{ox} = +13.2$$
$$2Fe^{2+} + Cl_2 = 2Fe^{3+} + 2Cl^- \qquad pE^\circ_{rex} = +9.8$$

(from Eq. 181)

Since the pE° value for the final redox reaction is positive, it is feasible for the oxidation of ferrous iron to be effected by chlorine. Another example:

$$Cu^{2+} + 2e^- = Cu(s) \qquad pE^\circ_{red} = +6.0$$
$$2Cl^- = Cl_2(g) + 2e^- \qquad pE^\circ_{ox} = +23.0$$
$$Cu^{2+} + 2Cl^- = Cl_2(g) + Cu(s) \qquad pE^\circ_{rex} = -17.0$$

Since the pE° value for this redox reaction is negative, it is not feasible for oxidation of the chloride anion to be effected by a cupric cation. It should be noted from Table 3-12 that, by convention, the reduction reactions that lie above the hydrogen reactions (Eq. 171) carry a negative sign, whereas those that lie below hydrogen carry a positive value.

Electron Energy Level Diagrams. Since Eq. 174 shows that the pE° value represents an expression of the free energy change of a reduction reaction, an electron energy level diagram may be constructed. This is conceptually analogous to the proton energy level shown previously. Figure 3-19 gives this diagram for the transfer of 1 mole of electrons from reductant to oxidant. The diagram is constructed so that, by convention, the electrons are transferred from H_2 to the oxidant. The energy gained from this transfer increases as one proceeds from top to bottom. Above the H_2,H^+ line, the feasible reaction would be:

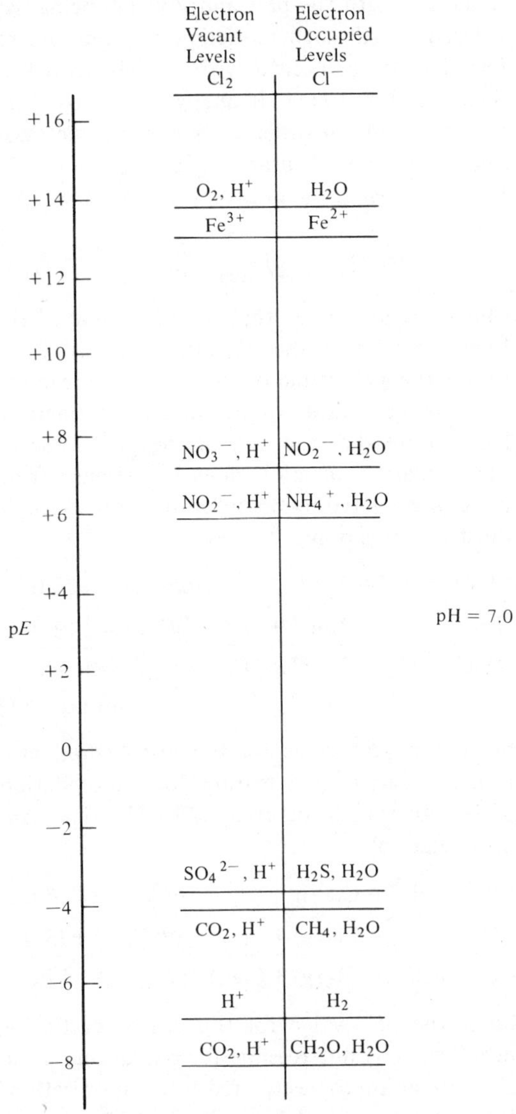

Fig. 3-19 Electron energy level diagram.

$$Fe(s) + 2H^+ = Fe^{2+} + H_2(g) \qquad (182)$$

whereas below the H_2, H^+ line it would be:

$$2Fe^{3+} + H_2(g) = 2Fe^{2+} + 2H^+ \qquad (172)$$

Another feature of this diagram shows the water oxidized energy level with the H_2O, O_2, H^+ line. It is noted that the Cl^-, $Cl_2(g)$ line lies below the water oxidized line which makes chlorine thermodynamically unstable in water. In the absence of such reductants as Fe^{2+}, the H_2O molecule would serve as a source of electrons for $Cl_2(aq)$. This accounts for the unstability of standard chlorine solutions.

Another conceptual aspect of Fig. 3-19 is that the right half of the diagram represents electron-occupied levels or reductants, whereas the left half represents electron-unoccupied levels or oxidants. Hypothetically, if one adds a reductant (e.g., Fe(s)) to a system containing several oxidants, the lowest unoccupied electron level would be populated first. For example, the Fe(s) would react first with $Cl_2(g)$, second with O_2, third with Fe^{2+}, and so forth. It follows that an oxidant must be chosen from an electron-unoccupied level lower than the reductant in order for a redox reaction to be thermodynamically feasible.

Nernst–Peters Equation. A more commonly used expression for oxidation-reduction reactions is V, the electrode potential. (Needless to state, symbolism in oxidation-reduction chemistry is confusing. The symbol E is used extensively in the technical literature. This chapter considers E and V, for all intents and purposes, interchangeable.) In the IUPAC convention, the reduction free energy may be expressed:

$$\Delta G^\circ = -nFV^\circ \qquad (183)$$

where n = number of moles of electrons transferred, F = the Faraday (96,484 absolute coulombs), and V° = the electrode potential in volts. The sign conventions for V are the same as for pE (sign invariant, $+V$ denotes feasible reaction, etc.). Substitution of Eq. 183 into Eq. 174 yields:

$$pE^\circ = \frac{nFV^\circ}{2.3\,RT} \qquad (184)$$

or at 25°C:

$$pE^\circ = V^\circ/0.059 \qquad (185)$$

which is the relation between pE° and V°.

Frequently, it is necessary to operate oxidation-reduction reactions at conditions other than standard (i.e., 1 mole of electrons). The nonstandard state values of V may be computed from the well-known Nernst-Peters equation:

$$V = V^\circ + \frac{2.3\,RT}{nF}\log\frac{[\text{oxid}]}{[\text{red}]} \qquad (186)$$

where [oxid] and [red] are the nonstandard state concentrations of the oxidant and reductant, respectively.

pE-pH Diagrams. Many oxidation-reduction reactions are influenced by the $[H^+]$, or as more commonly expressed, the pH value. This may be seen in the "water oxidation" reaction in Table 3-12 (Eq. 179) where 4 moles of protons, 4 moles of electrons, and 1 mole of $O_2(g)$ are involved. Thus, it may be readily seen that if one increases the $[H^+]$, there would be a concomitant increase in the $[e^-]$ in order for the water molecules to be oxidized. Another aspect of this reaction is that it represents the oxidative stability boundary for water (illustrated below). The Fe(0)-Fe(+II)-Fe(+III) system may be cited as a classical example of the influence of pH values on pE values. Several (but not all) iron reduction reactions are cited in Table 3-13.

The concept of pE-pH diagrams illustrates the effect of pH on the pE values of the iron system. Such a diagram is seen in Fig. 3-20, where pE is plotted versus pH. Please note that the pE ordinate is compressed for illustrative purposes, whereas a correct plot would

TABLE 3-12. Some Reduction Reactions and their Equilibrium Constants (25°C)[a]

Equation No.	Reaction	log K_{red}	pE°	V(volts)
(175)	$2H_2O + 2e^- = H_2 + 2OH^-$	−28	−14	−0.828
(176)	$Zn^{2+} + 2e^- = Zn(s)$	−26	−13	−0.76
(177)	$Fe^{2+} + 2e^- = Fe(s)$	−14.9	− 7.5	−0.44
(171)	$2H^+ + 2e^- = H_2(g)$	0.0	0.0	0.00
(178)	$Cu^{2+} + 2e^- = Cu(s)$	+12.0	+ 6.0	+ .34
(165)	$Fe^{3+} + e^- = Fe^{2+}$	+13.2	+13.2	+0.77
(179)	$O_2(g) + 4H^+ + 4e^- = 2H_2O$	+83.1	+20.75	+1.229
(180)	$Cl_2(g) + 2e^- = 2Cl^-$	+46	+23	+1.36

[a]Signs in accord with the IUPAC Convention.

TABLE 3-13. Some Iron and Nitrogen Reduction Reactions

Equation No.	Reaction	log K_{red}	log K_{eq}
(187)	$Fe(OH)^{2+} + H^+ + e^- = Fe^{2+} + H_2O$	15.2	–
(188)	$Fe(OH)_3 + 3H^+ + e^- = Fe^{2+} + 3H_2O$	17.8	–
(189)	$Fe^{3+} + H_2O = Fe(OH)^{2+} + H^+$	–	−2.2
(190)	$Fe(OH)^{2+} + 2H_2O = Fe(OH)_3(s) + 2H^+$	–	−2.41
(191)	$NO_2^- + 8H^+ + 6e^- = NH_4^+ + 2H_2O$	91.0	–
(192)	$NO_3^- + 2H^+ + 2e^- = NO_2^- + H_2O$	28.2	–
(193)	$NO_3^- + 10H^+ + 8e^- = NH_4^+ + 3H_2O$	119.0	–
(194)	$N_2(g) + 8H^+ + 6e^- = 2NH_4^+$	28.1	–

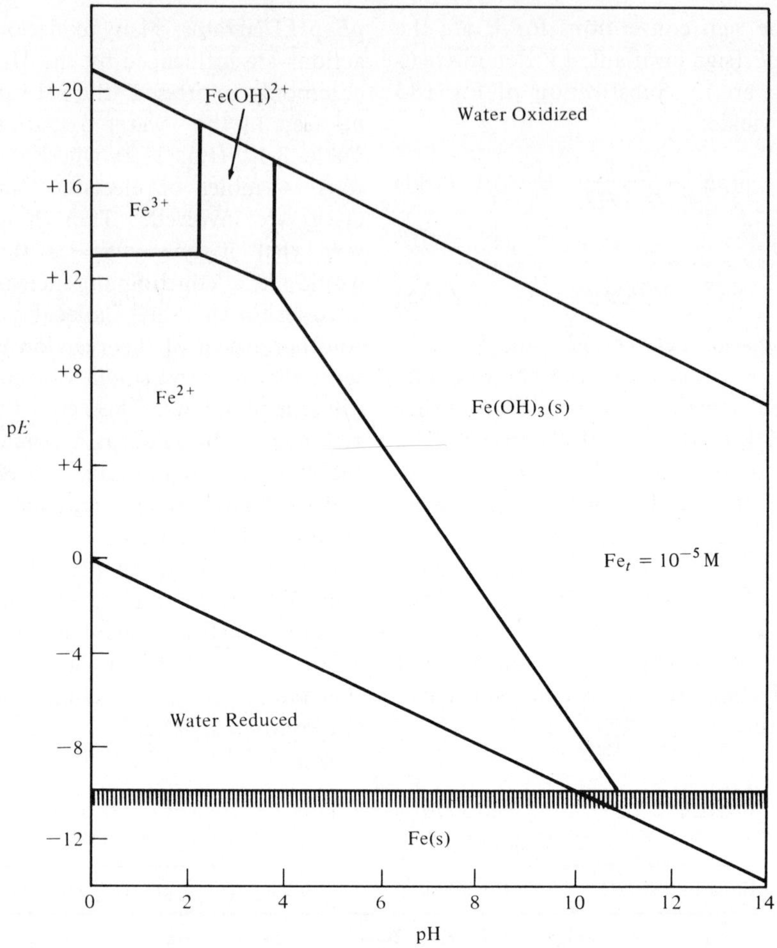

Fig. 3-20 p*E*–pH diagram for the iron system.

have one p*E* unit for every pH unit. Construction of the p*E*–pH diagram in Fig. 3-20 goes something like this. First the water oxidized (Eq. 179, Table 3-12) and water reduced (Eq. 175, Table 3-12) boundaries are drawn from their linear forms:

$$pE = 20.7 - pH \text{ (water oxidized)} \quad (195)$$

and,

$$pE = - pH \text{ (water reduced)} \quad (196)$$

Water is stable toward oxidation-reduction at pH-p*E* values between these two boundaries. The Fe(0)-Fe(+II) boundary is established from Eq. 177 (Table 3-12) and from the assumption that $[Fe]_t$ of the soluble species is 10^{-5} M. This boundary is set at a p*E* value of −9.95, below which Fe(s) is the stable species and above which $Fe^{2+}(aq)$ is the stable species. Since this reaction is the oxidative portion of the corrosion reaction for metallic iron, it may be combined with Eq. 175 (Table 3-12) as the reductive reaction. This shows that metallic iron pipes are thermodynamically unstable in water; i.e., they have the inherent tendency to corrode. The Fe(+II)-Fe(+III) boundary is computed from Eq. 165, Table 3-12, where the p*E* value is +13.0. At p*E* values > +13.0, Fe(+III) is the predominant species, whereas at p*E* values < +13.0, Fe(+II) is the predominant species. This reaction is independent of $[H^+]$ over the pH range of 0.0 to 2.2. Above this pH value of 2.2, Fe^{3+} enters into an acid-base equilibrium (Eq. 189; Table 3-13), whereupon $Fe(OH)^{2+}$ is the predominant Fe(+III) species. The pH boundary for Fe^{3+} and $Fe(OH)^{2+}$ may be readily computed from the equilibrium constant for this reaction. $Fe(OH)^{2+}$ is the predominate Fe(+III) species over the pH range 2.2 to 3.7, whereupon Eq. 190, Table 3-13 becomes operative. The pH boundary for $Fe(OH)^{2+}$–$Fe(OH)_3(s)$ may be computed from the equilibrium constant for this reaction.

The influence of pH on pE of the Fe(+II)–Fe(+III) system may be seen from Eqs. 187 and 188, Table 3-13. The linear form of Eq. 187 is:

$$pE = 15.2 - pH \quad (197)$$

and for Eq. 188:

$$pE = 22.8 - 3\ pH \quad (198)$$

These two equations may be used to calculate the pH–pE boundaries for Fe(+II)–$Fe(OH)^{2+}$ and Fe(+II)–$Fe(OH)_3(s)$, respectively. The influence of pH value on the pE value of Eq. 188 is seen readily from Eq. 198 and Fig. 3-20, where for every 1-unit increase of pH there is a 3-unit decrease in pE. This reaction is especially significant because it is employed in the removal of iron from natural waters. It is, of course, the oxidation of Fe^{2+} to Fe^{3+} with the subsequent precipitation of $Fe(OH)_3(s)$. This diagram also shows that the pE value required for the oxidation is lowered as the pH value is increased. This interpretation has a very practical operational application in the removal of iron from natural waters.

Please note that the Fe(0)–Fe(+II)–Fe(+III) system portrayed in Fig. 3-20 has been simplified greatly for illustrative purposes. It is, however, conceptually sound. For more comprehensive iron systems, the reader is directed to Refs. 1 and 2.

The pE–pH diagrams may be used to predict the stability or instability of wastewater constituents discharged or injected into natural aquatic environments. Highly-treated domestic sewage has been injected into the Magothy aquifer beneath Long Island, N.Y., which serves as a major source of potable water for many communities, for purposes of groundwater recharge.[22] One of the wastewater constituents is ammonia-N, which averages 32 mg/l. Since the ammonia-N is in a reduced oxidation state, the possibility of oxidation to nitrate-N should be considered.

Figure 3-21 shows the pE–pH diagram that was constructed from the four reduction reactions for nitrogen given as Eqs. 191, 192, 193, and 194 in Table 3-13. The linear forms of these reactions are, respectively:

$$pE = 15.2 - 1.33\ pH \quad (199)$$

$$pE = 14.1 - pH \quad (200)$$

$$pE = 14.9 - 1.25\ pH \quad (201)$$

$$pE = 4.68 - 1.33\ pH \quad (202)$$

The pE–pH values of the native Magothy water are seen plotted in Fig. 3-21 for eight observation wells. All of these points fell into the $N_2(g)$ stability domain. Consequently, oxidation of the injection waters' NH_3 (ammonia-N) to NO_3^- (nitrate-N) should not occur in the Magothy water. The diagram, however, does suggest an oxidation to $N_2(g)$. That this oxidation does not occur to a significant extent may be due to kinetic reasons.[2]

Kinetics of Fe(+II) Oxidation. In the removal of Fe(+II) from natural waters, the treatment process of aeration is employed frequently. The chemistry involves the oxidation of Fe(+II) to Fe(+III) with the latter precipitated as the hydrous oxide. After investigation of the stoichiometry and kinetics of this oxidative reaction,[23] the stoichiometric reaction was reported as:

$$Fe^{2+} + \tfrac{1}{4}O_2 + 2OH^- + \tfrac{1}{2}H_2O = Fe(OH)_3(s) \quad (203)$$

The kinetics, however, was followed by this equation:

$$-\frac{d\,[Fe^{2+}]}{dt} = k\,[Fe^{2+}]\,P_{O_2}\,[OH^-]^2 \quad (204)$$

where the rate of disappearance of the ferrous iron was first order with respect to $[Fe^{2+}]$ and the partial pressure of oxygen and second order with respect to $[OH^-]$. The rate constant, k, was evaluated as $8.0 \pm 2.5 \times 10^{13}$ liter^2/mole^2/atm/min. Thus, it may be seen that the rate of oxygenation of ferrous iron increases with an increase of P_{O_2} and hydroxyl ion concentration.

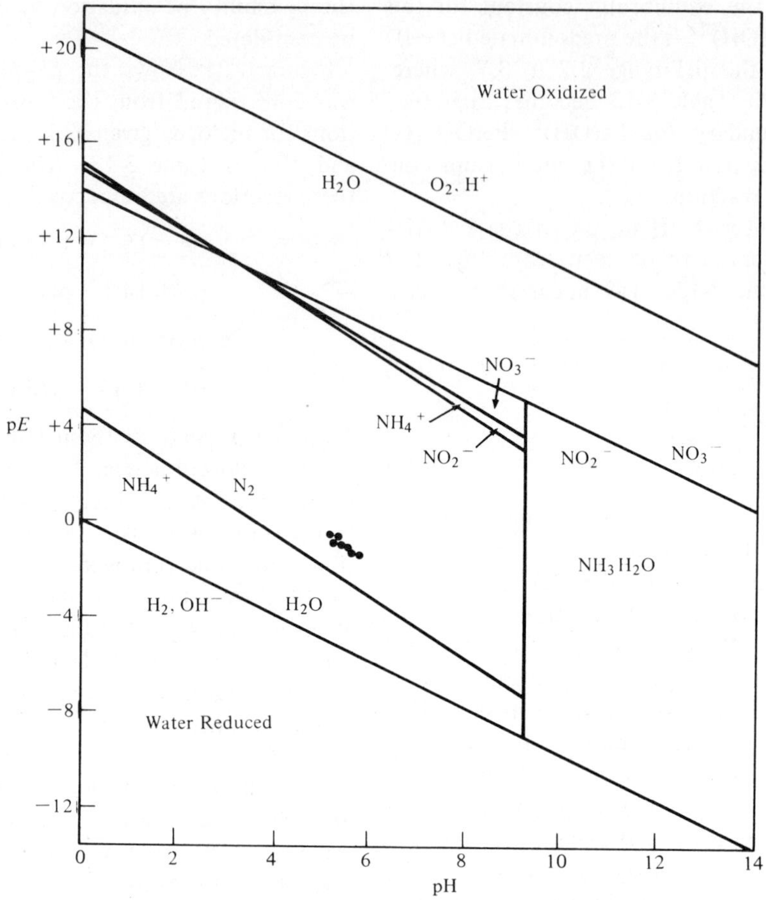

Fig. 3-21 p*E*–pH diagram for the nitrogen system.

Some very practical operational and design information may be derived from this kinetic study. Equation 204 may be forced into pseudo first-order kinetics by holding P_{O_2} and $[OH^-]$ constant:

$$K' = kP_{O_2}\ [OH^-]^2 \qquad (205)$$

The K' values in Table 3-14 were computed from those reported by Stumm and Lee.[23] If first-order kinetics are observed, then one may calculate the time required, for example, to reduce the initial concentration of Fe^{2+} by 50 percent from the equation:

$$t_{50} = \frac{0.691}{K'} \qquad (206)$$

These t_{50} values are given in Table 3-14 for the systems of Stumm and Lee. The effects of P_{O_2} and $[OH^-]$ are clearly seen. Moreover, these t_{50} values may be interpreted as the detention times required of the reactor vessels employed in the oxygenation of Fe^{2+}. These times are, of course, the 50 percent values. If greater removals are required, on a percentage basis, then t_{90} and t_{99} equations must be employed:

$$t_{90} = \frac{2.303}{K'} \qquad (207)$$

$$t_{99} = \frac{4.606}{K'} \qquad (208)$$

Kinetics of Oxidation of Organic Compounds. Many organic compounds are found in surface waters and account for the majority of taste and odor problems. When the raw water is treated and delivered as potable water, one of the major concerns is the organoleptic quality

TABLE 3-14. Pseudo First-Order Rate Constants and t_{50} Values for the Oxygenation of Fe(+II)[a]

P_{O_2} (atm)	$[OH^-]$ ($\times 10^8$)	K' (min^{-1})[b]	t_{50} (min)
0.107	1.86	3.34×10^{-3}	206.5
0.107	2.51	4.77×10^{-3}	144.5
0.153	3.16	1.23×10^{-2}	56.1
0.153	4.56	2.53×10^{-2}	27.3
0.153	5.02	2.61×10^{-2}	26.5
0.174	7.25	9.06×10^{-1}	7.6
0.184	9.12	1.29×10^{-1}	5.4
0.184	12.0	2.46×10^{-1}	2.8
0.195	18.2	4.84×10^{-1}	1.4
0.195	22.4	9.43×10^{-1}	0.7

[a]Ref. 23.
[b]Correction by Singer.[28] 20.5°C, alkalinity = 2.9 to 3.9×10^{-2} eq/l.

of the delivered water. Removal of organic compounds is accomplished usually by adsorption on carbon surfaces or by chemical oxidation. If the latter treatment is chosen, the oxidant is usually selected from the following: $KMnO_4$, Cl_2, or ClO_2.

There is much confusion in the application of inorganic oxidants for the removal of organic compounds. There have been many reports in the literature concerning the efficiency of these oxidative processes. Inefficiency may result from: (a) improper dosages of the oxidant; i.e., less than the stoichiometric quantities are added, (b) insufficient detection or reaction times allowed for the oxidation to occur, and (c) improper adjustment of such environmental factors as the pH value.

The kinetics of chemical oxidation of two aquatic herbicides, diquat and paraquat, have been investigated[24] in an effort to determine the optimum oxidative conditions. First, a thermodynamic approach was used to calculate the feasibility of $KMnO_4$, for example, to oxidize diquat:

$$3(C_{12}H_{12}N_2)^{2+} + 52MnO_4^- + 52H^+ = 36CO_2 + 6NH_4^+ + 52MnO_2(s) + 32H_2O \quad (209)$$

The free energy change, ΔG°, for this reaction was calculated to be -5781 kcal/mole. Thus, it was concluded that it was feasible to use $KMnO_4$ in the oxidation of diquat. Equation 209 reveals a rather complex stoichiometry (17.3 moles $KMnO_4$ per mole diquat) for the reaction that was verified by the appropriate experiment. Subsequent kinetic experiments yielded data that could be fitted to an integrated form of second-order equation:

$$\frac{2.303}{C^\circ_{ox} - C^\circ_{diq}} \log\left(\frac{C^\circ_{diq}\, C_{ox}}{C^\circ_{ox}\, C_{diq}}\right) = K_{ob}t \quad (210)$$

where t is the time in minutes, C°_{ox} is the initial oxidant concentration, C°_{diq} is the initial diquat concentration, C_{ox} is the oxidant concentration at time t, C_{diq} is the diquat concentration at time t, and K_{ob} is the observed rate constant in units of liter/mole/min. That the rate of $KMnO_4$ oxidation of diquat is pH dependent is seen in Table 3-15. These kinetic data suggest that alkaline pH conditions are more favorable for the rapid oxidation of diquat than are neutral or slightly acid values.

In order to calculate the detention times for the diquat-$KMnO_4$ oxidation, it was necessary to force the reaction into a first-order situation. Since the reaction apparently follows second-order kinetics, the rate of oxidation is proportional to the oxidant and herbicide concentrations in solution:

$$-\frac{dC_{diq}}{dt} = K_{ob}\, C_{ox}\, C_{diq} \quad (211)$$

The assumption is made that the oxidants' concentration is present in excess of the diquat concentration. Consequently, a first-order rate

TABLE 3-15. Oxidation of Diquat by $KMnO_4$

pH	k_{ob}[a]	k'[b] (min^{-1})	t_{50} (hr)
5.1	0.123	1.23×10^{-4}	94.1
6.0	0.087	8.69×10^{-5}	133.8
6.9	0.186	1.86×10^{-4}	62.0
8.0	1.18	1.18×10^{-3}	9.8
9.1	14.8	1.48×10^{-2}	0.78
10.0	54.6	5.46×10^{-2}	0.21

[a]I = 0.02 M, temp = 20° ± 0.2°C, $C^\circ_{diq} = 4.16 \times 10^{-5}$, and $C^\circ_{ox} = 1 \times 10^{-3}$ M. Unit of K_{ob} is liter/mole/min.
[b]First-order rate constant.

constant may be computed from $K_{ob}C^{\circ}_{ox}$. Then one may use Eq. 206 to calculate t_{50} values for the diquat-$KMnO_4$ reaction. These data are seen in Table 3-15. It should be obvious that an alkaline pH value between 8.0 and 9.1 is necessary for the rapid removal of diquat by $KMnO_4$ oxidation under detention times afforded by most water treatment plants.

It should be noted that diquat is not a typical organic pollutant of surface waters, although the above should be considered as the proper approach for the removal of organics from water. Kinetic studies yield data that should be utilized in the design of treatment units.

GEOCHEMICAL CONTROLS OF WATER QUALITY

In discussing the various chemical impurities in natural waters, the factors affecting their concentrations in different aquatic environments, and their significance and effect on man, a discussion of their immediate source should be included—that of the process of geochemical weathering. Weathering is perhaps the single most familiar chemical process to anyone who has studied the earth.

Although weathering processes are extremely slow, it is, at best, nature's way of seeking equilibrium in a reaction between the atmosphere, lithosphere, and hydrosphere; the end product being an oxidized residuum of the rocks weathered and chemical solutions containing the soluble cations and anions of the original rocks. This has been expressed as follows:[25]

$$\text{rock} + \text{oxygen} + \text{water} + p\int_{\text{mountains}}^{\text{sea level}} \longrightarrow \text{saprolite} \longrightarrow \text{soil} + \text{soluble inorganic and organic ions or compounds.} \qquad (212)$$

In the following paragraphs, the concern will be primarily with one of these end products—the soluble ions in solution—how they became dissolved, and some of the basic chemical reactions involved in the process.

Factors Affecting the Weathering Process The climate surrounding a region influences in a major way the manner of weathering that occurs. For example, in the polar regions the predominant form is mechanical, where rocks are degraded by freeze-thaw action. However, in a warm, moist environment such as the tropics, there is not a marked extreme between summer and winter and chemical weathering is at a maximum.

The amount of physical or mechanical weathering that occurs also plays a major role in geochemical weathering rates, especially in temperate regions. The process begins in joints and fractures which may be present in the rocks. These joints often tend to be initiated during tectonic disturbances (faulting), upon cooling from a magmatic stage, or upon removal of overburden during the erosion process. Nonetheless, it facilitates the production of many small particles of the once compact rock and, consequently, greatly increases the surface area to react with solutions.

In igneous and metamorphic rocks, the individual grain sizes influence the manner in which it disintegrates. A coarsely-granular rock will disintegrate more rapidly than a finer grained one due to the effects of thermal expansion of the individual minerals. Metamorphic rocks, on the other hand, most often have a preferred orientation of individual minerals, thereby making them more susceptible to disintegration. A granite, which most often has a random orientation to its mineral constituents, will not be affected in this manner. Sedimentary rocks generally contain grains within a narrow size range and tend to be formed of the more resistant minerals (e.g., quartz). The cement bonds of these individual grains control the final product.

It should be noted here also that the individual minerals within an igneous, metamorphic, or sedimentary rock generally follow a selective sequence of alteration as shown in Table 3-16. This sequence bears a striking resemblance to Bowen's "Reaction Principle"[26] where the minerals formed at the highest temperatures (olivine and calcic plagioclase) are those most susceptible to weathering processes. Silica, or quartz, being the last to crystallize from the

TABLE 3-16. The Common Minerals of Igneous Rocks Arranged According to General Rates of Weathering

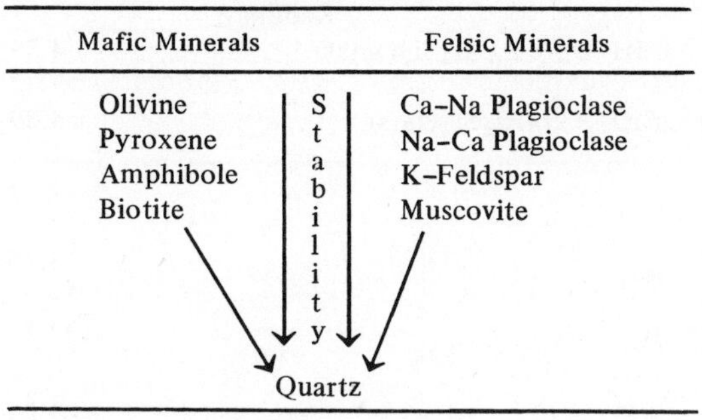

magma, is generally the most resistant to weathering. Table 3-16, although a generalization, is applicable to many rocks found in a majority of weathering environments. However, since the rate of weathering is dependent upon many factors including straight chemical susceptibility, these generalizations cannot cover all cases.

Perhaps the most important factor of all is that of the reactant nature of the hydrosphere. Rain can be considered as the principal source of water, but ultimately the water which occurs in the interstices of rocks and soil is the true reacting electrolyte.[25]

All water that comes in contact with the lithosphere (soil and rocks of the earth's crust) is slightly acidic. This is due in nature to the delicate equilibrium existing between the partial pressure of CO_2 in the atmosphere and soil–air and the buffering capacity of the carbonate system. In fact, all geochemical weathering can be considered as an "acid attack" originating at the atmosphere–lithosphere contact.

The amount and rate of the weathering process is in direct relation to the amount of water available to react with the rocks and the length of time it takes this water to drain through them. As an example, groundwater contained in some types of metamorphic rocks contains higher amounts of dissolved solids and certain species of solutes than water contained in unconsolidated Coastal Plain sediments. This is due primarily to the constricting nature of the crystalline rocks on the movement of water, as the joints and fractures controlling that movement are very often not continuous or interconnected. The velocity of groundwater under these conditions is dependent on: the hydraulic gradient of the water table (from point of recharge to point of discharge), the permeability (a function of the interconnection of joints and fractures), and the porosity (a function of the number of joints and fractures). Thus, it is conceivable that some groundwaters tapped by a well under these conditions may yield water that has been in the rocks for thousands of years.

Once leached from the parent rocks, those ions that become dissolved in the process eventually find their way into the hydrologic cycle. At this point, their effect on man becomes evident, as he relies on the hydrologic cycle to support many of his physical needs.

Weathering of Alumino-Silicates. Certainly the more important weathering reactions are those of the alumino-silicates. They comprise a major portion of the igneous and metamorphic rocks within the continents which are in contact with groundwaters and streams. Table 3-17 provides a list of the more common rock-forming minerals and examples of their typical weathering reactions. Tables 3-1 and 3-18 give the thermodynamic data for these minerals and their ions.

The behavior of silicate minerals attacked by acid is very important. The presence of ions in solution derived from them alters the composition of that solution and its ability to react with unaltered minerals. Thus, the final product of the weathering process may also be an intermediate stage. For example, kaolinite, montmorillonite, and mica are possible products. However, mica has been identified as an intermediate product of the weathering of K-feldspar.[2] Silica is slightly soluble at all pH values, whereas alumina is only soluble below pH 4 and above pH 8.5. Since the normal weathering reactions occur between pH values of 5.0 and 8.5, silica is continuously leached out in small quantities and leaves the alumina to accumulate in the clayey residuum. This explains the presence of dissolved silica in almost all natural waters, whereas aluminum is

TABLE 3-17. Some Geochemical Weathering Reactions at 25°C

Equation No.	Reaction	log K_{eq}
(213)	$NaAlSi_3O_8(s) + H_2CO_3 + 4\frac{1}{2}H_2O = Na^+ + HCO_3^- + 2H_4SiO_4 + \frac{1}{2}H_4Al_2Si_2O_9(s)$ Albite … Kaolinite	-5.87
(214)	$CaAl_2Si_2O_8(s) + 2H_2CO_3 + H_2O = Ca^{2+} + 2HCO_3^- + H_4Al_2Si_2O_9(s)$ Anorthite	4.39
(215)	$2KAlSi_3O_8(s) + 2H^+ + 9H_2O = 2K^+ + 4H_4SiO_4 + H_4Al_2Si_2O_9(s)$ Orthoclase	-4.39
(216)	$SiO_2(s) + 2H_2O = H_4SiO_4$ Amorphous silica	-2.77
(217)	$SiO_2(s) + 2H_2O = H_4SiO_4$ Quartz	-3.76
(218)	$FeS_2(s) + \frac{7}{2}O_2 + H_2O = Fe^{2+} + 2SO_4^{2-} + 2H^+$ Pyrite	192.2
(219)	$Fe_2O_3(s) + 6H^+ = 2Fe^{3+} + 3H_2O$ Hematite	-1.9
(220)	$FeOOH \cdot H_2O(s) + 3H^+ = Fe^{3+} + 3H_2O$ Limonite	–
(221)	$BaSO_4(s) = Ba^{2+} + SO_4^{2-}$ Barite	-10.0
(222)	$CaSO_4(s) = Ca^{2+} + SO_4^{2-}$ Gypsum	-5.3
(223)	$NaCl(s) = Na^+ + Cl^-$ Halite	+1.57
(224)	$FeCO_3(s) + H_2CO_3 = Fe^{2+} + 2HCO_3^-$ Siderite	-6.65
(225)	$Mg_2SiO_4(s) + 4H_2O = 2Mg^{2+} + 3OH^- + H_4SiO_4$ Forsterite	-27.61
(226)	$Fe_2SiO_4(s) + \frac{1}{2}O_2 + 2H_2O = Fe_2O_3(s) + H_4SiO_4$ Fayalite	34.8
(227)	$MgCO_3(s) + H_2CO_3 = Mg^{2+} + 2HCO_3^-$ Magnesite	-4.03

TABLE 3-18. Thermodynamic Data for Selected Geochemical Substances at 25°C[a]

Substance	$\Delta H°$ kcal/mole	$\Delta G°$ kcal/mole	$\Delta S°$ cal/mole-deg
Albite (low)	-937.1	-883.99	50.2
Anorthite	-1009.3	-955.6	48.45
Barite	-352.1	-325.3	31.6
Fayalite	-353.5	-329.7	35.45
Forsterite	-520.4	-491.9	22.75
Gypsum	-484.0	-430.1	46.36
Halite	-982.6	-91.8	17.24
Hematite	-197.3	-177.7	20.89
Kaolinite	-979.5	-902.9	48.53
Magnesite	-266.1	-246.1	15.7
Orthoclase	-933.3	-883.1	63.3
Pyrite[b]	-40.5	-37.8	12.7
α-Quartz	-217.65	-204.6	9.9
Siderite	-177.8	-161.0	25.1
Silica Glass	-215.9	-203.3	11.3
Silicic[b]	-349.1	-312.9	78.0

[a]After Ref. 5.
[b]After Ref. 26.

virtually nonexistent as a dissolved ion species.[11] The concentrations of silica normally range from less than 5 mg/l to 30 mg/l. The higher values are usually indicative of rock type and the temperature of the water. (Silica has an increased solubility with higher temperatures.) Aluminum concentrations normally are less than 1.0 mg/l. See water E in Appendix Table 3A-1 for a typical silica content.

Other Ion Species Present in Natural Waters:

Iron. Iron is present in almost all natural waters due to its abundance in a wide variety of crystalline rocks and soils. Literature concerning iron concentrations in water is especially voluminous because of the effects it has on water supplies when concentrations exceed a few tenths of a milligram per liter.[11]

The principal minerals containing iron include the amphiboles (hornblende), pyroxenes (augite), biotite, olivine, and magnetite. Table 3-17 lists several of these minerals and their weathering products.

Pyrite (FeS_2), an abundant accessory mineral in metamorphic and sedimentary rocks, is an important ferrous iron source for natural waters. In highly industrialized areas, especially the coal regions of Pennsylvania, acid mine drainage greatly increases the solubility of the ferrous species, and thus the total concentration of iron.

One of the problems that arises during the development of groundwater sources is the oxidation of the ferrous (Fe^{2+}) iron to the ferric or Fe^{3+} state. Groundwater at depth normally is very low in dissolved oxygen and has a low *Eh* value. Once pumping has started, the water takes up oxygen and the *Eh* rises. When this happens, precipitation of ferric hydroxide occurs. This can be represented as: $Fe^{2+} + 3H_2O = Fe(OH)_3 + 3H^+ + e^-$. The liberation of 3 hydrogen ions for each oxidized ferrous ion can drastically change the pH of the water, resulting in possible acid corrosion of the well screen and casing over a period of time, especially if the original HCO_3^- content of the water was relatively low.[27] When sampling, this type of water is clear when first drawn, then becomes cloudy, and eventually turns brown as the $Fe(OH)_3$ precipitates out.

Well screen encrustation can also occur from the precipitation of the ferric iron species, especially when there are large fluctuations in pumping. Biological activity can also cause encrustation problems when oxygenation around the well screen occurs (Hem in Ref. 1).

Iron concentrations have a very wide range in all geologic environments, as can be expected. Concentrations range from less than 0.1 to 50 mg/l: thus, iron removal is to be considered one of the more common treatments of water supplies. See water D in Appendix Table 3A-1 for an example of an iron-bearing water.

Manganese. Manganese occurrence in natural waters creates many of the same problems as that of iron. Small concentrations can cause objectionable staining of plumbing fixtures, clothes, etc.

The occurrence of manganese is widespread in the rocks and soil of the earth, although somewhat less abundant than iron. The most common forms are hydroxides and oxides in which the oxidation state of the element is +2, +3, or +4.[11] The more common oxides are pyrolusite (MnO_2), braunite (Mn_2O_3), and hausmannite (Mn_3O_4). Rhodochrosite ($MnCO_3$) and rhodonite ($MnSiO_3$) are also fairly common manganese-bearing minerals.

The solubility of manganese in natural waters is a function of the pH and *Eh*. The divalent ion Mn^{2+} in concentrations between 1.0 and 10 mg/l would be stable in a pH and *Eh* range of surface water exposed to the atmosphere. A slight increase in pH, however, could cause the precipitation of $MnO_2(s)$ ($Mn^{2+} + 2H_2O = MnO_2(s) + 4H^+ + 2e^-$). Bacteria are thought to influence the rates of manganese oxidation in a manner similar to the way they affect oxidation of ferrous iron.[11]

Concentrations of manganese are usually less than 1.0 mg/l in natural waters. There are some occurrences where concentrations have exceeded several milligrams per liter, especially in streams that have received acid mine drainage. Groundwaters reputed to carry large concentrations of manganese are usually from thermal springs associated with manganese oxide deposits.[11]

Calcium and Magnesium. Calcium is perhaps the most prevalent cation in natural waters. It is an essential constituent of many igneous and metamorphic rock minerals, especially the solid solution series of albite ($NaAlSi_3O_8$) and anorthite ($CaAl_2Si_2O_8$), pyroxene and amphibole. The dissolution of the albite-anorthite series is represented by Eqs. 213 and 214 in Table 3-17. Since the pH of such a reaction is slightly basic, given enough contact time calcium carbonate could precipitate from the solution.

Calcium also occurs as a carbonate precipitate in sedimentary environments. Limestone, principally $CaCO_3$ (calcite), is the common rock type, with dolomite $CaMg(CO_3)_2$ being less prevalent.

Magnesium is also very common as a dissolved cation species in water. The principal sources of magnesium are the ferro-magnesian minerals in igneous and metamorphic rocks such as olivine (Mg_2SiO_4–Fe_2SiO_4), biotite, augite, and hornblende. Magnesium also occurs in a variety of magnesium carbonates and hydroxycarbonates, most being considerably more soluble than calcite.

The occurrence of dolomite, the calcium-magnesium carbonate, is a perplexing problem to geologists. There is not much evidence to support the idea that dolomite formed as a primary precipitate.[26] It is presently thought that dolomitization occurred when magnesium ions invaded the calcite lattice of limestone shortly after its precipitation. Seawater, saturated with respect to magnesium, would have to be in continuous contact with the lime sediments in a reflux action. In any event, the details of the formation of dolomite remains obscure.

Concentrations of magnesium occur in a wide range on the weathering environment. Together with calcium, it is a primary source of hardness in water causing scaling problems in some industrial applications. When present in excess, the soluble sulfate $MgSO_4$ causes some laxative effects on consumers. See waters A, B, and C in Appendix Table 3A-1, for typical calcium and magnesium contents in groundwaters.

Sodium and Potassium. Sodium is present in a variety of rock types, primarily in igneous rocks where it exists in slightly more abundance than potassium. Once in solution, sodium tends to remain in that status unless acted upon by the process of ion exchange, where the sodium ions are exchanged for other cations on mineral surfaces such as clays.[26] The source of sodium, as previously mentioned, is varied. The dissolution of albite (Eq. 223, Table 3-17) is the principal weathering reaction involving the presence of sodium in natural waters. The solution of halite (Eq. 223, Table 3-17) is a less common source of sodium. Brines occurring in evaporite deposits normally exhibit sodium concentrations in excess of 1000 mg/l, although the natural range would be from 1–30 mg/l. See water F in Appendix Table 3A-1 for an example of a water containing "high" content of sodium and potassium.

Potassium, although slighly less abundant than sodium in igneous and metamorphic rocks, occurs in a much narrower range in natural waters. Potassium feldspar, $KAlSi_3O_8$, is very resistant to acid attack and yields kaolinite and leached silica much more slowly than the other feldspars (Eq. 215, Table 3-17). Potassium also enjoys a rather special function in the weathering process—it forms the adsorbed cation in mixed-layer clays such as illite, which also limits its presence as a dissolved cation in natural water. Concentrations of potassium in excess of 10 mg/l in fresh water are very unusual except in water very high in dissolved solids or in water from a thermal spring.

ACKNOWLEDGMENT

Paper of the Journal Series, New Jersey Agricultural Experiment Station, Rutgers, The State University, Department of Environmental Sciences, New Brunswick, New Jersey.

REFERENCES

1. Faust, S. D., and Hunter, J. V., eds., *Principles and Applications of Water Chemistry*, John Wiley and Sons, Inc., New York (1967).
2. Stumm, W., and Morgan, J. J., *Aquatic Chemistry*, John Wiley and Sons, New York (1970).
3. Wasser, J., *Basic Chemical Thermodynamics*, W. A. Benjamin, Inc., New York (1966).
4. Everdell, M. H., *Introduction to Chemical Thermodynamics*, W. W. Norton and Co., New York (1965).
5. Robie, R. A., and Waldbaum, D. R., U.S.G.S. Bulletin 1259 (1968).
6. Latimer, W. M., *Oxidation Potentials*, Prentice-Hall, Inc., Englewood Cliffs, N.J. (1952).
7. Morris, J. C., "Disinfection," *Water Purification and Wastewater Treatment and Disposals*, Vol. 2, John Wiley and Sons, Inc., New York (1968).
8. Harned, H. S., and Davies, R., Jr., *Journal American Chemical Society* **65**, 2030 (1943).
9. Harned, H. S., and Scholes, S. R., *Journal American Chemical Society* **63**, 1706 (1941).
10. *Standard Methods for the Examination of Water and Waste-water*, 13th ed., APHA, AWWA, WPCF, New York (1971).
11. Hem, J. D., "Study and Interpretation of the Chemical Characteristics of Natural Water," USGS Water Supply Paper 1473, 2nd ed. (1970).
12. Langmuir, D., *Geochim et Cosmochim Acta* **32**, 835 (1968).
13. Langelier, W. F., *Journal AWWA* **28**, 1500 (1936); *Ibid.*, **38**, 169 (1946); *Ibid.*, **38**, 179 (1946).
14. Flentje, M. E., *Journal AWWA* **53**, 1461 (1961).
15. Nordell, E., *Water Treatment for Industrial and Other Uses*, Reinhold Publishing Co., New York (1961).

16. Powell, S., *Water Conditioning for Industry*, McGraw-Hill, New York (1954).
17. Hem, J., "Form and Stability of Aluminum Hydroxide Complexes in Dilute Solution," USGS Water Supply Paper 1827-A (1967).
18. Freiser, H., and Fernando, Q., *Ionic Equilibria in Analytical Chemistry*, John Wiley and Sons, Inc., New York (1966).
19. Stumm, W., and Lee, G. F., *Rev. suisse d'Hydrologie*, **22**, 295 (1960).
20. Leckie, J., and Stumm, W., "Phosphate Precipitation," Water Resources Symposium No. 3; "Water Quality Improvement by Physical and Chemical Processes," University of Texas Press, Austin (1970).
21. Licht, T. S., and deBethune, A. J., *Journal of Chemical Education* **34**, 433 (1957).
22. Faust, S. D., and Vecchioli, J., "Chemical Problems Associated with the Injection of Highly Treated Sewage into a Deep Sand Aquifer," *Journal AWWA* **66**, 371 (1974).
23. Stumm, W., and Lee, G. F., *Industrial and Engineering Chemistry* **53**, 143 (1961).
24. Gomaa, H., and Faust, S. D., *Journal Agricultural Food Chemistry* **19**, 302 (1971).
25. Carrol, D., *Rock Weathering*, Plenum Press, New York (1970).
26. Krauskopf, K., *Introduction to Geochemistry*, McGraw-Hill, New York (1967).
27. Hem, J., "Restraints on Dissolved Ferrous Iron Imposed by Bicarbonate, Redox Potential and pH," USGS Water Supply Paper 1459-B (1960).
28. Stumm, W., and Singer, P. C., *Proceedings ASCE, Sanitary Engineering Division* **SA5**, 120 (1966).

APPENDIX

TABLE 3A-1. Quality of Selected Groundwaters Throughout the United States[a]

Substance[b]	A	B	C	D	E	F
Silica	8.4	15.0	8.4	41.0	99.0	22.0
Iron (total)	0.04	0.01	–	8.1	0.04	0.20
Manganese	–	0.01	–	0.0	–	–
Calcium	46.0	96.0	40.0	45.0	2.4	2.5
Magnesium	4.2	19.0	22.0	17.0	1.4	2.1
Sodium	1.5	18.0	0.4	63.0 (Sodium + Potassium)	100.0	182.0 (Sodium + Potassium)
Potassium	0.8	1.5	1.2		2.9	
Bicarbonate	146.0	133.0	213.0	208.0	111.0	412.0
Sulfate	4.0	208.0	4.9	29.0	130.0	3.5
Chloride	3.5	25.0	2.0	83.0	10.0	9.5
Fluoride	0.0	0.4	0.0	0.2	22.0	1.7
Nitrate	7.3	0.4	4.8	0.6	0.5	0.6
Dissolved solids	139.0	468.0	180.0	405.0	–	452.0
Hardness as $CaCO_3$	132.0	318.0	190.0	182.0	12.0	15.0
Specific conductance	250.0	690.0	326.0	638.0	449.0	718.0
pH	7.0	7.8	7.4	6.7	9.2	8.7
Color	5.0	3.0	5.0	23.0	–	1.0
Aluminum	–	–	1.4	–	–	–

[a]After Ref. 11.
[b]All expressed as mg/l except specific conductance (micromhos at 25°C), pH, and color.
Water A–Big Spring, Huntsville, Alabama. Limestone. 61°F.
Water B–Industrial well, Williamanset, Massachusetts, 120 ft deep. Chicopee shale. 54°F.
Water C–Spring 2.5 mi N.W. of Jefferson City, Tennessee. Knox dolomite. 58°F.
Water D–Elizabeth City, North Carolina, 30–80 ft deep. Coastal Plain deposits.
Water E–Owyhee County, Idaho, 800 ft deep. 122°F.
Water F–Wake County, North Carolina, 184 ft deep, Coastal Plain sediments.

4

Principles and Applications of Bacteriology to Polluted Water

H. Heukelekian

Consultant

BACTERIA IN THE AQUATIC ENVIRONMENT

Of all the microscopic organisms in the aquatic environment, bacteria are at the base of a complex series of reactions in transforming and stabilizing both organic and inorganic impurities. Their activities precede those of the higher forms of life such as protozoa, rotifers, nematodes, worms, insect larvae, and algae. Separating bacterial activity from that of higher forms of life is arbitrary, however, since in the ecosystem, the roles of various biota are interlocked and interrelated. Products of one group serve as food to the higher forms of life, giving rise to the "food chain."

Within a given group of bacteria, there are specializations of function and the ability of certain species to extend their range of activities by adaptation. The biochemical activities within a given species for a specific substrate and environmental condition constitute a series of complex chemical reactions which go through manifold intermediate steps and side reactions. The basic processes of bacteria in natural self-purification and wastewater treatment, however, are qualitatively the same. They differ only in intensity with respect to time and area required to stabilize pollution loads. It is therefore possible to present a unified viewpoint of basic bacterial processes as they relate to purification of polluted water.

Morphology of Bacteria. Bacteria are unicellular or multicellular plants devoid of chlorophyll. They multiply by binary fission which differentiates them from yeasts and fungi. The absence of chlorophyll differentiates them from algae, and has very significant and far-reaching implications related to polluted water purification.

Bacterial Forms. Bacteria have only a few outstanding structural characteristics. They are

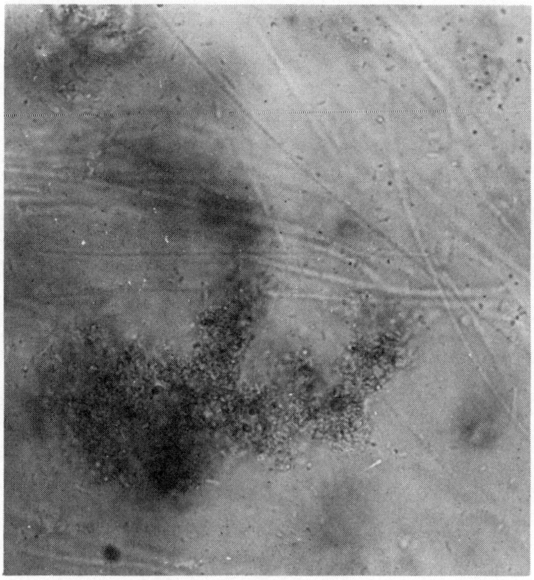

Fig. 4-1 *Sphaerotilus.*

either rod-shaped, spherical, or spiral. They may exist as individual cells or may form chains and other kinds of aggregates (see Figs. 4-1 and 4-2).

Identification of numerous bacterial species based on morphology and cell groupings alone is of limited value. Their characterization is, therefore, based largely on their physiological and biochemical reactions.

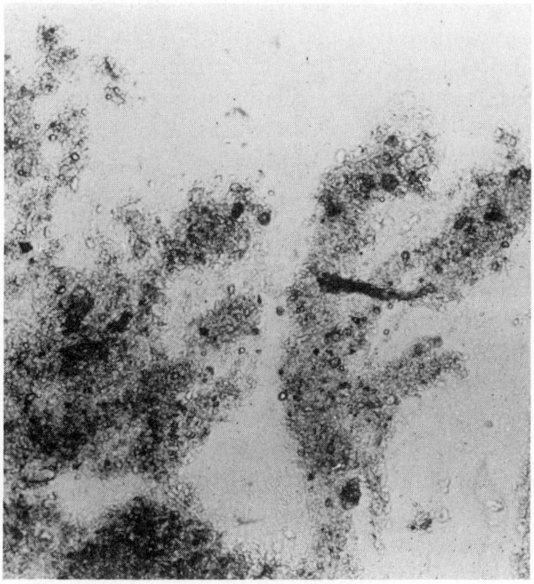

Fig. 4-2 *Zoogleal* bacteria.

Despite the lack of morphological differentiation, it would be wrong to assume that bacteria are simple forms of life, since all the vital activities are carried on within a single cell without the specialization of cells into tissues and organs as in higher forms of life.

Cell Structure and Function. The bacterial cell contains a nucleus which serves as the carrier of the genetic code and is the information center for the direction of specific cellular synthesis. The nucleus is surrounded by the cytoplasm which is the primary site of synthetic and functional activities. Within the cytoplasm are found reserve or storage material such as glycogen, lipids, and poly-β-hydroxybutyric acid. The presence, amount, and nature of such reserve materials are determined by environmental conditions.

A specialized part of the cytoplasm at the periphery, the cytoplasmic membrane, acts as a selective semipermeable membrane which guides the transport of materials from the substrate into the cell and the excretion of metabolic products in the reverse direction. It is the most important aspect of cell structure since it permits only the passage of nutrients of molecular and diffusible forms. Particulate materials in the substrate cannot be utilized unless they are first hydrolyzed by extracellular enzymes. The rate of their hydrolysis, therefore, determines their assimilation.

The cytoplasmic membrane is surrounded by a cell wall which gives the cell physical protection and structural rigidity. In many bacteria, the cell wall itself is covered by a slime layer which is a polysaccharide–a metabolic product secreted by the cell.

Cell Composition. A bacterial cell, a jelly-like mass, is high in water content, varying from 75 to 95 percent. The dry matter contains 50 percent carbon and 5 to 10 percent nitrogen. Ash content on a dry basis varies from 2 to 15 percent. The most abundant element in the ash is phosphorus (10 to 40 percent), with smaller concentrations of calcium, magnesium, sulfur, iron, silicon, potassium, sodium, and chlorine. The organic matter is primarily proteins and nucleic acids (60 percent), carbohydrates, polysaccharides, and fats.

The wide variations in composition of the bacterial cell are related to (a) composition of substrate, (b) species of bacteria, (c) age of culture, and (d) luxury consumption. This latter factor is invoked for the removal of relatively high amounts of phosphorus from sewage by activated sludge.

Bacterial Characteristics. Some bacteria are provided with flagella which serve as the organ of motility for free-swimming forms. Bacteria also exhibit Brownian movement; others grow attached to a substratum.

Certain bacteria produce spores which are resistant forms and are thus protected from unfavorable environmental conditions. Some spores can survive for many years (20–40) under laboratory conditions. Upon the return of favorable conditions, they germinate into the vegetative-growing forms. The significance of spores lies in the ineffectiveness of some common methods of disinfection for destroying bacteria.

Size and Propagation. Bacteria commonly vary from 0.5 μ to 3.0 μ in size. The significance of their small size lies in the fact that they expose large surface area per unit volume. Since all biochemical reactions are surface phenomena, this explains the ability of bacteria to grow rapidly and to bring about profound changes in the environment. The generation time of bacteria is about 20 to 30 min under favorable conditions.

Growth Factors. Substrate composition is an important consideration in the growth of bacteria. There are generally high enough concentrations of carbonaceous material in sewage, some organic industrial wastes, and in polluted waters that carbon is not a limiting growth factor. Nitrogen is likewise usually not a limiting factor in domestic sewage which contains 20 to 30 mg/l of NH_3 (ammonia-N) and 10 to 20 mg/l of organic nitrogen. However, in certain industrial wastes such as pulp mill effluents, nitrogen and phosphorus are frequently nutritionally limiting. In order to achieve successful biological treatment in such instances, it is important to supply these nutrients by the addition of compounds of nitrogen and phosphorus to yield ratios of 5 parts of nitrogen and 1 part of phosphorus to 100 parts of BOD_5.

The quantity of nutrients required is dependent upon the rapidity of biological oxidation. For example, in the case of oxidation in ponds requiring 20 to 30 days to provide 75 percent reduction in BOD_5, the addition of these elements to wastes deficient in them fails to accelerate the reaction. However, when the same reduction is achieved in 2 hr as with activated sludge, their presence is imperative.

Bacterial Aggregations. The phenomenon of capsule formation by individual cells leads to the bacterial aggregation which is a very important characteristic of growth in the aquatic environment. The merging of capsules of individual cells leads to the formation of attached slime layers on wetted surfaces, such as in trickling filters and stream beds, or floc formation in activated sludge. Numerous bacteria thus become imbedded in a common gelatinous matrix which is referred to as a zoogleal type of growth. This characteristic, often attributed to a bacterial species known as *Zooglea ramigera*, is also found in many other types of bacteria.

The significance of bacterial aggregations lies in the fact that concentration of bacteria per unit volume is much greater than would be possible if they were individually dispersed. Such aggregations, whether attached or free-floating, also make possible the adsorption of the particulate matter on the gelatinous surface for more expeditious hydrolysis by enzymes within the aggregate. As a result of the great concentration of bacteria in the aggregate and the enhanced surface adsorption of soluble matter and concentration of particulates, the rate of reaction is greatly accelerated. This is illustrated by the long detention time (days) required in oxidation ponds and in stream self-purification as compared with the high rate of purification obtained in trickling filters and activated sludge.

One additional form of bacterial aggregation is that of certain filamentous bacteria, such as *Sphaerotilus natans*. (see Fig. 4-1). The distinguishing characteristic of this type of growth, as compared with chains formed by bacterial rods, is that the cells in true filamentous bacteria are encased in a sheath.

METABOLISM OF BACTERIA

Food is required by bacteria both for growth and energy. Since they do not contain chlorophyll, they cannot utilize the sun as a source of energy, nor is electrical or thermal energy of use to them.

Higher temperature causes a more rapid functioning of the cell, but it cannot economize on food by utilizing the heat of the surrounding medium. A starving cell is not benefited by an increase of temperature which instead leads to more rapid death. Energy cannot be transmitted from one cell to another but must be produced within the cell.

The essential elements for food are those that are found in the cell and consist of the following: H_2, O_2, C, N, S, P, K, Na, Mg, Mn, Ca, and Fe. Carbon, which is the most abundant element of the cell, may be obtained from organic material or from carbon dioxide, depending on the type of organism. Some species are omnivorous and versatile while others utilize only certain specific compounds.

Metabolism is, therefore, defined as the sum total of all the biochemical reactions of an organism whereby food is utilized, growth and multiplication take place, and energy is made available. The definition includes all aerobic and anaerobic reactions, organic and inorganic sources of food, with and without the formation of gas.

It is comprised of (a) *catabolism*, or degradation, dissimilation, and energy-liberation reactions, and of (b) *anabolism*, or synthesis, assimilation, and energy-utilizing reactions.

Energy is thus required for growth, maintenance, repair, and locomotion. Relatively smaller amounts of energy are used for synthetic purposes, most of it being utilized for maintenance and repair.

The utilization of energy is greater for aerobic bacteria, which convert 40 to 50 percent of an organic substance to cell material. Anaerobic bacteria convert only 5 to 10 percent. The lower the percentage of energy utilized, the greater the amount of organic matter that has to be transformed to yield the same amount of protoplasm. Thus, under aerobic conditions, the same amount of food material decomposed will result in a greater mass of bacterial growth than under anaerobic conditions.

Bacterial metabolism can be divided into two main categories: organic and inorganic. In organic metabolism, bacteria require preformed organic materials as a source of both carbon and energy to bring the synthesis about. Bacteria with this kind of metabolism are referred to as heterotrophic bacteria.

There are also bacteria which obtain carbon from carbon dioxide and energy from reduced forms of inorganic materials. Such bacteria are known as autotrophs.

Organic Metabolism. In the decomposition of organic matter with the production of intermediate and end products, cell mass is formed even though the amount of growth represents only a small fraction of the organic matter attacked. Thus, in the aerobic decomposition of sugar, the reaction should be represented thus:

$$C_6H_{12}O_6 + 6O_2 \longrightarrow 6CO_2 + 6H_2O + 674 \text{ cal} \longrightarrow \text{protoplasm} \quad (1)$$

The destruction of organic matter is referred to as catabolism or dissimilation, while the synthesis and growth are referred to as anabolism. Catabolism and anabolism are coupled reactions. The concept is that of a common compound which traps the energy and forms the base in the synthesis of protoplasm.

Catabolism. Catabolic energy-liberation reactions take place by removal or transport of the hydrogen atom (or electron) from the organic molecule which becomes dehydrogenated (or oxidized). The liberated hydrogen then combines with a suitable hydrogen acceptor which becomes reduced.

Depending on the nature of the hydrogen acceptor, catabolism can be classified as:

1. Oxidative dissimilation or respiration when the hydrogen acceptor is molecular oxygen under strictly aerobic conditions;
2. Intermolecular respiration with an oxygen-rich compound serving as hydrogen ac-

ceptor, in the absence of molecular oxygen.

Nitrate, sulfate, and carbon dioxide are known to serve as hydrogen acceptors. The succession of the acceptors is: oxygen, nitrate, sulfate, and carbon dioxide.

Intermolecular Respiration. The dehydrogenation and hydrogenation steps with nitrate serving as a hydrogen acceptor take place when the energy liberation from the former step is greater than from the latter. The overall reaction is represented as follows:

$$5CH_3COOH + 8NO_3 \longrightarrow 10CO_2 + 4N_2 + 6H_2O + 8(OH) \quad (2)$$

This process, in which nitrate is reduced to nitrogen gas, is referred to as denitrification.

The reaction in which sulfate serves as hydrogen acceptor can be represented as:

$$2CH_3CHOHCOOH + SO_4 \longrightarrow 2CH_3COOH + 2CO_2 + H_2S + 2(OH) \quad (3)$$

The reaction in which CO_2 serves as the hydrogen acceptor can be represented as:

$$2CH_3CH_2OH + CO_2 \longrightarrow 2CH_3COOH + CH_4 \quad (4)$$

This is one mechanism of methane fermentation.

Intramolecular Respiration. Methane is also produced by intramolecular respiration or fermentation without the mediation of external hydrogen acceptors, by mere transport of hydrogen from one part of the molecule which is oxidized to another part which is reduced:

$$CH_3COOH \longrightarrow CH_4 + CO_2 \quad (5)$$

Another example of intramolecular respiration is the alcoholic fermentation of glucose:

$$C_6H_{12}O_6 \longrightarrow 2C_2H_5OH + 2CO_2 \quad (6)$$

There is a question as to when the reactions cease to be aerobic and become anaerobic. Oxidative dissimilation in the presence of molecular oxygen is unquestionably aerobic. Since anaerobic conditions are generally associated with the production of odors, then dissimilation in which nitrate serves as hydrogen acceptor can be considered aerobic, because the product of reduction is nitrogen gas. The bacteria that utilize the oxygen of nitrate prefer to utilize molecular oxygen if it is present. They can therefore be regarded as facultative organisms. Various types of bacteria such as *Pseudomonas*, *Spirillum*, *Bacillus*, etc., have the ability to utilize these two sources of oxygen. Furthermore, in the presence of nitrates, sulfates are not reduced to hydrogen sulfide.

The reaction involving the reduction of sulfate must, without question, be considered anaerobic. The degree of anaerobiosis is greater and the level of redox potential required is very low when sulfates and carbon dioxide serve as hydrogen acceptors. It has been shown that oxygen inhibits the growth of methane bacteria.

Endogenous Respiration. Endogenous respiration takes place at the expense of stored food material in the cell. This type of reaction takes place whether exogenous food is present or not, but its manifestations become pronounced in its absence. Endogenous respiration in the absence of exogenous food ultimately results in the death of the cell by self-digestion or autolysis. The cell becomes disrupted and the contents are dispersed into the substrate to be utilized by other microorganisms.

Thermogenesis. Part of the energy liberated which is not used for growth, repair, maintenance, or locomotion is dissipated to the surrounding medium as heat. The amount of heat generated is manifested in compost and manure piles, which under suitable conditions attain temperatures of 160°F (71°C) or higher. The essential requirements are aerobic conditions and proper moisture content. In sewage treatment processes such as activated sludge or sludge digestion, no increase in temperature of the substrate can be detected, even if it were produced, due to the large volume of water in relation to the concentration of organic matter.

Products of Metabolism. The products of metabolism vary with the material decomposed, with the organisms responsible for the decomposition, and with the environmental conditions. Products of decomposition of nitrogenous organic matter are different from those of carbonaceous materials. Different organisms break down the same materials to varying degrees, giving different end products. The same compound subjected to decomposition under aerobic and anaerobic conditions gives rise to different products. The main products of aerobic decomposition of carbonaceous organic materials are carbon dioxide and water. Under anaerobic conditions, additional products are formed consisting of methane, hydrogen, organic acids, and alcohols. Proteins under aerobic conditions produce carbon dioxide, water, ammonia, and amino acids, while under anaerobic conditions additional compounds having objectionable odors such as hydrogen sulfide, indole, skatole, and mercaptans are produced.

Enzymes. Both catabolic and anabolic reactions are catalyzed by enzymes. Enzymes in catabolic reactions bring about a stepwise decomposition of organic materials and produce intermediate compounds from which synthesis of new organic material can be started. The other function of enzymes is to make a part of the energy liberated available for synthesis.

Enzymes are defined as organic catalysts produced by living organisms which accelerate reactions under normal temperature conditions to rates which could otherwise take place only at high temperatures. They form an unstable compound with a specific substrate which is then broken down to end products. The enzyme itself becomes free for additional reaction. They are highly specific and act on definite structural configurations.

The rate of reaction depends on the concentration of the substrate and of the enzyme. With increased concentration of substrate, the enzyme systems may become saturated and limiting. Theoretically, they are not destroyed in the process, but actually there is need for additional production of enzyme.

Enzymes which are produced by the cell and diffuse out into the surrounding medium are classified as exoenzymes. They are hydrolytic in their mode of action, and break down complex organic molecules and particles into simpler forms which can then diffuse through the cytoplasmic membrane and become available as sources of energy and carbon. Such hydrolytic enzymes can be obtained by filtering the culture free of bacteria. The filtrate containing the enzyme has been shown in some cases to bring about the reaction.

Enzymes are classified according to the specific substances on which they act such as:

1. Amylase or diastase which converts starch and glycogen into maltose (disaccharide).
2. Cellulase which hydrolyzes cellulose into hemicellulose.
3. Cytase which converts hemicellulose to disaccharide.
4. Lipase which converts fat to fatty acids and glycerol.
5. Proteases which convert proteins to peptides.
6. Peptidases which convert peptides to amino acids.
7. Amidases which convert amino acids to ammonia.
8. Urease which converts urea to ammonia.

In addition, there are endoenzymes which are responsible for the energy-liberating catabolic reactions within the cell and enzymes responsible for the anabolic or synthetic reactions.

Of the great variety of enzymes produced by bacteria, the particular one and the amount formed at any given time are largely determined by the control mechanisms responsible for their production. Respiratory and synthetic enzymes which are essential for the maintenance and growth of bacteria are known as constitutive enzymes and are always formed irrespective of the composition of the substrate. Many other enzymes are synthesized only in response to the presence of a specific compound in the environment. Such enzymes are known as induced or adaptive enzymes. Induced enzymes are therefore very important in biological processes which occur in polluted waters since chemical composition of wastewater varies greatly.

Implications and Applications of Organic Metabolism. Organic metabolism has far-reaching and manifold implications and applications to the transformation of organic matter in polluted waters and to its stabilization in natural and artificial purification processes.

Organic matter in wastewater is one of the important ingredients which interfere with the beneficial uses of waters. Organics consist of numerous compounds of various composition and in different states of subdivision. Decomposition and stabilization is brought about most expeditiously and economically as a result of biochemical activities of bacteria and other forms of life. Organic materials are utilized as food and as sources of energy by a variety of bacterial types working in sequence to effect partial or complete degradation.

The condition of the environment of the substrate (sewage or polluted water) determines the types of bacteria and the type of metabolism. Among the many environmental factors, perhaps the presence or absence of oxygen has the most profound effect on the pathways and the products formed.

Under aerobic conditions (in the presence of molecular oxygen and nitrates), the oxidation reactions are complete and the main end products are carbon dioxide, ammonia, and water. Bacterial growths are also inevitably produced, the amount of which is less than the quantity of matter consumed. It therefore becomes necessary to waste sludge in activated sludge. In trickling filter processes, excess growth is sloughed off.

It is not implied that the quantity of sludge to be disposed of from these processes is all due to biological growth. Some of it is contributed by nonbiodegradable and incompletely utilized organic matter. Therefore, claims made of a biological process which would not produce any sludge are not based on sound principle. The residual organic matter, including the bacterial growth, can be further decomposed, albeit slowly, but never completely. Bacterial cells become autolyzed under starvation conditions and the organic matter they contribute to the surrounding medium serves as food for other microorganisms. In addition, the organic matter, including bacteria, serves as food to higher forms of life, particularly filter-feeding protozoa. This results in a further decrease of the organic matter accompanied by a small increase in biological growth.

Under anaerobic conditions, the energy liberation and utilization of organic matter are not complete, a result of which is the formation of a series of intermediate products. The anaerobic process is primarily one of transformation of matter, and the only such process that goes beyond mere transformation, i.e., from the original organic matter to its complete destruction, is in methane fermentation which converts the intermediate products formed to methane and carbon dioxide. The quantity of bacterial growths produced is less than under aerobic conditions.

The decomposition of these complex and numerous organic compounds, under varying environmental conditions, is brought about by the mobilization of those bacteria present in the polluted water which find the specific compounds in the substrate suitable as sources of food which they utilize through the production of constitutive and inductive enzymes. They multiply rapidly and become the dominant forms, each species following the other in sequence, depending upon the niche they can occupy in the series of transformation of the organic matter. Very little is known as to the nature of the proliferating bacteria in biological treatment processes, but somewhat more is known about some of the chemical transformations. The latter is especially true with respect to the end products, as well as with some of the intermediates which accumulate in the substrate.

The organic and inorganic nitrogen contents in the substrate have important implications. Sewage contains both organic and inorganic nitrogen (ammonia) and therefore nitrogen does not become a limiting factor for the synthesis of bacterial cells. In the treatment process, the settleable solids with their nitrogen content are recovered in the sludge. The organic nitrogen fraction of the nonsettleable solids subjected to biological treatment goes through a series of hydrolytic reactions which result in the formation of ammonia, adding to the ammonia content of the effluent. Part of the ammonia is assimilated and the rest may be oxidized to nitrate by biological treatment processes.

The organic nitrogen content of the raw sludge produced from domestic sewage, constituting about 4 percent of the dry matter, undergoes a similar process of hydrolysis during anaerobic digestion which increases the ammonia content of the supernatant liquor. Whether decomposition will result in the production of an excess amount of ammonia, or wehther the organic nitrogen content will be limiting for bacterial metabolism will depend on the percent nitrogen content of the organic matter. At about 2 to 2.5 percent nitrogen (on dry matter basis) there will be neither an excess of ammonia wasted to the substrate nor a deficiency for bacterial assimilation. It follows that for the decomposition of organic matter with lower percentages of nitrogen, artificial additions of nitrogen compounds will have to be made to support biological activity.

Inorganic Metabolism. Certain bacteria are capable of deriving their energy from the oxidation of inorganic compounds and their carbon for cell synthesis from carbon dioxide (autotrophs), in contrast with those that get their energy and carbon from organic materials (heterotrophs). The bacterial heterotrophic form of nutrition is, therefore, similar to that of animals, being dependent on preformed organic material produced by green plants. Autotrophic nutrition of bacteria, on the other hand, is comparable to that of green plants. The difference is in the source of energy required. Autotrophs can therefore develop in the dark as well as in an environment that is free of organic matter, and are able to produce organic matter in the form of cell substance from inorganic materials, specificially carbon dioxide and minerals. The source of energy necessarily has to be a reduced form of inorganic matter. These are ammonia (NH_3), nitrogen dioxide (NO_2), reduced sulfur compounds, and ferrous iron.

Nitrification. Bacteria which derive their energy from reduced nitrogen compounds are called nitrifying bacteria and the process is referred to as nitrification. Two highly specialized bacteria are involved in the overall conversion of ammonia nitrogen to nitrate in two sequential steps: in the first step, ammonia is oxidized to nitrate by *Nitrosomonas*, and in the second, nitrite is oxidized to nitrate by *Nitrobacter* asrepresented by the following equations:

$$(NH_4)_2CO_3 + 3O_2 \longrightarrow 2HNO_2 + CO_2 + 3H_2O + 147.2 \text{ cal} \quad (7)$$

$$2HNO_2 + O_2 \longrightarrow 2HNO_3 + 37.6 \text{ cal} \quad (8)$$

$$CO_2 + H_2O \longrightarrow HCOH + O_2 - 134 \text{ cal} \quad (9)$$

Equation 9 represents the generalized assimilation reaction using the energy liberated in Eqs. 7 and 8 for the reduction of carbon dioxide to the first organic compound formed as represented by formaldehyde. The oxidation of ammonia to nitrite appears to be the slower of the two reactions because normally nitrite is present only in low concentrations.

In the production of nitrite from ammonia, 3.43 parts of oxygen are utilized for each part of nitrogen oxidized. In the production of nitrate from nitrite, an additional amount of 1.14 part oxygen is utilized for each part of nitrite nitrogen oxidized. The energy liberated in these reactions is utilized to reduce and assimilate the carbon dioxide for growth. Essential considerations for the reactions are:

1. Aerobic conditions. It has been demonstrated that nitrification in the activated sludge process is minimal at 0.5 to 1.0 mg/l of dissolved oxygen and increases with increased oxygen concentration.
2. Presence of ammonia. Sewage contains an ample supply of ammonia nitrogen so that nitrification does not become limited by this factor.
3. Ample supply of carbon dioxide. This is generally present in adequate amounts.
4. pH value of the substrate. The optimum pH value is on the alkaline side of neutrality (pH values 7–9). It should be pointed out that the products of nitrification are mineral acids and they can consequently affect the pH values when relatively large

concentrations of nitrates are produced. The pH values attained are determined by the relative concentrations of nitrate nitrogen produced and the alkalinity of the substrate. Each part of ammonia nitrogen oxidized to nitrate reduces the alkalinity by 7 parts. Thus, in poorly buffered substrates, pH value can become unfavorable for further nitrification.

5. Temperature. This has an important effect on nitrification similar to its effect on other biochemical reactions.
6. Generation time. This aspect of nitrifying bacteria is longer than for most heterotrophic bacteria. It has been estimated by various investigators to be as high as 24 to 30 hr. Nitrification in sewage, therefore, becomes pronounced after a prolonged period, during which the few nitrifiers present in the substrate multiply following an autocatalytic reaction. This is manifested by the onset of nitrification in the BOD test after about 10 days of incubation, or similarly after 10 days of aeration of domestic sewage. Once a nitrifying population is established, however, the reaction can be rapid provided that the organisms are in a captive form attached to solid surfaces, as in trickling filters. In activated sludge, nitrification takes place only when the loss of the nitrifying bacteria in the wasted sludge washout rate is smaller than the rate by which the organisms are able to replace the loss by multiplication.
7. Effects of organics. Nitrifying bacteria do not utilize organic matter as a source of energy or carbon, and yet organic matter has an indirect effect on them. Heterotrophic bacteria, with their rapid rate of growth utilizing oxygen at a high rate, can starve the nitrifiers of their essential oxygen supply. The phenomenon of nitrification at the lower depths of the conventional trickling filters and towards the effluent end of the aeration tank in the activated sludge process can be attributed to this factor.

Soil scientists have established that NH_3 and nitrifying bacteria are concentrated on the surface of clay, thus giving greater opportunity for contact between the reactants and greatly accelerating the reaction rate. A similar phenomenon can take place in biological wastewater treatment processes.

Applications of Nitrification to Water Pollution Problems. Nitrification has been receiving considerable attention recently in this country as a means of maintaining dissolved oxygen in the receiving waters and for controlling eutrophication by reducing the nitrates produced to nitrogen gas under anaerobic conditions.

The oxygen demand exerted in the conversion of ammonia nitrogen to nitrate is 4.57 parts per part of nitrogen. Thus, secondary effluent with 20-30 mg/l of carbonaceous BOD_5 containing 20-25 mg/l of ammonia nitrogen could exert a potential additional oxygen demand of 90-115 mg/l. Whether this additional oxygen demand is exerted in the receiving stream or not would depend, among other factors, on the passage time in the stream and the relative wetted surface area per unit volume. The reason for the long passage time required is due to the long generation time of these organisms.

Effluents from biological treatment processes in which only a part of the ammonia is oxidized can seed the receiving water with nitrifying bacteria and accelerate nitrification. From the standpoint of dissolved oxygen content of the stream, discharge of partially nitrified effluent could have a greater effect than non-nitrified effluent.

The effect of large surface area per unit volume of water—e.g., shallow depths—follows from the fact that under these conditions the nitrifying bacteria, which grow attached, can bring about rapid nitrification as has been demonstrated in certain streams. The effect of bottom growths in deep streams would be relatively small.

Biological treatment processes are capable of producing nitrates under certain conditions. Conventional trickling filters are known to produce nitrates in the lower sections. Nitrification, however, is not complete and a part of the ammonia is discharged in the effluent.

With the advent of high-rate trickling filters in this country, nitrification was sacrificed for efficiency. Conventional filters may now be added to the existing high-rate filters in locations where effective nitrification is required either for the maintenance of dissolved oxygen in the receiving water or for the purpose of controlling eutrophication by the reduction of nitrate to nitrogen gas.

Conventionally-loaded activated sludge plants are known to bring about partial conversion of ammonia to nitrates. In order to induce more complete nitrification, it is necessary to decrease the sludge loading factor (increase the sludge age) and thus reduce the sludge growth rate to prevent loss of nitrifying organisms in the excess sludge. This is due to rapid heterotrophic bacterial utilization of organic material as compared with the slower growth rate of nitrifying bacteria. When both the removal of organic matter and the oxidation of ammonia are undertaken in the same aeration tank, deflocculation of the sludge results due to autooxidation of the heterotrophs under starvation conditions, and produces a turbid effluent containing small floc particles.

This problem can be solved by separating these processes into two systems: the first part designed to yield effective removal of the BOD and the second for nitrification. Each aeration tank would utilize a clarifier and facilities for recycling sludge separately (see Fig. 4-3).

Oxidation of Reduced Sulfur Compounds. The autotrophic bacteria which oxidize reduced sulfur compounds are the filamentous gliding bacteria *Beggiatoa* and *Thiothrix* and the single-cell *Bacterium thiobacillus.* *Beggiatoa* and *Thiothrix* both obtain their energy from the oxidation of hydrogen sulfide to elemental sulfur which is deposited within their cells. Upon the exhaustion of hydrogen sulfide, the organisms oxidize the deposited sulfur to sulfuric acid.

The growth of these organisms requires an environment which produces hydrogen sulfide under anaerobic conditions and at the same time requires aerobic conditions for oxidation. This situation is encountered in conventional trickling filters with a heavy accumulation of slime when the growth next to the stone becomes anaerobic. The hydrogen sulfide produced diffuses to the surface of the slime, contacts the atmosphere in the voids, and thus permits the growth of these organisms. They form a thin white growth covering the surface of the slime.

Thiobacillus thioparus is an autotrophic bacterium which obtains its energy by the oxidation of elemental sulfur, sulfide, thiosulfate, and tetrathionate.

Thiobacillus thiooxidans oxidizes elemental sulfur to sulfuric acid and can actually grow in a medium of 5 to 7 percent sulfuric acid. It utilizes carbon dioxide as its source of carbon. The role played by *Thiobacilli* in sewer corrosion will be discussed in conjunction with sulfate reduction.

Iron Bacteria. Iron bacteria cause deposition of ferric hydrate, reduce the water-carrying capacity of pipes, form slime growths, create turbidity and discolorations, and produce odors and tastes in water. They are aerobic, mostly filamentous, and are widely scattered in stagnant waters, wells, springs, streams, reservoirs, and water pipes. They are found only in waters containing iron in solution. The accumulation of iron on their cells, which may be 500 times the weight of cells, imparts a brown to rust red color.

Iron bacteria are associated with corrosion but are not the agents of corrosion. The source of iron may be either the water supply or the solution of iron in one part of the pipe

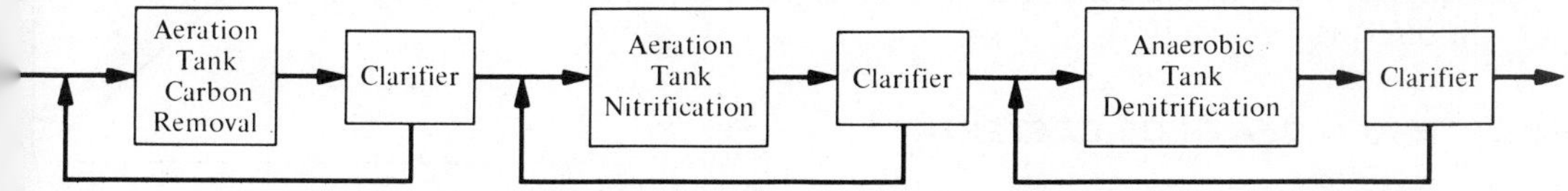

Fig. 4-3 Three-stage system for biological removal of carbon and nitrogen.

and its deposition in another part. Manganese is also oxidized by some iron bacteria.

Some iron bacteria are true autotrophs, obtaining their energy from the oxidation of ferrous iron to the ferric state. In other cases, the transformation of iron is incidental to the development of bacteria, resulting in a change of the environmental conditions. Under anaerobic conditions, iron is made soluble, while under aerobic conditions, iron is oxidized and precipitated.

A third type of iron transformation by these nonspecific bacteria is brought about by the decomposition of the organic radicle of an iron complex which results in the precipitation of the metal.

Some of the iron bacteria are:

1. *Siderocapsa*. These are rod or coccus forms growing in small colonies surrounded by capsular material with ferric hydrate surrounding the capsule. This type utilizes organic matter and hence is a nonspecific iron bacterium.
2. *Clonothrix, Leptothrix*, and *Crenothrix*. These filamentous bacteria formerly were considered as different types, but are now considered as special growth forms of *Sphaerotilus natans*. If the environment contains salts of iron and manganese, the hydroxides of these metals accumulate on their filaments. They are not autotrophic iron bacteria.
3. *Gallionella*. The only true autotrophic iron bacterium, it grows in a twisted ribbon form. The autotrophic oxidation requires a delicate balance of the environment to permit the existence of reduced iron in an otherwise aerobic environment.
4. *Ferrobacillus ferrooxidans*. An obligate autotrophic iron-oxidizing bacteria, it has been isolated from the drainage of bituminous coal mines containing iron. It is a motile rod-forming organism appearing singly or in pairs. It grows under acid conditions (pH value of 2.5).

GROWTH AND DEATH OF BACTERIA

Growth Phases. As a result of metabolism, bacterial cells increase in size and divide in half, thus multiplying in numbers. Bacteria cannot be studied as individuals but only as populations. The increase in size of an individual cell cannot be measured gravimetrically, since it would take 5×10^9 bacteria to produce 1 mg of dry matter. It is possible, however, to obtain a total cell count or total viable cell count by cultural or turbidimetric methods. It is also possible to harvest the cell crop in a culture medium by centrifuging and weighing the solids.

Growth and reproduction are not in direct proportion to time but according to a geometric progression. One bacterial cell with a generation time of 20 min could, if uninhibited growth could take place, produce in 48 hr 2.2×10^{43} cells weighing 2.2×10^{31} gm, 1000 times heavier than the earth.[1] Obviously, growth would be checked long before it reached such quantities either because of the exhaustion of food or the accumulation of metabolic products. After growth stops, death follows, decreasing the numbers. These events can best be illustrated by plotting the logarithm of the numbers against time (see Fig. 4-4). A typical growth curve has the following four phases:

1. lag phase
2. logarithmic or exponential growth phase
3. maximum stationary phase
4. death phase

Lag Phase. When a suitable medium is inoculated with a pure culture of bacteria, generally a certain period of time elapses before a constant growth phase is established. The lag phase generally occurs when the inoculum is taken from an old culture or when the inoculation is

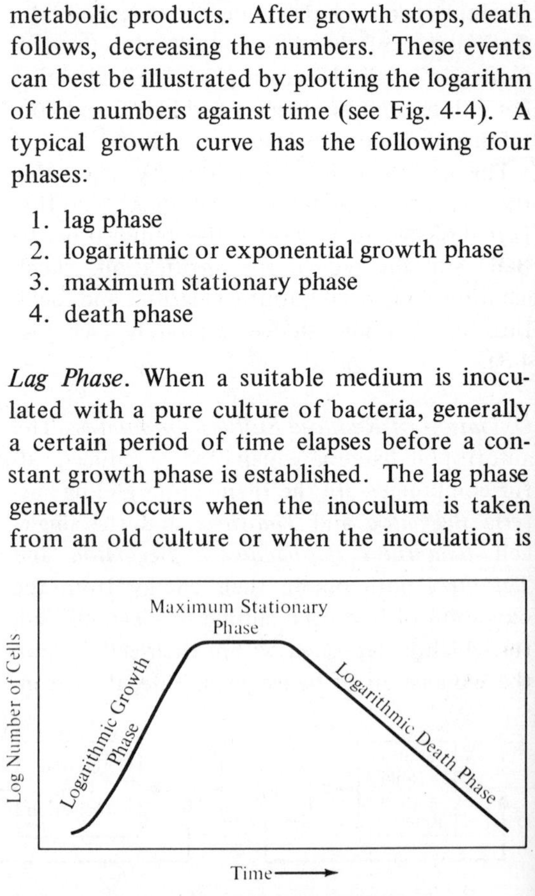

Fig. 4-4 Bacterial growth phases.

made into a medium of a different composition than the one in which the organisms were previously grown. The lag period is shorter with larger inocula. The lag under these conditions can be ascribed to the time required for the mutant forms that may be present in the inoculum in small numbers to multiply in the medium into which they were transferred and produce the requisite enzymes.

Exponential or Logarithmic Growth Phase. In the exponential or logarithmic growth phase, the rate of growth is constant. The actual rate depends upon the nature of the organism and environmental conditions such as pH value, temperature, composition and concentration of the substrate, and the presence or absence of oxygen. This phase in batch cultures lasts only for a relatively short time because of the unfavorable conditions brought about in the medium by the growth of the bacteria or by the exhaustion of food.

Biological sewage treatment processes are not affected by these factors. The rate of multiplication and populations are kept at a high level compatible with food concentration because of the continuous supply of food in the influent and the discharge of effluent which prevents the accumulation of inhibitory compounds.

Maximum Stationary Phase. In this phase, the numbers of bacteria in batch cultures remain constant at a plateau level since the multiplication and death rates are equal. Length of the stationary phase depends on the susceptibility of the bacteria to the unfavorable environmental conditions prevailing. With increasing concentration of food supply, the maximum number attained increases but not in direct proportion.

Actual populations and the rate of growth are also dependent on the size of bacterium. Smaller organisms achieve the greater numbers with a given food concentration, but the amount of growth produced is essentially the same as with the larger size organisms.

There is a minimum food concentration below which growth does not take place. An even lower limiting food concentration, however, will permit growth when exposed to an inert surface area.

It has been suggested that physical crowding (biological space) may be another factor which could explain the maximum stationary phase of population.

It should be pointed out that bacteria in the stationary phase metabolize and utilize oxygen at a low rate; thus, the highest rate of activity takes place in the logarithmic phase of growth.

Maximum numbers are attained and maintained in continuous-flow cultures, provided that environmental conditions do not change and the composition and concentration of nutrients remain constant. In sewage treatment processes, variations of these factors as well as flow occur commonly, so that ideal equilibrium conditions are not attained; hence, a dynamic equilibrium condition prevails.

Logarithmic Death Phase. In the logarithmic death phase, the viable count of bacteria decreases at a generally constant exponential rate. Deviations from the exponential order of decrease are encountered which are attributed to variations in individual resistance within a pure culture. In mixed populations, there are even greater variations in resistance of different species.

In the absence of food, endogenous respiration results in the death of most of the cells, although a small number remain at the end of the death phase; seldom are sterile conditions produced.

Measurement of Activity. The numbers of a mixed population of bacteria as determined by cultural methods cannot represent the actual numbers since no single culture medium can support the growth of all the bacteria. Furthermore, the numbers in a biological slime or activated sludge cannot be dispersed completely to individual cells to allow counting of the maximum numbers present. Numbers in themselves can also be misleading since at the maximum stationary phase the metabolic rate is at a minimum. A further question may be raised as to whether the numbers or the mass of growth is more important, since there are differences in the size of bacterial cells and much greater differences in size between bacteria and protozoa and metazoa. The most meaningful determination, then, is measuring the activity rate by using the change in a chemical parameter or oxygen utilization rate.

To establish the metabolism, physiology, nutritional requirements, metabolic products, and influence of factors affecting growth, it is necessary to make pure culture laboratory studies with artificial media. The various factors must be analyzed separately, but it should be remembered that they do not operate separately. They interact continuously, producing unexpected results.

Caution should be used in applying the results of pure cultures to a natural environment where many species of bacteria and other forms of life such as protozoa, rotifers, crustacea, worms, insects, and algae form a complex biological community. The nature and concentration of nutrients in the environment are entirely different from those in culture media. Additionally, the environmental conditions are subject to changes and are not constant. Under any condition, laboratory findings should be correlated with field observations. Attempts should also be made to reproduce as closely as possible the natural environment in the laboratory.

This can be done sequentially by: (1) replacing pure culture in artificial media with a progressively increasing complex population, (2) using the sterilized natural substrate inoculated with cultures isolated from the environment, (3) progressively increasing the multiplicity of bacterial types, and (4) ending with the unsterilized natural substrate with its population.

In mixed populations, biological interactions can take any of the following relationships:

1. *Symbiosis*; a mutually beneficial relationship between two kinds of organisms.
2. *Neutralism*; two types of organisms coexisting without either beneficial or harmful interaction.
3. *Commensalism*; one organism benefiting and the other being unaffected. An example of this kind of relationship might be the lowering of redox porential by one type of bacteria to a level where another, such as the sulfate-splitting bacteria, can function. Another is the need of liquefying bacteria to convert complex organic matter to lower volatile acids which the methane-producing bacteria can utilize in anaerobic sludge digestion. There are other similar relationships.
4. *Mutualism*; an interaction between two types of organisms leading to the production of a new metabolic product which would not have been produced by either one alone, or the growth of two organisms together which otherwise could not grow separately. Examples are the production of vitamin B_{12} by bacteria, essential for the growth of algae, and the relationship of bacteria and algae in oxidation ponds.
5. *Competition*; an interaction occurring between two organisms utilizing the same nutrient which is present in limited amounts, with the successful competitor being the one that has the faster growth rate. The inability of nitrifying bacteria to compete with the faster-growing heterotrophic bacteria in situations where oxygen supply is limited is an example of this category.
6. *Antibiotics*; production of growth-inhibiting substances by certain organisms under natural environmental conditions. Such an effect has been invoked to explain, without positive proof, the disappearance of coliform bacteria in seawater. Production of antibiotics requires special media and organsims growing under controlled conditions. It is doubtful whether these conditions exist in the sea to produce a sufficient concentration to explain the rapid death of coliforms.
7. *Parasitism*; a small organism consuming a larger one by means of secreted lytic enzymes or by actual contact between the two organisms. *Bdellovibrio*, which has been discovered to exist in sewage, is such an example.
8. *Predation*; an interaction limited to organisms which can engulf particles including bacteria. Protozoa and metazoa are examples of this kind.
9. *Viruses*; obligate intracellular parasites able to multiply only within the host cells. They cannot be cultivated in arti-

ficial culture media and pass through filters which retain bacteria. They are infective agents of plants and animal tissues and cells; those which attack bacterial cells are known as bacteriophages. These infective agents are not living organisms since they can be crystallized and are a molecular complex. They vary in size from 10 to 300 $\mu\mu$. The host range may be wide or may be limited to single species. The phage is adsorbed on the surface of a bacterial host cell. It penetrates it and multiplies within the cell, finally resulting in the lysis of the cell.

Viruses and phages are excreted from the intestinal tract of man. It is well established that infectious hepatitis is caused by a virus transmitted by drinking contaminated water or by ingestion of raw shellfish grown in polluted waters. This virus has not been isolated yet, but other enteric viruses such as polio, coxsackie, and echo viruses have been cultivated. The latter have not been proven to be transmitted by water. Viruses withstand inactivation by complete water treatment processes including the usual chlorination practices.

Bacteriophages are known to be present in sewage derived from intestinal excreta of human beings. They were first discovered by D'Herelle who assigned to them an important role in the destruction of cholera bacteria in Indian rivers. Others since then have attributed to them a function in bacterial self-purification.

Bacteriophage activity can be readily demonstrated by adding a bacteria-free filtrate into a bacterial culture which results in the destruction of susceptible organisms. However, lytic action of bacteriophages occurs only in a culture of multiplying bacteria, a condition that does not generally exist in polluted streams for coliform or other enteric bacteria. Bacteriophages further are readily adsorbed on inert particles which prevents their lytic activity. It is therefore not likely that bacteriophages play an important role in the reduction of bacteria in polluted waters. Their presence in sewage has been definitely established but their role under natural conditions is questionable.

Applications of the Growth Phases to Polluted Waters and Their Treatment. The numbers of bacteria in stored sewage samples undergo an initial increase followed by a decrease. Similarly, samples of well or surface waters stored in bottles at the same temperature as in their original environment manifest increases in total plate counts as well as exhibiting coliform bacteria counts several hundredfold over the original values in a few days. This phenomenon necessitates refrigeration and determination of numbers within a few hours after sampling. This increase in stored samples is due to the effect of change in the surface area to volume relationship from the original environment. Explanation of the phenomenon lies in the concentration of bacteria and food on the surfaces of the container which thus enhances the growth of the organisms on a substrate in which food supply is a limiting factor for growth.

By contrast, intestinal bacteria discharged into a stream generally undergo a decrease in number. A small initial increase below a point of discharge may or may not be due to an actual multiplication. The death rate of bacteria in streams is generally logarithmic. It has to be postulated that bacteria involved in the decomposition of organic matter in the stream must be increasing in numbers. The concentration of organic matter in streams is considered high when it exceeds a BOD of about 10 mg/l, which is a far cry from culture media with 5000 mg/l of readily-available organic matter such as in lactose or nutrient broth. The death rate of intestinal organisms should, therefore, be attributed to their inability to grow in the new environment into which they are discharged because of the contrast in the composition and concentration of food supply between the intestinal canal and the aquatic environment. The death rate in streams has been observed to be higher in the summer than in the winter, which is what would be expected under conditions of limited food supply, resulting in accelerated deterioration of the cell by endogenous respiration.

The natural bacterial self-purification process brings about substantial reduction of coliform bacteria in fresh waters, amounting to about 90 percent in 2 days with only 1 percent remaining after 4-5 days passage time.[2] The death rate of these bacteria is even greater in seawater.[3] Studies have shown a decrease of 90 percent in 2.5 hr.[4]

The death rate of intestinal pathogenic bacteria is even greater than that of coliform bacteria because of their more rigid food and environmental requirements. Therefore, a safety factor exists from the standpoint of public health in that by the time coliform numbers have been greatly reduced either by treatment or by natural self-purification, the pathogenic bacteria which may have been present could be eliminated.

Effect of Temperature on Bacterial Growths. The phases of growth are influenced to a great extent by temperature. There are three temperature limits; e.g., a minimum, optimum, and maximum. The position of these points differs greatly for different species of bacteria. The minimum of some bacteria may be above the maximum for others. Bacteria are grouped, therefore, according to temperature as follows:

1. Cryophilic or psychrophilic, with a temperature optimum below 20°C.
2. Mesophilic, which have an optimum between 20 and 45°C.
3. Thermophilic, whose optimum temperature is above 45°C.

Cryophiles are capable of growing at temperatures as low as 0°C but cannot develop at 30°C. Coliform bacteria as well as enteric pathogens have their optimum at 37°C. Coliforms do not grow at temperatures of 10°C even under otherwise favorable conditions. Some thermophiles do not grow at 30°C, but may develop at temperatures as high as 70 to 80°C.

Bacteria also manifest wide variation in resistance to extreme temperatures. Spores are more resistant to heat than vegetative forms. Some spores may not be completely killed when exposed to 100°C for 1 hr or more. Vegetative cells of most bacteria are killed by exposure to moist heat of 55 to 58°C.

The general van't Hoff or Arrhenius law of the doubling of the rate of chemical reactions with every 10°C increase in temperature applies as well to bacteria within the temperature range of growth.

In polluted waters, the death rate of intestinal bacteria is affected by the temperature of the water. Biological treatment processes are subject to seasonal changes of temperature in the wastewater. Only in anaerobic digestion has it been feasible to artificially control the temperature of the tank. The optimum temperature for digestion is at 35°C. Digestion in the thermophilic range has not found general application.

Effect of pH Value. The pH value of a bacterial cell is believed to be approximately neutral. Nevertheless, bacteria grow well within a range of pH values varying from 6.0 to 9.0, a thousandfold difference in hydrogen ion concentration. The cell membrane apparently is not highly permeable to hydrogen and hydroxyl ions. The growth and death rates are, nevertheless, greatly affected by pH values.

Some bacteria, such as the methane-producing ones, have a very narrow optimum pH value at above 7.0; at a pH value of 6.0 they become completely inactive. Others have a wide range of optimum pH values ranging from 6.5 to 8.5. The minimum pH value for many bacteria is at about a pH value of 4.0 to 4.5. The case of *Thiobacillus thiooxidans* referred to earlier which grows well at a pH value of 1.0 is a biological phenomenon.

Low or high pH values may be present in the original substrate or may be produced as a result of biochemical reactions. It has been generally observed that, within limits, bacterial growth in an acid medium tends to result in an increase in pH value, and under alkaline conditions the pH value tends to move towards the neutral point.

The biological treatment of domestic sewage does not offer any problems with pH control. Anaerobic digestion of sludge, however, develops acid conditions which may be caused by overloading, failure of temperature control, or toxic industrial wastes. Associated symptoms of this condition are a change in gas composition and a decrease in gas production. Lime addition generally helps correct the symptoms, especially if begun soon after the pH value shows signs of decreasing. Industrial wastes which are outside the proper range of pH value will need to have the pH value corrected.

NUMBERS AND TYPES OF BACTERIA IN SEWAGE

The numbers of bacteria in sewage vary considerably from place to place and from time to time. Raw sewage contains approximately 1 million bacteria/ml. Coliform bacteria are present in great numbers, with the daily per capita contribution of coliforms estimated to be about 300 billion. The calculated number with a sewage flow of 100 gpd/person would be 750,000/ml, which closely corresponds to the actual determined numbers.

The origin of bacteria in sewage is human and animal excreta, soil, and air. The types of bacteria that have been detected are: *Achromabacteria, Flavobacteria, Pseudomonas, Zooglea ramigera, Sphaerotilus, Vibrio, Alcaligenes, Beggiatoa, Nitrosomonas, Nitrobacter*, and *Sporovibrio desulfuricans*.

The coliform group is comprised of two important species, *Escherichia coli* and *Aerobacteria aerogenes*, and can be useful in detecting sewage pollution. The presence of coliform can be detected in as low as 1 bacterium/100 ml, which is equal to the detection of about 1 part of sewage in 100 million parts of water. They are, therefore, very sensitive indicators of bacterial contamination. Coliform organisms, however, serve only as presumptive evidence of sewage pollution since coliforms, especially the *Aerobacter*, are widespread in nature and are not restricted to the intestines of warm-blooded animals. The evidence of contamination with sewage is increased by the determination of fecal coliforms. At the present time, however, it is not possible to differentiate between human and animal fecal coliforms.

The U.S. Public Health Service (USPHS) standard[5] for the bacteriological quality of drinking water for interstate carriers, which has been accepted by all state regulatory agencies, specifies that coliform levels are not to exceed 1/100 ml of water. Such a coliform level is readily attainable and economically feasible with the present technology of water treatment. The use of coliforms rather than fecal coliforms is justified for drinking water since it provides an additional factor of safety. Numbers higher than this standard suggest either inadequate treatment, recontamination in the distribution system, or aftergrowths. There is a question, however, whether the present bacterial standard for drinking water affords protection against enteroviruses. Gastroenteritis cases of unknown epidemiological origin may be associated with viruses in water meeting the present bacteriological standards.

It has been established that a raw water supply containing coliforms not in excess of 5000/100 ml can, with modern water treatment processes, produce a potable water meeting the bacteriological standard.[6]

For outdoor bathing beaches, there are no generally accepted coliform standards. The recommended levels vary from 50 to 3000/100 ml in different states. There is a current trend to use fecal coliforms rather than total coliforms as the indicator organism at a level of 200/100 ml for swimming.[4] There is little epidemiological evidence that swimming in waters containing coliform or fecal coliform levels higher than those prescribed have caused enteric disease among the bathers. The common symptoms of disease caused by swimming in polluted waters are eye, ear, throat, and nose infections. In extreme cases of pollution with raw sewage, aesthetic considerations should be governing.

USPHS requirements for harvesting shellfish set a coliform level not exceeding 70/100 ml.

(U.S. drinking water standards and water quality criteria according to use are in revision.)

Other bacteria such as *Enterobacteria* and *Streptococcus* are also being used as indicator organisms. The detection of these organisms is significant from the public health standpoint in regard to the presence of waterborne pathogenic organisms which are also inhabitants of the intestinal track. The bacterium causing typhoid fever is called *Salmonella typhosa*, and the closely-related disease of paratyphoid fever is caused by other species of *Salmonella*. Bacillary dysentery is caused by *Shigilla*. Cholera is caused by the *Vibrio comma*. The nonbacterial waterborne diseases are amoebic dysentery caused by a protozoan known as *Endamoeba histolytica*, and infectious hepatitis caused by a virus.

Coliform and the other pathogenic and nonpathogenic intestinal bacteria enumerated above do not generally multiply after leaving their host organisms. The food and temperature conditions in the aquatic environment are not favorable for their survival. They therefore do not play an important role in the stabilization of organic matter either in polluted streams or in artificial biological treatment processes of wastewater.

Biological sewage treatment processes decrease the numbers of bacteria by 80-90 percent.

SEEDING

It is a remarkable fact that no biological sewage treatment process requires artificial inoculation. It follows that all the various bacteria necessary for the biochemical reactions under aerobic and anaerobic conditions are present in sewage. Since coliforms do not play an important role in this respect, then other types of bacteria must find access to the sewer in addition to those contributed by sewage. Bacteria capable of bringing about the diverse reactions culminating in the stabilization of the organic matter–nitrification, denitrification, methane fermentation, sulfate reduction, production of sulfuric acid from hydrogen sulfide, and a myriad of other reactions–must be present in the sewage since these processes are known to take place.

The establishment of these processes must be the result of the selection of only those types of bacteria that find the particular environment conducive for their growth. It is important, therefore, to create and maintain the optimum environmental condition for each treatment process. It then becomes a matter of time for the appropriate population to build up while unnecessary species are eliminated.

In addition to the mechanism for selecting out of a heterogenous population one suitable for a particular environment and substrate, there is the adaptation phenomenon. Bacterial cultures are known to adapt themselves to new and adverse environments. Cultures become drug-resistant when grown in the presence of drugs, or can be trained to use new sources of carbon or nitrogen. Such adaptations can be attributed to mutant forms present in the culture which are capable of growing under the particular conditions. It results in the selection of the few mutant cells, their multiplication, and the elimination of the rest of the cells which cannot develop. In addition, there is the mechanism of adaptation through the production of inducible enzymes discussed earlier. This latter type of adaptation can only be transient. The acquired characteristic will be lost when the chemical that caused the induction disappears from the environment.

There are a number of examples in water pollution literature where adaptation to a toxic substance has been demonstrated by using the techniques of repeated passages of cultures over increasing concentrations of the inhibitory substance. By this procedure, it was possible, for instance, to increase the oxidation of formaldehyde from an initial concentration of 45 mg/l to 1750 mg/l, while a higher initial addition of 175 mg/l did not utilize any oxygen during the 7-day incubation period.[7] When dealing with a heterogeneous culture at the start (sewage), it is not possible to determine which of the three mechanisms were responsible for the adaptation noted.

"Acclimatization" and "adaptation," as used in the water pollution field, are general terms indicating the development of an efficient flora from a mixed population suitable for a particular environmental condition and waste. Certain species in the mixed population find the conditions and food best suited for their development and with continuous or repeated feeding they become the predominant forms. Initially, with a limited number of these organisms in the inoculum, an extended development period may be encountered which, upon repeated passages of the seed, becomes shorter and shorter.

It is important to bear in mind that such a process of adaptation or acclimatization to inhibitory substances has definite limits.

Microorganisms exist in nature which are capable of degrading all naturally-occurring organic molecules. Certain synthetic organic substances produced by man such as fibers, plastics, some pesticides, and detergents are not biodegradable.

The dominant role played by bacteria in the decomposition of organic matter can be attributed to their ubiquitous distribution, their rapid rate of multiplication, and their great adaptability to changing environment. With a constant mutation rate and a high multiplication rate, the chances of a successful mutation occurring are greater.

Certain bacteria are capable of extending the range of their activity or use alternate pathways. Some have a restricted metabolic activity, such as the nitrifying and sulfur-oxidizing bacteria, while others, like the Pseudomonads, can utilize some 200 different organic compounds.

When an activity of a bacteria makes the environment less favorable for itself, another type generally exists for which the altered environment is not as unfavorable, and the latter replaces and succeeds the former type.

ANTISEPTICS AND DISINFECTANTS

Substances which kill or inhibit bacterial activity are classified as follows:

1. An antiseptic is a substance which inhibits the growth of microorganisms.
2. A bactericide is a substance which kills bacteria.
3. A bacteriostatic agent is a substance which inhibits the growth of bacteria.
4. A disinfectant is a germicide which kills pathogenic organisms. It does not bring about sterilization with the destruction of all living forms unless it is used in very high concentrations. In fact, the differences between these terms is primarily a function of concentration.

The factors which influence disinfection are:

1. The concentration and type of the disinfectant.
2. Higher temperature which increases the effectiveness of the disinfectant.
3. The nature of the microorganisms.
4. The time of contact.
5. The pH value of the substrate.

The disinfectant most commonly used in water pollution control for the destruction of pathogenic microorganisms is chlorine. When gaseous chlorine is added to water or wastewater it forms hypochlorous acid as follows:

$$Cl_2 + H_2O \longrightarrow HOCl + HCl \qquad (10)$$

The hypochlorous acid dissociates to hydrogen and hypochlorite ions. The concentration of hypochlorite increases at higher pH values. Of the two species of chlorine, the undissociated hypochlorous acid is the more active disinfectant.

Chlorine combines with a number of other substances in water or wastewater. It reacts with ammonia to form mono- and dichloramines. The disinfecting power of chloramines is much lower than that of hypochlorous acid. Chlorine also combines with organic compounds forming products with very low disinfecting capabilities. It further reacts with reducing substances such as ferrous iron and hydrogen sulfide to form products which have no disinfecting power. All these reactions take place more or less simultaneously, depending on the nature and concentration of the substances in the substrate, thus competing with and reducing the bactericidal activity.

Unless very high dosages of chlorine are added to satisfy all these interfering reactions, the chlorine in wastewater is all in the form of chloramines and not as free residual chlorine. Since residual chlorine is used to control the disinfection, it is imperative that adequate contact time be provided to bring about satisfactory disinfection. This contact time is generally 20 to 30 min.

There is an inverse relationship between the residual chlorine and the contact time necessary for a given degree of disinfection. Residual chlorine concentrations varying from 0.2 to 1.0

mg/l are generally used. It is possible to obtain 99.90 to 99.99 percent destruction of coliform bacteria under these conditions.

Actual effectiveness of disinfection should be determined by the numbers of coliform bacteria left in the effluent. Regulations differ in this respect. Some regulatory agencies require a coliform bacteria concentration as low as 1/ml, or about a millionfold reduction of the numbers in raw sewage. In many instances, receiving waters contain coliform bacteria in excess of this number. This is not necessarily attributable to pollution by sewage and may more likely be contributed by surface runoff.

It seems that with effective chlorination, the chances of any pathogenic intestinal bacteria surviving is remote. However,the inactivation of viruses and the destruction of the cysts of *Endamoeba histolytica* cannot be considered assured unless breakpoint chlorination is practiced where the active agent is free residual chlorine.

DENITRIFICATION

Reference has been made earlier to the phenomenon of nitrate serving as the hydrogen acceptor in the absence of molecular oxygen, resulting in the liberation of nitrogen gas. The process is referred to as denitrification or dissimilatory nitrate reduction. It should be differentiated from the reduction of nitrate when it serves as a source of nitrogen for the synthesis of protein.

Denitrification can be brought about by a number of different bacteria such as *Pseudomonas*, *Spirillum*, and *Bacillus*. The organisms are facultative and they can use either molecular oxygen or the oxygen of the nitrate, the latter being reduced only in the absence of dissolved oxygen. This reaction takes place only in the presence of oxidizable organic matter.

Denitrification has a number of applications in water pollution control. Since it requires the presence of nitrates, and domestic raw sewage does not normally contain any nitrates, denitrification can take place only if it is preceded by nitrification or when nitrates are added. Since sulfate reduction does not take place in the presence of nitrates, nitrates have been used to prevent the production of foul odors in situations where the dissolved oxygen supply has been exhausted. Canneries treating wastes in lagoons have resorted in the past to the addition of nitrates to supplement the limited supply of oxygen derived from the atmosphere. The development of mechanical surface aerators has reduced the use of nitrates except in emergency conditions for odor control.

Denitrification has been implicated in the blanket rising in the activated sludge process. This should be differentiated from "bulking" because it constitutes the rising of sludge after it has settled in the secondary clarifier. It has been associated with partial nitrification in the aeration tank. The other condition which is necessary for the phenomenon is the incomplete stabilization of organic matter in the sludge. This organic matter becomes dehydrogenated (oxidized) in the bottom of the settling tank at the expense of nitrate oxygen after the dissolved oxygen, which might be present in the interstitial water, becomes rapidly exhausted. The resulting production of nitrogen gas buoys up the sludge. The blanket sludge rising does not take place under overloaded conditions which preclude the possibility of nitrate production. On the other extreme, the phenomenon does not occur in underloaded systems because of the complete stabilization of the organic matter. The other factor that influences the rising of the sludge is the length of time the sludge remains in the bottom of the clarifier.

Denitrification is also an important method of reducing the nitrogen content of the effluent. It presupposes prior conversion of ammonia to nitrates in biological treatment by activated sludge or the trickling filter processes. To induce the reduction of nitrates to nitrogen gas, the system has to be under anaerobic conditions and have a source of available organic substrate to serve as hydrogen donor. The effluent from the preceding oxidation process cannot serve this purpose. Thus, methyl alcohol or some other low-cost source of organic matter is added. Alternatively, organic matter in sewage could serve the same purpose.

A three-stage system of treatment is preferable to treatment in one or even two stages, because it can be designed according to the optimum requirements of each stage. The first stage is designed for the removal of carbonaceous matter by aeration. The settled effluent is aerated in the second tank. The effluent from this stage is treated under anaerobic conditions. Each stage is provided with its own clarifier and the sludge is recycled within each stage (see Fig. 4-4).

SULFATE REDUCTION

In the absence of dissolved oxygen and nitrates, sulfates serve as the hydrogen acceptor which is reduced to hydrogen sulfide. Factors influencing the reduction of sulfate are:

1. Anaerobic conditions.
2. The presence of specific sulfate-splitting bacteria. There are only a few types of bacteria capable of utilizing sulfates as a hydrogen acceptor. *Sporovibrio desulfuricans* is recognized as one of the bacteria capable of bringing about the reaction. It is present in fresh sewage in limited numbers.
3. A very low redox potential. Anaerobic reducing conditions way below those required for denitrification are necessary.
4. Organic matter as a source of hydrogen donor. Not much is known as to the nature of organic compounds that are utilized by these bacteria, although lactate is used for growing them in artificial culture media. Undoubtedly, a number of other organic substances and organic acids present in sewage or produced under anaerobic conditions can be utilized.
5. Optimum temperature. The sulfate reduction takes place both in mesophilic and thermophilic temperatures. The optimum is between 30–35°C within the mesophilic range, and 50–55°C in the thermophilic range.
6. A pH range of 6.0 to 9.0 with an optimum at around the neutral point.
7. The presence of sulfate.

Applications of Sulfate Reduction. The reduction of sulfate is of importance from the standpoint of odor nuisances and it is also related to corrosion of concrete sewers.

The essential requirements for the production of hydrogen sulfide exist in sewers with the necessary organic matter contained in the sewage itself. Hydrogen sulfide production is more pronounced in stronger sewage with a BOD in excess of 80 mg/l. It appears that below this value surface reaeration in gravity sewers provides dissolved oxygen to oxidize the hydrogen sulfide. In force mains, the reduction can take place at an even lower BOD value.

Sulfates are always present in domestic sewage and are derived from the water supply. They are supplemented with sulfate in urine. Their content varies with the hardness of the water supply and the degree of dilution of sewage. In a soft water area, sewage contains about 15–20 mg/l of sulfate sulfur. Such a quantity is more than enough to create odor nuisances if the other conditions are favorable for reduction.

Industrial wastes and seawater infiltration can increase the sulfate content of sewage. Odor production is pronounced at higher temperatures during the summer and during all seasons in southern latitudes.

Sewage becomes septic with longer travel time which is a function of the length of the sewer and of the gradient affecting the velocity of flow.

Except in unusual cases, the seat of sulfide production is primarily in the bottom deposits and slimes attached to the walls of sewers. It appears that the sulfate-reducing organisms find the optimum conditions of organic matter and redox potential in these environments. Sulfates in the sewage flowing past such locations are reduced and the sulfide produced passes into the flow. The emission of sulfide to the sewer atmosphere depends primarily upon turbulence in such areas as drop manholes and meter chambers. Once liberated into the sewer atmosphere, hydrogen sulfide can escape through manhole covers and become a source of odor nuisance in the neighborhood.

It also becomes responsible for corrosion in gravity sewers. The hydrogen sulfide is absorbed by the moisture on the exposed surfaces of the sewer. It undergoes a series of chemical transformations, ending in the oxidation of the sulfur to sulfuric acid by the autotrophic *Thiobacillus thiooxidans*. The sulfuric acid reacts with lime in the concrete turning it into a soft putty-like mass. The pH value of the corrosion products can go as low as 1.0. In the course of time, if this reaction is allowed to go unchecked, the corrosion penetrates to the aggregates and results in the failure of the sewer.

Measures to control the reduction of sulfate to hydrogen sulfide fall either into preventive or corrective categories. Not much can be done from the standpoint of controlling the temperature of the sewage except by eliminating the discharge of high-temperature industrial wastes. However, deposition and long passage times can be prevented by designing sewers with sufficient grade to provide high-flow velocity. Also, prevention of turbulence in manholes will reduce the rate of evolution of hydrogen sulfide into the sewer atmosphere.

One of the effective and practical methods of controlling sulfate reduction is a program of sewer cleaning. It has been found that scouring velocities during high flows diminishes sulfide production for a period of time.

Chlorine, when added as a preventive measure, is effective in delaying septic action and reducing biological activity. Use of chlorine as a remedial measure applied intermittently in high dosages destroys the sulfide present. Either method of control by chlorination is expensive for high flows. The remedial method requires high dosages to react not only with the hydrogen sulfide present, but also with all the reducing substances associated with septic sewage. Forced aeration of the sewage in the summer is not very practical because of shallow depths and low oxygen transfer rates. Forced-draft sewer ventilation is highly effective in preventing corrosion by drying up the exposed surfaces and sweeping out the hydrogen sulfide from the sewer atmosphere.

Adequate measures for the control of odors emitted into the atmosphere should be taken by discharging the exhaust air from stacks or scrubbing the gases before discharge. Natural ventilation through perforated manhole covers is not effective or desirable. The addition of sodium nitrate would effectively control hydrogen sulfide production, but at a high cost. Discharge of gaseous ammonia into the sewer atmosphere can neutralize the acidity in the sewer walls. Use of corrosion-resistant materials and protective coatings would naturally increase the life of sewers. The use of clay liners is not entirely satisfactory because of the failure of the jointing material. Better paints and resinous coatings are needed to adequately solve this problem.

METHANE FERMENTATION

Several pure cultures of methane bacteria have been isolated. The essential physiological characteristics of these organisms are:

1. They are obligate anaerobes and are highly sensitive to oxygen.
2. They require carbon dioxide as a hydrogen acceptor.
3. Their development is very slow.
4. They are sensitive to the pH value.
5. They can utilize only the lower volatile acids and their alcohols. The substrates known to be utilized are: formic, acetic, butyric, propionic, valeric, and caproic acids and their corresponding alcohols. (The substrates are not only limited to these acids and alcohols, but each methane bacteria species utilizes a specific substrate.)
6. They have a low redox potential.

The formic acid fermenting bacterium utilizes hydrogen and carbon dioxide in the production of methane. Fermentation of ethyl alcohol proceeds with the hydrogenation of carbon dioxide and the production of methane and acetic acid. Similarly, butyl alcohol is dehydrogenated with the reduction of carbon dioxide, producing butyric acid and methane. Butyric acid, in turn, is further converted by similar dehydrogenation and hydrogenation reactions with the utilization of carbon dioxide and production of acetic acid and methane. Volatile acids and alcohols with 4 and 5 carbon atoms are also fermented by the same mechanism. The general reaction may be presented as follows:

$$4H_2A + CO_2 \longrightarrow 4A + CH_4 + 2H_2O \quad (11)$$

where H_2A represents the fermentable organic compound and A the dehydrogenated product.

The acetic acid into which all these fermentations are funneled is fermented by two different mechanisms. First:

$$CH_3COOH + 2H_2O + CO_2 \longrightarrow 2CO_2 + CH_4 + 2H_2O \quad (12)$$

The reduction of 1 mole of carbon dioxide is accompanied with the production of 2 moles of carbon dioxide from the acetic acid.

In the second reaction mechanism in the fermentation of acetic acid, carbon dioxide does not serve as hydrogen acceptor and fermentation proceeds by intramolecular dehydrogenation and hydrogenation. Thus:

$$CH_3COOH \longrightarrow CH_4 + CO_2 \quad (13)$$

The methane is derived from the methyl group and the carbon dioxide from the carboxyl group.

Applications to Sludge Digestion. The essential requirements for growth of methane bacteria control the sludge digestion process. There is no problem of excluding oxygen from digestion tanks, and low redox potentials can readily be reached as a result of digestion of the organic matter in sewage solids. Neither are carbon dioxide requirements likely to be a limiting factor because digestion produces carbon dioxide, part of which reacts with the ammonia produced to form bicarbonates.

The critical factor in digestion is the maintenance of the pH value. Methane bacteria are extremely sensitive to slight pH variations on either side of neutrality.

The complex organic materials in sewage sludge consisting of carbohydrates, fats, and proteins are not directly available to methane bacteria. Before they can be utilized, these compounds have to be first transformed to lower volatile acids and alcohols by a process commonly referred to as liquefaction. The accumulation of these intermediate acid products inhibits the activity of the methane bacteria. High concentration of ammonia produced from the digestion of nitrogenous organic matter, at times when the solids content of the sludge is high, similarly inhibits the methane bacteria due to the presence of undissociated ammonia molecules at high pH values.

Bacteria responsible for the liquefaction are not sensitive to the lower pH values. In addition, the rate of development of methane bacteria is slower than that of the liquefying bacteria. The combined effect of these factors makes it difficult to initiate methane fermentation in sewage sludge. It requires the continuous adjustment of the pH value with lime. In the course of time, methane bacteria develop slowly and destroy the accumulated acids. Also, the digestion of nitrogenous matter produces ammonia. As a result of these two factors, a gradual shift of the pH value toward neutrality takes place, resulting in an acceleration of methane fermentation.

Once the optimum pH value is obtained and a large population of methane bacteria becomes established (sludge is ripened), an equilibrium is maintained between the rate of acid production and subsequent conversion of the acids to methane and carbon dioxide. Digestion proceeds satisfactorily as long as the environmental conditions are favorable to the methane bacteria. Ripe sludge serves both as a large pool of methane bacteria and as an ideal buffering medium.

It is essential that nothing be permitted to slow down the rate of fermentation of the acids by methane bacteria since it constitutes the more sensitive and slower reaction. The maintenance of the optimum temperature is one of the requisites in this respect. The optimum temperature in the mesophilic range is 95°F (35°C). Apparently there are thermophilic methane bacteria because sludge digestion can also take place at 130°F (55°C).

NUISANCE ORGANISMS

Nuisances are created in the aquatic environment by excessive masses of heterotrophic bacteria and phototrophic growths.

Growths of Photosynthetic Nuisance Organisms–Eutrophication. The term eutrophication signifies the enrichment of the environment with nutrients, producing excessive masses of aquatic growths. The greatest quantities of chemical elements found in biological growths (bacteria, microscopic animals, or algae) are carbon, nitrogen, and phosphorus.

The important difference between bacteria and algae is in the source of carbon. Heterotrophic bacteria, as pointed out in an earlier section, utilize organic matter both as a source of carbon and energy. Algae, on the other hand, require inorganic carbon in the form of carbon dioxide and bicarbonate, and utilize solar energy for photosynthesis. Certain algae, however, are known to grow in the dark and utilize specific soluble organic compounds. Many algae also have an absolute requirement for certain vitamins such as B_{12}. Minor elements in very small concentrations are needed which do not normally become growth-limiting factors in surface waters.

Hence, there are four important factors: light, carbon, nitrogen, and phosphorus. Of the three elements, carbon is in the highest percentage in biological growths, with nitrogen and phosphorus in descending order.

The important sources of nitrogen in surface waters are sewage effluents and surface runoff from fertilized agricultural lands, animal feed lots, and bottom deposits. The types of nitrogen compounds in surface waters are organic, ammonia, and nitrates. When the latter two forms of nitrogen are present, ammonia nitrogen is utilized first since amino acids and proteins are synthesized from it more readily than from nitrates. Organic matter from various sources in bottom deposits contributes additional amounts of nitrogen as a result of benthic activity. Another factor is the ability of certain algae (blue-green algae) to utilize elemental nitrogen from the atmosphere.

The decrease of the nitrogen content of surface waters by treatment of sewage is more feasible than by controlling the runoff or contribution from the bottom deposits. Some of the nitrogen present in the sewage is removed by conventional treatment methods, by sedimentation, and by biological methods which remove roughly one-third of the nitrogen.

Conversion of ammonia to nitrate does not offer a solution to the eutrophication problem. Nitrification followed by denitrification, however, results in the conversion to nitrogen gas. Nitrogen from sewage can also be removed by the growth of algae in oxidation ponds, but discharge of such an effluent without prior removal of algae by flocculation and filtration is objectionable. The growth of algae being subject to climatic conditions and changes in diurnal solar radiation, oxidation ponds do not offer a positive method of control of nitrogen in the effluent. Volatilization of ammonia at high pH values is feasible but subject to some technical problems.

The primary sources of phosphorus are also sewage effluents, drainage from agricultural lands, and bottom deposits. The sources of phosphorus in sewage are human excreta and detergents in the form of phosphates. Phosphorus can be both organic and inorganic, either in solution or suspension. Both forms of phosphorus, in either state of subdivision, are interconvertible.

Biological waste treatment methods remove some phosphorus. More substantial removals by the activated sludge process have been reported in some cases. More positive and effective removal is obtained by lime treatment or precipitation with iron or aluminum salts.

The above discussion has assumed that nitrogen and/or phosphorus are the only nutrients involved in eutrophication, although the role played by carbon should also be taken into consideration. Carbon, either as carbon dioxide or as bicarbonate, can be augmented by the decomposition of organic matter.

Answers to the following questions in regard to eutrophication have not been fully established:

1. Which of the nutrient elements should be controlled or should be removed from sewage effluents?
2. What is the limiting concentration of either nitrogen or phosphorus?
3. Will such removals result in effective control of eutrophication?

It appears that phosphorus plays a major role in eutrophication and is the key element. Its removal is more effectively and readily accomplished than the removal of nitrogen, although nitrification followed by denitrification has additional side benefits from the standpoint of removing the oxygen demand of the ammonia in the receiving water. An additional reason favoring the removal of phosphorus is that some algae can fix atmospheric nitrogen, making it available to the organisms.

The limiting phosphorus level for algal growth has been stated to be about 0.1 mg/l. Whether effective removal of phosphorus from sewage effluents can control eutrophication is not certain because of the other contributing sources. It should be admitted, however, that the reduction of the concentration of phosphorus should reduce the intensity of algal blooms.

Excessive algal blooms are objectionable for the following reasons:

1. They deteriorate the aesthetic quality of the water.
2. They decrease the oxygen content of water.
3. They produce taste and odors in drinking water.
4. They accelerate the aging of lakes by the deposition of organic matter causing choking of the lake.
5. They interfere with fish production and harvesting.

The production of oxygen by algae ceases during the night and is reduced under cloudy conditions. Living algae utilize oxygen for their respiration throughout the day and night, so the depletion of oxygen due to respiration can become very pronounced during the night. During the day, with bright sunshine, the dissolved oxygen can become supersaturated, part of it escaping to the atmosphere. The pH value can go up as high as 9.0 or above. At night, the dissolved oxygen decreases to a point where the lack of it can cause fish kills. Aside from the diurnal effects of living algae, there is the oxygen demand of the algae after their death and decomposition. They settle to the bottom, increase the accumulation of the organic matter, and cover up the spawning grounds of fish. All these effects are more pronounced in lakes and reservoirs than in streams.

Algae are either free-floating (planktonic) or attached to a surface. If attached to an aquatic weed, they are referred to as epiphytic growths; attached to the bottom, they are called benthic. They are either unicellular or multicellular and filamentous.

In addition to the algae, there are the aquatic weeds which contribute heavy growths in shallow waters near margins of the lake or stream. They may be rooted, submerged, emergent, or free-floating.

A moderate level of nutrients in water gives rise to biota and constitutes the base of a food chain which ends with fish food in the form of adult and larval stages of insects, worms, etc. Workers in Oregon[8] have demonstrated that the addition of small amounts of sucrose to an experimental section of a stream increased the *Sphaerotilus* growth which, in turn, served as food to chironomous larvae and increased the trout population tenfold as compared with a control section of the stream.

Growth of Heterotropic Nuisance Organisms. Growths of heterotrophic organisms in organically enriched waters have been referred to as "sewage fungus." This is a misnomer because the dominant organism is only occasionally a fungus. It is a microbial community consisting mostly of *Sphaerotilus* and *Zoogleal* bacteria. *Leptomitus* and *Geotrichum* fungi and the sulfur bacterium, *Beggiatoa*, are found less frequently and in smaller numbers. Flagellates and ciliates are frequently found in this community, but do not contribute materially to the bulk of the growth. *Carchesium* is found occasionally and could add, to some extent, to the bulk. Such slime infestations grow attached to surfaces in a flowing stream and produce heavy, slimy, cottony growths varying in color from white to grey and brown.

Sphaerotilus require an available source of carbon such as carbohydrates (mono- and disaccharides and pentoses), alcohols, and organic acids. The nitrogen source could be amino acids, ammonia, and nitrate. Vitamin B_{12} is also required. The organism grows attached in flowing waters and extracts from it the nutrients required for its growth at low concentrations. Velocity of flow is an important factor because excessively high rates of flow prevent attachment and can cause breaking of the growths. Below such a velocity, however, higher flows furnish greater amounts of food independent of the concentration. *Sphaerotilus* growth has not been reported in lakes and reservoirs. The organism can grow in a wide range of dissolved oxygen and temperature (10–40°C). The pH value is variously reported to range from 6.8 to 9.0.

One objection to the massive growth of *Sphaerotilus* is that it makes the stream unsightly. In addition, growths, either in their attached or floating stages, lower the dissolved oxygen level. After settling, they form organic deposits and their decomposition exerts additional oxygen demand. It is believed by some that the sludge deposit can smother the spawning grounds of fish. Commercial and sport fishermen complain that their nets or lines become fouled by these growths, and that cleaning is difficult and net life is reduced.

The reliable method of controlling such growths is effective and more or less complete wastewater treatment by biological methods to remove the organic matter which creates extensive growths. In addition to sanitary sewage, many organic industrial wastes such as dairy and fermentation wastes, spent pulping liquors, etc. are known to cause such infestations.

Sphaerotilus is also associated with bulking of activated sludge, resulting in a slow rate of sedimentation. The actual cause of bulking, however, is not well established. This occurrence is dealt with more extensively in Chapter 12.

REFERENCES

1. Stanier, R. Y., Douderoff, M., and Adelberg, E. A., *The Microbial World*, Prentice-Hall, Inc., Englewood Cliffs, N.J. (1970).
2. Phelps, E. B., *Stream Sanitation*, John Wiley and Sons, Inc., New York (1944).
3. Carter, H. H., Carpenter, J. H., and Whaley, R. C., "The Bactericidal Effect of Seawater Under Natural Conditions," *Journal WPCF* **39**, 1184 (1967).
4. Federal Water Pollution Control Administration, "Report of the Committee on Water Quality Criteria," FWPCA, U.S. Dept. of the Interior (1968).
5. U.S. Public Health Service Drinking Water Standards, U.S. Dept. of Health, Education, and Welfare, USPHS Pub. 956 (1962).
6. Streeter, H. W., "Standards of Water Quality," *Journal AWWA* **31**, 1479 (1939).
7. Gellman, I., and Heukelekian H., "Biological Oxidation of Formaldehyde," *Sewage and Industrial Wastes* **22**, 10, 1321 (1950).
8. Warren, C. E., Wales, J. H., Davis, G. E., and Douderoff, M., "Trout Reduction in an Experimental Stream Enriched with Sucrose," *Journal of Wildlife Management* **28**, 617 (1964).

5

Toward a Conceptual System for Water Pollution Biology

Charles E. Warren
Department of Fisheries and Wildlife
Oregon State University
Corvallis, Oregon

THE OBJECTIVES, POSSIBILITIES, AND CONSTRAINTS OF WATER POLLUTION BIOLOGY

The Nature of the Problem. Charles Darwin once wrote: "Looking back, I think it was more difficult to see what the problems were than to solve them." Darwin, of course, was referring to problems primarily biological in nature, his brilliant, though partial, solution of which was the greatest single advance in biological thought, indeed one of the greatest advances in man's total understanding. Darwin achieved his solution through an impressive display of problem analysis and statement, careful observation, development of conceptual systems, and logic—the primary means of science.

Now the problem to be considered here, water pollution, is not simply scientific in the usual sense. It has its scientific aspects, but in addition it has environmental, natural resource, technological, economic, and sociological facets, and its solution requires management of the affairs of men. This is not to say that these facets are without scientific problems amenable to the scientific approach. It is only to say that as individual and interacting phenomena they go beyond the realm in which we can expect only scientific solutions. Men, as they behaviorally interact with each other and their environment, present much too complex a problem. Still, in regard to problem analysis, even here science offers a sensible though constrained way to proceed. It is just that we have learned not to expect too much of science, in the ordinary sense, when confronted with total problems of this sort.

A historic view is usually helpful in analyzing the nature of a problem, especially in determining what its elements are by analysis of its development through time as these elements entered to complex the problem, to lose their independent identity, and finally to make their separate recognition difficult. Even if the elements are separately recognizable, they do not operate separately in the system that generates the problem. The system we now vaguely perceive as groups of men interacting with each other and with their physical, biological, and cultural environments began millions of years ago through biological evolution. It has continued its development primarily through cultural evolution, which includes the development of most of the activities and works of man. It is the very inertia of this cultural evolution, which appears to lack certain necessary controls, that has led to the water pollution problem and which some believe may lead to circumstances for which man is not biologically, sociologically, or psychologically suited.[1]

In any event, although man is biological in his very nature, some of his activities have taken on geological force, resulting in geological and climatological changes in his physical environment. From these and other activities, all-pervading sociological changes have come. In the complex of man and his environment, particular groups of men have certain needs and desires. The development of this complex has made the satisfaction of some of these needs and desires more possible, but it has simultaneously made the satisfaction of other needs and desires less possible. It is the generation of this conflict of interest that has led to problems like water pollution.

Before we can usefully conceptualize any part of the water pollution problem so as to effectively apply any field of endeavor (biology, in this case) to the solution of that part, we must conceptualize the total system generating the problem and recognize that some solutions of this one part of the problem may hinder more than help to solve the total problem, because they aggravate in unmanageable ways other parts of the problem. The system generating water pollution problems has broad environmental, natural resource, biological, technological, economic, and sociological dimensions. Any problems generated may be mainly in one or another of these dimensions, but it is always an outcome of the entire system, and its correction requires adjustment throughout the entire system. We have scientific, management, and educational resources and capabilities for dealing in some fashion with each of these dimensions and the system as a whole. Much of what we do is initially mobilized through political activity. But too often, if not usually, we emphasize one dimension at a time, thus deemphasizing the other dimensions, and the system is distorted in another way leading to other serious problems. And too often we place too much confidence and emphasis on only one of our capabilities—scientific, management, educational, or political. In either case, we can expect to be disappointed with the results. The water pollution problem, comprising its environmental, natural resource, technological, economic, and sociological parts, is generated by a complex system having these dimensions and will be solved only to the extent that we can appropriately adjust the system by balanced application of our human capabilities.

The Need for Goals and Objectives. Failure to make satisfactory progress in resolving a problem, even after the total problem and its parts as well as the system generating it have been adequately defined, can result from another sort of initial inadequacy of approach: failure to define suitable goals and objectives for the total endeavor to solve the problem as a whole as well as for the endeavors directed toward various parts of the problem. Too often, the generation and existence of activity are taken as indications that progress is being made toward the solution of a problem, whatever the orientation of the activity, indeed whether it has any orientation at all. Carefully stated goals and objectives provide means of appropriately directing and relating effort to the problem at hand and measuring progress toward its solution.

A goal is understood here to be an aspirational statement of purpose. It should be general and it should realistically relate our capabilities to the problem we intend to solve. It should not be so general as "to solve the water pollution problem." Rather, it should give some indication of what would constitute a satisfactory solution. A goal may have motivating potential and may act to integrate the relevant aspirations and activities of individuals and groups. This is needed in general problems such as water pollution that involve diverse interests. In order to be useful in solving a water pollution problem, a general goal must subsume goals relevant to each of the parts of the problem.

An objective can be taken as one possible translation of a goal into operational terms; it is less general, more definite, and more amenable to realization. Development of objectives that should lead toward attainment of previously stated goals constitutes the next step in a logical approach to problem solution. It should be remembered, however, that important elements of the problem and the goal are generally omitted when a goal is translated into an objective. Thus, there usually are complementary objectives and often there are alternative ones, all of which should be examined for possible inclusion.

In approaching the total problem and each of its parts, we should have a hierarchical system of goals and objectives. They should be relevant to the problem, they should be realistic, and they should be the most effective among the choices available. Goals and objectives constitute statements of the relations between our efforts and the problems we seek to solve, they serve to guide our efforts, and they provide a measure for assessing our progress.

The Need for Conceptual Systems. We can take a conceptual system to be a group of concepts together with their interrelationships. A concept can be defined as a generalization about something. Now we can think of conceptual systems as having different levels of development or maturity; the more explicitly stated the generalizations and their interrelationships, and the more nearly they reflect what we take to be reality, the greater the maturity of the conceptual system.

The scientific deductive systems—the theories—employed by theoretical physicists[2] probably represent conceptual systems of the highest level of maturity. Not all conceptual systems have been or can be deductively represented. At the level of extreme immaturity, there are conceptual systems—call them vague points of view—in which the generalizations and their interrelationships are not clearly stated, understood, or even consciously recognized.

It is extremely important, however, in a scientific sense, that we realize we are psychologically and intellectually unable to approach a problem without having a conceptual system—be it a vague point of view, a scientific deductive system, or a system between these in maturity. Because we cannot escape this and because the conceptual system we consciously or unconsciously employ will influence, if not determine, the extent of our success or failure, it is vital to our problem solving process.

At the one extreme, there is scientific danger in the vaguely perceived point of view; as such, it does not help us to detect our biases and it does not permit us to clearly identify our assumptions, which have much to do with the relevance of our findings. More mature conceptual systems are more indicative. In addition, they improve our explanatory and predictive powers, lead to more fruitful hypotheses, and better show the relevance of our findings to the more encompassing system of which the system we study is but a part.

There should be a more or less explicit conceptual system for the overall system encompassing the part or parts we intend to study, as well as a conceptual system for each part. These systems should be internally consistent and externally adequate to the remainder of the more encompassing conceptual system. The same problem or system can be approached with different conceptual systems. Beyond consistency and adequacy, the primary criterion of choice should be utility or fruitfulness, so leaving matters of truthfulness to the epistemologists.

The title of this chapter suggests a search for a conceptual system for water pollution biology. If this is taken to mean a single, unified conceptual system, the search is probably hopeless. There are conceptual systems more or less adequate for dealing with some parts of the biological problem, but for these parts there are alternatives, and for other parts conceptual systems need further development. Beyond this, the history of science teaches that the greatest of conceptual systems fall from the weight of their inadequacies, but only after better ones have been developed to replace them.

The Possibilities and Constraints of Water Pollution Biology. The foregoing statement of the complex problems we are considering is perhaps a tedious exposition of the obvious. It was made, however, to underscore the very real constraints on biological solutions of water pollution problems. Biology as a science cannot solve sociological problems; no science can. This would be asking too much.

There is also another way in which we should not ask too much of biology or any other science. We cannot expect anything like full explanations of natural phenomena of a complex nature, nor can we generally expect very reliable prediction of events in systems as complex as biological ones. Regarding biology in general, and water pollution biology in particular, the bases of these difficulties will be considered in later sections.

On the other hand, neither should we ask too little of a science. This is particularly important when a scientific contribution is necessary to the solution of problems of human welfare. First, we should expect a clear statement of scientific goals and objectives that are an essential part of the solution of the overall problem, and we should expect a candid statement of the possible utility and probable limits of utility of the knowledge sought. We should then expect the pursuit of these goals and objectives to be as effective as possible. In this, only relatively mature conceptual systems, not vague points of view, will suffice. Only they can exploit the knowledge and methods of biology so as to lead to the fullest explanations possible and to the most reliable predictions. Only such conceptual systems, those that are parts of the overall conceptual system for the water pollution problem, can make clear the relation of the biological contribution to the total effort and the possibilities and limits of this contribution.

THE NATURE OF BIOLOGICAL SYSTEMS

Organismic and Cybernetic Characteristics of Biological Systems. The work of biology is the explanation of biological systems; this and prediction of what these systems will do under certain at least partially specified conditions, which presupposes at least some explanation. The work of water pollution biology is the explanation and prediction of pollutional changes in aquatic systems. Any explanation whatsoever depends on some frame of reference, some conceptual system, however immature or mature this system may be. But it is more than this. We cannot begin to study a natural system —we cannot begin to seek answers to questions, solutions to problems—without some idea of the nature of that natural system. Our idea of the nature of a natural system represents what we know, or think we know, about the system. Such ideas are obviously subject to change with increase in knowledge, but we always have them, even when we first approach a problem. Hopefully, the idea is a reasonable approximation of the system and does not unnecesarily constrain the search for appropriate explanations. I have already given some indication of what I take to be the nature of the system producing the water pollution problem. However, here we are concerned primarily with the biological part of the system and how it relates to the whole. So, before proceeding, I must give some indication of what may be the nature of the biological systems involved.

It seems to me that organismic biology and cybernetics, taken together, provide a conceptual system adequate for an initial view of the nature of biological systems in terms of individual organisms, biological populations, and biological communities. I believe they promise much more than this, but first things first. The organismic view of biological systems arose mainly during the first quarter of this century in response to inadequacies of the vitalistic and classical mechanistic views of these systems. The vitalistic view fell into disfavor, not so much because of the lack of evidence for the existence of an essential "vital spirit," but more because it was conceptually and methodologically inadequate to further the explanation of life.[3] Early supporters of the mechanistic view promised too much and provided too little by way of a conceptual system adequate for encompassing life and guiding our search for understanding. We may analyze life into its particular processes, one at a time, but we must somehow explain how, taken together in all their interactions, these processes produce life systems having properties quite unlike other physical-chemical systems.

Borrowing from both earlier views of life, organismic biology came to emphasize the essential unity of the whole living organism and the directiveness or purposiveness of the parts as well as the whole. In any time slice, the parts function to maintain the essential stability and integrity of the whole. But through time, the whole tends towards greater complexity and stability. In all this there is a conservation, integration, and utilization of information.

Still, many organismic biologists took the position that the whole was somehow more than the sum of its parts, a statement that raises not only semantic problems but also conceptual and methodological problems not unlike those of vitalism. A more satisfying and useful view was taken by a few:[4] the whole is more than the sum of its parts *studied in isolation*. A part, or a process, studied in isolation or away from its environment or interactions with other parts and processes of the body is no longer the same. We will return to this methodological problem from which we cannot escape, but cybernetics offers a satisfying though partial solution to the conceptual problem.

The publication of *Cybernetics: or Control and Communication in the Animal and the Machine*[5] formally announced a new system of scientific thought of vast possibilities. In biology, these possibilities are even now only vaguely perceived. But this much seems clear: beginning with its genetic information system, each organism develops and operates as a network of untold numbers of interlinking information and control systems. In principle, at least, such a system appears to be sufficient to explain the organismic characteristics of life. Such a system could lead the parts to make and maintain the whole, and would make study of parts in isolation insufficient. Little, almost nothing, is known of this complex of systems in organisms, and we will probably never know it all. However, we believe it is there, and for now it will serve admirably to guide our thinking.[6]

We think of the individual living being as an organism. It is from such an organism that we abstract our ideas of the nature of organismic systems and characteristics. In seeking a conceptual system for water pollution biology, it will be useful for us to think also of biological populations and biological communities as organismic systems—not "superorganisms," but as systems having certain organismic characteristics.

Concepts of Environment. To define the environment of a system, we must first define the system. The one cannot exist without the other, neither conceptually nor in the real world. The two systems, the organismic system and the environmental system composed of identities interacting with each other and with the organismic system, together form a more encompassing system.

The concept of organism and the concept of environment are the primary concepts of biology. Failure to adequately define these concepts and recognize the interrelations between them and between the real world systems they represent has often led to conceptual and methodological difficulties in biology. Reduction of these difficulties by appropriate development and utilization of organismic and environmental concepts will increase the effectiveness of our application of biology to water pollution problems. The difficulties are understandable; dealing with the concepts and the systems they represent is not easy.

It is easiest to deal with the concept of the environment of the individual organism, because the individual organism comprises a more discrete and unified system than does either the population or the community. Even so, it is difficult to conceptualize and demonstrate what identities compose the environment of the individual. It was concluded in one of the very few careful and logical analyses of the concept of environment that the environment of the individual organism could most usefully be defined as the sum of all phenomena that immediately and directly are operationally significant to the organism, with the exception, of course, of characteristics of the organism.[7] This was termed the "operational environment" of the individual. Other phenomena were considered to be part of the "history" of the environment. Thus, the conceptual and methodological difficulty of including nearly everything in the environment of the individual was avoided. However, at least some very important operational phenomena are one or more steps removed from the organism, and I think it necessary to include these in the concept of operational environment. It may then be useful to distinguish between the "directly operational environment" and the "indirectly operational environment."

When we move to the population level of biological organization, we find the concept of environment generally even more poorly handled. This is because of failure to recognize the population as a discrete, unified organismic system, so that we can distinguish between it and its environment. Some biologists[8] prefer not to define the environment of a population because, as they see it, the population is part of its own environment, the individuals composing the population are a part of the environment of one another. But so also are the parts of the body of the individual organism[9]; yet this does not prevent these same biologists from defining the individual as a discrete and unified system so as to be able to distinguish and define its environment. Admittedly, a population of organisms is not so discrete and unified a system as an individual organism, but, as will be explained in a later section, it is a reasonably unified system and is generally viewed as such. Moreover, we are most interested in populations, with which we work in nature. We need a concept of the environment of a population, whether our work be simply ecology or the solution of water pollution problems.

Though not generally explicitly stated as such, the concept of "niche," so important in ecological thought, is equivalent to the operational environment of the population, if the population is conceptualized as an organismic system. The niche of a population is the complete set in space and time of conditions and relationships within which the population can and must survive and reproduce if it is to persist.[10]

Conceptualization of the environment of the biological community presents even greater difficulties. This is because a biological community is even less discrete and unified than a population, and because biological communities modify the conditions of their existence. However, lower levels of biological organization do this also. Biological communities are not merely chance agglomerations of populations of plants and animals, as one extreme view holds. One does not need to view the community as a "superorganism" in every way analogous to the individual organism, only at a higher level of organization, to recognize that a community is composed of interacting populations forming a system having some organismic characteristics. If we are willing to view the community in this latter organismic sense, we can then think of the "habitat" or the environment of the community as being the primary physical and chemical conditions of existence of the community as a whole.

Water is, of course, the medium of life when we are concerned with the aquatic environments of individual organisms, populations, and communities. Untold numbers of species over geological time have evolved the fitness necessary for their success in various aquatic environments. On the other hand, water itself was, in a sense, preadapted with its many remarkable properties making it uniquely suitable for life.[11] Water, as it naturally occurs and as man alters it, determines in large part the possible distribution and abundance of the different kinds of aquatic life. From a limnological point of view, a large share of what is known of this has been set forth in general reference works.[12,13,14,15]

Then, of course, there is the knowledge of pollutional changes in the aquatic environment[16,17,18,19,20] This raises one final problem for this section: what is pollutional change? More generally, what is the distinction between environmental change and environmental degradation? Environments continuously change naturally, and organisms successfully adapt to this. Perhaps we can usefully consider environmental degradation to be change to which individuals and populations cannot adapt. But man changes environments in ways to which other organisms cannot adapt when he plows land, cuts trees, and builds cities, as well as when he discharges his wastes. Are all these degradations of the environments of other species to be considered pollution?[21] Recognizing the desirability, even the need, of maintaining as much of the beauty and diversity of life as humanly possible, I am inclined to believe that we must still define pollution primarily in terms of degradation of the human environment, which, after all, includes an aesthetic and ethical consideration of all life.

Before going on to consider briefly the nature of biological explanation and then finally to consider the objectives, possibilities, and constraints of particular biological approaches as they can be integrated into conceptual and methodological systems for dealing more effectively with water pollution problems, I shall first indicate something of the nature of the individual organism, the population, and the community. To some of my readers, this must seem a circuitous and unnecessary way to get to matters most concerning them, but I think it really is not. Were we to have moved immediately to consideration of particular biological approaches to water pollution problems, I would have found it necessary to justify in detail the positions I will take for each of these approaches with regard to their possibilities and constraints. There are, however, general matters regarding the possibilities and constraints of all biological studies, and it is more orderly and efficient to first clarify these. More important, we cannot consider ways of integrating biological studies into water pollution control efforts without first having some conceptual framework for doing this.

Individual Organisms. An individual organism represents the lowest level of biological organization that can generally maintain and reproduce itself in nature in a biologically meaningful way. As such, it represents the elementary unit of ecological systems. Guided by its initial genetic program, the individual organism changes throughout its development and life in response to its operational environment. This is only to say that, at any point in time, the organism is the product of the complex of interactions between its genotype—its total genetic complement—and the sequence of environmental conditions throughout its development and life up to that point in time. Were this individual to have had a slightly different genetic complement, or were its environmental history to have been slightly different, the individual would then in some way be different in its performance characteristics. If the genetic and environmental differences were great, the individual would be greatly different. No two individual organisms, then, are likely to be identical in their characteristics, no matter how closely they may be related. We will have occasion in later sections to return to the differences between individuals, but now we must consider the conceptual and methodological difficulties deriving from differences in an individual that result from differences in its environmental history.

When we bring an organism into the laboratory to learn how particular environmental conditions might affect that organism, we make that organism different from what it would have been had it remained in its natural environment. The effects of the environmental conditions we study in the laboratory may for this reason be different from what they are in nature. Also, different ways of maintaining the organism in the laboratory may lead to different effects of the same set of experimental conditions. Which then are the effects most relevant to conditions in nature, where our real interest lies?

The individual organism, more clearly than other organismic systems, exhibits the organismic characteristics we have already considered. It should, because these organismic characteristics have been abstracted primarily from the development, operation, and function of the individual organism. When we study the physiology of the individual organism, we seek to learn how its parts operate to maintain its organismic characteristics. When we study the survival, reproduction, growth, and behavior of an individual organism as a whole, we seek to learn how it operates to maintain its function of insuring the persistence of its population.

Populations. We can think of a population as a group of interacting and interbreeding individuals that has relatively little reproductive contact with other groups of the same species. Biological species are composed of such local populations. Each of these populations adapts genetically to changing environmental conditions, its individual organisms acquiring, through evolution, morphological, physiological, and behavioral characteristics more or less distinguishing them from individuals of other populations of the same species. The gene pool of a population, in interaction and dynamic balance with the environment of the population, tends to insure the distinctiveness of the characteristics of the individuals as a group.[22] In responding to each other and the environment in distinctive ways, the individuals tend to maintain the integrity of the gene pool.

Thus, a population tends to have a certain integrity, owing to maintenance of its distinctiveness and cohesion. It can reasonably be viewed as an organismic system exhibiting, in some degree, the organismic characteristics we have considered. A population is a group of interacting individuals, with such interactions tending to maintain the stability of the population at any time. By means of natural selection, the population through time adapts genetically to changing environmental conditions. The individual organisms, the parts, function to maintain the population, primarily through the success of their reproduction.

The population in nature is the persisting, evolving unit of greatest biological and utilitarian importance. It is the population nature and man seek to maintain for their own different reasons. When we study individual organisms in the laboratory or in nature, our ultimate reason is to explain the operation of populations in terms of how their parts function to maintain the whole. The population in nature, however, is a part of a community of plants and animals in which it serves a function and apart from which it cannot persist. This too must be taken into account.

Communities. The nature of the biological community has continued to be argued by persons holding different positions on a wide spectrum of viewpoints.[23,24,25] At the one extreme, some have argued that the community is little more than a group of plant and animal species associated only because they have similar environmental requirements. At the other extreme, there are those who argue that the community is a highly integrated biological system, so much so that it can be viewed as a "superorganism" having characteristics quite analogous to those of individual organisms, only at a higher level of biological organization. Such extremes of view are more or less reasonably arguable only when there is a paucity of truly relevant information. Existing information, I believe, makes neither of these extreme views very tenable.

Populations of plants and animals are able to persist together in an area in part because they have the capacities to survive and reproduce under the physical and chemical conditions generally prevailing. Beyond this, in nutritional and other ways, these populations and their combinations need one another. A system, though perhaps a loose one, is formed by these populations and the relations between them. Such a system does in some ways exhibit certain of the organismic characteristics we have abstracted from our knowledge of individual organisms. The populations, the parts of the community, operate in ways that tend to maintain the whole. Through time, by a process known as "succession," the community passes through a series of successional stages or seral communities toward some generally more complex and stable "climax" community. Community succession, then, is perhaps analogous to the development of an individual organism.

However we may choose to view the closeness of dependency or coupling between the populations of a community, it is important that we recognize that there is necessary coupling, that some regulation occurs through this, and that, to this extent, the community is a system exhibiting at least some organismic and cybernetic characteristics. This recognition is important in our biological studies if we are to remain aware that individual organisms and populations studied in isolation (uncoupled from their biological communities) are different than they are in nature. Only such awareness helps us to properly interpret our findings and see their relevance to the real world.

Ecological Systems. Organismic systems–individual organisms, populations, and communities–do not and cannot exist without operational environments, although the operational environments may be laboratory ones unlike those in nature, an important matter to which we will return. Thus, we are compelled to study an organismic system and its environment, not an organismic system or its environment. The information obtainable on either part of the combined system depends on the nature of the operational environment as well as on the nature of the organismic part of the system. We need, then, an expression to represent the concept of the system formed by an organismic identity together and interacting with its operational environment.

The expression *ecological system* best identifies and pictures the system composed of an organismic identity and its operational environment. Thus, I take an ecological system to be an individual organism together with its operational environment, a population together with its niche, or a community together with its habitat, according to the case being considered. Ecologists study and are primarily interested in ecological systems as wholes, whatever level of biological organization may be their focus or whatever scientific discipline they may employ. It is this that most distinguishes ecologists from other biologists and scientists, for ecologists may by discipline be geneticists, physiologists, behaviorists, demographers, biogeographers, evolutionists, or even physical scientists or mathematicians.

The recently popular term *ecosystem* was originally coined by Tansley[26] to represent the concept of a biological community together with its habitat. I think it most useful to take ecosystem to be a special case (to be but one kind) of ecological system. This usage may cause some semantic difficulty–though I believe not much–because the term ecosystem is so obviously formed by combining the terms ecological and system. However, we need a concept for the general case as well as one for the special case, and terms to represent them.

What, then, are the sorts of interactions between organismic systems and their operational environments that justify our considering them together to form systems, ecological systems? An organismic system as a whole maintains a certain stability in the face of environmental change through interactive adjustment of the operation of its parts. To do this, however, the organismic system must receive energy, materials, and information (signals) from its operational environment. Energy, materials, and information also flow from the organismic system to the environment, modifying the environment in the process. To me, it seems not only justified but necessary to view the organismic system and its operational environment as together forming a system having cybernetic characteristics.

In beginning this section, I mentioned that an organismic system always has an operational environment, but that when we necessarily utilize laboratory studies of individual organisms, populations, and communities, the operational environment is then different from the one in nature. The responses of the organismic part of the system will then be different from what they would have been in nature. It is important, both conceptually and methodologically, that we bear this in mind if we are to understand the meaning of our results and their relevance to the system and problem of concern in nature.

THE NATURE OF BIOLOGICAL EXPLANATION

Explanation in Terms of Development, Operation, Function, and Evolution. Explanation of phenomena is the primary objective of science. Prediction of events is both an objective and a test of science. The best predictions are based on an understanding of the systems involved and on our ability to explain how they operate at any given time and how they change through time. It has been said that any proper answer to a "why" question may be considered an explanation of a sort.[2] It is further stated that what is demanded of the answer to a why question is intellectual satisfaction of one kind or another and that this can be provided, partially or completely, in different ways.

A reasonably complete explanation of a given biological system would contain considerable information about how the system developed to its state at any given time, how it operates at that time, what its function is in the more encompassing system of which it is a part, and how, over a longer time span, it evolved through a series of systems to come to have its own characteristics. Developmental, operational, and evolutionary explanations tend to be of the sort usually termed causal. I think it is better, however, for philosophical, conceptual, methodological, and interpretive reasons, to avoid at least the simpler ideas of cause and effect sequences. With publication of David Hume's essay "Of the Idea of Necessary Connection"[27] in 1748 and John Stuart Mill's essay "Invariable and Unconditional Antecedents"[28] in 1843, it became more apparent that we are unable to unconditionally separate cause and effect sequences—which we see primarily as sequences in time—from the nexus of causal relations leading to a particular event. We deal with systems in which untold numbers of conditions are prerequisites to a given event. Better we seek operational explanations than cause and effect sequences.

Developmental and evolutionary explanations tend to be operational explanations of a historic sort: how systems have operated through time to lead to their present states. Developmental explanations pertain to the ontogeny of individual organisms and the succession of communities. Evolutionary explanations pertain to how individual organisms, populations, species, and communities, over long periods of time, have acquired the characteristics by which we now recognize them. We also need explanations pertaining to how, at any given time, an organismic system operates to maintain itself. It is this kind of explanation that I distinguish simply as an operational explanation. Such explanations must usually be in terms of how parts, through their interactive processes, maintain the whole. Thus, operational explanations for one level of biological organization tend to be in terms of the next lower level, in the sequence: cells, tissues, organs, individual organisms, populations, and communities.

Because of the apparent directiveness or purposiveness of biological systems, one other sort of explanation is necessary. Even after we have explained how a system operates, we may still be inclined to ask: why should it do this? what is it good for? what is its purpose? or, better, what is its function? Appropriate answers to questions of this sort are functional explanations. Functional explanations tend to be in terms of the role that a system at one level of biological organization plays in the maintenance of the next higher level of organization. For example, the function of the individual organism is to contribute to the persistence of its population.

The Biological Indeterminacy Difficulty. Before we can go on to explain some of the approaches scientists can utilize in solving certain biological problems, specifically water pollution problems, two other aspects of the structure of biological explanations must be considered: the nature of needed generalizations and systems of generalizations, which we will deal with in the next two sections.

Generalizations and systems of generalizations should be of a form adequate to minimize, insofar as possible, what I will term the *biological indeterminacy difficulty*, which is one aspect of the problem of wholes and parts. The problem of wholes and parts and the problem of distinguishing between reality, observation, and concept are, I believe, the most general difficulties of scientific thought. It is the former that needs consideration here, the latter being more generally recognized.

On the basis of certain possible interpretations of the Heisenberg uncertainty relations,[29,30,4] the need to recognize a "biological indeterminacy principle" was suggested.[4] One interpretation of these relations leads to the conclusion that systems of elementary physical particles are, in themselves, of an indeterministic nature, and thus, in principle, there can be no deterministic law; there is no ultimate causality. The invalidity as well as the validity of this interpretation have been argued with high authority[3] and I am unqualified to comment.

Another plausible interpretation of these relations—one that I believe to be of very great importance in determining the possible extent of biological explanation—is that such systems are only methodologically indeterminable, not that they are in themselves of an indeterministic nature.[3] It is this alternative interpretation that Bertalanffy[4] accepts and briefly develops as a "principle," though I think it is better termed a "difficulty."

There are really two parts to this difficulty. The first of these is the near impossibility of making operational measurements on a system without interference that, at least to some extent, compromises the meaning of the measurements. The second part of the difficulty is the very great magnitude of the task of determining the operation and interaction—down to the ultimate elements and processes—of all the subsystems forming high-order systems like biological ones, even if it were possible to make measurements on these subsystems without interference.

The first part of the difficulty can be illustrated with one interpretation of the Heisenberg uncertainty relations: that it is impossible to determine both the position and momentum of a particle because of instrumental interference. Generalizing this idea to include biological systems, we interfere with the system when we change its environment—as when we bring an animal into the laboratory or when we study a heart in isolation from the body—as well as when we make measurements within the system by physical or chemical means. Thus, we are uncertain about the meaning of our results and their relevance to nature.

However well we minimize this first part of the biological indeterminacy difficulty, the second part remains. When we seek to explain the operation of an organismic system (individual organism, population, or community) in terms of its parts, we are compelled to explain how all these parts interact to maintain the whole system. Then we must explain the operation of each part in terms of the interactions of its subparts, and so on to the ultimate elements and processes. But there are too many of these. An organismic system may in principle be explainable in operational terms, but the magnitude of the task and conceptual and methodological difficulties make full operational explanation quite improbable. Still, as has been recognized,[4] the nature of the biological generalizations we use can minimize this difficulty. Satisfactory explanation requires systems of generalizations, and we can develop systems of generalizations that further minimize the difficulty.

The Nature of Needed Generalizations. A generalization is a more or less general proposition concerning relationships between concepts of two or more identities or phenomena. Such generalizations are always in some degree hypothetical, unless they are trivial. A scientific hypothesis is a generalization amenable to refutation or to confirmation by means of sense experience or a valid system of deductive logic, but preferably by both.

What is a law? There is general agreement that it includes a generalization characterized by some degree of universality—by invariability and generality. However, "there is no agreement as to whether or not it includes anything else," particularly pertaining to causality. Braithwaite[2] takes the parsimonious view that it does not, and goes on to suggest that it is the place of a law in a hierarchical system of generalizations that in some degree makes it logically necessary. In agreement with this, Campbell[31] suggests that it is the peculiar importance and place of laws in scientific systems of thought that most distinguish them, beyond universality, from other scientific hypotheses.

No generalization can be fully universal in application to natural systems without restriction in the form of initial and boundary conditions. This is only to say that a generalization representing a natural relationship is only useful as a part of a system of generalizations representing a natural system of relationships. Application even of such a system of generalizations requires specified initial and boundary conditions because it represents but a part of an even more encompassing natural system.

From ideas briefly and probably too hastily advanced in this and the previous two sections, what can tentatively be said about the nature of generalizations needed in biological explanation? Because of the heirarchical and organismic nature of biological systems, heirarchical systems of generalizations representing the natural systems are required for developmental, operational, functional, and evolutionary explanations.

First, then, needed generalizations must be of a sort permitting them to be fitted into hierarchical systems of generalizations at one level in a way that they complement generalizations on the same level and explain, in operational or functional terms, generalizations on the next lower and higher levels. Second, each needed generalization should subsume, without losing its utility, such biological indeterminacy as may exist in parts of the biological system covered by the generalization, usually parts at lower levels of biological organization. Third, and obviously, needed generalization should be confirmed by experience and, insofar as possible, deductive logic.

Systems of Generalizations. Explanation of complex systems, like the organismic systems and ecological systems we will be considering, requires complex systems of generalizations. We are able to explain a system only to the extent that our conceptual construct of that system reflects its important operational characteristics. If the natural system is hierarchically organized, as organismic and ecological systems tend to be, then our conceptual system should have a hierarchical structure. When a part of a system, such as some levels of biological organization, tends to be reticulate in the relationships among its subparts, that part of the conceptual system should reflect this reticulateness.

Some of the most impressive achievements of physics have been explanations in the form of scientific deductive systems. Such a system, a theory, is composed of a set of hypotheses on different levels; those on the highest level occur only as premises in the system, those on the lowest level only as conclusions, and those on intermediate levels as conclusions from higher levels and premises for lower levels.[2] The possibilities of such conceptual systems have been generally recognized and often realized since Aristotle's time. There are other forms of explanation, however, and, even when achieved, a deductive explanation may be inadequate.[3] Moreover, it is not clear that such explanation of biological systems is generally possible, at least in our present state of knowledge. Still, efforts such as an axiomatization of Darwin's theory of evolution[32] are likely to clarify our views of the systems we would explain.

However, mature systems of generalizations, deductive or otherwise, do not generally come with one stroke, much less the first stroke. They are based on knowledge acquired by all the method and art of science and are probably most generally developed as a more or less final ordering process. Even when we start an investigation, we have some system of generalizations, be it so immature as a vague point of view. Nevertheless, the more mature we can make our systems of generalizations—the more explicit and the more reflective of nature—the more useful they become for explanation and prediction. A system of generalizations sets the conditions for any included generalization and its premises set the conditions for the system as a whole. Our confidence in any one of the generalizations is very much determined by its relations to other generalizations in the system and our confidence in them. Our ability to explain nature is a measure of the adequacy of our systems of generalizations.

PHYSIOLOGICAL OPERATION OF INDIVIDUAL ORGANISMS

Maintenance of the Individual as a Whole. The most universal generalizations that we have in biology are probably those derivable from the organismic view of biological systems: that the organismic system at a particular time tends to maintain a dynamic stability; that through time it tends toward greater organization and stability; and that these are accomplished by the operation of parts functioning to maintain the whole, with cybernetic (information and control) mechanisms being of paramount importance in this. Although originally abstracted from the individual organism and its development, these generalizations have, with differing degrees of emphasis, been found to apply to all levels of biological organization. We will, in later sections, consider application of these generalizations to higher levels of organization. In this section we must briefly touch on how physiological systems operate to maintain the individual organism and just what this has to do with water pollution biology.

The individual organism generally exists in an environment that is fluctuating, factor by factor, some of which may be changing in a more persistent way, often because of man's activities. The ability of the organism to exist depends entirely on its ability to maintain its internal conditions within some limits, which are in part dependent on the previous environmental history of the organism. To maintain some degree of internal stability in the face of a change in some environmental factor, we must suppose there exists within the organism a cybernetic system of the sort diagrammed in Fig. 5-1. The response of this system, though, may very well depend on the levels of other environmental factors, and the organism may also respond to changes in these by means of other cybernetic systems. We can perhaps view the organism as a complex of untold numbers of such cybernetic systems interlinked. Only in this way would organismic characteristics appear explainable. However, in spite of the volumes published,[33,34] few if any such cybernetic systems have been satisfactorily explained.[6] There is sufficient evidence that the systems are there, but the biological indeterminacy problem confounds us.

What has this to do with water pollution biology? First, by means yet not understood, the organism maintains remarkable stability in the face of environmental change within some limits. This internal stability is, however, brought about by adjustment or change in the parts of the organism. Thus, every change in the physiology of the organism is not to be viewed with suspicion, i.e., viewed as evidence of pollutional change in the environment. Second, whenever we measure an overall response of the organism, such as death, the level of the lethal factor leading to death will in part be dependent on the organism's previous experience with that factor and perhaps others. An organism that has been acclimated to high temperature may survive at higher temperatures than it otherwise would (Fig. 5-2). An organism that has been acclimated to low dissolved oxygen concentration may survive longer at lower than normal oxygen concentrations (Fig. 5-3). Organisms may also respond to toxic substances in this way.[35] We must take this into account in the design, conduct, and interpretation of our experiments, not only at lethal levels but at levels influencing reproduction, growth, and other responses as well. The ability of organisms to acclimate to environmental change is evidence of the

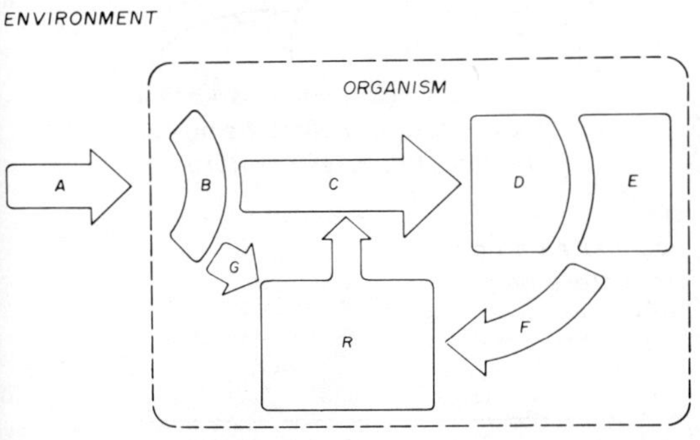

Fig. 5-1 Representation of the elements involved in homeostatic or adaptive responses: *A*, affector; *B*, receptor; *C*, transmitter; *D*, effector; *E*, output; *F*, feedback transmitter; *G*, forward transmitter; *R*, regulator. After Adolph.[102]

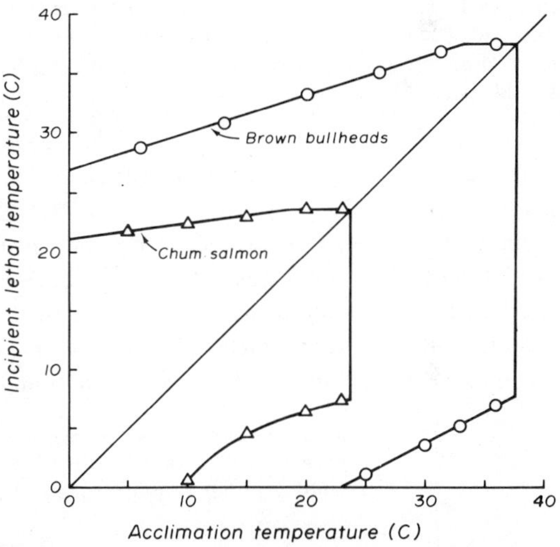

Fig. 5-2 The zones of tolerance of brown bullheads (*Ameiurus nebulosus*) and chum salmon (*Oncorhynchus keta*) as delimited by incipient lethal temperature and influenced by acclimation temperature. After Brett.[103]

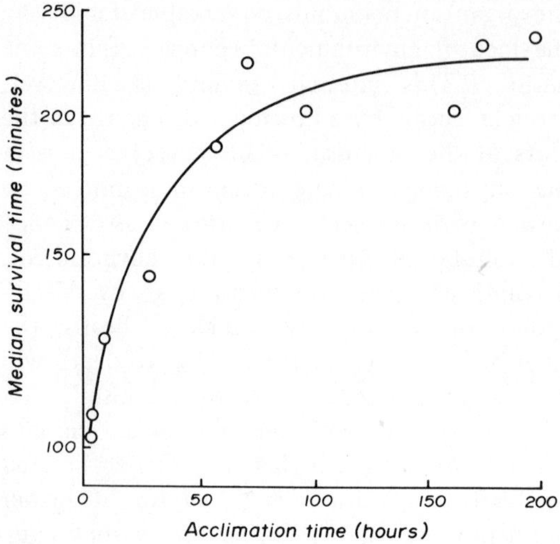

Fig. 5-3 Increases in the median survival time of juvenile brook trout (*Salvelinus fontinalis*) exposed to a lethal level of dissolved oxygen (1.06 mg/l) after the trout had been acclimated for different periods of time at a dissolved oxygen concentration of 3.8 mg/l. After Shepard.[104]

existence of homeostatic mechanisms of a cybernetic sort.

Temperature has a profound effect on the rates of biological processes. An animal acclimated to a high temperature will respond very differently to low temperatures than it would if it had been acclimated to a low temperature (Fig. 5-4). In a sense, it is a different animal. Across a wide range of temperatures, though, the metabolic rate of an animal may change relatively little if the animal is acclimated to each temperature at which it is tested. In nature, animals appear to acclimate rather rapidly to temperature change in the higher range, and, thus, they are able to maintain the rates of their biological processes relatively constant. All this must be taken into account, not only when we study temperature effects, but also when we study the effects of toxicants, low oxygen levels, and other environmental factors.

Knowledge of Physiological Systems. A few days browsing in a good library might lead one to conclude that knowledge of the physiology of aquatic organisms is very extensive indeed. In a sense, this conclusion would be correct. In tens of journals, thousands of articles are published each year. Review articles appear regularly, and multivolume sets of books are devoted to the physiology of just the crustaceans,[36,37] the mollusca,[38,39] and the fishes.[40] But when we would apply some of this knowledge to explaining or predicting the effects of some environmental change on the distribution and abundance of almost any par-

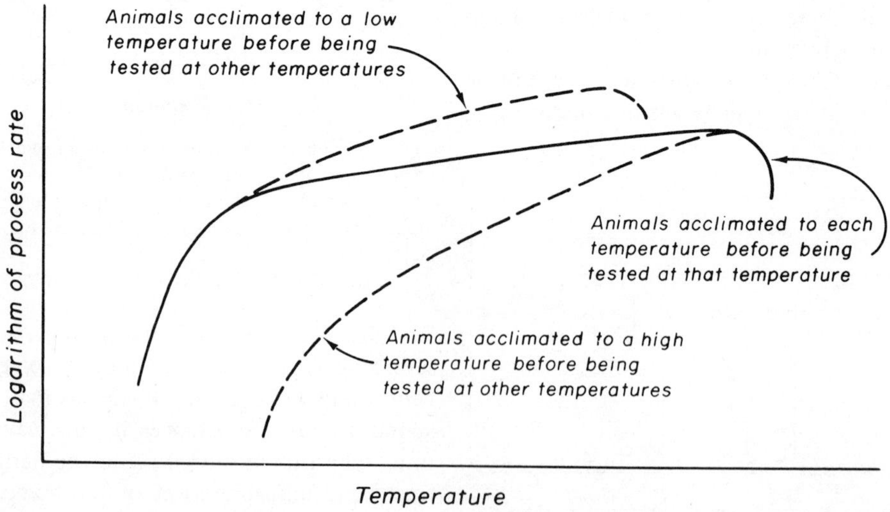

Fig. 5-4 Hypothetical curves showing the relationships between the rates of biological processes and temperature when animals are cold adapted, when they are warm adapted, and when they are perhaps more naturally adapted to the temperatures at which the rates are measured. After Bullock.[105]

ticular species, we find the relevant knowledge entirely insufficient to the task. Partly this is because our conceptual systems are generally inadequate to relate the behavior of physiological systems to the performance of the individual as a whole and the success of its population in nature, a matter to which we shall return in the next and later sections. This is an important reason for our failure, but there is another one. Our knowledge of the physiology of aquatic organisms is spread thinly over hundreds—probably thousands—of species, and for no single species or even group of closely related species is physiological knowledge sufficient to serve as a basis for adequately defining its overall environmental requirements. For economic and scientific manpower reasons as well as purely scientific ones, this state of affairs will probably persist into the distant future. Realistically, I think we cannot expect otherwise. Our knowledge of human physiology, extensive as it is, has been repeatedly shown to be inadequate to deal with environmental health problems, and few are the organisms we can hope to know as well as we do man.

Is this to say that physiological studies are all but hopeless, all but useless? Not in the least. It is that we must give much more attention to the objectives, possibilities, and constraints of physiological studies. If we would know the effects of an environmental change on the survival, reproduction, growth, or behavior of an organism, we should measure survival, reproduction, growth, or behavior, not study physiology. But if we would explain the mechanisms by which the environmental change affected growth, for example—and this is sometimes very important for predictive purposes—then we should study the physiological changes leading to the change in growth. No measurements of growth can ever explain the internal changes of the organism that lead to growth changes. We wish to predict, insofar as possible, the effects of environmental changes on the growth and other responses of living organisms, but we cannot study the growth and other responses of all important species under all possible conditions. Our predictions will be more reliable if they are also based on some understanding of the physiological reasons for changes in responses of whole organisms. This is only to say that the strongest explanations and the most reliable predictions are based on systems of generalizations, not isolated generalizations, and that these systems of generalizations should move from the whole living organism upward to its population in a community context and downward to physiological and even biochemical mechanisms. Available information will always be in some degree inadequate and unsatisfactory, however, primarily because of the biological indeterminacy problem. Still, as we will see in the next section, there are physiological concepts and measurements that help to subsume this indeterminacy.

Concepts of Physiological Scope. Except perhaps to suggest the rather extreme environmental conditions under which a species is unlikely to persist in nature, physiological studies have not generally been very successfully used to explain or predict the distribution and abundance of populations in nature. Partly this is because distribution and abundance in nature are determined not only by the physiological characteristics of organisms but also by predation, competition, and other ecological phenomena not included in physiological studies, and partly it is because physiological studies are rarely sufficient to adequately explain or predict the influence of environmental factors on the survival, reproduction, growth, and behavior even of organisms in the laboratory. These kinds of performance of the whole organism are most immediately relevant to its success in nature.

Physiological studies cannot explain the overall effects of, say, competition and food availability in nature, but they can tell us something about how natural stresses may unfavorably influence the physiological capacity of an organism to be successful in nature. That is, they can provide the kinds of generalizations that fit into systems of generalizations. Physiological studies also cannot fully explain the effects of, say, food availability on growth and reproduction, even in the laboratory, but they can make good use of concepts and measurements in subsuming much physiological indeterminacy, and can go far in explaining growth and reproduction in physiological terms.

Fry's concept of *metabolic scope for activity*, developed in his classic "Effects of Environment on Animal Activity,"[41] is one of the truly great contributions to the development of this kind of physiological thought and methodology. Fry considered the metabolic capacity of an animal to face the exigencies of its environment to be the difference between its standard or bare maintenance metabolic rate and its active or maximum possible metabolic rate under a given set of physical and chemical conditions. A change in this set of conditions could either reduce the active rate possible or raise the standard rate or do both of these, and so reduce the metabolic scope for activity and the energy the animal could make available to face the ecological realities of life in nature.

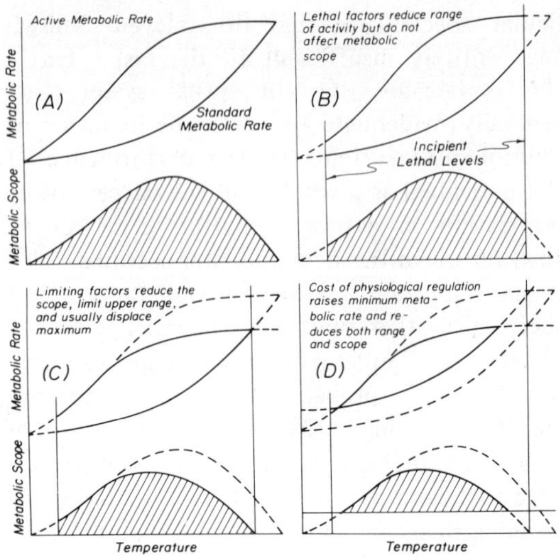

Fig. 5-5 The influences of a controlling factor such as temperature (A), lethal levels of the controlling factor (B), a limiting factor such as dissolved oxygen (C), and the cost of additional physiological regulation (D) on the active and standard metabolic rates and on the range and metabolic scope for activity of a hypothetical poikilothermic animal such as a fish. After Fry.[41]

In Fry's scheme, an increase in temperature—a controlling factor—would increase both the active and standard metabolic rate; this results, over a range in temperature, in first an increase in metabolic scope and then a decrease (Fig. 5-5A). If dissolved oxygen, a limiting factor, was in short supply, the active metabolic rate could increase only to some level and no further with increasing temperature or metabolic requirements. This would reduce the maximum possible metabolic scope and shift it toward lower temperatures (Fig. 5-5C). Physiological stress, such as that occasioned by the presence of some kinds of toxicants, would raise the standard metabolic rate at all temperatures and so reduce metabolic scope (Fig. 5-5D). Temperature, acting as a lethal factor at extremes, could reduce the range of temperatures over which life and metabolic activity would be possible (Fig. 5-5B). In ways like these, interacting environmental factors determine the metabolic scope for the activity of animals—the amount of energy they can make available under a given set of physical and chemical conditions to obtain food, escape predators, reproduce, and otherwise persist in nature.

Fry's scheme brilliantly shows us a way to conceptualize and evaluate how interacting physical and chemical factors determine the metabolic capacity of the organism to exist in nature. If this capacity is so reduced as to be insufficient for the organism to maintain necessary activities, we cannot expect it to be successful in nature. Generalizations derived from this sort of a concept are useful in systems of physiological and ecological generalizations. Generalizations at any level that subsume indeterminacy at lower levels must employ parameters that subsume such indeterminacy. Metabolic rate, the rate of oxygen consumption of the whole living organism, is such a parameter. It is what is known as a *sufficient parameter*:[42] a parameter measuring a functionally important outcome of all the processes leading to that outcome.

Organisms in nature face problems that we cannot adequately model in the laboratory. Thus, in the laboratory, we uncouple organisms from some of their natural environmental cybernetic and demand relations. Laboratory studies can tell us something of the physiological capacity of organisms. They cannot tell us where the organisms will be in nature. However, the concept of metabolic scope provides a way of determining how much "loading" by natural phenomena the organism can stand and still be able to persist. We will find it useful to extend this concept to cover higher levels of performance of the living organism when in later sections we consider studies of reproduction and growth.

PERFORMANCE OF INDIVIDUAL ORGANISMS

Performance in Terms of Survival, Reproduction, Growth, and Behavior. It is the performance of individual organisms as adaptable, integrated, living wholes in their environments as they find them that ultimately determines the distribution, abundance, and success or failure of their populations in nature. This overall response of living organisms to their environments that determines the distribution and abundance of their populations has been called *biopocrisis*.[43] Aristotle (*c*. 336 B.C.) saw that: "The habits of animals are all connected with either breeding and the rearing of young, or with the procuring a due supply of food; and these habits are modified so as to suit cold and heat and the variations of the seasons." Organisms have evolved in ways, then, that not only determine the distribution and abundance of their populations but also tend to insure the persistence of their populations, even in changing environments.

We can sometimes measure the survival, the reproduction, the growth, and the movements of organisms in nature, and so help to explain the distribution and abundance of their populations. It is difficult, though, to relate these responses to particular environmental factors, even when environmental measurements are made, for we cannot be sure which factor in the complex of changing factors is determining a particular response. Indeed, several factors are usually involved in complex ways. Often we bring a few organisms into the laboratory in a retreat from the face of complexity so we can study individual responses to particular factors in ways intended to explain what happens in nature. This can be helpful. But, here again, we have uncoupled these individual organisms from the cybernetic and demand relations they would have maintained in their natural environments. Because of this, the organisms we study and what we learn will in some degree be different from the organisms and their state of affairs in nature. We have gained some control and we have lost much reality. Still, this procedure is scientifically necessary, but we must never forget the severe constraints that it places on the meaning and relevance of our results.

Nevertheless, of the studies that we can conduct in the laboratory, measurements of the survival, reproduction, and growth of whole living organisms under particular sets of environmental conditions are the most immediately relevant to explanation and prediction of the influence of environmental factors on the distribution and abundance of their populations in nature. Only by experiment can we study changes in environmental conditions beyond those naturally existing but that man may one day cause.

For populations of valuable species to persist in nature, it is not enough that environmental conditions be barely adequate for the individuals in these populations merely to survive for a short or even an indefinite period of time. Conditions must also be adequate for the successful reproduction and growth of these individuals. Thus, in recent years, there have been more studies on the influence of environmental conditions on the reproduction, growth, and even behavior of organisms in addition to traditional tolerance studies. Water quality criteria for the protection of populations of fish and other valuable organisms will increasingly emphasize the maintenance of conditions suitable for reproduction, growth, and natural movements.

We will consider in following sections procedures for evaluating the tolerance of aquatic organisms for pollutional changes in their environments and influences of these changes on reproduction, growth, and behavior. Usually, different experiments are employed to evaluate different responses. In recent years, however, procedures have been developed for exposing groups of fish and other aquatic organisms to toxic and other conditions throughout their life cycle, including development, growth, and reproduction, from one generation to the next, under relatively simple laboratory conditions.[44,45] In this way, safe limits of environmental change for the entire life history of a particular species can perhaps be established. Still, for a species of interest to be successful in nature, conditions must also be satisfactory for its food organisms and the organisms these food organisms depend on for food, which are a part of the ever necessary food webs of aquatic communities. To help obtain the information needed not only on direct but also the indirect effects of environmental changes on valuable aquatic species, laboratory stream ecosystems as well as experimental channels and streams have been developed.[46,47,20]

Research of this sort is yet far from routine, and we cannot now reasonably expect waste dischargers to successfully adopt such procedures for the continual monitoring of wastes. Frequently now, industry is being required to maintain test animals such as fish in dilutions of effluents. Studies of the growth of animals can now be routinely accomplished, require little more effort than long-term tolerance studies, yield information more valuable both to industry and regulatory agencies, and are more likely to protect the resource.

Tolerance of Lethal Conditions. The tolerance of organisms for lethal conditions sets the extreme limits on their distribution in nature. The abundance of organisms in an area, however, is generally more closely related to their success of reproduction. More carefully stated, in accordance with the suitability of the environment, organisms become more or less abundant before reproduction rates are balanced by death rates. However, organisms can move into an area or persist for some time in an area where their reproduction has never been or no longer is possible. Studies of the tolerance of lethal conditions adult and juvenile organisms exhibit thus can tell us only a part of what we need to know about the suitability of possibile environments.

What organismic and environmental dimensions are really entailed in the concept of tolerance for lethal conditions? First of all, a lethal condition is a characteristic of some environment; tolerance is a characteristic of some organism. Still, an environment can only be lethal to some organism and an organism can only be tolerant of an environmental condition. Second, there is a time dimension, which is a function of the level of the lethal condition: an organism can tolerate for some time period a lethal condition, but this it cannot do indefinitely or the condition would not be lethal.

A lethal condition generally derives from two or more environmental factors interacting. Some single factor may indeed be the lethal agent, but the level of that factor leading to death in a given period of time will depend on the levels of other environmental factors. The tolerance of the organism is not independent of the environment but is derived from the interaction of the organism's genetic constitution with conditions throughout the organism's environmental history, with laboratory acclimation generally being only the most recent part of this. Here again we see an instance of an important organismic generalization: that the environmental history of an organismic system determines its character at any time; that information accumulates and is programmed for later use.

Our tolerance experiments should be designed so as to permit us to determine the *level of a lethal factor* that can be tolerated by a given percentage of the organisms for a given period of time, and, conversely, the *length of time* a given percentage of the organisms can tolerate a given level of a lethal factor. Lethal factor level and survival time are the primary variables of interest in tolerance experiments. Both the level of the lethal identity and exposure time are involved in any harm coming to the organism. Because different individual organisms differ in tolerance, we must work with groups of organisms and determine the percentages of mortality or survival under the various conditions of interest.

The detailed methodology of experiments to determine the tolerance of organisms for lethal conditions is reasonably well developed[48,49,100] and its explanation here is probably not necessary. Short-term tolerance tests, sometimes called *acute toxicity bioassays*, are designed to determine percentages of mortality at different levels of a lethal factor for different exposure times, and to determine survival or resistance times for various percentages of the test animals at particular levels of the lethal factor. For reasons of precision and convenience, *median tolerance limits* (TL_m's) and *median survival times* are generally estimated, as illustrated in Figs. 5-6 and 5-7.

Median tolerance limits can be determined for any time period in which the lethal agent is sufficiently effective, but usually they are determined for 24, 48, or 96-hr exposures and noted as 24-hr TL_m's, 48-hr TL_m's, or 96-hr TL_m's. Other equivalent notation is preferred by some: thus, TL50 or LC50 (median lethal concentration) for TL_m. Such latter notation is convenient for designating concentrations leading to the death of different percentages of the test organisms, e.g., LC10, LC20, LC80, LC90.

Probably the most generally interesting and useful expression of information of this sort is the graphical relationship between level of a lethal agent and median survival time. Such relationships assume different forms and sometimes conform well to simple mathematical formulas (Fig. 5-8). Within the limits of the data, we can estimate the length of time organisms can be expected to survive at a given lethal level, or, if we wish, the level of a lethal agent the organism can tolerate for a given period of time. Beyond the usual limits of such data, we can sometimes estimate the *incipient lethal level*, or *lethal threshold*, the level of the lethal agent that 50 percent of the organisms could tolerate indefinitely. Values obtained by such extrapolation are attended by worrisome uncertainties, because this is an important level to know. Long-term tolerance experiments appropriate for determining incipient lethal levels have become a pressing need, which fortunately has begun to be recognized and met.

In beginning this section, I noted that a lethal condition generally derives from two or more environmental factors interacting, that even when some particular identity is primarily responsible for death, the level of that identity

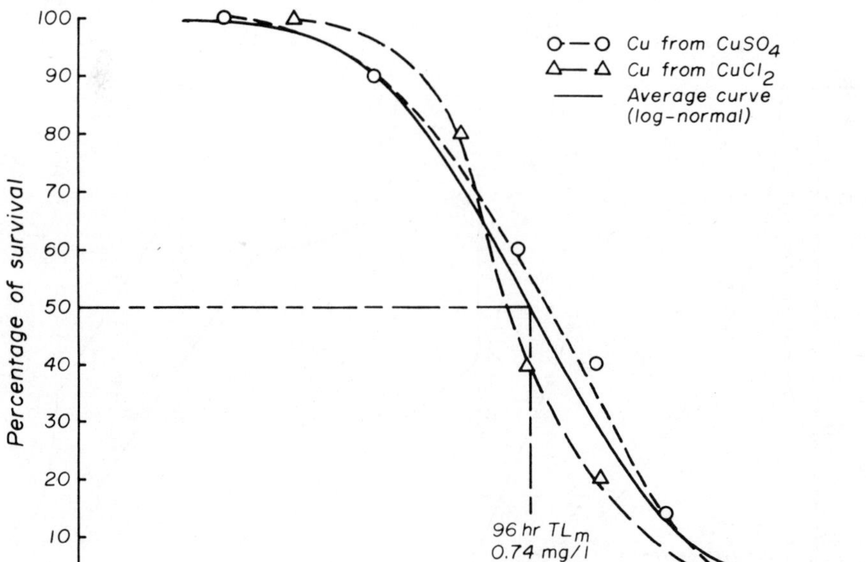

Fig. 5-6 Relationships between percentage of survival of bluegill (*Lepomis macrochirus*) for 96 hr and concentration of copper ion from copper sulfate and copper chloride, and the average theoretical log-normal relationship from which the 96-hr TL_m is estimated. After Trama[106] and Warren.[20]

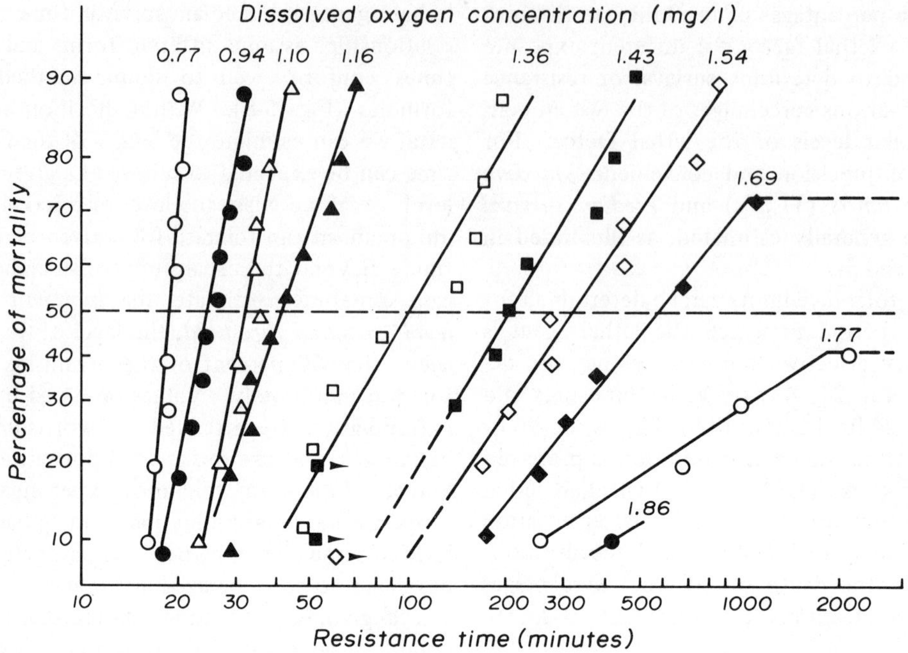

Fig. 5-7 Relationships between percentage of mortality and resistance or survival time of juvenile brook trout at 10 lethal levels of dissolved oxygen after the trout had been acclimated to a dissolved oxygen concentration of 10.5 mg/l. Median survival time at each of the 8 lowest concentrations can be estimated as the time at which just 50 percent mortality existed. After Shepard.[104]

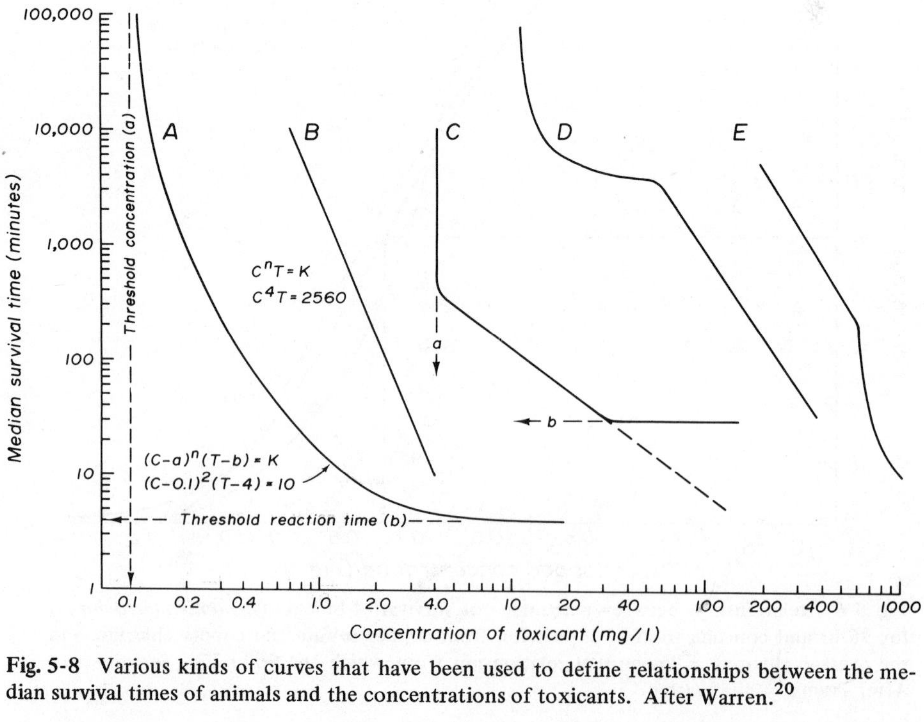

Fig. 5-8 Various kinds of curves that have been used to define relationships between the median survival times of animals and the concentrations of toxicants. After Warren.[20]

causing death will usually depend on the levels of other environmental factors. Thus, the median tolerance limits of rainbow trout (*Salmo gairdneri*) for ammonia vary widely with pH and total alkalinity, even when the latter factors are well within tolerance limits (Fig. 5-9). Temperature, salinity, and dissolved oxygen interact to determine the tolerance of American lobsters (*Homarus americanus*) for any one of these factors (Fig. 5-10).

Two or more toxic substances may interact to determine the overall toxicity of dilutions of their mixture. Some toxic substances will interact so as to result in a mixture less toxic than would be expected if the interaction were strictly additive; other toxic substances interact in ways leading to greater toxicity than would be expected were the interaction strictly additive (Fig. 5-11). Predictions of the toxicity of mixtures of toxic substances on the basis of the toxicity of the individual toxicants would be extremely valuable now that so many different toxicants are being introduced into natural waters. Because of the very nature of the problem, however, no general solution can soon, if ever, be expected. Still, much pertinent and good work is in progress, and partial solutions have been proposed, even if not sufficiently tested.[50,51,52,20]

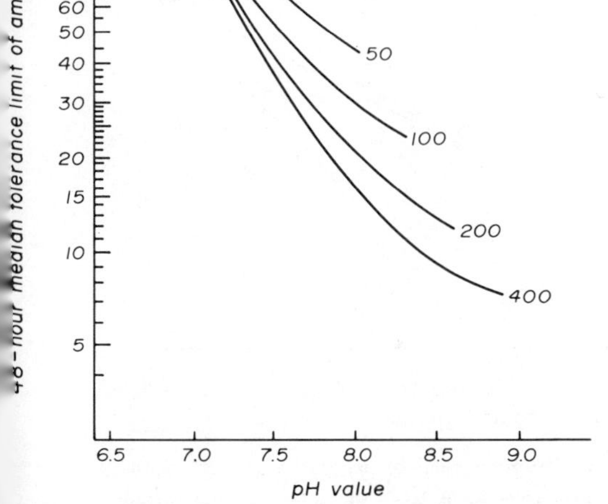

Fig. 5-9 Influence of pH and bicarbonate alkalinity on the toxicity of ammonia to rainbow trout (*Salmo gairdneri*) at 20°C. After Lloyd.[107]

This brings us to the problem of estimation of the concentration of toxicants that could be expected to be "safe" in natural waters—safe in the sense that they would in no way harm valuable species of aquatic life. Perhaps some decimal fraction of the 96-hr median tolerance limit of a species for a toxicant could safely be permitted in natural waters. Such a fraction has been generally referred to as an *application factor*. However, the effects of a given toxicant at low concentrations on all life history stages of an organism cannot be expected to be so simply related to concentrations acutely toxic to one life history stage. Under the leadership of Dr. Donald Mount at the National Water Quality Laboratory, Duluth, Minnesota, important studies on the levels of toxicants not influencing survival, reproduction, and growth of aquatic species are progressing.[45] Ways of similarly dealing with the problem of estimating safe concentrations of mixtures of toxicants are also under consideration there. Beyond this, the idea of safe concentration implies that not only direct but indirect effects of toxicants on species of interest will be controlled. We must, then, also be concerned about the necessary food webs in the biological communities of which our species of interest are a part. Laboratory streams and more natural experimental streams are proving of value in this.[20,47]

These matters go far beyond the problem of tolerance of lethal conditions, and we will discuss them more extensively in later sections. Still, in concluding this section, we can with confidence advance one more generalization: measurements of the tolerance of organisms for lethal conditions, though necessary, are insufficient for establishing ranges of environmental conditions suitable for aquatic life. This is a patently obvious generalization, yet one sometimes momentarily forgotten in our hurried efforts to resolve water pollution problems.

Reproduction and Development. The primary function of the individual organism is to con-

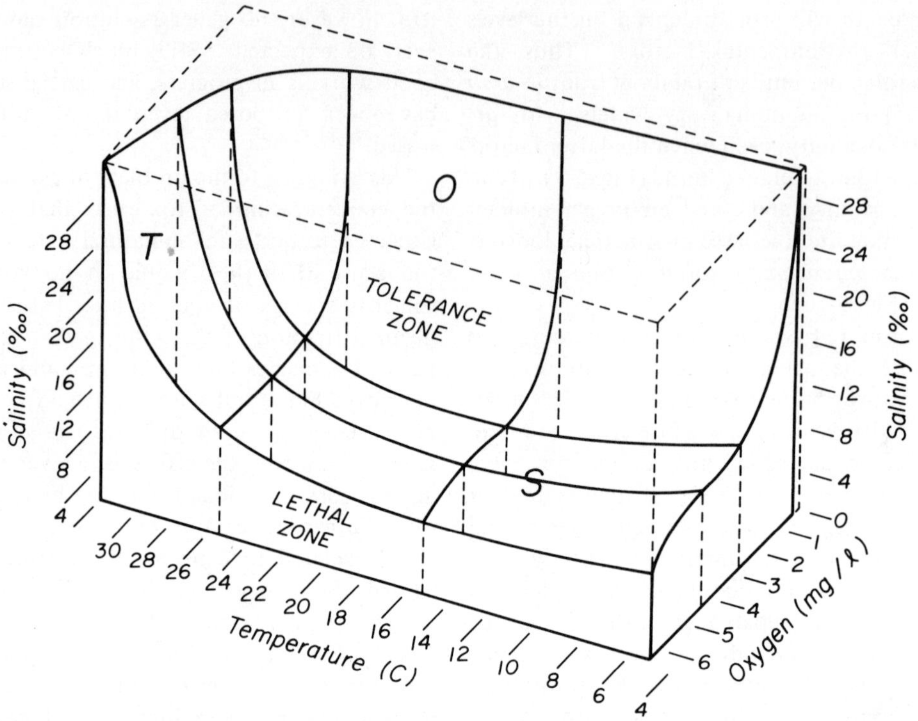

Fig. 5-10 Boundaries between the tolerance zone and the lethal zone, based on 48-hr median tolerance limits for the American lobster as determined by interactions of temperature, salinity, and dissolved oxygen conditions. In the regions marked T, S, or O, temperature, salinity, or oxygen, respectively, are primarily responsible for mortality. After McLeese.[108]

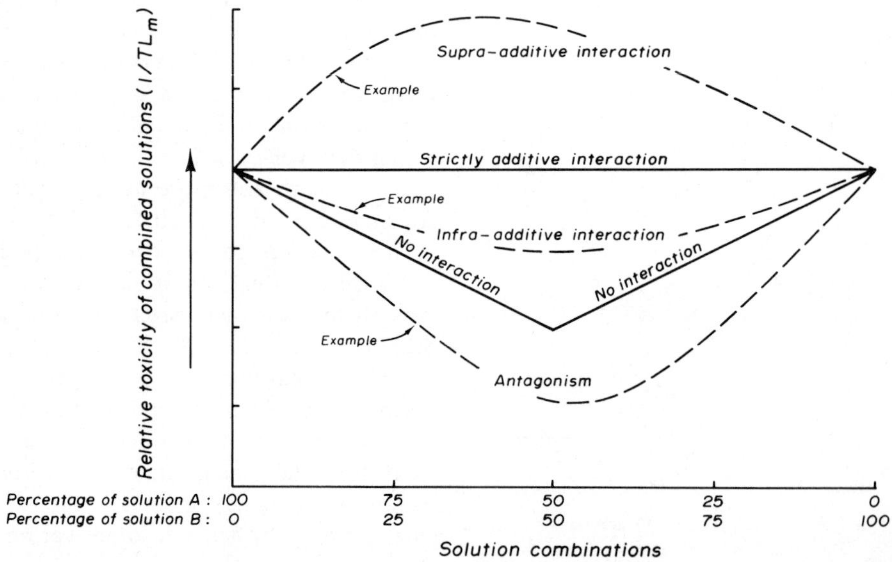

Fig. 5-11 Possible kinds of interactions between two hypothetical toxicants, A and B. Solutions of these toxicants are assumed to have been prepared so as to have equal toxicity when tested separately. For each kind of interaction there are shown the changes in relative toxicity (reciprocal of median tolerance limit) of the combined solutions that will occur when the two solutions are combined in different proportions. After Warren.[20]

tribute to the persistence of its population and kind. By far the most important part of this contribution is reproduction. For most species, it does not matter much what happens to individuals once they have made their reproductive contribution. The *fitness* of an individual organism in an evolutionary sense is thus measured primarily in terms of its contribution to the genetic characteristics of successive generations of its population. Still, to reproduce, an individual organism must survive for some time. Successful reproduction involves the survival and development of progeny themselves able to reproduce. Reproduction is successful to the extent of the numbers of progeny produced that are suited for prevailing environmental conditions.

The development of reproductive products within the parents, parental reproductive behavior, and the embryonic and larval development of progeny to more stable juvenile and adult stages are biological processes particularly and peculiarly sensitive to environmental conditions. The environmental requirements of the adults producing sex products, of the sex products within the adults, and of each successive embryonic and larval stage (whether within or external to the parent) are different because each developing individual is different, not only at every grossly recognizable stage, but from hour to hour and day to day. This, of course, presents a difficult problem when we try to determine environmental conditions suitable for the reproduction and development of species.

For a species present in a particular water throughout all its life history stages, we could perhaps estimate the highest concentration of a toxic substance that would in no deleterious way affect any phase of the overall reproductive process, were such a substance to be present throughout the life history of a species. This is the primary objective of recent research,[44,45,54] though information on toxicant concentrations not affecting survival and growth during the long periods of exposure is also obtained.

Species of fish are exposed to constant concentrations of toxicants in rather simple laboratory apparatus throughout the period of sex product development, reproductive behavior, embryonic and larval development, juvenile growth, and maturation to reproduction of the second generation if the generation length is no more than perhaps one year. For trout and other species requiring two or more years to mature, exposures have been maintained during the period of sex product development through early juvenile growth. Descriptions of the rather sophisticated procedures that have been developed for particular species of fish as well as for some fish food organisms can be obtained from the National Water Quality Laboratory in Duluth, Minnesota.

For species likely to be exposed to potentially harmful environmental conditions during only a part of their reproductive life cycle, it is generally more pertinent to model the periods and conditions of laboratory exposure to resemble those expected to prevail in nature. Anadromous salmonids, for example, would generally be exposed to toxic conditions prevailing in fresh water only during the period from the last part of sex product maturation through early juvenile growth. Sometimes, only one phase of the reproductive life history of a species will be in the part of the environment most likely to be degraded. Thus, dissolved oxygen concentrations may be more critical in streambed gravels where salmonid embryos and larvae develop than in the overlying water. Experiments to determine the influence of oxygen concentration and any interacting factors such as current velocity and water temperature may then be useful (Fig. 5-12).

Procedures have been developed for determining the influence of salinity, temperature, turbidity, toxic substances, and other environmental conditions on the reproduction and development of oysters, clams, mussels, and other invertebrates.[55,56,57,58,59,60] Here again, exposure to potentially harmful conditions may involve all or only a part of the reproductive life history (Table 5-1). As always, laboratory techniques must be adapted to the life history characteristics of the species of interest and to appropriate environmental conditions of concern.

The success of reproduction we observe in the laboratory at particular levels of particular environmental factors or combinations of these may or may not closely approximate reproduc-

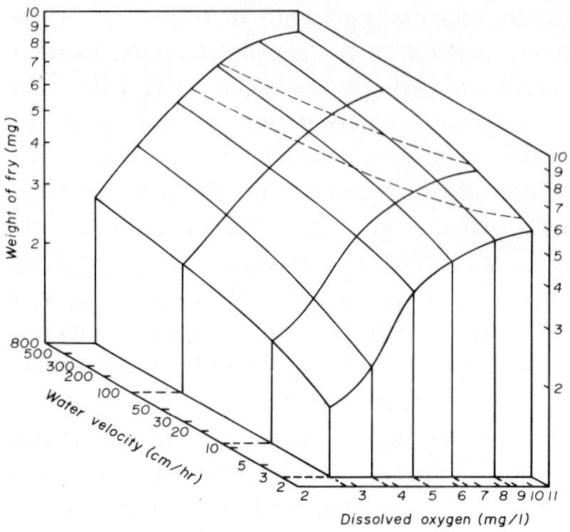

Fig. 5-12 A three-dimensional diagram showing the influence of both oxygen concentration and water velocity on the mean dry weights of newly-hatched coho salmon fry (*Oncorhynchus kisutch*) developing from embryos reared throughout development at the various combinations of oxygen concentration and water velocity represented by the intersections of the curves. After Shumway, Warren, and Doudoroff.[109]

tive success in nature, even at the same levels of these factors. If our laboratory procedures are good, however, reproductive success in the laboratory is likely to be greater than those that would prevail in nature. This is because in the laboratory we have eliminated or otherwise controlled environmental factors that in nature tend to reduce the success of reproduction and development. We generally eliminate the more ecological (as contrasted to physiological) demands the environment places on organisms, such as interaction with predators, competitors, or disease organisms. Still, it is reproductive success in nature that is our primary concern. We must, then, have a concept that usefully relates our laboratory experience to the natural reality. *Physiological scope for reproduction* would be such a concept. This would experimentally be defined as the changes in reproductive success caused by particular physical and chemical conditions of interest when the experimental environment was otherwise optimal. Across some range of temperature, physiological scope for reproduction would vary from zero to a maximum and again to zero. The addition of a toxic substance could reduce this scope at all temperatures and shift the maximum to lower temperatures. For given levels of temperature and toxicant, the maximum possible success of reproduction would be determined. In nature, factors not considered in the laboratory would tend to reduce reproductive scope, reproduction being less successful at optimal temperatures and particular concentrations of toxicant and being successful at all only over a narrower range of temperatures, much as indicated in Fig. 5-5 for metabolic scope for activity.

To relate this concept even more closely to the success of populations in nature, it can be extended to cover not only the reproduction but the survival of all life history stages. The concept of *innate capacity for increase in numbers*,[8] which involves both reproduction and survival,[20] is exceedingly important to ecologi-

TABLE 5-1. Mean Numbers of Olympia Oyster Larvae Released per Adult, Percentages of Abnormal Larvae, and Mean Larval Size, When Adults Were Held at Different Concentrations of Sulfite Waste Liquor Prior to and during Larval Development[a]

Concentration of SWL (mg/l)	Mean Number of Larvae per Adult	Abnormal Larvae (percent)	Mean Length of Larvae (microns)
0	36,408	0.22	157.8
0	48,585	0.15	158.6
2	32,071	0.19	154.9
4	31,367	1.05	153.3
8	31,319	9.24	150.9
16	31,634	65.46	139.2
32	4,826	97.22	161.2
64	910	92.48	159.2

[a]After Ref. 58.

cal thought. Application of the concept through laboratory experiments and their interpretation defines the physiological scope for increase in numbers of a species under environmental conditions of interest. We can only determine such a physiological scope for species able to complete their life histories in the laboratory: it has been done with *Daphnia* and with the common guppy,[61,62] and it can with great benefit be done with many other species. Even when such experiments are impractical, we need the concept to keep pertinent knowledge on reproduction in proper perspective to maintain a sound conceptual basis for the generalizations we must develop.

Bioenergetics and Growth. Growth is a universal biological process of great importance. To survive to reproduce, plants and animals generally must grow in some period of their life. Growth of individual organisms is the unitary process leading to the production of their populations and the possibility of yield to man. Growth is an integrated response of the whole living organism to its environment. The nature of the organism (as determined by its genetic characteristics and its environmental history) as well as conditions of its environment will determine its growth at any time. Properly studied, growth can be a very valuable parameter of the suitability of environmental conditions. Because it is a response of the whole living organism, growth as a process, as a measurement, and as a concept subsumes much biological indeterminacy. As a measurement, growth thus has the characteristics of a sufficient parameter. As a process, it presents many problems and possible levels of explanation.

To live and to grow, an organism must obtain from its environment energy and materials. One of the necessary conditions of the environment is, then, the availability of an energy and material resource of suitable quantity and quality. As we will see in later sections, this is dependent on other characteristics of the environment. As will be briefly explained in this section, some of these and still other characteristics of the environment will directly affect the ability of the organism to obtain energy and materials and to utilize them for its own maintenance and growth.

A change in the quality or quantity of food an animal consumes or a change in environmental conditions during a given food intake will alter the fates of this food within the organism and so alter growth—the outcome of primary interest. Samuel Brody's *Bioenergetics and Growth*,[63] is still by far the best treatment of the problems and methods of attaining understanding of the growth complex.

Figure 5-13 diagrammatically represents, in energetic terms, the fates of the food an animal consumes. Such a representation can never be fully satisfactory, for it must deal with defined concepts and parameters of identities and processes that in the animal body are not discrete. Nevertheless, it is useful both conceptually and methodologically. Of the food an animal consumes (Fig. 5-13), some is assimilated through the gut wall and the remainder passes out as fecal material. An animal cannot metabolize ammonia produced through deamination of amino acids, and this, sometimes in a conjugation product, is passed out as a nitrogenous waste. From the metabolizable materials (and those that might be incorporated in the body by a growth), some energy is lost as heat through deamination and similar processes, and some is used in the processes of digestion, movement, and incorporation of food materials: these two categories are together measured as specific dynamic action (SDA). The remaining energy and materials are utilized for standard (bare maintenance) metabolism and the metabolic cost of muscular activity beyond this, or they appear as growth.

To provide part of an adequate conceptual and methodological framework for the study of growth, we have developed, based on previous works,[63,117,41] the following bioenergetic equations:[53,20]

$$Q_c - Q_w = Q_g + Q_r \qquad (1)$$

where

Q_c = energy value of food consumed,
Q_w = energy value of waste products in feces, in urine, and lost through gills and skin,
Q_g = total change in energy value of materials of body (growth),

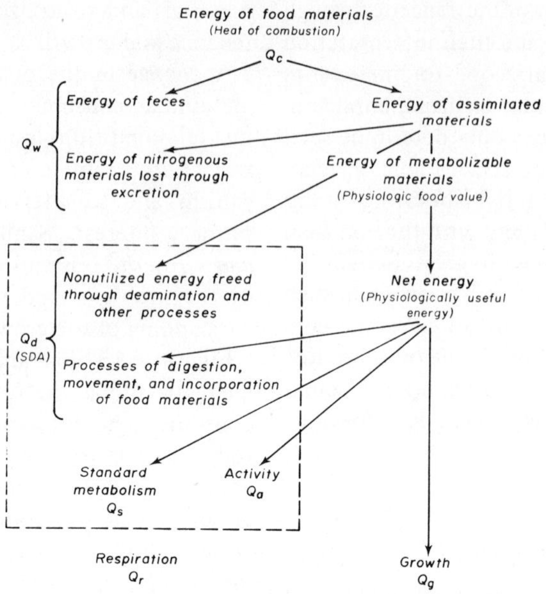

Fig. 5-13 The fates or categories of losses and uses of the energy and materials in the food an animal consumes. After Warren and Davis.[53]

Q_r = energy metabolically degraded with or without benefit to the organism.

Now if we let

$$Q_r = Q_s + Q_d + Q_a \tag{2}$$

where

Q_s = energy equivalent to that degraded in the course of metabolism of unfed and resting animals (standard metabolism),

Q_d = additional energy degraded in the course of digestion, assimilation, and storage of materials consumed (SDA),

Q_a = additional energy degraded in the course of muscular activity over and above that of the resting animal

then, substituting, we can represent the overall energy budget of an animal for any given period of time, in terms that are convenient for our purposes, as follows:

$$Q_c - Q_w = Q_g + Q_s + Q_d + Q_a \tag{3}$$

Equation 3 can be used to describe the uses and losses of energy and materials in all the food an individual animal consumes over an experimentally convenient period of time, say a week or a month, under a given set of environmental conditions.

However, the growth of which an animal is capable at any time is not only a function of food availability and other environmental conditions; it is also a function of the body size of the animal. To remove, insofar as possible, effects of changes in body size, we developed from Eq. 3 the following equation:[53,20]

$$A_c - A_w = A_g + A_s + A_d + A_a \tag{4}$$

where the Qs of Eq. 3 and the As of Eq. 4 are related by

$$Q_i = A_i W^x t \tag{5}$$

and

$$A_i = \frac{Q^i}{W^x t} \tag{6}$$

where

W = mean body weight or energy value,
t = time in days,
x = some mean power of W,
$i = c, w, g, s, d$, or a.

Space here will not permit me to explain these terms and equations and the assumptions upon

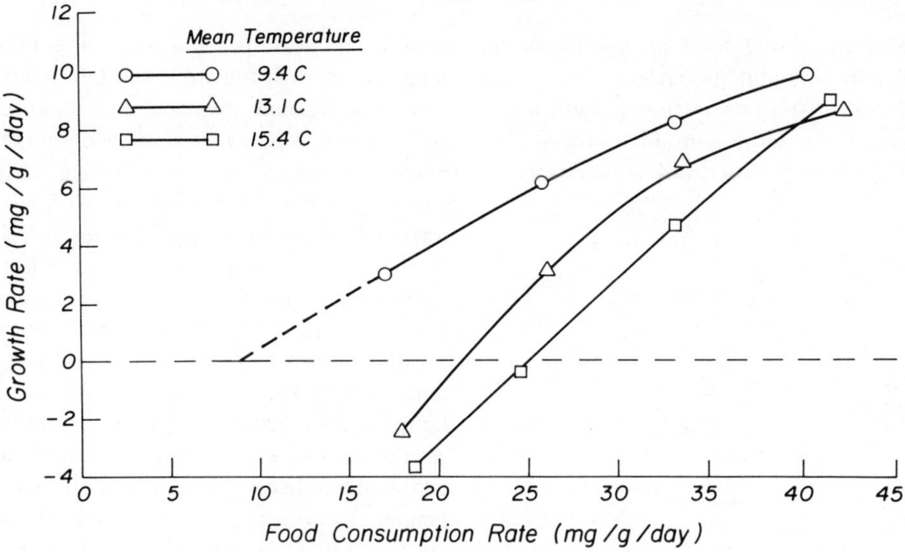

Fig. 5-14 Relationships between rates of food consumption and growth of juvenile coho salmon kept in aquaria from January 31 to March 2, 1970, at temperatures fluctuating about means of 9.4, 13.1, and 15.4°C. After Everson.[110]

which they are based. Neither will it permit me to explain the methods and difficulties of determination. These explanations, however, are developed fully in Ref. 20. The above will perhaps be sufficient to make clear the first part of a conceptual framework adequate for determining the effects of environment on growth.

We must now add dimensions of the environment to our conceptual framework. A single set of values for Eq. 3 or Eq. 4 represents an individual animal under only one set of environmental conditions. We need a set of values for each set of environmental conditions of concern because changes in these conditions will change the values. Measurement and interpretation of these changes in values help us to explain and predict the effects of changes in environmental conditions or growth.

First, let us consider a simple example (Fig. 5-14). At a given temperature and under other specified environmental conditions, an animal fed less than a *maintenance ration* will lose body mass. At the maintenance ration, it will neither gain nor lose mass. With further increases in food consumption rate (calories of food consumed per kilocalorie of body mass per day), its growth rate (calories of body mass gained per kilocalorie of body mass per day) will increase until the maximum consumption rate is reached (Fig. 5-14).

Were we to repeat this experiment at a higher temperature with a poikilothermic or cold-blooded animal, we would usually find the maintenance ration to be increased, primarily because of an increase in the standard metabolic rate. We would also usually find that at intermediate consumption rates, growth was lower at the higher temperature. If food is not limited, though, the animal at the higher temperature may eat more and thus attain higher consumption and growth rates than the animal at the lower temperature. Thus, the effect of temperature on growth is very much dependent on food availability.

In nature, food is generally limited. Experiments in which food is not limited cannot tell us much about the effects of temperature or other environmental conditions on growth in nature.

We can make use of Fry's[41] system of classifying environmental factors (controlling, limiting, lethal, etc.) and his concept of metabolic scope for activity (Fig. 5-5) to develop a conceptual system for simultaneously relating several environmental factors to the possible growth of an animal. To do this, it will be helpful to think of a *scope for growth:* the difference between the energy value of all the food an animal would and could consume and the energy value of all uses and losses of food other than growth under a particular set of environmental conditions.[53,20] Figure 5-15, in terms of Eq. 4, illustrates how

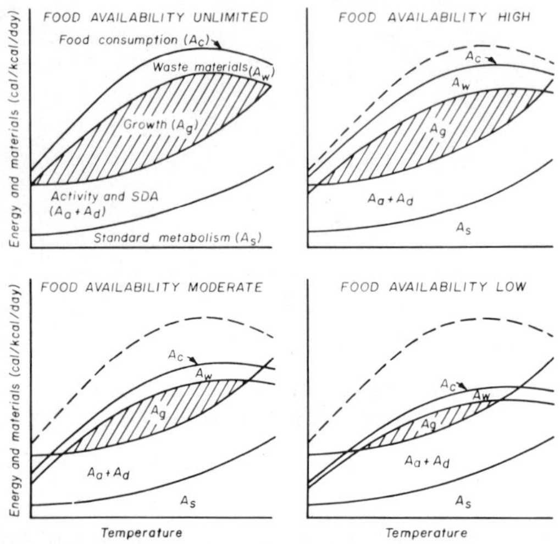

Fig. 5-15 Theoretical effects of temperature change on the food consumption, energy budget, and scope for growth of a hypothetical poikilothermic animal having food available in different amounts. After Warren.[20]

changes in food availability and temperature might in nature be expected to determine the scope for growth of a poikilothermic animal. Temperature, acting as a controlling factor, leads first to an increase in food consumption and then a slight decrease near the upper tolerance limit with increasing temperature at any one level of food availability. Standard metabolic rate also increases with temperature. Various simplifying assumptions, for ease of diagrammatic representation, have been made so that the combined costs of activity and specific dynamic action are constant across the levels of food availability and temperature shown in Fig. 5-15.[20] Nevertheless, the general conclusions that can be drawn have been shown by experience to be essentially correct. At any given level of food availability, scope for growth first increases and then decreases with increasing temperature. Decreases in food availability decrease the scope for growth at all temperatures, narrow the range of temperature over which growth is possible, and shift the maximum scope for growth toward lower temperatures.

How additional environmental factors, interacting with food availability and temperature, might be expected to influence scope for growth can now be considered. At low concentrations, dissolved oxygen can act as a limiting factor. Under oxygen stress, fish will limit their food consumption rate. At a particular temperature and level of food availability, this could reduce the scope for growth and would tend to shift the maximum scope toward lower temperatures, where standard metabolic rates would be lower. Some toxicants such as pentachlorophenol, which is believed to act mainly by uncoupling oxidative phosphorylation, tend to increase the standard metabolic rate at all temperatures and would thus further limit scope for growth. Scope for growth, then, is a concept that helps us to understand how interacting environmental factors may influence the growth of fish in nature.

Some toxicants (e.g., pentachlorophenol) act mainly to increase metabolic rate and thus the maintenance ration, the growth change being much like that caused by an increase in temperature (Fig. 5-14). If food is not limited, the animal can often compensate so as to grow well by eating more, but usually food is limited and growth is reduced. Toxic materials having more general effects may reduce the amount of food the fish will eat—even when food is available in excess—as well as increase maintenance metabolism. In this case, the animal cannot compensate and growth is reduced under all circumstances (Fig. 5-16). These are some of the reasons laboratory studies of growth must consider different food consumption rates.

Food availability in nature varies, and it is difficult to know how much animals are consuming. This is another reason we should consider possible effects of environmental changes at different food consumption rates in the laboratory. Food availability in nature involves questions of food density, distribution, particle size, and quality. Sometimes, but not always, it is possible to relate the growth rate of fish in nature to food density alone (Figs. 5-17 and 5-18). When this can be done, it is very useful, as we shall see in later sections.

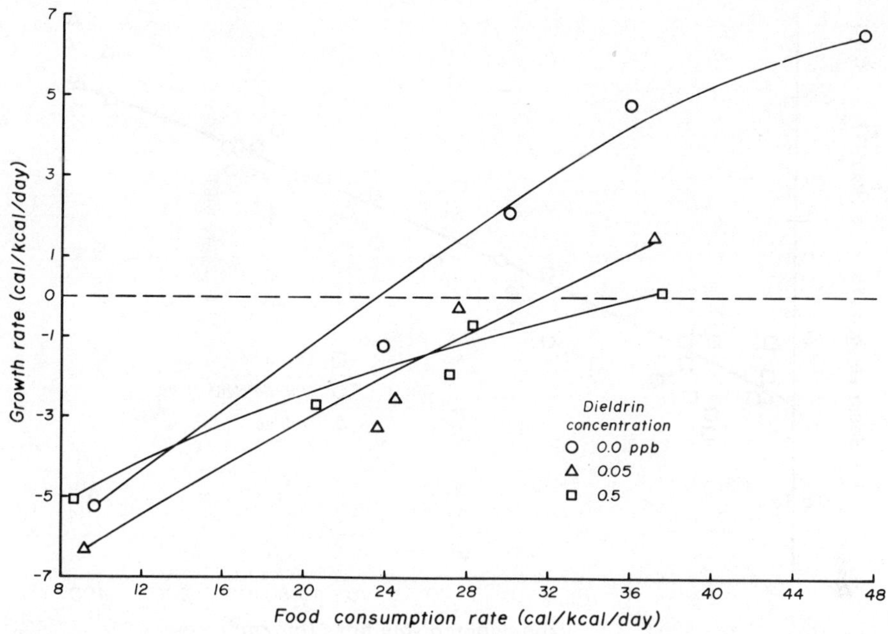

Fig. 5-16 Relationships between food consumption and growth of sculpins (*Cottus perplexus*) held at different concentrations of dieldrin. Unpublished data of Robert Brocksen and George Chadwick, Oregon State University. After Warren.[20]

Animals in nature will tend to compensate so as to maintain, insofar as possible, their natural characteristics and patterns in the face of environmental change. They will eat more, if they can, to grow well when environmental conditions tend to reduce growth.

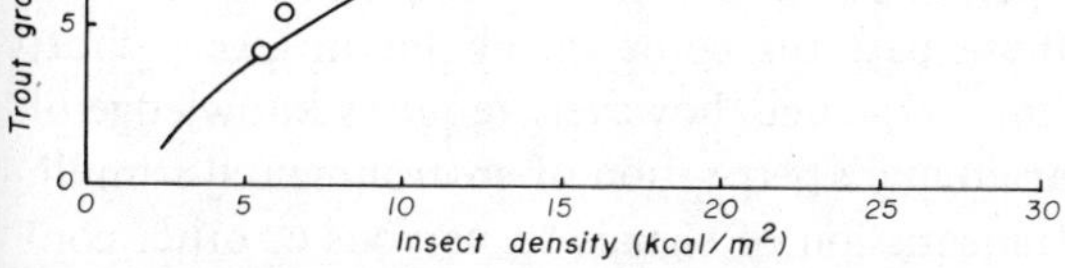

Fig. 5-17 Growth of cutthroat trout (*Salmo clarki*) in relation to the density of their food organisms in the benthos of the riffles of an experimental stream enriched with sucrose from October, 1965 through May, 1966. Trout growth rate and food organism density were estimated at monthly intervals. Unpublished data of John D. McIntyre, Oregon State University. After Warren.[20]

Animal Behavior. A satisfying explanation of animal behavior (as for other biological phenomena) should account for its development, operation, function, and evolution. Full explanation of any natural phenomenon is probably beyond human limits, but more than for other biological phenomena, explanation of animal behavior is likely to remain less than satisfactory. Adequate *functional* explanations we can expect to have; much less so for the development and operation of behavior. Confirmation of evolutionary explanations of behavior is probably beyond the pale, there being no fossil record.

Most of this discussion can perhaps best be directed toward the functions of animal behavior, because in this way I can most easily emphasize the great importance of behavior to the success of populations in nature, changed or unchanged.

Behavior, as we know it in the integrated movements of the whole living animal, develops through interaction of the genotype of the organism with its environment. No characteristic of an organism—morphological, physiological,

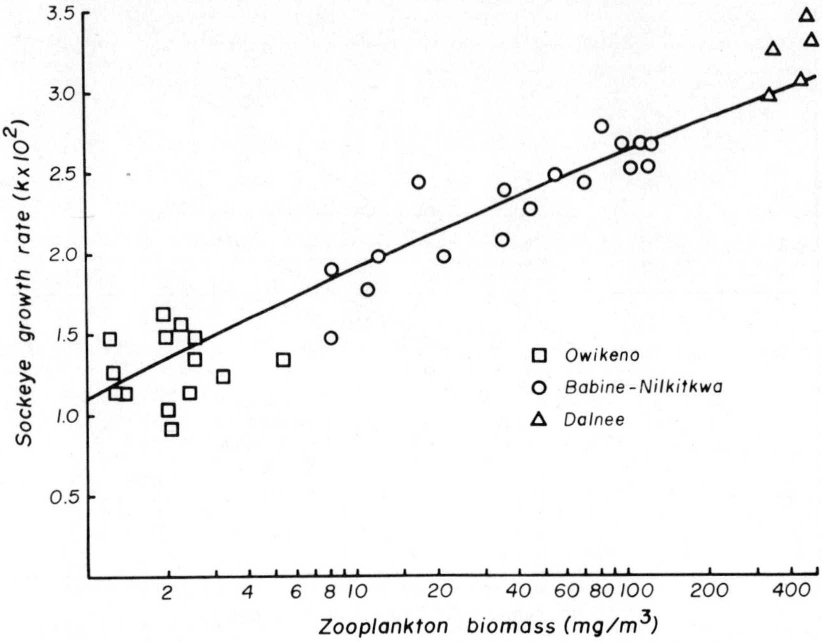

Fig. 5-18 Relationship between the daily percentage rate of growth of juvenile sockeye salmon and the density of their zooplankton food organisms in three lake systems. After Brocksen, Davis, and Warren,[111] and based on data of Johnson,[112] Ruggles,[113] and Krogius and Krokhin.[114]

or behavioral–is determined solely by its genes; no characteristic is determined solely by the organism's environment. It is never "either/or"; it is always "both."

Still, some behavioral patterns of animals are remarkably stable, whatever the environmental history of the animal. Such patterns early came to be known as instinctive or innate, in contrast to patterns learned through experience. Some behaviorists now believe the distinction between instinctive and learned behavior to be without merit;[64] they think the important matter is how behavior develops in the individual organism, not what category a particular pattern imperfectly fits. Undoubtedly, all behavior has heritable elements and individually acquired characteristics. Even so, there appear in animals patterns of behavior not much influenced by environmental history (they require little or no learning) and these patterns are very important to the persistence of the populations of the individuals having them. The development of other patterns is more dependent on the environmental history of the individual organism (it learns these patterns from its experiences) and such learned behavior may help the organism to better adapt to its environment.

The operation of behavior can be explained, in principle if not in actuality, at any level of biological organization from the whole organism down to nervous, hormonal, and genetic control systems. Full explanation requires the complete picture. Patterns of movement of an animal in relation to elements of its environment, including other organisms of the same and different kinds, are behavior in operation and can be described in order to provide one level of operational explanation. Explanation of how these patterns come into being in a very short time sequence, however, requires knowledge of an animal's perception of environmental stimuli, transmission of signals to nervous or other control centers, redispatching of information to bring about effective activities, and the mechanisms and processes integrating these activities so that they serve their functions. Some of this knowledge exists, but it accumulates slowly. If we recognize the biological indeterminacy problems, we will not expect ultimates in explanation. Still, as we learn more, we will come to better predict the effects of environmental change on animals we value.

The ability of a species of animal to behave as it does has evolved over long periods of time, much as have its morphological and physiological characteristics. These have not evolved separately but all together. Those behavioral characteristics that have favored the survival and reproduction of individuals of the species have persisted in the face of natural selection. We can suppose that nearly all the behavioral patterns we observe in animals have adaptive value, that somehow they make the persistence of its species more probable in its natural environment. The behavior of an animal can, within some limits, help it to adapt to changes man causes in its environment. If these changes go beyond the limits of behavioral adaptation, however, or if they cause pathological changes in an animal's behavior, the animal will not survive to grow and reproduce successfully, and its population will disappear just as surely as if the changes were immediately lethal.

Most if not all the activities of animals, except perhaps man, tend to insure the persistence of their kinds. This, then, is the function of animal behavior, whatever the pattern of behavior, so long as it is normal. For a population to persist, a sufficient proportion of its individuals must avoid harm, find suitable space and food, feed, migrate as necessary, and survive to reproduce successfully. Particular patterns of behavior function to fulfill these persistence requirements. Different patterns are involved in territorial behavior, reproductive behavior, feeding behavior, migratory behavior, and avoidance behavior.

Individuals of many species at some stages of their life histories have rather fixed space requirements; these individuals will defend such spaces as territories. Salmonids in streams do this by aggressive behavior involving intricate displays, and so tend to regulate the size of their populations and insure adequate food supplies.[65,66] In some species, reproductive behavior involves territorial behavior as well. Studies have led to descriptions of this and the complex behavior leading to mating and the care of young.[67,68] Successful reproduction of each kind of fish and other animal is only possible under certain environmental conditions.[69]

Each kind of animal has its own food preferences and habits. Its success depends not only on the availability of its food but also on its feeding behavior. This type of behavior within limits is adaptable to changing food opportunities, but it may be harmfully altered by toxic or other environmental changes. Some species—the anadromous salmonids are striking examples—migrate great distances to feed or to reproduce. Any interference with such migrations is a critical matter in the life of these animals and is often a concern of man, who with his dams and waste discharges has caused so much of this sort of interference. Some animals, including fish, have the ability to avoid certain harmful conditions if given the opportunity. Avoidance behavior may then lead to their survival, but it may also prevent their utilizing water areas otherwise suitable. I have reviewed these and other matters so important to the survival and reproduction of aquatic animals elsewhere,[20] so perhaps we should conclude this section with but a few final comments on a neglected area of water pollution biology.

The behavior of animals is difficult to study, either in the laboratory or in nature. This is probably the main reason so few attempts have been made to determine how pollutional conditions may affect the behavior of animals and thus their success in nature. Studies on the avoidance behavior of fish[16,70] have, however, indicated that fish in the laboratory are able to avoid certain harmful conditions. Such results are difficult to relate to nature because of reduction of spatial dimensions and consequent increase in steepness of environmental gradients in the laboratory experiments. Nevertheless, if we remain aware of such alteration and of other uncouplings of the natural environment when we interpret laboratory results—if we remember the biological indeterminacy problem—they may be helpful in explaining the effects of environmental changes on populations in nature. This is so for laboratory studies of other kinds of behavior, such as reproduction, feeding, and territorial behavior.

Sometimes we can find in nature circumstances that permit us to relate pollutional changes to changes in the behavior of fish and other animals. Effects of toxic substances on the migratory behavior of salmonids have been studied with reasonable success.[71] Unfortunately, the indeterminacy problem plagues us not only in the laboratory but also in nature, where conditions of interest vary and where other operating conditions are unknown. We sometimes are able to bring some experimental control over conditions in nature, as has been done with the influence of temperature and carbon dioxide concentration on the migratory behavior of fish.[72] Either way, we should not miss an opportunity.

The foregoing comments are intended only as warnings, not to discourage serious attempts to study how pollutional changes may harmfully alter the behavior of aquatic animals. These are needed and needed badly. The success or the failure of aquatic populations is as dependent on the behavior of the individuals as on any other of their characteristics.

Animals are behaviorally adapted in remarkably complex and sensitive ways to their natural environments. Change the behavior or the environment much and the behavior may no longer be appropriate or sufficient. Adequate explanation and prediction of the effects of environmental changes on animal populations cannot ignore behavioral responses.

POPULATION BIOLOGY, PRODUCTION, AND YIELD

The Biology of Populations. As organismic systems, populations lack certain obvious distinctiveness and cohesiveness possessed by individual organisms: they generally have no clear structural boundaries and are more loosely organized. Yet, in a sense, populations have a cohesiveness through time no individual can possess: with death individuals disintegrate, whereas populations may persist for many thousands of years by reproducing and adapting to environmental changes no individual could tolerate. Even so, populations disappear when environmental changes go beyond their capacity to adapt. This may be the natural course of events, but, since the advent of man, it has often resulted from the changes he has wrought so extensively and so rapidly.

The capacity of a population to maintain itself—as with any organismic system—resides in the characteristics and interrelationships of its parts, its individual organisms. To maintain itself, a population must replace through reproduction individuals lost to death, must regulate or be regulated so as to become neither too large nor too small for the area it occupies, and must in each generation and through many generations adapt to short and long-term environmental changes. If the individuals of a population are able to respond to the environment physiologically and behaviorally to maintain themselves and, on the average, to just replace themselves through reproduction, the population will persist. Over a few generations, a population will replace, regulate, and adapt itself according to existing capacities of its individuals and the extent of environmental change. Over many generations, a population can adapt to environmental changes too great for any single generation, because variation in the genetic capacities of its individuals and natural selection will lead to later generations of individuals more suited to the new environmental conditions.

Any population at any time has the advantage of the adaptive capacities of its individuals, but through time it has an additional adaptive advantage: through evolution, successive generations of individuals come to have capacities to face new environmental conditions.

In earlier sections, we have considered the capacities of individual organisms. We must now briefly consider some of the outcomes of these capacities in terms of the biology of populations, for it is maintenance of populations, not individuals, that concerns us in water pollution control.

A population will increase in numbers so long as its overall reproduction rate is greater than its overall death rate. As population numbers increase in a given area, there is a tendency for reproduction rates to decrease and death rates to increase, until, on the average, they balance, and the population attains an equilibrium that may or may not be stable. Were this area to have provided a more suitable environment, the population would have attained a higher numerical density before reproduction and death rates balanced. On the other hand, were the suitability of this area to have lessened, say by pollutional degradation of the environment, then the balance between reproduction and death would have been reached at a lower numerical density. If the degradation were sufficient to prevent reproduction from balancing death at any density, the population would then be eliminated.

The tendency for reproduction rates to decrease and death rates to increase as populations increase in numerical density leads to a certain regulation of population numbers. Intraspecific and interspecific competition for resources such as food and space are perhaps generally most important in this, but predation and parasitism may also be very important. This sort of regulation of population numbers does not lead to very constant numbers. Rather, such regulation as it generally exists merely tends to prevent populations from increasing to levels leading to the destruction of their resources and to prevent populations from decreasing to extinction for other reasons.

Generally, then, populations fluctuate according to their characteristics and the characteristics of their environment, sometimes greatly, sometimes not so much, sometimes regularly, sometimes irregularly. All this can occur apart from any effects of man's activities.

The density-dependent interactions of particular populations and species and their environments often lead to characteristic patterns of population change. Figure 5-19 illustrates for hypothetical species various relationships that may exist between the numbers of eggs produced by a parent stock and the numbers of eggs produced by the next generation. At low densities, parent stocks tend to more than replace themselves; they less than replace themselves at high densities, sometimes very much less. The populations fluctuate in nature, sometimes in ways rather like those that can be determined from such idealized reproduction curves as shown in Fig. 5-19.[73] Out of such relationships comes a kind of regulation and stability, even though it may be an oscillating one.

We are often surprised when aquatic populations disappear because of man's activities, but we should not be. It is more amazing that so many populations persist in spite of the changes man has caused in their environments. But even this should not surprise us. The populations now thriving in environments changed by man are composed of individuals that are genetically adapted to those environments. These individuals are genetically, morphologically, physiologically, and behaviorally different from their ancestors, most of which would not have had the capacity to survive to reproduce under the new conditions. In each generation, there are some individuals genetically more suited than

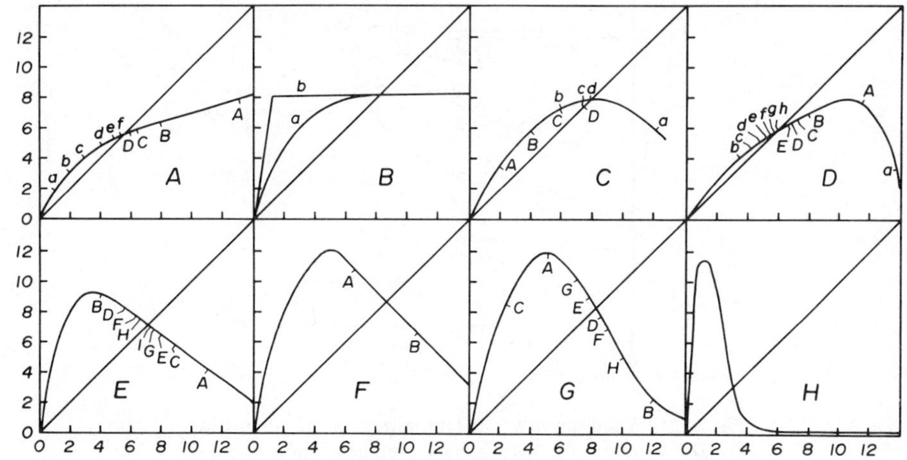

Fig. 5-19 Hypothetical reproduction curves whose abscissas represent the number of eggs produced by a parent stock in a given year and whose ordinates represent the number of eggs later produced by the progeny of that year. After Ricker.[73]

their fellows to existing environmental conditions, and these individuals contribute relatively more progeny to the next generation and more to the genetic characteristics of successive generations. Thus, differences in the genetic characteristics of individuals and natural selection of those most suited to reproduce under existing environmental conditions tend to insure that populations persist as environments change, by processes something like those illustrated in Fig. 5-20. There are limits to this, however, as exemplified by the all too often disappearance of populations and even species.

Why have I again burdened the reader with matters that may seem only of interest to biologists? It is because biological investigations of water pollution have not generally considered these important matters. They are difficult to study; they are difficult to take into account. Nevertheless, we cannot explain and predict the effects of pollution on aquatic life without knowing how populations of interest respond to their environments under controlled and uncontrolled conditions. We may study the effects of pollutional changes on the survival and reproduction of animals in the laboratory, but it is in nature that this concerns us. We need the laboratory studies, but we also need knowledge of the vital statistics of natural populations under different conditions, and there are routine methods for obtaining these.[74] We need to apply what we know about the ways natural populations fluctuate and are regulated to better interpret what we learn about changes in population numbers under pollutional conditions.

Beyond this, the responses to changes in environmental conditions we determine with samples of organisms from a population at one time are not likely to be the responses of the individuals of that population some generations later. The procedures of modern quantitative genetics[75] should make it possible for us to predict the capacity of at least some important populations to adapt genetically to environmental change.

The Production of Populations. In the management of any biological resource, man is interested in the amount or *biomass* of that resource

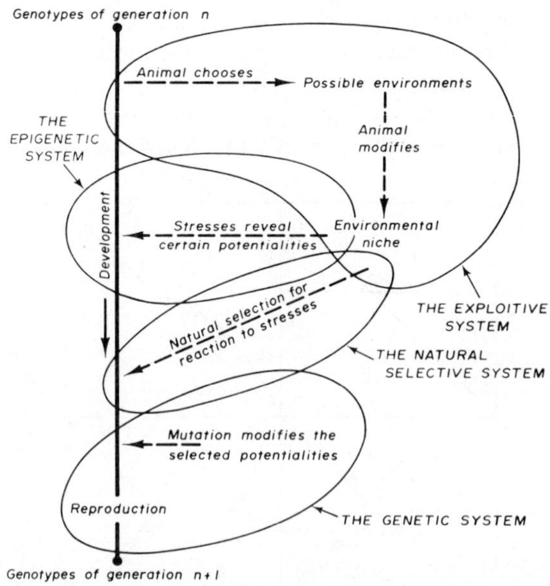

Fig. 5-20 Waddington's conceptualization of an evolutionary system. Changes in gene frequency in a population between successive generations depend upon the operation of four subsystems: the exploitive, the epigenetic, the natural selective, and the genetic. After Waddington.[115]

existing in an area at any time, for this is a matter of both practical and biological importance. But man is generally more interested in the ability of a biological resource to replace its biomass should man remove any part of it, as well as in its ability to accumulate more biomass should rates of removal be reduced. The total elaboration of new biological material, whether it is removed by man and other organisms or whether it accumulates as new biomass, is generally termed the *production* of the population constituting the resource. Thus, as defined by biologists, production is not what man realizes from a resource—this we will call *yield*—but rather is the total elaboration of biological material, whatever its fate (with one minor exception that we will consider later).

The fundamental unitary process involved in production is the growth of the individual organisms forming the population. Production of a population is only the total growth of all the individuals in the population over the period of interest. Thus, for any period of time, production, P (gm/day), is the product of the mean growth rate of the individuals, G (gm/kg/day), and the mean biomass, $\overline{B}$ (kg), of the population:

$$P = G\overline{B} \tag{7}$$

For convenience, biomass (kg/m^2) and production (kg/m^2/day) are generally expressed per unit area. Because populations are usually composed of different age classes of individuals having different growth rates and making up different proportions of the biomass, production must generally be determined for each age class separately and then summed to determine the total production of the population. Because growth rates and biomasses change over the year, production determinations should be made over shorter periods of time and summed over the total time period of interest.

Though conceptually simple, determination of the production of animals is methodologically difficult. Methods for determining the production of fish are generally practical and satisfactory.[76] This is not so for fish food organisms and other invertebrates.[77]

Table 5-2 illustrates arithmetic procedures for calculating the production of one age class of fish for a year, and it also includes data necessary for an equivalent graphical procedure illustrated in Fig. 5-21. Both procedures take into account negative production resulting from negative growth rates, the one fate of elaborated materials that properly must be excluded from production estimates.

Production is clearly a function of biomass (Eq. 7). In aquatic systems in which nutrient resources are limiting, growth rate is also a

TABLE 5-2. Computation of Production of a Single Age Group of Fish for 12 Months on the Basis of Hypothetical Data[a]

Date	Mean Weight ($\overline{w}$, gm)	Growth Rate (G)	Stock Number (N)	Stock Biomass (B, kg)	Mean Biomass ($\overline{B}$, kg)	Production (P, kg)
May 1	1.5	0.29	8000	12	10.5	3.0
June 1	2.0	0.22	4500	9	8.8	1.9
July 1	2.5	0.34	3500	8.7	9.6	3.3
Aug 1	3.5	0.26	3000	10.5	10.8	2.8
Sept 1	4.5	0.37	2500	11.2	11.6	4.3
Oct 1	6.5	0.06	2000	13.0	13.0	0.7
Nov 1	6.9	0.00	1900	13.1	12.4	0.0
Dec 1	6.9	−0.01	1700	11.7	10.9	−0.1
Jan 1	6.8	−0.03	1500	10.2	9.7	−0.3
Feb 1	6.6	0.00	1400	9.2	8.9	0.0
Mar 1	6.6	0.04	1300	8.6	8.0	0.3
Apr 1	6.9	0.07	1100	7.5	7.2	0.5
May 1	7.4		1000	7.0		
Annual production (ΣP)						16.4

[a]After Ref. 76.

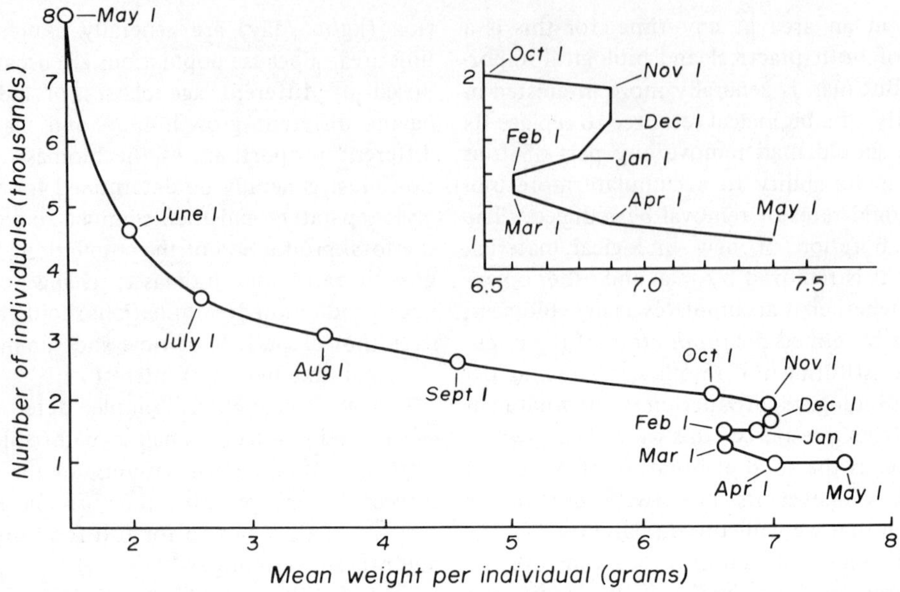

Fig. 5-21 Graphical estimation of the production of a single age group of fish for 12 months on the basis of hypothetical data given in Table 5-2. Production between any two dates is represented by the area under the curve between those dates. When negative production occurs, as from December 1 to January 1, this must be subtracted from the sum of production values for all other months in order to estimate annual production. All areas must be determined between the segment of the curve and the abscissal axis. After Chapman.[76]

function of biomass—a negative function so long as the capacity of the system to produce the product of interest remains constant. In such a system, increases in the biomass of a population decrease the availability of nutrient materials for each individual, and individual growth rates will decline as shown in Fig. 5-22. Thus, production—the product of growth rate and biomass—will increase from low levels at low biomasses to a maximum at intermediate biomasses and then decline with further increases in biomass.

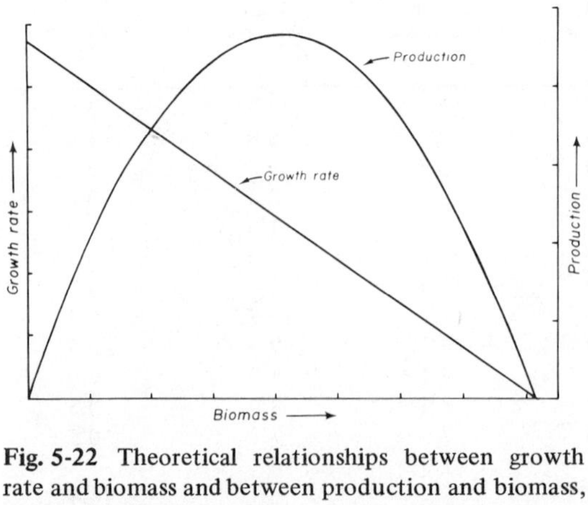

Fig. 5-22 Theoretical relationships between growth rate and biomass and between production and biomass, production being the product of growth rate and biomass. After Brocksen, Davis, and Warren.[78]

Definition of such biomass-production curves is not always possible in natural systems because sufficiently wide ranges of biomass may not occur. However, these relationships, or useful parts of them, can often be defined (Fig. 5-23). Even when this is not possible, it remains very important in the interpretation of existing production data to bear in mind that different biomasses would have led to different production values according to these fundamental relationships.[20]

Sometimes it is possible to determine the biological basis of a biomass-production relationship, and our explanatory and predictive powers are thereby increased. Increase in the biomass of the product of interest may lead to a measurable decline in the biomass or density of its food resource (Fig. 5-24). This decline in food availability may then lead to a decline in the growth rates of the individuals in the population of interest (Fig. 5-25). Thus, the forms

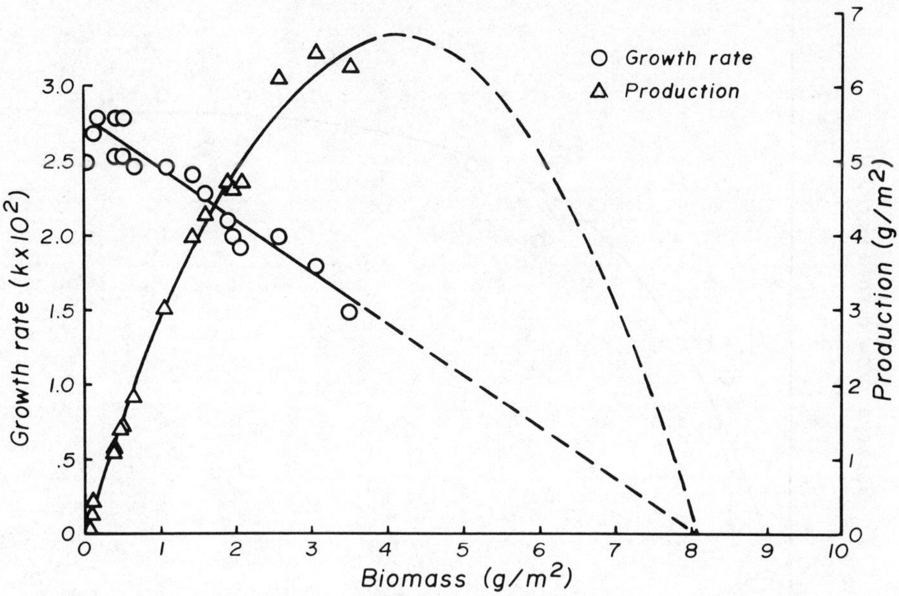

Fig. 5-23 Relationships of mean daily growth rate and production to biomass of juvenile sockeye salmon in the Babine-Nilkitkwa lake system, British Columbia, during 4-month periods from June through September, 1956 and 1957. After Brocksen, Davis, and Warren,[78] based on data of Johnson.[112]

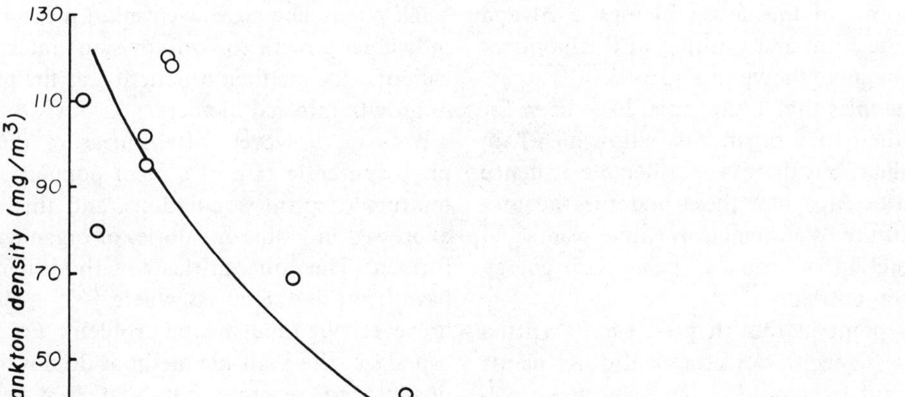

Fig. 5-24 Relationship between mean biomass of juvenile sockeye salmon and mean density or biomass of the zooplankton upon which they fed in the Babine-Nilkitkwa lake system, British Columbia, during 4-month periods from June through September, 1956 and 1957. After Brocksen, Davis, and Warren,[78] based on data of Johnson.[112]

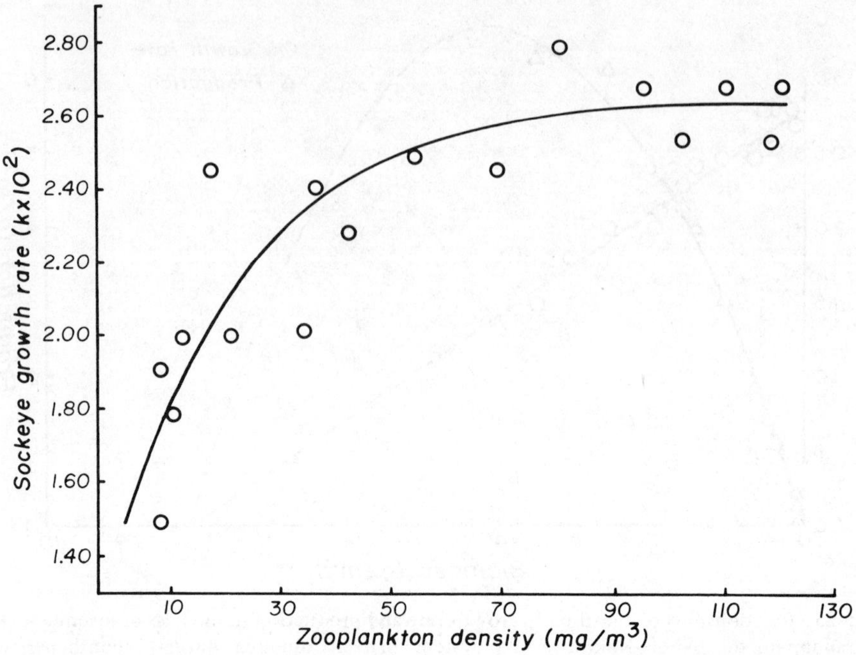

Fig. 5-25 Relationship between mean daily growth rate of juvenile sockeye salmon and mean density or biomass of the zooplankton upon which they fed in the Babine-Nilkitkwa lake system, British Columbia, during 4-month periods from June through September, 1956 and 1957. After Brocksen, Davis, and Warren,[78] based on data of Johnson.[112]

and positions of the curves in Figs. 5-24 and 5-25 fix the form and position of the biomass-production curve shown in Fig. 5-23.

The examples that I have used have been for fish and their food organisms (with which I am most familiar) but there is considerable evidence that relationships like these underlie the production of the food organisms eating plants and the production of plants utilizing solar energy and mineral nutrients.[78,20]

To this point in this chapter, I have written little that is specific for algae and other plants so important in aquatic systems and water pollution biology. Nevertheless, nearly everything that I have written, excepting that on animal behavior, is generally applicable to problems of plant as well as animal biology. The fundamental difference between autotrophic plants and heterotrophic animals is their manner of nutrition. Plants, through photosynthesis, utilize solar energy and mineral nutrients for their maintenance and growth. The growth of individual plants leads to the *net production* of their populations. Unfortunately with very small plants like algae, we cannot measure their individual growth (or cell division) rates, so we cannot calculate their production as the product of growth rate and biomass.

We can, however, often measure the total photosynthetic rate of a plant population or a mixture of plant populations, and this can be expressed in grams or calories of organic matter formed. The rather intricate methods necessary have been described elsewhere.[79,80] There is, however, one fundamental problem. For plankton algae, the available methods do not make it possible to separate out that part of total photosynthesis (gross production) that results in cell elaboration (net production). We can generally know only *gross production*—the sum of net production and respiration—because measurement of respiration of just the plant part of a biological community has not been satisfactorily accomplished.

The Yield of Populations. Man's utilization of a population of shore birds or a virgin forest may be purely aesthetic, and many populations are justifiably protected for this reason. But man also needs food and fiber, and this entails removal of animals and plants from their population.

The amount man removes from a population is the yield of that population to man. The yield of populations is dependent on social and economic factors as well as on biological factors. Declines in yields of fish populations have sometimes occurred during periods of increasing water pollution and have been attributed to water quality degradation, often probably correctly, but sometimes incorrectly. Overfishing can so deplete populations as to reduce their yield, and there are still other considerations.

Those of us concerned about the possible effects of water quality changes on aquatic resources need to be somewhat familiar with the phenomena determining yield if we are to interpret as correctly as possible such information as may be available. Only a few of the fundamental concepts of the theory of exploitation, which is rather highly developed in the field of fisheries, are presented here.[81,74] However, the following may still be of some value to uninitiated readers.

A population is in equilibrium when increments to its biomass through production are just balanced by decrements resulting from natural mortality and removal by man (yield); its biomass will not change so long as there is no emigration or immigration. If production increments are greater than mortality and removal decrements, then biomass will increase; if smaller, biomass will decrease. A *biomass equilibrium* can occur at any level of biomass consistent with the capacity of an ecosystem to support a population. It need not be the *maximum maintainable biomass.*

An *equilibrium yield* exists for each level of biomass; it is the yield that would lead neither to an increase nor to a decrease in biomass. Equilibrium yield is a curvilinear function of biomass (Fig. 5-26B), much as production is a curvilinear function of biomass (Fig. 5-22). At very low biomasses and at biomasses near the maximum maintainable biomass, equilibrium yield is near zero. *Maximum equilibrium yield* occurs at some intermediate biomass.

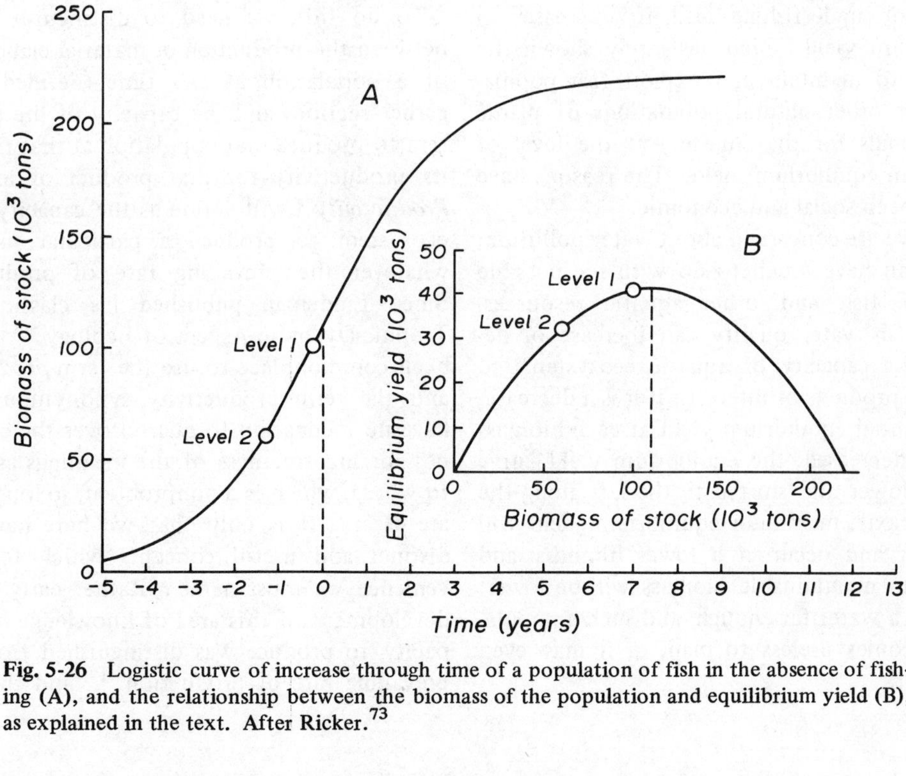

Fig. 5-26 Logistic curve of increase through time of a population of fish in the absence of fishing (A), and the relationship between the biomass of the population and equilibrium yield (B), as explained in the text. After Ricker.[73]

Figure 5-26 illustrates the relationship between population growth in biomass and the curve of equilibrium yield, with an example having some factual basis in the North Sea fisheries. Unexploited populations tend, through time, to increase in biomass in a way that can be described by an asymmetrical sigmoid curve, such as the *logistic curve* shown in Fig. 5-26A. The rate of biomass increase tends to decline with increase in population biomass, until, at the maximum maintainable biomass, population biomass increases no more. This is generally supposed to result from decreases in reproduction and growth and increases in mortality related to the increasing density or biomass of the population.

If at the point in time when the biomass of the population illustrated in Fig. 5-26 reached 100,000 tons (Level 1) man should begin to remove 40,000 tons annually (the equilibrium yield for that biomass), the biomass would no longer increase, nor would it decrease. Similarly, at the biomass of 60,000 tons (Level 2), annual removal of the equilibrium yield of 32,000 tons would lead to no further changes in biomass. Thus, the equilibrium yield of a population is very much dependent upon its biomass. Overfishing or underfishing lead to decreases in equilibrium yield. Man has rarely shown the wisdom to maintain and exploit fish populations (or other natural populations of plants and animals for that matter) at the level of maximum equilibrium yield. The reasons have usually been social and economic.

Here we are concerned about water pollution, which can have much to do with the possible yield of fish and other aquatic resources. Changes in water quality can increase or decrease the capacity of aquatic ecosystems to produce products of interest. If it is a decrease, the potential equilibrium yield at each biomass will be decreased; the equilibrium yield curve will be lower and shifted to the left along the biomass axis; maximum equilibrium yield will be lower and occur at a lower biomass; and maximum maintainable biomass will be lower. Degrade a water far enough, and such a population becomes useless to man, or it may even disappear.

PRODUCTIVITY OF ECOSYSTEMS: THE CAPACITY TO PRODUCE

The Meaning of Productivity. In the previous two sections on production and yield, we have considered certain relationships between the biomass of a population and the relative growth rate of its individuals (Figs. 5-22 and 5-23), its biomass and its production rate (Figs. 5-22 and 5-23), and its biomass and its potential equilibrium yield (Fig. 5-26). Everything previously said and the relationships illustrated by the figures will hold only so long as the capacity of the ecosystem to produce the product of interest remains constant. From season to season, however, even from month to month, the capacity of ecosystems to produce a particular product does not remain constant.

What then can be the value of the foregoing discussion and illustrations? They do have value, a great deal of value, but only as a part of a much broader conceptual framework; one that incorporates the concept of capacity to produce, that predicts the growth, production, and yield outcome of changes in this capacity, and that provides a way of explaining changes in outcome and capacity to produce.

To do this, we need to distinguish clearly between the production or material elaboration of a population at any time (defined in an earlier section) and the capacity of the ecosystem to produce that population at that time, or its productivity for that product of interest. *Productivity* I will define as the capacity of an ecosystem to produce a particular product, whatever the prevailing rate of production. Since Lindeman published his classic "The Trophic-Dynamic Aspect of Ecology,"[82] it has been commonplace to use the term *production* and the term productivity synonymously. I have no inclination to quarrel over the correctness or incorrectness of the meanings assigned to words, which is a nonproblem, so long as we are clear. It is only that we here have two distinct and useful concepts which for convenience we must name. Rather early in the development of this area of knowledge, the capacity to produce was distinguished from the prevailing rate of production,[117] and the concepts were defined and named much as I have done above.

What are the dimensions of—what is involved in—the concept of productivity? First, let us cover the whole matter with a general statement: any change in the rate of production that is not caused by a change in the biomass or age composition of the population of interest is caused by a change in the productivity of the ecosystem for that population. If production rate changes and there is no change either in the biomass or in the age composition of the population, we can say that productivity has changed.

How does such a statement logically follow from the foregoing discussion and what is its possible utility? In beginning this section, I stated that all we considered in the previous two sections was valid only so long as the capacity of an ecosystem to produce a product of interest did not change. This is to say that, with increasing population biomass, the relative growth rates of individuals will decline and production will lie along a single hump-shaped curve only so long as productivity remains constant. Therefore, production changes in a predictable manner even if productivity does not change. A change in the age composition of a population may also change production at the same level of productivity, but the production rate can change for reasons other than changes in population biomass or age composition; so can potential equilibrium yield. We must then determine the other phenomena that could change production, for these are the dimensions of productivity.

We will deal with this matter more concretely in the following two sections, but now I can generally indicate the kinds of phenomena that determine productivity and thus the production of a population having a given biomass and age composition. Starting at the top of the system of trophic or nutritive relations, we can say that a species competing for resources (i.e., food or space) with the species or population of interest will tend to reduce the productivity of the ecosystem for that population. In the food web of the population, any competitive relations that reduce the flow of energy and materials to the population of interest reduce the productivity of the system for that population. Furthermore, changes in solar radiation and changes in the introduction or regeneration of mineral nutrients may change productivity; so will changes in water quality parameters such as temperature, dissolved oxygen, and toxic substances. Our problem now is to develop a system of concepts subsumed by and adequate for dealing with such a high level abstraction as productivity.

Relations Between Productivity and Production. Before considering the ideas and relations subsumed by the concept of productivity, it will probably be helpful to the reader for me to be more concrete as to the outcomes in growth, production, and biomass relationships that might be expected were productivity to change in any of its dimensions. Examples from three lake ecosystems producing juvenile sockeye salmon (*Oncorhynchus nerka*) will suffice. Different ecosystems of the same sort producing the same species of interest provide a valid example of the outcomes of changes in productivity, if the systems differ in productivity, for the argument is the same for ecosystems differing in productivity as for a single ecosystem having productivity changing through time.

Figure 5-23 illustrates the decline in growth rate and the increase in production (which would have begun to decline with further increases in biomass) occurring with increases in biomass of juvenile sockeye salmon in the Babine-Nilkitkwa lake system of British Columbia during time periods having relatively constant and equal productivity. The growth rate and production relations abstracted from the data are represented as one of three sets of relations in Fig. 5-27. The other two sets of relations are based on data from two other ecosystems producing juvenile sockeye salmon—Owikeno Lake in British Columbia and Lake Dalnee in Kamchatka. These two ecosystems differ greatly in their capacity to produce these fish (productivity). Thus, it can be seen that growth rate and production increase with increasing productivity at a given biomass. Growth rate declines more steeply with increasing biomass in the less productive systems.

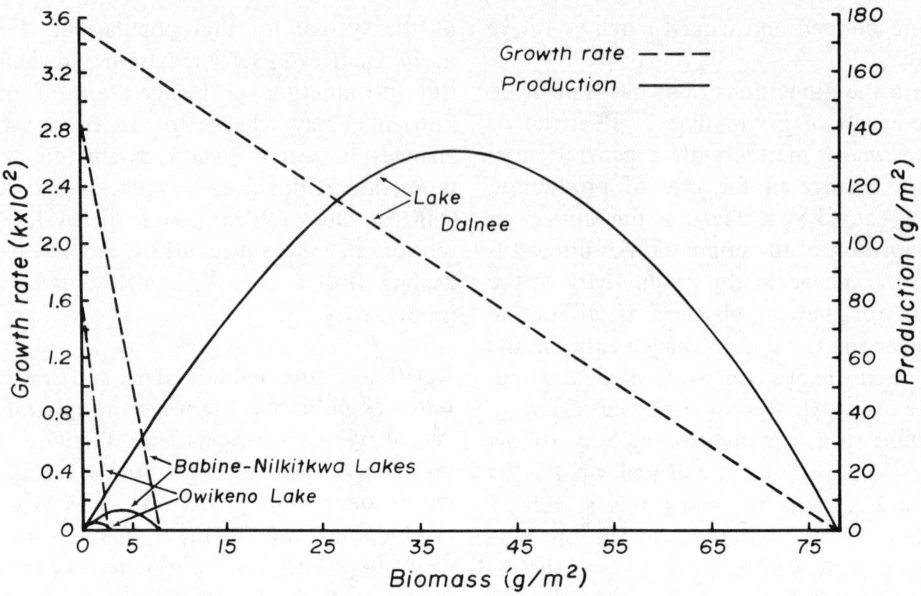

Fig. 5-27 Comparison of relationships of the biomass of juvenile sockeye salmon to their growth rate and production in Owikeno Lake, the Babine-Nilkitkwa system, and Lake Dalnee. After Brocksen, Davis, and Warren,[78] based on data of Ruggles,[113] Johnson,[112] and Krogius and Krokhin.[114]

Production is higher at given biomasses and across a wider range of biomasses in the more productive systems (Fig. 5-27).

In an earlier section, I defined for the Babine-Nilkitkwa system the relationships between sockeye salmon density and the density of their food organisms (Fig. 5-24) and between food organism density and slamon growth rate (Fig. 5-25) that determined the form and position of the relationships between growth rate, production, and biomass of salmon (Fig. 5-23), all at one level of productivity. Now I must explain how differences in productivity determine the forms and positions of relationships between the salmon and their food organisms and thus the forms and positions of the salmon production curves for the three lake systems.

Changes in the Physical and Biological Bases of Productivity. I have suggested that, during any period of constant productivity, the relationship between the biomass of a consumer and the density or biomass of its food organisms tends to be inverse, that the density of food tends to decline with increasing consumer biomass (Fig. 5-24). The form and position of such inverse relationships will be a function of the productivity of the ecosystem for the product of interest. Thus, as illustrated in Fig. 5-28, the midpoints of curves defining the inverse relationships between a consumer and its food organisms will tend to move upward and to the right with increases in productivity. This is to say that, during periods of changing productivity, the relationship between the biomass of a consumer and the biomass of its food organisms will tend to be direct. Because periods of constant productivity can be rather short, a given inverse relationship may not be fully defined by available data before a change in productivity tends to shift the system to a level described by a higher or lower inverse relationship. Thus, there are many possible curves along which data points could move during periods of changing productivity as the system shifts from one level to another, not simply one direct relationship between consumer and food organism biomasses as shown in Fig. 5-28. It is only during periods of constant productivity that changes in consumer biomass will tend to be inversely related to

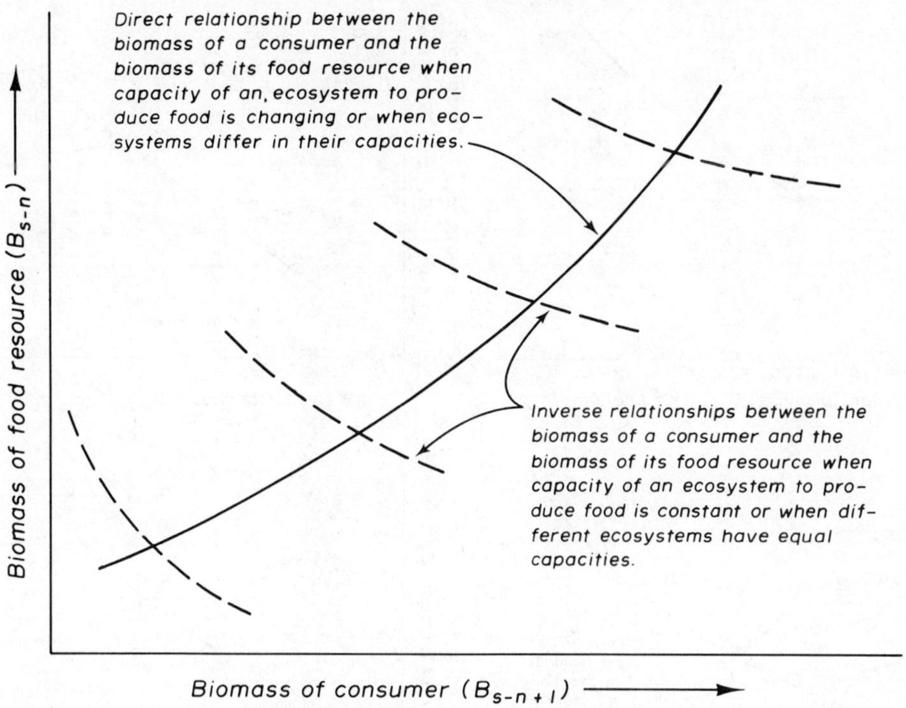

Fig. 5-28 Theoretical relationships between the biomass of a consumer and the density or biomass of its food resource. After Brocksen, Davis, and Warren[78] and Warren.[20]

changes in food biomass, whereas during periods of changing productivity, changes in consumer biomass will tend to be directly related to changes in food biomass.

In principle, the relationships illustrated in Fig. 5-28 hold not only at one step, say that between a fish population and its food organisms, but they hold also at each previous step along the trophic pathway leading from solar energy and mineral nutrients to the fish. The conceptual system we have been considering was developed over a number of years at our laboratory mainly on the basis of laboratory stream communities,[111,83,20] but has been applied to natural communities[78,20] such as those we have considered for sockeye salmon.

Figure 5-29, based on labroatory stream community studies having cutthroat trout (*Salmo clarki*) as the product of interest, illustrates how systems of direct and inverse relationships may be definable at each step along a trophic pathway leading to a product of interest. Such relationships are not always clearly definable—phenomena we do not yet know how to take into account are sometimes operating—but, over a period of several years, we have found relationships like these can often be defined, and we have been able to explain effects of pollutional changes not otherwise explainable.

I can now briefly suggest how various dimensions of productivity can operate in terms of their partial outcomes, as illustrated in Fig. 5-29. The inverse relationships shown for the four trophic steps in Fig. 5-29A, B, C, and D are sequentially related; that is, a relationship at one step roughly corresponds to a relationship at the same level on the next higher or lower trophic step. The level in any sequence was, in these experiments, determined primarily by the level of light energy input (Fig. 5-29A), one of the dimensions of productivity. An increase in the rate of introduction of mineral nutrients at high light levels would have similarly increased the level of productivity. Were another species of fish introduced into the streams, it could compete with the trout for food and so lower the level of insect biomass at any trout biomass, thus decreasing the productivity for trout (Fig. 5-29C). If a snail, not utilizable by the trout for food, were to have been introduced, it

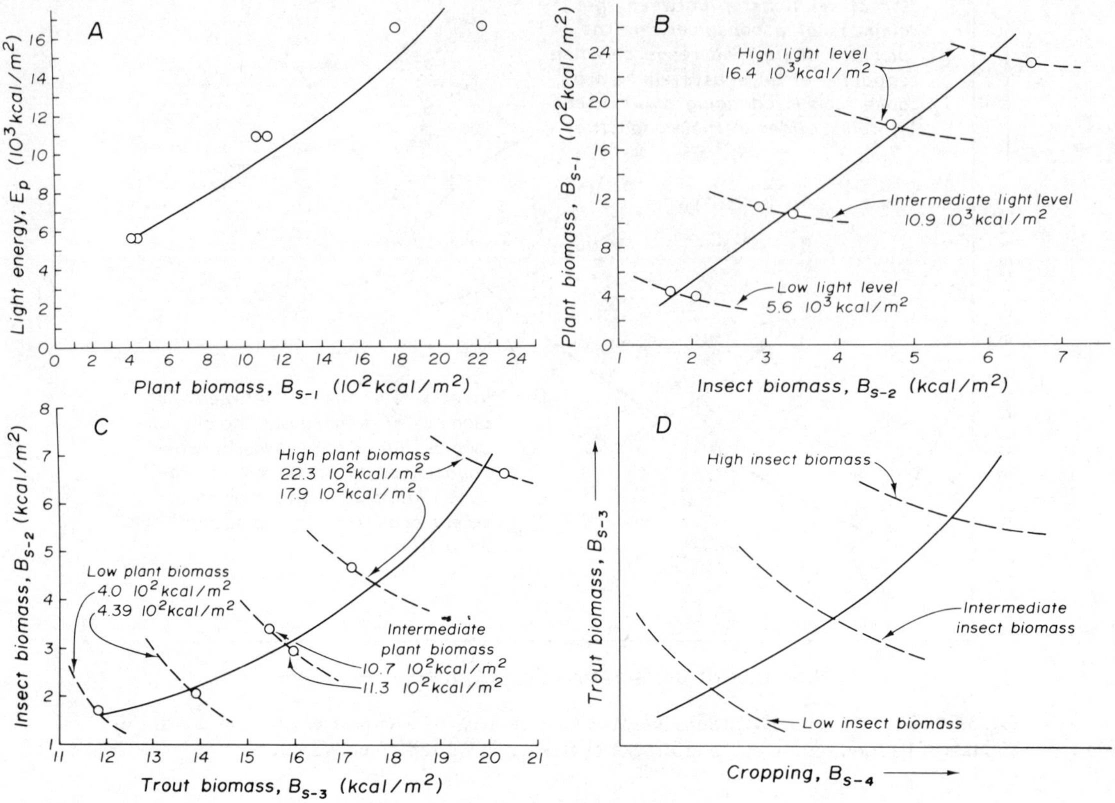

Fig. 5-29 Density relationships between successive steps along the pathway of energy and material transfer leading from light energy to the production of cutthroat trout in laboratory stream communities during a spring 1968 experiment. Only the solid lines defining relations between systems having different productive capacities are reasonably well defined. The broken lines only suggest the probable inverse relations that would exist at each level of productive capacity were more streams tested. After Brocksen and Warren[83] and Warren.[20]

would reduce the level of algae at any level of insect biomass, reduce the production of insects at that biomass, and so the productivity of the system for trout. Changes in temperature or the introduction of toxic substances could also change the productivity, and this would be indicated by upward or downward shifts in the positions of the inverse relationships.

Before we can define the production curve of a product of interest at any level of productivity, we must determine that other relationship between a consumer and its food, the relationship between the biomass or availability of the food and the relative growth rate of the consumer. Figure 5-30 does this for the cutthroat trout and the insects upon which they fed in the laboratory stream experiments we have been considering.

With the relationship between the growth rate of the trout and the biomass of its food (Fig. 5-30) and with any one of the inverse relationships between the biomass of the trout and the biomass of the insects (Fig. 5-29C), we can define one of the trout biomass-production curves shown in Fig. 5-31. If we do this with a higher inverse relationship, we define a higher and wider production curve at a higher level of productivity. On the other hand, if the relationship between insect biomass and trout growth rate were higher and to the left of the one shown in Fig. 5-30, this would define a higher production curve, even for the same inverse relationship. Thus, a change in temperature or the presence of toxic substances directly affecting the growth of the trout—and so their growth relationship—would change the production curve, even were the inverse relationship to remain unchanged. In

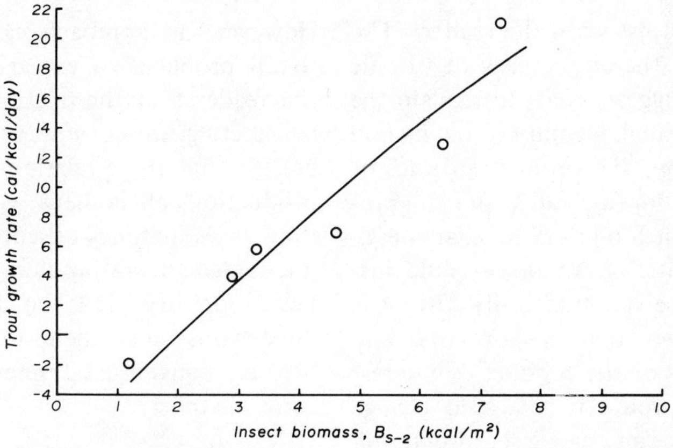

Fig. 5-30 Relationship between the growth rate of cutthroat trout and the density or biomass of the insects upon which they fed in laboratory streams having different productive capacities during a spring 1968 experiment. Data of Robert W. Brocksen, Oregon State University. After Warren.[20]

these and other ways, then, various environmental factors, as they interact in the trophic or nutritional complex leading to the production of products of interest, determine the productivity of ecosystems for these products.

Before leaving this section, there are two mat-

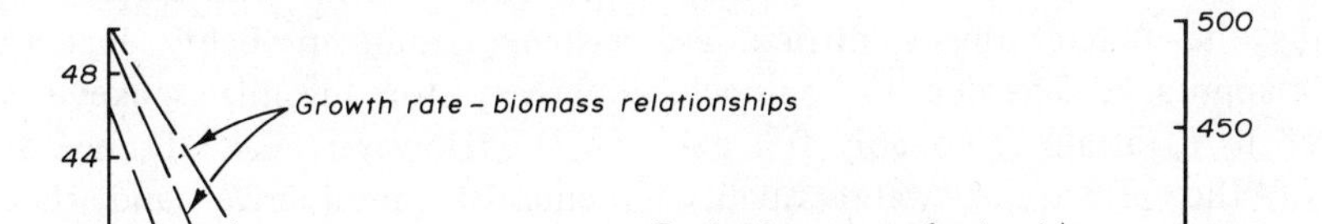

Fig. 5-31 Approximations of biomass-production curves for cutthroat trout in the laboratory stream ecosystems having 5 different basic capacities to support trout production in the spring 1968 experiment. The biomass-growth rate curves and the biomass-production curves were generated from relationships presented in Figs. 5-29C and 5-30. After Warren.[20]

ters about which I must warn the reader. The first has to do with the classic view of trophic relations[82] which suggests that, to explain the production of an animal, we must know its rate of food consumption, the production rates of its prey, their consumption rates, and the production rates of their food back to solar energy and mineral nutrients. Even if we could just once determine these simultaneously (30 years of effort has shown that we cannot), any shift in the biomass of the product of interest or any change in productivity would change all these rates. Thus, any single set of determinations, even if these could be made, would have little explanatory value. The classic approach, still followed by many, is utterly inadequate because of the biological indeterminacy difficulty.[20]

It was the need for an alternative approach that led us to develop the conceptual framework I have briefly explained. By making use of density relationships to explain the production of products of interest, this approach largely avoids the indeterminacy difficulties of the classic approach. The density approach has permitted us to attain reasonably full explanations of the effects of water quality changes on the productivity of aquatic ecosystems in work conducted at our laboratory. However, no approach can be fully adequate to all problems of explanation. I must warn the reader that the relationships we have been considering are not always clearly definable. I believe that these relationships underlie all the production phenomena we observe, but that they are sometimes obscured from our view by phenomena operating along dimensions we have yet to identify. It is the identification of these dimensions and their incorporation into a broader conceptual framework that now concerns me most.

Changes in Productivity, Production, and Yield. Environmental changes, those that result from man's activities as well as those that do not, lead to changes in the productivity of aquatic ecosystems. Some of these changes increase the capacity of systems to produce products of interest; others decrease such capacity. The introduction of additional mineral nutrients into the Owikeno and Babine-Nilkitkwa lake systems would probably increase their productivity for juvenile sockeye salmon (Fig. 5-27). However, Lake Dalnee is already tremendously productive, and the addition of nutrients into that system would be more likely to reduce its productivity for the juve-

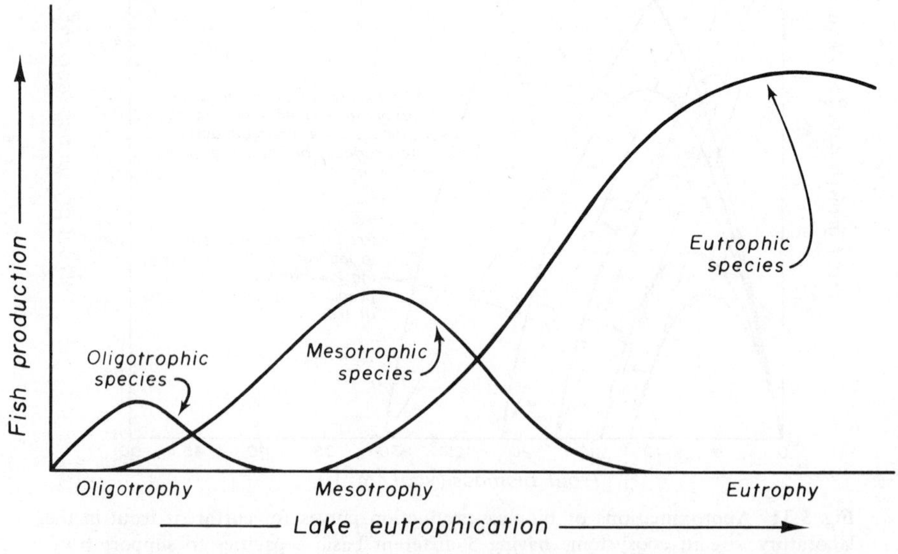

Fig. 5-32 Changes in the production of oligotrophic, mesotrophic, and eutrophic species of fish that would theoretically take place as a lake passed from the condition of oligotrophy to the condition of eutrophy through the process of eutrophication. After Warren.[20]

nile salmon. This might, however, increase its productivity for some other species of fish.

In terms of eutrophication, the Owikeno and Babine-Nilkitkwa systems are oligotrophic; Lake Dalnee is mesotrophic. Were Lake Dalnee to become eutrophic, it would no longer be an ecosystem suitable for sockeye salmon. Figure 5-32 illustrates how, through the process of eutrophication, the production of certain species first increases and then decreases. They may be replaced by species whose production reaches higher levels, but man may not value so highly the replacing species. He is then likely to judge the lake degraded, perhaps as a result of his own activities, perhaps by natural processes.

Increases in the productivity of a system for a product of interest increase the potential equilibrium yield of that product; decreases in productivity decrease equilibrium yield and at least one value of an aquatic system to man. In an earlier section I explained how at one level of productivity equilibrium yield is a function of the biomass of the population of interest (Fig. 5-26). A decrease in productivity will decrease the equilibrium yield at each biomass and shift to the left along the biomass axis the equilibrium yield curve. Both theory and some empirical evidence support this contention, but clear definition of these relations is not often possible with natural populations. With laboratory populations, such definition is quite possible. Figure 5-33 illustrates the different equilibrium yield curves for guppy populations fed at different rates, in order to simulate different levels of productivity. It is no chance thing that this family of equilibrium yield curves resembles the family of production curves for cutthroat trout shown in Fig. 5-31.

SUCCESSION, DIVERSITY, STABILITY, AND UTILITY OF COMMUNITIES

Beyond the Interest in Particular Populations. All too often, man does not see beyond his immediate interest in particular populations: a treasured fish population is declining, mosquitos are making summer evenings unbearable, the lake is being spoiled by blue-green algae. Such singularity of attention is understandable because it is human. We are concerned about immediate and concrete prob-

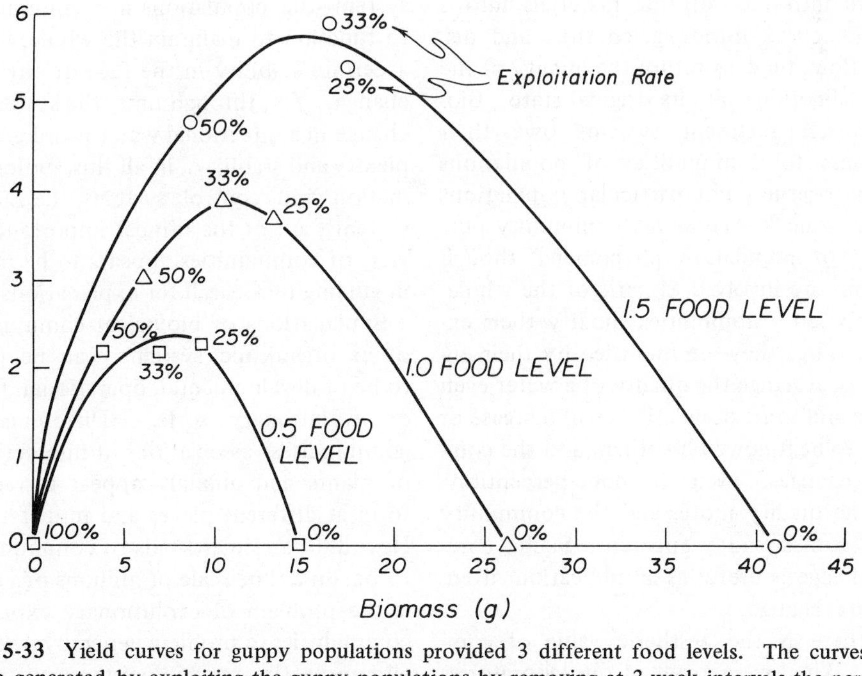

Fig. 5-33 Yield curves for guppy populations provided 3 different food levels. The curves were generated by exploiting the guppy populations by removing at 3-week intervals the percentages of biomass indicated beside the points along the curves. After Silliman.[116]

lems, developing and persisting sometimes because of our own activities, sometimes in spite of our efforts.

Population phenomena always have a community context. The distribution and abundance of populations are as much dependent on the characteristics of their communities as on physical and chemical conditions of their environments. Population problems may be a consequence of community alterations. Even when they are not, changes in populations that lead to problems cannot be understood apart from their community setting. One can understand unawareness of this on the part of a vacationer enjoying his summers less. Not so understandable are population research and management efforts that are apparently ignoring the community contexts of population problems. However, this has to do with concern over populations, concern over the proper orientation of population studies. What about concern for communities as wholes?

More than particular populations (excepting man), communities modify the chemical and physical properties of their environments. The quality of the water of lakes and streams is influenced by their aquatic communities. Where we introduce organic materials into a stream, its communities, given time and distance of flow, tend to return the quality of the water to something like its original state. Biological waste treatment systems owe their effectiveness to communities of populations operating together, not particular populations operating alone.[84] These are community phenomena, not population phenomena, though populations are involved as parts of the whole.

Not only do communities modify their environments, but they are modified by their environments. Change the quality of a water even very little and some populations will increase or decrease, to be followed by others, and the community changes, even if not perceptibly. Change the quality more, and the community change becomes readily apparent. Such a community change is useful as an indication of environmental change.

Then there is the aesthetic value of communities. When we get out of our laboratories and classrooms, when we get out of our offices and shops, few indeed among us are insensitive to the array of nature about us. We never outgrow our awe of a forest, a marsh, or a tidepool, all teeming with the life of their communities. This is important to us, and it should not be forgotten or neglected.

Beyond all this, man's life on earth became possible and will remain possible only because of the performance of biological communities. We don't know how much these communities can be changed without making life for man more difficult than it already is. It is not radical, indeed it is very conservative, to suggest that we should change these communities no more than is absolutely necessary.

The Succession, Diversity, and Stability of Communities. Early in this chapter I noted that a biological community is a system exhibiting organismic and cybernetic characteristics. This is not to say that a community is some kind of "superorganism"; this connotes far too much. Rather, it is to say that some important generalizations biologists have abstracted from their studies of individual organisms appear to apply also to biological communities. Organismic systems have a certain unity. The parts of the system—the populations in a community—tend to function to maintain the whole, to maintain a certain stability in the face of environmental change. Yet, through time, the system tends to change in a directional way toward greater complexity and stability. In all this, biological information and control systems, i.e., cybernetic systems, are of the utmost importance. Such a view of communities appears to be most useful in guiding our search for explanations.

Explanations of biological communities, like other organismic systems, can be considered to be of developmental, operational, functional, or evolutionary sorts. Different kinds of communities—associations of interacting species of plants and animals—appear in recognizable form at different places and in different times. How did particular kinds of communities come to be, on a time scale of millions of years? This is the problem of evolutionary explanation of communities, a problem generally treated rather circumspectly by biologists until recently.[85] However, here we best limit ourselves to developmental, operational, and functional explanations.

One kind of community can develop at different places or in different times, this process requiring less time than the evolutionary one. It is known as *community succession*. Upon an area denuded of life, a few hardy species of plants and animals come to live, and a *pioneer community* is formed. Over a period of time, perhaps rather long, this pioneer community so changes its environment that it becomes more suitable for other species. Another community becomes established, only to be followed later by still another, until finally, sometimes, a *climax community* becomes established—one that is in equilibrium with its environment and, hence, is more stable. Each of these communities in the series from pioneer community to climax community is called a seral stage or *seral community*, early ones paving the way for later ones.

This may be clearer to the reader if we consider a specific kind of succession, one we have already touched on in an earlier section. *Hydrarch succession* is the process by which an oligotrophic, or nutrient poor, lake becomes a eutrophic, or nutrient rich, lake; finally to perhaps become a swamp, a meadow, and then a forest.

Different seral communities are associated with each of these stages. A lake formed by geological activity will first be colonized by species of algae able to persist in nutrient poor water, and a few species of animals will come to feed on these plants. Over thousands of years, if the exchange flow of water is not too great, geochemical and biological activity will lead to an increase in the mineral nutrient content of the water, and more species in greater abundance will live there. Over more thousands of years, as the lake becomes richer, it may also become shallower with deposition of organic and mineral debris. Much life—almost too much from man's viewpoint—will then live in the lake. Oxygen at times may be depleted by natural processes; the lake has become eutrophic and some species man values no longer can live there.

Eutrophication is sometimes a natural process, but when man introduces mineral or organic nutrients into lakes, even very large lakes, the process is hurried. The time scale that would have been one of thousands of years becomes one of tens of years, and we end up with such situations as have occurred in Lake Zürich or Lake Erie.[86]

Rivers and streams are very different from lakes, perhaps most importantly because they are very open systems as compared to lakes. Lakes pass through succession and become eutrophic because through time they can accumulate organic and mineral materials. At any one location, rivers and streams can do very little of this, at least in reaches where velocities become sufficient for much scouring. Thus, it has been suggested that the communities of small, rapidly flowing streams are kept in pioneer seral stages by the continual downstream export of organisms and materials which prevents any persistent directional changes in conditions at any one location.[87] However, these organisms and materials, exported downstream, become available for colonization and utilization in downstream communities. Thus, progressing downstream, there is an accumulation of organic and inorganic materials, making ways of life of other species and formation of different communities possible. This downstream progression of communities, becoming generally more diverse, stable, and of greater biomass, was considered to be analogous to the succession through time that occurs in lake and terrestrial communities. The introduction of nutrient materials is, then, not so permanently damaging to a stream as to a lake because of this openness of stream systems.

So much now for developmental or successional explanations. What about operational explanations of community phenomena? Whatever characteristics communities have in the way of integrity, stability, and overall performance they have because of the ways their populations interact among themselves and with the conditions of the habitat. Symbiotic relations, competition for resources, and predation are important among these interactions which exhibit cybernetic characteristics.

We are perhaps most aware of the trophic interactions, the wonderfully complex food webs of communities. We are barely aware of other kinds of information and control systems, such as chemical information and control systems involving metabolites that inhibit, stimulate, or otherwise regulate the relations between populations. However, we are not as well informed about trophic interactions as some would have us believe. It is not just that we cannot determine all the rates of energy and material transfer in ecological systems, it is even that we often do not know who eats whom in these aquatic communities. The reader perhaps will recall that it was primarily the indeterminateness of such systems in dynamic terms that led me to suggest in an earlier section that we would better seek explanation of the productivity of ecosystems in terms of density relationships.

Functional explanations are generally rendered in terms of the role or purpose of a part in relation to a whole, the function of a subsystem in a more encompassing system. Communities clearly are a part of "spaceship earth."[88] They function in ways maintaining the characteristics of the atmosphere, the soils, and the waters of this system. This is a necessary holistic view, and we need to know more about the significance of communities in this respect. While we retain this grand view and exploit it however possible, what more restricted views might it subsume?

Already I have noted that in community succession there appear seral communities, each persisting awhile as it paves the way for the next seral community, which is usually more complex and stable. The whole series of seral communities is known as a *sere*. It is the sere that most fully exhibits organismic characteristics, each seral community functioning in the development and maintenance of the sere. For individual man, this is a very long view of an organismic system. For man and his concern for the earth, it may be a rather realistic and important view. The most important function of a particular community is probably the maintenance of the sere, of which it is only one part, one stage of seral development.

We are confronted with a very extensive and, according to our point of view, a hopelessly or beautifully complex system when we view a biological community. Each of its individual organisms and each of its populations displays much biological indeterminacy. Full explanation of communities in terms of the operation of each of their suborganismic, organismic, and superorganismic parts is well beyond the pale. Nevertheless, one path we must follow is from the part to the whole. But we cannot wait. We must make such measurements of the performance of communities as wholes as we have adequate conceptual systems to exploit. We need generalizations about communities that subsume their indeterminacy. The organismic generalizations at a very high level do this, but they are not enough.

We need generalizations of an operational sort that encompass concepts we can evaluate with observations of phenomena occurring over rather short periods of time. We have some of these. There is the concept of the community biomass, the total amount of biological material present in a community at any time. There is also the concept of community respiration, the rate of metabolic utilization and degradation of organic materials by the community as a whole. The community itself provides much or all of this organic material through total photosynthesis or gross primary production. Also, there is the diversity of the community, a concept involving the number of kinds of populations and their balance in the community. These concepts have been related by important generalizations, and there are sufficient parameters for evaluating them. These generalizations have much to do with the biology of polluted waters.

For aquatic systems, we can usually obtain reasonably good estimates of total photosynthetic and respiration rates of the entire communities. The ratio of photosynthetic rate to respiration rate (P/R ratio) for yearly periods is an indication as to whether a community is fixing energy in excess of, equal to, or less than its rate of energy utilization.[89] If a community has a P/R ratio in excess of 1, then it is either accumulating organic material or exporting material. If this ratio is less than 1, then the community is either importing organic material or utilizing organic materials accumulated by earlier communities. Figure 5-34 illustrates the approximate positions of dif-

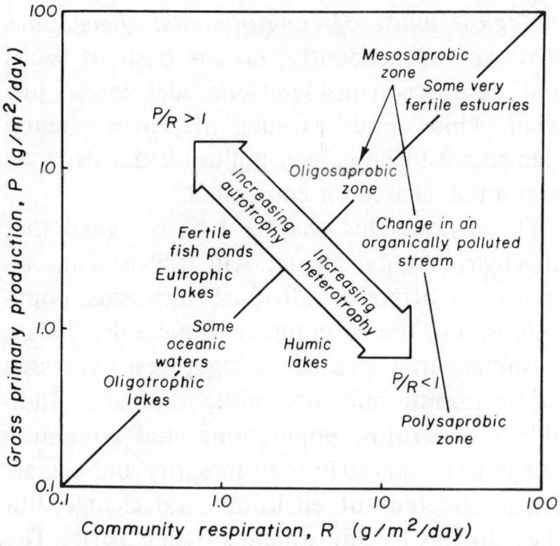

Fig. 5-34 Possible coordinates of gross primary production and community respiration for different ecosystems, some tending toward autotrophy, others tending toward heterotrophy. After Odum.[86]

ferent kinds of aquatic communities on coordinate axes for photosynthesis and respiration.

During lake succession through which eutrophic conditions develop, photosynthesis is generally greater than respiration and organic materials accumulate. When man introduces organic materials into a stream, respiration generally is first greater than photosynthesis, then becomes less than photosynthesis, and finally tends to balance photosynthesis as the stream passes from the polysaprobic through the mesosaprobic and into the oligosaprobic zones (Fig. 5-34).

It has been suggested[10] that aquatic systems of low maturity (low complexity or diversity) tend to have higher ratios of photosynthesis to biomass (P/B ratios) than do more mature systems. Not having relatively high biomasses to maintain, such low maturity systems may be capable of yielding to man greater amounts of fish and other products that interest him. As communities pass through succession, they tend to become more mature, their diversity increases, and their P/B ratio tends to decrease. Biomass and respiration are directly related, so these generalizations say important things about relations between photosynthesis, respiration, biomass, diversity, and succession.

I have not as yet clarified what biologists mean by *community diversity*. The concept of diversity involves the idea of an association of many different kinds of things, but it also involves the idea of lack of dominance of any particular kind. A community having many rare species and only one very abundant species would not give the impression of much diversity; it would exhibit *monotony*. Theoretically, *maximum diversity* would exist if there were a great many species, each represented by but one individual. Perhaps in most communities there are a few species that are quite abundant, many species that are quite rare, and many other species that are neither very abundant nor very rare. Such communities are neither very monotonous nor very diverse; they are intermediate in diversity.

Many different kinds of indices of diversity have been proposed and used,[90,20] including the application of information theory to the development of diversity indices.[91] Perhaps the most useful indices are those based on the Shannon equation.[92,93] In this approach, diversity is equated with the uncertainty that exists as to the species of an individual we might select at random from a community. The more species in a community and the more nearly equal their abundances, the greater would be the uncertainty as to the species of an individual we might select; hence, equating uncertainty and diversity, the greater the uncertainty the greater the diversity.[94]

It is generally recognized that more diverse communities are more stable. Environmental stability favors the development of diversity and diversity tends to insure community stability in the face of changing environments. However, the operational relations are neither clear nor agreed upon by biologists. Most would agree that it is not desirable to reduce the diversity of communities, except as truly necessary, say in agriculture. Even here, problems develop because of reduced community complexity. These are undesirable consequences. This happens when man introduces his wastes into natural waters. Changes in diversity can be used as indices of environmental changes, as is sometimes done in water pollution biology.

Biological Indices of Environmental Change. The whole subject of biological indices of community types and environmental conditions has long intrigued biologists, certainly before pioneers of the field of water pollution biology[95,96,97] in the early part of this century first began to apply the biological index concept to environmental problems caused by man. Apart from public health concerns, water pollution biology was mainly the search for biological indices until acute toxicity bioassays became popular. Now, of course, water pollution biology incorporates much more, and necessarily so.

We are still left with the task of rational incorporation of biological index and acute toxicity studies into the broader framework, because this has not as yet been accomplished. These two kinds of studies became the inadequate foundation of water pollution biology, and even today they too much dominate our thinking. They have a place, an important place, but it should be more in proportion to their possible utility. The approaches and methods of biological index studies have been reviewed,[98,17,20] and I am not so presumptuous or foolish as to suppose that I either know enough or have thought enough to be able to properly place biological index studies in the broader context. Perhaps, though, I can suggest some of the important considerations.

First, I think, we must distinguish clearly between the problems of measuring or indexing environmental changes and the problem of deciding just what environmental changes are to be considered as pollution. The first problem is primarily scientific and, as such, is amenable to rational solution. The second problem is primarily a social one that we hope is amenable to a satisfactory solution.

I do not believe that we should continue to use the expression "biological index of pollution," because this confounds these problems.[20] Appropriate biological indices can detect very slight environmental changes. It is not helpful to equate any environmental change, however small or whatever its meaning to man, with the idea of pollution. The concept and the expression "biological index of pollution" tend to do this. It is better that we speak of *biological index of environmental change* and that we independently, on the basis of social and resource considerations, determine just what kinds and extents of environmental change are to be judged pollution that is to be prevented, abated, or controlled.

Throughout this chapter I have argued that an environmental change will lead to some response by affected individual organisms, populations, and the community as a whole. This is a fundamental idea of biology, best expressed in organismic and cybernetic thought. Individual organisms, populations, and communities tend to maintain their integrity and character in the face of environmental change, but they do this by adjustment of their parts. The environmental change may be so slight as to elicit a response we are incapable of detecting, but there will be a response. According to our interest, knowledge, and techniques, but mostly according to our interest, we can detect almost any response. It is much more a question of how useful it would be to detect a response and how this can best be done.

Nearly all kinds of biological indices are subsumed by one or another of three basic concepts: the indicator organism concept, the concept of community composition, and the concept of community diversity. "The *indicator organism* concept in its purest form, a form often discussed and little used, is simply that the presence of a particular species is indicative of the existence of certain environmental conditions that interest man, whereas its absence is indicative of the absence of those conditions."[20]

The use of the coliform index for fecal contamination and of *Oscillatoria rubescens* for eutrophication are examples. Even these are fraught with difficulty, and generally there are even greater difficulties with the use of this concept. For sound biological and sampling reasons, this concept has fallen into disfavor.

Community composition is generally taken to mean the species present in a community and their relative abundances. Almost any environmental change will alter the relative abundances of species in a community; more appreciable changes will lead to the disappearance of some species and the appearance of others. The accumulation and tabulation of the species and their abundances in communities affords, for the experienced biologist, an extensive and information-rich picture of a community, one that alters perceptibly with environmental change.[17] Much knowledge and a considerable amount of subjective analysis are required in the interpretation of such pictures, and no simple representation, say by a number or a diagram, is at all adequate.

If a simple representation is desired or needed, we are well advised to go to indices of *community diversity*. Of these, the ones based on information theory are probably the best in that they make fullest use of available numerical data, are most sensitive, and are least compromised by sampling problems.[94] However, in dropping the identity, relative abundance, and what biologists know of particular species, much information is lost. If the practical problem is of sufficient moment, we would best combine analysis of community composition and community diversity in making biological evaluations of the effects and extent of environmental change.

There are, then, biological indices of environmental change. They are most assuredly not the biological answer to our water pollution problems. Nevertheless, they can be useful. We should be very clear as to our water pollution control objective, in any instance. We should then realistically determine whether or not biological index studies could help us in attaining this objective. Brushing over the possibilities and constraints of such studies will not help us in this. If it becomes clear that the use of biological indices of environmental change can substantially aid us in attaining a particular objective, then we should select for this purpose the procedures most appropriate for exploiting this concept.

MAKING BIOLOGICAL DECISIONS AND RECOMMENDATIONS

The Problem of Making Biological Decisions and Recommendations. The making of biological decisions and recommendations pertaining to resource utilization is difficult at best. Ecological systems—organismic systems and their environments—are complex at whatever level we choose to isolate them. When they are incorporated into human ecosystems and receive technological inputs and are expected to yield social benefits, then the systems involved are more complex, the difficulty of making decisions and recommendations becomes greater, an the cost to man of erring may become great.

Evaluation of what is good for man in a socio-cultural sense is beyond biology as a discipline. But evaluation of the performance of biological systems under different conditions is a task of biology that is important in attaining and maintaining man's well-being. Questions concerning levels of dissolved oxygen or levels of temperature to maintain in water are important and difficult, both biologically and sociologically. Knowing this, biologists are on the one hand anxious, yet on the other hand hesitant, to make decisions and recommendations. Thus, with insufficient knowledge often the justification, biologists are inclined to qualify their remarks in ways rendering their conclusions obscure and of little value for making decisions and recommendations. For few if any very complex systems, biological or physical, will knowledge ever be sufficient for full explanation. This is a reality of all scientific endeavors, not just biology. Any scientific explanation has initial and boundary conditions that will constrain the application of that explanation. When these are explicitly stated, the relevance of an explanation, its possibilities and constraints, should become clear. The realm and reliability of decisions and recommendations based on explanations can then be more nearly defined.

What kinds of decisions are to be made? First of all, it is helpful to recognize that there is an implicit if not explicit hierarchy of decisions that determine the value and reliability of any recommendation we would make. To begin, the biologist, with the help of others, must decide what the total system of interest may be, i.e., the nature of the social-biological problem. Then there is the decision as to how to define usefully the biological part of the problem. Next, a conceptual system adequate for the solution of the biological problem must be selected, and it should be a conceptual system relatable to the social part of the problem. Full explanation of the biological part of the problem will not be possible. Thus, there are the decisions of what investigation to pursue first and how to pursue it. Because the entire biological problem will not be solved, there are decisions pertaining to interpretations, in the largest context, of information obtained only on parts of the system of interest. From all this must follow decisions on the resource utilization recommendations to be made.

All this can be done, whatever may be our level of understanding of the systems involved. It is done and it must be done. The only question is how to do it best. Because circumstances differ, there is no simple answer to this question other than to say that formal recognition of the investigative processes involved makes clearer the decisions and assumptions to be made and thus the realm and reliability of scientific and administrative recommendations.

Statement of the Problem Needing a Solution. More often than not, when we fail to solve a social or scientific problem it is because we were not clear as to just what the problem was. The solution of social problems ultimately requires appropriate social action. The efficacy of such action, however, is according to the base of understanding of the nature of the problem. In resource matters, this involves explanation of the social-biological system of concern. Explanation is the work of science; it is the scientific part of the solution of a resource problem. Both the social and the scientific effort are facilitated by a statement of the problem that is not only clear but also useful.

The need for clarity is too self-evident to pursue here. How can we know whether or not we have solved an ambiguously stated problem? But what is a useful statement of a resource problem? A resource problem involves a hierarchy of problems or questions. At the highest level, the problems should be stated so as to clearly identify the social-biological system of concern. If properly stated, these should be useful in identifying the subsumed social and biological problems requiring solutions. Clear statement of the overall problem and any subsumed problems should suggest overall and partial conceptual systems of value in the effort. To be useful and not mislead further effort, the statement of the problem should make clear the initial and boundary conditions, a matter of importance in selecting appropriate conceptual systems and in all later steps in problem resolution. The statement of the problem and the selection of conceptual systems will aid in identifying the most appropriate courses of investigation.

Max Planck[99] wrote of *phantom problems*, problems for which there exist no indisputable methods of resolution or that are totally devoid of meaning. It is a psychologically and intellectually severe test to attempt to envision what a solution to a particular problem might be, indeed, whether or not there is any solution, but one that may save costly and pointless effort.

Usefully stated problems will not usually cover the entire realm of related social or scientific interest. All-encompassing statements are scientifically unmanageable. This makes it particularly important that the statement of a problem aid us in identifying the boundary conditions of any solution we might expect and the relevance of this solution to explanation and management of the resource system of concern.

Choice of a Conceptual System. The solution of any problem, social or scientific, requires some sort of conceptual system. Any social activity is based on some view of the nature of things. Any scientific explanation whatsoever presupposes some sort of conceptual system. It may be remembered that, in an early section of this chapter, I defined a conceptual system as a system of generalizations relating concepts. It may also be remembered that I suggested that conceptual systems differ in maturity, the most mature ones being most explicit as to the meaning of incorporated concepts and the nature of the relationships between them. Maturity of conceptual systems also increases as these come nearer to reflecting what we take to be reality, our only test of this being observation, with all its imperfections.

We may go about our daily activities or we may advance scientific explanations and be unaware of the conceptual systems underlying them, but they are always there. There may be social and scientific danger in proceeding without explicit conceptual systems, because then we are unaware of the assumptions and boundary conditions that limit the relevance of any social activity or scientific explanation. Thus, an explicit conceptual system helps us to know how much to expect from given social and scientific activities as well as to select the most appropriate activities and guide these efficiently toward the solution of problems.

Corresponding to the hierarchy of problems we would solve, there is needed a hierarchy of conceptual systems. And within itself, each conceptual system is likely to be hierarchical. To solve resource problems both socially and scientifically, there needs to be an all-encompassing, social-biological conceptual system, one corresponding as nearly as possible to what we believe at any one time to be the nature of the system of concern. Within this overall conceptual system, subsumed conceptual systems are necessary to orient social activity and to permit the scientific explanation on which such activity would best be based.

Let us suppose that some overall social-biological conceptual system has been selected as being the best possible for the solution of some resource utilization problem. Boundary conditions, defined by necessary assumptions, will constrain the realm and utility of this highest level conceptual system and all lower level conceptual systems derived from and subsumed by this system.

Some single biological problem needing resolution as a part of the solution to the overall resource problem will also require a conceptual system for its solution. It is to this level that the various biological conceptual systems sketched in this chapter pertain. Some of these relate to the performance of individual organisms, some to the biology or production of populations, and some to changes in communities. If the biological problem is at a high level in the hierarchy of scientific problems involved in the resource matter, it will subsume other biological problems, each requiring a conceptual system which will be useful to the extent that it relates to the other individual conceptual systems and the overall system of which they are parts.

Several conceptual systems may be suggested by any well-stated problem, but they are not likely to be equally fruitful. Choice among them should be made on the basis of their relevance to the overall social-biological conceptual system and the solution of the resource problem as well as on the basis of their probable scientific efficacy.

Narrowing of the Effort with Explicit Assumptions. If I am not wrong, most problems of resource utilization involve so many social and scientific problems as to make difficult, if not impossible, anything like a full solution. There is the complex hierarchy of problems, the solution of which would require a corresponding hierarchy of conceptual systems. In any period of time in which social action may be necessary, we are likely to be unable to solve many of these problems, whatever the construct of conceptual systems we may develop. Would, then, it be better to worry less about such constructs and do what little we can, problem by problem? Not in the least, I think. If our capabilities are insufficient, as they generally are, for full social solution or scientific explanation of a problem, then we would best apportion available effort among the most critical parts of the problem. It is here that the overall and subsumed conceptual systems become necessary. How do we conceive of the interrelating parts of the problem and what are their relative importances? Here again, the boundary conditions of the possible explanations of the parts of the problem become of concern. They tell us something of the utility of action or knowledge pertaining to any part.

Boundary conditions of possible explanations are defined by the assumptions necessary to exploit any particular conceptual system. The ease with which we can reach an explanation depends on the nature and number of assumptions we choose to make, but these assumptions also limit the realm and reliability of the explanation. If we must narrow our effort to solve a problem, we must increase the assumptions we are willing to make. Only mature conceptual systems permit us to choose those assumptions that will most reduce our effort and least reduce the relevance of our results.

Conclusions and Recommendations. Biologists are asked to make recommendations on the sorts of research that should be pursued and on the water quality requirements of organisms. Administrators are asked to make decisions and effect action based on such recommendations that will tend to insure wise resource utilization. Biologists, other scientists, and administrators, in view of the nature of resource problems, will never know enough to do this in a fully satisfactory manner. Yet, as scientists, we come to recognize and accept the limits of the scientific method; we may seek but we do not expect full explanation of natural phenomena. We always must put boundary conditions on our explanations. In the social realm, then, why do we hesitate to make firm recommendations limited by explicit assumptions?

If we would adopt such a pattern with attendant conceptual systems, then administrators could evaluate the risks of given action programs. Should the risks be judged unacceptable, yet the problem important, then more research could be ordered and supported, its justification clear. The removal of risks is in large part the validation of assumptions, an everyday affair in science.

REFERENCES

1. Dubos, R., *Man Adapting*, Yale University Press, New Haven (1965).
2. Braithwaite, R. B., *Scientific Explanation; A Study of the Function of Theory, Probability and Law in Science*, Cambridge University Press, Cambridge (1953).
3. Nagel, E., *The Structure of Science; Problems in the Logic of Scientific Explanation*, Harcourt, Brace and World, New York (1961).
4. Bertalanffy, L. von, *Modern Theories of Development: An Introduction to Theoretical Biology*, Oxford University Press, London (1933).
5. Wiener, N., *Cybernetics; or, Control and Communication in the Animal and the Machine*, 2nd ed., M.I.T. Press, Cambridge, Massachusetts, and John Wiley and Sons, Inc., New York (1961).
6. Bayliss, L. E., *Living Control Systems*, Freeman, San Francisco (1966).
7. Mason, H. L., and Langenheim, J. H., "Language Analysis and the Concept of Environment," *Ecology* 38, 325–340 (1957).
8. Andrewartha, H. G., and Birch, L. C., *The Distribution and Abundance of Animals*, The University of Chicago Press, Chicago (1954).
9. Whitehead, A. N., *Science and the Modern World*, Lowell lectures, Macmillan, New York (1925).
10. Hutchison, G. E., "Concluding Remarks; Population Studies: Animal Ecology and Demography," *Cold Spring Harbor Symposia on Quantitative Biology* **22**, 415–427 (1957).
11. Henderson, L. J., *The Fitness of the Environment*, Macmillan, New York (1913).
12. Hutchinson, G. E., *A Treatise on Limnology*, Vol. 1, Geography, Physics, and Chemistry, John Wiley and Sons, Inc., New York (1957).
13. Hutchinson, G. E., *A Treatise on Limnology*, Vol. 2, Introduction to Lake Biology and the Limnoplankton, John Wiley and Sons, Inc., New York (1967).
14. Hynes, H. B. N., *The Ecology of Running Waters*, University of Toronto Press, Toronto (1970).
15. Ruttner, F., *Fundamentals of Limnology*, (translated by D. G. Frey and F. E. J. Fry) University of Toronto Press, Toronto (1953).
16. Jones, J. R. E., *Fish and River Pollution*, Butterworth, London (1964).
17. Hynes, H. B. N., *The Biology of Polluted Waters*, Liverpool University Press, Liverpool (1960).
18. Klein, L., "Causes and Effects," *River Pollution*, Butterworth, London (1962).
19. Wilber, C. G., *The Biological Aspects of Water Pollution*, Charles C. Thomas, Springfield, Illinois (1969).
20. Warren, C. E., *Biology and Water Pollution Control*, W. B. Saunders Co., Philadelphia (1971).
21. McKee, J. E., "Parameters of Marine Pollution—An Overall Evaluation," *Pollution and Marine Ecology*, Wiley-Interscience, New York, pp. 259–266 (1967).
22. Mayr, E., *Animal Species and Evolution*, The Belknap Press of Harvard University Press, Cambridge, Massachusetts (1963).
23. Whittaker, R. H., *Communities and Ecosystems*, Macmillan, London (1970).
24. Odum, E. P., *Fundamentals of Ecology*, 3rd ed., W. B. Saunders Co., Philadelphia (1971).
25. Krebs, C. J., *Ecology: The Experimental Analysis of Distribution and Abundance*, Harper and Row, New York (1972).
26. Tansley, A. G., "The Use and Abuse of Vegetational Concepts and Terms," *Ecology* **16**, 284–307 (1935).
27. Hume, D., "An Enquiry Concerning Human Understanding," (1748); (reprinted 1910 in *English Philosophers of the Seventeenth and Eighteenth Centuries*, The Harvard Classics, Vol. 37, pp. 305–445, Collier and Son Company, New York).

28. Mill, J. S., *A System of Logic, Ratiocinative and Inductive* (1843); (reprinted 1900, 8th ed., Harper and Brothers, New York).
29. Heisenberg, W., "Uber quantentheoretische Umdeutung kinematischer und mechanischer Beziehungen," *Zeitschrift für Physik* 33, 879–893 (1925).
30. Heisenberg, W., *Die physikalischen Prinzipien der Quantentheorie* (translated by C. Ekart and F. C. Hoyt, *The Physical Principles of the Quantum Theory*) Dover Publications, Inc. (1930).
31. Campbell, N. R., *Physics, The Elements*, Cambridge University Press, Cambridge (1920).
32. Williams, M. B., "Deducing the Consequences of Evolution: A Mathematical Model," *Journal of Theoretical Biology* 29, 343–385 (1970).
33. Hughes, G. M. (ed.), *Homeostasis and Feedback Mechanisms*, Symposia of the Society for Experimental Biology 18, Academic Press, Inc., New York (1964).
34. Dill, D. B., Adolph, E. F., and Wilber, C. G. (eds.), "Adaptation to the Environment," *Handbook of Physiology*, Sec. 4, American Physiological Society, Washington, D.C. (1964).
35. Sprague, J. B., "Review Paper: Measurement of Pollutant Toxicity to Fish. III. Sublethal Effects and "Safe" Concentrations," *Water Research*, 5, 245–266 (1970).
36. Waterman, T. H. (ed.), *The Physiology of Crustacea*, Vol. 1, Metabolism and Growth, Academic Press, Inc., New York (1960).
37. Waterman, T. H. (ed.), *The Physiology of Crustacea*, Vol. 2, Sense Organs, Integration, and Behavior, Academic Press, Inc., New York (1961).
38. Wilbur, K. M., and Yonge, C. M., *Physiology of Mollusca*, Vol. 1, Academic Press, Inc., New York (1964).
39. Wilbur, K. M., and Yonge, C. M., *Physiology of Mollusca*, Vol. 2, Academic Press, Inc., New York (1964).
40. Hoar, W. S., and Randall, D. J. (eds.), *Fish Physiology*, Academic Press, Inc., New York (1969–71).
41. Fry, F. E. J., "Effects of the Environment on Animal Activity," University of Toronto Biological Studies Series 55, Ontario Fisheries Research Laboratory Publication 68 (1947).
42. Levins, R., "The Strategy of Model Building in Population Biology," *American Scientist* 54, 421–431 (1966).
43. Huntsman, A. G., "Method in Ecology-Biapocrisis," *Ecology* 29, 30–42 (1948).
44. Mount, D. I., and Stephan, C. E., "A Method for Establishing Acceptable Toxicant Limits for Fish–Malathion and the Butoxyethanol Ester of 2,4-D," *Transactions of the American Fisheries Society* 96, 185–193 (1967).
45. Mount, D. I., "Chronic Toxicity of Copper to Fathead Minnows (*Pimephales promelas*, Rafinesque)," *Water Research* 2, 215–223 (1968).
46. Warren, C. E., *et al.*, "Trout Production In an Experimental Stream Enriched with Sucrose," *Journal of Wildlife Management* 28, 617–660 (1964).
47. Warren, C. E., and Davis, G. E., "Laboratory Stream Research: Objectives, Possibilities, and Constraints," *Annual Review of Ecology and Systematics* 2, 111–144 (1971).
48. Doudoroff, P., *et al.*, "Bio-assay Methods for the Evaluation of Acute Toxicity of Industrial Wastes to Fish," *Sewage and Industrial Wastes* 23, 1381–1397 (1951).
49. *Standard Methods for the Examination of Water and Wastewater*, 13th ed., APHA, AWWA, WPCF, New York (1971).
50. Herbert, D. W. M., "Pollution and Fisheries," *Ecology and the Industrial Society*, pp. 173–195 (1965); British Ecological Society Symposium 5, (1964).
51. Brown, V. M., "The Calculation of the Acute Toxicity of Mixtures of Poisons to Rainbow Trout," *Water Research* 2, 723–733 (1968).
52. Sprague, J. B., "Review Paper: Measure of Pollutant Toxicity to Fish. II. Utilizing and Applying Bioassay Results," *Water Research* 4, 3–32 (1970).
53. Warren, C. E., and Davis, G. E., "Laboratory Studies on the Feeding, Bioenergetics, and Growth of Fishes," *The Biological Basis of Freshwater Fish Production*, Blackwell Scientific Publications, Oxford, pp. 175–214 (1967).
54. McKim, J. M., and Benoit, D. A., "Effects of Long-term Exposures to Copper on Survival, Growth, and Reproduction of Brook Trout (*Salvelinus fontinalis*)," *Journal of the Fisheries Research Board of Canada* 28, 655–662 (1971).
55. Davis, H. C., "Survival and Growth of Clam and Oyster Larvae at Different Salinities," *Biological Bulletin* 114, 296–307 (1958).
56. Davis, H. C., "Effects of Some Pesticides on Eggs and Larvae of Oysters (*Crassostrea virginica*) and Clams (*Venus mercenaria*)," *Commercial Fisheries Review* 23 (12), 8–23 (1961).
57. Dimick, R. E., and Breese, W. P., "Bay Mussel Embryo Bioassay," *Pacific Northwest Industrial Waste Conference* 12, 165–175 (1965).
58. Woelke , C. E., "Preliminary Report of Laboratory Studies on the Relationship Between Fresh Sulfite Waste Liquor and the Reproduction Cycle of the Olympic Oyster, *Ostrea lurida*," Reports on Sulfite Waste Liquor in a Marine Environment and its Effects on Oyster Larvae, pp. 107–148, Washington Department of Fisheries Research Bulletin 6 (1960).
59. Woelke, C. E., "Environmental Requirements of Marine Invertebrates," *Biological Problems in Water Pollution, Transactions of the 3rd Seminar, 1962, R. A. Taft Sanitary Engineering Center*, U.S. Department of Health, Education, and Welfare, pp. 67–77 (1965).
60. Woelke, C. E., "Application of Shellfish Bioassay Results to the Puget Sound Pulp Mill Pollution

Problem," *Northwest Science* **42**, 125–133 (1968).

61. Frank, P. W., *et al.*, "Vital Statistics of Laboratory Cultures of Daphnia Pulex De Geer as Related to Density," *Physiological Zoology* **30**, 287–305 (1957).
62. Roelofs, T. D., "Effects of Dieldrin on the Intrinsic Rate of Increase of the Guppy, *Poecilia reticulata* Peters," Ph.D. thesis, Oregon State University, Corvallis (1971).
63. Brody, S., *Bioenergetics and Growth*, Reinhold Publishing Co., New York (1945).
64. Hinde, R. A., *Animal Behaviour: A Synthesis of Ethology and Comparative Psychology*, McGraw-Hill, New York (1966).
65. Kalleberg, H., "Observations in a Stream Tank of Territoriality and Competition in Juvenile Salmon and Trout (*Salmo salar* L. and *S. trutta* L.)," Fishery Board of Sweden, Institute of Freshwater Research (Drottningholm) Report 39, pp. 55–98 (1958).
66. Chapman, D. W., "Aggressive Behavior in Juvenile Coho Salmon as a Cause of Emigration," *Journal of The Fisheries Research Board of Canada* **19**, 1047–1080 (1962).
67. Breder, C. M., Jr., "The Reproductive Habits of the North American Sunfishes (family Centrarchidae)," *Zoologica* **21**, 1 (1936).
68. Fabricius, E., and Gustafson, K. J., "Further Aquarium Observations on the Spawning Behaviour of the Char, *Salmo alpinus L.*," Fishery Board of Sweden, Institute of Freshwater Research (Drottningholm) Report 35, pp. 58–104 (1954).
69. Fabricius, E., "Heterogeneous Stimulus Summation in the Release of Spawning Activities in Fish," Fishery Board of Sweden, Institute of Freshwater Research (Drottningholm) Report 31, pp. 57–99 (1950).
70. Höglund, L. B., "The Reactions of Fish in Concentration Gradients," Fishery Board of Sweden, Institute of Freshwater Research (Drottningholm) Report 43 (1961).
71. Sprague, J. B., Elson, P. F., and Saunders, R. L., "Sublethal Copper-Zinc Pollution in a Salmon River–A Field and Laboratory Study," *International Journal Air and Water Pollution* **9**, 531–543 (1965).
72. Collins, G. B., "Factors Influencing the Orientation of Migrating Anadromous Fishes," Fishery Bulletin 73 (Woods Hole Oceanographic Institution Contribution 585); *Fishery Bulletin of the U.S. Fish and Wildlife Service* **52**, 375–396 (1952).
73. Ricker, W. E., "Stock and Recruitment," *Journal of the Fisheries Research Board of Canada* **11**, 559–623 (1954).
74. Ricker, W. E., "Handbook of Computations for Biological Statistics of Fish Populations," *Fisheries Research Board of Canada Bulletin* **119** (1958).
75. Falconer, D. S., *Introduction to Quantitative Genetics*, Oliver and Boyd, London (1960).
76. Chapman, D. W., "Production," *Methods for Assessment of Fish Production in Fresh Waters*, International Biological Programme Handbook 3, Blackwell Scientific Publications, Oxford, pp. 182–196 (1968).
77. Edmondson, W. T., and Winberg, G. G. (eds.), *A Manual on Methods for the Assessment of Secondary Productivity in Fresh Waters*, IBP Handbook 17, Blackwell Scientific Publications, Oxford (1971).
78. Brocksen, R. W., Davis, G. E., and Warren, C. E., "Analysis of Trophic Processes on the Basis of Density-Dependent Functions," *Marine Food Chains*, University of California Press, Berkeley, pp. 468–498 (1970).
79. Strickland, J. D. H., "Measuring the Production of Marine Phytoplankton," *Fisheries Research Board of Canada Bulletin* **122** (1960).
80. Goldman, C. R. (ed.), "Primary Productivity in Aquatic Environments," *Proceedings of an International Biological Programme Symposium, 1965*, University of California Press, Berkeley (1966).
81. Beverton, R. J. H., and Holt, S. J., "On the Dynamics of Exploited Fish Populations," Ministry of Agriculture, Fisheries and Food, Fishery Investigations Series 2, 19, pp. 3–533, London (1957).
82. Lindeman, R. L., "The Trophic-Dynamic Aspect of Ecology," *Ecology* **23**, 399–418 (1942).
83. Brocksen, R. W., and Warren, C. E., "Density Dependence of Trophic Processes in Laboratory Stream Communities," (unpublished).
84. Hawkes, H. A., *The Ecology of Waste Water Treatment*, Pergamon Press, Oxford (1963).
85. Whittaker, R. H., and Woodwell, G. M., "Evolution of Natural Communities," *Ecosystem Structure and Function, Proceedings of the 31st Annual Biology Colloquium*, Oregon State University Press, Corvallis, pp. 137–159 (1972).
86. *Eutrophication: Causes, Consequences, and Correctives*, National Academy of Sciences, Washington, D.C. (1969).
87. Margalef, R., "Ideas for a Synthetic Approach to the Ecology of Running Waters," *Internationale Revue der Gesamten Hydrobiobgie* **45**, 133–153 (1960).
88. Ward, B., *Spaceship Earth*, Columbia University Press, New York (1966).
89. Odum, H. T., "Primary Production in Flowing Waters," *Limnology and Oceanography* **1**, 102–117 (1956).
90. Pielou, E. C., *An Introduction to Mathematical Ecology*, Wiley-Interscience, New York (1969).
91. Margalef, R., "Información y Diversidad Especifica en las Comunidades de Organismos (Information and Specific Diversity in Communities of Organisms)," *Investigación Pesquera* **3**, 99–106 (1956).

92. Shannon, C. E., "A Mathematical Theory of Communication," *Bell System Technical Journal* **27**, 379–423, 623–656 (1948).
93. Brillouin, L., *Science and Information Theory*, 2nd ed., Academic Press Inc., New York (1969).
94. Wilhm, J. L., and Dorris, T. C., "Biological Parameters for Water Quality Criteria," *Bioscience* **18**, 477–481 (1968).
95. Kolkwitz, R., and Marsson, M., "Ökologie der pflanzlichen Saprobien," *Berichte der Deutschen Botanischen Gesellschaft* **26a**, 505–519 (1908) (translated, "Ecology of Plant Saprobia," *Biology of Water Pollution*, FWPCA, U.S. Dept. of the Interior (1967)).
96. Kolkwitz, R., and Marsson, M., "Ökologie der tierischen Saprobien," Beiträge zur lehre von der biologischen Gewässerbeurteilung, *Internationale Revue der Gesanten Hydrobiologie und Hydrogeographie* **2**, 126–152 (1909) (translated, "Ecology of Animal Saprobia," *Biology of Water Pollution*, FWPCA, U.S. Dept. of the Interior (1967)).
97. Richardson, R. E., "The Bottom Fauna of the Middle Illinois River, 1913–1925: Its Distribution, Abundance, Valuation, and Index Value in the Study of Stream Pollution," *Bulletin of the Illinois State National History Survey* **17**, 387–475 (1928).
98. Hawkes, H. A., "Biological Aspects of River Pollution," *River Pollution*, Butterworth, London, pp. 311–432 (1962).
99. Planck, M., *Scientific Autobiography and Other Papers* (translated from German by Frank Gaynor) Philosophical Library, New York (1949).
100. Sprague, J. B., "Review Paper: Measurement of Pollutant Toxicity to Fish. I. Bioassay Methods for Acute Toxicity. *Water Research* **3**, 793–821 (1969).
101. Margalef, R., "On Certain Unifying Principles in Ecology," *American Naturalist* **97**, 357–374 (1963).
102. Adolph, E. F., "Perspectives of Adaptation: Some General Properties," *Handbook of Physiology*, Sec. 4, American Physiological Society, Washington, D.C., pp. 27–35 (1964).
103. Brett, J. R., "Some Principles in the Thermal Requirements of Fishes," *Quarterly Review of Biology* **31**, 75–87 (1956).
104. Shepard, M. P., "Resistance and Tolerance of Young Speckled Trout (*Salvelinus fontinalis*) to Oxygen Lack, with Special Reference to Low Oxygen Acclimation," *Journal of the Fisheries Research Board of Canada* **12**, 387–446 (1955).
105. Bullock, T. H., "Compensation for Temperature in the Metabolism and Activity of Poikilotherms," *Biological Reviews* **30**, 311–342 (1955).
106. Trama, F. B., "The Acute Toxicity of Copper to the Common Bluegill (*Lepomis macrochirus* Rafinesque)," The Academy of Natural Sciences of Philadelphia, *Notulae Naturae* **257** (1954).
107. Lloyd, R., "The Toxicity of Ammonia to Rainbow Trout (*Salmo gairdnerii* Richardson)," *Water Waste Treatment Journal* 8, 278–279 (1961).
108. McLeese, D. W., "Effects of Temperature, Salinity and Oxygen on the Survival of the American Lobster," *Journal of the Fisheries Research Board of Canada* **13**, 247–272 (1956).
109. Shumway, D. L., Warren, C. E., and Doudoroff, P., "Influence of Oxygen Concentration and Water Movement on the Growth of Steelhead Trout and Coho Salmon Embryos," *Transactions of the American Fisheries Society* **93**, 342–356 (1964).
110. Everson, L. B., "Growth and Food Consumption of Juvenile Coho Salmon Exposed to Natural and Elevated Fluctuating Temperatures," M.S. thesis, Oregon State University, Corvallis (1973).
111. Brocksen, R. W., Davis, G. E., and Warren, C. E., "Competition, Food Consumption, and Production of Sculpins and Trout in Laboratory Stream Communities," *Journal of Wildlife Management* **32**, 51–75 (1968).
112. Johnson, W. E., "Aspects of the Ecology of a Pelagic, Zooplankton-Eating Fish," *Verhandlungen Internationale Vereinigung für Theoretische und Angewandte Limologie* **14**, 727–731 (1961).
113. Ruggles, C. P., "Juvenile Sockeye Salmon Studies in Owikeno Lake, British Columbia," *Canadian Fish Culturist* **36**, 3–21 (1965).
114. Krogius, F. V., and Krokhin, E. M., "Oburozhainosti molodi Krasnoi (*Oncorhynchus nerka* Walb.)," Izvestiia Tikhookeanskovo Nauchno-Issledovatelskovo Instituta Rybnovo Khoziaistva i Okeonografii **28**, 3–27 (1948) (translated by R. E. Foerster, "On the Production of Young Sockeye Salmon (*Oncorhynchus nerka* Walb.)," *Fisheries Research Board of Canada Translation Series* **109** (1958)).
115. Waddington, C. H., "Evolutionary Adaptation," *The Evolution of Life*, University of Chicago Press, Chicago, pp. 381–402 (1960).
116. Silliman, R. P., "Interaction of Food Level and Exploitation in Experimental Fish Populations," *U.S. Fish and Wildlife Service Fishery Bulletin* **66**, 425–439 (1968).
117. Ivlev, V. S., "Biologicheskaya produktivnost' vodoemov," *Uspekhi Sovremennoi Biologii* **19**, 98–120 (1945); translated by W. E. Ricker, "The Biological Productivity of Waters," *Journal of the Fisheries Research Board of Canada* **23**, 1707–1759 (1966).

6

Stream Analysis—Forecasting Waste Assimilation Capacity

C. J. Velz
Consultant
Longboat Key, Florida

THE SCIENCE AND ART OF STREAM ANALYSIS

The science and art of stream analysis has attained a mature stage of development. It is possible to forecast, with reasonable reliability, the waste assimilation capacity and resultant water quality of a stream. This can be done for any combination of residual waste loadings within the range of hydrologic and climatologic variations expected and under the impact of man-made river developments and uses. This approach is termed the rational method of stream analysis.

Rigorous stream analysis is the basis for assuring satisfactory quantity and quality of streamflow for intended uses. The successful design has as its foundation the efficient use of the waste assimilation capacity of the stream without waste or abuse. Assimilative capacity of natural waters is based on the determination of the allowable load placed on self-purification to ensure the desired water quality under the prescribed probability of natural drought or regulated streamflow.

DYNAMIC CHARACTER AND MULTIPLE USE OF WATER

Two unique characteristics of water are that it is a dynamic resource and inherently a multiple-use resource. Unlike a fixed mineral deposit, water varies in the amount that is available at any location from day to day and season to season. In this context water is dynamic, but observed over a period of years, streamflow, although variable, is orderly, following the laws of chance occurrence unless the natural setting has been altered.

Statistical analyses of the runoff records determine the range of streamflow available and the drought severities that can be expected. Man must either design his activities within this range of variation and decide on a level of risk of water shortage during droughts or develop the water resource by such practice as drought control to regulate variations between the extremes of flood and drought.

In turn, the natural waste assimilation capacity of the stream and the associated water quality are not fixed quantities but rather ranges in potential which parallel the natural variations in streamflow. In addition, man-made river developments and water uses, depending on how they are planned and operated, can have an effect that is either beneficial or detrimental.

The multiple uses of water by cities or private or governmental agencies also have an impact on other uses. As the urban-industrial complex grows, competition for the limited water resources intensifies.

THE NATURE OF STREAM SELF-PURIFICATION

Any system of stream pollution control is dependent on natural self-purification, the ability of the stream to assimilate wastes and restore its own quality, which is, to an extent, different for each type of waste. After discharge to the stream, liquid waste products from the four major sources–urban, industrial, agricultural, and natural–to a large extent lose their identity and are blended into a heterogeneous mixture. However, the river in its self-purification processes reclassifies the waste products as to type, principally into five broad classifications: (1) inorganic, (2) microbial, (3) radioactive, (4) thermal, and (5) organic.

Limitations of this chapter restrict full development of quantitative analysis to the organic type. However, the following thumbnail sketches of the nature of self-purification of other types illustrates the complexity of the phenomena. More detailed information on them may be found in Ref. 1.

Stable chemical wastes undergo little or no change in their journey down a river. Here the primary factor is dilution, and self-purification is almost wholly dependent on streamflow along the course.

In the case of bacterial wastes, streamflow, time of exposure (measured in hours or days), and water temperature are the three major determinants of self-purification. The decline in the number of organisms is a function of the latter two. The death rate increases with water temperature and length of exposure to unfavorable conditions in the stream environment.

Radioactive self-purification, like microbial self-purification, is subject to a die-off, except that the process is physical rather than biological, and thus is uninfluenced by temperature. The two primary factors involved are time of passage along the course of the stream and streamflow. For radioactive wastes with long half-lives of months or years, self-purification by decay in the relatively short time of passage along the stream is insignificant, and dilution by tributary runoff along the course is the principal factor.

Thermal self-purification (that is, heat dissipation) is a special case. In addition to streamflow and stream surface area, meteorologic factors of air temperature, wind velocity, relative humidity, and solar radiation are involved. Four principal surface mechanisms function in heat transfer and temperature change as a body of water receiving hot effluents is exposed to the atmosphere; heat is lost through evaporation, convection, and radiation, and is gained through solar radiation.

The most general type of waste is unstable organic matter from various sources, principally communities and industries. Here self-purification is a biochemical process of decay. Oxygen dissolved in the river water is utilized in the stabilization of the organic matter which is carried out by a chain of biological processes. The stabilization is a time–temperature function, and utilization of the dissolved oxygen increases as the temperature rises. Nature replenishes the depleted oxygen through the complex phenomenon of reaeration from the atmosphere, but unfortunately the oxygen saturation capacity of water declines as the temperature rises. Thus, streamflow, time of passage down the river, water temperature, and reaeration are the four major factors that govern organic stream self-purification.

The river waste assimilation potential controlled by self-purification depends on its size as defined by its hydrologic setting, climatic variables, and stream behavior through each reach of the stream.

The behavior depends on the configuration of the drainage area and the physical characteristics of the channel along its course. Where the channel is shallow and steep, the time of passage is rapid and short, with good reaeration; where the river is deep and meandering, with natural pools, the time of passage is slow and long, with poor reaeration. Also, the time of passage down the river varies inversely with runoff, being slowest and longest during the drought season.

Except in the case of stable inorganic matter, the decline in residual waste along the course of the stream is nearly logarithmic in character. Thus, the stream is rather quickly restored to its natural equilibrium state.

THE DUAL OBJECTIVES OF STREAM SANITATION

The variable nature of the primary factors controlling stream self-purification offers an opportunity for devising a rational system of stream sanitation, with two primary objectives: (1) the satisfactory disposal of the water-carried wastes and (2) the enhancement of satisfactory stream water quantity and quality for intended multiple uses. These two objectives are interrelated, but in an efficient sanitation system they follow quite different lines of attack, the former being primarily defensive and the latter primarily offensive.

Each river system, because of its dynamic hydrologic and meteorologic setting and pattern of water resource developments and uses, is unique. Hence, no arbitrary blanket solution applied wholesale can be expected to meet the varied needs of a technological society. Stream sanitation practice must, therefore, be specifically designed for each particular river, using that combination of defense and offense that ensures maximum efficiency at minimal cost.

For convenience, lines of defense and lines of offense are considered separately here, but it should be emphasized that efficient combination of both is the essence of good design and practice.

Multiple Lines of Defense. The satisfactory disposal of water-borne waste products is oriented toward the protection of the stream against overburden of its self-purification capacity (or the enhanced capacity provided by lines of offensive action) to maintain desired water quality. Effective defense is comprehensive; it is directed against all sources of man-made wastes—urban, industrial, and agricultural—and involves several lines of action.

Integrated Service Areas. Lines of defense are most effectively developed for an integrated service area encompassing, if necessary, several political units and including industrial as well as commercial and residential areas, with services extending to the urban fringes.

Segregation for Treatment or Special Disposal. Though the defense should be arranged for an integrated service area, it is necessary to segregate the wastewater in a separate sewerage system independent of the community storm water drainage system. This is desirable, since dilution of sewage prior to treatment complicates the problem. Furthermore, where combined sewers are utilized, treatment is bypassed with almost every rain and the combined flow of raw sewage and rainwater is discharged directly to the river through many overflows.

Similarly, segregation of industrial wastewater from rainwater and relatively clean plant cooling water should be practiced. Also in some instances, extremely concentrated industrial wastes are segregated for special disposal by evaporation and burning or by pumping into deep disposal wells.

In-Plant Practices. A number of in-plant lines of defense can be practiced to reduce an industrial waste load at its source: elimination of certain harmful products (such as mercury), efficient use of raw materials, recovery systems and water reuse, slug and peak discharge equalization, neutralization, process change or modification, industrial production tailored to waste assimilation variation of the receiving stream, and continuous wastewater surveillance and control.

Planning and Zoning. Many pollution problems stem from initial mislocation in relation to waste assimilation capacity available and in relation to other water uses. Where waste effluent outlets are placed is as much a concern as the magnitude of the residual waste discharge. Plant site location must consider satisfactory waste effluent disposal as well as water supply.

Community land use planning, which takes into account the interrelation of land and water, can enhance site location. Good zoning practices can also give guidance to direction of growth and avoid conflicting uses.

Regional treatment plants (serving two or more communities) with effluent outfalls strategically located to favor assimilation and to prevent adverse effects on other water uses should be considered.

Improved Agricultural Practices. Modern agricultural practices are a significant source of pollution. The trend toward livestock feedlots and associated on-site meat processing facilities concentrates organic waste loads which were heretofore spread over larger land areas. Significant residual organic matter, fertilizers, pesticides, and herbicides reach the stream through land drain systems, landwash, and erosion.

Elimination of certain hazardous chemicals may be necessary, as well as more efficient use of chemicals and improved irrigation practices. Better erosion control is essential for preventing widespread loss of soil to streams.

Environmental Control. Urban wash flushed to storm water drains and natural drainage channels may aggregate from 5 to 10 percent of the total community sewage. Strict control of sewer connections is essential, but in the final analysis the only effective approach is a comprehensive system of control of the community environment, including efficient refuse collection and disposal, control of litter, street cleaning, and public and private sanitation practices.

Wastewater Treatment. There is a popular misconception that treatment is all that is involved or necessary in satisfactory waste disposal and water quality control. While treatment will always play a major role, it is a mistake to promote the misconception that it is the cure-all. Treatment is a percentage removal; some residual fraction always remains. Wastewater treatment alone, therefore, is not the solution; a much more sophisticated system is required if we are to meet today's and tomorrow's needs.

With the wide diversity in natural capacities of rivers throughout the country and the widely unequal distribution of the waste burden, enhancement of stream quality by low flow augmentation and other offensive strategies should be considered. (Wastewater treatment systems are discussed elsewhere in this volume.)

Multiple Lines of Offense. The enhancement of satisfactory stream water quantity and quality for intended uses constitutes the second part of the dual objective of stream sanitation.

Positive developments are required to increase self-purification capacity and to permit fuller utilization of the water resource, as well as to provide protective or corrective measures against the detrimental impact of other water resource uses and developments.

Drought Control. The size of a river from the standpoint of quantity of streamflow may vary over a wide range, from flood to drought. Waste assimilation capacity and water quality fluctuate with these variations in quantity. If the total flow of a river could be equalized, a much higher assured flow would be continuously available and these characteristics would be greatly enhanced. When inefficient use is made of the total flow available, uncontrolled drought flow results and there is a drastic decline in the quantity and quality of the water.

Drought control is a logical complement to flood control. The floodwaters are retained for release to augment the dry-weather streamflow. Various types of river developments are adaptable to this practice; storage reservoirs and regulated release, multipurpose river developments, and pumped storage.

It is seldom economically feasible to completely equalize the streamflow, but the practice of drought control ensures continuous flow several times greater than that available under the natural fluctuations of runoff. It not only increases waste assimilation capacity and water quality but also benefits all water users along the stream.

Storage of Wastes and Regulated Release. Where it is not feasible to store water to augment the streamflow, a higher proportion of the total annual waste assimilation capacity can be made available by storing wastewater in lagoons and regulating its release in accord with the natural variation in streamflow. This is particularly adaptable to large, isolated industries where land area is available for construction of the required lagoon capacity. The lagoons serve a dual purpose as waste storage and as a treatment system. Reductions in BOD of 50 to 90 percent are attainable during the storage period.

Planning River Developments to Minimize Adverse Impact on Waste Assimilation Capacity. Almost every river development has the potential to damage or to enhance waste assimilation through alteration of the parameters that affect self-purification. The principal impacts are associated with the following: flood control works, hydroelectric plants, navigation works, thermoelectric plants, and diversion works. Steps which may be taken to minimize their adverse effects and to utilize their potential for augmenting waste assimilation capacity are discussed more fully later.

THE FIVE MAJOR PHASES OF STREAM ANALYSIS

Stream analysis usually involves five major phases:

1. Determination of waste loadings.
2. Definition of hydrologic and climatologic factors.
3. Adoption and verification of self-purification factors.
4. Forecasts of expected stream conditions.
5. Evaluation of the impact of river developments.

Determination of Waste Loadings. The details of waste loading determination are presented in other chapters. Only those aspects of waste loadings essential for meaningful stream analysis procedure involving organic wastes are discussed here.

The best single measure of organic waste is the biochemical oxygen demand (BOD). Special emphasis is given to the determination of BOD (particularly with respect to immediate demand), differentiation between the colloidal and dissolved fraction and the settleable solids fraction, and the development of the BOD rate curve.

Since BOD is the dynamic chemical manifestation of a chain of organisms acting on the unstable organic matter, it is to be expected that considerable variation is inherent in test results among samples. This does not invalidate the BOD test; however, it does emphasize the necessity of examining a sufficient number of samples to derive a statistically reliable representative mean value.

The Onshore Survey. The determination of waste loads should be made from an onshore survey and not derived from river sampling. After waste effluents are discharged into the stream, the identity of the source is obscured, and the difficulty of obtaining representative samples due to streaking and variation in streamflow leads to large errors. The onshore survey involves location and measurement of volume and characteristics of all (municipal, industrial, and agricultural) significant wastewater effluent outlets to the stream.

Since wastes are a reflection of man's activity, it is to be expected that variation in loading occurs during the day, from day to day, and from season to season. At best, therefore, the intensive onshore measurement is but a reflection of the load at a point in time or for a short interval of a few days. If adequate records are available (e.g., the daily operating reports of treatment works), these variations can be defined in statistical terms.

In small communities, the day-to-day variation in BOD is rather large, with a coefficient of variation (standard deviation as a percentage of the mean) of 30 to 40 percent; for large cities the coefficient is usually least, 10 to 20 percent.

In an industrial plant, many factors also induce variation in the waste loading, both in volume and characteristics. Generally the variation in the daily waste production under a consistent production schedule follows a normal probability distribution.

The degree of variation in industrial waste as reflected by the coefficient of variation is usually on the order of magnitude of that for municipal wastewaters, greatest with small industries and least with large industries that operate 24 hr a day, 7 days a week.

It is commonly observed that variations of the effluent BOD of biologically treated wastes form skewed distributions, usually log-normal.

Projecting Waste Loadings. In stream analysis, consideration must be given not only to the current situation, but also to expected future conditions 20 to 50 years hence. This presents the more difficult task of projecting the trend in waste loadings. For meaningful analysis, the projections must extend beyond area totals to assignments to specific service area waste effluent outlet locations. Future waste loadings are primarily a function of population and industrial and economic growth.

The urban organic waste load is closely related to population, and thus the projection of population growth based upon the historical trend provides a good basis for estimating future municipal loads. Industrial waste loads are more difficult to project because of the great variation among different industries and the changing technology of production.

Various methods of population projection range from simple graphical projections to mathematical models that include apportionment, rates of increase, ratios of smaller units to larger units, density per unit area, cohort survival or net migration and net births and deaths, and the logistic theory of growth.

The Bureau of the Census employs the cohort-survival approach, which, since it commences with the present population structure and specific birth and death rates and migration, is generally recognized as the most reliable method for short-range projections. This advantage, however, soon diminishes when applied to long-range projections.

The logistic method has the advantage of an underlying theory of growth that can be dealt with on a mathematical basis and is particularly applicable to long-range projections. In addition to projection in terms of numbers of persons, it provides a means of defining characteristics of growth that are useful in comparison among communities and in defining significant changes in patterns of growth.

Economic base studies are made to arrive at a measure of economic opportunity and as a basis for projecting future growth potentials. Three general techniques of regional economic base analyses in common use are the export-base theory, the interindustry (or intersectional) flow method, and the theory of proportionate share by economic sector.

Combinations of these methods are possible and can lead to sophisticated regional economic mathematical models, but these do not necessarily provide more reliable forecasts, particularly where social and political decisions concerning the development of resources may change the past historical trends. Thus, economic base projections are not to be taken as absolute; they merely provide guidelines and serve as a reference frame within which population and industrial projections are made.

The conversions of projections of future organic loads are based on 0.24 lb of ultimate biochemical demand (BOD) per capita per day as the basic population equivalent. This is considered as representative of a normal community composed of residential and commercial areas with a moderate mix of industries but without large wet-process industries contributing heavy organic waste loads.

In any specific situation, the total organic load comprises three fractions: (1) the normal community waste, (2) the large wet-process industries, and (3) urban wash. If the community is normal without special organic waste industries, the BOD conversions in population equivalents are directly proportional to the projected populations; to this is added the uncontrollable urban wash, which may be estimated at 5 to 10 percent of the BOD load of the projected population. If large wet-process industries are also involved, these are converted based on: (1) a higher per capita unit oxygen demand, or (2) population equivalents per unit product, or (3) population equivalents based on economic projection of employment by the industrial sector.

Hydrologic and Climatologic Factors. The degree to which waste assimilation capacity can be quantitatively determined is dependent to a large extent upon the detail with which the hydrology and climatology of the specific drainage basin are defined. Since these phenomena are highly variable in time and place, many of the parameters are defined in statistical terms from analyses of long-term records. Three variables are particularly pertinent: climate, stream runoff, and the physical characteristics of the river channel.

Climatological Patterns of Variation. The climatic setting establishes the special differences among drainage basins and subdivisions. Physiographic land forms, geology, topography, vegetative cover, land use, rainfall, and temperature all play roles.

Although physiography in a given basin is reasonably stable, the climatologic factors—precipitation, temperature, wind velocity, vapor pressure, and solar radiation—are highly variable in time and place, and induce variation in stream runoff and in waste assimilation capacity. The first two variables, precipitation and temperature, are of special concern in the evaluation of organic self-purification.

Precipitation. Precipitation patterns not only vary radically from basin to basin, but there are also variations from year to year and through the seasons. These variations can be defined by statistical analysis of long-term records of weather stations in a basin. For example, the analysis of a 51-year record in the Roanoke Basin shows the average annual geographical distribution pattern on the basin ranging from 50 in. in the Coastal Plains near the mouth, to 40 in. in the Upper Piedmont Plateau, and increasing up the eastern slope of the Blue Ridge Mountains to 46 in. at the rim of the headwaters.

Statistical analysis of annual average depths over the entire basin show considerable variation from year to year. The annual values follow a well-defined normal distribution as a straight line on normal probability paper about an average annual depth of 42.1 in. with a standard deviation of 6.1 in. Thus, in the long run, for 50 percent of the years the annual average basin depth can be expected to be within the range of 38.0 to 46.2 in. and for 90 percent of the years within the range 32.1 to 52.1 in.

The seasonal pattern of rainfall through the months of the year for the Roanoke Basin likewise is determined from statistical analysis of the 51-year record of depths by month as an average on the drainage area based on isohyets for each month. The variation for specific months, however, is skewed, plotting as a good fit on log-normal probability paper. From this, a most probable mean and variation ($\sigma_{\log}$) about the mean is readily defined.

Although these illustrations pertain to a specific basin, they are representative of the form of a general pattern and indicate a method of analysis useful in defining relevant characteristics for any particular basin with adequate precipitation data.

Air and Water Temperature. The temperature of the water is a major factor in organic waste assimilation capacity because it plays a triple role in affecting the rate of stabilization, the dissolved oxygen saturation capacity, and the rate of reaeration. As the water temperature rises, the rate of oxidation of organic matter increases, placing a greater demand on the dissolved oxygen. However, as the temperature rises, the dissolved oxygen saturation capacity of water declines. In reaeration, temperature plays opposing roles; an increase in temperature increases the rate of oxygen diffusion within the body of water, but the increase in temperature restricts solution from the atmosphere by reducing the saturation capacity.

Long-term records of water temperature are usually not available. However, the surface water temperature of many streams parallels reasonably the monthly mean air temperature; where observed water temperature records are for short periods or unavailable, the patterns developed from statistical analyses of long records of monthly mean air temperature serve as a reasonable approximation for many practical applications. This generalization, however, must be qualified depending upon unusual hydrologic settings of particular basins such as streams fed by large cold water springs.

If no water temperature data are available and the problem warrants (or if waste heat from steam electric power generation is involved), water temperatures can be computed from meteorological records of air temperature, wind velocity, relative humidity, and solar radiation.[1]

Analysis of air and water temperature records in drainage basins throughout the United States indicates that a well-defined seasonal pattern prevails; in northern latitudes the seasonal peak occurs in June–July, and in the more southern latitudes the peak rises and shifts toward July–August. The variation from year to year in monthly means is consistently normal, from which a most probable value and any desired confidence range can be defined. The once-in-10-year monthly average at the seasonal peak can be expected to exceed the most probable or normal monthly average by 5 to 7°F, and the once-in-10-year weekly average can be expected to exceed this by about an additional 5°F.

Other Climatological Factors. Long-term records of other climatological factors such as wind velocity, relative humidity, and solar radiation also consistently describe normal probability distributions from which the most probable value and any confidence range can be obtained, and in turn well-defined seasonal patterns can be constructed.

Quantitative Characterization of Streamflow. No phase of stream analysis is more important and more difficult than determining runoff from location to location along the course of the stream. Streamflow is determined from the drainage area tributary to locations along the course and from records of streamflow at a number of key locations in the drainage basin.

Some drainage basins are homogeneous in character, and streamflow along the course is directly proportional to the tributary drainage area. Streamflow then is readily obtained as drainage area in square miles multiplied by the yield per square mile as measured at the key stream gauging station or as weighted yields at more than one station.

Other drainage basins are composed of heterogeneous tributary areas of widely divergent yields. Estimating streamflow along the course, therefore, is as much an art as it is a science and depends on intimate knowledge of the drainage basin, its physiographic regions, land forms, vegetative cover, geologic and groundwater formation, and rainfall distribution as well as other climatologic factors. The estimation must also be tempered by experience and professional judgment.

While highly dynamic, streamflow variation is actually quite orderly and can be defined in useful statistical terms and graphical forms, each of which reflects elements of the hydrologic setting.

The Monthly Hydrograph–Seasonal Pattern. Plotting monthly average streamflow throughout the year provides the monthly hydrograph. The characteristics of this are influenced primarily by climate, the geographic and seasonal distribution of precipitation, temperature, evaporation, and transpiration. From statistical analysis of long-term monthly average streamflow records, a seasonal pattern emerges such as that illustrated for the Trinity River (Fig. 6-1) in the Gulf Coastal Plains.

In developing the seasonal pattern, each month of the year is analyzed independently; the variation in monthly average is generally logarithmically normal and is readily defined by plotting on log-normal probability paper. From this, the most probable monthly average and any confidence range, such as that expected for 80 percent of the years, can be obtained graphically.

Since the probability distributions are also logarithmically normal, the seasonal pattern is plotted on semilog paper as shown in Fig. 6-1. Curve A is the most probable or normal monthly average expected, and the shaded belt about the most probable curve is the range within which the monthly average streamflow for specific months of the year can be expected

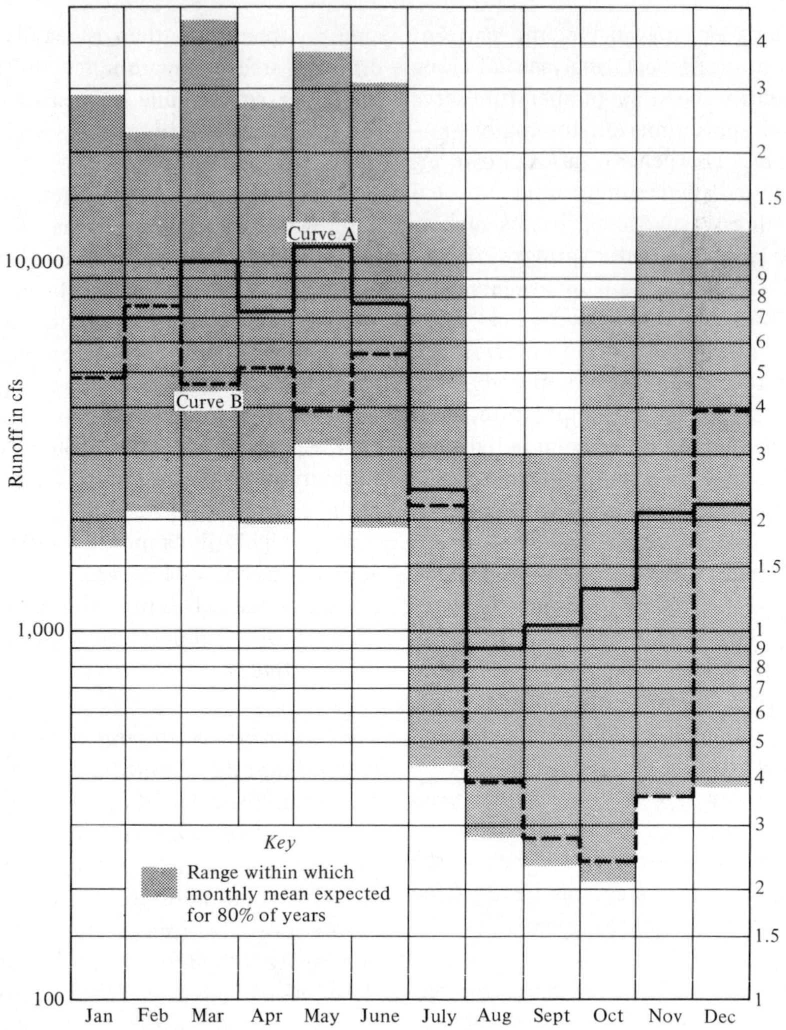

Fig. 6-1 Seasonal pattern of runoff, Trinity River at Romayer, Texas (based on 30-year record).

for 80 percent of the years; for 10 percent of the years the monthly average can be expected to be above the upper limit of the shaded belt, and for 10 percent of the years it can be expected to be below the lower limit.

The seasonal pattern for any specific year may deviate from the normal pattern since not all months through a year will experience streamflow with the same probability of occurrence; it can be expected, however, that 80 percent of the yearly patterns will fall within the frame of the shaded belt. Curve B shows the monthly hydrograph for the specific year 1939, generally recognized in this basin as a year of severe drought.

It is apparent from the monthly hydrograph that the lower Trinity River is subject not only to marked seasonal differences between the winter–spring wet period and the summer–fall drought period, but that within these periods wide deviations from the normal pattern can be expected. For example, while the normal monthly average streamflow during the drought season is expected in the range of 1000 to 2000 ft^3/sec once in 10 years, it can be expected that the monthly average will decline to 200 to 300

ft^3/sec. However, although subject to such extreme lows, it is also evident that in many years the warm-weather runoff of the Trinity will be very high, with a once-in-10-year monthly average range of 3000 to over 10,000 ft^3/sec.

There is risk in relying on the natural streamflow of a river such as the Trinity. While the average flow is high, it is not always available unless drought control is practiced to equalize the extremes of the spring high runoff season and the warm-weather drought.

The point to emphazize here is the value of defining not only the normal seasonal pattern but also the degree of variation from normal which can be expected. Not all rivers show such high degree of variation, and the monthly hydrograph seasonal pattern with its expected band of variation is a valuable device in characterizing the streamflow.

The Daily Hydrograph. Plotting the daily average streamflow establishes the daily hydrograph which primarily reflects the flashy or stable character of the flow, principally the physiographic elements—shape of the drainage basin, rugged or flat topography, geology, soil, and natural storage in lakes, ponds, and swamps. Steep slopes of impervious soil formation with lack of natural storage give rise to a flashy stream with a daily hydrograph punctuated by sudden rises and rapid declines with each occurrence of rainfall. Intervening periods between rains are usually periods of extremely poor yield. Such a flashy stream is illustrated by the daily hydrograph of the Jackson River (Fig. 6-2).

Conversely, where the drainage basin is relatively flat with pervious soil and good storage in groundwater, lakes, and swamps, the daily hydrograph is reasonably stable, less influenced by rainfall and drought, with much higher yield.

The daily hydrograph to some extent also reflects man-made influences, such as weekly operating patterns of hydroelectric plants without reregulation.

Since water quality fluctuates radically with sudden changes in streamflow, the daily hydrograph is invaluable in the interpretation of river sampling. Sampling during an erratic hydrograph gives a false reflection of stream conditions. The ideal time for sampling is during a steady runoff pattern, but sampling should not commence until the hydrograph has attained stability for a period at least equal to the time of passage through the critical reach to remove the influence of any preceding instability induced by a freshet. A critical analysis of recorded daily hydrographs aids in anticipating a period of stability in which to sample; if stability does not occur, it is useless to undertake sampling, as the observed stream conditions would be of little value in stream analysis.

The Instantaneous Hydrograph. The 24-hr average streamflow as reported in Ref. 2 does not reflect moment-to-moment variations in runoff. The variations are detected only at stream gauges with continuous automatic recorders, records of which can be obtained upon request from the U.S. Geological Survey. The primary purpose of the instantaneous hydrograph is to detect man-made influence on streamflow such as that induced by hydroelectric plants. The diurnal cycle during weekdays and the weekend shutdown of typical hydroelectric operating practices are shown in Fig. 6-3 for the Roanoke River at the Roanoke Rapids gauge below hydropower plants without reregulation to equalize flow.

Such pulsation patterns persist for many miles downstream, as illustrated by the instantaneous hydrograph of gauge height at three locations on the Pee Dee River (Fig. 6-4) below a hydroelectric plant. The extent and degree of influence depend upon channel characteristics, and if no reregulation is provided, drastic pulsations in streamflow extend many miles downstream. Weekend shutdowns in some locations have been observed to virtually shut off flow of the river with little or no release for as long as three days. Obviously, such induced curtailments and pulsations in streamflow are extremely detrimental to waste assimilation capacity and resultant water quality. The detrimental impact of such practices and the need for corrective action are discussed more fully later.

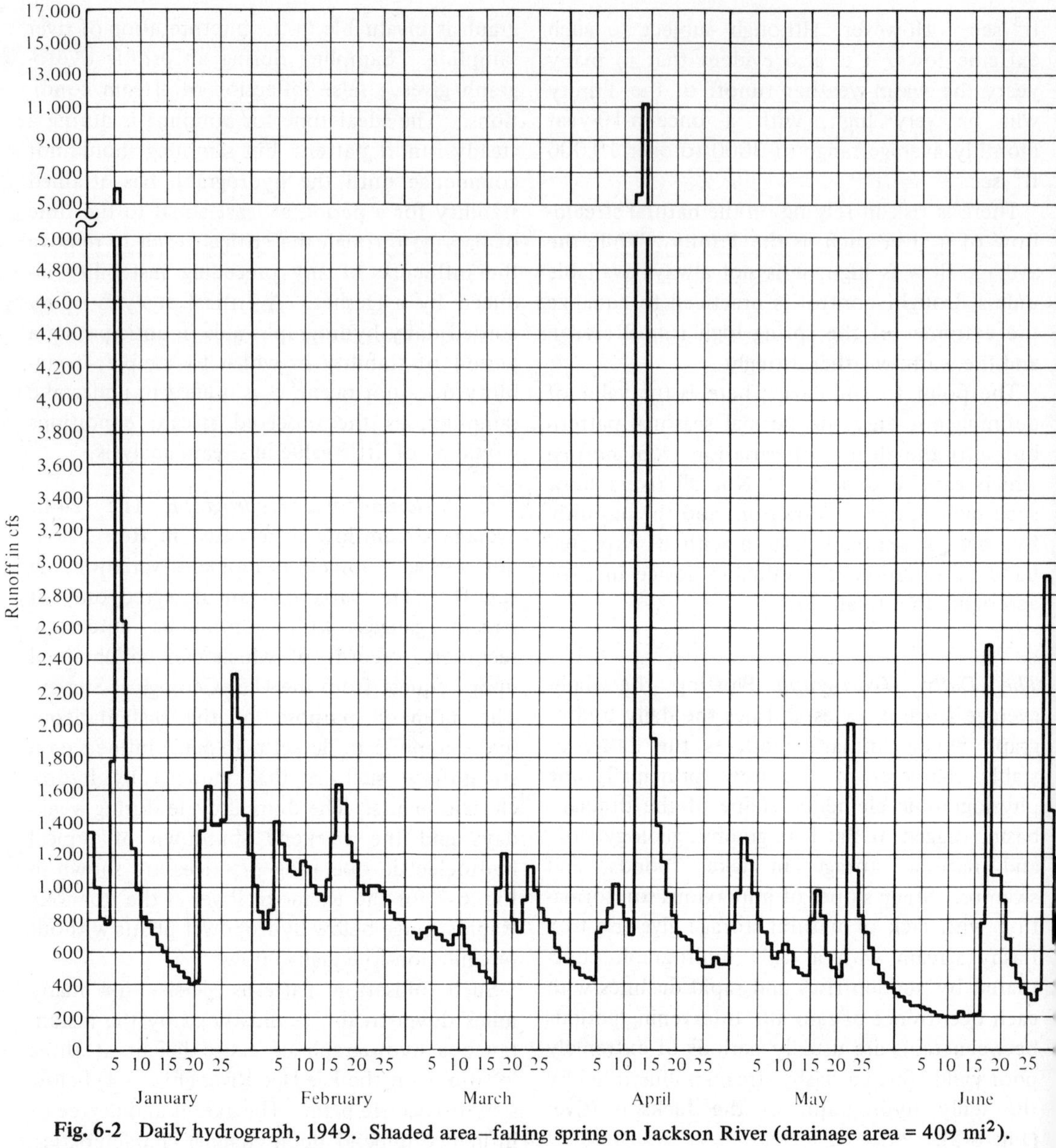

Fig. 6-2 Daily hydrograph, 1949. Shaded area—falling spring on Jackson River (drainage area = 409 mi^2).

Drought Flow Probabilities and Indices. Natural streamflow, as has been repeatedly stressed, varies between extremes of floods and droughts. Any level of development carries a risk of unsatisfactory stream conditions related to the probability of drought severity. The degree of community and industrial development, the downstream water quality objectives, the required degree of wastewater treatment, and the selection of defensive and offensive strategies rest on the definition of drought flow probability.

A fundamental approach to the statistical evaluation of the drought flow characteristics of streams stems from the theory of extreme values. It is well known that extremes of hydrologic phenomena, such as floods and droughts, do not follow a normal symmetrical distribution but are skewed. A standard skewed distribution based on the theory of

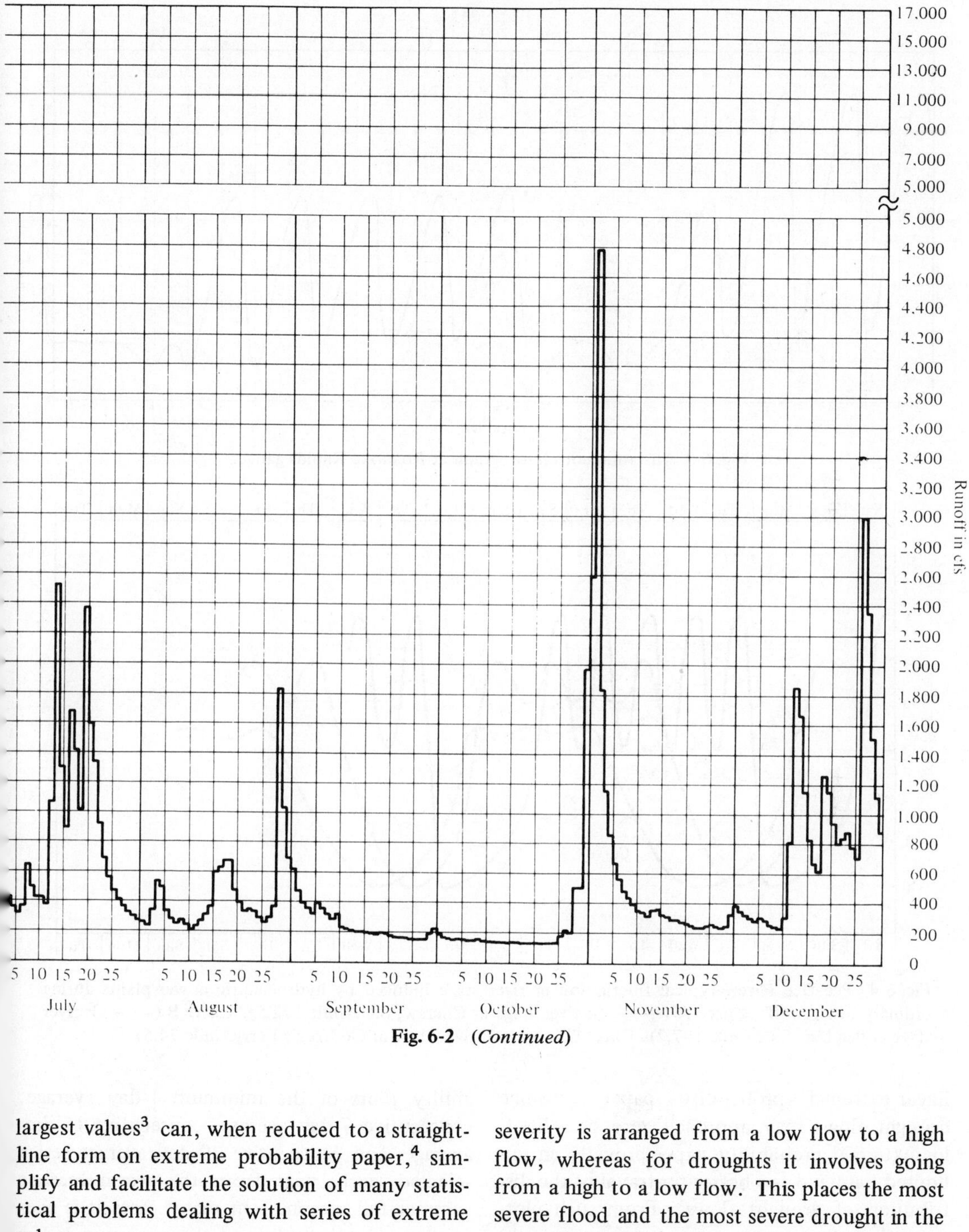

Fig. 6-2 *(Continued)*

largest values[3] can, when reduced to a straight-line form on extreme probability paper,[4] simplify and facilitate the solution of many statistical problems dealing with series of extreme values.

The extreme probability grid initially developed for flood flows is adapted to droughts by the simple technique of dealing with values in their order of severity rather than in their order of magnitude. For flood flows, the order of severity is arranged from a low flow to a high flow, whereas for droughts it involves going from a high to a low flow. This places the most severe flood and the most severe drought in the same position in terms of the probability of severity. Thus, the extreme probability grid is equally adaptabe to minima as well as maxima series arranged in order of severity.

In dealing with floods there is no upper limit to flow, and flood series are best plotted on

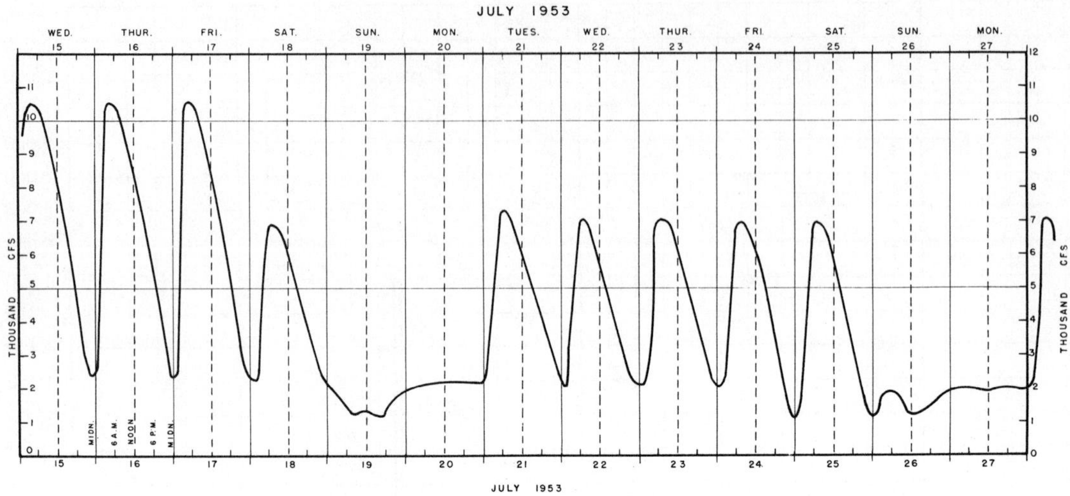

Fig. 6-3 Instantaneous hydrograph at Roanoke Rapids gauge.

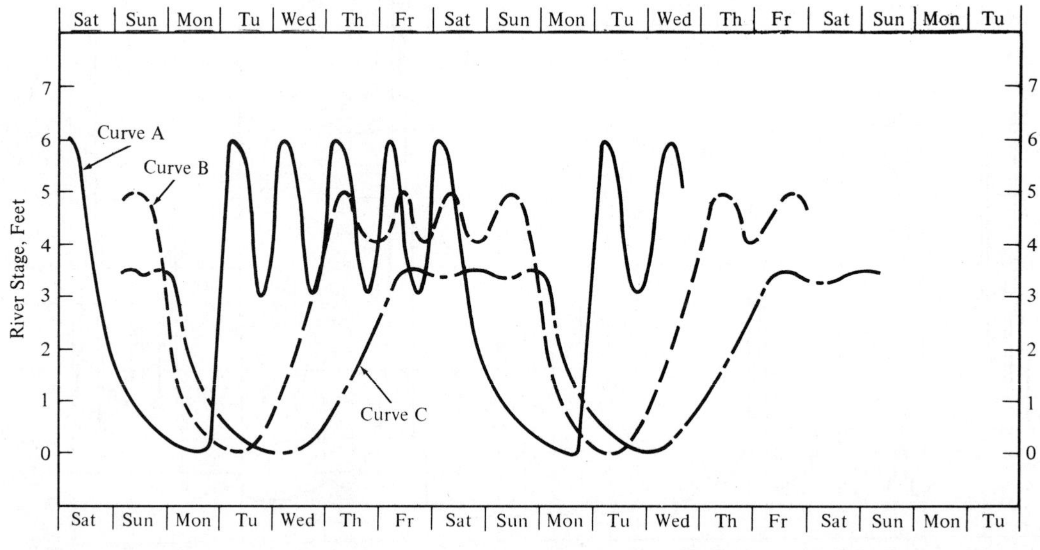

Fig. 6-4 Pee Dee River–typical fluctuation in river stage induced by hydroelectric power plants during ordinary low runoff. Curve A (———): river stage at Cheraw, S.C. (mile 172.5). Curve B (– – – –): river stage at Pee Dee, S.C. (mile 107.2). Curve C (– – - – –): river stage at Godfrey's Ferry (mile 74.5).

linear-extremal probability paper. Since drought flows may approach zero as a limit, log-extremal probability paper provides an unlimited base and is a better portrayal of the distribution of drought flows. From such plottings, drought severities expected for any probability of occurrence can readily be determined, such as once in 5, 10, or 20 years.

It is desirable to define the drought flow reference frame illustrated in Fig. 6-5. This is readily developed from log-extremal probability plots of the minimum 1-day average, consecutive 7-day average, consecutive 15-day average, and consecutive 30-day average series selected each year from the warm-weather summer–fall season. (In many cases it is adequate to develop plots of the minimum 1-day average, consecutive 7-day average, and the calendar monthly average series, which are more easily identified in stream discharge records.)

As one shortens the period over which the

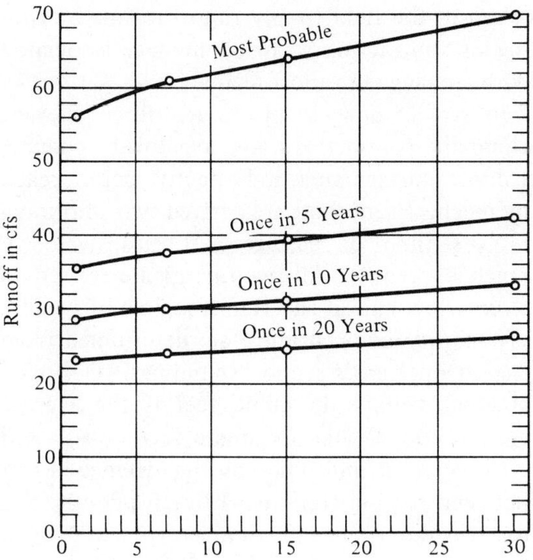

Fig. 6-5 Battle Creek at Battle Creek—drought duration vs. severity.

discharge is averaged, drought severity increases. Conversely, extremely low flows do not prevail for extended periods. From the standpoint of pollution control, consecutive low flow over a period equal to the time of passage through the critical reaches of the stream is the significant interval. This will vary from stream to stream and may range from 3 days to 2 weeks or more, depending on the physical characteristics of the channel and the distribution of pollution loads along the course. Drought flow reference frames afford a convenient means for interpolating severities as an average for any number of consecutive days appropriate to the situation.

In reaching decisions about pollution control requirements, water quality objectives, and levels of community and industrial development, it is obviously essential to take a position at some point within the drought flow reference frame. This position must be defined in terms of both coordinates—horizontally by the consecutive number of days over which flow will be averaged and vertically by the drought flow severity within the range from the most probable to, for example, that expected once in 20 years. If the decision is made on a consecutive 30-day average, it must be recognized that more severe droughts will surely be experienced for the shorter period averages. The 7- or 15-day average may be considered more appropriate. If the decision is made at, for instance, once-in-3-years severity, then it must be recognized that on the average in the long run drought flow will be more severe and streamflow will not be adequate once in 3 years. If this frequency of less than desired streamflow is unacceptable, it is necessary to restrict activity to a lower flow, such as a severity that is expected once in 10 years.

On the other hand, to be so conservative as to take a position at, let us say, the minimal 1-day average expected once in 20 years, development may be drastically restricted. One must take a reasonable position within the natural drought flow reference frame and evaluate the consequences of that position or one must consider alternatives of relieving the restriction by drought control.

Two indices, the yield and the variability ratio, afford a basis for comparing drought flow characteristics within and among drainage basins. They assist in estimating drought flow at locations in the system other than at stream gauging stations and in developing the drought flow along the entire course of the main stream.

The yield is simply the drought flow expressed as a unit per square mile of tributary drainage area. This affords comparison among basins of different size. If the basin is homogeneous, the yield may be applied to drainage areas tributary to locations other than the gauging station.

Some streams show wide variation from year to year in the severity of drought flow, whereas others are relatively stable. This variation in drought severity is measured by a simple index termed variability ratio. In order to present a useful index free of the mathematical complications of deviations in the skewed extremal distribution, the concept of a simple ratio is employed. It answers directly the question, what proportion of the normal drought flow is the rare drought flow? The variability ratio in a standardized form is defined as the ratio of the once-in-10-years drought flow to the most probable drought flow. Thus, a variability ratio of unity would indicate perfect stability with no variation; a small ratio would indicate a high degree of variation in drought flow from year to year, with great risk of occasional extremely severe low flow.

In comparing the drought potential of streams, both indices—the yield and the variability ratio—contribute to quantitative characterization. The yield defines the level or magnitude of flow; the variability ratio defines the degree of variation, or stability, and the degree of risk of severe deviation below normal.

Extensive studies of the drought flow characteristics of streams throughout the United States disclose that the theory of extreme values is consistently applicable and that drought flow indices are usable devices.

Quantitative Characterization of the Stream Channel.

Physical Factors. A phase of hydrology that is essential to quantitative computation of waste assimilation capacity is the detailed characterization of the stream channel. Rivers along their course vary radically in configuration, cross-sectional shape, depth, channel bed, and hydraulic gradient.

These features are reflected in cross sectioning at frequent intervals along the channel. In some instances cross section soundings are available; if not, they are best obtained by echo-sounding equipment or by rod or sounding lead. Cross sections should be taken at approximately 500-ft intervals along the course and more frequently at irregular configurations. Frequency of cross sectioning is more significant than precise measurements at any given section. Location of sections by survey traverse is not necessary since good maps or aerial photos are usually available on which sections can be marked.

If maps or photos are to a reasonable scale, river widths can be scaled; if not, widths are obtained by stadia. Cross section sounding should be taken at reasonably low-water conditions and a steady hydrograph, and be related to the current streamflow at a nearby gauge.

In addition to the discharge rating curves at key gauging stations, it may be necessary to establish a number of auxiliary river stage gauges. These permit adjusting soundings as taken in the field to any runoff regime by adding or subtracting a change in stage associated with a change in runoff regime.

In stream analysis procedure, three relevant channel parameters are occupied channel volume, surface area, and effective depth, reach by reach. From these are derived two additional factors—time of passage and mean velocity reach by reach. All five factors are related to streamflow and at any runoff regime of specific probability of occurrence are determined from the channel cross section soundings. Occupied channel volume in cubic feet is the average cross-sectional area in square feet of adjacent cross sections multiplied by the distance in feet between section stations. Effective depth for a reach is determined as occupied volume divided by surface area, where surface area is taken as the average of a number of scaled (or measured) widths multiplied by the distance from the beginning to the end of the reach in question.

Time of Passage. Time of passage along the course of the stream, as discussed previously, is one of the primary factors in self-purification. A distinction is made between the traverse of the hydraulic crest, the time at which changes in the rate of discharge are noted at locations downriver, and the mass movement of the body of water. It is the latter that is the measure of the time of passage of wastes from reach to reach along the course and which is the primary concern.

The time of passage is determined at any specific runoff regime as the occupied channel volume reach by reach divided by the tributary runoff to each successive reach. Occupied channel volume is obtained from the channel cross section soundings as outlined above; runoff along the course of the channel at any location is determined by the tributary drainage area and the appropriate yield as outlined heretofore. With occupied channel volume expressed in cubic feet and runoff in cubic feet per second, time of passage is in seconds. It is customary, however, to convert the time of passage to days. The summation of the increments of time from cross section to cross section provides the cumulative time along the course, whose plot against river mile locations provides the time of passage curve. Since channel parameters change with change in runoff regime, time of passage is tied to the probability of each runoff regime. Accordingly, the pattern of time of passage is represented by a nest of curves for the range in streamflow under study, usually the runoff identified by the drought flow reference frame.

Figure 6-6 shows time of passage curves for River A based on detailed channel cross sections. In this instance, data for each cross section sounding were punched on cards and a program was written for computer adaptation to adjust sounding for each runoff and to execute the extensive computations.

Time of passage as occupied channel volume displaced in blocklike manner by runoff, it is recognized, does not strictly hold, since a certain degree of short-circuiting and longitudinal dispersion takes place. However, where channel configuration is reasonably regular, occupied channel volume displaced by runoff is a valid measure of the mean time of waste passage along the course. This is verified by close agreement with dye tracer results and confirmed by analyses of resultant stream conditions.

In some instances it may be necessary to limit occupied channel volume to the active flow-through volume, such as elimination of wide shallow flats or isolated bays. In channels of great depth subject to stratification, the active flow-through depth may be limited to the upper 15 to 20 ft, and volume below this depth from the standpoint of time of passage is virtually dead area.

Sludge Deposit and Scour. In addition to providing the important factor of time, the time of passage curve affords a ready means of identifying potential zones of sludge deposit. Since the cumulative time of passage is plotted against distance along the river course, the tangential slope at a specific mile-point location on the time of passage curve, or slope through a reach, is the measure of velocity.

For example, the time of passage curves in Fig. 6-6 for the range in drought flow on River A show the velocity slopes in feet per second. A mean channel velocity of 0.6 ft/sec is the critical velocity at which deposits of organic settleable solids occur and accumulate. By translating the 0.6-ft/sec slope along the time of passage curves, four potential sludge deposit areas are located, as shown in Fig. 6-6. There is a good theoretical basis for this method of identifying the potential zones of sludge deposit, and field observations confirm these locations.

Fresh organic sludge deposits undergoing digestion and gasification may be readily scoured at velocities between 0.6 and 1.0 ft/sec. For resettled or wholly digested material for which active gasification has subsided, higher velocities, such as those associated with freshets and floods, may be required to scour deposits.

Adoption and Verification of Self-Purification Factors. In magnitude and complexity, the organic constituents of the end products of community and industrial activities comprise the most significant problem in satisfactory waste disposal. Usually, if the waste assimilation capacity of the stream is adequate to handle the organic effluent residual, the self-purification capacity is adequate for the other classifications of waste as well.

The Normal Rate of Biochemical Oxidation. A generalized expression has been developed[5] that serves as a basis for the quantitative determination of the liability fraction in organic self-purification: "the rate of biochemical oxidation of organic matter is proportional to the remaining concentration of unoxidized substance, measured in terms of oxidizability." Phelp's law then is expressed in differential form as:

$$\frac{-dL}{dt} = KL$$

which may be integrated to:

$$\log_e \frac{L_t}{L} = -Kt$$

or

$$\log \frac{L_t}{L} = -0.434Kt = -kt,$$

$$\frac{L_t}{L} = 10^{-kt} \tag{1}$$

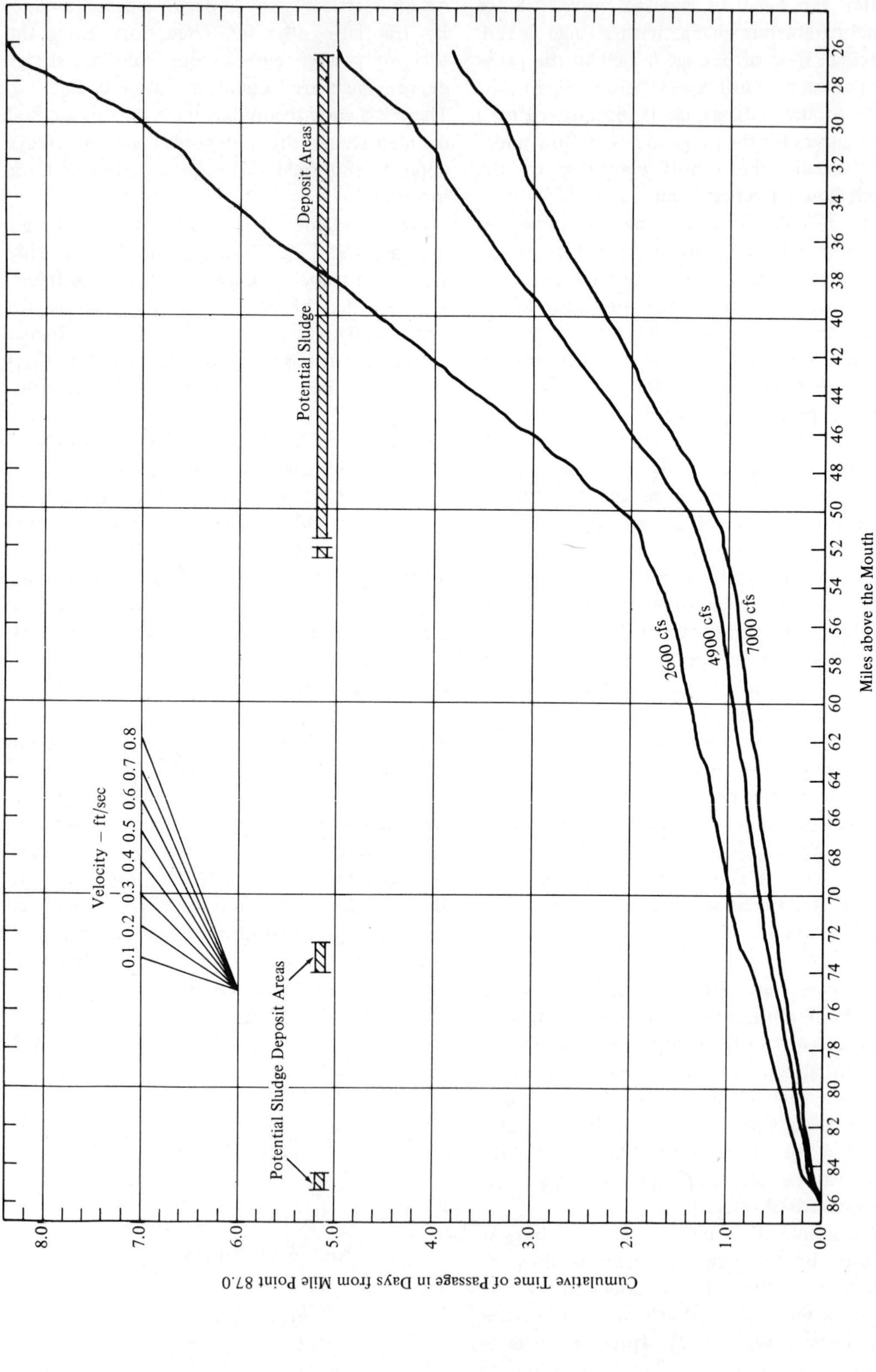

Fig. 6-6 River "A"—passage of time.

where L is the initial oxidizability at time zero (the initial total outstanding organic debt) and L_t represents the corresponding value at time t. The term L_t/L represents the fraction of the oxidizable organic debt remaining, and $(1 - L_t/L)$ is the fraction that has been oxidized, or amortized. The rate of reaction, the constant k, represents the rate at which the amortization proceeds. If time t is taken in days, k is found experimentally to have a value of 0.1 at 20°C.

Phelps' monomolecular reaction formulation is generally referred to as the normal rate of satisfaction of BOD. The normal amortization schedule expressed in percent of the initial outstanding BOD debt may be expressed as:

$$\log Y = 2.0 - kt \tag{2}$$

which represents a straight line relation as shown in Fig. 6-7. The reaction rate k is the slope of the line $\Delta Y'/\Delta X$. This is a convenient form for graphical integration of multiple sources of organic waste load along the course of a stream, as will be illustrated later.

Half-life and the Standard BOD. In the normal amortization schedule, two periods are of special interest–3 days and 5 days. At the end of 3 days, 50 percent of the debt is paid and 50 percent remains outstanding; thus, for the normal decay rate the half-life of the organic debt is 3 days. During each succeeding 3-day intervals the remaining debt is successively reduced by 50 percent.

The debt paid at the end of 5 days is referred to as the standard biochemical oxygen demand. This derives from laboratory procedure prescribed in *Standard Methods*,[6] wherein the amount of oxygen utilized in a 5-day period of incubation at 20°C is determined. At the normal rate in 5 days, about 32 percent of the debt remains unpaid and 68 percent of the total debt has been amortized. Hence, the normal standard BOD represents only 68 percent of the total. Thus, if the reaction rate is normal (k = 0.1), the standard 5-day BOD value is divided by 0.68 to convert to the total initial BOD debt that ultimately must be satisfied.

In expressing BOD, confusion can be avoided

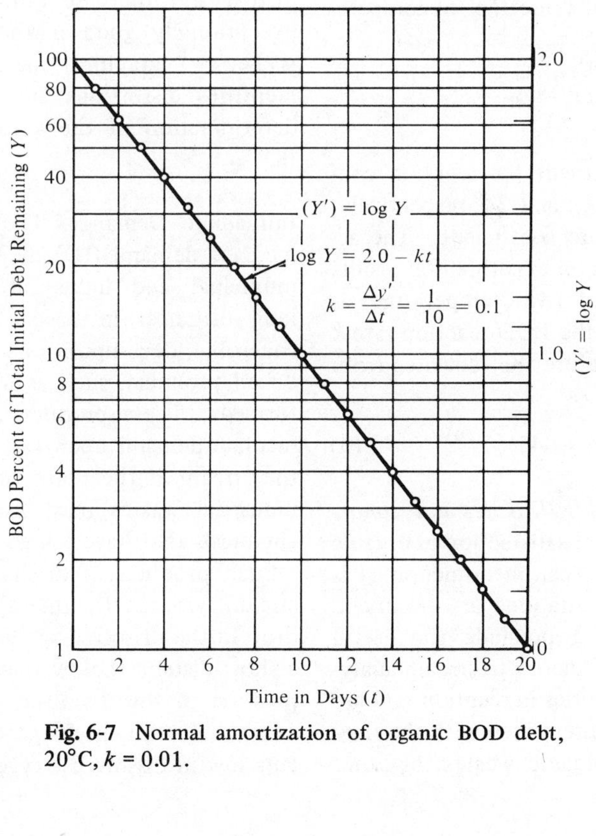

Fig. 6-7 Normal amortization of organic BOD debt, 20°C, k = 0.01.

by indicating the time of incubation by a subscript, with BOD_5 representing the standard 5-day 20°C value, BOD_3 representing the half-life 20°C value, and BOD_U representing the total initial debt, or the ultimate demand. If the temperature of incubation deviates from the standard 20°C, the specific temperature should also be indicated.

In addition to the period of incubation, the specific reaction rate must be known in order to determine the proportion of the total BOD that is exercised in the period of incubation. Usually, mixed organic wastes of the urban complex and many industrial wastes approximate the normal reaction rate, $k = 0.1$. Effluents of biological treatment systems and some industrial effluents may deviate significantly from normal, necessitating determination of the specific reaction rate before converting the standard 5-day (BOD_5) value to the total ultimate debt (BOD_u).

Effect of Temperature. Increasing the temperature above 20°C increases the reaction rate k, and decreasing temperature slows the reaction. A general expression of the effect of temperature is:

$$\frac{k_1}{k_2} = \theta^{(T_1 - T_2)}$$

where k_1 and k_2 represent the reaction constants at temperature T_1 and T_2, respectively, and θ is the temperature coefficient. The accepted coefficient based on experimental results over the usual range of river temperatures is taken as 1.047. Thus, the BOD reaction rate k at any working temperature that deviates from 20°C is:

$$k_T = k_{20°C} \times 1.047^{(T-20)} \quad (3)$$

Deviations from Normal BOD Rates and Abnormalities. Since the rate of satisfaction of organic debt is a complex biological phenomenon, it is not unexpected that variations are observed. These variations do not preclude the useful application of the basic theory to stream analysis. At times, abnormalities in reaction rate are lost when the waste is intermixed with the river water or with other organic wastes; in other instances they persist and must be taken into account.

Five significant abnormalities which may be encountered in biochemical oxidation of organic matter and which should be investigated are: abnormal specific BOD reaction rate, immediate demand, nitrification, sludge deposits, and biological extraction and accumulation.

Specific BOD Reaction Rate. Where the problem warrants and foreknowledge indicates a BOD reaction rate above or below normal, it is desirable to determine the specific reaction rate of major sources of organic wastes and of river water after intermixture of waste effluents. Generally the laboratory method is employed to determine the specific reaction rate for raw waste, effluents, and river waters.

There are various analytical procedures for interpreting the laboratory results, but the rigidity of analytical methods may lead to erroneous rates because of inability to take into account such factors as immediate demand, initial lag, or individual BOD values that are decidedly erratic. Such factors in many cases distort an otherwise quite normal rate. An exceptionally good method of interpretation is Lee's[7,8] graphical method, which visually identifies distortions and permits a graphical determination of the k rate and the ultimate BOD_U.

Immediate Demand. The abnormality of immediate demand (ID) may be associated with municipal and industrial wastes transported long distances in sewers and outfalls where anaerobic decomposition takes place and reduced products, such as hydrogen sulfide, are formed. These products can exercise an immediate demand upon the oxygen resources of the stream at the point of discharge. Certain industrial wastes may also contain inorganic chemicals that have a high demand for oxygen.

Such products, if discharged directly to the stream without treatment, can cause a sharp drop in the dissolved oxygen level at or within a short distance below the outlet. A sharp depression in the dissolved oxygen level is also associated with discharge of wastewater effluents low in dissolved oxygen when the flow of waste relative to the streamflow is large. Large volumes of condenser water used by thermoelectric plants, by virtue of rise in temperature, can significantly depress the amount of dissolved oxygen, particularly during the drought season.

Nitrification. The abnormality of nitrification usually is not manifest in the natural stream environment except under unusual conditions such as where biologically-treated effluents high in ammonia are discharged to small streams and environmental factors of temperature and pH are optimal. Unlike the carbonaceous reaction which is dominated by the rather rugged group of heterotrophic organisms, oxidation of nitrogenous material is carried out by highly-specialized autotrophic groups of organisms which are more restricted in numbers and much more sensitive to environmental conditions of the stream.

Biological treatment effluents, particularly of activated sludge systems, may contain adequate nitrifying bacteria, and if the effluent volume constitutes a large proportion of the streamflow and the stream temperature is optimal (25-30°C), nitrification can take place in the stream. This can depress the dissolved oxygen level substantially below the level accounted for by the residual carbonaceous BOD of the effluent. In such situations, the determination of the ammonia of the effluent is a measure of the oxygen demand. The stoichiometric relation of oxygen to ammonia in conversion to nitrite and then to nitrate is 4.57 to 1. Thus, 1 mg/l of ammonia nitrogen is equivalent to 4.57 mg/l of BOD. The reaction is not strictly stoichiometric due to carbon dioxide fixation, probably reducing the ratio from 4.57 to between 4.3 and 4.4. For application to stream conditions, the nitrogenous BOD reaction rate determination of the effluent intermixed with the streamflow, incubated at the optimal temperature (28°C) employing the jug technique, provides an approximation of the expected river reaction rate. (See Chapter 4 for a detailed discussion of nitrification.)

Sludge Deposits. In dealing with the abnormality of sludge deposits, a distinction is made between fresh organic deposits arising from municipal and industrial wastes and the old compacted accumulations of partially stabilized organic residue and river muds from surface wash after flood periods. The latter are referred to as benthal deposits and the former as sludge deposits, the primary concern in stream analysis.

With sludge deposits, active decomposition of a semianaerobic nature takes place with end products leaching into the overflowing stream and utilizing dissolved oxygen. Gasification in the deposits results in sludge boils, and on occasion large floating masses rise to the surface, inducing a process of refloating and redepositing until active decomposition is practically completed. If sludge deposits are excessive, the dissolved oxygen of the overflowing stream can be completely exhausted. As result, gas may rise and break at the water surface, releasing hydrogen sulfide and other gases to the atmosphere.

The effect of sludge deposits on the dissolved oxygen of the stream and measurement of that effect depend upon the relationship between accumulation of the deposits and the rate of demand from the accumulation. It is the BOD content (not the solids bulk) of the deposit that requires evaluation. A measure can be approximated by the laboratory standard BOD of the total waste sample and of the supernatant after 1- or 2-hr settling, the difference constituting the fraction of BOD subject to deposit.

The quantitative determination of deoxygenation involves the interrelation of several factors: (1) the BOD of sources of waste and fraction subject to deposit, (2) the amortization of these BOD fractions in transit down the stream to the sludge deposit area, (3) the net daily addition of BOD to the sludge pile, (4) the extent of accumulation in the sludge pile, and (5) the amortization of the BOD in the accumulation.

The second and third factors are determined from integration of the BOD of the sludge deposit fraction of waste sources down the river at the time of passage for the specific runoff and temperature to the sludge deposit area. The integration is at the normal aerobic rate adjusted to the specific temperature; the amortization in transit is the difference between the initial BOD and the residual BOD at the sludge deposit area; the daily deposit is the residual arriving at the deposit area.

The fourth factor, the extent of the accumulation in the sludge pile, depends upon the prevailing and preceding hydrograph, which predicts the interval over which depositing can take place without interruption or scour from a freshet. Accumulation also depends upon the rate of satisfaction within the accumulation. With daily additions of fresh BOD to the pile and daily amortization from the pile, an equilibrium accumulation is reached when these are in balance. Under these equilibrium conditions, the demand upon the dissolved oxygen of the stream is equivalent to the total BOD of the daily addition to the sludge pile.

The rate of BOD satisfaction of fresh sludge deposits undergoing active decomposition with end products leaching to the overflowing stream follows Phelps' law of constant rate, but the rate is slower, represented as:

$$\frac{S_t}{S} = 10^{-k^1 t} \tag{4}$$

where S is the initial oxidizability of the sludge at time zero, S_t is the corresponding value at time t, S_t/S is the fraction remaining, and $(1 - S_t/S)$ is the fraction amortized. The reaction rate of sludge decomposition k^1 at 25°C is found to range from 0.01 to 0.05, and for most practical stream conditions may be taken as 0.03.

The biological decomposition rate is influenced by temperature, and for the usual temperature ranges encountered in streams the decomposition rate may be taken as:

$$k_T^1 = k_{25°\mathrm{C}}^1 \times 1.047^{(T-25)} \tag{5}$$

The amount of accumulation under conditions at less than equilibrium, the extent of accumulation, and the demand from that accumulation for any finite period of continuous oxidation and depositing can be obtained from the relation:

$$S_{t'} = \frac{D}{2.3k^1}\,(1 - 10^{-k^1 t^1}) \tag{6}$$

where D is the rate of depositing and S_t is the accumulated storage. The time t^1 in this instance is not the time of passage downstream—it is the interval in days over which continuous depositing can take place and accumulate without scour or interruption.

Since theoretically an infinite time is required to attain equilibrium, the general equation reduces to the specific equilibrium form:

$$S_{t'} = \frac{D}{2.3k^1} = S_e \tag{7}$$

where S_e is the equilibrium accumulation. Even though an infinite time is theoretically required, it can be seen from these relationships that in 40 to 50 days equilibrium is practically attained.

In practical stream application, these theoretical conditions seldom hold, and accumulation and demand from it depends upon the preceding daily hydrograph. If the hydrograph remains low (velocities in the deposit areas are 0.6 ft/sec or less) for 40 to 50 days, it may be assumed the sludge pile reached equilibrium; if a scouring freshet intervenes, the entire accumulation may be flushed, or if velocity does not scour but prevents depositing for a limited period, less than equilibrium conditions prevail. Any number of conditions of accumulation can be investigated by resorting to graphical integration.

In the graphical method of integrating the accumulation, daily deposit is considered as a slug added at the beginning of each day rather than as a continuous contribution throughout the day. Each day's addition to the residual accumulation of prior days is then amortized through the day on semilog paper at the appropriate k^1 slope for the temperature involved. The level of accumulation is thus directly available at any period, and the demand exercised on any particular day is the difference between the accumulation at the beginning of the day and that at the end of the day.

Biological Extraction and Accumulation. The abnormality of biological extraction and accumulation is usually associated with infestation of filamentous growths, such as *Sphaerotilus*. (See Chapter 4.) In shallow streams the biological growth attached to the stream bed extracts organic matter from the stream flowing above, performing in a manner analogous to the trickling filter. Other rivers, depending on channel characteristics, may perform as natural activated sludge systems because the growth may be dispersed as biological floc throughout the flowing stream. In both instances, the effect on the stream is removal of BOD far in excess of the normal satisfaction in the time of passage through the reach, and under equilibrium accumulation a drastic demand is exerted on dissolved oxygen.

The abnormal BOD removal from the stream actually comprises two simultaneous independent processes–normal biochemical satisfaction and biological extraction, with the latter dominating. The BOD removal is not oxidized at the instant of extraction but is merely transferred from the flowing stream by biophysical contact with the biological floc; part is stored in the biomass and part is synthesized to additional biomass by anabolic reactions that do not require oxygen. Satisfaction of the extracted BOD then takes place slowly within the accumulated floc mass, usually as a sludge deposit. If extraction and accumulation of deposit along the course continue uninterrupted, an equilibrium is approached, and then the daily BOD satisfied from the accumulation equals the BOD extracted and added. At equilibrium, an abnormal demand is exercised on the dissolved oxygen of the stream. This is akin to the usual sludge deposit discussed above, except that the amount of the deposit is a function of the rate of BOD extraction from the flowing stream, and usually the zone of deposit is spread along a substantial reach of river rather than concentrated in localized areas.

The parameters that govern biological extraction rates are not sufficiently precise to permit predicting a specific rate of extraction for a particular river situation. Reliance must be placed on the stream survey to determine the time of passage and corresponding BOD residuals from which b is obtained. Laboratory investigation of waste sources and mixed river water defines the specific amortization rate of organics in transit, k. From these, the biological extraction rate, b^1, is then obtained ($b^1 = b - k$).

The rate of biological extraction, b^1, varies widely among rivers, ranging from 0.2 to 1.0 or more, depending on opportunity for contact between the biological mass and the water mass. Deep, slow-moving streams afford little or no opportunity for biological extraction. A shallow stream with a high roughness coefficient favors attached growth.

In quantitative determinations under equilibrium conditions, computation may be dealt with in the normal manner, substituting b for k in BOD integration. At less than equilibrium, partial accumulation is involved with much less consequent effect upon dissolved oxygen, and solutions follow concepts developed for sludge deposits.

Oxygen Resources. For practical application of stream analysis, the assets or income side of the dissolved oxygen ledger may be classified in five groups: (1) dissolved oxygen from increments of stream runoff, (2) reaeration, (3) photosynthesis, (4) stored dissolved oxygen, and (5) negative dissolved oxygen. The first two of these are the most important.

Photosynthesis is the process by which aquatic plants, principally algae, produce pure oxygen, but it is not dependable, nor always available, nor particularly amenable to quantitative determination. When present in large numbers, algae may add substantially to temporary income, but the income is available only during daylight hours because respiration at night institutes an outgo. Moreover, when dependable assets are most needed, during hot, dry weather, large masses of algae may die and the decomposing plants become a liability (see Chapter 4).

Oxygen stored in the form of nitrates is in the nature of reserve assets. This oxygen is available as a last resource to prevent complete bankruptcy when the dissolved oxygen supply is exhausted.

Negative dissolved oxygen is, in a sense, operation in the red. It is a reflection of more current obligations to be met than cash and reserves available. The result is hydrogen sulfide, which, if it remains dissolved in the flowing stream, may place an extra burden on the resources downstream. On the other hand, if hydrogen sulfide escapes to the atmosphere, the debt is paid by oxygen of the air rather than of the river water. Such a loss of negative dissolved oxygen in a sense can be considered as bankruptcy proceedings, where at least the red is washed from the books and the river starts again from scratch.

Oxygen Resource of Stream Runoff. Streamflow is the "bank" in which dissolved oxygen is contained. Since streamflow is a dynamic resource varying from season to season and from day to day, the oxygen bank is not a fixed size but flexible, shrinking or expanding as variations occur in runoff. These variations in the natural setting, as previously discussed, can be defined in statistical terms.

In addition to variation in runoff expected at a given location, the size of the oxygen bank is enlarged as it proceeds from headwaters to mouth down the river channel, acquiring increments of tributary drainage area en route. As noted earlier, estimating streamflow from section to section along the course of the river is one of the most difficult tasks, requiring experience and professional judgment.

Man-made river developments can add or detract from the oxygen bank by changes in streamflow, and these potential changes also must be taken into account.

The temperature of the streamflow is also a factor, as the size of the oxygen bank shrinks during the warm weather season and expands in the cold weather season because of the influence of temperature upon the oxygen saturation capacity of water.

Oxygen Resources of Reaeration. Reaeration is governed by two basic laws: the law of solution and the law of diffusion. The rate of solution is proportional to the dissolved oxygen deficit; the rate of diffusion through water and between two points is proportional to the difference in concentration between these two points. On these two laws and the experimental determination of the coefficient of diffusion, Phelps developed the mathematical formulation for the quantitative determination of reaeration from the atmosphere.[9] The solutions are involved, and recourse is necessary to extensive expansion by means of a Fourier series which leads to:

$$D = 100 - 81.06\left(0.779^c + \frac{0.105^c}{9} + \frac{0.0019^c}{25} + \cdots\right) \quad (8)$$

$$c = \frac{at\pi^2}{L^2}$$

where t is time of exposure, L is depth, and a is the diffusion coefficient for a specific temperature, with the influence of temperature:

$$a_T = 1.42 \times 1.1^{(T-20)} \quad (9)$$

To facilitate application of Phelps' formulation, a standard reaeration curve has been developed[10] as shown in Fig. 6-8. Curve A is based upon expansion of the series to as many as 500 terms to establish 5 points on the curve. With the advent of high speed computers, it has become possible to extend the expansion to a great many terms from which a refinement has been developed as Curve B,[11] which is in practical agreement with a recently developed direct solution.[12]

The application of the standard reaeration curve greatly facilitates computation of reaeration and readily takes into account measurable stream channel factors of occupied channel volume, effective depth, and water temperature. Mix interval is not directly measurable in the stream but has been empirically developed from comparisons of computed and observed dissolved oxygen profiles on a number of streams for which other basic data were well defined. Mix interval is related to stream velocity and particularly stream depth, as shown in Fig. 6-9. Streams are divided into three classes: those of moderated depth and velocity (Curve A), those of shallow depth and high velocity (Curve C), and tidal estuaries subject to ebb and flood translation (Curve B). Solution on the standard reaeration curve (Fig. 6-8) is arranged to function as a slide rule.

For example, for a stream depth of 10 ft, water temperature of 25°C, and mix interval of 24.2 min, dissolved oxygen as percentage of saturation absorbed per mix interval at zero initial dissolved oxygen is quickly obtained as follows: slide the control of scale 3 to coincide with 10 ft on scale 2; opposite 25°C on scale 3

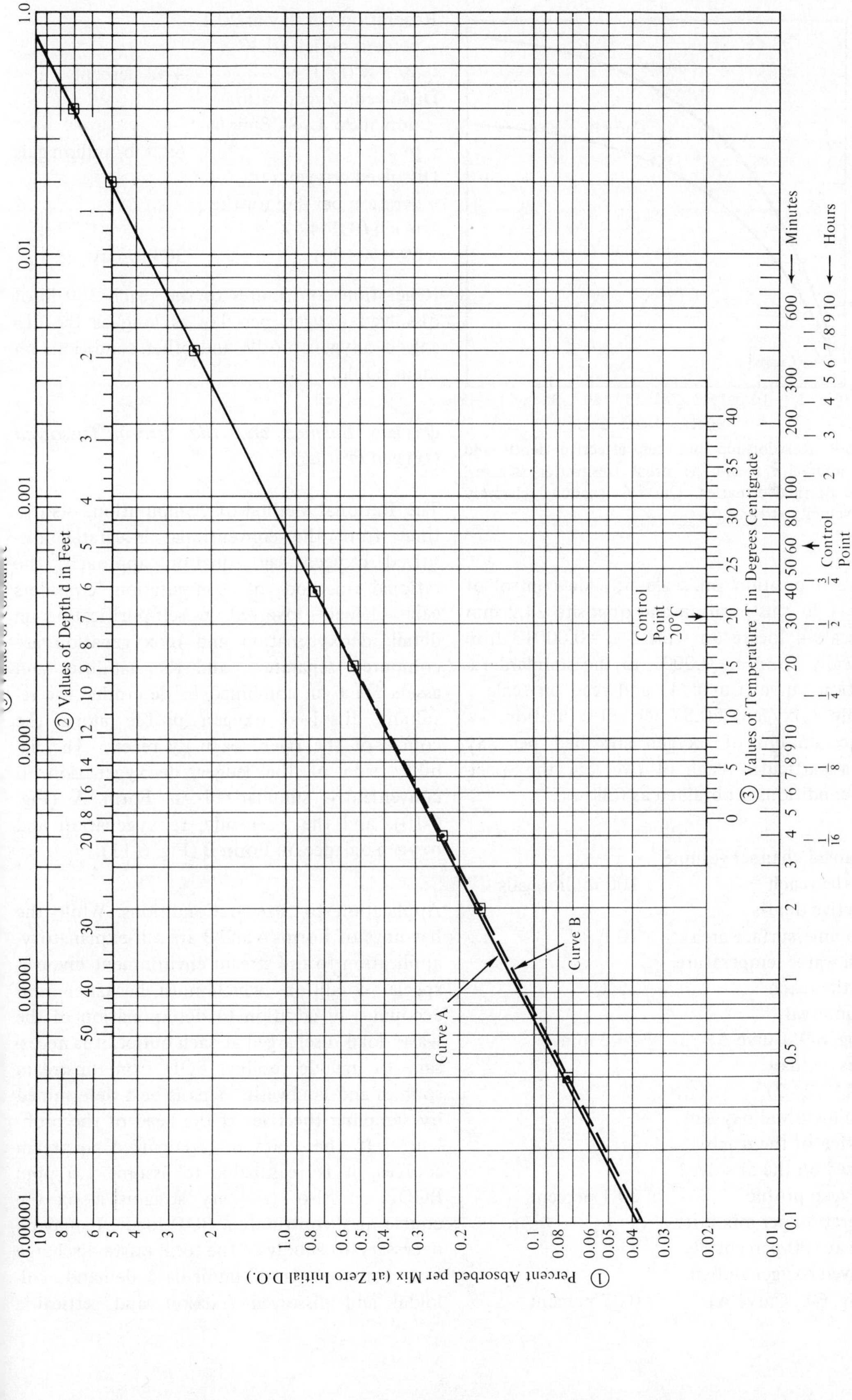

Fig. 6-8 Standard reoxygenation curve.

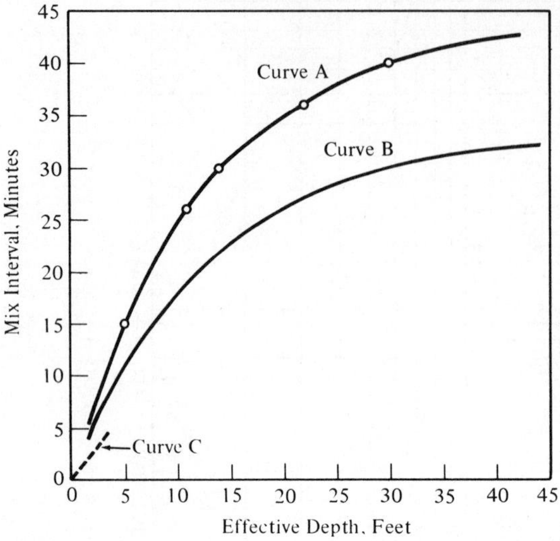

Fig. 6-9 Relationship between effective depth and mix interval. Curve A: usual freshwater streams. Curve B: tidal estuaries. Curve C: shallow, relatively high velocity streams.

locate a position on scale 5; slide control of scale 4 to this position and opposite 24.2 min on scale 4 locate on scale 5 $c = 0.0001$; drop vertically from $c = 0.0001$ to the standard reaeration curve (Curve A) and read on scale 1 dissolved oxygen of 0.37 percent saturation.

The amount of oxygen absorbed per day for a particular reach of river for the specified conditions is obtained as follows:

Occupied channel volume of the reach	100 million gals
Effective depth (volume/surface area)	10 ft
Mean water temperature of the reach	25°C
Mix interval (Fig. 6-9, Curve A)	24.2 min
Mixes per day (1440/24.2)	59.5
Mean dissolved oxygen deficit of the reach based on the dissolved oxygen profile	20.1 percent
Reaeration per mix interval at 100 percent dissolved oxygen deficit (Fig. 6-8, Curve A)	0.37 percent
Reaeration per day at 20.1 percent deficit (0.37 X 59.5 X 0.201)	4.42 percent
Dissolved oxygen saturation at 25°C, 8.38 mg/l or	69.9 lb/million gals
Dissolved oxygen reaeration per day for the reach (0.0442 X 69.9 X 100)	309 lb/day

Reaeration contributes to the reach 309 lb of dissolved oxygen per day as long as the dissolved oxygen profile and other conditions remain stable.

Oxygen Balance and the Stream Dissolved Oxygen Profile.

The Rational Method of Computation. As distinct from the conventional integrated dissolved oxygen sag equation approach, the rational method of computation considers each element involved in self-purification in detail, deoxygenation and reoxygenation are computed separately, and the liabilities and assets are then combined to determine the resultant dissolved oxygen profile along the course of the river, reach by reach. The liability side of the ledger, deoxygenation, is conveniently summarized in Form A (Fig. 6-10), and the asset side, reoxygenation and oxygen balance, in Form B (Fig. 6-11).

Application to Stream Conditions. While the headings of Forms A and B are self-explanatory, application to the stream environment involves special conditions which must be taken into account. In addition to determination of the waste load discharged at each outlet, it is necessary to include residual BOD from upstream sources and landwash. This is best determined by sampling the river at the head of the problem. If there are no recognized upstream sources, it is reasonable to assume 1.0 ppm BOD_5. Also, tributary streams along the course may contribute a BOD load. It may be necessary to subdivide the total waste discharge into components of immediate demand, colloidal and dissolved fraction, and settleable

FORM A: DEOXYGENATION (page 1)

Computation No. ____________

Computer ____________

Date ____________

Stream: ____________

Runoff: ____________ Ft³/Sec at Reference Gauge ____________

Drought Probability: ____________

Special Conditions: ____________

____

BOD Loading: Residual at Mile ____ from Upstream Sources and Land Wash:

____

____

Waste Loads:

River Mile (1)	Source (2)	Type of Treatment (3)	BOD Population Equivalents (PE)* Discharged to Stream			
			Immediate Demand (4)	Colloidal and Dissolved Fraction (5)	Settleable Solids Fraction (6)	Total (7)
	Upstream Residual					

$$\text{*Population Equivalent (PE)} = \frac{\text{BOD}_u \text{ lb/day}}{0.24}$$

Fig. 6-10 Form A: deoxygenation.

FORM A: DEOXYGENATION (page 2)

Computation No. ______________

				BOD in Population Equivalents (PE)				
Station River Mile (1)	Mean Water Temp. °C (2)	k at Mean Temp. (3)	Time Passage from Initial Station, Days (4)	Discharged at Station (5)	Residual from Station Upstream (6)	Total at Station (7)	Satisfied in Stream Between Stations (8)	Total Satisfied Above Station (9)

Fig. 6-10 *(Continued)*

solids fraction in order to deal with abnormalities. (If treatment precedes discharge, the effluents usually do not involve abnormalities.)

The integration of the BOD loads along the course of the stream is summarized on page 2 of Form A (Fig. 6-10). The time of passage is specific for each runoff regime and is determined as discussed under "Hydrologic and Climatologic Factors" above (Fig. 6-6). The integration of the BOD loading from station to

FORM B: REOXYGENATION AND OXYGEN BALANCE

Computation No. ______________

Computer ______________

Date ______________

Station River Mile (1)	Runoff at Station, ft^2/sec (2)	Mean Temperature, °C (3)	Effective Depth, ft (4)	Frequency of Turnover, min./mix. (5)	River Volume, million gals (6)	Total Demand Satisfied above Station, PE (Form A, col. 9) (7)	Dissolved Oxygen in Population Equivalents (PE): Added by Runoff between Stations (8)	Total Added by Runoff above Station (9)	Total Runoff at Station at Saturation (10)	Net at Station (9) minus (7) (11)	Added by Reaeration Between Stations (12)	Total Added by Reaeration above Stations (13)	Net Oxygen Balance at Station PE (11) plus (13) (14)	Percent Saturation at Station (14) Divided by (10) (15)

Fig. 6-11 Form B: reoxygenation and oxygen balance.

station for the time of passage at the reaction rate k (adjusted to the mean water temperature) may be graphical, as illustrated in Fig. 6-12, or analytical, which is facilitated by Eq. 2, substituting for Y, the population equivalents remaining. Any abnormalities which may be involved are computed separately and are added to the total satisfied.

In compiling Form B, (Fig. 6-11) the division of the river channel into a number of reaches for reaeration computation and oxygen balance is contingent on the desired refinement in the dissolved oxygen profile and such factors as time of passage, channel changes, junction of tributaries, sources of immediate demand, and location of sludge deposit zones. Since the dissolved oxygen profile is curvilinear, reaches should be short enough for the profile to be adequately defined as straight lines through reaches. A primary consideration in the selection of reaches is reasonable uniformity of effective depth—reaches that combine shallow and deep channel stretches should be avoided.

Runoff tributary to each reach along the river course is determined from the hydrologic study as previously discussed; likewise for the appropriate temperature for each

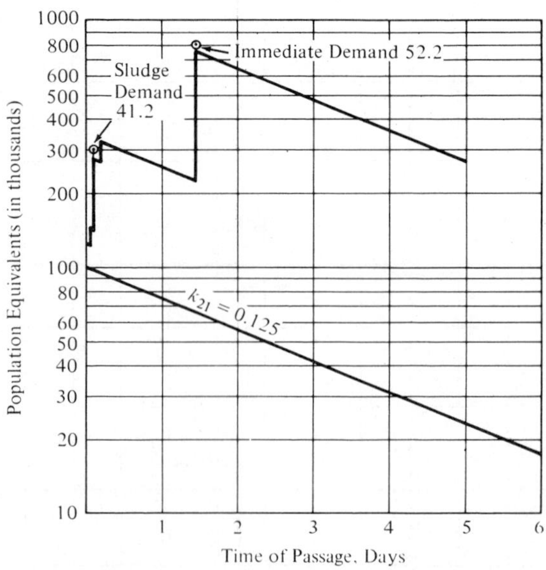

Fig. 6-12 River "A" integration of BOD loading (reduced scale).

reach. If steam power plants are involved, a water temperature profile must be computed separately.[1] Frequency of turnover, or mix interval, is a function of effective depth, as noted earlier, and is obtained for each reach from the appropriate curve of Fig. 6-9. Summation curves of occupied channel volume and surface area developed from the channel cross section soundings provide a ready means of determining the effective depth and river volume applicable for each reaeration reach. The total demand satisfied above the station (column 7) is obtained directly from Form A, column 9, or a graphical plot of column 9 if interpolations are necessary. Columns 8, 9, and 10 of Form B are the oxygen assets of runoff, a function of saturation value for the prevailing temperature and the extent to which increments are below saturation as they enter the respective reaches. Seldom are natural unpolluted streams fully saturated with oxygen, and in the absence of field survey data, 85 percent of saturation is a reasonable value to assume. Column 11 is a balance between assets of runoff and total liability, neglecting reaeration for the moment.

Reaeration is now computed, reach by reach, commencing at the head of the problem and proceeding downstream. Since the subdivision into reaches is made to provide a straight line profile in dissolved oxygen from the upper to the lower end of the reach, the dissolved oxygen deficit is obtained as the average at the beginning and end of each reach. Before a computation of reaeration for the reach can be made, it is necessary to know the dissolved oxygen level at the end of the reach.

We deal with this dilemma by successive approximations. Assume a dissolved oxygen value at the lower end, which with the previously determined value at the head of the reach provides a dissolved oxygen deficit necessary for the first reaeration calculation; make successive approximations until the assumed value is in close agreement with the computed one. We start at the head of the first reach usually with foreknowledge of the stream's condition as it arrives. With the knowledge of the net asset at the end of the reach (column 11 of Form B), a reasonable estimate of the dissolved oxygen at the downstream end can be made as the first approximation. The result of the first computation of reaeration discloses whether the approximation was too low or too high, and this constitutes the basis for a closer approximation in the second try. With experience, about three approximations usually lead to agreement within 0.1 percent of saturation, which is an adequate tolerance. The second and subsequent computations can be made quickly, since all other conditions remain constant and reaeration is adjusted in proportion to change in oxygen deficit.

Applying the rational method to River A, Fig. 6-12 is the graphical integration of the BOD loading which includes an immediate demand and a sludge deposit demand at equilibrium accumulation. Figure 6-13 is the resultant computed dissolved oxygen profile along River A.

Comparison of Computed and Observed Dissolved Oxygen Profiles. As verification of normal self-purification or abnormalities, a comparison is made between computed and observed dissolved oxygen profiles under identical conditions where feasible.

For reliable verification purposes, observed stream conditions entail executing a carefully designed river sampling survey under conditions of known waste loads and stable river hydrology. Intensive sampling over a short period of known stable conditions is superior to grab samples taken over long periods in which hydrology or loads may vary. Even under supposedly stable conditions, there is considerable variation among samples; hence, a sufficient body of data should be obtained to ensure statistically significant observed conditions. Sampling should also be sufficiently extensive to describe the entire profile, the decline, the critical reach, and the recovery. In making comparisons, it is more significant to consider the overall shape of the dissolved oxygen profile along extensive reaches of river than to make precise verifications at a particular sampling station or a limited reach.

Referring again to River A, Fig. 6-13 shows a comparison of the computed dissolved oxygen profile and results of a 10-day sampling survey. Curve A is the computed profile at a steady runoff of 4900 ft^3/sec (reference gauge) and a steady waste loading for which Fig. 6-12 represents the BOD integration. The results of the 10-day river sampling survey are super-

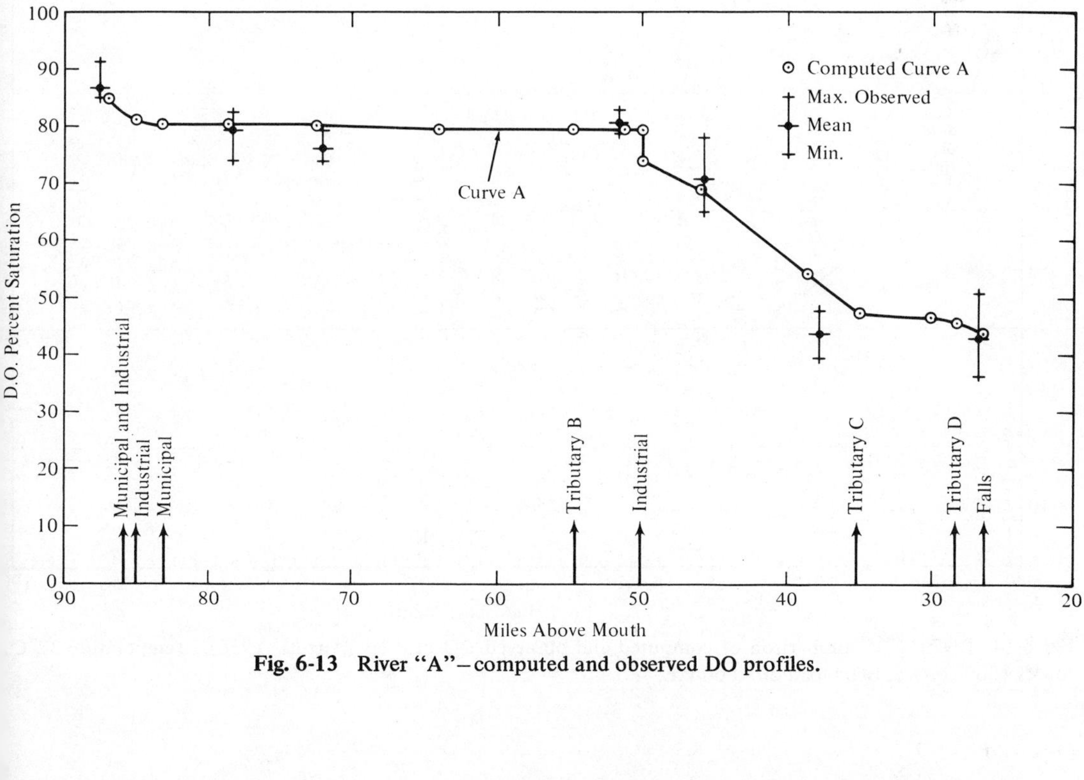

Fig. 6-13 River "A"—computed and observed DO profiles.

imposed on the computed profile. The conditions during sampling were not ideal but were reasonably stable with runoff varying between 5240 and 4600 ft^3/sec; the waste loadings likewise were subject to normal variations during the survey period. Considering these variations, there is reasonable agreement between the computed and observed dissolved oxygen profiles, particularly in the overall shape. This verifies the self-purification factors adopted, including the abnormalities in immediate demand and sludge deposits.

Figure 6-14 (Curve A) illustrates verification of normal self-purification in a relatively large river (River C) below a major industrial source of waste effluent, free of settleable solids subject to deposit entering at mile 238, and a relatively minor municipal and industrial load of 21,500 population equivalents at mile 206. Excellent channel cross section soundings and a number of river sampling traverses covering the reach (mile 238 to mile 210) were available. A period of stable, low flow (1000 ft^3/sec) and fairly steady waste effluent loadings prevailed during August, with a high natural river water temperature of 32°C. The observed dissolved oxygen profiles based on 7 river sampling traverses are shown in Fig. 6-14 by key; the independently computed dissolved oxygen profile for these conditions, based on normal BOD satisfaction adjusted to 32°C, is shown as Curve A, which is in close agreement with the observed profile.

Figure 6-15 represents an intensive, well-planned survey carried out by the Michigan Water Resources Commission in which 18 samples were taken over a 2-day period preceded by and during a stable hydrograph and a normal waste loading. Normal distributions of dissolved oxygen were developed for each sampling station, with the mean dissolved oxygen profile as shown by Curve A and a 90 percent confidence range in individual dissolved oxygen values shown as the shaded band. Although there is variation in the results of individual samples, there is reasonable stability in the mean profile based on the 18 samples. Curve B of Fig. 6-15 shows the independently-computed profile for the conditions that prevailed during the intensive sampling period, indicating quite good agreement in shape and magnitude with the observed profile.

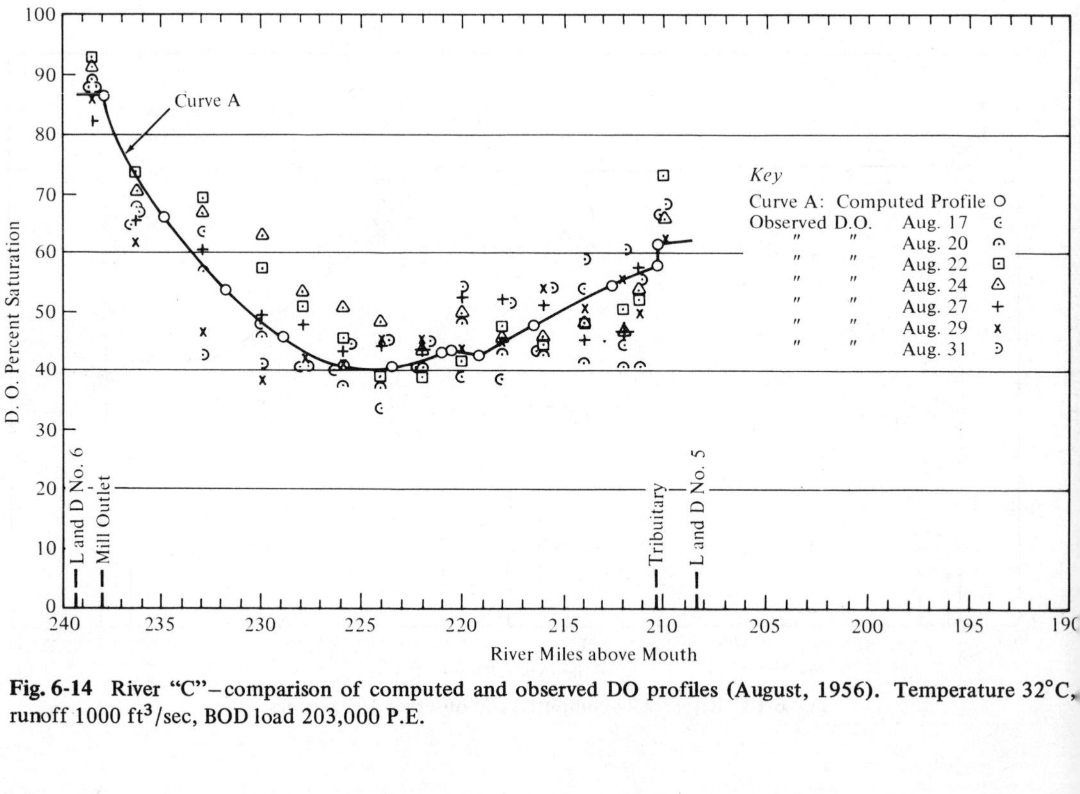

Fig. 6-14 River "C"–comparison of computed and observed DO profiles (August, 1956). Temperature 32°C, runoff 1000 ft^3/sec, BOD load 203,000 P.E.

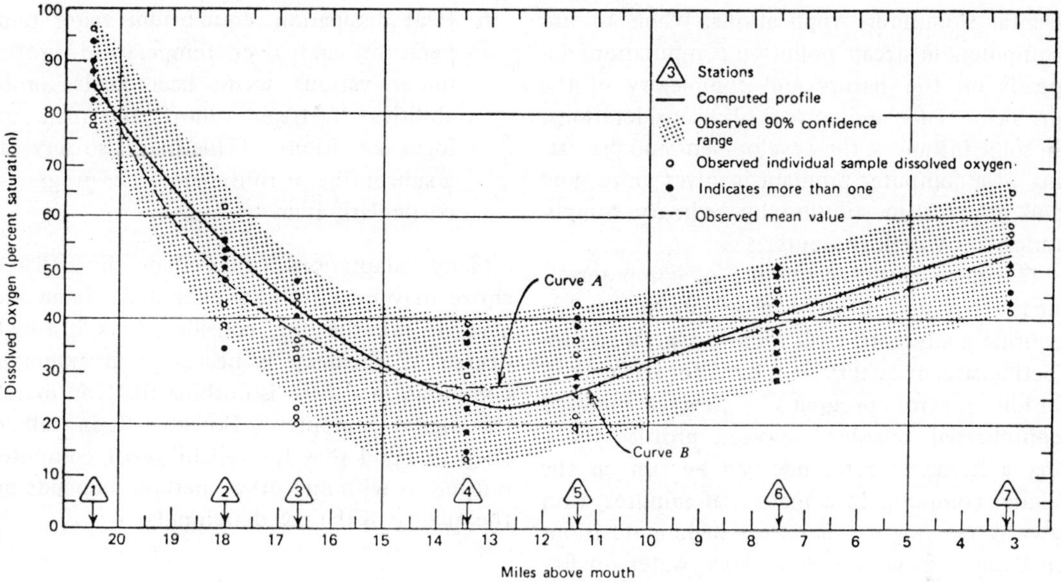

Fig. 6-15 River "H"–comparison of computed and observed DO profiles.

Limitations of the Simplified DO Sag Equation. The difficulties inherent in the mathematical solution of Phelps' original formulation for reaeration (prior to the availability of the graphical solution) led to a simplification[13] commonly referred to as the DO sag equation:

$$D = \frac{k_1 La}{k_2 - k_1}(10^{-k_1 t} - 10^{-k_2 t}) + Da \times 10^{-k_2 t} \quad (10)$$

In the customary units, D is the oxygen saturation deficit in parts per million at time t and Da is the initial deficit, La is the initial BOD in parts per million in the streamflow, t is the time of passage in days, and k_1 and k_2 are deoxygenation and reaeration rates based on common logarithms and the day as the time unit.

In this equation, the relevant physical and dynamic stream characteristics are lumped into a single reaeration coefficient, k_2. The gain in simplicity, however, is a loss in generality.

Lacking a rational basis for determination of the reaeration coefficient k_2 from measurable stream parameters, there has been a tendency to misuse it in the sag equation as though it were a constant along an entire course of a river. It is known, however, from the fundamental formulation to vary markedly with channel depth, which varies radically from reach to reach. As a result, detailed channel cross section soundings along the course have been ignored, and the sag equation has been employed without benefit of measured physical characteristics of the channel and without development of time of passage curves. The dissolved oxygen profile thus defined is unrelated to geographical location along the course and often is found to be distinctly at variance with actual observed stream conditions.

More recently, theoretical and empirical methods have been proposed[14,15] for estimating the reaeration coefficient based upon channel parameters including stream depth and velocity. Thus, if the reaeration coefficient is changed reach by reach based upon good channel cross section soundings (which is essential to a meaningful dissolved oxygen profile), the sag equation quickly loses its simplicity, and extensive computations ensue.

The advantages of the rational method of stream analysis over the sag equation is evidenced in less burdensome computations and more flexibility in taking into account characteristics of waste loads, abnormalities in self-purification, and detailed characteristics of the stream channel.

Digital Computer Applications. When to use computers in stream pollution computations depends on the nature and complexity of the problem. Time and cost are also considerations. In some instances, the development and pretesting of a computer program involves more time and effort than solution by ordinary computation or by desk computer.

The most advantageous use of computers is in the repeated use of a program, such as the routine analysis of river monitoring, or in projecting stream quality as a basis for stream pollution control programs. Several hundred complicated dissolved oxygen profiles forming a frame of reference can be run on the digital computer in a matter of minutes, thus greatly refining the basis for making decisions in design, water use allocations, water quality criteria, and water resource management. The computer is an indispensable tool in the systems analysis approach to comprehensive river basin studies.

Programs adapted to the digital computer have been developed,[16] tested, and found highly effective tools in the following problem areas:

1. Selection and preparation of stream runoff records for drought and flood flow analyses.
2. Selection and preparation of meteorologic records (temperature, wind velocity, vapor pressure, or solar radiation).
3. Reduction of river channel cross section data along extensive reaches of a stream and determination of occupied channel volume, time of passage, effective depth, and other parameters under various runoff probabilities. (This may also serve as a subroutine in other programs.)
4. Deoxygenation, reoxygenation, and dissolved oxygen balance profiles under various organic waste loadings and ranges in stream runoff both in freshwater streams and in estuaries.
5. Integration of organic sludge deposit accumulation and demand under various waste loadings and runoff probabilities. (This may also serve as a subroutine in item 4.)
6. Heat dissipation, equilibrium water temperature, and river temperature profile under various waste heat loads, probabilities of stream runoff, and meteorologic conditions. (This may also serve as a subroutine or routine in other programs, particularly item 4.)

Many adaptations other than those listed above may be profitably exploited. Time and cost are not the only considerations; if a high degree of accuracy is necessary in extensive computations, there is nothing that can match the digital computer. However, it is well to keep in mind that the reliability of computer output, as with any other method, depends on the quality of the raw data input.

Forecasts of Expected Stream Conditions. With self-purification characteristics fundamentally defined by the rational method of analysis and verified, it is possible to forecast, in quantitative terms, the stream conditions expected for various organic waste loadings over the whole range of natural or regulated streamflow. These forecasts define the range of waste assimilation capacity available, from which the relation between streamflow and allowable waste loading can be established to maintain any prescribed level of water quality. Such a reference frame affords the basis for designing an effective offensive and defensive system for satisfactory waste disposal and maintenance of satisfactory stream water quality. A basis is thus available for the evaluation of specific alternatives and combinations, such as water quality objectives and criteria, degree of wastewater treatment, low-flow augmentation, direction of urban-industrial growth, plant site location, land and water use planning and zoning, multipurpose river development, and the effects of river developments and water uses.

Waste Assimilation Potential Under Natural Hydrologic Variations. Waste assimilation potential is a dynamic variable following the patterns of variation in streamflow and water temperature. The potential is very great during the cool-weather high runoff season and declines drastically during the warm-weather drought flow season. Evaluation of the waste assimilation potential of a river necessitates defining the entire range of variation through the seasons and over a period of years.

An extensive series of dissolved oxygen profiles computed by the rational method constitutes the framework for determining waste assimilation capacity for the recorded range of streamflow and seasonal temperature (facilitated by adaptation of a digital computer). This is achieved by assigning arbitrary waste loads at selected runoffs and temperatures and determining the critical dissolved oxygen at the low point in each profile. From these critical dissolved oxygen values, control curves can be developed from which the allowable waste loading can be readily obtained for any runoff and temperature to maintain any prescribed critical dissolved oxygen level.

Total Annual Waste Assimilation Capacity and Seasonal Pattern. If full advantage is taken of the allowable waste loading (to maintain a prescribed critical dissolved oxygen level) each month according to the recorded monthly mean streamflow and water temperature, the total annual waste assimilation capacity can be determined with the aid of the control curves. If this is determined over the years of record, the distribution of annual values is defined (generally normal) from which the mean annual assimilation capacity and the standard deviation are obtained. The expected variation from year to year depends upon the hydrologic setting; in one river system with a high degree of variation in hydrology, the coefficient of variation (standard deviation as percentage of the mean) was found to be 45 percent.

An analysis of the monthly assimilation capacities, considering each month of the year separately, discloses distributions that are logarithmically normal. From these the seasonal pattern is developed. Figure 6-16 is such a seasonal pattern developed from a de-

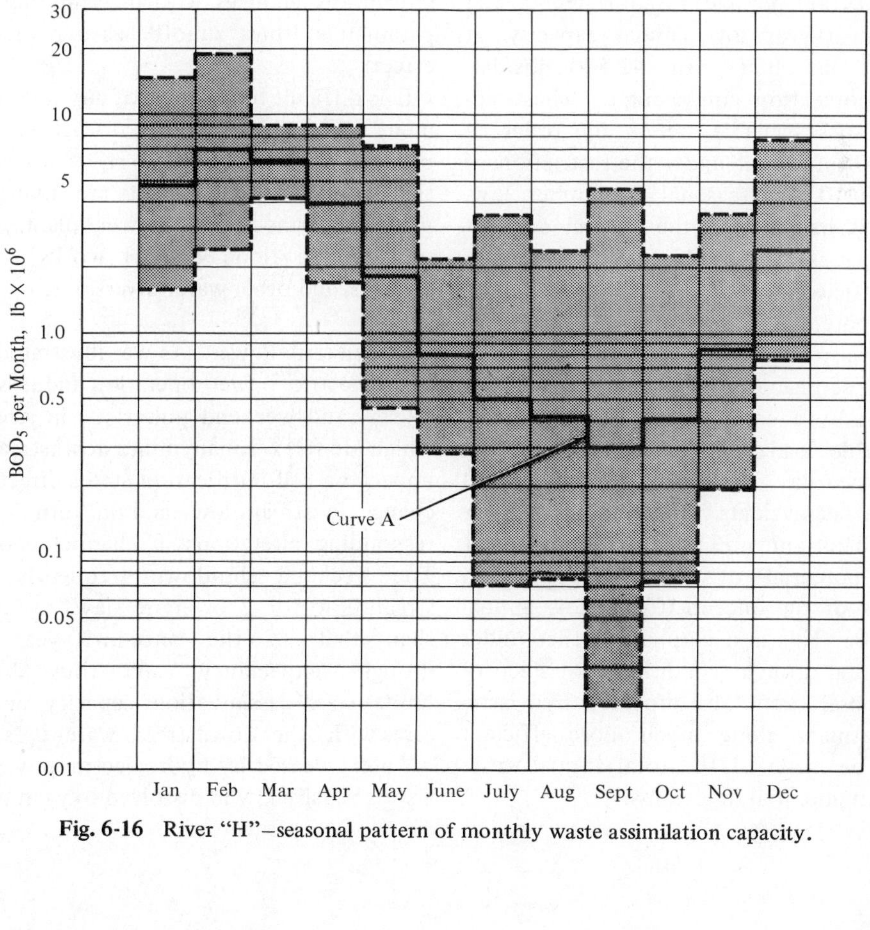

Fig. 6-16 River "H"—seasonal pattern of monthly waste assimilation capacity.

tailed study of River J to maintain a critical dissolved oxygen level of 4 ppm, assuming normal amortization and an effluent free of settleable solids subject to deposit. Curve A is the most probable monthly waste assimilation capacity, and the shaded band about Curve A is the range within which monthly capacity is expected for 80 percent of the years, with 10 percent of the years above the upper limit and 10 percent of the years below the lower limit. It is noted that there is high capacity in the winter–spring season and low capacity during the summer–fall season, but with wide variation and occasionally extremely low capacity during severe drought years. From the seasonal pattern, September is the critical month with the once-in-10-year low monthly capacity of 0.02×10^6 lbs BOD_5 or 667 lbs/day (4080 population equivalents).

An analysis of the total annual waste assimilation capacities for River J for the years of record discloses a most probable annual total of 36×10^6 lbs BOD_5 or averaging 98,500 lbs/day (605,000 population equivalents), and the once-in-10-year low annual capacity of 15.5×10^6 lbs BOD_5 or 42,500 lbs/day (260,000 population equivalents). These annual capacities available are on the order of 150 to 160 times that under the restriction of the once-in-10-year seasonal September low. If design is based upon the natural drought conditions, it is obvious that this represents a highly inefficient use of the natural waste assimilation potential of the river.

Even if waste treatment is extended beyond removal of settleable solids to an efficiency of 90 percent by a complete biological system, the allowable load (free of settleable solids) produced per day could not exceed 40,800 population equivalents under the 10-year drought. This compares with the annual most probable potential of 605,000 population equivalents or the once-in-10-year low annual of 260,000. This also emphasizes that under severe natural drought conditions, satisfactory waste disposal cannot be provided by wastewater treatment alone—much more efficient use must be made of the total annual waste assimilation potential of streams.

The Impact of River Developments on Waste Assimilation Capacity. It is easy to grasp the significance of the impact of waste disposal on other water resource uses, but all too seldom is adequate attention given to the effect of other water uses and river developments on the complex problem of waste disposal. Depending upon how they are planned and operated, other water uses and river developments can be either detrimental or beneficial to the waste assimilation capacity of the stream and hence to water quality.

Modification of the Parameters Influencing Self-Purification. The factors that influence self-purification of streams may be consolidated into three primary elements: (1) streamflow, (2) water temperature, and (3) channel characteristics. In addition to depth, the latter involves a number of interrelated parameters—surface area, volume, time of passage, and velocity—that are usually altered through change in depth of channel. Every change in streamflow induces a change in the channel parameters; thus, runoff has a compounding effect.

It is difficult to conceive of any river development that does not change at least one of these primary factors. Typical river developments that have significant impacts are: hydroelectric plants, thermoelectric power plants, navigation works, flood control works, irrigation systems, and other water diversions.

Hydroelectric Power. As was illustrated earlier, hydroelectric power operation induces radical diurnal and weekend pulsations in streamflow which are felt for many miles downstream. The impact on self-purification stems directly from change in streamflow, and in turn from corresponding alterations in channel parameters. The weekend shutdown frequently curtails streamflow for 2 or more days to levels less than that of the once-in-10-year natural drought expectancy, and thus drastically limits waste assimilation capacity and interferes with other downstream water uses.

Water released by hydroelectric power plants also is usually low in dissolved oxygen by virtue of deep withdrawal from the reservoir which, in the critical season, is usually highly stratified. Such oxygen-depleted water in a sense is a form of water pollution. The water, however, is usually cooler, but this asset is offset by lack of dissolved oxygen.

Reregulation reservoirs should be constructed below existing proposed hydroelectric power plants to iron out diurnal and weekend pulsations in discharge and to equalize the streamflow. Since most proposed hydroelectric power projects now are planned as large-storage types, such regulation would not only remove these fluctuations but would also ensure passing downstream the benefits of streamflow augmentation (i.e., water released from storage during the low runoff months).

The construction of a hydroelectric dam radically changes the channel characteristics. The increase in channel depth from a high dam may extend for miles upstream, replacing the natural channel by a slack-water pool. Time of passage is greatly increased, placing a heavy burden of amortization on a short reach of stream, and with the increase in depth the reaeration rate is greatly reduced. The effect for organic waste disposal is a dissolved oxygen profile sharply contracted from that of the natural channel. The velocity in the slack-water pool is so reduced that any natural or introduced suspended solids are certain to deposit and accumulate as sludge; this adds an additional burden on dissolved oxygen. Unless compensated by low-flow augmentation, these detrimental effects on upstream waste assimilation capacity are so drastic as to preclude adjacent industrial and urban growth.

Pumped-storage hydroelectric plants as discussed heretofore include reregulation and offer several other beneficial aspects.

Thermoelectric Power. Stream temperature superelevation resulting from the discharge of waste heat from the condenser circulating water is the principal effect of thermoelectric power plants. Temperature superelevation increases the rate of amortization of organic matter and lowers the dissolved oxygen saturation value, and can thereby result in a sharp decline in the dissolved oxygen profile. Fortunately, heat dissipates rather quickly and returns the temperature profile to the equilibrium temperature of the meteorological environment. If temperature superelevation is not excessive, it may not interfere with aquatic life over extended reaches, and may actually be stimulating to desirable species.

With the trend toward concentration of large power plants, there are but few rivers of sufficient size to supply the large quantities of condenser water; large off-channel reservoirs, either in headwaters or at pumped-storage developments, are preferred. For onstream location, upland storage that will augment streamflow to ensure high sustained runoff during the critical dry weather season provides required condenser water and lowers temperature superelevation. The increased waste assimilation capacity from flow augmentation may more than compensate for the superelevation.

If water resources are not adequate, recourse may be made to cooling ponds or cooling towers for recycling condenser water and dissipating waste heat. However, substantial water loss results from evaporative cooling, and make-up water is essential. Also, cooling towers have aesthetic disadvantages, and vapor plume, fog, and icing in winter are decidedly undesirable near urban areas.

Navigation Works. Navigation works affect stream self-purification by alteration in streamflow and by changes in the channel. In most current navigation projects the effect is detrimental. Dredging deepens the channel depth, thereby slowing the time of passage and diminishing reaeration. This places a greater burden on shorter reaches of stream. The recent trend toward locks and dams produces deep slack-water pools extending many miles upstream. This drastically reduces waste assimilation capacity and produces steep dissolved oxygen declines. The operation of locks induces irregular pulses in streamflow, at times with little or no releases. The slow velocity through pools is certain to permit depositing of any suspended matter, which, if organic in nature, can further reduce the dissolved oxygen profile.

Some navigation developments include reservoir storage to augment flow during the dry-weather season. Depending upon the magnitude and manner of such releases from storage, this may offset to some extent the loss in natural self-purification through deep pool reaches.

Flood Control Works. If designed to provide drought control, flood control developments can be a distinct asset. Levees and local protection works such as dredging deepen the channel and result in adverse effects upon waste assimilation capacity. Conventional flood control reservoirs (excluding pumped storage) are limited in organic waste assimilation capacity in the same manner as the deep slack-water pools of navigation works. However, usually flood control reservoirs are located in undeveloped headwaters areas, and, if of the multipurpose type, may serve as water-based recreation areas and provide for other beneficial water uses.

Irrigation and Other Diversions. Diversions without return or compensation from storage have a direct impact on streamflow, and, depending on the magnitude and the season, may be seriously detrimental to waste assimilation capacity.

In arid regions where prior appropriation is in effect, practically the entire streamflow may be diverted for agricultural irrigation during the warm-weather growing season. Most of the water is consumed by transpiration and evaporation, and any that returns by drainage is usually of poor quality and may contain residual fertilizers, herbicides, and pesticides.

Diversions for public water supply without compensation from storage, likewise, are a direct drain on streamflow.

In some cases proposals are made to divert freshwater streams from developed estuarial harbors. Such diversions result in sea-water intrusion with little or no net seaward movement and thus can drastically limit waste assimilation capacity of an estuary.

EFFICIENT USE OF STREAM WASTE ASSIMILATION CAPACITY

There are two obvious approaches to more efficient use of natural waste assimilation capacity: drought control through streamflow regulation, and regulated waste effluent release tailored to variations in waste assimilation capacity associated with the seasonal patterns of hydrology.

Drought Control. We have seen that under natural hydrologic variations the total waste assimilation capacity is relatively great, in some instances on the order of 100 times greater than that available under uncontrolled drought streamflow. Community and industrial development and the social and economic potentials of a river basin are drastically restricted if geared to the minimum waste assimilation capacity during uncontrolled drought. We cannot increase the total natural waste assimilation capacity, but we can, with moderate expenditure, remove the severe restrictions of uncontrolled drought and make much more efficient use of what is available.

Among the promising approaches to more efficient use of natural waste assimilation potential through drought control are:

1. Headwaters storage reservoirs and regulated release.
2. Incremental storage reservoirs along the river course and regulated release.
3. Multipurpose river developments, including augmentation of low streamflow.
4. Adaptation of large-scale pumped storage.

Drought control inherently provides benefits not only to waste assimilation capacity but to all existing and potential water uses along the course of the river.

Headwaters vs. Incremental Storage. The conventional method of augmenting low streamflow has been the construction of reservoirs in natural upland valleys remote from urban-industrial development. The locations must be suitable for a dam and have sufficient tributary drainage area to provide the necessary runoff to supply the required storage capacity.

Headwaters storage with regulated release can greatly enhance downstream waste assimilation capacity and improve water quality of the stream. However, a feature of low-flow augmentation from headwaters is that the increased flow greatly improves the upper reaches but shortens the time of passage, carrying a higher residual BOD downstream and kaleidoscoping loading to create a secondary downstream depression. As headwaters augmentation increases, more residual BOD is carried downstream, until a point is reached where augmentation brings about the desired improvement along the lower reach. Such an unbalanced dissolved oxygen profile is inherent in augmentation from headwaters, particularly where the lower river receives a succession of BOD loads along the course.

A more efficient use of low-flow augmentation is possible if reservoir sites are available along the lower river to provide increments of streamflow at points of need to maintain dissolved oxygen. In such cases, less storage capacity is needed to maintain prescribed water quality. However, as development continues along the lower main stem of rivers, incremental storage sites with adequate drainage areas are either preempted or too costly. Pumped storage offers an alternative.

Pumped Storage and Multipurpose Development–The Dual Power-Water System. Basically, pumped storage functions as an energy accumulator that stores low value off-peak energy (usually generated by steam plants) by using it to pump water from a lower pool, located in or just off the river channel, to a higher off-channel reservoir from which it can be returned to generate power during peak periods. The upper reservoir includes storage capacity requirements for both drought control augmentation water (which can be used for other purposes) and for power cycling; the lower pool, in addition to functioning in the conventional manner for recycling water for energy storage and generation, serves as a reregulation reservoir for release of water to the river to augment flow as needed.

Large-scale pumped storage, in addition to peak power and reserve, provides the opportunity to locate fossil- or nuclear-fueled steam plants on the upper reservoir site and insures adequate condenser water supply and large water surface area for dissipation of waste heat.

The multipurpose water resource phases of the dual power-water system include: efficient use of water in stream quality control, reservoir water quality improvement, community and industrial water supply, water-based recreation, and flood protection.

Pumped-storage reservoirs can be located along the lower main stem of rivers near points of need because no drainage area is needed to supply the water since it is pumped from the river. This location of the reservoir affords opportunity to control water quality almost instantly, whereas from headwaters reservoirs it may require days in travel to the point of need.

Pumped-storage reservoirs are less costly because less storage capacity is needed and extensive floodwater spillways are not required. The recycling operation to the upper reservoirs prevents stratification (so prevalent in conventional reservoirs) and insures cool augmentation water that is low in BOD and practically fully saturated with dissolved oxygen. The pumped storage reservoir also provides a high quality nearby raw water supply for community and industrial needs.

If operated in conjunction with other conventional or pumped-storage reservoirs, an excellent water-based recreation area can be made available near high concentrations of population. If additional storage capacity is provided in the reservoir, pumped storage can be employed to decapitate flood peaks by virtue of the very fast-acting pumping equipment.

Lagoon Waste Storage and Regulated Effluent Release. A second approach to more efficient use of natural waste assimilation potential inherent in all rivers is the storage of wastes and the regulated release of effluent tailored to the natural variation in waste assimilation capacity. This approach is particularly adaptable to industry located in areas where sufficient land is available.

The system can be divided into two types—seasonal storage and release, and storage with continuous release. In the former, all waste is stored during the season of critically low streamflow and the lagoons are emptied during the high runoff season; in the latter, a more refined practice involving continuous storage and continuous release is employed, taking fuller advantage of the annual waste assimilation capacity.

The Dual System—Treatment and Storage Release Regulator. The lagoons serve a dual purpose as a treatment system and as a storage release regulator. Lagoons provided in series serve as sedimentation basins, storage, and final oxidation ponds. A number of factors influence the treatment with efficiencies ranging from 50 to 90 percent reduction in BOD.

The design of the storage release regulation system requires the definition of the waste assimilation potential under natural hydrologic variations in streamflow and water temperature as previously illustrated for River J (Fig. 6-16). The allowable BOD release to the stream is determined from such control curves. There remains the determination of the required size of lagoons in relation to the industrial production level within the framework of the total annual waste assimilation capacity and the prescribed water quality criteria.

Since waste storage requirement is a function of advection to and from the lagoons, the principle of the mass diagram is applicable for determining lagoon capacity. The cumulative monthly allowable lagoon waste effluent discharge to maintain the prescribed critical dissolved oxygen level is plotted through the years of record. The industrial production rate (or rates) is plotted tangentially on the mass curve expressed as BOD of lagoon effluent, thus providing the maximum storage ordinates required each year expressed in terms of BOD of lagoon effluent. The annual storage requirement values usually form a normal distribution from which a most probable and any desired confidence range in storage capacity are defined. A balance then can be evaluated between the size of the industrial production and the required storage capacity at any level of probability of risk of inadequate lagoon capacity.

From application of such analysis to River J, it was determined that with a lagoon storage of 30,800 acre-ft, the natural waste assimilation capacity can support an industrial production of 1200 tons of product/day and insure a 4 ppm critical dissolved oxygen level for 90 percent of the years. In these computations, it was conservatively assumed that a 50 percent reduction in BOD was provided by the lagoon system. The industrial production level of 1200 tons/day is more than 20 times that possible during a once-in-10-year drought (including complete biological treatment of the waste with effluent discharged daily without benefit of regulated release). This again illustrates that reliance upon wastewater treatment as the sole defense is not the solution to pollution control and that much more efficient use of the range in natural waste assimilation capacity inherent in lines of offensive strategies is necessary.

BEHAVIOR OF ESTUARIES

Estuaries as considered here are restricted to the lower reaches of rivers discharging into the ocean. Here ebb and flood tides occur and salinity gradients extend through brackish reaches separating fresh water from seawater. Large wide bays and coastal bodies of water with relatively small tributary landwater drainage cannot be generalized, and each must be considered as a unique case.

Approaches to Evaluation of Self-Purification in Estuaries. The complexity of estuaries has led to several approaches to evaluation of self-purification. These can be roughly grouped into five schools: (1) tidal prism, (2) mathematical model, (3) experimental model, (4) statistical, and (5) rational.

The early concept of behavior of estuaries centered on the belief that the waters of the tidal prism are completely available for dilution and that wastes are flushed to sea on each ebb tide, not to be returned on the following flood. On this fallacy it was postulated that estuaries possessed great capacity to assimilate wastes. Actually, experience with deterioration of harbors has disproved the tidal prism theory, and it is now well established that in the majority of polluted estuaries wastes carried toward the sea on the ebb tide actually return on the flood tide, oscillating back and forth with but slow net movement toward the sea.

Currently a great deal of effort is being devoted to solutions through elaborate mathematical models that apply hydrodynamic theory developed for atmospheric diffusion to tidal phenomena. A great many independent variables are involved, only a few of which are measurable, and in the end, a number of quite arbitrary assumptions must be made in application of the mathematical models to a specific estuary.

Experimental models of specific estuaries are sometimes constructed to verify theoretical models. In addition to high cost, the limitation of hydraulic models is the difficulty of incorporating the features of the specific prototype. The usual procedure is to employ stable dyes. Although such investigations are useful in predicting the fate of stable, nondegradable waste products, there is risk in projecting these findings to the fate of unstable, degradable bacterial, organic, and radioactive wastes.

In the statistical approach, no attempt is made to define and formulate the fundamental relationships involved. Reliance is placed on statistical analyses of observed water quality end results from a given waste loading, correlating with recorded variables such as landwater runoff, seasonal water temperature, and tide. The statistical approach is helpful in interpolation within the frame of observed results, but, since it is not based on self-purification fundamentals, it cannot be employed in analyses of other conditions such as interception of waste loads for treatment with new locations of effluent discharge, channel changes, storage of upland runoff, or low-flow augmentation.

The rational approach divides the complex problem into rational parts and applies to each its primary relevant factors. The same basic principles of self-purification developed for application to inland rivers will apply to estuaries, and it is necessary only to consider modifications associated with tidal translation and seawater intrusion.

An appreciation of the complexities encountered in dealing with estuarial problems is gained from the nature of variation in tides and tidal currents. However, any approach which attempts to incorporate all of these vagaries quickly leads to insurmountable confusion. Much of this confusion can be avoided, without loss of significance, by reducing analysis to simple mean tide conditions. It is then possible to apply to estuaries the basic principles developed for fresh water self-purification.

Tides and Tidal Currents. There are some 20 or 30 short- and long-term periodic forces that are usually taken into consideration in predicting tides and preparation of the tide tables supplied by the U.S. Coast and Geodetic Survey. For most estuarial problems, we need be concerned only with three primary tide cycles: semidaily, the moon's phase, and the moon's declination.

For most areas of the world, the semidaily tide produces two highs and two lows in a lunar day of 24.841 hr. The moon passes a given meridian about 50 min later each day; hence, high and low water likewise occur about 50 min later each day. Assuming for a moment that the two tidal cycles in a lunar day are equal, the interval for a complete rise and fall in one cycle would be 12.42 hr. Thus, in observing (or sampling) a given location at the same hour each day, all phases of a complete cycle of a rising and falling tide are reflected in 12.42/0.841, or approximately 15 days.

The moon's phase cycle induces day-to-day variations in tide, depending on the position of the moon relative to the earth and the sun. Twice in the lunar month ($29\frac{1}{2}$ days) spring tides occur at times of the new and full moon; then high water rises higher and low water falls lower than usual. Neap tides occur when the moon is in its first and third quarters; then tide does not rise as high or fall as low as usual.

The moon's declination cycle ($27\frac{1}{3}$ days) induces diurnal inequality in the morning and afternoon tides of a day. When the moon is on the equator, the inequality is minimal. When it is in either of the tropics, inequality is maximal. Such diurnal inequality is a particularly complicating factor in interpretation of estuarial sampling.

Although all the periodic variables that influence tides are felt all over the world, geographical location determines to a great extent which factor or factors will dominate. Hence, the specific characteristics of tide vary radically from place to place.

Other nonperiodic factors such as terrestrial features greatly influence local tides and tidal currents. The most significant nonperiodic factor that affects self-purification in estuaries is landwater runoff.

The periodic vertical rise and fall of the tides induces corresponding horizontal tidal currents —flood with rising and ebb with falling tide. These alternating tidal currents are the major factors in the initial distribution of waste effluents. Where both diurnal inequality and landwater runoff are minimal, the ebb and flood translations are nearly equal; as runoff increases, ebb exceeds flood translation and there is a net movement toward the sea.

Net Time of Passage. The time of passage down the course of an inland stream from reach to reach is expressed as occupied channel volume displaced by runoff. This expression of the time of passage implies blocklike movement of the body of water with practically complete vertical and transverse mixing throughout the cross sections but no appreciable longitudinal mixing. This simplified concept is equally applicable to most estuaries with only minor modification.

Blocklike movement of the water mass is characteristic of estuaries, and extensive longitudinal mixing does not occur; the steep water quality gradients observed in polluted estuaries confirm these similarities to inland streamflow. If longitudinal mixing dominated, steep gradients would be leveled and the observed water quality would show more uniform conditions over long reaches. True, with reversal of tidal currents some longitudinal mixing does occur in short reaches, but not with the completeness assumed in some theoretical approaches. The more acceptable concept, therefore, is blocklike displacement by landwater runoff.

Since the occupied channel volume at mean tide remains quite fixed, it is axiomatic that, with the continuous addtion of landwater runoff from the tributary drainage area, an equivalent amount must find its way into the ocean; otherwise the water level would continue to rise. Though a seaward movement takes place in steplike action as net excess of ebb over flood translation, conversion to mean tide is equivalent to elimination of these oscillations, and movement through the estuary may be considered as a continuous forward net time of passage similar to that in a freshwater stream.

Depending on the configuration of the estuary, only minor modification is necessary in computing the net time of passage. When a well-defined channel without bays leads directly to the ocean, no modification need be made except to convert channel soundings to mean tide. In other instances, isolated bays or auxiliary channels are interconnected to the main channel, and the portions that interchange with the main channel are included as part of the total occupied channel volume.

Initial Distribution of Waste Loads. Waste discharges from outlets along the course of an estuary lose identity with respect to their onshore locations because of tidal translation and overlapping distribution from adjacent outlets. The waste contributed in any given reach is not measured as sharp increments at individual outlets but rather as a continuous increment expressed as a smooth distribution curve integrated by ebb and flood translation.

The nature of the initial distribution of wastes might best be visualized by considering the tidal estuary as a stationary body of water and the waste outlets as rotating through an arc up and downstream, spreading the waste over a reach of the estuary upstream of the outlet equal to the flood translation and downstream equal to the ebb translation, with overlapping sprays from outlets within their ranges of arc.

Employing this concept of distribution of waste effluent and blocklike movement of the water mass with ebb and flood translation about equal (which is usually the case during drought streamflow), a very complex phenomenon is reduced to a simplified solution by employing the principle of moving average on a mass curve of summation of the various waste loads along the course of the esturary. It can be shown that, from any single outlet, waste effluent is distributed over twice the distance of the tidal translation, identifiable a distance upstream equal to the flood translation and downstream equal to the ebb translation.

The amount of waste at any fixed geographical position is represented by the mean value through a tidal cycle passing such a fixed channel cross section, proportional to the time the blocklike movement is opposite the section. This places the maximum proportion of the waste discharged at a cross section position opposite the outlet, with proportionally less at other cross sections within twice the translation, symmetrically distributed above and below the outlet in the form of a bell-shaped sine curve. If this curve is integrated, commencing at the upstream end, the cumulative waste load shows that 50 percent of the total is distributed above the outlet and 50 percent below. It is further found that only 15 percent is distributed above a location one-half the flood translation and 15 percent below one-half the ebb translation, with 70 percent within the central portion about the outlet. Through this central portion, the distribution curve is approximately linear, a convenient feature permitting application of the moving average on the mass curve of wastes as a close approximation of the initial waste load distribution from overlappings of multiple waste discharges along the course of the estuary.

Figure 6-17 shows such a mass curve and moving average trace applied to Estuary A. Curve A is the plot of waste loads as the cumulative mass curve commencing with the upstream residual loads and progressively adding successive loads at the mile locations of each outlet. In this estuary, the mean tide translation for the estuary reach involved was approximately 7 mi or a total translation of 14 mi over the ebb and flood. Curve B is the trace of the moving average values through the mass curve (Curve A), produced by applying a 15-point moving average over 14 mi to the ordinates of Curve A. Commencing 14 mi

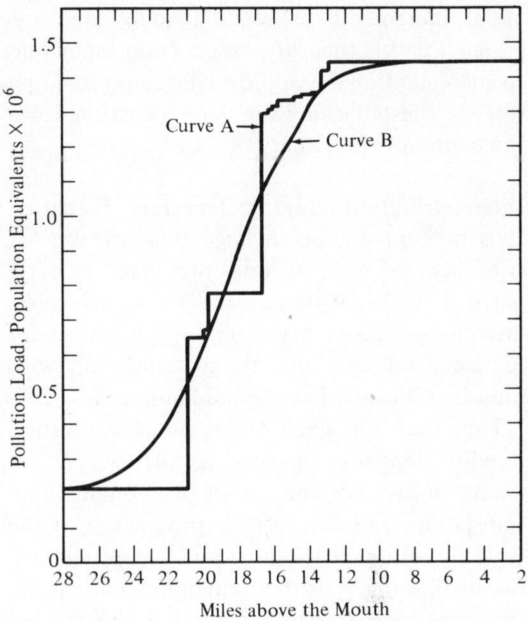

Fig. 6-17 Estuary "A"–distribution of pollution loads by tides. Curve A: mass curve of sources of pollution without tidal oscillation. Curve B: integrated distribution by ebb and flood oscillation.

above the first waste outlet, the average of the 15 ordinates is placed at the midpoint and regularly at 1-mi intervals, dropping an upstream ordinate and adding a downstream ordinate. This smoothed Curve B then approximates the initial integrated distribution of overlappings of the multiple outlets effected by ebb and flood translation.

The individual sources of waste now lose their identity and are replaced by increments of waste loading mile by mile along the course, not as specific contributions at each outlet but as the differential between any two successive mile points along the integrated distribution Curve B.

These differential increments replace the original sources and are dealt with as if they were the initial waste loading introduced along the estuary. With this modification in the expression of waste loading, deoxygenation can now be computed in the same manner as for inland, nontidal streams. The time of passage to the location of each increment is readily determined from the net time of passage curve for the particular runoff regime. The differential increments of waste loading are integrated for this time of passage from increment to increment at the appropriate deoxygenation rate, thus establishing the deoxygenation down the course of the estuary.

Interdistribution Among Tributary Estuaries. It is not uncommon in large tidal streams for tributary systems to join the main estuary, particularly in its lower reaches. In such cases, interchange takes place among the tributary branches on ebb and flood translation with interdistribution of assets and liabilities.

The basis for determining interdistribution usually depends on two factors which are readily measured: the extent of ebb and flood translation, and the relative proportion of the total floodflow at the junction that extends up the main estuary to that which extends up the tributary estuary. On the ebb tide, the channel contents of the main estuary and its tributary are combined and mixed laterally so that on the flood tide the combined contents are returned and redistributed to each in proportion to their respective tidal flows. The proportion of each estuary's liabilities or assets that is thus mixed can be estimated from the length of time during the ebb period that the liability or asset from a particular section of one estuary is commingled with that of a corresponding section of the other.

Seawater Intrusion–The Salinity Gradient. A brackish reach is formed in the lower reaches of rivers which discharge into the ocean by intrusion of seawater upstream counter to the landwater flow. The extent of the intrusion is well beyond the flood tide translation and depends primarily on the characteristics of the local tide, the configuration of the channel, and the tributary landwater runoff. The tongue of seawater intrusion reduced to mean tide conditions usually extends in a well-defined salinity gradient from a location near the mouth to a location of about 10 percent seawater, and then diminishes at a more gradual rate to a location of 100 percent landwater.

The mean tide salinity gradient comes to an equilibrium position during a steady hydrologic regime; during freshets or periods of marked change in the rate of runoff, the gradient is unstable and erratic and landwater tends to flow near the surface and seawater near the bottom. Steady conditions of equilibrium usually prevail during the low-runoff drought season, and it is then observed that the salinity distribution is quite uniform throughout the channel cross section. The stable condition at this period permits evaluation of the critical waste assimilation capacity of the estuary.

Basically, seawater intrudes on landwater by virtue of the difference in specific gravity of seawater (1.03 approximately). The seawater intrusion will extend upstream until the hydraulic head achieves a superelevation sufficient to counteract this difference in specific gravity and the additional head necessary to permit landwater flow to continue to the sea. Channel configurations, particularly mean cross section, mean depth, and bed profile, are also involved. The greater the landwater flow, the steeper is the hydraulic gradient, and thus the superelevation to offset the specific gravity difference is achieved in a shorter reach, and the salinity gradient of seawater intrusion slides downstream. Conversely, as landwater flow diminishes, the hydraulic gradient flattens and the salinity gradient slides upstream until again the required superelevation is attained.

Theoretical approximations of the slope and position of the salinity gradient can be made, but, in most cases extensive and intensive estuary sampling in the brackish reach under stable hydrologic conditions is necessary to establish the gradient at one or more stable landwater runoff regimes. Figure 6-18 represents salinity gradients developed from such intensive sampling. Under stable hydrologic conditions, a quite definite mean tide salinity gradient is established. The mean tide gradient remains reasonably constant at different levels of stable landwater runoff but decidedly changes its position along the course, shifting seaward at higher stable runoff and landward at lower stable runoff.

Replacement Seawater Flow–Relation to Landwater Runoff. By reducing to mean tide conditions, the relation between landwater and seawater exchanges necessary to maintain the salinity gradient may be resolved into two opposite net flows. One is equivalent to the

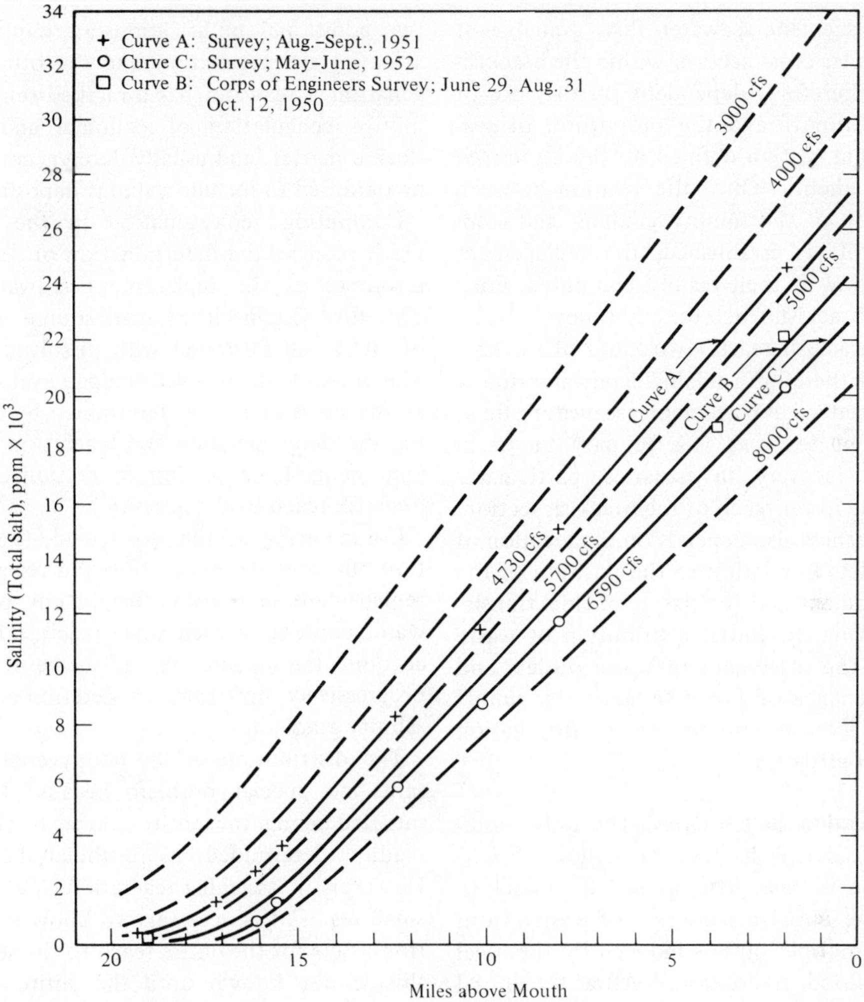

Fig. 6-18 Savannah Harbor salinity gradient–mean through tidal cycle at various fresh water runoffs, Clyo gauge.

landwater moving seaward and the other is equivalent to the seawater moving upstream and discharging into the downward landwater flow at such intervals and in such amounts as to provide the observed salinity gradient. Thus, this hypothetical seawater river is large near the ocean but gradually diminishes in volume as it proceeds upstream, vanishing at the upper end of the brackish reach. The magnitude of the landwater runoff is the key factor since it controls both the position of the salinity gradient and the amount of seawater needed to maintain the salinity gradient.

With stable equilibrium conditions, it follows that there must be a loss of landwater through the brackish reach to the sea equivalent to the landwater discharge to the brackish reach. Since under these conditions landwater and seawater are well intermixed throughout the cross section, the loss of landwater passing any section in the brackish reach must take with it seawater in proportion to the landwater and seawater mixture at that section. In turn, this seawater loss must be replaced by other seawater in equal quantity flowing upstream to maintain the stability of the salinity gradient at that cross section. This economy results in a large replacement seawater flow near the outlet into the ocean where the proportion of seawater to landwater is great, progressively decreasing upstream along the salinity gradient to zero at the upper end of the brackish reach.

The replacement seawater flow available at any particular cross section within the brackish reach is, therefore, dependent on the rate of landwater runoff and the proportion of seawater at that section defined by the associated salinity gradient. Once the relation between the position of the salinity gradient and landwater runoff is established, the replacement seawater flow is then readily computed along the entire brackish reach for any runoff.

In the brackish reach, two kinds of valuable dilution are therefore available: landwater runoff and associated replacement seawater flow. Replacement seawater is a primary factor in the sharp recovery in estuaries, particularly through the lower reach of the brackish section. Its effectiveness also depends on the location of waste outlets in relation to the position of the salinity gradient and the nature of tidal translation affecting the initial distribution of waste loading. The placement of waste outlets and the maintenance of adequate landwater runoff are both crucial to the waste assimilation capacity of estuaries.

Self-Purification in Estuaries. The only modification necessary in the calculation of self-purification in estuaries above the brackish reach is the initial distribution of wastes from single or multiple outlets induced by the local ebb and flood translation. After the initial waste load distribution, ebb and flood action can be eliminated by reducing to mean tide conditions; in effect, the estuary is dealt with as a continuously seaward-moving stream. Deoxygenation is dealt with as in inland streams. The differential increments of BOD loading taken from the moving average mass curve are integrated at the net time of passage for the particular runoff regime from increment to increment and summarized as usual on Form A (Fig. 6-10). Reoxygenation requires no special modification except the mix interval is shortened. Curve B of Fig. 6-9 is found to fit most estuaries. The oxygen balance and computed dissolved oxygen profile are summarized as for freshwater streams on Form B (Fig. 6-11).

In computing organic waste assimilation capacity in the brackish reach of estuaries, only two additional modifications are required: one affecting deoxygenation and the other reoxygenation. Seawater has a marked tendency to induce coagulation of colloidal and nonsettleable matter, and usually deoxygenation must be modified to include a sludge deposit factor.

Computing reoxygenation in the brackish reach requires the determination of the oxygen resources of the replacement seawater flow. This flow is considered at its source to be free of BOD and saturated with dissolved oxygen. The amount of dissolved oxygen available in it at any cross section is determined by distributing the deoxygenation and reaeration by small increments from section to section along the brackish reach to the ocean.

Commencing at the ocean end, the calculation allocates deoxygenation and reaeration to replacement seawater in proportion to the seawater content in each small reach, taking into account the decline in replacement seawater progressively upstream, as determined by the salinity gradient.

The distribution of the deoxygenation pre= sents no special problem because the BOD satisfied along the entire course to the sea is readily determined as outlined heretofore. However, in assigning reaeration assets to each small reach, it is necessary to know the reaeration assets for the entire reach to the sea. Since this is not known until the entire dissolved oxygen profile has been computed, it is necessary to utilize the procedure of successive approximations.

Initially assume a complete reaeration summation to the ocean; solve for resultant dissolved oxygen assets of replacement seawater section by section; compute the entire dissolved oxygen profile to the sea; and finally check the computed reaeration summation against that assumed. Successive approximations are made until what is computed reasonably approximates that assumed. Here, the adaptation of the digital computer is advantageous.

Verification—Comparison of Computed and Observed Dissolved Oxygen Profiles. Figure 6-19 shows a comparison of a computed dissolved oxygen profile for Estuary D and observed estuary samplings taken under the same

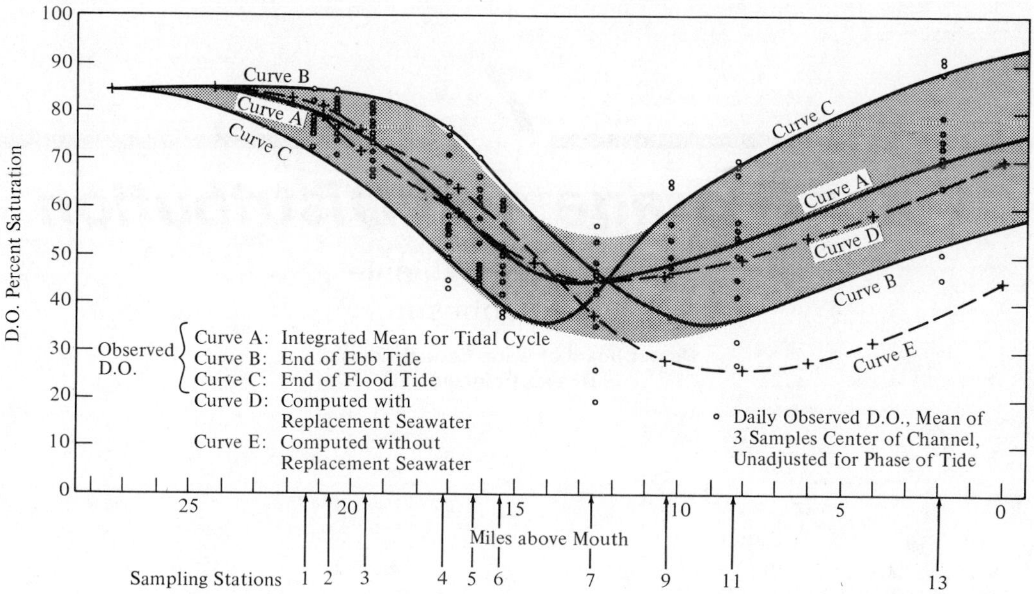

Fig. 6-19 Tidal estuary "D"–comparison of computed and observed DO profiles showing range through tidal cycle.

conditions. Curve A is the mean of observed sample results integrated through the tide cycle, and Curves B and C represent the observed end of the ebb and flood tide taken from smoothed plots of the tide cycle. The computed mean tide dissolved oxygen profile, Curve D, fits well within the observed dissolved oxygen and is in close agreement with the observed mean tide profile, Curve A. Curve E is the computed dissolved oxygen profile without taking replacement seawater into account in order to show the significance of this important asset in the recovery of the dissolved oxygen profile. Curve E, it is noted, is decidedly below the observed dissolved oxygen profile Curve A and the computed profile Curve D which took into account the dissolved oxygen assets of replacement seawater.

REFERENCES

1. Velz, C. J., *Applied Stream Sanitation*, John Wiley and Sons, Inc., New York (1970).
2. *Surface Water Supply of the U.S.*, USGS Series Water Supply Papers.
3. Gumbel, E. J., *Annals of Mathematical Statistics* **12,** 163 (1941).
4. Powell, R. W., *Civil Engineering* **13,** 105 (1943).
5. Phelps, E. B., *Stream Sanitation*, John Wiley and Sons, Inc., New York (1944).
6. *Standard Methods for the Examination of Water and Wastewater*, 12th ed., APHA, Inc., New York (1965).
7. Lee, J. D., *Sewage and Industrial Wastes* **23,** 2-164 (1951).
8. Velz, C. J., *Applied Stream Sanitation*, John Wiley and Sons, Inc., New York, pp. 148–154 (1970).
9. Black, W. M., and Phelps, E. B., "Report to the Board of Estimate and Apportionment, New York City, Appendix," pp. 88–100 (March 23, 1911).
10. Velz, C. J., *Transactions ASCE* **104,** 560 (1939); *Sewage Works Journal* **19,** 629 (1947).
11. Gannon, J. J., and Downs, T. D., *Water and Sewage Works* **110,** 3, 114 (1963).
12. Deininger, R. A., *Some Notes on Reaeration of Rivers*, Professional Paper, Dept. of Environmental Health, University of Michigan (July, 1964).
13. Streeter, H. W., and Phelps, E. B., *Public Health Bulletin* **146** (1925).
14. O'Connor, D. J., and Dobbins, W. E., *Journal ASCE* **SA6,** 1115 (1956).
15. Churchill, M. A., Elmore, H. L., and Buckingham, R. A., *Journal ASCE* **SA4,** 3199 (1962).
16. Gannon, J. J., and Downs, T. D., *Water and Sewage Works* **110,** 3, 114 (1963); **110,** 4, 157 (1963).

7

Water Storage and Distribution

Robert C. McWhinnie
Paul R. Johnson
Denver Board of Water Commissioners
Denver, Colorado

THE POTABLE WATER SYSTEM

A potable water system is basically an arterial network of piping and appurtenances, operated to deliver water of high quality to all points within it at predetermined pressures and rates of flow. Through pumping and repumping, valving and pressure regulation, and storage at optimum elevations within the system, pressure can be maintained. Through an arrangement of adequately-sized mains and service piping, and utilization of remote control monitoring, an adequate flow to meet all consumption demands, including emergencies, can be delivered to any point within the system from a central control station.

Elements of the system include transmission and distribution piping, valving patterns to allow bypassing and repairs, maintenance of pressure zones with regulation valves, high service and other pumps, potable water storage with quality control, and automatic operational controls. These elements will be discussed more fully under "Engineering Considerations," below.

Changes in the system network should be anticipated. When merging networks through new connections and extensions, hydraulic analyses of optimum flows within the total system or zone will point out weak areas and permit better design. Many mathematical techniques and analytical tools are available for network analysis.

In recent years the computer has made daily water system hydraulic analysis possible in most large cities. The use of monitoring equipment and transmittal of data to a recorder and computer allows rapid and precise adjustment to the most complex distribution system flows. However, the methods and mathematical tables are available to anyone interested in analyzing a small system or part of a large one by means of a calculator.

Design conditions can be estimated from past

flow and demand records and from the estimated growth rate of the community served by the system, or by an extension thereof. Determining the maximum flow that can be accommodated by existing pipe sizes will establish another design parameter. Working with known plant efficiencies and known available alternatives, an analysis of any system can be performed to estimate the optimum results that can be expected.

Damage to the system can occur through uncontrolled surges and water hammer due to improper phasing of pump operations, closure of valves, etc. Construction methods are as important as operating procedures to protect the system from these hydraulic transients. Leakage, cross connections, and faulty sanitary operation procedures impair water quality and system efficiency. These are reviewed below from the viewpoint of their impact on the potable water system and its successful operation.

ELEMENTS OF THE SYSTEM

Transmission and Distribution Piping. The size or function of a pipe classifying it as a transmission pipe or a distribution pipe is a matter to be individually determined according to the design situation. In general, transmission pipes operate under higher pressures and have a limited number of branch pipes or house service taps. Pressure reducing valves may be required to take water from these mains to lateral connections. Transmission pipes contribute the primary network for flow analysis in a system. They normally receive flows only from other transmission pipes or from a pump.

Distribution piping is the large subnetwork of secondary lines which supply water for consumer use and for fire flows in residential, commercial, and industrial zones. High-rise buildings may require increased pressure in the distribution main or may install supplementary pumping and storage within the building to supply the required pressure and volume to the upper floors.

In most cases, the required rate of fire flow exceeds the consumer demand in a residential zone though not necessarily in other zones. Design of the system is based upon producing the needed high flow at any hydrant, and upon maintaining the minimum main sizes that allow the design flow. The peak hour flow plus the fire flow is not generally used, as the probability of both occurring simultaneously is small. A reasonable safety factor taking increased future use into account should nonetheless be maintained. A pipe of 8 in. diameter is commonly considered a minimum size for residential fire flow.

System design calling for closed loops rather than dead ends doubles the amount of water available to a hydrant. Pumping through an auxiliary fire engine pumper boosts the pressure for the fire stream, although sufficient quantity of water must be available at the hydrant to support the stream.

Auxiliary water distribution systems have been constructed in the past for fire fighting, sometimes utilizing nonpotable water from such sources as lakes and harbors in order to avoid the costs of delivering potable water to a fire. In general, however, such systems are used over a limited area and are more expensive than a single potable system with enlarged mains. The auxiliary system requires a completely independent pipe and pumping system and, as a rule, is rarely used. The danger of cross connections is always present as are other health hazards associated with substandard water. Moreover, this system doubles plant and pipe maintenance costs and requires large storage reserves for emergencies—reserves which are built into any well-designed potable system.

Pipes and appurtenances making up the system vary with the zone they serve and the demands they must meet. The pipes are under continuous stress from such external loads as earth fill, traffic, and construction, while carrying a full flow of water under pressure. Internal loads are likewise not static. Pump surges, water hammer, valve operations, breaks, and similar causes impose extra impact loads beyond the operating pressure of the system, throughout the life of the conduit. It is thus essential to select a pipe able to withstand all probable maximum external and internal forces.

Materials now used in the manufacture of

pipes are metals, cement, and plastics. Among the metals are steel, cast iron, ductile iron, and, for small lines, copper and galvanized iron.

Cementitious materials go into the manufacture of concrete pipe and asbestos-cement pipe. Since Portland cement is weak in tensile strength, these two pipe types must be reinforced with material designed to take the loads in tension.

In the earliest manufacture of concrete pressure pipes, a steel cylinder was wrapped with tensioned wire, lined centrifugally with cement mortar, and covered with concrete. Now the wire is wrapped about the concrete encasement holding the steel cylinder and then coated with concrete. Noncylindrical concrete pipe with reinforcing bar cages of encased steel are used where the extra strength of prestressed steel is not required.

Asbestos-cement pipe incorporates the fibers of asbestos (to carry tensile loads) and finely-ground silica. Fittings at critical stress points are usually of cast iron or steel.

Plastic pipe of smaller size has been used for a number of years. Of late, increasingly larger pipe sizes have been manufactured from this versatile material. Some of the plastics suitable for water piping have been reported upon in detail in AWWA journals. Varying in type or composition, the plastics may be utilized for particular problems according to type. Principal plastics used include polyvinyl chloride (PVC), polyethylene (PE), and the acrylonitrile-butadiene-styrene (ABS) plastics. A fourth type, growing in popularity, is reinforced-plastic-mortar (RPM) pipe which is composed of a sand-resin base reinforced with strands of fiber glass.

Each of the materials used in pipe manufacture demonstrates advantages and disadvantages when size, weight, construction costs, and economic life are measured against the stresses the pipe will encounter at its proposed location. The risk of a broken main in a congested traffic zone, or a transmission line break endangering valuable real estate, and similar liabilities should encourage very careful selection of the pipe to be installed.

A principal cause of water main breaks is improper design resulting from insufficient knowledge of soil conditions. Specifications may not apply for both flexible and inflexible piping in trenching, bedding, and backfill. To prevent future breaks, it is important in the installation of pipe to recognize the characteristics and moisture stresses of the soil, to protect against shearing and beam stresses, and to ensure proper bedding and handling during construction. Inspection of the pipe at the construction site to determine if defects are present and an inspection warranty against defects from the manufacturer are good insurance against subsequent breaks.

Known causes of main breaks after installation include internal stresses of surge and water hammer, external stresses of soil movement, trench settling, and traffic impact loads. Insufficient protection against corrosion both on the inside and outside of the pipe can lead to eventual failure, as can temperature.

These common loads and stresses will be present regardless of pipe type. To withstand them, the pipe must meet established test criteria for:

a. Backfill load at incrementally increasing depths, both with the conduit under pressure and empty.

Backfill loading formulas follow the following general equation for rigid pipes:

$$W = CwB^2 \tag{1}$$

where

W = the vertical load per linear foot,
w = weight per cubic foot of backfill,
B = trench width at the top of the pipe,
C = a coefficient relating to height and width of backfill, width of pipe, and shearing forces on prisms of backfill lying between the pipe width projection and the trench wall.

For flexible pipe use:

$$W = CwBD \tag{2}$$

where D = the pipe diameter in feet. Values for C can be determined from Fig. 7-1.

Rigid pipe will include cast iron, ductile iron, concrete, and asbestos-cement. Flexible pipes include plastics and the thin-walled steels.

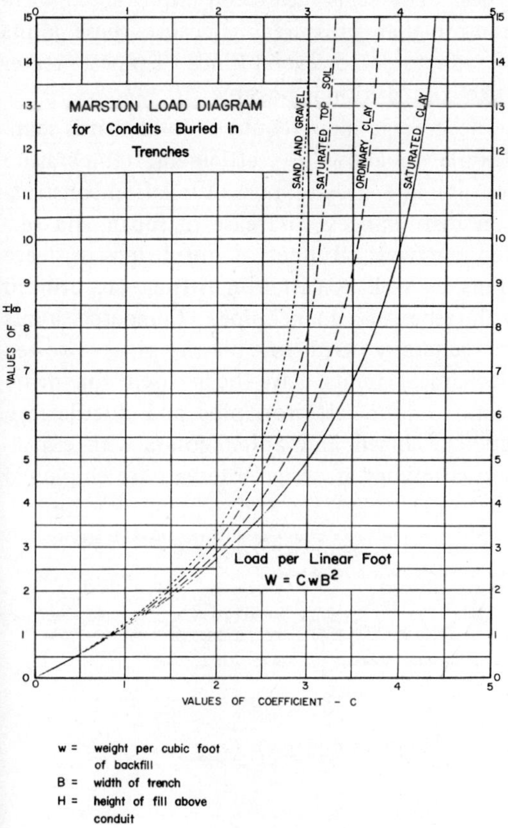

Fig. 7-1 Marston load diagram for conduits buried in trenches. (*Courtesy of Denver Board of Water Commissioners.*)

Specifications and manuals of the ASTM and the AWWA should be consulted to determine the exact characteristics of a pipe if there is any question.

b. Bursting pressure.

In a theoretical ring section of unit width, the force in tension P to be resisted by the pipe can be found by this equation:

$$P = whr \tag{3}$$

where

w = unit weight of water (62.43 lb/ft³),
h = head in vertical linear feet,
r = radius of pipe in feet.

The force P, in pounds, is measured over the area of the unit width of the section multiplied by the thickness of the pipe, or pounds per square inch (psi). The transient pressures of surge and water hammer must be added to the tension load. Each pipe material has its determined safety factor indicating how it will withstand common transient pressures.

c. Stresses from dynamic forces at bends which vary with the size of pipe, radius length, degree of bend, and flow velocity in the pipe.

The general equation for determining the tensile force on the pipe bend is

$$P_b = \frac{wAV^2}{g} \tag{4}$$

where

w = unit weight of water,
A = area of pipe opening in square feet,
V = velocity of flow in feet per second,
g = acceleration of gravity (32.17 ft/sec²).

The total force to be resisted by a thrust block or tied joints resultant to a bend is given by the equation:

$$P_c = 2A\left(\frac{wV^2}{g} + P\right)\sin\frac{\theta}{2} \tag{5}$$

where

θ = degree of bend,
P = force in tension (see Eq. 3).

Pipe joints are a vulnerable part of the main. Longitudinal forces will loosen them, especially at bends. A common cause of bend failure in a well-designed main with proper "kick-blocks" to withstand the calculated forces at the bend is a lateral shifting of earth due to digging near the bend after the main is in operation. Disturbance of the homogeneity of the supporting earth weakens structural support.

This problem can be overcome at times by tying the joints of the pipe on either side of the bend. This is done by clamping longitudinal steel rods to the body of the pipe at each joint. The rods and clamps run back from the bend to a point where the total weight of the tied length of main plus the skin friction of the main against the tamped backfill equals the internal forces at the bend. Similarly, shorter longitudinal rods can be extended back to a reinforced concrete anchor structure which straddles the main and extends several feet into the ground below the main. A thick expansion strip separates the pipe and the concrete at the straddle point.

High stresses from hydraulic transient pressure waves can loosen seals and start leaks that may go undetected for years. Some systems, it is estimated, lose as much as 40 percent of their water through leaks and other wastage during periods of high pressure and maximum use.

Each pipe material dictates the type of joint best suited for it. "Stab joints" with plastic seals which seat when the bell of the first pipe receives the spigot of the second to the proper depth are now commonly used in several pipe types. The seals have been improved to withstand higher pressures. Because these joints are adaptable to several kinds of pipes, greater price competition is possible.

The average main requires a joint that is semi-flexible to allow for settlement, temperature and pressure stresses, post-installation construction disturbances, and ease of repair and section removal. Rubber or butyl rings in sleeve joints as well as cast iron fittings are utilized with asbestos-cement pipe. Cemented joints are generally used for plastic pipe. Bolted mechanical joints have been used for many years to join cast iron pipe and ductile iron pipe. Flanged and bolted joints with expansion couplings at design intervals are employed

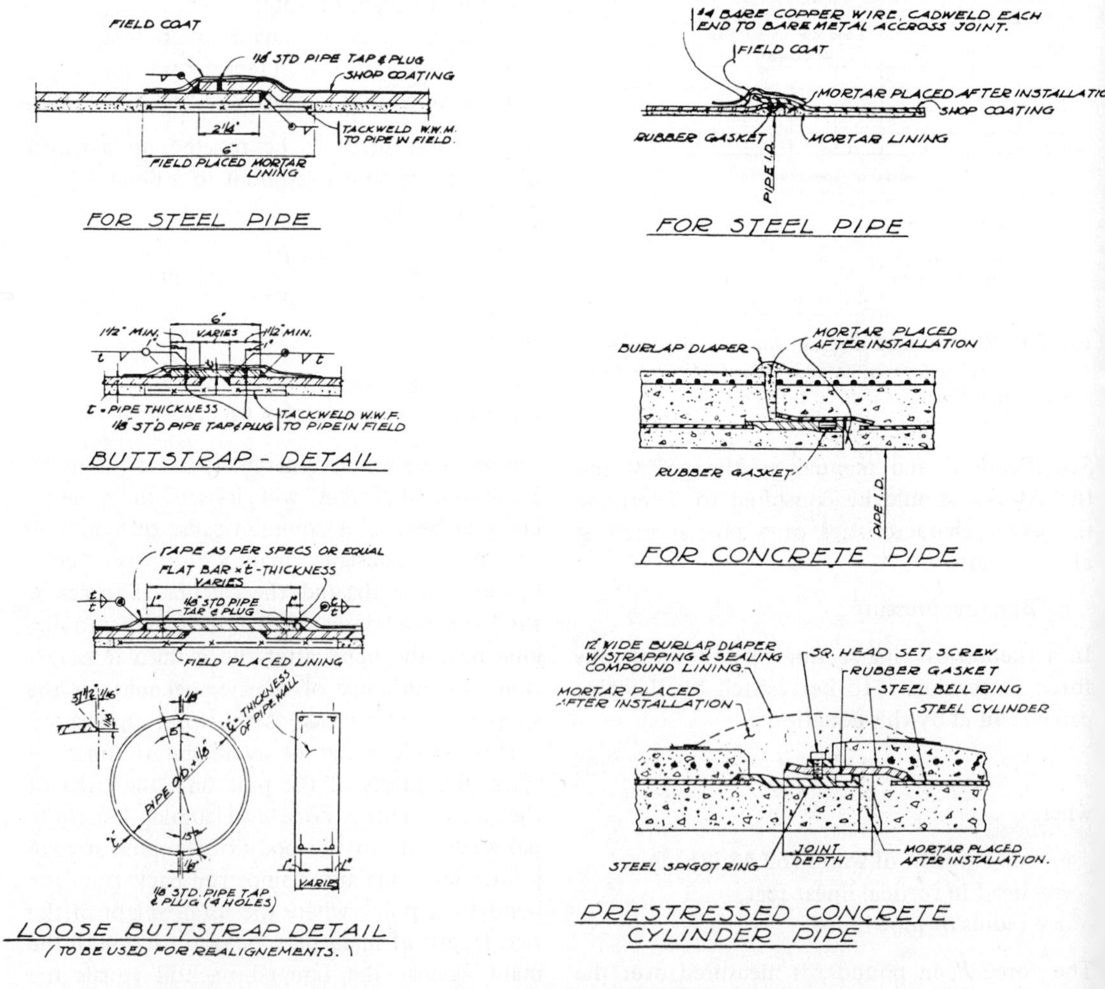

Fig. 7-2 Pipe joints and connections. (*Courtesy of Denver Board of Water Commissioners.*)

on steel pipe and are adapted to other varieties of pipe. There is no limit to the versatile use of joints. Overall, however, size, weight, pressure stresses, velocity of flow expected, and the cost and economic life of the pipe are all considerations in selecting the pipe material and joints. Some types of joints for pipes of various materials are shown in Fig. 7-2.

Each year pipe manufacturers are offering a wider selection of improved products. To evaluate the pipes available, reference should be made to standard tests and specifications of the ASTM and the AWWA. Since these standards change with time, the latest edition should be obtained.

Valves used in the system include numerous line valves which are arranged to close off short sections of a main for repairs and send flow around the section. Commonly, the line valves are located at the end of blocks or near the intersection of loops and are arranged to give the most flexibility to flow routing. There may be three or four valves in a street intersection controlled through valve boxes. These are usually gate valves which are illustrated in Fig. 7-3. They range in size from 4 to 18 in. in diameter although larger sizes may be procured on special order. In recent years the butterfly valve has been promoted as a replacement for the more complex gate valve and has been used with various seating seal materials. Its utilization in waterworks is bound to increase as it becomes more familiar to operators and designers (see Fig. 7-4).

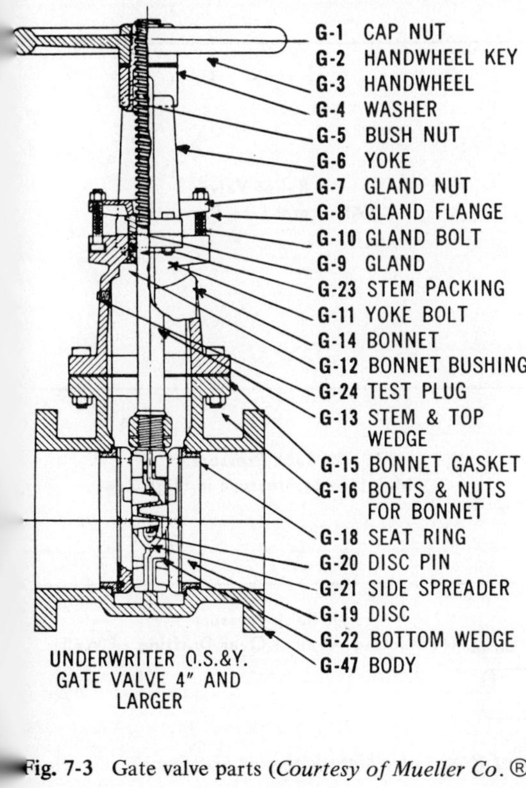

Fig. 7-3 Gate valve parts (*Courtesy of Mueller Co.* ®)

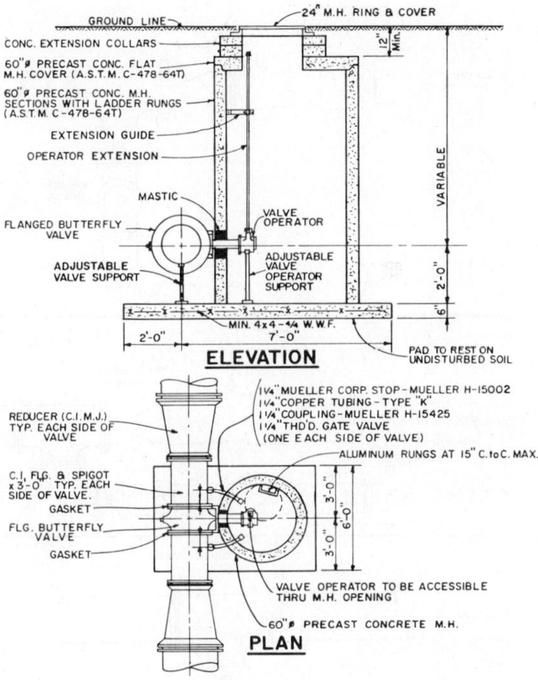

Fig. 7-4 Standard butterfly valve installation in cast iron pipe 16-in. diameter and larger. (*Courtesy of the Denver Municipal Water Works.*)

It is not always necessary that the valve and line size be the same. Depending upon the hydraulic flow characteristics of the valve, its size may be decreased by one pipe size without appreciable headloss. However, velocity of the flow through the valve (10 ft/sec being a suggested design limit) plus the extra costs of reducers and other fittings may nullify the savings of a smaller valve, especially in the small-diameter pipes (4 to 12 in).

In addition to line valves (which are predominantly gate valves), a system incorporates many other valves to ensure smooth flow throughout the network, transient wave supression, automatic control, and other objectives. Such valves include surge control valves, pressure control valves, altitude control valves, check valves, plug valves, glove valves, and others. Many of them are located in pump stations. Altitude control valves are used in inlet pipes to storage reservoirs or tanks. Pressure control valves are used at points of differential pressure in the system. These are covered in greater detail in "Pumping and Pressure Control Valves," below.

Air valves and vacuum valves are installed at points in the main where low pressure, low flow, or the separation of air from water causes air pockets, or where the evacuation of water from a pipe section causes a partial vacuum. These

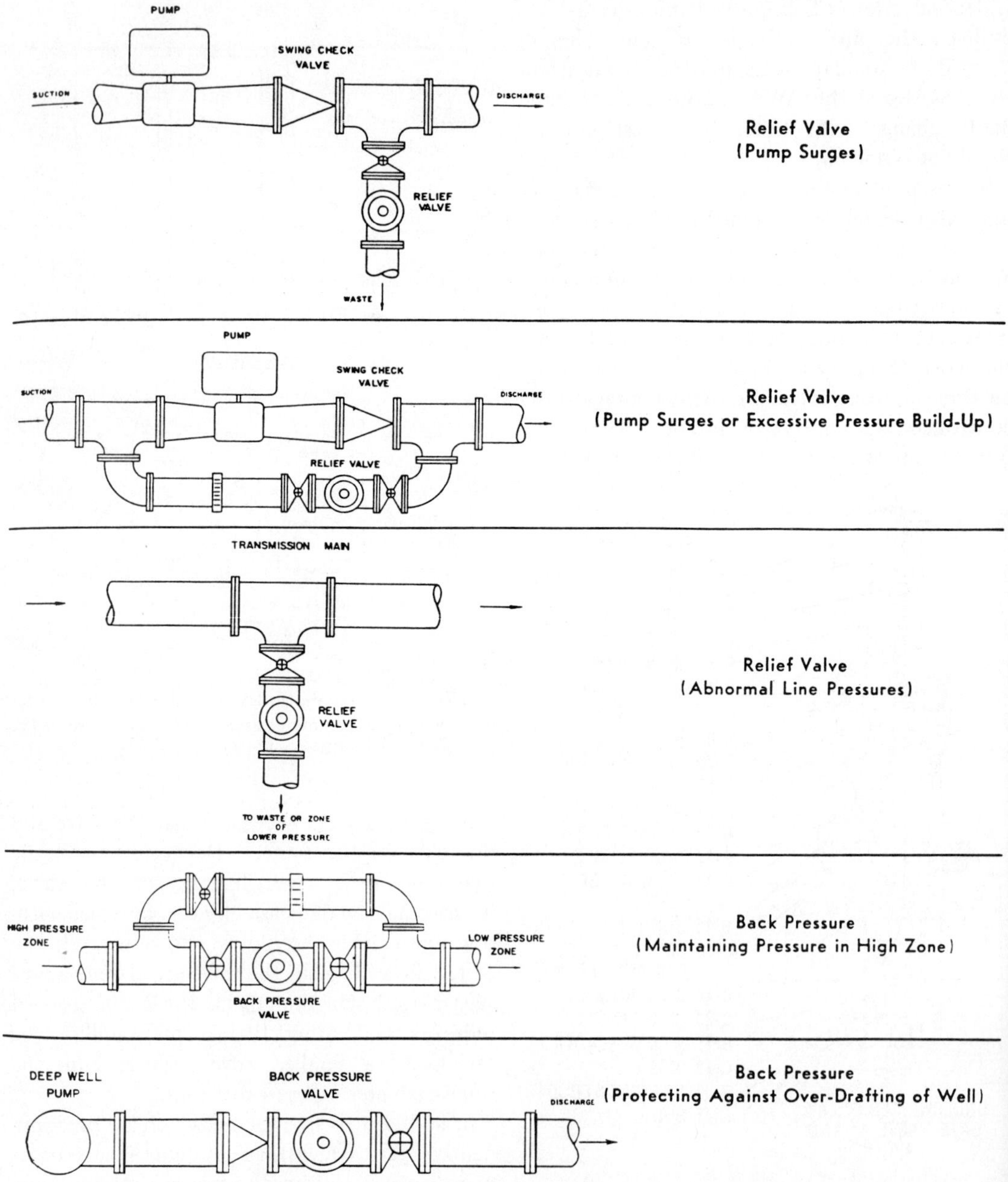

Fig. 7-5 Typical piping arrangements. (*Courtesy of Ross Valve Manufacturing Co., Inc.*)

occurrences arise more frequently in large conduits flowing at changing rates through extreme grade changes, but the valves may also be required in a distribution system with hilly streets, dead ends, and the like.

Typical piping arrangements, including placement of several types of valves serving different purposes, are illustrated in Fig. 7-5. A ball valve is shown in Fig. 7-6.

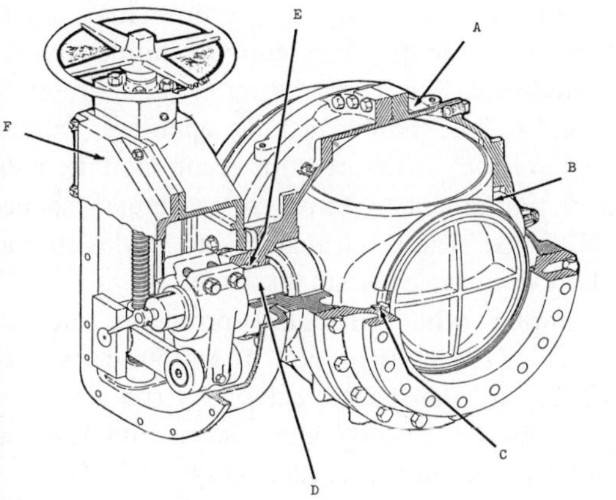

Fig. 7-6 Ball valve (*Courtesy of Bingham-Willamette Co.*) A: body, B: plug or ball, C: seats, D: shaft, E: stuffing box, F: torque unit to rotate ball.

Distribution Water Storage. The most economical force to move water is gravity. If gravity can be utilized in the distribution system to maintain adequate flows or to provide storage water under pressure during heavy demand periods, the load on the electrical pumping system will be reduced. Designers therefore try to establish storage points at elevations which will gain the maximum pressure head for the system. If the full pressure required for the system is not obtainable, partial pressure to the suction side of the pumping facility will increase efficiency. Since it is unlikely that the filter plant and clear water storage will be located in mountains above the distribution system, it will be necessary to pump to storage during off-peak hours and combine pumping and gravity flow during peak hours.

Treated water in the clear well of a filtration plant or in the manifold pipe of a well field is usually pumped to the distribution system storage through the same transmission lines that supply the distribution lines. To conserve energy, the pumping is done at a median rate during hours of low demand. Without storage, pumping facilities would have to match the hourly fluctuations of demand at increased energy levels while the consumer load is at a peak for both water and electric power. Charges for peak hour power rates are substantially higher than those for median rates. Further, pumps may be unable to meet an unusually high demand for prolonged periods. Fire demand peaks, irrigation water peaks, power demand peaks, power failure, and similar demands occurring simultaneously in multiple combinations could seriously reduce the water supply.

It is sometimes impossible to achieve a design that incorporates enough natural elevation for storage to benefit from the force of gravity. In such cases, there remain the options of pumping from ground level storage reservoirs or of creating gravity with elevated water tanks.

Sizing of the distribution reservoir will depend upon the zone it serves, the size of the zone mains, the capacity required for fire demand, and other criteria. A rule of thumb for determining storage water amounts is to calculate a maximum day's flow and add 10 hr fire flow. For best hydraulic efficiency, the ground level reservoir or storage tank should be located near the center of the zone it serves. Since it is rare that a distribution reservoir can be placed at the optimum point for hydraulic efficiency, the location and size of mains delivering water from the reservoir must minimize the head loss implicit in long runs. Fire demands are covered below.

Construction of ground level reservoirs involves the use of an impermeable liner, a cover, and control appurtenances, often including automatic valves and pumping facilities. As treated water is a valuable commodity that should not be lost in transit and as chemically-treated water is a perishable product, the water must be kept secure, clean, and free of any contamination while in storage.

A number of materials are available for reservoir lining. Reinforced concrete slabs with copper or plastic seals and expansion joints, asphaltic paving with coal tar surfacing to prevent bituminous leaching, jointed asphaltic panels, membrane linings of plastics, butyl rubber, and other substances are all in use at present. Gunite and pressure-jetted concrete are often employed. Combinations of sandwiched materials are also available in prefabricated sheets or may be constructed.

The structural stability of the underlying earth or fill is most important. One leak can do a great amount of damage by undermining the lining. Detection of leaks, structural strength, and ready repairability of the lining material are prime considerations in the long life of the reservoir. By laying a gravel base on an impervious membrane over which the liner is installed, seepage may be carried to a drainage sump for monitoring. A clay base helps to seal leaks.

The slope of the sides is a most important factor, affecting area, preparation of the reservoir base, installation of the lining, and ease of subsequent maintenance and repair operations. Asphaltic paving requires more gentle slopes than paving with reinforced concrete or other more rigid lining materials. In estimating the durability of a proposed lining, the weight of maintenance equipment and workers on the finished surface and other wear and tear should influence design and selection.

The shape and area of the reservoir may govern the selection of a lining. For example, if roof support columns are to be spaced across the floor of the reservoir, selection of a method for sealing the floor at each column will become both a construction problem and a cost consideration. If the lining and the roof are structurally independent, there will be a wider choice of suitable linings, but roof costs may increase.

The principal function of a distribution reservoir roof is to keep the water clean. Installation of a light metal or wooden roof therefore normally suffices. However, in some cases, tennis courts, grassed parks, and even parking lots have been built atop reservoir roofs. Also, covering a concrete reservoir roof with earth and grass has the advantage of keeping the stored water cool in summer and unfrozen in winter. To meet the loads imposed on these double-duty roofs, prestressed concrete tee beams, steel beam and girder construction, reinforced concrete floor slabs, and other devices have been employed.

Structural members of the roof will be subject continuously to conditions of very high humidity. Materials and finishes must therefore be resistant to rust and corrosion as well as biological growths. Roofing will inhibit the growth of algae and other plants which are dependent upon sunlight. It will not, however, eliminate fungi and other plants which do not require direct sunlight. Spores may enter through air vents. The reservoir should thus be periodically emptied and cleaned. Architectural design should include openings, ramps, and related structures to accommodate men and equipment. Exposed structural shapes should be readily cleanable, and grades should allow drainage of waste discharge.

Among other structural problems encountered in the construction of a ground level reservoir are the stresses caused by the action of groundwater against sides, floors, and foundations when the reservoir is empty. To control the groundwater level, the use of drainage tile or other procedures may be required. Differential settlement of the earth over the reservoir area, especially in fill, can cause leaks, column foundation problems, and other difficulties. Expansion and contraction of an exposed roof places stresses on all structural supporting members.

These stresses can be extensive on days of extreme temperatures. Movement can be taken by proper expansion joints, rocker or roller supports, etc. The coefficient of expansion of the roof material will determine the total movement to be expected laterally from fixed points in the structural system of the roof.

The term *elevated storage* generally refers to water tanks supported in the air by structural members or located on hills at a height permitting development of a pressure head sufficient to keep the mains in the zone served operating at design pressure. The water in the tank supplies the needed input to the system during periods of high demand, supplementing the pumped flow in the transmission lines. These tanks often "ride on the system," acting collectively as a pressure control, surge suppressor, and dissipator of hydraulic transients, as well as a reservoir to maintain volume. Required storage volumes are discussed later in this chapter.

The surface of the water in the storage tank is at atmospheric pressure. If the tank overfills, the overflow is wasted to a storm sewer or another available wasteway. It is not good practice to connect the overflow pipe to a sanitary sewer because of public health concerns and the risk of flooding the sewer and neighborhood basements connected to it.

A small tank on the ground is called a standpipe. Often the height exceeds the diameter. Pressure is produced by the height of fluctuating water levels above the main. The basic weaknesses of standpipes as components of a system are their lack of great height and the "dead area" of water in the lower part, below the effective zone of fluctuation. For example, a 50-ft water level will give only 22 psi pressure in the main. The limit set by the height of the standpipe will be the limit of pressure for the mains and the zone. These considerations have led designers to limit standpipe use in laying out new systems.

If it is possible to utilize the natural elevation of a hill within a reasonable distance of the zone being served by an elevated tank, the elimination or reduction of the supporting structure may permit a cost savings, or an increase in tank size may be realized for the same cost. If the distance from the tank to the zone is excessive, account must be taken of the added cost of increased main sizes or construction of additional mains to handle the maximum flows to the center of the zone.

Steel is the material most commonly used in fabricating tanks to be elevated on a column structure. Many shapes are in use, the design being determined by the amount of water to be stored and structural factors. As a great fluctuation in water level is not desirable, most tanks are limited in height and are enlarged more horizontally than vertically as the design volume increases. Small tanks are often cylindrical with domed or conical tops. Oblate spheroids, spheres, and combinations with hemi-ellipsoidal bottoms and domed tops are produced in all sizes by prefabricating companies.

Selection of the tank and supporting structure is based on the need to contain and elevate the greatest amount of water at optimum design efficiency and structural safety. Sizes may range from a 40,000-gal tank with a diameter of 20 ft and a fluctuation of 17 ft to a 2,000,000-gal tank with a diameter of 80 ft and a fluctuation of 35 ft.

Cleaning, painting, and protection from freezing are important maintenance concerns with a structure of this type. Heating elements to keep float controls from freezing, insulated riser pipes, underground vaults for valving and controls, etc., are essential to the design of tanks elevated by a supporting column structure. Rust, corrosion, growths, and the like also call for constant vigilance both inside and out.

On a hill location, the foundation for a tank at ground level should be constructed on undisturbed earth or in an excavation rather than on a fill. This will help avoid differential settlement problems which can be considerable with the large area and weight per square foot involved.

Larger volumes of water can be stored in these tanks than can be supported by column structures. Welded steel, reinforced concrete, and prestressed concrete are commonly used in the construction of storage tanks of varying diameter. There is no theoretical limit to the size of such a tank, although a limit of 5,000,000-gal has been found to be economical. Where more storage is required, the construction of two or more tanks per installation will allow greater flexibility in operation and better maintenance.

Pumping and Pressure Control Valves. Pumping begins first at the source of the treated water, at the clear well of the filter plant, or at the manifold pipe of a well field. Sufficient water is pumped to bring the pressure of the transmission lines up to operating level. During hours of low demand, enough water will be pumped to serve the needs of the distribution system and to replenish storage reservoirs in anticipation of heavy demands later in the day.

At the second echelon of pumping are the high service pumps used to lift water and boost pressure. They can deliver storage water from low-level reservoirs to transmission mains during hours of high demand and may double

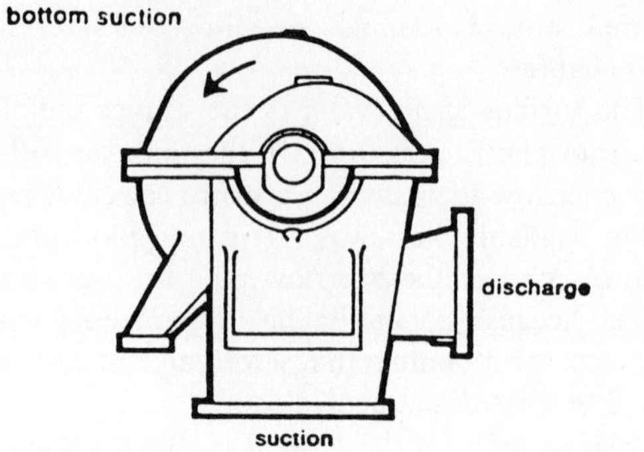

Water taken by negative suction can be pumped to a high pressure line.

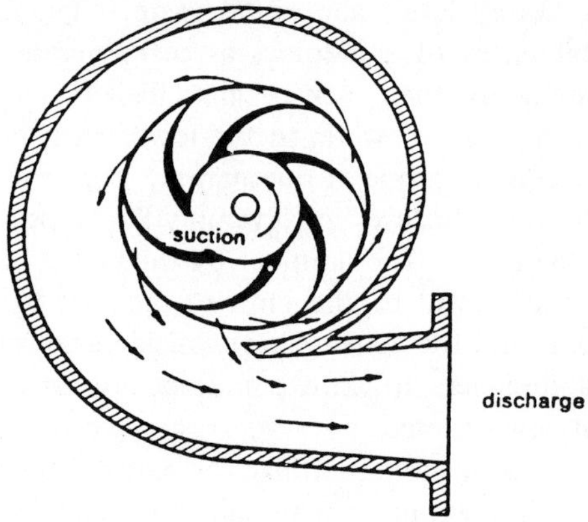

Water taken at center of the impeller is thrown by centrifugal force to a volute section.

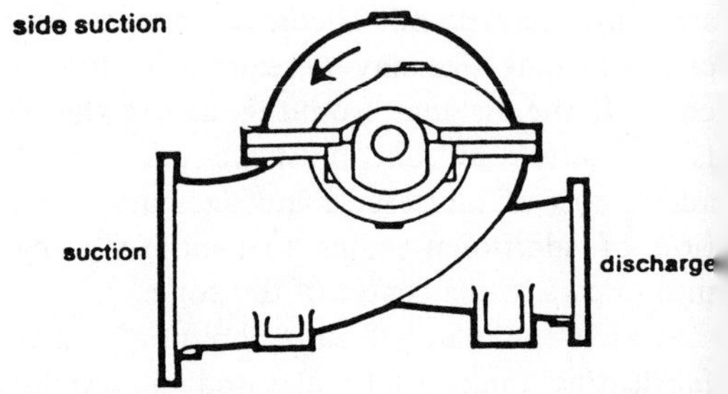

On-line movement of flows can be boosted to higher pressures.

Fig. 7-7 Centrifugal pump functions. (*Courtesy of Worthington Engineered Pumps, Inc.*)

as pumps for elevated storage. Through the design of controls, valving, and piping, they may be employed alternately as individual pumps or as syncronized pumps to deliver peak capacity flows. The necessity of lifting water in the system to zones at various elevations and maintaining the zone pressures may require the constant use of booster or high-service pumps as on-line facilities. Otherwise, they may only be required during periods of high demand.

Once water is lifted to a higher zone, pressure is maintained by altitude valves or by pressure control valves which prevent backflow to the pump or to a pressure zone at a lower elevation. Exceptions can be made, if desired, setting specified differential pressure limits in the controls.

Pumps most commonly found in potable water pumping stations are of the centrifugal type illustrated in Figs. 7-7 and 7-8. This pump usually utilizes a cast iron or bronze impeller on a drive shaft which spins at high velocity. The impeller receives the water near its center of rotation and thrusts it centrifugally to its outer perimeter, forcing the water out under pressure. This function can be performed by a single impeller or in several stages to increase the pressure to the pressure head required by the lift. A single-stage centrifugal pump is limited to a pressure head of about 300 ft or 130 psi, minus losses due to individual pump efficiency. Where greater pressures are required, multistage centrifugal pumps can be used, or single-stage pumps used in series.

An efficiency and horsepower curve for centrifugal pumps is shown in Fig. 7-9. The centrifugal pump comes in a wide selection of sizes. It is simple, compact, and relatively inexpensive. Vertical models take water from a vertical pipe on the suction side and pump it to a horizontal line. On-line horizontal pumps boost the pressure and flow in one direction.

In order to specify a pump, the following service conditions must be determined: (1) the quantity of water to be pumped, in gallons per minute or cubic feet per second; (2) the total head to be pumped against, including pipe and valve losses, etc.; (3) the suction head or lift; (4) the maximum and minimum operating heads; (5) the minimum shut-off head; (6) the pump suction diameter and discharge diameter; (7) maximum revolutions per minute and minimum horsepower of motor; and (8) pump efficiency characteristics at specified quantities of flow and pressure heads to be met.

Pumps other than the centrifugal type used in

Fig. 7-8 Double volute centrifugal pump. (*Courtesy of Worthington Engineered Pumps, Inc.*)

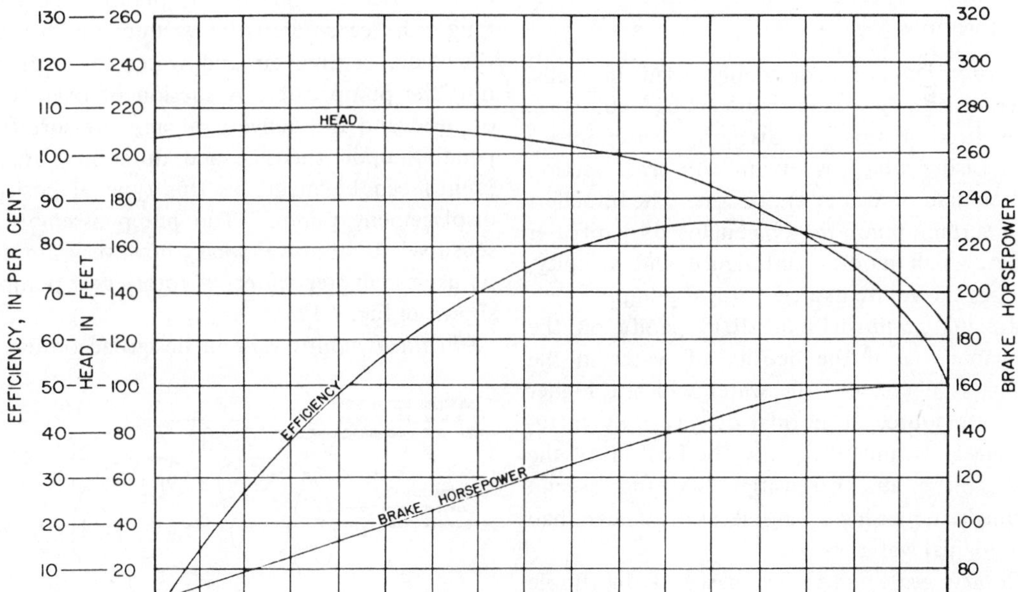

Fig. 7-9 Efficiency and horsepower curve for centrifugal pump. (*Courtesy of Denver Board of Water Commissioners.*)

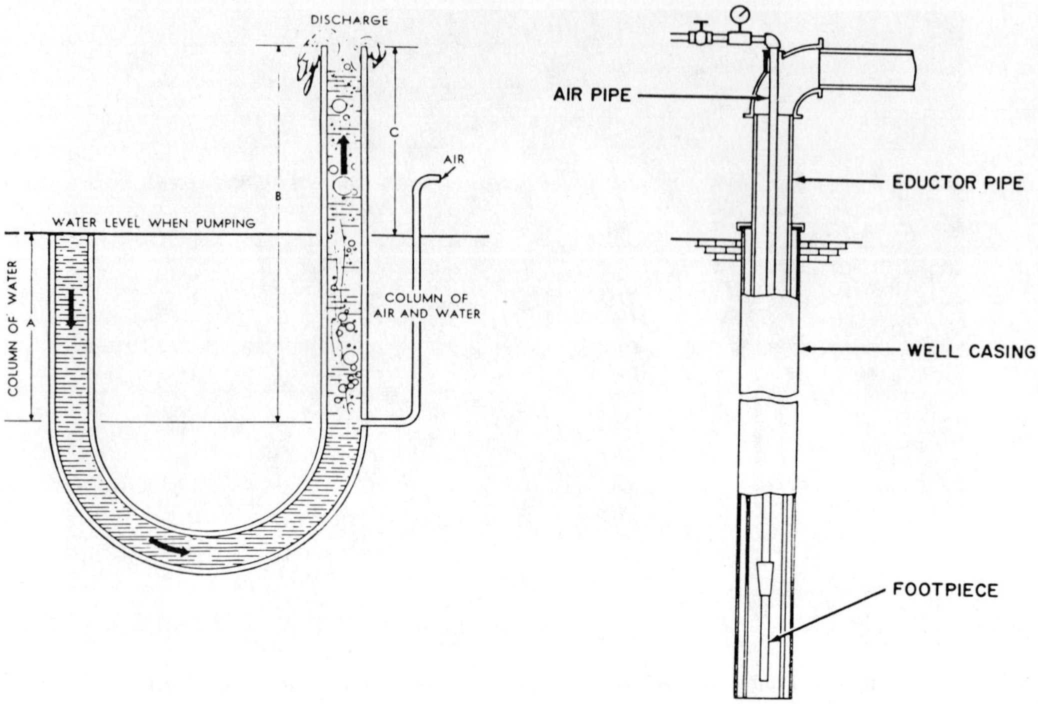

Fig. 7-10 Theory and assembly of an air lift pump. (*Courtesy of the Department of the Army.*)

waterworks operations fall into the following classifications:

Turbine pumps, commonly used in wells, operate in the same manner as multistage centrifugal pumps but discharge water into a well casing which is of uniform cross section rather than of volute expansion. The impellers of a turbine pump may be enclosed by diffuser vanes which modify turbulence and velocity, as does the volute in a centrifugal pump.

Air lift pumps (Fig. 7-10) operate on the principle that if the density of water in the casing is decreased, the water will rise higher than its source at normal density. By introducing a stream of air into the bottom of the casing, the air and water mix and ascend through the casing to an elevation higher than the original water level.

Rotary gear pumps are based upon simple gears in a case, both of which are machined to close tolerances which will allow no backflow to the suction side of the pump. As the gears turn, a positive displacement of water is carried between the case and the gear body to the discharge side of the pump. The spinning gears cause a partial vacuum on the suction side, causing air and water to be drawn into the pump case. A suction of over 20 ft of head and the delivery of any pressure (depending upon the size and durability of the pump) can be made by this type of positive displacement pump. The pump assembly is sensitive to abrasive wear, however, limiting its use to silt-free water. A rotary gear pump is shown in Fig. 7-11.

Chemical pumps vary in design according to

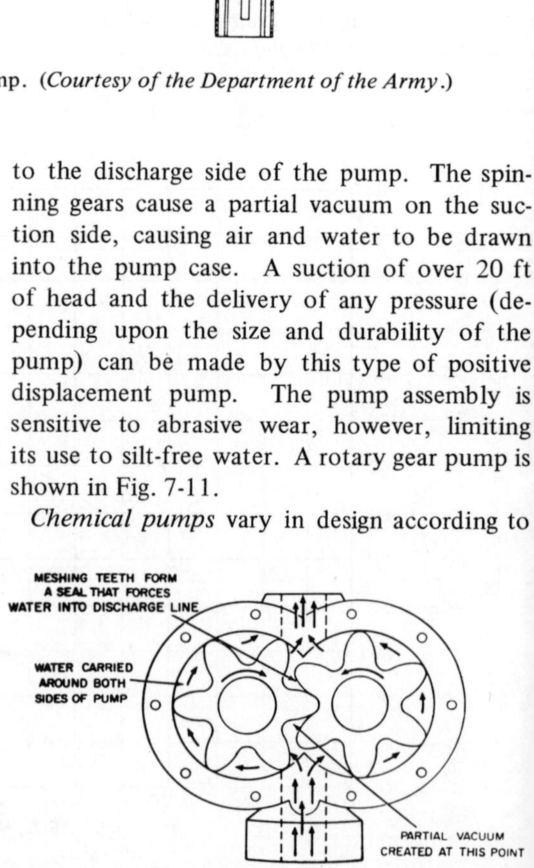

Fig. 7-11 Rotary gear pump. (*Courtesy of the Department of the Army.*)

the nature of the material being pumped, i.e., its density, viscosity, etc. Chemical slurries can be handled by pumps utilizing the principle of the Archimedes screw, propelling the mass forward by metal or plastic vanes or shaft conformations. Extrusion through a rubber tube—by rotating intermittently-spaced rollers on a wheel to depress the tube and force the "slug" forward—and other discrete patented methods are also in use. These must be positive displacement pumps which are not subject to wear from abrasion and which are easy to disassemble and clean.

Sump pumps must be able to pass abrasive silts and other foreign material. Open impellers on centrifugal pumps, jet or siphon-action pumps using a clean, separate water source, and other sewage-handling pumps are available to pump this type of wastewater.

Pressure reducing valves (PRVs) or pressure

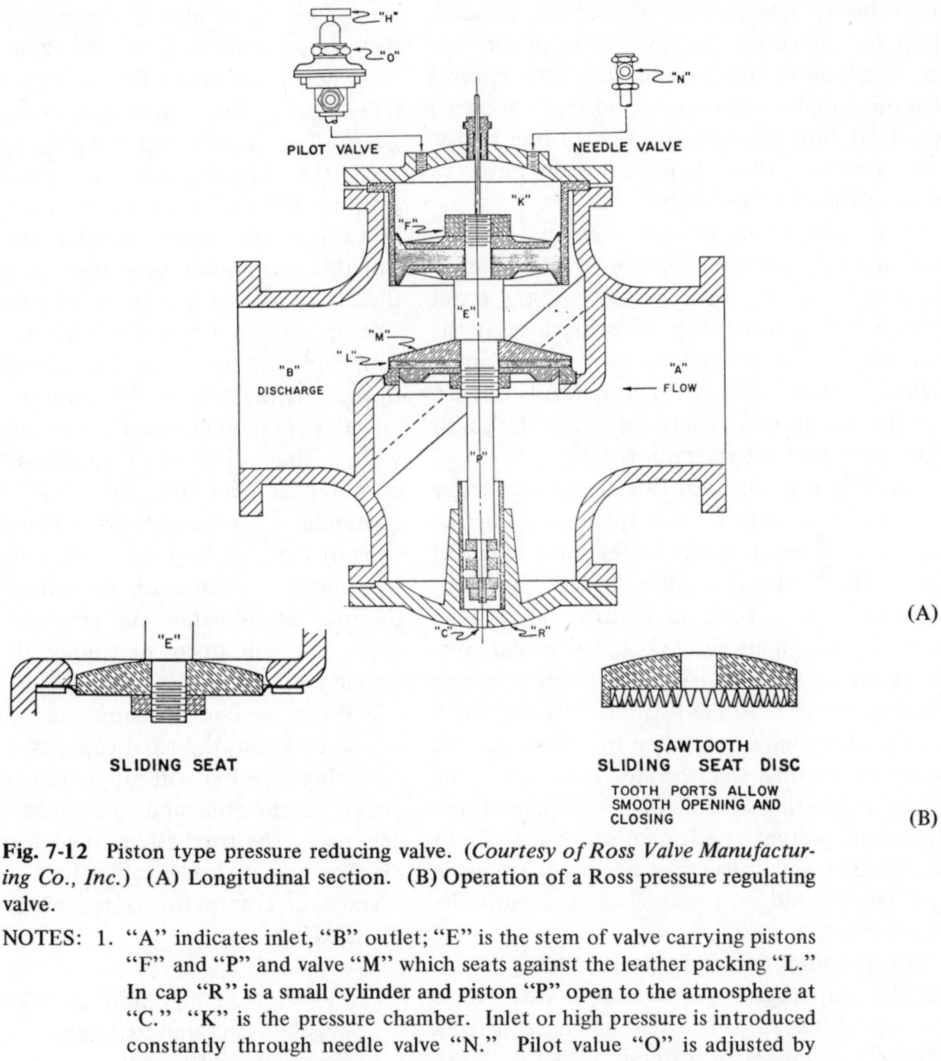

Fig. 7-12 Piston type pressure reducing valve. (*Courtesy of Ross Valve Manufacturing Co., Inc.*) (A) Longitudinal section. (B) Operation of a Ross pressure regulating valve.

NOTES: 1. "A" indicates inlet, "B" outlet; "E" is the stem of valve carrying pistons "F" and "P" and valve "M" which seats against the leather packing "L." In cap "R" is a small cylinder and piston "P" open to the atmosphere at "C." "K" is the pressure chamber. Inlet or high pressure is introduced constantly through needle valve "N." Pilot value "O" is adjusted by turning handwheel "H." Turning "H" out reduces the pressure, turning it in increases the pressure. Needle value "N" is on the inlet side and is set to control or time the movement of "M." Pilot "O" is on the outlet side.

2. To set valve to desired outlet pressure, place pressure gauge on aircock on outlet side of valve and adjust to desired pressure by slowly turning handwheel "H."

control valves found most commonly in waterworks are of the piston type with two chambers, the piston closing on its valve seat between them (see Figs. 7-12A and 7-12B). The piston is opened, allowing flow, or closed, depending upon which side has the higher pressure. Small-diameter bypass lines connecting the two chambers of the valve have hydraulic controls which can be set at the desired pressure balance between the two chambers. Additional controls can be interposed to allow the valve to reverse flow should the pressure drop too much on the lower side or increase too much on the higher side. The valve control will open under the preset conditions, allowing water to flow through the bypass line to the other chamber, thereby activating the piston to move in the direction desired.

In the case of an altitude valve, whose purpose is to close the intake line to a reservoir or elevated tank, the static pressure of the water storage will activate the valve to close as the reservoir reaches its full elevation. A solenoid switch activated by the valve controls can signal the monitoring panel of a pump station or filter plant that the reservoir is full.

A simple mechanical float control can also be used to activate a valve to close an intake line or to open it as the water level rises and falls. This device is sensitive to ice, however—more so than a pressure control valve in an underground piping vault (Fig. 7-13)—and must be protected from freezing. An electric heating element attached to the float can afford protection by keeping a hole open in the surface ice in which the float can operate. In addition, the space above the water may be insulated and heated to prevent the formation of ice. Slush ice in lines and valves can be quite damaging and costly, and it can collect and stop the flow entirely.

For pressure reduction in smaller pipes, a simpler diaphragm-type pressure control valve can be employed. The diaphragm is usually adjusted by means of a fine-threaded bolt and calibrated springs. Some models have chamber bypass lines and controls adjusted and monitored by needle valves and pressure gauges. They produce the same result in small lines that the piston-type valve does in large mains. Pressure on the diaphragm exceeding the set limit causes a throttling of flow and a drop in pressure downstream to the set limit. These valves are usually designed for only one direction of flow but, like the larger valves, may be used in reversed pairs or in batteries to serve local needs.

In designing a pressure reduction point in the distribution system, a number of questions arise. Important ones are: will the pressure reducing valve (PRV) act in concert with others? should the others be located at the same point as the PRV or at a distance on the same line? This may occur when similar feeds are located a few blocks apart into a common pressure zone. Two unbalanced installations can affect each other, causing "hunting" or rapid alternate opening and closing of the piston valves which in turn creates water hammer and line noise. For this reason, one must exercise great care in the sizing and setting of differential pressures. The operating range of the PRV point will have to include the lowest as well as the highest fire flows. As the average PRV will not cover this range, a compound installation is usually made with a large PRV and a small one in parallel on a bypass main line. The small PRV is set to operate at a few pounds pressure higher (downstream) than the large one. Thus, the small one will operate continuously until the downstream pressure drops below the set limit, when the large one will open to supply the increased demand.

If PRVs feeding the same zone are installed in combination, the total capacity of the PRVs operating together will supply the high flow demands to the zone and to adjacent zones if so designed. The total effect of adjacent zones on fire flows must also be calculated.

The basic criteria for sizing a piston-type PRV are as follows:

1. Flows: maximum, minimum, and fire flow, in gallons per minute (gpm)
2. Pressures: inlet, outlet, and differential, in feet of head
3. Velocity: in feet per second
4. Friction loss: in the PRV, pipe, fittings, and gate valves, in feet
5. PRV size estimate: use maximum allow-

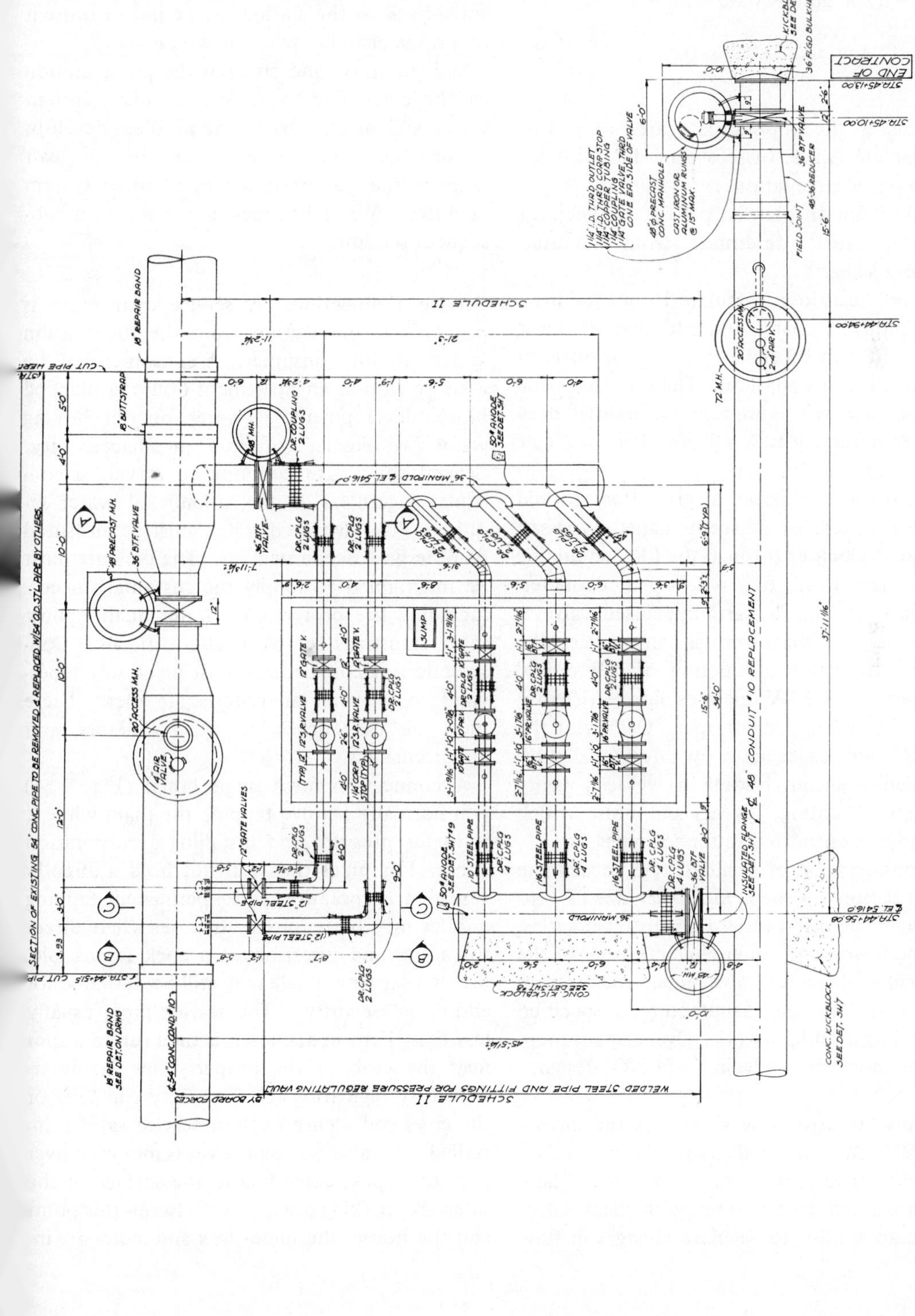

Fig. 7-13 PRV vault piping plan. *(Courtesy of Denver Board of Water Commissioners.)*

able velocity through PRV, 20 ft/sec (given–1200 gpm):

$$Q = AV; \quad A = Q/V;$$

$$\text{gpm} \times 0.002228 = \text{ft}^3/\text{sec}$$

$$Q = 1200 \text{ gpm} \times 0.002228 = 2.6736 \text{ ft}^3/\text{sec}$$

$$A = 2.6736/20 = 0.1337 \text{ ft}^2$$

This is the approximate area of a 5-in. diameter pipe. Use a 6-in. PRV for the first trial calculation.

6. Minimum differential pressure to keep the PRV open: determine from manufacturer's charts.
7. Total head loss: calculate from PRV manufacturer's charts and add that of pipes, bends, valves, reducers, etc. throughout the PRV installation. These should total less than the minimum differential pressure necessary to keep the PRV open (see above).
8. Velocities: velocities at given flows should be checked for adjacent mains, bypasses, etc. Velocities through the PRV should be calculated at full wide-open diameter. Flow through a partially closed valve is similar to that through an orifice, the velocity being a function of differential pressure. A PRV operates more efficiently at 20 to 100 percent open. Oversizing the PRV will cause it to operate in a slightly open position, leading to velocity problems, "hunting," wear, and noise which can be transmitted to service pipes.
9. Consideration of smaller PRV: recalculate all factors for the next *smaller* size PRV to see if the differential pressures, velocities, line losses, and other characteristics of the smaller diameter valve, pipe, and fixtures are within the manufacturer's specified limits and might work. Oversizing is more common than undersizing in PRV design.

Pressure sensitive valves such as the piston-type PRV can be utilized as surge control valves by slight valve design modification. These valves are often used in series with check valves at a pump station to regulate changes in flow rate, which can be brought on by back surges, hydraulic transient waves, and similar causes. The valve controls can be set to read the pressure wave preceeding a surge. Slow-closing check valves together with the surge control valve can then close the line and bypass the surge back to the station intake line or route it to a surge chamber where it is dampened.

Sudden starts and stops at the pump station can be controlled to avoid hydraulic transient waves and surges. When the problem develops outside the pump station, it can often be dealt with at the pump station or at other system facilities. Water hammer is covered in a subsequent section.

Service Connections. A service connection is primarily a connection from the distribution system to the consumer. A consumer may be a single house, an apartment house, a planned block development, a water district buying water "wholesale," or even an adjacent city. Several things are common to serving a consumer regardless of the amount of water required. The delivered water must be measured or otherwise accounted for. The pressure must be maintained to supply the customer's needs, including fire protection. Water quality must meet standards set by health officials. Construction of connections must be closely supervised to see that all criteria are met. There must be easy access to valves, meters, and appurtenances.

A connection for a single house (Fig. 7-14) will normally involve tapping the main while it is under pressure and installing a corporation cock. For pipes that will not hold a directly threaded corporation cock, permanent tapping saddles may be required. In cases where direct threading and insertion of a cock is possible, use of a tapping saddle may still be desirable for additional security. The service pipe, usually less than 1 in. in diameter, is then run to a spot near the curb or the property line (as determined by regulations of the utility company or the city) and a curb cock or service valve is installed. A valve box and cover is mounted over the curb cock, extending to the surface of the sidewalk, parking area, etc. Between this point and the house, the meter box and meter are in-

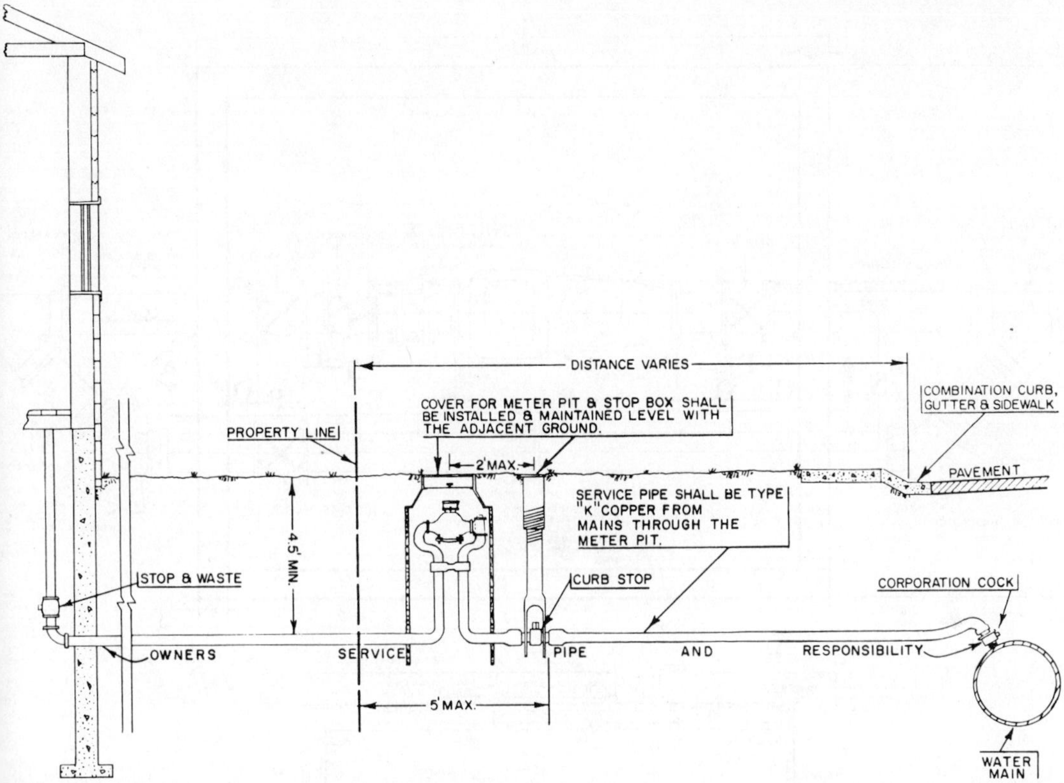

Fig. 7-14 Typical service line unit meter for a single residence. (*Courtesy of Denver Board of Water Commissioners.*)

stalled. The depth and design of these are dictated by the climate, local standards, and anticipated maintenance problems. If desired, meters can be placed inside the house and provided with outside connections for reading. If it is necessary to install a pressure reducing valve for the service line, it can also be situated in the basement of the house. The responsibility of the water utility often ends at the main or at the curb cock. Meter and pressure reducing valve maintenance are then carried out at the consumer's expense. Copper lines are commonly used between the main and the curb cock. The consumer has the option of using whatever equipment the building code will permit on his property.

Consumers of large amounts of water such as apartment houses or planned block developments require proportionately larger connections of a similar type. In fact, fire protection demands may necessitate the installation of even larger diameter pipes and appurtenances than the population differential would otherwise require. For example, a high-rise apartment with fire hose outlets on each floor requires piping equal to the possible fire demand on several floors, depending upon the local fire code, building water storage capacity, and related considerations.

Planned block developments, which may include several buildings and span a large area, are becoming more prevalent for residential housing and commercial buildings. A private intrablock street system with fire hydrants will need service lines which are, in effect, main extensions. These service connections demand compound meters (see Fig. 7-15) which allow a fire flow or a bypass valve in the meter vault that will admit the water called for by a fire. Ready access to the vault or other location of an automatic valve control is of prime importance. Elevated water storage can be used to respond to unusual water demands from these private developments.

Serving an entire water district or neighboring city may require several points of supply,

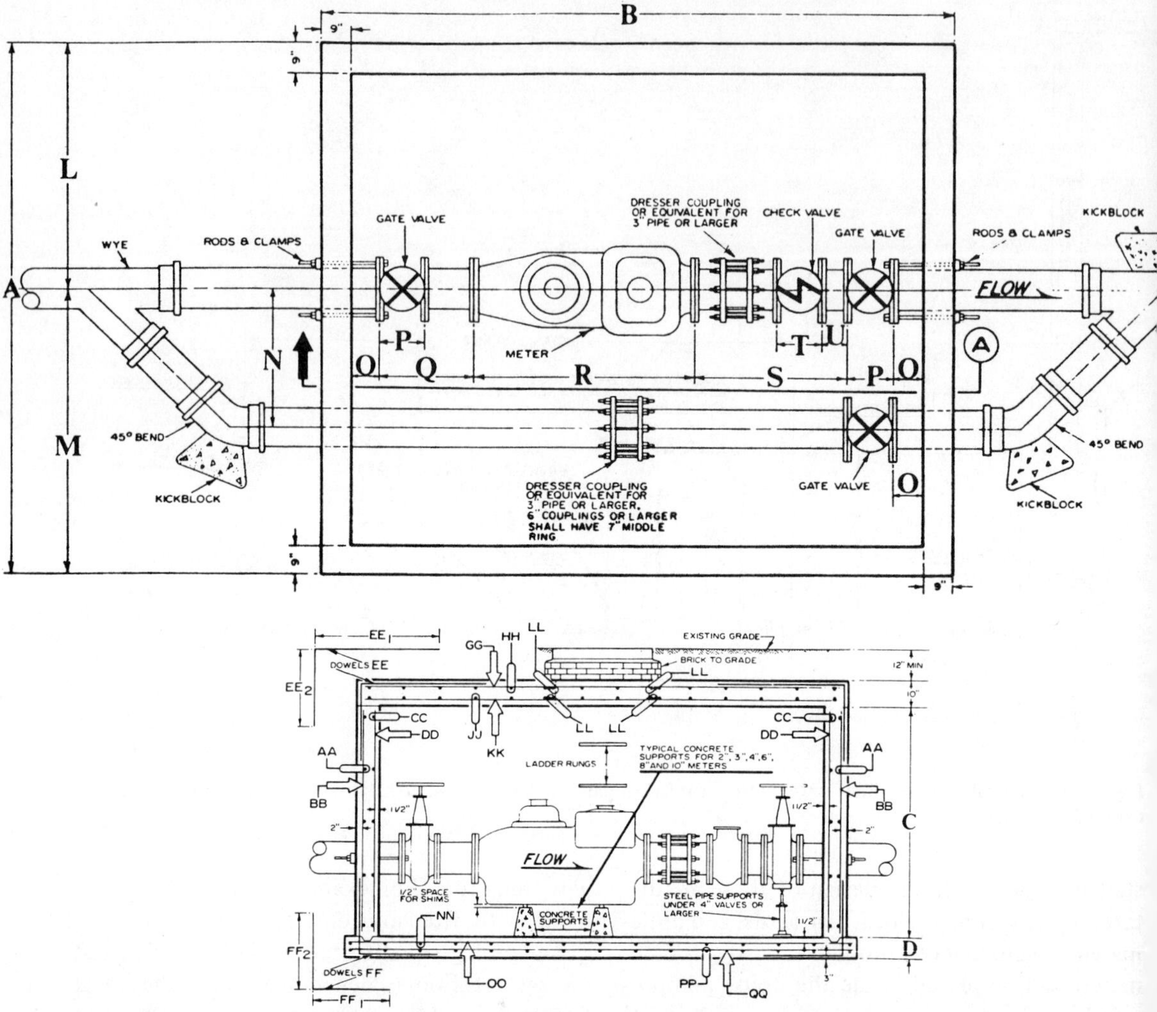

Fig. 7-15 Compound meter vault (above). (*Courtesy of Denver Board of Water Commissioners*.) Typical piping plan section through meter vault (below).

metering, pressure control, and, perhaps, water storage. A metering point in these instances is usually a "point of no return." If the flow is to be reversed and returned to the originating system, equipment costs and measurement problems are compounded. If the receiving consumer has other sources of water supply, a return flow may be complicated by legal and public health problems. In such a case, the service connections would in all likelihood include double check valves on all lines.

Because of the turbulence created by pressure reducing valves, a metering vault should be situated upstream from them. If the upstream line pressure exceeds 150 psi, however, the meters should be located far downstream. The costs for meters to handle pressures greater than 150 psi are much higher since they are classified as special equipment.

Automatic Operational Control. In order to assure efficient operation of the water distribution system day and night in all weather without expending great time and labor to open and close valves, check water levels and pressures, start and stop pumps, and the like, automatic controls are utilized. These controls may vary from a simple switch to start a pump by remote control to a computer data collection and logging system which can keep the pressures balanced between several zones by anticipating needs and activating inputs to the distribution system where needed.

Choice of equipment for an automatic control system depends, of course, upon the size and the complexity of the distribution system. Maintaining pressures and volumes of water at various points in the distribution system may require inputs from several sources and these inputs will need monitoring. Automatic controls measure and record positions of valves, rates of flow, and differential pressures. To assure efficient performance of these tasks at the least cost is the major problem.

Distance is the first and usually the most costly consideration in setting up automatic controls. Stringing wires to a remote point or burying them with a conduit and operating them may be simple or may constitute a major undertaking. Radio controls may cover the distance easily but because they are subject to interference, power outages, and other hazards, they may require reinforcement by auxiliary equipment to guarantee performance. Repeating and checking codes may have to be sent on the assigned band to verify information as it is transmitted. Receiving equipment will have to translate the codes and activate the necessary inputs to the distribution system controls. False information must be detected and discarded. The services of skilled technicians may be required, thus adding to the cost of a proposed signal system.

Utilizing telephone lines is a common solution to handle signals from remote points to a control center. Here, as in all of the other options, an engineering and economic study must be conducted to see if the costs are warranted.

The general methods of transmitting signals, each with its own technical and economic assets, are the *private wire*, the *leased wire* (provided by the telephone company or other franchised agency), and *radio systems*, of which there are many. Radio equipment with frequency modulation (FM), amplitude modulation (AM), or microwave equipment are among the transmitting and receiving systems with a capacity to handle signals from remote points.

A combination of methods can be used to relay signals. For example, a signal can be sent via private wire or radio to a substation where telephone lines are available to carry it through the congested urban area to a distribution system control center. Local conditions such as costs, service, and interference dictate which signal system to install.

Remote controls are used most often to start and stop pumps. To avoid surges and water hammer, an "on or off" status, modified by time delays, is applied. Pressure-sustaining and reservoir level controls, however, might be classed first as monitoring functions, with optional actions available to activate or depress inputs to the distribution system. Automatic opening of valves is usually limited by control overrides for specific conditions expected in the system. For example, during certain hours each day it may not be good policy to pump to storage. The level control or altitude valve for an elevated tank might be overridden during those hours to prevent the pump from drawing water from the transmission main.

Under changing conditions of demand, an operator at a distribution control center may be preferable to an entirely automatic system. If the number of controls to be monitored and recorded throughout the day becomes too burdensome for him to make efficient judgments, the installation of totalizing equipment and computers may be necessary.

The impulse initiated by a transmitter and translated by a receiver acts to move a stylus in recorders. Concurrently, this impulse can turn on a pump or light switch, set off an alarm, or activate the pilot valve of a pressure regulating valve. The transmitter may be activitated by the movement of a pressure-sensitive diaphragm and connecting linkage, the movement of an external arm on a pressure regulating valve indicating how much the valve is open, the movement of a hydraulic accumulator, the change in an electrical rheostat, or the position of a mercury column coded by intermittent impulses.

In the event of a line surge, the overriding controls must themselves be overridden. For example, when the line to a reservoir is open, surges are absorbed by the inertia of the storage water. When a level control valve is closed, however, the potential for damaging surges is at its highest. Controls then must sense a surge and reopen the valve. Surge detectors must act in seconds to be effective, and thus cannot await a decision by the operator. Alternatively, valves can be opened to route surges to surge chambers or back to the suction side of a pump station.

All system inputs, whether by automatic control or an operator, are continually scanned at short time intervals of 15 min or so. Any operational deficiencies can then be interpreted from the records and charts. Unusual recordings can be verified by checking nearby points with similar recordings at the time indicated. It is important to establish key points at which to record the total effect distribution zones have on one another, the coordination of inputs, and optinum equipment utilization. Recording is covered further in the following section.

ENGINEERING CONSIDERATIONS

Service Criteria. Pressure ranges within a distribution system are governed by the needs of the zone being served and the safety precautions deemed necessary to avoid overstressing distribution pipes, service lines, or building water fixtures. Excessive pressure will cause wastage by fixtures taking more water than is necessary to operate them. It will also start or increase leakage in mains.

With the increase in use of appliances such as automatic washing machines, dishwashers, and garbage disposals, and of such modern conveniences as multiple bathrooms and lawn sprinkling systems, the demand for water is on the rise in residential zones. Air-conditioning systems and other water-using equipment have similarly increased water demands in other zones. Water pressures have risen concurrently, especially in older areas where undersized mains and high head losses from corrosion are most frequent. These systems require the highest pressure to deliver the volume demanded.

The average water main is designed to handle stresses of 150 psi, with a reasonable safety factor to withstand surge pressure, water hammer, and other hydraulic transient forces. The safety factors vary with each pipe material, but elasticity plays an important part in flexible conduit, while strength and inertial properties are paramount in rigid conduit. There is usually a selection available locally to answer each conduit design problem. Experience has shown, however, that a junction between rigid and flexible conduit can be vulnerable to unusual stresses. If one "gives" elastically and the other one does not, cracks and leaks can result, usually in the rigid conduit.

Transmission mains can carry up to 250 psi as a routine pressure. Standard pipe and fittings are available in most materials. Higher pressures can be accommodated by designing the conduit to the required specifications, and for this purpose steel and prestressed concrete are commonly used. In general, however, the need for pressures higher than 150 psi in a distribution system is limited to the transmission lines supplying pressure zones at higher elevations.

Working pressure for residential zones normally lies between 40 and 60 psi. Other zone pressures can be regulated up or down to satisfy the needs of the consumers. Downtown buildings, high-rise apartments, and similar structures may be furnished water from a private pressure system installed to serve the building. A pressure head of 100 ft is equal to 43 psi in the main. High-rise buildings generally average 10 ft/floor. Plumbing head losses should be added in to calculate the total pressure needed to serve the top floor. Few plumbing fixtures will work satisfactorily with less than 20 psi pressure. Automatic home appliances should be supplied with a minimum pressure of 40 psi for best operation. Thus, an automatic washer operating on the tenth floor of an apartment building would need about 90 psi. For normal purposes, such pressure is undesirably high. Private pressure-boosting systems are usually more efficient in supplying upper floors.

Designing water service to fight fires demands knowledge of each neighborhood, each pressure zone, the building construction materials allowed or proposed, the type of business in the service area, and the type of stored materials that would determine fire potential. The

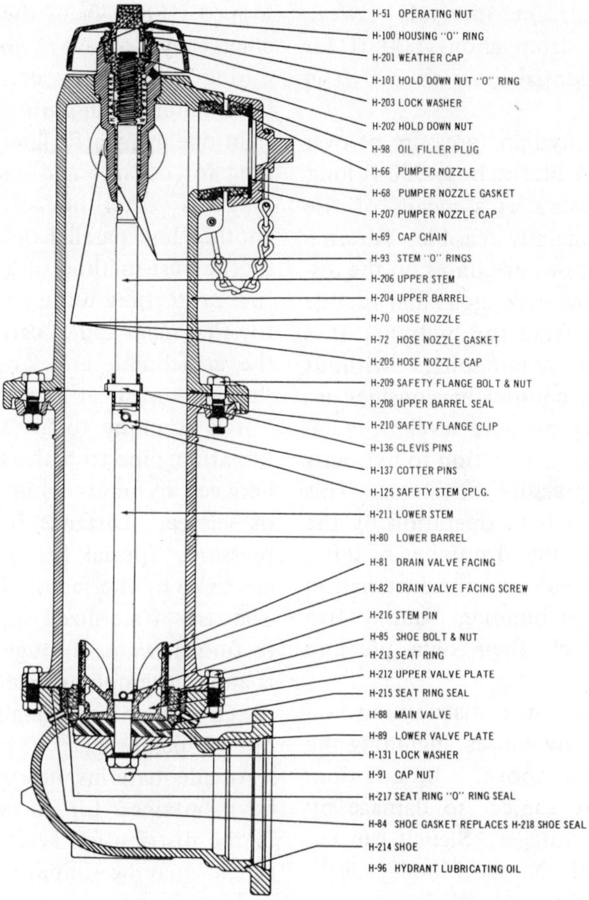

Fig. 7-16 Fire hydrants (*Courtesy of Mueller Co.* ®)

American Insurance Association* has established standards to rate communities by fire preparedness in relation to insurance risk. Their latest publications should be reviewed in designing a new system or upgrading an old one.

Minimum fire flows in residential zones start at 500 gpm, leaving a reserve flow and a residual pressure of 20 psi for other neighborhood requirements during the fire. The fire flow will average about 50 percent higher than the average daily flow of the main. A minimum main size of 8 in. diameter is commonly used to handle residential fire flow specifications. Looping the mains (Fig. 7-16) to leave no dead ends will provide twice the water to a hydrant, allowing a drop in main size to 6-in. diameter for sparsely built-up residential areas. On streets in densely populated areas and for long runs of main without cross connections, a pipe with a minimum diameter of 12 in. should be utilized.

As the density of housing units in a residential area becomes greater, the fire flow demands increase sharply. For mains supplying closely-spaced single houses or small apartment houses a flow of 1000 gpm is recommended. For buildings such as large apartments and hotels, 3000 gpm is suitable. Three-story buildings in densely built-up areas demand 6000 gpm.

Spacing of fire hydrants should allow up to 70 percent of the total fire flow to be concentrated on any one building. The distance from the hydrant to the fire is, of course, important. For each 100 ft of hose, an increase in pressure of about 20 percent will be necessary to

*Successor to the National Board of Fire Underwriters.

provide the same fire stream. Intervals between hydrants should vary from about 600 ft for sparsely built-up residential areas to 200 ft in the densest areas.

As it would take a hydrant pressure of over 90 psi to operate a $2\frac{1}{2}$-in. fire hose 300 ft long, maintaining high pressure as a means of fire preparedness is not generally feasible. Alternative methods include booster pumps on the distribution system in high-risk areas and mobile pumps drawing water from the hydrant. It is recommended** that communities without extensive fire-fighting equipment consider increasing their working pressure level from 60 to 75 psi to allow direct connection to hydrants without interposing pressure boosters. This solution allows quicker hose operation by the fire department for a limited number of lines. The greater pressure also permits installation of sprinkler systems in buildings four to five stories high which lack their own pressure systems.

Interruption of service in a distribution system can be due to many causes, including the pipe failures discussed above. In addition, pump installations are subject to damage by line surges and power outages. Signals can fail to function at critical times. Floods, earthquakes, wind, and other natural forces, plus human activities such as construction, explosion, and fire can all result in the interruption of service in varying degrees. To minimize interruption, system components most likely to be impaired should be reinforced and system design and emergency preparedness procedures should be geared to continue operations under unusual conditions.

In the layout of a pipe network, looping and placement of valves to allow the bypassing of any section out of operation is, of course, elementary design. In the event no parallel mains exist—as in the case of a boulevard business route extending out from the city with no depth of subdivision laterally or a main serving an outlying development—special measures can be undertaken to ensure service. One solution is the installation of a storage tank, elevated or at ground level with pumps, to be removed when parallel lines are installed and looped. Flow from the tank can serve during emergency outages and maintain pressure during normal service. Installation of dual lines behind each curb of the boulevard, rather than one in a traffic lane, is a good way of planning for outages and extensions. If the planning is after-the-fact, however, installing another line parallel behind a curb might still be the best method of guaranteeing service and fire protection during interruption of flow in the first main and of carrying increased flows as the area builds up. Loops between the main should be made at intervals.

It is common to use fire hoses or aluminum irrigation pipe to make temporary connections between hydrants while bypassing a main out of service. Distance between hydrants, flow pressures, special fittings, and street crossings are some of the factors in emergency planning. The use of sterilized water trucks with pumps to furnish water through a hydrant connection is another temporary method of maintaining reduced service to a small area while a main is out of operation.

An adequate inventory of repair parts should be maintained for all types and sizes of pipe in the distribution system. Planning for emergencies may be simplified by standardizing new and repair pipe and fittings in the system so far as past practices will allow.

The installation of replacement pumps and valves in an emergency is often quite difficult because of earlier selections of pipe and fittings. Sizes that are odd as compared to present stock, especially in outside diameters, are often found in old cast iron pipe and "tailor-made" connections. They should be identified, and their repair during emergencies should be anticipated.

Avoiding damage to mains by street excavation machinery is an important preventative measure. Accurate mapping of mains and a permit system can forestall inadvertent puncture by a backhoe "feeling its way" or joint separation caused by the digging of a trench behind a thrust block, removing earth support.

Signals, electric valve operators, pumps, and many other system components are dependent upon a steady power source. Loss of the source should be foreseen as a probable interruption of service and steps should be taken to provide

**By the American Water Works Association.

standby service. This service might be in the form of diesel engine generators, ready to start in a few seconds, other separate electrical supplies for signals, or manual controls to open valves to elevated storage tanks. Each electrical control normally operates in coordination with others in the distribution system. Therefore, in design of standby service, individual and total effects must be studied.

Analysis of the System–Evaluation. To analyze the amount and location of existing and future peak water demands, existing flows must be measured and the future use in *each zone* must be calculated. Pressure readings, taken at available taps and hydrants during peak hours, will help determine the existing demand, system deficiencies, etc. Local master plans and zoning regulations for areas to be developed will provide an estimate of proposed construction and population growth upon which to base future water facility construction requirements.

In these determinations, loss factors must be added. In transporting water to any zone there will be a system loss, or water which cannot be accounted for. This loss is demonstrated by the difference in quantities of water leaving the filter plant or pump station and the water metered at consumers' sites of use. Losses can be attributed to unmeasured flows used for fire fighting, filling of tank trucks at hydrants, blowoffs to clean mains, other maintenance and construction uses, and leaks.

Leaks can lose a very high percentage of the water in a system, expecially under conditions of high pressure. Losses exceeding 30 percent have been calculated by cities with extensive old systems. It may thus be economical to set up zone or block metering to detect unusual flows if incomplete metering in the distribution system prohibits a comprehensive check for losses. Areas where old and undersized mains handle the distribution flow under higher than normal pressures should be given particular attention. Sewer infiltration and wet basements are indicative of water line losses.

Past daily demand records plotted into selected daily, monthly, and annual demand curves for zones and for the system as a whole will provide a data base for many studies. Growth trends can be interpreted and projected at a determined rate into the future. Cycles of high demand years can be detected. Unit increases in the amount of consumption per capita per year can be formulated.

Planning for system balance with increased storage and/or increased pumping capacity against the expected demand will then be possible. Optimum results can be foreseen by studying the use of the existing system with added storage, added filtration capacity, enlarged pumping input, and other expanded facilities against the expected incremental growth in various parts of the system. The combination which will produce the maximum water input at the least cost can then be incorporated into preliminary plans.

Flow measurements and C-factor tests on the transmission and distribution mains most affected by future plans should be recorded for use as design data. If it is not possible to determine the C-factor through Pitot tubes and gauges at stations along the pipe or by visual inspection, it will be necessary to use a Hazen and Williams "C" value from hydraulic tables to compute flows and losses. Information on the age of the pipe, material, and lining can be correlated with known local pipe corrosion characteristics to produce a reasonable estimate of the "C" value. If the present velocity of the flow in the pipe under study is approaching 7 ft/sec and its rated flow is being consumed during peak hours, installation of additional conduit should be planned for future expansion.

Design conditions should respond to all information on existing pipe in an area of growth, including its limitations and ultimate capacity. Estimated conditions include the expected increase in water service to specified points, sizing of facilities to handle the maximum day without overtaxing any component of the system, and producing enough water from all facilities, including storage, to handle the maximum hour on the maximum day. Tank filling is normally done daily in low demand hours and does not require added capacity or larger mains.

A fire flow of 10 hr plus 24-hr storage, available to any point in the system covered by the local fire code, is an often-used design criterion to which the known or estimated water losses of the system are added as a percentage. The expected demand for a zone or for the entire system can then be totalled.

Provisions for expansions or extensions of the system, including material and installation costs of pipe, pump stations, storage reservoirs, controls, special valves, etc., plus land and right-of-way costs, are next translated into a capital improvement program. Financing costs, costs of a bond issue and election, or other costs incurred in the administration of a capital improvement program to plan, design, and construct water distribution facilities may also have to be considered.

The use of the Hazen and Williams "C" factor, or the molecular drag created by the inside surface of the pipe within the cross section of the flow, might be discussed a bit more at this point. Over the years, or centuries, formulas have been produced by noted hydraulic researchers and engineers which are now in wide use. Most of them have been developed through the use of experimental apparatus and the results have been formulated into mathematical language. The application of more than one of these formulas to the solution of a single problem may produce slightly different results. Some are more suited to one type of problem than another. Open channel flow or gravity flow in conduits or closed pipe is perhaps more easily calculated by use of Manning or Scobey equations. Pressure piping problems are generally better handled by Darcy-Weisbach and Hazen-Williams formulas and tables. These six founders of hydraulic science are only a few of a great number. They are, however, among those whose work is most prominent in text and tables for engineering design.

The amount of water that can be forced through a pipe will depend upon the pressure or head in vertical feet (translated to pounds per square inch), the diameter and length of the pipe, the roughness of the inside of the pipe, and the number of constrictions or enlargements such as bends, reducers, expanders, valves, and so forth in the length of pipe.

The velocity head ($h_v = v^2/2g$) is the amount of pressure necessary to initiate and sustain the velocity. This is measured in feet lost, as are all constrictions and enlargements which cause energy losses. Tables showing pipe fittings with their rated losses in fractions of $v^2/2g$ are readily available for the designer from equipment manufacturers and other sources. The constant $2g$ is equal to 64.4 (ft/sec^2), or twice the acceleration of gravity.

The quantity of water flowing in a length of untapped pipe is constant after it reaches equilibrium ($Q = AV$). The area and the velocity will change proportionally to each other to accommodate the quantity. That is, if the pipe area decreases as the flow goes through a constriction such as a valve or a venturi meter, the velocity will increase at that point. If the constriction is too great or a turbulent condition is started, the Q may drop to a new equilibrium throughout the length of pipe. The velocity will likewise drop throughout the rest of the length, but will remain proportionately high at the area of constriction. An example of this is the increase in velocity at a pressure reducing valve, where an average velocity of about 4 ft/sec will increase to 20 ft/sec through the valve as the valve closes the main down to a lesser flow or gradually shuts it off.

Before the pipe flow reaches equilibrium, the inside surface of the pipe will exert an influence which will slow down the cross-sectional velocity. This velocity is greatest at the center of the pipe and near zero at the pipe surface. However, the surface roughness influence is across the entire cross section. This influence, the head loss from it, and the resultant velocity and quantity of water that will be transported have long been the subject of hydraulic research and mathematical analysis.

Selected formulas and roughness coefficients applicable to water pipe are listed below. Further information on this subject can be found in texts on hydraulic analysis.

Pipe Formulas For Water Pipe

Nomenclature

V = velocity in feet per second
Q = quantity of water in cubic feet per second
H_f = loss of head in feet per 1000 feet of pipe
h_f = loss of head in length L of pipe

L = length of pipe in feet
n = coefficient of roughness for Manning formula
C = coefficient for Hazen-Williams formula
C_s = coefficient for Scobey concrete pipe formula
f = friction factor in Darcy-Weisbach formula
D = diameter of pipe in feet
S = hydraulic slope = h_f/L
R = hydraulic radius of pipe = $D/4$ (= A/P for other conduit)
g = acceleration of gravity = 32.2 ft/sec^2
A = cross-sectional area in square feet
P = wetted perimeter in feet

Manning formula

$$V = \left(\frac{1.486}{n}\right)(R^{2/3} S^{1/2})$$

$$= \left(\frac{0.590}{n}\right)(D^{2/3} S^{1/2}) \quad (6)$$

$$Q = \left(\frac{0.463}{n}\right)(D^{8/3} S^{1/2})$$

$$= \left(\frac{0.0147}{n}\right)(D^{8/3} H_f^{1/2}) \quad (7)$$

$$H_f = \left(\frac{4660}{D^{16/3}}\right)(nQ)^2 \quad (8)$$

n values for pipe materials:

Asbestos-cement	0.010 to 0.012
Concrete and mortar lined	0.011 to 0.017
New Cast iron (coated)	0.011 to 0.013
Ductile iron (coated)	0.011 to 0.013
Welded steel (coated)	0.011 to 0.013
Galvanized iron	0.011 to 0.013
Plastics	0.009 to 0.011

Scobey formula for concrete pipe

$$V = 4.73\, C_s D^{5/8} H_f^{1/2} \quad (9)$$

$$Q = 3.71\, C_s D^{21/8} H_f^{1/2} \quad (10)$$

$$H_f = 0.0448 \frac{V^2}{C_s^{1/2} D^{5/4}}$$

$$= \frac{0.0725}{D^{21/4}} (Q/C_s)^2 \quad (11)$$

Values of C_s for concrete water pipe vary from 0.370 for smooth concrete or cement-mortar lining with no rough finished joints to 0.345 for average or aged concrete. For gunite or similar finishes, tunnel linings, etc., use 0.310.

Darcy–Weisbach formula

This formula involves the introduction of critical velocity and viscosity into flow studies. The friction factor f is dimensionless and refers to a velocity head loss per diameter length of pipe. That is, if $f = 0.004$, then 0.004 of a velocity head is lost for each diameter length. If the pipe was 1000 diameters long, h_f would equal 4 diameters. The basic formula is:

$$h_f = f \frac{L}{D} \frac{V^2}{2g} \qquad f = \frac{64}{N_R} \quad (12)$$

$$Q = 0.1994\,(D^{5/2} H_f^{1/2})/f^{1/2} \quad (13)$$

$$H_f = 25.16\, f Q^2/D^5 \quad (14)$$

Note: N_R is the Reynolds number which varies with pipe size, viscosity, and turbulence of flow. An N_R range of 2000 to 3000 for water continually moving in pipes is the limit most waterworks problems might fall into. Below 2000 is viscous flow, above 3000 begins turbulent flow. Reference to a more detailed text on hydraulic analysis is recommended for problems in turbulent flow.

Hazen–Williams formula

$$V = 1.318\, C R^{0.63} S^{0.54}$$

$$= 0.550\, C D^{2.63} S^{0.54} \quad (15)$$

$$Q = 0.432\, C D^{2.63} S^{0.54}$$

$$= 0.0104\, C D^{2.63} H_f^{0.54} \quad (16)$$

$$H_f = 4960/D^{4.87} (Q/C)^{1.85} \quad (17)$$

C values for pipe materials:

Asbestos-cement	120
Concrete and mortar lined	110
New cast iron (coated)	110
Ductile iron (coated)	110
Welded steel (coated)	110
Galvanized iron	110
Plastics	130
Old pipe, average wear	100
Rough to very rough	80–90

Analysis of the System—Techniques. The water distribution system is a series of pipes and tanks through which water flows from the sources to the users. The wide range in quantity of water used for fire fighting, industry, and domestic purposes causes a similarly wide range in the rates of flow through the pipes. For purposes of analysis, the system must be studied under various conditions ranging from static or no flow to maximum flow.

Static conditions are simple to analyze since the only variable is the elevation. As the elevation changes, so does the pressure head on the system. Accordingly, the low elevation boundary can be established based upon the maximum acceptable static pressure.

Dynamic conditions in which the flow of water through pipes induces a head loss is a more complex analysis problem. As described earlier, distribution systems may include numerous looped pipe systems, several sources, and a withdrawal at each home or business.

Obviously, the best analysis would be based upon an accurate simulation of the actual system. This is complicated by a large number of variables, including the physical characteristics of the pipes, ranges of flow, and location and number of users. Therefore, analysis techniques have been developed to reduce the number of variables and maintain the reliability of the simulation of the system.

The first step in simulation is to plot the physical layout of the system showing elevation, length, size, age, and material of the pipes; location of the sources and major withdrawals; location of any unusual features such as hydrants, tanks, and valves other than full-open; and the known physical conditions of the system such as rates of flow and pressures. Next is the tabulation of addresses and consumption for each customer. The computation of fire flows based upon the requirements of the American Fire Insurance Institute and the projection of the demands for other uses will complete the basic data for the analysis. However, it should also rely upon field observations of the relationships between pressures and flows.

Many techniques have been developed for the actual analysis including the graphical method, physical simulation of the water system via an analog method, and the iterative or digital method. Availability of computers to handle the repetitive calculations required of the latter method have made it the predominant procedure in use today. The graphical methods are so laborious and time consuming that they are seldom used. The analog simulation is limited both by the accuracy of the physical components and the number of components required to simulate a large system. The iterative method most commonly used is that developed by Hardy Cross. This method assumes flows and carries out repetitive calculations of head losses and correction of the assumed flows until a hydraulic balance is obtained.

Another helpful tool in any analysis is the equivalent pipe calculation which combines a number of pipes into one equivalent for further study. Simply put, this is the determination of the size, length, and roughness of one pipe that will be equivalent to two or more pipes at the same rate of flow. A nomograph, based upon the Hazen–Williams formula, is usually used for the calculation.

For example, the pipe equivalent to a 12-in. main, 2000 ft long and an 8-in. main, 1000 ft long connected in parallel is calculated as follows: assume a 3-ft head loss through each, and note that the summation of the head loss around any loop should be zero. Thus, the 12-in. main will carry 590 gpm with a 3-ft head loss and the 8-in. main, which is only one-half as long, will carry 300 gpm. The equivalent pipe could be assumed to be one that would carry 890 gpm with a 3-ft head loss, which, for a 2000 ft length, is a 14-in. pipe.

Another handy tool is the method of sections developed by Hazen. This method assumes that the flows through a network are predominantly in one direction. Next, the system is sectioned with lines drawn perpendicular to the flow at any point in which there is significant change in the total flow. Based upon the allowable head loss across a section, the flow of each pipe cut by the section can be computed. The total of these flows, when tested against the required flows, will indicate if the present system is adequate or if additional pipes are required.

The Hardy Cross method, like all others, is more easily understood if the system is schematically drawn, and summing of the loads on the system at pipe junctions is a common practice. The first step in this analysis is to assume the flows and direction of flow in each pipe and calculate the head loss in each. Assuming that clockwise flows are positive and counterclockwise flows negative, compute the total head loss in each loop. A flow correction for each loop is then computed by the formula Q = algebraic summation of the head loss around the loop divided by the constant, 1.85, times head loss divided by the flow. The flow in each pipe is adjusted by the above calculation for that loop, and pipes in more than one loop will be adjusted for each loop. The above procedure is repeated until all of the loops are in balance to the accuracy required.

The ease with which computers carry out the repetitive calculations necessary in the Hardy Cross analysis permits its use for the simplest or the most complex system. Nearly every computer manufacturer or service bureau has a Hardy Cross program available.

Figures 7-17A and 7-17B illustrate a typical Hardy Cross loop and a sample Hardy Cross computation.

The final steps in the analysis are a careful study of the results of the computations and a test for reason. The system is analyzed to find a historical condition for which field data is available against which to compare. The comparison should point out any inaccuracies in the assumed physical characteristics or simulation of the system. Thus, the field and simulated conditions should compare closely before any conditions are analyzed.

The results of the analysis are plotted on the original system layout and carefully studied. The sensitivity of various possible alternative changes should be tested as well as the logic of the reaction of the system to the various changes. Any method of analysis, either by hand or by computer, is only as good as the basic data, and the results should be tested and retested for accuracy.

Hydraulic Design and Layout. In graphs of distribution systems, the *pressures* at each pump station, elevated tank, and pressure regulating valve might be shown as rising bars with values at specific hours, with accompanying bars showing the *rate of flow* produced at the same time. The capacities of pumps and pipes can be compared to the actual performance or *measured flow* at a specific hour to determine the efficiency of input to the system at that point. Cumulative inputs at various parts of the system can be measured as a resultant at any point. This is done by reading the pressure at a hydrant or other checkpoint and determining the flow there, by measurement or calculation.

Unusually high pressures or velocities required to deliver the flow demanded may indicate undersized pipes, the need to revise pipe layout, or the need to change pressure regulation or type of pumps. It may be necessary to extend arterial transmission lines further into an area (intersecting existing mains), or to install new interconnecting pipe to develop new patterns of looped main flow. These situations can occur when a new subdivision or another developed service area is interposed between parts of an existing system with inadequate flow patterns and main sizes for the added area.

Storage reservoir input is limited to the amount of water that can be replaced during hours of low demand. Normally during this period, pumping for storage water replacement is not limited. However, pipes on the suction side of the pump station may also be supplying a growing service area. In this case, future demands on the supply pipe on critical days should be evaluated. Storage tanks located in isolated areas away from major transmission mains will be most affected by replacement problems during a run of hot summer days and nights.

Considering all of the limitations on the distribution system's storage water input together with available rates of pumping for primary and high-service pumps, one more stress should be added. A major fire should be superimposed in a critical area of the system for theoretical calculation. Once the figures are assembled to represent optimum needs for flows at critical points of the system, field checking at these points of stress during existing maximum flows should be undertaken to determine the percent of capacity available for future consideration.

In estimating the need for storage, a rule of

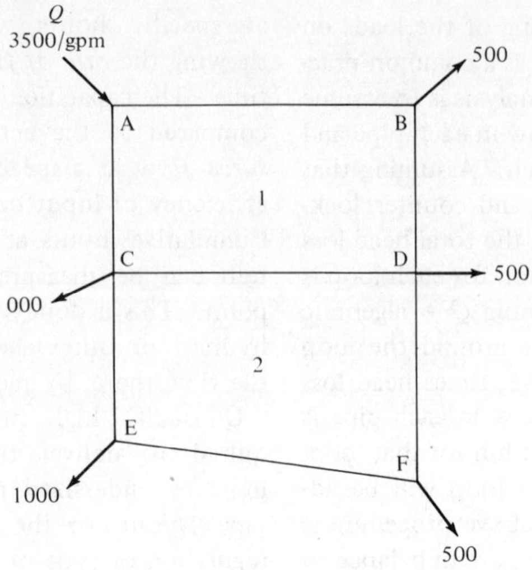

(A)

			Trial 1				Trial 2			
Line	Diameter, inches	Length, feet	Assumed Q, gpm	H_f, feet/1000	ΣH_f, feet/ 1000	H_f/Q	Assumed Q, gpm	H_f, feet/1000	ΣH_f, feet/ 1000	H_f/Q
AB	12	2000	2000	16.9	+33.8	0.0169	1600	10.0	+20.0	0.0125
BD	12	1500	1500	9.0	+13.5	0.0090	1100	5.0	+ 7.5	0.0068
DF	12	1500	800	2.8	+ 4.2	0.0053	400	0.8	+ 1.2	0.0030
FE	12	2000	300	0.5	+ 1.0	0.0033	100	0.09	− 0.18	0.0018
EC	12	1000	700	2.2	− 2.2	0.0031	1100	5.0	− 5.0	0.0045
CA	12	1500	1500	9.0	−13.5	0.0090	1900	14.2	−21.3	0.0112
					+36.8	0.0466			+2.5	0.0398

Trial 1 $\Delta Q = 36.8/1.85\ (0.0466) = 427$ gpm (use 400 difference in Trial 2, Assumed Q)
Trial 2 $\Delta Q = 2.5/1.85\ (0.0398) = 33$ gpm (OK for balance)

Line	Diameter, inches	Length, feet	Assumed Q, gpm	H_f, feet/1000	ΣH_f, feet/ 1000	H_f/Q	Assumed Q, gpm	H_f, feet/1000	ΣH_f, feet/ 1000	H_f/Q
AB	12	2000	2000	16.9	+33.8	0.0169	1600	10.0	+20.0	0.0125
BD	12	1500	1500	9.0	+13.5	0.0090	1100	5.0	+ 7.5	0.0068
DC	8	2000	200	1.5	+ 3.0	0.0150	200	1.5	− 3.0	0.0150
CA	12	1500	1500	9.0	−13.5	0.0090	1900	14.2	−21.3	0.0112
					+36.8	0.0499			+3.2	0.0455

Trial 1 $\Delta Q = 36.8/1.85\ (0.0499) = 399$ gpm (use 400 difference in Trial 2, Assumed Q)
Trial 2 $\Delta Q = 3.2/1.85\ (0.0455) = 39$ gpm (OK for balance) (B)

Fig. 7-17 Typical Hardy Cross loop (A) and sample computation (B).

thumb applied is 1 day's supply plus a 12-hr fire flow. Since system designs are discrete, tailor-made to local needs, and vary widely in type and capacities of pumps, mains, and other facilities, storage needs may exceed the rule, or may not be considered important. So long as the distribution system can deliver enough water to supply all of the demand at peak hours and maintain an adequate reserve capacity for emergencies, the needs of the system are met.

Estimating the contingencies which might handicap the system is the job of the designer. Power failure, faulty equipment, conduit damage from natural disasters, and other factors all affect reserve capacity. The conservation of electric energy during hours of great demand is another consideration. In any case, adequate storage at critical points is an important asset and should be given due attention in any distribution layout.

Pumps in the distribution system are usually limited to on-line centrifugal type equipment and submerged well pumps where groundwater is the source of supply. As discussed previously, several other types of pumps may be used for specific purposes in preparing the water for distribution, but in the system itself these two are predominant. A battery of well pumps located in a well field may both supply the initial motive force and maintain pressure for water delivery to consumers.

If pressure losses, friction losses due to great distances, overstressing of individual well capacities, or similar problems of imbalance occur, secondary pumping facilities are usually installed. These facilities are similar in design to pump stations lifting treated water from the clear well of filter plants. The suction side pressures may vary markedly, however. Well water which is often chlorinated at the well head is run to a large manifold pipe which acts as the suction side supply. In either case, the pump station must be designed to deliver treated water into a transmission main with the least amount of shock or reaction when the water is needed.

Great ingenuity in design of automatic pump controls is sometimes required to accomplish this in response to an incremental drop in pressure. For example, several large pumps with fixed speeds may be quite adequate to handle up to the maximum flow, but as each comes on line it may throw more force into the main than is required to replace the incremental drop and can cause a surge. Therefore, several sizes of pumps may be used, singly, in pairs, in alternating pairs, in trios, etc., to bring the pressure and flow to various levels without exceeding the demand.

The use of multiple-speed and variable-speed motors to solve critical surge problems is well known. These motors are often much larger than necessary for an average flow and may, therefore, be more costly to install and operate. For specific problems in critical distribution zones, however, they are excellent. Fixed-speed motors and pumps of several sizes are more commonly installed for distribution system pumping. Surge relief valves are installed concurrently to handle expected line surges. High service pumps and reservoir filling pumps are very often dampened by pressure reducing valves or altitude valves. These pumps may be installed at isolated locations where they are less complicated than an on-line pump station, or they may be installed as part of the pump station to add their input to a general transmission line. In other words, storage pumping and high service capacity needs are sometimes handled by the general pump station capacity.

The elements of a distribution system layout which were covered in earlier sections may be summarized as follows. The transmission mains are designed to deliver water to each distribution zone of the system. These mains may have to carry higher pressures than the distribution mains they supply. Pressure reducing valves may be required to reduce the pressure between a transmission main and a distribution zone. These valves are also used to maintain pressure differentials between zones, often where they are at considerably different elevations. When the differential pressure exceeds the design setting between the two zones, water can be released from the higher pressure to the lower. Transmission mains usually supply storage reservoirs and tanks, but a distribution main may serve this purpose in isolated areas. Altitude valves or a water level control will shut off the flow when a storage structure is full. These structures are needed when the transmission main is inadequate for peak flows. Long runs of a main should be interconnected with others to form loops, if possible. If not possible, the diameter of the pipe should be increased to insure adequate peak flows.

In laying out distribution mains, it is necessary to include valving which will make it possible to isolate any section for repair work or other emergencies. Arrangements should be planned to bypass a closed section or deliver water from another area. Looping will do away with a number of problems and will deliver twice the volume of water for fire flows or other heavy drafts.

The vulnerability of river crossings, pipe hung

on bridges, shallow buried pipe, and pipe subject to vibrations from traffic or construction should be recognized and emergency planning done before damage occurs. Weather in the form of extreme heat or, more often, freezing and heaving of soil is the prime cause of damage. The placement of line valves to permit alternate routing of water to bypass potential outage points is a measure of good design.

A negative pressure may be created when a main is drained through a blow-off valve. This pressure will increase as the slope of the main above the blow-off valve increases, or the drop in head increases as the main empties. The use of vacuum relief valves will avoid the intrusion of groundwater into an emptying main in this situation.

Air valves are placed at high points in mains where air can collect and saturate the water. It is good practice to install outlets at high points in mains to allow future installation of an air valve if there is a question of the need for one at the time of design. These outlets are relatively inexpensive and will also provide auxiliary openings for measuring pressure and flow.

In sizing mains, the ultimate use and the maximum flows should be the criteria, of course. However, since these figures may not be fully determined at the time of design, a safety factor is often employed. One can be found in the velocities required to deliver design flows for given quantities. The range of velocities are generally held between 2 and 7 ft/sec. Valves may cause brief jumps which can generally be ignored in hydraulic calculations if they are not excessive. The safety factor of using the slower velocities–i.e., the larger diameter pipe–should be investigated within the limitations of standard pipe sizes.

In order to size and select pumps and motors for a pumping station, certain factors must be kept in mind. No motor should run continuously day and night; 50 percent of the time will be sufficient for most equipment. The efficiency of pump motors over a period of time will average around 70 percent. However, 50 percent operation at 70 percent efficiency does not necessarily mean that 3 times as many pumps as the capacity of the pump station will be required. Since peak flows occur only once each day, with secondary peaks at separate times, an economical pumping schedule which alternates the pumps can be worked out. Adequate coverage should be available from other pumps in the station to compensate for a unit undergoing repair; this deficiency may also be made up by increasing the output of other pump stations. Extra capacity designed into the system for emergencies and storage water may also be put to more use temporarily.

Diesel-engined pumps, installed as part of a disaster operations plan, may be made available for use if their piping and valving are designed to act in concert with the system, rather than as a separate alternate. Diesel-driven standby generators are of no use, however, if the pump itself is out of service. Selection of standby equipment for civil disaster preparedness might well include the criterion of standby aid for operations emergencies which occur much more frequently than disasters.

The subject of hydraulic transients has been intensively studied and reported. Surges and water hammer are two common occurrences in any distribution system. They are not the same thing. A surge is a comparatively slow-acting rise or fall in water pressure. This can usually be sensed by controls and remedied by surge relief valves or "surge tanks" which may be any vessel that can receive the volume of water propelled by the surge. A reservoir or elevated tank riding on the system and not valved off from the main can act as a surge structure. In areas which generate or receive constant surges, a special tank may be installed to abate surge force. These tanks are partially filled with water and are airtight. The air may be compressed to provide a cushion of design "hardness" to dampen the surge. The elasticity of the air counterbalances the force of water.

One problem with this type of tank is the solubility of air in water. Air in solution can be carried into the main. Repressurizing by an air compressor is usually necessary. The insertion of a rubberlike sack in the tank to hold air has been successful in smaller tanks. To calculate volumes for surge tank design, the rise of water level in a tank can be determined from the length of pipe involved in the surge, the flow in cubic feet per second before valving off or after starting a pump, and the pipe and tank areas proposed:

$$H = Q \left(\frac{L}{agA}\right)^{1/2} \quad (18)$$

where

H = rise in feet,
Q = cubic feet per second,
L = length of pipe in feet,
a = cross-sectional area of pipe in square feet,
A = cross-sectional area of surge tank in square feet,
g = acceleration of gravity = 32.2 ft/sec^2.

Water hammer is a dynamic pressure wave, elastic in nature, which travels near the speed of sound in water. A hammering impact at destructive magnitudes can be created which can loosen joints, damage valves or pumps, and split pipe. If brought to a sudden halt through rapid closing of a valve, pump power failure, or similar action, a column of water traveling at a high velocity can set up a rebounding water hammer wave that can damage not only the main, but also the branching service pipes.

The reason for this wave is basic in physics. Water is compressible to a small degree. That is, there is space between the molecules of water which will compress if the column is suddenly stopped because of the inertial force in the moving column. Secondly, the pipe is elastic to a degree, depending upon its material, thickness, and other qualities. The combination of these two physical stresses sets up the elastic wave which will rebound back and forth through the confined section of pipe until damped out by friction and time.

Most studies on water hammer have been done under controlled conditions with unmixed pipe materials–i.e., the pipe is usually of one material throughout the length of the section under study. This does not represent the conditions usually found in the field where, for example, a pipe may be cement-morter lined steel part way and prestressed concrete or asbestos-cement pipe in the balance of the section. For this reason, it is possible to simplify the calculations for field conditions by using more general values for the M_r coefficient. This is the ratio of the modulus of elasticity of water to that of the pipe material. This coefficient varies from 0.01 for steel to 0.4 for lead, with other pipe materials lying between the extremes. An intermediate value of 0.2 can generally be used.

In order to avoid water hammer, enough time must be used to close a valve to exceed the time required for the water hammer cycle. The cycle time can be calculated as follows:

$$T = \frac{2L}{a} \qquad a = \frac{4660}{(1 + M_r t_r)^{1/2}} \quad (19)$$

where

T = time in seconds,
L = length of pipe section in feet,
a = velocity of the cycle in feet per second,
Velocity of sound in water = 4660 ft/sec,
M_r = ratio of moduli of elasticity, water to pipe material,
t_r = ratio of pipe diameter to wall thickness of pipe.

The total feet of pressure developed by water hammer can be determined as follows:

$$h = \frac{av}{g} \quad (20)$$

where

h = shock load of water hammer in feet of water,
v = drop in velocity flowing through the pipe section,
g = acceleration of gravity = 32.2 ft/sec^2.

Construction. Construction requirements of a distribution system are wide in scope. Maintenance construction and new extensions of service may be almost continuous operations. The laying of conduit with valves and appurtenances is the most frequent construction which requires design, scheduling, contract administration, and inspection during installation. However, planning and construction of pump stations, storage tanks and reservoirs, reinforced concrete vaults, and other structures may be necessary in conjunction with conduit extension. Monitoring flow and pressures in new areas of service will require installation of controls, signals, and recording equipment, and the proper housing for them.

Construction standards, therefore, depend upon several sources of information. Excavation and backfill standards must consider all problems known to exist in the area with regard to soil characteristics, rock removal, groundwater in excavations, etc. Soil stresses arising in excavation for construction of underground structures such as service vaults, basements, and foundations for buildings are usually described in local building codes and in design files on successful construction in the area. Special conditions, such as the use of explosives in excavation, require rigid adherence to safety codes and expert methods of handling.

Trenching problems are a function of the stability of the earth and the ease of excavation. Once dug to required depth, the stability of the sides and bottom may become critical, usually decreasing with the length of time the trench is open. Trenches may be dug through fine or coarse-grained soil, shales, rock lenses, gumbo, clay, or even sand with a tendency to turn to quicksand as groundwater flows upward through it. Each involves special methods of handling, equipment requirements, and cost factors. Construction experience and proven engineering design standards in the area should be researched for input into plans and estimates for new construction.

Dewatering a trench or other excavation is a common construction necessity. The rate of inflow of groundwater and the stability of the sides of the excavation will determine the extent of the dewatering operation. For a small flow in the bottom of a cut in an average conglomerated soil or reasonably tight soil, a sump hole of a few cubic feet will suffice to collect the water at a low point. In a trench, an offset is dug for the sump. Hose connections are installed from the sump to a portable gasoline-driven pump and to a suitable discharge point away from the work site. Since the pump must be capable of handling mud and small rock without damage, a diaphragm or open-impeller centrifugal type is frequently employed. Use of a sand or pea gravel pipe bedding will make a wet trench easier to work.

In more critical soils which will displace, undercut, or shear off because of groundwater flow, it may be necessary to install well points to lower the water table. This can be done before or after excavation begins. Well points are installed in gravelly soil by jetting them into place. In more dense soils, it may be necessary to drive a pipe or well casing first. The dense soil is then removed from the driven pipe and the well points are inserted and packed with gravel or sand.

Each well point is then connected to a header pipe by a flexible hose with a valve for individual regulation. The header runs to a self-priming centrifugal pump with a discharge line to a drainage way. Spacing of well points depends upon the density of soil and the amount of flow produced per connection. Spacing generally is at 5 to 10 ft. Flows may vary from 2 to 20 gpm per connection, but average about 3 to 4. The depth of the well points generally exceeds the excavation to intercept the underground flow.

Backfill requirements will vary with the location of the excavation, expected surface loads, soil characteristics, and many other factors. Tamped backfill around a newly installed pipe is necessary to protect it from external stresses, future settlement or earth movement, or other disturbance. Minimum backfill standards above the pipe usually require soil compaction to near its original density. Streets and highways require the most care in backfill and have the most precise specifications. Pipe crossings of canals, ditches, and other waterways require careful compaction and restoration during backfill to avoid creating a new transverse waterway.

Backfill around new buildings, vaults, and other structures must be placed in a manner which will avoid a lateral thrust on reinforced concrete walls, which do not reach their design strength for nearly a month. Trucks driven over backfill next to a new building can exert a lateral thrust that will crack a green concrete wall. This can lead to leakage and possible failure of the structure.

The swell and shrinkage factors of soils often determine their suitability as backfill. A soil having a large percentage of swell may not be desirable. Select backfill may be brought in to replace it rather than spending excessive time compacting it to an acceptable density—i.e., material costs may be preferable to labor costs. For example, in city streets which are busy traffic arteries, a select sand and gravel mix for the full depth of a pipe trench may be specified. This will allow fast repaving without settlement.

The use of sand or pea gravel is often specified at 6 in. for pipe bedding and fill and for each side of the pipe. The sand or pea gravel is then hand (and foot) tamped and sometimes vibrated or jetted to optimum density. The soil from the excavation or hauled-in select material is placed over this and compacted by machine tamping in lifts of 6 to 12 in. Density tests are made on the materials to see that they reach required compaction, usually as close as possible to the original density of the soil at the site. Both overcompaction and undercompaction can cause structural problems or pipe damage. Tests include those discussed in the U.S. Bureau of Reclamation's *Earth Manual* and in Standard Proctor Maximum Density Tests for earth material, as well as those performed by portable nuclear test kits with probes to gauge moisture and density of the soil by reading reflected gamma rays emitted by the probe.

Inspection of pipe installation to see that all joints are fitted and finished properly according to specifications is the best insurance for a trouble-free pipe network. Tied joints, anchored joints, thrust blocks, and other structures must be effective for several years even though the street is cut and recut for new utilities. When securing a bend or other point of stress, disturbances and unusual excavations nearby should be expected. The placement and protection of anchorage and ties are most important design and construction functions. Tie rods and other exposed metal should be treated or wrapped to prevent rust and decomposition.

Most pipe is now lined or coated inside to prevent corrosion and to improve the friction factor in conduit flow. There are some exceptions, such as plastics and asbestos-cement, which have naturally smooth and corrosion-resistant finishes developed in manufacturing. All require careful fitting and sometimes finishing of joints inside to maintain an unbroken surface.

Technical associations representing individual pipe materials, such as cast iron, steel, prestressed concrete, asbestos-cement, plastics, and so forth, all provide technical literature and specifications for obtaining optimum results in the installation of each type of pipe. The American Water Works Association and the American Society for Testing Materials both prepare specifications for water pipe materials and installation.

Workmanship in pipe assembly, finishing, and protection will be the final responsibility of the contractor and construction inspector. Installation is the point at which a pipe failure can begin.

Steel and reinforced concrete structures, general building construction, the installation of electric systems, plumbing, and gas, and the handling of compressed air and chemicals are all subject to building and safety codes and construction specifications. These specifications will provide pertinent design data and code requirements for construction. Inspectors from city building departments are often available to conduct inspections for an expanding utility department.

Quality Control. Water in any pipe of a distribution system is expected to be clean and potable. This means that it is free of harmful bacteria and viruses, contains no sediment from corrosion or other source, has no biological growths and a minimum of entrapped air, and has a reasonably fresh turnover in flow. The system is further expected to be sealed tight against intrusion by dust, insects, birds, mammals, or other contaminants or pollutants.

Water is normally at its best as it leaves the filter plant or the well house. It will not improve, but may deteriorate in quality or lose its disinfecting agent. Thus, booster chlorination stations or the addition of clarifying chemicals at storage reservoirs may be needed. Generally, it is better as a long-range policy to attempt to remove the sources of deterioration rather than to introduce more chemicals intransit. Finding these sources and preventing future intrusion of contaminants are continuing and important functions of waterworks operation.

The first source to investigate is the network of pipes which carry pressure and flow day and night. Wear from corrosion, velocity, pressure, and time, and aided by chemicals, growths, bacteria, and improper operation and maintenance can be expected, especially in older pipe. Old, unlined cast iron or steel pipes can be expected to be a constant source of rust and sediment. Iron bacteria in water will attack an exposed pipe surface and strive to reduce it to its elements. Old cement-mortar linings may crack or spall, adding debris and exposing underlying metal to corrosion. New concrete and asbestos-cement pipe and cement-mortar linings may elevate the alkaline content of water which is especially noticeable in areas of slow velocities. Improperly installed field joints for concrete pipes or pipes with cement-mortar lining, finished by hand, may be a source of alkaline debris. New copper pipes for house connections can impart a taste and an unwanted disinfectant quality to water standing in them overnight. All of these pipes, however, are excellent water conduits once in use if properly maintained to preserve their optimum flow and strength. Replacement or relining of old pipe will remove these sources of contamination from the system. Good construction inspection procedures will diminish potential contaminants in new pipe installations.

The contaminants mentioned above are distinguished from pollutants in that they are not infectious or generally harmful to humans in great dilutions. Pollutants can be. These may be growths which aid infectious agents or are in themselves infectious or harmful. The impact of a contaminant or pollutant in a system or the use of chemicals to overcome it should be gauged to some extent by the vulnerability of babies and small children. Their health and small tolerances should be kept in mind.

Open reservoirs can be the recipient of wind-blown dirt, fertilizer, weed seed, spores, etc. Water birds are as amenable to landing on finished water reservoirs as they are on sewage-laden ponds, streams, and other polluted water bodies. Wires or netting strung across the reservoir to deter water birds can invite land birds to rest and to nest. Rodents, snakes, and insects are attracted by open water and build an ecosystem around it. Insects include the gnatlike Chironomidae fly, whose larvae is called the "bloodworm," which feeds on algae often found in open reservoirs. Once an infestation has taken place, a constant effort must be maintained to eradicate the pest by destroying eggs and larvae. This can be done through heavy chlorination or the use of copper sulfate or DDT. Emptying and cleaning the reservoir is often necessary to insure complete eradication. If allowed to multiply, the larvae will infest the distribution pipes as well as the reservoir. Water quality can suffer appreciably, from both the cause and the cure. The use of open reservoirs will undoubtedly be discontinued due to their negative impact on water quality maintenance.

Roofed reservoirs or storage tanks still present problems in handling stormwater, corrosion, infestation, and biological growths, but to a lesser degree and in a better controlled environment. A reservoir roof should be watertight and structurally stable in order to exclude all stormwater runoff. It should be vented and screened or filtered to prevent entry of windblown dust, yet allow adequate ventilation. It must be impervious to spores, coliform organisms, the smallest flies, and any other form of insect or animal life. The site must be graded and drained to prevent stormwater pollution or groundwater infiltration. Coating or sealing materials used in the reservoir must be chemically stable, imparting no taste or odor.

Storage tanks must be similarly kept free of infestation. Although nutrients, plants, etc. available to insects on the ground are not so readily available in an elevated tank, algae and fungi will grow and attract flying insects. Wear and corrosion, especially at the water line, is often a maintenance problem which will require cleaning and coating tank surfaces periodically with nontoxic materials. Coal-tar enamel paints and a number of epoxy-bonded plastic paints have been used and reported upon in journals of the waterworks industry.

The presence of coliform bacteria in the distribution system is not an unusual occurrence. In themselves, they may be relatively harmless, but their presence indicates the possibility that other bacteria and viruses may also have entered the system. Coliform group fecal bacteria are a good indicator in laboratory tests since the group is rapidly and easily identified.

Its absence does not guarantee the absence of other bacteria and viruses, however. Biological growth in the mains should be a warning of improper initial disinfection of water, aftergrowth in the main from a seeded source of pollution, or the intrusion of pollutants from back siphonage, cross connections, negative pressures allowing groundwater to enter through a leak, or careless construction connections.

As discussed previously, when a section of a main is valved off and drained through a blowoff valve, a negative pressure can be developed. If this phenomenon permits the entry of groundwater, a thorough disinfection is required. This is usually accomplished by introducing chlorine (hypochlorite tablets) in powerful concentrations, discharging the solution, and refilling without creating the initial pressure problems. This procedure theoretically solves the problem although experience in local conditions may modify the theory. In any case, care must be taken to leave the main sterile.

Back siphonage from a consumer's piping system can enter a main in which the pressure has been dropped. Backflow through industrial hoses, sprinkling systems, sanitary facilities, etc. without proper plumbing safeguards, such as vacuum breakers and backflow prevention valves, is a potential pollution threat. Cross connections with private water supplies, cooling systems, etc. can introduce questionable water into the public distribution system. Local plumbing codes and ordinances should be reviewed to determine the extent of protection they provide. Rigid standards of enforcement for plumbing inspection should be supported.

It is important to know the physical and chemical characteristics of water at any point in the system at any time. Loss of residual chlorine or a drop in dissolved oxygen can indicate the intrusion of a pollutant. Changes in temperature, turbidity, alkalinity, or other characteristics should be known to the plant operator as soon as possible. The laboratory customarily knows the precise condition of water as it leaves the filter plant, but most frequently depends upon check points, random sampling, and telephone calls to establish the distribution system's water quality patterns. An active manual sampling routine provides good coverage, but is time-consuming and expensive. In recent years, there has been an increasing effort to introduce automatic monitoring.

Because of research grant programs and development of new monitoring equipment, it is possible to detect more finitely the changes in water as it flows through a conduit of the distribution system. Initial progress is indicated in the measurement of these parameters:

Temperature	Dissolved oxygen
pH (alkalinity)	Conductivity
Hardness	Free residual chlorine
Nitrates	Chlorides
Fluorides	Metals (Cu, Cd, Pb)
Corrosion	

The use of ion-selective electrodes, polarization, the nephelometer (for turbidity), and other very sensitive electronic and membrane devices will allow us to monitor water quality in the future as we monitor pressure and flow today.

Design considerations in water quality control, beyond those mentioned above, include the protection of valves, appurtenances, and electrical equipment from flooding through proper location and drainage. Air relief valves, electric operators, and monitoring equipment are especially susceptible to storm and groundwater damage. This means that underground vaults must be protected by adequate sump pumps, and manholes where vulnerable equipment is installed should be drained to a wasteway or a French (sand) drain. Tight-bottomed sumps which can be pumped with a portable sump pump can be used in standard valve vaults and manholes. Open-bottomed sumps will let groundwater into the structure.

Water mains should be located at least 10 ft away from and 5 ft above sewer mains, if possible. It is common practice to establish fixed zones on opposite sides of streets in order to locate water mains away from sanitary or storm sewers. Groundwater surrounding a sewer line can become quite polluted, especially along old lines. Where sewers must cross above a water line, or closer than 5 ft below it, it is good practice to require that the sewer be constructed of cement-mortar lined cast iron pipe. This pipe should have leaded or mechanical joints and extend for a distance of 10 ft or more on each side of the water line.

Storage reservoirs must be protected from groundwater infiltration, with special care taken in areas where groundwater can flow from the vicinity of sanitary sewers. The connection of drain lines from waterworks structures to sanitary sewers is generally poor policy. The danger of pollution from a backed-up sewer is always present plus the risk of basement flooding mentioned earlier in this chapter.

Improper design of dead-end mains, such as those often found in cul-de-sacs or other limited accessways, can lead to water quality problems. For example, if the pipe ascends in grade, air saturation can cause "milky" water. If it descends, sedimentation may collect. According to the circumstance and location, the installation of an air relief valve, a blow-off valve, or fire hydrant to flush the line should be considered for dead-end mains. Similarly, the location of required air valves, vacuum relief valves, and the flushing or draining of sections should be anticipated throughout the total run of the distribution system section under investigation. Problems can be anticipated from the critical high and low points on profile drawings, and the means of isolating sections by line valves. Flanged outlets may be installed for the future if the need is not yet imperative.

Velocity is an important factor in water quality. Very slow-moving water will encourage the build up of sediment at the invert of the pipe. Testing of hydrants, reversal of flows, flushing, and other activities will stir up the sediment which may stay in suspension for several hours. Design velocities should take into account the fact that particles will settle out in flows of 2 ft/sec or less. Sometimes a zero velocity is impossible to avoid during several hours of the day. Looping and designing for increased velocities for other parts of the day can help reduce this problem.

Operational considerations should include recognition of the various expected sources of contamination and pollution. These include sudden velocity changes which might stir up sediment. Hydrant use by the fire department or others, or their periodic testing may generate complaints of murky or bad tasting water. A full flushing program concurrent with testing or subsequent to other uses which engender complaints might prove feasible. Flushing may require a great deal of water, so it must be timed at hours of low demand. Quality should be checked frequently in areas where trouble is anticipated, such as low velocity dead ends which may require scheduled flushing.

If reports on a main are consistently poor, showing high sediment or biological growths, the introduction of additional chlorine daily should be considered as a temporary measure. Depending upon the pollution source, a permanent solution may include installation of more blow-off points, heavy disinfection with a strong chlorine solution for several hours with the area out of service, or replacement of part of the conduit.

Storage tanks and reservoirs should be sampled and tested frequently for insects and windblown dust and debris. Special tests for suspected bacteria or other growth might be made routinely. Visual inspections can be made as the water level drops.

BIBLIOGRAPHY

General Publications, Texts, Handbooks, and Manuals (refer to the AWWA Yearbook for complete list of publications and recommended texts)

Manual M-8, A Training Course in Water Distribution
Manual M-9, Installation of Concrete Pipe
Manual M-11, Steel Pipe Design and Installation
Index to all AWWA Journals (by subject)
AWWA Standards and Handbooks (see Yearbook)

Manufacturers and Materials Associations Books and Data

Handbook of Cast-Iron Pipe (several printed by individual pipe companies)

Concrete Pipe Handbook (see American Concrete Institute, Portland Cement Association, American Concrete Pipe Association, and individual pipe companies)

Ductile-Iron Pipe Data (see individual companies manufacturing or distributing pipe)

Prestressed Concrete Pipe Data (see AWWA Specification C303-70, individual pipe companies)

Asbestos-Cement Pipe (see individual asbestos-cement pipe manufacturers' literature and AWWA specifications)

Plastic Pipe Data (see manufacturers' data on each type of plastic material)

Asphalt in Hydraulic Structures, The Asphalt Institute

Textbooks and Handbooks

Davis and Sorenson, eds., *Handbook of Applied Hydraulics*, 3rd ed., McGraw-Hill, New York (1969).

King and Brater, *Handbook of Hydraulics*, 5th ed., McGraw-Hill, New York (1963).

Abbett, R. W., *American Civil Engineering Practice*, Vol. II, John Wiley and Sons, Inc., New York (1956).

Addison, H., *A Treatise on Applied Hydraulics*, 5th ed., Chapman and Hall Ltd., London (1964).

Rich, G. R., *Hydraulic Transients*, 2nd ed., Engineering Societies Monographs, Dover Publications, Inc. (1963).

Arches, Continuous Frames, Columns and Conduits, Selected Papers of Hardy Cross, Univ. of Illinois Press, Urbana (1963).

Ground Water and Wells, 1st ed., Edward E. Johnson, Inc., St. Paul, Minn. (1966).

Public Water Supply Treatment Technology, AWWA Research Foundation, 1973 Report for Water Resources Research, U.S. Department of the Interior, AWWA (1973).

Water Quality and Treatment, 3rd ed., Prepared by the AWWA, McGraw-Hill, New York (1971).

Peurifoy, R. L., *Construction Planning*, Equipment and Methods, 2nd ed., McGraw-Hill, New York (1970).

Hicks, T. G., *Pump Selection and Application*, McGraw-Hill, New York (1957).

Articles

Transmission and Distribution Piping:

"Comparison and Evaluation of the Design Criteria for Pipe of Various Materials," *Journal AWWA*, pp. 81–114 (Jan., 1963).

Bottles, D. A., "Use of Plastic Pipe," *Journal AWWA*, pp. 55–59, (Jan., 1970).

Several articles on asbestos-cement and plastic pipe design, *Journal AWWA*, pp. 441–456 (July, 1972).

Romanoff, M., "Performance of Ductile-Iron Pipe in Soils," *Journal AWWA*, pp; 645–655 (June, 1968).

Morris, R. E., Jr., *Journal AWWA*, pp. 782–798 (July, 1967).

Distribution Water Storage:

Hoedeman, K. A., "Practical Aspects of Tank Design," *Journal AWWA*, pp. 169–172 (March, 1972).

Wormald, L. W., "Water Storage Tank Failure Due to Freezing and Pressurization," *Journal AWWA*, pp. 173–175 (March, 1972).

Prazer, S. J., "Reservoir Renovation: Butyl-Rubber Lining and Floating Cover," *Journal AWWA*, pp. 151–154 (Feb., 1973).

Lackey, J. M., "The Use of Asphalt in Hydraulic Works," *Journal AWWA*, pp. 401–408 (April, 1950).

Kenmir, R. C., "Concrete Reservoir Design," *Journal AWWA*, pp. 1181–1194 (Oct., 1968).

Automatic Controls and Valves:

End, W. F., Jr., "Automation and the Automatic Valve," *Journal AWWA*, pp. 301–308 (March, 1957).

Lescovich, J. E., "Locating and Sizing Air Release Valves," *Journal AWWA*, pp. 457–461 (July, 1972).

Weaver, D. L., "Surge Control," *Journal AWWA*, pp. 462–466 (July, 1972).

"Telemetering: Five Utilities' Experiences," *Journal AWWA*, pp. 491–507, (Aug., 1971).

Freeston, R. C., "Feasibility of Remote Controls for Distribution Systems," *Journal AWWA*, pp. 723–728 (Nov., 1970).

Lee, R. L., "Controls for a Medium-Sized Water Utility," *Journal AWWA*, pp. 527–530 (August, 1971).

Computer in Waterworks:

Van Dyke, R. P., "Computerized Operation of Distribution Systems," *Journal AWWA*, pp. 234–240 (April, 1970).

Becher, Bizjak, and Schulz, "Computer Techniques for Water-Distribution Analysis," *Journal AWWA*, pp. 410–417 (July, 1972).

Loehr, R. C., "Analog and Digital Computers in Distribution System Analysis," *Journal AWWA*, pp. 1001–1008 (Aug., 1964).

McIlroy, M. S., "Direct-Reading Electric Analyzer for Pipeline Networks," *Journal AWWA*, pp. 347–365 (April, 1950).

8

Developing Groundwater Resources

G. F. Briggs
Technical Director, Johnson Division
Universal Oil Products Company

AVAILABILITY OF GROUNDWATER vs. SURFACE WATER

Because surface waters are visible and because such tremendous amounts of money have been spent in building dams, levees, artificial reservoirs, aqueducts, and irrigation canals, it is a natural tendency to think of surface water as the major source for the world's needs. Actually, less than 3 percent of the earth's fluid fresh water available at any given time occurs in streams and lakes. The other more than 97 percent—an estimated 8×10^{12} acre-ft—is underground.

Not all the water beneath the earth's surface can be recovered from the water-bearing formations in which it is found. Some lies in rock formations so deep that pumping costs alone rule out its recovery. Other water-bearing formations resist recovery to varying degrees and may even defy pumping efforts.

While comparative figures of the amounts of water available from surface and ground sources thus cannot be used as true indicators of actual water resources, they do point to the fact that the available supply of groundwater is many times greater than the available supply of surface water.

HYDROLOGIC PRINCIPLES

Subsurface Distribution of Water. Understanding of groundwater occurrence requires a study of the vertical distribution of water in subsurface geologic formations. The outer part of the earth's crust is normally porous to a greater or lesser depth. This part is called the zone of rock fracture. The pores or openings in this portion of the lithosphere may be partially or completely filled with water.

The upper stratum, where the openings are only partly filled with water, is called the zone of aeration. Immediately below this, where all the openings are completely filled with water, is the zone of saturation (Fig. 8-1).[1]

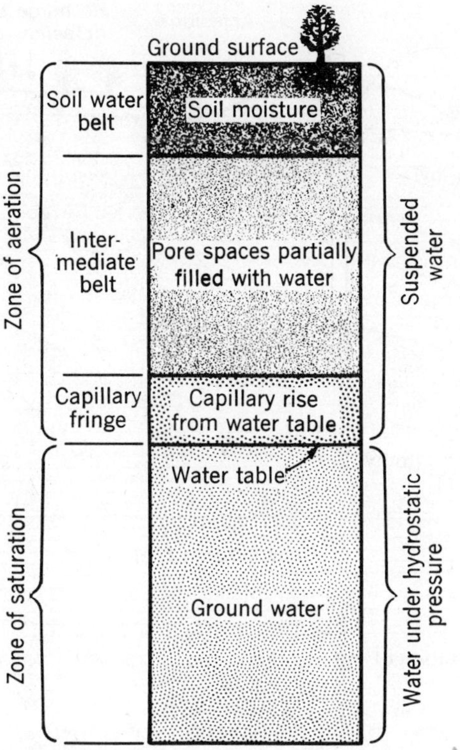

Fig. 8-1 Groundwater phase of the hydrologic cycle.*

*All figures in this chapter courtesy of the Johnson Division, Universal Oil Products Company.

Groundwater. Water in the zone of saturation is the only part of all subsurface water which is properly referred to as groundwater (Fig. 8-1). The saturated zone may be viewed as a huge natural reservoir (or system of reservoirs) whose capacity is the total volume of the pores or openings in the rocks that are filled with water. Groundwater may be found in one continuous body or in several separate strata.

The thickness of the zone of saturation varies from a few feet to many hundreds of feet. Factors that determine its thickness are: the local geology, the availability of pores or openings in the geologic formations, the recharge, and the movement of water within the zone from areas of recharge toward points or areas of discharge.

Formations or strata within the saturated zone from which groundwater can be obtained for beneficial use are called aquifers. To qualify as an aquifer, a geologic unit must contain pores or open spaces which are filled with water, and these openings must be large enough to permit water to move through them toward wells and springs at a sufficient rate for these outlets to serve as practical sources of water supply. "Waterbearing formation" and "groundwater reservoir" are other terms often used in place of the word aquifer.

The upper surface of the zone of saturation is called the water table. The shape of the water table is controlled partly by the topography of the land and tends to follow, in a general way, the shape of the land surface (Fig. 8-2).

The interrelation of surface water and groundwater is shown in Fig. 8-3.

Water Table Conditions. Groundwater in some aquifers occurs under water table conditions. This means that the upper limit of the aquifer is defined by the water table itself (Fig. 8-2). At the water table (the top of the saturated portion of the geologic formation), the water in the pores of the aquifer is at atmospheric pressure as if it were in an open tank. Under this condition, the aquifer itself is referred to as a water-table aquifer. The terms "unconfined aquifer" and "unconfined groundwater" are also used with reference to this form of groundwater occurrence.

The hydraulic head at any level within a water-table aquifer is equal to the depth below static level and may be expressed as head in feet of water. For example, groundwater at a depth of 50 ft below the water table is under a static head of 50 ft.

When a well is drilled in a water-table aquifer, the static water level in the well stands at the same elevation as the water table.

In some cases, a local zone of saturation may exist at some level above the main water table. This situation can occur where an impervious stratum within the zone of aeration interrupts percolation and causes groundwater to accumulate in a limited area above that stratum. The upper surface of the groundwater in such a case is called a perched water table.

The water table is not stationary but periodically moves up and down—rising when more water is added to the saturated zone by vertical

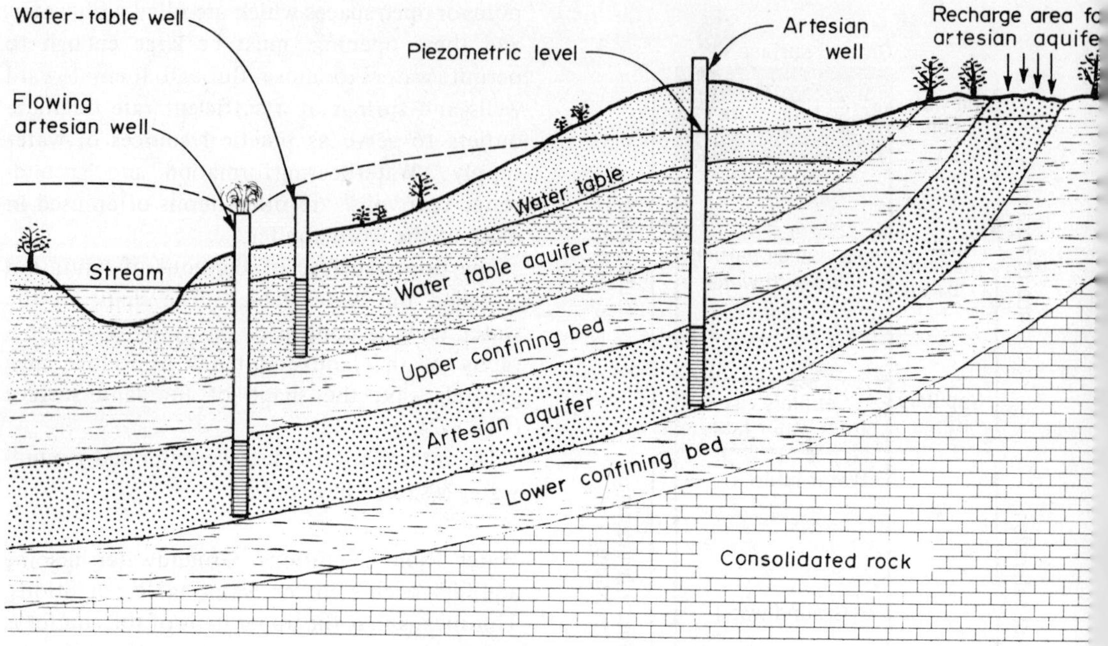

Fig. 8-2 Subsurface distribution of water.

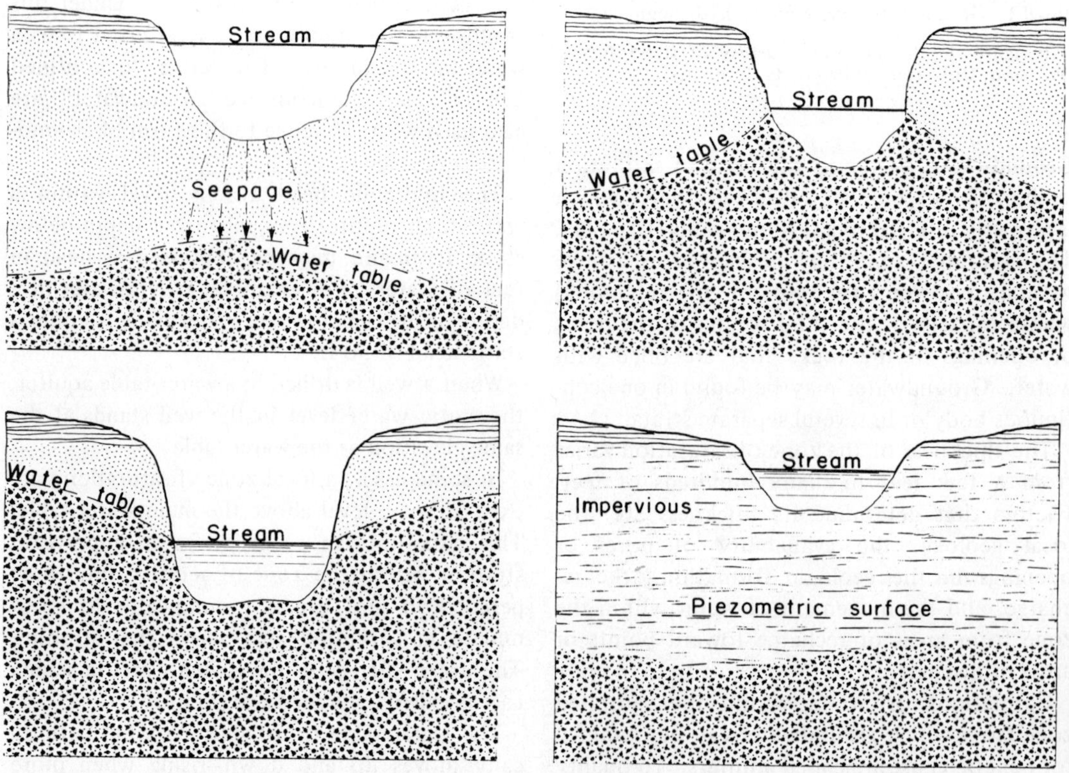

Fig. 8-3 Interrelations between groundwater and surface water.

percolation, and dropping during drought periods when previously stored water flows out toward springs, streams, wells, and other points of groundwater discharge.

Artesian Conditions. The zone of saturation may include both permeable and impermeable layers of earth materials. The permeable layers are aquifers. Where an aquifer is found between impermeable layers above and below it, both the aquifer and the water it contains are said to be confined. Because of the presence of the upper confining layer, the water of the aquifer is not open to atmospheric pressure. It thus occurs within the pores of the aquifer at pressures greater than atmospheric. Groundwater in such a situation is said to occur under artesian conditions, and the aquifer is called an artesian aquifer (Fig. 8-2).[2]

Aquifer Functions. An aquifer performs two important functions–a storage function and a conduit function. It stores water, serving as a reservoir, and it acts as a pipeline to transmit water. The openings or pores in a water-bearing formation similarly serve both as storage spaces and as a network of conduits. The groundwater is constantly moving over extensive distances from areas of recharge to areas of discharge. Movement is very slow, with velocities measured in feet per day or even feet per year. As a consequence of this and of the great volume represented by its porosity, an aquifer detains enormous quantities of water in transient storage.[3]

Two properties of an aquifer related to its storage capacity are its porosity and its specific yield.

Porosity. The porosity of a water-bearing formation is that part of its volume which consists of openings or pores–the proportion of its volume not occupied by solid material. Porosity is an index of how much groundwater can be stored in the saturated material. For example, if 1 ft^3 of sand contains 0.30 ft^3 of open spaces or pores, we say that its porosity is 30 percent.

While porosity represents the amount of water an aquifer will hold, it does not indicate how much water the porous material will yield. When water is drained from a saturated material by gravity force, only part of the total volume stored in its pores is released. The quantity of water that a unit volume retains when subjected to gravity drainage is called its specific retention. Both specific yield and specific retention are expressed as decimal fractions or percentages.

If 0.10 ft^3 of water is drained from 1 ft^3 of saturated sand, the specific yield of the sand is 0.10, or 10 percent (Fig. 8-4). Assuming that the porosity of the sand is 30 percent, its specific retention is 0.20, or 20 percent.

A water-table aquifer extending over an area of 20 mi^2 and with an average thickness of 40 ft occupies a total volume of 22.3×10^9 ft^3. With a porosity of 25 percent, this groundwater reservoir would have 5.6×10^9 ft^3 of groundwater stored within its pores. If the specific yield of the material were 10 percent and the upper 5 ft of the aquifer were drained by lowering the water table 5 ft, the total yield would be about 280×10^6 ft^3 of water, or about 2.1×10^9 gal (U.S.). This quantity would supply 4 wells pumping 700 gpm continuously, 12 hr each day, for 1042 days or

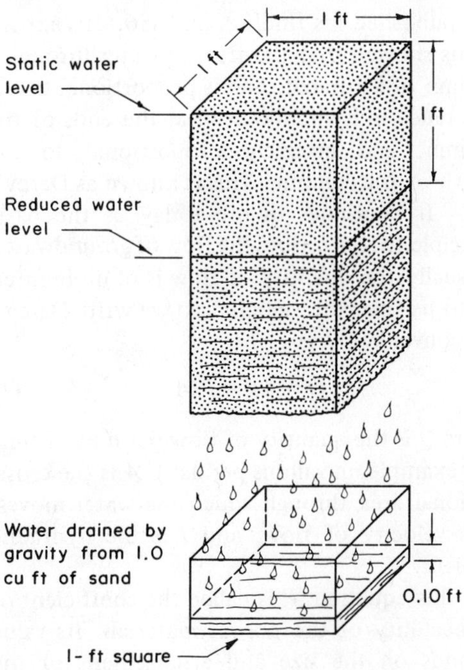

Fig. 8-4 Physical concept of the specific yield of sand. Its value here is 0.10 ft^3/ft^3 of aquifer.

almost 3 years. It would be sustained by the groundwater stored in the upper 5 ft of the aquifer in the absence of any replenishment during the 3-year period.

This simple example shows how the storage function of an aquifer makes it possible to use groundwater at a constant rate, even though replenishment to the aquifer may be intermittent and irregular. In this respect, groundwater reservoirs serve much more effectively than do surface water reservoirs because of their enormous capacity.

Permeability. The property of a water-bearing formation which is related to its pipeline or conduit function is called its permeability. Permeability is defined as the capacity of a porous medium for transmitting water. Movement of water from one point to another in the material takes place whenever a difference in pressure or head occurs between two points. Permeability may be measured in the laboratory by noting the amount of water that will flow through a sample of sand in a certain time and under a given difference in head.

Henri Darcy, a French engineer, investigated the flow of water through beds of filter sand and published his findings in 1856. His experiments showed that the flow of water through a column of saturated sand is proportional to the difference in hydraulic head at the ends of the column, and inversely proportional to the length of the column. This is known as Darcy's law. It continues in use today as the basic principle that describes the flow of groundwater.

Usually the quantity of flow is of more interest to us than the velocity, so we write Darcy's law conveniently as

$$Q = AV = PIA \tag{1}$$

where Q is the quantity of flow per unit of time (for example, in gallons per day), A is the cross-sectional area through which the water moves, V is velocity of flow, and I is the hydraulic gradient.

In this equation, P is called the coefficient of permeability of the porous material. Its value depends on the size and arrangement of the particles in the unconsolidated formation and on the size and character of the surfaces of the crevices, fractures, or solution openings. It may change with any variation in these characteristics. The coefficient of permeability is the quantity of water that will flow through a unit cross-sectional area of a porous material per unit of time under a hydraulic gradient of 1.00 (100 percent) at a specified temperature.[4]

For convenient use in well problems, P is expressed as the flow in gallons per day (gpd) through a cross section of 1 ft^2 of water-bearing material under a hydraulic gradient of 1.00 and at a temperature of 60°F. The permeability unit with these dimensions is called a Meinzer unit after the late Dr. O. E. Meinzer of the U.S. Geological Survey (Fig. 8-5). The term "permeability" is often used to denote the coefficient of permeability, and this is the sense in which it is used in subsequent discussions.

Calculating Flow. Darcy's law shows that flow in a saturated sand varies directly with the hydraulic gradient. If the hydraulic gradient (head loss per foot of travel) is doubled, the rate of flow in a given sand is also doubled.

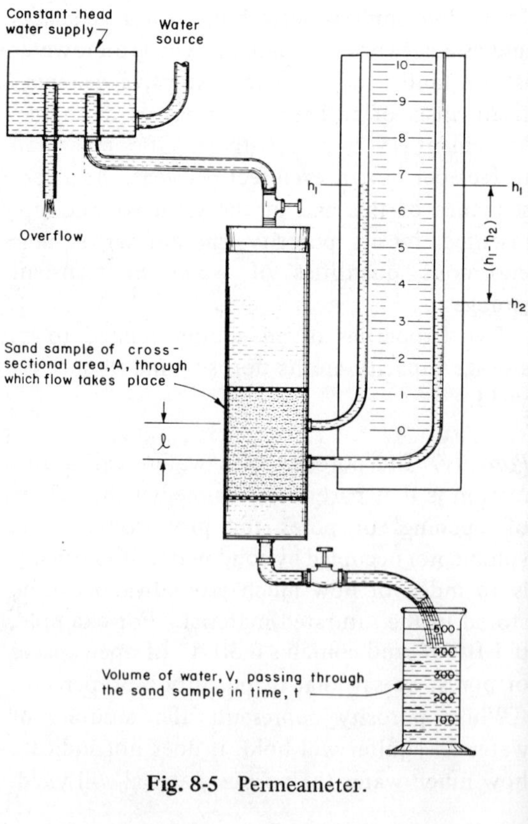

Fig. 8-5 Permeameter.

This means that to double the rate of flow in a given sand requires doubling the hydraulic gradient.

The type of fluid flow described by Darcy's law is known as laminar flow. Water flow in pipes can be laminar in character at low velocities, but more commonly takes place under conditions described as turbulent flow. In turbulent flow, the flow rate does not increase in direct proportion with change in hydraulic gradient–doubling the hydraulic gradient may only increase the flow rate by about $1\frac{1}{2}$ times.

The slope of the water table or the slope of the piezometric surface is the hydraulic gradient under which groundwater movement takes place (Fig. 8-6). The total flow through any vertical section of an aquifer can be calculated if we know the thickness of the aquifer, its width, its average coefficient of permeability, and the hydraulic gradient at the section in question. The flow, q, through each foot of width of an aquifer is:

$$q = PmI \qquad (2)$$

where P is the average coefficient of permeability of the material from top to bottom of the aquifer, m is the thickness of the aquifer in feet, and I is the hydraulic gradient.

C. V. Theis, in 1935, pointed out the convenience of using the product of P and m as a single term to represent the water transmitting capability of the entire thickness of an aquifer.[5] He introduced the term "coefficient of transmissivity" for this, defining it as the rate of flow in gallons per day through a vertical section of an aquifer whose height is the thickness of the aquifer and whose width is 1 ft when the hydraulic gradient is 1.00. The temperature assumed in this definition of the coefficient is the temperature of the groundwater in the aquifer.

When the coefficient of transmissivity, T, is introduced in the Darcy equation, the flow through any vertical section of an aquifer is expressed as

$$Q = TIW \qquad (3)$$

where T is the coefficient of transmissivity, I is the hydraulic gradient, and W is the width of the section through which the flow occurs (see

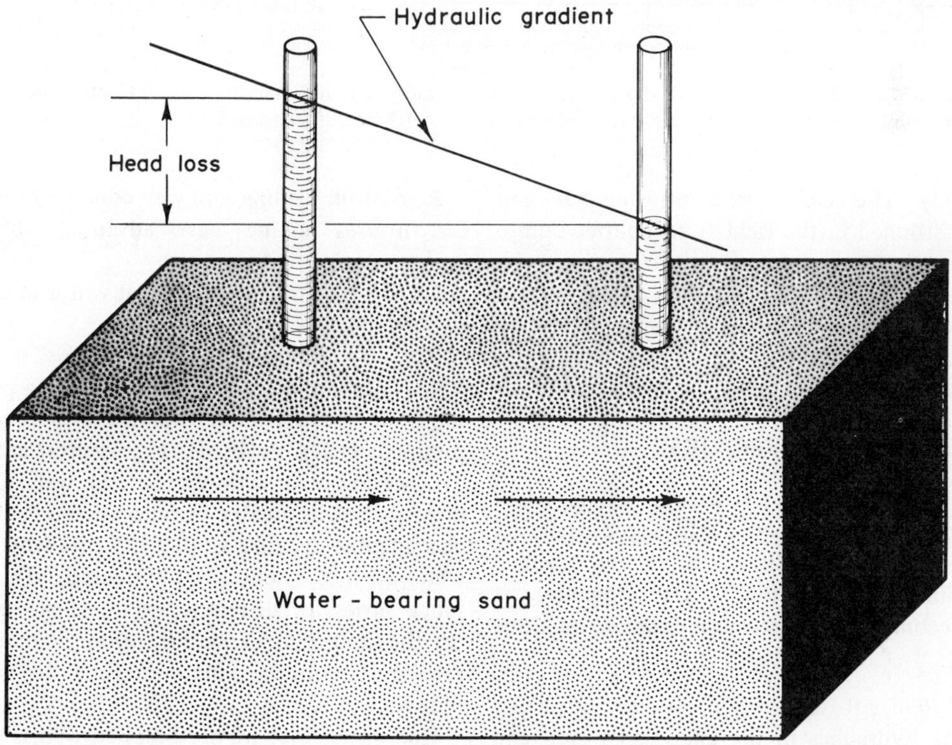

Fig. 8-6 Hydraulic gradient is head loss divided by the distance between the two points.

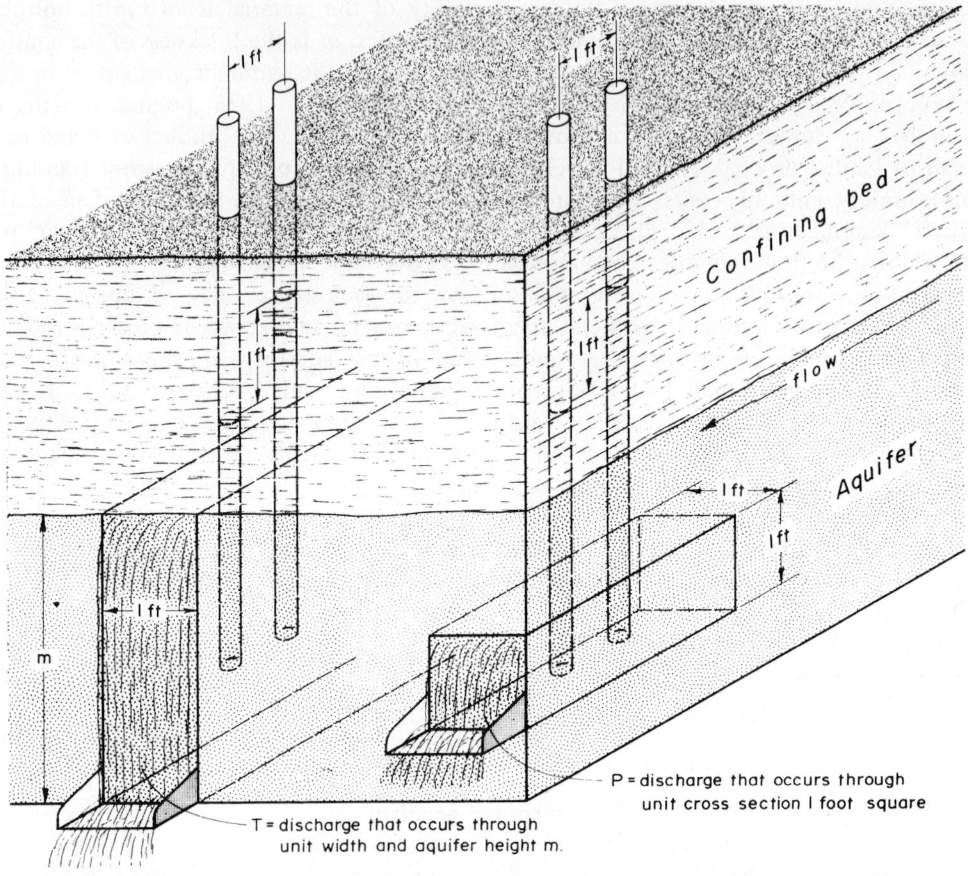

Fig. 8-7 Graphical concepts of the coefficients of permeability and transmissivity. Coefficient of permeability multiplied by the aquifer thickness equals coefficient of transmissivity.

Fig. 8-7). The coefficient of transmissivity can be determined in the field from aquifer pumping tests. This practical fact overcomes the problem of getting reliable values of the average coefficient of permeability from laboratory tests.

WATER WELL HYDRAULICS

A well is a hydraulic structure which, when properly designed and constructed, permits the economic withdrawal of water from a water-bearing formation. How adequately it will accomplish this purpose depends on three things:

1. Intelligent application of the principles of hydraulics in the analysis of well and aquifer performance.
2. Skill in drilling and well construction that insures taking best advantage of the geologic conditions.
3. Selection of materials that will insure long life.

The Nature of Converging Flow. When pumping is started, the water level in the vicinity of a pumped well is lowered. The greatest amount of lowering or drawdown is at the well. Drawdown is less at greater distances from the well, and at some distance a point is found where the water level is essentially unaffected. This distance varies for different wells. It also varies for the same well, depending upon the pumping rate and upon the length of time the well is pumped.

The water level in the pumped well is lower than at any place in the water-bearing forma tion surrounding it, so water moves from the formation into the well to replace that being withdrawn by the pump. The force or pressure that drives water toward the well is the head represented by the difference between the water level inside the well and the water level at any place outside the well.

Water flows through the aquifer from every direction toward the well. As the water moves closer and closer to the well, it must move through successive cylindrical sections that are smaller and smaller in area. Accordingly, the velocity of the water increases as it approaches the well. In Fig. 8-8, A_1 represents the area of a cylindrical surface 100 ft from the center of the well, and A_2 represents the area of a similar surface 50 ft from the center of the well. It is readily seen that A_1 is twice A_2. However, the same quantity of water is flowing toward the pumped well through areas A_1 and A_2, so the velocity V_2 must be twice V_1.

Darcy's law shows that in flow through porous media the hydraulic gradient varies directly with the velocity. With increasing velocity, the hydraulic gradient increases as flow converges toward a well. As a result, the lowered water surface develops a continually steeper slope toward the well. The form of this surface resembles a cone-shaped depression known as

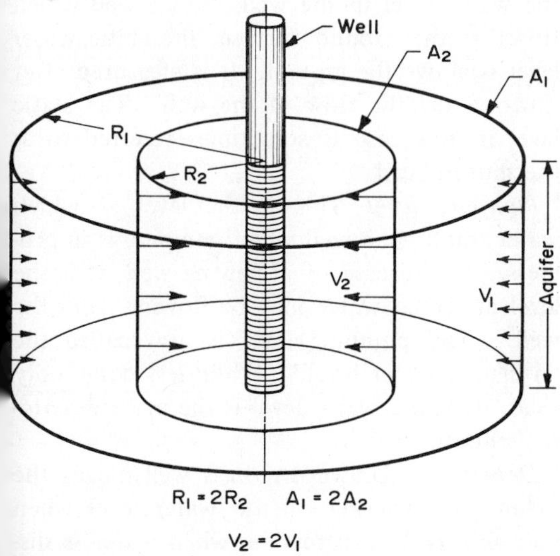

Fig. 8-8 Flow converges toward a well as it passes through imaginary cylindrical surfaces that are successively smaller as distance to the well decreases.

the cone of depression. Any well, when pumped, is surrounded by such a cone. Each differs in size and shape depending upon the pumping rate, length of pumping period, aquifer characteristics, slope of the water table, and recharge within the zone of influence of the well.

Cones of Depression. Figure 8-9 shows two cones of depression around pumped wells which illustrate how the transmissivity of the aquifer affects the shape of the cone. In a formation with low transmissivity, the cone is deep and has a small base with steep sides. In a formation with high transmissivity, the cone is shallow and has a large base with flat-sided slopes. The amount the original water level or pressure surface is lowered below on the base of the cone and within the well itself is called the drawdown at that point.

Figure 8-10 shows how the drawdown is distributed within the cone of depression on one side of a pumping well. The curve shows the levels at which water would be found in observation wells drilled at various distances from the pumped well. In a water-table aquifer, it represents the water surface within the aquifer as the well is being pumped. In an artesian aquifer, it represents the hydrostatic pressure in the aquifer. The difference between the water level indicated by the curve and the static water level is drawdown at any given point.

Head loss is a better term to describe the difference in head, or pressure, that is required to cause a certain rate of flow from one point to another in an aquifer. It represents the force required to overcome the resistance to flow. The head losses from point to point along the pumping water level curve in Fig. 8-10 are the changes in drawdown between these points.

Suppose, for example, the well is being pumped at a constant rate of 200 gpm. The total drawdown of 20 ft in the well is the head required to move 200 gpm through the aquifer (within the zone of influence of the well) and into the well.

At a distance of 20 ft from the well the drawdown is about 6 ft. This indicates that 6 ft of head are required to force 200 gpm through the formation from the outer limit of the

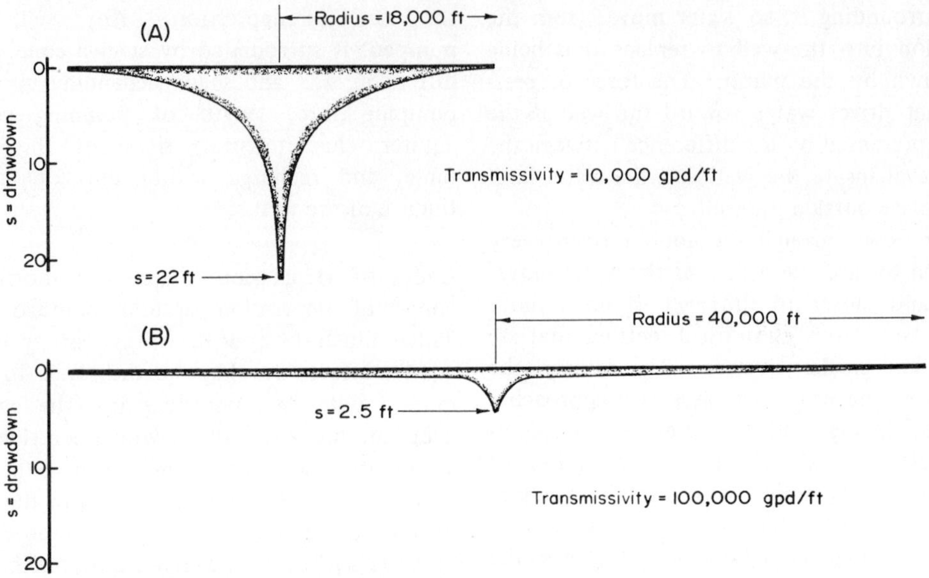

Fig. 8-9 Effect of differing coefficients of transmissivity upon the shape, depth, and extent of the cone of depression (pumping rate and other factors being the same in both cases).

cone of depression to within 20 ft of the well. It requires another 6 ft of head to move the 200 gpm from a distance of 20 ft to within about 5 ft of the well. At this point the drawdown is about 12 ft. The remainder of the total drawdown or head loss is used in pushing the water through the last 5 ft of the formation and through the well screen.

These figures show clearly the increasing head loss per unit of distance along the flow path as the flow in the aquifer converges toward the well.

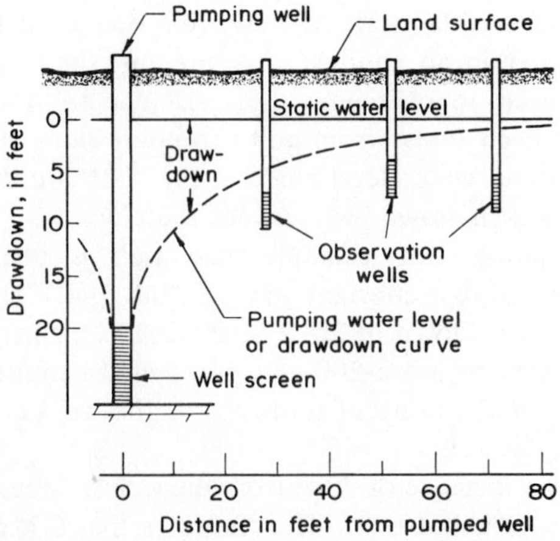

Fig. 8-10 Variation in drawdown with distance from a pumped well.

Definition of Terms. Important terms used in discussing hydraulics of wells are:

Static water level. This is the level at which water stands in a well when no water is being taken from the aquifer either by pumping or by free flow. It is generally expressed as the distance from the ground surface (or from a measuring point near the ground surface) to the water level in the well. For a well which flows at the ground surface, the static water level is above the ground. It is measured after shutting off the flow of the well. The static level in this case is sometimes referred to as the shut-in head.

Pumping level. This is the level at which water stands in a well when pumping is in progress. In the case of a flowing well, it is the level at which water may be flowing from the well. The pumping level is also called the "dynamic water level." While less commonly used, dynamic water level is the more descriptive phrase.

Drawdown. Drawdown in a well means the extent of lowering of the water level when pumping is in progress or when water is discharging from a flowing well. Drawdown is the difference, measured in feet, between the static water level and the pumping level. This represents the head, in feet of water, that causes water to flow through the aquifer toward a well at a given rate.

Residual drawdown. After pumping is stopped, water levels rise and approach the static water level observed before pumping started. During such a recovery period, the distance that the water level is found to be below the initial static water level is called residual drawdown.

Well yield. Yield is the volume of water per unit of time discharged from a well, either by pumping or by free flow. It is measured commonly as the pumping rate in gallons per minute (gpm). Other units employed are gallons per hour (gph) for small yields, and cubic feet per second (ft^3/sec) for large yields.

Specific capacity. Specific capacity of a well is its yield per unit of drawdown, usually expressed as gallons per minute per foot of drawdown. Dividing the yield by the drawdown, each measured at the same time, gives the value of the specific capacity. For instance, if the pumping rate is 160 gpm and the drawdown is found to be 20 ft, the specific capacity of the well is 8 gpm/ft of drawdown at the time the measurements are taken.

The terms static water level, dynamic water level, drawdown, and residual drawdown apply similarly to measurements made in a pumped well or in other nearby wells used as observation wells.

Radius of influence (R) is the distance from the center of the well to the limit of the cone of depression. It is larger for cones of depression surrounding artesian wells than for those around water-table wells.

Coefficient of storage (S) of an aquifer is the volume of water released from storage, or taken into storage, per unit of surface area of the aquifer per unit change in head. In water-table aquifers, S is the same as the specific yield of the material unwatered during pumping. In artesian aquifers, S is the result of two elastic effects—compression of the aquifer and expansion of the contained water—when the head or pressure is reduced during pumping. The coefficient of storage is a dimensionless term. Values for S for water-table aquifers range from 0.01 to 0.25; values for artesian aquifers range from 0.00001 to 0.001.

Coefficient of transmissivity (T) of an aquifer is the rate at which water will flow through a vertical strip of the aquifer 1 ft wide and extending through the full saturated thickness under a hydraulic gradient of 1.00 or 100 percent (see Fig. 8-7).

Values of the coefficient of transmissivity range from less than 1000 to over 1,000,000 gpd/ft. An aquifer whose transmissivity is less than 1000 can supply only enough water for domestic wells and the like. Where the transmissivity is on the order of 10,000 or more, well yield can be adequate for industrial, municipal, or irrigation purposes.

The coefficients of transmissivity and storage are especially important because they define the hydraulic characteristics of a water-bearing formation. The coefficient of transmissivity indicates how much water will move through the formation and the coefficient of storage indicates how much can be removed by pumping or draining. If these two coefficients can be determined for a particular aquifer, predictions of great significance can usually be made. Some of these are:

1. Specific capacity of wells of different diameters.
2. Drawdown in the aquifer at various distances from a pumped well.
3. The drawdown in a well at any time after pumping commences.

Water Supplied from Storage. When water is pumped from a well, the quantity discharged initially is derived from aquifer storage immediately surrounding the well. As pumping continues, more water must be derived from storage at greater and greater distances from the well bore. This means that the cone of depression must expand, so that the radius of influence of the well increases. Drawdown also increases to provide the additional head required to move the water from a greater distance. The cone expands and deepens at a decreasing rate with time, however, because with each additional foot of horizontal expansion, a larger volume of stored water is available than from the preceding one.[6]

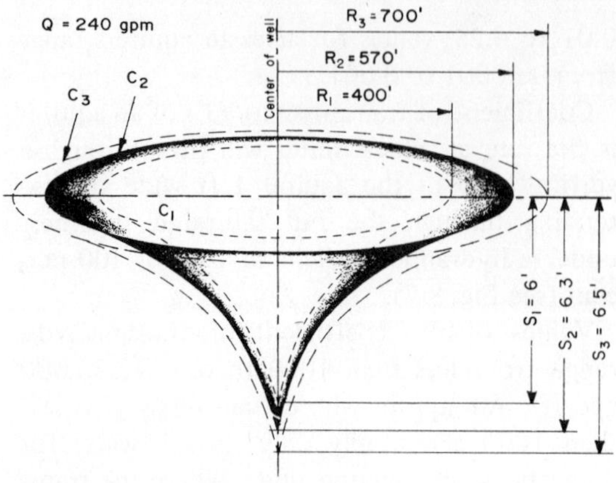

Fig. 8-11 Changes in extent and depth of cone of depression after equal intervals of time, assuming a constant pumping rate.

Figure 8-11 illustrates how the cone of depression expands during equal intervals of time. Assume that after 1 hr of pumping the radius of the cone may be 400 ft and its depth 6 ft at the well bore. At the end of the second hour, the cone's radius may have expanded to 570 ft and its depth increased to 6.3 ft. In the second hour, therefore, the cone has extended outward an additional 170 ft and deepened by additional 0.3 ft. The third hour of pumping produces an additional radial expansion of 130 ft and an increase in depth of 0.2 ft. Calculation of the volume of each of the cones would show that C_2 has twice the volume of C_1, and C_3 has 3 times the volume of C_1. This occurs because the same volume of water is discharged from the well during each 1-hr interval.

After some hours it requires precise measurements to see how much the depth of cone (drawdown) is increasing during limited intervals of pumping. This often leads observers to conclude that the cone has reached a stabilized position as pumping continues. The fact is, however, that the cone continues to enlarge and deepen until aquifer recharge is received equaling the pumpage. Recharge may occur in one or more of the following situations:

1. The cone enlarges until it intercepts enough of the natural discharge from the aquifer to equal the pumping rate.
2. The cone enlarges until it intercepts a body of surface water from which enough water enters the aquifer to equal the pumping rate.
3. The cone enlarges until there is enough vertical recharge from precipitation within the radius of influence to equal the pumping rate.
4. The cone enlarges until there is sufficient leakage through overlying or underlying formations to equal the pumping rate.

When the cone has stopped expanding due to one or more of the above reasons, a condition of equilibrium exists. There is no further increase in drawdown with increase in time of pumping. In some wells, equilibrium occurs within a few hours after pumping begins; in others, it does not occur even though the length of the pumping period may be extended for years.

Equilibrium Well Formulas. Well-discharge formulas for equilibrium conditions were derived many years ago by various investigators. There are two basic formulas—one for artesian conditions and the other for water table conditions.[7] Both assume recharge at the periphery of the cone of depression. Figure 8-12 shows a vertical section of a well constructed in a water-table aquifer. The formula for the water-table well is:

$$Q = \frac{P(H^2 - h^2)}{1055 \log R/r} \tag{4}$$

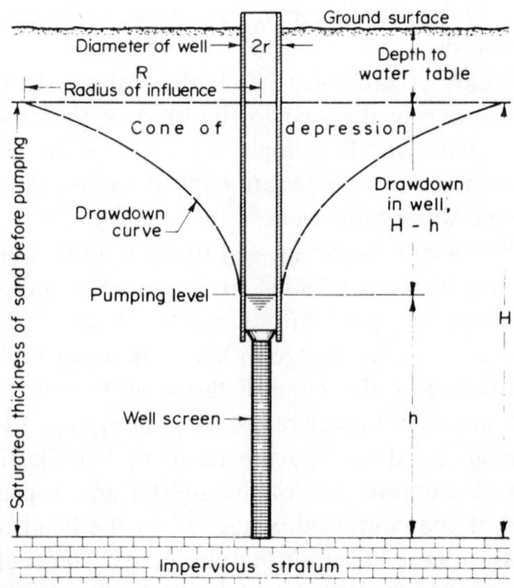

Fig. 8-12 Diagram of a water-table well showing the significance of various terms used in the equilibrium formula.

where

Q = well yield or pumping rate in gallons per minute,
P = permeability of the water-bearing sand in gallons per day per square foot,
H = saturated thickness of the aquifer before pumping in feet,
h = depth of water in the well while pumping in feet,
R = radius of the cone of depression in feet,
r = radius of the well in feet.

Figure 8-13 is a vertical section of a well pumping from an artesian aquifer. The formula for a well operating under artesian conditions is:

$$Q = \frac{Pm(H - h)}{528 \log R/r} \qquad (5)$$

where

m = thickness of the aquifer in feet,
H = static head at the bottom of aquifer in feet of water.

All other terms are as defined for Eq. 4.

Derivations of the foregoing formulas are based on the following simplifying assumptions:

1. The water-bearing materials are of uniform permeability within the radius of influence of the well.
2. The aquifer is not stratified.
3. For a water-table aquifer, the saturated thickness is constant before pumping starts; for an artesian aquifer, the aquifer thickness is constant.
4. The pumping well is 100 percent efficient.
5. The pumping well penetrates to the bottom of the aquifer.
6. The water table and the piezometric surface are horizontal surfaces.
7. Laminar flow exists within the radius of influence of the well.
8. The cone of depression has reached equilibrium so that both drawdown and radius of influence of the well do not change with continued time of pumping at a given rate.

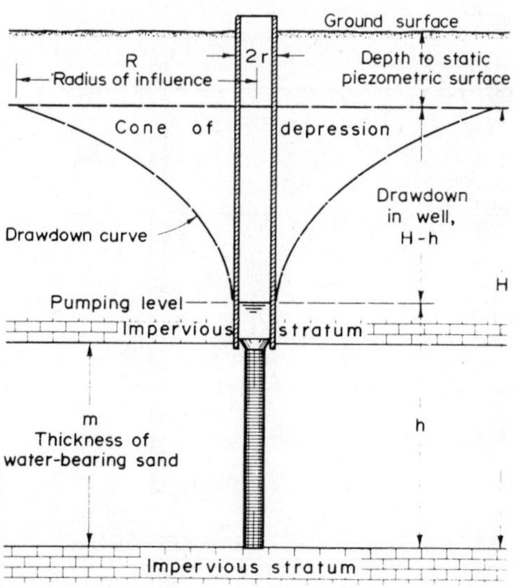

Fig. 8-13 Diagram of a well in an artesian aquifer showing the significance of various terms used in the equilibrium formula.

These assumptions appear to limit severely the use of the two formulas. In reality, however, they do not.

Uniform permeability is seldom found in a real aquifer, but the average permeability as determined from aquifer pumping tests has proved to be reliable for predicting well performance. For artesian wells where most of the aquifer thickness is penetrated and screened, the assumption of no stratification is not an important limitation. For water-table aquifers, where drawdown reduces the saturated thickness considerably, the situation can be handled when the stratification is known and taken into account in applying the formula.

Assumption of constant thickness is not a serious limitation because variation in aquifer thickness within the cone of depression in most actual situations is relatively small. Where changes in thickness are important, they can be taken into account, as will be explained later.

The assumption that the well is 100 percent efficient can cause the calculated well yield to be seriously in error in cases where the real well may be inefficient as the result of improper design or construction. Factors that contribute to inefficiency are discussed later.

Assumption that the water table or the piezometric surface is horizontal before pumping is never fulfilled. The slope or hydraulic gradient, however, is usually very flat and the effect on calculation of well yield is negligible in most cases. Slope of the water table or the piezometric surface does cause distortion of the cone of depression, making it more elliptical than circular.

Flow at all places in the aquifer is considered to be laminar. Some investigators have theorized that turbulent flow near a well could result in relatively high head losses. Actual laboratory and field tests, however, show that some departure from laminar flow causes additional head loss of only minor proportion.

Determining Aquifer Permeability. Equations 4 and 5 can be used to calculate well yield if P, H, m, and R are known. The well log provides values of H and m, R is usually estimated, and P must be determined from laboratory or field tests. For a water-table aquifer, the formula for calculating P is:

$$P = \frac{1055Q \log r_2/r_1}{(h_2^2 - h_1^2)} \tag{6}$$

where

P = permeability in gallons per day per square foot,
Q = pumping rate in gallons per minute,
r_1 = distance to the nearest observation well in feet,
r_2 = distance to the farthest observation well in feet,
h_2 = saturated thickness in feet at the site of the farthest observation well,
h_1 = saturated thickness in feet at the site of the nearest observation well.

Figure 8-14 shows a sectional view of a pumping test layout in a water table formation for the purpose of determining the permeability of the formation. All of the pertinent factors are easily measured in this kind of test and the permeability of the aquifer can be accurately determined.

For artesian conditions, the formula for calculating the permeability is:

$$P = \frac{528Q \log r_2 r_1}{m(s_1 - s_2)} \tag{7}$$

where all terms except the following are the same as for Eq. 6:

m = thickness of the aquifer in feet,
s_2 = drawdown in feet at the site of the nearest observation well,
s_1 = head in feet at the site of the nearest observation well.

Besides providing accurate means for calculating the average permeability of the aquifer, the equilibrium formulas are useful for studying the relationship of various factors to each other

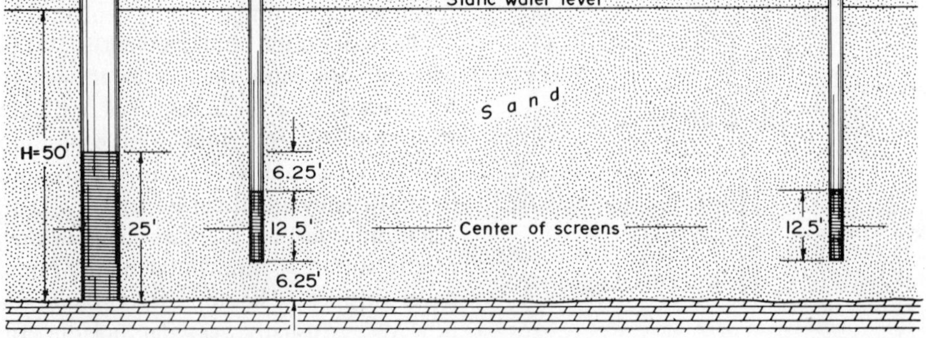

Fig. 8-14 Typical arrangement of pumped well and observation wells to obtain field data needed for calculating the coefficient of permeability from well discharge formulas.

and to well yield. They show, for example, that the well yield is directly proportional to the permeability, all other things being equal. A formation with double the permeability of another should have double the yield.

For an artesian aquifer, Eq. 5 indicates that the yield is directly proportional to the formation thickness, all other things being equal.

Relation of Well Size to Yield. How does the radius or diameter of a well affect the yield? Many persons assume that doubling the diameter of a well doubles its yield. This is far from true. With all other factors constant, Eq. 4 shows that Q varies as $K/\log R/r$, where K represents all the constant terms. From this relationship, the theoretical increase in yield that results from enlarging the well diameter can be calculated. Table 8-1 shows the yield ratios when $R = 400$ ft—a typical value for water table conditions. It indicates that if a 6-in. well will yield 100 gpm with a certain drawdown, a 12-in. well constructed at the same spot will yield 110 gpm with the same drawdown; an 18-in. well will yield 117 gpm. A 48-in. well will yield 137 gpm, or 37 percent more water at the same drawdown.

These ratios apply both to specific capacity and total yield. For example, if a 12-in. well is producing 20 gpm/ft of drawdown, then a 24-in. well in the same location would produce 111 percent as much or 22.2 gpm/ft of drawdown. The table shows that doubling the diameter of a water-table well will increase its yield about 11 percent.

Rules for selecting the proper diameter under various situations are given in the discussion on well design. The decision depends on a study of each case and on the cost factors involved.

TABLE 8-1. Well Diameter vs. Yield Ratio, in Percent

Well Diameters						
6 in.	12 in.	18 in.	24 in.	30 in.	36 in.	48 in.
100	110	117	122	127	131	137
–	100	106	111	116	119	125
–	–	100	104	108	112	117
–	–	–	100	104	107	112
–	–	–	–	100	103	108
–	–	–	–	–	100	105

Relation of Drawdown to Yield. Equation 5 for a well operating under artesian conditions shows that the yield is directly proportional to the drawdown, $H - h$, as long as the drawdown does not exceed the distance from the static piezometric surface to the top of the aquifer. If the drawdown exceeds this amount, then m is reduced and the proportionality no longer holds true. Theoretically, this means that if the drawdown is doubled, the yield is doubled. Stated another way, the specific capacity of the well is constant at any pumping rate as long as the aquifer is not unwatered.

For a well in a water-table aquifer, the part of the formation within the cone of depression is actually unwatered during the pumping. This influences the ratio of drawdown to yield. When the drawdown is doubled, the well yield is less than doubled. The specific capacity decreases with increased drawdown; in fact, it decreases directly in proportion to the drawdown.

Nonequilibrium Well Formula. Theis developed the nonequilibrium well formula in 1935.[5] The Theis equation was the first to take account of the effect of time of pumping on well yield. Its derivation was a major advance in the field of groundwater hydraulics. By use of this concept, the drawdown can be predicted at any time after pumping begins.

The transmissivity and average permeability can be determined from the early stages of a pumping test rather than only after water levels in observation wells have virtually stabilized or reached equilibrium. The aquifer coefficients can be determined from the time-drawdown measurements in a single observation well rather than having to use two observation wells as is required in the use of Eq. 6 or 7.

Derivation of the Theis equation is based on the following assumptions:

1. The water-bearing formation is uniform in character and permeability in both horizontal and vertical directions.
2. The formation has uniform thickness.
3. The formation has infinite areal extent.

4. The formation receives no recharge from any source.
5. The pumped well penetrates and receives water from the full thickness of the water-bearing formation.
6. The water removed from storage is discharged instantaneously with lowering of the head.

These assumptions are essentially the same as those for the equilibrium formulas except that, as noted above, the water levels within the cone of depression need not to have stabilized or reached equilibrium.

In its simplest form, the Theis equation is:

$$s = \frac{114.6Q}{T} W(u) \qquad (8)$$

where

s = drawdown in feet at any point in the vicinity of a well discharging at a constant rate,
Q = pumping rate in gallons per minute,
T = coefficient of transmissivity of the aquifer in gallons per day per foot,
$W(u)$ is read "well function of u," in which $u = 1.87r^2S/Tt$, where
r = distance in feet from the center of the pumped well to any point on the cone of depression,
S = coefficient of storage (dimensionless),
T = coefficient of transmissivity in gallons per day per foot,
t = time since pumping started in days.

Table 8-2 shows values of $W(u)$ corresponding to values of u. If T and S are known, values for these and other terms can be substituted in the equation to obtain a desired unknown.

For example, suppose the transmissivity of an artesian aquifer is 50,000 gpd/ft and the coefficient of storage is 5×10^{-4}—a typical value for artesian conditions. What would be the specific capacity, Q/s, of a 12-in. well after 1 day of continuous pumping?

In this case, $r = 0.5$ ft and $t = 1$ day. First, we calculate u:

$$u = \frac{1.87r^2S}{Tt} = \frac{1.87 \times (0.5)^2 \times 5 \times 10^{-4}}{50{,}000 \times 1}$$
$$= 4.7 \times 10^{-9}$$

Next, the value of $W(u)$ corresponding to this value of u is read from Table 8-2. $W(u)$ in this case is 18.60. Rearranging Eq. 8 we have:

$$\frac{Q}{s} = \frac{T}{114.6W(u)} = \frac{50{,}000}{114.6 \times 18.60}$$
$$= 23.5 \text{ gpm/ft of drawdown}$$

This calculation assumes that the well is 100 percent efficient. The result can be considered as the theoretical specific capacity of the well.

Having calculated the specific capacity, the drawdown in the pumped well for any pumping rate is easily determined. If the pumping rate is 200 gpm, then

$$s = \frac{Q}{Q/s} = \frac{200}{23.5} = 8.5 \text{ ft}$$

assuming that the pumped well is 100 percent efficient.

If a well is discharging 200 gpm, what would be the drawdown 1000 ft from the pumped well after 1 day of pumping?

$$u = \frac{1.87 \times 1000^2 \times 5 \times 10^{-4}}{50{,}000 \times 1} = 1.9 \times 10^{-2}$$

For this value of u, Table 8-2 shows that $W(u)$ = 3.4. Substituting in Eq. 8 we have

$$s = \frac{114.6 \times 200 \times 3.4}{50{,}000} = 1.6 \text{ ft}$$

This calculation is independent of the efficiency of the pumped well.

Determining T and S. Use of the Theis equation to determine T and S from a field pumping test requires measurements of drawdown in at least one observation well.[8] These should be made at proper intervals of time after the instant that pumping starts. Measurements in more than one observation well can also be used.

A direct calculation from Eq. 8 is not possible. Theis worked out a graphical means of solution, however, which makes it possible to find T and S if values of the other terms are known.

The method involves matching a curve plotted from specific pumping test data with what is called a "type curve." The type curve is pre-

TABLE 8-2. Values of $W(u)$ Corresponding to Values of u for Use in the Theis Equation (Values of u are $N \times 10^n$)

N	$N \times 10^{-15}$	$N \times 10^{-14}$	$N \times 10^{-13}$	$N \times 10^{-12}$	$N \times 10^{-11}$	$N \times 10^{-10}$	$N \times 10^{-9}$	$N \times 10^{-8}$
1.0	33.96	31.66	29.36	27.05	24.75	22.45	20.15	17.84
1.5	33.56	31.25	28.95	26.65	24.35	22.04	19.74	17.44
2.0	33.27	30.97	28.66	26.36	24.06	21.76	19.45	17.15
2.5	33.05	30.74	28.44	26.14	23.83	21.53	19.23	16.93
3.0	32.86	30.56	28.26	25.96	23.65	21.35	19.05	16.75
3.5	32.71	30.41	28.10	25.80	23.50	21.20	18.89	16.59
4.0	32.57	30.27	27.97	25.67	23.36	21.06	18.76	16.46
4.5	32.46	30.15	27.85	25.55	23.25	20.94	18.64	16.34
5.0	32.35	30.05	27.75	25.44	23.14	20.84	18.54	16.23
5.5	32.26	29.95	27.65	25.35	23.05	20.74	18.44	16.14
6.0	32.17	29.87	27.56	25.26	22.96	20.66	18.35	16.05
6.5	32.09	29.79	27.48	25.18	22.88	20.58	18.27	15.97
7.0	32.02	29.71	27.41	25.11	22.81	20.50	18.20	15.90
7.5	31.95	29.64	27.34	25.04	22.74	20.43	18.13	15.83
8.0	31.88	29.58	27.28	24.97	22.67	20.37	18.07	15.76
8.5	31.82	29.52	27.22	24.91	22.61	20.31	18.01	15.70
9.0	31.76	29.46	27.16	24.86	22.55	20.25	17.95	15.65
9.5	31.71	29.41	27.11	24.80	22.50	20.20	17.89	15.59

N	$N \times 10^{-7}$	$N \times 10^{-6}$	$N \times 10^{-5}$	$N \times 10^{-4}$	$N \times 10^{-3}$	$N \times 10^{-2}$	$N \times 10^{-1}$
1.0	15.54	13.24	10.94	8.633	6.332	4.038	1.823
1.5	15.14	12.83	10.53	8.228	5.927	3.637	1.465
2.0	14.85	12.55	10.24	7.940	5.639	3.355	1.223
2.5	14.62	12.32	10.02	7.717	5.417	3.137	1.044
3.0	14.44	12.14	9.837	7.535	5.235	2.959	0.906
3.5	14.29	11.99	9.683	7.381	5.081	2.810	0.794
4.0	14.15	11.85	9.550	7.247	4.948	2.681	0.702
4.5	14.04	11.73	9.432	7.130	4.831	2.568	0.625
5.0	13.93	11.63	9.326	7.024	4.726	2.468	0.560
5.5	13.84	11.53	9.231	6.929	4.631	2.378	0.503
6.0	13.75	11.45	9.144	6.842	4.545	2.295	0.454
6.5	13.67	11.37	9.064	6.762	4.465	2.220	0.411
7.0	13.60	11.29	8.990	6.688	4.392	2.151	0.374
7.5	13.53	11.22	8.921	6.619	4.323	2.087	0.340
8.0	13.46	11.16	8.856	6.555	4.259	2.027	0.311
8.5	13.40	11.10	8.796	6.494	4.199	1.971	0.284
9.0	13.34	11.04	8.739	6.437	4.142	1.919	0.260
9.5	13.29	10.99	8.685	6.383	4.089	1.870	0.239

pared by plotting values of $W(u)$ against $1/u$ on graph paper with logarithmic scales, as shown in Fig. 8-15. Values from Table 8-2 are used for this. Data from a pumping test are plotted on similar graph paper with logarithmic scales and cycles identical to those of the type curve.

Table 8-3 shows some data from a pumping test. Drawdown was measured at frequent intervals in an observation well 400 ft from a well that was pumped at a constant rate of 500 gpm. Since the measurements show that the water level is still dropping after 240 min of pumping, analysis of the test data requires use of the Theis nonequilibrium concept.

For solution by the nonequilibrium equation, the data are plotted on logarithmic graph paper as shown in Fig. 8-16A. Drawdown is plotted on the vertical axis and time since pumping started on the horizontal axis. This graph is then superimposed on the type curve sheet so that the plotted points fall on or fit some portion of the type curve. In finding the position of best fit, the axes of both graphs must be kept parallel.

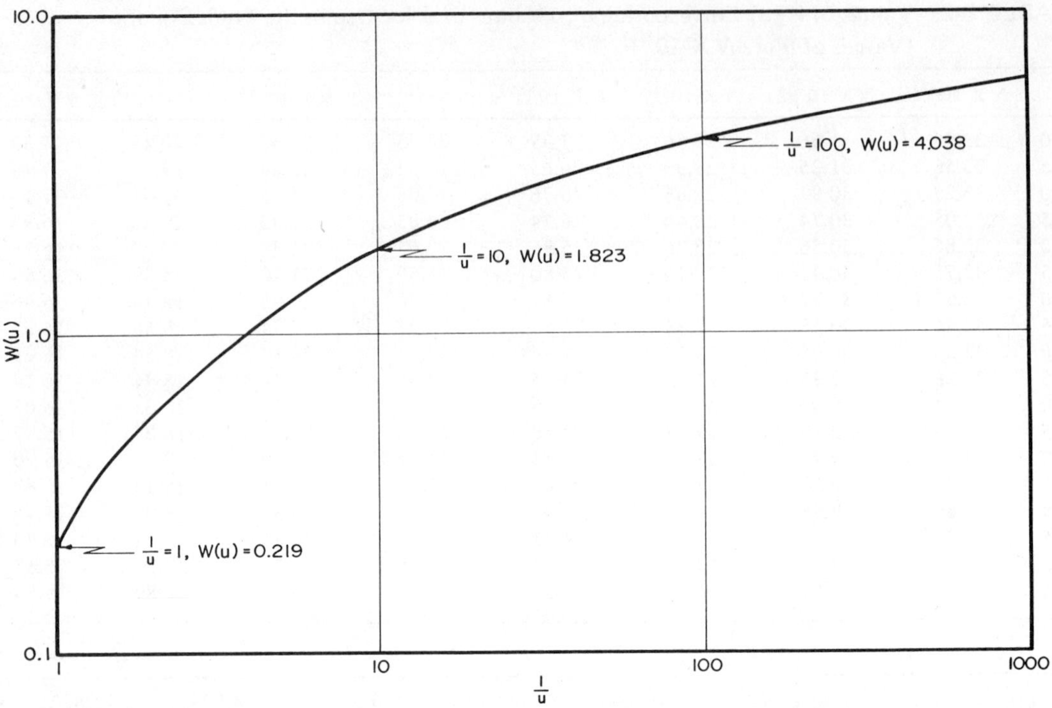

Fig. 8.15 Type curve for use in solving Theis nonequilibrium equation graphically with values of $W(u)$ (well function of u) corresponding to values of $1/u$. The curve is plotted on a logarithmic scale.

TABLE 8-3. Drawdown Measurements in Observation Well

Time Since Pump Started, minutes	Drawdown, s, feet
1.0	0.16
1.5	0.27
2.0	0.38
2.5	0.46
3.0	0.53
4.0	0.67
5.0	0.77
6	0.87
8	0.99
10	1.12
12	1.21
14	1.30
18	1.43
24	1.58
30	1.70
40	1.88
50	2.00
60	2.11
80	2.24
100	2.38
120	2.49
150	2.62
180	2.72
210	2.81
240	2.88

Once a good matching position is found, a match point is selected. The match point can be any point in the overlap area of the curve sheets. It is most convenient to select a match point where the coordinates on the type curve are known in advance. This simplifies the computations.

In Fig. 8-16B, the match point is shown as the point on the type curve where $1/u = 100$ and $W(u) = 4.038$. At the corresponding point on the time-drawdown diagram, $s = 2.3$ ft and $t = 83$ min or 83/1440-day.

Substituting in Eq. 8 we have

$$T = \frac{114.6Q}{s} W(u)$$

$$= \frac{114.6 \times 500}{2.3} \times 4.038$$

$$= 100{,}000 \text{ gpd/ft}$$

After determining T, we can calculate S from the following relationship:

$$S = \frac{uTt}{1.87r^2}$$

The value of r in this example is 400 ft, the

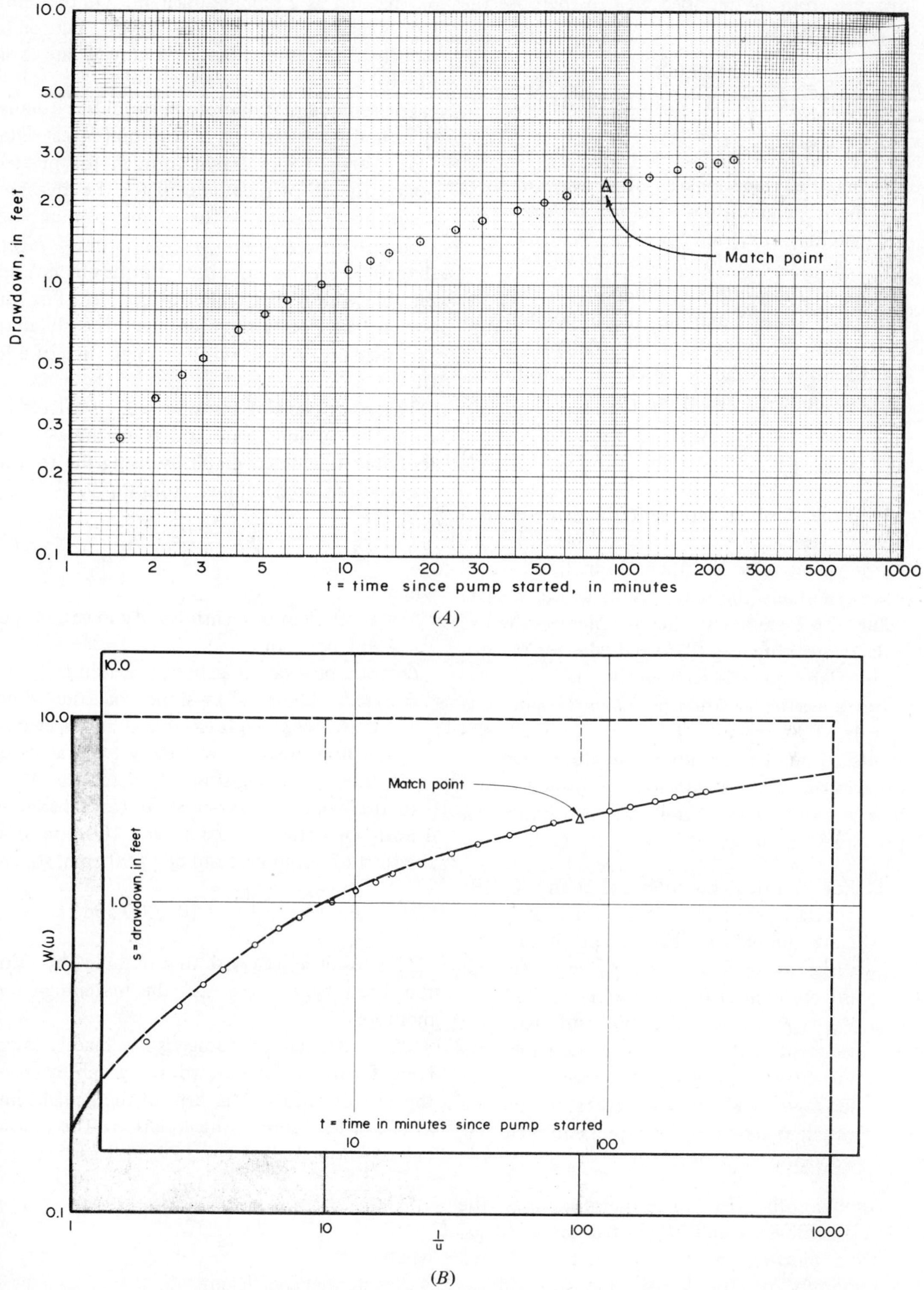

Fig. 8-16 (*A*) Data given in Table 8-11 plotted on a logarithmic scale define a curve similar in shape to that of the type curve shown in Fig. 8-15. (*B*) Diagram of plotted points representing the pumping test data superimposed on the type curve in position where the points fall on the type curve. Match point chosen here is for $1/u = 100$.

distance from the pumped well to the observation well. Thus,

$$S = \frac{1 \times 100{,}000}{100 \times 1.87 \times 400^2} \times \frac{83}{1440}$$

$$= 1.9 \times 10^{-4}$$

Detailed explanations of other ways of solving the nonequilibrium equation by means of the matching-curve technique are available.[8]

Modified Nonequilibrium Equation. In working with the Theis equation, Jacob pointed out that when the value of u is sufficiently small, the equation can be modified to the following form without significant error:

$$s = \frac{264Q}{T}\left(\log \frac{0.3Tt}{r^2S}\right) \qquad (9)$$

where the symbols represent the same quantities as in Eq. 8.

For values of u less than about 0.05, Eq. 9 gives essentially the same results as Eq. 8. The value of u becomes smaller as t increases and as r decreases; thus Eq. 9 is valid where t is sufficiently large and r is sufficiently small.

For a particular situation when the pumping rate is held constant, Q, T, and S are all constants. Equation 9 shows, therefore, that the drawdown, s, varies with $\log t/r^2$ provided that u is less than 0.05. From this, two important principles can be stated:[9]

1. For a particular aquifer and at any specific point (where r is constant) the terms s and t are the only variables in Eq. 9. Thus, s varies as $\log K_1 t$, where K_1 represents all the constant terms in the expression.
2. For a particular formation and at a given value of t, the terms s and r are the only variables in Eq. 9. In this case, s varies as $\log K_2/r^2$, where K_2 represents all the constant terms in the expression including a specific value of t.

Applying the first of these principles, the time-drawdown graph is plotted on semilogarithmic plotting paper.[8] Time, t, is plotted horizontally on the logarithmic scale; drawdown, s, is plotted vertically on the arithmetic scale. Figure 8-17 shows the data from Table 8-3 plotted as a semilog diagram. On the semilog diagram, most of the points fall on a straight line rather than on a curved line as in Fig. 8-16.

The semilog plot, therefore, simplifies calculations and interpretation of the aquifer test data.

The straight line drawn through the plotted points fits all of them except for the measurements made during the first 10 min of pumping. During the first 10 min, the value of u is larger than 0.05, so the modified formula is not applicable within that region of the test. For this reason, the points representing the early measurements do not lie on the straight line of best fit.

The coefficient of transmissivity is calculated from the pumping rate and from the slope of the time-drawdown graph by using the following relationship developed from Eq. 9:

$$T = \frac{264Q}{\Delta s} \qquad (10)$$

where

T = coefficient of transmissivity in gallons per day per foot,

Q = pumping rate in gallons per minute,

Δs (read "delta s") = slope of the time-drawdown graph expressed as the change in drawdown between any two values of time on the log scale whose ratio is 10.

In the example, $\Delta s = 1.30$ ft (the change in drawdown between 10 min and 100 min after the start of pumping), and $Q = 500$ gpm, so:

$$T = \frac{264 \times 500}{1.30} = 101{,}000 \text{ gpd/ft}$$

This result agrees with that obtained by using the Theis type curve and the matching-curve method.

The coefficient of storage is also readily calculated from the time-drawdown graph by using the zero drawdown intercept of the straight line as one of the terms in the equation. The following equation, derived from Eq. 9, is used:

$$S = \frac{0.3Tt_0}{r^2} \qquad (11)$$

where

S = storage coefficient,

T = coefficient of transmissivity in gallons per day per foot,

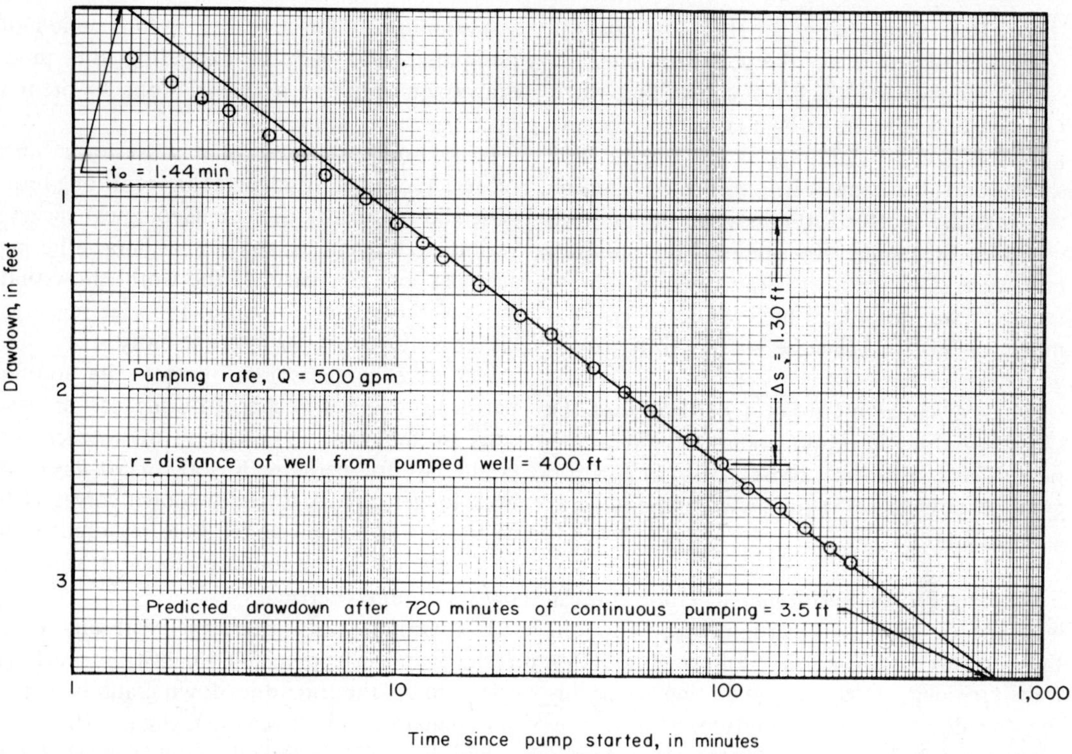

Fig. 8-17 When data from Table 8-3 are plotted with semilogarithmic coordinates, most of the plotted points fall on a straight line. From this graph, Δ_s and t_o are determined as indicated.

t_0 = intercept of the straight line at zero drawdown in days,

r = distance in feet from the pumped well to the observation well where drawdown measurements were made.

In our example, $t_0 = 1.44$ min, or $1.44/1440 = 0.001$ day; $T = 101{,}000$ gpd/ft; and $r = 400$ ft.

$$S = \frac{0.3 \times 101{,}000 \times 0.001}{400 \times 400} = 1.9 \times 10^{-4}$$

This result also agrees with the calculation from the Theis type curve.

The data used in this example are from an observation well located 400 ft from a production well pumped at a rate of 500 gpm for 240 min. If measurements had been taken in another observation well which was farther from the pumped well, and if the data were plotted in Fig. 8-17, the points would fall along a straight line parallel to the one shown but above it. If measurements in an observation well closer to the pumped well—or in the pumped well itself—were plotted in Fig. 8-17, they would fall on a straight line parallel to the one shown but below it. The values of T and S calculated from all three sets of data would be the same.

In addition to its use for calculating the aquifer constants, the time-drawdown diagram provides a graphical means of predicting future drawdown. The straight line in Fig. 8-17 may be extended to the right to indicate the drawdown that would occur in the observation well (400 ft from the pumped well) after any period of continuous pumping at 500 gpm.

In the example, the predicted drawdown after 12 hr or 720 min of continuous pumping is indicated by extrapolating the time-drawdown curve. Figure 8-17 shows that for $t = 720$ min, $s = 3.5$ ft.

It is good practice to pump an artesian well for a period of 24 hr to get data for the time-drawdown curve. A water-table well should be pumped for 3 days. The 240-min pumping period in the illustrative example is too short for most situations. A longer test provides data that define more reliably the slope and position of the line of best fit for the plotted points.

For a particular well, Eq. 9 shows that when t is constant, the drawdown, s, is directly proportional to the pumping rate. Figure 8-17 shows that the drawdown in the observation well was 2.5 ft after 120 min when the pumped well was discharging 500 gpm. If the pumping rate were 1000 gpm, the drawdown in the observation well after 120 min would be twice as much, or 5.0 ft. Likewise, at $Q = 1500$ gpm, s would be 7.5 ft.

For a given aquifer, Eq. 10 shows that the slope, s, is directly proportional to the pumping rate. The test shows s to be 1.30 ft at $Q = 500$ gpm. If Q were 1000 gpm, s would be twice as great or 2.60 ft.

Estimating Future Performance. Good use can be made of these two relationships. Applying them to test data from one or more observation wells provides a means of predicting accurately the drawdown in the observation wells at rates of pumping different from the test rate. Applying them to drawdown data from the pumped well provides a way of making similar calculations, within reason, for its performance, provided that it is an efficient well and that the saturated thickness of the aquifer is not too greatly reduced at a higher assumed pumping rate.

The foregoing discussion and examples of the use of the Theis equation and the Jacob modification adhere to the basic assumptions employed in deriving these mathematical expressions. At this point, it is helpful to review some of these assumptions and to see the effects of some departures from them upon the time-drawdown relationship. Real aquifers do not conform fully to the assumed conditions, so limits of application must be judged for those cases where the differences are significant.

The assumption that the aquifer receives no recharge during the pumping period is one of the six fundamental conditions upon which the nonequilibrium equations are based. This means that all water discharged from a well is assumed to be taken from storage in the aquifer. This fundamental concept makes it possible to calculate the transmissivity from the time-drawdown data employing Eq. 10.

Assuming no recharge during the period of pumping also permits extending the time-drawdown graph at its initial slope to predict drawdowns at future times.

It is known, however, that most formations receive recharge. Hydrographs from long-term observation wells by the U.S. Geological Survey, various state agencies, and similar data-gathering agencies in other parts of the world show that most water-bearing formations receive recharge either continuously or intermittently.

Where recharge is intermittent, the aquifer performs without recharge during certain seasons of the year. The period of no recharge could last for perhaps 30, 60, or 90 days, or even longer. Thus, time-drawdown curves such as those that are presented here represent well performance corresponding to a season or period of no recharge.

If a pumping test is performed carefully while recharge in an appreciable quantity is reaching the aquifer, the time-drawdown graph from the pumping test will reflect the recharge effect.

Figure 8-18 is a time-drawdown graph for a pumped well operating under conditions of no recharge. The well was pumped at a constant rate of 350 gpm and the drawdown measurements were obtained at intervals of time during a pumping period of 360 min. The points plotted on semilogarithmic paper define a straight line with a slope, or s value, of 9.3 ft. As has already been seen, the future drawdown in this well for any period of continuous pumping at 350 gpm can be quickly estimated by extending the straight line as desired. The drawdown corresponding to 5000 min of continuous pumping—about $3\frac{1}{2}$ days—is 73.3 ft in this case.

The anticipated pumping level and the pump setting required to provide adequate submergence for the pump bowls are easily determined by this method. Any desired safety factor can be applied to the calculated pumping level to take into account changes in well performance resulting from possible incrustation or from interference effects of other wells that might be constructed nearby at a future time.

Continuous pumping means operating 24 hr a day with no opportunity for recovery of water level. A well that is pumped for only a part of the full 24 hr each day will not show the same cumulative drawdown after 7, 30, or 90 days as

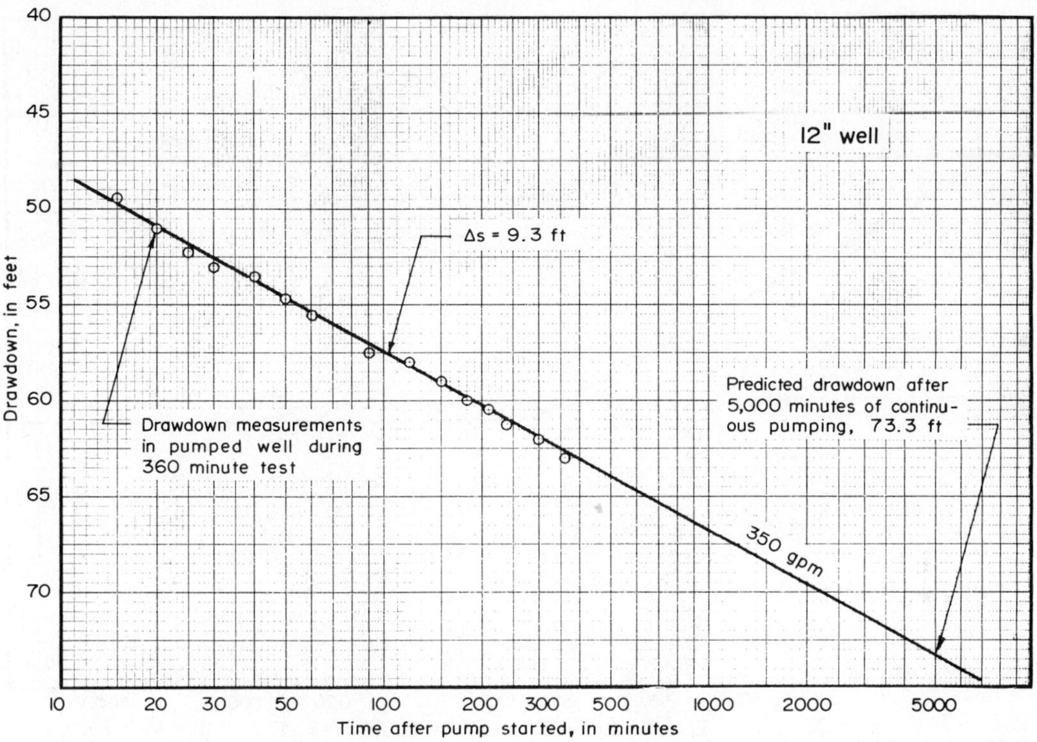

Fig. 8-18 Time-drawdown curve for the pumped well can be extrapolated to predict drawdown for a period of continuous pumping beyond the test period.

would be indicated by a time-drawdown graph such as Fig. 8-18. Obviously, a well that is pumped on a cycle of 12 hr on and 12 hr off benefits from the recovery of water level during the 12-hr idle period.

On the other hand, unless some recharge takes place while the pump is stopped, the water level will not fully recover to the original static level. Each time pumping is resumed, the drawdown starts from a new level slightly lower than that which prevailed at the start of the previous pumping period.

In the absence of recharge, the drawdown at the end of 90 days of intermittent pumping might possibly be about equivalent to the drawdown at the end of 45 days of continuous pumping. This would be merely a rule of thumb equivalent; it does not represent a definite mathematical relationship for intermittent pumping.

When recharge to the aquifer within the zone of influence of the pumping well equals the rate of discharge of the well, the drawdown will stabilize. No further lowering of the water levels will occur as pumping continues at a constant rate. The time-drawdown graph then becomes horizontal, as in Fig. 8-19.

The first part of the curve in Fig. 8-19 shows that the cone of depression was continually enlarging during the first 240 min of pumping. At about this time, the cone of depression, or area of influence of the well, encountered a source of recharge. The rate of recharge was sufficient to equal the rate of pumping, resulting in an equilibrium of water levels throughout the area of influence. The second part of the graph in Fig. 8-19 represents a change to equilibrium conditions.

It sometimes happens that the recharge within the cone of depression takes place at a rate less than the rate of pumping of the well. While this changes the slope of the time-drawdown curve, it may not become horizontal. The slope becomes flatter than the original slope and indicates that the cone of depression is enlarging more slowly than during the first part of the pumping period.

The future drawdown in the well may be pre-

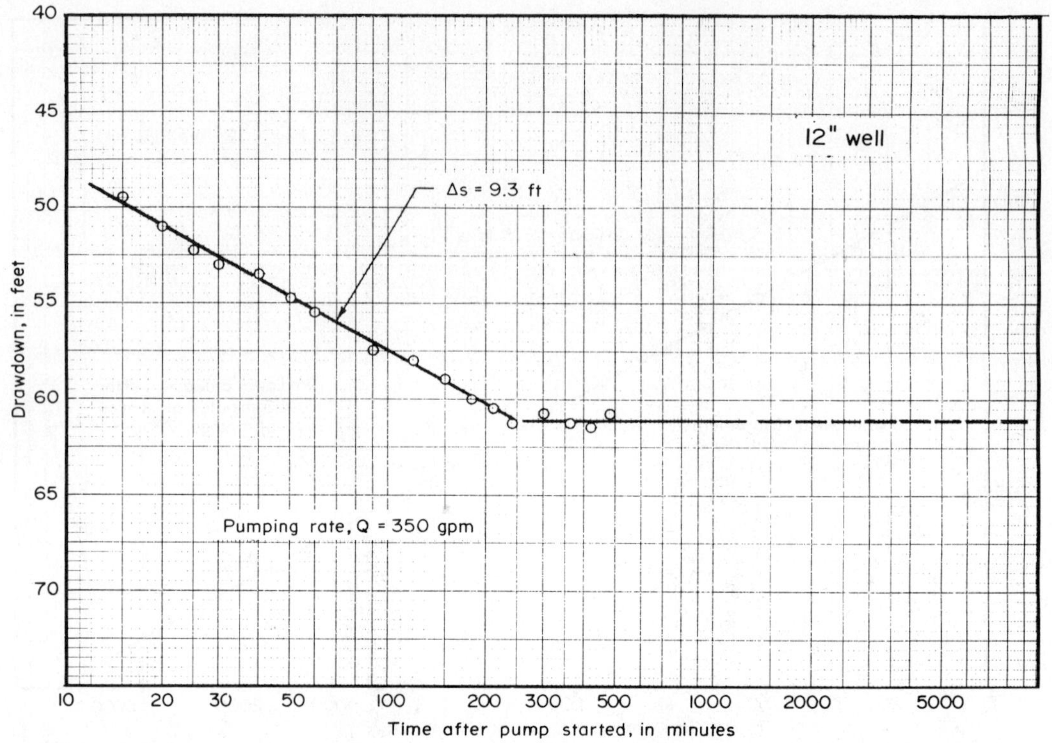

Fig. 8-19 When recharge to the aquifer occurs within the zone of influence of the well, the slope of the time-drawdown curve becomes flatter. The horizontal leg in this case indicates that recharge equals well discharge, and that equilibrium conditions were reached after 240 min of pumping.

dicted by extending the straight line at the slope of the second leg of the graph and reading the drawdown indicated by the extrapolated curve for any assumed period of continuous pumping.

The future drawdown for any period of continuous pumping must be determined graphically, using the method described here, when any condition causes a change in the slope of the drawdown graph during the pumping test. If the slope does not change, the future drawdown can be computed from the Theis formula or it may be found by the graphical method. If the slope changes, however, the Theis formula cannot be applied to any time period after the change from the initial slope.

It should be pointed out in passing that calculation of the transmissivity, T, of the water-bearing formation must be made from the value of s corresponding to the slope of the first part of the time-drawdown graph. Beyond the point where a change in slope occurs, a numerical value that may represent the slope of the second part of the graph is of no significance in analyzing the pumping test data. No attempt should be made to use any such value in either the Theis nonequilibrium or modified nonequilibrium formulas.

Recharge from a River. Development of equilibrium conditions that stabilize the cone of depression around a pumping well may occur under several general situations. One of these is where an aquifer is recharged from a river or lake. Figure 8-20 illustrates such a situation when equilibrium has been reached.

During the early part of the pumping period, the cone of depression does not extend to the river and no recharge effect is evident. When the cone of depression spreads beneath an area of the river bed, a hydraulic gradient develops between the groundwater in the aquifer and the water in the river. River water then percolates downward through the pervious stream bed

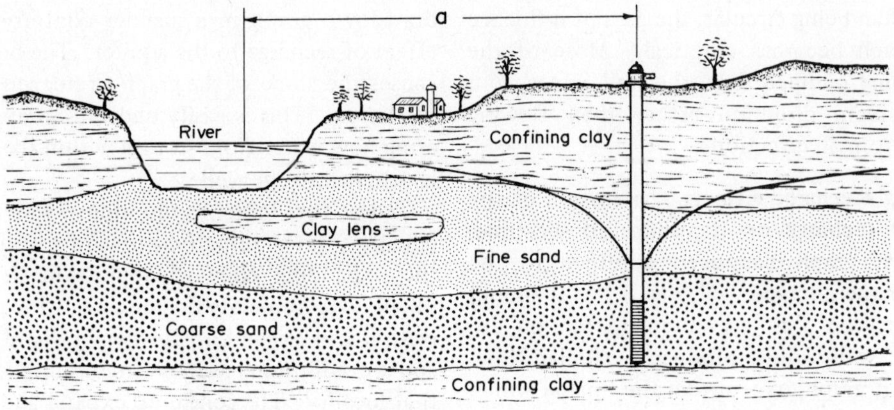

Fig. 8-20 Cone of depression expanding beneath the river bed creates a hydraulic gradient between the river and the aquifer. This can induce recharge to the aquifer from the river.

under the influence of the hydraulic gradient, if the stream bed is hydraulically connected with the aquifer. The river thus recharges the aquifer at a rate which increases as the cone of depression enlarges.

When the rate of recharge to the aquifer equals the rate of discharge from the well, the cone of depression and the pumping level become stable. The horizontal part of the graph in Fig. 8-19 corresponds to this condition and to the situation shown in Fig. 8-20.

From the extension of the graph in Fig. 8-19, it is seen that the predicted drawdown after 5000 min of continuous pumping would be 61.5 ft. The recharge to the aquifer in this case reduces the drawdown 11.8 ft from that shown in Fig. 8-18 after the same period of continuous pumping.

Recharge from Vertical Percolation. Another situation leading to the establishment of equilibrium conditions is one where vertical recharge occurs throughout the whole area of influence around the pumping well. An example of this would be where a well is completed in a water-table aquifer in which all the material in the zone of aeration, from the ground surface to the water table, is permeable sand. Assume that as the cone of depression is enlarging, infiltration occurs as the result of rain falling within the radius of influence of the well. When the quantity of water percolating to the water table within the circle of influence of the well equals the discharge from the well, the cone of depression ceases spreading and pumping levels within the well and the aquifer reach equilibrium.

Perhaps the more common type of recharge that brings about equilibrium conditions is vertical leakage of water from saturated layers of material above an aquifer tapped by a production well. Overlying strata in the zone of saturation often have considerably lower permeability than the deeper material in which the well is completed. The difference in permeability of the upper and lower strata may be such that the upper material may be considered as a confining layer. When the area of the circle of influence covers tens of thousands of square feet, the total vertical seepage from the upper material—even though it is of low permeability—can equal the well discharge and bring about equilibrium in the pumping levels.

Effect of a Sloping Water Table. The groundwater flow within an aquifer, toward and past a well, produces another situation which can bring about virtual stabilization of the pumping level. Groundwater moves because of the hydraulic gradient that has developed between areas of recharge and areas of discharge. The gradient is represented by the slope of the water table or piezometric surface. Most natural slopes of the water table or piezometric surface are relatively flat and do not affect appreciably the well performance curves. A relatively steep slope, however, causes distortion of the cone of depression around the well.

Rather than being circular, the area of influence of the well becomes elliptical. More of the water being pumped from the well comes from the upstream direction rather than flowing equally from all directions.

Detecting Recharge Effects. Discussion of the effect of recharge on the shape of the time-drawdown graph has been confined to the graph for the pumped well, Fig. 8-19. It should be understood that measurements in an observation well would produce a similar time-drawdown graph, except that the time at which the change in slope occurs would differ, depending upon the relative distance to the source of recharge.

Data from observation wells are usually more reliable and accurate than data from the pumped well, so the time-drawdown graphs from observation wells are most often relied upon to reveal the performance of the aquifer.

Theoretically, the two graphs—pumped well and observation well—should show the same value of Δs.

Assuming that the value of Δs from time-drawdown graphs for both an observation well and the pumped well is 9.3 ft, the same as shown in Fig. 8-19, the transmissivity of the water bearing formation would be:

$$T = \frac{264 \times 350}{9.3} = 10{,}000 \text{ gpd/ft}$$

It should be evident from this discussion that accurate and frequent measurements of drawdown at the start of a constant rate pumping test are extremely important. Sufficient dependable data must be obtained during the early part of the test to reflect clearly both the well performance and the aquifer performance before recharge effects or other outside influences which invalidate the basic nonequilibrium theory come into play.

Impervious Boundaries. Relatively few aquifers conform to the basic assumption of infinite extent in all directions from the pumped well. In many localities, definite geologic and hydraulic boundaries limit aquifers to areas ranging from a few to many square miles. This is especially true in glaciated regions.

An impervious boundary affects the time-drawdown graph in a manner contrary to the effect of recharge to the aquifer. The boundary causes the slope of the graph to steepen instead of flatten. This is easily understood by considering how water from storage within the aquifer is supplied to the well.

Under the basic assumptions, water flows equally from all directions toward a well. When the expanding cone of depression strikes an impervious boundary on one side of the pumped well, it can expand no further in that direction and no additional water can be supplied from that locality. The cone must expand and deepen more rapidly in all other directions to maintain the yield of the well. The effect on the semilogarithmic time-drawdown graph is to steepen its slope, as shown in Fig. 8-21.

Care must be used when analyzing aquifer test data where either nearby boundaries or recharge sources are suspected. Aquifer coefficients must be calculated from the early part of the test data if boundary effects are apparent. In no case should calculations be made from the slope of any part of the time-drawdown graph that reflects the effect of a boundary. The latest established slope of the time-drawdown graph should be extended when it is desired to predict water levels after longer terms of continuous pumping.

It is also pertinent to note the reason for the previous suggestion that the duration of pumping tests be at least 1 day for artesian wells and at least 3 days for water-table wells. Figure 8-21 shows evidence of a boundary effect 100 min after the pump was started. If the test had been conducted for only 100 min, the boundary effect would not have been revealed. Extending the initial slope in such a case would indicate a drawdown of 62 ft after 7 days (10,000 min) of pumping at 250 gpm, whereas the correct estimate of future drawdown is 73 ft, as determined from the extension of the second leg of the graph.

In an artesian aquifer, enlargement of the cone of depression during a 24-hr pumping period is usually extensive enough to encounter boundaries that could appreciably affect drawdown predictions from the semilog diagram. In a water-table aquifer, the cone of depression expands more slowly, so a longer pumping period

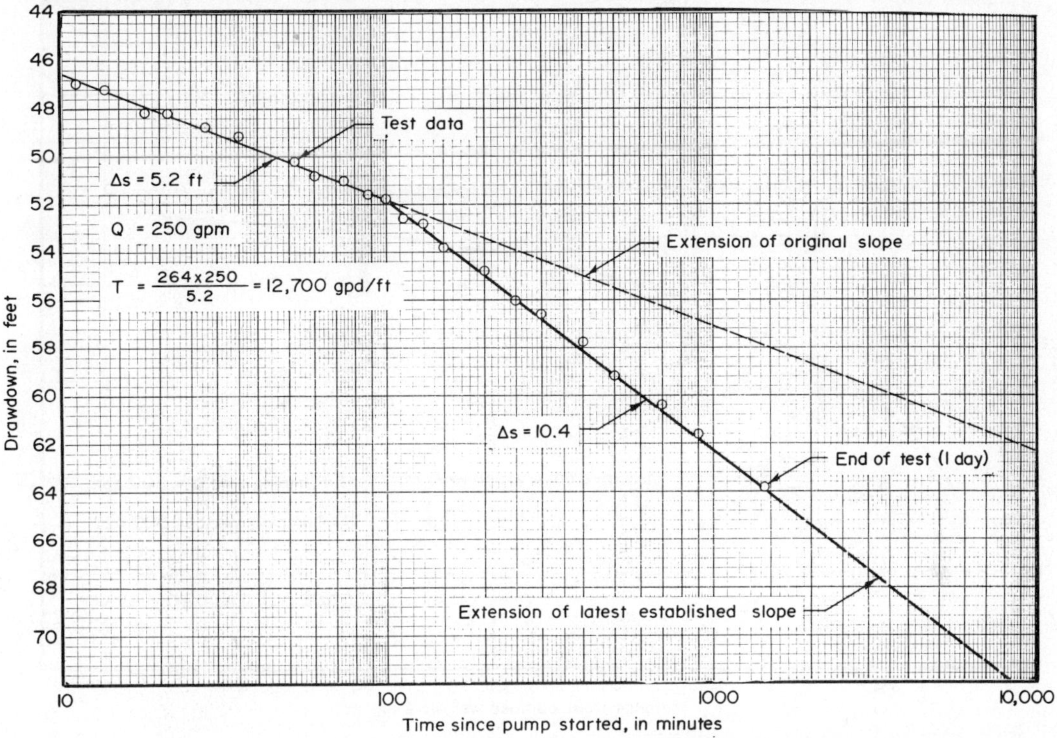

Fig. 8-21 Steepening of the time-drawdown curve indicates the presence of a hydraulic barrier. Here the expanding cone of depression encountered one or more impermeable boundaries after about 100 min of pumping.

is required to obtain evidence of boundaries that may exist.

Distance-Drawdown Diagram. If drawdown is measured simultaneously in three or more observation wells, a semilog distance-drawdown graph can be constructed. Three observation wells—each at a different distance from the pumped well—are needed to construct an accurate distance-drawdown graph.

If these values are plotted on ordinary graph paper, as in Fig. 8-22, the familiar trace of the cone of depression near a pumped well is obtained. This is similar to the curves in Figs. 8-9, 12, 13, and 20.

If the drawdowns in the same three wells are plotted on a semilog diagram, the drawdown curve becomes a straight line as in Fig. 8-23. The vertical scale represents drawdown; the horizontal scale represents the logarithm of the distance from the pumped well.

Points representing drawdowns in distant observation wells would fall a little below the straight line shown in Fig. 8-23, because at some distance from the pumped well the value of u is larger than 0.05. Where u is greater than 0.05, the straight line relationship of s to log r is no longer strictly true.

The semilog plot of the cone of depression—the distance-drawdown diagram—simplifies applications of the distance-drawdown relationship. The straight line may be extended to the right to determine the effect of the pumping at any distance from the pumped well. As an example, Fig. 8-23 shows that when the test well was pumped at 200 gpm for 500 min, it would have produced a drawdown of 2.6 ft in another well 300 ft away.

A simple transformation of Eq. 9 permits calculation of the transmissivity from the distance-drawdown diagram.[8] The slope of the straight line is used in a manner similar to the procedure used for the time-drawdown diagram. The equation for this is:

$$T = \frac{528Q}{\Delta s} \qquad (12)$$

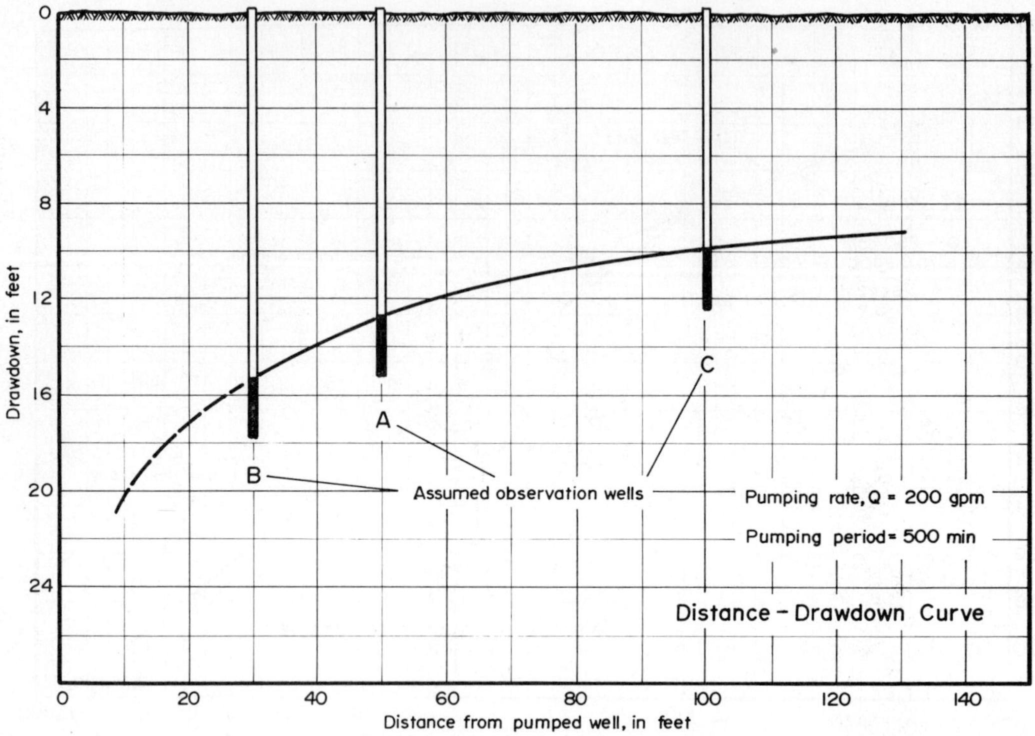

Fig. 8-22 Plotting drawdowns for three observation wells defines a trace of the cone of depression.

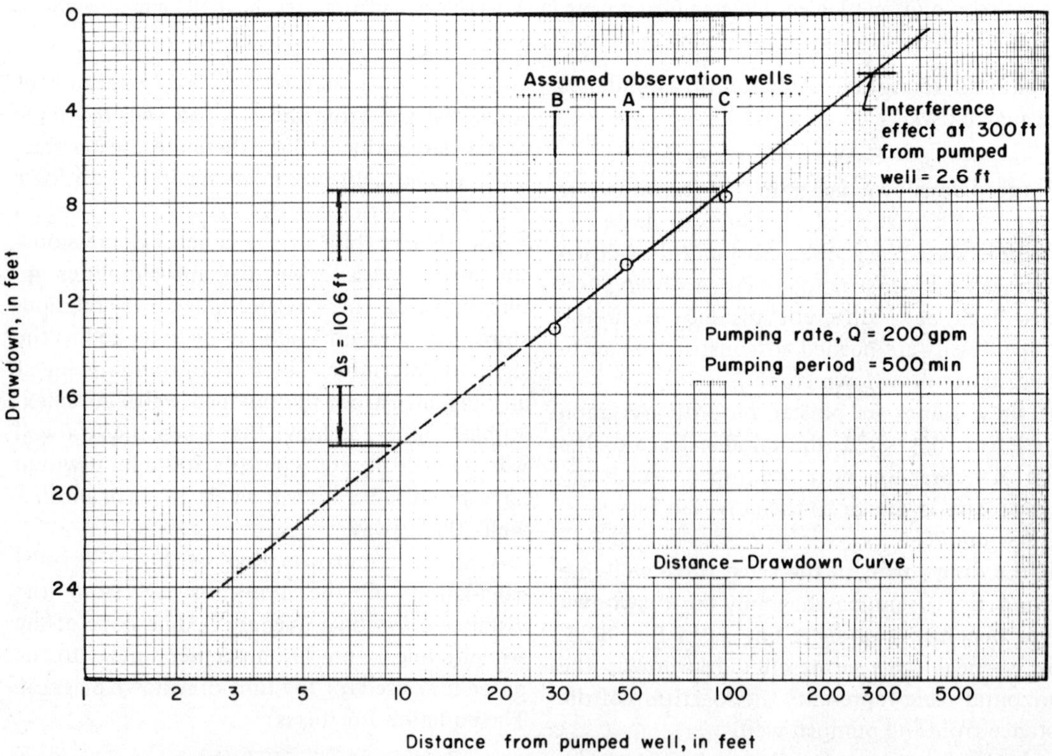

Fig. 8-23 Trace of the cone of depression plotted on semilog coordinates becomes a straight line. Drawdown in each well was measured at 500 min after starting the pumping test.

where

T = coefficient of transmissivity in gallons per day per foot,
Q = pumping rate in gallons per minute,
Δs = slope of the distance-drawdown graph expressed as the change in drawdown in feet between any two values of distance on the log scale whose ratio is 10.

For the example shown in Fig. 8-17,

$$T = \frac{528 \times 200}{10.6} = 10{,}000 \text{ gpd/ft}$$

The coefficient of storage can be obtained from the distance-drawdown diagram[9] using the following equation:

$$S = \frac{0.3Tt}{r_0^2} \qquad (13)$$

where

S = coefficient of storage,
T = coefficient of transmissivity in gallons per day per foot,
t = time since pumping started in days,
r_0 = intercept at zero drawdown of the extended straight line in feet.

From Fig. 8-23, the value of r_0 is 450 ft; T is 10,000 gpd/ft; and t is 500/1440 or 0.347-day, so:

$$S = \frac{0.3 \times 10{,}000 \times 0.347}{450 \times 450} = 5.1 \times 10^{-3}$$

It is seen, then, that aquifer coefficients may be calculated from two relationships revealed by an aquifer test:

1. Rate of lowering of water level with time at any place within the cone of depression—time-drawdown diagram.
2. Shape and position of the cone of depression at a given time—distance-drawdown diagram.

These calculations are independent of one another, so the result from one may be used as a check on the other.

It may be noted that the numerical constant of Eq. 12 is exactly twice that of Eq. 10. This is so because t appears in Eq. 9 to the first power, whereas r appears as the square or to the second power.

Since log r^2 is the same as 2 log r, it follows that the value of Δs for the distance-drawdown graph is just twice the Δs for the time-drawdown graph. This ratio of 2 for the slopes of the two straight lines—for a given aquifer and a given pumping rate—is a fixed relationship.

This means that when the value of Δs from a time-drawdown diagram is determined, it is known at once that the slope of the distance-drawdown graph, with the well pumping at the same rate, should be twice as great. This permits the construction of a distance-drawdown graph from a time-drawdown graph plotted with measurements from a single observation well. Where only one observation well is used, however, independent calculations of aquifer constants are not possible.

The following example explains how distance-drawdown graphs can be constructed. Figure 8-24 shows a semilog time-drawdown graph of data from observation well A which is 50 ft from the pumped well as indicated in Figs. 8-22 and 8-23. The Δs from the time-drawdown graph is 5.3 ft, exactly one-half the value of Δs from the distance-drawdown graph of Fig. 8-23.

If it is desired to construct a distance-drawdown graph after 300 min of pumping have elapsed, it is first noted from Fig. 8-24 that the drawdown in well A at 300 min is 9.4 ft. This measurement is plotted at the 50-ft point on a new diagram as shown in Fig. 8-25. A straight line with $\Delta s = 2 \times 5.3 = 10.6$ ft is then drawn through this point. Thus, a distance-drawdown graph is obtained which represents the cone of depression after 300 min of pumping at 200 gpm.[10]

Evaluating Interference Effects. The interference or drawdown in another well a distance 200 ft from the site of the pumped well in this case is obtained directly from Fig. 8-25. At the end of 300 min of pumping, the drawdown 200 ft away is 2.5 ft. If the pumping rate were 400 gpm, the drawdown in the aquifer at this point would be twice as great, or 5.0 ft. This would be the mutual interference between two wells 200 ft apart each pumping at 400 gpm.

The interference effects of several wells spaced at various distances can be evaluated in the same way. The total effect in any one well

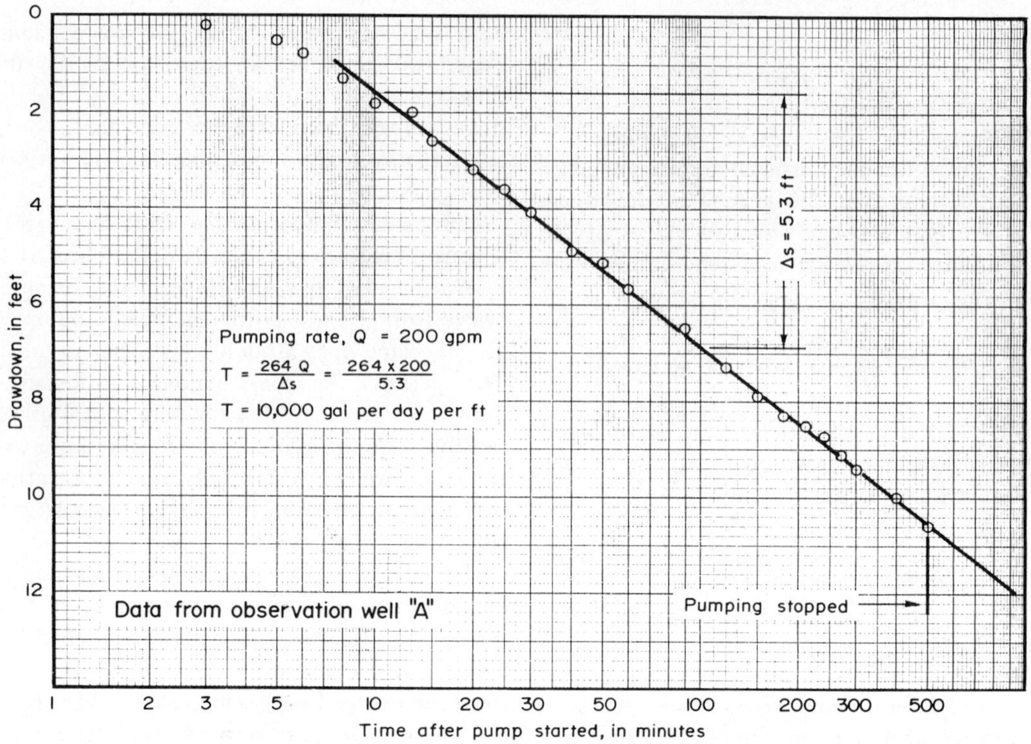

Fig. 8-24 Time-drawdown curve for observation well A situated 50 ft from a well being test-pumped.

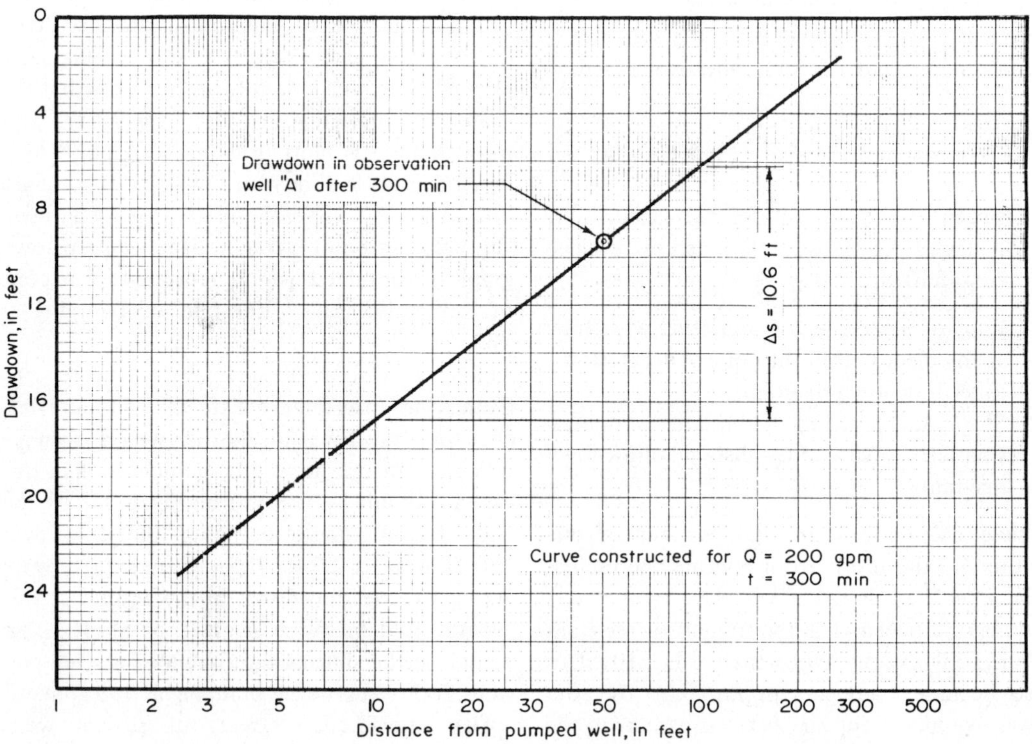

Fig. 8-25 A distance-drawdown curve constructed from the time-drawdown data, Fig. 8-24, represents the position of the cone of depression after 300 min of pumping and 200 gpm.

is the sum of the influences produced by all the others of the group.

In this discussion, the calculations of well interference are for conditions as they would exist in the aquifer after pumping for 300 min, since Fig. 8-25 was constructed on this basis.

If interference data for a different period of continuous pumping are desired, one must construct another distance-drawdown graph, starting with another point taken from the original test data. The drawdown in observation well A after 1000 min of pumping, shown by Fig. 8-24 to be 12.3 ft, might be used. The graph showing conditions after 1000 min of pumping at 200 gpm in Fig. 8-26 is parallel to the line for a pumping period of 300 min.

To provide a clear picture of interference in a well field where there are several wells pumping whose cones of influence overlap, Fig. 8-27 should be studied carefully. In Fig. 8-27A, the profiles of two wells that are 2000 ft apart are shown each pumping continuously at a rate of 500 gpm for 10 min. Note that at the end of 10 min the cones of depression have not reached one another. In Fig. 8-27B, the extent of the cones of depression after 2 days of continuous pumping is shown. If either of the two wells were pumped alone, then each cone of depression would appear as shown by the dashed profile for the well being pumped. However, when both wells are pumping, the net result is depicted by Fig. 8-27C, which combines the drawdowns at any point within the zone of influence of both wells.

A time-drawdown graph for each of the two wells in Fig. 8-27 would appear as shown in Fig. 8-28. The correct transmissivity of the aquifer would be obtained only from the initial slope of this curve. The secondary slope—about twice as steep—shows the effect of the other well. The net result of the second well, when pumping at the same rate, is mathematically equivalent to a formation boundary.

Pumped Well Efficiency. The efficiency of the pumped well can be estimated from the distance-drawdown graph in some cases. To do this, the straight line representing the profile of

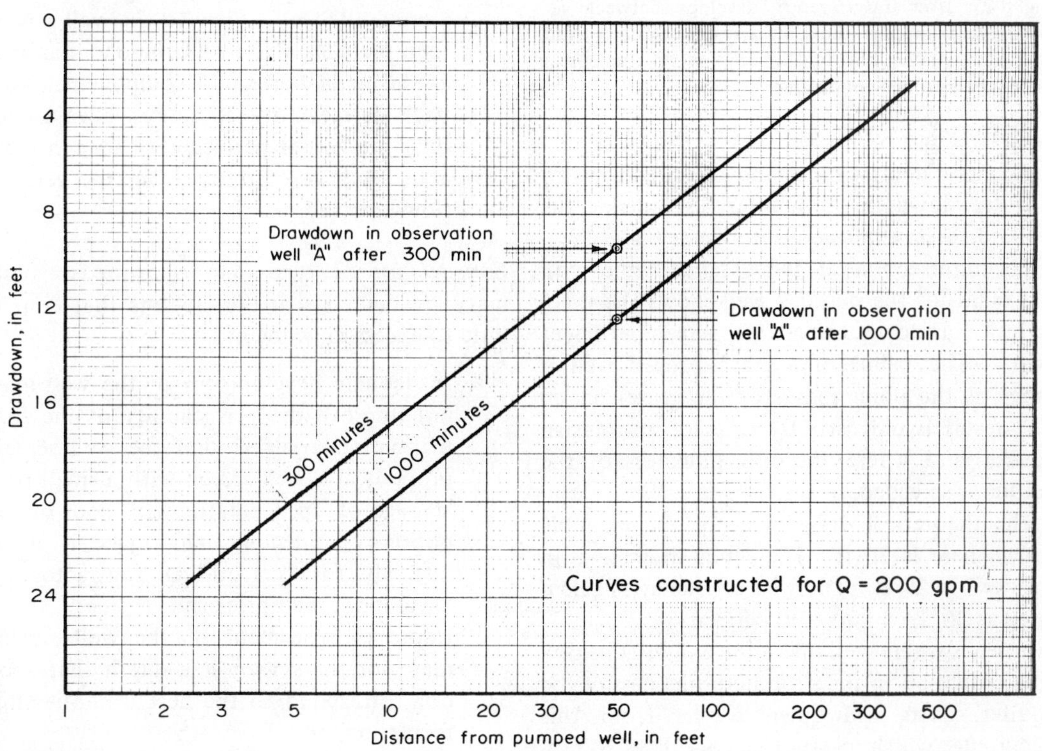

Fig. 8-26 The distance-drawdown diagram constructed to show the position of the cone of depression after pumping for 300 min and 1000 min, respectively, at a constant rate of 200 gpm.

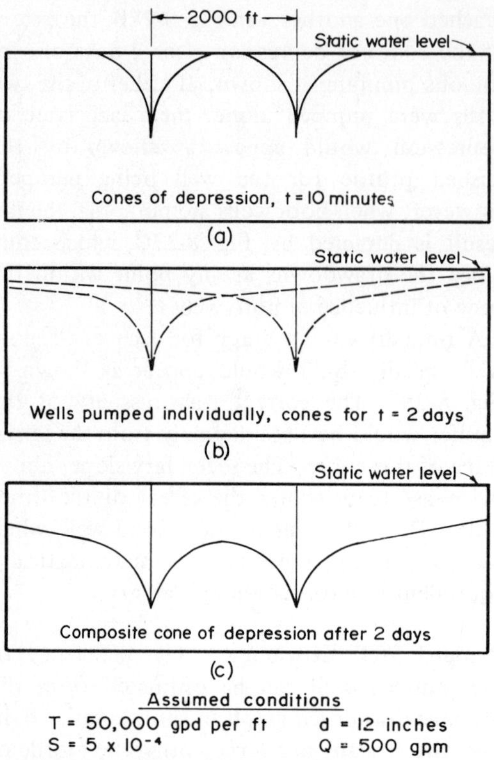

Fig. 8-27 How interference develops between adjacent wells tapping the same artesian aquifer. Composite cone is for both wells pumping simultaneously under the assumed conditions.

the cone of depression can be extended to show the drawdown in the aquifer just outside the well. Such an extension is drawn in Fig. 8-29.

The intersection of the extended line with the radius of the pumped well shows the theoretical drawdown for a 100 percent efficient well. (The radius of a gravel-packed well is taken as the outer radius of the gravel pack.) The result is only strictly true for an artesian aquifer and where the full thickness of the aquifer is screened.

The theoretical drawdown for the example shown here is 26 ft. The actual drawdown in the well is 29 ft. The well efficiency is calculated as $26/29 \times 100 = 90$ percent.

In cases where well operation causes considerable reduction in the saturated thickness of the aquifer, extra drawdown results from this occurrence which may not be considered as representing a defect in the construction of the well. This would be true in water-table wells. For such wells, a correction factor for reduction in saturated thickness must first be applied before calculating the efficiency of the well itself.

Design Factors. The factors that contribute to excess drawdown in wells can be grouped within two classes. One group includes those related primarily to choices made in the design of the well. The other group includes items related to construction operations. The following is a summary of the design factors that contribute to excess drawdown:

1. Choosing a well screen with insufficient open area creates excessive entrance velocities which result in greater than normal entrance losses.
2. Poor distribution of screen openings results in excessive convergence of flow near the individual openings. In some cases, this has caused the drawdown to be twice as much as it should have been.
3. Insufficient length of well screen, resulting in partial penetration of the aquifer, distorts the flow pattern for some distance around the well. Flow to the well screen then includes a vertical component as well as the main horizontal component. Vertical permeability is almost always less than horizontal permeability so considerable head losses result from the vertical flow involved.

Construction Factors. The following is a summary of the construction factors that contribute to excess drawdown:

1. Inadequate development of the well may not improve the permeability of the formation materials around the well screen properly. A well screen with a small percentage of open area or with poor distribution of openings may make it most difficult for the drilling contractor to develop the well properly.
2. Improper placement of the well screen may put the screen at a depth that does not correspond to the best water-bearing formation.

Even though extra drawdown may result, a reduced length of well screen may be used in many instances because of other design consid-

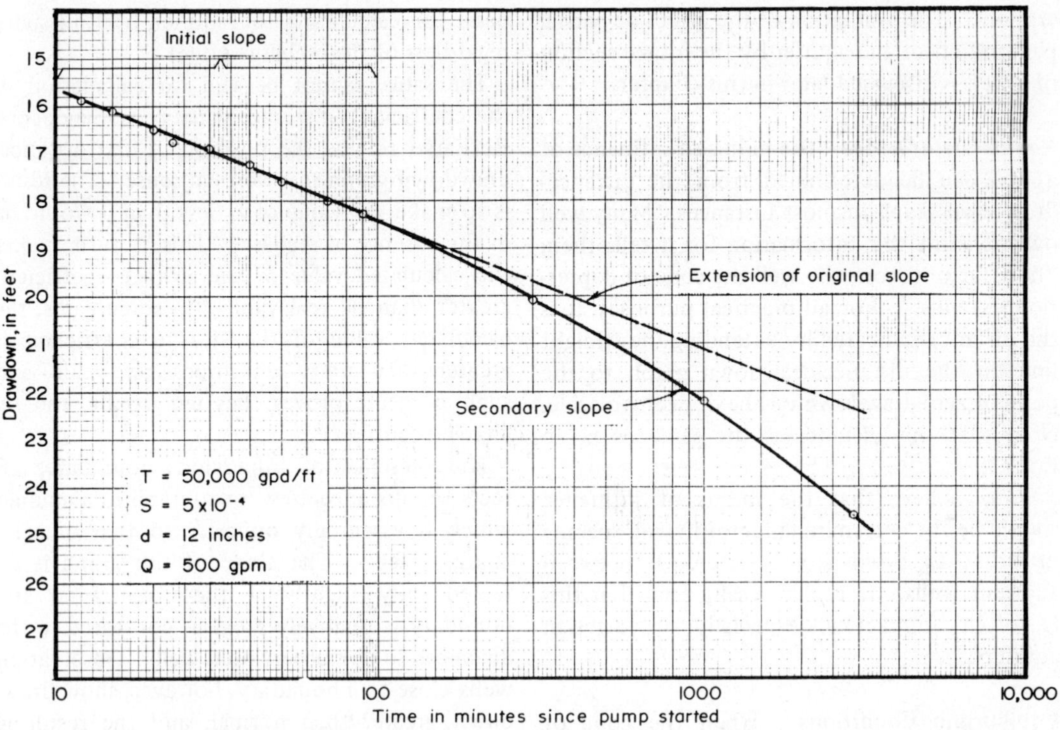

Fig. 8-28 A time-drawdown diagram for either well in Fig. 8-27 shows the effect of well interference beginning at the time each cone of depression extends to the site of the other well.

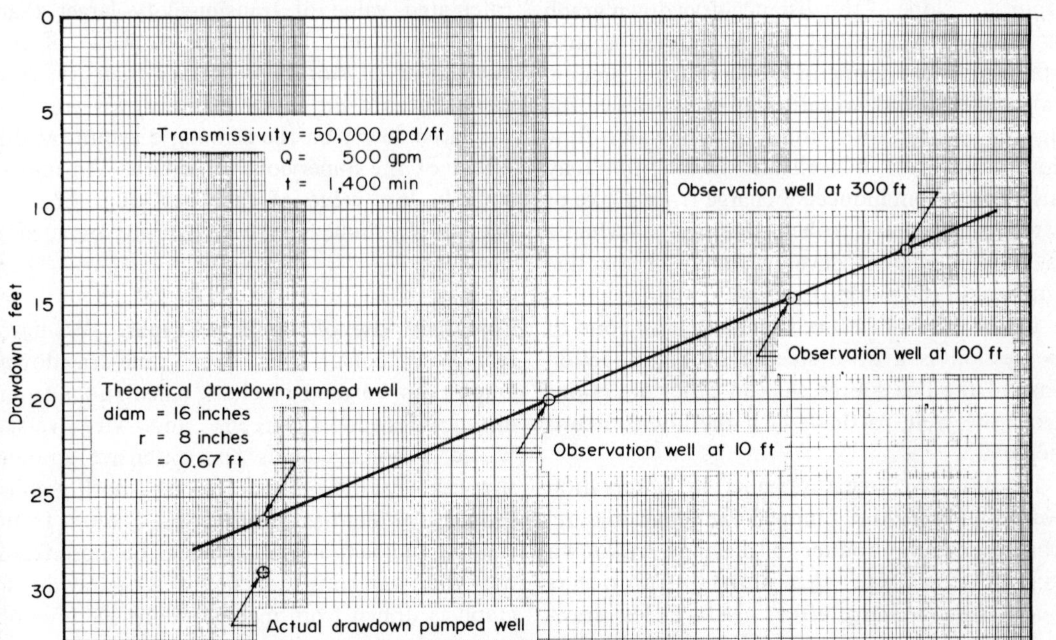

Fig. 8-29 The theoretical drawdown for a pumped well can be compared with the actual drawdown by extrapolating the distance-drawdown curve to a point where the radius of the well (outer face of the well) is indicated on the abscissa.

erations. The evaluation of the effect of partial penetration of the aquifer by the intake section of a well is discussed later in this chapter.

Radius of Influence. The radius of influence of a well can be determined from the distance-drawdown graph in most instances. Many well owners want this information for it tells them "how far out from the well the cone of depression extends." For all practical purposes, it is the distance indicated by extending the straight line on the distance-drawdown graph to the point of zero drawdown on the drawdown scale. Note that this distance is the same as r_0 in Eq. 13.

Some believe that the radius of influence should be the determining factor in well spacing in order to avoid any interference between adjacent wells. It can be easily seen that this is often impractical—especially in artesian aquifers.

Equilibrium Conditions. When the cone of depression tends to become stable and approach equilibrium during a pumping test, the effect upon the slope of the distance-drawdown graph is usually slight. The cone of depression may stabilize as the result of recharge from precipitation, leakage from saturated strata above the aquifer proper, or natural groundwater flow toward and past the well site. Equilibrium may also result from induced recharge from a lake or stream—considered to be a line source—when a hydraulic gradient develops between the line source and the pumped well.

In the case of induced recharge, the cone of depression is a little steeper in the direction toward the line source than in other directions from the well, particularly if the line source is nearby.

In this case, measurements in observation wells between the pumped well and the recharge source produce a distance-drawdown graph that is steeper than normal. This occurs because the recharge effect results in less drawdown in the observation well nearest the line source than in other observation wells on the opposite side of the pumped well. Except for this situation, however, the slope of the distance-drawdown graph provides a reliable basis for calculating the aquifer transmissivity regardless of the recharge effect.

The same cannot be said for calculation of the coefficient of storage when recharge occurs. Recharge affects the position of the distance-drawdown graph in a vertical sense. The result is to make the value of r_0 less than it would be in the absence of recharge. This, in turn, makes the calculated value of the storage coefficient greater than its real value. In some cases, the calculated value of the storage coefficient is greater than unity—an impossible value—and such a result proves, beyond doubt, the recharge occurrence.

The presence of impervious boundaries affects the distance-drawdown graph in a manner which is essentially opposite to that of a recharge effect. The slope of the graph is affected only slightly if the distance to any boundary is relatively great as compared to the distances to the observation wells. Observation wells close to a boundary, however, show drawdown greater than normal, and the resulting distance-drawdown graph is a little flatter than would be the case otherwise. This leads to a calculated value of transmissivity larger than the true value. The calculated value of the coefficient of storage is smaller than the true value because the distance-drawdown graph is displaced lower on the semilog diagram by the effect of the impervious boundary. An unreasonably low value of the coefficient of storage usually indicates the existence of such a boundary.

Combined Use of Semilog Graphs. We have seen how calculations from time-drawdown graphs can be used to check calculations from distance-drawdown graphs and vice versa. Where a homogeneous and extensive aquifer supplies a well from storage only, both graphs should give identical results. Calculations from the one diagram are independent of those from the other, assuming that the distance-drawdown graph is drawn from measurements in three or more observation wells.

If a recharge effect appears during the pumping test, or if the expanding cone of depression encounters an impervious boundary, the effects upon the two graphs are quite different as

TABLE 8-4. Comparisons of Recharge and Boundary Effects on Semilog Diagrams

Recharge Effect During Pumping Test	
Time-Drawdown Graph	**Distance-Drawdown Graph**
1. Slope of graph becomes flatter. If transmissivity is calculated on the basis of the flatter slope, it will be higher than the true value.	1. Slope of straight line remains almost unchanged. Aquifer transmissivity calculated from the graph is usually close to its true value.
2. Extending straight line of flatter slope results in an erroneous value of t_0, making it too low. A calculation using this figure gives a value for the storage coefficient that is smaller than the correct one.	2. Straight line is displaced upward. Extension to zero drawdown gives a value of r_0 which, when used to compute the storage coefficient, results in a value higher than the correct one.
Boundary Effect During Pumping Test	
Time-Drawdown Graph	**Distance-Drawdown Graph**
1. Slope of graph becomes steeper. If transmissivity is calculated on the basis of the steeper slope, it will be lower than the true value.	1. Slope of straight line remains almost unchanged. Aquifer transmissivity calculated from the graph is usually close to its true value.
2. Extending line of steeper slope results in erroneous value of t_0 which is too high. A calculation using this figure gives a storage coefficient that is larger than its correct value.	2. Straight line is displaced downward. Extension to zero drawdown gives erroneous value of r_0, which makes calculated value of storage coefficient smaller than the correct one.

noted above. Table 8-4 summarizes these contrasts.

Knowledge of the differing effects can be used advantageously in aquifer test interpretation.

Effect of Partial Penetration. Up to this point, most of our discussion of well performance has assumed that flow toward the well is entirely radial. Figure 8-8 illustrates this condition—all flow lines are horizontal with no vertical component. The flow is radial in this diagram because it pictures (1) an artesian aquifer, (2) a well with a well screen whose length is the same as the aquifer thickness, and (3) a pumping condition in which the formation near the well is not unwatered.

Where strictly radial flow does not exist, the flow pattern and drawdowns differ somewhat from those calculated by the methods presented thus far. Figure 8-30 shows an artesian aquifer with only the upper 50 percent screened. Here the arrows represent some typical flow lines—paths of water particles as they move through the formation to the intake portion of the well. The departure from radial flow is evident.

Water in the lower part of the aquifer must move along the curved lines to reach the well screen openings. In doing so, this water must take paths that are longer than strictly radial flow lines. The result of the longer flow paths is to increase the drawdown in the well relative to that in the aquifer at any given horizontal distance from the well. For a given yield, therefore, the drawdown in the pumping well is greater if the aquifer thickness is partially screened rather than completely screened.

The mathematical analysis of partial penetration problems is difficult for homogeneous aquifers and equations for stratified aquifers are almost impossible to develop. The curves in Fig. 8-31 provide a simplified means for estimating results from partially penetrating wells in artesian aquifers that are reasonably homogeneous. This diagram was developed from an equation derived by Kozeny.[11]

The Kozeny equation is not valid when the aquifer thickness is small and the percent penetration and the well radius are large. The curves in Fig. 8-31 represent conditions within the valid range of the formula.

To use the curves, the screen length must be expressed as a percentage of the aquifer thickness. After locating this value on the horizontal scale, move upward along a vertical line until intersecting the curve that represents the ratio

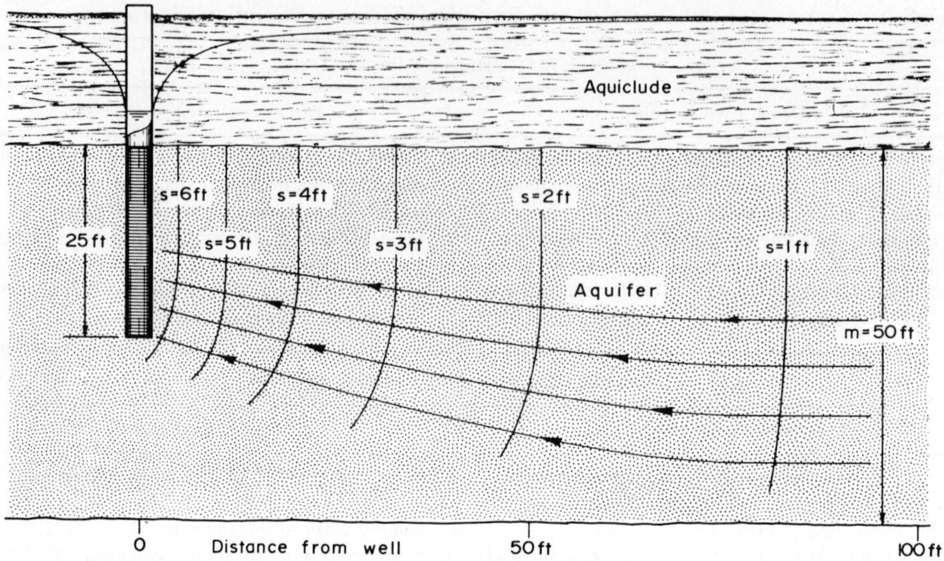

Fig. 8-30 When a well partially penetrates an aquifer, flow lines in the aquifer deviate somewhat from the radial flow pattern associated with a fully penetrating well.

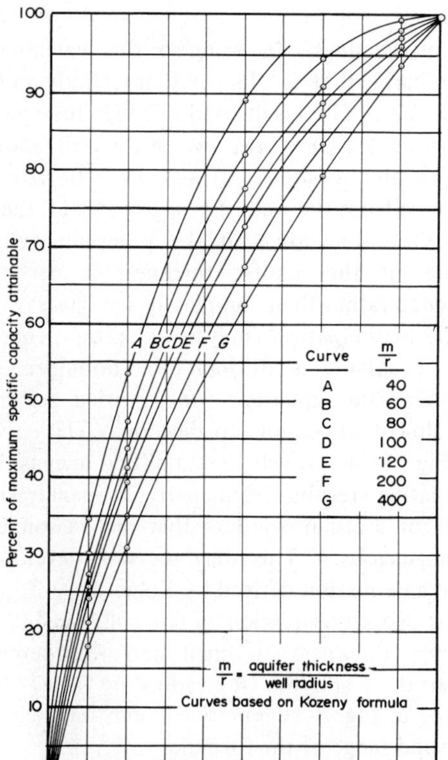

Fig. 8-31 Relationships of partial penetration and attainable specific capacities for wells in homogeneous artesian aquifers.

of aquifer thickness to well radius for the case in question. Move horizontally and read the percentage value of the vertical scale. This result represents the specific capacity of the partially penetrating well expressed as a percentage of the specific capacity that could be obtained with a fully penetrating well. The specific capacity of a fully penetrating well is referred to here as its maximum possible specific capacity.

In a water-table well, the problem of partial penetration must always be reckoned with because pumping the well dewaters the upper part of the water-bearing sand. This reduces the saturated thickness and necessarily shortens the intake portion of the well. A water-table well is often pumped so that its drawdown is a large proportion of the aquifer thickness. This results in a greatly distorted flow pattern as compared to strictly radial flow.

Using Water Level Recovery Data. When pumping of a well is stopped, water levels in the well and the aquifer recover from their depressed condition toward the original prepumping levels. The rate at which recovery occurs provides a means for calculating the coefficients of transmissivity and storage. The time-recovery record, therefore, is an important part of an aquifer test.

The time-drawdown measurements during the pumping period and the time-recovery measurements during the recovery period provide two distinct sets of information from a single aquifer test. Values obtained from analysis of the recovery record serve to check calculations based on the pumping record.

Where at least one observation well within a reasonable distance of the pumped well is available, the water level recovery data from that well fully reflect the hydraulic characteristics of the aquifer. Where no observation well is available, the water level recovery data from the pumped well can be used for limited calculations of aquifer capabilities.

One observation well should be provided, if at all possible. Two or more are desirable. In all cases, water levels should be measured in the pumped well and in each observation well.

During the pumping period, pumping must be done at a constant rate. Recovery data can be analyzed as described in the following paragraphs only when this condition is met. Recovery measurements following a variable rate test, such as a step-drawdown test, cannot be utilized.

The exact time of starting and stopping the pump must be recorded, noting also any unavoidable changes in pumping rate and the exact time at which each occurs. The recovery curves that reflect the aquifer response show the change in water level with time. An essential part of the record, therefore, is the exact time at which each measurement is made. All wells do not have to be measured simultaneously, but the time intervals between readings should be about the same for each one.

At each well, the elevation of the measuring point—often the top of the well casing—should be determined. An electric sounding device or a weighted steel tape are needed to measure the depth to water accurately. Readings taken with an air line and pressure gauge are not reliable enough for this type of test work.

Figure 8-32 shows the manner in which water levels in a well change with time. The left half of the diagram corresponds to the pumping period and the right half to the recovery period after pumping ceases. The figure shows that the recovery curve is practically an inverted image of the drawdown curve. The exact shape of each of the curves is influenced by the physical characteristics of the aquifer. The analysis which reveals these characteristics amounts to a detailed examination of the time-recovery curve.[12]

The points actually plotted on the right-hand half of Fig. 8-32 represent the residual drawdown in the well during the recovery period.

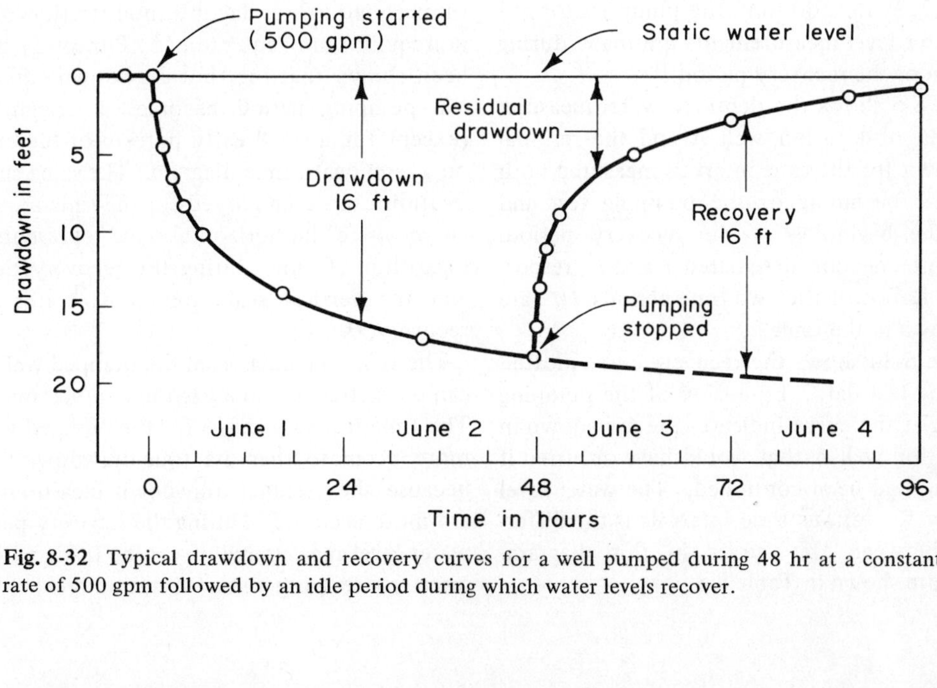

Fig. 8-32 Typical drawdown and recovery curves for a well pumped during 48 hr at a constant rate of 500 gpm followed by an idle period during which water levels recover.

Each one is the difference between the original static water level and the depth to water at a given instant during the recovery period. Each value, then, is the remaining drawdown—the distance the water level has yet to rise to reach the initial static level.

The actual recovery in the water level—the distance that the water rises after pumping ceases—must be expressed with reference to the pumping water level. The hydraulic theory of well and aquifer performance pictures the water level changes during the recovery period as being the result of an imaginary recharge well. If such a well injects water into the aquifer at the same rate as the real well pumps it out, with both wells working simultaneously after a given instant, the recovery curve will be that shown in Fig. 8-32. The rise in water level due to the imaginary recharge well is the vertical distance between an extension of the time-drawdown curve and the curve representing water levels during the recovery period.

Recovery, then, means the difference between the measured water level in an observation well at a given time after stopping the pump and the level at which the water would have been found had pumping continued until that instant.

Assume that a 6-in. test well and observation well 50 ft away are available for an aquifer test as shown in Fig. 8-33. After pumping the well at 200 gpm for 500 min, the pump is stopped and water level measurements are made during 400 min of the recovery period.[16]

Table 8-5 shows the depth to water measurements in observation well A and the residual drawdown for the time intervals measured both from the beginning of the pumping test and from the beginning of the recovery period. These intervals are designated t and t', respectively. Ratios of the two time periods, t/t', are also shown in the table.

Figure 8-34 shows the recovery curve plotted from the test data. Extension of the pumping portion of the curve indicates the drawdown in observation well A that would have occurred if pumping had been continued. The water level recovery for various time intervals is the differences between the curves in this diagram. Values are shown in Table 8-5.

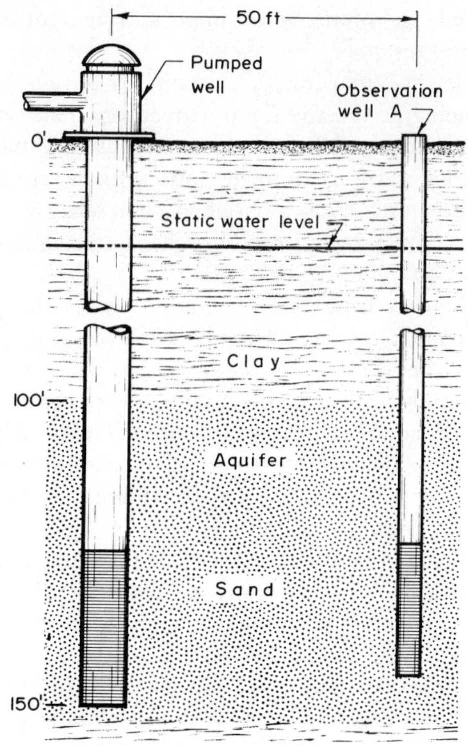

Fig. 8-33 Producing well and observation well A used in performing a typical aquifer test.

The curve as plotted in Fig. 8-34 does not lend itself to mathematical analysis. It can be simplified for analysis, however, in either of two ways—Theis' corollary to the nonequilibrium equation[5] or Jacob's modification of the nonequilibrium equation.[12] Previously it has been shown that the time-drawdown curve for the pumping period becomes a straight line (except for a small early portion of the curve) on a semilogarithmic diagram. The same simplification can be employed for the time-recovery curve where the horizontal scale represents the logarithm of time during the recovery period and the vertical scale represents water level recovery, $(s - s')$.

The recovery data from the pumped well also can be plotted and analyzed in a similar manner. The time-recovery curve for the pumped well is more accurate than its time-drawdown curve because the residual drawdown measurements are more accurate. During the recovery period, water level measurements can be made without interference from pump vibration and without

TABLE 8-5. Residual Drawdown and Calculated Recovery

Time Since Pump Test Started, t, minutes	Time Since Pump Stopped, t', minutes	Ratio, t/t'	Depth to Water,[b] feet	Residual Draw-down,[b] s', feet	Drawdown, s, from pumping Curve,[a] feet	Calculated Recovery, $(s - s')$, feet
500	0	–	18.60	10.60	10.60	0.00
501	1	501	18.40	10.40	10.60	0.20
502	2	251	18.50	10.50	10.60	0.10
503	3	168	18.40	10.40	10.61	0.21
504	4	126	18.09	10.09	10.61	0.52
506	6	84	17.72	9.72	10.62	0.90
508	8	64	17.22	9.22	10.63	1.41
510	10	51	16.64	8.64	10.64	2.00
520	20	26	15.27	7.27	10.67	3.40
530	30	17.7	14.50	6.50	10.70	4.20
540	40	13.5	13.63	5.63	10.73	5.10
560	60	9.35	12.95	4.95	10.80	5.85
590	90	6.55	12.01	4.01	10.96	6.95
650	150	4.33	10.80	2.80	11.15	8.35
710	210	3.38	10.70	2.70	11.35	8.65
770	270	2.85	10.06	2.06	11.56	9.50
830	330	2.51	9.96	1.96	11.76	9.80
890	390	2.28	9.60	1.60	11.95	10.35

[a] Average pumping rate during preceding pumping period was 200 gpm.
[b] Static water level–8.00 ft.

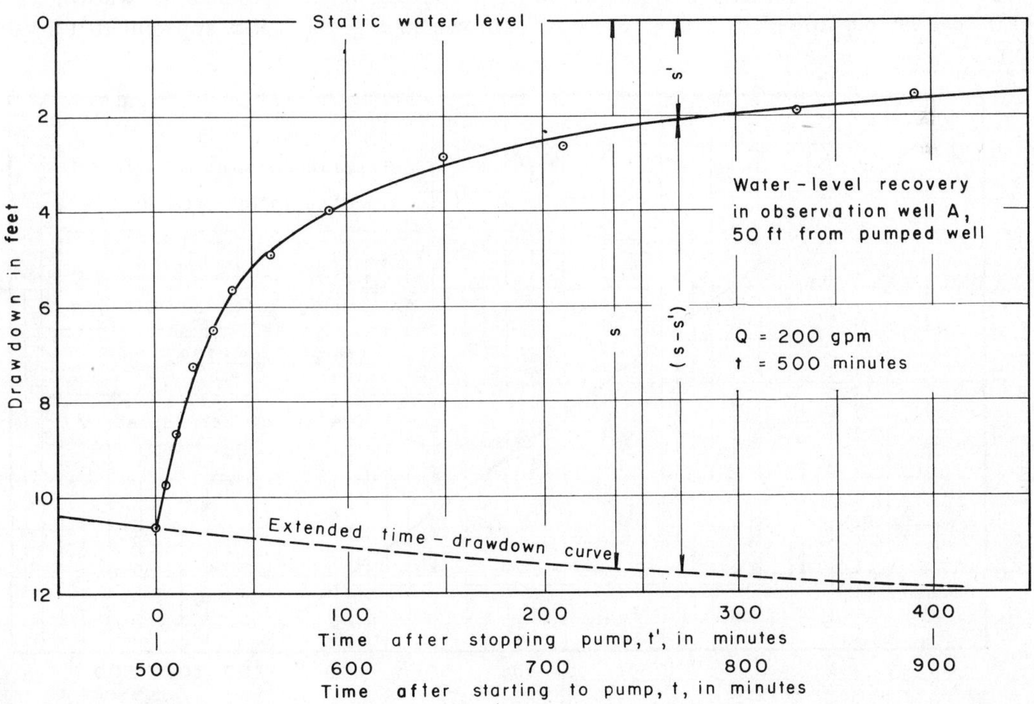

Fig. 8-34 Residual-drawdown curve from observation well A, with extended time-drawdown curve, plotted on arithmetic scales. This shows how calculated recovery is determined at any instant during the recovery period. The producing well was pumped at 200 gpm for 500 min.

being affected by momentary variations in the pumping rate.

A convenient method of plotting the data permits direct use of the residual drawdown without calculating the recovery from an extension of the time-drawdown curve. It can be shown that the residual drawdown is related to the logarithm of the ratio t/t' as follows:[8,11]

$$s' = \frac{264Q}{T} \log t/t' \qquad (14)$$

This equation shows that when values of s' are plotted against corresponding values of the ratio t/t' on semilogarithmic graph paper, a straight line can be drawn through the plotted points. Figure 8-35 shows the data from Table 8-5 plotted on a semilog diagram with values of s' indicated on the vertical arithmetic scale and values of t/t' on the horizontal logarithmic scale. The transmissivity is then calculated from the following formula:

$$T = \frac{264Q}{\Delta s'} \qquad (15)$$

where

T = coefficient of transmissivity in gallons per day per foot

Q = pumping rate in gallons per minute,

$\Delta s'$ = change in residual drawdown per logarithmic cycle of values of t/t' in feet.

Note from Fig. 8-35 that time during the recovery period increases toward the left in this method of plotting, whereas on the time-drawdown curves time increases toward the right.

If no observation well is available, the recovery data from the pumped well usually provide the best basis for calculating the transmissivity of the aquifer. The residual drawdown curve, illustrated by Fig. 8-35, should always be used in such a case.

Determining Storage Coefficient. If measurements are made in at least one observation well during the recovery period, the storage coefficient can be calculated from a semilog plot of $(s - s')$ against $\log t'$. The residual drawdown curve cannot be used for determinining the storage coefficient, even though that curve is valid for calculating the transmissivity.

Referring to the residual drawdown curve shown in Fig. 8-35, it is apparent that a value

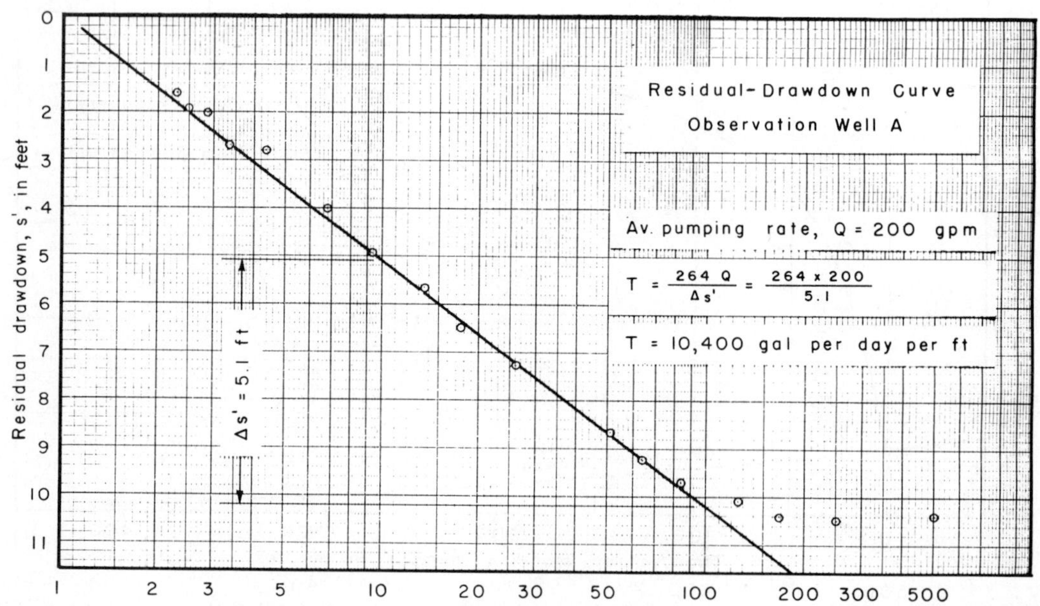

Fig. 8-35 Residual drawdown plotted against the ratio t/t' becomes a straight line on semilog graph paper. This plotting permits calculating the transmissivity as shown. Time during the recovery period increases toward the left in this diagram.

for t'_0 cannot be obtained from that diagram. The horizontal scale is a ratio which is a dimensionless number. The intercept of this curve at zero drawdown has an entirely different significance on this graph.

To discuss this point, it is necessary to review the basic assumptions which were used in developing the equations for analyzing test data from both the pumping period and the recovery period of an aquifer test.

When the aquifer conforms to these assumptions, the residual drawdown curve when extended to the left, as in Fig. 8-35, should pass through the zero-drawdown point where the ratio t/t' becomes unity. The ratio t/t' approaches unity as the lengths of the recovery period is extended.

After a long period of recovery, the water level throughout the aquifer tends to return to the original static level, with the residual drawdown approaching zero as t/t' approaches unity. Theoretically, therefore, the residual drawdown curve should pass through the upper left-hand corner of the diagram as in Fig. 8-35.

A study of residual drawdown curves from actual aquifer tests reveals that the curve does not always pass through this point, which may be called the origin of the diagram. When the curve fails to pass through the origin, it is concluded that the aquifer conditions do not conform to the assumed idealized conditions.

Three ways in which the conditions differ from the theoretical aquifer may be indicated by the residual drawdown curve. If the graph indicates zero drawdown at a value of $t/t' = 2$ or more, it is concluded that some recharge took place during the pumping period. The result of the recharge is to bring about full recovery to the original static level during a relatively short recovery period—long before t/t' approaches unity. The upper curve of Fig. 8-36 shows this situation.

A different condition is indicated when the curve to the left shows a residual drawdown of several inches or more as t/t' approaches unity. This situation would occur in the case of an aquifer of limited extent with no recharge. The lowest curve of Fig. 8-36 illustrates this kind of result.

The third condition which can account for a little displacement of the residual drawdown curve results from a variation in the value of the storage coefficient, S. In theory, the storage coefficient is assumed to be constant. In prac-

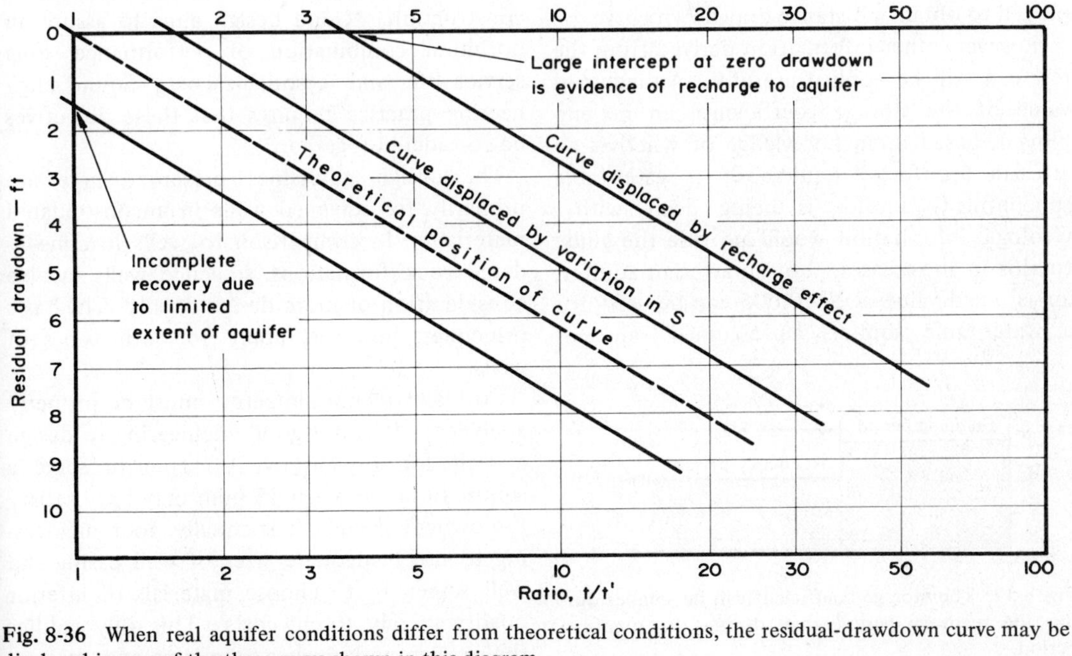

Fig. 8-36 When real aquifer conditions differ from theoretical conditions, the residual-drawdown curve may be displaced in any of the three ways shown in this diagram.

tice, S probably varies and is apt to be greater during the pumping period than during the subsequent recovery.[12]

The value of S for a confined artesian aquifer depends upon the elastic properties of the formation. If the bed is not perfectly elastic, it does not rebound vertically at the same rate that it is compressed as a result of the drawdown during the preceding pumping.

During pumping from a water-table aquifer, air occupies the voids in the sand within the cone of depression since that part of the formation is actually dewatered. The volume of water drained per cubic foot of the formation is the value of S. When pumping is stopped, the rising water table may trap some of the air as bubbles in the voids of the sand. Thus, a slightly smaller volume of water will refill the dewatered portions of the formation, resulting in a correspondingly lower value of S during recovery. Figure 8-37 pictures the situation in its simplest form, showing the value of S during the recovery period as two-thirds its value during the pumping period. The effect of the variation in S on the residual drawdown graph is shown by the middle curve of Fig. 8-36.

While analysis of residual drawdown data is useful in interpreting the results of constant rate of pumping tests, these data cannot be employed to obtain a distance-drawdown curve.

However, other information derived from the test can still be applied usefully. An assumed value of the storage coefficient can be employed, based upon knowledge of whether an artesian (confined) aquifer or a water-table (unconfined) aquifer is being dealt with.[7] Geologic information would provide the guide to this in most cases. For an artesian aquifer, an assumed value of $S = 0.005$ can be used; for a water-table aquifer, an assumed value of

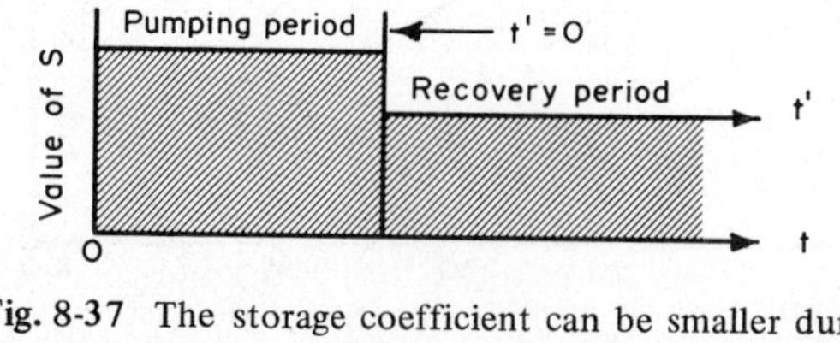

Fig. 8-37 The storage coefficient can be smaller during the recovery period than during the pumping period.

$S = 0.10$ is utilized. The resulting calculations will be less accurate than if the real values were known, but are close enough for most situations.

While most aquifers do not conform to all the theoretical conditions assumed by the author of the Theis concept, experience in applying the equations and their graphic relationships has been good. Where the aquifer is not uniform, as assumed, hydraulic interconnection throughout the geologic formation results in continuous adjustment of flow between local regions of differing permeability. The cone of depression then tends to deepen and spread in a manner which reflects the average and overall values of the transmissivity and storage characteristics of the aquifer.

The values of T and S must be considered as average values for an area. Accordingly, we must expect some variations in well yield at individual well sites when predicting future performance from the results of aquifer tests.

WATER WELL DESIGN

Designing a water well involves selecting the proper dimensional factors for the well structure and choosing the materials to be used in its construction. Good design aims to assure an optimum combination of performance, long service life, and reasonable cost. Sound engineering practice requires that these objectives be considered together.

The design guidelines presented here are primarily for screened wells in unconsolidated materials. In comparison to wells in consolidated rock formations, screened wells involve consideration of more design details. The basic principles, however, apply to both types of wells.

Cost and technical factors must be properly analyzed. It is not good engineering to design a well for a yield of 300 gpm to serve a suburban home when 15 gpm may fully satisfy the owner's needs. It is equally poor engineering to use inadequate sizes of well casing and well screen or to choose materials of inferior quality merely to cut costs. This only saddles the owner with higher pumping and maintenance costs, as well as reduced useful life of the well. Any additional investment for a properly designed, efficient well will, in the long run, usually produce maximum economy.

Most of the discussion that follows relates to the design of municipal, industrial, and irrigation wells. Such wells are usually designed to obtain the highest yield available from the aquifer, and the highest efficiency in terms of specific capacity. These factors bear directly upon operating costs. Another important cost factor in these types of installations is the economic loss that may result from interruption of service in large water supply developments. Good design minimizes this danger by building into the water well the features that will assure long and trouble-free life.

The special design factors that relate to smaller wells, such as domestic and farm wells and small commercial wells, are discussed separately.

It is helpful to consider the well structure as consisting of two main elements. One element is the part of the well that serves as a housing for the pumping equipment and as a vertical conduit through which water flows upward from the aquifer to the level where it enters the pump. This is commonly the cased portion of the well, although some of its length may be uncased where the well is constructed in consolidated rock materials.

The other main element is the intake portion of the well. Since this is where water enters the well from the aquifer, the design of this element requires careful consideration of the hydraulic factors that influence well performance. This applies particularly to a well that derives water from an unconsolidated, water-bearing sand. In such a case, a well screen is employed, and it functions as the intake portion of the well structure.

A properly constructed well screen allows water to enter the well freely at low velocity, prevents sand from entering with the water, and serves as the structural retainer to support the loose formation material.

In a consolidated rock aquifer, the intake portion of the well is usually an open borehole drilled into the aquifer to an adequate depth. The yield of such a well varies with the number, size, and continuity of the openings in the rock that are encountered in drilling the open borehole in the aquifer.

Well Diameter. Choice of proper well diameter is important because it affects significantly the cost of the structure. The well may or may not be the same diameter from top to bottom. After starting construction with casing of a given size, drilling conditions or other factors sometimes make it necessary to reduce the diameter at some depth and complete the lower portion of the well in a smaller size.

The well diameter must be chosen to satisfy two requirements:

1. The well casing must be large enough to accomodate the pump with proper clearance for installation and efficient operation.
2. The diameter of the intake section of the well must be such as will assure good hydraulic efficiency of the well.

In choosing the size of the casing, the controlling factor is usually the size of the pump that is expected to be required for the desired or potential yield of the well. The diameter of the well casing should be two sizes larger than the nominal diameter of the pump. Under no circumstances should it be less than one nominal size larger than the pump bowls.

Table 8-6 shows casing sizes recommended for various ranges in well yield or pumping rate.[13] To construct the table, the bowl size of the most efficient vertical turbine pump that would be used to pump a given quantity of water was first determined. The optimum casing size was then selected as being two nominal pipe sizes larger than the best bowl size. Velocity and head losses resulting from the water moving vertically from the bottom of the well through the casing to the pump intake also had to be taken into account. For the pipe sizes and pumping rates shown in the table, these head losses will be small.

Values of theoretical specific capacity given in Table 8-7 for a range of aquifer characteristics can be used to estimate potential well yield. The theoretical specific capacity for a given case must be reduced by applying a factor

TABLE 8-6. Recommended Well Diameters

Anticipated Well Yield, gpm	Nominal Size of Pump Bowls, inches	Optimum Size of Well Casing, inches	Smallest Size of Well Casing, inches
Less than 100	4	6 ID	5 ID
75 to 175	5	8 ID	6 ID
150 to 400	6	10 ID	8 ID
350 to 650	8	12 ID	10 ID
600 to 900	10	14 OD	12 ID
850 to 1300	12	16 OD	14 OD
1200 to 1800	14	20 OD	16 OD
1600 to 3000	16	24 OD	20 OD

representing well efficiency. Multiplying this result by the allowable or anticipated drawdown gives the estimated well yield.

If the casing size is selected according to Table 8-6, there will be adequate clearance for the vertical turbine pump, the pump shaft will be plumb, and binding will not occur even if the casing is slightly out of line and not exactly plumb. Clearance will be fully adequate for a submersible pump. Also, if the pump is set below any screened section, there will be sufficient area around the bowls to allow water to pass downward with minimum head loss to the pump intake.

In deep wells where the pumping levels are expected to be relatively high, the well diameter can be reduced at some point below the lowest anticipated pump setting. This is done in many wells which tap artesian aquifers where the artesian head is relatively high.

Well Depth. The expected depth of a well is usually determined from the log of a test hole, from logs of other nearby wells in the same aquifer, or during the drilling of the production well. Generally, a well is completed to the bottom of the aquifer. This is desirable for two reasons:

1. More of the aquifer thickness can be utilized as the intake portion of the well, resulting in higher specific capacity.
2. More drawdown can be made available, permitting greater well yield.

One exception to this rule is made where the well screen is centered between the top and bottom of the aquifer—a practice sometimes followed to make more efficient use of a given length of screen in a uniform artesian aquifer.

Another situation that requires departure from the basic rule is where water of poor quality is found in the lower part of an aquifer. In such a case, the well should be completed to a depth which will avoid the undesirable water and obtain the better quality water that is available. Well depth in crystalline rocks may also be limited by cost factors.[14]

Well Screen Length. The optimum length of the well screen is chosen in relation to the thickness of the aquifer, available drawdown, and stratification of the aquifer. Rules applying to four typical situations are discussed in the following four subsections.

Homogeneous Artesian Aquifers. In this type of aquifer, 70 to 80 percent of the thickness of the water-bearing sand should be screened—assuming that the pumping water level is not expected to be below the top of the aquifer. Good design practice dictates that the maximum available drawdown in an artesian well should be the distance from the static water level to the top of the aquifer.

If the aquifer is less than 25 ft thick, the screening of 70 percent is satisfactory. If it is between 25 and 50 ft thick, 75 percent of the thickness should be screened. If it is more than 50 ft thick, 80 percent of the thickness should be screened. Screen lengths corresponding to these rules make it possible to obtain about 90 percent or more of the maximum specific capacity that could be obtained by screening the entire aquifer thickness.

TABLE 8-7. Theoretical Specific Capacity (Q/s) for 100 Percent Efficient Water Wells (gpm/ft of drawdown)

Aquifer Transmissivity, gpd/ft	Artesian or Confined Aquifer (Storage coefficient = 5×10^{-4})						Nonartesian or Water-Table Aquifer (Storage coefficient = 1×10^{-1})					
	6-in. Diam. Well			12-in. Diam. Well			6-in. Diam. Well			12-in. Diam. Well		
	12 hr	1 day	10 days	12 hr	1 day	10 days	12 hr	1 day	10 days	12 hr	1 day	10 days
1000	0.57	0.54	0.47	0.62	0.59	0.51	0.86	0.81	0.67	1.00	0.93	0.75
2000	1.08	1.04	0.91	1.19	1.13	0.99	1.62	1.52	1.27	1.86	1.73	1.41
3000	1.59	1.52	1.34	1.74	1.66	1.45	2.34	2.20	1.85	2.58	2.50	2.05
4000	2.08	2.01	1.77	2.27	2.17	1.90	3.04	2.86	2.42	3.46	3.27	2.69
5000	2.56	2.47	2.18	2.79	2.68	2.35	3.71	3.5	2.96	4.2	3.95	3.26
10,000	4.94	4.64	4.21	5.35	5.12	4.52	7.01	6.7	5.7	7.9	7.5	6.2
15,000	7.24	6.98	6.21	7.85	7.54	6.65	10.3	9.7	8.3	11.5	10.8	9.1
20,000	9.5	9.16	8.17	10.30	9.90	8.75	13.3	12.6	10.8	14.9	14.0	11.8
30,000	13.9	13.42	12.00	15.06	14.48	12.82	19.4	18.4	15.9	21.6	20.5	17.4
40,000	18.3	17.60	15.80	19.78	19.00	16.90	25.4	24.	20.8	28.1	26.7	22.7
50,000	22.6	21.85	19.58	24.30	23.50	20.90	31.1	29.5	25.5	34.6	32.6	27.9
60,000	25.9	25.95	23.30	29.00	27.85	24.82	37.1	35.0	30.4	40.9	38.8	33.1
70,000	31.1	30.05	26.95	33.40	32.25	28.70	42.6	40.6	35.2	47.0	44.2	38.4
80,000	35.4	34.10	30.70	37.95	36.60	32.65	48.2	46.0	39.8	53.5	50.6	44.4
100,000	43.6	42.10	38.00	46.80	45.10	40.30	59.3	56.7	49.4	65.3	62.2	53.4
125,000	53.8	52.20	47.00	57.00	56.00	50.00	73.2	69.8	60.8	80.2	76.5	65.8
150,000	64.4	62.10	56.00	69.00	66.50	59.60	86.7	82.8	72.4	95.3	90.7	78.1
175,000	73.7	71.80	65.00	79.60	76.90	68.90	99.5	95.7	83.5	110.0	104.5	90.4
200,000	84.3	81.80	73.70	90.70	87.50	78.50	113.	108.	95.	124.	118.	102.

Best results are obtained by centering the screen section in the aquifer or by dividing the screen into sections of equal length and interspacing them with sections of pipe.

Stratified Artesian Aquifers. In this type of aquifer, it is obviously best to screen the most permeable strata. Determination of the more permeable or productive strata can usually be made by one or more of the following techniques:

1. Permeability tests are performed on samples that represent the respective strata of the water-bearing formation.
2. Sieve analyses are made of samples which are representative of respective strata of the aquifer. Comparison of the grain size curves can indicate the relative permeability of each sample. If the samples do not differ greatly in uniformity, relative permeabilities may be estimated by comparing the squares of the effective sizes.
3. Visual inspection and comparison are made of samples representing each strata. The relative permeability of each is estimated from the observed coarseness and cleanliness (lack of silt and clay) of the material.

The three techniques are listed in the order of their reliability, which is also the order of cost—that is, the first is both most reliable and most costly. The economic factors governing the work dictate how much expense can be justified to determine most accurately the favorable portions of the aquifer. It is highly recommended that at least sieve analyses be performed on formation samples from any industrial, municipal, or irrigation well.

Homogeneous Water-Table Aquifers. Theory and experience have shown that screening the bottom one-third of the aquifer provides the optimum design for this situation. In some wells, however, the bottom half of the aquifer may be screened in order to obtain higher specific capacity. Higher efficiency is more desirable than higher total yield in some cases.

For water-table wells, selection of screen length is something of a compromise between two factors. On the one hand, higher specific capacity is obtained by using as long a screen as possible. On the other hand, more available drawdown results from using as short a screen as possible. These two conflicting aims are satisfied, in part, by using an efficient well screen.

Available drawdown is the distance between the static water level and the top of the screen or a point a little above the top of the screen. A water-table well is usually pumped so that the pumping water level is maintained a few feet above the top of the screen. The well screen is positioned in the lower or bottom portion of the aquifer, since the upper part of the aquifer is necessarily unwatered to cause water to move into the well.

Figure 8-38 shows a properly designed water-table well in an homogeneous formation. This well was designed to pump 1500 gpm continuously. Note that the screen length is about one-third the saturated thickness of the homogeneous sand.

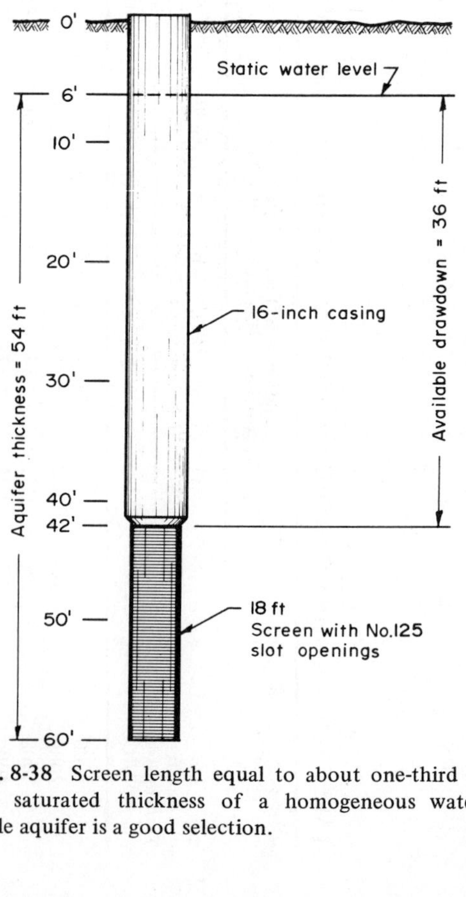

Fig. 8-38 Screen length equal to about one-third of the saturated thickness of a homogeneous water-table aquifer is a good selection.

Stratified Water-Table Aquifers. The principles of design that apply to the nonhomogeneous or stratified artesian aquifer apply also to the design of a well in this type of water-table aquifer. The only variation in the case of the water-table aquifer is to position the screen or screen sections in the lower portions of the most permeable strata in order to provide for maximum available drawdown.

Well Screen Slot Openings. For naturally developed wells, well screen slot openings are selected from a study of sieve analysis data for samples representing the water-bearing formation. A sand analysis curve is plotted for each sample. For a homogeneous formation that consists of fine uniform sand, the size of screen openings (often called the slot size) is selected as the size that will retain from 40 to 60 percent of the sand.

To determine the correct slot opening, it is only necessary to select the point on the graph where the 40 percent (or 60 percent) line intersects the sand analysis curve and then determine the screen opening from the horizontal scale. Figure 8-39 is an example of a grain size curve representing a uniform sand.[15] For a screen to be installed in this sand, the proper slot size would be either 0.016 in. if the screen is to retain 40 percent of the sand, or 0.012 in. if it is to retain 60 percent.

The 40 percent size is usually chosen when the groundwater is not particularly corrosive and when there is little doubt as to the reliability of the sand sample. On the other hand, the 60 percent size is selected if the water is corrosive or if there is some doubt as to the reliability of the sand sample. The more conservative choice is wiser where the water is corrosive, since enlargement of the openings of only a few thousandths of an inch due to corrosion could cause the well to produce sand.

For a homogeneous formation that consists of coarse sand and gravel, the designer has greater latitude in selecting the slot openings. The grain size curve for sand and gravel is flatter than the curve for fine sand (Fig. 8-40). A variation of a few thousandths of an inch in the slot size in such a case permits only a small difference in the amount of material that will pass through the well screen during development of the well. If the openings chosen retain only 40 percent, more material is brought through the screen during the development process. This increases the time required to develop the well.

Offsetting the cost of extra development work, however, are the advantages gained from the open area of the screen that results from the larger slot size. Where the water is incrusting, longer service life can be expected before plugging reduces the well yield. Larger slot size also makes it possible to develop a thicker zone in the material surrounding the screen.

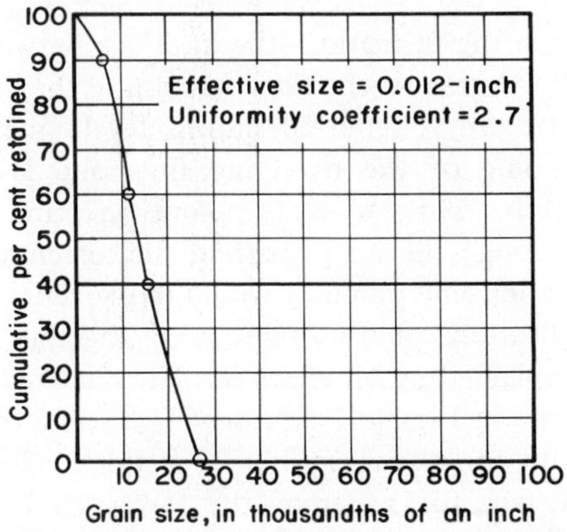

Fig. 8-39 The effective size of sand is the grain size corresponding to 90 percent retention. Uniformity coefficient is the 40 percent size divided by the effective size.

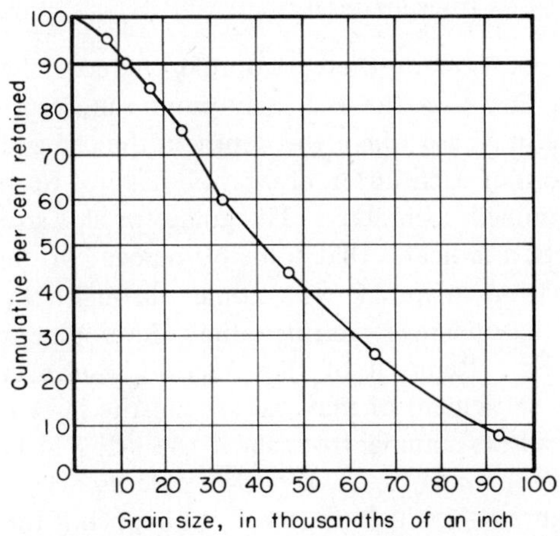

Fig. 8-40 From this grain-size curve, No. 50 (0.050-in.), or slightly smaller openings would be selected for a well screen for a naturally developed well.

This generally increases the specific capacity of the well, making the well more efficient.

A more conservative selection of slot size is called for when there is some doubt about the reliability of the samples, where the aquifer is thin and overlain by fine-grained loose material, or when development time is at a premium. Under these conditions, slot sizes that will retain 50 to 60 percent of the aquifer material are preferred.

Nonhomogeneous formations—stratified aquifers—occur more generally in nature. When dealing with a nonhomogeneous formation, slot openings of different sections of the well screen are chosen according to the gradation of the materials of the different strata. Each section of screen is made with openings to fit the material of each individual stratum according to the criteria already described for fine uniform sand and for coarser sand and gravel. Two additional rules apply, however, when selecting the openings for a multiple-slot screen. They are:

Rule 1. If fine material overlies coarse material, extend not less than 2 ft of the screen with the slot size designed for the fine material down into the coarse stratum below.

Rule 2. If fine material overlies coarse material, the slot size for the screen section to be installed in the coarse stratum should not be more than double the slot size for the overlying finer material.

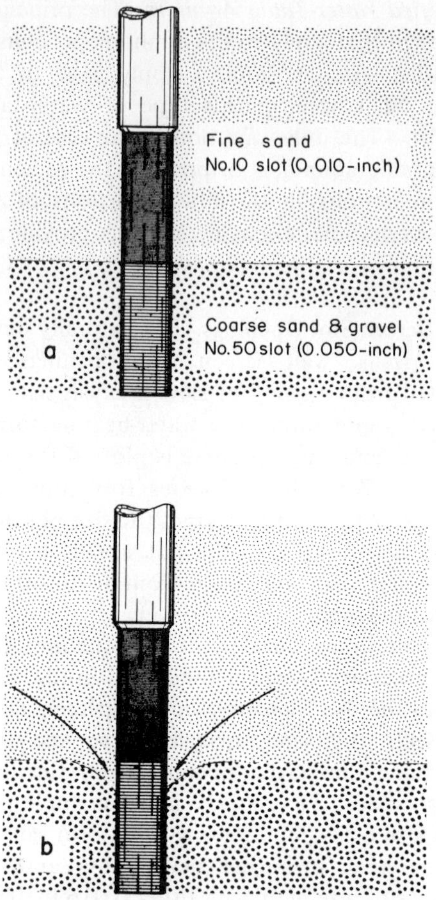

Fig. 8-41 Screen in lower part of stratified aquifer (a) should be shorter than the thickness of the coarser sand to avoid (b), otherwise fine sand may enter the upper part of the screen after developing the well.

Application of these two rules reduces the possibility that the well may pump sand especially in a case where the depth to the top and bottom of each different stratum has not been determined accurately. The guide for slot size selection indicates that some 60 percent of the formation material will come through the screen openings during the development process. Removal of this fraction results in some settlement of material around the screen. This allows material from above to settle a little also.

Figure 8-41 illustrates what can happen if the first of the above rules is not applied. Here the section of the screen with openings selected to fit the coarser sand begins at the depth of the charge in the formation—the interface between the two strata. As the finer fraction of the bottom stratum is removed during development, a slumping of the overlying fine sand layer can occur. This can easily permit fine sand to pass through the lower part of the screen and continuous sand pumping would then occur.

Application of the two rules together is best explained by an example. The sand analysis curves in Fig. 8-42 represent two strata that make up an artesian aquifer 25 ft thick. Good design calls for screening the lower 18 ft of the formation, which means that more than half the length of the screen is to be set in the most permeable part of the aquifer. To readily evaluate the situation, it is convenient to

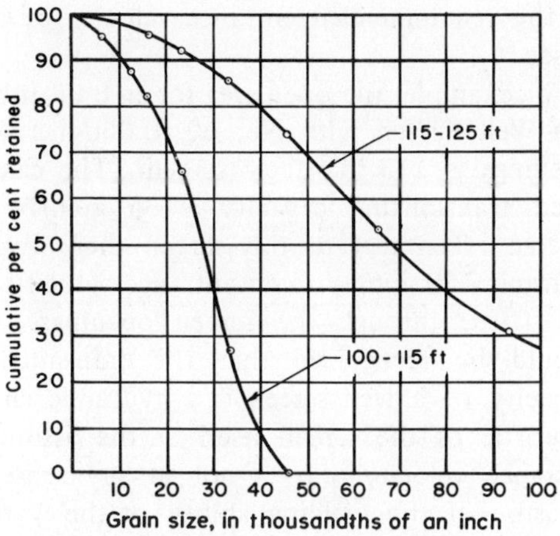

Fig. 8-42 Grain-size curves representing fine and coarse portions of a stratified sand aquifier.

assemble the available information in a design table as illustrated in Table 8-8.

The depth and thickness of each stratum, the effective grain size of each sample, and the 40 percent size of each sample are put into the table first. The square of the effective size is then put down to indicate the relative permeability of each sample. A range of slot sizes, above and below the 40 percent size, is then written in. These sizes should be values that could be considered for a screen in each stratum by itself, without considering the existence of other strata above or below.

Slot sizes for a screen to be used in a formation consisting of only two strata can be selected readily without making a design table. The table is extremely useful, however, where a large number of samples of material of different gradations are to be compared.

A screen 18 ft long, installed in the lower part of the aquifer, puts the top of the screen at 107 ft. Selection of the screen openings begins at the top. Average conditions are assumed in this example, so the design is based on slot sizes that retain 40 percent of the material. The screen section for the upper stratum should have No. 30 slot openings and the section for the lower stratum should have No. 80 slot openings.

Applying Rule 1, the finer openings must extend into the coarser stratum at least 2 ft. This puts the lower limit of the section with No. 30 slot openings at 117 ft. The upper 10 ft of the screen, therefore, is to be No. 30 slot openings, set from 107 to 117 ft.

If it were not for Rule 2, No. 80 slot openings would be chosen for the lower 8 ft of the screen. Since 80 is more than twice 30, this would violate Rule 2. A transition section at least 1 ft long with No. 60 slot openings should be provided between the No. 30 slot and the No. 80 slot sections. The completed selection of screen openings is as follows:

10 ft of No. 30 slot openings, from 107 to 117 ft.

1 ft of No. 60 slot openings, from 117 to 118 ft.

7 ft of No. 80 slot openings, from 118 to 125 ft.

The use of the proper screen openings to fit each stratum assists in attaining the highest specific capacity possible and greatly reduces the possibility of pumping sand with the water.

Well Screen Diameter. From the above analysis of well hydraulics, it has been seen that the diameter of the intake portion of a well can be varied somewhat without greatly affecting specific capacity or well yield.

Screen diameter is selected to satisfy an essential basic principle—enough total area of screen openings must be provided so that the entrance velocity of the water will not exceed

TABLE 8-8. Design Table for Screen Slot Size

Depths, feet	Thickness, feet	Effective Grain Size, thousandths of an inch	Effective Size Squared	POSSIBLE SCREEN OPENINGS, THOUSANDTHS OF AN INCH		
				Minimum	40 Percent Retention	Maximum
100 to 115	15	10	100	28	30	30
115 to 125	10	26	676	68	80	95

the design standard. Screen diameter is a factor that can be varied after the length of the screen and size of the screen openings have been selected.

Screen length depends upon the thickness of the water-bearing sand; screen openings depend upon the gradation of the sand. To a large extent, the natural characteristics of the aquifer fix these dimensions, leaving the diameter as a factor that can be varied.

Laboratory tests and field experience show that if the screen entrance velocity is equal to less than 0.1 ft/sec, the following will result:

1. Friction losses in the screen openings will be negligible.
2. Rate of incrustation will be at a minimum.
3. Rate of corrosion will be at a minimum.

The entrance velocity is calculated by dividing the expected or desired yield of the well by the total area of openings in the screen. If the figure is greater than 0.1 ft/sec, the screen diameter should be increased to provide enough open area so that the entrance velocity is less than 0.1 ft/sec.

If, on the other hand, the calculated entrance velocity is less than 0.1 ft/sec—for example, 0.05 ft/sec—the screen diameter may be reduced somewhat. A well screen with a greater percentage of open area offers a decided cost advantage where different types of screens are compared at the same entrance velocity. Screen diameter should be as large as possible, however, to keep the entrance velocity safely below the 0.1 ft/sec limit.

The foregoing statements assume that the pump will be set above the well screen (the usual case) and that the head loss associated with the vertical flow of water upward through the screen will be small (which is also usually true).

Screen Transmitting Capacity. Most screen manufacturers provide tables that show the open area per foot of screen for each size of screen and for various widths of slot openings. The transmitting capacity of a well screen, expressed in gallons per minute per foot of screen, is calculated readily from the open area figures at the recommended entrance velocity of 0.1 ft/sec.

For example, the open area for an 8 in. Johnson Well Screen with No. 60 (0.060-in.) slot openings is 113 in.^{2}/ft of screen. The calculated transmitting capacity is 35 gpm/ft of screen. Ten feet of this screen then would transmit 350 gpm with an entrance velocity of 0.1 ft/sec through the screen openings. It should be recognized that the transmitting capacity of a well screen is a hydraulic characteristic of the screen itself at the assumed entrance velocity. It is not in any sense a measure of the yielding ability of the water-bearing formation in which the screen might be installed.

Figure 8-43 illustrates an example where use of an efficient well screen permitted choosing a screen diameter a few inches smaller than the well casing. The well taps the water-bearing sand and gravel of the Chicot formation in sourthern Louisiana. The expected yield of 1800 gpm or more required the use of 20-in. OD casing to provide adequate space for the pump. A 12-in. pipe-size well screen provided enough open area, however, so that the entrance velocity would not exceed 0.1 ft/sec even if the well were pumped at 3000 gpm.

When it is necessary to set the pump within the screen, or to set it within pipe that connects two screen sections, the screen diameter to the lowest anticipated pump setting should be selected from Table 8-6. Any screen section below the pump setting position, however, can be of smaller diameter, provided the conditions for entrance velocity are met.

Selection of Material. Three things govern the choice of metal of which the screen should be fabricated. These are:

1. Mineral content of the water.
2. Presence of bacterial slimes.
3. Strength requirements of the screen.

The mineral content of the groundwater is obtained from a chemical analysis. A study of the analysis usually can reveal if the groundwater is corrosive or incrusting. In some cases, it may cause both corrosion and incrustation.

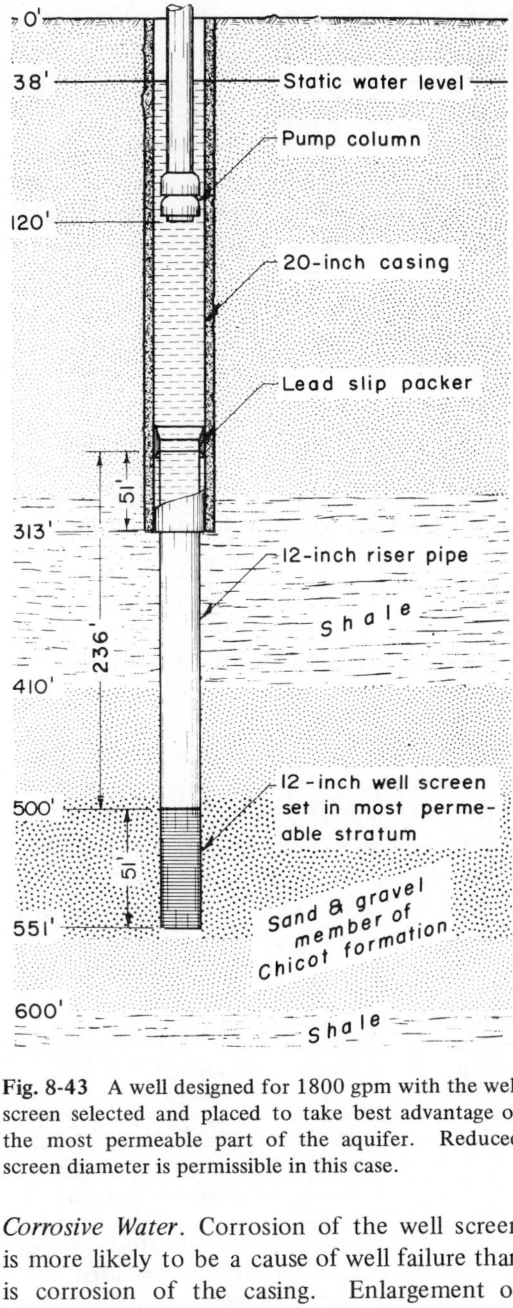

Fig. 8-43 A well designed for 1800 gpm with the well screen selected and placed to take best advantage of the most permeable part of the aquifer. Reduced screen diameter is permissible in this case.

Corrosive Water. Corrosion of the well screen is more likely to be a cause of well failure than is corrosion of the casing. Enlargement of screen openings resulting from removal of only a few thousandths of an inch of metal can permit an excessive amount of sand to enter the well. On the other hand, corrosion could remove $\frac{1}{8}$ to $\frac{1}{4}$ in. of the casing wall and still leave enough wall thickness to prevent collapse of the well or entrance of undesirable water. It is important, therefore, to use a well screen fabricated of a corrosion-resistant metal.

The following list of indicators of corrosive water can help the well designer recognize potentially corrosive conditions:[19]

1. Low pH. If the pH is less than 7.0, the water is acidic and corrosive conditions are indicated.
2. Dissolved oxygen. If dissolved oxygen exceeds 2 ppm, corrosive water is indicated. Dissolved oxygen is likely to be found in shallow water-table wells. It is difficult to determine dissolved oxygen accurately.
3. Hydrogen sulfide. Presence of hydrogen sulfide can be detected readily from its characteristic rotten egg odor. Less than 1 ppm can cause severe corrosion and this amount can be detected by odor or taste.
4. Total dissolved solids. If total dissolved minerals exceed 1000 ppm, electrical conductivity of the water is great enough to cause serious electrolytic corrosion. To avoid this, well screens made of a single, corrosion-resistant metal are necessary.
5. Carbon dioxide. If free carbon dioxide exceeds 50 ppm, corrosive water is indicated.
6. Chlorides. If chlorides exceed 500 ppm, corrosion can be expected.

Combinations of any of these corrosive agents appear to increase the corrosive effect, and any two of them almost always cause more corrosion than the sum of their individual effects.

Incrusting Water. Incrusting water tends to deposit minerals on the screen surface and in the pores of the formation just outside the screen.[20] Such deposits plug both the screen openings and the formation.

Mineral deposits from incrusting-type groundwater are often removed by putting strong acid into the well to dissolve the deposits. Well screens used in this type of water should be made of corrosion-resistant metals to withstand the corrosive effect of the acid treatment.[21]

A group of organisms, commonly referred to as iron bacteria, are found in some groundwaters. Although they are noninjurious to health, they are nuisance organisms that cause plugging of the pores in the water-bearing formation and the openings of the well screen. The bacteria produce accumulations of a slimy material of jellylike consistency, and they oxidize and precipitate dissolved iron and manganese. Because of the combined effect of the growth of the organisms and the precipitated minerals, almost complete plugging of the well can occur in a short time. Cases are reported where reduction of 75 percent in well yield has occurred in periods of 3 months to a year.[22]

An effective treatment for this condition is a relatively strong solution of chlorine which destroys the organisms. Hydrochloric acid treatment is often used following the chlorine treatment for the purpose of dissolving the precipitated iron and manganese, thus making it possible to remove them from the zone surrounding the well by pumping. The corrosiveness of strong chlorine solutions, plus the fact that the treatment must be repeated periodically, makes it necessary to use a well screen fabricated of a corrosion-resistant metal to withstand the effects of the chemical treatments. This justifies choosing the proper material.

Screen Strength. Choice of the well screen metal sometimes is based on strength requirements. The two loads imposed on the screen which are to be considered most frequently are column load and collapse pressure. Where a long screen supports a considerable weight of pipe, it functions as a slender column. Where earth pressure and caving materials squeeze the screen, it must have good collapse resistance.

The screen's resistance to both column and collapse loading pressures is directly proportional to the modulus of elasticity of the material used in fabricating the screen. It follows from this, therefore, that the screen made of stainless steel, whose modulus of elasticity is 30×10^6 psi, has twice the strength of one made of a copper alloy such as Everdur whose modulus is 15×10^6 psi, when screens with identical sizes of vertical and horizontal members are being compared. For strength requirements then, a screen might be fabricated of stainless steel rather than Everdur bronze.

It is poor practice to overdesign and to provide unnecessary excess strength because this reduces the open area of the screen. The purpose of the screen is to permit water to enter the well with as little friction loss as possible. The screen should have adequate strength to resist the forces to which it will be subjected and maximum open area consistent with the strength requirements.

Well screens are manufactured from any of several metals or alloys. Table 8-9 provides a guide for metal selection, indicating relative cost and corrosion resistance.

TABLE 8-9. Metals for Well Screens and Suggested Applications

Metal or Alloy	Nominal Composition	Cost Factor	Suggested Applications
Monel	70 percent nickel 30 percent copper	1.8	High sodium chloride combined with dissolved oxygen as in sea water. Usually not needed for potable groundwater.
Stainless steel	74 percent steel 18 percent chromium 8 percent nickel	1.0	Hydrogen sulfide. Dissolved oxygen. Carbon dioxide. Iron bacteria. Excellent strength.
Everdur	96 percent copper 3 percent silicon 1 percent manganese	1.0	High total hardness. High sodium chloride where dissolved oxygen is absent. High iron. Extremely resistant to acid treatment.
Galvanized steel	99.35–99.72 percent iron 0.09–0.15 percent carbon 0.20–0.50 percent manganese	0.5	Not corrosion-resistant. Generally used in temporary wells, such as test wells or wells for dewatering. Satisfactory in some ares of southwestern U.S. where waters are noncorrosive and nonincrusting.

Gravel-Pack Design. The artificially gravel-packed well differs from the naturally developed well in that the zone immediately surrounding the well screen is made more permeable by removing the formation material and replacing it with artificially-graded coarser material. In the naturally developed well, the fine material in the formation surrounding the screen is removed by development to create a more permeable zone. In either case, the net hydraulic result is an increase in the effective diameter of the well.[17]

In the naturally developed well, it has been shown that the width of the slot openings of the screen is chosen to retain about 40 percent of the natural formation and to permit 60 percent to pass through the openings during the development process. In the gravel-packed well, an artificially-graded gravel is chosen that will retain essentially all of the formation material; a well screen opening then is selected of a size that will retain the gravel.

Some geologic conditions favor artificially gravel-packed wells and this design should be considered in those situations. Some of these are discussed in the following paragraphs.

Fine Uniform Sand. In this type of formation, gravel packing should be considered because larger slot openings in the screen may be used. The open area of the screen is, accordingly, greater. Low percentage of open area is characteristic of screens with small slot openings. If the slot opening selected on the basis of a naturally developed well is smaller than 0.010 in. (No. 10 slot), the gravel pack alternate should be considered.

Examples include the Eutaw formation in Alabama, the Raritan sand in New Jersey, the Sparta sand in Louisiana, and aquifers of the Indus Plains in West Pakistan.

Thick Artesian Aquifers. In this type of aquifer where a long screen is required and where the pump is to be set above the screened section of the well, a smaller diameter screen can be centered in the borehole and the annular space filled with gravel. This is preferable to using a shorter screen of a diameter about the same as that of the borehole.

Loosely Cemented Sandstone. Many excellent sandstone aquifers are poorly cemented. The Dakota sandstone in North Dakota and South Dakota, the Jordan sandstone in Minnesota, the Garber and Elk City sandstones in Oklahoma, and the Bocacatu formation in Brazil are examples. If a well is finished as an open hole in such aquifers, some sand particles slough from the walls of the hole and this results in a sand-pumping well.

Another reason for gravel packing a sandstone aquifer is that the formation material usually provides little or no lateral support for the screen. The formation does not readily slump or cave against the screen during development as do unconsolidated sands. Some void spaces may remain between the screen and the wall of the hole. This leaves the possibility that a section of the sandstone could break off and fall against the screen and damage it at any time. Loose granular material inserted between the screen and the wall of the hole accommodates itself to all the irregularities of the borehole. It supports the wall of the hole and at the same time gives lateral support to the screen. Figure 8-44 shows the construction details of a gravel-packed well finished in the Dakota and Lakota sandstones.

Extensively Laminated Formations. Some aquifers consist of thin alternating fine, medium, and coarse layers. Examples include the Magothy formations of Long Island, New York, some coastal plain deposits of North Carolina, the Ogallala formation of the High Plains region, and some aquifers in Turkey and Chile. It is often difficult to determine precisely the position and thickness of each individual stratum and to choose the proper length of each section of a multiple slot screen corresponding to the stratification. It is usually more desirable in such a case to reduce the chances of error by designing the well for artificial gravel packing.

The grading of the gravel pack should be based on the layer of finest material in the water-bearing section. A gravel pack selected in this manner does not restrict the flow from the layers of coarsest material because the perme-

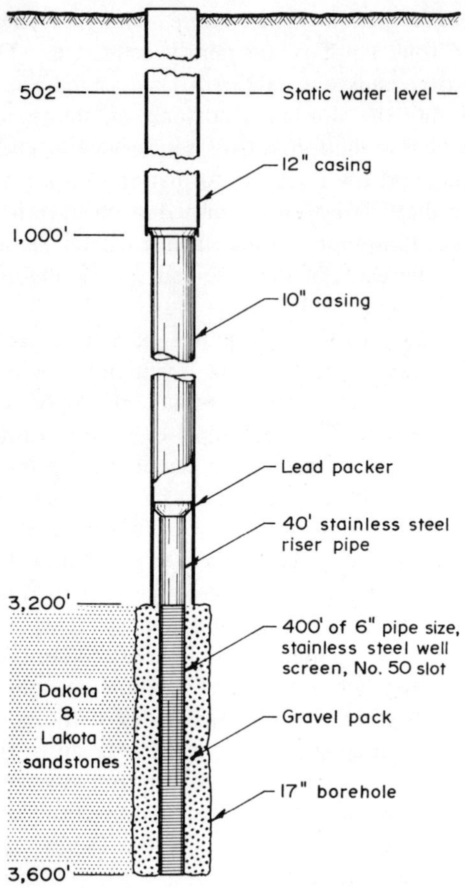

Fig. 8-44 Gravel-pack design is usually preferred for wells completed in sandstone aquifers, such as the Dakota and Lakota formations in South Dakota.

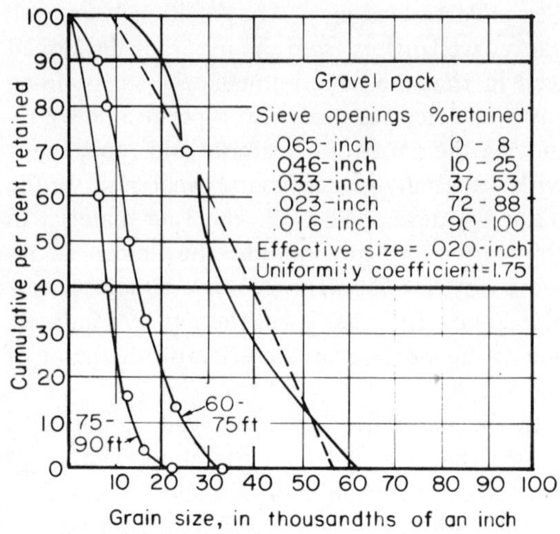

Fig. 8-45 Grain-size curves for aquifer sand and the corresponding curve for properly selected gravel-pack material.

ability of the pack would still be several times the permeability of the coarsest stratum.

Gravel-Pack Gradation. The following are logical steps in designing an artificial gravel pack:

1. Plot sieve analysis curves for all strata comprising the aquifer. Determine the stratum composed of the finest sand and select the grading of the gravel pack on the basis of the seive analysis of this material. Figure 8-45 shows the grading of two samples of typical water-bearing material that make up an aquifer 30 ft thick. The finest material lies between 75 and 90 ft. The design of the gravel pack in this example will be based on this stratum. (In some instances, it is good practice to disregard unfavorable strata and use blank tube or pipe at these places between sections of screen positioned in the better parts of an aquifer.)
2. Multiply the 70 percent size of the sand by a factor between 4 and 6.* Use 4 as the multiplier if the formation is fine and uniform; use 6 if it is coarser and nonuniform. Place the result of this multiplication on the graph as the 70 percent size of the gravel. In Fig. 8-45, 0.005 in. is the 70 percent size of the sand between 75 and 90 ft. Using 5 as the multiplier, we have $5 \times 0.005 = 0.025$ in., the 70 percent size of the gravel. This is the first point on the curve that represents the grading of the artificial gravel-pack material.
3. Through the initial point on the gravel-pack curve, draw a smooth curve representing a material with a uniformity coefficient of 2.5 or less. This must be done by trial and error. In Fig. 8-45, the curve drawn as a solid line has a uniformity coefficient of about 1.75. It could have been

*Use a factor between 6 and 9 where the formation sand is nonuniform in gradation and includes silt, and where the aquifer is sandstone.

drawn somewhat differently, as shown by the dashed line, which has a uniformity coefficient of 2.47. It is better practice to draw the gravel-pack curve so that it is as uniform (low uniformity coefficient) as practical. The material indicated by the solid-line curve is more desirable, therefore, than the material indicated by the dashed-line curve.

4. Prepare specifications for the gravel-pack material by first selecting four or five sieve sizes that cover the spread of the curve and then setting down a permissible range for the percent retained on each of the selected sieves. This permissible range may be 8 percentage points below and above the percent retained at any point on the curve.

   When selecting gravel-pack material, the designer should keep in mind local sources of filter sand for rapid sand filters. Firms that produce these materials have large stocks of clean, uniformly-graded sands and gravels that readily fit the requirements for gravel packing water wells.

5. As a final step, select a size of the well screen openings that will retain 90 percent or more of the gravel-pack material. In the example here, the correct size of the slot openings is 0.020 in.

If the well designer follows the foregoing steps carefully, sand-pumping wells can be avoided because the design is based on the proper ratio between the grain size of the formation and that of the gravel pack. A pack having such a proper ratio of grain sizes provides mechanical retention of the formation sand and prevents the sand from moving into the gravel envelope and into the well itself.

Gravel-pack material should be clean, with well-rounded grains that are smooth and uniform. These characteristics increase the permeability and porosity of the pack material. With uniform material, less hydraulic separation of the particles occurs while the material is being placed or allowed to settle through a considerable depth of water.

Gravel-pack material consisting mostly of siliceous rather than calcareous particles is preferred. Up to 5 percent calcareous material is a common allowable limit.

Thickness of the Gravel Pack. Since the design theory of gravel-pack gradation is based on mechanical retention of the formation particles, a limited thickness of only two or three grain diameters is all that is actually needed to retain and control the formation sand. Laboratory tests made by the Johnson Division, Universal Oil Products, show that a pack with a thickness of only a fraction of an inch successfully retains the formation particles regardless of the velocity of water tending to carry the particles through the gravel pack. It is impractical, however, to place a gravel pack only a fraction of an inch thick in a well and expect the material to completely surround the well screen. To insure that an envelope of gravel will surround the entire screen, therefore, a thickness of 3 in. is the minimum that is considered practical for installation in the field.

Under most conditions, the thickness of the gravel pack should not exceed about 8 in. A thicker envelope does not materially increase the yield of the well, and thickness does nothing to reduce the possibility of sand pumping because the controlling factor is the ratio of the grain size of the pack material to the formation material. Too thick a gravel pack can make final development of the well more difficult.

Formation Stabilizer. Introduction of clean, coarse sand or a sand-gravel mixture to fill the annular space around a well screen installed in a well drilled by the hydraulic rotary method assists greatly in completing such a well by natural development.

The term "formation stabilizer" has been coined to describe the fill material used for this purpose. This term distinguishes the material from a more uniform material used for artificial gravel packing. The formation stabilizer may include a wider range of grain sizes.

In drilling through the water-bearing formation by the rotary method, it is necessary to make the diameter of the drilled hole somewhat larger than the outside diameter of the well screen. This provides enough clearance to permit installation of the screen without interference. The diameter of the drilled hole is usually about 4 in. larger than the well screen, which leaves a 2-in. annular space around the screen. It is of benefit to fill this small space around the screen, at least partially, to prevent silt and clay materials above the aquifer from caving or slumping when the development work is started. Proper development of the well is more easily carried out if such caving can be avoided.

A sand mixture of a grading about the same as, or slightly coarser than, the water-bearing formation itself serves best as a formation stabilizer.

Size of the well screen openings should be selected to permit proper natural development of the aquifer materials around the screen just as if no formation stabilizer were to be used. The finer particles of the formation stabilizer, together with the fines from the aquifer, will be brought through the screen openings during the development process.

It is not necessary to provide centering devices to center the well screen in the drilled hole as is usually done for gravel packing. The fact that the formation stabilizer may not entirely surround the screen is relatively immaterial. With the screen openings chosen to permit full development of the natural formation, a complete envelope of the added material is not essential in bringing the well to a sand-free condition.

Placing Formation Stabilizer. The quantity of formation stabilizer to be placed should be sufficient to fill the space around the screen and well casing to a level 30 ft or more above the top of the screen. Considerable settlement of the material will occur since part of the stabilizer will be brought through the screen openings during the development process.

Extraction of the finer particles from the formation stabilizer while removing fines from the natural formation at the same time helps significantly to break the mud cake from the wall of the drilled hole in the screened section of the well. Development work shifts and agitates particles of the formation stabilizer. Some settlement takes place and this movement of the material tends to remove the mud wall that was formed during the drilling operation.

Removal of the finer particles from both the formation stabilizer and the water-bearing formation increases the permeability of the envelope of coarser material which is developed around the well screen.

Based on tests of samples of the aquifer material, a well screen with No. 40 slot openings was used in one formation. From the left-hand curve in Fig. 8-46, it is noted that the No. 40 slot would allow about 60 percent of the formation stabilizer to pass through the screen openings during the development process.

The permeability of the formation stabilizer represented by this curve is 1700 gpd/ft^2. After removal of most of the particles smaller than 0.040 in., the grading was changed to that shown by the right-hand curve in Fig. 8-46. This increased the permeability to 15,000 gpd/ft^2—about 9 times the original value.

The formation stabilizer represented by Fig. 8-46 is a typical concrete or mortar sand. Such material is readily available at low cost in most areas. This type of coarse sand would be suitable for use under a fairly broad range of formation conditions. It is suitable as a fill material where the water-bearing sand would require slot openings in the well screen as small as No. 20 (0.020 in.). It would also be satis-

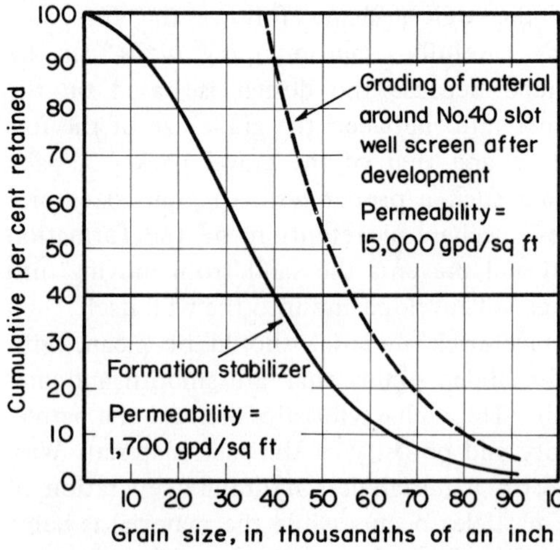

Fig. 8-46 Grain-size curve of formation stabilizer and corresponding curve for the coarser fraction of this material which remains in place around the well screen after development of the well.

factory where the natural aquifer materials are of such grading as to permit the use of No. 50 (0.50 in.) slot openings.

Design of Small Wells. Many of the design features already discussed for large capacity and highly efficient industrial, municipal, and irrigation wells apply to domestic and farm wells. The selection of the well screen openings, metal of fabrication, and entrance velocity requirements are especially valid.

Thousands of wells are drilled every year for suburban homes and farms where the total water requirements may be from 5 to 30 gpm. For these requirements, the use of long screens in relatively thick aquifers would be uneconomical. The farmer and the home owner, however, need a dependable water supply that can be obtained at a reasonable drawdown. Here it is necessary to make a compromise between well cost and well efficiency. Definite rules for choosing the screen length cannot be laid down, but a study of several examples of how short screens are used in specific cases illustrates the principles involved.

For the situation shown in Fig. 8-47A, the length of the screen for a small well should be at least one-third the thickness of the coarse sand stratum. For the situations shown in Fig. 8-47B and 8-47D, the screen length should be nearly as great as the entire thickness of the lower stratum of coarse sand. If a screen of

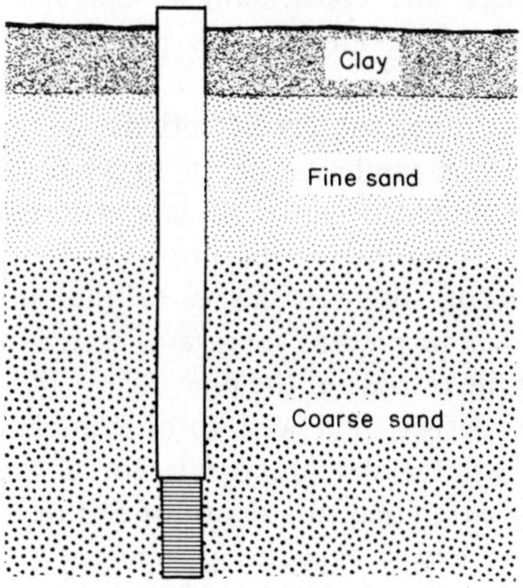

(A) Coarse part of formation is thick

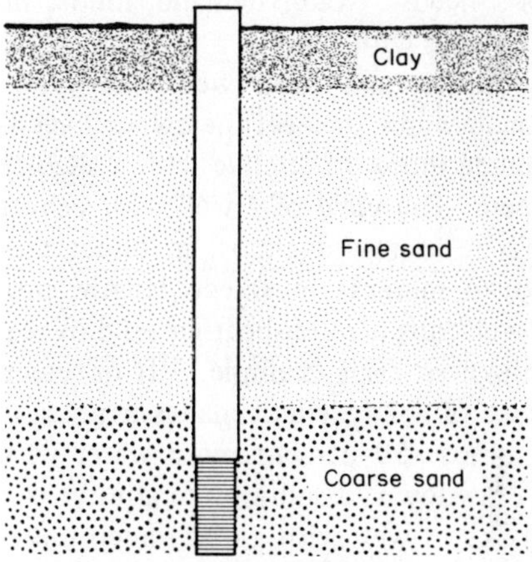

(B) Coarse part of formation is thin

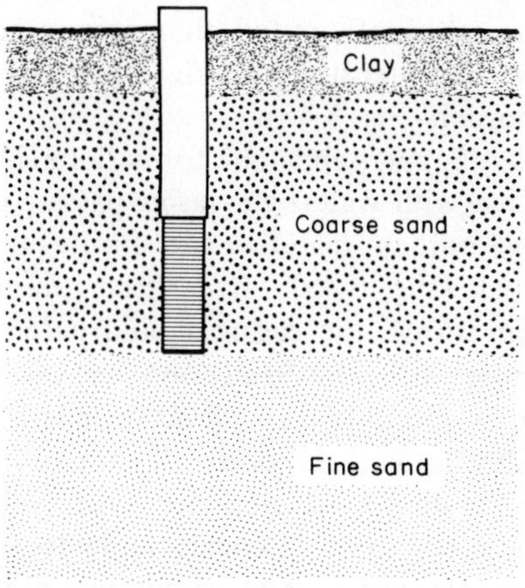

(D) Coarse material above fine sand

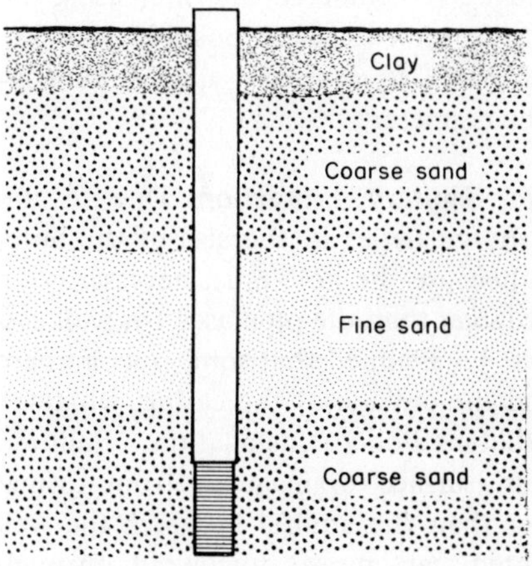

(C) Alternate layers of coarse and fine sand

Fig. 8-47 Suggested positioning of well screens in stratified water-bearing sand formations.

this length does not provide sufficient open area for the desired yield, the screen may have to be extended a short distance into the finer sand above.

A small well, constructed in a formation such as that shown in Fig. 8-47C, should be finished only as deep as the bottom of the coarse sand stratum. The length of the screen should be about one-half the thickness of the coarse sand. Ordinarily, it would not be purposeful to extend the screen deeper into the fine sand.

Though it has been pointed out that it may not be economical to use long screens in small wells, it should be emphasized that it is important to equip such wells with screens of the correct length to insure sufficient capacity to meet the owner's needs. Water demand almost invariably increases with time.

Many suburban and farm wells that yield barely enough water to meet the owner's needs could be made more productive with a slightly longer screen. The yield of a well with a short screen may be increased almost in direct proportion to an increase in screen length, provided the well taps a water-bearing sand of reasonable thickness. For example, 3 ft of screen used in place of a 2-ft length makes it possible to pump almost 50 percent more water with the same drawdown, assuming the aquifer is 7 or 8 ft thick. A well with a 4-ft screen under these circumstances will yield about one-third more water than the same well with a 3-ft screen.

It should be remembered that increasing the length of the screen, where possible, will improve well yield to a greater degree than increasing the diameter.

Design for Sanitary Protection. Due to unavoidable irregularities in the size of the borehole, and because the hole must necessarily be somewhat larger than the pipe used for well casing, it must be assumed that some openings are left around the outside of the casing, whatever may be the method of installation. Polluted water from surface drainage or from formations other than the aquifer in which the well is completed can move downward through these spaces. This can result in contamination of the water being pumped from the well. As a general rule, then, a means for sealing an opening outside the casing must be planned.

Watertight construction of the cased portion of a well should be carried into an impervious subsurface formation which caps the aquifer, or it should be carried to a safe depth below the anticipated pumping level. If these and certain other rules are followed, the water supply from the aquifer is likely to remain free from harmful bacteria because proper construction preserves the natural factors that normally protect most good aquifers—particularly water-bearing sand formations—from pollution.

In addition to the hydraulic features already discussed in detail, the design of a water well to supply potable water should include those measures for constructional operations that provide continuing sanitary protection for the supply.

Two important considerations are involved. One is the minimum depth to which the well casing should be installed as a watertight conduit. The other is the provision for positive sealing of the well casing in the subsurface materials to prevent seepage of water vertically along the outside of the pipe.

The design of a well to be completed in a sand aquifer should provide for a watertight casing extending to a depth of 5 ft or more below the lowest expected pumping level. Where the pumping level is less than 25 ft from the surface, the casing should extend 10 ft below the pumping level. Reasonable exceptions to these general rules may be made where a water-bearing sand of limited thickness is confined under an extensive clay or other impervious overburden (Fig. 8-48).

Wells tapping sandstone formations should be designed with watertight casing passing through all overburden material and firmly seated in the rock (Fig. 8-49). If creviced rock overlies the sandstone, the casing should extend 15 ft or more into the sandstone and should be sealed by grouting throughout its length.[18]

Wells tapping creviced or fractured rock aquifers should be designed with watertight casing extending to a considerable depth below the ground surface and below any outcrop of the rock formations that may occur within the general vicinity. The objective is to tap the

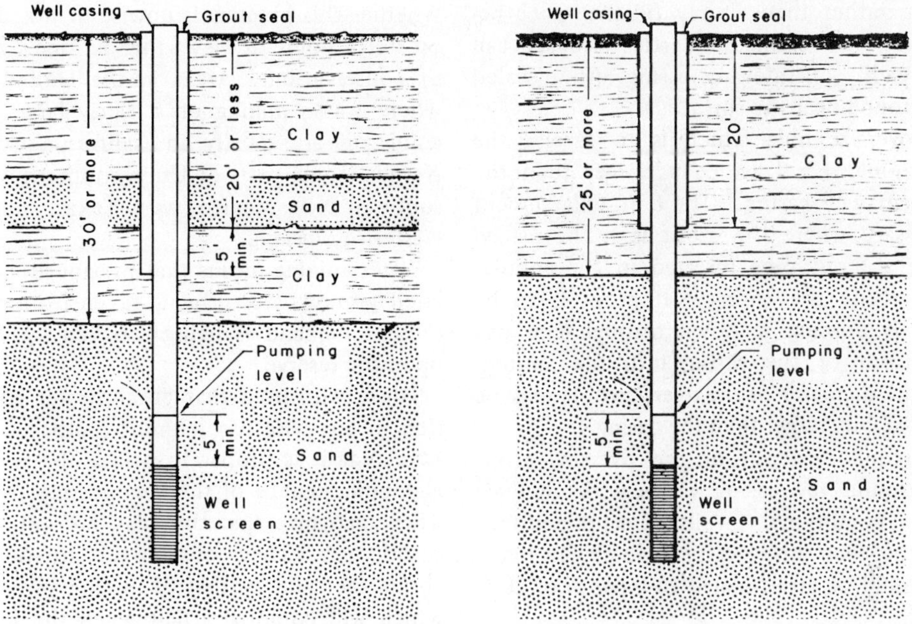

Fig. 8-48 Oversized borehole needed for grouting the upper 20 ft of casing in clay overburden.

deeper part of the aquifer. The greater percolation distance provides more opportunity for natural protective features to safeguard the quality of the supply. Rock formations with extensive systems of large openings, however,

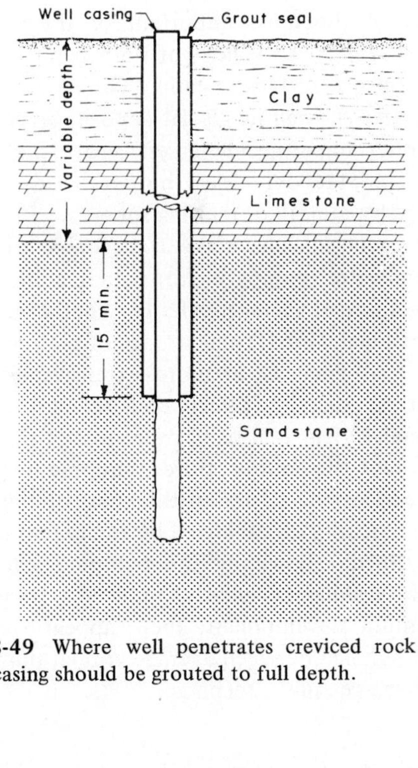

Fig. 8-49 Where well penetrates creviced rock, the well casing should be grouted to full depth.

provide little chance for the natural purification that normally takes place where groundwater moves through unconsolidated materials.

Provision should be made in the design of a well for grouting the well casing in the ground from the surface down to an adequate depth. This normally means planning an oversize borehole to the proper depth to provide an annulus into which fluid grout can be placed around the pipe.

Experience has shown, time after time, that failure to seal the space around the casing is largely responsible when samples of water from wells show the presence of coliform bacteria. Although proper precautions may have been applied at the ground surface and a sufficient depth of casing employed, pollution can occur if the upper section of the casing is not carefully sealed. Subsequent sealing of the casing, when water samples have indicated a contaminated supply, has been shown to effectively clear up the condition in almost all cases for wells completed in water-bearing sand.

The oversize borehole may be an uncased hole in stable formations such as stiff clay, shale, or rock, or it may be lined with a temporary casing of proper diameter. It is important to remove the temporary casing when grouting rather than simply fill the space between the two casings. Vertical seepage can occur readily outside any casing left unsealed in the ground near the surface.

The oversize hole which is to receive the grout should be 2 to 6 in. larger than the permanent well casing. The 6-in. enlargement is necessary where the grout is to be placed through a grout pipe inserted in the annular space outside the casing. If grout is to be placed by forcing it down through and out the bottom of the casing into the annular space, the diameter of the oversize hole may be as little as 2 in. larger than the casing.

The required depth of the oversize grout space around the casing varies with geologic and site conditions. Wells in sand aquifers, where the near surface strata are clayey materials, should be grout-sealed to a depth of 20 ft or more.

Wells completed in sandstone aquifers where the overburden consists of unconsolidated materials should have the casing grouted to a minimum depth of 25 ft. If the overlying formation is creviced or fractured rock, the casing should be grouted to its full depth.

Full-depth grouting of the casing should also be planned for wells completed in creviced or fractured rock aquifers.

LIMITS TO DEVELOPMENT

Theoretically speaking, groundwater is a renewable resource since its origin is the water that falls as rain and snow on the land surfaces. The time factor of replenishment to some groundwater reservoirs, such as deep artesian aquifers, must be measured in terms of one or more generations of man. With respect to current beneficial use that can be made of such water, therefore, its mode of occurrence and time required for renewal put certain groundwater bodies in the nonrenewable classification.

In some local areas, groundwater is being pumped out of water-bearing formations faster than natural recharge is replenishing these formations. Water tables in such areas are gradually dropping as a result of overdraft on the water previously accumulated as storage. Whether this practice is unwise or not depends on one's point of view—that is, one's concept of conservation. Areas in the United States where water mining on a fairly large scale is going on are mostly in California, southern Nevada, some parts of Utah, southern Arizona, southern New Mexico, west Texas, and southwestern Louisiana.[5]

Water mining means that previously accumulated groundwater is being taken from storage. Present demand is being satisfied by drawing upon the reserve.

Those who hold that groundwater conservation means wise use without undue waste believe that it is proper to utilize the water for economic benefit of the area as long as it lasts. They recognize that this may be only 40, 50, or 60 years. They consider the situation to be about the same as that of mining an ore body which will come to a halt when the supply is exhausted.

Safe Yield of Water-Table Aquifers. Water-table aquifers that are recharged by direct infiltration of precipitation and by influent seepage from freshwater streams may be drawn upon safely up to the limit of natural or induced recharge. Such aquifers are usually not of great thickness nor of great areal extent. Many of the most important of them are in alluvial fill in the valleys of both perennial and intermittent surface streams. Others are of the character of shallow glacial deposits where clay or other impermeable layers above the water table provide only local barriers to direct recharge by rainfall or melting snow.

Three conditions make it desirable to limit groundwater pumpage from such aquifers to the volume of average recharge, whether natural or artificial. First, total storage of groundwater is not as great as in more extensive artesian aquifers; second, direct recharge to sustain perennial yield is at hand over most of the areal extent of the aquifer; third, pumping necessarily reduces the saturated thickness of the sand in the vicinity of each well by the extent of the drawdown. Lowering the pumping level too much results in greatly reduced efficiency. Optimum development, then, appears to call for a balance between average annual pumpage and average annual recharge.

Under such a plan, considerable fluctuations in the water table may occur. Steady pumping during 2 or 3 years of below-normal rainfall would remove quantities of water from normal groundwater storage, causing water levels to go down. These levels would go up again during above-normal recharge when the amount of water being added to storage exceeds pumpage during a series of wet years.[2]

Coastal Aquifers. When an aquifer is exposed below sea level along the coastline of a land body, the upper part of the water-bearing formation usually contains fresh water while the lower part is saturated with salt water. The body of fresh groundwater may be considered as floating on seawater within the aquifer.[25]

Because fresh and salt waters have different densities and because the fresh water table is above sea level, the boundary zone between the two waters is maintained in hydraulic balance. Roughly speaking, fresh water extends to a depth about 40 times the height that the water table is found above mean sea level. This relationship is known as the Ghyben-Herzberg principle.

Since the existence of fresh water to any appreciable depth requires that the water table always be above sea level, it follows that there is also always a freshwater gradient towards the sea. Under the influence of this hydraulic gradient, fresh water discharges into the sea. Part of the aquifer exposed below sea level is a natural outflow zone for fresh water. A hydraulic gradient and a seaward freshwater movement is essential to maintain the position of the contact zone of fresh water and salt water at some depth below the land surface. The position and fluctuation of this contact zone depend upon a hydrodynamic balance of the two waters. Figure 8-50 shows the shape of the contact zone in a simplified fashion.[23,24]

If fresh water is taken from a coastal aquifer by pumping from wells, the hydrodynamic balance changes—outflow of fresh water in a seaward direction from the wells is reduced, and the water table is lowered. As a result, the salt water body moves inland some distance—an occurrence called salt water intrusion.

If only a part of the normal groundwater flow is pumped out and if the wells are a reasonable distance inland from the ocean beach, the salt water in the aquifer can be held far enough seaward by the remaining freshwater flow so that the wells can always yield fresh water.[26]

Aquifer Potential. The foregoing discussion, together with other details of groundwater occurrence and hydraulics of wells presented earlier in this chapter, brings up a basic question about a groundwater supply. How long can one pump at a given rate from a water-bearing formation?

The answer to this question depends upon the amount of water in storage, which is fixed, and the amount of recharge, which varies. It also depends upon the economics of the situation.

Before man arrives on the scene to start a

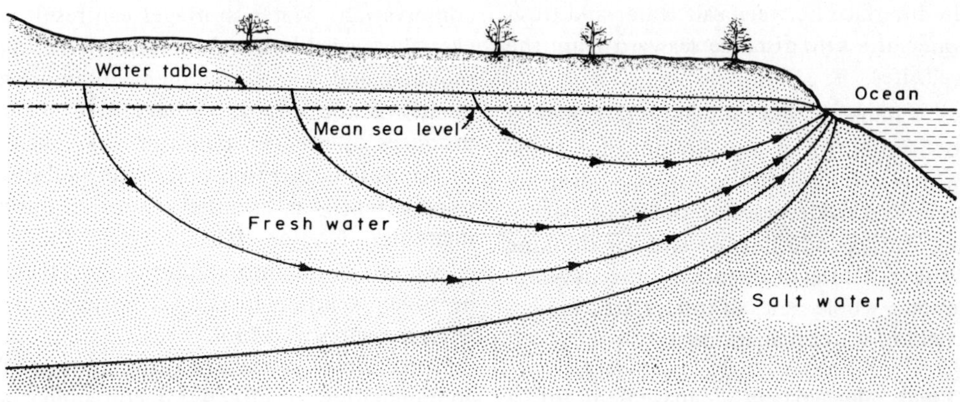

Fig. 8-50 Vertical section shows the elements of the hydrodynamic balance between fresh water and salt water in a coastal aquifer where uniform recharge of fresh water maintains the water table at a constant level. Pumping from wells can change the balance and cause salt water to move inland.

groundwater development, the water movement and storage in an aquifer is in hydrologic balance—average inflow equals average outflow. Man's water wells subtract water from some part of the natural system. In the use of wells, he can make total outflow exceed inflow, or he can maintain a balance by diverting from its normal course only an amount that does not exceed inflow.

The Optimum Development Concept. It is sound practice to look upon optimum development of groundwater in three ways, each specifically related to a type of hydrologic system and mode of occurrence of the resource. Controlled use to achieve optimum development is consistent with sound values in resource conservation and in economic benefits. Hydrologic systems and the corresponding proportion of total supply which may be taken from them by water wells fall into the following broad classifications:

1. For aquifers readily recharged from the surface, average pumpage may equal average recharge, but should not exceed it.
2. For deep artesian aquifers with little or no natural recharge near the areas where the water is desired, average pumpage may exceed recharge—but this must be managed, based on quantitative estimates of how much of the total groundwater storage is being used.
3. In coastal aquifers or on oceanic islands, average pumpage must be considerably less than average recharge in order to maintain the interface between salt water and fresh water at a safe distance seaward from the well sites.

These statements are probably too general and too simple for many groundwater situations which are admittedly complex. Nevertheless, they do provide practical points of view in planning for groundwater use and for thinking about the potential influence of the development upon a whole series of interrelated factors in the field of water management.

Well Field Management. Management of a groundwater development logically begins in the exploratory stage before the water supply facilities are constructed.[27] A geologic and hydrologic study, including test drilling and test pumping, is normally a part of the exploratory work. All records of each test well, including construction details, description of material penetrated, and quality of water, should be carefully collected. An accurate map of test well locations should be prepared and a simple identification system should be adopted that can be maintained over the years.

One of the most important items of information to be compiled is a record of fluctuations of groundwater levels. These constitute a permanent record of groundwater conditions before new facilities are constructed.

After production wells are in operation, a regular program of observation should be continued. Water levels should be measured periodically in all wells, whether pumping or idle. Water samples should be collected for chemical analysis, detailed records of well performance should be maintained, and a running tabulation of pumping should be recorded. Such records are of great value in handling well operation and for analyzing the effect of pumping upon aquifer performance over the long term. Problems arising from excessive pumping, contamination, salt water intrusion, water level decline, and unusual changes in temperature can be attacked constructively when adequate data are available concerning the performance history of the well field.

Conservation of Water Quality. Pollution or contamination is one of the problems in water conservation. Water shortages can result wherever the available supply is not of the quality required and where treatment to improve its quality is too costly. While there are many instances of needless pollution that ought not to occur, it is also true that water pollution to some extent is an essential feature of our legitimate use of water. It is imperative that we do all we can to minimize pollution, but there will be some pollution in natural water no matter what measures we adopt.

Protection of sources of drinking water from contamination, or effective decontamination or disinfection of such supply, is paramount. In this top priority—water that is necessary for life—the distinctive advantages of groundwater are of special importance.

In the overall picture, the more valuable water becomes, the more conflicts of interest arise over its use and management, for we face the universal fact that water is a common denominator of human activity.

REFERENCES

1. Meinzer, O. E., "Occurrence of Ground Water in the United States," USGS Water Supply Paper 489 (1923).
2. Thomas, H. E., "Underground Sources of Our Water," *Water*, U. S. Dept. of Agriculture Yearbook, pp. 62–77 (1955).
3. Thomas, H. E., "Ground-Water Regions of the United States—Their Storage Facilities," Interior and Insular Affairs Committee Report, U. S. Congress, pp. 3–13 (1952).
4. Meinzer, O. E. (ed.), *Hydrology*, McGraw-Hill, New York (1942).
5. Theis, C. V., "The Relation Between the Lowering of the Piezometric Surface and the Rate and Duration of Discharge of a Well Using Ground Water Storage," *Transactions of the American Geophysical Union*, pp. 519–524 (1935).
6. Muskat, M., "The Flow of Homogeneous Fluids Through Porous Media," J. W. Edwards, Inc., Ann Arbor, Michigan, pp. 121–286 (1946).
7. Theim, G., *Hydrologische Methoden*, Leipzig, Germany, p. 56 (1906).
8. Brown, R. H., "Selected Procedures for Analyzing Aquifer Test Data," *Journal AWWA* **45** (8), 844–866 (1953).
9. Cooper, H. H., Jr., and Jacob, C. E., "A Generalized Graphical Method for Evaluating Formation Constants and Summarizing Well Field History," *Transactions of the American Geophysical Union* **27**, 516–524 (1946).
10. *Ground Water and Wells*, Johnson Division UOP, St. Paul, Minnesota (1966).
11. Kozeny, J., *Theorie and Berechnung der Brunnen*, Vol. 28, Wasserkraft u. Wasser Wirtschaft, p. 101 (1933).
12. Jacob, C. E., "Recovery Method for Determining the Coefficient of Transmissibility," USGS Water Supply Paper 1536I (1963).
13. *Hydraulic Handbook*, Fairbanks, Morse and Co., Kansas City, Kansas, pp. 28–32 (1959).
14. Davis, S. N., and Turk, L. J., "Best Well Depth in Crystalline Rocks," *Johnson Drillers Journal* **41** (4) (1969).
15. Ahrens, T. P., "Water Well Engineering," *Water Well Journal* **XII** (11, 12) (1958).
16. Smith, H. F., "Gravel Packing Water Wells," *Water Well Journal* **VIII** (1, 2) (1958).
17. Smith, R. C., "Relation of Screen Design of Mechanically Efficient Wells," *Journal AWWA* **55** (5), 609–614 (1963).
18. "Rules for Well Construction and Pump Installation," Wisconsin Administrative Code, State Board of Health, Madison (1956).
19. Leopold, L. B., "Water and the Conservation Movement," USGS Circular 402 (1958).
20. Thomas, H. E., *The Conservation of Ground Water*, The Conservation Foundation, McGraw-Hill, New York (1957).
21. Geraghty, J., "Water Newsletter," Vol. 4, No. 9, Water Information Center, Inc., Port Washington, N. Y. (1962).
22. Banks, H. O., "Priorities for Water Use," *Proceedings National Conference on Water Pollution*, USPHS, pp. 153–156, 179, 181 (1960).
23. Bear, J., and Todd, D. K., "Transition Zone Between Fresh and Salt Waters in Coastal Aquifers," Hydraulic Laboratory, University of California, Berkeley (1960).
24. Todd, D. K., *Ground Water Hydrology*, John Wiley and Sons, Inc., New York (1960).
25. Sayre, A. N., and Smith, W. O., "Retention of Water in Silts and Sands," USGS Professional Paper 450-C (1962).
26. Hazen, A., "Some Physical Properties of Sands and Gravels," Massachusetts State Board of Health, 24th Annual Report (1893).
27. Mogg, J. L., "The Effect of Aquifer Turbulence on Well Drawdown," *Proceedings ASCE, Hydraulics Division*, pp. 99–112, (Nov., 1959).

9

Preparing Potable Water

James C. Vaughn
(Deceased)
Former Engineer of Water Purification
City of Chicago

DRINKING WATER QUALITY REQUIREMENTS

In developing a municipal water supply, there are two basic considerations: how much water is available, and what are its characteristics? This chapter will deal with the various aspects of the latter and the treatment practices utilized to prepare raw water for potable purposes.

This involves consideration of two sets of parameters: (1) requirements of state boards of health which usually reflect the U.S. Public Health Service Drinking Water Standards,[1]* and (2) a list of goals which have been expressed by the American Water Works Association's Committee on Water Quality Goals.[2] The drinking water standards are set primarily at levels that will ensure the consumer a water that is reasonably pleasant and completely safe to drink. Some of the standards of mineral content are such that the majority of water supplies may meet them.

The "water quality goals" are set at a level that will produce what the AWWA committee agreed upon as an "ideal" water. The relative stringency of the goals can be illustrated by the fact that they set a limit of 80 ppm (mg/l) hardness while the standards have no requirements for the hardness of a water supply. There are other differences in the prescribed levels of chlorides and sulfates. There are, for example, many water supplies, particularly in the southwestern part of the United States, that could never meet the recommended goals, but many of them could meet the drinking water standards.

WATER TREATMENT PLANTS

After the characteristics of a water supply are determined, the level of the various parameters that the water utility desires to meet must be

*In revision.

ascertained. Three limitations then prevail, which, in order of importance, are money, space, and time.

The treatment of water is always a continuous process. If it were done in "batch" quantities, the product of every water utility would always be ideal. But that type of treatment is, of course, impossible. Depending upon how much money and space are available, a plant is designed that will provide adequate treatment and allow time for the various treatment steps to be effective.

Water treatment plants may be divided, for the purposes of this discussion, into two general classes. The first class includes those that are known simply as filtration plants. These plants perform three primary functions: (1) clarification by coagulation, sedimentation, and filtration, (2) sterilization, and (3) removal of objectionable tastes and odors. The second class of water plants are those that may perform the above functions but whose primary purpose is softening of the water.

Screening. Screening here refers to the removal from the water supply of objects or solid materials which would interfere with pumping or other treatment processes. This is separate and apart from filtration or the use of microstrainers.

A rough screening may be accomplished by building intake structures with such small apertures as to keep out objectionable objects. When this is done, it is necessary to make the intake structure adequately extensive so that there is no significant loss of head through the apertures. The city of Hammond, Indiana has a water intake structure which consists of a 40-ft square wooden piling around the end of the intake pipe which is surrounded by an inverted cone of stone rip-rap. The base of this cone is probably 100 ft across. This provides sufficient apertures to eliminate any loss of head that otherwise might be present.

The apertures are also sufficiently small so that they will screen out ice. Since building this structure in 1923, there has been no shortage of water at Hammond caused by ice clogging at the intake. Where there is danger of ice clogging at the intake cribs, for example, in Chicago, it is necessary to raise the screens above the water intake ports during the winter season.

Each of Chicago's two water filtration plants has two intakes—one shore and one crib. The crib intakes are located on top of shafts 16 ft in diameter which go down to transmission tunnels about 180 ft below mean lake level. The center line of the sluice gates in each shaft is about 18 ft below lake level. The two crib structures are about 2 mi from the shoreline.

The shore intakes are located at the perimeter of the plant structures (Fig. 9-1). They consist of port gates about 7 ft by 9 ft and are in about 25 ft of water. The bottom lines of these gates are about 22 ft below lake level.

Most of the intakes along the western shore of Lake Michigan consist of pipes laid along the bottom of the lake. They end in a pipe cross with elbows turned upward which gives three

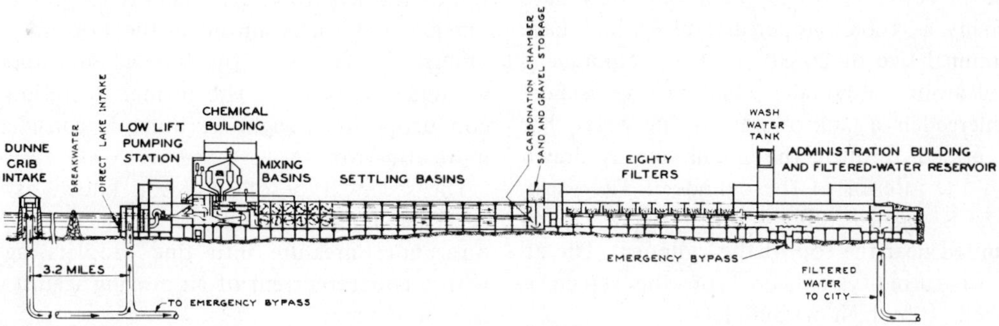

Fig. 9-1 Sectional view of flow pattern of Chicago's South Water Filtration Plant. (By permission of the Department of Water and Sewers, City of Chicago.)

apertures to cut down the head loss. This cross in some cases is covered with a wooden structure or stone rip-rap. This type of intake is frequently subject to ice troubles.

Many intake structures along rivers and lakes consist of nothing more than a port at the shoreline protected by a bar screen. Where metallic screens are used, it is necessary to size the mesh so as to block the passage of undesirable objects such as blocks of wood and/or fish. Again, the limitation of mesh size is the loss of head across the screen.

Most screens range in mesh size from $\frac{1}{2}$ to $1\frac{1}{2}$ in. The loss of head of a given size mesh may be overcome by having more screen area. This increases the size of the structure which is subject to a cost and space limitation.

Many water plants have fixed screens, but some of the larger plants have revolving screens from which the blocked objects are removed by high pressure jets. Now that the return of water plant wastes to water sources is considered objectionable, these objects must be sent through a comminutor and ground up before being added to the settled sediment.

In many plants, the screens remove a large amount of filamentous algae. The labor required to remove this material from fixed screens is considerable. It is no problem where there are revolving screens cleaned by high pressure jets. This material will also have to be sent through a comminutor before adding it to the settled sediment.

Straining. In this context, the word "straining" refers to the passage of water through a fine mesh membrane for the purpose of removing algae or turbidity. This membrane may have as many as 160,000 apertures/in.^{2} which have a nominal size of 20-30 μ. The membrane is fitted around a cylinder which revolves almost submerged in a tank of water. The water that has passed through the membrane is drawn from the interior of the cylinder. The washwater is carried off by means of a trough mounted near the top of the cylinder. The algae or turbidity collected on the screen is washed off by high pressure jets.

As a roughing process prior to further treatment, microstrainers can often be effectively used in reducing the amount of sedimentation time or the amount of final filter area. Some of the major installations are at Denver, Colorado, Kenosha, Wisconsin, and Montreal, Quebec. The Denver reports show approximately 90 percent removal of microorganisms and a 20 percent reduction of turbidity while using about 3 percent washwater.

Taste and Odor Control. Objectionable tastes and odors in surface water supplies may originate from a number of sources. These are industrial wastes, sewage plant effluents, and natural sources such as algae and decaying organic matter. Even well supplies may have unpleasant tastes and odors due to the presence of such materials as hydrogen sulfide, sulfur, and iron bacteria. In treating water, there are several methods to control the various taste- and odor-producing substances. Some of them, such as the addition of ozone, take place just prior to coagulation, but more often they are performed in conjunction with coagulation. Several substances used for this purpose are also utilized for disinfection and are discussed in that context in ensuing pages.

Aeration. Aeration is most helpful in removing dissolved gases such as hydrogen sulfide or entrained materials such as volatile hydrocarbons as well as remaining septic odors. There are two general types of aerators that are quite successful. One is the coke-tray aerator in which water is trickled over racks of lumps of coke, a porous carbonaceous material. There is a countercurrent of air going against the water stream which may be induced by a fan at the top of the structure. The fan discharges to the outside and sucks air in at the bottom. In others, the fan is at the bottom and applies air under pressure. The former operation, if conducted in a slight vacuum, is considered more effective.

The second type of aerator is known as the spray-jet aerator. Here the water is sprayed out under pressure into fine droplets, again with a countercurrent of air moving against the stream of water.

Aerators as a class have one limitation. It is difficult to reduce the substances of interest below their solubility level in water under the temperature and pressure conditions which exist in the aerating mechanism.

Oxidants. Superchlorination refers to the treatment of water with quantities of chlorine in excess of the amount required for sterilization. Any dosage in excess of 10 mg/l may be considered superchlorination. Chlorine is a comparatively mild oxidizing agent and is effective in reducing the odors of hydrogen sulfide and septic odors from sewage, as well as those from phenols not mixed with other organic contaminants. Feeding large quantities of chlorine involves a number of problems, however. Quite often, subsequent dechlorination with some such agent as sulfur dioxide, sodium sulfite, or activated carbon is required before the water is delivered to the consumers.

Chlorine dioxide is a stronger oxidizing agent than chlorine. It is extremely rapid in its reaction but is equally unstable; therefore, it is usually generated at the point of application. It is prepared by the introduction of sodium chlorite solution into the chlorinator discharge line ($2NaClO_2 + Cl_2 \rightarrow 2NaCl + 2ClO_2$). A ratio of chlorine to sodium chlorite of 1:2 is recommended. It is most effective in oxidizing phenolic matter and algal growths and is unaffected by pH values between 6 and 10.

Potassium permanganate is a mild oxidizing agent which is possibly less effective than chlorine dioxide in removing tastes and odors due to industrial wastes. It will oxidize those materials that are readily oxidized up to a certain point. Beyond this point, it is usually necessary to use activated carbon. It is quite effective in removing iron and manganese.

Potassium permanganate is usually made up in a 10 percent solution and fed into the water at the beginning of the mixing period. The degree to which it is carried through the treatment process may be observed by the existence of the characteristic pink color in the settling basin. It is highly undesirable that this color be carried through the filters into the finished water. There is always the hazard of overtreatment which will leave undesirable levels of manganese with the inherent problem of manganese stains.

Ozone is a form of oxygen usually carrying the formula O_3. In the process of breaking down O_3 to O_2, it releases an atom of nascent oxygen. This is a very active oxidant. In reducing tastes and odors, it is particularly effective on odors from algae decomposition, septic sewage, phenols, and the readily oxidizable organic contaminants. Ozone is produced by passing dry air through a closed chamber through which a high-tension electric current is also passing. Since ozone is unstable, it is applied to the water immediately after generation. It is also quite expensive.

Adsorbing Agents. A number of clays are used as adsorbents in the treatment of water supplies. These are Fuller's Earth, Bentonite, and similar materials of various names. These have a limited adsorption capacity but under special circumstances can be quite effective. For instance, when it was necessary to remove film oil from the surface of the raw water supply at a small plant with a raw water channel roughly 3 ft wide, dry Fuller's Earth was distributed over the surface of the channel by means of a metal cone. The clay adsorbed the film oil and at a later point of treatment was coagulated with chlorinated copperas (itself a good oil adsorbent). Clays may also be used to adsorb the surface active agents of detergents.

Settled Sediment. The settled sediment from a sedimentation basin of a water filtration plant has a limited adsorbing capacity. In experiments with this problem, it was found that recycling settled sediment was effective in reducing surface active agents of detergents through about three recycles. After that point, it began to be a source of contamination. It would never be desirable to use this material except in an emergency.

Activated Carbon. Activated carbon is produced by passing a carbonaceous material such as coke, charcoal, or paper char through a heating mechanism at high temperatures against a countercurrent of steam and flue gases. It exists in two forms: powdered and granulated. Powdered carbon is ground to extreme fineness, 92 percent passing through a screen of

300 mesh/in. It must possess the properties of easy wetability, prolonged suspension, and effective odor absorption capacity.

It is usually shipped in bulk (to the large users) and is wet down upon receipt into a slurry. It is fed by an appropriate proportioning device at about 1 lb carbon/gal slurry.

Granular carbon is available in several mesh sizes: 8-20 mesh and 12-40 mesh. It may, of course, be obtained in any desired mesh at some cost.

It is customary to use the granular carbon either in pressure filters or in beds on top of the regular sand filters. Since the time of contact of the carbon with the odor-producing substances is very short when the water is passed through a bed of granular carbon, there has been a tendency to make the beds very deep (5 to 10 ft). This, of course, increases the time of contact and if the odor level is sufficiently low, a thick bed will do an effective job.

In feeding the powdered carbon, the more effective method is known as the "split treatment." In this method, part of the carbon is added at one point in the treatment and the remainder is added at another point, say, just ahead of the filter bed. In the removal of phenolic compounds from the water, it is more desirable to feed a good portion of the powdered carbon ahead of chlorination, as chlorophenols have a very low rate of adsorption in addition to producing a very obnoxious taste and odor.

Chlorine-Ammonia Treatment. This is primarily a method of odor prevention. When phenols occur at a comparatively high level (e.g., 0.05 mg/l), it would take an enormous amount of activated carbon to absorb them. Quite often, chlorophenolic odor can be prevented by first treating the water with ammonia. This results in a formation of aminophenols which are tasteless and odorless. When the chlorine is added, the product of the reaction is then chloroaminophenol. These compounds are also tasteless and odorless. There are a number of water supplies heavily polluted with industrial wastes which require the pre-ammonia treatment.

Coagulation. For many years, particularly in England, waterworks engineers attempted filtration without any pretreatment of the water with the exception of the use of impounding reservoirs in which the heavier particles of turbidity or sediment were settled. In the latter part of the nineteenth century, other engineers discovered that the quality of the filtered water could be improved by coagulating the suspended particles and then allowing time for the larger and heavier coagulated particles to settle out prior to filtration.

The use of alum as a coagulating agent was known to the Egyptians as far back as the year 2000 B.C. The ancient Chinese added alum to turbid water, allowed the precipitated particles to settle, and then filtered the supernatant water through a cloth. However, it was in the years 1895–97 that George W. Fuller clearly established the fact that alum is a most satisfactory coagulating agent.[3] It was the work of Hyatt, Jewel, and Fuller that established the use of coagulants in relation to rapid sand filtration (Fig. 9-2).[4]

The removal of the heavier particles of clay and sand in turbid water presents no problem. Given time in an impounding reservoir, these particles will readily settle, leaving the finely-divided colloidal particles in suspension.

There are two general theories of coagulation of these colloids.[5] One is that the coagulating agent, for example, aluminum or iron sulfate, reacts with the alkalinity present in the water to precipitate the gelatinous aluminum hydroxide or ferric hydroxide. These hydroxides, in settling, enmesh the colloidal particles.

The other theory is that the colloidal particles carry an electrostatic charge which is neutralized by interaction with the coagulant, and once neutral, is readily agglomerated to larger particles that will settle.[6] This theory is based, in general, on the colloidal particle being surrounded by a double layer of electrical charge. The inner layer is firmly attached to the surface of the particle. The outer layer is partially diffused into the liquid media surrounding the particle which, in this discussion, is water.

The difference between the two layers is known as the plane of shear, and the degree of this plane is known as the "zeta potential."

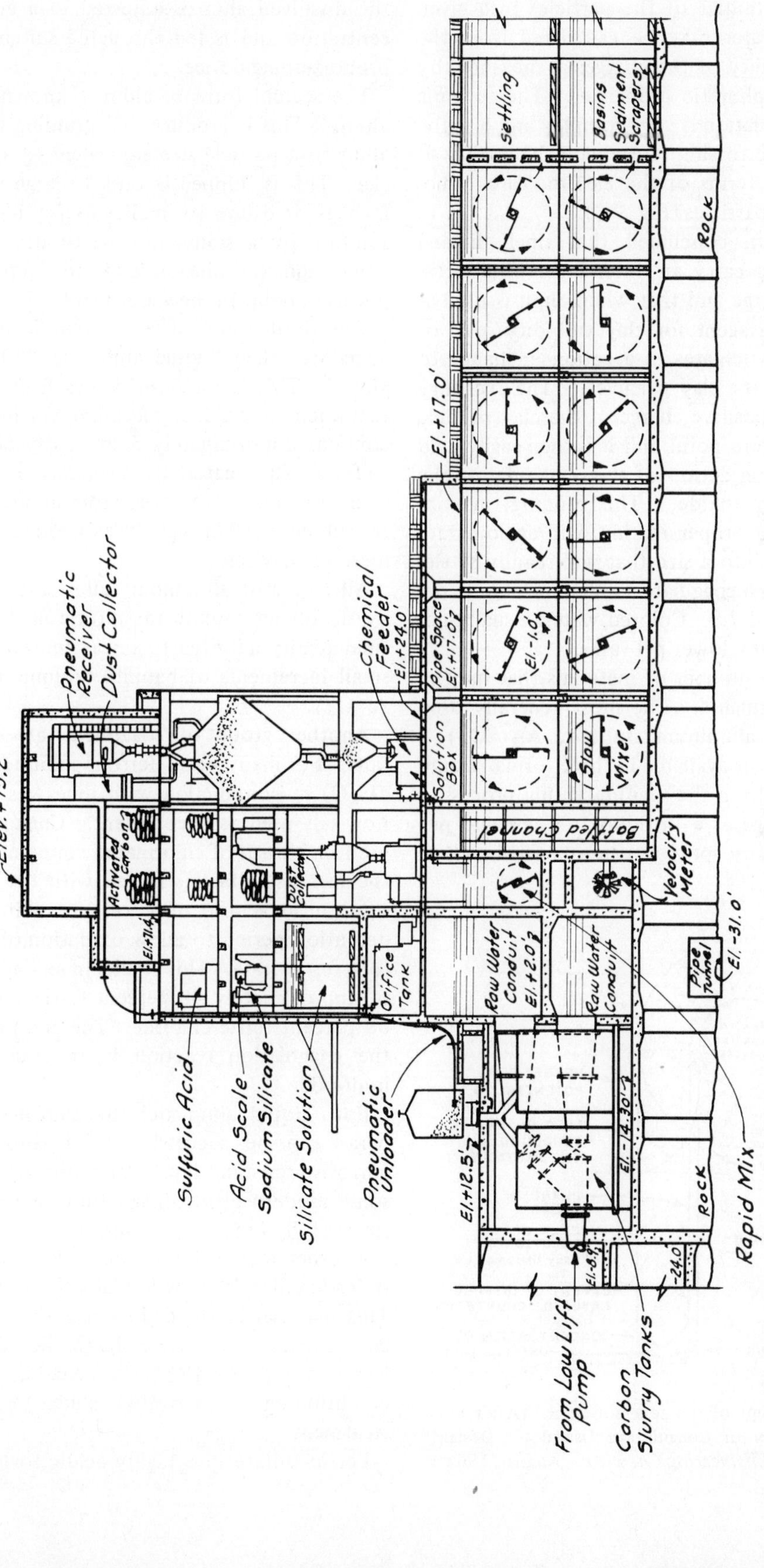

Fig. 9-2 Typical section through the chemical building and mixing basins–Chicago's South Water Filtration Plant. (By permission of the Department of Water and Sewers, City of Chicago.)

The zeta potential of the particles in a given sample of water may be calculated from the electric mobility of the charges as measured by their electrophoretic properties. The formula for this calculation is very complex and it is the usual practice to simply evaluate the quality of the water in terms of the electrophoretic mobility of the particles (Fig. 9-3).

It has been concluded that finely-divided clay particles carry an outer layer of positive electrical charge and that when alum is used as a coagulating agent for clay turbidity, part of the alum participates in a cation exchange of charges with the clay particles.[7] The exchange reduces the positive charge of the clay particle to the near zero point and it is then easily held by the binding action of the remainder of the aluminum hydroxide. This binding, in conjunction with proper mixing, will agglomerate the particles into a size that will readily settle.

Most waters coagulate best at pH levels between 6.0 and 7.8. Colored waters usually coagulate best at a lower pH value.

There are a number of materials that may be used as coagulants. The most generally used material is aluminum sulfate, $Al_2(SO_4)_3 \cdot 14H_2O$. This is available in three forms. Lump alum comes from the manufacturing process in pieces as large as a breadbox. These are put into solution by spray jets in a large bin where

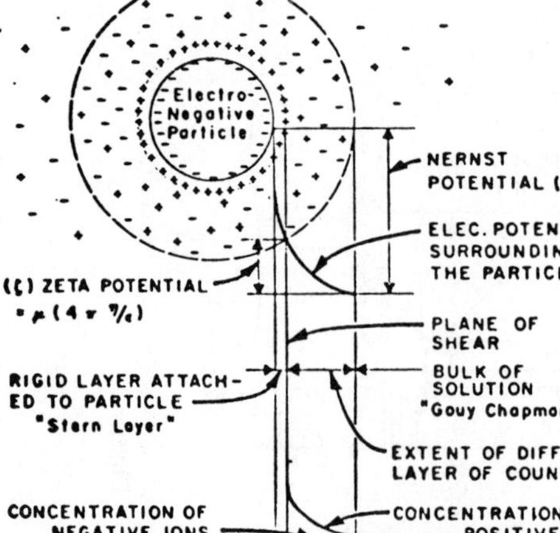

Fig. 9-3 Concept of the zeta potential. (After Priesing, "A Theory of Coagulation Useful for Design," *Industrial and Engineering Chemistry*, August, 1962.)

the dissolved alum is adjusted to a given concentration and is fed through a suitable liquid proportioning device.

The second form of alum is known as "rice alum." This is produced by grinding the lump alum to a particle size equivalent to a grain of rice. This is shipped in bulk to large users and in bags or drums to smaller users. It is put in solution by a standard type of dry chemical feeder and the solution is fed to the proper application point in the water plant.

The third form of alum that is now produced is called "liquid alum" or "50 percent alum." This latter term comes from the fact that each gallon of liquid alum weighing 10 lb contains approximately 5 lb of dry aluminum sulfate. This material is generally available in areas not too far removed from an alum plant. It can be fed through in noncorrosive liquid metering devices.

All forms of aluminum sulfate are produced by dissolving bauxite in sulfuric acid with the final acidity adjusted to a near neutral point by small increments of bauxite (aluminum oxide, Al_2O_3).

Another group of coagulants are the iron salts. For many years, ferrous sulfate ($FeSO_4 \cdot 7H_2O$) in conjunction with lime was the only iron salt used as a coagulant. Currently, it is used in a form of chlorinated copperas wherein the ferrous sulfate is mixed with the chlorine solution at a ratio of 1:7.8 parts with a short detention period to allow oxidation of the iron to a ferric state. This results in an end product of approximately 40 percent ferric sulfate and 60 percent ferric chloride. The end product in the coagulation reaction is, of course, ferric hydroxide.

Both liquid alum and the chlorinated copperas can be successfully fed through any corrosive-resistant liquid metering device. In some modern plants, magnetic flow meters are successfully used for this purpose.

Another iron salt used in some water plants is ferric chloride, now available in liquid form. This material in the dry form is very deliquescent and very corrosive. In the liquid form it is still very corrosive. Its principal use is in the conditioning of activated sludge in sewage treatment.

Ferric sulfate in a highly acidic form is used

in some plants. It requires corrosive-resistant materials for storage and feeding.

Another coagulant with a somewhat limited use is sodium aluminate ($Na_2O \cdot Al_2O_3$). It is most effective on soft or colored waters. It is alkaline in reaction which helps to raise the pH. It works best at a pH between 6.0 and 7.0. Colored waters are coagulated at low pH values with alum and sulfuric acid, and then sodium aluminate and alum are added to produce the final coagulation.

Some of the clays are used as coagulating agents. They are usually fed in the dry form directly into the water, thereupon swelling into reasonably large particles that enmesh the suspended sediment during the settling process. The principal use of the clay is to provide nuclei around which the precipitated coagulant may form.

Flocculation. Traditionally, coagulation has been defined as addition of the proper dosage of a coagulating chemical to water and mixing it with the water until such time as precipitated iron or aluminum hydroxide is formed (this is known as "floc"). However, as investigations have revealed more and more information about mixing, flocculation is treated as a separate subject.

The primary purpose of flocculation, which is sometimes called rapid mix, is to thoroughly diffuse the coagulating chemical throughout the water (Fig. 9-4). The secondary purpose is to put enough energy into the water so that the proper dispersion of power will take place. To accomplish these objectives, a certain amount of time is required. In some water plants, a detention time of 20 sec or less is frequently found, and mixing units providing 1-2 hp for each ft^3/sec of flow are within the range of desirability. The power input is usually expressed as the temporal mean velocity gradient G (which is the dimension of 1/time):

$$G = (P/\mu V)^{1/2}$$

Here G is the power dissipated in the water (foot-pounds per second), V is the volume of the tank or basin of water to which the power is applied (cubic feet), and μ is the absolute viscosity of the water (pounds per second per square foot). It has been shown[6] that when mixing at G values, 2 min is insufficient to produce complete flocculation. A more appropriate time is 5 min. However, mixing at excessive G values (10,000/sec or more) for as long a time as 2 min seems to delay substantially or retard floc formation (Fig. 9-5).

There are many ways to accomplish rapid

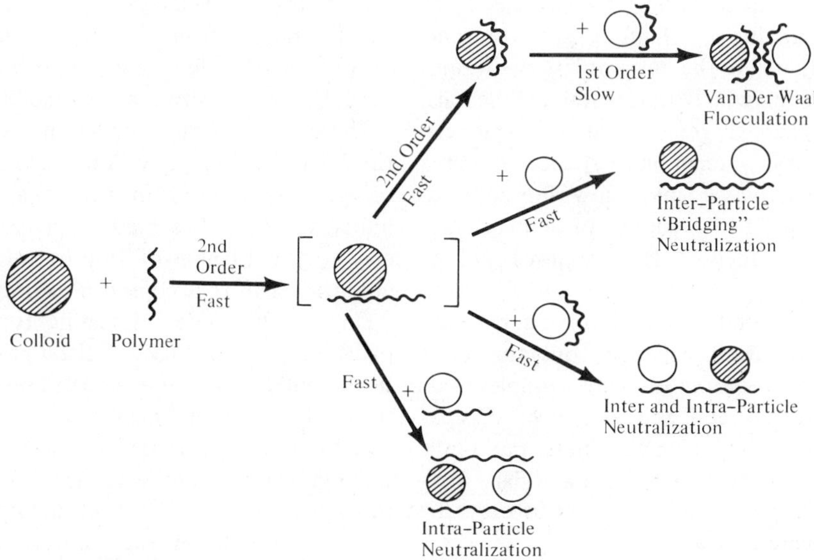

Fig. 9-4 Proposed flocculation mechanism. High salt concentrations distort the polymer molecule. (After Priesing, "A Theory of Coagulation Useful for Design," *Industrial and Engineering Chemistry*, August, 1962.)

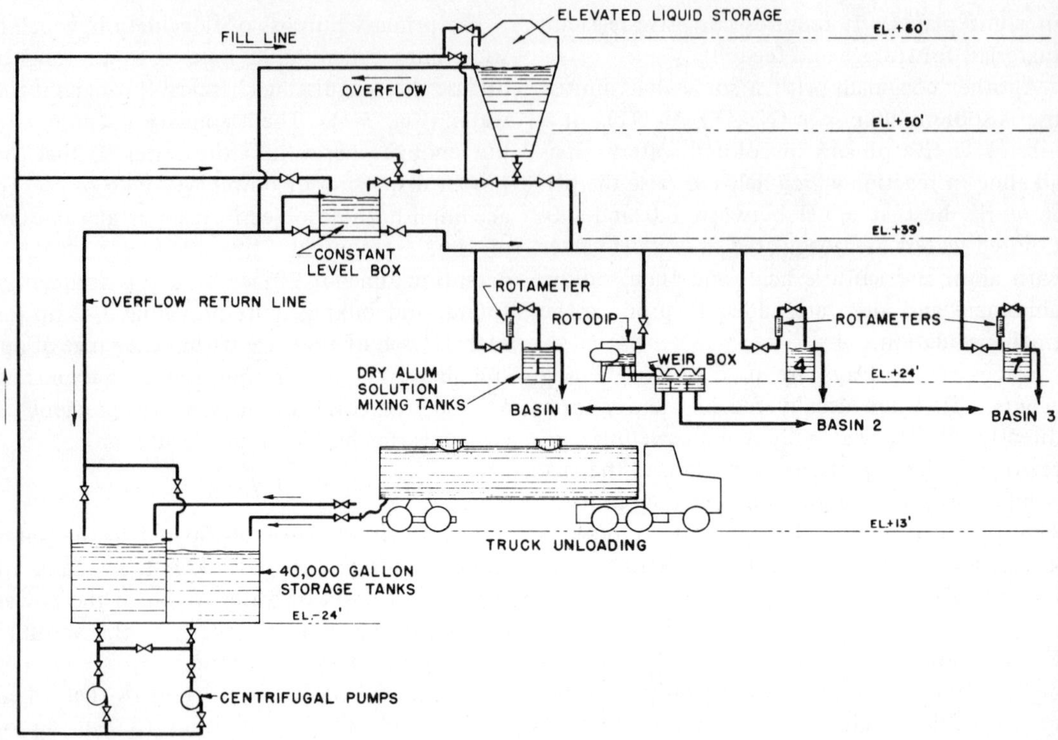

Fig. 9-5 Liquid alum handling system—Chicago's South Water Filtration Plant. (By permission of the Department of Water and Sewers, City of Chicago.)

mixing. There are flash mixers, orifice mixers, and mixing by adding the coagulants through a pump. (The latter is usually insufficient.) Walking-beam flocculators are available in which the mixing mechanism goes up and down, and there are the more or less traditional rapid mixers in which horizontal paddles are turned in a channel, generally at comparatively high speeds, say for example, 3 ft/sec paddle-tip velocity. The water may flow horizontally along the axis of the paddle or it may flow across the channels. This requires proper baffling.

Some plants produce rapid mixing by long chemical application channels, but this does not necessarily accomplish both complete coagulant diffusion and proper power dissipation. Some modern plants without rapid-mix installations have had to provide auxiliary mixing in the form of turbine mixers in order to obtain adequate flocculation.

The traditional form of flocculation system employs one rapid-mix channel and four or more slow-mixing channels with dividing baffles so arranged that the water passes from one channel to the next around the end of a baffle. The paddle-tip velocity of the slower mixer usually ranges from 1.0 to 1.6 ft/sec. The purpose of the slow mixer is to build the floc particles up to a size that will facilitate settling.

There are also upflow clarifiers where the flocculated water passes up through a blanket of precipitated floc. In waters having low turbidity values, it is sometimes necessary to add clay to this blanket of floc in order to give it enough weight to accomplish its purpose.

One of the more interesting types of filter plant involves the use of the upflow clarifier in the middle of a large circle. The water flows from the centrally-located clarifier across a doughnutlike space to a circular ring channel surrounding the entire device. Generally in this ring there is the traditional sand filter which accomplishes the third step of clarification. In plants with a design filtration rate of 2 gpm/ft^2, the retention time in the rapid and slow-mixing channels ranges from 30 to 45 min.[8]

Sedimentation. Sedimentation is the second step in the process of clarifying water in those plants where coagulation is used. The basins used for sedimentation are sometimes called flocculation basins because, in some designs, the floc continues to form after the water enters the sedimentation basin.

Sedimentation basins may come in a variety of shapes and sizes. The most common form is rectangular. Some of the rectangular basins are double-decked with straight-line flow on both levels. In other double-decked rectangular basins, the water flows out on either the upper or lower level and back to the filters on the alternate level. Those basins where the flow is out on the upper level permit the installation of sediment scrapers that do not have to be completely immersed in the water. Since the outward path is above the water level, power requirements for the scraper are reduced. A large part of the sediment settles out in the first third of the sedimentation basin. Therefore, scraper requirements are limited to that portion or, at most, the first half of the basin. The upflow clarifiers have scrapers that operate continuously. This means that, if the floc blanket is working as it should, the sediment settles in the designed area.

Radial-type sedimentation basins, as exemplified in the upflow clarifier type, work best at the design flows. Very high flows result in tangential currents which short circuit the settling time. Short circuiting is also a problem of the rectangular type. When the water enters the sedimentation basin from around the end of a baffled-channel flocculation basin, it tends to either short circuit through the near ports or the entrance to the sedimentation basin. At the same time, it builds up a head velocity against the far wall of the basin. This can result in a short-circuiting flow through the basin in the form of a spiral or a "corkscrew" movement. This, of course, interferes with the sedimentation process.

Proper baffling of the entrance to the sedimentation basin can correct these tangential flows. This correction is aided by the addition of "training" walls in the first part of the basin. These are usually vertical walls that are constructed parallel to the longitudinal direction of the basin.

In the double-decked basins of the "out and back" flow type, there is often short circuiting as the water flows from the out level to the back level. This can be corrected with properly designed and located patented baffles.[9]

The size of the sedimentation basin is usually chosen on the basis of the desired settling time at the design flow.[10] This settling time ranges from 2 to 8 hr. For a plant with a filtration rate designed at 2 gpm/ft^2, this settling time is usually about $3\frac{1}{2}$ hr. In a study of twelve plants, it was concluded that short filter runs were due primarily to the lack of adequate settling time.[11]

The depth of settling basins usually ranges from 8 to 12 ft for each level of flow, although there are some plants in which the depth is greater. In Chicago's two water filtration plants, the maximum depth is approximately 15 ft. If the sedimentation basins are shallow and the sediment is not regularly removed at higher flows, the water tends to "scour" the beds of settled sediment, thus increasing the turbidity rather than permitting the settling of the suspended matter. A good settling basin design permits the settling out at high flows of a minimum of 70 percent of the flocculated sediment. At lower flows, the percent settled out may at times reach 90 percent.

The time it takes floc to settle is inversely proportional to both the length and the depth of the sedimentation basin. (If this logic is carried to its extreme, the ideal sedimentation basin would have no length and no depth.)

How well floc settles is proportional to its density rather than its size. Carbon dioxide, a by-product of the coagulating reaction, can be entrained in floc and can make it buoyant as well as bulky. Treating the water with an alkali in sufficient quantities to neutralize the free CO_2 will make the floc more dense. The use of activated sodium silicate as a coagulant aid is another mechanism for giving the floc greater density and better settling characteristics.

A recent innovation in sedimentation basins to enhance floc settling and keep it free of interference by currents and short circuiting is the "tubular settler." In this basin, the water flows upward through a bank of tubes placed at an angle. These may be at any angle from obtuse to acute. The flow settles out within the tubes and the floc slides to the bottom of the tubes to a sediment draw-off device (Fig. 9-6).[12]

Filtration. In filtration, water is passed through a filter medium in order to remove the particulate matter not previously removed by sedimentation. There are several functions involved. One is simply the building up of a floc layer on the surface of the filter which in itself acts as a filter medium. The second is the penetration of the floc into the bed where tubes of penetration and further branch tubes are developed through which the water flows, leaving behind the particles of suspended matter. More particle reduction is achieved when the water is mixed with the filter medium, and the sediment along with the medium are retained on a filter membrane. Another function of the filter is to remove the particulate matter by electrostatic exchange. Here the charge on the filter medium neutralizes the charge of the floc, thereby permitting the floc to be removed. In the backwashing action, this electrostatically neutral material is removed and the charge on the filter media is replaced.

Slow Sand Filters. Slow sand filters are those where the action is primarily biological and the straining effect is secondary. These filters were quite common around the turn of the century when there was an abundance of land and labor for their construction and maintenance.

In addition to the layer of sand, the slow sand filter usually contains a layer of gravel up to 12 in. in thickness surrounding tile underdrains which lead to a main channel for the filtered water. The sand layer may in the beginning be as thick as 48 in. It is usually quite fine sand, with a 5 mm effective size or less.

The yield of a slow sand filter is expressed in millions of gallons per acre which means that the yield per minute per square foot is comparatively low. As the surface becomes clogged

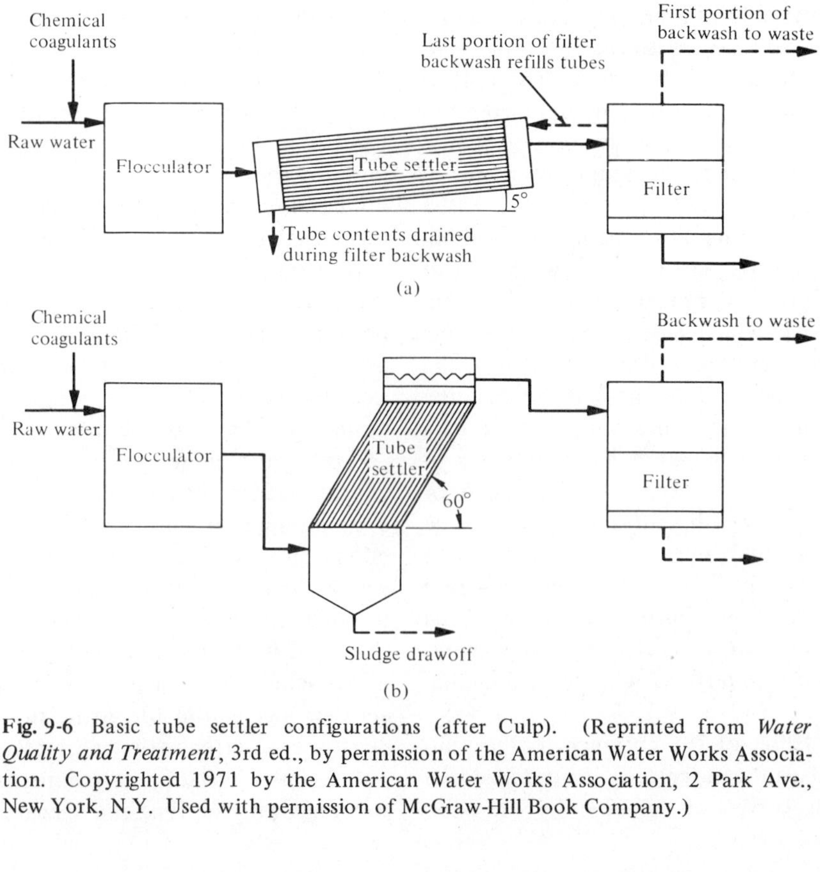

Fig. 9-6 Basic tube settler configurations (after Culp). (Reprinted from *Water Quality and Treatment*, 3rd ed., by permission of the American Water Works Association. Copyrighted 1971 by the American Water Works Association, 2 Park Ave., New York, N.Y. Used with permission of McGraw-Hill Book Company.)

with particulate matter, the yield drops below the economic level. At that time, the filter is taken out of service and the top inch of sand is scraped off and transferred to cleaning equipment. Starting out with 48 in. of sand, after twenty-four cleanings the thickness of the sand bed is down to 24 in. Cleaned sand is then added to build the layer back up to 48 in.

The only pretreatment ever used on water to be slow sand filtered is primary settling in an impounding reservoir or filtering through a coarse rapid sand or "roughing" type filter where the large particles of sediment are removed. This reduces the load on the slow sand filter and provides much longer periods of service between cleanings.

The effect of the slow sand filter is to remove bacteria by biological action and to remove most of the turbidity by straining. It is not particularly effective in removing colloidal turbidity.

Rapid Sand Filters. A typical rapid sand filter has a layer of sand from 4 to 36 in. in thickness, the sand ranging in effective size from 5.0 to 1.0 mm. In most rapid sand filters, the sand does not exceed 7 mm effective size.

The sand layer is supported by a gravel layer from 12 to 24 in. in thickness which usually consists of five grades of gravel ranging from $2\frac{1}{2}$ in. for the coarser grade down to $\frac{1}{10}$ in. for the finest grade.

The purpose of the gravel is threefold. In addition to supporting the sand layer, it acts as a collecting mechanism for the filtered water and as a distributing mechanism for the backwash water. The coarser gravel is placed on the bottom between the underdrains, which may be of cast iron, vitrified clay blocks, or precast concrete. It has been found that the finer grades of gravel ($\frac{1}{8}$ to $\frac{1}{12}$ in.) tend to mound during the backwash period, resulting in the loss of sand in the filter area adjacent to the mound.[8,13] To prevent this, a revised grading in the gravel has been developed. In this regrading, the $\frac{1}{8}$- to $\frac{1}{4}$-in. gravel layer is increased from $3\frac{1}{2}$ to 4 in. Above that, 2 in. of $\frac{1}{4}$- to -in. gravel are placed; above that, 2 in. of $\frac{1}{2}$- to 1-in. gravel; and above that, 5 in. of 1- to 2-in. gravel. A total of forty-five filters at the South Water Filtration Plant in Chicago are constructed in this manner (Fig. 9-7).[12] There has been no mounding of gravel with the subsequent loss of sand in these filters.

There are perforations in the underdrains, usually on the lower side. These perforations may be from 3 to 12 in. apart and may range in size from $\frac{3}{8}$ to $\frac{1}{2}$ in. in diameter. In the case of cast iron underdrains, the perforations are usually lined with a thin brass ferrule to prevent erosion.

The underdrains may be from 6 to 12 in. or more apart within a given filter. The drains lead to a collecting channel which flows into the clearwell through a rate controller. This channel and the underdrain pipes are also part of the filter water collection system and of the backwash water distributing system, the backwash line being "T'd" into the underdrain system behind the rate controller.

The rapid sand filter is kept clean by two systems of washing. The backwash water is fed into the underdrains from a tank under a fixed head (if a pump is used, it should be equipped with a bypass so that the head of the water forced into the underdrains may not exceed a certain fixed value).

In most filter plants, a surface water system is activated prior to turning on the backwash water. There are two types of these systems. One is the fixed-jet system. In Chicago's two water filtration plants, the Baylis fixed-jet system is used. It consists of a header pipe to which water laterals are attached 38 in. apart. In each lateral on the horizontal portion, 1-in. nipples extend down to about 2 in. below the sand surface. Each nipple is equipped with a pipe cap drilled in five places with $\frac{1}{8}$-in. holes. One hole is in the center pointing directly down. The other four holes are in the corner of the cap angle at 37° and drilled on the four diametrically-opposed sides of the cap. In practice, the surface wash system is operated for 1 min under about 8 psi pressure before the backwash water is turned on. The rate of rise of the backwash water is usually about 15 gpm/ft^2 of filter area. In Chicago, the rate of rise is changed with the change in temperature of the water so that the lifting power remains

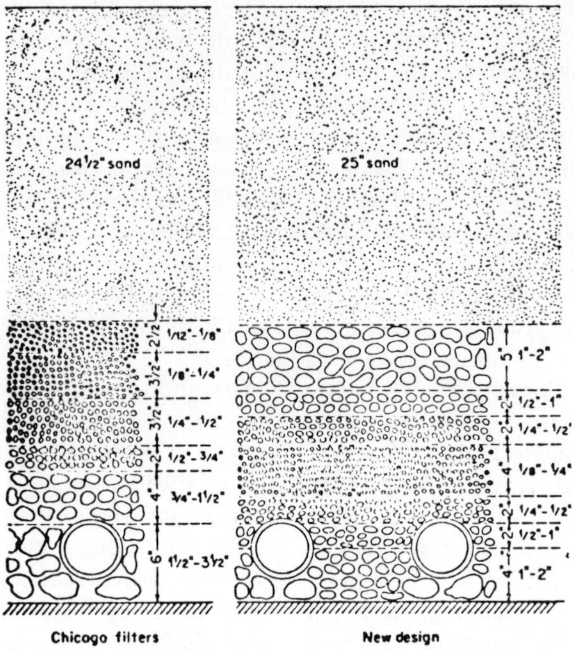

Fig. 9-7 Gravel depths in Baylis' old and new design of filter. (Reprinted from *Water Quality and Treatment*, 3rd ed., by permission of the American Water Works Association. Copyrighted 1971 by the American Water Works Association, 2 Park Ave., New York, N.Y. 10016. Used with permission of McGraw-Hill Book Company.)

constant for the entire year. With the 24-in. sand layer, a 4-in. sand expansion is thus obtained.

The second type of surface wash system is a revolving type. The water is fed to a central box and flows outward through the two ends of the feeder pipe which is mounted on a swivel joint. The jets on either end are angled in opposite directions so that the water causes the main feeder pipe to rotate by the jet action of the nozzles. In order to reach the corners with this system, it is customary to put a jet nozzle in either end of the revolving pipe, which is sometimes called a filter sweep. Care must be taken so that this jet does not come too close to the side walls of the filter box. There have been cases where this end jet had a sand blasting action on the side wall of the filter, wearing away the concrete down to the reinforcing steel rods.

The traditional design rate of operation for a rapid sand filter is 2 gpm/ft^2. However, such filters may be operated at rates as high as 5 gpm/ft^2 and still produce a satisfactory quality of effluent. This depends, of course, on adequate pretreatment of the water and adequate sizing of the hydraulic features of the filters such as piping and controls. The entire filter washing procedure can readily be automated.

Ultrahigh-Rate Filters. In recent years, filters have been built that permit rates of filtration from 5 to 15 gpm/ft^2. These are usually "coarse and fine" filters in which the coarse layer, usually anthracite coal, acts as a roughing filter in permitting fine particulate matter to penetrate deeply into the filter for removal by a fine filter sand.

This process is usually aided by the addition of polyelectrolytes. These polyelectrolytes consist of long-chain organic molecules that connect two or more of the sediment particles producing a "bridging" action which makes the particles sufficiently large that they are easily caught on the coarse layer of the filter. In some of these plants, no settling time at all is used. In all of them, the entire dependence is placed on the efficiency of the polyelectrolyte treatment.

Pressure Filters. The pressure filter is usually a rapid sand filter completely enclosed in a steel tank so that the water may be applied to the filter under pressure. It normally consists of the regulation sand layer on top of gravel with the proper arrangement of underdrains.

Pressure filters usually are not equipped with surface wash systems. They are not in general use, being limited to industrial applications and to certain areas where the topography permits the operation of the pressure filter under a gravity head (Fig. 9-8).[12]

Diatomite Filters. Diatomite filters consist of an industrial-type filter with a filter membrane arranged usually in a series of leaflets. The membrane is precoated with a thin layer of diatomaceous earth ground to a certain particle size. This precoated layer is not sufficient to stand up very long under operating conditions. As a result, more diatomaceous earth is added to the applied stream of water, and the suspended sediment mixed with the diatomaceous earth is caught on the filter membrane. This layer of sediment and diatomaceous earth is then removed by backwashing. It is necessary to pre-

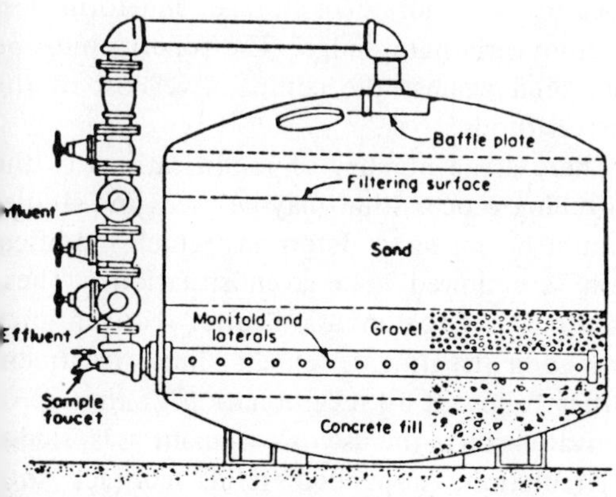

g. 9-8 Cross sections of typical pressure filter. eprinted from *Water Quality and Treatment*, 3rd ., by permission of the American Water Works sociation. Copyrighted 1971 by the American ater Works Association, 2 Park Ave., New York, Y. 10016. Used with permission of McGraw-Hill ok Company.)

coat the membrane again and continue to add diatomaceous earth to the applied stream. This is a fairly expensive operation and is usually limited to industry and the smaller municipal operations. The largest diatomaceous earth filter plant known to the author has a capacity of about 2.6 mgd.

Multimedia Filters. Multimedia filters have two or more grades of filter medium of varying specific gravities so that in the backwashing period the several layers remain more or less intact. The simplest form of multimedia filters is the anthracite-sand combination. The anthracite has a specific gravity of 1.5 while the sand layer has a specific gravity of 2.65. There is little problem of intermixing. However, in the use of the surface wash system, the anthracite layer has to be expanded to a level 2 in. above the surface wash nozzles.

Most of the multimedia filters consist of three layers of filter material of different specific gravities–the top layer being the coarsest in effective size, for example, 1.0 mm; the middle layer being of intermediate size, for example, 0.5 mm effective size; and the bottom layer being of very fine media, for example, 0.2 mm effective size. If the specific gravities differ sufficiently, there is little intermixing of the three layers during the backwash period. In these filters, the coarser layers act as roughing filters, putting a much smaller load on the finest layer. This type of filter produces a satisfactory quality of effluent at a much higher rate than the rapid sand filter.

Household Filters. There is an appreciable market for household filters that are usually attached to the kitchen tap. These are normally small-scale pressure filters without equipment for backwashing.

Softening. The hardness of water is that property which causes it to form curds when soap is used with it. Some waters are very hard and the consumption of soap by these waters is commensurately high. There are other adverse effects such as "bathtub ring," deterioration of fabrics, and, in some cases, stains. Therefore, it becomes desirable where the hardness is considered high to soften the water. This also reduces detergent consumption, scale formation, and corrosion.

However, there is some indication of a broad correlation between the hardness of water consumed and cardiovascular disease. For example, the state of Arizona, which has on the average the hardest water in the continental United States, has the lowest number of cardiovascular deaths per 100,000 persons. By contrast, the state of South Carolina, which has on the average the softest water in the nation, has the largest number of cardiovascular deaths per 100,000 persons.

The value assigned to an "ideal" water is 80 ppm (mg/l). Waters ranging in hardness from 80 to 150 mg/l are considered "moderately" hard waters. Usually, the additional soap required by such waters costs less than softening these waters to 80 mg/l. The hardness of water from Lake Michigan, the source of Chicago's supply, is 136 ± 6 mg/l. Softening such a water would have to be considered a luxury rather than a necessity.

Some of the factors affecting the softening process are temperature, reaction zone efficiencies, time, and the catalytic effect of retained slurries.

Lime Soda. The hardness of most water supplies is caused by the presence of calcium and magnesium compounds in solution. Sometimes strontium, ferrous iron, and manganese may contribute to a small part of the hardness. The elements are usually present as bicarbonates and sulfates, although in some waters there are some chlorides and nitrates.

The bicarbonates of calcium and magnesium are known as "carbonate hardness," and the calcium and magnesium sulfates, chlorides, or nitrates are called "noncarbonate hardness." The carbonate hardness of an untreated water can be ascertained by making an alkalinity determination. In some waters, there is sodium alkalinity present, in which case the total alkalinity would exceed the total hardness. The chemistry of water softening is illustrated in the following equations:

$$CO_2 + Ca(OH)_2 = CaCO_3 \downarrow H_2O \qquad (1)$$

$$Ca(HCO_3)_2 + Ca(OH)_2 = CaCO_3 \downarrow + 2H_2O \qquad (2)$$

$$Mg(HCO_3)_2 + Ca(OH)_2 = CaCO_3 \downarrow + MgCO_3 + 2H_2O \qquad (3)$$

$$MgCO_3 + Ca(OH)_2 = CaCO_3 \downarrow + Mg(OH)_2 \downarrow \qquad (4)$$

$$2NaHCO_3 + Ca(OH)_2 = CaCO_3 \downarrow + Na_2CO_3 + 2H_2O \qquad (5)$$

$$MgSO_4 + Ca(OH)_2 = Mg(OH)_2 \downarrow + CaSO_4 \qquad (6)$$

$$CaSO_4 + Na_2CO_3 = CaCO_3 \downarrow + Na_2SO_4 \qquad (7)$$

In Eq. 1, carbon dioxide is shown as reacting with lime to produce insoluble calcium carbonate. Carbon dioxide is not of itself hardness, but it will consume some lime. Therefore, this must be taken into consideration in the total lime dosage. In Eq. 5, the sodium bicarbonate shown is not hardness, but if present it will also react with lime to produce insoluble calcium carbonate. In Eqs. 2, 3, and 4, the removal of bicarbonate hardness by lime is shown. The removal of magnesium noncarbonate hardness, usually magnesium sulfate, by lime is shown in Eq. 6.

An alternative method to the lime-soda process is the use of caustic soda instead of sodium carbonate. Liquid caustic soda (50 percent NaOH) is usually used. It is easy to handle does not deteriorate in storage, and forms less calcium carbonate sludge. The benefits must be measured against the comparative cost of the two methods.

There are a number of modifications of the softening process that may be used, and studie should be made to determine which modification is required in a given situation. Thes methods are carbonation after conventiona lime-soda treatment, excess lime treatmen either followed by recarbonation or the use c activated silica, the use of coagulant aids, sludg recirculation, suspended solids contact sof eners, the use of polyphosphates to inhib calcium carbonate scale formation, and th substitution of cation-exchange resins for sod ash to remove noncarbonate hardness.

In a water low in magnesium hardnes just enough lime to precipitate the carbona hardness may be used. If the water contains more than 40 mg/l of magnesium hardness (as $CaCO_3$), it is usually desirable to resort to excess lime treatment to obtain the required magnesium reduction. It is undesirable to pump highly caustic water to consumers. Therefore, the excess lime remaining in solution after settling out precipitated magnesium hydroxide and calcium carbonate in the primary basin must be neutralized. The easiest methods are to convert it to calcium carbonate by recarbonation or by the use of bypassed raw water which is called "split treatment." When excess lime treatment is followed by recarbonation, the water must again be flocculated and settled (Fig. 9-9).

Split treatment consists of treating a large portion of hard water with all the required lime, then after settling the precipitated magnesium hydroxide and calcium carbonate, adding the balance of hard water and again mixing and settling. Adding all the lime required to only a portion of the water results in excess lime treatment for the fraction of untreated water, with the result that all of the magnesium is removed.

There are a number of plants using what is called "three-stage" treatment. These plants usually reburn the lime-softening sludge which provides a cheap supply of lime for treatment, and, in some cases, produces enough excess lime for the plant to supply other softening plants that do not reburn the sludge. In this process, primary treatment is subdivided and only the lime required by the free carbon dioxide and carbonate alkalinity of the water is added. After settling, the practically pure calcium carbonate sludge can be pumped to the lime-reburning plant.

The effluent of the first stage passes to the second stage where more lime is added to provide the hydroxide alkalinity needed for the desired reduction in magnesium hardness. The sludge from the second stage not only contains hydroxide but some calcium carbonate as well, which precipitates when the calcium hydroxide is added. These sludge particles are quite light and may be weighted by recirculating a small amount of calcium carbonate sludge from the third stage back to the second stage quick-mixed basins. This is a very economical coagulant aid. The second stage also may be recarbonated using waste gases from the lime-reburning plant for this purpose. The effluent from the second stage goes to the third and final stage where it is recarbonated and mixed with the balance of the raw water if split treatment is used.

Where salt is plentiful and cheap, it is sometimes preferable to remove noncarbonate hard-

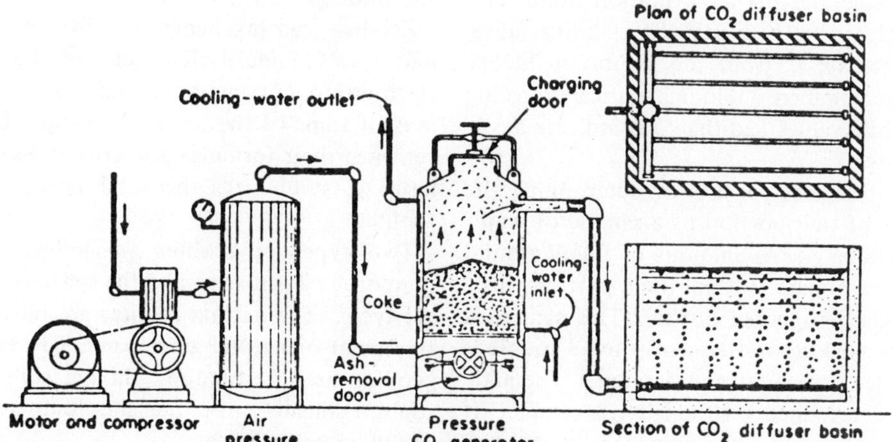

Fig. 9-9 Recarbonation system (National Lime Association). (Reprinted from *Water Quality and Treatment*, 3rd ed., by permission of the American Water Works Association. Copyrighted 1971 by the American Water Works Association, 2 Park Ave., New York, N.Y. Used with permission of McGraw-Hill Book Company.)

ness by means of cation exchange rather than by use of soda ash. In this method, the water is first treated with lime to convert all free carbon dioxide and noncarbonate alkalinity to normal hardness. After mixing and flocculating, the precipitated calcium carbonate is settled out. A very small amount of magnesium is also removed in this stage, but most of the magnesium remains in solution. After settling, the lime-softened water is carbonated with carbon dioxide gas to convert most of the normal carbonates to bicarbonates. The pH should be reduced at least to the pH of the water after lime softening to prevent the deposition of calcium carbonate in the cation-exchange material. After carbonation, the water flows to sand filters and then, depending on the hardness itself, is passed through cation-exchange softeners.

One of the more useful tools in establishing the chemical stability of treated waters is the Langelier index, which is the difference between the actual pH and the calculated pH at the saturation point. A positive value for the index indicates a lack of excess carbon dioxide and a "plating" characteristic of the water. A negative value for this difference indicates an excess of carbon dioxide, or scale-dissolving qualities.

Another index is known as Ryznar's stability index. This involves the expression $2pH_s - pH$, or twice the pH at saturation minus the actual pH. This is called the stability index to differentiate it from the saturation index. Values greater than 7.0 indicate a scale-forming water, while values less than 7.0 indicate a corrosive water.

The marble test consists of determining the alkalinity of one portion of a sample of water before adding excess amounts of finely-ground calcium carbonate to another portion in a 300 ml glass-stoppered bottle. This sample is mixed by shaking it frequently for 3 hr. It is then settled overnight and filtered. The alkalinity is determined on the filtered sample. If alkalinity is increased, the water is not saturated with calcium carbonate and would not be stable. If the alkalinity is decreased, the water is saturated and would be scale-forming.

The Enslow stability index is simply a device for a continuous marble test. A quantity of fine calcium carbonate is contained in a revolving bulb or in a separatory funnel and two glass tubes containing 6-in. deep beds of crushed marble which is screened to pass a 14-in. mesh sieve and held on a 30-in. mesh sieve. The crushed marble is supported by a 6-in. bed of small silica gravel. The water tested is passed continually up through the calcium carbonate in the funnel and then through the two tubes in series. The alkalinity before and after passage through the treatment device is evaluated in the same method as the marble test.

Ion Exchange. Ion exchange takes place when in a given medium one ion is displaced by another. When this is applied to water treatment, it means that there is a reversible exchange of ions between a liquid and a solid. This does not make any particular change in the solid structure. The solid body must have ions of its own to exchange for others. Cation exchange, or base exchange, is the displacement of one positive ion or cation by another positive ion. In natural waters, the cations usually found are calcium, magnesium, sodium, hydrogen, iron, and manganese. Anion exchange, or acid exchange, occurs when one negative ion or anion is displaced by another negative ion. The anions usually found in natural waters are chloride, sulfate, nitrate, carbonate, hydroxide, and fluoride.

Zeolites can exchange monovalent sodium ions for: (1) multivalent ions of the alkaline earth group, (2) ammonia, and (3) the divalent ions of some of the metals in water. For convenience their formulas are written Na_2Z. The letter Z symbolizes the solid portion of the zeolite.

Two types of zeolites are known: natural nonporous green sand and the synthetic porous gel type. The natural zeolites are obtained by treatment of natural green sand. The synthetic zeolites are prepared by mixing solutions of sodium silicate and either aluminum sulfate or sodium aluminate.

Resinous cation exchangers are synthetic organic polymers generally of a sulfonate polystyrene type. They are produced in an effective size and uniformity coefficient comparable to filter sand. Resinous anion exchangers are organic polymers consisting of amines and quaternary ammonium compounds. They may be regenerated by the use of any caustic. Fluoride exchangers are anion exchangers that consist of a processed tricalcium phosphate, which removes fluoride by ion exchange. Iron and manganese are removed during the normal softening cycle or by zeolites treated with manganese sulfate and potassium permanganate. In this case the iron and manganese are removed by contact oxidation.

There are three methods for water softening by ion exchange. They are sodium cycle, split stream, and demineralization. The principal one is the sodium-cycle operation. Softening water by this method consists essentially of passing hard water through a bed of cation exchangers. There the hardness components of the water, calcium and magnesium ions, are removed from solution and replaced by sodium. The exchange rate is so rapid that any water can be easily and completely softened at normal flow rates regardless of hardness.

The reactions which occur in cation-exchange water softening and in regeneration can be shown by chemical equations. The following equations, in which R represents an anionic component of the ion exchanger, indicate the reactions which take place:

$$Ca(HCO_3)_2 + Na_2R = CaR + 2NaHCO_3 \quad (8)$$

$$\underset{\text{Carbonate hardness}}{Mg(HCO_3)_2} + Na_2R = MgR + \underset{\text{Soluble nonhardness}}{2NaHCO_3} \quad (9)$$

$$CaSO_4 + Na_2R = CaR + Na_2SO_4 \quad (10)$$

$$CaCl_2 + Na_2R = CaR + 2NaCl \quad (11)$$

$$MgSO_4 + Na_2R = MgR + Na_2SO_4 \quad (12)$$

$$\underset{\text{Noncarbonate hardness}}{MgCl_2} + Na_2R = MgR + \underset{\text{Soluble nonhardness}}{2NaCl} \quad (13)$$

When the exhausted exchanger is regenerated, essentially the reverse action takes place:

$$CaR + NaCl = CaCl_2 + Na_2R \quad (14)$$

$$MgR + 2NaCl = MgCl_2 + Na_2R \quad (15)$$

Only a small amount of iron, manganese, and aluminum will be removed from the exchanger during regeneration. These elements will readily form an insoluble hydroxide and will be permanently locked within the structure. This neutral substance may cause a loss in softening capacity of the resins. In split-stream softening, a portion of the water is passed through a conventional salt-regenerated exchanger. The remaining water is passed through a resinous exchanger which has been regenerated with mineral acid. The net result is complete removal of these metals and destruction of the bicarbonate, thus leaving only free carbon dioxide:

$$Ca(HCO_3)_2 + H_2R = CaR + 2H_2O + 2CO_2 \quad (16)$$

$$Mg(HCO_3)_2 + H_2R = MgR + 2H_2O + 2CO_2 \quad (17)$$

$$2NaHCO_3 + H_2R = Na_2R + 2H_2O + 2CO_2 \quad (18)$$

When sulfates and chlorides are present, the corresponding mineral acids are formed by the exchange:

$$CaSO_4 + H_2R = CaR + H_2SO_4 \quad (19)$$

$$MgSO_4 + H_2R = MgR + H_2SO_4 \quad (20)$$

$$2NaCl + H_2R = Na_2R + 2HCl \quad (21)$$

The ratio of the two streams on remixing is adjusted so that the free mineral acidity in one is enough to neutralize and leave whatever degree of alkalinity is desired in the mixed effluent.

In demineralization, both cation and anion exchanges in single or dual tanks are utilized to completely remove all metal and nonmetal ions from solution. This method is usually only used in laboratories or industrial processes where a demineralized water is required.

An important aspect of softening by ion exchange is the softening regeneration cycle. In order for softening to be accomplished, the ion exchanger is usually retained in a vessel where the flow of water and regenerants going to the unit can be controlled. Softeners are usually operated with either pressure or gravity flow. Pressure vessels for softeners are identical in structure and function to those which regulate pressure filters, whereas gravity softeners may be simply concrete, steel, or wooden staves.

The pressure softeners operate down-flow or up-flow. The head loss due to down-flow may be overcome by passing the water upward through the bed during the softening cycle. Many large municipal water softening plants use gravity softeners with softening up-flow and brining down-flow.

On completion of the softening cycle, the softeners should be backwashed to loosen the ion-exchange resin. This should flush out any particulate matter which may have accumulated on the bed surface and make it more receptive to the regeneration cycle by preventing channeling or brining through the bed. In regenerating the pressure softener, brine is introduced to the bed at a controlled rate. There are a number of different techniques for accomplishing this.

In some small domestic softeners, the tank is drained down several inches and rock salt is dropped onto the top of the resin. Water is passed at a controlled rate through the salt to produce a brine of the proper concentration. After the brine has been applied to the softener, all the spent brine must be removed before the unit is returned to service. Because of the greater density of the brine, it moves on down the bed in a piston fashion. If rinsing is continued at this rate until the chlorides have dropped to an acceptable level, the time required is excessive. The best way to accomplish complete removal of the chlorides is to increase the rinse rate fivefold for the balance of the cycle.

In evaluating the performance characteristics of the cation exchanger, it is good practice to measure its performance in gallons per minute per cubic foot. Another mechanism of evaluation is to express the pounds of salt per cubic foot of exchange required for complete regeneration. Experience has indicated that the optimum concentration of brine used in regeneration should lie between 10 and 15 percent.

Seawater may be used in regeneration of cation exchangers but it should be free of solids and/or soluble iron and manganese. It should also be free of slime-producing organisms. Some plants using seawater find it desirable to sterilize it with chlorine and then filter it through a sand filter in order to prevent contamination of the resin.

One of the great problems in ion-exchange softening is the disposal of the spent brine from a regeneration cycle. These waste products consist principally of the chlorides of calcium, magnesium, and sodium. There may also be some iron, manganese, and aluminum present. It is no longer considered desirable to discharge this brine into waterways. The discharge of such brines into the ocean may be permissible but also may create an undesirable situation at the point of discharge.

The possibility of using evaporation ponds, uncontrolled dilution, controlled dilution, and brine disposal wells has been suggested. Evaporation ponds will only work in areas where the average daily temperature is sufficiently high. The level of natural dilution which will prevent damage to aquatic life occurs when the discharge of waste brine into a waterway amounts to only a very small percentage of the total flow. Controlled dilution is the storage of waste brines so that the discharge rate may be kept below the harmful level. Brine disposal wells have been used successfully in oil well areas. Since this brine comes out of the ground with the oil in the first place, returning it to the formation by means of brine disposal wells should not change the geology of the situation. It is anticipated that other uses of brine disposal wells may have only a limited application before the geologic formation becomes hopelessly plugged.

Disinfection. Disinfection is the destruction of harmful bacteria and other organisms which may be present in a water supply. It has been known from the beginning of civilization that boiling water makes it generally safe to drink. Modern practice is to treat the water with various chemical disinfecting agents. The germ theory of disease was not expressed until 1876 by Koch and Pasteur. It was some years later that this concept was generally accepted and practices to prevent the transmission of disease by water were undertaken.

Chlorination. The most widely used chemical in the field of disinfection is, of course, chlorine. The earliest experiments with electrically-produced chlorine gas were conducted in 1896 during the Louisville experiments.[3] In 1908, sodium hypochlorite was used to disinfect the 40 mgd Boonton Reservoir at Jersey City, New Jersey. The first use of a dry chlorine feeder was by Major C. R. Darnall of the U.S. Medical Corps, Fort Meyer, Virginia. Chlorination was started in Chicago in 1912 at the Bubbly Creek plant of the Union Stockyards.

Chlorine is a greenish-yellow gas at ordinary temperatures and pressures. Under 95 psi pressure it becomes a liquid. These properties make it highly desirable as a disinfection agent because its application is easily controlled. When chlorine is added to chemically-pure water, a mixture of hypochlorous acid (HOCl) and hydrochloric acid (HCl) is formed:

$$Cl_2 + H_2O = HOCl + H^+ + Cl^- \qquad (22)$$

At the pH level of most water supplies (above 4.0), the reaction goes to the right, leaving little chlorine in solution. Hypochlorous acid is weakly dissociated at pH values below 6.0. Above this level it dissociates almost completely as follows:

$$HOCl = H^+ + OCl^- \qquad (23)$$

A modern theory (which has some substantiation) of the action of chlorine in destroying bacteria is that there is a chemical reaction between the HOCl produced and an enzyme system. (The reactions are presumed to involve irreversible HOCl oxidation of sulfhydryl-containing enzymes.)

Most of the early-type chlorinators draw chlorine from the gas side of the container. It is put into solution in the chlorinator itself after being drawn through a metering orifice under a vacuum. The chlorine solution is then transmitted to the point of application through corrosive-resistant lines, rubber-lined steel being the preferred material.

In modern chlorination practice, chlorine is taken from the liquid side of the container. Pressure of the gas in the container is allowed to force the liquid into an evaporator where it is converted into a gas. This, in turn, is drawn through a regulating orifice under vacuum and is transmitted to the point of application by means of a vacuum created by a water eductor. The water eductor then puts the chlorine into solution at that point.

There are a few large water systems in the continental United States that use chlorine in tank cars. Among them are the systems of Minneapolis, Philadelphia, and Tampa. Most of the other larger water plants use chlorine in ton containers. The smaller water plants find it desirable to feed the chlorine from 100- to 150-lb cylinders. The municipal code of some cities, Chicago for example, forbids the use of bulk chlorine in congested population areas.

Chlorine under control is an effective chemical but its use can also be hazardous. There are a number of safety practices that should be followed regardless of the type of container employed in the chlorination process. Most manufacturers of chlorine are members of the Chlorine Institute, a cooperative organization for developing standardized equipment and safety practices. Each manufacturer has manuals outlining the materials that should be used and the practices that should be followed. A user can readily obtain copies of these booklets.

Superchlorination. Superchlorination is the application of a much greater quantity of chlorine than is used in normal chlorination. These dosages may range as high as 10 mg/l or more.

Superchlorination is sometimes required to destroy the organic material in the water in order to satisfy the chlorine demand of the water. If an abnormal residual of chlorine is left after superchlorination, then the water must be dechlorinated before it can be delivered to the consumer.

Most of the pathogenic bacteria can be destroyed by the use of chlorine in normal dosages. Superchlorination, however, if carried far enough (for example, above 10 mg/l application), may inactivate some of the virus organisms as well. Some of the materials used in dechlorination following superchlorination are sodium thiosulfate, sodium sulfite, sulfur dioxide, and activated carbon.

Chlorine Dioxide. Chlorine dioxide is a comparatively powerful oxidizing agent produced by the action of chlorine with sodium chlorite. (See section on taste and odor control.) It is usually generated at the point of application. It is said to be a more effective bactericide and sporicide than chlorine although it is not considered effective in destroying virus organisms.

Potassium Permanganate. Potassium permanganate is a mild oxidizing agent ranking somewhere between superchlorination and chlorine dioxide. It is considered quite an effective bactericide but, again, is not considered effective in destroying viruses.

It is usually prepared in 10 percent solution which is fed into the water ahead of filtration through an appropriate liquid metering device and applied at the beginning of the mixing period. Its progress through the mixing and settling process may be readily followed by observing the characteristic pink color which should not be allowed to pass through the filters. Care must be taken to avoid an excess of permanganate to check this. There is always a danger of overtreating and creating manganese problems in the finished water.

Ozone. Ozone has the formula O_3. It is prepared by passing dry air through a high-tension electric current. It is unstable and is usually applied to the water immediately after generation.

Ozone is a most effective sterilizing agent and also is capable of inactivating hepatitis and poliomyelitis viruses. Its effectiveness is assumed to be due to the fact that in breaking down O_3 to O_2 it releases an atom of nacent oxygen which is a very active oxidizing agent.

It is largely used in Europe, particularly in France. The Paris Suburban Water Company has a plant at Choisey-Le Roi which has a very extensive ozone installation to treat water from the Seine River. The purpose of the ozone treatment, in addition to reducing objectionable tastes and odors as discussed above, is primarily virus inactivation.

Ultraviolet Light. When this treatment method is used, the water is usually passed under the lights in a very shallow layer so that all of it is exposed to the ultraviolet rays. This requires an elaborate installation to treat reasonably large volumes of water. It is not in general use in public water supplies and is limited mostly to small, special-purpose supplies.

Iodine. Iodine reacts much more slowly than chlorine but over a long period of time is a somewhat more effective chemical. It is relatively insoluble in water itself and must be brought into solution by use of potassium iodide. This adds to the complexity of disinfection by iodine. Therefore, it is not extensively used in public water supplies. Its principal use is in swimming pools.

Silver. On rare occasions, silver has been used in the sterilization of water. A form of activated silver known as Katadyn sand filter has been on the market since 1929. Another way of introducing the metal into the water is by passing an electric current through it into the water. This process has also been used principally in swimming pools.

Bromine. Bromine is a highly corrosive liquid, is reddish brown in color, has very irritating fumes, burns the skin, and is slightly soluble in water. It is sometimes used in the sterilization of water but it must be handled with great care and with corrosive-resistant equipment throughout the process. It is not a very important factor in the purification of water.

Hypochlorites. Hypochlorites have been used for many years as a sterilizing agent. Long before the use of chlorine, chlorinated lime was employed.

There are two forms of hypochlorites available today. One is calcium hypochlorite which has at least 70 percent available chlorine and is employed in many of the smaller installations where continuous control of the chlorination is not practical. It is usually made up in a certain strength by dissolving it in water in a stoneware crock, and is then fed in a proper dosage by a proportioning liquid feed pump.

Sodium hypochlorite is a readily soluble compound. Most of the chlorine bleaches now available on the domestic market are 3 to 5 percent available chlorine solutions of sodium hypochlorite. Some fairly large installations are using sodium hypochlorite as a sterilizing agent because of the hazard of chlorine usage to either the equipment or the surrounding population.

Chlorine–Ammonia Treatment. Assuming the chlorine in water to be present as HOCl, the reaction on adding ammonia (NH_3) is as follows:

$$NH_3 + HOCl = \underset{\text{monochloramines}}{NH_2Cl} + H_2O \qquad (24)$$

At pH values below 7.0, there would be some dichloramines present as shown:

$$NH_2Cl + HOCl = NHCl_2 + H_2O \qquad (25)$$

In most water supplies where the chlorine-ammonia treatment is used, the combined chlorine is present as dichloramines. The dichloramines are colorless, tasteless, and are more persistent in water distribution systems than is free chlorine. Many water supply systems use this combination as posttreatment for this reason. Given ample time of contact (more than 45 min), the chloramines are more effective bactericides than free chlorine.

WATER PLANT WASTE DISPOSAL

Settled Sediment. The disposal of water plant wastes has become a worldwide problem. Solutions to each problem are peculiar to local situations and must be evolved by a competent consultant.

The disposal of the sediment or sludge from the softening process is particularly difficult because of the large quantities of it. For those larger installations in which the softening sludge is comparatively pure calcium carbonate, reburning the sludge to produce fresh lime for use in the softening process can be profitable. Some water utilities lagoon the softening sludge and, when it is finally dried, sell it for agricultural lime.

Discharge of these settled sediments to water sources is now forbidden by both federal and state regulations. In Illinois, all new softening plants must be designed with a system of disposal of settled sediment other than to the source before their plans for construction are approved. Older plants have a limited period in which to arrange for appropriate disposal.

In filtration plants which are not softening plants, the most convenient method of disposal has been back to the source.

Some water plants attempt to recover chemicals, such as aluminum sulfate, by treatment of the sediments. Usually the concentration of the alum in such cases is so low as to make the process unprofitable.

A great deal of experimental work has been undertaken to find methods of concentrating these sediments. Percolation through sand beds, lagooning, the use of polyelectrolytes, vacuum filtration, and percolation have resulted in no better than 40 percent dried solids. The handling of dried sediment again produces a prohibitively expensive operation.

In several cities, water utilities are now putting sediment into sanitary sewers and it is treated with other wastes at a sewage treatment plant. Within limitations, where it is possible this is the most desirable method of disposal.

Wash Water Disposal. Returning filtration plant wash water to the source is also frowned upon by regulatory bodies. In the softening plants, the filters are backwashed with softened water; therefore, it becomes economically profitable to return the backwash water to the primary treatment basin. In the standard filtration plant, the wash water may be treated to reduce the solids content and then added to the regular plant product at an after-treatment point. Again, in most filtration plants the percentage of solids in the backwash water is so low as to make this unnecessary. By far the better means of disposal is to return the wash water to the intake basin and let it go through the treatment process again. Some care must be taken, though, since most wash water contains chlorine or chloramines and/or fluorides. The wash water is added in increments, usually 10 percent or less, which will not interfere with chlorination and fluoridation.

HOUSEHOLD SOFTENING AND CONDITIONING

In hard water areas where there are no municipal water treatment plants, it is the usual practice to install household softeners. These are nearly always of the ion-exchange type. They may be regenerated in place or they may be serviced with regenerated containers, or softeners can be obtained that have a built-in automatic regeneration facility. The choice of method is purely economic.

Some softener service organizations urge the householder to put the softener on the entire water system. This means that water used for lawn sprinkling, washing cars, and other uses as well as for drinking is all softened water. Such a practice adds appreciably to the cost of regeneration by requiring more service time.

There are some household softeners that consist of a small device that fits on the kitchen tap and contains an ion-exchange resin which will soften the water until the activity of the resin bed is exhausted. There is no mechanism for regenerating or backwashing this type of softener and it is difficult for the average householder to determine when the resin bed is exhausted. If the replacement of the charged material is neglected, the filter may become a source of contamination by those materials it is designed to remove.

From time to time, various devices appear on the market that are alleged by their advertisers to perform the functions of water treatment magically. Some of these consist of nothing more than a cast iron unit through which the water is passed. Others consist of alternate layers of magnesium and brass which are supposed to act like magnesium rods hung in hot water tanks. In one case, such a device was on test for a period of 6 years with water flowing continuously through it. It had no effect whatsoever on the water and the magnesium strips had lost almost none of their weight.

There is no such thing as magic water treatment. Nearly all of the treatment processes involve chemical reactions of one kind or another. The products of these chemical reactions are well known and their application is well understood.

REFERENCES

1. U.S. Public Health Service Drinking Water Standards, USPHS, U.S. Dept. of Health, Education, and Welfare (revised) (1962).
2. "Quality Goals for Potable Water," *Journal AWWA* **60,** 1317 (1968).
3. Fuller, G., *Experimental Work at Louisville, Kentucky Water System*, (c. 1900).
4. Baker, M. N., *The Quest for Pure Water*, AWWA, New York (1948).
5. "A Theory of Coagulation Useful for Design," *Industrial and Engineering Chemistry* (Aug. 1962).
6. "State of the Art of Coagulation," *Journal AWWA* **63,** 99 (1971).
7. Langelier, W. F., "The Mechanics of Flocculation," *Journal AWWA* **41,** 163 (1945).
8. Babbitt, H. E., Doland, S. S., and Cleasby, J. L., *Water Supply Engineering*, 6th ed., McGraw-Hill, New York (1962).
9. Patent of A. A. Hirsch, former Superintendent, Shreveport, Louisiana Filtration Plant.
10. *Manual of Water Treatment Plant Design*, ASCE, AWWA, CSSE (1969).
11. Gamet, M. B., and Rademacher, J. M., "Study of Short Filter Runs with Lake Michigan Water," *Journal AWWA* **52** (1), 137–152 (1960).
12. *Manual of Water Quality and Treatment*, 3rd ed., AWWA (1971).
13. Baylis, J. R., "Washing and Maintenance of Filters," *Journal AWWA* **46,** 176 (1954).

10

Water in Industry

Robert Parlante
Closter, New Jersey

INTRODUCTION

Industry uses water to generate steam, to cool and heat, and to fill the varied and voluminous needs of process and general plant services. Estimating the true economic impact of these water needs involves determining not only the required flows and the available supply, but also quality requirements and treatment costs. It also involves the consideration of the resultant wastewater volume and available receiving waters, and the cost of treating those wastewaters and their ultimate pollution effects. In order to improve the economy of industrial water use, it becomes imperative to reduce wastewater volumes, recover water or by-products from wastewater, and reuse wastewater, where possible.

Reduction, Recovery, Reuse. Recovering wastewater for reuse is not an uncommon practice in industry. For example, closed recirculating cooling water systems, with proper treatment, have replaced once-through cooling water use in many applications because it is less expensive to reuse water, even taking into account the additional equipment and chemicals required. In many industrial plants, the water used once in a particular application is still of a higher qaulity than the original raw water. Therefore, it can be cascaded or reused in another application requiring a lesser quality. A complete survey of plant usage becomes essential in order to apply this cascade principle to water reuse. It is also important to define water quality criteria required for any given usage.

BASIC TREATMENT STEPS

To select the most practical and economical method of water treatment, it is vital to know what impurities are present in the water supplies. Their effects on a treatment process must be understood and a method should be selected for their reduction or removal.

There is a wide range of possible water treatment processes. Selection of any one application will depend primarily on volume of water to be treated and degree of purification required. This selection is best left to a qualified water consultant, to be made only after all possible source controls have been implemented and the potential for reduced flows, recovery, and reuse has been evaluated.

Whether the water source is a river, lake, well, or municipal supply, impurities normally encountered fall into four general categories: suspended matter, dissolved solids, dissolved gases, and microorganisms. In turn, these four general impurities contribute color, taste, and odor.

Treatment will probably involve one or more of the fundamental processes. These processes along with the basic chemicals are the tools of treating industrial plant water.

Coagulation. Coagulation is the process of forming suspended matter and color in the water into a bound floc. This increases the size of solids suspended to the point where they can settle. Since most uses require a water that is at least clear and colorless, this process is frequently the first step in treatment applications.

The degree of coagulation depends on (1) how often the floc collides with itself and (2) the tendency of the floc to stick together after it collides. The frequency of collision is controlled by mixing. Floc adhesion is enhanced by added coagulants.

Coagulants are chemicals which produce one or more of the following actions:

1. Neutralization of the repulsive forces around the suspended solids. This enables the van der Waals cohesive force to hold the particles together once they collide.
2. Precipitation of sticky flocs such as metal hydroxides. This floc entangles the suspended solids and color.
3. Bridging the suspended solids by natural or synthetic long-chain high molecular weight polymers. These polymers are believed to act by absorption and the "bridge" effect created tends to take up particles missed by (1) and (2). It also strengthens flocs produced by them.

Chemicals for Coagulation. The most commonly used coagulants are aluminum sulfate, ferric sulfate, and ferrous sulfate. They are acidic in nature and react with the natural or added alkalinity of the water to produce sulfates of calcium, magnesium, or sodium and a gelatinous precipitate. The precipitate is assumed to be a hydroxide, but actually it is a complex formula of varying composition.

Most of the colloidal particles of turbidity and color carry a negative charge, whereas the metallic hydroxide formed carries a positive charge. Neutralization of particle ionic charges, as well as mechanical entanglement, are involved in coagulation by aluminum and iron salts.

For each coagulant there is an optimum pH zone. It is desirable to control the pH to the point of maximum precipitation of the coagulant, equivalent to the minimum solubility of the floc. The best pH for this precipitation is dependent not only on the coagulant, but also upon the mineral characteristics of the water. For example, alum will coagulate best at a pH of about 5.5 for waters with very low dissolved solids content. However, for most natural waters, the ideal pH is higher and may range from 5.7 to 7.8. The range for iron coagulants is usually above pH 8.4.

When aluminum or iron salts are added to water, they lower the pH. Therefore, it may be necessary to increase the pH to the optimum level by introducing an alkali such as lime, soda ash, or caustic soda.

The temperature of the water also affects the process of coagulation. At temperatures near the freezing point, increased mixing time as well as increased coagulant dosages may be required. In general, successful coagulation involves three main factors:

1. The presence of a minimum quantity of aluminum or iron ions to form an insoluble floc.
2. The presence of a strong anion such as sulfate or chloride.
3. The control of pH within a definite range.

It is not possible to predict from the analysis of a water sample which coagulant will be the best to use, the quantities that will be required, or the optimum pH for control of the process. For best coagulation, it is necessary to run laboratory tests with different coagulants at varied concentrations and pH levels. Laboratory stirring devices should be used in these tests to provide uniform test conditions. While previous practical experience with a given water will aid in selecting coagulant dosage and optimum pH, in order to consistently secure the required results it is usually necessary to change coagulant dosage and pH control to compensate for varying raw water characteristics.

Alum–$Al_2(SO_4)_3$. The most widely used coagulant is aluminum sulfate or filter alum. It reacts with natural or added alkalinity as follows:

$$Al_2(SO_4)_3 + 3Ca(HCO_3)_2 = 2Al(OH)_3 + 3CaSO_4 + 6CO_2 \quad (1)$$

$$Al_2(SO_4)_3 + 3Na_2CO_3 + H_2O = 2Al(OH)_3 + 3Na_2SO_4 + 3CO_2 \quad (2)$$

$$Al_2(SO_4)_3 + 3Ca(OH)_2 = 2Al(OH)_3 + 3CaSO_4 \quad (3)$$

Ferrous Sulfate (Copperas)–$FeSO_4 \cdot 7H_2O$. The natural alkalinity of most waters is not sufficient to react with copperas to form the desired ferric hydroxide floc. Lime or another alkali must be added to produce this result. The first reaction with lime produces ferrous hydroxide which in turn is oxidized by the dissolved oxygen in the water to form ferric hydroxide. This oxidation occurs only at higher pH values and copperas is a suitable coagulant only in the alkaline pH range above 8.4. Chlorine may also be used to oxidize the ferrous hydroxide:

$$FeSO_4 + Ca(OH)_2 = Fe(OH)_2 + CaSO_4 \quad (4)$$

$$4Fe(OH)_2 + H_2O = 4Fe(OH)_3 \quad (5)$$

Ferric Sulfate (Ferri-Floc)–$Fe_2(SO_4)_3$. Ferric sulfate also reacts with alkali and precipitates ferric hydroxide as is the case with copperas. However, since the iron is in the ferric state, dissolved oxygen is not required for oxidation:

$$Fe_2(SO_4)_3 + 3Ca(HCO_3)_2 = 2Fe(OH)_3 + 3CaSO_4 + 6CO_2 \quad (6)$$

$$Fe_2(SO_4)_3 + 3Ca(OH)_2 = 2Fe(OH)_3 + 3CaSO_4 \quad (7)$$

Alkalies. Coagulant feeding alone is usually not sufficient for proper coagulation because the pH must be adjusted to its optimum value. The most common alkalies used for pH adjustment are:

1. Calcium hydroxide ($Ca(OH)_2$), commonly called hydrated lime.
2. Sodium carbonate (Na_2CO_3), commonly called soda ash.
3. Sodium hydroxide (NaOH), commonly called caustic soda.

Clays. Waters which are low in turbidity are more difficult to coagulate than waters which are highly turbid. Adding clay to the water to be treated increases turbidity and results in better settling flocs. The use of clay also tends to broaden the pH range required for coagulants.

Polyelectrolytes. Polyelectrolytes are high molecular weight water-soluble polymers classified as anionic, cationic, or nonionic. Polymers whose functional groups in a water solution give positively charged particles are known as cationic. Polymers that dissociate to form negatively-charged ions are called anionic. Polymers in the nonionic group provide both positive and negative charges in water.

As described previously, the pH of a water is usually a very critical factor. However, the use of a polyelectrolyte not only broadens the pH range over which satisfactory flocculation could occur, but may also reduce the quantity of primary coagulant required.

In many cases, the use of polyelectrolytes has been justified on the basis of improved water quality rather than cost savings. These polymers may also prove advantageous in permitting high flow rates through existing equipment. The jar test is the most dependable tool for establishing both the dosage and the most effective polymer for the job. The usual dosage is less than 1 ppm.

Coagulant Dosages. The following can be used for estimating the magnitude of chemical feeding requirements, but do not necessarily represent the maximum dosages to be encountered:

A. For Coagulation of Suspended Matter:
   1. *Aluminum sulfate.* 2 grains/gal (gpg) with sufficient alkali for a resultant pH of 6.7-7.0. If the alkalinity and carbon dioxide are such that the addition of 2 gpg will not depress the pH to 5.7-7.0 or lower, then additional aluminum sulfate can be used. For large plants where the aluminum sulfate dosage can become excessive, it may be economical to obtain some pH reduction with sulfuric acid.
   2. *Ferric sulfate.* 1.5 gpg as ferri-floc with sufficient alkali for a resultant pH greater than 9.0.
   3. *Ferrous sulfate.* 3 gpg as copperas with sufficient alkali for a resultant pH of 9.0 or greater. Equipment for feeding chlorine for oxidation to ferric hydroxide is normally used. Exact chlorine dosage cannot be determined, as the amount of chlorine required depends on chlorine demand, oxygen content, and temperature of raw water.
   4. *Clay.* When the turbidity of the raw water is below 100 mg/1, a dosage of 2 gpg clay is typically used.

B. For Color Removal:
   1. Aluminum sulfate is almost exclusively used. A sufficient amount is added to depress the pH to 5.6, but in no case is the dosage less than 2.5 gpg. If 2.5 gpg depresses the pH below 5.6, an alkali is used to adjust the pH to 5.6. Equipment for feeding chlorine is usually required for optimum color removal.

Coagulation Equipment. The type of equipment most widely used for coagulation and clarification is a solids contact-type unit (Fig. 10-1). This type of unit operates on the principle of crystallization from a supersaturated solution and a more efficient reaction in the presence of the same type of precipitate.

The equipment is divided into two zones by

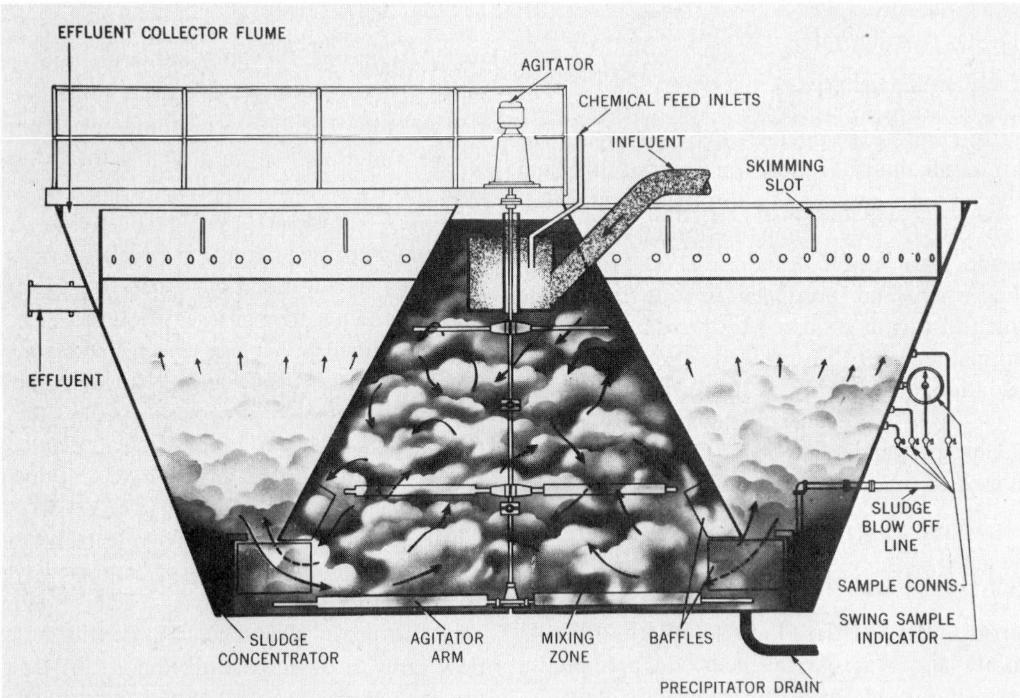

Fig. 10-1 Section of solids contact unit shows thorough mixing of chemicals with influent and previously formed sludge. (*Courtesy of Permutit Co.*)

means of sloping baffles. In one portion, the mixing takes place; in the other portion, the separation of the sludge floc from the water occurs.

The mixing zone is provided with agitation that brings the incoming water chemical mixture into intimate contact with previously precipitated and coagulated materials. This mixing and sludge recirculation is gentle enough so as to not break up the sludge. In this manner, equilibrium is quickly established. The newly formed solids do no separate individually as small particles; instead, the fresh precipitates deposit on the sludge floc already present.

Between the mixing zone and the sludge separation zone is a constricted area designed to produce a uniform flow from one zone to another. It is through this area that the sludge recirculation occurs to replace the sludge removed from the mixing zone.

The separation zone is designed so that the upward velocity of the water–sludge floc mixture decreases so that there is a point when this upward velocity will no longer support a particle of sludge floc. At this point, a definite line of separation exists between the sludge bed and the clarified water which has been filtered through the bed of its own sludge. The sludge floc left behind further combines with other particles to grow larger and larger and drop further in the blanket. In this way, the particle eventually drops into a concentrator section for eventual disposal.

The blow-off is usually allowed to concentrate in a semiquiescent zone within the unit to reduce water losses. The concentrated material is then blown off by a controlled cycle to maintain a fixed balance of solids in the unit. This balance is determined by the type of precipitate, the flow rate through the unit, and the temperature of the water which affects viscosity and the buoyancy tendency.

The water, after it leaves the sludge blanket, continues to flow upwards to the collector system. The effluent is uniformly collected from the whole surface of the unit. It is uniformity of distribution and collection and the internal baffling that permits this type of equipment to approach the theoretical detention time for which they are designed.

Tube Settlers. In recent years, equipment manufacturers have begun to market tube settling devices of various configurations for increasing the capacity of existing clarification equipment, as well as for optimizing new plant designs. The tube concept is based upon a shallow settling basin and a short detention time within that basin. For example, a particle settling at a rate of 1 in./min requires 120 min to fall to the bottom of a conventional clarifier of 10 ft depth. If the basin were 2 in. deep, the particle would only require 2 min to fall to the bottom.

Tubes are usually assembled into a bank and installed at an angle of about 60°. Under this condition, as water flows upward through the tube, solids settle rapidly and drain continuously from the bottom of the tube tank. Tube settlers usually permit higher through-put rates than conventional clarification equipment for the same given space. Figure 10-2 shows a typical tube installation in a conventional horizontal-type clarifier.

Cold Lime-Soda Softening. Lime-soda softening is the process by which the calcium and magnesium hardness are chemically precipitated from the water through the use of lime ($Ca(OH)_2$) and soda ash (Na_2CO_3). The process may be varied according to the composition of the raw water and the end use requirements.

In some cases, only partial treatment is required. For instance, municipal water supplies are usually softened only to about 85 ppm of total hardness. Many industrial and process waters only require a reduction of calcium alkalinity to about 35 ppm. Carbonated beverage waters usually require only a reduction of total alkalinity to 50 ppm.

In other cases, a greater degree of hardness removal is necessary. This may involve more complete treatment in which the total calcium and magnesium contents, whether present as bicarbonate hardness, noncarbonate hardness, or a mixture of both, are reduced as far as possible without the use of excess chemicals. In other cases, the softening process may involve excess chemical treatment to reduce the hard-

Fig. 10-2 Typical tube settler installation in a conventional horizontal type clarifier. (*Courtesy of Permutit Co.*)

ness to even lower values than are obtained in the complete treatment.

Chemicals for Softening. A water of virtually any chemical composition can be softened by various combinations of these chemicals:

1. *Coagulants.* The most widely used coagulant is alum (aluminum sulfate, $Al_2(SO_4)_3 \cdot 18H_2O$). Other coagulants that may be used are copperas (ferrous sulfate, $FeSO_4 \cdot 7H_2O$) and ferri-floc (ferric sulfate, $Fe_2(SO_4)_3$).
2. *Hydrated lime* ($Ca(OH)_2$). This compound is usually used in 93 percent purity in the cold lime-soda process. In systems where large quantities are required, unslaked lime (CaO) may be slaked to hydrated lime before use, as follows:

$$CaO + H_2O = Ca(OH)_2 \qquad (8)$$

3. *Soda ash* (Na_2CO_3). This is used in the process at the usual specification of 59 percent Na_2O, which is equivalent to 99.2 percent Na_2CO_3.
4. *Calcium chloride* ($CaCl_2$). This may be needed in waters where alkalinity is substantially in excess of hardness.

Chemistry Involved. Each of these chemicals does a specific job once it is added to the water, thoroughly mixed, and permitted to react. To do the job, the dosages must be accurately calculated from the water analysis. Therefore, interpreting the water analysis is important.

It is convenient to picture the bicarbonate ion (HCO_3^-) as tending to bind first to the calcium and second to the magnesium. Then, for example, when lime is added, it reacts first with the calcium, but continued addition of lime would react with the magnesium. In general, alkalinity is first calculated as being bound to the calcium and is called *calcium alkalinity*. If there is any alkalinity left over, it is calculated as being bound to the magnesium and is termed *magnesium alkalinity*. With a small number of natural waters, an excess of alkalinity may be present above the calcium and magnesium alkalinities. In such cases, it is calculated as being bound to sodium and is termed *sodium alkalinity*. If there is less alkalinity than the total of calcium and magnesium hardness present, then the amount of hardness in excess of alkalinity is called noncarbonate hardness. Noncarbonate hardness generally means hardness associated with sulfate or chloride ions. See Fig. 10-3 for a definition of terms used in describing the chemical composition of water.

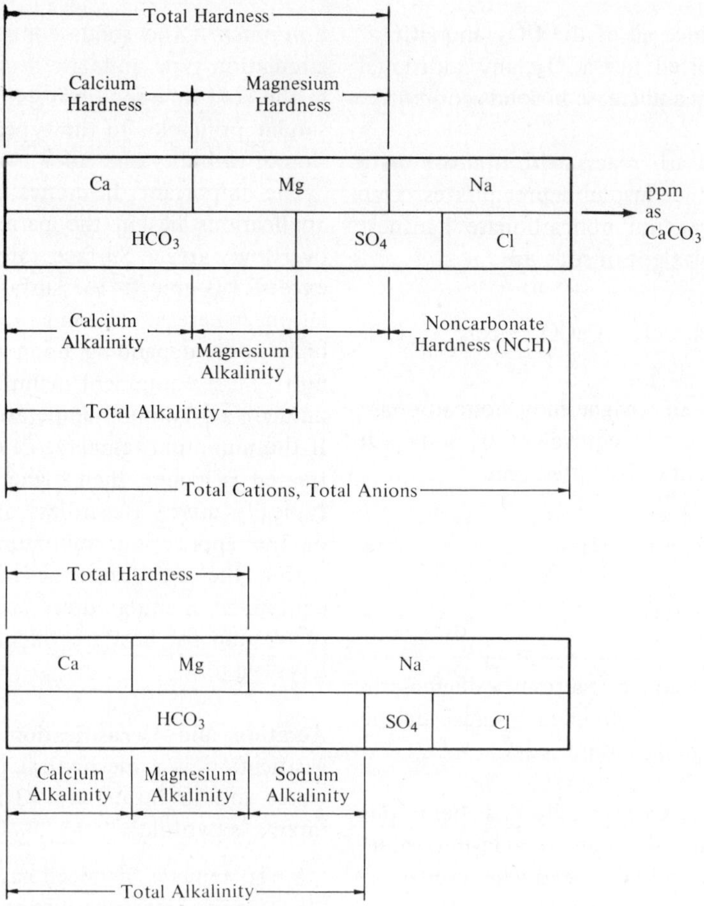

Fig. 10-3 Definition of terms used in describing chemical composition of water. (*Courtesy of Permutit Co.*)

Coagulant. When hardness is precipitated, it is so finely divided that the use of small dosages of coagulant is necessary to produce flocculation and settling. The following dosages are typical:

Alum	20 ppm
Copperas	20 ppm
Ferri-floc	10 ppm

As the coagulants react with the alkalinity present, a fluffy, gelatinous precipitate or floc is formed, which grows in size until it is large enough to settle. This floc has the ability to attract or entangle particles of suspended matter and permits them to be physically separated from the water. The coagulant reactions are the same as those for turbidity and color removal.

Lime. Hydrated lime reacts to chemically precipitate the carbonate hardness present in water to produce insoluble calcium carbonate ($CaCO_3$) and magnesium hydroxide ($Mg(OH)_2$). The required dosage will vary with the alkalinity in the raw water. The lime will also react with any carbon dioxide (CO_2) present and then with the bicarbonate ion (HCO_3^-). Quantities required can be calculated from the chemical equations of the reaction:

$$CO_2 + Ca(OH)_2 = CaCO_3 + H_2O \quad (9)$$

$$Ca(HCO_3)_2 + Ca(OH)_2 = 2CaCO_3 + 2H_2O \quad (10)$$

$$Mg(HCO_3)_2 + 2Ca(OH)_2 = 2CaCO_3 + Mg(OH)_2 + 2H_2O \quad (11)$$

Note that twice as much lime is required to precipitate magnesium than is required to precipitate calcium. Once all of the CO_2 and HCO_3^- present is converted to $CaCO_3$, any additional lime will result in a decrease in solids content.

Soda Ash. Soda ash reacts with noncarbonate hardness to produce insoluble precipitates. One equivalent of calcium noncarbonate hardness requires one equivalent of soda ash:

$$Ca^{SO_4}_{Cl_2} + Na_2CO_3 = CaCO_3 + Na_2{}^{SO_4}_{Cl_2} \quad (12)$$

One equivalent of magnesium noncarbonate hardness requires one equivalent of soda ash plus one equivalent of hydrated lime:

$$Mg^{SO_4}_{Cl_2} + Na_2CO_3 + Ca(OH)_2$$

$$= Mg(OH)_2 + CaCO_3 + Na_2{}^{SO_4}_{Cl_2} \quad (13)$$

Note that in the above reactions a decrease in solids content is not obtained because of the production of soluble sodium salts.

Calcium Chloride. Occasionally this chemical is required to reduce alkalinity to a desired value. Calcium chloride will be needed whenever:

1. Magnesium alkalinity exceeds 15 ppm and (magnesium hardness) - (magnesium alkalinity) = less than 40 ppm, or;
2. Sodium alkalinity is present. (In general, natural waters do not contain sodium alkalinity unless magnesium alkalinity is present.)

The reason it is needed is that magnesium carbonate ($MgCO_3$) or soda ash (Na_2CO_3) are soluble and will not precipitate from water with lime or soda ash. They will, however, react with calcium chloride ($CaCl_2$) as follows:

$$MgCO_3 + CaCl_2 = MgCl_2 + CaCO_3 \quad (14)$$

$$Na_2CO_3 + CaCl_2 = 2NaCl + CaCO_3 \quad (15)$$

Note that when reducing magnesium carbonate alkalinity, there is no reduction in hardness. The magnesium chloride ($MgCl_2$) formed will require an additional equivalent of soda ash and lime. See reactions under "Soda Ash."

Equipment. The solids contact and sludge recirculation type units are most commonly used in the cold lime-soda process. They operate on similar principles to the types used in coagulation of turbidity and color.

The important difference between the two applications lies in the permissible surface or overflow rates. Surface rates do not usually exceed 1.0 gpm/ft^2 of surface area for coagulation, whereas softening applications permit high rates, depending upon the design of the unit. Most equipment manufacturers rate their units on a minimum temperature of 32°F (0°C). If the minimum temperature of the water to be treated is higher, then higher surface rates are typically used. Regardless of the temperature or the application, minimum detention time within the unit must be maintained. Most equipment manufacturers specify a minimum of 60 min for best treatment. Figure 10-4 is a typical unit.

Aeration and Degasification. Aeration or degasification is a mechanical process of mixing water and air intimately. The purpose of this mixing is twofold:

1. To remove dissolved gases such as carbon dioxide by substituting air and thereby establishing a state of equilibrium between

Fig. 10-4 Cutaway view of a vertical solids contact unit. (*Courtesy of Permutit Co.*)

the gas present in the water and those present in the atmosphere. This process is usually thought of as degasification.

2. To dissolve air in the water to produce a chemical reaction such as oxidation of the soluble ferrous and manganous bicarbonates to the insoluble hydroxides. This is usually thought of in terms of aeration.

Degasification Theory. The removal of gases from water by aeration follows Henry's law which, in brief, states that the solubility of a gas in water is directly proportional to its partial pressure in the surrounding atmosphere. At 20°C and at sea level, water in equilibrium with the atmosphere will contain approximately 15.8 ppm nitrogen, 9.4 ppm oxygen, 0.5 ppm carbon dioxide, and no measurable amount of hydrogen sulfide.

Both carbon dioxide and hydrogen sulfide are relatively soluble in water. At 20°C and at sea level, carbon dioxide is soluble to the extent of approximately 1700 ppm, while hydrogen sulfide solubility is approximately 3500 ppm. The partial pressure of each of these gases in a normal atmosphere, however, is practically zero. Consequently, the establishment of a state of equilibrium between water and air by means of aeration will result in the saturation of the water with nitrogen and oxygen and the practically complete elimination of carbon dioxide and hydrogen sulfide from the water.

The basic requirement for removing gases by aeration is the establishment of a state of equilibrium between the gases in the water and those present in the surrounding atmosphere. The removal of gases by aeration is favored by an increase in temperature, aeration time, volume of air in contact with the water, and surface area of water exposed to the air. The efficiency of aeration is greater where the concentration of the gas to be removed is high in the water and low in the atmosphere. For example, aeration efficiency is higher with a water containing 100 ppm carbon dioxide than with one containing only 10 ppm.

Aeration Theory. The aeration of iron- and manganese-bearing waters is common practice.

Ferrous bicarbonate is soluble and colorless when in solution, while ferric hydroxide has a negligible solubility (less than 0.1 ppm). As result, upon oxidizing the ferrous bicarbonate, the iron content of the water will be precipitated out as ferric hydroxide. Oxidation is rapid and complete at pH values of 7.0 or higher. At a somewhat lower pH, the oxidation proceeds at a much slower rate. The reaction is as follows:

$$2Fe(HCO_3)_2 + H_2O + \tfrac{1}{2}O_2 = 2Fe(OH)_3 + 4CO_2 \quad (16)$$

Theoretically, therefore, 1 ppm dissolved oxygen will oxidize 7 ppm ferrous iron, expressed as Fe.

The oxidation of manganous bicarbonate, which is not commonly found in water, is similar to that of iron, except the pH should be above 10. The reaction and the theoretical quantity of oxygen required is the same; however, the time required for complete oxidation is longer. Where there is both iron and manganese in solution, it is also necessary to aerate above pH 10. At lower values, the iron will precipitate, but the manganese will remain in solution. If the pH of the water is not sufficiently high, lime, caustic soda, or soda ash can be added to increase the pH. In some cases where iron alone is being removed, it may not be necessary to add any alkali because aeration partially reduces the carbon dioxide content as well, and the pH would increase automatically. However, the increase is not always sufficient for some iron removal jobs, and alkali must then be added.

Equipment. The type of aerator used depends on a number of factors: (1) the carbon dioxide content, (2) the degree of removal required, (3) whether or not iron, manganese, or hydrogen sulfide removal are required, and (4) the volume of water to be handled.

The most commonly used equipment design is the forced-draft wood slat or packed tower type. The typical forced-draft degasifier (Fig. 10-5) is a closed round or square structure containing a series of closely-stacked, superimposed wood-slat trays. The water enters the top of the aerator where a tray distributes the water evenly over the topmost of them. The slats in

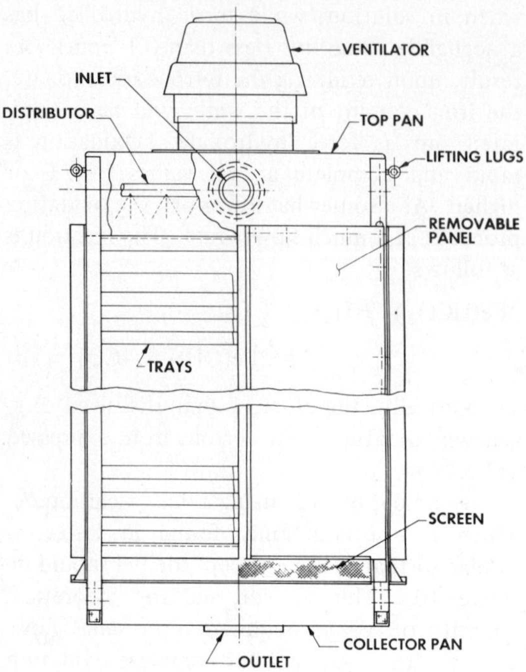

Fig. 10-5 Basic components of a slat tray degasifier. (*Courtesy of Permutit Co.*)

the trays are staggered so that the films of water raining over the edges of the slats in one tray fall on the centers of the slats in the tray just below. At the bottom of the aerator, the aerated water flows through an air seal to a catch basin or reservoir.

Air is supplied to the bottom of the aerator by a blower which blows it upward through the aerator, countercurrent to the downward flow of the water at a rate of about 1 ft^3 air/min/gpm of flow. In the upper part of the aerator, air, plus its content of gas removed from the water, is discharged through a number of baffled air stacks to an air space above the distributor tray, from which an air outlet discharges it to the atmosphere.

The high efficiency of this type of aerator is due to this controlled countercurrent flow of air and water, plus the action of the large number of slat trays in presenting fresh surfaces of films and drops of water. Typical application rates to this type of degasifier design are as high as 30–40 gpm/ft^2 of degasifier cross-sectional area.

Depending upon the height of the degasifier and the number of trays used, the raw water carbon dioxide concentration could be reduced to 5 or 10 ppm.

Raw water alkalinity, temperature, and iron and hydrogen sulfide concentrations are important factors to consider in the design of a wood slat degasifier.

Packed Tower. The packed tower design is very efficient and is only used to reduce carbon dioxide to extremely low levels. It is not normally used for iron removal or where iron in the raw water exceeds 0.5 ppm because the iron precipitates on the tower packing. The available area of the rings, as well as the efficiency of the unit, are subsequently affected.

Removal of Iron, Manganese, and Hydrogen Sulfide. In ground and surface water supplies, iron generally exists as divalent ferrous salts or as organic complexes; manganese, with few exceptions, exists in the divalent manganous state. The divalent salts and complexes of both are soluble, and the most common soluble form is the bicarbonate.

Iron and manganese are chemical constituents of shale and sandstone found in the soil. When reducing conditions are present, insoluble iron and manganese are converted from their insoluble higher valence state to their soluble divalent state. This is brought about in nature by the decomposition of organic materials in soil which consume oxygen, thereby creating reducing conditions and releasing carbon dioxide. As water trickles through soil, the carbon dioxide is dissolved, forming carbonic acid. This in turn lowers the pH, thereby increasing the reducing conditions.

If water exposed to these conditions remains underground and excluded from air contact (for example, in a well supply), the iron and manganese remain in solution. Even in pumping the water out of the well to a point of treatment, the iron is a colloidal precipitate and a part of turbidity.

Organic complexes of iron and manganese are formed by the combination of iron and manganese ions with negatively-charged organic units. These complexes can be visualized as having the iron or manganese in the center, surrounded by organic units. Generally, they are more difficult to oxidize than inorganic compounds because of this organic "protective shell." These complexes and chelates are commonly referred to as "organically-bound" iron and manganese, and are most common in surface supplies.

Hydrogen sulfide in a water supply is usually due to bacterial breakdown of organic matter containing sulfur. The presence of hydrogen sulfide is more common in well supplies than in surface supplies because these bacteria work best in the absence of oxygen.

Hydrogen sulfide, as noted above, is very soluble in water and imparts a characteristic taste and "rotten-egg" odor. If the pH is high, this odor may be slight since most of the sulfide may be present as alkaline sulfide rather than as hydrogen sulfide. The following tabulation gives some approximate values relating hydrogen sulfide to total sulfide content:

Percentage of Total Sulfide Present as H_2S

pH	Percent H_2S
5.0	98
7.0	33
9.0	0.5

Choosing the method of treatment for removal depends greatly on the ionic states of the iron and manganese and the concentration of hydrogen sulfide.

Insoluble Iron and Manganese. When iron and manganese are already insoluble, no oxidation is necessary. Simple coagulation (if softening is not required in a solids contact unit) followed by filtration will generally remove them satisfactorily. If lime-soda softening is utilized, the insoluble iron will be physically removed in the sludge blanket. It is also possible that filtration alone might do the job if the content of insoluble iron is not too high.

Soluble Iron and Manganese. The following methods can be applied to remove soluble iron and manganese:

1. *Oxidation by air, hypochlorites, chlorine, or chlorine dioxide followed by filtration.* Air oxidation is usually applied through the use of a wood slat degasifier.

   Hypochlorites are applied as the sodium or calcium salt which in turn hydrolyzes to hypochlorous acid, a stronger oxidant than molecular oxygen in air. Therefore, soluble iron and manganese are more rapidly and completely oxidized. This process, however, offers certain limitations since hypochlorous acid does not efficiently oxidize organically-bound iron and manganese.

   Chlorine, like hypochlorites, must first react with water-forming hypochlorous acid, the active oxidizing agent. It is thus subject to the same limitations as hypochlorites.

   Chlorine dioxide is a strong oxidizing agent which will rapidly oxidize soluble iron and manganese. However, it is seldom used because the cost of chlorine dioxide formed by reacting sodium chlorite with chlorine in aqueous media is generally prohibitive.
2. *Lime and lime-soda softening.* Where either lime or lime-soda softening is used, the pH of the treated water is sufficiently elevated to provide the softening reactions, and iron and manganese are precipitated as hydroxides. Aeration is sometimes provided prior to softening in order to supply any oxygen needed for oxidation.
3. *Ion exchange.* With certain waters, sodium exchange is used to remove soluble iron and manganese. Removal of iron by cation exchange is similar to and simultaneous with removal of calcium and manganese. In order to avoid fouling the exchanger bed with precipitated hydroxides, it is necessary for the iron and manganese to be in the soluble state. Air must be excluded from contact with the sodium cation exchange, at all times. Iron and manganese bicarbonate, carbonate, or sulfate can also be removed by a hydrogen cation exchanger as part of a demineralizer system.
4. *Manganese zeolite–potassium permanganate.* This topic is covered in detail below.

Organic Iron and Manganese. Generally, organic iron and manganese are removed by alum coagulation followed by settling and filtration. Manganese zeolite-potassium permanganate are also effective in removing these compounds.

Combination of Insoluble, Soluble, and Organic Iron and Manganese. The only effective means of removing combinations of these substances is with manganese zeolite-potassium permanganate.

Hydrogen Sulfide. Complete reduction of hydrogen sulfide is not achieved through aeration alone and the amount removed is not always predictable. At higher pH values, the sulfide content will be present mostly as alkaline sulfides. Carbon dioxide is less soluble than hydrogen sulfide and is removed by aeration more rapidly. As the carbon dioxide is removed, the pH rises and the equilibrium between alkaline sulfides and hydrogen sulfide gas is shifted so that the reaction proceeds in the wrong direction for the most complete removal of hydrogen sulfide. To some degree, this problem can be overcome by feeding acid to maintain a pH range of 4.0 to 5.0. Therefore, essentially all of the hydrogen sulfide would be in a gaseous form and could be removed by degasification.

The absorption of oxygen by the water during aeration also assists hydrogen sulfide removal by oxidizing part of the hydrogen and alkaline sulfides to free sulfur. This reaction is not as dependent upon pH as straight degasification, but long detention periods following aeration are required. Usually, air oxidation alone is not as effective as pH-reduction degasification.

Aeration-settling and aeration combined with pH reduction cannot, however, produce a "zero" hydrogen sulfide content effluent. This can be done by chlorination alone or in combination with aeration. The chlorination reaction is as follows:

$$H_2S + Cl_2 = 2HCl + S \tag{17}$$

In the above reaction, hydrogen sulfide is oxidized to elemental sulfur. This fine colloidal precipitate can then be removed by coagulation and/or filtration. Theoretically, 2.1 ppm of chlorine are required for every 1 ppm of hydrogen sulfide. Consequently, it is usually more economical to remove the bulk of the hydrogen sulfide by aeration and use chlorination only to oxidize residual hydrogen sulfide.

Manganese Zeolite–Potassium Permanganate.

Regenerative Batch System. The regenerative batch system uses manganese zeolite as both the oxidizing source and the filter media. The manganese zeolite is made from processed greensand zeolite by alternate treatments with manganese sulfate and potassium permanganate. The following is a simplified version of its preparation:

$$\underset{\text{Greensand zeolite}}{NaZ} + Mn^{2+} = MnZ + 2Na^{+} \tag{18}$$

$$MnZ + KMnO_4 = \underset{\text{Manganese zeolite}}{ZMnO_2} + K^{+} \tag{19}$$

This results in the higher oxides of manganese in and on the granules of greensand.

In the batch manganese zeolite process, no softening of the water or hardness removal occurs. The bed oxidizes ferrous or manganous bicarbonate to the higher insoluble oxides and simultaneously filters them out of the water. This system can also operate in conjunction with other iron removal methods such as aeration. In this case, the manganese zeolite oxidizes only residual iron and manganese. The oxidation reaction is as follows:

$$ZMnO_2 + \begin{matrix} Fe^{2+} \\ Mn^{2+} \end{matrix} = ZMn_2O_3 + \begin{matrix} Fe^{3+} \\ Mn^{3+} \\ Mn^{3+} \end{matrix} \tag{20}$$

The capacity of manganese zeolite is usually about 0.09 lb iron or manganese/ft^3 of material. Once the oxidizing capacity of the bed is exhausted, soluble iron and manganese leakage occurs. To prevent this, the bed is batch regenerated with potassium permanganate at predetermined intervals. This is normally done by treating the bed with $\frac{1}{2}$ percent solution of potassium permanganate for several hours; the usual amount required is 0.15-0.20 lb permanganate/ft^3 manganese zeolite:

$$\underset{\substack{\text{Exhausted}\\\text{manganese}\\\text{zeolite}}}{ZMn_2O_3} + KMnO_4 = \underset{\substack{\text{Regenerated}\\\text{manganese}\\\text{zeolite}}}{ZMnO_2} \tag{21}$$

In addition to regeneration, the zeolite bed is backwashed at intervals when the accumulated deposits in the bed build up pressure losses of 7 to 9 psi.

In standard practice, the use of manganese zeolite in treating large flows is limited to raw waters with a 1 ppm maximum iron or manganese concentration. However, at lower rates it is possible to treat waters containing somewhat higher amounts of iron or manganese. Conversely, lower levels of iron and manganese permit slightly higher rates.

The reasons for the 1 ppm limit are:

1. Potassium permanganate is expensive and is uneconomical at high flows and iron/manganese concentrations. Besides, some excess potassium permanganate is required for regeneration which in turn is wasted.
2. Flow rates must be reduced to obtain good removals in waters with greater than 1 ppm iron or manganese concentration. This also increases the operating cost.
3. Experience has shown that the zeolite bed clogs rapidly with a high iron content. This interferes with efficient results.

Although the batch manganese zeolite process is still in use today, it is rapidly being replaced by the continuous permanganate feed system.

Continuous Permanganate Feed. The continuous potassium permanganate feed system also uses a bed of manganese zeolite, such as is employed in the batch regeneration process. However, instead of the bed being regenerated with potassium permanganate, the permanganate is fed continuously in-line upstream of the manganese zeolite unit. In this manner, iron, manganese, or hydrogen sulfide in the water is oxided to an insoluble state before it reaches the manganese zeolite bed.

The bed, therefore, operates basically as a filter media. When the iron and manganese deposits have built up, the unit is backwashed as is an ordinary filter. The backwash is best carried out with treated filtered water since raw water containing iron and manganese would precipitate particles which would tend to clog the underdrain system.

In addition to its basic function as a filter bed, the manganese zeolite also serves as a "buffer." It oxides any residual dissolved iron and manganese when the permanganate dosage is slightly low, and removes any excess unreacted permanganate by partial regeneration of the manganese zeolite bed when the dosage is high.

The continuous feed method has many advantages over the batch regeneration system. First, since the zeolite acts as a filter instead of an oxidizing media, the runs are longer because the bed's oxidizing capacity tends to exhaust before the pressure drop buildup becomes excessive. Since the bed is never batch regenerated but only intermittently backwashed, down time is at a minimum. Finally, the potassium permanganate has a higher oxidizing potential than MnO_4, the oxidizing agent in a conventionally-regenerated manganese zeolite filter. Therefore, organic iron, some tastes and odors, and manganese at lower pH values can be treated more effectively. These usually are more difficult to handle in a conventional system.

Equipment. A conventional pressure filter type vessel with a bed of 6 in. of anthracite on top of 18 in. of manganese zeolite is the usual set up, although automatic gravity filters have been used (Fig. 10-6).

Reaction Time. Potassium permanganate requires only a 20-60 sec contact time before the water enters the filter for complete oxidation. As a rule, the potassium permanganate is fed approximately 20 ft upstream from the bed as a 1-2 percent solution. The soluble iron, manganese, and hydrogen sulfide are rapidly oxidized to an insoluble state and primarily filtered out in the anthracite layer. The reaction is influenced by the pH of the raw water, particularly in the case of manganese removal. For this reason, where raw water to be treated is relatively low in pH, it is generally desirable to feed an alkali along with the potassium permanganate in order to raise the pH to about 7.5-8.0 in order to insure optimum performance. As all waters vary somewhat in their reaction, there will be other occasions when alkali may not be necessary, and similarly there will be other

Fig. 10-6 Manganese zeolite system using continuous potassium permanganate feed. (*Courtesy of Permutit Co.*)

occasions where a somewhat different pH will be required.

Control. When the unit is first put into operation, it runs for about 10 hr without the addition of potassium permanganate upstream. This is necessary to partially exhaust the manganese zeolite which is usually supplied fully regenerated.

Control of the potassium permanganate dosage is simple and is handled visually. When potassium permanganate is added to water which contains soluble iron and manganese, precipitates are formed and the water remains colorless until all the iron and manganese are oxidized. At this point, any excess potassium permanganate imparts a pink color to the water since no agents for reaction remain. As more excess is added, the color becomes more intense.

Two samples are required to determine if the dosage is correct. One sample should be taken at the tap at the inlet of the unit (Sample 1). Another sample should be taken directly above the manganese zeolite bed (Sample 2). The following color comparison is usually made:

a. Sample 1. The sample drawn from the line before water enters the unit should have a definite pink color.
b. Sample 2. The sample drawn from the tap immediately above the filter bed should be just on the borderline between colorless and a very faint pink color.
c. The effluent from the filter unit should remain colorless at all times.

If the potassium permanganate feed rate is too low, the sample from the inlet line will be only faintly pink or will be colorless. The sample above the bed will show colorless also. If the potassium permanganate feed rate is too high, then a definite pink color both in the raw water sample and in the sample taken just above the bed will be observed.

If the feed rate is incorrect, results may not be noticed immediately. However, an overdose of potassium permanganate for an extended length of time would result in exhaustion of the bed and eventual leakage of iron and manganese to the effluent.

Operating Results. The operation is generally efficient, and most system designs will reduce combined iron and manganese to 0.3 ppm. Hydrogen sulfide will normally be reduced to less than 0.5 ppm if its raw water concentration does not exceed 10 ppm. Typical potassium permanganate usage is about 2 ppm potassium permanganate per 1 ppm of combined iron and manganese, and 4 ppm potassium permanganate per 1 ppm hydrogen sulfide.

Filtration. Filtration is employed in water treatment to remove or reduce suspended solids which may be present initially in the raw water or may be the result of coagulation or a precipitation process.

For example, in lime-soda softening, calcium and magnesium salts are precipitated and final clarification by filtration is necessary before use. In general, most water treatment processes involving coagulation and precipitation are followed by filtration for complete removal of suspended impurities.

Theory Involved. With proper sedimentation following chemical coagulation or softening treatment, the heavier particles are removed prior to filtration. Only the smaller and lighter particles of floc reach the filters. When a freshly-backwashed filter is first placed in operation, many of the finely-coagulated particles penetrate into the filter bed through the many voids in the bed surface. As the particles lodge between grains of filter medium, flow is restricted. Coagulated particles then build up on the surface of the filter bed. With continued filtration, the rate of flow increases through the larger unclogged passages and decreases through the smaller and partially-clogged passages. As the flow penetrates the bed further, the water spreads out and velocity decreases with the lodging of coagulated particles at points of low velocity.

Penetration of the conventional filter medium by coagulated particles normally does not extend deeper than 2-4 in. Most of the filtration occurs at the surface or in the first 1 or 2 in. of the bed. This coagulated mat acts as a fine filter for smaller particles. It is necessary in selecting the size of the filter medium to provide sufficient coarseness so that the top few inches of the bed are penetrated. With no penetration of the bed by coagulated material, the loss of head would increase rapidly and filter runs would be short.

In an effort to increase the dirt-holding capacity of the filter medium, two important modifications have developed over the years. The first is the dual media filter consisting of about 6–12 in. of silica sand with a 12–24 in. overlay of anthracite coal. This provides a roughing anthracite filter on top of the sand layer which in turn acts as a polishing filter to insure an effluent of low turbidity. This type of filter media design can be very successful under steady-state operation.

More recently, mixed media filter beds have become popular because of their resistance to particle breakthrough caused by surges and changes in flow rate. Mixed media usually consist of intermixed sand, anthracite, and garnet to provide a tapered void gradation within the bed. The coarse, low-density anthracite remains at the top while the small, high-density garnet remains at the bottom.

Filter Media Grades. The size of filter media is important because it should:

1. Prevent suspended matter from passing through.
2. Hold suspended matter as loosely as possible in order to provide easy backwash.
3. Be able to hold as large a volume of suspended matter as possible without clogging.

To determine whether a filter media meets a definite standard which permits the above conditions, two parameters, *effective size* and *uniformity coefficient*, are used. These measurements are made by performing a screen analysis of the filter media using a series of U.S. Standard Sieves. The results (cumulative percent retained on each screen) are plotted against sieve opening (millimeters).

The effective size is the size at which 10 percent of the filter media by weight is smaller and 90 percent is coarser–in other words, the size which accumulates 90 percent of the sample. Therefore, effective size is the minimum size of the bulk of the filter media. However, this does not indicate particle size variation in the bulk.

To assure that variation is not too great, the second measurement must be made. This is the size at which 60 percent is finer and 40 percent is coarser. This size divided by the effective size is the uniformity coefficient.

Types of Media. Basically there are four common types of filtering media used within the water treatment industry: silica sand, anthracite, calcite, and garnet.

Silica sand is the most widely used, although it is not used in situations where silica leached from the sand could be detrimental. Effective sizes typically range from 0.45 to 0.6 mm.

Anthracite, similar in size to fine sand, is equally as effective. Extended filter runs, less backwash water requirements, and higher filtration rates are possible with anthracite. Also, its sharp, annular particles produce larger voids than sand, and therefore have more dirt-holding capacity. Effective size range is 0.65 to 1.0 mm, with a typical uniformity coefficient of 1.85. Anthracite is generally employed in those cases where sand imparts silica due to the action of high alkalinities and heat.

Calcite is processed $CaCO_3$ and neutralizes part of the carbon dioxide content when the pH of the raw water is under 6.8; at the same time it serves as a filter medium. It works by the following reaction:

$$CaCO_3 + CO_2 + H_2O = Ca(HCO_3)_2 \quad (22)$$

Equilibrium under the recommended conditions is reached at about pH 7.0 or slightly higher, with some carbon dioxide remaining in the water. It will be noted that the calcite is consumed in this reaction and calcium bicarbonate goes into solution. Replacement is made whenever the bed depth decreases by 10 percent.

The use of garnet for filtration is relatively new and it is usually mixed with sand and anthracite to form mixed media beds. A typical mixed media filter consists of 10 percent garnet, 30 percent sand, and 60 percent anthracite coal.

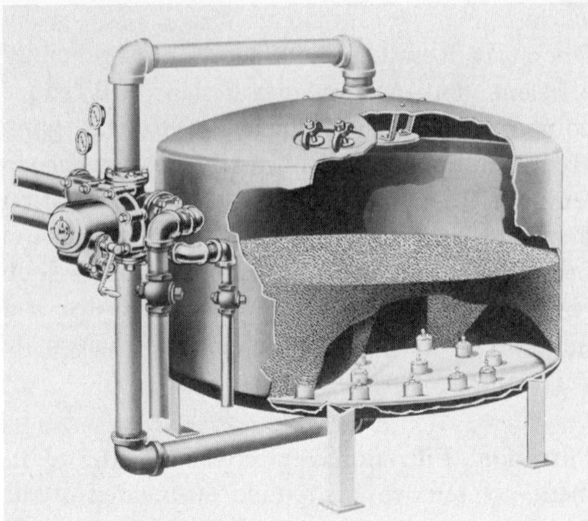

Fig. 10-7 Cutaway view of a pressure filter. (*Courtesy of Permutit Co.*)

Filtration Equipment. The two most widely used filter equipment designs in industrial applications are the pressure type and the automatic gravity type.

Pressure Filters. A pressure filter is a closed cylindrical steel tank containing a bed of granular filter medium over a collector system at the bottom (Fig. 10-7).

Periodically, the flow is reversed and the filter is backwashed to carry away the dirt which accumulates on and within the filter bed. This backwashing is done at least once per week, and if possible, once per day. In any case, backwashing should be carried out before the pressure drop through the unit exceeds 5 psi.

Pressure filters are used where the raw or coagulated water is supplied under pressure. Since the effluent from these filters is also under pressure, the necessity of repumping to the point of service is eliminated.

Automatic Gravity Filters. These filter designs automatically conduct all of the filter operation cycle—filtering, backwashing, rinsing, and return to service—with one three-way valve and, in one particular design, with no valves (Fig. 10-8).

In the valveless design, water from a constant level source at least 25 ft above grade enters through the inlet pipe into the upper part of the filter bed compartment and filters downward through the filter bed. Passing through

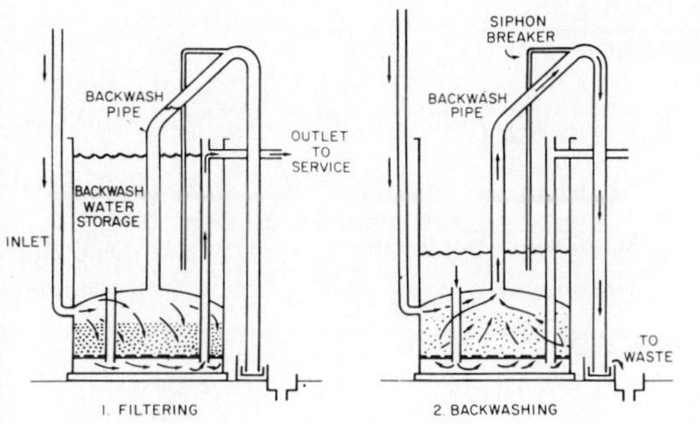

Fig. 10-8 Operation of an automatic gravity type filter. (*Courtesy of Permutit Co.*)

the strainers into the collector chamber, the filtered water then rises from the chamber up through the outlet to service. During the filter run, the accumulated floc on the filter bed slowly builds up a backpressure which causes a gradual rise of the water in the backwash pipe. When this reaches a predetermined level, usually 4 to 5 ft above the level of the filtered water outlet, a self-actuated primer system rapidly exhausts air from the backwash pipe and starts the siphoning action that backwashes the filter.

Since the backwash pipe carries approximately 20-21 gpm/ft^2 of filter bed area and only 2-3 gpm/ft^2 flows through the inlet pipe, the remainder is drawn from the backwash storage compartment through the ducts leading from it to the collector chamber, and from there up through the filter bed, thus backwashing it. The backwashing, therefore, starts at a higher rate than it ends because of the diminishing level in the backwash storage compartment. It starts at 18–20 gpm/ft^2 and decreases to 8–10 gpm/ft^2; the average rate is 13–15 gpm/ft^2. This cleanses the filter bed and carries off accumulated floc through a backwash regulator to waste. This steadily diminishing backwashing rate aids greatly in hydraulically regrading and evenly settling the filter bed. When the level of the water in the backwash storage compartment reaches the siphon breaker, this admits air to the backwash piping, breaks the siphon, and stops the backwashing operation.

At the end of the backwashing operation, the incoming water from the inlet pipe filters through the filter bed and, rising through the ducts from the collector chamber, fills the backwash storage compartment with filtered rinse water for the next backwashing operation. When this compartment is filled to the level of the filtered water outlet, the water then ceases flowing into this compartment and flows through the filtered water outlet, thus returning the filter unit to service.

Activated Carbon Filters. Activated carbon is widely used in industrial applications for removing tastes, odor, and color from water, as well as for dechlorinating water supplies. Granular activated carbon is generally used in equipment which is similar to conventional filters. Bed depths are about 2-3 ft and application rates range from 2 to 6 gpm/ft^2 of cross-sectional bed area. Periodic backwashing of the activated carbon is necessary when the head loss across the bed reaches a predetermined level.

Activated carbon is normally employed when organics causing taste or odor are low in concentration. Although carbon has a limited adsorption capacity for organics, it can be steam-regenerated to restore a portion of the original adsorption capacity. Because of carbon's limited capacity, it is usually less expensive to remove the bulk of taste and odor by means of a chemical coagulation process followed by carbon filtration for polishing treatment.

The removal of residual chlorine from water supplies with activated carbon is a chemical reaction which converts the chlorine to a chloride ion. Carbon capacity for dechlorination is high, and bed replacement is usually on the order of every 1 to 2 years.

Ion Exchange Softening.

Softening Operation. The ion exchange water softening operation is typical of almost all ion exchange processes in that it is a "batchwise" operation. A vessel containing a bed of resin, usually about 30 in. deep, serves as a water softener. As hard water percolates downward through the resin, the ion-exchange beds are continually bombarded by the calcium and magnesium ions, which gradually replace the sodium on the resin skeleton (Fig. 10-9).

The most widely used softening exchangers are the synthetic resins. Their cation-exchanging

THE SOFTENING PROCESS

SOFTENING REACTION

$$\left.\begin{matrix}Mg\\Ca\end{matrix}\right\}\left\{\begin{matrix}(HCO_3)_2\\SO_4\\Cl_2\end{matrix}\right. + Na_2Q = \left.\begin{matrix}Ca\\Mg\end{matrix}\right\}Q + Na_2\left\{\begin{matrix}(HCO_3)_2\\SO_4\\Cl_2\end{matrix}\right.$$

Calcium and/or Magnesium } { Bicarbonate Sulfate and/or Chloride (*soluble*) + Sodium Cation Exchange Resin (*insoluble*) = Calcium and/or Magnesium } Cation Exchange Resin (*insoluble*) + Sodium { Bicarbonate Sulfate and/or Chloride (*soluble*)

REGENERATION REACTION

$$\left.\begin{matrix}Ca\\Mg\end{matrix}\right\}Q + 2\,NaCl = Na_2Q + \left.\begin{matrix}Ca\\Mg\end{matrix}\right\}Cl_2$$

Calcium and/or Magnesium } Cation Exchange Resin (*insoluble*) + Sodium Chloride (*soluble*) = Sodium Cation Exchange Resin (*insoluble*) + Calcium and/or Magnesium } Chloride (*soluble*)

Fig. 10-9 The softening process. (*Courtesy of Permutit Co.*)

properties are principally due to sulfonic groups introduced by a sulfonation step during the manufacture of the resin. These exchangers are usually styrene-base products in the form of small spherical beads which are typically amber in color.

The bed of ion exchange is somewhat like a storage battery. It can continue to supply sodium ions only until its charge is exhausted. It must then be regenerated. This is done by washing the resin bed with a relatively strong solution of ordinary salt. Excess salt is rinsed out, and the softener is then ready for another cycle of operation. There are various other steps in the regeneration process designed to improve performance and produce acceptable efficiency. Since the resin acts as a filter, an important step is an upflow wash to rid the bed of debris accumulated during the treating process.

When the ion-exchange bed is "exhausted," the upper resin bed layers have a high hardness content and very little sodium-form resin remains, even at the bottom layer. After backwashing, the ionic matter is redistributed throughout the bed, so that there is substantially the same composition at each bed level. After regeneration, brine has converted the upper layers completely to the sodium form, since all of the brine used for regeneration passes downward and the amount contacting the top few inches is considerably greater than is needed. The contaminants from the upper layers are pushed downward, and at the end of the brine regeneration there is still hardness in the lower layer.

When the softener is put back into operation, hard water passing downward through the top layers becomes completely softened. This softened water is high in sodium, acts as a regenerating chemical, and displaces off a certain amount of hardness. This so-called "leakage" (re-exchange) is relatively low compared to the

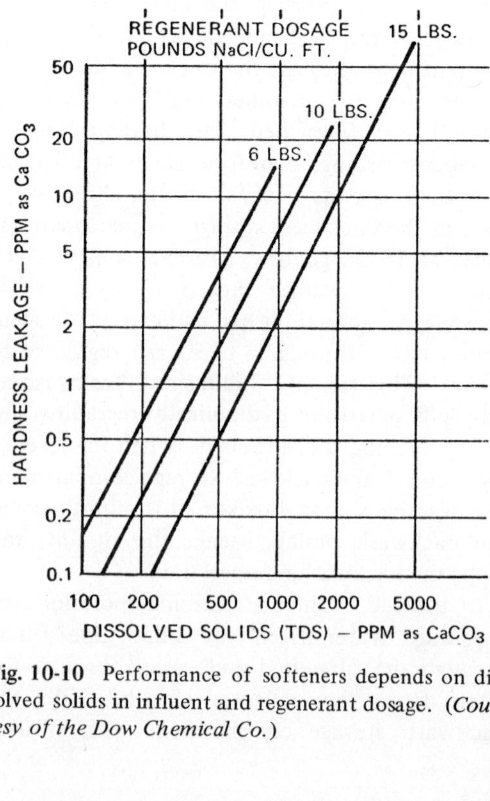

Fig. 10-10 Performance of softeners depends on dissolved solids in influent and regenerant dosage. (*Courtesy of the Dow Chemical Co.*)

raw water hardness, but is often troublesome. This contamination is directly proportional to the solids content of the raw water, as shown in Fig. 10-10.

Toward the end of the run, there is a certain amount of slippage of incoming raw water hardness through the lower layers into the outlet. The quality of finished water from the beginning to the end of a fixed column run is shown by Fig. 10-11, illustrating that the contaminants found in the effluent are the sum of those ions released by "re-exchange" and those resulting from true "slippage." Since the bulk of the contamination is from re-exchange, the quality of finished water normally continues to improve during an operating run up to the point where the unit is within about 5–10 percent of exhaustion.[1]

Some Aspects of Design. The design of equipment using ion-exchange materials to process water is the result of a series of compromises between quality of the finished water, chemical operating costs (related to both regeneration efficiency and resin losses), and gadgetry which could be added to the system to improve performance.

For example, a softener can be designed to reduce hardness below the level of detection by normal chemical analysis. To obtain such quality, however, might require an elaborate regeneration system. It is important to know what quality is really necessary in order to evaluate the extras which must be added to the system to produce it.

The usual softening operation is not a highly efficient process. For example, 1 lb of salt should theoretically displace 0.85 lb of calcium (as $CaCO_3$). This is equivalent to 6000 grains or 6 kilograins (kgr). In a typical plant, the ion-exchange material may be regenerated at a level of 6 lb of salt/ft^3 and produce a capacity of 20 kgr/ft^3. The salt consumption is 0.3 lb/kgr, as against a theoretical consumption of about 0.17 lb/kgr. The total capacity available in this type of cation exchanger is about 40 kgr/ft^3. If the softener uses only 20 kgr/ft^3, this means that about half of the ion exchange centers are not being converted to the sodium form by brine regeneration. This accounts for the re-exchange phenomenon explained earlier as a prominent factor in effluent quality.

To minimize leakage, a salt regenerant is applied at 10–15 percent sodium chloride with adequate contact time allowed to obtain maximum regeneration for the salt dosage used. Upflow wash or rinse with hard water should be avoided after regeneration. For extremely low leakages, recycle of rinse water is sometimes used. In any case, rinse volume will be greater at low salt dosages.

Introduction of continuous type (actually intermittent) high flow rate softeners for hardness reduction has greatly reduced salt requirements to produce "soft" water. By pulsing resin in a direction counter to the flow of water and regenerant, efficiencies approach theoretical values. By resin pulse control counter to the flow of regenerant and water, water can be softened while operating a high capacity resin at a rating as low as 8 kgr/ft^3 with the leakage value maintained at a constant level. Cyclic column operations usually operate at a resin capacity in excess of 20 kgr/ft^3.[1]

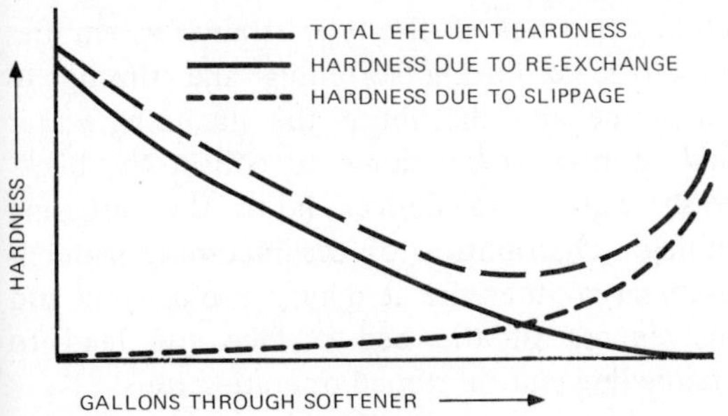

Fig. 10-11 Effluent contamination as a summation of re-exchange, heredity effect, and true slippage. (*Courtesy of the Dow Chemical Co.*)

Leakage. Hardness leakage is a result of the effect of dissolved solids on hardness left on the bed (particularly in the lower bed area) from the preceding regeneration. A typical strong acid resin has about 42 kgr/ft^3 total capacity. When regenerated with 6 lb sodium chloride/ft^3, maximum capacity is 24 kgr/ft^3, leaving 18 kgr hardness scattered throughout the bed. With 15 lb salt/ft^3, maximum capacity is 32 kgr/ft^3, leaving 10 kgr in the bed. During the softening cycle, dilute sodium salts formed by the ion-exchange process act as a regenerant.

The effectiveness of this dilute regenerant is directly related to the concentration. Hardness appears in the effluent due to this regenerating effect on calcium and magnesium ions left following the preceding regenerations. Figure 10-10 shows this effect when a typical strong acid cation exchanger is regenerated at 6, 10, and 15 lb salt/ft^3.[1]

Softening Brackish Waters. The technology of hardness (calcium and magnesium) removal from water has been extended to softening of brackish waters and even to the removal of a few parts per million of calcium from saturated brine. Techniques change as the salinity of the solution increases. Resins can do the job effectively by using beds in series and counterflow regeneration with large excesses of salt regenerant applied to the second unit series. The series technique can be used up to about 8000 ppm dissolved solids.

Leakage increases with dissolved solids concentration, and salt wells having as little as 0.75 percent (7500 ppm) sodium chloride are somewhat effective as regenerants for strong acid cation resin. At the 5000–8000 ppm range of dissolved solids, chelate type resin can replace the second strong acid cation bed in the system. This resin effectively removes all calcium and magnesium present in the effluent of the first bed. Regeneration of the chelate resin is done in two steps: hydrochloric acid is applied first to rid the bed of hardness, and sodium hydroxide is then used to neutralize the bed. Small amounts of calcium, magnesium, and other multivalent cations can be removed from brine at saturation concentration (26 percent) with chelate-type resin.[1]

Equipment. The basic components of a batchwise sodium cation exchange water softener are:

1. The shell.
2. Underdrain system.
3. Cation-exchanger bed.
4. Inlet distribution system.
5. Brining system.
6. Controls, including multiport-type valve.

Shells. The pressure type softener is a steel cylinder with heads usually dished to a radius equal to the diameter of the softener and metal thicknesses designed to withstand whatever pressure is required. Standard softener shells are designed for a working pressure of 75 psi, although higher working pressure units can be furnished. The tanks are generally unlined carbon steel, but corrosion-proof linings are available, if needed.

Underdrain System. The pressure softener, which is the most widely used, typically contains a double-dish type underdrain. The function of the underdrain is to collect the softened water, distribute the backwash water, and remove salt/rinse water during the brining and rinsing operation. Other designs which have been used are deflector plate and header-lateral types. However, some of these systems require gravel and sand subfills.

Cation-Exchanger Bed. Lying on top of the underdrain system is the cation-exchanger bed which, in the industrial sizes of water softeners, is not less than 2 ft and seldom over 8 ft in thickness. As the bed is leveled and hydraulically graded by the backwashing operations, the coarsest particles are in the bottom of the bed and the finest at the top, which assures an even resistance and therefore an even flow of water through the bed. Above the bed is a rising space so that the cation-exchanger bed may expand on backwashing, thus separating the particles and releasing any insoluble matter which may have collected in the bed.

Inlet Distribution System. In the upper portion of the shell, slightly below the top of its straight side, is the wash water collector. This is usually a series of disc-type strainer assemblies which serve during softening and rinsing to introduce and distribute the incoming water and, during backwashing, to collect the backwash water and deliver it to the outgoing piping. Distribution of the incoming water is necessary; otherwise it may cause drifting and unevenness of the bed surface and lead to channeling and shortened operating runs.

Brining System. A short distance above the surface of the exchanger bed is the brine distributing system, which serves to introduce and uniformly distribute the diluted brine over the resin bed so that all parts of the bed will come in even contact with it.

Since common salt varies but little in solubility over ordinary ranges in temperature, the necessity of weighing the salt dosages is eliminated. A saturated solution of salt is used and is measured volumetrically. The saturated brine is picked up, diluted to the proper strength, and introduced into the softener by means of a hydraulic ejector. In those cases where brine is delivered to the softener by a pump instead of an ejector, provisions are made to dilute it to the proper strength before it enters the softener by means of a line introducing water into the delivery line from the pump.

Salt storage is also part of the brine system. The simplest consists of a plastic or galvanized steel tank which contains enough salt for several regenerations. Water is passed through the salt, dissolves it, and produces saturated brine. In larger softener installations, elevated tanks or concrete basins are used to store carload quantities of salt which in turn are used to make saturated brine.

Controls: Multiport Valve. Water can be controlled during service or regeneration by individual valves or a single multiport valve. Manual or automatic, the multiple settings of most multiport valves simplify the operations required in regeneration.

Basic Demineralization by Ion Exchange.[2] Demineralization, or deionization as it is often called, is the process of removing dissolved ionic impurities from water. These positively-charged cations and negatively-charged anions allow conductance of electricity and are called electrolytes. Electrical conductivity is thus a measure of water purity, with low conductivity corresponding to high purity. Weakly-ionized contaminants such as silica contribute only slightly to the conductivity of water.

Ion-exchange demineralization is basically a two-step process. Cations (positively-charged ions) are removed in one step by an ion exchanger known as a cation exchanger. The anions (negatively-charged ions) are removed by a second ion exchanger known as an anion exchanger. Thus, when a water containing sulfates, chlorides, or nitrates of calcium, magnesium, sodium, or potassium is passed through a hydrogen cation exchanger, the metallic cations are removed and held by the ion-exchange resin, which gives up an equivalent amount of hydrogen ions in exchange. Since an equivalent amount of the anions of sulfate, chloride, or nitrate are also present, the net product is a dilute solution of sulfuric, hydrochloric, or nitric acid. Using calcium sulfate as an example and RSO_3H as the symbol for the complex cation-exchange resin, the reaction is as follows:

$$CaSO_4 + 2RSO_3H = (RSO_3)_2Ca + H_2SO_4$$

$$\underset{\text{(soluble)}}{\text{Calcium sulfate}} + \underset{\text{(insoluble)}}{\text{Hydrogen cation exchanger}}$$

$$= \underset{\text{(insoluble)}}{\text{Calcium cation exchanger}} + \underset{\text{(soluble)}}{\text{Sulfuric acid}} \qquad (23)$$

These strongly ionized acids—sulfuric, hydrochloric, or nitric—are then removed by an anion exchanger. Using the symbol R_4N to represent the complex anion-exchange resin, the reaction is as follows:

$$H_2SO_4 + 2R_4NOH = (R_4N)_2SO_4 + 2H_2O$$

$$\underset{\text{(soluble)}}{\text{Sulfuric acid}} + \underset{\text{(insoluble)}}{\text{Highly basic anion exchanger}}$$

$$= \underset{\text{(insoluble)}}{\text{Anion exchanger sulfate form}} + \text{Water} \qquad (24)$$

Note that the net effect of the two steps shown in these reactions is the removal (exchange) of the soluble salt (calcium sulfate) on two insoluble ion exchangers.

Regeneration of the Cation Exchanger. When the useful exchange capacity of the hydrogen cation exchanger is exhausted, it is regenerated with an acid, which removes the accumulated metals such as calcium, magnesium, or sodium. Simultaneously, the cation exchanger takes on an equivalent amount of hydrogen ions in exchange, thus restoring itself to the original hydrogen form. The acid generally used for this regeneration is dilute sulfuric acid. Hydrochloric acid is also an excellent, but usually more expensive, regenerant.

Using a calcium form to illustrate an exhausted cation exchanger and the symbol

RSO_3H to represent the complex cation-exchange resin with sulfuric acid as the regenerant, the reaction is as follows:

$$(RSO_3)_2Ca + H_2SO_4 = 2RSO_3H + CaSO_4$$

Calcium cation exchanger (insoluble) + Sulfuric acid (soluble)

= Hydrogen cation exchanger (insoluble) + Calcium sulfate (soluble) (25)

Regeneration of the Anion Exchanger. When the useful exchange capacity of an anion exchanger is exhausted, it is regenerated with an alkali which removes the accumulated anions in the form of their soluble salts and restores the anion exchanger to its original state. Using the highly basic anion exchanger exhausted to sulfuric acid as an example, and the symbol R_4N to represent the complex anion-exchange resin, the regeneration reaction may be written as follows:

$$(R_4N)_2SO_4 + NaOH = 2R_4NOH + Na_2SO_4$$

Anion exchanger (insoluble) + Caustic soda (soluble)

= Anion exchanger (insoluble) + Sodium sulfate (soluble) (26)

Chloride, nitrate, bicarbonate, and silicate are removed in a similar manner by regeneration with caustic soda which removes the acids in the form of their sodium salts and simultaneously restores the anion exchanger to its original state.

Ion-Exchange Resins. There are many different types of ion exchangers available from resin manufacturers. The principal resin types have generally similar characteristics and can be categorized as follows:

Cation-Exchange Resins

1. Strongly acidic
   a. Sulfonated polystyrene type with divinylbenzene cross-linking.
   b. Carbonaceous type (sulfonated coal).
2. Weakly acidic–carboxylic acid type with a cross-linked acrylic matrix

Anion-Exchange Resins

1. Strongly basic
   a. Polystyrene type with quaternary ammonium functionality, Type I.
   b. Polystyrene type with quarternary ammonium functionality, Type II.
2. Weakly basic
   a. Aliphatic polyamine type.
   b. Epoxy-polyamine type.
   c. Polystyrene type with tertiary amine functionality.

Each of these major resin types has one or more characteristics that are necessary to the demineralizing process. The important features of these resin types are tabulated below.

Cation-Exchange Resins

1. Strongly acidic–capable of exchanging the H^+ ion for the cations of all salts: for example, the Na^+ ion in NaCl, $NaHCO_3$, Na_2SO_4, etc.
   a. Sulfonated polystyrene type–most common of all cation exchangers. It is made with varying percentages of divinylbenzene cross-linking (8 to 20 percent). The lower percent cross-linking results in the best chemical efficiency (best utilization of regenerating chemicals), but the lowest resin stability. Therefore, where abnormal oxidizing conditions exist, higher cross-linked resins may be justified. This sacrifices some chemical efficiency in the interest of longer resin life.
   b. Carbonaceous type–good resin stability, good chemical efficiency, particularly with waters containing substantial alkalinity. However, relatively low capacity (25 to 35 percent that of polystyrene type resins) limits its application.
2. Weakly acidic–capable of exchanging the H^+ ion only for the cations of the salts of weak acids: for example, it will exchange the H^+ ion for the Ca^{2+} in $Ca(HCO_3)_2$, but not in $CaSO_4$. Thus, it is rarely used as the only cation resin in a demineralizing process, but its extremely good chemical efficiency warrants its use in many cases as an adjunct to the less efficient strong

cation resins. Most weak acid resins in current use are of the same general type—the carboxylic type with a cross-linked acrylic matrix. Capacities are very high and chemical utilization of regenerants is extremely good.

Anion-Exchange Resins

1. Strongly basic—capable of exchanging the OH^- ion for the anions of all acids: for example, for the SO_4^{-2} ion in H_2SO_4, or for the HCO_3^- ion in H_2CO_3. Its most important characteristic is that it is capable of removing dissolved silica.
   a. Type I strongly basic anion resins—most stable of the normal strongly basic types, although it has slightly less chemical efficiency than the Type II anion resins. It is generally used for demineralizing surface waters because of better resistance to fouling by organic contaminants frequently present in surface water supplies.
   b. Type II strongly basic anion resins—more efficient in the use of chemical regenerants than the Type I resins, but they generally do not have the stability or the ability to reduce silica to the degree possible with Type I resins. It is more susceptible to organic fouling than the Type II resins.
2. Weakly basic—capable of removing strong acids only: for example, HCl, H_2SO_4, or HNO_3. (Note: in the case of resins of this type, the entire acid molecule is absorbed on the resin rather than just the exchange of the OH^- ions for the anion radical, as in the case of the strongly basic resins. However, the net result is the same, and frequently an ion-exchange reaction is written similarly for both strongly basic and weakly basic resins.) These weakly basic resins have high chemical efficiencies, but they are incapable of removing weak acids such as carbonic and silicic acid. Hence, their main application is as an adjunct to the less efficient strongly basic anion resins.
   a. Aliphatic polyamine type—good efficiency and stability, but not capable of being regenerated with strong alkalies such as sodium hydroxide. Usual regenerant is soda ash (Na_2CO_3).
   b. Epoxy-polyamine type—good efficiency and high capacity, but sometimes sensitive to organic fouling with some surface water supplies. Usual regenerant is sodium hydroxide.
   c. Polystyrene type with tertiary amine functionality—good efficiency, good capacity (but usually less than that with the epoxy-polyamine type), and good resistance to organic fouling. Usual regenerant is sodium hydroxide.

Basic Demineralizing Systems. A properly designed demineralizing system should take into consideration the possibility of using any combination of the major resin types for any given application. This requires an intimate familiarity with these resins and the modifications of each type that are currently available from all resin manufacturers. In addition, it is essential to be completely familiar with the process equipment costs for the ion-exchange vessels, components, and auxiliaries. This usually dictates a rather substantial economic study which involves a high degree of specialized knowledge and experience. The major process arrangements that should be considered and the major reasons for their use are illustrated in Fig. 10-12.

Cation Resin—Hydrogen Cycle.[3] The quality of a demineralized water is expressed in terms of "leakage" of ionic components from the ion exchange system. The leakage level experienced will depend primarily upon the specific design of the system in use. For example, a simple two-bed demineralizer composed of a bed of strong acid cation-exchange resin followed by a bed of strong base anion-exchange resin will allow a measurable leakage level. However, a system using a mixed bed of strong cation and strong anion resins gives virtually no leakage.

In the two-bed system during the cation exchange, hydrogen ions released into solution in the upper portion of the cation bed can re-exchange for unregenerated sodium ions in the bottom of the bed. Thus, there is a tendency

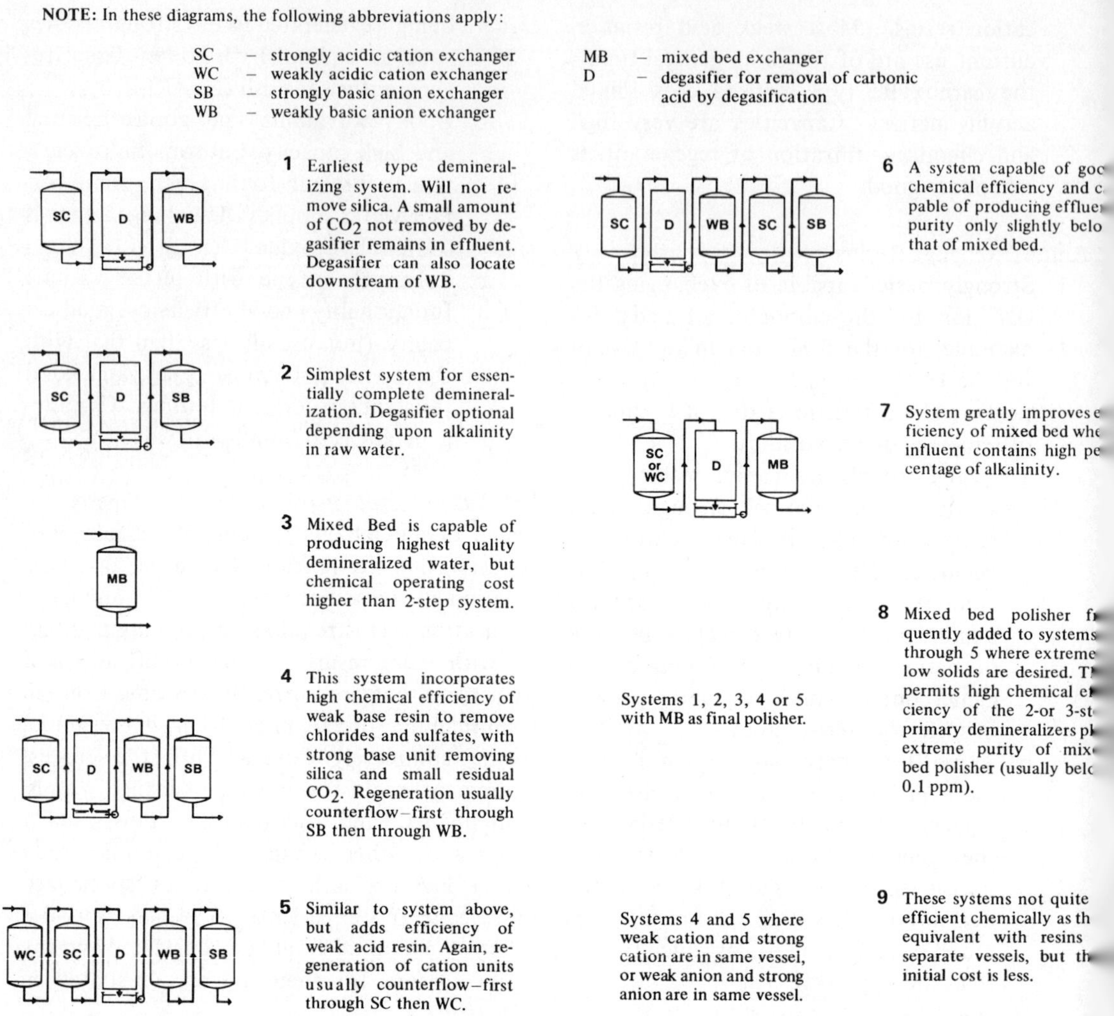

Fig. 10-12 Major demineralizing process arrangements. (*Courtesy of Permutit Co.*)

for a small amount of sodium to leak past the cation exchanger. In the anion exchanger, anions such as chloride exchange for hydroxide ions which are neutralized. Anions associated with the sodium leakage also exchange for hydroxide ions, thus leading to sodium hydroxide leakage. With a mixed bed, both cation- and anion-exchange reactions occur simultaneously, with the hydrogen ions released from the cation resin being almost immediately neutralized by hydroxide ions from the anion resin. Since hydrogen ions are not present to re-exchange with unregenerated sodium, leakage does not occur. In mixed bed operations, cation and anion resins are separated prior to regeneration. Thus, similar regeneration procedures apply for both mixed bed and separate bed resin units.

Cation Leakage. In a two-bed system, leakage of cations from the cation exchanger determines final water quality. This leakage is a function of the feedwater composition, the acid regenerant, the level of acid applied, and the regeneration procedure. The degree of removal of the various ions depends on cation resin selectivity. Since cation resins exhibit a much greater selectivity for divalent cations such as calcium and magnesium ions than for monovalent cations like sodium and potassium, leakage

TABLE 10-1. Cation Leakage in ppm as $CaCO_3$ for Dowex HCR-S

Total Dissolved Solids, ppm as $CaCO_3$	8 lb/ft³ H_2SO_4					
	At 0 Percent Alkalinity		At 50 Percent Alkalinity		At 75 Percent Alkalinity	
	50 percent Na	75 percent Na	50 percent Na	75 percent Na	50 percent Na	75 percent Na
50	1.2	5.5	0.74	3.4	0.42	1.9
100	1.8	10.3	1.1	6.4	0.63	3.6
250	4.1	29	2.5	18.0	1.4	10.1
500	7.2	59	4.5	36.6	2.5	21
1000	14.5	120	9.0	74.4	5.1	42

from the cation exchanger tends to be sodium rather than calcium or magnesium. Increased sodium concentration as a percentage of total cations in the feedwater will give higher sodium leakage from the cation exchanger.

The total dissolved solids (TDS) level and the alkalinity (carbonate and bicarbonate ions) in the feedwater also affect cation leakage. Waters high in dissolved solids exhibit higher leakage at fixed regeneration conditions while the presence of alkalinity in the water leads to reduced cation leakage. The effects of the sodium level, TDS, and alkalinity on cation leakage for a given regeneration level (8 lb/ft³ sulfuric acid) are demonstrated in Table 10-1.

For a given feedwater, leakage from a cation exchanger is most critically affected by the acid regeneration procedure and level. During downflow acid regeneration, the top of the bed is more completely converted to the hydrogen form than the bottom. Therefore, unless large quantities of acid are employed, the bottom of the bed will contain a band of sodium. During the following service cycle, influent cations are exchanged for hydrogen in the top of the bed. These hydrogen ions then exchange for sodium in the bottom of the bed, leading to leakage in the effluent. Cation leakage is reduced by increasing the acid regenerant level as is evidenced in Figs. 10-13 and 10-14 for standard 8 percent cross-linked resins. The band of sodium in the bottom of the bed is reduced by higher acid levels.

Cation Resins. Nearly all demineralization systems employ some type of strong acid cation-exchange resin. Specific resin selection depends on the system design and feedwater conditions. Under highly aggressive oxidative conditions, use of the maximum stability macroporous type cation exchanger is advisable. Although exhibiting lower capacities than the gel cation

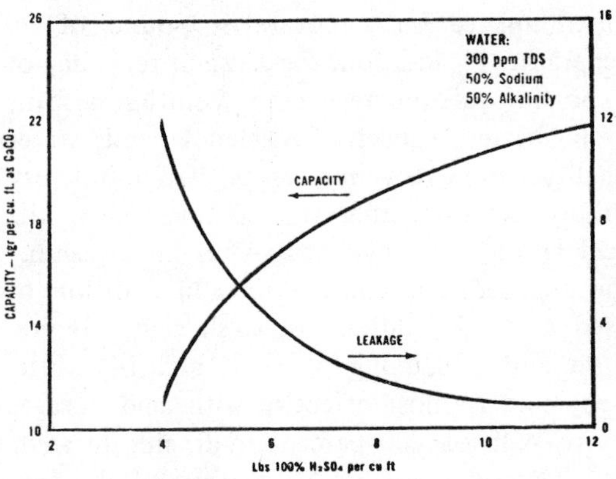

Fig. 10-13 Sulfuric acid regeneration–capacity and leakage vs. acid dosage. (*Courtesy of the Dow Chemical Co.*)

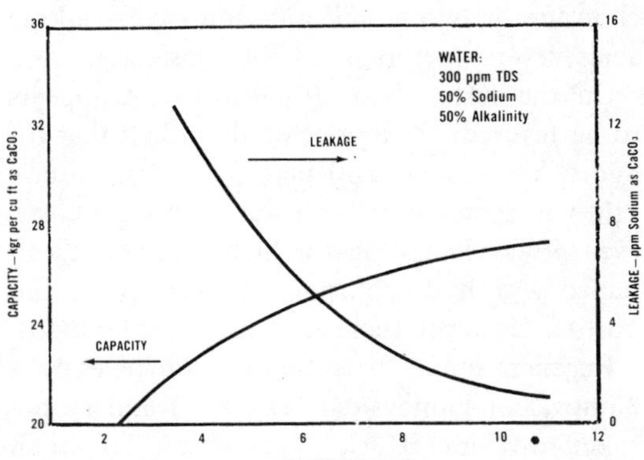

Fig. 10-14 Hydrochloric acid regeneration–capacity and leakage vs. acid dosage. (*Courtesy of the Dow Chemical Co.*)

resins, macroporous resins show superior resistance to degradation and also perform well in mixed beds.

Cation Regeneration. The operational characteristics and economics of the cation cycle depend primarily upon the regeneration step. Upon exhaustion of the resin, the acid is introduced to reconvert the resin to the hydrogen form for another exhaustion cycle.

Operating capacity for a strong acid resin depends on the regeneration level employed. For regeneration, a source of hydrogen ions is required. Standard commercially available strong mineral acids such as sulfuric, hydrochloric, and nitric acids are normally used. As is evidenced in Figs. 10-13 and 10-14, hydrochloric acid gives higher operating capacities than sulfuric acid at comparable regeneration levels. Because the resins are strongly acidic, acid utilization is incomplete: thus, the regenerant effluent contains unreacted acid. In some cases, spent regenerant acid can be recycled to increase utilization.

For acid regeneration, several variables must be considered—acid choice, concentration, dosage, flow rate, flow direction, and contact time. The choice of acids is influenced by availability and economic considerations. The most commonly used acids are sulfuric and hydrochloric, although nitric acid can be employed. For most large installations in most sections of the country, sulfuric acid has been employed. However, hydrochloric acid does exhibit some advantages and is often used. In choosing between sulfuric and hydrochloric acids, several factors should be considered.

On the surface, use of sulfuric acid appears to be favored. At a price of about half that for hydrochloric acid (100 percent basis), sulfuric acid will generally incur lower regenerant costs, even after accounting for higher capacities attained with hydrochloric acid (Figs. 10-13 and 10-14). However, there are other considerations.

Regeneration of the cation resin involves both monovalent-monovalent (H^+, Na^+) and monovalent-divalent (H^+, Ca^{2+}) exchange. To obtain the most favorable mass action effect, an acid concentration of at least 4–5 percent is desirable. However, with sulfuric acid as the regenerant, slightly soluble calcium sulfate is formed and may precipitate in the resin bed. It may leach out on a subsequent exhaustion cycle. Because of this factor, regeneration with reduced concentrations of sulfuric acid at selected flow rates is necessary. The regeneration procedures are thus more critical and there are more potential problems than with hydrochloric acid.

Regeneration with sulfuric acid is most effectively accomplished by gradually increasing the regenerant concentration, either stepwise or continuously. This method of application gives optimum contact time while still preventing calcium sulfate precipitation. By this method, the dosage is reduced to a minimum consistent with the exchange capacity and quality of effluent desired. The regenerant flow rate is governed by the acid concentration and the calcium in the exhausted bed of cation resin. Suggested flow rates are listed in Table 10-2 for various calcium levels.

For many installations, a low concentration (1–2 percent) sulfuric acid regeneration at high flow rates has been employed to avoid the precipitation problem. This method offers simplicity of design and is characterized by a short regeneration time. Compared to this procedure, the stepwise procedure outlined above gives higher acid utilization (Fig. 10-15) and higher capacity. Such improved acid utilization will mean improved operating costs.

Since none of the cations commonly occurring in water supplies form insoluble chloride salts, hydrochloric acid can be applied at any concentration and flow rate that will effect maximum cation removal per pound of regenerant. In addition, the reuse or recycling of a portion of the regenerant solution affords even better capacity efficiency and water quality. Acid recycling can be carried out very simply by collecting the volume below the acid lateral for recycling. After backwashing, the regeneration consists of: (1) addition of used acid, (2) addition of fresh acid, (3) prerinse and reclaiming of acid, and (4) rinse. Recylcing is most effective with acid dosages above 4 lb of 100 percent hydrochloric acid/ft^3. Recovery of 30–50 percent of the acid applied is feasible at regeneration levels in excess of 4 lb/ft^3.

The concentration of hydrochloric acid to be

TABLE 10-2. Boiler Water Concentration Limits Recommended by the American Boiler Manufacturer's Association with Respect to Operating Pressure to Assure Good Quality Steam

Pressure, psi	Total Solids, ppm	Total Alkalinity, ppm as $CaCO_3$	Suspended Solids, ppm	Silica,[a] ppm as SiO_2
0–300	3500	700	300	125
301–450	3000	600	250	90
451–600	2500	500	150	50
601–750	2000	400	100	35
751–900	1500	300	60	20
901–1000	1250	250	40	8
1001–1500	1000	200	20	2.5
1501–2000	750	150	10	1.0
Over 2000	500	100	5	0.5

[a]Silica limits based on limiting silica in steam to 0.02 to 0.03 ppm.

applied may vary depending upon the dominating cations in the exhausted exchanger bed. If sodium is the larger constituent, lower acid concentrations are used, i.e., 4–5 percent hydrochloric acid. When calcium and magnesium predominate, higher concentrations of hydrochloric acid (8–10 percent) are used. Therefore, to obtain maximum capacity per pound of hydrochloric acid when handling high hardness waters, the acid concentration should be as high as 10 percent hydrochloric acid and the contact time as long as practical.

Thus, the acid regenerant selection is not merely a question of acid costs but includes considerations for operational procedures. When overall operations are considered, hydrochloric acid may be more advantageous than sulfuric acid.

Other Cation Methods. For those conditions where the tendency for sodium leakage is great

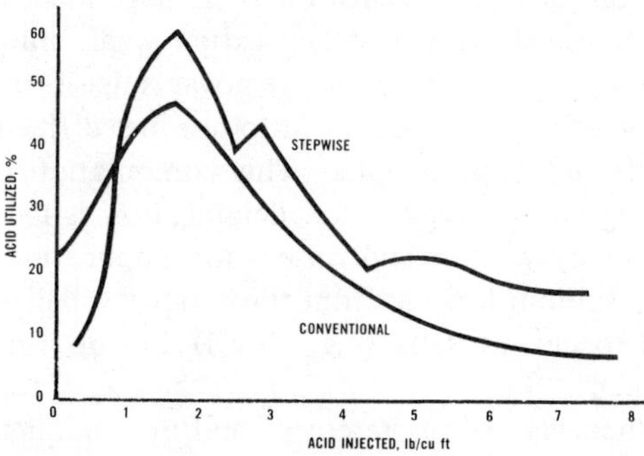

Fig. 10-15 Acid utilization for stepwise and conventional regeneration techniques. (*Courtesy of the Dow Chemical Co.*)

or where very low sodium leakage must be assured, upflow regeneration of the cation resin is practiced. Addition of acid at the bottom of the bed assures that the resin at the bottom is the most completely regenerated. Sodium leakage is thus reduced because any sodium not removed is retained in the upper regions of the bed and exchanges for hydrogen in the lower region in subsequent service cycles. Countercurrent regeneration methods can cause sufficient resin bed expansion to impair acid–resin contact. This yields a lower overall acid utilization and results in a lower operating capacity. However, special equipment designs and procedures that minimize bed expansion are available.

Several variations in the design of ion-exchange demineralizers are utilized. The use of multiple cation–anion systems improves final water quality, with mixed cation–anion beds giving the highest water quality. Use of a weakly acidic cation-exchange resin offers significant operating economies for many water treatment applications. These resins exhibit outstanding acid utilization efficiency and high hardness and alkalinity levels. They are used in series with strong acid exchangers and in some cases in layered single beds of strong acid–weak acid resins.

Anion Resin–Hydroxide Cycle.[4] In many demineralizer systems it is the operation of the cation cycle that determines final water quality. Leakage of sodium from the cation system of a two-bed demineralizer will hamper total effluent purity even with perfect operation of the anion unit. Thus, careful attention must always be given to the cation unit to minimize leakage.

However, operation of the anion cycle is no less critical to demineralizer performance. In fact, the economics of the anion cycle are more crucial since a more expensive regenerant is normally employed (caustic soda versus sulfuric acid for cation resin regeneration) and because anion resins are more expensive and inherently less stable than cation resins. In addition, even though cation leakage most markedly affects the electrical conductivity of the final water (a widely used measure of purity), leakage of small amounts of silica from the anion unit can also be a very serious problem, and in many applications silica content is continuously monitored along with electrical conductivity. Of further importance in considering the anion cycle is the possibility of organic fouling of the resin. Thus, it is quite evident that operation of the anion cycle must be closely controlled in order to assure efficient production of high quality water.

In the cation cycle, cations present are exchanged for hydrogen ions, thus generating the corresponding acid. It is the function of the anion cycle to remove this acidity to yield a demineralized water.

Choosing the Anion Resin System. Both strong base and weak base anion-exchange resins are widely used in demineralizer systems. Strong base resins are capable of reducing all strong and weak acids in the cation effluent.

Since the hydroxyl ions released are neutralized by the acids, a favorable acid-base equilibrium exists and there is no acid leakage. Even silica and carbon dioxide are reduced to very low levels. Regeneration of the strong base resins is merely the replacement of the exchanged anions with hydroxyl ions from an alkaline source. Strong base resins are highly dissociated and do not exhibit an especially high preference for hydroxyl ions. Regeneration is possible only because a relatively high concentration of caustic soda is used, thus giving an appreciable driving force to push the resin back into the hydroxide form. Due to this rather low affinity for hydroxyl ions, regeneration of strong base resins is not highly efficient and there remains a significant portion of nonutilized caustic soda in the spent regenerant.

Weak base anion resins efficiently remove strong acids like HCl and H_2SO_4 but do not absorb weak acids such as H_2CO_3 and H_2SiO_3. Weak base resins behave like weak organic bases such as aniline.

The weak base resin has no hydroxide form as does a strong base resin, and the entire acid molecule, and not just the specific ion, is adsorbed. Because the resin is weakly basic, its reactions are acid-base in nature and go virtually to completion. Therefore, the alkaline regenerant needs only to neutralize adsorbed acid, not to provide appreciable hydroxyl ion concentration for re-exchange. As a result, low-cost weakly basic reagents such as ammonia or sodium carbonate can be employed as regenerants. In addition, weak base resins are able to utilize nearly 80 percent of applied regenerant as opposed to about 40 percent for strong base resins.

Further adding to the advantages of weak base resins are their very high operating capacities for strong acids. For example, in treating the same stream of strong acids at equal regeneration levels (5 lb/ft^3), strong base anion resins give an operating capacity of about 15 kgr/ft^3, whereas weak base resins give about 30 kgr/ft^3.

Thus, there is considerable economic motivation to utilize weak base resins wherever possible. When silica and carbon dioxide removal are not necessary, a weak base resin can be used in a simple two-bed demineralizer. For example, on a strong acid water (i.e., no silica or carbon dioxide), qualities of 1-2 megohms can be achieved with a strong cation-weak anion system. If carbon dioxide removal is necessary, a degasifier can be employed to produce effluent levels of about 5 ppm. This combination of strong cation-weak anion deionization is ideal for many process and rinse water applications. Any sodium leakage from the cation exchanger will occur as salts (e.g., NaCl) in the final effluent.

Whenever silica removal and/or complete carbon dioxide removal is required, a strong base resin must be used. For smaller demineralization units, a strong cation-strong anion combination is the usual system. However, in larger systems with waters having appreciable free mineral acidity, use of a weak base-strong base resin combination is economically attractive. The weak base resin acts as the free mineral acidity adsorber with the following strong base resin retaining silica and carbon dioxide. Any sodium leakage will occur as alkalinity with this system. Spent caustic regenerant fractions from the strong base resin can be utilized as weak base resin regenerant, thus improving overall regenerant efficiency.

To further improve economics, layered beds of strong and weak base resins are being used. In this design, a layer of weak base resin is placed on top of the bottom layer of strong base resin. The service cycle is operated downflow so as to expose strong acids present to the weak base resin first. Regeneration should be upflow in order to maximize efficiency. Various manufacturers of industrial water treatment equipment have developed special techniques for design and regeneration of layered beds.

Carbon Dioxide Removal. Carbonate and bicarbonate ions commonly occur in water supplies and are collectively known as "alkalinity." On passing through the cation cycle, alkalinity is converted to carbonic acid. If water requirements dictate that this acid be removed, two general avenues are available: (1) mechanical, and (2) strong base resin exchange. Mechanical techniques such as degasification are economically favored over strong base exchange and are normally employed when significant carbon dioxide removal is required. The level of carbon dioxide remaining after mechanical treatment is normally a few parts per million, which is then reduced further with a following strong base exchanger. Mechanical carbon dioxide removal is employed after the cation cycle. If placed ahead of the weak base resin, the most efficient carbon dioxide removal is attained but the equipment must be acid-resistant. It can also be placed after the weak base unit.

Silica Removal. Silica is normally present in water supplies at maximum levels of only a few parts per million, but its complete removal is critical for some applications such as high pressure boilers. On passage through the cation unit, silica is converted to silicic acid (H_2SiO_3). This very weak acid is retained by strong base resins. Because it is the weakest acid of the adsorbed anions, silica breakthrough occurs before the other anions. Very careful consideration must always be given to silica removal and regeneration since improper procedures can lead to silica leakage problems.

As silica accumulates and ages in the anion resin, it can polymerize within the resin polymer structure. Thus, in the regeneration with caustic soda, it is not only necessary to achieve re-exchange of ions but also to dissolve this polymerized silica. Use of hot caustic soda solution is needed to assist the solution process. When silica is properly removed on regeneration, leakage problems are minimized.

APPLIED TECHNOLOGY—BOILER FEEDWATER TREATMENT

Treatment of boiler feedwater aims at preventing scale and sludge depositions, corrosion, and carry-over.

Boiler Scale Formation. Proper water treatment is fundamental to the satisfactory operation of any boiler, whether of a fire tube or water tube design. The need for proper water treatment becomes more apparent with an understanding of the mechanics of scale formation; i.e., how scale forms in a boiler and what happens when the scale forms.

At the prevailing temperatures in a steam boiler, both carbonate and noncarbonate hardness form scale. A breakdown of calcium and magnesium carbonate hardness is shown by the following equations:

$$\underset{\text{Calcium bicarbonate}}{Ca(HCO_3)_2} = \underset{\text{Calcium carbonate}}{CaCO_3} + \underset{\text{Water}}{H_2O} + \underset{\text{Carbon dioxide}}{CO_2} \tag{27}$$

$$\underset{\text{Magnesium bicarbonate}}{Mg(HCO_3)_2} = \underset{\text{Magnesium carbonate}}{MgCO_3} + \underset{\text{Water}}{H_2O} + \underset{\text{Carbon dioxide}}{CO_2} \tag{28}$$

$$MgCO_3 + H_2O = Mg(OH)_2 + CO_2 \quad (29)$$

Magnesium carbonate + Water = Magnesium hydroxide + Carbon dioxide

The carbon dioxide is released with the steam whereas calcium and magnesium carbonates and magnesium hydroxide deposit as scale.

Under higher temperatures and pressures, the solubilities of calcium and magnesium carbonates and magnesium hydroxide decrease. For example, at 210 psi and 392°F (200°C), the solubility of calcium carbonate is less than 5 ppm as $CaCO_3$ and magnesium hydroxide is only slightly over 1 ppm.

Calcium sulfate, the only scale-forming salt in the noncarbonate hardness, follows the same solubility pattern. At 322 psi, its solubility is equivalent to 40 ppm calcium carbonate.

Scale formation occurs when these salts crystallize from a solution which is locally supersaturated. Feedwater forms in a thin film along boiler tube surfaces and tends to become concentrated, especially if the water is moving at a low velocity. Consequently, the solubility of the calcium and magnesium salts is first exceeded in this thin film. The result is crystallization of scale directly on the heating surface, even when the solubility of the scale-forming salts has not been exceeded in the boiler water itself.

With a fire tube boiler, hot gases from combustion are within the tubes, with boiler water and scale deposits on the outside. A water tube boiler is reversed, and scale forms inside the tubes.

Scale creates a problem in boiler operation because it has a low degree of heat conductivity. It acts like a thin but effective film of insulation across the path of heat transfer from the hot flue gases to the boiler water. As more scale forms, the insulating effect increases. Unless prevented or cleaned off, it continues to increase until the safe maximum temperature of the tube metal is exceeded. Eventually, this rise in metal temperature causes tube failures and costly equipment down time.

Besides tube failures, overall efficiency is affected since scale can reduce the heat transfer rates by as much as 10 to 12 percent. A $\frac{1}{8}$ in. thickness of scale can cause 2 to 3 percent loss in boiler efficiency.

Another source of boiler scale is silica. Its thermal conductivity is very low and tube failures often occur with even very thin silica scales. However, in low and moderate pressure steam boilers, hard, glassy silica scale can usually be avoided by maintaining a proper alkalinity-to-silica ratio, thus keeping the silica in solution.

Preventive Measures. Scale formation can be avoided by removing or reducing calcium and magnesium before feedwater reaches the boiler (external treatment), or by adding chemicals such as phosphates, organics, colloids, etc. to the boiler water, the hardness causing a tendency to form sludge instead of scale (internal treatment).

Ion-Exchange Softening. Softening boiler feedwater by sodium cation exchange is the most widely used external treatment for low pressure boilers. Its chief advantages are complete removal of calcium and magnesium and simplicity of operation. The sodium salts produced are more soluble than calcium and magnesium salts, and, therefore, scale formation is minimized. However, one thing a softener cannot do is reduce the alkalinity of the total solids content of the water, as is sometimes necessary for efficient boiler operation.

If alkalinity reduction is required, the following approaches can be used:

1. Partial neutralization of alkalinity in the softener effluent with acid; or
2. Series operation of sodium cation exchange and chloride anion exchange; or
3. Split-stream hydrogen and sodium cation exchange.

Only the third approach produces a reduction in total solids content corresponding to a reduction in alkalinity.

Boiler Compounds. A great variety of boiler compounds are used for internal treatment. Among them are caustic soda, phosphates, and various types of organics which serve as reactive colloids. In general, the action of the boiler compound may be divided into two classes:

1. Reduction of hardness by chemical action. The sodium phosphates are the most commonly used chemicals for this purpose. When any of these phosphates enter a boiler, they convert to trisodium phosphate, a precipitate in flocculation. Magnesium salts are precipitated by caustic alkalinity as the extremely insoluble magnesium hydroxide.
2. Mechanical action on sludge with the intention of coating the precipitates to prevent their cementing together. Once sludge forms, it must be kept fluid since tricalcium phosphate and magnesium hydroxide can also adhere to heating surfaces. Organic materials, usually proprietary compounds, can maintain this fluid condition.

Even feedwater treated by external means will require internal boiler treatment to remove residual hardness, silica, oxygen, and other impurities. Internal treatment alone is usually applied when boilers operate at low pressure or where the feedwater is steam condensate. Moderate and high pressure systems require a combination of treatment.

Blowdown. Because all solids (dissolved and suspended) that enter or precipitate in a boiler remain in the drum and tubes, continued addition of make-up results in greater and greater concentration. This concentration continues until some limit is reached beyond which operation is unsatisfactory—i.e., a point at which priming and foaming occur. Priming is the spouting or surging of water into the steam outlet, while foaming is a buildup of bubbles until they reach the outlet. Either priming or foaming (or both) results in "carry-over" of solids and moisture in the steam. The specific concentration at which they take place varies according to the amount and type of dissolved solids and the presence or absence of suspended matter.

Table 10-2 shows the limits recommended by the American Boiler Manufacturer's Association with respect to operating pressure to assure good quality steam. It is obvious that as the boiler operating pressure increases, further removal of total solids in the feedwater is required for economical operation. For boilers with over 600 psi operating pressure, complete demineralization of the feedwater may be justified.

One commonly used treatment process for reduction of hardness and alkalinity is the combination of the hydrogen cation exchange effluent with (1) alkali neutralization, (2) raw water neutralization, or (3) blending with the sodium ion exchange effluent. These methods of operation make it possible to obtain a feedwater with varying degrees of hardness and any predetermined desired alkalinity. These processes have enough flexibility to meet most medium pressure boiler manufacturer's feedwater specifications.

Treatment of Hydrogen Cation Exchange Effluents. The effluents obtained in treating waters by the hydrogen cation exchange process then contain various amounts of carbon dioxide, sulfuric acid, nitric acid, and hydrochloric acid. While it is possible to rid the system of carbon dioxide by degasification, sulfuric and hydrochloric acids cannot be removed by this method. These acids must be neutralized; otherwise, severe acidic corrosion of the metal in the system will result. The sulfuric, hydrochloric, and nitric acids may be (1) removed by an anion exchanger, thereby demineralizing the water, or (2) neutralized with alkali from the effluent of a sodium cation exchanger, or in some cases with raw water.

The Hydrogen Cation Exchange Plus Caustic Soda Process. This method of operation includes the exchange units, degasifier, and caustic soda feed. The acid effluent from the exchanger is passed through a degasifier to remove free carbon dioxide. Caustic soda is then fed to the degasifier effluent to neutralize the sulfuric/hydrochloric/nitric acid. This method of neutralization finds some use with waters low in sulfates, chlorides, and nitrates. High levels of these anions may make this process uneconomical since the cost of caustic soda becomes excessive.

The alternate method of blending the hydrogen cycle exchange effluent with either sodium-exchange softened water or raw water is usually a more desirable approach. In each of these two cases, the amount of water to be blended is calculated with regard to the desired final alkalinity. The desired flow rate is equal to the total flow rate after blending, and the percentage of this total flow rate which must be treated in the hydrogen cation exchanger can be calculated by the following equation:

$$\% \text{HR} = 100 \times \frac{\text{Alk A}_{\text{raw}} - \text{Alk A}_{\text{mixed}}}{\text{Alk A}_{\text{raw}} + \text{ThMA}_{\text{raw}} - \text{ppm cation leakage}}$$

where

- % HR = percent of water to be treated in the hydrogen unit,
- $\text{Alk A}_{\text{raw}}$ = total alkalinity of the raw water as ppm $CaCO_3$,
- $\text{Alk A}_{\text{mixed}}$ = required alkalinity in the mixed or blended water as ppm $CaCO_3$,
- ThMA_{raw} = theoretical mineral acidity of the raw water (sum of Cl^-, SO_4^{2-}, and NO_3^-, all as ppm $CaCO_3$),
- Leakage = cation leakage from the hydrogen exchange unit.

The Hydrogen Cation Exchange Plus Raw Water Process. In cases where reduction of the alkalinity to a specified level and partial but not complete removal of the hardness are required, the effluent from the hydrogen cation exchanger is mixed with enough raw water so that its bicarbonate alkalinity neutralizes the acidity and furnishes whatever excess of alkalinity is desired. The mixture is then passed through a degasifier. The raw water is mixed with the hydrogen cation exchange effluent ahead of the degasifier so that the carbon dioxide liberated by the neutralization, as well as that liberated in the hydrogen cation exchanger, will be reduced to the tolerance required.

The uses for this process are rather limited. However, it has been applied in certain industrial applications where the raw water has a high sodium bicarbonate content and very low hardness.

The Hydrogen and Sodium Cation Exchange Process. The arrangement used in this method of treatment consists of (1) hydrogen cation exchange unit(s), (2) sodium ion exchange unit(s), and (3) a degasifier. As in the process of neutralization with raw water, the degasifier is used to decarbonate the neutralized effluent. The mixture of hydrogen cation exchange and sodium ion exchange softener effluents is decarbonated so as to remove the carbon dioxide formed in the neutralization as well as that formed by the action of the hydrogen cation exchanger. Using softened water to neutralize the hydrogen cation exchange effluent provides a final blend that not only is reduced in alkalinity content, but is also completely softened.

The flow of the effluent from each of the exchange units is adjusted so that its sodium bicarbonate content is sufficient to neutralize the acidity in the flow of the effluent from the hydrogen cation exchange unit(s) and also furnish whatever excess of alkalinity may be required. The percentage of water to be treated in the hydrogen unit(s) can be calculated using the equation given previously.

Corrosion. Boilers, economizers, feedwater heaters, and piping must be protected against general corrosion and pitting. Corrosion of boiler systems usually occurs when the boiler water is acidic or when dissolved oxygen or carbon dioxide is present.

Preventive Measures. Boiler feedwater is normally maintained at a high pH to prevent corrosion due to acidity. Carbon dioxide and oxygen can be removed by deaerating heaters and special ion-exchange resin systems. Chemical oxygen scavengers—commonly sodium sulfite or hydrazine—are typically used in conjunction with deaerating heaters to insure complete corrosion protection. These chemicals are usually fed to the deaerator storage compartment.

Dealkalizers. Alkalinity reduction is primarily used on feedwater to low and moderate pressure boilers. The ultimate goal is to reduce carbon dioxide in the steam, which is a major cause of condensate return line corrosion.

Since most boiler plants have deaerating heaters which remove free carbon dioxide, the amount of carbon dioxide in the steam is governed by the bound carbon dioxide in the feedwater.

Bound carbon dioxide is that carbon dioxide which can be formed from the breakdown of bicarbonate and carbonate alkalinity. At boiler operating temperatures, bicarbonate or carbonate alkalinity will decompose, liberating free carbon dioxide.

Basic Reactions. A strong base anion resin regenerated with salt (NaCl) would then operate on the chloride cycle. When water containing bicarbonate ions (HCO_3^-) passes through this chloride-cycle resin, the following reaction occurs:

$$HCO_3^- + R_4NCl = R_4NHCO_3 + Cl^- \quad (30)$$

The bicarbonate alkalinity in the water has been replaced by chlorine ions; in effect, there is dealkalization. In this process, all other anions excepting chlorides go through this exchange.

Unsoftened water is usually *not* passed through the anion unit operating on the chloride cycle. Precipitates of calcium carbonate or magnesium hydroxide can form in the bed. Therefore, dealkalizer is preceded by a cation exchange sodium softener to remove hardness.

Dealkalizer Regeneration. The cation exchange regeneration using sodium chloride is conventional. The anion exchanger regeneration reaction is as follows (shown for HCO_3^- only; same for SO_4^{2-}, NO_3^-, CO_3^{2-}):

$$R_4NHCO_3 + NaCl = R_4NCl + NaHCO_3 \quad (31)$$

Typical dealkalizer design practice is to use 3.5 lb sodium chloride/ft^3 of strong base anion resin. An alternate method of regeneration is to use 3.5 lb sodium chloride/ft^3 of anion plus 0.35 lb caustic soda/ft^3 of resin. The use of caustic soda plus salt on certain waters results in a higher exchange capacity. The effluent carbon dioxide content is somewhat different when using salt–caustic soda as compared to salt alone.

When using salt alone for regeneration of the resin, it appears that some of the bicarbonates exchanged by the resin are converted to carbonate in the process of being taken up. This results in some free carbon dioxide being liberated as follows:

$$HCO_3^- + R_4NCl \begin{cases} \nearrow R_4NHCO_3 + Cl^- \\ \searrow R_4NCO_3 + H_2O + CO_2 \end{cases} \quad (32)$$

The free carbon dioxide of the effluent will be a few parts per million higher than that in the raw water because of the additional carbon dioxide produced.

When caustic soda in addition to the salt is used for regeneration, hydroxyl exchange groups as well as chloride exchange groups are provided on the anion resin.

The result is that the free carbon dioxide in the raw water, as well as any carbon dioxide which might be liberated from the bicarbonates, is taken up by any hydroxyl exchange groups on the resin. Chloride exchange groups, similarily on the resin, will not take up any of this carbon dioxide.

$$H_2O + CO_2 = H_2CO_3 \text{ (either in raw water or from bicarbonate break down)} \quad (33)$$

$$H_2CO_3 + R_4NOH = R_4NHCO_3 + H_2O \quad (34)$$

The hydroxyl groups also play a role in determining the type of alkalinity leakage, but not the amount from the anion unit. Total alkalinity leakage is the same whether salt or salt–caustic soda is used. When salt is used alone, all alkalinity leakage is bicarbonate. When salt and caustic soda are used, it is a mixture of carbonates and bicarbonates. Presumably, this mixture occurs because some hydroxyl (OH^-) groups are displaced from the resin as in conventional demineralization by other anions, whether bicarbonate, sulfates, etc. This free hydroxyl ion in turn reacts with either carbon dioxide or bicarbonate in the water, forming carbonates. Most of these carbonates are taken up by ion exchange for chlorides or other hydroxyl ions in the resin bed, but some, of course, slip through and appear in the effluent.

Since it is possible for carbonates and/or hydroxyl ions to be present in the anion exchange system, it is desirable to feed only softened water to the dealkalizer to eliminate any possible precipitates of calcium carbonate or magnesium hydroxide.

Dealkalizer Effluent Quality. Most equipment manufacturers will guarantee that the effluent from a dealkalizer will insure a maximum concentration of bound carbon dioxide regardless of the actual form of the effluent alkalinity.

The free carbon dioxide content of the effluent will vary depending upon the type of regeneration cycle used. Salt regeneration alone usually gives an effluent carbon dioxide concentration which is 10–15 ppm greater than the influent. The effluent carbon dioxide is about zero when salt and caustic soda regeneration are used.

Total anions or cations are unchanged by the dealkalizer treatment. Alkalinity and sulfates which are removed are replaced by an equal amount of chlorides. The sulfate content of the effluent is reduced to near zero.

Silica reduction does take place during the early part of the dealkalizer run when salt and caustic regeneration are used. This usually amounts to about 70–80 percent reduction, but effluent silica content concentration in the latter part of the run will peak at 3–4 times the influent level. Overall average silica removal is about 50 percent.

Deaeration. Deaeration is the process of removing dissolved gases such as carbon dioxide and oxygen from water by:

1. Heating the water to the boiling temperature for the pressure under which the process is conducted (saturation conditions).
2. Using a suitable equipment design which provides for intimate mixing.
3. Providing a means of venting the gases released from the system.

The primary function of this process is to remove the noncondensable gases from make-up water used for boiler feed, and from return condensate. Removal of oxygen and free carbon dioxide protects piping, pumps, and boiler and condensate return lines from corrosive dissolved gases.

Since the solubilities of oxygen, carbon dioxide, and nitrogen are zero at the boiling point, water may be freed from dissolved air and carbon dioxide by boiling it and venting off the released gases.

Boiling may be accomplished at atmospheric, elevated, or reduced pressures. Since cold water will boil in a vacuum, deaeration of water at ordinary temperatures may be handled in that manner. The important factors are: (a) the boiling of the water—whether this takes place at ordinary temperatures or at or above 212°F (100°C) is immaterial; and (b) the venting off of the released gases.

The Chemistry of Corrosion. There are various theories to explain the corrosion of metals; however, the electrochemical process is probably the most widely accepted. It is quite complex, but the following is offered as a simplified version.

When iron or steel is brought in contact with water, its ferrous ion will enter the water, leaving electrons behind in the metal:

$$\text{Fe(metal)} = Fe^{2+} + 2e^- \qquad (35)$$

The point where the metal leaves and goes into solution is the *anodic area*. Eventually, an equilibrium condition can be reached between the metal ion concentration in the water and the concentration of electrons in the metal. Corrosion could stop at this point, provided the equilibrium condition were not upset.

Unfortunately, a number of factors upset the balance. Anything that will remove or neutralize either the metal ions in solution or the electrons in the metal, or both, will cause the corrosion process to continue.

The hydrogen ions (H^+) in the water which will react with and remove electrons are one factor. This upsets the equilibrium and more Fe must go into solution to fill the electron gap:

$$2H^+ + e^- = \underset{\text{Atomic hydrogen}}{H} \qquad (36)$$

$$2H = \underset{\text{Molecular hydrogen}}{H_2} \qquad (37)$$

Equation 37 will reach equilibrium very quickly because of the low solubility of molecular hydrogen in water. However, again the equilibrium can be upset by removing either the atomic hydrogen or the molecular hydrogen.

If equilibrium is maintained, the atomic hydrogen forms a thin coating or layer over the *cathodic area* (the area where the electrons leave the metal in order to react with hydrogen ions). This coat of atomic hydrogen prevents the electrons from leaving the metal, and thus stops the corrosion process. Unfortunately, the atomic hydrogen is easily removed by direct combination with other chemicals in the water or by forming molecular hydrogen which in turn reacts. The formation of the atomic hydrogen layer is called polarization.

Depolarization is the removal of the hydrogen layer. This is usually done by the removal of molecular hydrogen by dissolved molecular oxygen (O_2) (Eq. 38):

$$H_2 + \tfrac{1}{2} O_2 = H_2O \qquad (38)$$

The above is concerned only with accelerating corrosion at the cathode area. However, the anode area is just as active. For example, dissolved oxygen has a role in removing the ferrous (Fe^{2+}) ions in solution, thereby preventing equilibrium from being established. This creates more activity in the anode area. The Fe^{2+} reacts with OH^- ions present in the water:

$$Fe^{2+} + 2\,OH^- = Fe(OH)_2 \qquad (39)$$

$$4Fe(OH)_2 + O_2 - 2H_2O = 4\,Fe(OH)_3 \qquad (40)$$

Carbon dioxide, of itself, also abets the corrosion process, even in oxygen-free boiler systems. It creates a low pH and thus a source of hydrogen ions to react with electrons:

$$CO_2 + H_2O = H_2CO_3 = H^+ + HCO_3 \qquad (41)$$

Carbon dioxide can even accelerate oxygen corrosion. When both gases are present in the same solution, they will produce more corrosion than the same volumes of these gases would produce if considered individually.

Dalton's law states that total pressure of a mixture of several gases is equal to the sum of the pressure which each gas would exert were it alone present in the volume occupied by the mixture.

Henry's law states that the concentration of the dissolved gas in the solution is directly proportional to the partial pressure of the gas in the free space above the liquid. An exception to this would be where the gas has chemically united with the liquid.

In accordance with the basic principles contained in the above laws, it is evident that a dissolved gas can be removed from water by reducing the partial pressure of that gas in the surrounding atmosphere (regardless of the total pressure on the system). The simplest way to do this is to bubble another gas through the water or to spray the water into a countercurrent flow of another gas. In addition, the gas must be freely vented from the system. In this manner, the concentration of dissolved gas in the surrounding atmosphere is reduced by dilution with the scrubbing gas. The partial pressure of the dissolved gas in the surrounding atmosphere then decreases and, in accordance with Henry's law, the concentration of the dissolved gas in the water must decrease in direct proportion to its decrease in partial pressure in order to reestablish equilibrium. Practical application of this principle is the removal of carbon dioxide from water by blowing air countercurrent to falling streams of water as in a degasifier.

Oxygen can also be removed in this way, although air cannot be used as the other gas because of the high percentage of oxygen (20 percent) in the air. The partial pressure of the oxygen in the surrounding atmosphere could never be reduced sufficiently. Nitrogen could be used, but it is impractical.

The most desirable method for removing dissolved oxygen is to use steam as a scrubbing gas. It is accomplished by the following mechanism:

1. Mechanical separation of the gas particle results when saturation conditions (boiling) are achieved and when the vapor pressure of the water and the solution pressure of dissolved gases appreciably exceed the pressure imposed on the system. Under these conditions, all gases not chemically bound become insoluble.
2. The gas particle diffuses to the surrounding atmosphere where it is eventually vented out of the system.

Deaerating Heaters. There are two types of deaerating heaters: (1) the spray heater in which the water is sprayed into an atmosphere of steam and then passed through a scrubber, and (2) the tray heater in which water flows down over a series of trays while in an atmosphere of steam. Both types are capable of reducing oxygen to less than 0.005 cc/liter and titratable free carbon dioxide to zero.

Most equipment manufacturers offer several different designs of deaerators. They all perform the same basic function, but the configuration of each design varies to fit available space and storage requirements. For example, sometimes all the functions –spraying, preheating, scrubbing, condensing, and storage–are accomplished in one shell (Fig. 10-16). In other designs, the finished water is stored in a horizontal or vertical tank and is connected to another vessel where deaeration is first accomplished. This connection can be by means of a downcomer pipe, or the smaller vertical deaeration vessel can be an integral part of the larger horizontal storage tank, looking much like a tank car. If a larger vertical storage tank is used, a "potbelly" configuration results, with the smaller deaeration vessel sitting piggyback on top.

Typical Spray Operation. Incoming cold water first enters the spray valve compartment where stainless steel spray valves direct the flow in conical sheets through the stainless steel internal direct contact vent condenser into the steam atmosphere of the preheater section. Here most of the corrosive gases are removed before the water strikes any steel surfaces. This hot and partially-deaerated water then passes from the preheater section to the deaeration steam scrubber section.

Oxygen-free steam is introduced uniformly into this section through a steam distributor. The large excess of steam collides at high velocity with the heated and partially-deaerated water and violent scrubbing and mixing action occurs, even at very low load conditions. This action mechanically scrubs the last remaining traces of noncondensable gases out of the preheated water.

The deaerated water is discharged from the top of the steam scrubber section where the steam separates from it and passes into the preheater section to continue the cycle. Vented steam is condensed in the internal direct contact vent condenser with a small amount of steam being vented to the atmosphere to carry away the noncondensable gases released from

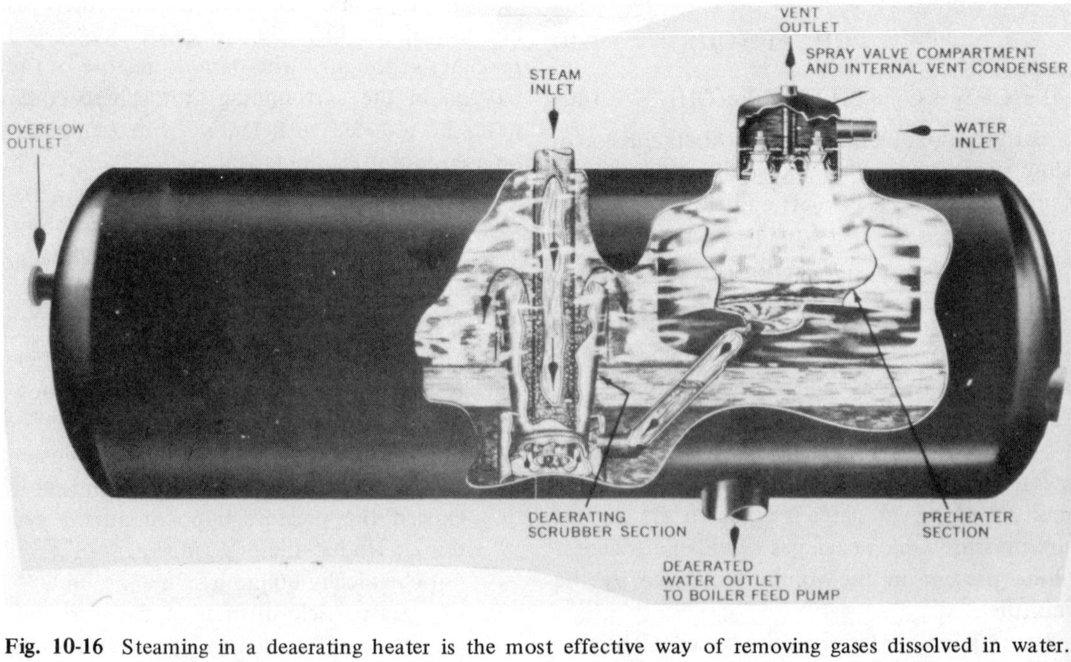

Fig. 10-16 Steaming in a deaerating heater is the most effective way of removing gases dissolved in water. (*Courtesy of Permutit Co.*)

the water. The internal direct contact vent condenser is a stainless steel section beneath the spray valve compartment where noncondensable gases are separated from the preheater section and concentrated before being discharged to the atmosphere.

Spray Design Parameters. The two important parameters in designing a spray deaerator are: (1) the amount of water to be deaerated, and (2) the deaerated water storage requirements. The amount of deaerated water affects the sizing of the spray valves and the type of scrubber to be used. For example, spray valves are ordinarily rated at about 50 gpm or 25,000 lb/hr. Therefore, the amount of water to be deaerated determines the number of spray valves.

The scrubber design is also affected by capacity. Up to 300,000 lb/hr capacity units frequently use a Venturi design scrubber. Above 300,000 lb/hr, a rectangular scrubber is used which incorporates a baffle design.

Storage requirements are usually 2 or 10 min and standard designs use either one or the other. The choice is usually set by the customer's consultant or the job specifications. Sometimes, the required storage is a period of other than 2 or 10 min. In these cases, the standard tank sizes are adjusted to accommodate the storage requirement.

Condensate Softening. Condensate softeners are generally used for treatment of condensate where boiler pressures are variable up to about 1200-1400 psi. Usually, steam from these plants is used for various industrial processes other than power generation.

The object of condensate softening is to remove both suspended and dissolved copper and iron as well as hardness so that the condensate can be reused as boiler feedwater. Total dissolved solids are generally not a factor in treating condensates from boilers with pressures up to 1200-1400 psi. This application is quite distinct from the use of condensate polishing demineralizers in large power stations where the water is returned to high pressure steam generators. In this case, mixed bed units are supplied to maintain very low concentrations of silica, iron, copper, and dissolved solids.

In contrast, water quality requirements for the low pressure boilers are much less stringent, materials of construction less restrictive, and contamination of the steam or condensate more likely. For example, the level of contaminants can range as follows:

Total dissolved solids	5-30 ppm
Hardness	up to 0.1 gr/gal
Iron and copper	0.02-1 ppm

Iron-copper concentrations, in particular, vary widely from one boiler system to another. Also, the percentage of total iron and copper which is suspended, rather than dissolved, is important because this may have some effect on equipment design.

Condensate softeners operate in one of two ways: (1) condensate is first passed through a filter (such as a precoat tubular type) to remove any suspended iron, the condensate then going to the softener, or (2) unfiltered condensate is sent directly to the softener where simultaneous ion exchange and filtration take place.

Both of these methods are used today, but the latter is becoming more generally accepted since the softener is acting as an ion exchanger as well as a filter. One problem, however, with passing unfiltered condensate through an exchanger is that the resin beds become coated with filtered-out iron. Because this coated iron is not alwyas removed from the resin by backwashing and regeneration, most equipment designs make provisions for chemical cleaning of the resin beds. Equipment utilized for this process differs from the conventional type mainly in that it must be able to handle a normal condensate temperature of 160 to 200°F.

The regeneration scheme plays a key role in the design of condensate softeners. There are two methods of regeneration, conventional in-place or external. For external regeneration, the resin is sluiced out of the vessel and regenerated and/or chemically cleaned in a separate tank. Since service flow rates are usually around 25 gpm/ft^2 (conventional softeners operate at 8-10 gpm/ft^2), coarser grades of ion-exchange resins are normally used in this application to minimize pressure drop.

In those cases where cold water is used for backwashing and regeneration, higher cross-linked resin is specified. This is necessary because alternate use of hot and cold water could create problems with temperature shock, and too rapid contraction and expansion of the resin bed could lead to physical breakdown of the resin. Results of this type of treatment indicate that 80 to 90 percent removal of copper and iron can be expected when salt dosages of at least 10 lb/ft^3 are used.

APPLIED TECHNOLOGY—RECIRCULATING COOLING WATER SYSTEM

Treatment of cooling water is required to avoid scale, corrosion, algae and slime, and cooling tower wood decay. There are two modes of operation for controlling these parameters. The first takes place in an acidic environment and represents the dominant technology for cooling water systems. The mode of operation whereby the critical parameters are controlled in an alkaline environment represents recent technology geared to meeting today's increasingly stringent requirements for pollution control.

Scale. There are five common types of scale which may form in cooling systems. There are calcium carbonate, calcium sulfate, calcium phosphate, silica, and magnesium silicate. The scale that forms in most parts of the United States, if cooling water is not treated, is predominantly calcium carbonate. Although this type of scale can be removed from cooling system components such as tube bundles through the use of acid cleaning, repeated applications will normally shorten the life of the component. Calcium sulfate and calcium phosphate scale will form only where water treatment is being attempted but improperly controlled. These two types of scale are unaffected by acid cleaning and must be removed from components by mechanical means. Silica and magnesium silicate scale occur only in water which has a high silica content, or a high silica to hardness ratio.[5]

Corrosion. Corrosion rates well under 1 mm/yr will be consistently obtained where the cooling water is treated properly. It is not uncommon for untreated cooling water to cause corrosion rates of 20-25 mm/yr.[5]

Algae and Slime. Microbiological organisms can grow in cooling water systems under most adverse conditions and can seriously impair the efficiency of heat transfer as well as the structural strength of the cooling tower if the tower is wooden. Biocidal treatment is required in all cooling water systems to minimize these problems.[5]

Acidic Mode of Treatment. Concentrated sulfuric acid, when used with the proper bleed-off rate, will prevent the formation of scale of any kind with almost any type of water. Acid is added in an amount to make the water slightly corrosive to keep the metal surfaces clean; then an effective corrosion inhibitor is added to protect the metal against the corrosive water.

Three points of control are required in order to maintain cooling water in a scale-free and minimum corrosion condition:

1. The pH or acidity of the water must be controlled by adding acid.
2. The total dissolved solids level must be controlled by bleeding a portion of the circulating water to waste.
3. An effective corrosion inhibitor, typically chromate-based, must be selected and controlled at the recommended concentration.

The most critical of these three points of control is the pH of the recirculating water. Damage to the cooling system will be caused by too much or too little acid feed.

The total dissolved solids level is the next most critical. This level should be maintained as high as possible to save water, but not so high that scale will form by exceeding the solubility of certain components.

The least critical of the three control points is the inhibitor level. Recommendations as to the allowable inhibitor level, made by manufacturers, cover a wide range varying by as much as a factor of three from the lowest to the highest allowable level. For economy reasons, the inhibitor level should be maintained at the lowest effective level.[5]

Acid. Technical grade concentrated sulfuric acid, 66° Baume, is by far the most highly concentrated and least expensive of the acids commonly used for pH control in closed cooling water systems. Hydrochloric acid is required in rare instances where the sulfate content in the make-up water is so high that adding additional sulfate in the form of sulfuric acid is uneconomical because of the high blowdown rates required to maintain the sulfates in the cooling water at a given level.[5]

Corrosion Inhibitors.[5] The chromate-based corrosion inhibitors supplied by a large number of chemical companies are composed of chromate in combination with one or more of the following: a polyphosphate, a metallic salt, and a dispersant or surfactant. The ratio of chromate as CrO_4 to the total quantity of inhibitor varies from as little as 1 part chromate to 4 parts total inhibitor to as much as 1 part chromate to 2 parts total inhibitor.

Chromate-based formulations have been the backbone of cooling water treatment for many years. By themselves, chromates inhibit very effectively at concentrations of 200-300 ppm. Losses from closed systems are usually minimal, so the high concentration of chromate is retained and remains effective and practical.

In open recirculating systems, chromate usually is combined with polyphosphate and zinc. The optimum pH range in such systems for best sclae protection and inhibitor performance is 6.0 to 6.8. These combinations perform synergistically so that excellent steel protection is achieved with chromate concentrations as low as 20-50 ppm. This concentration commonly provides corrosion rates of less than 1 mm/yr, and pitting is completely eliminated. Because of such effective treatment, steel can be used in cooling system equipment instead of more expensive corrosion-resistant alloys.

If chromate in moderate concentrations is used in a cooling system, the blowdown may need special treatment to make the effluent comply with environmental standards. Chromate (chromium in the +6 valence state) may be reduced by any of several reducing agents (such as sodium metabisulfite) to chromium in the +3 valence state. The soluble trivalent chromium sulfate can then be precipitated by neutralizing the acid waste with lime to a pH of 8.5.

Biocides.[5] Biocidal treatment is required for all recirculating cooling water systems. Chlorine is the most widely used biocide and is usually quite effective at a concentration of about 0.5 ppm. Chlorine can be fed directly as a gas with a gas chlorinator, in liquid form as a sodium hypochlorite solution, or as a solid in the form of "organic chlorine." Organic chlorine has the added advantage of slowly releasing chlorine into the water on-site, and therefore is least subject to aeration. Solid calcium hypochlorite is usually not a recommended source of chlorine because the calcium content may cause scaling. Best results can usually be obtained by shock treating periodically, whether the cooling tower appears to require treatment or not. One a week is a reasonable frequency of shock treatments. Experience may require applications more or less frequently.

It may be necessary to shock treat the system from time to time with an alternate biocide such as sodium pentachlorophenate at a level of about 200 ppm. The frequency should be determined by experience, but every other month would be a reasonable starting point.

Where pentachlorophenate is not allowed in the bleed from the cooling tower, or where it is not sufficiently effective, a host of exotic materials are available from water treatment chemical suppliers. Chemical companies should be consulted for their recommendations. The most effective combination of biocides can be selected in difficult cases if samples of the microorganism are taken and tested for sensitivity in a microbiological laboratory.

One very important requirement for any biocide added to a recirculating cooling water system is that it can be compatible with the corrosion inhibitor being used.

Alkaline Mode of Treatment.[6] While acid treatment has been the dominant technology for cooling water systems, alkaline treatments are gaining in usage and popularity. Alkaline cooling water treatment is not new, however.

For many years before acid chromate and phosphate treatments came into widespread use, cooling waters were treated with small amounts of polyphosphates and natural organic materials to condition calcium carbonate scale so that it could not form a hard deposit on heat transfer surfaces. This is called threshold treatment and products of this type are known as crystal-growth inhibitors. They distort the lattice of growing calcium carbonate crystals and prevent the formation of large adhering crystals.

Scale Control in Alkaline Water. Calcium carbonate is by far the most common scale-forming substance in alkaline cooling water. Sulfonated lignins, tannins, etc. are effective calcium carbonate crystal-growth inhibitors. Polyacrylates act as dispersants for crystalline sclaes. Polyphosphates act as crystal-growth inhibitors, and also, at sufficient dosage, chelate calcium ions to prevent scale formation. Polyphosphates are limited in use, however, because of their tendency to revert to orthophosphate at high temperature.

More recently, new scale-control chemicals have been developed that avoid the reversion problem of polyphosphates and are effective at higher pH and alkalinity values. These new chemicals have made possible today's widespread use of alkaline cooling water treatments. The polyol esters are a class of moderate molecular weight polyhydric alcohol phosphate esters, containing a carbon–oxygen–phosphorus linkage. The phosphonates, on the other hand, contain direct carbon–phosphorus bonds. Both classes of chemicals have demonstrated effective scale control under alkaline conditions in both laboratory and filed tests. It is absolutely essential in any application of these products to maintain proper chemical dosage.

In an alkaline system, scale formation can be prevented only by the presence of scale-control chemicals. If chemical dosage is lost, even for a brief time, some scaling will occur, resulting in reduced heat transfer. Once formed, scale cannot be removed easily by the same scale-control chemicals. Other means, such as mechanical cleaning or acidizing the system, usually are required.

Corrosion Control In Alkaline Systems. The same corrosion inhibitors are employed in both acid and alkaline cooling systems; i.e., chromates, phosphates, zinc, and some organic chemicals. While these are the most effective corrosion inhibitors presently available, the search continues for effective, nontoxic inhibitors for cooling water systems. Silicates are used occasionally in alkaline systems, and good results have been achieved with organic–inorganic combinations. However, neither the silicates nor the organic–inorganic blends are as effective in controlling corrosion as the synergized chromate products.

With careful control of tower concentrations, some cooling water systems can be treated solely for corrosion control (no scale control) at pH values usually less than 8. This is simply an extension of normal acid treatment programs, usually involving the synergized chromate products. Increasing the pH of water from 6.0 to 7.5 generally makes the water less corrosive, so that lower dosages of corrosion inhibitors can be used. Also, less acid is required, so that the treatment program becomes less expensive.

More commonly, the pH in alkaline cooling water is allowed to reach its equilibrium level, which often falls above 8.5. Scale control usually is required under these conditions, and scale-control chemicals are often combined with chromate-based inhibitors to provide a complete treatment program for such systems.

Fouling Control In Alkaline Systems. Deposit control in cooling water systems is essential for continued good heat exchange rates. However, control of deposits is often more difficult in alkaline systems than in acid systems. The make-up water may contain dissolved organic matter or suspended solids, both of which can contribute to fouling. It may be microbiologically contaminated and contain substantial BOD, particularly if the make-up water is part of a process effluent stream.

More substances are insoluble in alkaline water than in acid water. For example, inorganic deposits such as calcium carbonate and ferric hydroxide are almost completely insoluble above a pH of 8, whereas they are dissolved in acid water. Colloidal organic matter tends to coagulate and precipitate more easily in alkaline systems than under acid conditions. Microbiological deposits are sometimes more difficult to remove in alkaline systems.

There are numerous sources of fouling deposits in recirculating cooling water. The raw water may carry in colloidal organic matter, silt and dirt, soluble iron, and microbiological contamination. The air blowing into the tower may contain dirt, or be contaminated with reactive process gases such as hydrogen sulfide, sulfur dioxide, ammonia, etc. The recirculating water may deposit calcium carbonate, calcium sulfate, or magnesium silicate scale, or corrosion products such as iron oxide tubercles. The recirculating water is, of course, the major source of microbiological growths. Process leaks are a common source of hydrocarbons and sulfides. Sulfides can react with oxidizing biocides and zinc ion corrosion inhibitor, and may cause chromate reduction with additional fouling.

It is interesting to note that most of the commonly used scale-control chemicals are also good fouling dispersants. However, the particular dispersant selected for an alkaline system will depend upon the deposit composition. Natural organics, including sulfonated lignins and tannins, are most effective in dispersing hydrous oxides and organic deposits. Polyacrylates are excellent dispersants for silt, sand, hard iron oxides, and other inorganic deposits. Polyol esters and phosphonates are, as would be expected, dispersants for calcium carbonate and calcium sulfate scales. However, as was pointed out above, this scale-removal action takes place gradually. These products are not generally effective for rapid cleaning of scaled heat transfer surfaces. As a general rule, it is always better to prevent formation of insoluble deposits rather than to remove deposits already formed.

Dispersants plus scale-control chemicals make effective alkaline treatments for cooling towers in which the make-up matter contains suspended solids.

Microbiological Control. Microbiological deposits are a special case of cooling water fouling. Treatment often requires both biocides to kill microbiological growth and dispersants to loosen and float away deposits.

The most common biocide in both alkaline and acid systems is still chlorine. Chlorine is less effective at high pH (above 8.0) than it is below pH 7 because of the well-known hypochlorite equilibria. Nevertheless, it is still an effective biocide in alkaline systems, and is an indispensable part of many alkaline cooling water microbiological control programs. Chlorine is available as gas, hypochlorite solution, and slow-release hypochlorite.

There are some problems associated with chlorine use, however. In the presence of ammonia, it forms chloramines, which are less effective biocides than free chlorine, and represent a chlorine demand in the system. Chlorine can react with some organic materials, particularly phenolic products, to form "hard" chlorinated organic compounds which are nonbiodegradable and thus present possible pollution control problems. Original fears of tower lumber deterioration with chlorine in alkaline waters have not been substantiated in practice. Both users and suppliers have been concerned about this possibility and have been watching towers closely since alkaline treatments were introduced. It is now generally accepted that with proper control, cooling towers can be operated above pH 8.0 using chlorine without danger of delignification of the wood.

Just appearing on the horizon as a possible new biocide for alkaline systems is ozone, the most powerful oxidizing agent known. Ozone is ideal from a pollution control viewpoint since its only end product is free oxygen. However, ozone is unstable and expensive, and must be generated on site, at least at present.

Along with chlorine, nonoxidizing biocides are essential for good control with most alkaline cooling water systems. Chlorine alone cannot do the whole job. Nonoxidizing biocides are, in many cases, specific for different fouling organisms. Also, these biocides can be formulated with dispersants, scale inhibitors, and other cooling water treatment chemicals. Thus, the proper biocide program can be selected for each specific cooling tower.

Sidestream Filtration. When unclarified water is used as make-up to a cooling tower, sand, silt, or turbidity may enter the system in any number of ways. The suspended matter can be present in the tower make-up water, wind can deposit dust in the cooling system, and algae and slime growths can occur in the water and cause turbidity. Corrosion products and leaks can also contribute to the suspended matter. Not only does this suspended matter foul heat exchangers, but it impairs heat transfer rates and accelerates corrosion problems caused by silt deposits. Most important, it increases cleaning and maintenance costs.

Although any type of filter can be used to eliminate this foreign matter, the automatic gravity type filters are most suitable for this service. Filters are normally located near cooling towers which are sometimes remote from other plant facilities. Consequently, a filter that has minimum manpower requirements for operation and maintenance is ideal. In a cooling water system, the filter reduces the amount of suspended matter in the circulating water and maintains it at low levels by continuously filtering a small portion of the main cooling stream.

Figure 10-17 shows two possible locations for the sidestream filter in a cooling water system. If the filter is located on the cool side, which is usual case, a slight addition to the pumping capacity is required. This is so because the pump must have a capacity equal to the cooling water flow through the heat exchanger system plus the flow through the filter. The additional capacity is small since sidestream flow is usually a small percentage of the cooling water flow.

If the filter is located on the hot side, the additional pump capacity is not required. However, sufficient pressure must be available to permit flow to the sidestream filter; this will depend upon the local layout of any given plant. If sufficient pressure is not available, another pump would be required, making the hot side location undesirable.

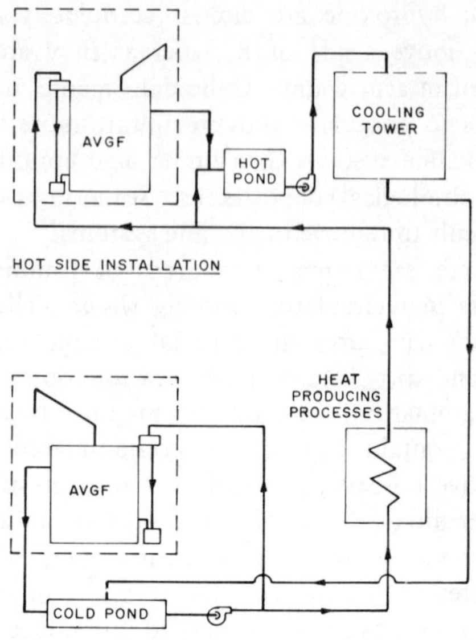

Fig. 10-17 Two locations for sidestream filter in cooling water system may be on hot side or cool side. (*Courtesy of Permutit Co.*)

Sidestream filters are usually sized for 2.0–2.5 percent of the cooling water recirculation rate or two to three turnovers of the total cooling system volume every 24 hr through the filter, whichever is smaller. The filtration rate is typically 2.5–3.0 gpm/ft^2 of filter bed area.

APPLIED TECHNOLOGY–ULTIMATE WATER

Ultrapure water or ultimate water is defined as water that is virtually free of all contaminants, both dissolved and suspended, organic and inorganic. It is widely used in manufacturing many electronic components, such as transistors or semiconductors, which are extremely sensitive to minute contamination. Even traces of dust, dirt, inorganic salts, or organic matter on these parts before they are encapsulated can cause malfunctions. Therefore, they are manufactured under conditions of cleanliness and sterility normally found only in a hospital or surgical room. A final step before these parts are sealed is a wash, or rinse, with extremely pure or ultimate water.

A common specification for ultimate water is

merely a requirement of water of 18 megohm resistance. This is an incomplete and unrealistic specification. Conductivity (actually megohms are a measure of resistivity) gives a measure of dissolved electrolytes only, and even then, it means nothing without a temperature reference.

The theoretical resistivity of absolutely pure water at 25°C (the usual reference temperature) is about 18 megohms. At 18°C, which is sometimes used as a reference temperature, pure water has a resistivity of about 22 megohms. Thus, temperature must be specified along with resistivity, and a purity greater than the theoretical value certainly is impossible.

In addition to extremely low dissolved electrolytes, which can be measured by conductivity instruments, ultimate water must be equally free from suspended matter. Even low concentrations of bacteria or algae in this water are extremely detrimental. Thus, a specification for "18 megohm water," which infers high quality demineralization, also means high quality filtration and sterilization for removal of suspended solids, organic matter, and bacteria control.

The basic process diagram shown in Fig. 10-18 is typical of the treatment systems that produce ultimate water for use in the manufacture of semiconductors and microcircuits. Note that the basic treatment consists of the following steps in order:

1. Sterilization and/or coagulation.
2. Filtration and dechlorination with activated carbon.
3. Primary demineralizing.
4. Mixed-bed polishing.
5. Upgrading and maintenance of high quality water in the ultimate water loop.

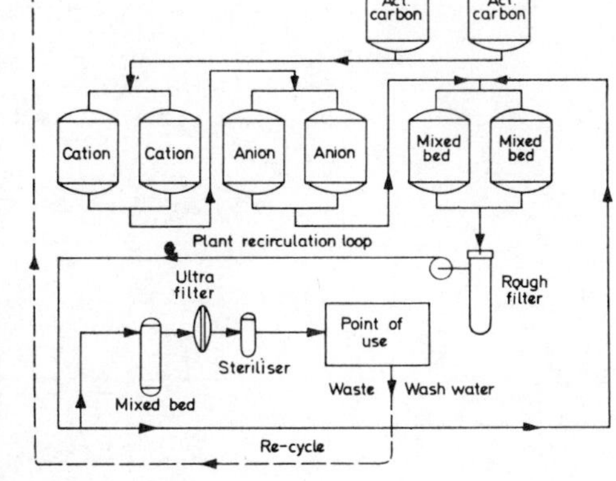

Fig. 10-18 Process for producing ultimate water. (*Courtesy of Permutit Co.*)

Sterilization must be included, using chlorine at high levels (approximately 10 ppm), with adequate contact time for reaction and destruction of organic matter. On most supplies, coagulation is also required, and sterilization can be accomplished in the coagulation unit. Sand filtration and activated carbon purifiers are invariably required. Because of the high chlorine dosages employed, both filter and purifier must be lined. In addition, the carbon purifier should be arranged for convenient periodic sterilization with steam.

Primary demineralizers may be either two-bed or mixed-bed type, depending on the flow rates and water analysis. For small flows (10 gpm range), mixed-bed demineralizers will usually prove to be the most economical choice.

Mixed-bed polishing can be accomplished at flow rates of up to 15 gpm/ft^2. This unit will normally produce an effluent of about 10 to 12 megohms in protracted service. (All mixed-bed units produce water of exceptional quality for the first few runs. Thus, this unit may well produce up to 16 megohm water for a short period of time, but 10 to 12 megohms is a more realistic design average.)

Upgrading and maintenance of quality in the ultimate water loops is an important factor in these plants. Here the quality of the water may be stepped up to 15 to 18 megohms by recycling the water around a pipe loop through non-regenerable mixed-bed units. This is essential, as this extremely high purity water is easily degraded by dissolving minute amounts of even the most inert piping materials. Takeoff lines from this loop to points of use must be very short, for the same reason. Where these takeoff lines cannot be kept short, individual, point-of-use demineralizers must be used at these points.

Cartridge type filters should be used downstream of the nonregenerable mixed-bed units in the ultimate water loop. These filters should be able to remove all particles larger than 5 to 25 μ.

Final polishing filters in the submicron range must be used as the finishing treatment. These are low capacity filters having a flow range of only a few gallons per minute, and therefore can be installed in the loop only if the flow around the loop is small. In most instances these filters must be installed at the point of use.

REFERENCES

1. *Idea Exchange* **1** (2) 1–4, Dow Chemical Co. (Oct., 1972).
2. *Idea Exchange* **2** (1) 1–3, Dow Chemical Co. (Jan., 1972).
3. *Idea Exchange* **2** (2) 1–4, Dow Chemical Co. (Apr., 1972).
4. *Idea Exchange* **2** (4) 1–3, Dow Chemical Co. (Oct., 1972).
5. *Uniloc Design Manual–Water Treatment Control Systems*, Universal Interloc, Inc., Santa Ana, California (1970).
6. Freedman, A. J., and Shannon, J. E., "Modern Alkaline Cooling Water Treatment," *Industrial Water Engineering*, pp. 32–34 (Jan., 1973).
7. "Demineralizing by Ion Exchange," Bulletin 5614, The Permutit Company, pp. 2–8 (1971).

11

Wastewater in the City

David V. Auld

Former Director
Department of Sanitary Engineering
District of Columbia

INTRODUCTION

The subjects covered in this chapter are for the most part those which first challenged the ingenuity and prompted the investigations and experiments of the early practitioners within the then embryonic field of sanitary engineering. Wastewater treatment and water quality control evolved as concerns subsequent to the initial problems of how much wastewater to anticipate and how to confine it and move it from the places in which it was to be generated. So it is that the following material has become reasonably well standardized and its use routine whenever collection system design is undertaken. It has been presented in one form or another in standard texts and manuals over the past 50 or more years. The storm runoff computations as given herein have been in use for 80 years.[1,2,3,4]

Within the limits of the handbook concept, and with the expectation that the serious designer will take advantage of the more penetrating treatments of these topics which are available elsewhere, the following presentations are confined to those highlights whose recognition would be essential to any general consideration of piping system problems.

SANITARY SEWAGE COLLECTION

Sources of Flows Carried by Sanitary Sewerage Systems. Sanitary sewerage systems are intended to convey the spent water of the community. This may be contributed from domestic, commercial, institutional, and industrial sources. Also, these systems usually receive limited quantities of surface and independently-developed process waters. Groundwater as the result of infiltration of seepage into structures, including piping, will almost always be present. The relative quantities of surface and process waters will be determined by the character and enforcement of restraining regulations. The quality of construction, both in the public and the tributary private systems and appurtenances, and the relationship of the groundwater table thereto, will determine the quantity of infiltration to be handled.

Domestic Waste–Population to be Served. The complexity of the methods employed for estimating the future population tributary to a wastewater system, or part thereof, should be governed by the extent of the territory to be served and the distance into the future on which the design is to be based. In sizing local collectors, lift stations, and the like, the kinds of land use to be anticipated from prevailing or proposed zoning should suffice as a basis. For new communities to be developed on open land, the architect's layout and housing assumptions should provide a dependable guide to future population. In established settlements, cities, or metropolitan regions, census data should be examined to determine past population numbers, distribution, and rates of change, either as growth or decline. This demographic history may not be extrapolated directly into dependable predictions except after giving due weight to the economic, political, and social forces which may be expected to influence future growth. In general, the rates of growth will be greater for the younger, developing communities. In the older cities, growth is slower, and in some cases may already have reached its ultimate under prevailing conditions.

Comparisons with the experience of other similar but older and larger communities may be helpful. Local and state planning agencies and public utility planning offices may be excellent sources of information. Where the area of interest may have been embraced within a study or program of regional scope (as for water resources, industrial, mineral, agricultural, or economic development) by an interstate or federal agency, any pertinent population projections should be given great weight.

The period of the population forecast used as a design basis for any given structure or system component should correspond with the time during which it is to be adequate for the intended service. This, the "period of design," will not be the same for all elements of a sewerage system. Local service mains, lateral collectors, and trunks and interceptors in existing or anticipated congested settings should suffice for the ultimate loadings. Other trunks, interceptors, and outfalls may be designed for from 25- to 50-year periods, depending on anticipated rate of growth, difficulty of providing increased future capacity, and financing. Pumping stations, if they are interim necessities, should be adequate just for near-term flows; if permanent, they should be designed for incremental expansion to ultimate loadings. However, the mechanical and electrical equipment which is subject to obsolescence may justify a design period of no more than 10 to 25 years.

Specific methods of mathematical and graphical population projection are presented in Chapter 2 for potable water requirements. These methods are identical to those used in establishing the population basis for wastewater quantity calculations.

Domestic Waste–Per Capita Production. The various forms of land use in a community will determine the quantities of wastewater to be produced per unit area. The measurement of flows generated within established nearby sections will give guidance as to those to be anticipated from areas of like development. Wide regional or climatic dissimilarities, however, will render any useful correspondence unlikely.

Residential occupancy may vary from an ultimate density of 2 to 3 persons/acre to over 1000/acre. The former rate occurs with single family houses on large lots, the latter with multistory apartments or condominiums. Single family dwellings on small lots may hold 15 to 35 persons/acre, and multifamily dwellings from 35 to 100/acre.

Some 70 percent of the water supplied to the community ordinarily becomes wastewater. The remainder is consumed by general municipal usage for fires, street cleaning, etc. Domestic consumption includes evaporation from refrigeration and heating, lawn watering, swimming pools, and the like. In arid areas, increased lawn and garden watering, pool usage, and cooling may result in only some 40 percent of the water supplied becoming wastewater.

Water usage as between various classes of users and the proportions thereof becoming wastewater vary substantially. The per capita usage in a separate single family household will be greater than that of the apartment dweller. This can be accounted for by more appliances, gardens, children, and possibly social activity in the single family home. Also, the higher valued residential properties have been found to produce increased per capita sewage flows (Fig. 11-1).[5]

The generally rising standard of living, the low cost of water, and the growing maturity of established communities all contribute to an upward trend in the individual rates of water usage and sewage production.

Some comparative average water consumptions and sewage flows for various localities are given in Table 11-1.[1]

Surface Water Effects. Surface runoff from precipitation and snowmelt may reach sanitary sewers. The discharge of roof, yard, driveway, and foundation drainage to sanitary sewer systems should not be permitted since such flows on occasion will greatly exceed the peak sanitary flows, and, if accepted, will dictate excessive sewer sizes. Surreptitious contributions from such sources can be revealed by gauging during and after rainfall.

If peculiar local conditions give no feasible alternative to the admission of flows from one or another of these surface sources, the quantities of stormwater to be handled should be determined by methods given in the section "Stormwater Runoff." The allowances to be provided for foundation drainage, if unavoidable, are conjectural. One published figure for Kansas City, Missouri is 1.25 gpm/house.

In some localities, drainage from areaways, depressed garage driveways, and window wells may be legally connected to the sanitary house lateral. The quantities of stormwater expected to reach the sanitary sewer system in the District of Columbia, based on a survey of permissible drainage sources and flow measurements, is given in Fig. 11-2.[6]

Flooded or ill-fitting manhole covers, broken pipes at stream crossings, and other system defects may admit stormwater. Since proper maintenance should prevent such occurrences, system design should not have to take them into account.

Groundwater Effects. Groundwater enters a sewerage system through pipe joints or breaks and cracked or porous masonry as it occurs in manholes, monolithic sections, etc. Poor jointing usually accounts for most such seepage or infiltration. Private plumbing connected to

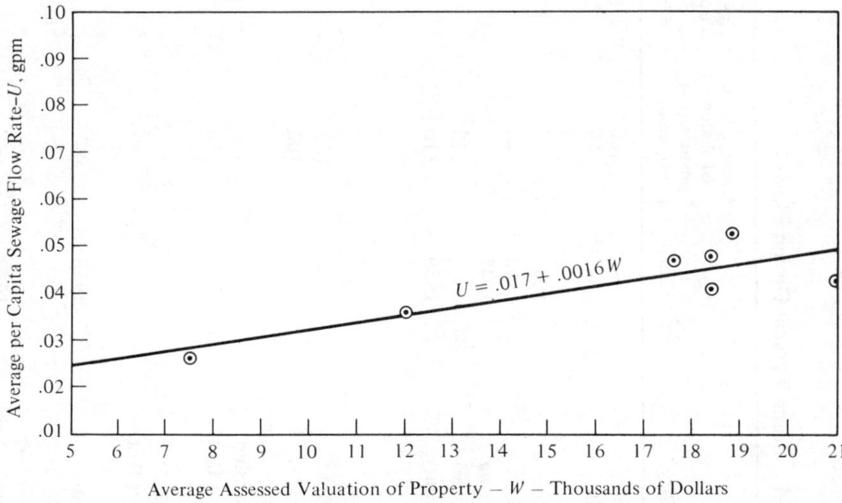

Fig. 11-1 Average per capita domestic wastewater flow vs. average market value of property. (*Courtesy of the Water Pollution Control Federation.*) *Source:* Geyer, J. C., and Lentz, J. J., *Journal WPCF* **38**, 1139 (1966).

TABLE 11-1. Some Typical Design Flows

City	Year	Average Rate of Water Consumption (gpd/person)	Population Served (1,000s)	Average Sewage Flow (gpd/person)	Sewer Design Basis (gpd/person)	Remarks
Baltimore, Md.	–	160	1,300	100	135 × factor	Factor 4 to 2
Berkeley, Calif.	–	76	113	60	92	
Boston, Mass.	–	145	801	140	150	Flowing half full
Cleveland, Ohio	1946	–	–	100	–	
Cranston, R.I.	1943	–	–	119	167	
Des Moines, Iowa	1949	–	–	100	200	
Grand Rapids, Mich.	–	178	200	190	200	
Greenville County, S.C.	1959	110	200	150	300	Service area includes city of Greenville. Sewers 24 in. and less designed to flow half full at 300 gpd/person; sewers larger than 24 in. designed to have 1-ft freeboard
Hagerstown, Md.	–	100	38	100	250	
Jefferson County, Ala.	–	102	500	100	300	
Johnson County, Kan.	1958					
Mission Township Main Sewer Dist.		70	70	60	1,350	Most houses have basements with exterior foundation drains
Indian Creek Main Sewer Dist.		70	30	60	675	Most houses have basements with interior foundation drains
Kansas City, Mo.	1958	–	500	60	675	For trucks and interceptors
					1,350	For laterals and sub-mains. Many houses have basements and exterior foundation drains
Lancaster County, Neb.	1962	167	148	92	400	Serves city of Lincoln
Las Vegas, Nev.	–	410	45	209	250	
Lincoln, Neb. (Lateral Dists.)	1964			60		For lateral sewers max flow by formula: Peak flow = $5 \times$ ave flow $\div$ (pop in 1,000s)$^{0.2}$

[illegible]	[illegible]	30	100	50	100	
Los Angeles, Calif.	1965	185	2,710	85	*	*85 gpd residential multiplied by peak factor
Los Angeles County Sanitation Dist.	1964	200	3,500	70**	–	**Domestic flow only, ranges from 50 to 90 gpd/person depending on cost of water, type of residence, etc. Domestic plus industrial averages 90 gpd/person
Greater Peoria, Ill.	1960	90	150	75	800 8,500	Based on 12 persons/acre for lateral and trunk sewers, respectively
Madison, Wis.	1937	–	–	–	300	Maximum hourly rate
Milwaukee, Wis.	1945	–	–	125	–	All in 12 hr 250 gpd/person rate
Memphis, Tenn.	–	125	450	100	100	
Orlando, Fla.	–	150	75	70	190	
Painesville, Ohio	1947	–	–	125	600	Includes infiltration and roof water
Rapid City, S.D.	–	122	40	121	125	
Rochester, N.Y.	1946	–	–	–	250	New York State Board of Health standard
Santa Monica, Calif.	–	137	75	92	92	
Shreveport, La.	1961	125	165	–	–	Sewer design 150 gpd/person plus 600 gpd/acre infiltration. Sewers 24 in. and less designed to flow half full; sewers larger than 24 in. designed to have 1-ft freeboard
St. Joseph, Mo.	1960	–	85	125	450	Main sewers
					350	Interceptors
Springfield, Mass.	1949	–	–	–	200	150 gpd/person was used on a special project
Toledo, Ohio	1946	–	–	–	160	
Washington, D.C. Suburban Sanitary	1946	–	–	100	2 to 3.3 × ave	
Wyoming, Mich.	1960	150	50	82†	400	†Calculated actual domestic sewage flow, not including infiltration or industrial flow

Note: gal × 3.785 = liters; gpd/acre × 0.00935 = m^3/day/hectare; ft × 0.3 = m; in. × 2.54 = cm.
Source: ASCE Manual No. 37.

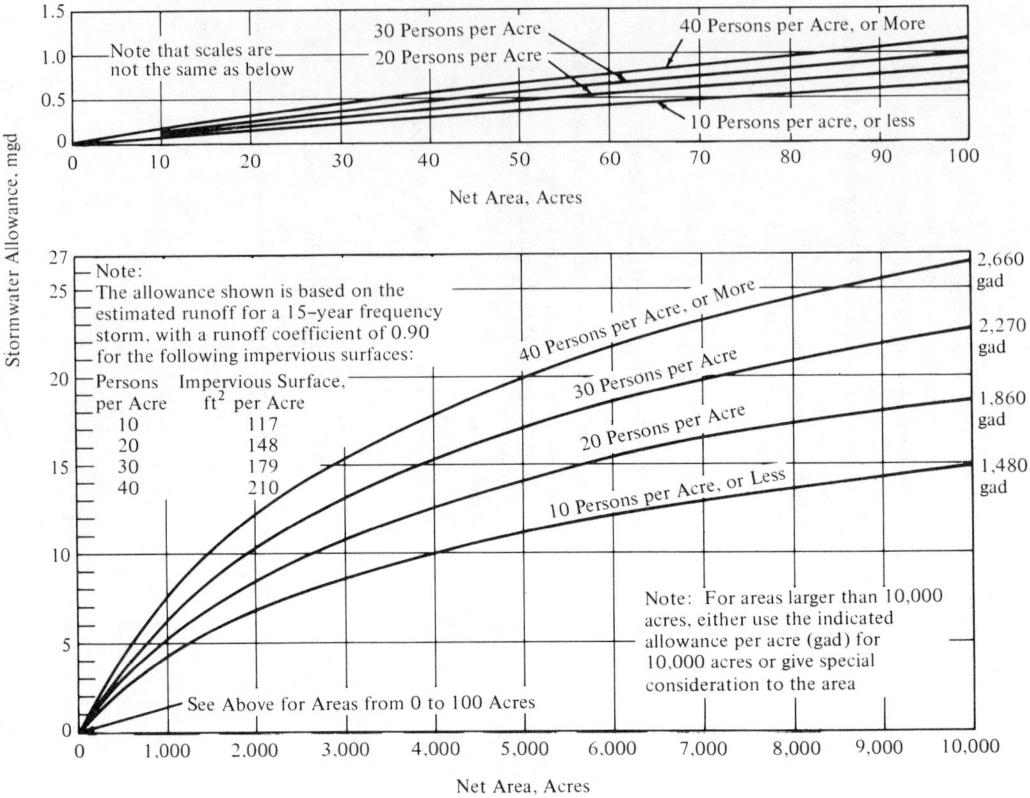

Fig. 11-2 Stormwater allowance for design of separate sewers. (*Courtesy of Department of Environmental Services, District of Columbia.*)

public sewers usually collects groundwater, and in some situations has been shown to contribute as much as 90 percent of the total infiltration in a given setting.[1] Only sound standards of construction and their enforcement can reduce this wasteful loading which may necessitate excessive pipe sizes and become an unwanted burden on pumping and treatment facilities.

The relation of the water table to the underground facilities will affect the quantities of infiltration. In dry seasons flows may diminish greatly, only to increase to maximum quantities with the return of rainfall.

In the design of sewage handling facilities, allowance must be made for infiltration as an addition to the peak sanitary flow and possibly other flows. Gauging and calculations in the District of Columbia gave an average rate of infiltration for the entire system of 730 gpd/acre; for the 2400 acres tributary to a major trunk line infiltration was 690 gpd/acre. Infiltration allowances, or limits, for new construction usually relate to the length and size of the conduits. Design allowances of several cities are shown in Table 11-2.[1]

Local conditions, such as water table relation-

TABLE 11-2. Variation of Infiltration Allowances among Cities

Number of Cities Reporting	Allowance (gpd/in. diam/mi)
4	1,500
4	1,000
1	800
2	700
1	600
63	500
11	450 to 300
16	250 to 150
21	100
5	500

Note: gpd/in. diam/mi × 0.000925 = m^3/day/cm diam/km.
Source: ASCE Manual No. 37.

ship, existing plumbing and pipeline defects, and intended quality of new construction will influence selection of the infiltration limits. If infiltration limits are to be related to tributary area alone, they may range from 500 to 5000 gpd/acre, with 2000 gpd/acre as a reasonable average. If limits are to be related to the length of sewer without regard to size, they may range from 5000 to 100,000 gpd/mi, with 30,000 as a reasonable average. For calculations based on infiltration per inch of diameter per mile, additional seepage of 100 gpd/manhole is customary.

Commercial, Industrial, and Institutional Effects. No firm rules as to flows to be expected from commercial, industrial, and institutional users can be offered. The waste production from identical areas may vary greatly; for example, laundries versus storage warehouses. Even within similar industries, differing pro-

TABLE 11-3. Sewer Capacity Allowances for Commercial and Industrial Areas

City	Year	Commercial Allowance (gpd/acre)[a]	Industrial Allowance (gpd/acre)
Baltimore, Md.	1949	6,750–13,500; 135 gpd/person, resident population	7,500 min
Berkeley, Calif.	–	–	50,000
Buffalo, N.Y.	–	60,000	–
Cincinnati, Ohio	–	40,000	–
Columbus, Ohio	1946	40,000; excess added to residential amount	–
Cranston, R.I.	1943	25,000	–
Dallas, Tex.	1960	30,000 added to domestic rate for downtown; 60,000 for tunnel relief sewers	–
Detroit, Mich.	–	50,000	–
Grand Rapids, Mich.	–	Offices, 40–50 gpd/person; hotels, 400–500 gpd/room; hospitals, 200 gpd/bed; schools, 200–300 gpd/room	250,000
Hagerstown, Md.	–	Hotels, 180–250 gpd/room; hospitals, 150 gpd/bed; schools, 120–150 gpd/room	–
Houston, Tex.	1960	Peak flows; Offices, 0.36 gpd/ft^2; retail, 0–20 gpd/ft^2; hotels, 0.93 gpd/ft^2	–
Las Vegas, Nev.	–	Resort hotels, 310–525 gpd/room; schools, 15 gpd/person	–
Los Angeles, Calif.[b]	1965	Commercial, 11,700: each; hospitals, 0.75 mgd; schools, 0.12 mgd; universities, 0.73 mgd	15,500
Los Angeles County Sanitation Dist.	1964	10,000 ave; 25,000 peak	–
Kansas City, Mo.	1958	5,000	10,000
Lincoln, Neb.	1962	7,000	–
Memphis, Tenn.	–	2,000	2,000
Milwaukee, Wis.	1945	60,500	–
St. Joseph, Mo.	1962	6,000	–
St. Louis, Mo.	1960	90,000 ave; 165,000 peak	–
Santa Monica, Calif.	–	Commercial, 9,700; hotels, 7,750	13,600
Shreveport, La.	–	3,000	–
Toledo, Ohio	1946	15,000 ave; 30,000 peak	–
Toronto, Ont.	1960	63,500, downtown sewers	–

Except as otherwise noted.
Values used for future planning; individual studies made for specific projects.
ote: gpd/acre × 0.00935 = m^3/day/hectare; gpd/ft^2 × 0.0408 = m^3/day/m^2; gal × 3.785 = liters.
ource: ASCE Manual No. 37.

cesses or methods may produce a variety of waste quantities. Therefore, if available, data from comparable developments, industries, or processes will provide the best guide to future needs. Table 11-3[1] shows the wide range of allowances made in a number of cities.

It will be noted that flows may be related to the numbers of workers, rooms, beds, or acres. Further information is given in Tables 11-4,[1] 11-5,[1] and 11-6.[1]

The strictly sanitary flows from industrial plants have been estimated at from 8 to 25 gal/person/shift. The forecast of flows to be expected from prisons, hospitals, homes for the elderly, schools and colleges, government office buildings, etc. will require special investigation of analogous institutions. Some approximation of these may be developed through use of the "fixture-unit method" described below.

TABLE 11-4. Average Commercial Flows

Type of Establishment	Avg Flow (gpd/person)
Stores, offices, and small businesses	12 to 25
Hotels	50 to 150
Motels	50 to 125
Drive-in theaters (3 persons per car)	8 to 10
Schools (no showers), 8-hr period	8 to 35
Schools (with showers), 8-hr period	17 to 25
Tourist and trailer camps	80 to 120
Recreational and summer camps	20 to 25

Note: gal × 3.785 = liters.
Source: ASCE Manual No. 37.

TABLE 11-5. Measured Flow from Apartment Projects, Houston, Texas

Stories per Structure	Dwelling Units per Acre	FLOW (GPD/ACRE)		
		Average	Peak Day	Peak Hour
2	16.6	3,430	3,690	14,900
2	20.3	4,340	5,880	15,000
15	80.3	8,720	10,200	20,000

Note: acre × 0.405 = hectare; gpd/acre × 0.00935 = m^3/day/hectare.
Source: ASCE Manual No. 37.

TABLE 11-6. Estimated Average Wastewater Flows in a Section of New York City

Contributor	Population	Water Consumption (gpd/person)	Wastewater Flow (mgd)
Residents	67,000	200	8.55
Workers	409,000	30	8.56
Transients	392,000	15	3.80

Note: gal × 3.785 = liters; mgd × 3,785 = m^3/day.
Source: ASCE Manual No. 37.

Flow Relationships. The rates of wastewater production by all classes of water users vary over a wide range during each day. Minimum flows occur between midnight and 4 AM. Maximum flows usually peak between 7 AM and 10 AM. The smaller the area and population tributary to a collector, the greater will be the relative variations during any day. Conversely, the larger the volume, the less pronounced will be the peaks.

The infiltration and possible stormwater components of the total flow are added to the sanitary waste and are not subject to diurnal fluctuations. Sewerage facilities must be designed to handle the peak sanitary flows plus infiltration and stormwater increments at rates reflecting their respective peaks, since these infiltration and stormwater increments could coincide with periods of maximum sewage production.

While maximum flows may be 6 or more times the average daily rate, a generally accepted rule of thumb for domestic sewage from smal areas provides:

Maximum daily flow = 2 × average daily flow
Maximum hourly flow = 3 × average daily flow
Minimum daily flow = $\frac{2}{3}$ average daily flow
Minimum hourly flow = $\frac{1}{3}$ average daily flow

A general relationship has been developed fo North American application as follows:[7]

$$\frac{Q_{max}}{Q_{ave}} = \frac{(18 + \sqrt{P})}{(4 + \sqrt{P})} \qquad (1$$

where Q_{max} is the maximum rate of domest flow, Q_{ave} is the average rate of domestic flov and P is population in thousands.

For conditions in the District of Columbi the ratios of peak sewage flows to average flo are shown in Fig. 11-3.[6] The ratio of extren flows to average daily flow in New England illustrated in Fig. 11-4.[1]

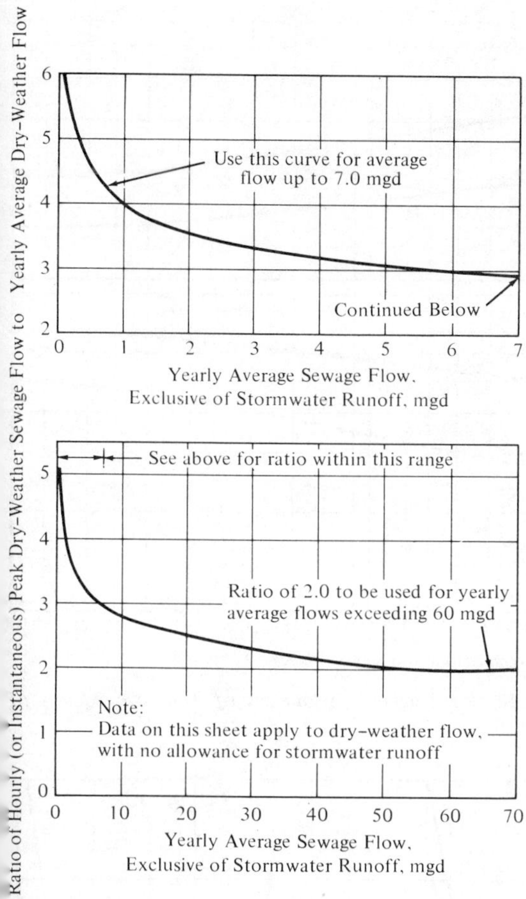

Fig. 11-3 Ratio of peak sewage flow to average flow exclusive of stormwater runoff. (*Courtesy of Department of Environmental Services, District of Columbia.*)

A Typical Computation. The tributary geographic area and population to be served by any component of a sewerage system will be determined by topography, land use, piping layout, and possibly limits of political jurisdiction. The sizing of that system component should be governed by the maximum flows to be expected during the period of design. Figure 11-5 illustrates typical design flows for separate sewers serving residential areas.[6] While the values may not correspond with those applicable to other cities, they are used here to illustrate the method by which similar tables may be developed.

The design hourly peak flow for a tributary area of 1000 acres having a population density of 30 persons/acre is developed assuming:

Domestic flows = 65 gpd/person
Commercial flows = 15 gpd/person
Infiltration = 700 gpd/acre

Average daily flow rates:

Domestic
1,000 X 30 = 30,000 at 65 = 1.95 mgd
Commercial
1,000 X 30 = 30,000 at 15 = 0.45 mgd
Infiltration
1,000 X 700 = 0.70 mgd
Yearly average dry-weather flow = 3.10 mgd

Using Fig. 11-3, it can be determined that for a yearly average dry-weather flow of 3.10 mgd, the peak hourly dry-weather flow has a ratio of 3.35.

Thus, the peak dry weather flow = 3.35 X 3.10 = 10.4 mgd.

Going to Fig. 11-2 for stormwater allowance for 1000 acres with 30 persons/acre, a flow of 6.4 mgd is indicated. The design hourly peak flow = 10.4 + 6.4 = 16.8 mgd. The per capita hourly peak flow becomes 16.8 mgd/30,000 = 560 gpd/person. In this example, with a yearly domestic usage of only 65 gpd/person, the peak hour (reflecting infiltration and stormwater) attains a rate of 560 gpd/person, a ratio of 8.65. The stormwater increment due only to areaways and depressed driveways is 38 percent of the total load. In the study cited, the minimum design peak for the whole or any part of the system was set at 300 gpd/person. The per capita flows used in a number of American cities as the basis of design are given in Table 11-1.

Fixture-Unit Method of Design. Institutional and some commercial peak sewage flows may be approximated by use of the fixture-unit method. The United States of America Standards Institute National Plumbing Code, USASI A40.8-1955, defines the "fixture-unit flow rate" as "the total discharge flow in gallons per minute of a single fixture divided by 7.5, which provides the flow rate of that particular plumbing fixture as a unit of flow." Table 11-7 gives the fixture-unit values assigned to the various fixtures usually encountered.

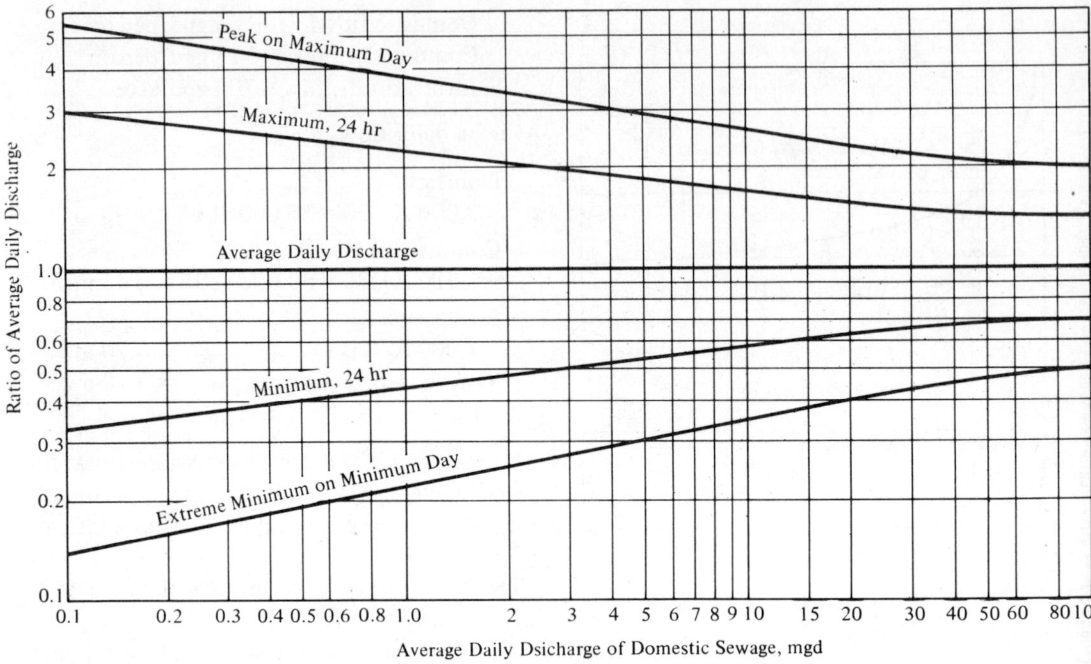

Fig. 11-4 Ratio of extreme flows to average daily flow in New England. (*Courtesy of American Society of Civil Engineers.*) *Source:* ASCE Manual No. 37.

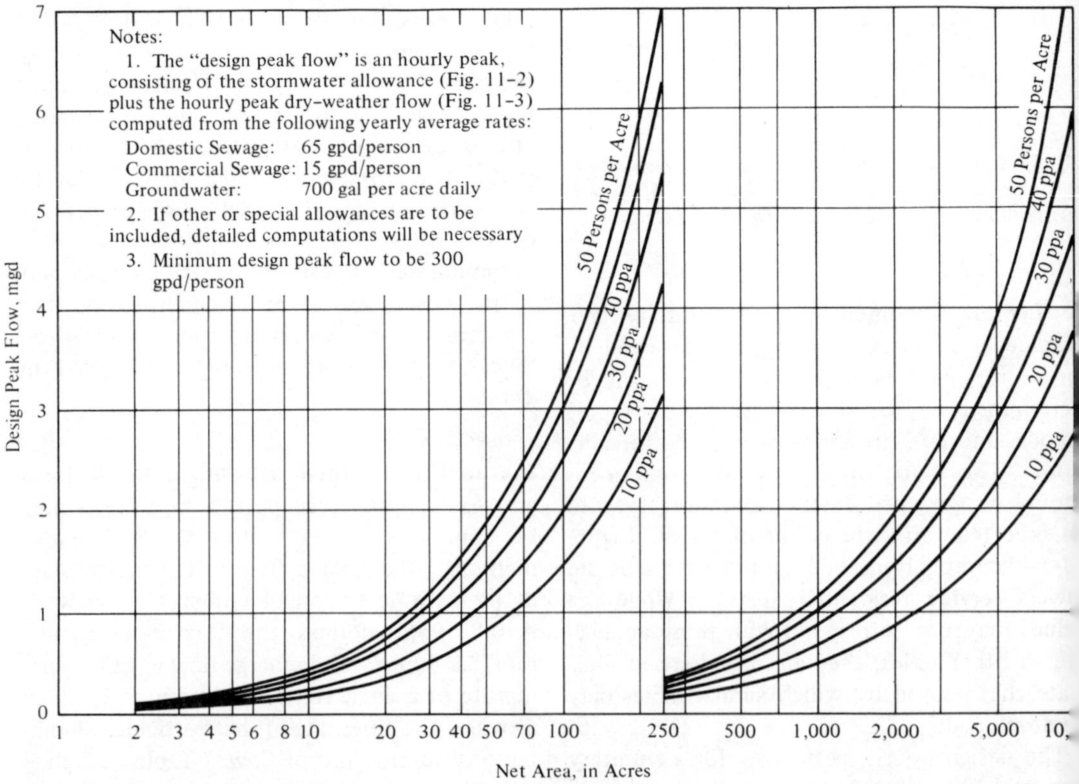

Fig. 11-5 Typical design flows for separate sewers serving residential areas. (*Courtesy of Department Environmental Services, District of Columbia.*)

TABLE 11-7. Fixture-Units per Fixture or Group[a]

Fixture Type	Fixture-Unit Value as Load Factors
1 bathroom group consisting of tank-operated water closet, lavatory, and bathtub or shower stall	6
Bathtub[b] (with or without overhead shower)	2
Bidet	3
Combination sink-and-tray	3
Combination sink-and-tray with food-disposal unit	4
Dental unit or cuspidor	1
Dental lavatory	1
Drinking fountain	$\frac{1}{2}$
Dishwasher, domestic	2
Floor drains	1
Kitchen sink, domestic	2
Kitchen sink, domestic, with food waste grinder	3
Lavatory	1
Lavatory	2
Lavatory, barber, beauty parlor	2
Lavatory, surgeon's	2
Laundry tray (1 or 2 compartments)	2
Shower stall, domestic	2
Showers (group) per head	3
Sinks	
Surgeon's	3
Flushing rim (with valve)	8
Service (trap standard)	3
Service (P trap)	2
Pot, scullery, etc.	4
Urinal, pedestal, syphon jet, blowout	8
Urinal, wall lip	4
Urinal stall, washout	4
Urinal trough (each 2-ft section)	2
ash sink (circular or multiple), each set of faucets	2
ater closet, tank-operated	4
ater closet, valve-operated	8

From United States of America Standards Institute ational Plumbing Code, USASI A40.8–1955.
A shower head over a bathtub does not increase the xture value.
ote: For a continuous or semicontinuous flow into drainage system, such as from a pump, pump ejector, r-conditioning equipment, or similar device, two xture units shall be allowed for each gpm of flow.
urce: ASCE Manual No. 37.

An apartment or single-family house with one throom, combination sink-and-tray with food-posal unit, and dishwasher totals 12 fixture-its. Households average between 3 and 4 persons each, so an allowance of between 3 and 4 fixture-units per capita would be reasonable.

For the facility or area under investigation, an estimate of the total fixture-units to be in service should be determined. It will be apparent, however, that all such units will never be in use at maximum rates simultaneously. Accordingly, the aggregate peak rates to be anticipated have been estimated on the basis of probability and have been plotted as shown in Fig. 11-6.[1]

While this method can be a valuable tool, it is not absolute. The designer should make every effort to compare the results obtained in this manner with estimates based on measurements where possible, or other conventional means as discussed elsewhere in this chapter.

STORMWATER RUNOFF

General. The source of all natural runoff is precipitation as rain, sleet, or snow. Considering the entire earth's surface, the total precipitation is equaled by the total evaporation. In any local setting, however, this equilibrium will not be apparent. Storm runoff is the portion of rainfall which is not lost by infiltration or evaporation and is not retained by surface irregularities or absorption. In most terrain it will flow to natural watercourses and augment the normal groundwater runoff or other waters carried therein.

The surfaces on which rain falls will govern the amount which runs off and the rapidity of its travel across such surfaces. Also, the total runoff and its rate from the same areas may vary from different storms of apparently similar intensities but having different patterns. The wide range of climate across the United States produces extensive differences in the amount and characteristics of rainfall in different localities. For each locality, experience will show characteristic rainfalls which may be correlated as to frequency, intensity, and duration, each of which must be a significant consideration in a stormwater system design.

The history of rainfall at any location as determined by use of rain gauges over an extended period will provide the most dependable basis for estimating stormwater production.

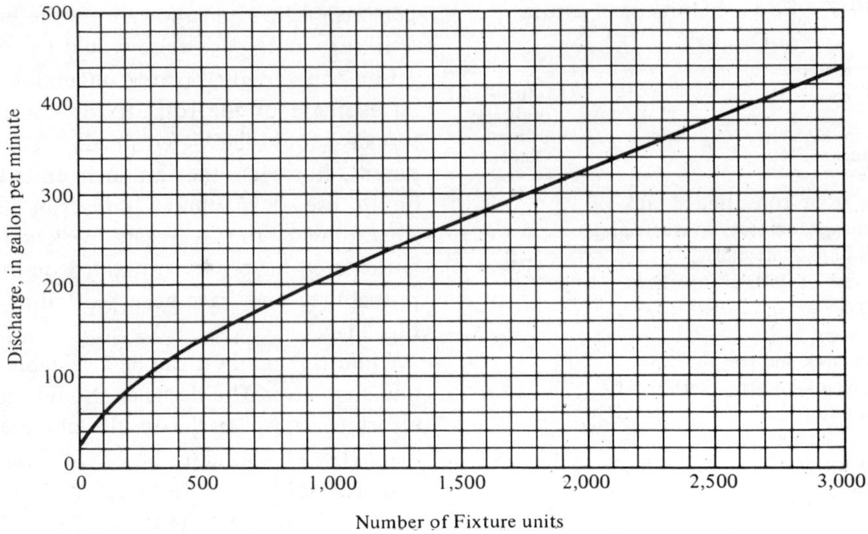

Fig. 11-6 Relation of peak discharge to fixture units. (*Courtesy of American Society of Civil Engineers.*) *Source:* ASCE Manual No. 37.

The U.S. Weather Service has accumulated records of this kind for much of the country.

Rainfall Intensity and Duration. The rate at which rain falls, expressed in inches per hour, is designated as "intensity." The most intense rainfall usually occurs near the beginning of a storm, but may develop at any time thereafter. The intensity is highest for short periods and decreases with the length of the period of observation. The patterns of the most intense storms experienced in Washington, D.C. between 1897 and 1945 are shown in Fig. 11-7.[6]

The curves usually plotted to represent the intensity-duration relationship portray the arithmetic mean rate of rainfall for any point of duration selected. Figure 11-8 is the typical form.[1]

Such curves will result from application of the equation:

$$i = \frac{A}{t + b} \quad \text{or} \quad i = \frac{A}{(t + b)^n} \tag{2}$$

where i is the average intensity, t is duration, A and b are abstract constants, and n is an exponent.[1] The plotting of data from rainfall observations will establish points in close agreement with an equation of this form. The Washington curve for 15-year frequency becomes:

$$i = \frac{66.6}{(t + 13.2)^{0.75}}$$

Emphasizing the variations between different localities, other values have been employed as follows:[8,9]

A. N. Talbot, Maximum storms in eastern United States $i = 360/t + 30$

A. N. Talbot, Ordinary storms in eastern United States $i = 105/t + 30$

W. W. Horner, St. Louis, Mo., *Engineering News*, August 29, 1910 $i = 56/(t + 5)^{0.85}$

The observed and calculated intensities usually relate to a point or place at which they occur. Since rainstorms are not uniform within themselves, having variable patterns and formations, the application of the point intensities t large areas will lead to excessive values. There fore, some reduction in the intensities to b anticipated in stormwater control facilities ma safely be assumed. The larger the tributar area, the larger the reduction, or the small the percentage of the point intensity which wi be applicable. Figure 11-9 shows the distribu tion of rainfall intensity with respect to th area covered and the duration of the storm developed for Washington, D.C. It should b noted that the longer the duration of the storm the more nearly the intensity, even in larg

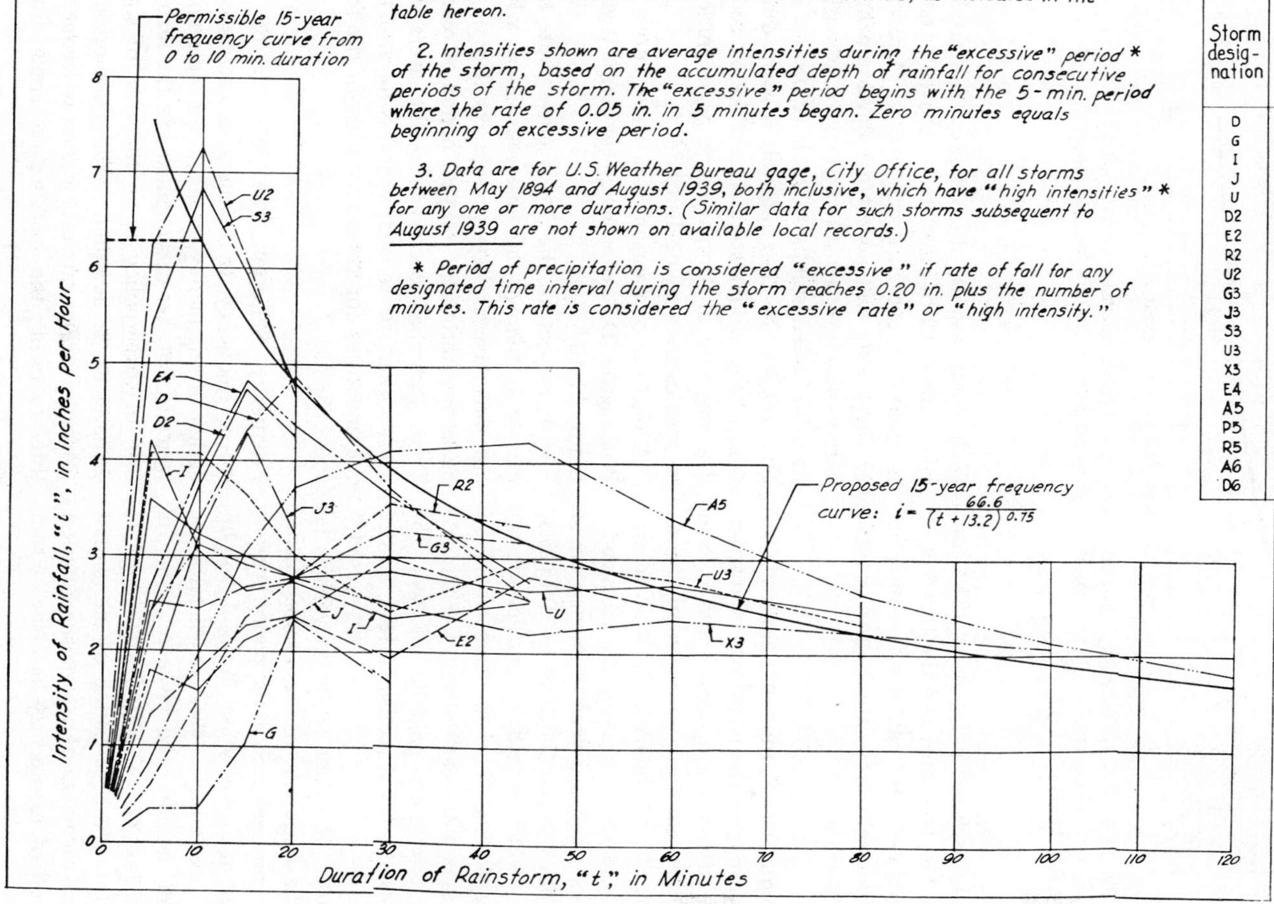

NOTES:

1. Letters and numbers denote the individual storms, as indicated in the table hereon.

2. Intensities shown are average intensities during the "excessive" period * of the storm, based on the accumulated depth of rainfall for consecutive periods of the storm. The "excessive" period begins with the 5-min. period where the rate of 0.05 in. in 5 minutes began. Zero minutes equals beginning of excessive period.

3. Data are for U.S. Weather Bureau gage, City Office, for all storms between May 1894 and August 1939, both inclusive, which have "high intensities" * for any one or more durations. (Similar data for such storms subsequent to August 1939 are not shown on available local records.)

* Period of precipitation is considered "excessive" if rate of fall for any designated time interval during the storm reaches 0.20 in. plus the number of minutes. This rate is considered the "excessive rate" or "high intensity."

TABLE

Storm designation	Date of storm	Amount of rainfall prior to beginning of excessive rate, inches
D	Aug. 10, 1897	Trace
G	Sept. 25, 26, 1899	0.52
I	June 2, 1900	0.04
J	June 8, 1900	Trace
U	July 5, 6, 1905	0.01
D2	Aug. 21, 1906	Trace
E2	Aug. 24, 1906	Trace
R2	July 14, 1912	0.02
U2	July 30, 1913	0.01
G3	June 14, 1917	0.02
J3	July 14, 1917	0.03
S3	July 15, 1921	0.01
U3	July 13, 1922	0.02
X3	Sept. 2, 1922	0.66
E4	Sept. 2, 1926	0.02
A5	Sept. 12, 1934	0.08
P5	Sept. 4, 1939	---- See Note 3
R5	July 23, 1940	---- See Note 3
A6	Aug. 2, 1944	---- See Note 3
D6	July 2, 1945	---- See Note 3

Fig. 11-7 Intensities of rainfall based on cumulative amounts for most intense storms. (*Courtesy of Department of Environmental Services, District of Columbia.*)

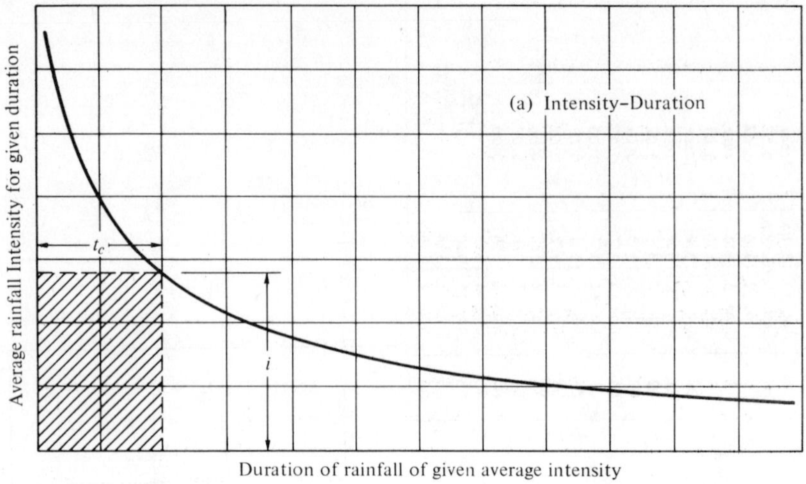

Fig. 11-8 Rainfall-intensity duration curve; i = average intensity of rainfall, t_c = time of concentration. (*Courtesy of American Society of Civil Engineers.*) *Source:* ASCE Manual No. 37.

areas, will approach the point value. The configuration of the land area and the location of the stormwater control facilities will require the exercise of judgment in the application of these intensity reductions.

Rainfall Frequency. Storms of moderate or nominal intensity are the usual case. Occasionally a more intense storm occurs, and at even more infrequent intervals even greater storms are experienced. Mathematically at least, the greater the intensity the lesser the frequency. Thus, a storm with a 2-year frequency could be equaled or exceeded 50 times in 100 years, and a storm which would be equaled or exceeded only 4 times in 100 years would be designated as a "25-year" storm. While improbable, it is entirely possible for two "25-year" storms to occur consecutively in the same season.

For purposes of design, it is convenient to determine the frequency or the number of occurrences in any selected time period that a rainfall of given intensity and duration will be equaled or exceeded. Weather Service or other rainfall data for the place or point of interest for the entire period of record is essential. From this, a listing of all rainfall intensities for each step of progressive duration for the period of record may be made. From such compilations, the number of occasions when any level of intensity for a given duration is equaled or exceeded will emerge. A plotting of these data will permit the development of curves which relate successions of intensity–duration relationships to their respective frequencies, or the time intervals in years during which they would not normally be exceeded. The majority of storms will be of insignificant intensities and may be disregarded.

A plotting of data and calculated curves for Washington, D.C. over a 60-year period is given in Fig. 11-10.[6]

With respect to the 2-year curve, note how closely it follows the observed values, which are indicated as having been equaled on 30 occasions during the 60-year period of record. Thus, they had occurred on the average of every other, or second, year. Note also that the equations for these curves are of the form shown by Eq. 2.

Frequency Selection. The selection of the storm frequency to be used as the basis of design of any facility intended to handle stormwater is a matter of judgment and the recognition of local policy as to the risks of storm damage to be accepted. The assumption of storms of higher intensity and lesser frequency, thus requiring the handling of larger flows,

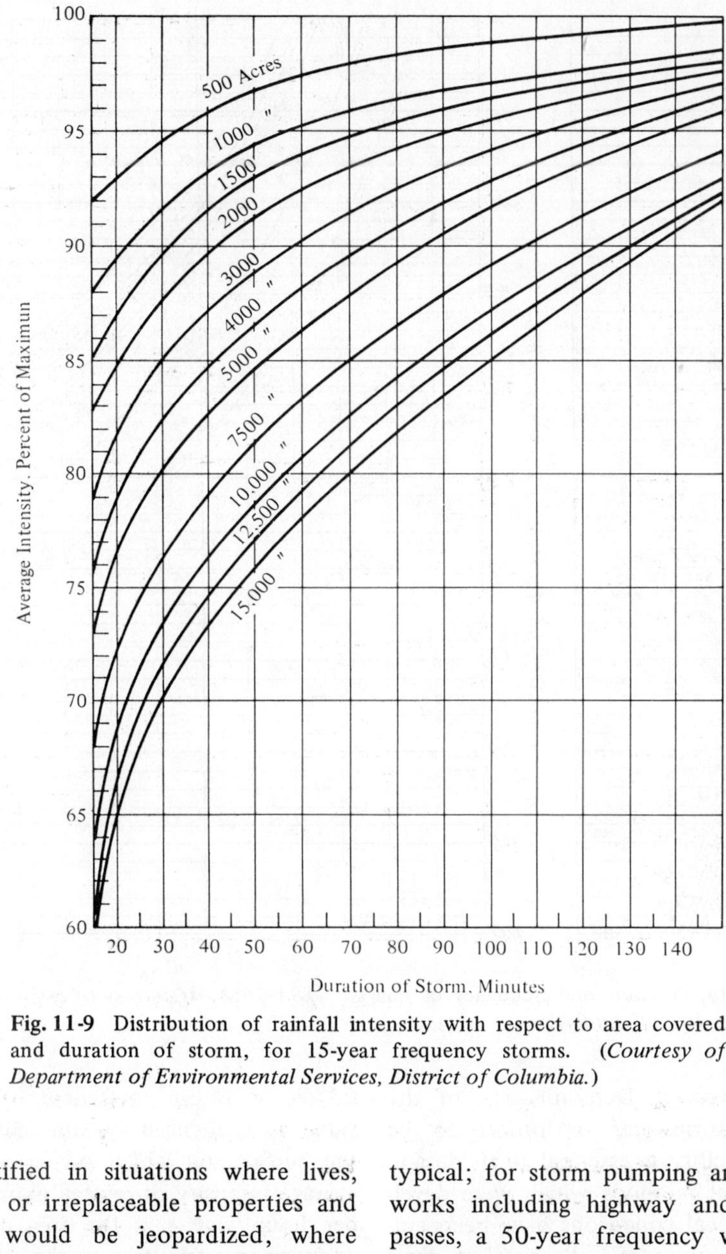

Fig. 11-9 Distribution of rainfall intensity with respect to area covered and duration of storm, for 15-year frequency storms. *(Courtesy of Department of Environmental Services, District of Columbia.)*

would be justified in situations where lives, public health, or irreplaceable properties and art treasures would be jeopardized, where public transportation, power, communications, water or waste treatment works, etc. would be seriously disrupted, or where the later reinforcement of such planned stormwater facilities would be unduly expensive, if not impossible.

For residential areas, from a 2- to 5-year frequency is customary, except in a combined system, in which case not less than 15 to 25 years are required to reduce likelihood of basement flooding. For commercial and high-value districts, frequencies from 10 to 50 years are typical; for storm pumping and flood control works including highway and railway underpasses, a 50-year frequency value is used for both pumps and discharge piping.

In undertaking any given design, a comparison of the effect on the pipe sizes of a 5-year frequency value and of a 15-year frequency value, for example, may show the difference to be so slight that the cost savings would not justify the increased flood hazard under the circumstances. This may not be of concern in speculative development, but should be an important consideration in setting municipal policy.

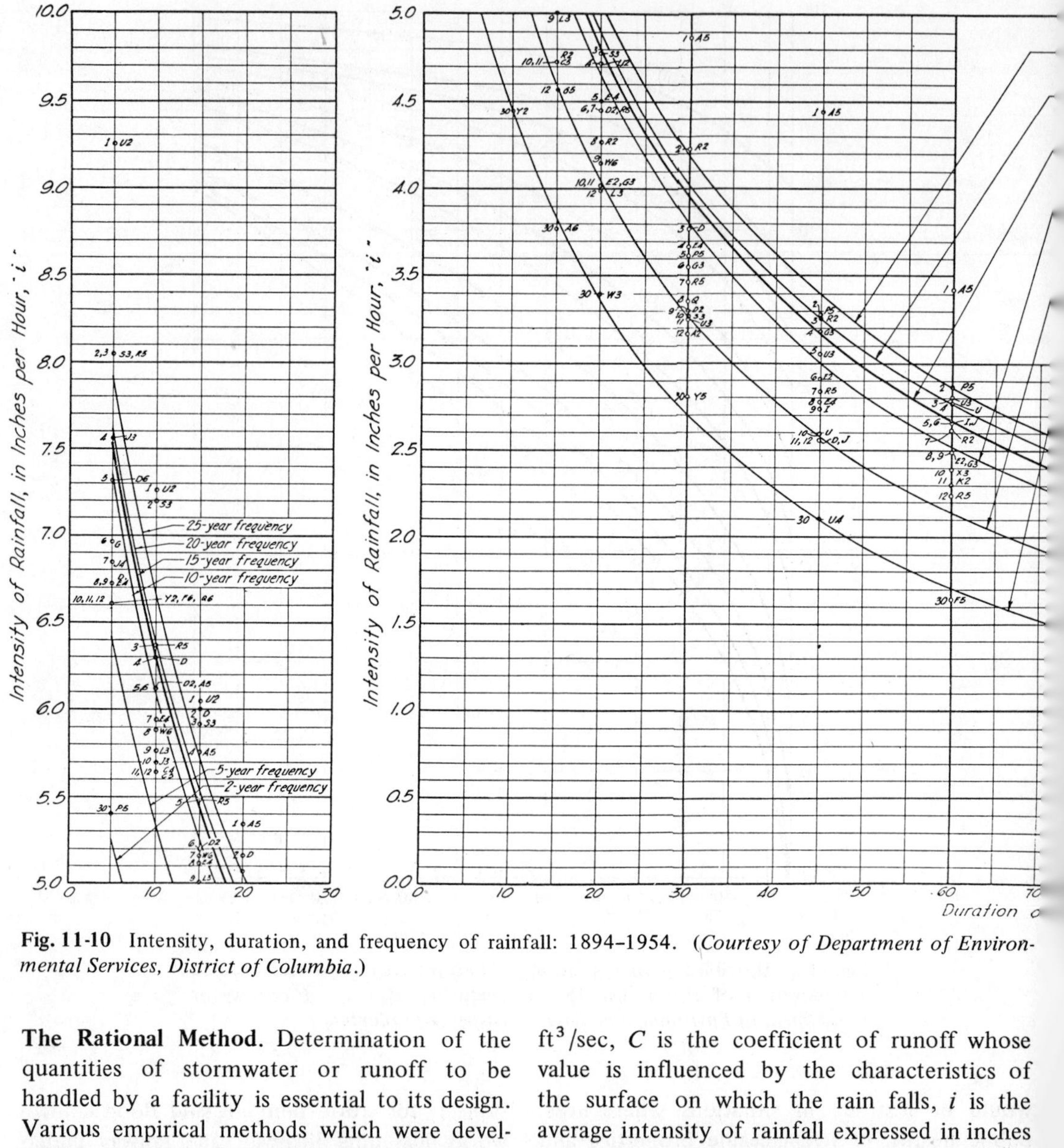

Fig. 11-10 Intensity, duration, and frequency of rainfall: 1894–1954. (*Courtesy of Department of Environmental Services, District of Columbia.*)

The Rational Method. Determination of the quantities of stormwater or runoff to be handled by a facility is essential to its design. Various empirical methods which were developed to meet local conditions have been employed for this purpose. Because of their unique applications, they are not suited to general use.

One method, however, known as the "rational method," has enjoyed widespread use in the United States and the United Kingdom for many years. The runoff to be handled is related to rainfall intensity by the formula:

$$Q = CiA \tag{3}$$

where Q is the rate of runoff expressed in ft^3/sec, C is the coefficient of runoff whose value is influenced by the characteristics of the surface on which the rain falls, i is the average intensity of rainfall expressed in inches per hour, and A is the area in acres of the drainage area tributary to the facility. By convenient coincidence, and using conventional English units, a runoff rate of about 1 ft^3/sec is produced by the discharge of 100 percent ($C = 1$) of the rain falling on 1 acre at the rate of 1 in./hr.

Use of the formula is generally limited to urban areas not exceeding 5 mi^2. While "correct" for larger applications, the uncertainties in selection of the appropriate coefficient and intensity values generally result in substantial over-design.

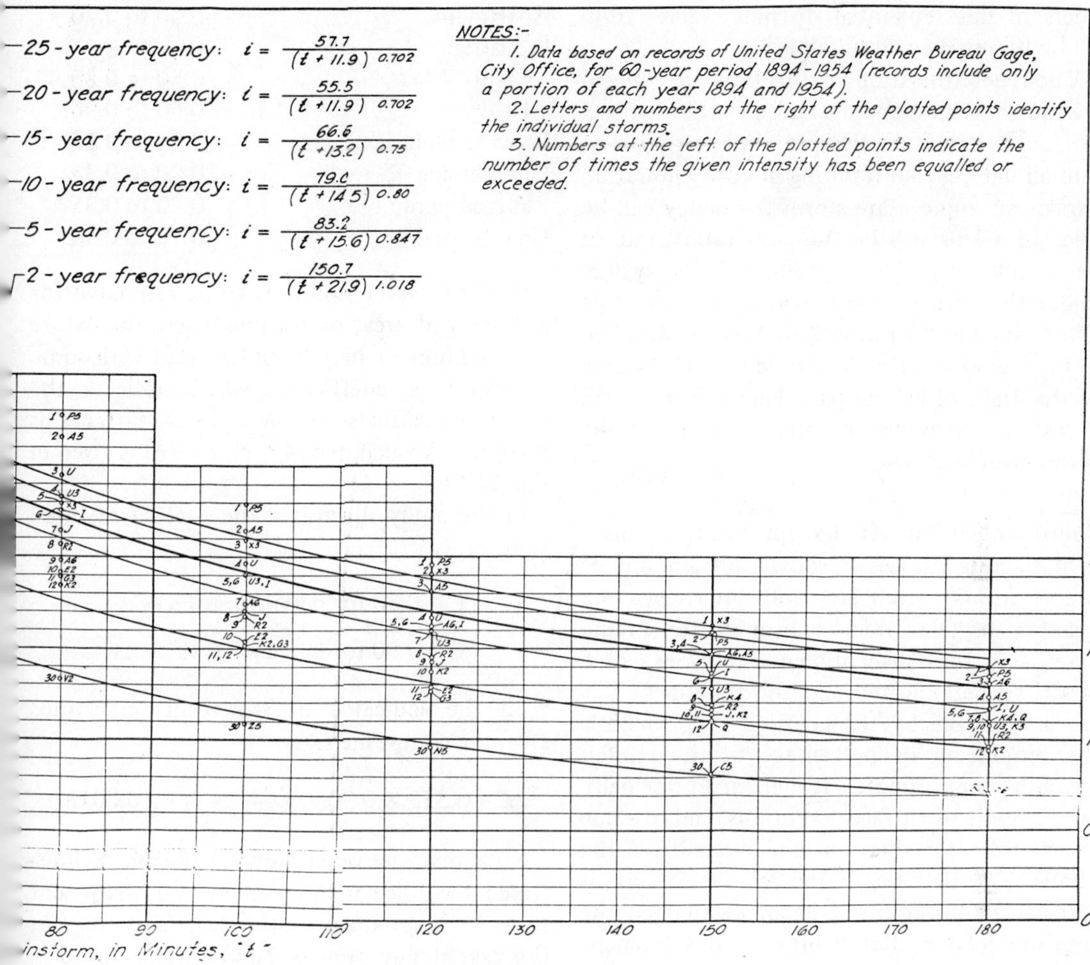

Fig. 11-10 *(Continued)*

The area tributary to a stormwater system may be determined from surveys, maps, or aerial photography. The topography and present or proposed street layout will determine the general outline of the system from its most remote inlet to its point of discharge. Even though built progressively over a span of years in possibly noncontiguous segments, the ultimate loading to be handled must be envisioned for each part when designed. For each location in the system where design is to be undertaken, a "time of concentration" must be determined. This will be the time for the runoff from the most remote part of the drainage area to reach the inlet intended to receive it, and then flow through the system to that point under design. Obviously, the time of concentration will be peculiar to each element, as will be the tributary area. Water falling on the most remote place may reach the point of design before water from another even closer area because of topography and system layout.

The time for water to flow over the surface to the first inlet is called the "inlet time." This will vary with slope and character of surface, rate of absorption into the soil, paving, etc., and catchment in depressions. Higher intensities tend to reduce the inlet time.

Inlet time for central city conditions with much paving and closely-spaced inlets could be assumed at not more than 5 min. With flat slopes in medium development, values from 10 to 15 min are commonly used. With infrequent inlets in flat residential districts, values from 20 to 30 min may be suitable.[2]

Upon determination of the time of concentration, a corresponding rainfall intensity will be found based on the prior investigation of rainfall data. From resulting intensity-duration curves, an appropriate storm frequency can be selected. This will be the peak runoff rate to be handled by that segment of the system under the design, thus becoming the value of i to be used in the rational method of computation. The value of A to be used will be the area of the drainage area or part thereof from which flows to be handled by that segment of the system will originate.

Coefficient of Runoff. Except during a coincident melting of ice or snow, the amount of runoff from a given area will not exceed the quantity of rain which falls upon it. In the absence of these exceptional conditions, some loss is to be expected and the amount of this loss is reflected by the coefficient C, which is the proportion or percentage of the rainfall which becomes runoff. It will vary principally as the result of surface conditions, and to some extent from the duration and intensity of the rainfall. Evaluation of the relative effects of surface conditions must be predicated on the land use to prevail after ultimate development of the watershed, unless the facility is intended for interim service only.

The principal loss of rainfall is from infiltration into permeable soil. Lesser losses result from evaporation, transpiration, and retention in depressions or areas with vegetation or forest cover. These losses are of less significance in an urban setting.

Average coefficients for various kinds of land use are frequently employed:[1]

Description of Area	Runoff Coefficients
Business	
Downtown	0.70 to 0.95
Neighborhood	0.50 to 0.70
Residential	
Single-family	0.30 to 0.50
Multi-units, detached	0.40 to 0.60
Multi-units, attached	0.60 to 0.75
Residential (suburban)	0.25 to 0.40
Apartment	0.50 to 0.70
Industrial	
Light	0.50 to 0.80
Heavy	0.60 to 0.90
Parks, cemeteries	0.10 to 0.25
Playgrounds	0.20 to 0.35
Railroad yard	0.20 to 0.35
Unimproved	0.10 to 0.30

If storm water runoff is to be estimated for a developed area, or for one where the nature of development may be anticipated with some assurance, a coefficient which reflects the composite effects of the various surfaces is feasible. An example of such an area is given in Fig. 11-11.[6]

In the study illustrated the coefficients used were:

$$C = 0.30 \text{ for pervious surfaces}$$

$$C = 0.90 \text{ for impervious surfaces}$$

With the indicated 35.3 percent impervious area, the composite is:

$$C = 0.353 \times 0.90 + 0.647 \times 0.3 = 0.6018$$

Note that the open land on the block illustrated was filled with an assumed development so as to approximate ultimate conditions. In this case an inlet time of 10 min was assumed.

Coefficients for various surface types fall into the following ranges:[1]

Character of Surface	Runoff Coefficients
Pavement	
Asphaltic and concrete	0.70 to 0.95
Brick	0.70 to 0.85
Roofs	0.75 to 0.95
Lawns, sandy soil	
Flat, 2 percent	0.50 to 0.10
Average, 2 to 7 percent	0.10 to 0.15
Steep, 7 percent	0.15 to 0.20
Lawns, heavy soil	
Flat, 2 percent	0.13 to 0.17
Average, 2 to 7 percent	0.18 to 0.22
Steep, 7 percent	0.25 to 0.35

The listed values are for 5- to 10-year frequency storms. More intense, less frequent storms would require some upward adjustment

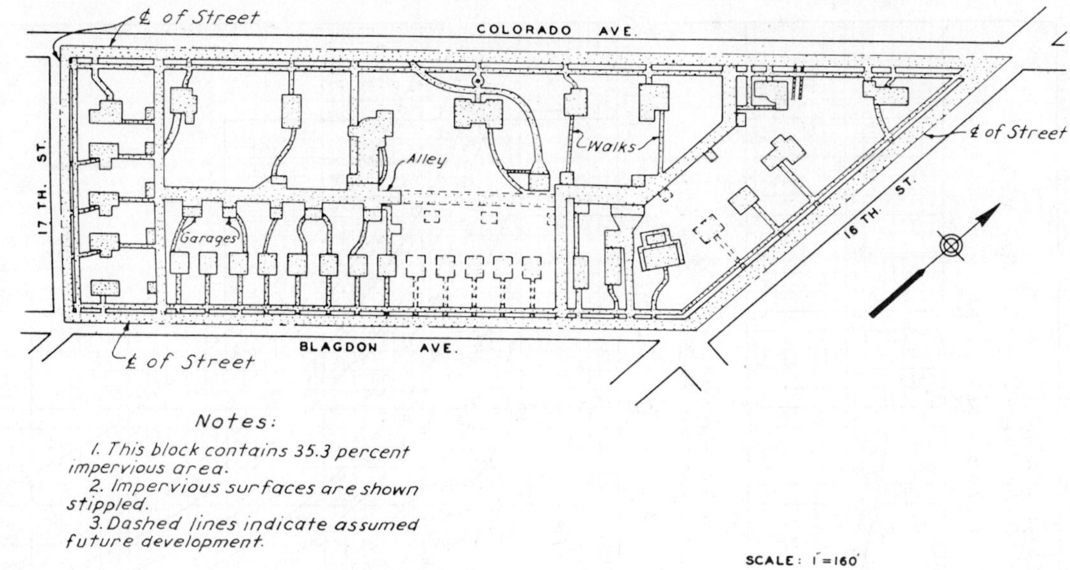

Fig. 11-11 Plan showing impervious surfaces. *(Courtesy of Department of Environmental Services, District of Columbia.)*

It may be found convenient to develop runoff curves which show the product *Ci* for various ratios of pervious to impervious surfaces, based on the design frequency which has been selected. Such curves are shown in Fig. 11-12.[6]

It will be apparent that selection of the coefficient to be used in the rational method is the most uncertain element in the procedure. Quite apart from accurate assumptions as to the nature and extent of the surfaces and their effects on runoff is the further influence of the form and timing of the storm itself. If the "design" storm has been preceded by other rainfall so that absorption and percolation are reduced and areas of surface retention are already filled, or even worse, the ground is frozen, the proportion of runoff will be greater than the conventional coefficient values would indicate. Also, the longer the storm lasts, the fewer the losses as the ground and other surfaces become saturated and percolation is reduced. A reflection of these effects is given in Fig. 11-13.[1,9]

The coefficients indicated above should not be used directly. The time of concentration and the establishment of design flows probably will not start with the beginning of rainfall; the critical loading may well occur somewhat later.

SEPARATE AND COMBINED SEWER SYSTEMS

Description–Separate Sewer Systems. A sanitary sewer is intended to receive and convey sewage from residences, places of business, institutions, and possibly industrial establishments without the admixture of stormwater, except in limited quantities under unusual circumstances. It carries the wastes described under "Sanitary Sewage Collection." In urban settings it should extend to the limits of land development to provide service throughout its tributary area. Flows are conducted to the point of disposal, which under modern practice is a treatment facility.

A stormwater sewer or drain is intended to receive and convey primarily surface water as storm runoff. In practice, such drains may properly be used for disposal of clean process water (as from air-conditioning, condensing, etc.), filtered water reservoir and tank overflows, swimming pool and footing drainage, and subsoil and groundwater seepage. They are designed to collect water accumulating in gutters, streams, and so forth, ideally while the volume remains manageable. Such drains are rarely extended to the extremities of their tributary areas and ordinarily discharge to the

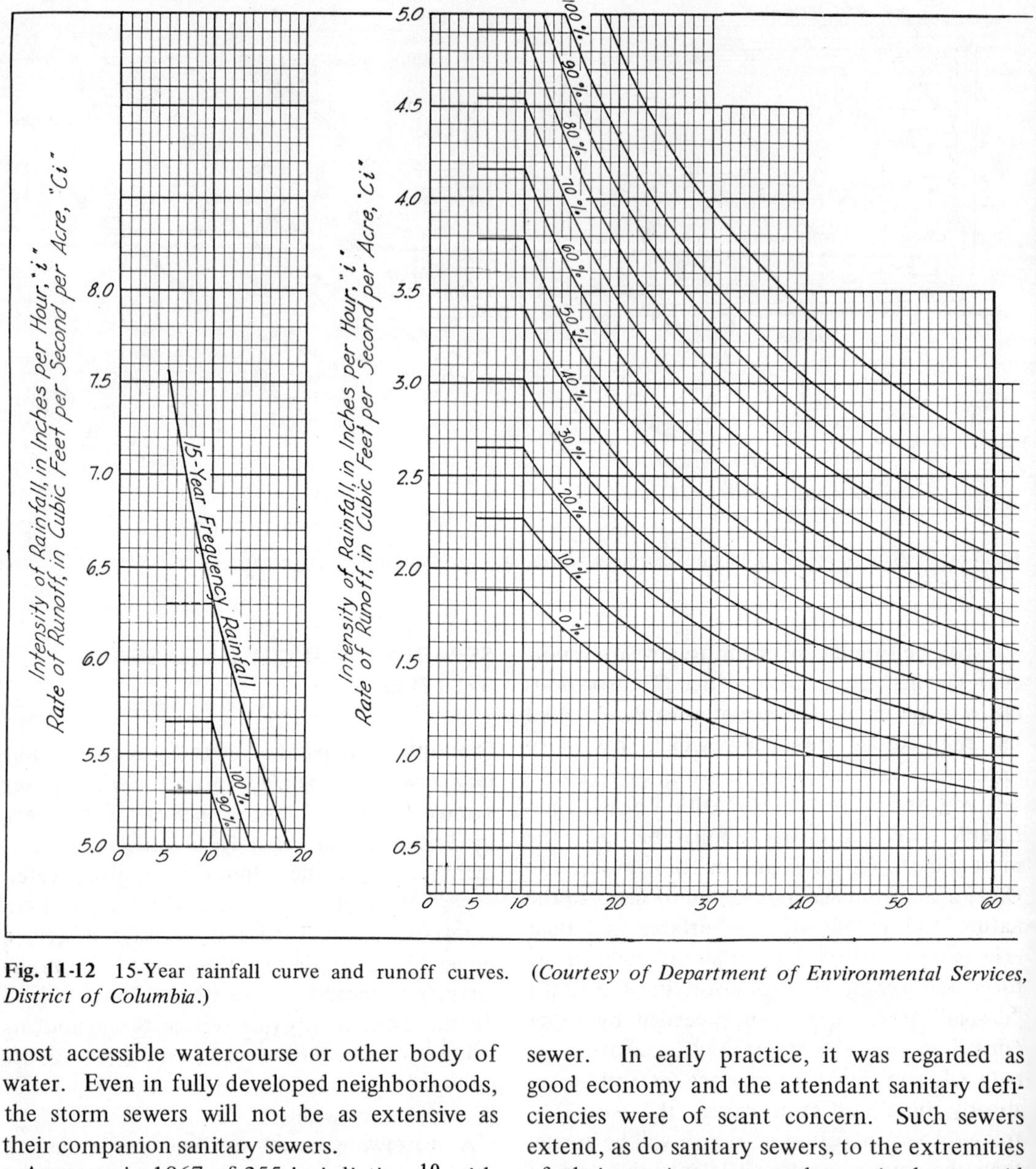

Fig. 11-12 15-Year rainfall curve and runoff curves. *(Courtesy of Department of Environmental Services, District of Columbia.)*

most accessible watercourse or other body of water. Even in fully developed neighborhoods, the storm sewers will not be as extensive as their companion sanitary sewers.

A survey in 1967 of 255 jurisdictions[10] with separate sewer systems showed the relative lengths of the two systems as follows:

System Type	Length (miles)	Population (millions)	Area Served (acres)
Sanitary	44,286	15.4	2,088,000
Storm	17,151	9.5	1,329,000

Description–Combined Sewer Systems. A combined sewer is intended to perform the functions of both a sanitary sewer and a storm sewer. In early practice, it was regarded as good economy and the attendant sanitary deficiencies were of scant concern. Such sewers extend, as do sanitary sewers, to the extremities of their service areas and are sized to avoid danger of backup into basements during times of peak storm runoff. Since flow of the sanitary component is continuous, the combined sewer system will extend to the point of disposal for the sanitary wastes, probably the treatment facility. Inasmuch as the quantity of storm flow may be from 50 to 100 times greater than the sanitary flow, it is impracticable to attempt to carry any significant proportion of these peaks beyond the points of convenient release to receiving waters. Also, treatment and

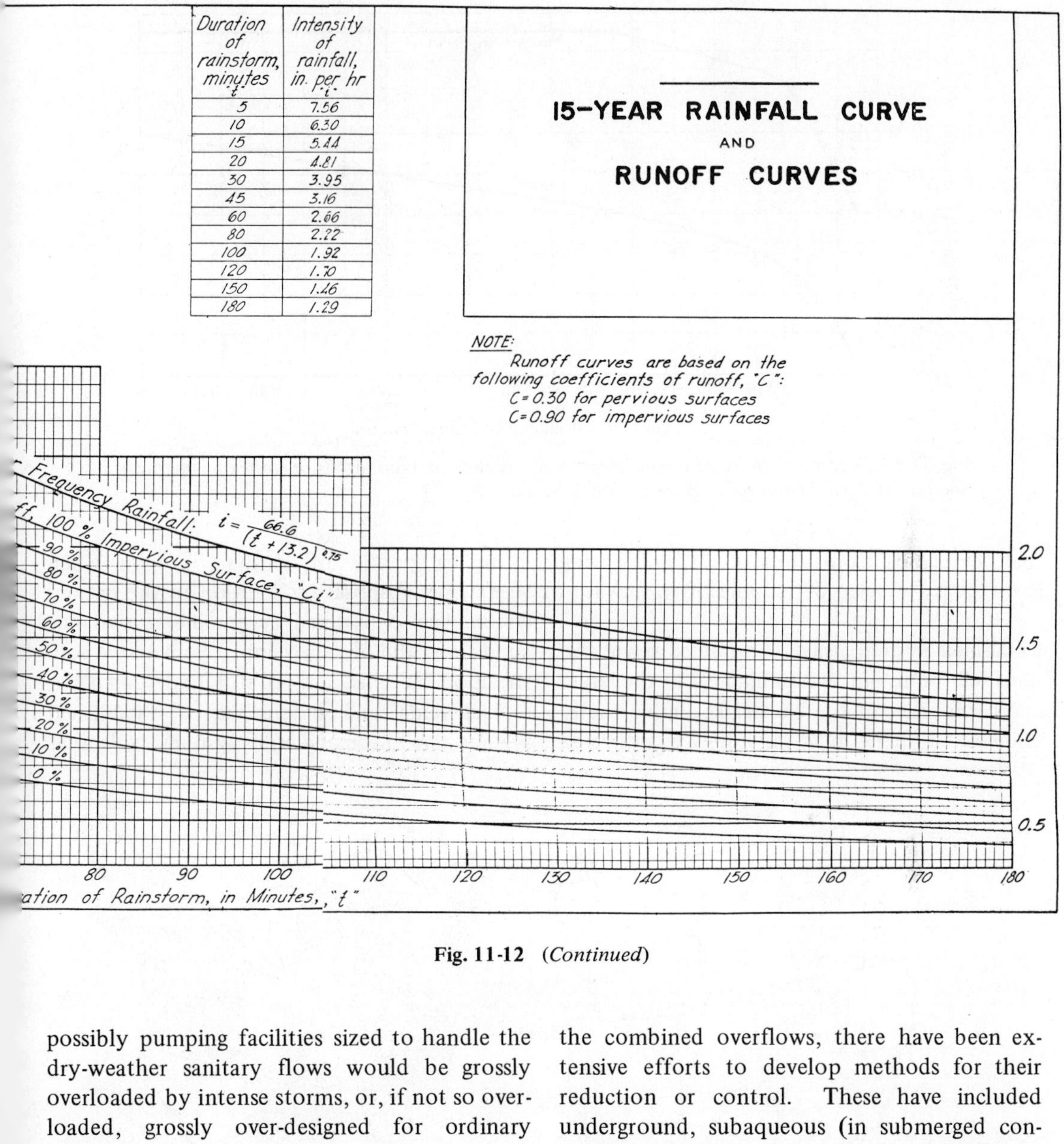

Fig. 11-12 (*Continued*)

possibly pumping facilities sized to handle the dry-weather sanitary flows would be grossly overloaded by intense storms, or, if not so overloaded, grossly over-designed for ordinary conditions.

The flows retained after relief at the overflow points are usually intercepted by an express connection to the treatment facility. This line is known as an interceptor. Since the arteries of the combined system will operate independently within their respective drainage areas, the interceptor is usually located so as to carry all dry-weather flow plus some predetermined proportion of the storm flows from the several arterial components.

Recognizing the polluting characteristics of the combined overflows, there have been extensive efforts to develop methods for their reduction or control. These have included underground, subaqueous (in submerged containers), or surface retention for some part of the peak flows from normal rainfalls, with later release to the treatment facility. In large systems, use may be made of the storage capacity of the major sewers to hold back part of these mixtures, so a greater proportion of the total may be treated. Also, special tanks for settling and disinfection before release are used. The most costly of these efforts is the entire separation of the combined system into dual systems of sanitary and storm sewers, including the alteration of all plumbing on private property so

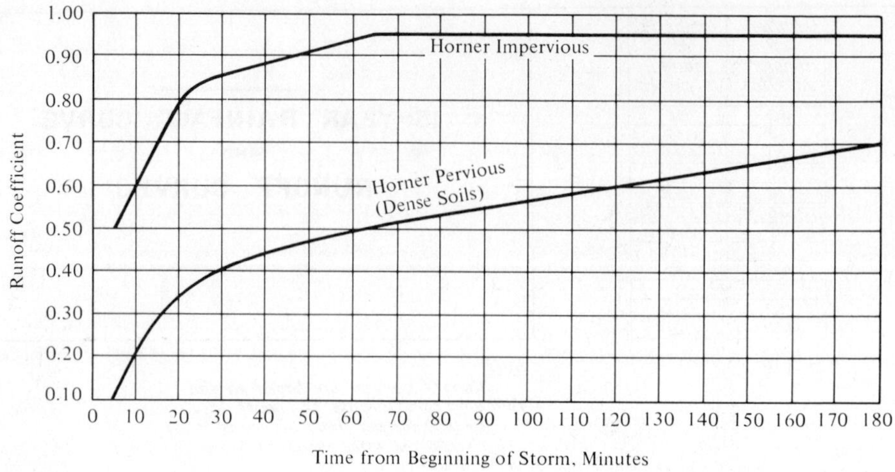

Fig. 11-13 Variation in runoff coefficient with period of rainfall. *(Courtesy of American Society of Civil Engineers.)* *Source:* ASCE Manual No. 37.

that the runoff from roofs, yards, and paving is no longer mingled with domestic wastewater.

Combined System Usage. Combined sewer systems will be found predominantly in the larger and older cities of the Northeast and northerly Midwest. For the most part, they were started during a period of national growth when wastewater disposal by dilution in the natural waters was considered acceptable if nuisance conditions did not develop. Some new construction of this sort is still undertaken for replacements and extensions of established systems. While many such systems evolved as

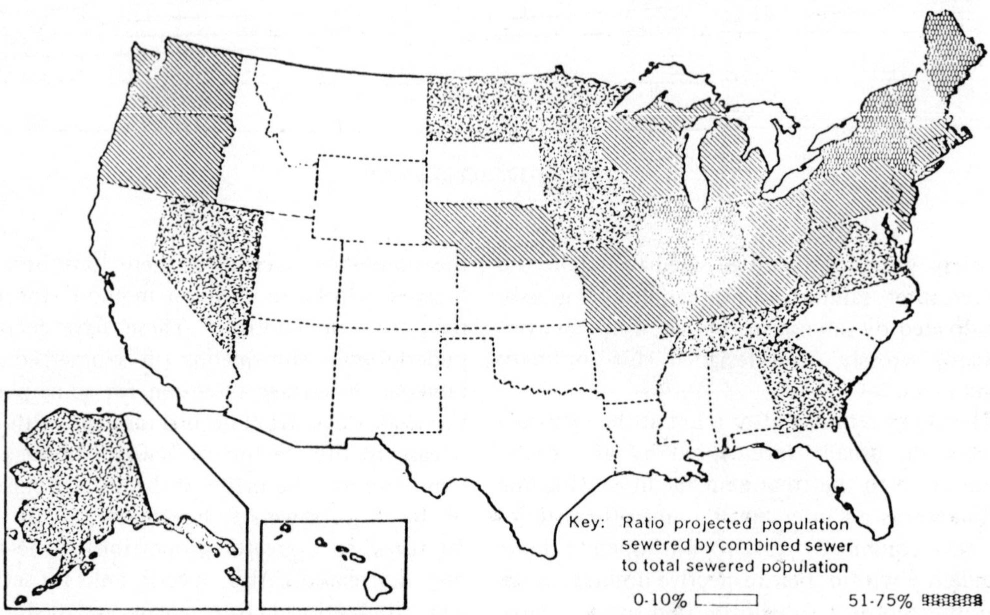

Fig. 11-14. Relative use of combined sewers compared to total sewered population by states. *(Courtesy of Federal Water Pollution Control Administration, U.S. Department of the Interior.)*

expedients to permit abandonment of private cesspools by connecting houses to storm drains, for example, some were designed according to the best practices of their day. The general national distribution of combined systems is shown in Fig. 11-14.[10]

The relationship of the projected 1329 jurisdictions in the United States having combined sewers (serving 3,025,000 acres) to the total national population in 1967 was:[10]

Combined systems	36,236,000
Separate systems	89,534,000
Unsewered	74,230,000
Total population	200,000,000

SYSTEM SIZING AND DESIGN

General. Determination of the quantities of wastewater and storm runoff and types of systems employed are described in earlier sections. All of these systems have flows characteristic of those which occur in piping that is intended to act as an open channel (the water surface exposed to the atmosphere).

The quantities to be handled vary over wide ranges, especially in storm lines which may be empty most of the time and then suddenly carry maximum flows. Lower limits of suitable working velocities for piping in these services must be observed since substantial quantities of suspended and floating materials must be transported without deposition, if possible. Upper velocity limits must be observed to protect the pipe from erosion.

Gravity flows are to be preferred in all such systems although pumping through pressure conduits may become necessary where suitable grades are not available. The grade or energy available in the system should be conserved, provided excessive depths do not result.

Hydraulic Principles. Over time, flows in sewers are unsteady, that is, variable in their rate of discharge. By virtue of their variability, they are also nonuniform–i.e., the velocity or depth or both are not the same from point to point along an open channel at all times. However, except for situations of sequential changes in conduit size, grade, shape, etc., it is customary to assume uniform flows in calculating the various hydraulic elements of a pipe or channel section.

In any conduit, the quantity of water passing any given cross section will be equal to the product of the area of the water at the cross section and the mean velocity of the water at that section:

$$Q\text{ (quantity)} = A\text{ (area)} \times V\text{ (mean velocity)} \tag{4}$$

If area is expressed in square feet and mean velocity in feet per second, the quantity will be cubic feet per second.

The word "head" in relation to water indicates one or more aspects of the energy which it may possess. If subjected to pressure from a pump or piston, it may have a "pressure head." Its elevation above any datum gives it a "potential head" with respect to that datum. Water in motion possesses a "velocity head." This is expressed by the term $V^2/2g$, where the velocity is in feet per second, and g is the gravitational constant. Each of these forms of head is a linear quantity and may be expressed in feet.

$$H\text{ (velocity head)} = V^2/2g \tag{5}$$

Theoretically, H is the distance a body would fall in a vacuum to attain a velocity V.

It has been found that resistance to the flow of fluids in conduits is in part due to friction, part due to the form or shape of the conduit, and part due to separation of the flow from the boundary of the conduit. For turbulent flow which prevails in sewers, the loss due to friction will vary approximately as the second power of the velocity. Hydraulic friction is independent of pressure, and ordinarily in sewer design temperatures are ignored as having negligible influence. The resistance to flow may be expressed:

$$H = \frac{kV^n}{2g} \tag{6}$$

The quantity k will have some relationship with the roughness of the conduit as well as its size, and both it and n, an exponent, need to be determined experimentally. The value of n varies from 1.75 to 2.0.

To relate k to a channel or pipe, some expression suitable for all shapes and sizes is necessary. The hydraulic radius, a linear term, meets this requirement:

R (hydraulic radius)

$$= \frac{\text{Area of water cross section}}{\text{Length of wetted perimeter}} \quad (7)$$

This equation is equally applicable to open or closed conduits and to pressure or atmospheric flows. For partly-filled channels or pipes, the area will be just the area of the cross section of the water at that section, and the wetted perimeter will be only the wetted part of the pipe or channel. The surface open to the atmosphere is not included in the perimeter since it offers no resistance to flow.

Experiment and analysis have shown the general relationship:

$$k = \frac{fl}{4R} \quad (8)$$

where f is a friction coefficient, and l is the length of the conduit for which loss is to be found. Using n as 2, for convenience, and substituting for k, the expression for resistance to flow becomes:

$$H = \left(\frac{fl}{4R}\right)\left(\frac{V^2}{2g}\right) \quad (9)$$

$H/l = s$ (slope), the rate of friction loss or head per unit of length. Substituting and transposing:

$$V^2 = \frac{8g}{f \times Rs} \quad (10)$$

If $8g/f$ is designated as a constant for any given channel, and for convenience is called c^2, then:

$$V = cR^{1/2}s^{1/2} \quad (11)$$

This is the Chezy formula developed in 1775. It is the basis for the Manning and Kutter formulas which are those most widely used in sewer design. The Manning and Kutter formulas attempt to provide a means of evaluating c in the Chezy formula. Each is suited to closed or open channel flows and conduits of any shape. The coefficients applicable to them have been extensively investigated.

For the Manning formula, the value of c becomes:

$$c = \frac{1.486}{n} \times R^{1/6}, \text{ and}$$

$$V = [1.486/n \times R] \times R^{1/2}s^{1/2} = \frac{1.486}{n} R^{2/3}s^{1/2} \quad (12)$$

The Kutter formula is:

$$V = \frac{\dfrac{1.81}{n} + 41.67 + \dfrac{0.0028}{S}}{\dfrac{n}{R^{1/2}}\left(41.67 + \dfrac{0.0028}{S}\right)} R^{1/2}s^{1/2} \quad (13)$$

The constants given above are suited to English units of measure. The values of R and s will relate to the specific conduit section, depth of flow, and slope of the hydraulic gradient. These, in effect, are "knowns," or must be assumed if velocity is being sought. In uniform atmospheric flow, if the depth is constant in the direction of flow (in a conduit of fixed section), the water surface then has the same slope as the conduit. If a desired velocity is established and the conduit size is fixed, then the necessary slope to produce the desired velocity will result, and so forth. The inclusion of the slope, s, in the Kutter coefficient adds to its complexity, and the validity of its influence is now in doubt. The Manning formula is generally preferred for its ease of use without sacrifice in the accuracy of results compared to the Kutter formula.

The selection of n is of critical importance. The values of n are substantially the same for both the Manning and Kutter formulas. Note that this n is not the same as the exponent n which first appears in the basic expression for resistance to flow (Eq. 6).

The value of n is governed by the material of the conduit, its size and shape, and the depth of flow therein. It will also be greatly influenced by the quality and condition of the conduit. Rough and lipping joints, sudden breaks in line and grade, obstructions such as roots, deposits and coatings such as grease, large amounts of transported solids or large-sized transported solids, and flows from laterals all will increase the value to be used. Table 11-8 provides typical values of n suitable for use in either the Manning or Kutter formulas applicable to sewer design and Fig. 11-15 is an alignment chart for solution of the Manning formula.[1]

While it is indicated that the chart is to be used for pipe flow, it may also be employed for other shapes of closed conduits and open channels by using a value 4 times the hydraulic radius on the diameter scale; the discharge or Q scale will not be valid in this situation, but the velocity or V scale will be correct.

TABLE 11-8. Values of Friction Formula Coefficients

Conduit Material	Manning n ($ft^{1/6}$)
Closed conduits	
Asbestos-cement pipe	0.011–0.015
Brick	0.013–0.017
Cast iron pipe	
Cement-lined and seal coated	0.011–0.015
Concrete (monolithic)	
Smooth forms	0.012–0.014
Rough forms	0.015–0.017
Concrete pipe	0.011–0.015
Corrugated-metal pipe ($\frac{1}{2}$-in. × $2\frac{2}{3}$-in. corrugations)	
Plain	0.022–0.026
Paved invert	0.018–0.022
Spun asphalt lined	0.011–0.015
Plastic pipe (smooth)	0.011–0.015
Polyethylene pipe	0.009
Vitrified clay	
Pipes	0.011–0.015
Liner plates	0.013–0.017
Open channels	
Lined channels	
Asphalt	0.013–0.017
Brick	0.012–0.018
Concrete	0.011–0.020
Rubble or riprap	0.020–0.035
Excavated or dredged	
Earth, straight and uniform	0.020–0.030
Earth, winding, fairly uniform	0.025–0.040
Rock	0.030–0.045
Unmaintained	0.050–0.14
Natural channels (minor streams, top width at flood stage < 100 ft)	–
Fairly regular section	0.03–0.07
Irregular section with pools	0.04–0.10

Source: ASCE Manual No. 37.

A diagram via the Kutter formula giving the discharge of circular sewers flowing full, having a value of $n = 0.013$, is shown in Fig. 11-16.[11]

Inasmuch as sewers flow less than full most of the time, their behavior under such open channel conditions takes on importance. The variations due to the different depths of flow can best be illustrated by plotting what are called the "hydraulic elements" as ratios to the respective values which prevail at full flow. For such a plotting, the shape of the conduit must be selected and the slope must be considered as fixed. This will then leave for the Manning formula the elements R (hydraulic radius), V (velocity), n (roughness coefficient), Q (quantity), and A (area of the flow). In Fig. 11-17, the values for the full section carry the subscript "f".[1]

In sewer design, use of the lowest values should ordinarily be avoided. For conduits in good to fair condition, values in the middle range will be safer. While it has been customary to use lower values for larger conduits, this practice cannot be supported by hydraulic theory, and in fact, the reverse is true,[1] except for extremely rough interiors. For partly-filled conduits, the value of n becomes greater than the value when full by possibly up to 25 percent. This variation, while frequently ignored in practice, is illustrated in Fig. 11-17 along with its influence on the hydraulic elements of the conduit section.

The interesting properties of the circular section when n is constant are that: from half full to full depth the velocity will be equal to or greater than that at full depth; the discharge at half depth will be exactly half that at full depth; and at about 93 percent of full depth the discharge will be almost 10 percent greater than that at full depth. The introduction of a variable n value reduces the zone of higher velocity to the upper 80 percent, and discharges exceeding those when full become negligible.[12] The actual change in n through varying depths, while detected by numerous investigators, has not been established to the extent that it can be evaluated with confidence. In the words of the ASCE Manual on the design and construction of sewers, "Until more information and better analyses are available, the decision to

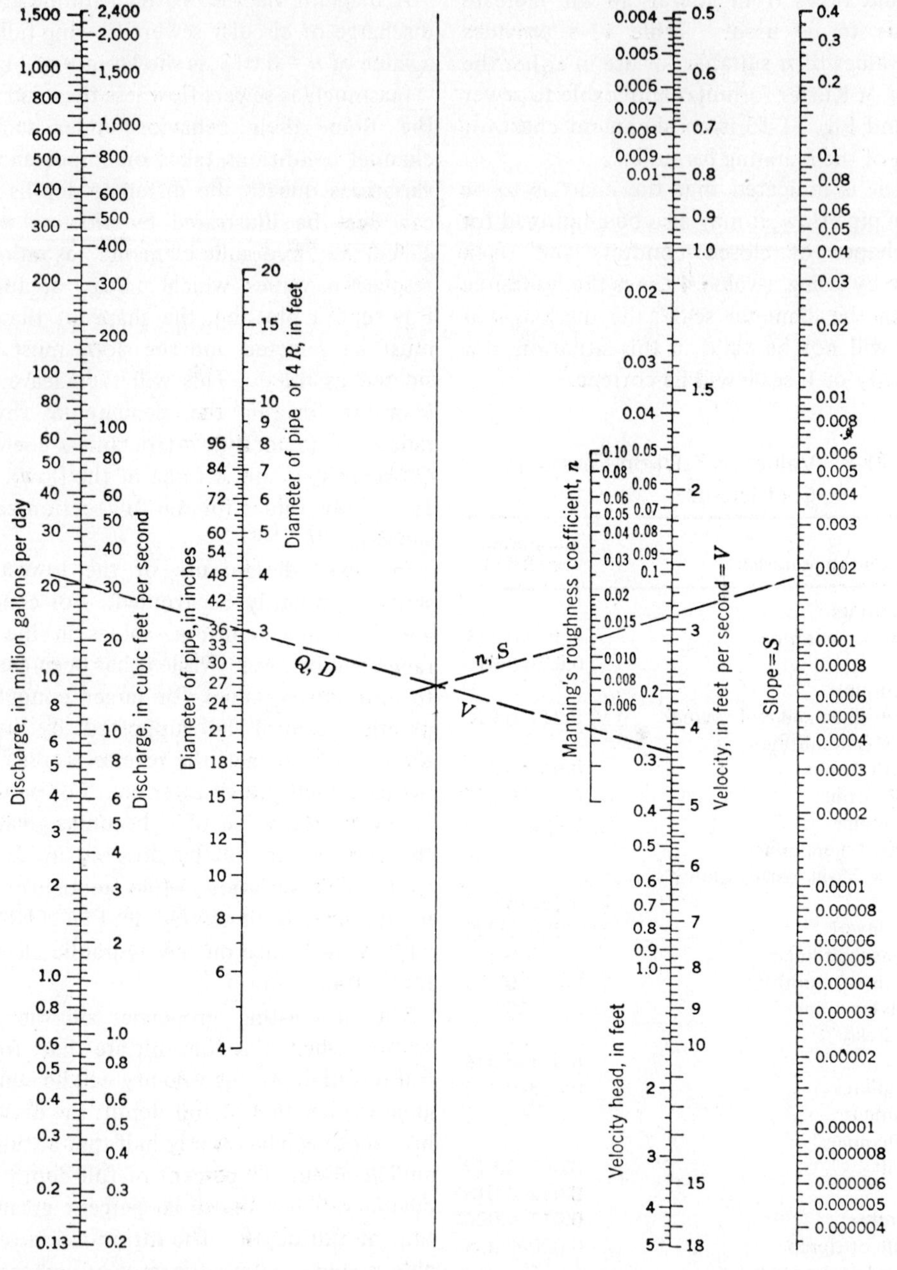

Fig. 11-15 Alignment chart for Manning formula for pipe flow. *(Courtesy of American Society of Civil Engineers.) Source:* ASCE Manual No. 37.

use a constant or variable n must be left to the individual designer."[1]

EXAMPLE: find the velocity and discharge of a 15 in. diameter pipe running 0.60 full, laid on a grade of 10 ft per 1000 ft; assume constant $n = 0.013$.

From Fig. 11-16, for pipe running full: $V = 5.45$ ft/sec and $Q = 6.7$ ft^3/sec.

From Fig. 11-8, at 0.60 full: $V/V_f = 1.09$; $Q/Q_f = 0.68$; $V = 1.09 \times 5.45 = 5.94$ ft/sec; $Q = 0.68 \times 6.7 = 4.56$ ft^3/sec.

Self-Cleaning Velocities. The maintenance of velocities adequate to prevent the deposition of materials carried in suspension must be considered. The variations in sewage flow over

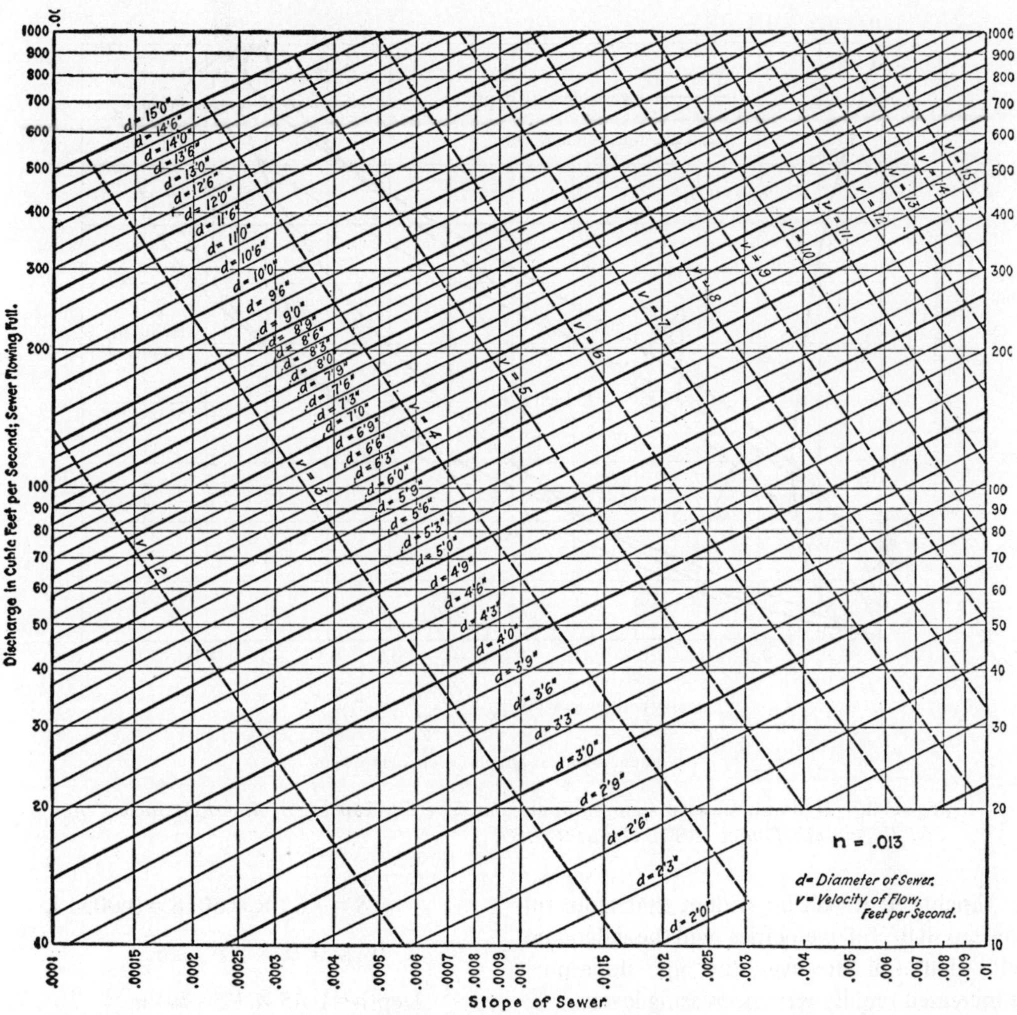

Fig. 11-16 Discharge of circular sewers. *Source:* Metcalf and Eddy, Inc., *Sewerage and Sewage Disposal*, McGraw-Hill Book Co., N.Y. (1922). Used with permission of McGraw-Hill Book Co.

each day and over the life of the sewer as the number of connections increases with the years are of concern. In the early life of a sewer, adequate scouring velocities would be desirable at average flows, even if not feasible at minimum flows. The extremities of combined sewers which are sized for stormwater probably will not maintain such velocities during dry weather. However, the attainment of scouring velocities during moderate storms should be attempted, both for such combined sewers as well as separate storm sewers.

With determination of desired scouring velocities with flow at full depth, the slopes necessary to maintain this condition at flows less than at full depth are related to the several hydraulic elements as follows:

$$\frac{S}{S_f} = \frac{R_f}{R}; \quad \frac{V_s}{V_f} = \frac{n_f}{n}\left(\frac{R}{R_f}\right)$$

$$\frac{Q_s}{Q_f} = \frac{n_f A}{n A_f}\left(\frac{R}{R_f}\right) \qquad (14)$$

In these, the subscript "f" indicates the full condition, V_s is the velocity for the self-cleaning condition, and Q_s is the discharge at the self-cleaning velocity.

These relationships have been developed for a circular conduit as shown in Fig. 11-18.[1]

From the curve for S/S_f, which is unaffected

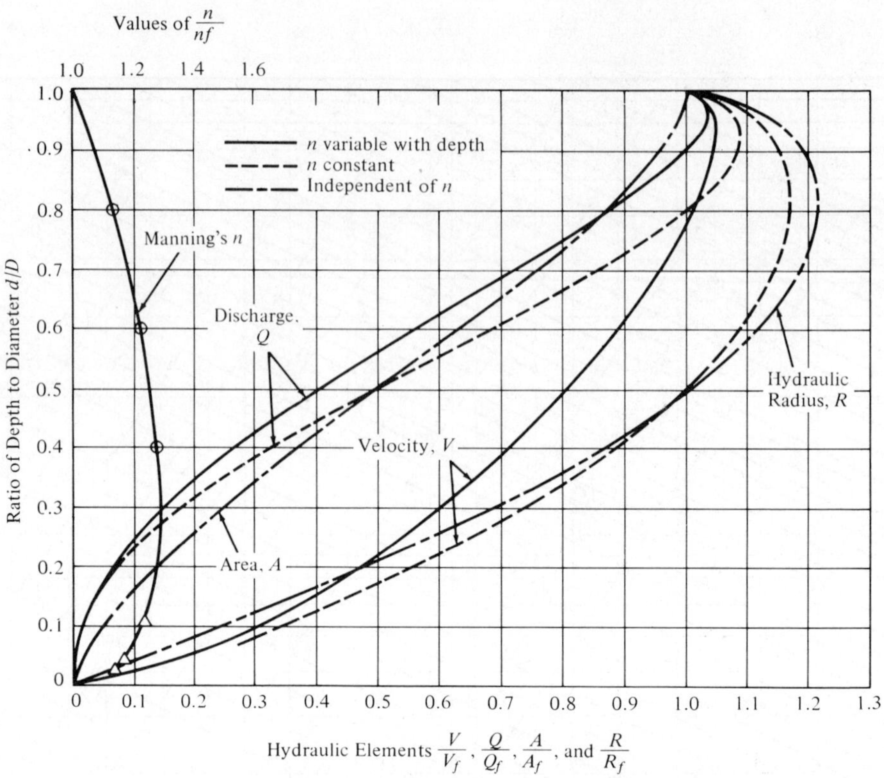

Hydraulic–elements graph for circular sewers.

Fig. 11-17 Hydraulic-elements graph for circular sewers. (*Courtesy of American Society of Civil Engineers.*) *Source:* ASCE Manual No. 37.

by variable n, it will be evident that from full flow to half full velocities will be adequate. Below half full, however, the pipe slope must be increased rapidly with decreasing levels.

EXAMPLE: for a 12-in. pipe which has an adequate velocity of 2 ft/sec when flowing full, find the equivalent self-cleaning velocity and necessary slope when the pipe is discharging only 0.60 ft^3/sec. Use $n = 0.013$, but assume variable n with changing depth of flow.

According to Fig. 11-16, the 12-in. pipe when flowing full attains a discharge $Q_f = 1.68$, which requires a slope of $S_f = 0.0018$. Given $Q_s = 0.60$,

$$\frac{Q_s}{Q_f} = \frac{0.60}{1.68} = 0.357$$

Using Fig. 11-19 for this value of Q_s/Q_f:

$$\frac{S}{S_f} = 1.57; \quad \frac{V_s}{V_f} = 0.78; \quad \text{Depth } \frac{d}{D} = 0.46;$$

$$S = 1.57 \times 0.0018 = 0.00314;$$

$$V_s = 0.78 \times 2 = 1.56;$$

$$\text{Depth} = 0.45 \times 12 = 5.4 \text{ in.}$$

A minimum velocity of 2 ft/sec for sanitary sewers running full or half full and 3 ft/sec for storm sewers is considered good practice. However, higher values may be necessary if dense and coarse materials are expected to be carried in the flowage.

Nonuniform Flows. At transitions in size, shape, or grade, and at junctions between two or more conduits, the stream or streams will undergo changes in velocity, shape, or depth, or all of these. The energy lost will be primarily a function of the change in velocity heads between the incoming and outgoing flows. Under normal conditions, velocities in sewers are so low that perhaps little measurable energy or head is lost. For subcritical flow in an open

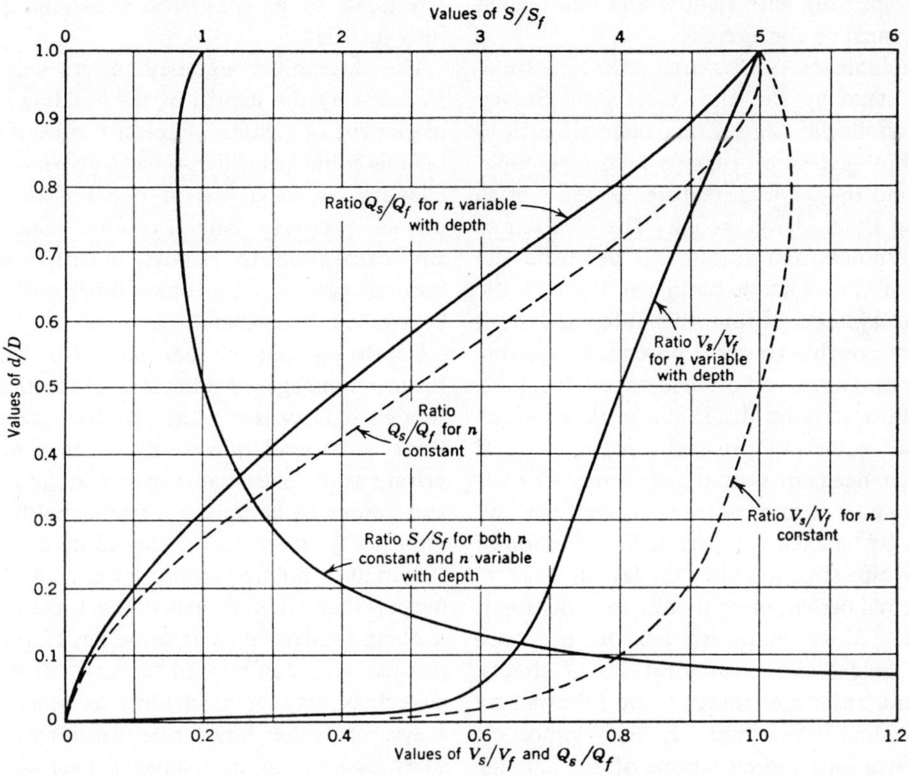

Fig. 11-18 Hydraulic elements of circular sewers that possess self-cleansing properties at all depths. (*Courtesy of American Society of Civil Engineers.*) *Source:* ASCE Manual No. 37.

channel transition, the loss of energy may be expressed as:[1]

$$H_L = K\left(\frac{V^2}{2g}\right) \tag{15}$$

The change in velocity head above and below the transition is represented by $(V^2/2g)$. If the transition has wall deflections below 12.5° and is well designed, K will approximate 0.10 for channel contractions and 0.20 for channel expansions.

As in transitions, losses in junctions where flows from one or more sewers join do not ordinarily require special consideration unless velocities are 6 to 8 ft/sec. The angle of entry with respect to the main line should be no greater than from 30 to 45°, and the junction should be sized so that the velocities of the merging streams are about equal at full flow. (Note: discussion of critical depth, the development of drawdown and backwater curves, hydraulic jump, etc. is considered beyond the scope of this chapter.)

Layout. To the maximum extent possible, any system of sewers or drains should conform to the topography of the area it is to serve. In "new towns," subdivisions, or extensions of established communities, the street arrangement and grading will necessarily follow in large part, or be responsive to, the ground surfaces. The sewer lines would preferably fall within these streets or related alleys. Ready accessibility to such lines is essential, even after their construction. If the street layout does not extend to the site of the disposal facility or point of discharge, the sewers leading to those locations should fall within the stream valleys. Maximum use should be made of gravity flow and, except when physically, economically, or ecologically imperative, pumping should be avoided. Not only costs but continuity of service, especially with remote and unattended stations, must be considered.

For preliminary studies and estimates, topographic maps by the U.S. Geological Survey may be available. Any other material such as aerial photographs or surveys recording topography and the physical features of the area to be served should be gathered. The location of the high points and ridges will delineate the outline of the drainage basin tributary to the proposed system. From this information, it should be possible to select potentially suitable sites for treatment or other disposal facilities. Taking into account the subbasin flows which must be gathered progressively, a general system arrangement should take form. The use of appropirate assumptions as to land use and flows to be handled will permit the selection of tentative pipe sizes suited to the layout.

For actual design, more precise information is required. Most important are the proposed finished grades and exact locations of streets, alleys, and any other routes to be followed by the individual sewer lines. Equally important are the locations and elevations of any existing or future structures or piping to be served. Investigation of the soils to be encountered, other piping or utilities (present or future), proposed structures to be avoided (such as possible underpasses, rail lines, etc.), and intended land use of tributary areas is necessary. Field surveys should be conducted to tie in intended development with existing conditions and establish bench marks and other points for both vertical and horizontal control so that profiles along the sewer routes may be plotted. The development of contour maps of critical locations and the plotting of the street system and other physical features affecting design are also required. Selection of the lower terminus of the system should be made only after a thorough investigation of the type of treatment or other method of disposal to be employed and the suitability of the site for this operation.

In residential areas, sanitary sewers are generally located close to the center of the street to be accessible to both sides. Major thoroughfares, however, would generally justify placement of the sewers outside the roadways and under or behind the sidewalks. Such dual sewers in other congested situations may be fully justified.

The desirability of alley sewers will be influenced by the depths of the building lots and placement of houses. Excessive runs of private laterals could be burdensome to owners. Service connections to sewers in rights-of-way across private property where routine maintenance and excavation to remove obstructions may become necessary are operational problems to be avoided if possible.

Depths should be adequate for basement fixture drainage of abutting properties, even though it is system policy to discourage cellar floor and foundation drains. The contour of private land, placement of houses, and lengths and grades of laterals (preferably with a minimum of $\frac{1}{4}$ in. per foot) should be considered. Generally, a public sanitary sewer should not be deeper than 10.5 ft below the finished street grade in residential areas, although 12 ft may be justified in commercial areas. Stormwater collectors may be as shallow as nearby water mains and other interfering utilities will allow. They should be in roadways near the curbs where the stormwater inlets are located.

Manholes are used at breaks in line or grade, at junctions with other sewers or large building laterals, and on sewer terminals. Manholes should be spaced from 300 to 400 ft apart on sewers too small to permit entrance. On larger lines they may be 500 ft or more apart.

House connections should be made through wyes or tees on the sewer, not by breaking out part of its wall and cementing a saddle or stub in place as was once customary, since this practice can result in the serious infiltration of groundwater.

In general, it is ideal if the sewer line can use a grade or slope which is the same as that of the overlying street or ground surface. In this way, both excessive and insufficient depths may be avoided. If placement is too deep, the cost to build both the public sewer and connecting house laterals will be too great; if too shallow, the intended service may not be provided. Observance of maximum and minimum velocities to be expected, the maximum use of the minimum pipe sizes, the fullest utilization of each segment, and the arrangement of the several segments with respect to each other will all be considerations as pipeline grades are selected. Sewers are never placed in the same trenches with water mains.

Sewers can be laid on long radius curves (100 ft or 200 ft and upward)[1] to keep their positions in curving streets. This presupposes, however, that modern pipe-cleaning apparatus will be used for maintenance, not the older bucket and cable method. The individual pipes may be "broken" at the joints, the degree of deflection depending on the particular pipe being used, or, preferably, beveled pipe cast for the desired radius may be used. Monolithic sewers readily adapt to curved alignments. Also, for larger pipelines, it is often desirable to use monolithic construction for shorter radius changes in direction.

Ranges of Flows and Velocities

Sanitary Sewers. For small sewers in system extremities at the beginning of the design period, the minimum daily flow will be $\frac{2}{3}$ the average daily flow, and the extreme minimum flow will be $\frac{1}{2}$ the minimum daily flow, or $\frac{1}{3}$ the average daily flow.[2] The extreme minimum would be critical for self-cleaning velocity. At the end of the design period, the average daily flow will be 2 times the average daily flow at the beginning, the maximum daily will be 2 times this average daily, and the extreme maximum will be $1\frac{1}{2}$ times the maximum daily. The range over the life of the sewer has a ratio of 18:1. The extreme maximum will be critical for design of piping and pumps. For large sewers and extensively developed tributary areas, the ratio of extremes would be 4:1.

Pipe slopes or grades are set so that when flowing full or half full, the velocity will be a minimum of 2 ft/sec. Some designers prefer 2.5 ft/sec. Fine sand will be transported at 1.0 ft/sec and fine gravel at about 2.0 ft/sec. Since the pipes probably will not run half full during the early years, it is wise to check conditions to be expected during times of the daily maximum, as these early peaks may suffice to resuspend deposits. Grade adjustment to facilitate this should be considered. Notwithstanding the desirability of maintaining adequate velocities, it may be necessary to build on flatter slopes to avoid an excessively deep and costly system.

Velocity is influenced very little by pipe size for any given flow and slope. Table 11-9 shows the slopes necessary for velocities of 2 ft/sec for various flows without regard to pipe size.[1]

The maximum velocities for sanitary sewers will be governed by the amount and hardness of the grit to be carried and the abrasion resistance of the sewer invert material. Vitrified clay pipe and monolithic structures with vitrified brick invert linings may withstand a velocity of 10 ft/sec continuously. Softer materials, including sewer brick, probably should be limited to 5 to 8 ft/sec, and concrete to 8 to 10 ft/sec.

Sanitary sewers should not be designed to run full. Some air space is desirable for ventilation and the supression of sulfide generation. The sewers at the extremities of the system should be sized to run about half full when handling their design peaks. The larger lines will be less subject to wide flow fluctuations since the greater tributary population and mixture of flows from commercial businesses and other sources will tend to damp out the effects of individual contributors. In these larger lines, deeper flows at design peaks are justified,

TABLE 11-9. Slopes Necessary for Velocities of 2 ft/sec[a]

Q ft³/sec	Slope (per thousand)
0.1	9.2
0.2	6.1
0.3	4.8
0.4	4.1
0.6	3.22
0.8	2.73
1.0	2.39
1.5	1.89
2.0	1.59
3.0	1.26
4.0	1.06

[a]For circular pipes of any size when the indicated flow is not less than 0.1 percent nor more than 95 percent of full-pipe capacity.
Source: ASCE Manual No. 37.

especially if subject to the effects of surface runoff from areaways.

Combined Sewers. Combined service sewers must be sized to carry the sanitary flows plus the peak storm flows. The relative quantities of these flows will vary with each design situation. If all surface runoff is to be accommodated, the sanitary portion could be insignificant. If only limited runoff is admitted—for instance, only that from roofs—the sanitary-runoff ratio could be near unity, especially if infiltration is also present in large quantities.

A minimum velocity of 3.5 ft/sec when flowing full is desirable to decrease the deposition of putrescible materials and to scour deposits of sand and grit. Although designed essentially as a storm sewer, the depth must correspond with that of a sanitary sewer in the sections which provide service to contiguous property. In the reaches upstream of the stormwater relief outlets and interceptors, the peak dry-weather flows may range from 2 to 7 times the average. Since some stormwater allowance is included in computations of peak sanitary flows where areaways, etc. are admissible, there should be no duplication of such areas in assigning the total storm runoff to the combined sewer.

Interceptors should be designed for nothing less than the peak dry-weather sanitary flow plus infiltration. The customary provision of capacity for several times the dry-weather flow, known as "dilutions," has been intended to carry off the initial slug of material which has accumulated in the sewers and has been resuspended by the storm flow. While such early flows have been found to be heavily polluting, it would appear that any significant subsequent quality improvement is likely to occur only during prolonged storms.[13]

Investigations on the effects of higher dilution ratios show disappointing benefits. It has been indicated that from 80 to 160 dilutions may be required to prevent all overflows; studies also show that for Boston 5 to 6 overflows per month will occur with interceptors just adequate for the dry-weather peak flows.[14] In Detroit, it was found that with 10 dilutions there are 3 overflows per month.[15] The Board of Engineers reporting on the District of Columbia system in 1957 developed curves for the frequency and duration of overflows relating to the number of dilutions (Figs. 11-19 and 11-20).[16] Overflow durations of 400 hr/yr with 2 dilution interceptors would be reduced to 60 hr/yr with 10 dilutions.

Storm Sewers. There is no lower limit for flows in storm sewers—they will be dry much of the time. They should scour clean when in use

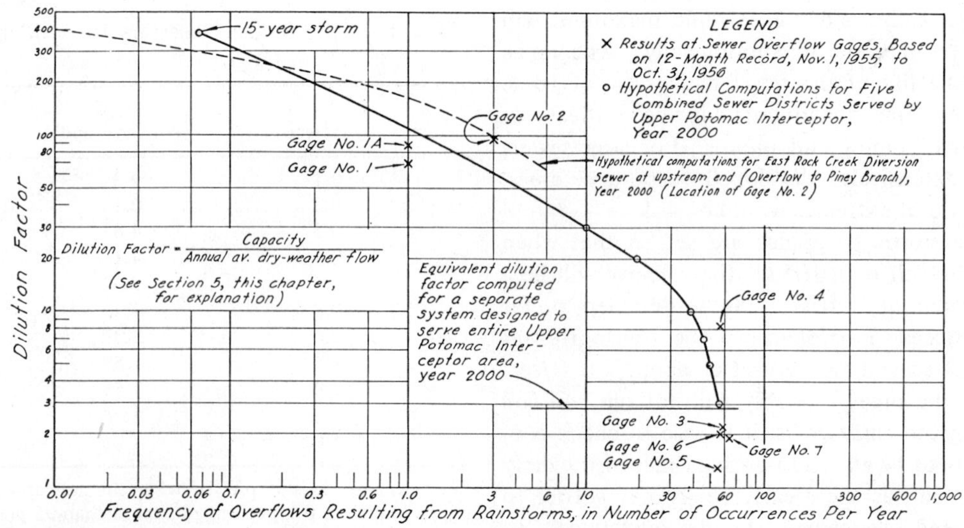

Fig. 11-19 Frequency of sewer overflows to surface streams. (*Courtesy of Department of Environmental Services, District of Columbia.*)

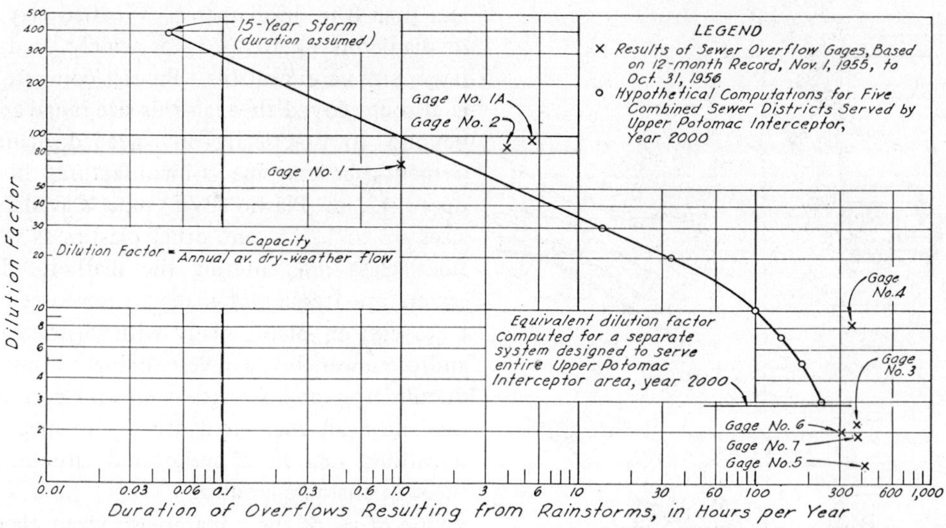

Fig. 11-20 Duration of sewer overflows to surface streams. (*Courtesy of Department of Environmental Services, District of Columbia.*)

without the necessity of carrying a design storm, such as a 5- to 10-year event. Sudden changes in velocity should be avoided. There is no customary upper velocity limit—20 ft/sec could be satisfactory. Since ventilation and sulfide attack are not problems, storm drains should be designed for running full. Some surcharge may be acceptable providing the surface water inlets continue to function as intended.

Building Connections. The pipe conveying wastewater from a building to a public sewer is called the house sewer, lateral, or building connection. It may be 4, 5, or 6 in. or more in diameter and is usually required to be of cast or ductile iron where it passes through or beneath the building wall. For the remainder of the distance to the public sewer, local building regulations will govern. The pipe is usually made of vitrified clay, but may also be cement-asbestos, plastic, cast iron, or concrete. Laterals to storm sewers are not necessarily of the same materials as the sanitary laterals.

Connection to the public sewer should be through a special fitting built originally into or inserted in the line. Typical service connections are shown in Figs. 11-21, 11-22, and 11-23.[1,17]

The normal minimum slope for laterals 6 in. and smaller is $\frac{1}{4}$ in. per foot, which is approximately 2 percent or 20 ft per 1000 ft. Since service connections have been shown to be serious sources of groundwater infiltration into sewer systems, they should be constructed to the highest standards and receive capable inspection before backfilling.

Pipe Sewer Details and Manholes. For sanitary service, the smallest public sewer should not be

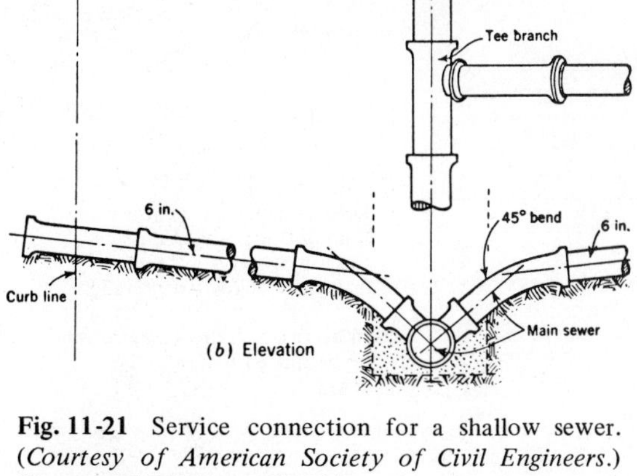

Fig. 11-21 Service connection for a shallow sewer. (*Courtesy of American Society of Civil Engineers.*) *Source:* ASCE Manual No. 37.

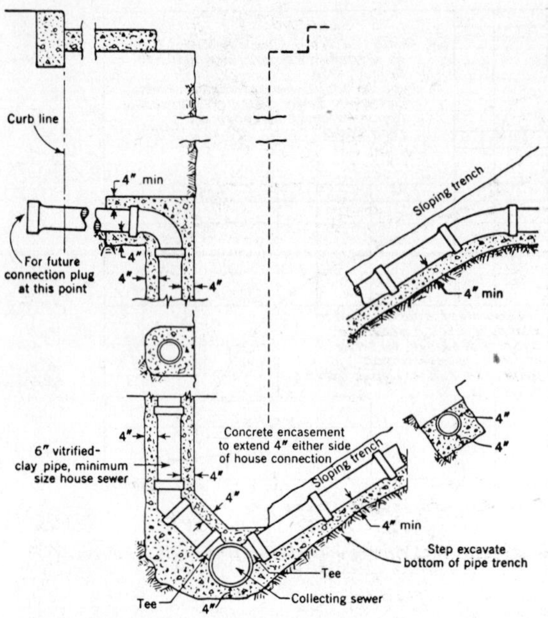

Fig. 11-22 Service connection for a deep sewer. (*Courtesy of American Society of Civil Engineers.*) *Source:* ASCE Manual No. 37.

less than 8 in. in diameter. Vitrified clay pipe in diameters up to 24 in. is widely used, and larger sizes are available. Precast concrete pipe is also employed through this size range and on beyond to practically any needed diameter. Cement-asbestos pipe is manufactured in sizes up to 42 in. Plastic (PVC) pipe is available in sizes up to 12 in., and other plastics as well as fiber glass are entering the market. Large sewers are frequently made of reinforced concrete cast in place, many with vitrified brick and/or sewer brick invert linings. Cast and ductile iron and smooth wall and corrugated steel have all been used, usually in special applications–i.e., in above-ground situations or through unstable ground.

Pipe of all of these materials except clay and corrugated steel with its conventional jointing could be suited to use in pressure lines and force mains. The tightness of the joints is essential, and modern developments have gone far in the reduction of infiltration. The early

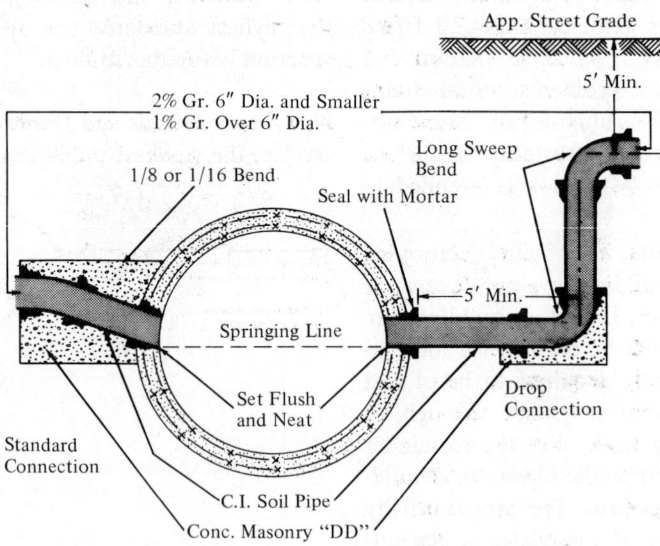

Notes:

1) Drop connections to storm sewer shall not be permitted unless the top of sewer is 7 feet or greater below street grade.
2) In no case will drop connections be permitted to a sanitary sewer.
3) Materials and workmanship shall be in accordance with the D. C. Plumbing Code.
4) After a house sewer has been connected and before any part has been covered, the plumbing shall be inspected and approved by a D. C. Plumbing Inspector.

Fig. 11-23 House sewer connections–reinforced concrete pipe or masonry sewers. (*Courtesy of Department of Environmental Services, District of Columbia.*)

dry packing with jute, sealing with cememt mortar or dry-formed packing, and the use of hot-poured sealant have given inferior results to those now available through use of preformed rubber sealing rings, "O" rings, and plant-assembled compression joints of PVC. The selection of pipe material and type of joint will depend on external loading, soil conditions, the service intended, and infiltration tolerance. For systems leading to treatment facilities, maximum advantage should be taken of the newer methods of jointing.

The minimum storm sewer size is usually 12 in. While any standard pipe may be used, precast concrete is the most widely used in the urban setting.

The method of pipe laying will be governed by the strength and flexibility of the pipe, soil conditions, and external loading. The load-carrying ability is increased with the support provided by shaping the subgrade to conform with the pipe. A complete cradle of concrete supporting the pipe up to or above its haunches will provide even greater strength. Pipe should not rest directly on an irregular rock subgrade or other rigid point of support. A cushion of gravel, concrete, or graded borrow material should be provided. The pipe should be deep enough not to be affected by traffic shocks, with a minimum earth cover of not less than 2 ft. Where groundwater is present in sufficient quantity, an underdrain may be used to improve working conditions during construction.

Manholes are to provide access to the sewer. They should be large enough to admit personnel and equipment for maintenance work and be not less than 4 ft at the bottom, round or square. The channel across the invert should have a "U" shape in section and preferably have a depth equal to the pipe diameter. When of this depth, the bottom is said to be "fully developed" and this insures the minimum energy loss. The grade of the pipe may be continued through the manhole: however, if two or more pipes enter, bringing flows together, some allowance for changes in velocity heads of the streams would be in order. A rule of thumb has been cited without endorsement[2] which would allow a drop of 0.10 ft in a through manhole; 0.2 ft in the presence of one lateral or bend; and 0.3 ft for two laterals. If a change of direction occurs in the manhole, the curve of the channel should have a radius of no less than $1\frac{1}{2}$ diameters.

Where the outlet pipe is larger than the incoming pipe or pipes, the springing lines may be kept on the same grade with sizes under 24 in. Above this, the tops may preferably be kept together if they are expected to run full.

Manholes may be of brick, concrete poured in place, or precast concrete. The latter has gained in popularity with increased labor costs. Joints between the precast segments may be sealed with rubber gaskets. The top openings are protected by cast iron covers set in frames which usually rest on several courses of brick. This will facilitate adjustment to the paving grade. The clear openings are a minimum of about 24 in. Watertightness, support of traffic loading, and difficulty of removal by vandals are prime elements in their design.

Some examples of sewer placement and laying, and manhole details are shown in Figs. 11-24 and 11-25.[2,17]

Manholes are installed where any substantial difference in grade between the incoming and the outlet pipes occurs. A typical form of "outside control" for this situation is shown in Fig. 11-26.[17]

For very deep drops, special designs become necessary to dissipate the excess energy. Baffles, water cushions, and "morning glory" shapings to produce vortical flow have all been used.

Inverted Siphons. Where streams, depressed highways, or other structures intersect the grade line of a conduit and it may not be adjusted to maintain open channel flow, it is customary to depress the conduit so as to pass beneath the obstruction. In so doing, the depressed section or siphon will be surcharged or submerged below the hydraulic gradient. Flow will be induced by the difference in levels at the two ends of the siphon.

Considering the difficulty of access, it is important to prevent the deposition of materials in the depressed conduit as much as possible. This becomes difficult in view of the nonuniformity of sewage flow. The development of scouring velocities of 2 to 3 ft/sec is usually attempted. To accomplish this, several barrels of differing sizes are built, and flows are controlled by weirs on the approach end. The smallest would be sized for the daily maximum

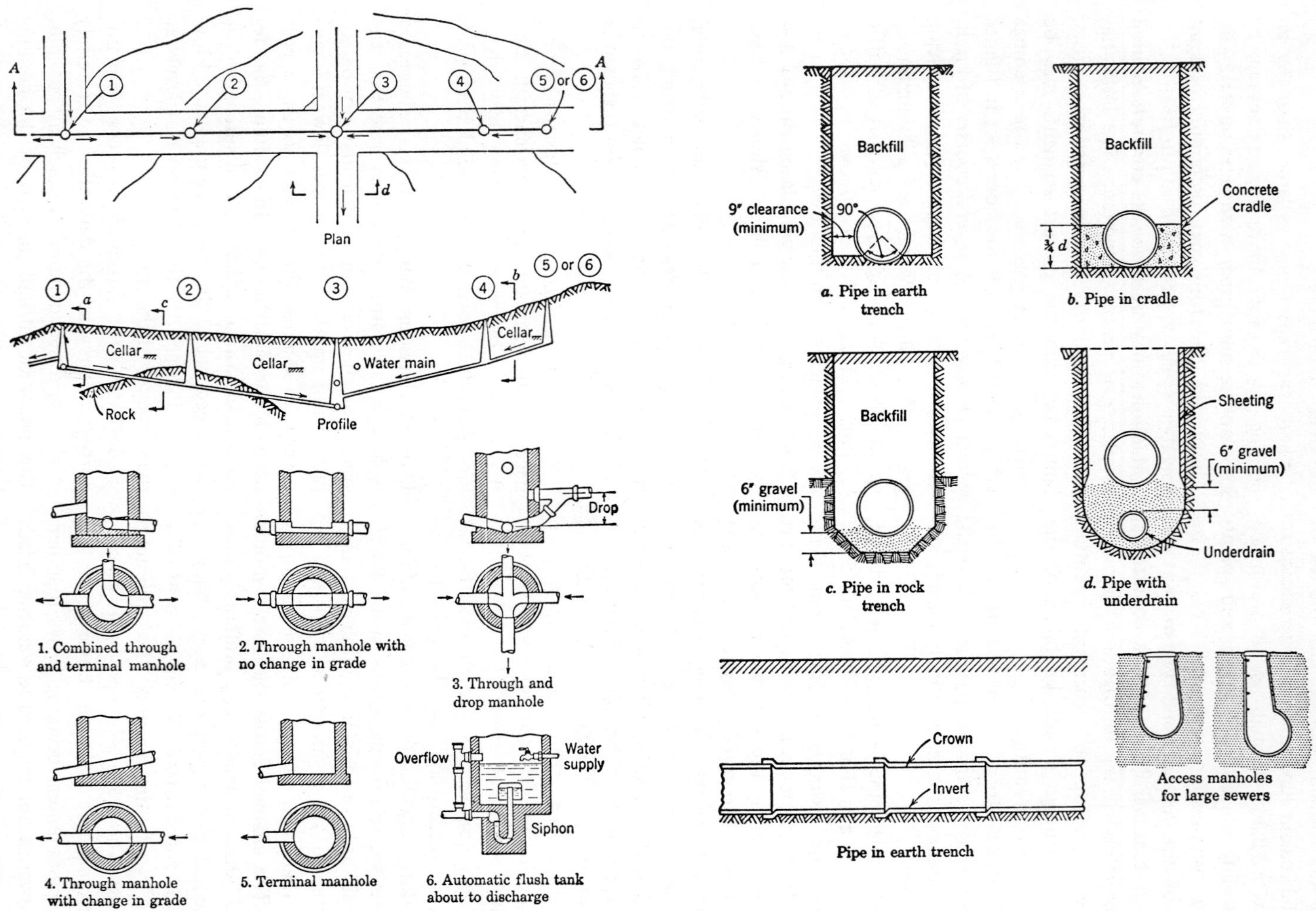

Fig. 11-24 Plan, profile and constructional details of sanitary sewers. (*Courtesy of John Wiley & Sons.*) *Source:* Fair, G. M., and Geyer, J. C., *Elements of Water Supply and Waste Water Disposal*, John Wiley & Sons, N.Y. (1958).

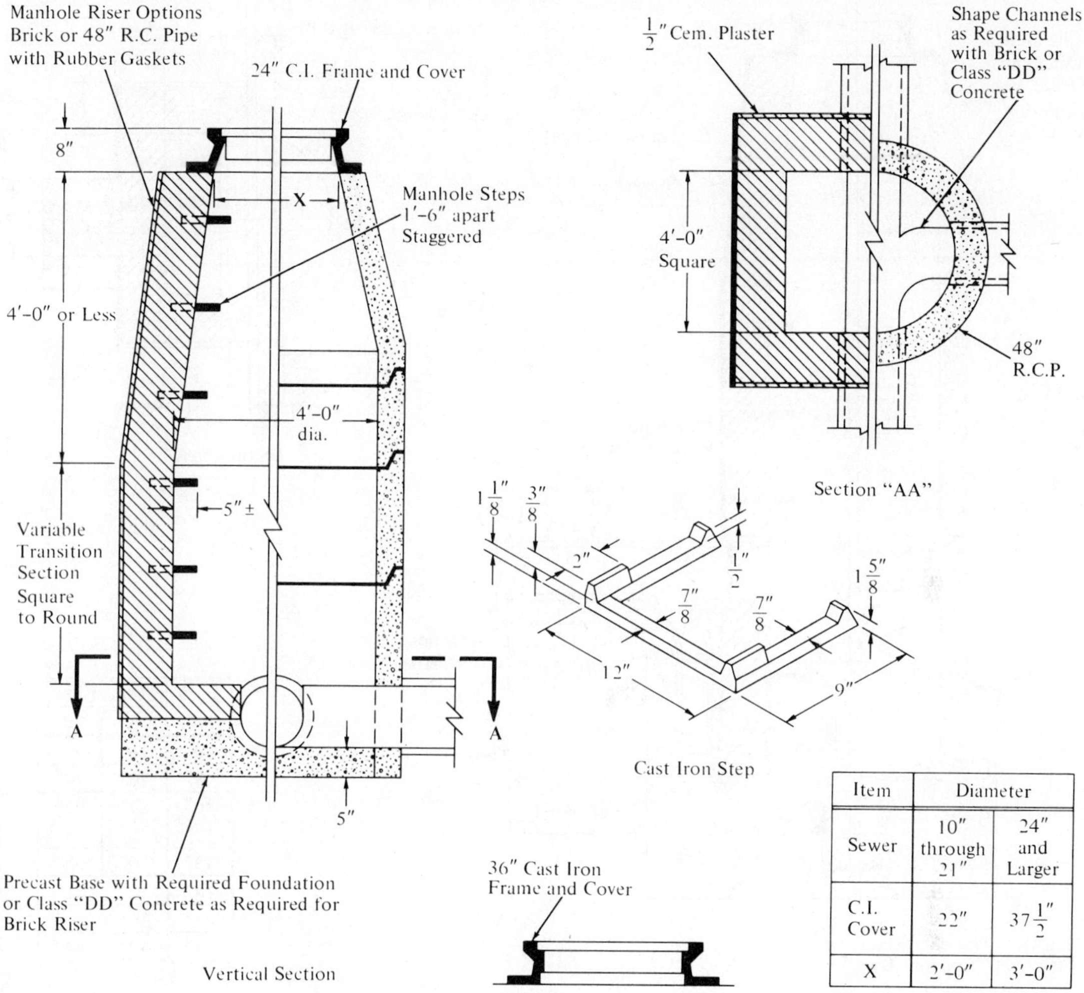

Item	Diameter	
Sewer	10" through 21"	24" and Larger
C.I. Cover	22"	37 1/2"
X	2'-0"	3'-0"

Unless otherwise specified, when manhole depth exceeds 12′ the thickness of the brick riser shall be increased to 13″ below 12′ or reinforced concrete riser

Fig. 11-25 Manholes on sewers 36-in. and smaller. *(Courtesy of Department of Environmental Services, District of Columbia.)*

during the early years, and the other or others would, in turn, be sized to accommodate progressively the additional flow increments as they developed in that particular system. Considering that all siphon pipes will stand full even when not flowing, it is certain that suspended materials will settle during periods of inaction. Details of a twin-barrel siphon used by the Greater Peoria Sanitary District are shown in Fig. 11-27.[1]

Since the need for cleaning is likely to be frequent, especially on combined systems subject to street scour, there should be no less than two siphons at any such crossing so that one may be kept in service while the other is cleaned.

The buildup of air at the chamber on the approach end, which is induced by flow, requires relief. Odors released by the sewage flowing over the weirs make such relief objectionable. Flows downstream of the siphon draw air into the siphon outlet. A closed air conduit connecting the upstream and downstream chambers, called an "air jump," relieves this problem. Since the jumper will fill with

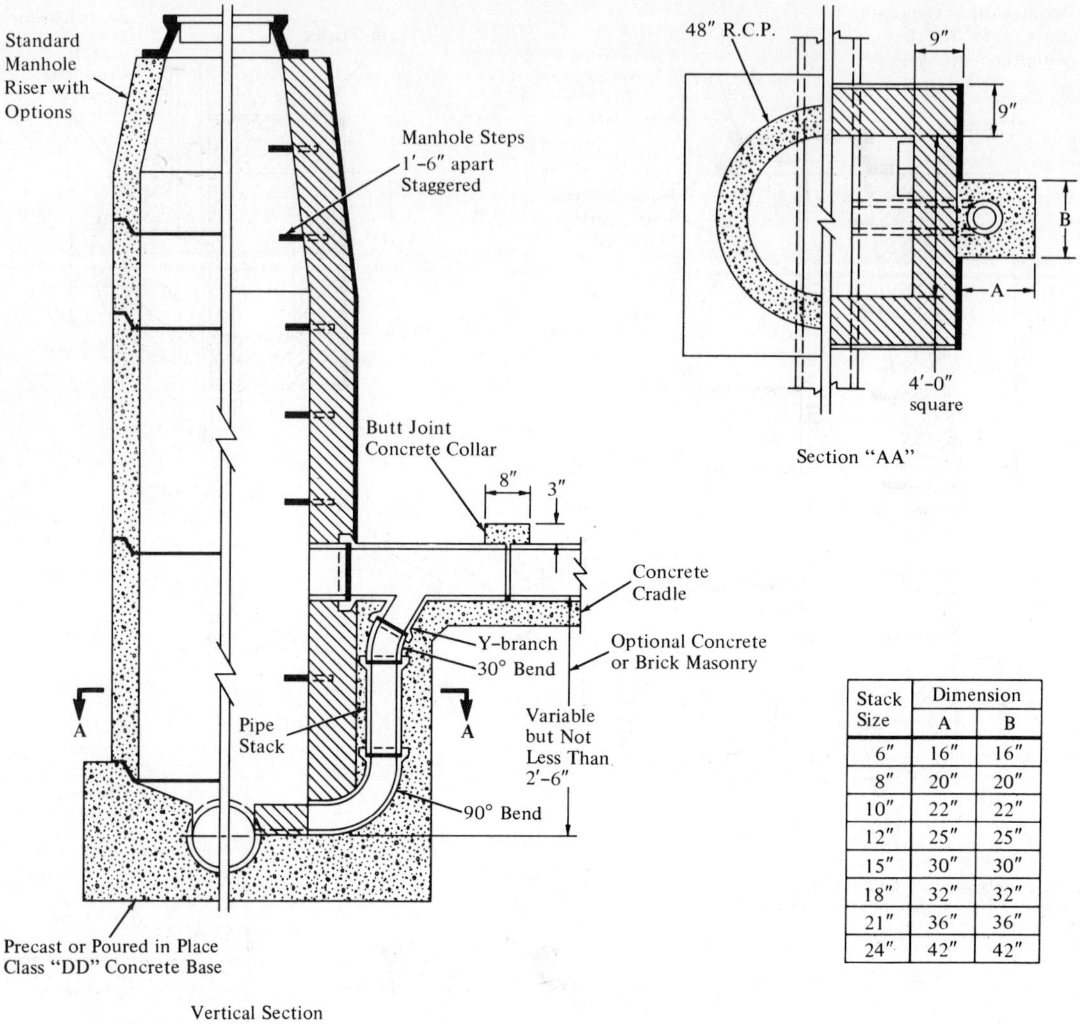

Stack Size	Dimension A	Dimension B
6″	16″	16″
8″	20″	20″
10″	22″	22″
12″	25″	25″
15″	30″	30″
18″	32″	32″
21″	36″	36″
24″	42″	42″

Fig. 11-26 Drop manhole. (*Courtesy of Department of Environmental Services, District of Columbia.*)

condensate or possibly sewage at times, some means of bleeding or pumping must be provided. Sulfide generation is likely in siphons as in all other pipes entirely full of sewage.

Stormwater Inlets. Surface runoff from improved streets reaches the stormwater drains by way of stormwater inlets. These are distinguished by the locations of their inlet openings. Thus, those with horizontal openings in the paving surface are gutter inlets; those with vertical openings flush with the face of the curb are curb or side inlets. Either may be approached by or enveloped by a slight depression in the paving surface which will increase the quantity of entrapped water. Combination of curb and gutter inlets, or their use in sequence, are matters of judgment influenced by the quantities of water to be handled and the grades of the pavement both along the curb line and along the crown normal to the curb.

Connections from inlets should be 10 in. or more in diameter and will usually terminate in manholes over the storm drain.

Both gutter and curb inlet structures may be provided with catchment chambers to retain the grit in the runoff. When so equipped, the side inlets are called catch basins. Where discharge is to combined sewers, the outlet pipe may be built so as to retain a water seal to prevent the escape of odors. However, since cleaning after storms and mosquito control during

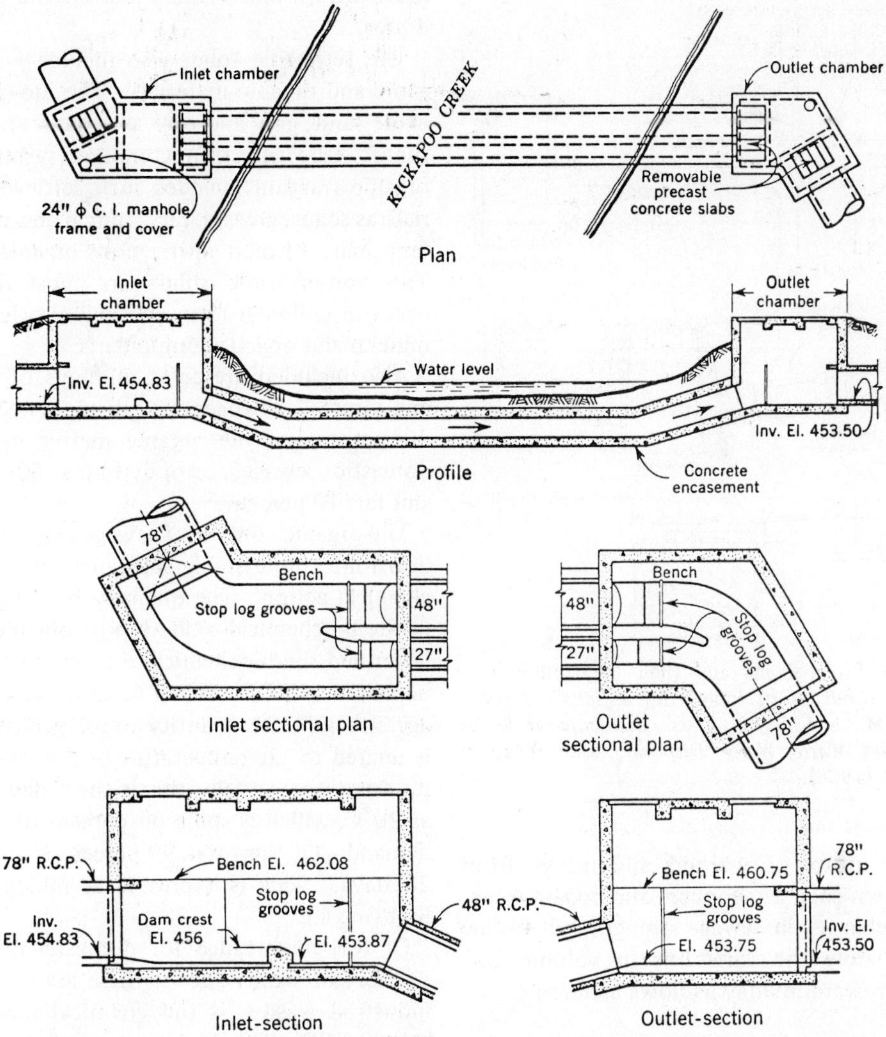

Fig. 11-27 Twin-barrel siphon. (*Courtesy of American Society of Civil Engineers.*) Source: ASCE Manual No. 37.

the summers become burdensome, use of the water-sealed catch basin is diminishing. On the other hand, the grit retained in these, including abrasives used for ice control, is material which the combined sewers and the treatment facilities to which they lead may well do without. Street inlets and their connection to a manhole are illustrated in Fig. 11-28.[2]

URBAN WASTEWATER CHARACTERISTICS

Municipal Sewage–Sources. The principal sources of municipal sewage are generally households and activities related to support of urban life such as business places, restaurants, hospitals, and governmental institutions. The proportions of these with respect to each other will vary from place to place. Some communities may have disproportionately large numbers of visitors and tourists, while others have concentrations of industries or manufacturing activities whose wastes could materially alter the normal municipal composite. Wide seasonal fluctuations in quantity and possibly in constituents are not exceptional with such overbalancing influences. Weekend flows may be materially less due to the absence of the

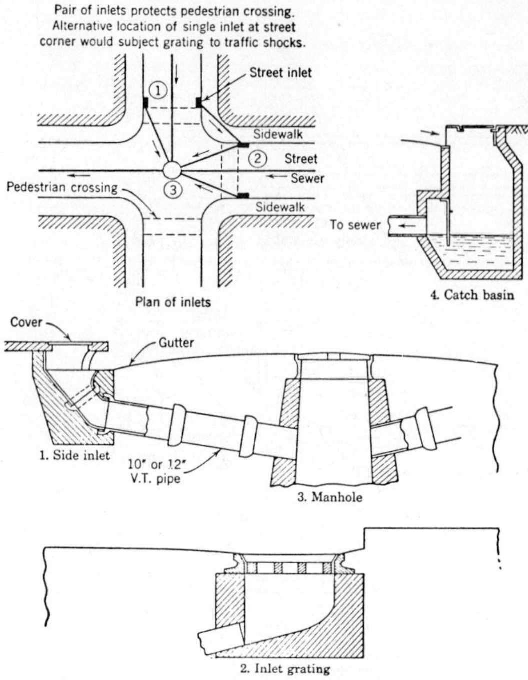

Fig. 11-28 Street inlets and their connection to a manhole. (*Courtesy of John Wiley & Sons.*) *Source:* Fair, G. M., and Geyer, J. C., *Elements of Water Supply and Waste Water Disposal*, John Wiley & Sons, N.Y. (1958).

weekday influx of workers and others from out-of-town such as shoppers and businessmen. Diurnal changes in sewage strength will follow approximately the variations in volume—i.e., stronger toward midday as flows increase.

Municipal Sewage—Composition. Inasmuch as sewage is the "spent water supply" of the community, its liquid portion is almost entirely water, which becomes an easy vehicle for the transport of wastes. Its solids content is comprised of those in solution or "dissolved solids" and those in suspension or "suspended solids." A portion of the suspended solids is "settleable" and the remainder is "nonsettleable."

The mineral dissolved solids include chlorides, carbonates, and silicates of sodium, potassium, calcium, and magnesium, some of which may have been present in the original water supply or may have entered the sewers as groundwater infiltration. The dissolved organics may be said to consist of urea and proteins, which include most nitrogenous organic matter, and carbohydrates.

The settleable solids also include both inorganic and organic materials. The inorganics include sand, grit, and clay as scour from surface runoff, especially from combined systems. The organic fraction includes such settleable materials as soaps, greases, tars, animal and vegetable fats, hair, oil, and nitrogenous organic matter. The nonsettleable solids are most likely to occur in colloidal form and will also have both mineral and organic components.

Also included are great numbers of bacteria and viruses. Nitrogenous substances constitute 40 percent of the organic matter in average domestic sewage, carbohydrates 50 percent, and fats 10 percent.[2]

The organic content of sewage may be broken down into more stable substances through biochemical action. The quantity of oxygen used in the biochemical oxidation of organic matter is termed the biochemical oxygen demand, abbreviated as BOD. When used as a test of sewage strength, the quantity of oxygen consumed is related to the temperature of the process and its duration. Usually this is the 5-day demand at 20°C, which is some 68 percent of the total demand. Of the total, 99 percent is satisfied in 20 days. This is expressed as milligrams per liter (mg/1).

Of less importance for domestic flows, but possibly of significance if these are mixed with industrial wastes, is the chemical oxygen demand, COD. This is a measure of the oxygen-consuming capacity of both the organic and inorganic matter as shown through a test with a chemical oxidant. There is no necessary correlation between this value and the BOD.

The total suspended solids and the putrescible matter as BOD are the indicators usually used to express sewage strength, since these are the components for which treatment plant capacity must be provided. The volumes of waste to be handled are also vital to plant design; however, the bacterial content does not constitute a significant element in considerations of ordinary plant loadings since they take up no space. However, disinfection facilities of satisfactory capacity are usually provided. The emulsification of grease and oils by synthetic detergents

TABLE 11-10. Average Per Capita Solids and BOD in Domestic Sewage
[grams/capita/day (1 gram per capita = 2.2 lb per 1000 population)]

State of Solids	Mineral	Organic	Total	5-Day, 20°C BOD
Suspended	25	65	90	42
Settleable	15	39	54	19
Nonsettleable	10	26	36	23
Dissolved	80	80	160	12
Total	105	145	250	54

Source: Fair, G. M., and Geyer, J. C., *Elements of Water Supply and Waste Water Disposal*, John Wiley & Sons (1958).

in the sewage may reduce the quantities of solids which settle in the primary sedimentation tanks.

Tables 11-10 and 11-11 give the average per capita solids and BOD, and the average composition, respectively, for domestic sewage.[2]

The average raw suspended solids over 17 years at the District of Columbia treatment plant (tributary population of 1.25 million) was 164 ppm, equivalent to 164 mg/1. Average BOD was 142 ppm. The tributary system was about 40 percent combined. As the basis of design for plant expansion, garbage additions to the sewage were assumed to average 25 ppm suspended solids and 16 ppm BOD.

The per capita strength shown in Table 11-11 is predicated on the average United States production of 100 gpd/person. Stormwater, groundwater, substantial additions of cooling water, etc. may result in higher volumes but correspondingly lower relative strengths.

The temperature of sewage is usually higher than that of the water supply because of additions of warm water from users. This difference will vary with each city. Mean annual temperatures from typical cities range from 53 to 69°F (11.6 to 20.5°C). With higher temperatures, bacterial activity increases, and any dissolved oxygen or oxygen of the nitrates and nitrates present may be consumed so that the sewage becomes septic and offensive through the production of hydrogen sulfide. Similar disagreeable conditions will result if sewage becomes stale in transit.

TABLE 11-11. Average Composition of Domestic Sewage, mg/l

State of Solids	Mineral	Organic	Total	5-Day, 20°C BOD
Suspended	65	170	235	110
Settleable	40	100	140	50
Nonsettleable	25	70	95	60
Dissolved	210	210	420	30
Total	275	380	655	140

Source: Fair, G. M., and Geyer, J. C., *Elements of Water Supply and Waste Water Disposal*, John Wiley & Sons (1958).

Organisms. Bacteria in great numbers and possibly enteric viruses are present in raw domestic sewage. While pathogenic bacteria may be present, their identification is not customary in usual practice and instead organisms of the coliform group are sought as indicators of fecal contamination. This group includes organisms from differing natural sources and habitats, including human and animal intestines. It is estimated that human feces contain a mean value of 1×10^8 coliform bacteria /gm.[19] While this will include a variety of types of coliforms, one of particular interest will be *Escherichia coli*, which has become known as the "fecal coliform." Another indicator of fecal material is the fecal streptococci which are present in the feces of all warm-blooded animals. The proportions of these and fecal coliforms in man and the lower animals differ widely. Thus, the fecal coliform to fecal streptococcus ratio for domestic sewages is about 4 to 1.[20] In contrast, this ratio for cats is typically 0.29, dogs 0.02, and rats 0.04. This difference becomes significant in the identification of possible sources of surface runoff. Coliforms are reported on the basis of the most probable number, MPN/ 100 ml.

Virology techniques have not been developed to the point where virus enumerations are a routine procedure in sewage analysis. From 1×10^4 to 1×10^5 infectious doses of infectious hepatitis may be contained in 1gm of human feces; however, only a small portion of the population is so infected. The relative scarcity of hosts of these and other enteric viruses account for the fact that domestic sewage normally contains some 10,000 times as many coliforms as viruses.[19]

Stormwater Runoff. Urban surface runoff is being recognized as a significant source of water pollution. Rains tend to scour the areas on which they fall and collect accumulations of many kinds—street litter, either in bulk or simply the washout of its soluble contents; debris consisting of remnants of household and commercial litter, lawn and garden litter, and leaves from contiguous buildings; abrasives and chemicals used for ice control; chemicals used as pesticides, herbicides, and fertilizers; droppings from animals and birds; and substances in stack emissions deposited from the air.

The BOD of the soluble dust and dirt fraction of street litter has been shown to average 5 mg/gram, amounting to 0.40 lb BOD/day/curb mile.[21] The sometimes high levels of particulate matter such as sand and surface scour from eroding land may cause shoaling at points of discharge. The use of salt and calcium chloride in winter produces high concentrations of these materials in subsequent runoffs. Single applications of salt average from 200 to 400 lb/lane mile, resulting in as much as 20 tons/lane mile/season in some localities. Thirteen states have reported damage to vegetation and twelve states have reported pollution caused by chemical intrusion of salt into surface and groundwater sources. In a

TABLE 11-12. Characteristics of Stormwater

City	BOD (mg/l)	Total Solids (mg/l)	Suspended Solids (mg/l)	Coliform (/l)	Chlorides (mg/l)	COD (mg/l)
East Bay Sanitary District						
Minimum	3	726	16	4	300	
Maximum	7,700		4,400	70,000	10,260	
Average	87	1,401	613	11,800	5,100	
Cincinnati, Ohio						
Maximum seasonal means	12	260				110
Average	17		227			111
Los Angeles County						
Average, 1962–63	161	2,909			199	
Washington, D. C.						
Catch-basin samples during storm						
Minimum	6		26		11	
Maximum	625		36,250		160	
Average	126		2,100		42	
Seattle, Washington	10			16,100		
Oxney, England	100[b]	2,045				
Moscow, U.S.S.R.	186–285	1,000–3,500[b]				
Leningrad, U.S.S.R.	36	14,541				
Stockholm, Sweden	17–80	30–8,000		40–200,000		18–3,100
Pretoria, South Africa						
Residential	30			240,000		29
Business	34			230,000		28
Detroit, Michigan	96–234	310–914	102–213[c]	930,000[b]		
Criteria for:[a]						
Potable water						
(to be filtered)				5,000	600[b]	10
(not to be filtered)				50		10
Body contact water				2,400	NA	

[a]New York State.
[b]Maximum.
[c]Mean.
Source: Federal Water Pollution Control Administration, *Water Pollution Aspects of Urban Runoff*, FWPCA, U. S. Dept. of the Interior.

controlled test of the Kennedy Expressway at Chicago, with 101,600 lb of salt applied during February and March 105,150 lb were discharged between February and April (obviously salt was tracked into the test area from contiguous sections). Other studies have indicated that runoff usually accounts for only about half of the salt applied.[21]

The characteristics of stormwater runoff in a number of settings are given in Tables 11-12[21] and 11-13.[13]

As the result of numerous observations during storms of varying intensities and durations, DeFilippe and Shih, concluded:[13]

"The concentrations of waste constituents increase in proportion to the discharge rate during the initial flushing period. The average organic and nutrient concentrations in separated storm sewer runoff were observed to be approximately one-third of those in combined sewer discharges. The quantity of waste materials carried by the initial flush of storm runoff is proportional to the length of the dry weather between storms. The bacterial counts vary with flow rates of sewer discharges, with a peak count in the initial flush of each storm. Fecal coliform occurrence in separated storm sewer discharges was observed to be approximately one-eighth of that in combined sewer discharges."

It is evident that storm runoff should not be concentrated in the vicinity of water-oriented recreational activities, even though the coliform content most probably is not of human origin. Also, its other constituents will add materially to the assimilative burden of the receiving waters.

Combined Sewage Characteristics. The dry-weather composition of flows in combined sewers would correspond to that of separate sanitary sewage. On the occasion of rainstorms, the surface runoff reaching the combined sewers would have the characteristics shown for stormwater. Typical of the mixture would be the constituents shown in Table 11-13. To the extent that the interceptors can convey combined flows to treatment during storms, there will be a sharp increase in grit to be removed. This will be most pronounced during a sudden thaw following the application of abrasives. Treatment plant design for systems with tributary combined sewers must provide grit-handling facilities of greater capacity than required for separate systems.

TABLE 11-13. Characteristics of Combined Wastewater and Separated Storm Flow

	COMBINED WASTEWATER		SEPARATED STORM FLOW	
Waste Constituents (mg/l)	Range	Mean	Range	Mean
COD	80–1,760	382	29–1,514	335
BOD	10–470	71	3–90[a]	19[a]
TS	120–2,900	883	338–14,600	2,166
TVS	40–1,500	344	12–1,004	302
SS	35–2,000	622	130–11,280	1,697
VSS	10–1,280	245	0–880	145
Settleable solids	0–1,308	229	0–7,640	687
Total phosphate	0.8–9.4	3.0	0.2–4.5	1.3
Total nitrogen	1.0–16.5	3.5	0.5–6.5	2.1
Orthophosphate	0.1–5.0	2.0	–	–
Ammonia nitrogen	0–4.7	1.5	–	–
pH	5.6–6.7	6.3	7.2–6.0	6.5
Bacteria (counts/100 ml)				
Total coliform	420,000–5,800,000	2,800,000	120,000–3,200,000	600,000
Fecal coliform	240,000–5,040,000	2,400,000	40,000–1,300,000	310,000
Fecal streptococcus	1,000–49,000	17,200	3,000–60,000	21,000

[a]Excluding sample from a June storm, which had a BOD of 660 mg/l.
Source: De Filippi, J. A., and Shih, C. S., *Journal WPCF* **43**, 2033 (1971).

It has been estimated that overflows from combined sewers may contain from 3 to 5 percent raw sewage and may represent some 5 percent of the total pollutional discharge into the nation's watercourses. During storm peaks, as much as 95 percent of the sanitary sewage in the sewer may overflow directly to the receiving streams.[22]

Continuous gauging of the combined portion of the sewer system in Washington between 1955 and 1958 showed that 3.2 percent of the sanitary flows were lost through overflows during storms.[23]

The polluting characteristics of combined overflows are greater than those of surface runoff unmixed with other wastewaters. Mean median bacterial counts of a separate stormwater system and a combined system are compared in Fig. 11-29.[24,27]

Here the mean median total coliform density for combined flows was 8 times that for separate storm flows, the fecal coliforms 33 times larger, and the fecal streptococci 4 times larger. The significant number of fecal organisms from the storm system were presumably of animal origin.

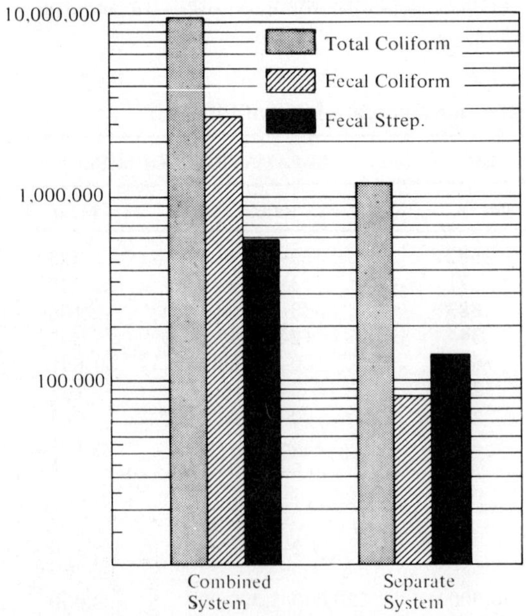

Fig. 11-29 Mean median bacterial counts per 100 ml; separate system (Ann Arbor) and combined system (Detroit). (*Courtesy of the Water Pollution Control Federation.*)

The high bacterial content of storm runoff causes the separation method of combined system elimination to become something less than an entirely satisfactory remedy unless the stormwater is disinfected.

Urban Peculiarities. Conventional treatment facilities can cope with the usual urban sewage, and, through the addition of selected processes, can produce an effluent meeting any reasonable standards. Even bacterial and viral content can be reduced to comparatively safe levels. Plant siting, however, becomes a most difficult problem because of the usual lack of space and the proximity of other establishments. The very size of the sewage load dictates a large facility, and the vast areas required for land disposal of sludge or effluent, as advocated by some, simply are not available in or near the average city. Transportation of sludge from the plant becomes complex and costly.

With the growth of fringe areas, it is not unusual for outlying jurisdictions to connect to the city system. These jurisdictions may produce a somewhat stronger sewage, and depending on distance, it may be stale on arrival for treatment. However, sewage from distant sources may reach the plant at off-peak periods for city flows and tend to even out the plant loading.

INDUSTRIAL WASTEWATER HANDLING

Sources and Characteristics. The broad categories of industry and their respective requirements are shown for 1964 in Table 11-14.[25]

Of the vast intake, approximating 49×10^{10} gal/yr, just under 1×10^{10} gal/yr (about 2 percent) was consumed, and much of this was by evaporation. Of the discharged water, some 44×10^{10} gal/yr (over 90 percent) was used for cooling and condensing, and little if any should have found its way into urban wastewater treatment systems. Perhaps no more than 4×10^{10} gal/yr (8 percent) of the total water withdrawn required some form of treatment. It is significant that 31 percent of the water used was recycled.

The approximate quantities and strength of

TABLE 11-14. Industrial and Investor-Owned Thermoelectric Plant Water Intake, Reuse, and Consumption, 1964

(From the 1963 census of manufactures, water use in manufacturing, and water resources activities in the United States—electric power in relation to the nation's water resources)

	WATER INTAKE (BILLION GAL/YR)							
	PURPOSE							
Industrial Group	Cooling and Condensing	Boiler Feed Sanitary Service, etc.	Process	Total	Water Recycled (billion gal/yr)	Gross Water Use, Including Recycling (billion gal/yr)	Water Consumed (billion gal/yr)	Water Discharged (billion gal/yr)
Food and kindred products	392	104	264	760	520	1,280	72	688
Textile mill products	24	17	106	147	163	310	13	134
Lumber and wood products	71	24	56	151	66	217	28	123
Paper and allied products	607	120	1,344	2,071	3,945	6,016	129	1,942
Chemicals and allied products	3,120	202	564	3,886	3,688	7,574	227	3,659
Petroleum and coal products	1,212	99	88	1,399	4,763	6,162	81	1,318
Leather and leather products	1	1	14	16	2	18	1	15
Primary metal industry	3,387	195	996	4,578	2,200	6,778	266	4,312
Subtotal	8,814	762	3,432	13,008	15,347	28,355	817	12,191
Other industries	571	197	271	1,039	1,207	2,246	71	968
Total industry	9,385	959	3,703	14,047	16,554	30,601	888	13,159
Thermoelectric plants	34,849	[a]	–	34,849	5,815	40,664	68	34,781
Total	44,234	959[b]	3,703	48,896	22,369	71,265	956	47,940

[a]Boiler feedwater use by thermoelectric plants estimated to be equivalent to industrial sanitary service, etc. water use.
[b]Total boiler feedwater.
Source: Federal Water Pollution Control Administration, "Report of the Committee on Water Quality Criteria," FWPCA, U. S. Dept. of the Interior (1968).

certain industrial wastes are given in Table 11-15.[2]

Industrial wastes may be related to domestic or combined wastes on the basis of their population equivalent. This would be the ratio of the quantity of the BOD, suspended solids, or other significant industrial waste component to the quantity of the corresponding element as produced on a per capita basis. For example, the production of 1 barrel of beer, generating 1000 gal of waste with a BOD of 1200 mg/l, would have a total BOD of 4542 gm. As shown in Table11-10, the daily per capita production of BOD is 54 gm., thus, the population equivalent is 4542/54 = 85.

Wastes from industry, including those shown, may contain metals, acids, caustics, organic chemicals, solids, grease, dyes, radio-isotopes, brines, etc., resulting from the many procedures employed for manufacture, extraction, processing, and so forth. These may be toxic or otherwise objectionable of themselves or may become toxic or objectionable when combined with other substances through synergistic action after discharge.

The sanitary wastes generated incidental to

TABLE 11-15. Approximate Quantities and Concentration of Industrial Wastewaters

Wastewater	Production Unit	Gallons per Unit	Suspended Solids (mg/l)	5-Day BOD (mg/l)	Special Characteristics
Beet sugar[a]	1 ton beets	3,000–4,000	800	450	Without Steffens' waste[b]
	–	–	–	700–2,000	With Steffens' waste
Brewery	1 barrel beer	300–1,000	250–650	500–1,200	–
Cannery	1 ton stock	2,500–8,000	200–3,000	300–4,000	–
Coal washing	1 ton coal	600–2,400	3,000–150,000	–	Fine coal
Distillery	1 bushel grain	45–55	20,000–40,000	15,000–20,000	Starchy
	100 gal molasses	200–300	–	20,000–30,000	High in potash
Gas and coke	1 ton coal	200–400	200–3,000	1,000–6,000	Phenols 1,000–5,000 mg/l in untreated liquor
Laundry	100 lb clothes	1,500	400–1,000	300–1,000	Alkaline
Malthouse[a]	100 bushel barley	800	–	400	–
Milk plant	1,000 lb milk	100–225	–	300–2,000	Lactose, sours readily
Oil-well brine	100 gal oil	4–2,800	–	–	Total solids 11,000 to 325,000 mg/l
Packing house, small	1 hog[c]	1,000–1,500	500–1,500	600–2,000	High in fats, proteins
large	1 hog[c]	500–800	400–1,000	350–1,000	–
Paper-making	1 ton paper	5,000–100,000	150–1,000	20–100	Fiber, clay
Paper pulp, ground-wood	1 ton pulp	0–12,000	500	20	–
sulfite	1 ton pulp	30,000–100,000	100–150	400	–
sulfite cooker-liquid	1 ton pulp	2,000–3,600	–	16,000–25,000	Lignin, sulfites
Tannery	100 lb hides	600–700	1,000–5,000	500–5,000	High in lime
Textile, cotton kier[d]	100 lb goods	120–250	–	1,000–1,600	High in caustic alkali
kier-liquor only	100 lb goods	30	–	4,000–8,000	–
cotton kier and bleach	100 lb goods	1,000–1,600	–	100–200	High in oxidizing agents
Flax retting[a]	1 ton straw	5,300	–	1,800–2,800	–
Fulling[a]	–	–	4,000	520	–
Silk boiling[a]	100 lb silk	850	100–500	700–1,000	–
Sulfur dye (dye-bath only)	100 lb goods	500–1,800	–	1,000–2,000	Sulfides
Wool-scouring, batch	100 lb wool	160–500	1,000–170,000	200–10,000	Wool grease
countercurrent	100 lb wool	40–100	1,000–170,000	200–10,000	Wool grease

[a] Data from one source only.
[b] The Steffens process precipitates tricalcium sucrate from beet juices that have yielded their crystallizable sugar. Steffens' waste is the liquor filtered from the sucrate.
[c] 1 steer = 2.5 hogs; 1 sheep or calf = 1 hog.
[d] A kier is a tub.
Source: Fair, G. M., and Geyer, J. C., *Elements of Water Supply and Waste Water Disposal*, John Wiley & Sons (1958).

industrial operations will probably be segregated and disposed of separately using a public system if available. As stated earlier, sanitary wastes will range from 8 to 25 gal/person/shift.

Disposal. With the growth of water demand from generally fixed sources and the national policy to upgrade water quality, there is great incentive to conserve water and reuse it where feasible. The current Environmental Protection Agency (EPA) system, which requires permits for industrial discharges and imposes effluent standards for toxic wastes from point sources, stimulates efforts to recover reagents, processing chemicals, and possible by-products of marketable value. This system also prompts consideration of disposal through irrigation, well injection, permanent storage, or connection to public sewer systems.

Treatment facilities for domestic and other municipal wastes are sensitive, especially their biological processes, and may not be expected to receive and treat significant quantities of many of the refractory substances contained in industrial wastes successfully. On the other hand, there may be clear advantages both for the industry and the public authority, whose treatment plant the industry could use for waste disposal. Some industrial wastes may be more easily treated after mixture with municipal sewage and at a possible savings to the industry. The addition of the industrial load could, with proper charges, become a benefit through economy of scale. Depending on relative size, the public facility may be able to provide superior technical skill in the treatment effort.

It has been projected by the Federal Water Pollution Control Administration, a predecessor agency to the EPA, that the discharge of industrial wastes into municipal sewers in the United States will increase by 1977 to represent about 50 percent of the total municipal wastewater, and have a population equivalent of 160 million.[26, 18] Estimated percentages of wastes discharged to municipal systems are shown for individual industries in Fig. 11-30.

Even where an industrial waste is unsuitable for introduction into a public system, it may be advantageous for the industry to remove only

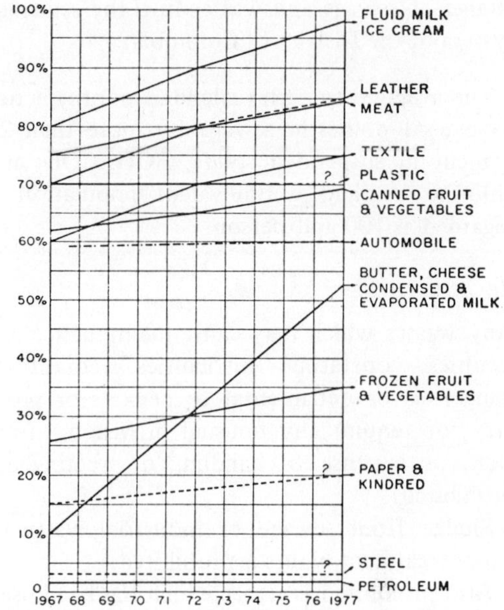

Fig. 11-30 Estimated percentages of industrial wastes discharged into municipal sewers in the U.S. (*Courtesy of the Water Pollution Control Federation.*) *Source:* Maystre, Y., and Geyer, J. C., *Journal WPCF* **42**, 1277 (1970).

the objectionable elements or characteristics and to discharge this partially-treated waste to a public facility where treatment would be completed. If the unacceptability in the public system is simply due to the volume of the waste, releases may be controlled at the source so that the reduced rates of flow of the hard-to-handle constituents may be more easily assimilated into the municipal wastes. Unacceptable concentrations of certain substances or of heat may be reduced through dilution.

The relative sizes of the industrial and the municipal loads and the specific constituents and characteristics of the former will justify analysis on a plant-by-plant basis. It is well for the public authority to anticipate such problems and determine as a matter of public policy what may and may not be introduced into the public sewerage system. The matter of charges for the conveyance and treatment of the industrial wastes must also be determined.

Elements of a Typical Regulation. The following is a digest of regulations concerning the discharge of sewage and wastes into the sewerage system of the District of Columbia:

The average strength (calendar month) is not to exceed domestic sewage by more than 20 percent in suspended solids, BOD, COD, and chlorine demand. Equivalent population is regarded as 100 gpd/person.

PROHIBITIONS:

Any wastes which may cause maintenance difficulties, operating difficulties, conditions dangerous or detrimental to persons or property, or require any unusual provision, attention, or expense of handling or treating are prohibited.

Sludge from sewage or industrial waste or water treatment plants is prohibited.

Also prohibited is water which has been used for cooling or heat transfer purposes without recirculation, and drainage associated with construction operations.

Any liquid or vapor over 150°F (65.5°C); emulsified oil or grease exceeding 100 ppm of ether-soluble matter; gasoline, naphtha, benzene, fuel oil, lubricating oil, or other flammable or explosive liquids, solids, or gases; wastes containing more than 10 ppm hydrogen sulfide, sulfur dioxide, or nitrous oxide; water or waste with a pH of less than 5.5 or higher than 9.0; toxic or poisonous substances in sufficient quantity to injure or interfere with any treatment process; any cyanides in excess of 2 ppm as CN from any outlet, with lesser amounts permitted only with continuous control; acid iron pickling wastes or plating wastes unless treated prior to discharge to system; exhaust gas, steam, malodorous gas, or any other substance which either singly or by interaction with other wastes is capable of creating a nuisance or hazard to life or of preventing entry into sewers; any radioactive toxic isotopes of over 100 days half-life (^{131}I and ^{32}P from hospitals not prohibited if diluted at source); highly colored dye wastes, spent tanning solutions, or similar wastes; and grease, oil, or other substances that will solidify or become discernibly viscous between 32 and 150°F (0 to 65.5°C) are all prohibited.

The 24-hr average concentration at the point of connection to the sewer shall not exceed:

Iron as Fe	1 ppm
Chromium as Cr (hexavalent)	5 ppm
Zinc as Zn	2 ppm
Copper as Cu	3 ppm
Phenols	0.05 ppm

and at no time shall the concentration exceed 3 times the 24-hr average concentration.

Other prohibitions relate to items not likely to originate with industry.

The foregoing may be so severe that industries would prefer complete in-house treatment of their wastes. The stated limits, however, provide protection to the public system and are suitable guides for comparison with any waste characteristics whose admission to a system may be under consideration.

Charges to Industrial Users. Charges to industrial users are usually based on the volume of the waste, the strength resulting from its suspended solids content, and the BOD which it exerts. Chlorine demand, grease content, and number of garbage grinders have been factors in some rate schedules.[18]

An equitable formula for the annual charge to an individual user is:[18]

$$c_i = v_o V_i + b_o B_i + s_o S_i \qquad (15)$$

where

c_i = charge to user i in dollars per year,
v_o = cost of transport and treatment chargeable to volume in dollars per gallon,
b_o = unit cost of treatment chargeable to BOD in dollars per pound,
s_o = unit cost of treatment (including sludge treatment) chargeable to SS in dollars per pound,
V_i = volume of wastewater from user i in gallons per year,
B_i = weight of BOD from user i in pounds per year,
S_i = weight of SS from user i in pounds per year.

Application requires the distribution of system costs for handling volume, treating solids, and satisfying oxygen demand, including maintenance, operation, and the cost of capital.

Determination of the SS and BOD contributed by the user may also require care. It should be noted that Eq. 15 provides the entire charge to the user, not a surcharge over some base rate or "nominal" system loading.

This general topic is thoroughly treated in Ref. 18 and in Chapter 15 of this volume.

SYSTEM PLANNING

General Considerations. Planning in some form is necessary for every wastewater disposal scheme. The nature of the undertaking will determine the extent and direction of the planning effort. Extension or improvement of an established system would entail a consideration of past performance and a review of a large amount of available physical data. While a system for a community yet to be built will depend upon information of a more general nature, modern water quality requirements are precipitating the planning of collection and disposal systems in rural areas where old private septic systems have failed. Also, new fringe subdivisions are being required to develop some form of community disposal.

Preliminary Concerns. The manner of disposal of the collected wastewater has become so important that it should be the first consideration. If state policy requires that there be no effluent, the terrain and ground cover are unsuited to irrigation or the like, and the geology makes underground disposal impracticable, no new community system or subdivision should be attempted unless the wastewater can be transported elsewhere for disposal. Without a firm, workable idea of how to dispose of effluent, it is folly to proceed further. In this first planning step, consultation with the state or regional agency having water quality control responsibility is essential at the outset.

If resolved, fact gathering may safely proceed. This should include a general reconnaissance of the site or setting. Taking into account the topography, a determination of the area to be served in the present, near future, and distant future should be possible. This requires the selection of a "design period." Topography will also determine the form of the collection system and the most likely place for effluent disposal. Climate will affect the generation of wastewater as pointed out previously. Street layouts, zonings, and land use patterns, and residential, commercial, industrial, and institutional developments and their locations require consideration. If not available, they must be hypothesized. An approach to system layout is also given elsewhere.

The system under study should be placed in perspective in relation to any other area or regional system, either existing or proposed. In general, the larger the system, the cheaper for both collection and disposal, and the broader the base upon which the cost of operating and maintenance forces and equipment may be spread. Smaller subdivisions and communities should look for opportunities to connect to, or become part of, larger systems. This may be done without the necessary sacrifice of local jurisdiction of local sewers and building connections.

At this stage in the planning studies, the general pattern should begin to emerge. If choices become apparent, their relative merits and faults must be weighed in general terms as to the economic and operational potentials and possible political and institutional consequences. The method of disposing of surface runoff should also be considered. As basic policy, little or none should be admitted to the sanitary sewers. The general outline of complementary stormwater facilities could well take form. Even if the study concerns the extension of an existing combined system, serious effort should be made to avoid introducing additional surface runoff by way of the new extension.

General system layout should now be attempted and estimates of populations and other sources of flows should be made. From these, tentative sizes may be fixed for key elements, including treatment facilities if necessary. It is not unlikely that more than a single layout or arrangement must be considered. While still not of the detail required for design, some estimates of needed capacities and likely pipeline sizes, lengths, and locations in several patterns should be made. With the assignment of cost estimates to these plans, a preliminary evaluation and comparison of alternatives should be possible. Costs for the initial, intermediate, and final stages of system development may be important in some studies. If pumping is extensive and multiple treatment facilities are being weighed against single facilities, then operating costs must also enter the comparison. Dependability, ease of maintenance, accessibility, need for rights-of-way and plant sites, aesthetics, social and ecological impact, etc. must all be taken into account. The cheapest in dollars may not be the "best."

The financial feasibility of the alternatives, if still undecided, should be examined and consideration should be given as to how and when the system will be financed. There could be a striking difference between a do-it-yourself treatment plant and the simple connection to another available system for disposal. Under one alternative, a nearby self-sufficient industrial facility could be attracted as a customer or participant in system financing if the industry was able to profitably abandon its independent treatment facility.

With the final selection of a preferred plan, the preliminary assumptions as to service to be provided, the growth pattern and rate to be anticipated, the size and character of flows to be accommodated, the disposal means to be used, and the general quality and type of construction to be employed should be carefully reviewed so they may become firm bases for future project development. Missing information on field conditions which could have a significant effect on cost estimates should be sought. Informal agreements with other persons, utilities, enforcement agencies, and political jurisdictions concerning possible permits, land access, or other matters which may have been made during the study should be confirmed as far as possible. The preliminary cost estimates may then be revised with confidence.

A report should be made which includes the following: a description of the proposed system (which might include drawings and maps showing its elements and its relation to the area, other systems, etc.), the expected phases of construction, the reasons for its selection, its purpose, cost estimates, description of difficulties, means of operation, and commitments to and by others. Also, in anticipation of future financing efforts, growth in demand, user charges, and revenues to be expected may be included.

Financing. Early American practice in the financing of municipal sewerage systems was to pay for capital and operating costs almost entirely from general taxes, usually on property. The use of a special benefit assessment against individual properties as sewer service was extended to them was also frequent. This assessment, intended to recover a substantial part of the cost of constructing the local service sewer, was usually predicated on the length of sewer which abutted the building lot or parcel.

It has now become a growing practice to cover the cost of operations by individual service charges, either as a flat rate or as some proportion of the water billing. These charges may also be used as the source of capital funds and contribute to the liquidation of debt previously contracted for system expansion. Benefit assessments may be used in conjunction with such service charges. An ad valorem tax on land can be justified, notwithstanding use of service charges and special assessments, to support the costs of investment in future system capacity and the costs of the nonpaying customers such as government and other public uses.

Bond issues for sewerage system improvements are general practice, and these are likely to be pledged on the property tax base, even though paid off by service charges. Statutory ceilings on municipal debts may prompt the creation of an independent authority or agency which may take jurisdiction of the sewerage system and enjoy financial independence of the municipality. Obligations of such authorities, depending as they do on revenues from service charges, are usually revenue bonds.

Real estate developers are being required to shoulder a growing proportion of the cost of sewer construction in new subdivisions. The entire subdivision sewer system may be required as a contribution to the local jurisdiction, including either temporary or permanent treatment facilities, all built to the specifications of the local governmental authority.

The federal government, under several programs, is able to offer grants to local government agencies in the states to assist in the cost of constructing pollution abatement facilities which may include certain sewer lines as well as treatment plants. Some states have similar grant programs. These may be employed in conjunction with the federal grants or independently.

New agencies at the state and interstate levels, the latter through interstate compacts, are authorized to provide disposal service to local communities, thus possibly providing economy of scale. These agencies may in some cases guarantee or underwrite local debts incurred for disposal and related works.

Storm and drainage systems are usually paid for by property taxes or bonds funded thereby. State and federal highway funds may also be important sources of funds for storm runoff control.

Study of the means of financing a proposed system or facility must include consideration of the administration under which it will operate. An independent authority may have many advantages both in dealing with the several governmental jurisdictions which may be involved and in developing better credit. An estimate of necessary customer charges and the resulting revenues will be essential. The outstanding obligations, those foreseen for other purposes, the life of the proposed obligations, and many other detailed concerns will influence the final decision.

Active Phases. After plan adoption and after agreements as to the financing and the respective responsibilities between agencies or other parties for construction and operation of the system are made, sites for key facilities should be selected and acquired or preempted if possible, and the street system or other routing committed for future availability. Coordination with all other agencies and utilities to be affected is essential. Permits should be sought to put structures on or under navigable waters and across federal, state, or other lands. Firm commitments should be acquired from state or federal regulatory agencies, and approaches to agencies having control of construction grant funds should be made.

With the first availability of funds, design of the elements necessary to provide initial service may begin with construction to follow as needed.

A new agency will need to establish regulations for system use, schedules of charges for service and assessments, penalties for nonpayment, enforcement authority, procedures for dealing with applicants for service, and publicizing of service. It will also need to devise immediately a system of recording all "as-built" work; complete locations should be recorded for everything installed.

REFERENCES

1. *Design and Construction of Sanitary and Storm Sewers*, ASCE Manual No. 37 (1969).
2. Fair, G. M., and Geyer, J. C., *Elements of Water Supply and Waste Water Disposal*, John Wiley & Sons, Inc., New York (1958).
3. Fair, G. M., Geyer, J. C., and Okun, D., *Water Supply and Waste Water Removal*, John Wiley & Sons, Inc., New York (1966).
4. Metcalf and Eddy, Inc., *Wastewater Engineering*, McGraw-Hill, New York (1972).
5. Geyer, J. C., and Lentz, J. J., "An Evaluation of the Problems of Sanitary Sewer System Design," *Journal WPCF* **38**, 1139 (1966).
6. District of Columbia, Dept. of Sanitary Engineering, Investigation of Sewerage System, Metcalf and Eddy (1955).
7. Harmon, W. G., "Forecasting Sewage Discharge at Toledo," *Engineering News Record* **80**, 1235.
8. H. E. Babbitt, "Sewerage and Sewage Treatment," 8th Ed., John Wiley and Sons, New York (1958).
9. Horner, W. W., "Modern Procedures in District Sewer Design," *Engineering News* **64**, 326 (1910).
10. Federal Water Pollution Control Administration, *Problems of Combined Sewer Facilities and Overflows*, FWPCA, U. S. Dept. of the Interior, WP-20-11 (1967).
11. Metcalf and Eddy, Inc., *Sewerage and Sewage Disposal*, McGraw-Hill, New York (1922).
12. Camp, T. R., "Design of Sewers to Facilitate Flow," *Sewage Works Journal* **18**, 3 (1946).
13. De Filippi, J. A., and Shih, C. S., "Characteristics of Separated Storm and Combined Sewer Flows," *Journal WPCF* **43**, 2033 (1971).
14. McKee, J. E., "Loss of Sanitary Sewage Through Storm Water Overflows," *Journal Boston Society of Civil Engineers* **34**, 55 (1947).
15. Palmer, C. L., "The Pollutional Effects of Storm-Water Overflows from Combined Sewers," *Sewage and Industrial Wastes* **22**, 154 (1950).
16. Board of Engineers' Report on Improvements to Sewerage System, District of Columbia (1957).
17. Dept. of Sanitary Engineering Office Manual, District of Columbia.

18. Maystre, Y., and Geyer, J. C., "Charges for Treating Industrial Wastewater in Municipal Plants," *Journal WPCF* **42**, 1277 (1970).
19. U. S. Public Health Service Drinking Water Standards, USPHS, U. S. Dept. of Health, Education, and Welfare (1962).
20. Geldreich, E. E., *et al.*, "The Bacteriological Aspects of Stormwater Pollution," *Journal WPCF* **40**, 1861 (1968).
21. Federal Water Pollution Control Administration, *Water Pollution Aspects of Urban Runoff*, FWPCA, U.S. Dept. of the Interior, WP-20-15 (1969).
22. U. S. Public Health Service, "A Preliminary Appraisal, Pollution Effects of Stormwater and Overflows from Combined Sewer Systems," USPHS (1967).
23. Johnson, C. F., "Equipment, Methods, and Results from Washington, D. C. Combined Sewer Overflow Studies," *Journal WPCF* **33**, 7 (1961).
24. Benzie, W. J., and Courchaine, R. J., "Discharges from Separate Storm Sewers and Combined Sewers," *Journal WPCF* **38**, 400 (1966).
25. Federal Water Pollution Control Administration, "Report of the Committee on Water Quality Criteria," FWPCA, U. S. Dept. of the Interior (1968).
26. Federal Water Pollution Control Administration, "The Cost of Clean Water," FWPCA, U. S. Dept. of the Interior (1968).
27. Burm, R. J., and Vaughn, R. D., "Bacteriological Comparison between Combined and Separate Discharges in Southeast Michigan," *Journal WPCF* **38**, 400 (1966).

12

Sanitary Sewage Treatment

A. J. Fischer

Consultant
Manhasset, New York

INTRODUCTION

The treatment of sanitary sewage is accomplished in a sequential series of unit operations, some common to water and sewage treatment, such as coarse screening and sedimentation, and others specific to waste treatment, such as biological treatment. Basic design considerations as well as the need for certain of the unit operations are generally dictated by governmental agencies. Most pollution regulatory agencies have established effluent standards, detention periods, and other criteria affecting design and operation. Considerable latitude is often given, however, in enforcing these standards depending on plant size, local conditions, and other factors.

Primary sewage treatment generally combines the unit operations of screening or communition, grit removal, and sedimentation. For secondary treatment, oxidation by alternate biological means is employed and effluent disinfection may be added in either case. Where it is desired to improve sedimentation efficiency, flocculation with chemicals may be employed ahead of sedimentation in the primary system.

Thus, the treatment of sanitary sewage may be regarded as a series of steps to remove solids of gradually decreasing size and/or specific gravity as follows:

1. Screens to remove or comminute large solid objects.
2. Grit chambers to remove high specific gravity solids.
3. Sedimentation to remove settleable suspended solids.
4. Sedimentation with chemicals to remove both settleable and nonsettleable suspended solids.
5. Biological oxidation (with final sedimentation) to remove residual suspended solids in addition to oxidizing dissolved organic solids.

PRIMARY TREATMENT OPERATIONS

Screening. Screening devices are used for the primary purpose of removing and/or comminuting large solid objects and rags that would injure subsequent equipment, clog piping, or wrap around pump impellers, interfering with their operation.

Bar Screens. Coarse Racks. Coarse racks are used chiefly in large plants to remove sizable objects that would injure mechanisms used for cleaning conventional bar screens. They are also used as protective devices ahead of fine screens and comminutors. In the very large plants, hand-cleaned cage-type units that can be lifted from the channel are frequently employed. They are often used in duplicate so that one unit may be kept in service while the other is cleaned. Alternately, mechanically-cleaned trash screens such as those used for protecting water intakes are employed. Hand-cleaned coarse racks or screens are also used in the medium to large plant range.

Clear openings between bars on coarse racks are generally 3 in. where the rack is mechanically cleaned, but may vary from 3 to 6 in. depending on plant size and the type of unit which follows. In large plants where the screen channel is relatively deep, the rack is generally vertical, or at a very slight angle from vertical. In the smaller plants where the racks are cleaned by hand, they may repose at an angle of from 30° to 60° from the vertical. Here the flatter surface is preferable as it facilitates the hand cleaning.

Standard Bar Screens. Standard bar screens are used for the same purpose as are coarse racks. The chief difference is the opening between bars. In the case of bar screens, the clear openings will vary from $\frac{1}{2}$ to 2 in., the most common opening being 1 in. Because of this comparatively narrow opening, bar screens are generally mechanically cleaned to prevent high head loss due to screenings accumulations (Fig. 12-1).

Bar screen slopes vary from 45° to 90° from

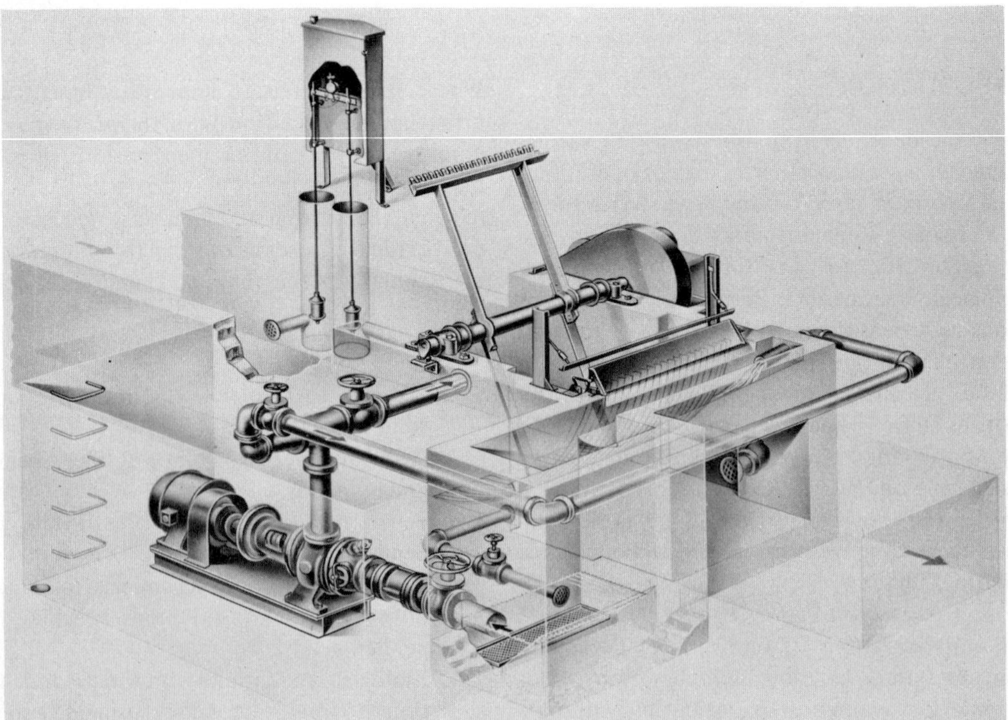

Fig. 12-1 Bar screen and disintegrator. (*Courtesy of Dorr-Oliver, Inc.*)

the horizontal, 60° being the most common angle. There are, however, screens with curved bars (forming a 90° arc) to conform with the circular path of the rotating cleaning mechanism. The screen channel should be designed so that the velocity through the screen should not exceed $2\frac{1}{2}$ to 3 ft/sec at (average) design flow and should not fall below about 1.0 ft/sec at minimum flow. This practice also applies to coarse racks.

Operating Controls. The operation of bar screen cleaning mechanisms may be controlled by: (1) manual switches, (2) loss-of-head switches, (3) time clocks, (4) high-level switches, or (5) overload switches. The time-clock control seems to be the most satisfactory. A selective timer, however, that will change the frequency of cleaning several times a day to conform with changes in screenings would appear to be desirable.

Screenings Volume. The volume of sewage screening obtained from coarse racks and standard bar screens varies greatly with the screen openings and the character of the sewage, normal variations running from 0.1 to 5.0 ft^3 wet screenings/million gal of sewage. With 1 in. openings, an average of about 1.0 ft^3/million gal appears to be a good general figure. Screenings volumes reported from various typical U.S. plants[1] are presented in Table 12-1.

Disposal of Screenings. Screenings from bar screens may be disposed of by: (1) grinding and returning to raw sewage, (2) incineration, (3) digestion, and (4) burial.

Where grinding is employed, two types of grinding devices are available: the cutter pump and the "hammer mill" type shredder. In both types, conveying liquid must be provided. Raw sewage or plant effluent may be used for this purpose. This method of handling screenings is highly satisfactory in both large and small plants.

Special incinerators are offered for sewage screenings. In general, such units have been found satisfactory for medium and large plants, but have not been found suitable for small ones. Attempts have been made to incinerate screenings along with sludge in conventional sludge incinerators. Difficulties are experienced in such cases unless the screenings are first ground. This also applies to digestion of the screenings either alone or mixed with raw sludge.

Screenings burial is a satisfactory method of disposal, especially in small plants, as long as land is available and precautions are taken to provide suitable disinfection and earth cover such as that employed in the landfill garbage disposal system.

Comminuting Screens. Comminuting screens are devices for screening solids in the sewage flow and cutting the screened solids without removing them from the flow (Fig. 12-2). There are several types of mechanisms for accomplishing this purpose. In one, the screening unit consists of a revolving vertically-slotted drum mounted in a specially-designed channel. In

TABLE 12-1. Bar Screen—Operating Results

		SCREENINGS—AVERAGE			
Plant	Opening Between Bars	Ft^3/million gal wet	Lbs/million gal wet	Percent moisture	Lbs/million gal dry
Plainfield, N.J.	$\frac{1}{2}$ in.	5.7	261	84	42
Aurora, Ill.	$\frac{1}{2}$ in.	1.4	77	86	71
Milwaukee, Wis.	$\frac{3}{4}$ in.	210	150	88	18
Baltimore, Md.	$\frac{7}{8}$ in.	1.9	100	81	19
Aurora, Ill.	$1\frac{1}{4}$ in.	0.8	52	81	9
Grand Rapids, Mich.	$1\frac{1}{2}$ in.	0.6	–	–	–
Champaign, Urbana, Ill.	$1\frac{9}{16}$ in.	1.0	–	–	–
Worcester, Mass.	2 in.	1.8	121	–	–

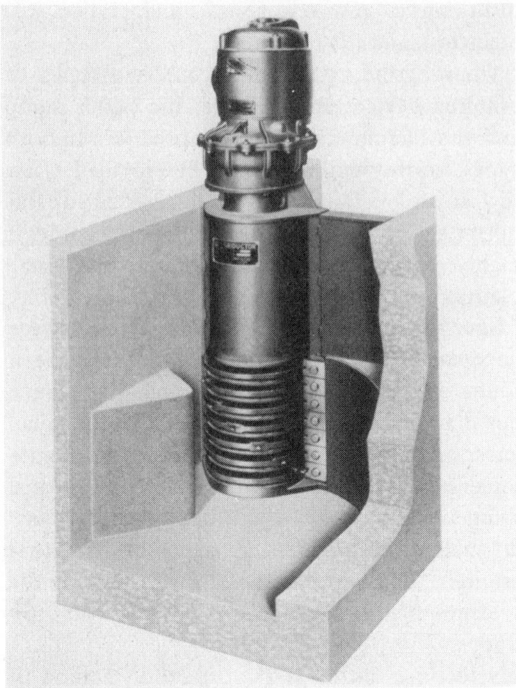

Fig. 12-2 Comminutor-type screening device. (*Courtest of Dorr-Oliver, Inc.*)

another, the screen consists of semicircular horizontal bars set in a horizontal channel. In a third type, a special flat horizontal bar screen is used.

All have distinctive moving types of cutters. Comminuting screens have openings $\frac{1}{4}$ to $\frac{3}{4}$ in., the smaller size being most commonly used with the drum and semicircular bar type units.

The advantage of comminuting screens is that they take the place of conventional bar screens equipped with separate grinders and accomplish the grinding operation below the sewage surface, eliminating fly and odor nuisances. A disadvantage is that they are subject to high wear from grit; hence, they are most frequently located in the system following grit chambers. This often causes operating troubles with grit-removal mechanisms. An added disadvantage with certain types is that a special channel construction is required.

Fine Screens. Fine screens currently in use in sewage treatment plants are of the disc and the drum types. Both use (as a screening medium) a perforated plate with round or slotted openings or woven wire cloth. Perforations are usually $\frac{1}{16}$ to $\frac{3}{4}$ in. across.

In the disc type, a round flat disc rotates on a slightly inclined axis. Sewage flows through the lower section and retained solids are removed by rotating brushes as they are carried above the sewage level on the inclined disc.

In one drum type, a horizontal cylindrical drum rotates transversely to the flow. Sewage passes through the screen and flows out at right angles to the next stage of treatment. In one variation, rotating brushes are mounted on top of the screen for the removal of retained solids. In another, the screening drum rotates rapidly, producing a cascading action that causes the retained solids to drop into a pit where they are removed by a bucket elevator with perforated buckets.

In another drum type, the drum rotates parallel to the flow into the interior at one end and out through the screen at right angles. Solids retained on the inside drum surface are removed by sprays and conveyed out by means of a screen conveyor.

Depending on the size of the screen openings and the character of the sewage, fine screens are capable of removing from 5 to 35 percent of the suspended solids. Normally, however, the removals will be in the range of 20 to 25 percent. Operating results reported from several typical plants are given in Table 12-2.

Other Screen Types. Other types of screens less widely used at present in the sewage treatment field are the vibrating screen for fine screening, the tangentially-fed curved screen for intermediate size screening, and the so-called microstrainer for very fine straining.

A variety of types of vibrating units are offered. All use a flat or slightly inclined screening deck of 20 to 200 mesh metal cloth. This screen is vibrated either by mechanical or electrical means. The vibration causes the liquid to pass through the screen and conveys the retained solids to the point of discharge.

The tangentially-fed curved screen consists of a series of spaced wedge-wire bars set perpendicular to the line of flow and arranged in a series of short chords that approximate a 60° arc of a circle. Raw sewage is fed tangentially at

TABLE 12-2. Fine Screen Operating Results[a]

Plant	Type Screen	Size Openings	SCREENINGS—AVERAGE Ft³/million gal wet	Lbs/million gal wet	Percent moisture	Lbs/million gal dry
Plainfield, N.J.	Disc	$\frac{1}{16}$ in.	21	1323	88	159
Long Beach, Cal.	Drum	$\frac{1}{16}$ in.	40	1960	80	392
So. Yonkers, N.Y.	Disc	$\frac{1}{16}$ in.	–	1000+	90	100+
Milwaukee, Wis.	Drum	$\frac{3}{32}$ in.	14	975	92	78

[a]If the average raw sewage suspended solids content is assumed to be 200 ppm, the removal obtained by the screens at these plants vary from 4.6 to 23.5 percent.

the upper end and retained screening discharge is at the lower end of the screen. This type of screen requires a high head, and discharges relatively wet screenings. It is, however, of relatively simple construction, is self-cleaning, and has no moving parts.

The microstrainer is a revolving drum-type screen with a special woven wire cloth having openings as small as 20 μ. Sewage flows through the screen and out through the interior at right angles. Retained solids are removed by water sprays into discharge troughs. This type of screen, because of its fine mesh surface, is capable of use for effluent polishing. It may also be used for treating stormwater overflow. Test results[2] on combined sewer overflow are presented in Table 12-3.

Grit Removal. Grit removal devices are used for the removal of sand, gravel, and other high specific gravity inorganic solids from sewage that would cause undue wear on pumps, sludge disposal systems, etc., clog underflow lines, and segregate in settling and sludge digestion tanks, thus reducing their effective capacity. Strictly speaking, grit also includes some high-gravity or fast-settling organics, but these are nonputrescible solids such as coffee grounds, pulverized egg shells, seeds, etc.

Grit Collectors. Most grit collector tanks are designed to separate solids of 2.65 specific gravity in the size range of 65 to 100 mesh (0.21 to 0.15 mm). To accomplish this requires settling in a tank having a surface area of such a size that the maximum overflow rate is in the range of 32,000 to 46,000 gpd/ft^2. These figures correspond to settling velocities of the particles of about 3.0 to 4.3 ft/min. An average overflow rate of 40,000 gpd/ft^2 is commonly used.

Tank widths and depths should be selected so that the velocity of flow should not exceed 0.75 ft/sec at design flow. Due allowance must be made in depth to accommodate grit accumulations in collection tanks that are cleaned mechanically on a continuous basis. Most modern plants use mechanically-cleaned units.

When flow velocities fall below 0.75 ft/sec due to a drop-off in flow or the use of greater tank depths than are required for mechanical reasons, a considerable amount of putrescible organic matter will settle with the grit. This is no longer of consequence as modern mechanical grit units are equipped with grit washers.

TABLE 12-3. Microstrainer Performance—Storm Sewer Overflow

Screening Rate (gpm/ft^2 at 1.0 rpm of screen)	Drum Speed (rpm)	SUSPENDED SOLIDS (PPM) Influent	Effluent	Percent Removal	Screen Openings (microns)
1.9	1.9	288	72	75	23
4.1	1.75	173	89	49	35
6.4	2.5	159	48	70	23
32.0	6.0	69	15	78	23

Grit collectors may be rectangular, square, or circular in shape. Rectangular units are the most common. It is important that inlet and outlet designs of such units minimize inlet turbulence, high bottom scour, and flow velocity changes with varying flows. Such disturbing factors may be lessened by the use of proper approach channels to even out the flow and the use of proportional flow devices such as the Sutro or Rettger weir or the Parshall flume. A rectangular outlet weir may be used to control velocity if the bottom sides of the collection tank are sloped inwardly. This design has the added advantage of requiring a narrow grit removal mechanism.

In large plants, especially those treating sewage from combined systems, it is desirable to install multiple units so that the number of units in service may be varied depending on flow variations and grit load.

In square tanks, greater care must be exercised in distributing the feed to avoid short circuiting with local high velocities. This may be accomplished by the use of a series of directional flow fins.

Relatively deep circular tanks with steep sloping bottoms have been used in small plants. Such units are fed at the center via a deep feed well and overflow at the periphery.

There are many varied mechanisms for removing settled grit from collection tanks. Among those used widely in rectangular tanks are the chain-mounted flight, bucket, screw, and reciprocating rake conveyors.

In square tanks, rotating mechanisms with outward raking blades are most commonly employed. Corner fills are used in such tanks so as to avoid the use of troublesome extension arms on these mechanisms.

In some designs, the conveying mechanism carries the grit up a ramp at the inlet to a suitable point of discharge. In others, the grit is raked to a sump where it is picked up by pumps, bucket elevators, or sloping deck screw or rake mechanisms.

Grit Washers. Where a clean grit is required (e.g., one low in putrescible solids), washing of the grit to remove relatively slow-settling organic matter is essential. Most modern grit chambers, especially those mechanically equipped, provide for such washing (Fig. 12-3). This may be accomplished by mechanical, hydraulic (air or water), or cyclonic means.

The oldest method of separating the light organics, which is still in use, is aeration of the collection tank to maintain a suitable internal velocity. In applying this method, care must be taken to avoid diffuser clogging, especially at the inlet end of the tank where the bulk of the heavier grit settles. Care must also be taken to avoid highly violent aeration so that fine grit is not resuspended.

Mechanical washing involves the vertical conveying of the grit above the liquid level by means of bucket, screw, or rake conveyors, or the jigging of the settled grit to loosen and separate the fine organics.

The use of water washing is accomplished by using water (or plant effluent) in conjunction with the vertical conveying means.

Cyclonic grit washing (Fig. 12-4) involves the pumping of the settled grit to a cyclone that comprises a cylindrical feed section and a conical section. The feed enters the feed section tangentially. Liquid containing the lighter organics exits through a center connection at the top. The washed grit discharges at the cone apex. Wash water may be added with the feed, if desired. When mounted on its side, the cyclone is self-regulating in that very little liquid discharges at the apex when no grit is present.

Further dewatering of the grit is required. This may be accomplished in a sump with a drainage connection in small plants or in a small rake or spiral clarifier in larger plants. The latter has the added advantage that additional washing may be obtained by adding wash water on the deck of the clarifier unit. To avoid clogging, the minimum diameter of a cyclone in this application should be 12 in.

Combination Grit Systems. In small plants where the amount of grit is relatively low, grit removal may be accomplished in the clarifier feed inlet or feed well. In another unit, preaeration, grit removal, and clarification are combined. Another system eliminates the grit collection tank and provides grit collection and sedimentation in a single clarifier. Dilute un-

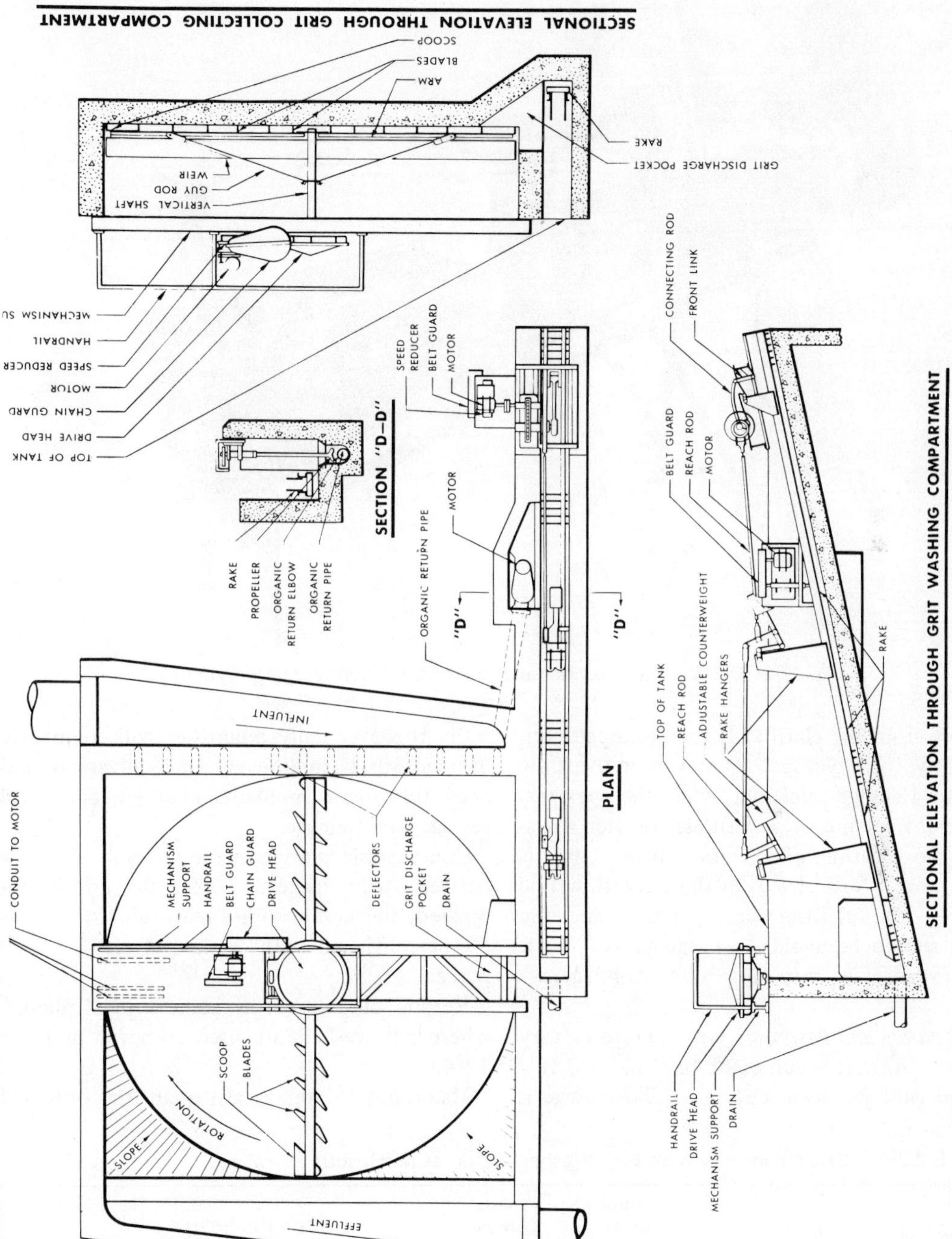

Fig. 12-3 Washing-type grit chamber. (*Courtesy of Dorr-Oliver, Inc.*)

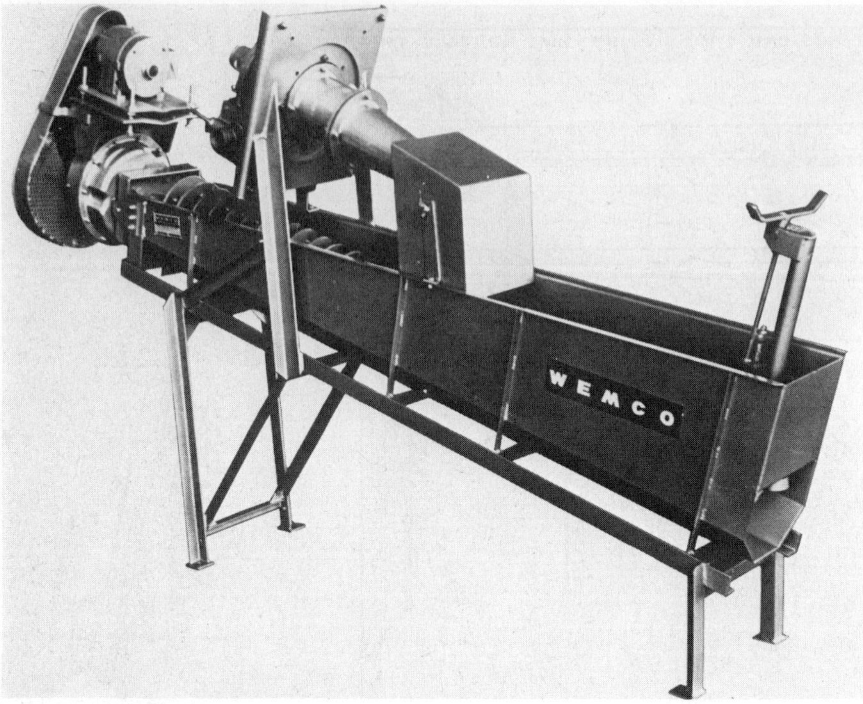

Fig. 12-4 Cyclonic grit separator and grit washer. (*Courtesy of WEMCO*.)

derflow from the clarifier is then pumped continuously to a cyclone. The cyclone overflow goes to a sludge thickener, while the apex discharges to a sump or a clarifier. In such a system, the location of the underflow pump is important. Vertical legs in the underflow line between the clarifier sludge pump and the pump should be avoided as segregation of grit in this vertical leg may cause a line stoppage.

Grit Volume and Disposal. Grit quantities vary greatly. Annual volumes of 0.20 to 12.0 ft^3/million gal have been reported. The average is 3.0. However daily quantities will frequently average 30 ft^3/million gal and will greatly exceed this figure in plants handling combined sewage and drainage.

Combustible matter in the grit will also vary between wide ranges. Where the grit is not washed, the combustibles may run from 40 to 55 percent, and with washing from 1 to 10 percent.

Results obtained from some typical plants[3] where grit washers are used are given in Table 12-4.

Maximum to average grit quantities obtained

TABLE 12-4. Grit Chambers With Grit Washers—Operating Results

Plant	Maximum Daily Flow (mgd)	Overflow Rate at Maximum Flow (gpd/ft^2)	Grit Volume at Maximum Flow (ft^3/million gal)	PERCENT OF GRIT RETAINED ON SCREENING			Percent of Putrescibles in Dry Grit
				48 Mesh	65 Mesh	100 Mesh	
Baltimore, Md.	243.0	22,500	14.6	66.1	88.0	97.4	3.1
Berea, Ohio	2.0	20,000	20.0	95.2	97.0	97.5	4.6
Birmingham, Ala.	16.0	31,200	0.7	89.6	95.3	97.5	3.5
Findlay, Ohio	10.0	25,000	0.8	88.8	95.5	98.1	2.6
Muskegon, Mich.	10.0	25,000	8.1	69.1	92.2	98.8	0.6
Sioux Falls, S. Dak.	9.0	23,000	15.0	84.4	95.2	98.5	0.4
Souderton, Pa.	0.8	8,000	18.0	97.5	98.5	99.0	1.2

TABLE 12-5. Variations in Grit Quantities

Plant	GRIT REMOVAL/DAY (ft^3/million gal) Average	Maximum	Maximum/Average
Cleveland, Ohio-Easterly	0.3	540.0	1800.0
Battle Creek, Mich.	18.6	170.0	9.2
Muskegon, Mich.	1.5	8.1	5.4
Baltimore, Md.	5.4	14.6	2.7
Green Bay, Wis.	7.0	7.6	1.1

at a number of plants[3] are presented in Table 12-5.

Adequately washed and drained grit is inoffensive and may be used as landfill. Grit that is not washed is highly putrescible and should be buried. In a few instances, unwashed grit is incinerated along with sludge cake. This method of treatment is not recommended because of undue wear on equipment.

Sedimentation. Sedimentation is used to remove settleable suspended solids from sewage not removed in screening and grit removal units.

Primary sedimentation is used as a means of reducing the suspended solids and biological oxygen demand load on the receiving stream, or for reducing such loads in intermediate treatment systems using chemical precipitation or in secondary systems employing biological treatment devices.

Secondary or final sedimentation is used for removing coagulated suspended solids produced in chemical or biological treatment.

Significant factors affecting the efficiency of sedimentation tanks, apart from detention or overflow rates, are inlet and outlet designs and adequate provisions for sludge and scum removal with minimum disturbance of the tank contents.

Tank shape has apparently no influence on tank effectiveness. Rectangular tanks are in use with lengths ranging from 30 to 300 ft and with widths of 12 to 107 ft. Length-to-width ratios vary from 1:1 (square tanks) to 7.5:1. The most generally used maximum lengths, however, are 100 to 120 ft, while maximum widths are 16 ft where single-flight raking mechanisms are used and 30 ft where double-flight units are used.

Circular tanks range in diameter from 12 to 200 ft, but few tanks are in use for primary sedimentation having diameters of more than 100 to 120 ft. The average size unit in use is in the 50 to 60 ft diameter range. Square tanks with circular sludge removal mechanisms are in the same size range as are circular units.

Tank side water depths for all types of tanks vary from 7 to 15 ft; the average is in the 10 to 12 ft range. Floor slopes toward the sludge outlet are generally $\frac{1}{8}$ in. per foot in the case of rectangular tanks and 1 in. per foot for round and square units.

Feed distribution at the tank inlet is of importance in reducing excessive streaming of the feed toward the tank outlet and so reducing sedimentation efficiency. In the case of rectangular or cross-flow square tanks, this may most readily be done by feeding over a weir behind a vertical baffle spaced about 2 to 3 ft from the tank wall and extending down about 2 to 3 ft below the water surface.

In circular tanks and center-fed square tanks, feeding is usually accomplished via a center feed well having an area of about 3 percent of the tank surface area and extending from above to about 4 ft below water level. The center well is fed via a horizontal pipe extending from the tank side in the case of small tanks or through an inverted siphon in the case of large tanks.

Outlet arrangements for most units are overflow weirs located at or near the tank wall opposite the inlet (the tank periphery in the case of circular or center-fed square tanks). Overflow weirs may be straight or V-notched. The latter are preferred as they provide far better distribution at times of low flow or where wind causes surface disturbance. A variation of the usual method of feeding and overflowing circular tanks is one where the feed is introduced peripherially behind a deep baffle and the overflow is taken off at the center.

The mechanism most commonly used for removing deposited sludge from the rectangular tanks is the endless chain drag, in which a pair of chains carry, at about 10 ft intervals, wood flights that span the tank width. The sludge is raked to hoppers or cross-collectors at the inlet end of the tank where the mechanism is driven by chains and sprockets from above. An alternate type consists of a carriage that spans the tank width and is moved on rails along the tank length by either cables driven from one end of the tank or by a drive unit mounted on the carriage. A sludge raking member is suspended from the carriage.

For circular tanks, the most common sludge removal mechanism consists of arms attached to a revolving center shaft or cage. Attached to the arms are plows or blades set at an angle so that as the unit is rotated, the deposited sludge is pushed toward a central discharge cone or hopper (Fig. 12-5). Other sludge removal mechanisms include a spiral scraper supported by the arms, and nozzles connected to the pipe arms whereby the sludge is removed by hydrostatic means. The latter type is more frequently used for activated sludge.

Skimming arrangements for the removal of grease and other floating material are generally a part of the sludge removal mechanism. Where skimming is employed (usually in most primary and some secondary settling units), a scum baffle must be used ahead of the effluent outlet.

In the case of chain conveyor-type mechanisms, skimming is accomplished by the wood flights on their return travel and the scum removed via a slotted pipe or tipping trough located at the surface near the outlet scum baffle. The slotted pipe or tipping trough may be operated manually or automatically. In circular tanks, a skimming blade is rotated by the center mechanism and rakes the scum up a ramp to a discharge trough.

Rectangular tanks have an advantage in that they may be "nested" and so require less ground area where multiple tanks are required. Also, common wall construction can be used and covers provided at lower cost. Their disadvantage is that where chain drag sludge removal mechanisms are used, maintenance costs are higher due to the multiplicity of underwater moving parts.

Primary Sedimentation. Suspended solids in raw sewage vary widely in character and composition. Therefore, their settling characteristics differ greatly from the patterns followed by such discrete particles as sand, for example, or

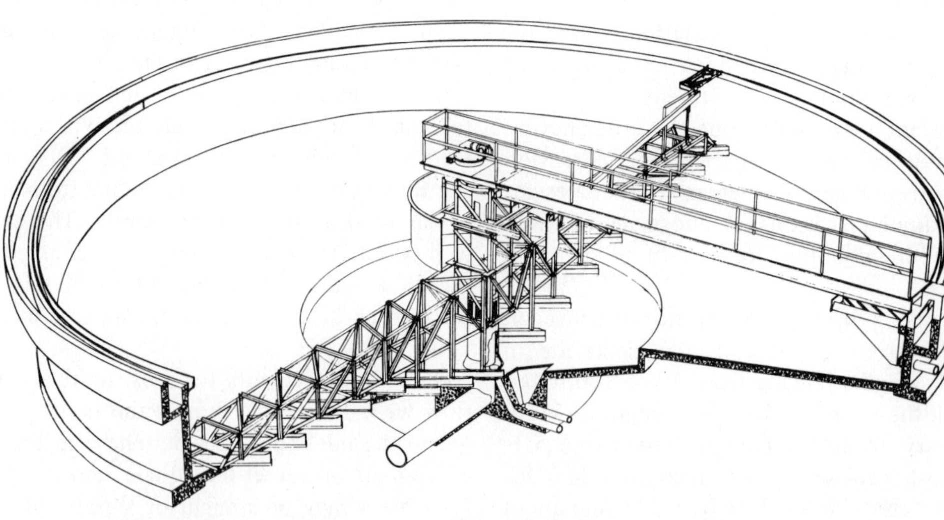

Fig. 12-5 Primary clarifier with sludge thickening zone. (*Courtesy of Envirotech*.)

flocculated material such as activated sludge. Tank surface area, although of major importance, is, therefore, not the dominant factor in tank design. Detention period also has an important role. This is due to the fact that autoflocculation of the finer suspended solids fraction occurs during the holding period.

Normally, minimum design detention periods for primary sedimentation tanks vary from about 2.0 hr for small and medium-sized plants to 1.0 hr for large plants. Design overflow rates for small plants usually vary from a maximum 600 gpd/ft^2 where secondary treatment is not employed to 900 gpd/ft^2 for medium-sized plants and for small and medium-sized plants employing secondary treatment. In large plants, usual maximum design overflow rates vary from 900 to 1200 gpd/ft^2, but may run considerably higher where local conditions warrant.

Thus, in small plants where the tank depth is 7 ft at 2 hr detention, the upflow rate will be 3.5 ft/hr and the overflow rate 630 gpd/ft^2. Where the tank depth is 12 ft at 1.0 hr detention, the upflow rate will be 12 ft/hr and the overflow rate 2160 gpd/ft^2.

There is no evidence that weir loading rates have any effect on primary tank efficiency. Settleable solid reduction generally exceeds 95 percent and total suspended solids (primary sedimentation) will usually average between 50 to 60 percent at a 2.0 hr detention and an overflow rate of 900 gpd/ft^2. A corresponding reduction in BOD will normally average between 30 to 35 percent. At 1.0 hr detention and 2160 gpd/ft^2 overflow rate, removal of about 35 to 40 percent of the suspended solids and 20 percent of the BOD might reasonably be expected. Typical performance results[4] of plants with rectangular and circular settling tanks having a detention period of 1.5 to 2.5 hr and a raw sewage suspended solids content of 200 to 300 ppm are presented in Table 12-6.

Sludge removal from primary settling units is usually intermittent, often with time clock control of the sludge pumps. Sludge solid concentrations will normally run from 4 to 6 percent, depending on the volatile matter content of the sludge and adequate pump control. However, where sludge thickeners are used following primary settling, a very thin underflow is usually pumped to them continuously. Sludge solids and volatile matter content from a number of circular settling tanks[4] are given in Table 12-7.

Imhoff tanks are a combination of primary sedimentation and sludge digestion units. The sedimentation compartment is above the digestion compartment and has a sloping floor so that the settling sludge drops continuously into the digestion compartment through a slot connecting both compartments. The sedimentation compartment floor normally has a vertical to horizontal slope of 1.4 to 1.0. Units may be longitudinal or round.

In longitudinal-type units, the sedimentation compartment may be considered as equivalent to separate rectangular settling units. Likewise, in round units the settling compartment may be considered as equivalent to separate circular sedimentation units.

Periodic flow reversal is essential in longitudinal units to prevent excessive sludge deposition in the digestion compartment at one end. Also, because of steep floor slopes, construction costs are high. In both longitudinal and round units, despite the steep sedimentation compartment floor slopes, solids adhere to the floor and give rise to septic conditions and scum formation. A modification of the round unit that avoids this is one where the floor slopes at only 1 in. per foot and a standard circular clarifier mechanism is used to rake deposited sludge to a central outlet connected to the lower digestion compartment. These units are normally supplied in sizes from 12 to 40 ft in diameter.

Flotation. Flotation has been used to assist in the removal of oil and grease from domestic sewage where such constituents are unusually high. The practice has been to add diffused air to partly emulsify the oil and grease and so assist their flotation, either in separating tanks preceding the primary clarifier or in the clarifier itself. No attempt is made in this situation to float a substantial portion of the suspended solids contained in the sewage.

The objective of the use of flotation as a substitute for primary clarification is the removal of the heavier portion of the suspended matter

TABLE 12-6. Primary Sedimentation Tank Data

				Suspended Solids (ppm)			BOD (ppm)		
Plant	Detention (hours)	Overflow Rate (gpd/ft^2)	Weir Rate (gpd/ft)	Influent	Effluent	Percent Removal	Influent	Effluent	Percent Removal
RECTANGULAR TANKS									
Rochester, N.Y.	1.56	914	41,000	233	184	21	260	206	21
Hammond, Ind.	1.32	1800	24,000	273	161	30	206	154	25
Abington, Pa.	2.02	885	44,400	237	145	39	198	141	29
Dallas, Tex.	2.00	1080	24,000	258	888	66	256	151	41
Roanoke, Va.	1.87	1010	120,000	230	76	67	190	93	51
Average	1.75	1138	50,680	246	133	46	222	149	33
CIRCULAR TANKS									
Buffalo, N.Y.	1.60	1690	67,200	209	114	46	138	107	23
Albuquerque, N.M.	2.21	995	19,800	254	91	61	282	150	45
Baltimore, Md.	1.64	1360	56,200	214	83	61	281	204	28
Mansfield, Ohio	2.38	905	14,700	208	87	58	227	139	39
Austin, Tex.	1.69	1257	23,900	263	95	64	285	152	46
Green Bay, Wis.	1.60	1120	14,600	256	128	50	219	156	29
Average	1.85	1224	32,750	234	100	57	233	151	35

TABLE 12-7. Primary Sludge Characteristics

Plant	Percent Solids	Percent Volatile Matter
Washington, D.C.	8.0	67.5
Buffalo, N.Y.	5.8	59.0
Albuquerque, N.M.	3.9	81.0
Yakima, Wash.	7.0	74.4
Appleton, Wis.	5.6	58.0
Baltimore, Md.	3.9	82.7
Mansfield, Ohio	4.2	76.0
Cedar Rapids, Iowa	5.5	81.2
Austin, Tex.	4.0	83.0
Denver, Colo.	5.4	76.0
Ypsilanti, Mich.	8.2	71.4
Monroe, Mich.	5.2	67.7
Average	5.6	73.2

as well as oil and grease. Here, air is introduced under pressure ahead of the flotation vessel, and the fine solids, buoyed up by fine bubbles which attach themselves to the solids, are caused to float in the flotation tank where they are skimmed off. Air may be added by diffusion through porous tubes or plates, by aspiration through an injector, or by mechanical means. The flotation vessel may be at atmospheric pressure or under a vacuum (Fig. 12-6).

By means of such systems it is possible to effect a suspended solid removal of 40 percent at an overflow rate of 7500 gpd/ft^2 and a detention period of about 20 min. BOD removals should average about 25 percent. Pilot plant tests at Racine, Wisconsin on an atmospheric flotation unit gave results[5] as presented in Table 12-8.

Pilot plant tests run at Cleveland, Ohio on a continuous flow vacuum flotation unit gave results[6] as shown in Table 12-9.

Results obtained[7,8] on plant units operating on domestic sewage are given in Table 12-10.

Fig. 12-6 Vacuator. (*Courtesy of Dorr-Oliver, Inc.*)

Sludge removal mechanisms for flotation tanks are substantially the same as for conventional sedimentation tank units. Due to the settling of heavier solids, it is essential that provisions be made for regularly removing settled sludge as well as floated solids.

Advantages claimed for vacuum flotation are that finer air bubbles, released when the aerated sewage is subjected to a vacuum of 9 in. of mercury, attach themselves to the solids and cause them to float better than larger air bubbles which readily become detached and lose their effectiveness. Disadvantages are that vacuum vessels are more costly to construct and are less accessible for repair and maintenance.

Flotation systems should not be used as the sole means of primary clarification, except

TABLE 12-8. Atmospheric Flotation Unit—Pilot Plant Performance

		SUSPENDED SOLIDS (ppm)			
Detention (minutes)	Recirculation (percent)	Influent	Effluent	Percent Removal	Float Solids (percent)
1.0	50	180	63	65	9.1
1.25	40	145	70	52	8.4
1.67	30	130	59	55	9.7
2.5	20	158	93	41	8.2

TABLE 12-9. Vacuum Flotation Tank—Pilot Plant Performance

Detention (minutes)	Overflow Rate (gpd/ft^2)	SUSPENDED SOLIDS (ppm) Influent	Effluent	Percent Removal	Scum (percent solids)	Grit[a] (percent solids)	Ratio Lbs Scum Solids to Lbs Grit Solids
22	5090	270	162	40	6.1	50.7	0.93
15	7640	250	150	40	5.6	51.7	0.92
11	10180	244	155	37	6.0	55.5	0.55
9	12700	278	210	25	6.0	49.8	0.67
7.5	15280	279	216	23	6.0	44.6	0.65

[a]Grit was obtained by washing the underflow from the flotation unit in a grit washer and recycling the fines to the system.

possibly where secondary treatment is provided and where the flotation unit can be used in emergencies as a short-period sedimentation unit.

Flocculation. Primary sedimentation removes only a portion of the settleable suspended solids contained in sewage. For example, where analytical results show that a 55 percent removal of suspended solids is obtained in continuous-flow tanks, laboratory settling under quiescent conditions for 1 hr may show a 75 percent removal. This would indicate that 20 percent of the suspended matter is non-settleable and that the removal of settleable suspended solids is actually 73.3 percent. This may be regarded as the sedimentation efficiency.

This efficiency may be increased in a number of ways, such as by increasing the period of sedimentation, or by mechanical or air flocculation with or without the addition of chemicals and with or without the addition of return sludge.

Preflocculation Without Chemicals. Prolonged sedimentation may increase the suspended solids removal to, say, 60 percent, giving a "sedimentation efficiency" of 80 percent due chiefly to autoflocculation and the deposition of finer, slow-settling solids. This same result or better results are attainable by means of mechanical or air flocculation at lower settling tank detentions. Also, BOD removals will be increased an additional 5 percent. Flocculation periods of 20 to 40 min are generally provided followed by conventional settling at an overflow rate of 800 gpd/ft^2.

Mechanical flocculation units should provide a gentle mixing action with velocities in the range of 0.9 to 1.2 ft/sec. Where air flocculation is employed, air diffusion at a rate of 0.10 to 0.30 ft^3/gal of sewage is commonly used.

Mechanical flocculation units may be of vari-

TABLE 12-10. Vacuum Flotation Tank—Plant Performance

Plant	Flow (mgd)	Detention (minutes)	Overflow Rate (gpd/ft^2)	SUSPENDED SOLIDS (ppm) Influent	Effluent	Percent Removal	BOD (ppm) Influent	Effluent	Percent Removal
Palo Alto, Cal.	2.1	53	2140	249	128	49	202	166	19
Santa, Maria, Cal.	1.3	59	1930	418	252	40	304	181	40

	PALO ALTO Scum	Sludge
Percent solids	4.2	4.9
Percent volatile matter	81.9	80.1

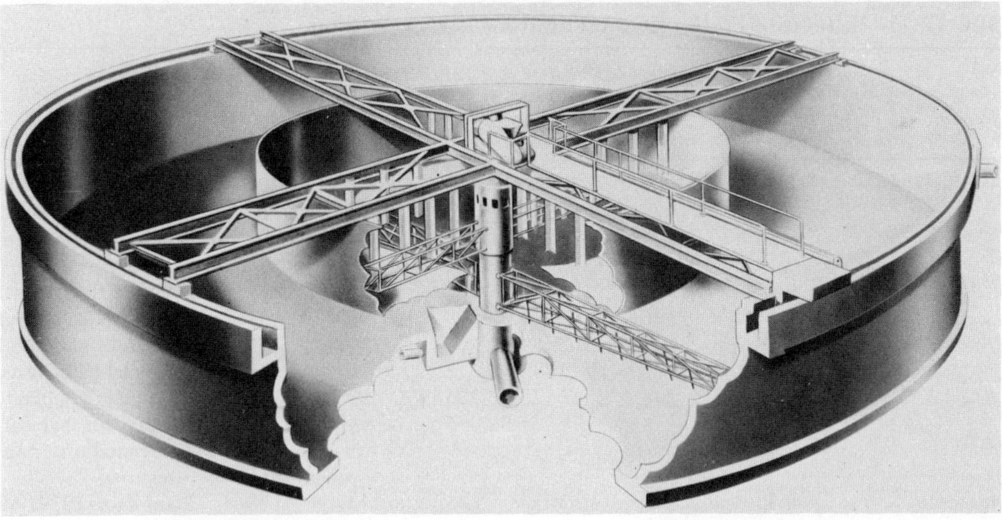

Fig. 12-7 Dorr-Oliver clariflocculator. (*Courtesy of Dorr-Oliver, Inc.*)

ous rotating types ranging from relatively small high-speed impellers or propellers with or without draft tubes to large low-speed paddles rotating on horizontal or vertical shafts. Other types provide an oscillating or up-and-down motion to promote particle contact. The low-speed types are preferred as they avoid floc breakup. For the same reason, air diffusion through porous media, perforated pipes, or ejectors is preferred to mechanical aeration.

Combination-type units where both the flocculation (or aeration) and the sedimentation units are contained in a single tank are available for both rectangular and round tanks (Figs. 12-7 and 12-8). Such units have the advantage that high-transfer velocities are avoided between the flocculation and the sedimentation units. Also, a more uniform feed to the settling compartment is obtained. Heavy sludge deposits in the flocculation zone are avoided by extending the settling tank sludge-raking mechanism beneath the flocculation unit. The disadvantage is that if the flocculation unit is taken out of service for repairs, the sedimenta-

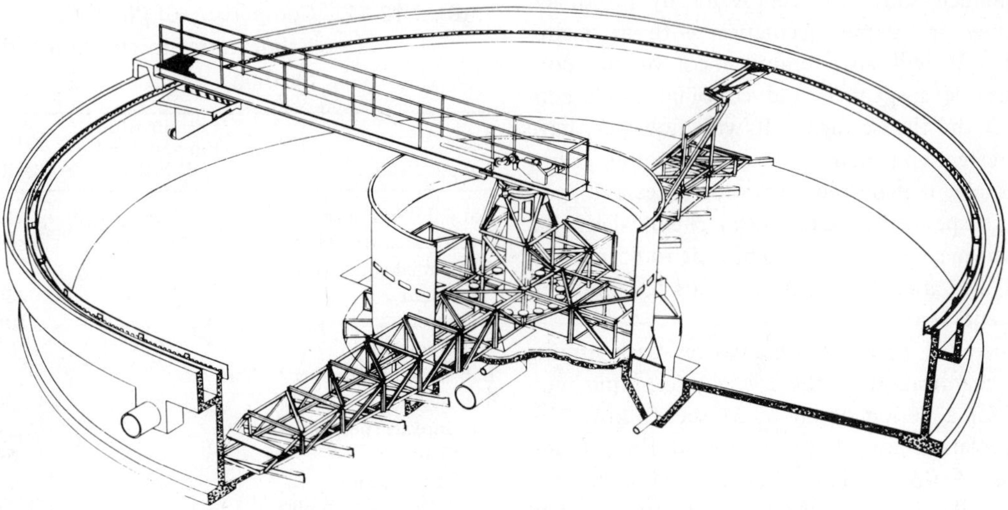

Fig. 12-8 Primary clarifier with preaeration. (*Courtesy of Envirotech.*)

TABLE 12-11. Effects of Preflocculation on Raw Sewage Settling

Plant	DETENTION (hours)		Clarifier Overflow Rate	SUSPENDED SOLIDS (ppm)			Remarks
	Flocculation	Settling	(gpd/ft^2)	Influent	Effluent	Percent Removal	
Ypsilanti, Mich.	0	2.3	790	155	82	97	Combination-type unit. Mechanical flocculation. Parallel plant tests.
	0.4	1.9	805	155	66	57	
Cedar Rapids, Iowa	0	1.9	660	208	126	39	Separate floc and settling units. Mechanical flocculation. Parallel plant tests.
	0.67	1.9	660	208	96	54	
Los Angeles, Cal.	0	2.0	–	286	91	68	Pilot plant. Separate units. Mechanical flocculation. Parallel tests.
	0.5	2.0	–	275	64	77	
Denver, Colo.	0.33	2.8	630	145	43	70	Plant data, 1939. Mechanical flocculation.
Ames, Iowa	0	–	–	219	83	62	Parallel plant tests. Air flocculation 0.075 ft^3 air/gal. Plant data, 1939.
	0.07	–	–	219	64	71	

tion unit is also deactivated. Table 12-11 gives plant scale test data[9,10] on commercial and pilot plant units.

Laboratory versus plant performance at Ypsilanti[11] in Michigan are shown in Table 12-12.

Flocculation with Chemicals (Chemical Precipitation). Flocculation with chemicals will remove most of that portion of settleable suspended solids not removable by prolonged settling or by preflocculation without chemicals. It will also remove most of the nonsettleable suspended and colloidal solids contained in the sewage. It will, however, have no effect on dissolved sewage solids. For example, if plain sedimentation gives a 55 percent suspended solids removal, chemical precipitation may increase the removals to 85 percent and also increase BOD removals to about 50 percent.

Chemical precipitation involves flash mixing the chemicals with the sewage; i.e., rapid mixing for a period of about 30 sec, followed by flocculation for 20 to 40 min at flocculation rates of 0.9 to 1.8 ft/sec depending on the type of chemical coagulant employed, and settling at overflow rates of 800 to 1200 gpd/ft^2 at average flow. Here, as the precipitated solids are flocculent in nature, sedimentation tank area rather than detention is the dominant factor in settling tank design.

Flash mixing may be accomplished by means of high-speed impellers or by aeration. Prolonged mixing should be avoided to prevent premature floc buildup and breakage.

Flocculation and sedimentation mechanisms

TABLE 12-12. Comparison of Plant vs. Laboratory Flocculation and Settling

	Combination Flocculator–Clarifier	Clarifier
Total detention (hours)	2.06	2.12
Overflow rate (gpd/ft^2)	880	861
Suspended solids (ppm)		
Influent	211	230
Effluent	77	106
Lab settled	61	63
Lab flocculation plus settled	36	38
Removals (percent)		
Plant	64	54
Lab settling	71	72
Lab flocculation plus settled	83	84

are of the same type used for preflocculation or preaeration without chemicals and for primary sedimentation, respectively.

A variety of chemicals have been employed for chemical precipitation. The most widely used have been lime, alum, ferrous sulfate, ferric chloride, chlorinated copperas, and, more recently, various organic polymers. Lime has frequently been used with alum and iron salts for pH value correction. In general, the choice of chemicals is dependent on the character of the sewage being treated and on the cost of the chemicals.

The use of chemicals generally results in the production of a voluminous sludge of relatively low solids content due to retention of fines and hydration of the coagulant. It is difficult to dewater when alum and iron salts are used as the coagulant. In general, the use of chemical precipitation of sanitary sewage has declined in recent years in favor of biological treatment methods which are more effective in removing BOD. It may be advantageously used for reducing loads on subsequent biological treatment steps. Some typical results[12,13,14,15] obtained by chemical precipitation are presented in Table 12-13.

Sludge Return as an Aid in Flocculation. The return of sludge for aiding flocculation has been practiced with and without chemical addition. When chemicals are not used, it has been necessary to avoid sludge septicity by aerating the sludge. When this aeration is carried out to the point of converting the sludge to activated sludge, excellent clarification is obtained with 20 to 40 min contact of the sludge with the raw sewage ahead of sedimentation. The contact may be by means of mechanical flocculation or by aeration.

In such cases, it is preferable to provide two sludge draw-off points, one for the heavier sludge to go to waste, the other for the lighter sludge to be returned to process. The amount of sludge return depends on the strength of the sewage and will usually vary between 15 and 30 percent by volume.

Sedimentation tank capacity is determined by area rather than detention because of the flocculent nature of the sludge. Overflow rates of 800 to 1200 gpd/ft^2 should be suitable with well-aerated sludge. Suspended solids removals of as high as 90 percent and BOD reductions of 60 to 70 percent are obtainable by this method, which is termed "bio-precipitation." It is also known as "contact aeration."

Where the return sludge is not kept fresh by aeration, its recycling may be detrimental rather than beneficial. The use of presedimentation is of considerable advantage in applying this process as it reduces the load on the sludge aeration stage. Laboratory tests[9] showing the effect of mechanical flocculation of raw sewage with activated sludge are shown in Table 12-14. Laboratory tests[16] showing the effect of the addition of raw settled sludge on settling ef-

TABLE 12-13. Chemical Precipitation—Plant Results

	SUSPENDED SOLIDS (ppm)			BOD (ppm)			CHEMICALS USED
Plant	Influent	Effluent	Percent Removal	Influent	Effluent	Percent Removal	70 lb Cl plus 480 lb
Birmingham, Ala. (Shades Valley)	124	10	92	86	12	86	$Fe_2(SO_4)_3 \cdot 7H_2O$
Palo Alto, Cal.	290	43	85	306	106	65	143 lb $FeCl_3$/million gal Spent sludge regenerated and recycled
New York, N.Y. (Coney Island)	132	10	92	132	41	69	900 lb lime, 161 lb $FeCl_3$, 385 lb paper/million gal
Dearborn, Mich	101	25	76	95	37	62	147 lb lime, 235 lb $FeCl_3$/million gal
Erie, Pa.	140	71	52	88	63	27	6.25 lb Purifloc A21/million gal
	151	125	18	92	76	17	No coagulant

TABLE 12-14. Effect of Addition of Activated Sludge on Raw Sewage Settling

Flocculation Time (minutes)	Settling Time (hours)	SUSPENDED SOLIDS (ppm)			Remarks
		Raw Sewage	Settled Sewage	Percent Removal	
0	1.0	368	166	55	No activated sludge added
30	0.5	368	103	72	No activated sludge added
0	1.0	497	132	64	Activated sludge added
30	0.5	497	44	88	Activated sludge added

ficiency with and without aeration are presented in Table 12-15. Plant scale results[17] at Denver, Colorado showing the effect of raw sludge recycling are shown in Table 12-16.

The use of sludge return where chemicals are added has been limited to cases where lime alone has been used as the coagulant. Here the sludge return will serve to reduce the amount of chemicals required and to aid in the absorption of the finely-divided sewage solids by the larger flocs.

Secondary Sedimentation. Secondary or final sedimentation tanks are similar in design to those used for primary settling. Two overall points of difference are that secondary units are not always provided with equipment for the removal of scum, and large tanks are often provided with greater overflow weir length so as to reduce the weir loading.

The latter is true especially where activated sludge is settled. This is accomplished in the case of rectangular tanks by adding more effluent troughs near the outlet end. In circular tanks, the outlet trough is located several feet from the peripheral wall and overflow weirs are provided on each side. Additional annular troughs with weirs may be added if required. Weir loadings are generally kept below 10,000 gpd/linear ft on large tanks (average 75 ft diameter or equivalent).

In recent years there has been an increased

TABLE 12-15. Effect of Addition of Settled Sludge on Settling Efficiency

Aeration Time (hours)	Settling Time (hours)	Sludge Added (ppm)	SUSPENDED SOLIDS (ppm)		
			Raw Sewage	Settled Sewage	Percent Removal
0	1	0	150	100	33
6	1	0	150	70	63
0	1	555	150	120	20
6	1	555	150	185	–
0	1	1020	150	160	–
6	1	1020	150	130	–

TABLE 12-16. Denver, Colorado—Effect of Raw Sludge Recycling on Primary Sedimentation[a]

Period	Sludge Return (percent	Flocculation Time (minutes)	Settling Time (hours)	Overflow Rate (gpd/ft^2)	SUSPENDED SOLIDS (ppm)			BOD (ppm)		
					Raw	Settled	Percent Removal	Raw	Settled	Percent Removal
1942	25	15	2.0	865	196	37	81	222	91	59
1944	25	16	2.1	825	200	47	77	240	113	53
1945	25	16	2.1	835	195	59	70	220	109	50

[a]15-min preaeration plus 1 hr sludge aeration ahead of mechanical flocculation with raw sewage. Air used = 0.30 ft^3/gal of sewage.

tendency to provide skimming arrangements on secondary tanks as considerable amounts of floating solids may appear in these units; e.g., flies from trickling filters, activated sludge flocs floated by gas bubbles, grease, etc.

The efficiency of the final sedimentation tanks depends chiefly on overflow rate. This is especially so in the case of activated sludge. In the case of trickling-filter effluents, autoflocculation and, therefore, detention periods do have some effect. This is not so with activated sludge.

Table 12-17[18] gives results of plant-scale parallel tests run at the Chicago Southwest plant showing weir-rate effect. Here the tests were run on 126 ft diameter tanks with peripheral overflow and with added concentric weir troughs. Overflow rates were 860 gpd/ft^2 (1120 gpd/ft^2 feed rate), mixed liquor solids were 3290 ppm, and the sludge index was 52.

In activated sludge plants, overflow rates depend on the percent solids in the mixed liquor fed to the sedimentation tank and on the sludge volume index. In general, with mixed liquor solids of less than 2500 mg/l (ppm) and a sludge volume index of under 100, overflow rates should be in the range of 800 to 1200 gpd/ft^2.

For settling high- or low-rate trickling filter effluents, overflow rates should be in the 800 to 1000 gpd/ft^2 range. Effluents from activated sludge settling units should normally contain from 10 to 20 mg/l suspended solids on the average. However, daily variations will be considerable. Even in well-operated plants, daily average effluent suspended solids may range from 1 to 70 mg/l. Settled trickling-filter effluents will normally average 30 to 40 mg/l suspended solids. Here again, daily averages will vary widely, with extremes running from 10 to 100 mg/l suspended solids.

TABLE 12-17. Effect of Weir Rate on Activated Sludge Settling

Weir Rate (gpd/ft)	Effluent Suspended Solids (ppm)
9,900	10.6
17,700	13.8
28,200	18.2

Sludge from activated sludge settling tanks should be drawn quickly and continuously. Its solid content will generally run between 0.5 to 1.5 percent, 1.0 percent being a good average. Waste sludge may be returned to the raw sewage (as is frequently done in small plants) or separately thickened before mixing with primary sludge prior to disposal.

Sludge from trickling-filter settling units is withdrawn continuously or intermittently and is generally returned to the raw sewage, or to separate thickeners. In the case of high-rate filters, continuous withdrawal is preferred. Where intermittent sludge withdrawal is practiced, its solid content will normally run from 3.0 to 4.5 percent. Where it is withdrawn continuously, the solids content is very low.

Combined Sedimentation Units. Both primary and secondary sedimentation may be carried out in a circular unit with a single sludge-raking mechanism by use of a vertical partition along the tank diameter extending from above the water level down to the rotating rake support arms. One compartment is used for primary settling and the other for secondary settling. Such units have been used to date only for small high-rate filter plants. Sizes range from 10 to 40 ft in diameter.

A disadvantage of this unit is that there can be "leakage" of primary sewage below the baffle directly into the secondary compartment. Also, primary sludge is in contact with secondary compartment liquor. Leakage may be minimized by recycling the final effluent back to the secondary compartment inlet so that a backflow beneath the dividing baffle is created.

BIOLOGICAL TREATMENT

Biological treatment provides a high degree of treatment because it achieves not only the highest reduction in suspended solids but also (due to the oxidation of dissolved solids) the greatest reduction in biochemical oxygen demand. It essentially consists of the contacting of settled sewage with microbiological growth in the presence of dissolved oxygen. This may be accomplished in several ways, such as: (1) passing the microbiological growth through the sewage as is done in the activated sludge process, or (2) passing the sewage through a fixed bed containing the growth as is done in trickling filters, contact beds, stabilization ponds, and intermittent sand filters. Of these, the ones in greatest use at present are the activated sludge process and trickling filters.

The Activated Sludge Process. In the conventional activated sludge process, sewage is mixed with sludge separated from a previously aerated sewage-sludge mixture. It is then aerated and settled. Part of the settled sludge is returned to the aeration tank. The excess sludge not returned is wasted. The sewage-sludge mixture in the aeration tank is termed "mixed liquor." The settled sludge recycled or wasted is termed "activated sludge."

A number of modifications of the conventional system have been developed. Some of these involve a change in the flow sheet, others involve differences in aerator detention periods, BOD loadings, and/or aerating tank solids concentrations. The general flowsheets involved are shown in Fig. 12-9.

In the conventional process, BOD loadings are usually in the range of 25 to 50 lb/day/100 lb of suspended solids in the aeration tank. On this basis, with mixed liquor solids say at 2000 mg/l and with the settled sewage BOD_5 at 130 mg/l, the required aeration period would be 5.0 hr at the lower loading and 2.5 hr at the higher rate, both with a sludge return of 25

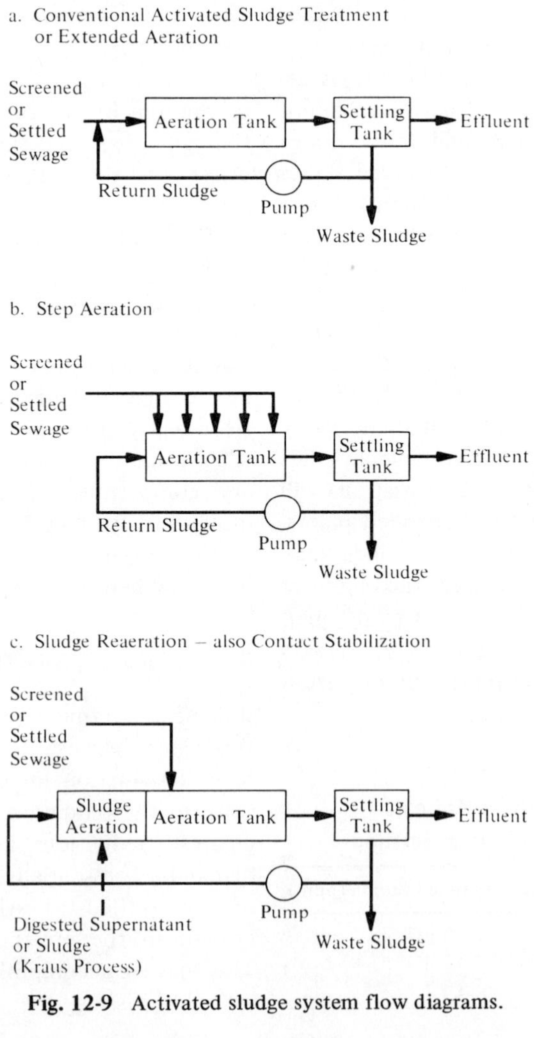

Fig. 12-9 Activated sludge system flow diagrams.

TABLE 12-18. Conventional Activated Sludge Plant Operation

	NO SLUDGE REAERATION							SEWAGE (ppm) TO AERATORS		PERCENT REMOVAL	
	Flow (mgd)	Aeration Period (hours)	Return Sludge (percent)	Mixed Liquor Solids (ppm)	Air (ft^3/gal)	BOD Removed (ft^3/lb)	BOD Loading (lbs/1000 ft^3)	Suspended Solids	BOD	Suspended Solids	BOD
Ann Arbor, Mich.	5.0	5.3	32.0	1840	1.05	770	39.0	128	182	90.7	89.6
Baltimore, Md.	10.6	9.3	21.5	940	1.24	728	28.8	92	217	91.3	93.6
Chicago, Ill. (North Side)	213.0	4.9	22.7	2000	0.48	800	19.0	88	77	88.5	92.5
Cleveland, Ohio (East)	86.9	6.2	24.9	2180	1.06	1600	17.4	146	92	90.8	86.9
Ft. Wayne, Ind.	14.9	7.7	30.5	2420	0.95	773	23.7	151	161	89.1	91.0
Gary, Ind.	19.0	4.4	33.2	1330	0.47	796	19.4	101	77	92.9	89.6
Hagerstown, Md.	3.9	6.7	31.5	1810	0.92	500	44.1	286	254	95.5	86.8
Hammond, Ind.	20.2	7.2	51.9	2200	0.66	655	19.5	152	135	88.8	91.5
Jackson, Mich.	8.0	5.8	15.5	2800	0.65	690	26.4	144	118	88.5	94.9
Springfield, Ill.	7.8	6.6	27.6	1680	0.70	710	23.5	120	132	90.9	90.2

percent. The lower loading would give a higher degree of nitrification and exhibit greater stability to shock loads. Stronger sewage would require a porportionately longer aeration period at the same loading. Air quantities, where difused air is used, would be about 600 ft^3/lb of BOD removed; e.g., with settled and final effluent BODs of 135 and 10 mg/l, respectively, air requirements would be 0.625 ft^3/gal of sewage.

Typical conventional activated sludge plant results[19] are presented in Table 12-18.

Another control index that has been applied to the conventional system has been that of "sludge age"; i.e., the average time a suspended solids particle remains in the aeration tank. It may be calculated by dividing the product of the aeration tank volume in million gallons and suspended solids in milligrams per liter by the product of the sewage flow in million gallons per day and its suspended solids content in milligrams per liter. In the examples cited above, the sludge age with a 5-hr aeration period would be 4 days. With half that aeration period it would be 2 days. The higher figure would be considered adequate for good performance. Lower figures would give questionable results.

With a fixed aeration tank volume, the only possible way to increase the sludge age in the regular conventional system is to increase the mixed liquor solids concentration. This in turn will have an adverse effect on the settling characteristics of the mixed liquor, thus affecting the overflow rate at which the final settling tank can effectively operate.

A means of increasing the sludge age without affecting the final clarifier feed is to use "step aeration." This is a modification wherein settled sewage is fed in steps along the aeration tank, all of the return sludge being fed at the head end. This method also permits reduction in required aeration tank volume.

Results[20] for step aeration at New York City plants are presented in Table 12-19.

Another possible way of reducing aeration tank volume in the conventional system is to employ sludge reaeration wherein the return sludge from the final settling tank is aerated before being mixed with the sewage flow. This system has been in use for many years and has been claimed to be of advantage where high-strength sewages were encountered or where shock loads such as from industrial wastes periodically entered the domestic sewage treatment plant. In most cases where it was used, however, sludge reaeration as originally practiced was considered to be of little or no advantage. In earlier plants, the sludge reaeration period was relatively short, the reaeration tank volume varying from 3 to 33 percent (averaging about 10 percent) of the total aerator volume. Table 12-20 shows typical results[19] obtained in older plants.

TABLE 12-19. Typical Step Aeration Results—New York City

Plant	Flow (mgd)	Return Sludge	Sludge Density Index	Air (ft^3/gal)	Aeration (hours)	Clarifier Overflow Rate (gpd/ft^3)	Sludge Age (days)	Aerator Suspended Solids (ppm)
Wards Island	50	28	0.9	0.43	2.5	870	3.4	950
Hunts Point	95	28	0.9	0.54	2.9	800	3.6	700
Tallmans Island	31	28	0.8	0.59	4.2	900	5.5	800

	FINAL EFFLUENT		PERCENT REMOVALS	
	Suspended Solids (ppm)	BOD (ppm)	Suspended Solids	BOD
Wards Island	6	6	95	94
Hunts Point	19	16	86	86
Tallmans Island	17	12	88	90

TABLE 12-20. Activated Sludge Plants with Sludge Reaeration

Plant	Flow (mgd)	AERATION PERIOD (hours) Mixed Liquor	Sludge	PERCENT SOLIDS Mixed Liquor	Return Sludge	Air (ft^3/lb BOD removed)	BOD Loading (lbs/1000 ft^3)	SEWAGE TO AERATOR (ppm) Suspended Solids	BOD	PERCENT REMOVAL Suspended Solids	BOD
Houston, Tex. N.	10.2	6.0	1.4	0.25	0.74	1350	38.0	298	208	78	85
Houston, Tex. S.	4.5	2.0	1.8	0.18	0.37	1760	45.0	235	178	78	79
Topeka, Kan.	5.6	5.9	4.2	0.125	0.45	1050	29.3	106	176	78	86

More recently, a modification of the sludge reaeration principle was developed wherein the volume of the reaeration unit was increased to 50 to 70 percent of the total aeration tank volume. This system is termed "extended reaeration" or contact stabilization. Table 12-21 gives pilot plant test results[21] obtained at Austin, Texas, where severe sludge bulking conditions occurred, causing frequent upsets in the conventional activated sludge system using short period sludge reaeration. Table 12-22 presents plant-scale data[22] obtained before and after the plant was converted to contact stabilization.

A further modification of the sludge reaeration or contact stabilization system is one in which nutrients are added to the reaeration or mixing tank. In the Kraus process, digester supernatant liquor or sludge is used as the source of nutrients. Table 12-23 shows the results of plant-scale tests[23] on the effect of digester sludge additions on sludge bulking at Peoria, Illinois.

The conventional activated sludge process,

TABLE 12-21. Biosorption Data—Austin, Texas Pilot Plant

DETENTION (minutes) Mixing	Sludge Reaeration	Settling	Clarifier Overflow Rate (gpd/ft^2)	SUSPENDED SOLIDS (ppm) Influent	Effluent	Percent Removal	Return Sludge	Mixed Liquor	BOD (ppm) Influent	Effluent	Percent Removal
14	96	65	910	226	14	94	5220	3250	264	20	93

TABLE 12-22. Austin, Texas Data—Conventional Activated Sludge vs. Biosorption

Treatment	Period	Flow (mgd)	Return Sludge (percent)	Air (ft^3/gal)	SUSPENDED SOLIDS (ppm) Raw	Primary Effluent	Final Effluent	Total Percent Removal	Return Sludge Solids (ppm)	Mixed Liquor Solids (ppm)	Sludge Index
Activated sludge[a]	1945	8.2	26.5	0.89	243	98	74	70	6080	1280	340
Biosorption	1955	7.7	41.3	1.59	226	–	18	92	6880	1900	115

Treatment	BOD (ppm) Raw	Primary Effluent	Final Effluent	Total Percent Removal	BOD Loading (lbs/1000 ft^3)	AERATION PERIOD (hours) Mixed Liquor	Sludge
Activated sludge[a]	249	151	91	63	–	–	–
Biosorption[b]	307	–	20	93	158	0.42	5.7

[a]Pool effluent due to sludge bulking. Sludge reaeration used.
[b]No primary settling used with biosorption.

TABLE 12-23. Effect of Addition of Digester Sludge to Reaeration Tank–Peoria, Illinois

Period	Primary Effluent (mgd)	Return Sludge (mgd)	Digester Sludge Added (mgd)	Percent Sludge	Air (ft^3/gal)	Mixed Suspended Solids (ppm)	Liquor Sludge Index	SUSPENDED SOLIDS (ppm)			BOD (ppm)			Lbs BOD/ft^3 Aeration Tank
								Primary Effluent	Final Effluent	Secondary Percent Removal	Primary Effluent	Final Effluent	Secondary Percent Removal	
1	11.7	5.0	0	–	1.68	1120	252	193	32	83	268	37	86	33.5
2	15.0	7.1	0.142	1.97	1.33	2090	120	133	15	89	275	19	93	43.7
3	15.8	5.2	0.125	2.03	1.14	1700	118	162	16	90	282	23	95	45.7

when compared to other biological processes, has the advantage of providing a high degree of treatment with minimum area requirements and head loss. Installation costs, however, are relatively high and a high level of supervision is required.

Modified aeration is a high-rate activated sludge process in which a short period of aeration is employed and where the mixed liquor solids are carried at a relatively low level as compared with the conventional process. BOD loadings are considerably higher, being on the order of 200 to 400 lb/100 lb of mixed liquor solids. Sludge age is likewise much shorter, running from 0.2 to 0.5 day. Mixed liquor solids may vary from 200 to 1000 mg/l and aeration periods from 1.5 to 3.0 hr. For a settled sewage BOD of 130 mg/l, a mixed liquor solids content of 400 mg/l, and a loading of 400 lb BOD/100 lb mixed liquor solid, an aeration period of 1.8 hr would suffice with 10 percent return sludge. The sludge age would be 0.25 day.

Air requirements for modified aeration will range from 0.3 to 0.5 ft^3/gal of sewage. Effluent quality is intermediate between primary sedimentation and conventional activated sludge, with effluent suspended solids between 30 and 50 mg/l and BODs between 30 and 60 mg/l being attainable.

Typical results[20] of treatment by modified aeration in New York City are presented in Table 12-24. Results[24] at Los Angeles and Miami are shown in Table 12-25.

Extended aeration is a form of activated sludge treatment in which the mixed liquor is aerated for a 24-hr period and where BOD loadings are in the range of 20 to 40 lb/1000 ft^3 of aerator capacity. The rate of return sludge is high and mixed liquor solids usually run between 3000 to 5000 mg/l. Taking 4000 mg/l as an average mixed liquor solids concentration, the BOD loading related to aerator solids is then 8 to 16 lb/100 lb. The volatile matter in the return sludge is much lower than in the conventional activated sludge process due to aerobic digestion. It normally runs about 50 percent versus about 70 percent for normal activated sludge.

Air quantities for extended aeration will vary

TABLE 12-24. Typical New York City Plant Results—Modified Aeration[a]

Plant	Aeration Period (hours)	Clarifier Overflow Rate (gpd/ft²)	Air (ft³/gal)	Sludge Age (days)	Mixed Liquor Solids (ppm)	FINAL EFFLUENT (ppm) Suspended Solids	BOD	PERCENT REMOVALS Suspended Solids	BOD
Wards Island	3.0	880	0.30	0.24	130	39	39	68	61
26th Ward	2.6	670	0.54	0.37	280	34	38	76	71
Bowery Bay	1.5	930	0.39	0.17	200	50	64	64	57
Owls Head	2.1	770	0.48	0.21	370	44	59	74	63
Jamaica	1.7	730	0.44	0.49	1000	38	35	81	76
Rockaway	2.1	850	0.38	0.44	470	38	32	67	64

[a]Primary settling is employed at Wards Island, 26th Ward, and Bowery Bay but not at the other plants.

TABLE 12-25. Modified Aeration Results at Los Angeles, California and Miami, Florida

Plant	Aeration Periods (hours)	BOD Loading (lbs/1000 ft³)	Air Requirements (ft³/lbs BOD removed)	SUSPENDED SOLIDS (ppm) Influent	Effluent	Percent Removal	BOD (ppm) Influent	Effluent	Percent Removal
Los Angeles	3.0	103	620	214	60	72	206	45	78
Miami	3.3	63	975	149	44	71	140	45	68

from 1000 to 1500 ft³/lb of BOD removed; i.e., about twice the quantities required for conventional activated sludge. Overall plant BOD removals are high (85 to 95 percent), but suspended solids removals are relatively low (70 to 75 percent) due to the discharge of finely-divided solids in the effluent that are end products of aerobic digestion.

Final settling tanks in extended aeration systems are usually designed at a very low overflow rate (400 gpd/ft²) and with a long detention time (4 hr) based on total daily flow. This assures that adequate settling capacity is available during peak daytime flows. Although the mixed liquor is relatively high in suspended solids content, it settles rapidly. The volume of waste sludge is very low.

Extended aeration plants, because of their high initial and operating costs, have in the past been generally restricted to small plants (up to 0.25 mgd). However, their range has more recently been greatly expanded to include plants up to 1.0 mgd. Their operation, due to the long aeration period, is relatively trouble-free.

Aeration tanks are generally rectangular in shape and have a water depth of 10 to 15 ft. Widths vary from about 10 to 32 ft. Length to width ratios of most units installed to date have been high, ostensibly to reduce short circuiting. In view of results obtained in step aeration and in newer "complete mixing" concepts, it is questionable whether the short circuiting theory is valid, so that tank size and shape may be considered as a matter of economics and convenience. Where the aeration tank capacity is in excess of 5000 ft³, two or more units are furnished.

Average results[25] from 27 plants in Florida are shown in Table 12-26. A summary of results[26] obtained from 50 plants by the U. S. Public Health Service is shown in Table 12-27.

Aeration may be accomplished by air diffusion or by mechanical means. Air may be diffused via porous diffusers (flat plates or tubes), perforated pipes, slotted plastic tubing, or ejectors (Fig. 12-10). In most such cases, the diffusers are arranged near one side of the tank so that a spiral flow pattern is obtained. They may be located near the tank floor or close to the water surface behind a baffle. With wide tanks, a double-spiral effect is produced by locating the diffusers along the tank center line.

TABLE 12-26. Average Aeration Results at 27 Florida Plants[c]

		SUSPENDED SOLIDS (ppm)			BOD (ppm)		
	Flow (gal/day)	Raw	Final	Percent Removal[a]	Raw	Final	Percent Removal[a]
Maximum	21,000	726	60	98	691	22	99
Minimum	3,000	72	4	55	103	2	91
Average[b]	10,200	282	26	94	118	7	94

[a]Maximum and minimum removals are for individual plants.
[b]Not a weighted average.
[c]These plants were designed for an aeration period of 24 hr with an air supply of 3000 ft^3/lb of applied BOD. In most cases, operation was at below plant design rates.

TABLE 12-27. Summary of Aeration Results at 50 Plants

			ACTUAL FLOW PERCENT OF DESIGN FLOW			AVERAGE PERCENT REMOVALS	
Type Facility Served	No.	Design Flow Range (gpd)	Maximum	Minimum	Average	Suspended Solids	BOD
Schools	15	4,700–63,000	103.7	20.3	39.6	60.3	87.1
Housing developments	7	15,000–100,000	105.0	16.8	57.3	69.4	91.6
Industry	6	6,900–100,000	131.5	31.2	79.7	62.7	88.1
Community	5	15,000–120,000	115.3	28.3	69.6	76.2	89.2
Institution	3	7,000–40,000	128.6	66.3	91.8	83.2	94.7
Motels and trailer parks	4	2,500–25,000	152.0	38.7	81.7	46.5	74.9
Shopping center	8	12,000–100,000	145.6	20.0	78.0	65.9	94.0
Miscellaneous	2	8,500–21,400	38.8	24.4	31.6	43.1	75.5
Total	50						
Average (all plants)						62.2	86.5
Maximum (all plants)						96.5	99.3
Minimum (all plants)						0	33.3
Median (all plants)						75.2	91.5

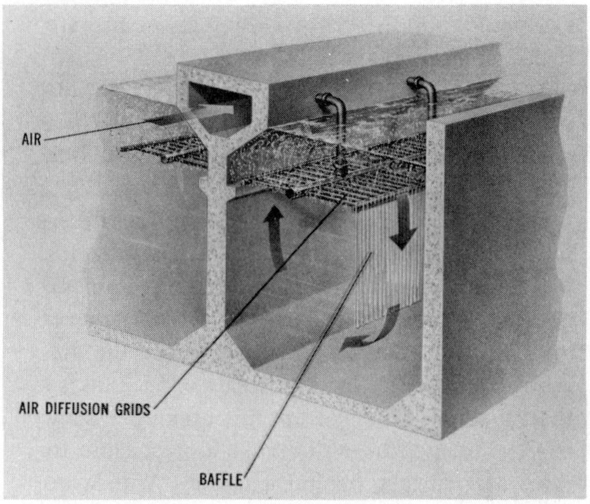

Fig. 12-10 Diffused air system using raised slotted tubing. (*Courtesy of Dorr-Oliver, Inc.*)

Other low patterns are obtainable by various other diffuser placements.

It is necessary to maintain a velocity of 1.0 to 1.5 ft/sec across the tank bottom so as to avoid solids deposition. To promote the spiral flow, a 45° fillet should be located at the tank bottom along the wall opposite that near which the diffusers are located.

The number of diffusers required depends on the specific type of unit supplied. Porous plates operate most efficiently at air rates of 1.5 to 3.0 $ft^3/min/ft^2$. They may be arranged uniformly along the tank length or tapered so as to allow the introduction of more air at the tank inlet. A preferred arrangement is one where assemblies of diffuser tubes may be mechanically raised above the liquid surface for servic ing. Air filters are required where porous-type diffusers are used.

A large variety of mechanical and surface aerators have been used in the activated sludge process in the past. The type that is in most general use at present is the surface impeller with a vertical draft tube having a conical top section. In this unit, the sewage is aerated by the revolving action of the impeller and is also caused to flow up or down a vertical central tube, thus causing agitation and contact between air and the sewage. These units are manufactured in various sizes.

The range of cone sizes is 2 to 12 ft and the drive units vary from 1 to 100 hp. The ratio of tank to cone size is usually 5-7 to 1. Tank depths vary from 7 to 30 ft. The average cone size is about 8 ft and the normal tank size is about 40 ft square by 16 ft depth. Multiple units may be installed in long tanks. The down-draft type has the advantage that it may be used in tanks with flat bottoms. The up-draft types require a hopper bottom. These aerators are capable of providing 4.5 to 5.5 lb of available oxygen per horsepower.

Most earlier designs of mechanical aerators had small impellers and drive units and so were limited in oxygenation capacity. As a result, they operated best with mixed liquor solids in the 800 to 1200 mg/l range. Therefore, longer aeration periods were required; e.g., twice the aeration was required for mechanical aeration than for air diffusion where a mixed liquor concentration of 2000 mg/l was maintained. More recent designs with higher power inputs are capable of providing greater oxygenation and can operate with higher mixed liquor solids than diffused air practice (Fig. 12-11).

Mechanical aerators have the advantage over air diffusion in that such auxiliary equipment as air filters, blowers, air piping, diffusers, etc. are not required. Due to the aeration limitations of the earlier designs, they were used mostly in small plants (up to 2.0 mgd capacity). The newer high-capacity units, are, however, adaptable for any plant size. Table 12-28 shows typical plant data[26] from earlier (before 1945) mechanical-type aeration plants.

The greatest difficulty encountered in activated sludge operation is that of sludge bulking.

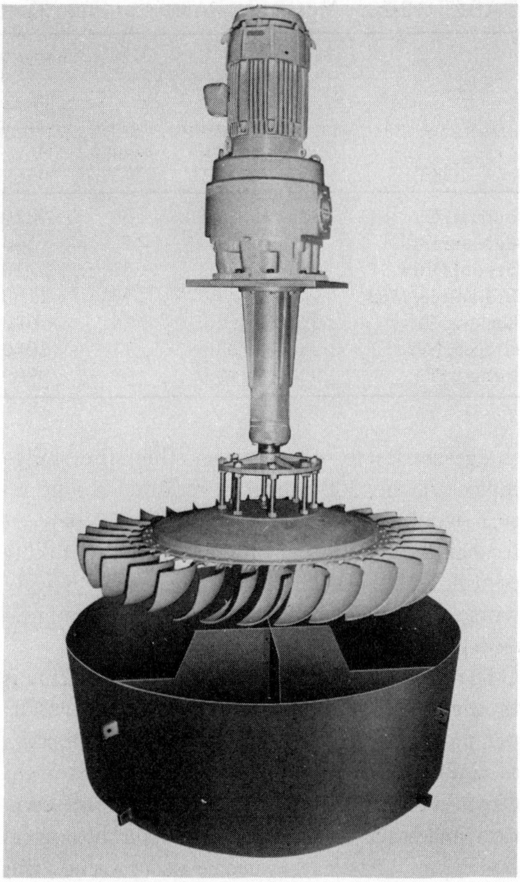

Fig. 12-11 Mechanical aerator. (*Courtesy of Clow.*)

This is defined as a condition wherein the volume of sludge per unit of weight increases to a point where the sludge in mixed liquor fails to separate from the liquor in the final sedimentation tank when operated at reasonable overflow rates (i.e., 800 to 1200 gpd/ft^2).

Tendency toward this condition may readily be monitored by regularly determining the Mohlman sludge index, which is the volume in milliliters occupied by 1 gm of sludge after settling aerator-mixed liquor for 30 minutes. This is done by settling 1 liter of mixed liquor for 30 min and noting the volume in milliliters occupied by 1 gm of sludge. For example, if sludge in mixed liquor containing 2000 mg/l settles to 200 ml, then the sludge index is 100.

To maintain the mixed liquor solids at 2000 mg/l, the return sludge needs to be only 20 percent of the sewage flow. If, however, the

TABLE 12-28. Mechanical Aeration Plant Data

Plant	Average Flow (mgd)	Aeration Period (hours)	Return Sludge (percent)	SUSPENDED SOLIDS (ppm) Mixed Liquor	Return Sludge	Sludge Index	BOD Loadings (lbs/1000 ft^3)	BOD (ppm) Primary Effluent	Final Effluent	Percent Removal
Bataria, Ill.	0.9	4.6	30	820	1520	56	17.5	70	9	87
Belvidere, Ill.	1.2	7.0	17	560	–	643	20.4	126	5	96
Bryon, Ohio	0.7	9.3	30	710	2040	603	9.2	67	10	85
E. Lansing, Mich.	1.8	4.8	33	1100	2700	53	12.7	40	10	75
Geneva, Ill.	0.9	6.0	14	620	2520	110	34.5	170	11	94
Greece, N.Y.	0.6	8.8	17	1040	4170	173	25.6	170	17	90
Radnor, Pa.	0.8	10.0	40	900	2000	155	11.0	104	7	93

sludge settles to, say 400 ml, then the sludge index will be 200, and the required sludge return would have to be increased to 50 percent to maintain stability in the aerator. As bulking continues, this condition becomes progressively worse until the return sludge and/or sedimentation tank capacities become overtaxed.

The most common cause of sludge bulking is the overloading of the aerators in terms of BOD per pound of mixed liquor solids. Introduction of industrial wastes that have a deleterious effect on biological growth, or insufficient aeration necessary to maintain favorable biological conditions in the aerator can also produce this effect.

To cope with normal load fluctuations of domestic sewage, it is customary to provide return sludge capacity of from 10 to 100 percent of the incoming sewage flow, and blower capacity (or aeration capacity equivalent) equal to 150 to 200 percent of normal requirements. Dissolved oxygen in the aerators should be maintained at a minimum of 2 to 3 mg/l in conventional systems. It should be noted that modified aeration requires only about half of the amount of air as conventional systems, while extended aeration, because of its low loading and oxygen requirements for biological oxidation, requires twice as much. Where bulking is caused by industrial wastes that either overload or poison the system, the only remedies are the elimination of such wastes from the system or the pretreatment of them.

Trickling Filters. Trickling filters are beds of stone, tile, or plastic media coated with a biological growth or film called zooglea. This forms when settled sewage is caused to flow through the bed, either continuously or intermittently. Contact with the zoogleal film after it becomes well-established causes biological oxidation of the sewage solids in much the same way as it is accomplished in the activated sludge process. The liquid discharged from the bed is then normally settled to remove solids sloughed from the bed. This sloughing may be intermittent or continuous.

Sewage was formerly dosed intermittently to the bed via fixed nozzles fed by dosing tanks equipped with automatic siphons. Practically all modern units are now supplied with rotary distributors (Fig. 12-12). In general, there are two types of trickling filters—low-rate or conventional units, and high-rate units.

Low-rate units are generally operated in the range of 25 to 100 gpd/ft^2 and at a BOD loading of 5 to 25 lb/1000 ft^3 of bed volume. Recirculation was not employed in the past, but is now quite common. Filter media depths average 6 ft. On this basis, a 1.0 mgd plant having a BOD of 130 mg/l in the settled sewage would require a filter bed 113 ft in diameter by 6 ft deep. Here the hydraulic loading would be 100 gpd/ft^2 and the BOD loading 18 lb/1000 ft^3. This plant could be expected to produce an effluent averaging 30 to 40 mg/l suspended solids and 20 to 40 mg/l BOD. These would represent monthly averages. Here, as in activated sludge treatment, however, daily and seasonal variations will be considerable, especially in view of the fact that low-rate filters usually slough periodically. High nitrification

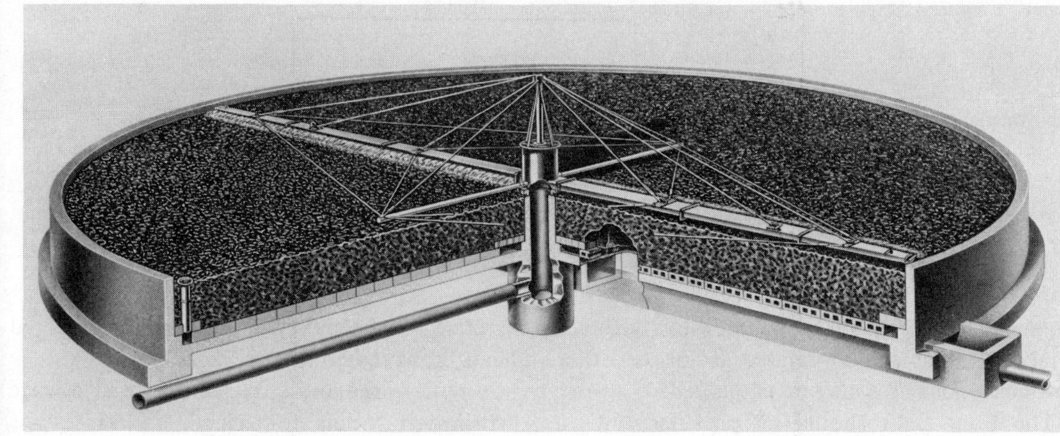

Fig. 12-12 Trickling filter with rotary distributor. (*Courtesy of Dorr-Oliver, Inc.*)

would be obtained with this type of plant. Typical low-rate filter performance data[27] are presented in Table 12-29. All plants have final settling tanks.

High-rate filters operate in the range of 200 to 1000 gpd/ft^2 and at BOD loadings of up to 300 lb/1000 ft^3 of filter media. There is some evidence that the high BOD loadings apply to strong sewages and that for normal sewages in the 130 to 180 mg/l BOD range (settled sewage), BOD loadings of 75 to 100 lb/1000 ft^3 will give the best overall results. Bed depths vary from 3 to 6 ft. Two-stage operation is common, with each stage being 3 ft deep. Recirculation is generally employed in high-rate units, with rates usually in the range of 0.5 to 3.0 gal/gal of incoming sewage, based on average daily flow. A typical single-stage flow sheet is shown in Fig. 12-13. In computing filter hydraulic capacity, the total applied flow (including recirculation) to the filters should be taken into consideration. Recirculated BOD, however, need not be added in computing organic loadings. Results [28,29] obtained at two army camps using flow diagram A in Fig. 12-9 are presented in Table 12-30.

Earlier filters with fixed-nozzle distribution systems were rectangular in plan, but since rotary distributors have come into common use, they are now usually circular. Low-rate filters are frequently designed to permit flooding so as to control filter flies. Watertight walls and valved outlets must be provided. Recent designs using high rates and recirculation avoid the necessity of flooding since continuous unloading of the filter occurs and flies, if any, are flushed down through the filter bed and so are prevented from accumulating at the bed surface.

As a result, thinner concrete walls may be

Table 12-29. Standard Trickling Filter Performance Data

Plant	Average Flow (mgd)	Depth of Media (feet)	Filter BOD Loading (lbs/1000 ft^3)	BOD (ppm) Primary Effluent	BOD (ppm) Filter Effluent	BOD (ppm) Final Effluent	BOD (ppm) Percent Removal Primary–Final
Akron, Ohio	47.9	10	13.5	207	67	46	78
Aurora, Ill.	7.8	6	4.4	72	22	14	80
Dayton, Ohio	34.4	7.5	11.8	137	46	33	76
Decatur, Ill.	11.0	6	18.9	129	32	25	80
Galesburg, Ill.	4.5	6	13.2	112	56	44	60
Plainfield, N.J.	4.3	6	25.0	320	26	13	96
Elgin, Ill.	4.6	8	6.1	94	46	41	56
Madison, Wis.	4.8	10	5.7	139	50	33	76

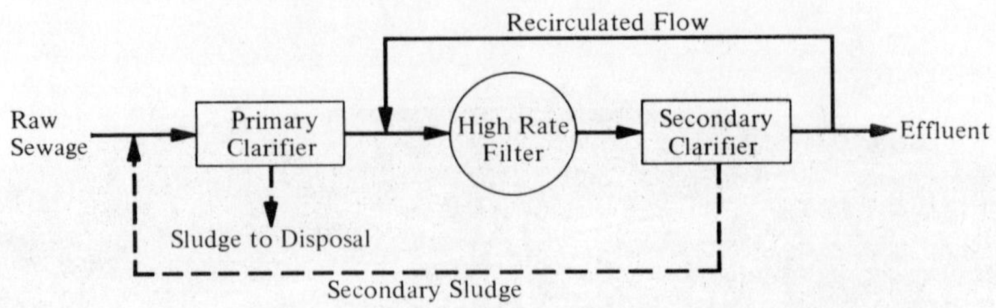

Fig. 12-13 Single-stage high-rate filter flowsheet.

used; or, where the filters are above ground, chain-link fencing, corrugated metal, or other low-cost enclosures may be employed.

The filter floor should be provided with a suitable drainage system to allow adequate ventilation and to permit free outflow of filter effluent at velocities sufficient to avoid solids deposition. These velocities should be 2 to 3 ft/sec. The floor should, therefore, be sloped toward a collection channel or other outlet means and should be provided with an adequate underdrain system. Forced ventilation by means of fans has been used in some filter installations in the past. It is now felt that such provisions are not justified. Since filters perform like cooling towers, the temperature differential between the sewage and the air causes circulation through it.

Filter media most commonly used are crushed rock screened to certain size specifications. The type of stone used depends largely on local availability and on resistance to disintegration during service. Trap rock and granite are the most desirable, while limestone is generally poor. For average conditions, stone 2-$\frac{1}{2}$ to 4 in. in size is highly suitable. In two-stage, high-rate plants, 1-$\frac{1}{2}$ to 2-$\frac{1}{2}$ in. stone may be used in the secondary filter.

In addition to crushed stone, blast furnace slag, redwood, and synthetic-fused or plastic material have been used as media. The plastic material has been in the form of corrugated sheets, honeycomb shapes, etc.

Currently, the most frequently used sewage distribution means for trickling filters is the reaction-type rotary distributor. It consists of two or more horizontal arms supported by a center column. Sewage enters through the base of the center column via an inverted siphon and discharges to the bed through slots or orifices in the trailing side of the arms. The reaction of the discharged flow causes the unit to rotate.

Velocities in the arms are usually held below 4 ft/sec at maximum flow so as to avoid excessive head loss. This maximum velocity and high head loss may also be controlled by the use of an overflow arrangement in the center column whereby two arms or compartments are in operation at times of low flow and four arms or compartments are in operation during high flows.

The speed of rotation of the distributor may be regulated by adjusting the size of the discharge slots or number of orifices in the arms. Reverse or braking orifices may also be used on the opposite sides of the arms. Clearance between the rotation arms and the bed media is usually 12 to 24 in. Rotary distributors are available for filters ranging from 20 to 200 ft in diameter.

Other types of distribution devices less frequently used than reaction-type distributors are

TABLE 12-30. Single-Stage Army Camp Plant Results

Army Camp	RECIRCULATION RATIOS		CLARIFIER DETENTION (hours)		BOD Loading[a] Based on Raw Sewage (lbs/1000 ft³/day)	SUSPENDED SOLIDS (ppm)			BOD (ppm)		
	R_1	R_2	Primary	Secondary		Raw	Final	Percent Removal	Raw	Final	Percent Removal
A	1.09	0.86	3.8	4.2	126	256	13	95	310	18	94
B	1.93	2.19	2.9	2.6	113	249	45	82	356	70	83

[a]Due to the recirculated flows, it is not possible to obtain BOD figures on primary settled sewage without recirculated filter effluent. In plant B, the filter loading based on the total flow applied to the filter was 100 lb/1000 ft^3/day.

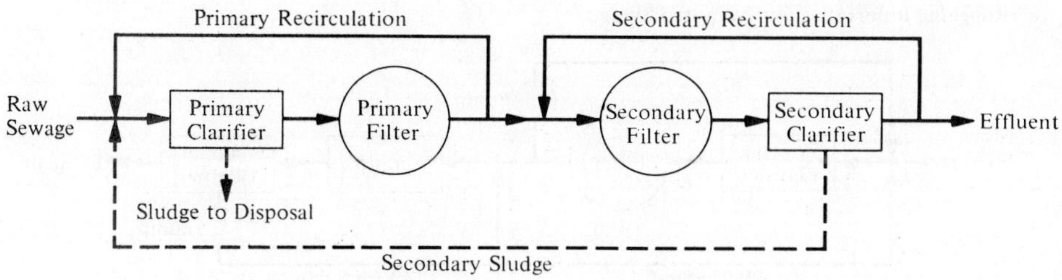

Fig. 12-14 Two-stage high-rate filter flowsheet.

motor-driven distributors and rotating discs for filters up to 35 ft in diameter. Dosing siphons are required where there are considerable fluctuations in flow; i.e., low-rate filters where recirculation is not employed.

Trickling filters are higher in costs than conventional activated sludge plants and require more area and operating head. Operating costs, on the other hand, are relatively low and operation is not affected by shock loads. Final effluents are inferior to conventional activated sludge.

Two-stage, high-rate filter plants usually operate with two 3-ft deep filters in series. The flow sheet most commonly used is shown in Figure 12-14.

This flow diagram is very similar to that used in single-stage operation with the filter divided in half. This system has definite advantages when very strong sewages are treated and will generally provide a better quality effluent when the sewage is of normal strength. Loadings, etc. for two-stage units are the same as for single-stage units having the same volume of filter media. Effluents averaging 20 to 30 mg/l BOD_5 and suspended solids are attainable. Results of two-stage plants using the flow diagram in Fig. 12-14 are presented in Table 12-31.

Trickling Filter–Aerator Combinations. Where a high degree of treatment must be attained and where shock loads and/or the presence of industrial wastes in sewage are apt to interfere with normal activated sludge operation, the use of trickling filters as "roughing filters" ahead of aeration will provide stable operation. The flow sheet used here is shown in Fig. 12-15A.

With 1.0 mgd sewage with 250 mg/l BOD and using a 1 : 1 recirculation rate, a 51 ft diameter by 3 ft filter would have a hydraulic loading of about 1000 gpd/ft^2 and a raw BOD loading of 337 lb/1000 ft^3 of media. At this loading, a 60 percent reduction of BOD in the primary system could reasonably be expected so that the BOD load to the activated sludge system would be 883 lb/day. At a loading of 50 lb/100 lb suspended solids in the aerator and with a mixed liquor concentration of 2000 mg/l, an aeration period of only 2 hr would be required.

Another trickling filter–aerator combination

TABLE 12-31. Two-Stage Plant Data—High-Rate Filters

Plant	RECIRCULATION RATIOS R_1	R_2	BOD Loading[a] Based on Raw Sewage (lbs/1000 ft^3/day)	SUSPENDED SOLIDS (ppm) Raw	Final Effluent	Percent Removal	BOD (ppm) Raw	Final Effluent	Percent Removal
Army Camp 1	0.87	1.57	117	250	31	92	212	24	89
Army Camp 2	1.30	1.14	136	187	38	80	281	70	75
3	1.33	0.32	145	232	24	90	255	37	85
4	0.53	0.07	86	228	17	92	149	12	92
5	1.12	1.12	123	196	26	87	240	44	81
Liberty, N.Y.	2.56	2.92	50	160	26	84	202	22	89

[a]Here again, BOD loadings are based on raw sewage figures, as the recirculated flows affect the primary settling performance.

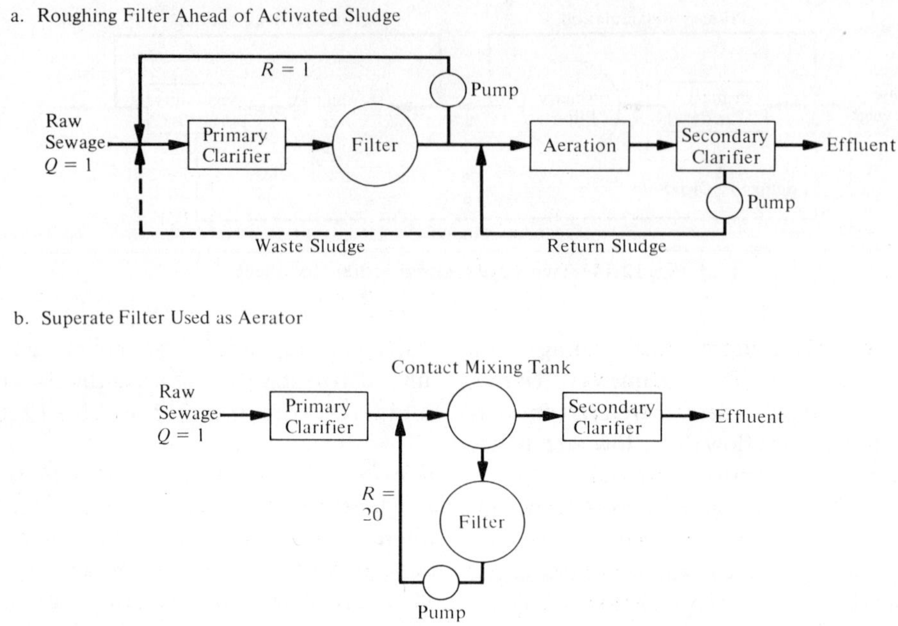

Fig. 12-15 High-rate filters used as roughing filter and aerator.

system in use has a filter with a coarse media depth of only 1 ft and uses very high recirculation ratios (20-25:1). The "superate filter" is used primarily as an aeration device, with further detention between the activated sludge floc and the sewage provided in a mixing tank as shown in the flow diagram in Fig. 12-15B. Average results of plants using the "superate filter" principle are shown in Table 12-32.

Contact Aeration. In this process, settled sewage is passed through an aeration tank containing media to which is attached a biological growth such as that found in a trickling filter bed. Air is diffused in the tank so as to maintain aerobic conditions. The media to which the growth adheres may be closely-spaced plastic sheets, wood slats, etc. The effluent from the contact unit is then settled. No sludge return is required, as the growth in the contact bed serves to supply the biological requirements in much the same way as in trickling filters.

The contact surface requirements are about 180 ft^3/lb of applied BOD; air requirements are about 1.3 ft^3/gal of sewage. Results are about the same as those from trickling filter plants. The use of contact aeration has been limited.

Operating results[30] at several military post treatment plants using two stages of contact aeration are given in Table 12-33.

Oxidation Ponds. Oxidation ponds, also called "stabilization ponds" or "lagoons," are shallow

TABLE 12-32. Plant Results—Trickling Filter Used as Aerator

Plant	Monterey, Cal.	Pismo Beach, Cal.
Flow (gpd)	72,000	170,000
Recirculation ratio	26.6:1	21:1
Mixed liquor suspended solids (ppm)	1740	713
Sludge index	89	160
BOD loading (lbs/100 lbs sludge)	47.1	20
Suspended solids (ppm)		
Raw	–	305
Primary effluent	78	85
Final effluent	20	23
Percent removal (secondary only)	74	73
BOD (ppm)		
Raw	–	287
Primary effluent	177	131
Final effluent	23	19
Percent removal (secondary only)	86	86

TABLE 12-33. Contact Aerator Operating Data

Post	Flow (mgd)	Aeration (ft^2 surface per lb/BOD)	Detention (hours)	Air (both stages) (ft^3/gal)	SUSPENDED SOLIDS (ppm) Influent	Final Effluent	Percent Removal	BOD (ppm) Influent	Final Effluent	Percent Removal
A. PRIMARY STAGE										
Greenville AFB	0.36	77.5	1.49	1.99	90	33	63	138	51	63
Hondo AFB	0.56	81.3	0.86	1.35	65	50	23	111	77	31
Kingsville Field	0.45	147	1.18	1.75	105	70	33	93	55	41
Roswell AFB	0.43	58.5	1.72	2.15	106	54	49	187	107	43
Rucker	1.6	92	3.5	4.21	141	69	51	197	78	60
B. SECONDARY STAGE										
Greenville AFB	188.5	1.35			33	19	42	51	37	28
Hondo AFB	117	0.86			50	31	38	77	43	44
Kingsville Field	250	1.18			70	45	36	55	51	7
Roswell AFB	102.5	0.95			54	30	44	107	52	51
Rucker	215	3.24			69	20	71	78	64	18

ponds in which sewage effluents are stored. During the storage period, natural biological action takes place with resulting reduction and oxidation of organic matter. The chief source of oxygen necessary for biological activity is photosynthesis; algae, in the presence of light, synthesize new organic matter from end products of decomposition and liberate oxygen. Bacteria present in the sewage then act to aerobically decompose the sewage organic matter. This action is accelerated by aeration either by use of air diffusers or by mechanical surface aeration (Fig. 12-16). Recirculation of pond effluent at a 1:1 ratio also gives improved results.

Depths and loadings of oxidation ponds depend to a great extent on climatic conditions. In cold climates where thick ice covers the surface for several months, 4 to 5 ft depths are commonly used with BOD loadings of 3 to 4 lb/acre-ft/day. Detention periods are at least 30 days. In warm climates, where there is little or no ice formation, depths from 2 to 4 ft and BOD loadings of 15 to 50 lb/acre-ft/day are used. On this basis, a 3-ft pond depth would provide a detention period of about 7 days at maximum loading for a primary effluent of 130 mg/l BOD. Where diffused air is used, pond depths of 10 ft are not uncommon.

Oxidation ponds are capable of reducing applied BODs by 85 to 95 percent, giving effluents comparable to those obtained in conventional activated sludge plants. They should, however, be designed for the lowest temperature conditions to produce such results year-round. The formula:

$$\text{acre-ft} = \text{lbs BOD/day} \times 0.00644 \times 1.072^{35-T}$$

where T = pond temperature in °C has been suggested as a method of determining pond volume requirements at any operating temperature. Using this formula, 1.0 mgd of primary effluent with a BOD of 130 mg/l would require a volume of 19.6 acre-ft at 20°C (68°F). At 0°C (32°F), the volume requirement would be 78 acre-ft.

Suspended solids removal in oxidation ponds should be about the same as in extended aeration systems; i.e., 70 to 75 percent.

The shape of oxidation ponds has no effect

Fig. 12-16 Oxidation pond with mechanical aerator. (*Courtesy of Clow.*)

on results. Due to the long detention periods, short circuiting should be minimal. However, it is advisable to use multiple inlets to distribute the settleable suspended solids as well as possible, and so avoid local pockets where anaerobic decomposition of sludge will occur. It is likewise desirable to arrange the outlets so that operation at different depths and complete drainage of individual ponds may be effected.

Due to their size, oxidation ponds are of earth construction. Where necessary, bottoms and sides should be sealed to prevent percolation.

Because of their large area requirements, the use of oxidation ponds is generally restricted to small plants and to nonresidential neighborhoods. They have the disadvantage of giving rise to odors, especially in colder climates, when their ice cover thaws. They do have the advantage, however, of giving a high degree of treatment at low cost if conservatively designed.

Results[29] of stabilization ponds at a U.S. Navy post are presented in Table 12-34. Here there were four series of connected ponds with a total area of 101 acres and a total volume of about 300 acre-ft. Results reported[31] on the

TABLE 12-34. Oxidation Pond Data—U.S. Naval Station, Shoemaker, California

	Average	Maximum	Minimum
Flow (mgd)	3.7	4.1	2.9
Suspended solids (ppm)			
Raw	254	310	205
Primary effluent	95	115	70
Pond effluent	54	140	25
BOD (ppm)			
Raw	236	260	180
Primary effluent	143	180	100
Pond effluent	34	50	20
Loading (lbs BOD/acre-ft)	14.5		
Pond detention (days)	26.5		

TABLE 12-35. Operation Results—Lagoon with Mechanical Aerator

Flow (gpd)	BOD Loading (lbs/acre/day)	Lbs BOD/Acre-ft/Day	BOD (ppm) Influent	Effluent	Percent Removal
19,000	270	86	240[a]	58	78.5

[a]Calculated on a 24-hr basis.

operation of an aerated lagoon are given in Table 12-35.

Intermittent Sand Filters. Intermittent sand filters are capable of producing very high quality effluents. When handling primary effluents, they are usually dosed at rates of 2.3 to 3.5 gpd/ft^2. Sand depths vary from 1.5 to 3 ft, with an average of $2\frac{1}{2}$ ft. For an average primary effluent with a BOD of 130 mg/l, the BOD loadings will then be equivalent to 25 to 38 lb BOD/1000 ft^3 of sand.

Effluents of intermittent sand filters will surpass in quality those from any other secondary treatment method and should average about 5 mg/l BOD or less, with suspended solids also under 5 mg/l. Effluent clarity should be excellent and stability and nitrification high because of the biological and straining action of the filter.

For flexible operation, there should be a minimum of three intermittent sand filters in an installation to permit alternate use, resting (for drying), and cleaning. Walls may be of earth or concrete, the latter being preferred to avoid the washing of earth onto the sand. Floors are usually of earth with sufficient slope to permit adequate drainage via a suitable underdrain system.

The sand filter media should have an effective size of 0.25 to 0.5 mm and should have a uniformity coefficient of 3.5 or less. It should be supported on a 5-in. layer of $1\frac{1}{2}$ to 2 in. stone topped with layers of grated aggregate ranging from 1 in. stone to pea gravel, the total coarse media depth being about 12 in.

Sewage may be discharged to the filters via pumps or a dosing tank. Distribution onto the bed is important in order to avoid local overloads that will result in ponding and oxygen depletion. Concrete troughs and metal or ceramic piping may be used for this purpose, with concrete aprons located at points of discharge so as to prevent sand scouring.

Because of their large area requirements and high initial and maintenance costs, intermittent sand filters are rarely used for handling flows over 0.25 mgd. Also, because they may give rise to odor nuisance, they must be located a considerable distance from dwellings. Typical results[32] of intermittent sand filters treating settled sewage are given in Table 12-36.

ODOR CONTROL AND DISINFECTION

Odor Control. Odor at sewage treatment plants may be caused by septic sewage entering the plant or may originate in the plant at almost every step in the treatment process. Good operation and general plant cleanliness will as a

TABLE 12-36. Typical Intermittent Sand Filter Data

Test	Sand Bed Effective Size (mm)	Depth (inches)	Dosing Rate (gpd/ft^2)	BOD_5 Loading (lbs/1000 ft^3 sand/day)	Suspended Solids (mg/l) Influent	Effluent	Percent Removal	BOD_5 (ppm) Influent	Effluent	Percent Removal	NO_3-N Effluent
1	0.25	18	2.3	18.0	87	3	97	142	3	98	32.4
2	0.25	18	3.45	30.7	105	8	92	161	3	98	19.3
3	0.25	18	4.0	31.8	86	6	93	143	2	99	28.7
4	0.25	30	2.3	10.8	87	4	95	142	2	99	38.1
5	0.25	30	3.45	18.4	105	9	91	161	3	98	20.8
6	0.25	30	4.0	19.1	86	10	88	143	3	98	24.6
7	1.04	30	4.0	18.1	93	15	84	136	25	82	11.8

rule suffice to obviate preventable odor nuisance. Other means are available in cases where the odors are caused by abnormal circumstances. Such means involve the use of chemicals or combustion.

Chemicals. Of the chemicals employed for odor control, chlorine is the most widely used and is most effective for preventing aerial nuisance due to incoming septic sewage, excessively long primary clarifer detentions, trickling filter clogging, and activated sludge bulking.[33,34] Generally, the application of about 5 mg/l of chlorine to raw sewage is sufficient to control odors only. However, this will leave no residual. It is more desirable to apply heavier doses so as to give a residual of about 0.2 mg/l in the effluent of the primary settling tank. This will prevent odors from emanating at the screening and grit removal stations and at the primary settling tank.

At small plants, sodium or calcium hypochlorite or chlorinated hydrocarbons marketed under a variety of trade names may be used instead of chlorine. They are fed as a liquid and are more costly than chlorine. These compounds are also of value in controlling odors where screenings are buried and where grit containing putrescible organic matter is used as landfill.

Ozone has been used for odor control to a limited extent, especially in connection with ventilating systems–i.e., where trapped gases from screen houses, covered grit removal units, sludge storage tanks, settling chambers, trickling filters, etc. are vented to the atmosphere.

Combustion. Odors may be eliminated by combustion in a ventilating system either by incineration at fairly high temperatures (1200 to 1500°F) or by catalytic oxidation at much lower temperatures (500 to 650°F). In the latter method, the heated gases are passed over a platinum–nickel catalyst where complete oxidation of the odor-producing organics takes place.

Disinfection. Disinfection involves the use of chemicals to remove substantially all of the bacteria in sewage effluents prior to their discharge to the receiving stream. The same chemicals that are used for odor control are also used for disinfection. Here again, chlorine is the most widely used while the use of hypochlorites is generally restricted to the smaller plants. Ozone is less effective on turbid effluents than on effluents of high clarity.

The amount of chlorine required for effective treatment depends on the quality of the sewage effluent. For a normal sewage, the following average quantities will be required:

Effluent From	mg/l
Primary sedimentation	8
Chemical precipitation	6
Trickling filter	5
Activated sludge	4

Such dosages should give an effluent residual chlorine content of at least 0.5 mg/l after a 15-min contact period.

Contact tanks providing good mixing are required where chlorine is applied for disinfection. Such units may be plain baffled concrete tanks and should provide a minimum detention of 15 min at maximum hourly flow. They may be of any desired shape but are usually rectangular in plan and with over-and-under or end-around baffles to minimize short circuiting. These should be provided with drains to permit periodic flushing to remove settled solids.

Chlorine is supplied as a liquid in 150-lb cylinders, in ton containers, and in tank cars. It may be fed as a gas (small plants) or as a solution in water. A large variety of manual control and automatic feeders is available. Chlorinator capacity for normal domestic sewage should be:

Treatment	Capacity (mg/l)
Primary	20
Chemical	15
Trickling filter	15
Activated sludge	8

Chlorine feeding equipment and chlorine supplies should be housed in separate areas. The chlorination room should be heated and well ventilated. It should be accessible only through an outside door.

Hypochlorites are supplied as a liquid or as a

TABLE 12-37. Effect of Mixing on Coliform Organism Residuals

Flow (gph)	With Mixing	Without Mixing
Contact time (minutes)	27,400	26,000
Chlorine dosage (ppm)	46	46
Chlorine residual	15	14.6
Initial coliforms (MPN/ml)	2.4	2.0
	307,500	359,000
Residual coliforms (MPN/ml)	2,135	80,750

dry powder. They may be fed by means of standard solution type or dry feeders.

Ozone is produced electrically by exposing dry air to a high voltage discharge in ozone generators. Several types are available, ranging in capacity from 0.5 to 63 lb/24 hr. Power requirements are 9 kw/lb ozone generated. As little as 0.5 mg/l will sterilize clear water. The dose requirement increases considerably, however, as turbidity and suspended solids content increases.

The effect of mixing in chlorine following primary settling is shown by the data presented in Table 12-37. Table 12-38 presents data on the chlorination of various types of sewage effluents.

PACKAGE TREATMENT PLANTS

Package plants were originally considered to be those plants that are completely fabricated at the factory and shipped assembled as an integral unit. They were generally of the complete treatment type and most of them used extended aeration or contact stabilization. A few high-rate trickling filter units have been installed at Air Force posts. Due to transportation limitations, such assembled units were necessarily limited in size, usually for treatment of flows up to 15,000 gpd in a single unit.

So-called package plants now include large combination-type units and assemblies of units requiring field erection and interconnection. Such plants are common for treating flows up to 1.0 mgd. The field for such units are: housing developments, motels, military posts, turnpike stations, large restaurants, shopping centers, institutions, schools, etc.

The design basis of package plants is generally more conservative than larger plants of the same type or of plants treating normal sewage flows. Reasons for this are: the raw sewage is usually stronger with a BOD of 300–500 mg/l; most of the flow arrives at the plant in an 8- to 12-hr period; and good operation supervision is often lacking.

Package-type extended aeration plants normally provide a minimum 24-hr aeration period and a final settling of 4 hr based on 24-hr flow. The aerator loading is, however, less than non-package units, being 15 to 20 lb BOD/1000 ft^3 of aerator capacity. A sludge storage tank having a capacity of 4 ft^3/person is generally provided. The air supply is usually 2500 to

TABLE 12-38. Sewage Effluent Chlorination Data

Type Plant	No. of Plants		Chlorine Contact Time (minutes)	Chlorine Dose (lbs/million gal)	Chlorine Residual (mg/l)	Effluent Suspended Solids	BOD	Coliform Bacteria in Effluent (count/100 ml)
Primary settling	1	Average	15	79	1.7	104	144	2900
		Range	15	79	1.7	104	144	
Contact aeration	1	Average	15	50	0.6	20	20	Under 10
		Range	15	50	0.6	20	20	
Low-rate trickling filter	5	Average	80	64	1.4	57	72	630
		Range	15–153	23–100	0.7–2.0	7–152	14–215	40–110
High-rate trickling filter	10	Average	72	56	1.3	47	40	210
		Range	10–260	7–150	0.5–2.5	17–150	17–150	6–550
Conventional activated sludge	15	Average	46	49	1.0	24	24	1220
		Range	15–180	17–134	0.2–2.5	7–43	7–50	100–5000
High-rate activated sludge	8	Average	39	41	1.2	43	41	237
		Range	15–63	20–56	0.2–4.0	24–57	15–56	1–735

3000 ft^3/lb of BOD loading. This includes air for sludge return, air lifts, etc. The raw sewage is usually bar-screened or comminuted ahead of aeration. Final effluent chlorination with a 15- to 30-min contact period is normal.

Package contact stabilization plants normally are designed with a mixing or contact period of 1.5 to 2.5 hr based on the 24-hr raw sewage flow plus 100 percent return sludge. The sludge reaerating tank is sized to provide a sludge detention period of 6 to 8 hr with the sludge volume equal to the sewage flow. The settling period is normally about 4 hr, with an overflow rate of 400 to 600 gpd/ft^2 based on raw sewage flow. Sludge digestion is carried out aerobically with a digester capacity equal to that of the sludge aeration tank or about 4 ft^3/person. The air supply is generally 2500 to 3000 ft^3/lb of BOD loading, divided about $\frac{1}{5}$ to the mixing or contact zone, $\frac{2}{5}$ to the sludge aeration zone, and $\frac{2}{5}$ to the digester. BOD loadings will vary from 15 to 30 lb/1000 ft^3 of total aeration capacity. This includes mixing, sludge aeration, and digestion. As in package extended aeration plants, no primary settling is provided, and the final effluent is chlorinated with a 15- to 30-min contact period.

High-rate trickling filter installations that would qualify as package-type plants under the broad interpretation have been installed at a

TABLE 12-39. Ohio Turnpike High-Rate Filter Plant Results

	Average	Range
Average flow (gpd)	34,000	24,000–41,000
Suspended solids (mg/l)		
Raw sewage	223	126–304
Final effluent	11	3–15
Percent removal	95	90–98
BOD (mg/l)		
Raw sewage	250	192–320
Final effluent	5	1–10
Percent removal	98	95–99
Loading (raw BOD/1000 ft^3) (total filter volume)	3.3	1.7–5.0

TABLE 12-40. Contact Aeration Plant Performance Data

BOD (ppm)								
Plant	Flow (mgd)	Percent of Design Flow	Raw	Primary Effluent	Intermediate Effluent	Final Effluent	Percent Overall Removal	Applied BOD Loading (lbs/1000 ft^2 plate/day)
Bossert Estates	0.066	59	259	214	100	47	82	4.6
Greenfield Village	0.15	135	162	136	89	42	74	7.2
Pemberton	0.1	55	235	199	22	24	90	4.1
Runnymede No. 1	0.15	67	327	156	27	7	98	4.9
Runnymede No. 2	0.326	92	403	314	130	42	90	13.3
Shore Crest	0.10	97	151	106	60	26	83	4.9
Swedesboro No. 2	0.083	71	204	332	134	15	93	9.3
Woodbury Hts.	0.065	100	406	336	123	42	89	14.2

SUSPENDED SOLIDS (ppm)					
Plant	Raw	Primary Effluent	Intermediate Effluent	Final Effluent	Percent Overall Removal
Bossert Estates	395	99	74	36	91
Greenfield Village	174	112	62	46	74
Pemberton	215	81	20	10	95
Runnymede No. 1	226	94	23	3	99
Runnymede No. 2	263	96	81	33	87
Shore Crest	159	96	54	35	78
Swedesboro No. 2	167	81	29	5	97
Woodbury Hts.	200	91	50	34	83

number of turnpike service areas. Both single-stage and two-stage systems are employed. The design flow for a number of two-stage plants on the Ohio Turnpike was taken at 100 gpm, with maximum daytime flows at 175 percent and peak flows at 350 percent of the average 24-hr flow. A raw sewage suspended solids content of 250 mg/l and a BOD of 500 mg/l were assumed. The design recirculation ratios were 1 : 1 in each stage when the raw sewage flow was 100 gpm. Plant unit design figures were as follows: primary clarifier 2.2 hr detention, 568 gpd/ft^2 overflow rate; primary filter hydraulic loading 159 gpd/ft^2, BOD loading 55.5 lb/1000 ft^3/day; secondary filter hydraulic loading 159 gpd/ft^2, BOD loading 14.4 lb/1000 ft^3/day; final clarifier detention 2.0 hr, overflow rate 840 gpd/ft^2; and chlorine contact detention 21 min. Combination type clarification–digestion units were used, the digester capacity being 4.7 ft^3/person. Filter recirculation was back to the primary clarifiers in the first stage, and directly back to the filter in the secondary stage. Operating results[35] over a year's time covering 8 plants of this type are presented in Table 12-39.

Package-type contact aeration plants have been used for treating sewage at small housing developments. In these plants, the usual basis for design is an average sewage flow of 100 gpd/person having a BOD of 200 mg/l. The applied (primary effluent) BOD loading is normally 5 lb/1000 ft^2 of contact plate area, using two stages of aeration preceded by primary settling for 2.5 hr, and 1.5 hr for intermediate and final settling. The detention period in each aerator stage is also 1.5 hr. The total air supply to both aerators is based on a minimum of 1.5 ft^3/gal of sewage, with about 60 percent going to the first stage. The contact plates are commonly of corrugated asbestos $\frac{1}{4}$ in. thick and spaced on $1\frac{1}{2}$-in. centers. Air is usually diffused through porous tubes. Final effluent is chlorinated with 30-min to $1\frac{1}{2}$-hr contact time. Sludge digestion tank capacity is usually 3 to 4 ft^3/person.

Results[36] of a number of housing development plants where BOD loadings were close to or above design loadings are presented in Table 12-40.

REFERENCES

1. Boruff, C. S., and Buswell, A. M., "The Anaerobic Stabilization of Sewage Screenings," *Sewage Works Journal* 4 (Nov., 1932).
2. Diaper, E. W., and Glover, G. E., "Micro Straining of Combined Sewer Overflows," *Journal WPCF* 43 (Oct., 1971).
3. Kivell, W. A., and Lund, N. B., "Grit Removal from Sewage," *Water Works and Sewerage* 87, 4(1940).
4. Joint Committee Report of ASCE and WPCF, "Sewage Treatment Plant Design," ASCE Manual of Practice No. 36 (1959) and WPCF Manual of Practice No. 8 (1957).
5. Hay, T. J., "Air Flotation Studies of Sanitary Sewage," *Sewage and Industrial Wastes* 28 (Jan., 1956).
6. Wirts, J. J., and Fischer, A. J., "Pilot Plant Results of Vacuator and Auxiliary Equipment," *Sewage Works Engineering* (Nov., 1946).
7. Logan, R. P., "Scum Removal by Vacuator at Palo Alto," *Sewage Works Journal* 21 (Sept., 1949).
8. Mays, T. J., "Vacuator Operation at Santa Maria," *Sewage and Industrial Wastes* 25 (Oct., 1953).
9. Fischer, A. J., "Developments and Trends in Sewage Treatment," *Engineering and Contract Record* (Canada) (Feb. 18, 1942).
10. Seidel, H. F., and Baumann, E. R., "Effect of Preaeration on the Primary Treatment of Sewage," *Journal WPCF* 33 (April, 1961).
11. Fischer, H. J., and Hillman, A., "Improved Sewage Clarification by Pre-Flocculation Without Chemicals," *Sewage Works Journal* 12 (March, 1940).
12. Committee Report of APHA Public Health Engineering Section, "Chemical Treatment of Sewage," *Sewage Works Journal* 7 (Nov., 1935).
13. "Instant Abatement," Dow Chemical Co. Bulletin (1967).
14. Cygan, P., "The Use of Polyelectrolytes as an Aid to Raw Sewage Sedimentation," presented at Pennsylvania WPCF Meeting (Aug. 3, 1966).
15. Hendon, H. H., "Experiences with Chemical Treatment of Sewage at Birmingham, Alabama," *Sewage Works Journal* 8 (March, 1936).
16. Heukelekian, H., "The Effect of Varying the Daily Charge of Sewage Solids on the Activated Sludge Process," *Sewage Works Journal* 7 (May, 1935).
17. Gunson, C. P., "Preaerated Raw Sludge Recirculation Increases Primary Sedimentation Efficiency at Denver," *Sewage Works Journal* 18 (Nov., 1946).
18. Anderson, N. E., "Design of Final Settling Tanks for Activated Sludge," *Sewage Works Journal* 17 (Jan., 1945).
19. Hazeltine, T. R., "Sludge Reaeration in the Activated Sludge Process–A Survey," *Journal WPCF* 33 (Sept., 1961).
20. Torpey, W. N., and Chasick, A. H., "Principles of

Activated Sludge Operation," *Sewage and Industrial Wastes* **27** (Nov., 1955).
21. Ullrich, A. H., and Smith, M. W., "The Biosorption Process of Sewage and Waste Treatment," *Sewage and Industrial Wastes* **23** (Oct., 1951).
22. Ullrich, A. H., and Smith, M. W., "Operating Experience with Activated Sludge–Biosorption at Austin, Texas," *Sewage and Industrial Wastes* **29** (April, 1957).
23. Kraus, L. S., "Digested Sludge–An Aid to the Activated Sludge Process," *Sewage Works Journal* **18** (Nov., 1946).
24. Sawyer, C. N., "Activated Sludge Modifications," *Journal WPCF* **32** (March, 1960).
25. Baker, R. H., "Package Aeration Plants in Florida," *Proceedings ASCE, Sanitary Engineering Division*, **SA6**, 88 (1962).
26. Dreier, D. E., "Summary of Experience in Mechanical Activated Sludge Plant Operation," *Sewage Works Journal* **17** (Jan., 1945).
27. Dreier, D. E., "Summary of Experience in the Operation of Standard Trickling Filters," *Sewage Works Journal* **18** (July, 1946).
28. Fischer, A. J., "The Present Status of Biofiltration," *Water Works and Sewerage* **89** (Oct., 1942).
29. Committee report, "Sewage Treatment at Military Installations," *Sewage Works Journal* **18** (Sept., 1946).
30. Nelson, F. G., "Spiral Contact Aeration in Biological Treatment," *Sewage and Industrial Wastes* **30** (July, 1958).
31. McKinney, R. E., and Edde, H., "Aerated Lagoon for Suburban Sewage Disposal," *Journal WPCF* **33** (Dec., 1961).
32. Furman, T. deS., Callaway, W. T., and Grantham, G. R., "Intermittent Sand Filters–Multiple Loadings," *Sewage and Industrial Wastes* **27** (March, 1955).
33. Heukelekian, H., and Day, R. V., "Disinfection of Sewage with Chlorine," *Sewage and Industrial Wastes* **23** (Feb., 1951).
34. Chambers, C. W., "Chlorination for Control of Bacteria and Viruses in Treatment Plant Effluents," *Journal WPCF* **43** (Feb., 1971).
35. Hauenstein, H. W., "Sewage Treatment for Turnpike Service Areas," *Water Works and Sewerage* (March, 1960).
36. Wilford, J., and Conlon, T. P., "Contact Aeration Sewage Treatment Plants in New Jersey," *Sewage and Industrial Wastes* **29** (Aug., 1957).

13

Industrial Wastewater Handling

C. Fred Gurnham
Betty A. Rose
Thomas K. Nedved
William T. Fetherston
Gurnham and Associates, Inc.
Chicago, Illinois

INTRODUCTION

The development of industrial wastewater treatment methods is historically more recent, but at the same time has been more rapid than the development of sanitary and municipal waste treatment. Large variations exist in the composition and the concentration of wastewaters from industry to industry, from plant to plant within the same industry, and even from day to day and from hour to hour in a single plant. This has challenged the wastewater engineer to make innovations in traditional treatment processes, to adapt physical, chemical, and biological material-processing techniques to the treatment of industrial wastewaters, and to develop entirely new technologies.

As greater emphasis is placed on water quality, public and political attention is ever more strongly focused on industry as a major water user and therefore as a major contributor to the pollutional burden. Industrial processing of almost every type and every material generally results in greater concentrations of specific pollutants than are found in domestic wastewater flows. Increased concentrations, together with the relative simplicity in identifying and targeting a single plant as a polluter in contrast to an entire community or area, have contributed to a frequently disproportionate emphasis on industrial wastes.

All environmental engineers, whether municipal or industrial, must be competent to design and operate facilities to treat any or all types of wastes. More stringent effluent criteria for municipal discharges and the demand for additional sanitary sewage treatment within the relatively narrow geographic boundaries of metropolitan areas offer an opportunity to transfer sophisticated industrial treatment technology to sanitary wastes. Space limitations serve also to intensify the growing trend toward treatment of combined municipal or regional domestic flows and industrial flows. The municipal engineer must evaluate the treatability of specific industrial wastes in his plant and establish pretreatment requirements. The increased and more sophisticated industrialization necessary to support a growing economy and rapidly developing metropolitan areas has multiplied the complexity of industrial waste flows and resulted in an increased demand for industrial waste treatment specialists.

Industrial waste treatment cannot realistically be separated from the larger problem of industrial waste management. Unlike the municipal sanitary official, the industrialist can exercise some degree of control over the quantity and quality of his wastewater flows by selection of alternate raw materials and/or manufacturing processes. The industrial waste treatment specialist must therefore have additional knowledge of the interrelationships between material and process alternatives, and of wastewater flows.

No two industries are alike in the nature and the quantity of the wastewaters they produce. Even within a single industry, the flows and compositions differ among individual establishments, sometimes by as much as a thousandfold, in spite of nominally identical raw materials and products. At any single plant, the flow and composition of the raw waste varies during the day, reflecting changes in production rate, shifts in products manufactured, clean-up operations, and the like. Seasonal variations, too, are characteristic of many industries, particularly those depending on natural raw materials such as agricultural products.

The variation in composition, concentration, and flow of industrial wastewaters from plants within the same industrial category precludes defining rigid design parameters for industrial waste treatment facilities. It also prevents an accurate prediction of efficiency for specific treatment techniques. The most effective industrial waste treatment systems are individually designed as a part of a total waste management program and tailored to handle unavoidable pollution control problems for a specific manufacturing facility. This chapter outlines the principles and procedures which are generally followed in solving specific industrial problems. Wastewater characteristics and treatment practices in individual major industries are then reviewed.

PRINCIPLES AND PROCEDURES FOR INDUSTRIAL WASTE MANAGEMENT AND TREATMENT

The basic systems engineering approach is the most scientific and logical method for developing a solution to a specific industrial pollution control problem. The first step in this approach, generally called the preliminary investigation, is the single most critical procedure and will be discussed in detail under the following major headings: waste categorization, plant survey, laboratory analyses, and other procedures. Selection of various technologies which will solve specific problems as identified in the preliminary investigation are subsequently covered under sections on in-plant waste controls, ultimate disposal, and waste treatment.

Preliminary Investigation.

Waste Categorization. The initial step in a preliminary investigation is to determine the various effluent standards and regulations which apply to the facility. The term "waste categorization" is related to the two federal systems for categorizing industrial wastes. One system, developed under Public Law 92-500, relates raw materials, wastewater characteristics, manufacturing processes, size and age of facilities, and other factors to the categorization of industries by the Standard Industrial Classification (SIC) of the U.S. Bureau of the Census. Federal effluent limitation guidelines and standards of performance for new sources are thus established by industrial category and subcategories, and state regulations may be expected to follow this pattern. It is therefore necessary to determine the category or subcategory which is applicable to a specific manufacturing facility or section of the facility. Multiple process or product plants may be covered by several sets of limitations and standards. These industrial regulations apply only to process wastewaters. Auxiliary or ancillary processes such as steam generation or cooling are covered by separate guidelines.

In any case, the effluent limitations and standards are tied to two levels of pollutant control and treatment technology. The first, which is required to be in operation by July 1, 1977, is defined as the best practicable control technology currently available; application of the second level, the best available technology economically achievable, is required by July 1, 1983. These technologies vary, of course, from industry to industry and from segment to segment of the same industry.

In addition to establishing effluent limitation guidelines, the federal regulatory documents also describe the characteristics of the wastewaters of the industry category. These profiles list significant pollution parameters and may serve as a source of information to aid in developing a detailed survey program.

The second federal system categorizes industrial wastes as compatible or incompatible with municipal treatment systems. On this basis, pretreatment standards applicable to industrial wastes entering such systems have been formulated to implement Public Law 92-500.

Compatible Pollutants. Compatible pollutants are those materials which the municipal facility is designed to remove or destroy. Most of the food manufacturing industries and a number of other organic industries produce raw wastes with some degree of similarity to municipal sewage. The parameters are the same, although the industrial effluent may range in strength from much weaker than sewage to a hundred or more times as strong. Such wastes may be mixed in common sewers and treated at joint treatment plants. Appropriate design allowance must, of course, be made for increased load and concentration.

Primary treatment, usually involving coarse screening and sedimentation, removes a substantial portion of the suspended solids, and will handle industrial settleable solids equally well, provided the equipment is not overburdened by an excessive load or by a settled sludge of abnormal characteristics.

Secondary treatment, typified by activated sludge and trickling filters but which may also include other aerobic biological processes, is designed to oxidize and destroy a major part of the organic matter in sewage. This matter is commonly measured and expressed as the biochemical oxygen demand (BOD or BOD_5), or by certain related but indirect measures such as chemical oxygen demand (COD) and total organic carbon (TOC). Biodegradable components of an industrial waste can be similarly removed, provided the necessary conditions for biological growth are not destroyed, as they might be by toxic components, extremes of temperature or pH, inadequacy of nitrogen and phosphorus nutrients, or simply an overload of organic matter.

Digestion, an anaerobic biological process, is commonly employed on both primary and secondary municipal sludges to reduce their volume and to further destroy or stabilize the organic material. Industrial wastes that are compatible with sewage in primary and secondary treatment commonly yield sludges that are compatible with sewage sludge digestion or are treatable in separate systems without sewage.

Disinfection is not generally required in the treatment of industrial wastes. However, with a few exceptions, the presence of industrial wastes in sewage does not interfere with the chlorination process. Certain reducing chemicals, such as sulfides, sulfites, and ferrous iron, increase the chlorine demand; however, these materials should have been removed or destroyed before the waste reaches the chlorination process, which is normally the last treatment before discharge.

Incompatible Pollutants. In contrast to food and other organic industrial wastes which are generally treatable by the same processes that are conventionally used for municipal sewage, the wastes from many industries contain pollutants that are incompatible with biological sewage treatment. The most serious type of incompatibility is interference with the operation of the treatment processes, such as the destruction or inhibition of the biomass by toxic substances. These include cyanide, heavy metals, acids, and petroleum-based oils and grease. These substances in weak concentrations represent another type of incompatibility; in this case, the pollutant neither affects nor is affected by the treatment process. It passes through the plant into the receiving stream unchanged except for some degree of dilution.

There are other parameters, in addition to the above types of incompatible pollutants, which are entirely prohibited from entering municipal sewers. The reasons are usually obvious. Forbidden parameters include: flammable and explosive materials, corrosive wastes, solid or viscous wastes that may cause obstruction and other interference with proper operation, and severely fluctuating waste flows or waste loads.

Many of the chemical and metallurgical industries produce pollutants that are incompatible with municipal sewers and sewage treatment. Heavy metals, cyanides, and extremes of pH are common in these wastes and are toxic to the treatment organisms. Other undesirable materials, either organic or inorganic, occur in many wastes and must be removed or inactivated before the waste can be discharged to a public sewer or treated in-house by biological techniques.

Final characterization of a pollutant as compatible or incompatible must be based on a study of the specific municipal system into which the industrial waste is to flow. The biomass of any biological treatment system may become acclimated to specific organic materials which do not appear on the federal list of compatible pollutants. Withdrawal of such materials from the municipal flow may cause an upset in the treatment system.

Advanced waste treatment systems or independent physical-chemical systems at the municipal plant, incorporating pH adjustments and chemical additions to treat sanitary wastes, also remove certain inorganic pollutants. Such removal capacities should be considered in determining acceptable concentrations of these inorganic materials in the influent to the specific municipal system.

Waste categorization, as the initial step in a preliminary investigation, serves first to develop a list of pollutants which can be expected to appear in the industrial waste flows. Such a listing is important in planning the plant survey. Secondly, categorization by compatible and incompatible pollutants points out the type of treatment processes which should be considered.

The Plant Survey. Again, because no two plants have identical wastewaters, it is not possible to delineate the wastewater problems of a specific plant except by an individual study of that plant. Background information or general data on similar plants is frequently available in the literature and can be very useful; however, a well-planned and carefully conducted survey is the only satisfactory procedure.

It is essential not only to establish parameters for the design, improvement, or expansion of a waste treatment system, but serves other purposes as well. It will determine if the plant is in compliance with local, state, and federal regulations; and, if it is not in compliance, the survey will reveal which parameters are in violation, how often, and to what degree. This information may be useful in preventing adverse publicity, or in showing how to counteract the noncomplying parameters by corrective measures. A series of properly managed surveys, over a period of time, is also a deterrent to or a defense against nuisance suits.

The primary goal of the wastewater survey is, however, to provide facts and data which are necessary to develop a complete waste management program. A waste management program includes more than a waste treatment system. Waste management begins with an effective water management program which can lead to cash savings, to increased convenience in providing utility services to the plant, and ultimately to a reduction in wastewaters to be treated. Survey data point out means for water conservation by recycling and reuse or, more simply, by curtailing actual usage. A literature study is especially useful here as it can show comparative figures for similar plants in the industry. A closely-related benefit may be the detection of material wastage and loss, discovered by any unusual concentration of product, raw material, or associated parameter in the wastewater.

The data acquired by the survey are needed by plant management in order to calculate or verify surcharge rates. They are also requisite to the execution of applications for discharge permits at the local, state, or federal level. Finally, as noted earlier, reliable and unbiased survey data provide the best defense against lawsuits by neighbors or by regulatory authorities.

Planning the Survey. A complete survey is designed to measure the quantity, and frequently the quality, of influent water as well as the quantity and quality of wastewater flows. Careful advance planning of the wastewater survey is essential if maximum benefits are to accrue. In preparing the survey program, the purpose must be thoroughly defined and understood. If it is to measure the degree of compliance or violation, the pertinent regulations must be specified. The type and quantity of data which the survey is expected to provide should be completely outlined, and provision made for obtaining each necessary item.

Commonly, the survey is conducted on the entire manufacturing plant as a unit. However, for some purposes, it may be limited to a specific area within the plant, such as a single department or small group of departments. It may be the area served by a single outfall. Whatever the extent of the survey area, the planning step should include acquisition and evaluation of maps showing boundaries, manufacturing equipment, water and wastewater sources, and sewer layouts in detail. In older establishments, sewer connections are frequently not known and it may be necessary to trace sewer lines by dye tests or equivalent methods.

The duration of the survey must be established in advance; the minimum period is usually 24 hr or one complete operating day. A thorough survey for design purposes requires from 5 to 14 days, and even this does not take into account seasonal or other long-range variations.

The wastewater survey should be correlated with production operations. It should be conducted during a period of "normal" plant operations, to the extent that there is any such period, unless there is a specific reason to the contrary. Wastes that occur during a clean-up shift are markedly different from production wastes and should be the subject of a separate survey, or they should be sampled separately and resulting data set apart in the survey log.

Plans should include measurement of water entering the survey area during the survey period for comparison purposes as well as to record its origin to indicate such things as metered city water, metered or otherwise measured pumped flow from wells or streams, recycled water, and stormwater flow if it is significant. Certain federal and state effluent regulations are based on the quantity of pollutants added to the water as it is used by industry. Background contamination is subtracted prior to determining the individual plant standards. Under those circumstances, it also becomes necessary to monitor or sample influent water quality during the survey period, although sampling need not be as frequent as wastewater sampling.

All waste streams must be measured and sampled at one or more points. These points must be selected to serve the stipulated purpose of the survey. A single outfall may suffice if the survey is being conducted solely to determine compliance with discharge regulations. If the survey is to establish design parameters, a number of branch flows should be evaluated in addition to the final discharge. In any case, all equipment for flow measurement and for monitoring and sampling must be installed and in operation before the survey begins.

Production data should be recorded for a period concurrent with the wastewater data because it may be desirable to report flow and pollutants in pounds per unit of production, as well as in the more conventional units of concentration. Furthermore, comparison of survey data with production data permits an estimate of the probable environmental effects of increased productivity.

In addition, consideration should be given to collecting data on upstream and downstream flow of the receptor, whether it is a sewer or moving stream. In certain areas, effluent requirements for discharges to surface watercourses vary according to the dilution ratio. Therefore, it is necessary to determine upstream flows. Discharge regulations may also require that specific stream quality be maintained; upstream and downstream sampling then becomes essential. A determination of the quality of upstream flow is also necessary in order to evaluate the synergistic effects of specific pollutants.

Finally, the planning stage must cover training of the survey crew, so that each man knows his responsibilities and duties and is prepared to carry them out.

Flow Measurement. An essential part of the wastewater survey is the measurement of flows at all strategic points. These include all water flows into the plant or survey area, all effluents leaving the area, and a selection of intermediate flows such as process streams. The intermediate flows may be deferred to a later survey; however, if one or more are known or are expected to be particularly significant, and if they can be included without overburdening the survey team, especially the laboratory, their inclusion increases the survey's value.

The methods of flow measurement are numerous and are thoroughly described in the literature. In addition to the conventional types of meters and techniques, industrial surveys frequently employ such procedures as timing the filling of a calibrated pail or other container, observing the rate of liquid level rise in a sump, or using an elapsed-time meter on a pump of known calibration.

The variability of flow rate is often equally as important as the 24-hr average flow and provision should be made to record it. This is necessary if proportional sample composites are to be prepared.

Any type of flow meter can be constructed to: (1) indicate flow, which requires personal observation at appropriate intervals; (2) record flow, which provides a continuous chart; or (3) integrate flow, which provides a total flow figure over the survey period or any portion of it. An automatic proportional sampler can also be regulated by a meter designed to integrate flow measurements.

Monitoring may be defined as the continuous measurement of pollutant parameters by automatic indicating or recording instruments. Such a procedure is superior to any type of sampling and laboratory analyses, but it is presently adaptable only to a limited number of parameters, and the initial or capital cost is relatively high. Costs may be offset, in part or in total, if the monitoring equipment can later be installed to control or monitor the waste treatment system.

The pollutional parameters most commonly monitored are temperature and pH. Recorders for one or both of these are widely used on effluent flows either to sewers or watercourses. They are frequently equipped with alarms or flow diverters which are activated by temperature or pH values that are in violation or tending toward it.

Other water quality parameters that can be monitored with modern instruments are electrical conductance (a measure of dissolved ionic salts), color, turbidity, and dissolved oxygen. Also available, though not yet widely used, are monitors for oil and grease and for specific ions. Others are continuously being developed, and the comparative efficiencies and costs of those presently available are being improved.

For the many significant parameters that cannot be economically or practically monitored, wastewater quality can be judged by taking samples from the main waste stream. These samples can be examined in the field for appearance, odor, temperature, and other obvious characteristics; with suitable kits, more elaborate field tests can be made. More sophisticated analyses are made in the laboratory. To fulfill any or all of the purposes of a survey, samples must be truly representative–i.e., the composition of a sample must be in close agreement with the composition of the stream from which it was taken.

The simplest type of sample is the "grab" sample. It is taken by dipping into the stream with an appropriate utensil. Acceptance of the sample as being representative of that portion of the stream which was invaded at the moment of sample withdrawal is a matter of judgment. The disadvantages and risks of accepting a single grab sample as being representative of the day's flow are obvious; however, there are circumstances where this is the best that can be done, and the results are not meaningless.

A composite sample, prepared by blending a series of grab samples, is of course far more realistic than a single grab sample. Such a composite might be made of 12 grab samples taken at 5-min intervals and would approximate the true average stream composition for that hour; a 24-hr composite might be made from 24 samples at hourly intervals, or 288 samples taken at 5-min intervals. Another technique for preparing a composite sample is to withdraw a small continuous stream from the main flow into the sample bottle. The only truly representative composite is made by collecting in a single vessel the entire waste stream flow, mixing thoroughly, and withdrawing a sample. This technique is rarely used, and is only possible on very small flows of stable wastes. Intermittent sampling, properly done, is an acceptable substitute.

When the waste flow varies with time during the survey or sampling period, as is almost universally the case, the composite sample is not truly representative unless each individual component sample is proportioned to the flow at the time it was taken. This can be done in several ways. For example, if samples are collected hourly, a portion of each which is proportional to the main stream flow is added to the aliquot. For example, if 10,000 gal pass the sample point from 7:00 to 8:00 AM, 100 ml of the 7:30 grab sample is placed in the composite bottle; if 12,000 gal pass from 8:00 to 9:00 AM, 120 ml of the 8:30 grab are placed in the composite bottle; if 25,000 gal pass from 9:00 to to 10:00 AM, 250 ml are placed in the composite bottle; and so forth for the whole 24-hr day. This method is frequently described as VvTc (volume variable, time constant).

A second method for collecting flow-proportioned composites reverses the volume–time relationships by collecting a fixed amount of sample whenever a specified number of gallons are measured; this method is VcTv (volume constant, time variable). The choice of either method is dependent on hydraulics of the site to be sampled, the characteristics of the waste, convenience, safety, budget, and whether the installation is to be permanent or temporary.

Sampling can be done manually or automatically. For a single short survey, the former is probably cheaper. If there is any intention, however, of extending the survey beyond about 24 hr, of repeating it, or of making further surveys in the same company, the automatic sampler will probably pay for itself in a short time.

Automatic samplers are available that provide composites proportional to flow, that provide a sequence of 24 separate samples for the 24-hr period (each a mini-composite of several proportional samples during its hour), and that provide other benefits such as refrigeration where necessary. The suppliers of automatic wastewater samplers have developed a line of variations to serve almost any survey need.

Preservation of samples until they are brought to the laboratory may be important. Refrigeration has been mentioned as a popular approach, and is recommended for samples to be tested for BOD. For some specific industrial components, chemical preservatives are placed in the sample bottle before the survey. Since some of these preservatives interfere with other desired analyses, the program must be planned in advance in order to provide everything necessary.

Sampling of the receiving stream may be difficult or even impossible under certain circumstances, and the survey team must exercise ingenuity to overcome the difficulties or to conduct a satisfactory survey in the enforced absence of a desirable sample. If a certain stream is totally inaccessible, its upstream tributaries can be sampled and blended either physically or mathematically. One difficulty which is practically impossible to overcome with full satisfaction is the sampling of a stratified stream with floating oil on the surface or sediment moving along the bottom. Again, ingenuity is called for. There is no way to include coarse materials such as rags, pieces of paper, and sticks in a representative sample; these must be accounted for by observation.

The sampling log, which includes observations regarding debris, changes in color or odor, times of accidental spills, or deviations from production schedules, is critical to the correct interpretation of the survey results.

Analyses. The analytical procedures needed for an industrial wastewater survey to determine compliance are well documented in *Standard Methods for the Examination of Water and Wastewater*[1] and in various federal procedural manuals. Some of the methods are controversial, especially for the very low concentrations that exist in some samples; in general, the analyst should conform to the methods that are employed by the cognizant regulatory authority in its own laboratories. Some methods of analysis are very arbitrary (the test for oil and grease is a good example), and the recommended procedures must be followed exactly.

Effluent discharge regulations are frequently written for groups of materials, such as those for oils, greases, or cyanides. The analytical methods described in *Standard Methods* or in federal manuals do not identify specific materials which are a part of the group. More precise analytical procedures may be necessary to develop the data on which the design of treatment or recovery systems can be based. For example, the standard analysis to determine violation of cyanide discharge standards measures the total cyanide, both simple and complex, in the waste stream. However, accepted treatment to destroy cyanide in wastewaters, namely chlorination, would have little or no effect on complex cyanides. Additional analyses are therefore essential for design purposes. Analytical procedures must therefore be selected according to the purpose of the survey.

If one goal of the survey is to provide design data for a treatment facility, specific laboratory tests of a functional rather than an analytical nature may be desirable. These are not routine, but provide design factors needed by the design engineer. Examples are: settleability tests for the removal of suspended solids, determination of the pH titration curve to evaluate neutralization of the waste, and biodegradability tests to explore the feasibility of biological treatment. There are other examples, the selection depending on the waste characteristics and the possible or probable treatment technologies.

The above functional tests need not be carried out in conjunction with the wastewater survey, although the sampling program provides a realistic way to acquire raw material for the tests. During the planning sessions, their inclusion may be decided, with due consideration for the heavy additional workload on the laboratory.

Another type of test, sometimes run in connection with the survey, is the determination of toxicity, or bioassay. This calls for highly-specialized techniques and trained personnel. The experimental organisms may be fish or other species, preferably native to the local area.

Evaluation of the Survey. The survey data should be critically examined to evaluate their adequacy for the intended purpose, and to find any discrepancies that defeat their credibility. Material balances produced from the survey data are often useful, but perfection must not be expected. Errors of 10 percent are about the minimum, and a 20 percent error is more common; even larger discrepancies do not destroy the validity of the survey, although they do suggest caution in its interpretation. The material balances may be on the total flow volume or on any single pollutant or group of pollutants.

The survey, if it is at all detailed within the plant, serves to pinpoint the major sources of pollution or material loss. It provides the data to calculate waste loadings from individual or groupings of manufacturing processes, and can be expressed as pounds per day of each specific pollutant or pounds per unit of production. Variations in waste load may be as significant as the daily average. Comparison of loadings with those of other plants in the industry, as known through the published literature or from private sources, may be of considerable interest.

The completed survey and its evaluation frequently lead to specific recommendations. The first recommendation may be simply that the survey be considered as only a trial run, and that it be repeated with correction of any apparent errors in planning, conduct, or type of laboratory analyses to be performed. The succeeding survey may be of a more specialized or limited nature, because the first survey succeeded in indicating the major waste sources. Almost always, an initial survey will include a number of items that later experience shows to be unnecessary, such as testing for pollutants that are found in insignificant amounts. Elimination of such work gives the ensuing survey a greater chance of success.

The successful survey will, of course, achieve the purposes for which it was intended. Specific recommendations can then be made to bring the effluents into full compliance, or to maintain compliance if no violations are discovered.

Control and Treatment Technologies.

In-Plant Waste Control. The logical first step in resolving a pollution problem is to attack the waste at its source—i.e., to prevent the formation of a waste rather than to depend upon treatment to remove or destroy it. Even if total elimination of the waste is not practical, any significant reduction of its total volume or strength produces a benefit in economics and convenience. An early goal of the wastewater survey should thus be to point out in-plant opportunities for waste reduction, so that only a minimum burden will reach the treatment or discharge facility.

In-plant reduction or elimination of pollutants can be achieved by process modification, by raw material substitution, by material recovery, by operational changes, or by segregation of wastes. The utilization of any or all of these in-plant control techniques is related to the recognition of water as a raw material, and wastewater as a secondary product of the manufacturing process. If these facts are accepted, then it is logical to apply the principles of quality control to the production of wastewater.

Process Modification and Raw Material Substitution. Because these two approaches to the in-plant control of wastes are frequently interrelated, they are discussed here as a single topic. The literature on production techniques and pollution control are full of successful case histories on the application of these control techniques. The opportunities to effect a reduction of pollution by process modifications appear limitless. However, the application of any specific technique to a given industrial facility must be evaluated for both technical and economic feasibility. Each facility must be studied individually.

Several widely accepted in-plant control techniques are presented to illustrate the general approach. One example which involves both process change and raw material substitution that has been extensively adopted for the purpose of pollution control is the paper industry's conversion of its sulfite pulping process from a calcium base to a magnesium or other soluble base. This change permits the recovery of materials that formerly reached the waste stream and required costly treatment. In another example, many steel plants have converted their pickling medium from the conventional sulfuric acid to hydrochloric acid. Hydrochloric acid lends itself to recovery and reuse instead of neutralization, settling, sludge handling, and disposal. In the same industry, the quantity of waste pickle liquor has been reduced by preliminary dry operations such as grit blasting. In any industry, conversion from wet to dry processing reduces the direct water pollutional load; however, the air pollution potential of dry processes may require installation of wet scrubbers, which in turn will produce liquid waste.

Material Recovery. A more common in-plant practice to reduce the pollution burden is to salvage reusable or recoverable materials from the waste stream. This must be done promptly after the waste is generated before it becomes mixed with other wastes and waste streams. It is simply good economics to salvage the material at its most concentrated point, and before it becomes diluted and contaminated with other substances. In the steel industry, iron-bearing dusts and sludges (such as mill scale) can be separated from their carrier stream and returned to the blast furnace. In wood pulping, chemicals can be salvaged by separating them from the pulp, concentrating and burning them, and reinforcing the residue with new chemicals for resue.

Where direct reuse is not practical, the waste material may nevertheless be salvaged and put through a regeneration process to render it suitable for another cycle. It is a common procedure, when activated carbon is used to purify sugar or chemicals, to regenerate that carbon by controlled heating and oxidizing. In a more complicated example, nickel can be removed from electroplating rinses by ion exchange, the exchange resin regenerated, and the concentrated nickel salt purified for reuse in the nickel plating bath. A still more complex procedure, but economical and widely used, is the salvage of silver from photograph-developing wastewaters; this can be accomplished by direct displacement in a steel wool cartridge or by electrodeposition in a flow-through unit in the waste stream. In either process, the recovered silver is not directly reusable; it is returned to the central processor for reforming into photographic materials.

The most widely practiced in-plant salvage of a useful material from the waste stream is the reuse of water itself. Water is an expensive raw material, and wastewater may be even more costly in its disposal; it is only logical, therefore, to reuse water in the plant as many times as possible. Reuse in the form of complete recycle requires partial purification as part of the cycle, but costs may be optimized if the used water moves through a succession of downgraded uses before it is finally regenerated for reuse or treated for discharge. Countercurrent rinsing of electroplated parts by gravity overflow from one tank to the next, for example, results in two or three uses of each drop of water with no deleterious effect on product quality.

Wastewater which has been contaminated only by the addition of heat, such as from its use as a coolant in a surface condenser, only needs cooling for complete recycle. Evaporative cooling of the recycling water is the most common method. This leads to a gradual increase of the natural mineral salts or other solids in the raw water, and must be compensated by bleed-off or blow-down of a small stream for discharge or treatment. Despite this requirement, the recycling of cooling water by means of cooling towers or spray ponds is widespread in industrial and commercial establishments.

When otherwise recyclable water is contaminated with materials rather than just heat, more complex but still only partial treatment is required within the cycle. Simple sedimentation may be adequate; at least one small steel mill treats its cooling water by settling for removal (and anticipated recovery) of mill scale, with simultaneous skimming and salvaging of free oil. The water from this treatment is of suitable quality for direct reuse in the mill. The evaporation of chromium plating rinse waters yields a concentrate that can be recharged to the plating bath and a distillate of high quality that can be used for several purposes in the plant.

Chemical or biological treatment is sometimes required to prepare the spent water for reuse, but this treatment is generally less extensive than would be required for discharge. In one full-scale but still somewhat experimental installation, barometric condenser wastewater, heavily contaminated with dissolved and suspended organic matter, is treated on a "biological cooling tower," which simultaneously cools the water and removes much of the organic content by aerobic biodegradation. The effluent is acceptable for reuse in the barometric condenser. Sludge withdrawl from the system provides a bleed stream which removes inorganic salts and nondegradable organics.

Some of the components of industrial wastewaters cannot be salvaged for reuse because they are no longer in a suitable condition for use as raw materials or intermediates, nor can they be used as products because of the risk of inferior quality. Nevertheless, they may be recovered as by-products, perhaps foreign to the main product line of the manufacturer, but still marketable. The most common example of by-product recovery is in the food industries, where spilled or damaged (not spoiled) food materials, considered to be unsuitable for human consumption, are prevented from entering the waste stream or recovered from that stream for processing as animal feed. Food wastes that are too badly degraded for use as animal food can be rendered or otherwise treated for the recovery of chemical by-products and fertilizer.

In the nonfood industries, the manufacture of by-products may be more complicated, and the end material not at all related to the principal industry product; nevertheless, by-product recovery is an accepted practice in many plants. Integrated steel mills, for example, produce coal-tar chemicals from the waste streams of their coke batteries. In times of fuel shortage, these same wastes may be burned as auxiliary fuel. In addition to the direct recovery and reuse of wood pulping chemicals themselves, chemical by-products can be recovered from spent pulping liquors that would otherwise be wasted. Turpentine and furfural are examples of such products; in a rather unusual but successful process, the flavoring material vanillin is obtained from spent sulfite liquor.

The salvage of reusable materials or the recovery of by-products from waste streams generally reduces their load of pollutional material, making it easier to treat and to dispose of them. In addition, this recovery is usually an economical advantage, although the economics is not always simple. Some by-products cost more to recover than their market value, but are economical because of the simultaneous reduction in waste treatment costs. On the other hand, vanillin is an example of a by-product that removes almost no pollutional burden from the waste stream; it is marketable at a profit, however, and this can be credited against the financial burden of treatment.

Operational Changes. In many industries, significant in-plant reduction of wastes can be accomplished by improvement of clean-up techniques. The cleaning of product, equipment, and work areas may be a continuous operation or it may be conducted batchwise on a nonproduction shift. Some units require cleaning very infrequently, perhaps once per year. Whatever the cleaning schedule and frequency, the resulting wastes differ from the normal production wastes, and place an additional and different load on the discharge facilities.

Consideration should be given to reducing the total flow. This can be accomplished by the use of high-pressure, low-volume jet or spray streams for washing. The amount of pollution in the waterborne waste can be reduced by preliminary dry cleaning operations. The characteristics of the waste can be improved by changing the cleansing agents used.

The same principles can be applied to product cleaning operations. The on-line cleaning and rinsing of electroplated parts in a conveyorized system can be carried out in dip tanks or, with significantly less water consumption, by spray nozzles or fog sprays. The sprays need not run continuously but can be automatically timed to operate for a few seconds for each piece of work. The same principles apply to the very different poultry dressing industry, and to many other manufacturing processes. Dry cleaning is practical and beneficial in such diverse operations as the cleaning of dough-mixing equipment in a bakery, the cleaning of metal products by sand blasting ahead of pickling, and the cleaning of floors in a meat processing plant. The more material that can be removed in solid form for disposal as solid waste, the simpler and less expensive is the treatment and disposal of the wastewater. In solid form, these materials may in some cases also be more easily processed for by-product recovery. The changing of cleansers is not always feasible, as an optimum has usually been selected based on effectiveness and cost; however, when the detergent in use is a major source of phosphorus, boron, phenols, or other specific polluting components, it may be in order to consider a change if this can be done without introducing another major pollution problem. The excessive use of cleaning agents should be avoided.

There are numerous simple internal measures that can be taken in every plant to reduce its wastewater problem. All of these may be described as common sense practices; they are commonly referred to as "good housekeeping." Many of them involve the elimination of accidental spills which can cause heavy pollutional loads, upsets in the treatment system, and major losses of costly material. Batch tanks that are filled by a manual operation can be equipped with a high-level alarm or more positive control to prevent over-topping. Material losses resulting from filling, mixing, boiling, or aerating of tanks can be prevented by increasing the freeboard or adding splash plates at strategic points. Proper maintenance and repair of leaks, whether occasional or regular (such as a leaking pump gland) are obvious, and can be important in combating both material losses and excessive pollutional loads.

Other simple housekeeping programs can be devised by the alert operator. Increasing the drainage time of wet materials (such as emptied milk cans, electroplated parts, and wet-processed textiles) before moving the material to the next step will prevent the transfer of liquids and the spread of a pollution problem. The drained liquid is returned to the process and does not enter the wastewater system.

Where spillage is inevitable or can be anticipated, containment is desirable. Containment areas may be small to collect and salvage drips from a single source or they may be diked areas to contain spills from an entire department or a group of tanks. If an outdoor tank farm is diked, provision must be made for the removal of stormwater from the diked area, either automatically or manually after each storm.

Segregation of Wastes. A final aspect of in-plant waste control is the collection of wastes, and particularly the decision of whether to keep them separate or to blend them. There are possible merits to each approach, and proper consideration should be given to both. Obviously, a waste which is scheduled for salvage, recovery, or reuse must be segregated from all other wastes that might interfere with that goal. Even simple dilution is undesirable, as it makes the recovery process more complex and more costly; mixing with other wastes may make the recovery totally impossible.

Similarly, wastes that require a specific treatment should not be mixed with others, except for a valid reason. For example, chromate wastes that are to be treated by chemical reduction should not be unnecessarily diluted as this will increase the chemical dosage required; they may, however, be mixed with simple acid wastes, as acidity is desirable for the reaction. Cyanide wastes intended for destruction by chlorination must not be allowed to mix with noncyanide heavy metal wastes, because the formation of complexes retards or stops the chlorination process.

On the other hand, blending of certain wastes presents no problem, and blending, in some cases, is a benefit. Continuing the example from metal finishing, it is quite proper to blend chromium wastes after reduction, cyanide wastes after chlorination, nonchromate noncyanide heavy metal wastes, pickle liquors and other acid wastes, and alkaline cleaners and other alkalies. The acid and alkaline materials counteract each other, at least in part, and the next step in treatment, addition of alkali to precipitate heavy metals, is appropriate for all these wastes. However, premature mixing can destroy the treatment operations.

One type of mixing which is almost universally beneficial when properly applied is the equalization of wastes. Industrial wastes, as formed, are usually highly variable in flow rate and in composition; this characteristic makes them difficult to treat in continuous systems and less acceptable in the receiving stream. The sudden peaks of both flow and concentration can be largely damped out by providing an equalization unit in the line ahead of treatment or discharge. For small flows, especially those with major peaks, it is often possible to contain the entire flow of a working day (or other realistic time period) in a tank or basin of reasonable size. Equalization occurs during filling. The contents are batch treated prior to discharge to the sewer or stream. Two tanks, operated in sequence, permit an extended discharge time, but this is not always necessary. In a common example, the rinse waters and an occasional dumped bath in a small electroplating tank are collected in a holding tank of perhaps 1000-gal capacity, mixed, treated during the evening, and allowed to settle during the night; the clear liquid is drawn off to the sewer before the following day's work shift starts. The small volume of sludge is left in the holding tank to accumulate until the weekend when it is removed.

Equalization may also be achieved in a flow-through tank with overflow which is equipped with agitators or mixing baffles. Only minor leveling of flow variations is accomplished in this manner, unless the surface area is large (as in a pond), but concentration peaks and random swings of pH can be smoothed quite satisfactorily. The flow-through tank with variable level can provide some improvement in flow leveling; this is an oversized tank which accepts the incoming wastes at whatever rate they are predicted. The tank is pumped out to treatment or discharge at a constant or controlled rate that approximates the average waste flow; the level varies, and the "constant" pumping rate must be controlled to prevent overflow or complete emptying.

Ultimate Disposal. It is not possible to destroy matter, only to convert it from one form to another. This applies to water, which can be changed from its normal liquid form into vapor, or broken down into its chemical elements, but it cannot be totally lost. Water which is taken into an industrial plant must, therefore, leave it as part of a product, as vapor, or as wastewater.

Total recycling of all wastewaters is not technically or economically feasible for every industry, or for all facilities within a specific industrial category. Based on the results of preliminary investigation, and particularly on an in-plant survey, it is possible to evaluate the various ultimate disposal alternatives. These include discharge to surface watercourses, to subsurface waters through underground wells or by application to land, or to municipal sewerage systems. It is necessary to explore the short- and long-term advantages and disadvantages of each disposal alternative prior to selection of the ultimate disposal site for the specific facility.

Whether treated or not, industrial wastewaters are returned to the natural environment principally by discharge to a natural watercourse—a stream, lake, or the ocean, or a ditch leading to one of these. In the United States, the discharge of polluted waters into natural waters is regulated by the federal, state, and local governments. These regulations are discussed elsewhere in this book, but it should be noted here that the wastes of industry are usually subjected to special limitations because of the varied and perhaps "unnatural" components. The degree of treatment required by law frequently exceeds that required for municipal wastewaters because the cost of industrial wastewater treatment is not a direct tax burden.

Until the decade of the 1970s, limits on waste discharges (industrial and other) were based on an allowable concentration for each pollutional parameter. The limit might be expressed as concentration in the effluent itself, or in the receiving body of water beyond a specified mixing zone. The relationship between these is equal to the inverse ratio of flow rates in the effluent and in the combined stream. Because most of the harmful effects of pollution are a function of its concentration, this traditional approach is a logical way to report and to control water quality in the natural stream.

In the early 1970s a new concept of pollution control developed and began to displace the concentration concept. The new procedure is to measure and control each parameter by units of total weight, for example, by pounds of cyanide discharged. Controls have been specified as pounds of a specified pollutant per unit of production; i.e., as pounds of cyanide per square foot of metal electroplated, or as pounds of BOD per 100 lb of raw milk converted to cheese, or as pounds of oil and grease per 250 lb of hog processed. The intent of this concept is to equalize the burden of treatment by relating it to production, regardless of the size of the facility.

A comparison of the results of the in-plant survey with existing and forthcoming local, state, and federal regulations will reveal the quantity and type of pollutants which must be removed prior to surface discharge. Preliminary cost estimates can be based on these comparisons.

Another important cost consideration is the disposal of solids in the form of sludge, which are a by-product of all wastewater treatment systems. Ultimate disposal of these sludges is perhaps a more critical problem than the disposal of the treated or untreated wastewaters. The alternatives for sludge disposal, including costs, must be considered in relationship to the various water disposal choices. A sludge handling problem may be the determining factor in selecting the optimum approach to wastewater processing.

Discharge to Subsurface Waters. Industrial wastewater, usually after adequate treatment, may be ultimately discharged to subsurface waters rather than to streams. However, pollution of these resources is as significant as pollution of surface waters, and is more long-lasting because of lower transport or flow rates. Since it is less obvious and less subject to control, the laws governing this procedure are apt to be based on prohibition rather than regulation because the subject is poorly understood.

Underground discharge may be direct, as in deep disposal wells. A high degree of clarification and disinfection is required, frequently with additional chemical treatment, in order to avoid possible plugging of the well or receiving stratum. Deep well disposal has been extensively used for toxic and strong organic wastes, for brines, and for several specific materials such as steel mill pickle liquors. Lack of information on the ultimate fate of such materials, once they have been released into the underground environment, has led to public aversion to this technology and to prohibition by regulatory agencies.

Alternatively, underground disposal is accomplished by spreading of the effluent on the land surface. This is done by spray, by other means of irrigation, or by ponding. Some portion of the water is lost by evaporation and some by transpiration by the cover crop; these phenomena, however, remove water only, not pollutants contained in the effluent, although some pollutants may be removed by oxidation, biological action, or volatilization. Much of the water and most of its components percolate into the soil, which adsorbs and holds some materials. If pure enough, the percolating water becomes a valuable source of groundwater recharge. The degree of purification is dependent on the depth and nature of the soil through which the water passes.

Land disposal areas frequently must be diked and tiled to prevent overflows or lateral seepage of unpurified waters. Monitoring wells are required to prevent contamination of aquifers.

Because some of the pollutant materials are absorbed by vegetation cover, caution must be exercised in using such wastewaters for the irrigation and fertilization of crops for human consumption. While the nutrients in such treated wastewaters may increase crop yields, plants may also absorb trace metals. For that reason, regulations have been established to control the quality of wastewaters dispersed on land surfaces. Extensive pretreatment is often required.

Discharge to Municipal Sewerage Systems. Certain industrial wastewaters may be discharged into a public sewer for treatment and release by the municipality; however, industrial wastes discharged to a municipal sewer must be compatible with the sewerage system as discussed previously under "Waste Categorization." They must not cause fire or explosion hazards in the sewers or treatment works, or create destructive corrosion. Solid or viscous wastes are banned if they could cause obstruction to the sewer flow or interfere with operation of the treatment plant. Excessive variations in flow or in any pollutional parameter are prohibited because of possible adverse effects on sewers and treatment. Beyond these four basic prohibitions, an industrial waste is "compatible" only if the treatment works has been designed to treat its specific pollutants, and does in fact remove them to a satisfactory degree. In general, then, a pollutant material is acceptable if it is not harmful to any part of the system and if it is substantially removed by the treatment provided in the system. The pollutant is not compatible if it simply passes through the system unchanged.

Federal regulations define compatible pollutants as BOD, suspended solids, pH, and fecal coliforms and list the following examples of additional acceptable parameters: COD, total organic carbon (TOC), phosphorus compounds, nitrogen compounds, and fats and oils of animal or vegetable origin.

In order to utilize a municipal sewer for the discharge of industrial wastes, pretreatment must customarily be provided. The major goal of pretreatment is to meet the prohibitions and limitations described above, plus any other regulations that may be established by the municipality or sanitary district. A second goal is to reduce the concentration of compatible pollutants in order to reduce surcharges or special fees to the industrial user.

When an industrial facility can pretreat its wastes to a condition acceptable to the sewer, and to a concentration that makes the surcharge tolerable, sewer disposal is generally preferable to discharge to a natural watercourse. Treatment requirements for the latter are usually far more stringent. An exception is high-volume wastes that require no treatment or only moderate treatment for discharge to the stream, such as uncontaminated or noncontact cooling waters. These need only be cooled to avoid thermal pollution; in the sewer they would present an unnecessary hydraulic burden on the conduits and treatment facilities.

Pretreatment techniques are generally the same as treatments for surface discharge, except that lower efficiencies may be acceptable. They are described in the treatment section of this chapter.

Sludge Disposal. During the purification of wastewaters for discharge to the natural environment or to municipal sewers, many pollutants are removed in the form of sludge. In addition, sludge contains chemicals which were used to remove the initial pollutants. Sludge derives from settling tanks and clarifiers, flotation units, bar racks and screens, filters, and other phase separation units. It may be a dirty liquid containing as little as 0.5 percent solids, or it may be soggy, wet press cake; whatever its form, suitable disposal is not easily achieved.

Many industrial plants dispose of their sludges, and sometimes of their entire wastes that are difficult to treat (such as spent electroplating baths) through a scavenger service. The problem of solid waste disposal is thus transferred, for a fee, to a specialist who is presumably better able to handle it. The scavenger has the same ultimate disposal options as the individual industrialist. However, because the volume of the material he handles is greater, the cost benefits from the recovery of oils, solvents, or metals will be larger.

Disposal by scavenger service is frequently the simplest and most economical solution to the sludge problem for the small manufacturer, and sometimes also for large plants. The legal problems of responsibility have not been fully resolved; in some cases the original sludge owner has been held liable for materials he employed the scavenger to remove and destroy.

Prior to the disposal of sludge, whether by scavenger or by in-house means, it is often desirable to condition the sludge and to reduce its volume. This may not be true if the sludge is to be discharged to a sewer—an unusual but occasionally acceptable practice—or spread in liquid form on land, but volume reduction is generally desirable. The water removed during the concentration of sludge is usually of very poor quality and is recycled to the raw waste stream or into the waste treatment process.

Sludge conditioning and concentration processes, including digestion, filtration, centrifugation, and drying are identical to those discussed in other sections of this book. However, the nature of industrial sludges may require certain process modifications. Because the techniques are similar, however, they will not be discussed in this chapter.

Recovery of valuable materials from sludge is a highly specialized operation, and is not generally practiced by the individual industrial installation. It is more common for the plant to segregate such material for collection by or sale to appropriate scavenger and recovery firms. Detailed discussions of recovery processes are beyond the range of this chapter.

Two basic processes are available to the industrialist or to the scavenger service for the ultimate disposal of sludge: incineration and land disposal. Disposal to the ocean, once a widespread practice in coastal areas, is no longer favorably regarded by regulatory authorities. However, disposal of certain fish processing wastes is being reviewed to determine whether such materials can be returned to the natural food chain by ocean disposal.

The incineration of sludges is a common practice in municipal sewage treatment and in large industrial establishments. It is particularly suitable for sludges of high organic content, although inorganic components receive the benefit of drying and sometimes of fixation into a less soluble form.

Auxiliary fuel is required to start the incineration process, and usually to maintain it, especially if off-gases require odor control by an afterburner. If the incinerator is properly designed, trash and similar solid wastes may be used as an auxiliary fuel supply. Incineration is not truly "ultimate" disposal, as there is inevitably some residue or ash, but it does provide a very substantial volume and weight reduction and may also result in the conversion of certain pollutional materials to less troublesome compounds. The residue must still be removed for disposal on land.

Land disposal of sludges or incinerator residue is practiced in several forms. If no significant toxic substances are present, organic sludges may have value as fertilizers or soil conditioners, and can be applied to croplands or woodlands by spraying, ploughing in, or similar techniques. Valueless or toxic sludges are dumped on undeveloped land and are usually covered with soil in a typical landfill operation. However, the disposal of toxic waste in this manner is subject to many restrictions in order to guard against later exposure or leaching of dangerous materials into nearby surface waters or groundwater.

Waste Treatment. An industrial plant faced with the problem of disposing of its process wastewaters should first utilize all reasonable in-plant measures to eliminate or reduce this waste at its source. Techniques for accomplishing this goal were outlined in an earlier section, and include water recycle and reuse, recovery of marketable or usable materials from the waste, and overall good housekeeping. The plant should then consider the alternatives available to it for discharge of that portion of the waste which cannot be bottled up or otherwise eliminated. These alternatives may include release to surface waters or to an underground stratum, discharge to a public sewer, disposal by a scavenger service, and perhaps others. Each of these has its benefits and problems, and whichever is chosen (except perhaps the scavenger service) it will probably require some form of treatment of the wastewater before it is allowed to leave the premises.

The treatment technology available is extremely varied, including all of the processes described elsewhere in this book for municipal sewage, plus other techniques that are specific for certain types of industrial wastewater. Selection of the type of treatment or series of treatment processes for a particular plant should depend primarily on the parameters of pollution that must be removed and the degree of reduction required. Obviously, there are additional considerations involved in this selection, such as the volume of waste to be treated, the pollution concentrations in the raw waste, and the variability of flow and composition. There also are factors such as climate, which affect some types of treatment, and, of course, the overall economics of the choice.

The reader is referred to earlier chapters for the theory and practice of individual treatment techniques. The discussions following relate to treatments specific to the industrial waste field and to modifications in conventional processes required for industrial wastes. The subject is discussed under four categories: phase separation, biological treatment, chemical treatment, and miscellaneous techniques. Each type of treatment process is designed to remove pollutants in a specific form. Therefore, selection is based not only on the quantity of the material to be removed, but also on the form in which it is present. The simplest distinction is that of solid versus liquid. However, more sophisticated determinations are necessary to insure optimum process efficiency.

Phase Separation. The most useful technique in the treatment of industrial wastes and other wastewaters, and the simplest in concept, is the physical separation of phases. Of greatest importance is the removal of solid-phase materials from the main body of liquid. The separation of a nonaqueous liquid phase, such as oil, from the water phase is also significant. Gas-liquid separation has more limited application, but is occasionally of value for specific wastes.

With many industrial wastewaters (and with municipal sewage), a substantial degree of purification is accomplished simply by separation of the solid phase that is initially present. Solids removal is also important to the effectiveness of subsequent treatment processes. Thus, solids separation is the basic step in almost every industrial waste treatment system.

Sedimentation, or the separation of phases by the influence of gravity, is utilized in primary settling tanks as the first or a very early step in wastewater treatment. It may be carried out in two stages; a short-time detention tank for the removal of heavy grit that settles rapidly, and the main unit which removes more slowly settling solids over a period of 2 hr or more. Oily droplets that are lighter than water float to the surface at the same time that heavier-than-water solids (and high specific gravity oils) settle to the bottom. Both the bottom sludge and the floating film or scum are removed by appropriate mechanisms, operated continuously or intermittently as dictated by the nature and quantity of the settled material.

The efficiencies of solids separation by settling vary from one industry to another, depending primarily on the difference in density between the particulate matter and the carrier liquid, the size of particulates, and the time allowed. Settling is adversely affected by eddy currents, which are made worse by improper design, overloaded facilities, temperature variations, and other velocity and density effects. The efficiency of settling can be improved by the agglomeration of small particles into larger pieces, as by chemical coagulation, and by increasing the holding time. The latter approach, of course, requires large vessels, and introduces the possible problem (in many industries) that the tank contents or the sludge may become anaerobic.

The same solids separation techniques are applicable for final polishing after biological or chemical treatment. Sedimentation in final clarifiers serves to remove residual solids that escaped the primary settling tanks plus, more important, the new solids created by chemical and biological reactions. Final settling may be a two-stage process: first, in conventional tanks with mechanical sludge removal, and second, in large polishing lagoons. The latter are suitable only when there is abundant land available.

In elaborate treatment systems, which may be required for complex or very concentrated wastes, additional clarifiers may be used at intermediate positions or on individual waste streams. On the other hand, primary settling can be omitted if the raw wastewater carries a light burden of suspended material, as is true in most dairies, some metal finishing operations, and a few other industries or individual plants.

The removal of free oil by sedimentation follows the same principles as solid particulate settling, and can be accomplished using the same equipment. However, special equipment, namely the American Petroleum Institute (API) separator, has been designed for wastewaters carrying a high concentration of floatable oils, such as those of petroleum refining. Emulsified and soluble oils, like dissolved solids, cannot be removed by settling and require prior treatment, usually by chemical additions.

Small flows of industrial wastes are often treated by batch settling on the fill-and-draw principle. This may be combined with chemical treatment, so that the solids initially present plus those created in the treatment step are all collected together in the batch treatment tank. It is common practice to employ two tanks and to use them alternately, one filling while the other is undergoing treatment and settling. It is also common, when the quantity of settleable solids is small, to run several batches through the tank, withdrawing the clear liquid or supernatant each time, but permitting the solids to accumulate for less frequent removal.

Sedimentation relies on gravity as its driving force, but its action can be speeded by the use of centrifuges. However, because of their high cost, both capital and operating, centrifuges are not often employed on whole wastes, either raw or treated. They are often practical for concentrating sludges, and are widely used when the volume of wash or waste fraction is small. Centrifuges require less space than conventional settling tanks, and their use in waste treatment is increasing.

All sedimentation techniques are dependent on a specific gravity differential between the particle or droplet to be removed and the supporting medium. If this difference is small, efficiencies are low except at unreasonably prolonged detention times. In many industries, it is possible to alleviate this problem by the use of the flotation process, in which air bubbles are caused to nucleate on the particles. This has the effect of lowering the specific gravity of the particle–bubble unit, so that it rises to the surface and is removed by skimming. Formation and attachment of the air bubbles is accomplished by pressurizing the waste with air to put more air into solution; when the pressure is released, the air comes out of solution by forming bubbles (nucleating) on all available surfaces, including the particulates to be removed.

Flotation can also be accomplished by applying a vacuum to the waste after saturating it with air at atmospheric pressure. This process has some advantages, but the bulky and expensive equipment usually gives the advantage to pressure flotation.

In pressure flotation, it is not necessary to pressurize the entire waste stream with air. Often, a side stream is adequate, and the side stream is reblended with the main stream as soon as it leaves the pressure relief valve or mechanism. Alternately, a portion of the effluent is air-pressurized and is then mixed with the raw waste flow. Each of these techniques has its advantages, and each is widely used.

The most common application of flotation is in the treatment of oily wastes from refineries, chemical plants, steel mills, and oil-handling installations. It works equally well on animal and vegetable oils from edible oil refineries, meat packing houses, and soaperies.

Screening is an even simpler technique than settling for the removal of solid particulate matter, but has the limitation that it separates only particles of substantial size. Coarse screens, usually in the form of bar racks, are almost universally employed at the inflow to the treatment works. These have large openings, typically 2 in., and serve to remove coarse solids that would cause trouble in the pumps and other equipment. In plants where such solids are practically nonexistent, a calculated risk may be taken by omitting this early screening.

Fine screens are used in canning and meat processing and in many other industries to remove particles larger than about 40 mesh. This removal lessens the burden on the settling tanks and their sludge-moving mechanisms. Even more important, the presence of coarse solids interferes with and retards the removal of fine solids in the sedimentation units.

The fine screens may be of a simple flow-through type, but these are difficult to keep clean. More common are the moving types, vibrating or rotating, and a fixed screen with a continuous liquid flush over the surface, provided by the waste itself.

Solid particulates that are too small to be removed on a fine screen can be separated by filtering. The base of the filter media is equivalent to a fine screen, consisting of cloth, paper, or a bed of solids; the true medium, however, is the accumulation of solids themselves on the web or bed. Filtration is rarely used on raw industrial wastes because of low flow-through capacity and the need for frequent cleaning. Filters are most applicable for final clarification of treated waste, particularly where space is not available for polishing lagoons.

It is almost always desirable to remove particulate solids at the earliest possible stage, often in a series of operations including, for example, bar rack, fine screen, grit chamber, and primary settling tank. However, there may be times when size reduction is preferable to removal. The comminutor, and underwater cutter or grinder, serves this purpose. The solids are then passed along with the flow to the next removal stage. Comminutors are more common in municipal sewage treatment systems, however, then in industrial facilities.

Biological Treatment. The biological waste treatment processes that have been developed for municipal sewage are, in many cases, suitable for wastewaters from the organic industries. With possible modification required because of a different concentration range, biological treatment is widely practiced in the food industries (including canning, dairy, meat, and poultry industries), pulp and paper manufacture, tanning, textile production, organic chemical manufacture, and other organic industries.

Biological treatment of wastes may be either aerobic or anaerobic. Each type of process has its own dominant strains of microorganisms and yields a different quality of product, although the basic purpose of both is to convert organic constituents of wastewater into gases, separable solids, and water.

Aerobic treatment requires the addition of oxygen, usually in the form of air but sometimes as oxygen gas or nitrate. Aerobic microorganisms are developed which convert the organic matter into carbon dioxide, water, and more organisms, with some production of sulfate and nitrate ions. In addition to the requirement for a continuing and adequate supply of oxygen, the process must be protected against pH outside the desirable operating range, temperature extremes, toxic substances such as metal ions, and sudden variations of any sort. Despite these necessary precautions, aerobic processing is a reasonably rugged technique and is in wide use.

The most common type of aerobic treatment is the activated sludge process, including several modifications of the basic process to adapt it to the usually higher BOD loads of industry. Aeration time may be extended from the conventional 2 to 8 hr to 24 hr or even several days for strong industrial wastes, such as those from pharmaceutical plants and distilleries. Other modifications are also designed to improve the gas–liquid contact and increase the oxygen transfer rate; the use of oxygen in place of air has this goal, as does the sludge stabilization or aerobic digestion process.

The trickling filter is an alternative to the activated sludge unit as an aerobic biological process. It is somewhat more tolerant of flow variations and shock loads than is activated sludge. The traditional rock filling of the filter has largely been displaced by various types of plastic media which provide better air–water contact and a higher degree of treatment. Because of the light weight of the plastic media, these units can be built as tall as 40 ft to permit longer contact time and to provide for increased oxygen supplies by forced-air ventilation.

The rotating biological contactor is a modern variation of the trickling filter. Large plastic discs or cylinders of a webbed structure rotate above the waste, dip into it for about 40 percent of their surface, and carry a film of waste into the air for aerobic action. Power costs are low and efficiency is high. Multiple stages are common. First used in sewage treatment, the rotating contactors are now being employed in several organic industries.

Where space is available, aerobic lagoons provide an inexpensive means of treating weak organic wastes, or of providing a polishing treatment subsequent to activated sludge or other biological processes. The major cost is in land, and even that is recoverable on termination of the operation.

In its simplest form, the aerobic lagoon or oxidation pond relies on its large surface and shallow depth, usually under 5 ft, to provide oxygen input by direct absorption and photosynthesis. For heavier organic loads, lagoons are operated at greater depths and the necessarily increased oxygen transfer is provided by floating or fixed surface aerators or fine-bubble compressed air injection. If a bottom deposit of settled organic solids forms, it becomes anaerobic; however, the main body of water can be kept in an aerobic condition by restricting the BOD inflow in relationship to available dissolved oxygen. Aerobic lagoons in series are commonly used to treat high-BOD wastes.

Anaerobic biological treatment is utilized only on strong organic wastes, including sludges. Organic matter is destroyed by microorganisms that do not require free oxygen to maintain their activity, but can extract the oxygen from sulfates, organic compounds, and water itself. The metabolic products, in addition to the growth of living cell tissue, are methane, hydrogen, hydrogen sulfide, ammonia, and lesser amounts of carbon dioxide. The gaseous products of this decomposition are flammable and odorous and must be collected and/or controlled in populated areas. Some degree of control can be achieved by covering the lagoons. In the meat industry, the cover may be self-provided by floating material, but this is commonly reinforced by additions of straw on the surface. If the level does not fluctuate excessively, a satisfactory seal is provided that does not permit the escape of odors. The liquid product of anaerobic digestion is obnoxious, especially if it is agitated; it is not suitable for discharge but must be transferred with proper care to the aerobic lagoon or other succeeding operation.

In order to treat the higher organic loadings of many industrial wastes, it is often necessary to recirculate biological process effluents through the process or to provide multiple-stage treatment. Multiple-stage treatment may consist of one process in series or a combination of several processes.

High-strength industrial wastewaters, although organic in nature, may be deficient in one or more of the nutrients necessary to support the biomass. Pilot plant studies of these wastes may be necessary to determine if nutrient additions are required. Additions of nitrogen and phosphorous are frequently essential to the efficient biological reduction of some vegetable and most fruit processing wastewaters.

Chemical Treatment. Chemical treatment has played a minor role in the treatment of municipal sewage, limited to an occasional coagulant aid. Since about 1970, independent physical-chemical (IPC) treatment has made a full-scale appearance as a competitor for biological sewage treatment, although many regulatory agencies still classify it as experimental. In contrast, many industries utilize chemical treatment technology for the optimum control of their wastewaters. This is especially true in the inorganic industries, but all industries employ chemical treatments to some degree. In addition to chemical flocculation, the processes of IPC treatment have many applications in industrial waste treatment, as do a number of chemical reactions that have no counterpart in municipal treatment.

The most universal chemical treatment is pH adjustment. Neither acidic nor alkaline wastes are acceptable for discharge to natural streams, to municipal or industrial treatment plants, or to an in-house biological pretreatment plant. The pH of wastes must be brought to near neutrality. A pH range of 5 to 10 is acceptable for some purposes, but tighter limits such as 6 to 8.5 are common for stream discharge.

Acid and alkaline wastes can neutralize each other to some extent. This is a major benefit derived from the blending and equalization of wastes, and is commonly practiced in diverse chemical manufacturing plants and other industries that produce both acid and alkaline wastes. Equalization also serves to dampen wide variations in pH within a single stream.

Self-neutralization has its limits, and almost always requires supplementary acid or alkali additions to achieve the desired range. Sulfuric acid is most commonly used to neutralize the excess alkali in an equalized waste, although other more readily available acids are satisfactory. In some areas, carbon dioxide in the form of flue gas can be used if proper contacting equipment is available. For acidic wastes, the operator has a choice of sodium hydroxide or carbonate, which are convenient to use, or lime, which is cheaper. Limestone is the least expensive reagent, but is not generally satisfactory for sulfate-bearing wastes because it becomes coated and inactive.

Automatic pH control is recommended in neutralization or any pH adjustment step (except in very small batch operations). Adequate mixing is required, and time must be allowed for slow reactions such as neutralization with lime or limestone, but automation assures an effluent within the allowable limits and prevents wasteful use of reagent.

Adjustment to a pH other than neutral is an intermediate step in some treatment systems. The removal of ammonia from a waste by air or steam stripping requires a high pH, achieved by the addition of lime or caustic soda. The high pH must be neutralized before final discharge. At the other end of the pH scale, sulfuric acid is used to produce a low pH for phenol stripping or for the now obsolete process of cyanide volatilization. Low pH is also required for chromate reduction in the electroplating industry. High pH is needed for cyanide destruction by chlorination, and moderately high pH serves to precipitate most of the heavy metals as hydroxides.

The removal of industrial waste pollutants by precipitation is a widely accepted practice. As noted above, heavy metal ions can be precipitated by raising the pH to 8 or 9. One exception is hexavalent chromium which must first be chemically reduced to the trivalent form. Certain metals that are complexed by cyanide can be precipitated only after the cyanide has been destroyed. The precipitation reaction is used for the removal of phosphate in the mining, fertilizer, and chemical industries; this calls for treatment with lime or alum or both. The removal of fluoride in the aluminum and chemical industries is accomplished by precipitation with lime.

Chemical coagulation is an indirect form of precipitation in which the pollutant may not itself react with the reagent, but is removed along with an induced precipitate. The principles of coagulation and flocculation are well established in water and sewage treatment practice; in industrial waste treatment they are applied to oily emulsions and to finely divided and nonsettleable solids such as pigments, paper fibers, meat and tannery effluents, and mineral beneficiation slimes.

Chemical oxidation is competitive with biological oxidation as a treatment technique. It is usually more costly, but it can provide a stronger action suited for more resistant industrial wastes. Air and oxygen are relatively weak oxidants which are very useful when used with microorganisms but of little value otherwise. Chlorine and ozone have broader applications for pollutants that are difficult to oxidize, such as ammonia, cyanides, phenolic compounds, dyestuffs, and other chemical wastes. Chlorine may be applied as gaseous chlorine or in solution form as hypochlorite. Hydrogen peroxide may be used to prevent septic conditions from developing, but it is more expensive than other oxidants. Electrolytic oxidation has been used to treat cyanides and other metal finishing wastes, and as a disinfectant and oxidizer in the presence of chlorides that can be converted to hypochlorite.

Chemical reduction has only one significant application: the reduction of hexavalent chromium from electroplating, anodizing, passivating, and other metal finishing wastes to the more treatable trivalent form. Any strong reducing agent can be used for this purpose; most common are sulfur dioxide, sodium metabisulfite, and other sulfite derivatives.

The selection of any chemical treatment system must be based on the quantity and quality of the influent, the cost and availability of the necessary chemicals, the safety of the process, and the quantity and quality of the waste sludges. Consideration must also be given to proper storage and handling of the necessary chemicals, and to energy requirements.

Pilot plant studies on a significant portion of the actual wastewater flow or on meticulously simulated wastes are almost mandatory in order to develop final design parameters. It is not uncommon for these pilot studies to be conducted by equipment or chemical suppliers. However, they must be carefully supervised by the plant owner or his consultant.

The need for process controls is greater than in biological treatment systems. If the quantity of chemicals introduced into the process is excessive, or if adequate reaction time is not provided, the chemical reaction may be carried beyond the optimum and result in the formation of other pollutants which are more difficult to treat.

Miscellaneous Treatments. Combinations of the above treatment techniques are more common than any single one alone. For example, as has been mentioned, aerobic biological treatment is invariably followed by physical treatment (usually sedimentation) to remove the growths of microorganisms. Sedimentation is sometimes followed by filtration to provide a more highly polished effluent; filtration alone would accomplish the same thing, but the heavier loading on the filters would require more frequent backwashing and down-time. Strong organic wastes sometimes receive an initial treatment in anaerobic lagoons, with further removals of pollutants in one or a series of aerobic lagoons. In the aerobic lagoon series, mechanical aeration is often used in the first pond where oxygen requirements are greatest, and the subsequent lagoons rely on natural aeration and photosynthesis to furnish sufficient oxygen.

The treatment processes which have been discussed under phase separation, biological treatment, and chemical treatment are all proven techniques which have been extensively practiced to treat industrial waste flows. Individual processes or combinations of them have resulted in removal efficiencies in excess of 80 or 90 percent. However, existing and forthcoming regulations may require complete (or at least 98 percent) removal.

In order to achieve such high degrees of purification, it will be necessary to consider the use of processes designed for product purification. The technical and economic feasibility of applying these techniques to wastewater treatment has not, in all cases, been fully studied. It can be stated, however, that the cost of achieving ultimate purity, or to remove the final 5 to 10 percent of pollutants, often equals or exceeds the cost to remove the first 90 to 95 percent.

The most extensively studied advanced waste treatment processes include the more sophisticated filtration processes such as rapid sand filtration, multi-media filtration, and microstrainers. Carbon adsorption is being used to remove soluble organic material in both industrial and municipal facilities. Other processes include ion exchange, reverse osmosis, electrodialysis, and coalescence, which are discussed in detail in other chapters in this volume.

The overall treatment system and the choice of processes used must be tailored to the specific application, wastewater, and receiving stream. The application of any of these advance techniques can only be justified after extensive, large-scale pilot plant studies.

If these costly treatment processes are used, the product water may be of such superior quality that total recycling is not only possible but economically mandatory.

Disinfection is or should be a universal final treatment for any wastewater that contains sanitary wastes. For most industrial process wastes that are free from human wastes, disinfection is not required. On the other hand, chlorination is commonly practiced on effluents from the meat, leather, and related industries, and is routinely required prior to the discharge of at least the first flush of stormwaters from their installations.

Chlorination usually follows all other treatment operations. It is not effective if the organisms requiring disinfection are encapsulated or enclosed within particulate matter. Prior treatment processes are also necessary to remove other chlorine-demanding components. Chlorine and chlorine compounds are the usual disinfecting agents; however, ozone and other materials have been used.

Cooling is a necessary treatment for wastes that do not meet the stream or sewer temperature standards. Direct heat exchange is rarely practical unless the waste is very hot; even then it rarely provides adequate treatment. Dilution with other cooling wastes is the simplest technology, but dilution with clean water is not acceptable to regulatory authorities. Evaporative cooling, as practiced in spray ponds or cooling towers, is common in industry; often the cooled waste is suitable for reuse in the plant. Under certain conditions, a cooling tower can be used simultaneously as a trickling filter for the destruction of biodegradable organic pollutants, but this practice is considered to be experimental and is not widely used.

Incineration of solid wastes, such as trash and garbage, and industrial organic refuse, such as cannery solids, is extensively practiced. It can also be used for organic sludges from other industries, preferably mixed with combustible trash to supply the heat needed for drying and to initiate combustion. It is also useful for waste liquids such as oils and solvents but is not generally practiced for waterborne wastes.

WASTEWATER CHARACTERISTICS AND TREATMENT PRACTICES IN MAJOR INDUSTRIES

It is axiomatic that every industrial plant is different from every other in the nature, volume, and variability of its wastes; in the technical and economic factors that govern its choice of treatment technologies; and very possibly in the standards its effluent must meet. Nevertheless, the plants within any single industry or subcategory of an industry usually have features in common.

The characteristics of some major industries are discussed below. These include the major production operations which produce wastewater, the volume and characteristics of the wastewaters, the methods for their control and treatment, and the limitations that have been imposed on their quality and quantity.

These discussions are based on the latest available information which commonly relates to exceptional plants. Under ideal circumstances, they may provide concrete guidelines for the solution to specific industrial problems. In most cases, they will serve as general guidelines on which to base individual plant surveys and suggestions as to the treatment processes which warrant intensive study.

Iron and Steel. The basic steel industry encompasses the manufacture of direct steel mill products, such as semifinished shapes (ingots, slabs, and bars), structural shapes, rails, bars, pipe and tubing, sheets, and plates, from iron and raw steel produced at the mill. In 1972, the industry manufactured 122×10^6 tons of steel shapes, consuming 93×10^6 tons of coal, 73×10^6 tons of iron ore, 32×10^6 tons of fluxes (limestone, lime, and fluorspar), and various other materials. The industry uses a large quantity of water in its operations, approximately 81 percent for cooling and the balance for process purposes. A typical fully-integrated steel mill averages 20,000 gal of water per ton of finished steel. The operations and products of such a mill are illustrated in Fig. 13-1.

Wastewater sources in the basic steel industry are material handling and storage, coke production, iron and raw steel manufacture, and finished steel production. These sources are discussed in greater detail below.

The raw materials, coal, iron ore, limestone, and fluxes, are received at the mill via rail, waterway, and highway and are stored in large piles. Rainfall runoff from these piles contains suspended solids and soluble constituents from the materials. Water draw-off from liquid fuel storage tanks also contains pollutants.

Coke operations center around the coking ovens wherein soft coal is subjected to high temperatures which convert the coal into the porous form of carbon called coke. The finished coke is pushed from the ovens into railroad cars and then quenched with a deluge of water. The quench waters are screened and settled to remove coke fines and then recycled back to the quench operation or discarded. Volatile matter driven off in the coking step consists of tars, ammonia, light oils, and fuel gases. It is cooled to condense out tars and some water, and is then either scrubbed with sulfuric acid or cold water to remove ammonia. The ammonia is recovered as crude ammonium sulfate crystals or as ammonia water. The water solution remaining from this operation is dilute ammonia liquor and is used to quench coke or is sewered. Gases from the ammonia recovery step are scrubbed with a petroleum solvent ("straw oil") to remove light oils. They are then used directly as fuel gas or passed through desulfurizing units prior to being burned.

Sources and Characteristics of Wastewaters from Iron and Steel Production. Wastewaters from these operations occur in the wet scrubbing of spent off-gases, the condensation of water of reaction, the quenching of slags from the furnace, and in steel-forming operations. Iron fines and dusts developed in the mill are mixed with cooler coke and flux and heated to form an agglomerated product called sinter. The combustion gases contain dusts which are removed by cyclones or in wet scrubbers. In the blast furnace, iron ore, coke, and limestone are fed to the top of the furnace while heated air is blown in at the bottom. The air oxidizes the carbon in the coke to form carbon monoxide; hydrogen in the coke is converted to water vapor. These gases reduce the iron ore to iron and carbon dioxide. The Brent reaction gases (containing carbon dioxide, carbon monoxide, hydrogen sulfide, water vapor, dusts, and hydrocarbons) leave the furnace as overhead product, while iron and molten slag are tapped off at the bottom. The overhead gases are processed through dust cleaning equipment, including wet scrubbers. Scrubber discharge waters contain suspended solids and soluble constituents from the gases.

Steel is made by melting iron from the blast furnace and scrap steel with flux, and treating the mix with oxygen or iron ore to control the carbon content. The furnace off-gases are frequently scrubbed with water to reduce dust discharges. Slags produced in the blast furnace and steel furnaces are often quenched with water sprays. Water draining from the slag piles is contaminated with suspended solids and dissolved matter.

Steel-forming operations develop various wastewaters such as:

1. Rolling mill wastewaters from direct contact cooling and lubrication with emulsified oils.
2. Spent baths and rinse waters from alkaline cleaning of steel shapes prior to further operations.
3. Spent pickling and rinse waters from the processing of steel plates and sheets through a sulfuric acid or hydrochloric acid bath to remove scale followed by a water rinse.
4. Plating rinse waters from the coating of steel with zinc, tin, or other metals by electroplating.

Other wastewaters from the typical steel mill include sanitary waste, powerhouse effluents, cooling tower blow-downs, and cooling waters.

Standard raw waste loads for typical mills, reported in an Environmental Protection Agency (EPA) study[2] of the basic steel industry, are

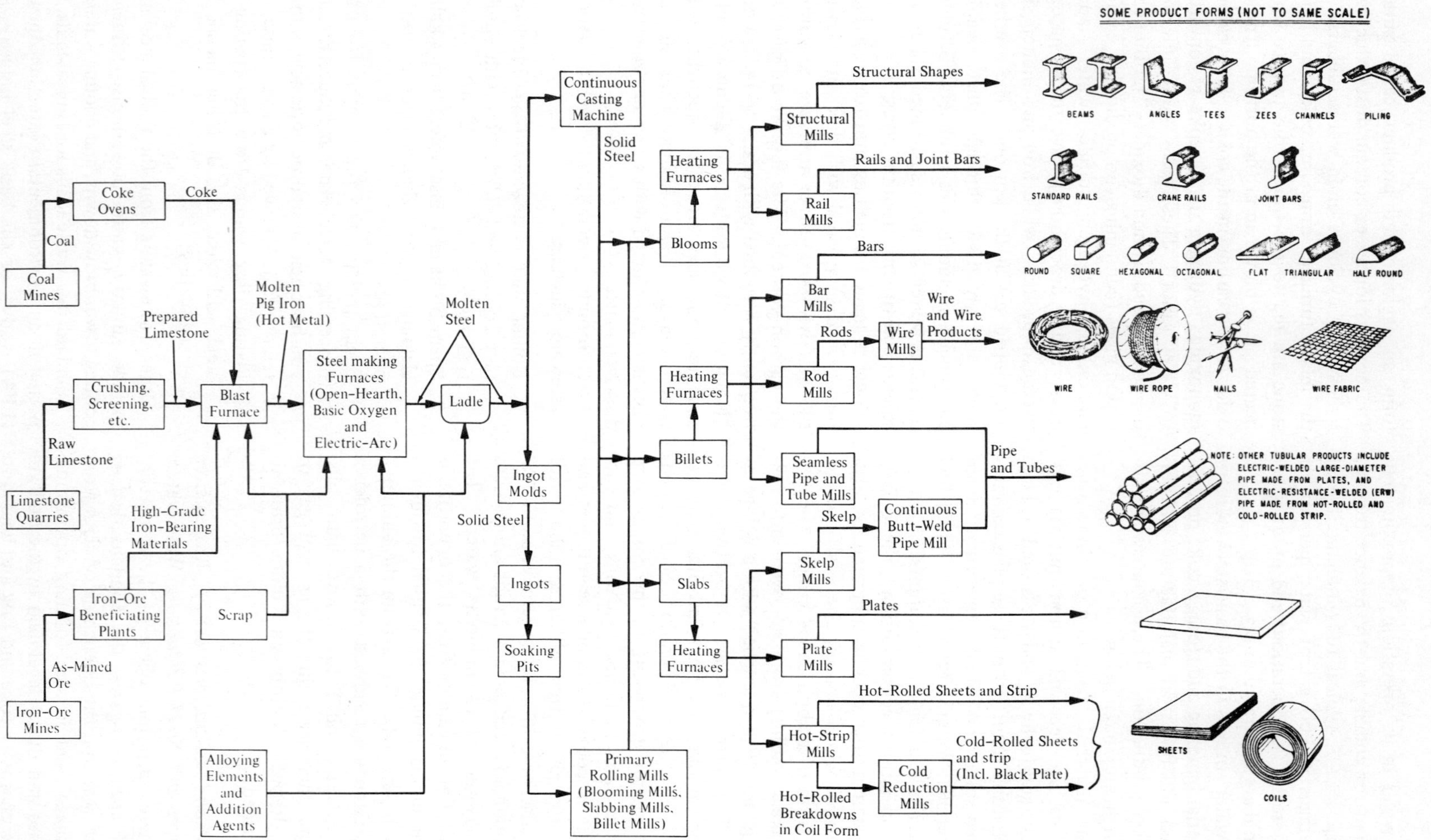

Fig. 13-1. Flow diagram of an integrated steel mill (From *The Making, Shaping, and Treating of Steel*, U.S. Steel Corp. 1971. Copyright 1971 by United States Steel Corporation.)

TABLE 13-1. Standard Raw Waste Loads of the Steel Industry (all values in units/ton product)

	COKE PLANT		BLAST FURNACE AREA		STEEL PRODUCTION AREA		HOT FORMING		COLD FORMING	
	Average	Range	Average	Range	Average	Range	Average	Range	Average	Range
Flow (gals)	3000	1400-4200	2200	1500-4000	4000	3000-7200	7500	430-17000	2200	460-5300
Suspended solids (lbs)			55	3-223	30	10-50	40	10-75	7	0.09-23.0
Oil (lbs)							3.5	0.02-7.35	9.0	0.15-16.0
Phenol (lbs)	0.5	0.1-0.9	0.02	0.002-0.15						
Cyanide (lbs)	0.3	0.0025-3.9	0.09	0.005-0.40					0.05	0.0009-0.10
Ammonia (lbs)	0.35	0.025-1.95	0.25	0.002-10						
Acidity (lbs)									10	0.9-55
Chlorides (lbs)									4	0.5-20
Sulfate (lbs)									4	1.4-22
Phosphate (lbs)									0.4	0.005-2.5
Total chromium (lbs)									0.50	0.0005-3.6
Hexavalent chromium (lbs)									0.40	0.01-2.0
Iron (lbs)									5	0.2-30
Tin (lbs)									0.7	0.005-4.5
Zinc (lbs)									1.0	0.005-2.7
BOD_5 (lbs)									2	0.085-5.0

presented in Table 13-1, in units of pollutant per ton of steel produced.

The principal significant parameters of pollution anticipated in steel mill effluents have been tabulated by EPA for various operations in the basic steel industry.[3] These parameters and their observed concentration ranges are:

Coke plant: 5-35 mg/l phenol; 0.1-55 mg/l cyanide; 1-78 mg/l ammonia; 7-600 mg/l BOD_5.

Blast furnace area: 0.1-8 mg/l phenol; 0.3-22 mg/l cyanide; 0.1-54 mg/l ammonia; 160-12,000 mg/l suspended solids.

Steel furnaces: 330-1700 mg/l suspended solids.

Hot forming: 160-1200 mg/l suspended solids; 0.3-120 mg/l oil.

Cold forming (including rolling, pickling, galvanizing, and tin plating): 0.05-60 mg/l cyanide; 3-275 mg/l BOD_5; 5-1250 mg/l suspended solids; 8-900 mg/l oil; 50-3000 mg/l acidity; 10-1750 mg/l iron; 1-200 mg/l total chromium; 0.5-100 hexavalent chromium; 0.3-150 mg/l zinc; 0.2-250 mg/l tin; 200-1000 mg/l chloride; 75-1200 mg/l sulfate.

Wastewater Treatment. Typical steel mill wastewater treatment involves the physical or chemical removal of pollutants, followed by polishing of the wastewaters in a biological system. The latter is typically owned by a municipality, although some mills have installed their own systems. Formerly, contaminated waters were diluted with cooling waters and the combined stream discharged to the nearest waterway. This practice is now forbidden by regulatory authorities, and treatment is the standard practice. Typical treatment methods in use are described in the following paragraphs.[4]

Phenols are removed from coke plant effluents by vapor (steam) stripping or by solvent extraction. Biological treatment is necessary to reduce the phenols to the low values called for in modern effluent standards.

Suspended solids and oils are removed in sedimentation basins ("scale pits") equipped with top and bottom skimmers. The addition of polyelectrolytes to improve solids separation is practiced with increasing frequency. Air flotation following an emulsion breaking process is utilized to recover lubricating oils for reuse. Lime treatment followed by coagulation-settling and mixed media filtration polishes the wastewaters of remaining suspended matter and precipitates and removes dissolved heavy metals.

Chromium is removed from tin mill and galvanized sheet wastewaters by reduction with ferrous sulfate (copperas) or sulfur dioxide.

The reduced chromium is then precipitated with lime.

Cyanide in tin mill and galvanized sheet wastewaters can be treated by alkaline chlorination. However, cyanide in coke and blast furnace wastewaters must be treated biologically because chlorination of these effluents forms chlorophenols that impart an objectionable taste to water and are resistant to further destruction.

Pickling rinse waters are subject to a variety of treatment and disposal methods. These include lime neutralization followed by discharge to a local waterway or sewer, lagooning of the sludge, deep-well injection, ion exchange, or regeneration.

Cooling waters and miscellaneous process wastewaters are processed through granular media filters to reduce the suspended solids content.

Biological treatment is necessary for many of the constituents of coke plant, blast furnace, and rolling mill wastewaters. It is accomplished by aerated lagoons, biological filters, or some variation of the activated sludge process. In many plants, the biological unit belongs to a nearby municipality.

EPA has established that the best practical technology currently available to meet the 1977 goals for steel mill effluents includes: vapor stripping or solvent extraction of coke plant effluents, the use of "fines free" limestone fluxes to make iron and steel (thus reducing suspended solids), neutralization of pickle liquors, emulsion breaking for oil-water mixtures, oil recirculation, settling of suspended solids, and air flotation units for recovering fine solids and oil.[5]

The best available technology economically achievable to meet the more stringent 1983 goals requires the addition of biological treatment for coke plant effluents and the recycling of effluents from iron and steel manufacturing units. Subsequent treatment, limited to "blow-down" wastewaters, requires coagulation settling units. Pickling wastewaters are to be treated by precipitation, flocculation, clarification, or regeneration of the acid for reuse, and steel forming wastewaters are to be treated by chemical or magnetic means. Heavy metal removal requires ion exchange, with biological treatment for residual oil.

Nonferrous Metals. The nonferrous metal industry nominally encompasses all metal production other than iron and steel, although by tradition such metals as nickel and chromium are usually considered in the ferrous metal sphere. For purposes of this section, the discussion of nonferrous metals is limited to primary and secondary production of aluminum, copper, lead, zinc, and combinations of these metals. For clarity of presentation, and because there is little similarity among them, the industries are discussed separately.

Primary Aluminum. Primary aluminum is one of the major metallurgical industries in the United States, and 4,120,000 tons of aluminum ingots were produced in 1972. A typical plant is illustrated in Fig. 13-2.

The bauxite ore as received at the alumina plant is treated with a caustic soda solution; or it is dried, mixed with lime and soda ash, calcined, and dissolved in water. The alkaline bauxite solution is then heated under pressure to solubilize the alumina. Digested slurry is processed through thickeners to remove insolubles, and the solution overflowing the thickeners is sent to crystallizers to recover crystalline alumina. Weak liquor from the crystallizers is fortified with fresh caustic soda and is recycled to the process to recover additional alumina. Slurry from the bottom of the thickener is washed countercurrently with water to remove additional caustic soda, then discharged as wastewater (called "red mud") to a holding basin. An alternative treatment is to mix wastewater, soda ash, and limestone with the washed slurry, and to calcine the mixture. The calcined mix is then dissolved in water and filtered. The filter cake, called "brown mud," is disposed of by slurrying with water and discharging to a holding basin.

"Red mud" slurry is the principal waste stream from a bauxite refinery, though other wastes exist, such as condensate streams, cooling waters, sanitary wastes, and barometric condenser waters. The barometric condensers are part of the evaporators used to remove

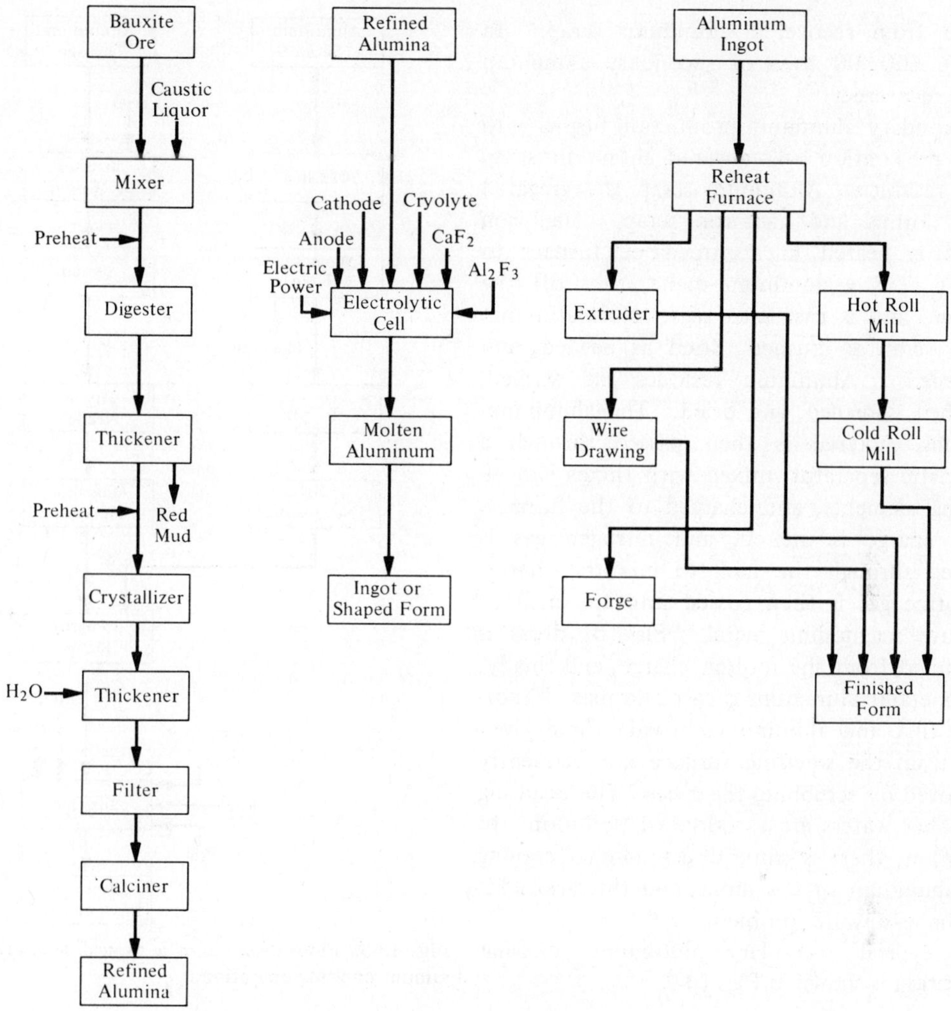

Fig. 13-2. Flow diagram of a typical primary aluminum plant.

excess salts from the recycle-caustic liquor. Normally the condensate is neutral, but "carry-over" may add alkaline values to the condensate, which will then need treatment.

Refined bauxite is mixed with cryolite (sodium aluminum fluoride), calcium fluoride, and aluminum fluoride in an electric cell. The charge is melted and an electric current is passed throughout the cell. The carbon in the electrodes reduces the alumina to aluminum metals and is oxidized to carbon monoxide and carbon dioxide. Off-gases from the cell are scrubbed with water to remove fluorides. The scrubber waters are a major source of pollution.

The next source of wastewater is the rolling mill, where aluminum ingots are heated in a furnace, rolled to an intermediate shape in a hot mill, and then finished in a cold mill. The mills are similar to those used in the steel industry. The rolled shapes are cooled by direct contact with water, or cooled and lubricated by spraying the rolled piece with an oil–water emulsion. The wastewaters from this step normally contain oils, suspended solids, and dissolved solids.

Secondary Aluminum Smelting. Approximately 13 percent of the aluminum produced in the United States is secondary aluminum, made from recovered aluminum scrap. In 1971, 600,000 tons of secondary aluminum were recovered.

Secondary aluminum production begins with feed preparation of recovered aluminum scrap and residues. Aluminum scrap is segregated into normal and high-iron scrap. High-iron scrap is heated in a sloped-bed furnace to 760°C. The aluminum melts, runs off the charge, and is cast into pigs. Normal aluminum scrap is crushed, dried as needed, and screened. Aluminum residues are washed, crushed, screened, and dried. The aluminum-bearing material is then passed through a magnetic separator, mixed with fluxes and alloying elements, and charged to the furnace. The charge is melted, and nitrogen gas is passed through the melt to mix the charge. Chlorine gas is then passed into the melt to remove magnesium metal. Slag or dross is skimmed from the molten charge, and finally, the metallic aluminum is cast into pigs. Vaporized flux and aluminum chloride fume given off from the smelting furnace are frequently removed by scrubbing the gases. The resulting scrubber waters are a source of pollution. In addition, there is some direct contact cooling of aluminum or the mold, and this also adds to the wastewater problem.

A typical secondary aluminum smelting operation is shown in Fig. 13-3.

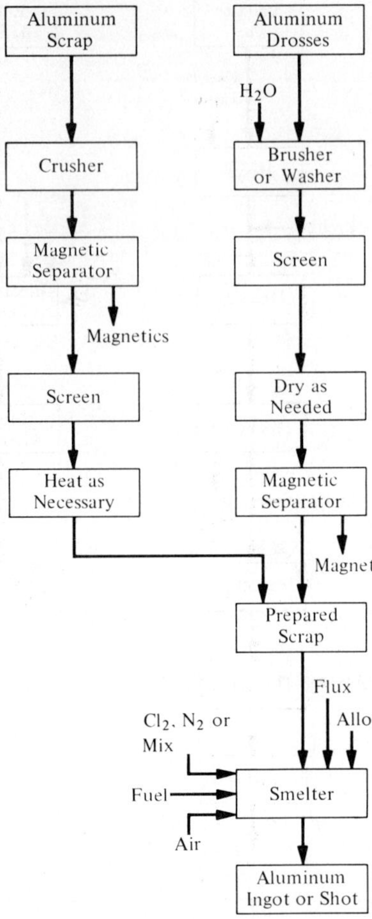

Fig. 13-3. Flow diagram of a typical secondary aluminum smelting operation.

Wastewater Characteristics. The principal wastewater parameters in the aluminum industry are suspended solids, dissolved solids, oils, and fluorides. Typical values for the various parameters are noted in Table 13-2.[6]

TABLE 13-2. Typical Nonferrous Metal Wastewater Characteristics (lbs/ton of product)

Parameter	Smelting	Secondary Smelting	Casting and Foundry	Fabrication
Volume (1000 gal/ton)	15–42	0.1–4	0.1–12	0.2–34
pH (units)	2.7–4.0	6.0–8.5	6.9–7.6	5.0–8.1
Total solids	–	–	3.7	4.3–18.5
Suspended solids	Trace–92	5–8300	–	0.8–5.0
Dissolved solids	64–70	80–130	6.4–16.2	2.2–300
Oil	–	–	1–1.6	0.1–15
COD	–	–	–	5–15
Chloride	–	–	–	1.3
Fluoride	11.4–158	–	–	–
Sulfate	12–25.2	–	–	–

The "red mud" waste stream from the bauxite refinery averages 3000 to 20,000 tons/day of slurry containing 15-20 percent solids. The pH approximates 12.5 and chemical analysis of the mud ranges as follows:

Fe_2O_3	30-60 percent
Al_2O_3	11-20 percent
SiO_2	2.5-14 percent
TiO_2	Trace-11 percent
CaO	5-10 percent
Na_2O	1.5-8 percent

Wastewater Treatment. "Red mud" wastewaters have not yet been satisfactorily treated. The common practice is to impound the waste in lagoons to permit the solids to settle out, and to allow solar evaporation of the water. The former practice of diluting the slurry and discharging it to a local waterway is now prohibited.

Suspended solids can be removed in a sedimentation basin. The use of coagulants and polyelectrolytes to enhance settling is recommended.

Oil- and grease-bearing wastes should be processed through an oil-water separator of the American Petroleum Institute (API) design. The oil is frequently emulsified and requires emulsion-breaking through the application of heat or pH changes. Oil separation can be improved somewhat by using inclined lamella separation devices or air flotation devices.

Fluorides can best be handled by precipitation using lime or calcium chloride. The concentration can be further reduced by passing the wastewater through a packed activated alumina or hydroxyapatite bed.

Neutralization of various streams is required to bring the wastewater pH into the 6.5 to 8.0 range.

The 1977 EPA effluent guidelines[7] based on best practical technology will require bauxite refineries to install lagoons to contain all process wastewaters. Primary smelting plants will have to provide fluoride, suspended solids, and oil and grease removal. Casting operations must be equipped with fume scrubbing, with neutralization, suspended solids removal, and elimination of direct-contact water cooling.

The major significant requirement in the 1983 guidelines for primary plants is an internal one—i.e., a water conservation program to reduce wastewater volume. In addition, secondary fluoride reduction must be practiced to reduce fluoride concentrations.

The 1983 requirements for secondary smelters involve a fumeless process to remove magnesium from the molten aluminum.[8,9]

Copper, Lead, and Zinc. In 1969, the United States production of these metals was as follows:

Copper	1,585,000 tons
Lead	639,000 tons
Zinc	1,041,000 tons

The ores from which they derive are principally sulfides which emit sulfur dioxide and large quantities of particulate matter during roasting in the presence of air. In general, the principal wastewater discharges from these plants are effluents from fume scrubbers installed on the roasting ovens and smelting units, and relatively infrequent equipment cleaning. However, since the primary smelting plants of these industries are usually located in relatively dry and remote locations, the wastewaters discharged are limited in volume because water is usually too valuable to discard.

A process flow diagram of copper, lead, and zinc smelting is presented in Fig. 13-4.

Copper. Copper ores are dried and roasted at 1200°F (650°C) to drive off volatile oxides and to convert most of the sulfide in the ores to sulfur. The roasted ore is mixed with a flux and charged to a reverberatory furnace where the charge is melted and separated into slag and a matte. The molten matte is charged to converters where air is blown through it to oxidize iron sulfides to oxides. These combine with silica to form a slag, and copper sulfide is converted to crude copper and sulfur dioxide. The crude copper is usually cast into anodes. The anodes are shipped to an electrolytic plant where the copper is placed in an electrolysis cell as the anode. The electrolyte is copper sulfate and the cathode is pure copper. An electric current is charged to the cell

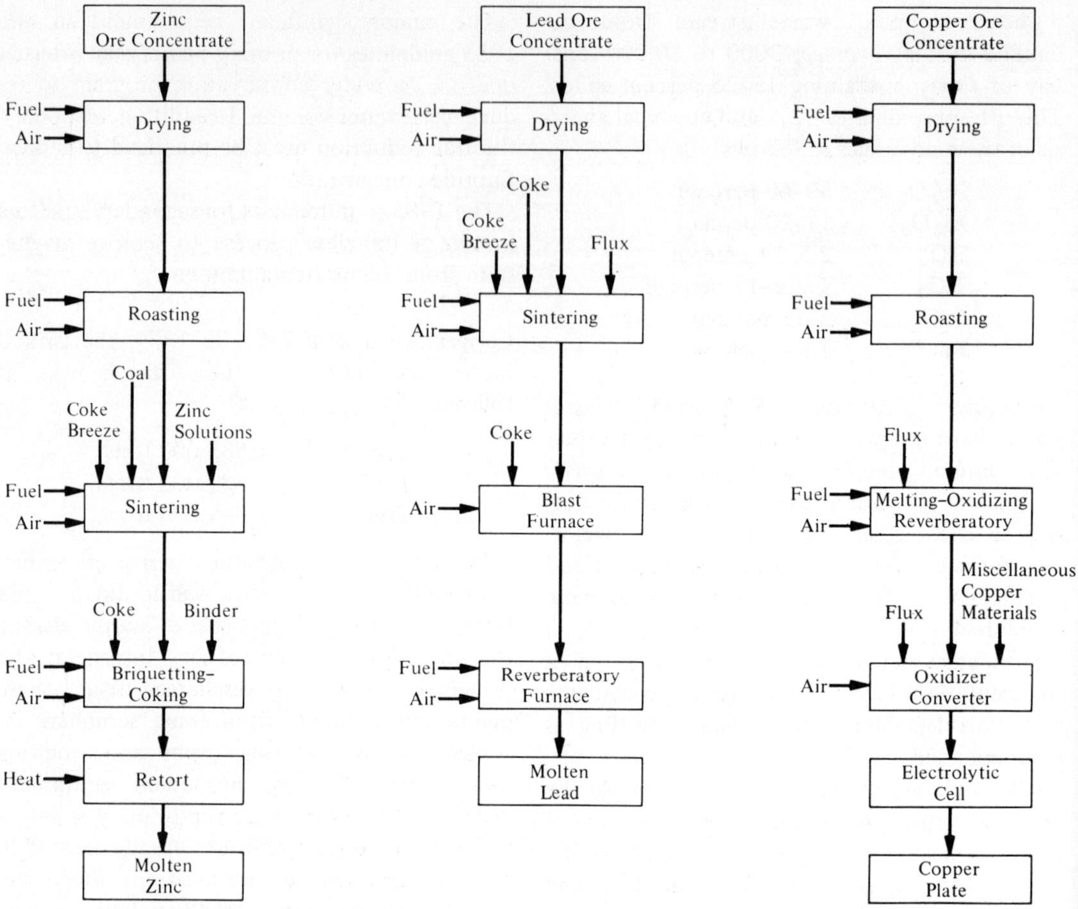

Fig. 13-4. Flow diagrams for typical zinc, lead, and copper smelting operations.

and pure copper plates on the copper cathode, while impurities fall to the bottom as sludge.

Wet scrubbing of off-gases from the roaster, reverberatory furnace, and oxidizing converter is a significant source of pollution, producing an effluent containing suspended solids, sulfur dioxide, and heavy metals. Spent electrolyte from the electrolysis plant is usually treated with iron (cementation process) to precipitate copper, leaving a dilute ferrous sulfate solution to be discarded or treated.

Lead. Lead ores are dried, mixed with coke and a flux, and sintered to lead oxide and sulfur dioxide. The sinter is charged to a blast furnace with coke, lime, and scrap iron, and air is supplied to the bottom of the furnace. The coke reduces the oxide to lead. The coke is oxidized to carbon monoxide and dioxide. The lead collects as a molten pool on the bottom of the furnace, topped by a matte of copper and iron sulfides. This is covered by a bed of slag. The matte and slag are tapped from the furnace and the lead is tapped off for collection. The matte is returned to the process for recovery while the slag is granulated with water. Lead from the blast furnace is charged to a reverberatory furnace where, under controlled temperature conditions and with the use of additives and air, the impurities in the lead are driven off, and the final lead product is drawn off and cast.

The principal waste streams are the slag granulation waters and flue gas scrubber effluents. The slag granulation waters are partially recycled, but a portion is discarded. Wet scrubbing of lead smelter gases, especially from the sintering and blast furnace operations,

TABLE 13.3 Primary Smelting Wastewaters (discharge in units/ton product)

	LEAD		ZINC				
	Slag Granulation	Wet Scrubber	Roaster Scrubber	Cathode Wash	Furnace Scrubber	Retort Residue	Retort Scrubber
Volume (1000 gal/ton)	4.5–80	0.2–2.3	0.2–4.0	2.5	0.6	0.5–2.0	2.0
Suspended solids (lbs)	600–800	–	–	–	–	1.7	–
Dissolved solids (lbs)	–	–	25–100	–	–	1.2	–
Sulfate (lbs)	–	–	10–50	–	–	–	–
Zinc (lbs)	–	–	0.7–3.0	–	–	0.2	–

produces an effluent containing sulfur dioxide, suspended solids, heavy metals, and other materials found in the ore.

Zinc. Zinc ores are received at the smelter and dried. The ore is first roasted to convert the sulfite ore to zinc oxide and sulfur dioxide. The roasted ore is mixed with coke, coal, and zinc solutions and charged to a sinter machine. The sinter is briquetted with coke and a binder and charged to a retort. The zinc distills out of the charge along with other volatiles, leaving impurities behind in the retort. The vaporized zinc is condensed to the liquid state on the condensers and is then cast into the final form. The off-gases are cleaned by passing through a scrubber.

Wastewaters arise through the scrubbing of off-gases from the roaster, furnace, and retort; from washing of electrostatic precipitators; and from sluicing the residue from the retort. Dissolved solids, sulfates, and zinc are the principal impurities. Table 13-3 presents typical raw waste loads emanating from the various processes of primary lead and zinc smelting.[10]

Secondary Smelting. A significant portion of the copper, lead, and zinc used in the United States is produced by processing scrap and metal-bearing materials other than ores (drosses, slags, solutions, chemical residues, etc). It is estimated that 45 percent of the copper, 60 percent of the lead, and 20 percent of the zinc production is by secondary processes.

The recovery operations for all three metals are similar. The scrap metal and metal-bearing materials are received and reduced in size. Lead is processed through a burner to remove impurities. The metal-bearing charge is then mixed with alloying materials and flux and charged to a reverberatory furnace to convert the charge to molten metal and a slag. Lead and copper are tapped off as molten metal; zinc can be tapped off or distilled out of the charge and recovered by condensation. Figure 13-5 illustrates the methods for secondary zinc and lead smelting.

The wastewaters produced in these processes are wet scrubber waters from washing of the flue gases, slag waters from sluicing or granulation, and clean-up wastewaters.

The principal parameters of pollution in the secondary nonferrous metals industry are solids and heavy metals. Typical values from the few such installations that have reported wastewater characteristics are shown in Table 13-4.[10,11]

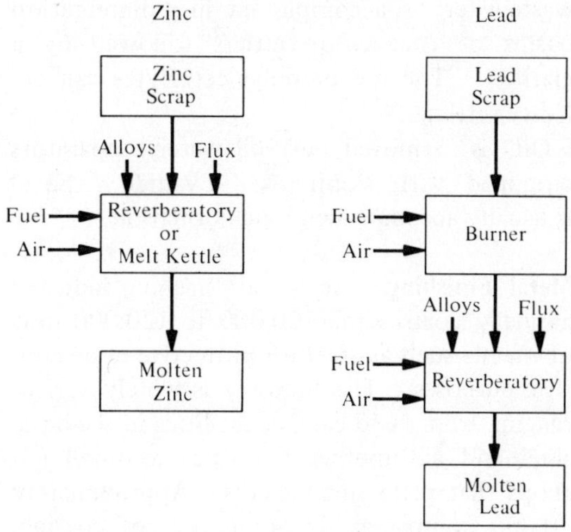

Fig. 13-5. Flow diagram for typical secondary lead and zinc smelting.

TABLE 13-4. Secondary Smelting Wastewaters (lbs/ton of product)

	Copper and Brass	Lead
Volume	0.5–37[a]	0.4[a]
Acidity ($CaCO_3$)	1.15–14.25	
Total solids	1.5–3.0	
Oil	0.16–2.14	
Sulfate		400
Cyanide	0–0.75	
Chromium	0.02–0.37	
Copper	0.01–2.19	
Lead		>22
Nickel	0.01–0.20	
Zinc	0.04–1.22	

[a]1000 gal.

Wastewater Treatment. The recommended wastewater treatment procedures are quite similar for the copper, lead, and zinc industries. The wastewaters are treated to remove suspended matter and to reduce heavy metal concentrations. They are neutralized to about the neutral pH range with lime as the preferred source of alkalinity. Lead wastes should be adjusted to above pH 7.0 to minimize the presence of soluble lead compounds. Zinc wastes should be adjusted to the pH 6.0 to 7.0 range to prevent solubilizing of the zinc. Copper should be adjusted to above 6.5 to maximize metal precipitation.

The removal of suspended solids from these wastewaters is accomplished in sedimentation basins or coagulation-settlers followed by a clarifier. The use of polyelectrolytes can enhance settling.

Oil is removed in oil-water separators equipped with skimmers. Cyanide removal is usually accomplished by chlorination.

Metal Finishing. The metal finishing industry annually coats some 100,000 to 120,000 tons of metals such as steel for protective or decorative purposes. The industry is widely varied, ranging from large captive facilities in the basic steel and automobile industries to small job shops that plate small pieces. Approximately 20,000 companies do some sort of coating; 3500 of these are job shops. The coatings include metals (principally copper, nickel, zinc, and tin), organics, and paints. Wastewaters arise in this industry through precleaning operations to remove scale, dirt, and oil, and they also arise from pickling operations and rinsing of the coated piece. Disposal of spent coating baths, spills, leaks, and sprays contribute only intermittently to the wastewater problem, but their impact is quite pronounced.

The coating process steps are shown in Fig. 13-6. They and the wastewaters they generate are described in more detail below.

Metal Cleaning and Preparation. The base metal surface to be coated is invariably covered with scale, rust, dirt, or oils and grease which

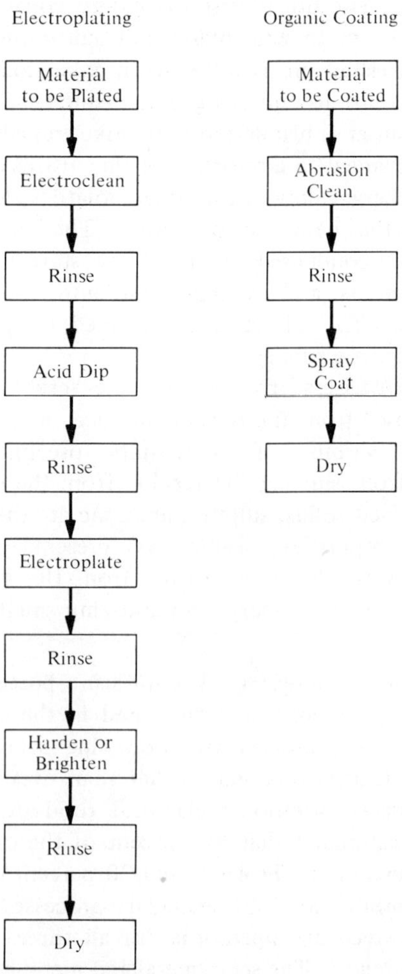

Fig. 13-6. Flow sheet for typical metal finishing processes.

must be removed prior to coating. The metal surface must be treated to ensure that the coating will adhere. Scale is removed ("stripped") by mechanical means (peening, wire brushing, or wet or dry burnishing) or by immersing the metal piece in a bath that can be acidic, alkaline, or of the electrocleaning variety. Dirt and oil can be removed in alkaline or phosphate baths, acidic baths, or by solvents. Surface preparation includes mechanical means or immersion in a bath that is usually acidic, thereby etching the surface. The various cleaning steps require water rinses to remove the bath materials from the base metal in order to stop further chemical or cleaning action. The rinse waters and spent baths contain suspended matter, oils and grease (frequently emulsified), soluble metals, and materials dissolved in the cleaning and preparation step.

Plating and Coating. The base metal piece, cleaned and prepared by the above steps, is coated by immersing or dipping the piece in a bath and sending an electric current between the piece and an immersed anode, or by immersing the piece in a molten metal bath, batch dipping, or spraying the coating onto the piece. The sources of wastewater in these processes are spills and leaks from the bath and extraneous spray that reaches the drains. The baths are rarely deliberately discarded as they are too valuable to lose. However, lost materials frequently get to the drains in clean-up operations, contributing suspended matter, heavy metals, acids, cyanides, and other pollutants to the wastewater discharges from the plating area.

Final Rinse. The coated piece is given a final rinse or rinses to remove all traces of the coating bath. This step invariably transfers part of the bath to the rinse and sends it to the drain. At this point, it contains all of the bath chemicals such as heavy metals, acids, and cyanides. The remaining operations are dry and contribute little to the wastewater problem. Sanitary and power plant wastes may be present, as is the case in any industry.

Wastewater Characteristics. The metal coating industry is so varied that a typical wastewater profile cannot be developed. However, wastewater characteristics of a plant plating copper, chromium, nickel, and zinc were reported by the EPA in 1973.[12] Standard wastewater loadings were given in milligrams per liter. A 1967 study of motor vehicle production reported on wastewater characteristics in this industry and presented the data in units per 100 vehicles produced.[13] These examples are shown in Table 13-5.

Waste Treatment. The principal wastewater treatment methods in use in this industry and their customary sequence are as follows:

1. Cyanide is destroyed by adjusting the pH to 11.0 with caustic soda and adding chlorine or hypochlorite. The reaction is time-dependent and the chlorination step should be allowed to continue for 1 hr.
2. The hexavalent chromium solution, preferably not mixed with any other wastes, is adjusted to pH 3.0. Sulfur dioxide or sodium bisulphite is added to reduce the hexavalent chromium to the trivalent form.
3. Heavy metals, including reduced or trivalent chromium and wastes after cyanide destruction, are precipitated by adjusting the pH to 10.0 with caustic soda or lime. This converts the heavy metals to hydroxides or oxides which come out of solutions as suspended solids.
4. Suspended solids are separated from solution by settling in batch tanks or in continuous settling basins. Polyelectrolytes can be added to improve the settling.
5. Separable oils and greases can be removed in batch tanks or in settling basins by skimming. Air flotation can be used to improve the separation.

There are other less widely used treatment methods in the industry to enhance wastewater renovation.[14] These include ion exchange to remove heavy metals for recovery and reuse, which is especially common for precious metal recovery. Cyanide can be destroyed with a mixture of formalin and hydrogen peroxide. This process has the advantage that no toxic by-product is produced. Evaporative

TABLE 13-5. Examples of Wastewater Characteristics from Metal Finishing

		MOTOR VEHICLE MANUFACTURE (lbs/100 vehicles)		
	Electroplating Wastewaters (mg/l)	Metal Cleaning	Coating and Painting	Electro-plating
Volume	–	5000[a]	2400[a]	5000[a]
BOD	–	4	6	–
Suspended solids	–	12	4	8
Oil	–	4	4	8
Hexavalent chromium	17	0.4	0.2	0.7
Trivalent chromium	0.05	–	–	–
Iron	–	0.2	0.2	0.2
Aluminum	–	–	1.8	0.2
Zinc	32	–	0.1	–
Copper	7	–	–	–
Nickel	2.5	–	–	–
Phosphate	34	2.1	2.1	–
Cyanide	50	–	–	1.8
Fluoride	–	0.4	0.4	–
Phenol	–	0.0018	0.0004	–

[a]Gal.

recovery has been practiced on a single batch system (principally chromium solution) but has not found favor in mixed metallic bath systems. Reverse osmosis has been tried but is still considered experimental in this industry.

The wastewater treatment problem is difficult because the solutions to be treated are relatively dilute, although volume reduction is possible through multiple water reuse. This can be achieved by countercurrent multiple-tank rinsing, where a piece is rinsed in one tank and then rinsed in a second one. The first tank may be concentrated for bath make-up. The second tank is usually extremely dilute and can be used in the first tank (countercurrent plan). Spray rinsing can reduce the volume of rinse water, as only the amount needed is used. Water recirculation after clean-up can materially reduce the amount of fresh make-up water needed.

The technology which the electroplating industry must apply to achieve the 1977 effluent limitations includes: cyanide destruction, hexavalent chromium reduction, and heavy metal precipitation followed by solids removal. The 1983 goal for water pollution control will require the plating shops to eliminate discharge of wastewaters. The "zero discharge" status will require the shops to install extensive water conservation measures to reduce wastewaters to a minimum volume, reverse osmosis treatment of this volume, followed by evaporation of the blow-down stream. Clean waters from the reverse osmosis operation will be recirculated.[12]

Petroleum Refining. In 1973, 247 refineries processed 13×10^6 barrels (42 U. S. gal per barrel) of crude oil per day. The products of refining are gasoline, aviation fuel, lube oil, other middle distillates, residual fuels, and asphalt, plus a large number of smaller volume products such as coke, wax, synthetic gas, and grease.

Petroleum refineries are usually large installations, with a complex series of operations as shown in Fig. 13-7, and there is a considerable variety of product operations among them. The processing operations that are principal sources of waste are discussed below.

Storage and Transportation. Crude oil is received at the refinery by pipeline, tank car, or waterway commerce and is stored prior to its consumption in the refinery. Losses are

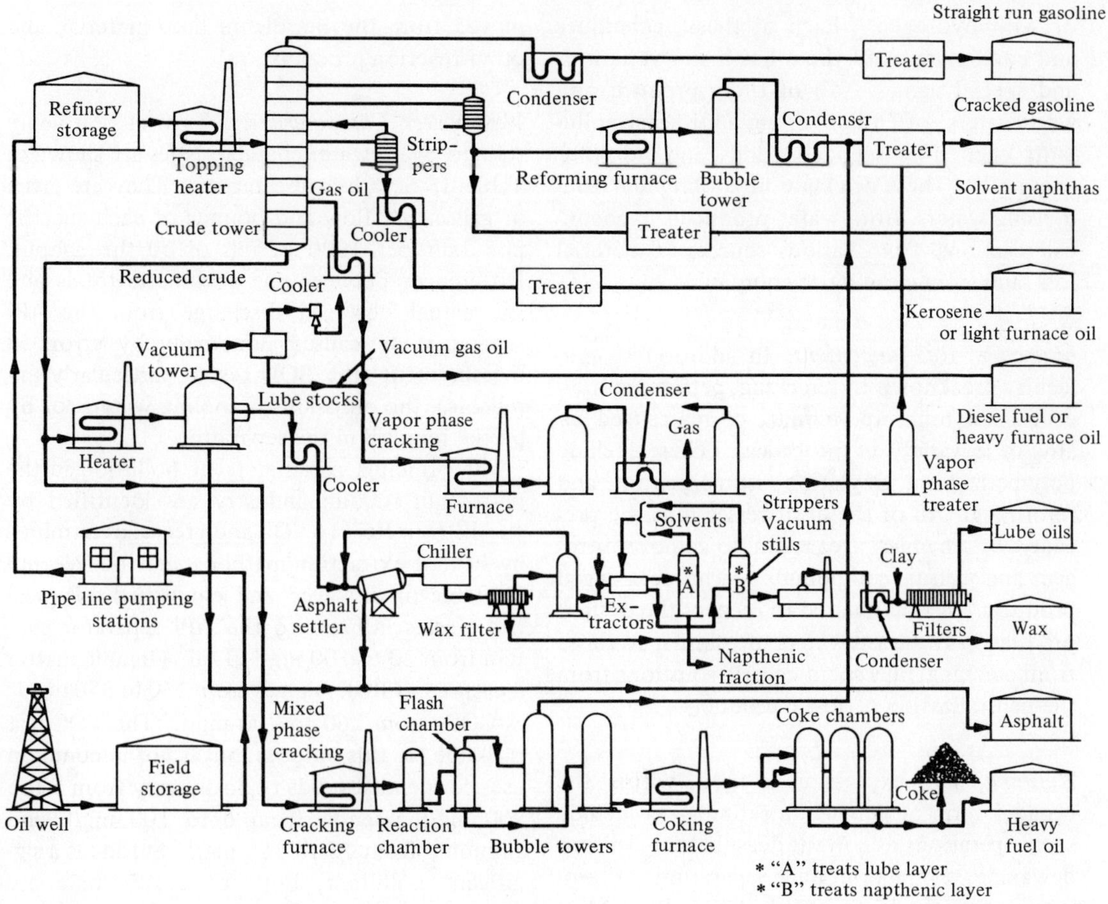

Fig. 13-7. Schematic drawing of a petroleum refinery (From *Chemical Process Industries*, McGraw-Hill Book Co. 1956, p. 839).

inevitable during transfer and during any time that tanks, pipelines, and equipment are cleaned. Water separates during transport (ballast water) and in storage tanks and must be removed; it is a highly contaminated wastewater. The storage of products is a similar source of pollution.

Crude Oil Desalting. At the wellhead, crude oil is produced in a mixture with brines. Even after separation in the field, the crude oil shipped to the refineries contains more salt than can be tolerated in the refining equipment. This is removed by wash water in the desalter, and the aqueous stream from this unit is contaminated with salt, oil, and smaller amounts of other materials.

Fractionation. The fractionation or separation of crude oil by distillation is accomplished in several stages, depending on the individual refinery's market demands and available raw material. The principal wastewater is cooling water, which is clean if it comes from surface coolers, except for occasional leaks. In barometric condensers and other water-oil contact coolers, and particularly in steam stripping operations, the wastewater contains a significant quantity of the oily fraction. There is also pollution of process waters by leaks and spills, and clean-up wastes are particularly troublesome.

Cracking. The chemical breakdown of large petroleum molecules into lighter fractions, called cracking, is done thermally, catalytically, or with hydrogen. Each of these techniques and combinations of them has many variations and special names. All of them produce oily wastewaters, often containing additional pollutants such as suspended solids and non-oily chemicals. These originate in cooling and condensing waters, from water produced in chemical reactions, from various sources of material loss, and from cleaning of equipment.

Molecular Rearrangement. In addition to molecular breakdown by cracking, petroleum molecules are built up, rebuilt, or rearranged by any of a variety of processes. These include polymerization, alkylation, isomerization, and reforming. All of these processes must be preceded by chemical treatment to remove nitrogen and sulfur compounds. The rearranged products are fractionated to recover the desired products. The wastewaters are similar to those from cracking, plus spent caustic solutions from the neutralization of acidic products.

Refining. This expression is used, within the broader term of petroleum refining, to include such operations as solvent deasphalting, solvent dewaxing, extraction and separation of aromatic compounds, recovery of butadiene from four-carbon fractions, and hydrotreating. The wastewaters are primarily spent caustic solutions, equipment washout waters, and various emulsions formed almost every time that water is brought into contact with oils, solvents, and similar materials.

Miscellaneous Processes. A number of more specific operations, not conducted at all refineries, are significant waste producers where they are practiced. Among these are grease production or the blending of petroleum oils and soaps to specific properties, asphalt production and compounding with catalysts and other substances, drying by use of clay-base filters, and sweetening for the removal of sulfur and nitrogen compounds. The wastewaters from these various processes are themselves varied, including, in addition to oily materials and heat, the various treatment chemicals and substances used in the process, materials removed from the petroleum base material, and other reaction products.

Wastewater Characteristics. Typical petroleum refinery wastewater characteristics are shown in Table 13-6 [15] by subcategory. They are given in gallons of flow and pounds of each specific pollutant per 1000 barrels of oil throughput. Differences between the arithmetic totals and the actual observed discharge from the API separator are caused principally by errors in measurement (the BOD test is particularly unreliable), the omission of cooling waters, or by partial removal in the separator.

The principal parameters of pollution in the petroleum refining industry are identified by the EPA in Ref. 16. Oil and grease, determined by hexane extraction methods, may be free and floatable or dissolved and emulsified. Typical raw wastes influent to the API separator contain from 20 to 100 mg/l of oil. Organic matter measured as BOD, ranges from 250 to 350 mg/1; as COD, from 260 to 700 mg/l. The TOC test is usable for this purpose, but is not in common use. Suspended solids typically vary from 50 to 200 mg/l, phenols from 6 to 100 mg/l, and ammonia from 15 to 150 mg/l. Sulfide is a significant pollutant, but no specific data are given in this reference.

TABLE 13-6. Wastewater Profile, Petroleum Refining Industry (per 1000 barrels oil throughput)

Source	Volume (gals)	BOD (lbs)	Phenols (lbs)	Sulfides (lbs)
Crude oil production and storage	400	1.0	–	–
Desalting	200	2.0	0.1	2.0
Fractionation	65,200	7.3	1.7	1.0
Cracking[a]	29,000	14.5	15.4	6.6
Molecular rearrangement[b]	7000	0.1	0.1	1.7
Refining[c]	35,800	53.9	5.2	0.7
Total	137,600	78.8	22.5	12.0
Discharge from API separator	150,000	204.5	16.9	5.6

[a]Includes thermal, catalytic, and hydrocracking.
[b]Includes reforming, polymerization, alkylation, and isomerization.
[c]Includes solvent refining, dewaxing, hydrotreating, deasphalting, drying, and sweetening.

Other pollutants, less significant or less common, include total dissolved solids, cyanides, pH, temperature, and heavy metals.

Waste Treatment. Many treatment techniques are available to and used by petroleum refineries, but the universal basic equipment is the gravity oil separator, specified in considerable detail by the American Petroleum Institute. These units remove the greater part of floating oil in refinery wastewaters, although they do not break emulsions or separate dissolved substances. Their effectiveness depends on temperature, oil density and droplet size, and the presence of other suspended matter. The parallel-plate separator is another device that is increasingly popular for the same purpose. Effluent from gravity separation contains from 5 to 35 mg/l of residual oil, principally in emulsified form.

Other oil separation technologies, which are also effective in reducing suspended solids and the pollutants carried with them, include dissolved air flotation and granular filters. The former, especially, is in widespread use, usually supplemented by the addition of chemical flocculating agents such as iron and aluminum salts plus organic polyelectrolytes. Flotation is commonly used following gravity separation for additional purification. Filters are most commonly employed near the end of the treatment sequence, following activated sludge or other biological or chemical treatment.

Biological treatment by the activated sludge process is capable of removing from 80 to 99 percent of the BOD and oil, although the reactions are slower than with municipal sewage. Higher concentrations of mixed liquor suspended solids are required, typically from 1500 to 3000 mg/l; the detention time is longer or the contact stabilization is modified. Trickling filters and aerated lagoons or oxidation ponds are also used for biological treatment of refinery wastewaters.

The breaking of oil emulsions is accomplished, together with the destruction of the oils, in biological treatment of the final refinery effluent. In individual process streams, chemical additions may be more convenient because they permit recovery of specific petroleum fractions at their source. Acidification is most common, although other chemicals are sometimes preferable in particular operations. Heating, although expensive, is used for high-boiling fractions in holding tanks equipped with steam coils.

EPA has recommended, as the best practical control and treatment technology (1977 goal), in-plant modifications plus treatment equivalent to single-stage biological treatment followed by granular-media filtration. Best available technology, the 1983 goal, specifies activated sludge in the above system, and adds adsorption on activated carbon.[16, 17]

Chemical Manufacture. The chemical industry is extremely complex. It is usually divided into two broad categories, inorganic and organic, though there is some overlapping of the two. Inorganic chemicals are principally made from minerals, including air and water, while organic chemicals are produced from animal, vegetable, or carbonaceous raw materials (originating from coal or petroleum). The petrochemical industry is an important branch of the organic chemical industry; its raw materials are derived from petroleum light front end fractions or natural gas.

The industry produces about 150 bulk chemicals and 31 classes of resins and synthetic fibers. Overall production approximated 270×10^6 tons in 1970.

Chemical processing steps are varied, ranging from physical changes to complex chemical rearrangements or synthesis. The majority of the manufacturing processes used industrially consist of some combination of physical and chemical steps, such as chemical reaction, size reduction, dissolution, crystallization, distillation, absorption, extraction, heating, cooling, and evaporation. Four typical operations are illustrated in Figs. 13-8 and 13-9.

The chemical industry is one of the largest water-using industries in the United States. Approximately 65 percent of the water is for cooling purposes, with the remainder used in processing. In addition, many chemical reactions generate water as a reaction product, and this becomes a part of the wastewater problem. About 20 percent of the chemical process plants have no wastewater discharge. In the

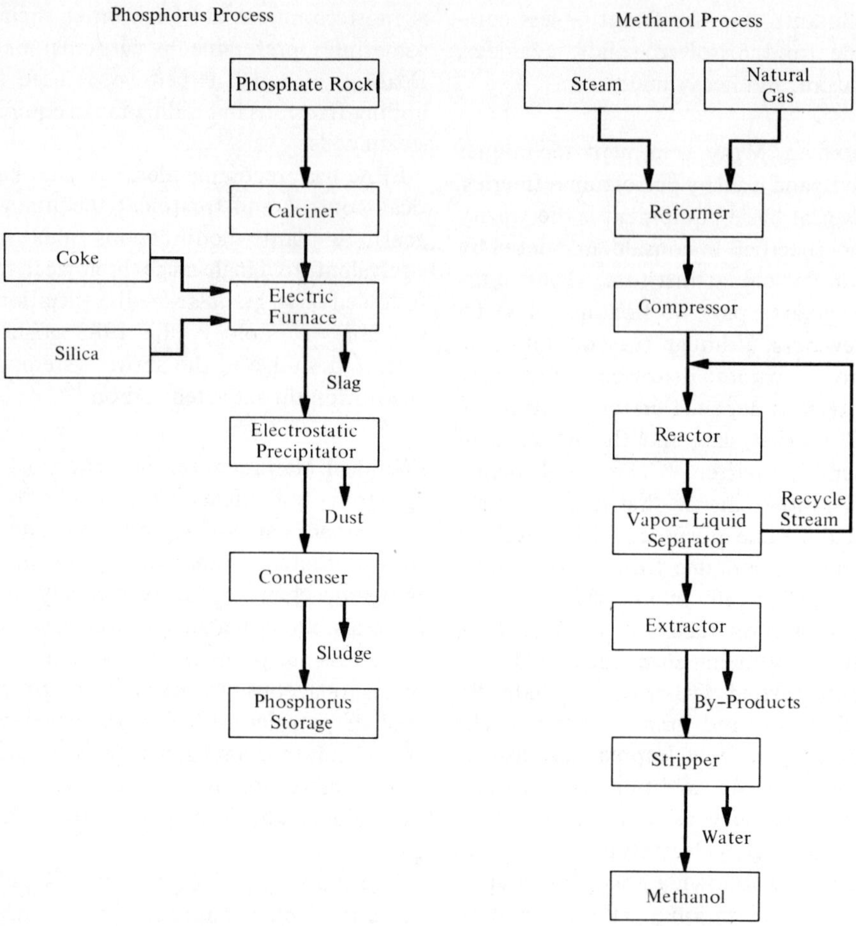

Fig. 13-8. Flow diagrams for typical chemical manufacturing.

others, process water is used as a solvent, carrier agent (steam is included in this category), direct contact coolant, and equipment and area clean-up media. In addition, rain can cause problems as it contacts spills or stored materials; after such contact, it must be treated.

Waste Characteristics. The wastewaters developed throughout the industry are extremely varied, and it is difficult to develop representative values for effluent parameters for such a complex industry. Instead, industrial waste profiles are tabulated in Table 13-7 for a few select processes, chosen to give an indication of the typical values encountered.[18-24]

Control and Treatment Technology. Chemical plant wastewaters range from innocuous to highly toxic. Most untreated wastewaters are in the high pollution load category. Applicable wastewater treatments include:

1. In-plant treatment to recover valuable materials and to recycle both water and materials back to the process. The actual treatment required is usually a standard chemical process unit operation handled as a normal process-improvement project.
2. Oil-water separation. This operation uses batch separation tanks, API separators, and various flotation devices. The design of continuous flow separators follows the outline given in the API manual. Batch treatment follows the same procedure, allowing a reasonable separation time.
3. Solids separation. This is effected in coagulation settling tanks, sedimentation

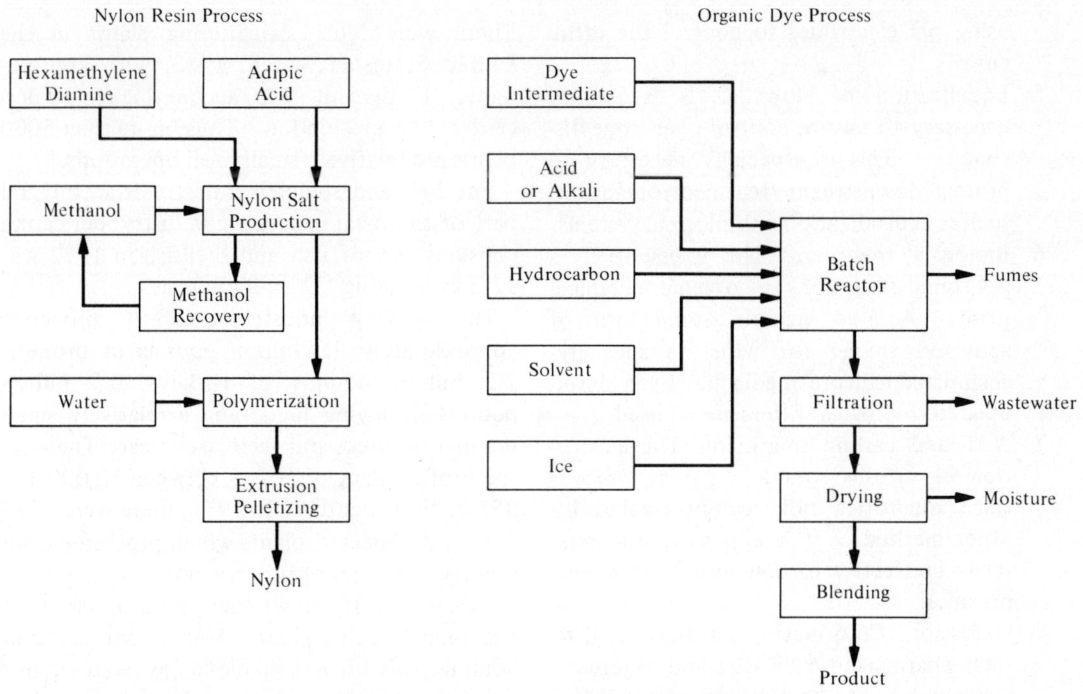

Fig. 13-9. Flow diagrams for typical chemical manufacturing.

basins, thickeners, filters, or centrifuges. Design is based on information available in waste control or chemical engineering texts on these types of equipment.

4. Neutralization. This step is required in most plants, as normal effluents are frequently acidic or alkaline. Design should be based on continuous neutralization,

TABLE 13-7. Chemical Industry Wastewater Profiles (mg/l except pH and volume)

	Nitroglycerine and Dynamite	Nylon	Adipic Acid	Fine Chemicals, Pharmaceuticals	Organic Dyes	Phosphorus
BOD	–	–	–	2000–10,000	325–620	–
COD	350	74,000–87,800	112,600–169,800	–	–	–
Total solids	–	–	–	1000–8000	–	–
Total dissolved solids	10,800	–	54,000–74,000	500–7000	–	–
Suspended solids	40	–	–	–	2500–5412	1000–5000
pH (units)	7	7.4–12.0	4.1–5.2	2.0–8.0	–	1.5–2.0
Alkalinity	–	22,500–40,700	1500–6000	–	–	–
Oil	–	152–367	–	–	–	–
SO_4	1000	0–194	142–160	–	3100	–
PO_4	–	5–78	1–4			
Cl	–	475–3340	726–901		600	
NH_3—N	1270		98–280			
NO_3—N	550					
Phosphorus						700–3000
Fluorine						500–2000
Wastewater Volume (10^6 gal)				0.4–0.7	1.5	0.14

using pH electrodes to control the effluent pH.

5. Equalization of flows. It is frequently necessary to ensure control over surge discharges. This is especially necessary to protect downstream treatment operations such as neutralization or biological systems.
6. Biological treatment. This is essentially a polishing step for an organic chemical plant. Aerated lagoons, some form of activated sludge for which a specially-acclimated culture media has been developed, or biological filters are all used.
7. Activated carbon treatment. The adsorption of various soluble organics removes trace quantities not readily treated by other methods. It is expensive and relatively ineffective for low molecular weight organics.
8. Oxidation. Chlorination has been used to treat pharmaceutical wastes and to remove taste and odor chemicals in many industries. Ozone is often more effective, but is limited by its high installed equipment cost.

It is possible at this writing that the 1977 control technology for the chemical industry might be as strict as zero discharge, depending on the industry segment involved. The techniques to reach this goal include segregation of process from cooling waters, water recycling, liquid-solid and liquid-liquid separations, flow equalization, neutralization, and biological treatment. The 1983 goal would intensify the water conservation schemes, and add more exotic treatment steps such as media filtration, adsorption, reverse osmosis, and evaporation.[18–23]

Meat, Fish, and Poultry. Meat-packing plants slaughter and process cattle, hogs, calves, and sheep for meat products and by-products. The plants in this industry perform any number of the operations shown in Fig. 13-10, ranging from a single process, such as a simple slaughterhouse, to the complete packing house.

In 1972, 59,223 million pounds (live weight killed) were processed, with beef representing 63 percent and hogs 34 percent of the total. There were 5991 slaughtering plants in the United States in March, 1973, with approximately 15 percent (900) accounting for 90 percent of the production. Thus, more than 5000 plants are relatively small local operations.

The fish and shellfish industry is an integral part of the meat processing industry; per capita consumption of fish and shellfish in 1972 was 12.2 lb, totaling 1.25 million tons.

The poultry industry presently processes approximately 10 billion pounds of broilers, 2.2 billion pounds of turkeys, 0.5 billion pounds of laying hens, and a relatively small number of ducks and geese each year. The average broiler plant processes between 50,000 and 95,000 birds per day. In 1971, there were 1237 federally-inspected plants which processed both poultry and meat or poultry alone.

Figure 13-10 shows the typical processes of a full-line packing plant. Animals are placed in holding pens for less than one day, watered but not fed, and protected from the elements. The animals are stunned mechanically (a blow to the head), electrically (shock), or chemically (carbon dioxide atmosphere). They are then stuck (cutting the jugular vein) and bled. The hides are removed, and the carcasses are eviscerated (heart, liver, tongue, and kidneys removed), washed, cut, and cooled.

The blood may be processed for pharmaceutical preparations, evaporated for animal feed, or sewered. The viscera (paunches, stomachs, and intestines) are emptied, and may be processed for tripe, chitterlings, sausage casings, surgical sutures, or mink or pet food. Hides may be demanured, washed, defleshed, brine cured, and sent to the tannery.

The cooled carcasses are cut for direct marketing of smaller sections or individual cuts. Trimmings are processed into products such as sausages and canned meats, or may be rendered for edible fats and solids. Cuts may be further processed to produce sausages, hams, bacon, canned meats, pickled meats, hamburger, corned beef, and smoked meat.

Fish and shellfish are harvested and taken directly to the processer or iced or frozen for later delivery. Preprocessing includes washing, thawing, beheading, de-icing, and similar preparations for butchering and eviscerating. Fresh

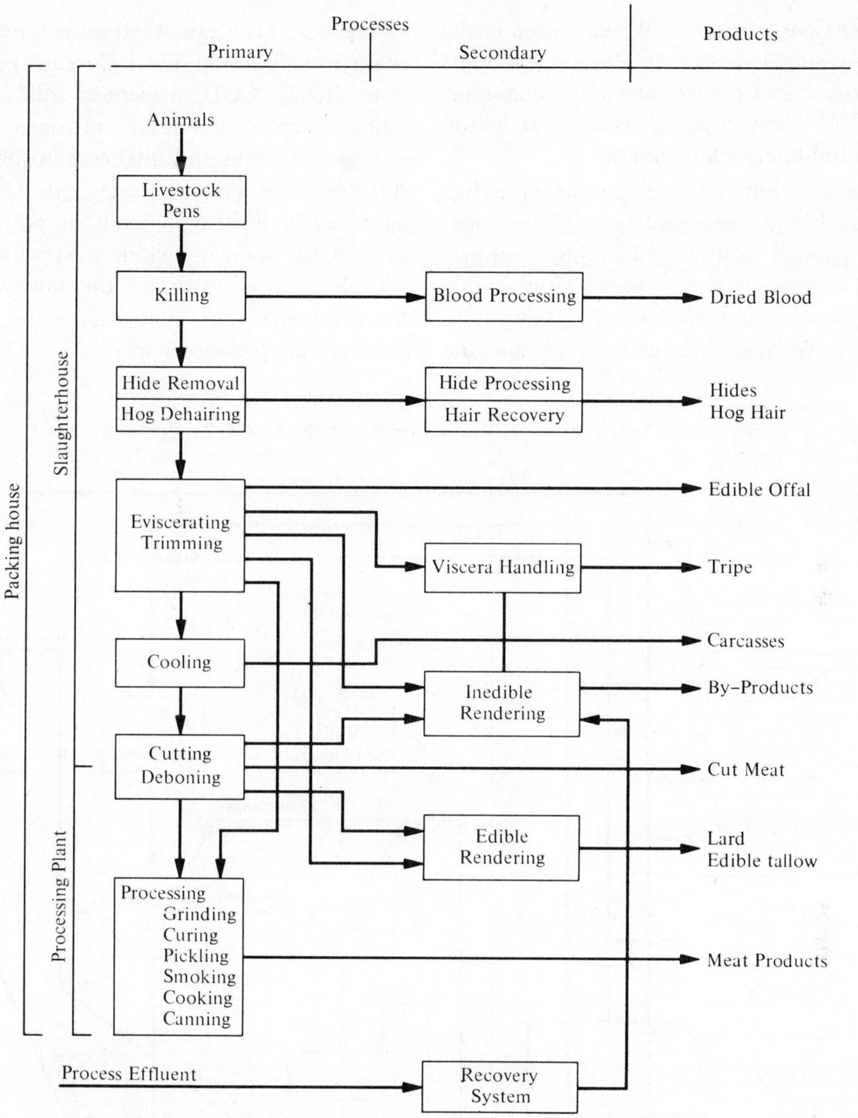

Fig. 13-10. Process diagram of a typical full-line packing plant.

fish and shellfish are packed, refrigerated, and shipped to retail. Prepared fish and shellfish are cooked, blanched, or precooked for removal of skin, bone, shell, gills, and other wastes, and are frozen, canned, pasteurized, or refrigerated.

Birds received at poultry plants are killed, bled, scalded, plucked, eviscerated, washed, and chilled. Most birds are frozen and shipped as eviscerated pack. Others are further processed by cutting, deboning, skinning, cooking, grinding, and breading. Pollutants consist of blood, protein, fat, cleaning chemicals, nutrients, manure, feed, and feathers. Product flow and processes are similar to those shown for cattle and hogs.

Rendering (wet or dry) separates fats and water from tissue for either edible (fats and tallows) or inedible (animal feed) products. Materials to be renderd, including trimmings, bones, offal, and condemned animals, are usually ground and heated under pressurized, atmospheric, or vacuum conditions. The fat phase is processed by screening or centrifugation and sent to refining. Nonfatty materials may be combined with other recovered materials and processed for animal feed.

Wastewater Characteristics. Water is used in the meat processing industry to cleanse products and to remove and convey unwanted materials. Figure 13-11 shows typical sources of wastewater in a full-line packing plant.

Wastewaters from the meat processing industry contain blood, meat and fatty tissue, meat extracts, paunch contents, bedding, manure, hair, dirt, condenser water, losses from edible and inedible rendering, curing and pickling solutions, preservatives, and caustic or alkaline detergents. The raw wastewaters are generally characterized using the following parameters: flow, BOD_5, COD, suspended solids, dissolved solids, grease, ammonia nitrogen, Kjeldahl nitrogen, nitrates and nitrites, phosphorus, and chloride. In general, hazardous toxic pollutants, such as heavy metals or pesticides, are not present in meat-packing wastewaters.

Table 13-8 summarizes the raw wastewater characteristics for slaughterhouses, packing houses, and processing plants[25]. The data in-

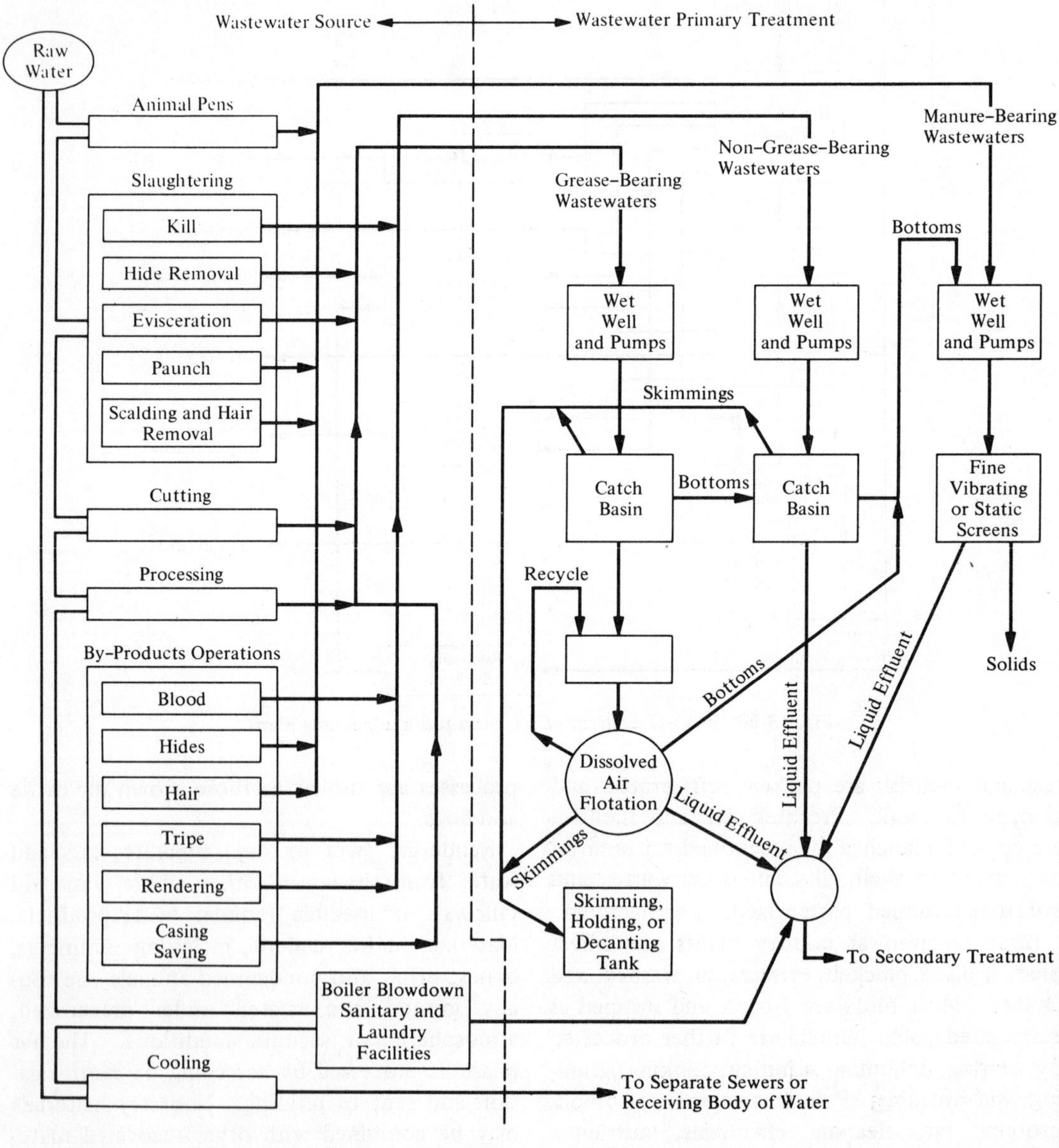

Fig. 13-11. Separate treatment of grease-bearing, nongrease-bearing, and manure-bearing wastewaters.

TABLE 13-8. Meat Packing Industry Raw Waste Load Characteristics (lbs/1000 lb lwk)

	SLAUGHTERHOUSES		PACKING HOUSES		PROCESSING PLANTS	
	Average	Range	Average	Range	Average	Range
Flow (gals)	696	290–887	1046	388–2330	1265	380–2900
BOD_5	5.8	1.7–8.4	12.1	3.4–27.3	5.7	0.8–16
Suspended solids	4.7	0.6–8.3	8.7	0.6–22.5	2.7	0.5–4.8
Grease	2.5	0.9–4.6	6.0	0.7–32.8	2.1	0.6–3.9
Kjeldahl-N	0.3	0.06–0.70	1.0	0.08–6.75	0.8	0.4–1.3
Chlorides	2.8	0.5–4.8	4.7	0.2–17.6	0.6	0.1–1.1
Phosphates	0.04	0.01–0.17	0.17	0.15–0.40	0.28	0.05–0.51

dicate the following typical compositions for these wastewaters:

Parameter[a]	Slaughter-house	Packing House	Processing Plant
BOD_5	1000	1400	550
TSS	810	1000	260
Grease	430	690	200
Kjeldahl-N	52	115	76
Chloride	480	540	57
Phosphate	7	20	27

[a] All values in mg/l.

Table 13-9 summarizes the raw wastewater characteristics for catfish, crab, shrimp, and tuna processing.[26] Table 13-10 summarizes the composition and waste loads from poultry processing plants.[27]

Control and Treatment Technologies. Various in-plant control techniques are used in meat packing and poultry and fish processing plants, or have been demonstrated as technically feasible. Holding pens in packing plants should be covered, dry-cleaned with only occasional washdown, and provided with separate sewers and manure pits. Bedding material and manure can be applied to farmland as fertilizer. Liquid wastes can also be applied directly to the land or treated in secondary systems.

Blood should not be sewered but collected separately and processed into animal feed or pharmaceuticals. Paunch materials and viscera should be processed by dry or semi-dry procedures. Water use and spillage from the rendering, hide processing, scald tank, and pickle and curing operations should be minimized, contained, and controlled. Good water conservation and clean-up practices should be employed to reduce water consumption. In poultry plants, improved blood recovery techniques, the transport of blood and lungs to rendering in tanks, more dry clean-up, automatic killing accompanied by shocking, automatic and dry evisceration, reuse of water, increased use of grease and grit removal systems, screens, and shell and tube condensers, more on-site rendering plants, and more prepared products will all reduce wastewater pollutional loads.[28]

In fish and shellfish processing plants, upgrading of solid waste to human foods or high grade animal feed components substantially reduces the wastewater pollutional loads.

Primary treatment in packing houses consists of flow equalization; screens to remove paunch manure, hog stomach contents, hog hair, stickwater solids, and hide-washing and curing materials; catch basins to recover grease (50 to 60 percent grease removal, 30 percent BOD, 40 to 50 percent suspended solids); and dissolved air flotation (60 percent removal of suspended solids, 90 percent with chemical addition, 80 to 90 percent grease).

The secondary treatment processes which predominate in the meat processing industry are anaerobic lagoons and anaerobic contact systems. These are economical and efficient, with low power costs and land requirements. Reductions of up to 97 percent BOD_5 and 95 percent suspended solids can be achieved in anaerobic lagoons with typical loadings of 15 to 20 lb of BOD/1000 ft^3, and with detention times of 5 to 10 days. Anaerobic contact systems consist of equalization tanks, digesters with mixing, air or vacuum gas stripping units, and sedimentation tanks. Reductions of 90 to

TABLE 13-9. Fish and Seafoods Industry Raw Waste Load Characteristics[29]

	CATFISH[a]		CRAB[b]		SHRIMP[c]		TUNA[d]	
Parameter	Mean	Range	Mean	Range	Mean	Range	Mean	Range
Flow rate, m^3/day	109	42–153	95	–	472	342–602	4490	450–11,700
(mgd)	(0.029)	(0.011–0.040)	(0.025)	–	(0.124)	(0.0905–0.159)	(1.19)	(0.119–309)
Flow ratio, l/kkg	19,000	12,000–29,000	15,000	–	60,000	47,000–73,000	15,700	5570–33,000
(gals/ton)	(4600)	(2900–6900)	(3560)	–	(14,000)	(11,000–18,000)	3765	(1335–7910)
Settleable solids, ml/l	0.6	0.3–1.2	90	40–150	75.8	33.4–117.8	3.0	0.2–5.9
Settleable solids ratio, l/kkg	10.6	1.6–27.6	1336	594–2227	4550	2000–7070	11.3	6.9–15.6
Screened solids, mg/l	305	117.9–476.6	–	–	–	–	–	–
Screened solids ratio, kg/kkg	6.7	3.5–10.2	–	–	–	–	–	–
Suspended solids, mg/l	445	274–756	177	87–279	968	652–1284	1541	1202–1769
Suspended solids ratio, kg/kkg	8.3	3.3–12.9	2.66	1.30–4.19	54	48–61	8.5	5.5–10.2
5-day BOD, mg/l	511	212–1006	439	288–618	2112	1310–2915	1527	421–2505
5-day BOD ratio, kg/kkg	8.3	5.0–16.2	6.58	4.32–9.28	116	96–137	13.8	13.4–14.1
20-day BOD, mg/l	–	–	–	–	2530	1900–3100	–	–
20-day BOD ratio, kg/kkg	–	–	–	–	152	114–186	–	–
COD, mg/l	891	496–1600	740	487–1040	3582	2233–4932	8301	1313–3942
COD ratio, kg/kkg	15.5	10.2–25.7	11.1	7.3–15.6	197	163–232	30.4	21.9–93.3
Grease and oil, mg/l	257	114–448	–	–	716	605–827	434	130–589
Grease and oil ratio, kg/kkg	4.5	2.9–7.2	–	–	42	39–44	4.2	3.2–5.0
Organic-N, mg/l	32.7	20.7–49.8	94	58–138	215	164–266	30	–
Organic-N ratio, kg/kkg	0.5	0.3–0.8	1.40	0.86–2.06	12.2	12.0–12.5	0.99	–
Ammonia-N, mg/l	0.8	0.5–1.4	6	4–9	6.9	4.4–9.5	7.6	2.2–13.0
Ammonia-N ratio, kg/kkg	0.01	0.006–0.02	0.09	0.06–0.14	0.38	0.32–0.45	0.072	0.072–0.072
pH	6.4	5.8–7.0	7.3	7.1–8.5	7.5	7.3–7.6	–	–

[a] 5 plants.
[b] 1 plant.
[c] 2 plants.
[d] 3 plants.

TABLE 13-10. Composition of Combined Poultry Plant Wastes[27]

	Range
BOD_5 (mg/l)	150–2400
COD (mg/l)	200–3200
Suspended solids (mg/l)	100–1500
Dissolved solids (mg/l)	200–2000
Volatile solids (mg/l)	250–2700
Total solids (mg/l)	350–3200
Suspended solids (percent of total solids)	20–50
Volatile solids (percent of total solids)	65–85
Settleable solids (mg/l)	1–20
Total alkalinity (mg/l)	40–350
Total nitrogen (mg/l)	15–300
pH	6.5–9.0

PROCESSING PLANT WASTE LOADS PER 1000 CHICKENS

	No.	Waste (gals)	BOD_5 (lbs)	Suspended Solids (lbs)
Flow-away	10	7000		
With blood recovery	5	–	25	13
All blood wasted	5	–	41	23
Nonflow-away	3	4500		
With blood recovery	1	–	23	12
All blood wasted	2	–	35	21

Source: Porges, R. and Struzeski, E. J., Jr., "Wastes from the Poultry Processing Industry," U. S. Public Health Service Technical Report W62-3 (1962).

97 percent in BOD_5 and suspended solids are achievable at temperatures of 90 to 95°F (32 to 35°C), loading between 0.15 to 0.20 lb/ft^3, and detention times between 3 to 12 hr.

Aerated lagoons have been applied to meat processing wastewaters to a limited extent; however, the number of installations is increasing because of stricter effluent limitations. The lagoons use fixed mechanical turbine-type aerators, floating propeller-type aerators, or diffused air systems; they are usually 8 to 15 ft deep and have detention times of 2 to 10 days. BOD_5 reductions range from 40 to 60 percent. These lagoons are usually followed by an aerobic lagoon to remove suspended solids and to provide additional treatment.[29]

Aerobic lagoons are large, shallow ponds (3 to 8 ft deep), loaded at a rate of 20 to 50 lb of BOD_5/acre/day, with detention times from 1 to 7 months. Trickling filters, activated sludge systems, and more recently, the rotating biological contactor, have found only limited application for meat processing wastewaters, primarily because of their inability to handle shock loads.

The pollution control technology available to the seafood processing industry has not been adequately demonstrated at the operational level. Most processors have little if any significant wastewater treatment. Bilge wastes from fishing boats, which contain substantial amounts of soluble components, as well as wastewaters from processing plants are often discharged directly into the adjacent harbor waters.

EPA has recommended, as best practical technology (the 1977 goal), in-plant solids and grease recovery systems followed by biological wastewater treatment. Best available technology (the 1983 goal) adds improved pretreatment, such as dissolved air flotation with pH control and chemical flocculation, an ammonia control step which may involve ammonia stripping or a nitrification–denitrification sequence, and a sand filter or equivalent following second treatment. Zero discharge technology is available using land disposal techniques after adequate treatment.[30,31,26]

Dairy Products. During the past two or three decades, dairy plants in the United States have decreased in number and increased in size. In 1970 there were approximately 5350 dairies, which processed about 55×10^6 tons of milk into fluid products (45.1 percent), butter (22.2 percent), natural cheese (17 percent), ice cream and other frozen products (11.4 percent), evaporated milk (2.85 percent), cottage cheese (1 percent), and dry milk (0.5 percent). Dairy plants receiving from 1000 to over 2 million pounds of milk and milk by-products per day are located throughout the country with regional concentrations in Minnesota, Wisconsin, New York, Iowa, and California.

Fluid Products. Most milk is processed by pasteurizing, homogenizing, vitaminizing, and bottling into fluid milk, cream, and related products (see Fig. 13-12). Raw milk is received in tank truck or 10-gal cans and is pumped

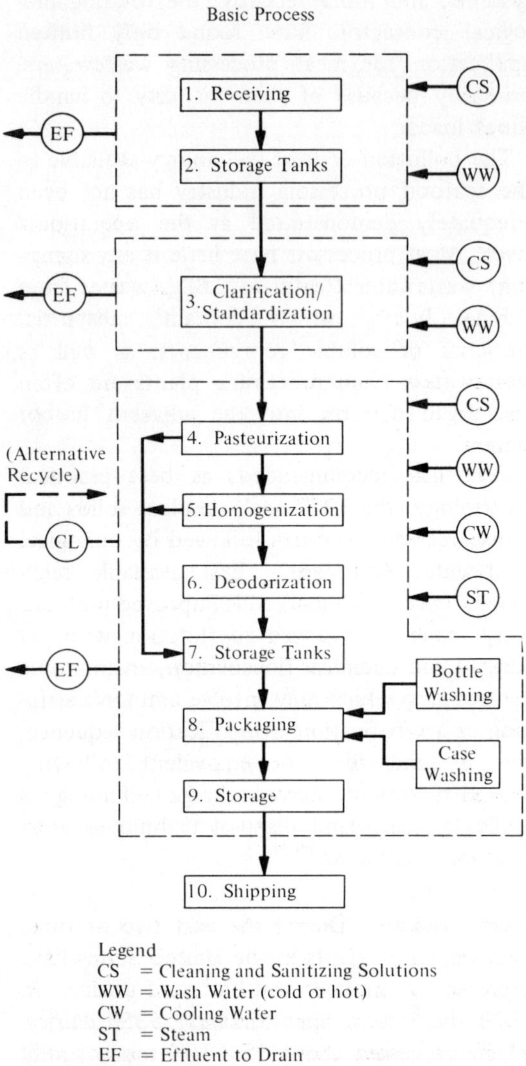

Fig. 13-12. Typical production processes and water and wastewater streams for fluid milk (From EPA *Development Document for Proposed Effluent Limitations Guidelines and Standards of Performance–Dairy Products Industry*, 1973).

from receiving to refrigerated storage. It is clarified by centrifugation or mechanical filtration, pasteurized, homogenized, deodorized by vacuum steam injection if necessary, cooled, stored until packaged or bottled into glass bottles, waxed paper cartons, plastic bottles, and plastic bags in cardboard boxes, and again cold stored until shipped.

Some raw milk is separated into cream and skim milk by centrifugation. Cream is blended, pasteurized, homogenized, cooled, stored, and shipped. Skim milk is used for the standardization of raw milk to a controlled percentage of butterfat or is processed to packaged skim milk, buttermilk, cheese dressing, or cottage cheese.

Creamery Butter. Raw milk is warmed to 90°F (32°C), centrifuged (skim milk goes to by-products use), cooled, stored, pasteurized, stored, churned at 45°F (7°C), drained (buttermilk goes to by-product use or is wasted), washed, drained, rewashed, drained, and standardized to 80 percent fat with water; salt, color, and flavoring are added, and the butter is "worked" to the desired consistency, wrapped, packaged, cold stored, and shipped.

Natural and Processed Cheese. Raw milk is heated, centrifuged (cream goes to by-product use), pasteurized, cooled, pumped to cheese vats, innoculated with culture, drained (whey goes to by-product use or is wasted), washed, cooled, salted, cut or milled, pressed into molds, incubated (aged), and packaged or processed. Processed cheese is stabilized, flavored, blended with additional ingredients, pasteurized, cooled, packaged, cold stored, and shipped.

Ice Cream and Frozen Desserts. Liquid cream, condensed milk, whole milk, corn syrup, cane sugar, stabilizers, emulsifiers, chocolate powders, and fruits are stored, liquefied with water or milk, blended, pasteurized, homogenized, cooled at 40°F (4°C), and refrigerated; fruits, flavoring, and nuts are prepared and added; and the product is packaged, hardened, stored, and shipped.

Condensed and Evaporated Milk. Whole milk, skim milk, cheese whey, buttermilk, liquid sugar, and liquid corn syrup are received and stored; separated, pasteurized, evaporated, cooled, and stored; packaged, sterilized, stored, and shipped. Some evaporated milk is spray dried, instantized, packaged, stored, and shipped. Liquid dairy and syrup ingredients and various dry ingredients are liquefied, assembled, pasteurized, pumped to a hot well, evaporated, homogenized, cooled, refrigerated, and stored; packaged, stored, and shipped.

Wastewater Sources and Characteristics. The major source of waste from the fluid milk process is whey which, if not converted to powder for human or animal feed (a break-even or slight profit process), becomes a difficult disposal problem because of the high protein and acid content. Skim milk and buttermilk from butter production create similar problems, with spillage and cleaning wastewaters a secondary source of wastewaters. Cheese whey constitutes the major disposal problem from cheese processing operations for the same reasons. Ice cream and frozen dessert processes produce spillage and equipment cleaning wastewaters. Significant wastes from the condensed and evaporated milk processes originate in miscellaneous spillages during normal processing and packaging operations and from cleaning equipment with soap and chemical cleaning solutions. Significant uncontaminated wastewaters emanate from once-through cooling and condensing water practices.

Typical dairy wastewater characteristics for each product manufactured are presented in Table 13-11.[32] The values are given in pounds (not gallons) of flow per pound of milk received and pounds of BOD_5/1000 lb of milk received.

The most important wastewater parameters for the dairy industry are BOD_5, COD, suspended solids, and pH. Others of lesser importance include temperature, phosphates, chlorides, and nitrogen. The BOD_5 concentrations of dairy raw wastewaters typically range from 1000 to 4000 mg/l; the suspended solids level is approximately 2000 mg/l; pH ranges from 6.0 to 9.0; temperature ranges from 46 to 100°F (7.8 to 37.8°C); phosphorus ranges from 12 to 210 mg/l (49 mg/l average); chloride ranges from 46 to 1930 mg/l (482 mg/l average); and nitrogen ranges from 1 to 13.2 mg/l (5.4 mg/l average).[2]

Control and Treatment Technology. In-plant control requires the management of water resources and waste materials in conjunction with the engineering aspects of plant, equipment, processes, and ancillary systems. Whey is a major challenge to the industry; at present, it is dried or condensed on bran for animal feed (lactose modification may increase use), used as a protein supplement in pet foods, and used in frozen desserts. With the exception of whey, dairy wastes are generally amenable to biological treatment. Three typical treatment schemes, activated sludge, trickling filter, and aerated lagoon, are shown in Fig. 13-13. Recommended design and operating parameters for these systems are presented in Table 13-12.

Other treatment systems successfully applied

TABLE 13-11. Dairy Plant Wastewater Characteristics

Products Manufactured	WASTEWATER VOLUME DISCHARGED (lbs/lbs milk received)		BOD_5 IN WASTEWATER DISCHARGE (lbs/1000 lb milk received)	
	Average	Range	Average	Range
Milk receiving	6.1	4.6–12.5	1.0	0.02–4.8
Milk	4.3	0.1–18.6	1.6	0.02–22.0
Cheese	2.3	0.3–5.7	1.85	0.2–4.1
Ice cream	2.8	0.8–5.6	5.8	1.9–21.4
Condensed milk	1.9	1.0–3.3	4.5	0.2–13.3
Butter	3.5	0.8–8.3	1.3	0.8–2.1
Powdered milk	2.95	0.8–11.5	7.8	0.02–12.3
Cottage cheese	6.0	0.8–12.4	34.2	1.3–71.2[a]
Cottage cheese and milk	1.8	0.5–7.2	3.5	0.7–8.6
Cottage cheese, ice cream, and milk	2.2	0.8–3.9	5.3	0.6–12.9
Mixed products	1.8	0.8–6.8	2.5	0.9–6.95
Average	3.1	0.1–18.6	4.4	0.2–71.2

[a]Includes cheese whey.

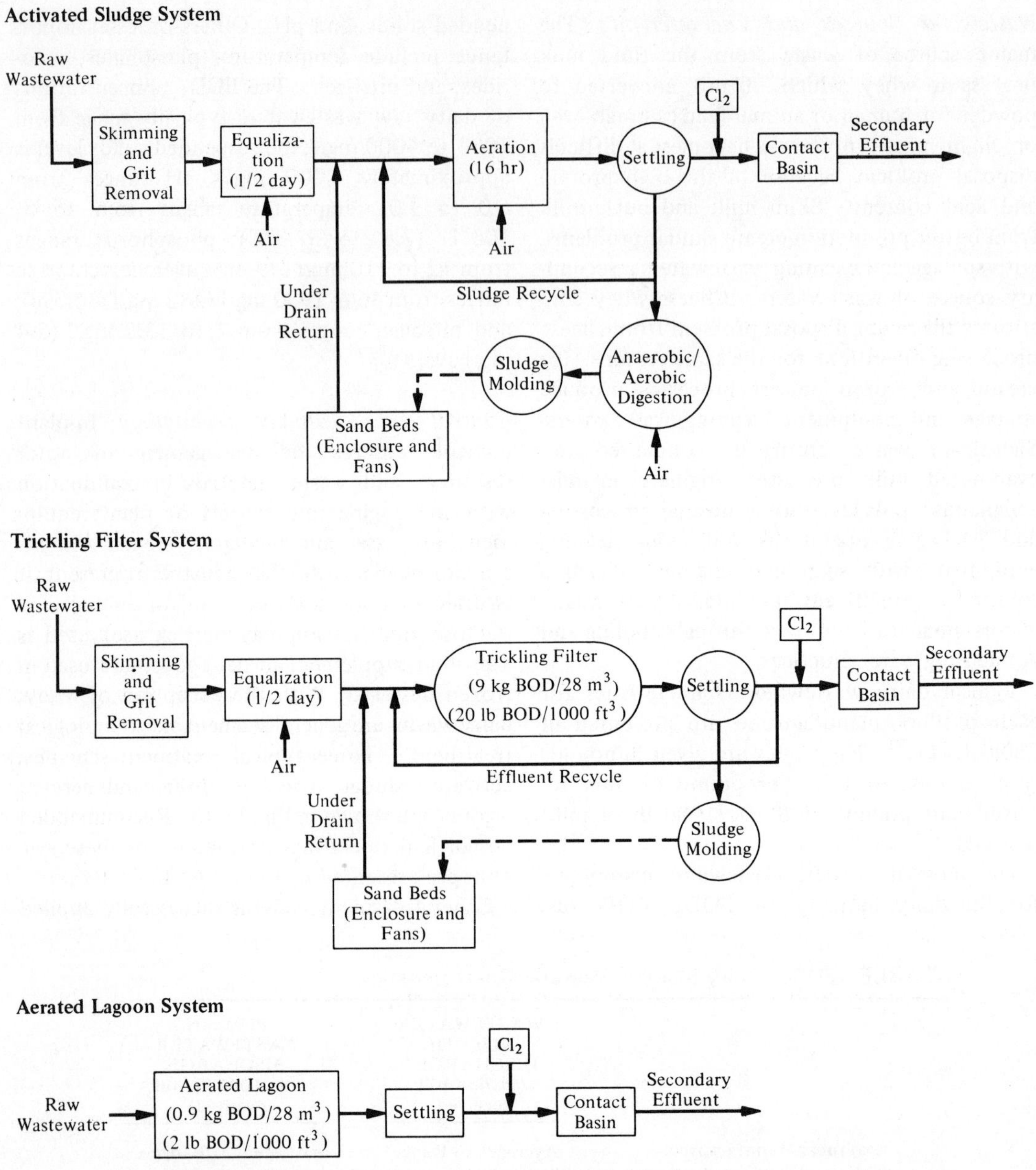

Fig. 13-13. Recommended treatment systems for dairy wastewater (From EPA *Development Document for Proposed Effluent Limitations Guidelines and Standards of Performance–Dairy Products Industry,* 1973.)

to dairy wastewaters include: stabilization ponds; spray, ridge and furrow, and truck irrigation; anaerobic digestion (septic tanks); and various combined systems.

Typical BOD_5 and suspended solids performance data for various treatment systems for wastewaters from various dairy operations are shown in Tables 13-13 and 13-14.

EPA has recommended, as best practicable control technology currently available (1977 goal), improved in-plant management programs and improved plant equipment followed by a biological treatment system (activated sludge, trickling filter, or aerated lagoon) and a sand filter. Where adequate land is available, a properly managed irrigation

TABLE 13-12. Recommended Design Parameters for Biological Treatment of Dairy Wastes[33]

Activated Sludge	Trickling Filter	Aerated Lagoon
1. Removal of floating substances.	1. Removal of floating substances.	1. Applied BOD_5 loading of 3.2 kg/100 m^3 (2 lb/1000 ft^3).
2. 12-hr equalization to buffer fluctuating BOD_5 and detergent loads. Diffused air supply to prevent acid fermentation.	2. 12-hr equalization to buffer fluctuating BOD_5 and detergent loads. Diffused air supply to prevent acid fermentation.	2. Air supply for sufficient oxygen dispersion.
3. Activated sludge tank to provide 36 hr retention.	3. Applied BOD_5 load of 32 kg/100 m^3 (20 lb/1000 ft^3).	3. Nutrient nitrogen and phosphorus addition if below BOD:N:P ratio of 100:5:1.
4. Microorganism population in the aerated tank to maintain a maximum loading of 0.5 kg BOD/kg volatile mixed liquor suspended solids.	4. Rock size of 6 to 9 cm (2.5 to 3.5 in.) or equivalent plastic media to allow proper ventilation and prevent clogging. Diffused air supply is helpful.[3]	4. Settling basin to sediment suspended solids.
5. Air supply of 60 m^3/kg (1000 ft^3/lb) BOD_5 applied.	5. 100 percent recycle of treated effluent.	5. Segregation of whey and cheese wash water from wastewater.
6. Nutrient nitrogen and phosphorus addition if below BOD:N:P ratio of 100:5:1.	6. Nutrient nitrogen and phosphorus addition if below BOD:N:P ratio of 100:5:1.	6. Reduction of milk waste concentration to a minimum through in-plant control.
7. Use of defoamers to prevent foam.	7. Steam injection of equalization tank if temperature drop impairs BOD removal.	7. Chlorination of final effluent.
8. Steam injection of equalization and aerated tanks if temperature drop impairs BOD removal efficiency.	8. Winter enclosure of filter in cold regions.	
9. Segregation of whey and cheese wash water from wastewater.	9. Segregation of whey and cheese wash water from wastewater.	
10. Reduction of milk waste concentration to a minimum through in-plant control.	10. Reduction of milk waste concentration to a minimum through in-plant control.	
11. Chlorination of final effluent.	11. Continuous dosing of filter to prevent drying up of slime.	
	12. Chlorination of final effluent.	

system is recommended. Best available control technology economically achievable (1983 goal) is essentially the same, with best performance and improved engineering resulting in BOD_5 reductions of 96 percent.[33,34]

Canning and Frozen Foods. The canning industry includes the canning, drying (dehydrating), or freezing of fruits, vegetables, juices, preserves, jams and jellies, and specialty products. Fruit and vegetable processing is a large industry in the United States with 1745 plants (1970). The activity level of these plants varies in response to the seasonal nature of the harvest.

Commercial canning processes depend upon the characteristics of the particular fruit or vegetable and the make-up of the final packaged product. However, certain operations which generate wastewater discharges are common to practically all production systems (see Fig. 13-14 for a typical apple processing operation). These operations are discussed below.

Receiving, Storage, and Washing. Fruits and vegetables are harvested from the fields, stored until needed, and delivered to the processing plant. Storage periods vary from 1 to 3 days for certain citrus fruits to almost a year for some potatoes. The produce is removed from storage as required and conveyed, sometimes by high-pressure water hoses, to the processing areas. It is washed with various combinations of high-pressure water sprays, simple water immersion, brush scrubbers, and rotary drum or

TABLE 13-13. Typical BOD and Suspended Solids Concentrations of Dairy Effluents[33]

Operation	Treatment System	Influent BOD (mg/l)	Influent SS (mg/l)	Effluent BOD (mg/l)	Effluent SS (mg/l)	Percent BOD Reduction	Percent TSS Reduction
Italian cheese	Anaerobic + activated sludge	827	376	14	32	98.3	91.5
Cottage cheese	Activated sludge	590	243	20	25	96.6	89.7
Fluid and cultured products	Activated sludge	1291	176	17	18	98.7	89.8
Yogurt and ricotta cheese	Aerated lagoon	637	503	24	29	96.2	94.2
Whey processing	Activated sludge	1373	–	46	–	96.6	–
Italian cheese	Aerated lagoon	1910	602	52	108	97.3	82.0
American, cheddar, and colby cheese	Anaerobic + bio-surf	1062	314	41	46	96.1	85.3
Fluid and cultured products	Activated sludge + aerated lagoon	1712	300	139	80	91.9	73.3
Average		1175	359	44	48	96.2	86.6

cylindrical tumblers to remove insecticide and pesticide residues, soil particles, insects, dried juices, and other foreign materials. In some cases, detergents or other chemicals are added to the wash water. Although internal recirculation of wash water is generally practiced, this source represents a significant fraction of the total wastewater volume.

Product Preparation. The cleaned raw produce is further processed by various combinations of sorting, peeling, coring, pitting, trimming, slicing, dicing, and juice extraction. Sorting is generally done by hand in order to remove inferior and undesirable material. In some subcategories of the industry, for example in apple canning, this inferior fruit can be redirected to another product line, such as apple juice, which can accept the lower quality raw material. Peeling methods include heat, mechanical peelers, caustic solutions, and water sprays. Potato peeling usually involves pretreatment with lye or high pressure steam, followed by water sprays, abrasive rolls, and

TABLE 13-14. Performance of Dairy Wastewater Treatment Plants[33]

	DATA FROM UNIDENTIFIED PLANT SOURCES[133]			DATA FROM IDENTIFIED PLANT SOURCES		
		PERCENT BOD_5 REDUCTION			PERCENT BOD_5 REDUCTION	
Type of Treatment	Number of Plants	Average	Range	Number of Plants	Average	Range
Activated sludge	100	84.0	24–99.6	3	97.3	96.6–98.7
Trickling filters	46	82.8	35–99.8	2	94.0	93.0–95.0
Aerated lagoons	2	96.5	95–98.0	4	96.2	95.2–97.3
	Average	83.8		Average	96.1	
Stabilization ponds	1	95.0	–	None	–	–
Combined systems	None	–	–	4	95.7	91.9–99.6
Anaerobic digestion	None	–	–	2	30.5	19.8–41.3
Sand filtration (of secondary effluent)	None	–	–	1	81.0	81.0– –

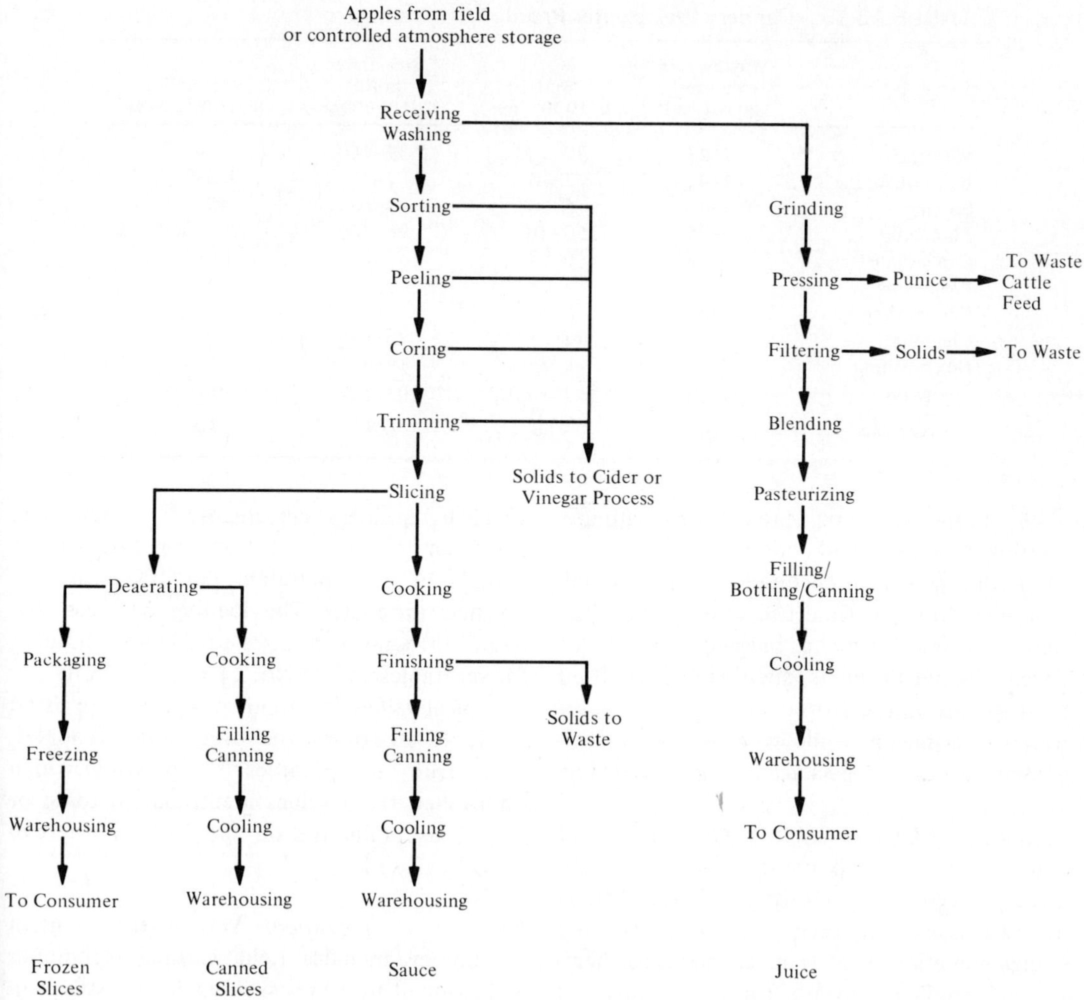

Fig. 13-14. Schematic drawing of a typical apple processing plant.

rubber rolls or rubber discs to remove the peel. Coring or pitting, when required, is accomplished by mechanical means.

Sliced fruits that are to be frozen or canned are deaerated by immersion in a brine under vacuum; they are then drained and either quick frozen or cooked for canning.

Potatoes and green vegetables are blanched to deactivate enzymes that cause undesirable flavor or color changes, to remove surface and intercellular air and gases, to remove excessive sugars and starches, to effect a degree of sterilization, to improve skin texture, to reduce color loss in subsequent operations, and to wilt leafy vegetables such as spinach to facilitate packaging. Blanch waters are the major source of soluble pollutants in fruit and vegetable processing wastewaters.

Raw juices are extracted in presses from precomminuted whole produce or cores, peelings, and trimmings. Seeds, pulp, and other large foreign particles are removed by screening. The juice is then further clarified by various filtering processes, frequently with the aid of diatomaceous earth, and is chilled, pasteurized, or concentrated by evaporation.

Finishing. Processed fruits and vegetables are cooked, dehydrated, or frozen. Canning operations include dispensing into jars or cans; topping with syrup, juice, or brine; exhaustion

TABLE 13-15. Cannery Wastewater Profile[36]

	Wastewater Volume (gals/case)	BOD (lb/1000 cases)	Suspended Solids (lb/1000 cases)	Total Solids (lb/1000 cases)
Washing	7–23	50–300	39–440	30–150
Belt conveying	1–4	2–10	7–40	3–9
Sorting	0.5–4	5–60	13–70	9–46
Blanching	6–25	150–400	96–400	200–400
Can exhausting	0–23	0–13	0–13	0–18
Processing	1–5	5–56	5–47	4–83
Can cooling	5–28	5–60	5–49	26–111
Clean-up	4.5–13	27–113	26–140	20–93
Box washing	2–5	8–23	14–40	16–40
Total	27–130	252–1035	205–1220	308–950
Average	70	680	780	720

of air by steam jets or heated boxes; capping, covering, or sealing; and cooling.

Dehydrating can be accomplished in a tunnel, conveyor, drum, or fluid bed drier. The dehydrated product can be bagged, canned, or frozen. Some products, such as French fried potatoes, are quick frozen in a tunnel freezer prior to packaging, while other products, such as strawberries, are packaged prior to freezing.

Wastewater Characteristics. Large volumes of water are used in all phases of the fruit and vegetable processing industry. These waters are used as cleaning agents to remove dirt and foreign materials, heat transfer media for heating and cooling, solvents for the removal of undesirable ingredients from the product, carriers for the incorporation of additives into the product, and conveyors for the transport and handling of raw materials and product. Typical fruit and vegetable processing wastewater characteristics are presented in Table 13-15 by industry subcategory.[35] These data are given in gallons of flow per case and as pounds of each pollutant per 1000 cases of finished product. The loading per case was based on equivalent cases of 24 No. 303 cans for vegetables and 24 No. $2\frac{1}{2}$ cans for fruit.

Typical values for minor pollutants are listed in Table 13-16 for the processing of apples, citrus fruit, and potatoes.[36] The study which established these values identified no toxic or hazardous pollutants on the basis of the evidence reviewed.

Wastewater Treatment. Wastewater control technology includes field washing techniques and in-plant measures such as dry caustic peeling, wash water recirculation, nonwater conveying systems, microwave fluidized beds, hot gas blanching, water recycling, and improved housekeeping and clean-up practices. Preliminary treatment systems include flow equalization and screening, plain sedimentation, hydro-

TABLE 13-16. Typical Wastewater Characteristics[36]

	Apples	Citrus Fruits	Potatoes
pH	6.6–11.2	7.1	0.3
Temperature (°C)	19	28	25
Alkalinity (mg/l as $CaCO_3$)	140–932	236	873
Total nitrogen (mg/l)	3.88	20.8	5.5
Total phosphorus (mg/l)	0.63	2.0	4.3
Total solids (mg/l)	–	3083	4227
Total dissolved solids (mg/l)	1337	2751	2340
Fecal coliforms (MPN/100 ml)	3–24,000	–	–
Pesticides (μg/l)	0.19	8.69	0.53

clones, and flotation to separate the solid wastes from the principal effluent wastewaters. Chemical treatments include pH adjustment, chlorination for odor control and disinfection, nutrient addition (cannery wastewaters are generally deficient in both nitrogen and phosphorous), and lime and alum coagulation and precipitation (reductions in BOD_5 as high as 89 percent have been reported). The standard primary treatment system in the food processing industry consists of a clarifier and rotary vacuum filter to treat the settled sludge. With chemical additions, average BOD_5 removals range from 25 to 40 percent, and average suspended solids removals range from 40 to 70 percent.

Biological treatment by the activated sludge process is capable of removing from 90 to 95 percent of the BOD_5 as long as loading rates are maintained in the range of 50 to 70 lb BOD_5/day/100 lb of sludge solids. Higher rates reduce removal efficiencies markedly, and result in less stable and bulking sludges. Contact stabilization and step aeration modifications, with nutrient addition, have produced treated effluents with BOD_5 concentrations of 50 mg/l or less.

Synthetic media trickling filters employing high liquor recirculation and forced air ventilation have been successfully applied to strong wastes in the 300 to 4000 mg/l BOD_5 range, with reductions of 70 to 77 percent.

Lagoons, ponds, or basins have been extensively used for treating food processing wastes where sufficient land is available. Anaerobic ponds have produced reductions of as much as 97 percent of the BOD_5 and 95 percent of the suspended solids. These are widely used as the first step in secondary treatment systems, or as pretreatment prior to discharge to a municipal system. Aerated lagoons are used in conjunction with other unit operations to remove 90 to 95 percent of the BOD_5 and to produce a high quality final effluent suitable for discharge directly to receiving waters. Aerobic lagoons with long detention times have attained up to 99 percent reduction in BOD_5, even though they are susceptible to operational difficulties such as algae growth, ice and snow cover, wave damage, short circuiting, rodent and weed problems, and odor. Ultimate disposal methods have included the use of percolation and evaporation lagoons. Spray irrigation has also been used where sufficient land with proper characteristics was available.

EPA has recommended, as best practical control technology (1977 goal), good housekeeping and water use practices, with treatment consisting of screening, primary settling, and biological waste treatment. Best available technology (the 1983 goal) specifies additional aeration, or ponding, and a sand filter, or equivalent, following secondary treatment.[37]

The Fermentation Industry. The fermentation industry includes the manufacture of malt, malt liquor, distilled spirits, and wine.

The objective of malting is to promote enzymatic changes in the starch and protein content of barley in order to form fermentable sugars and other substances necessary in the brewing of beer, malt, and ale. Barley grains are steeped by soaking in cold water for 2 to 3 days in large cylindrical tanks; moisture is absorbed during several water changes; and the dormant embryo in each kernel becomes activated. The barley is transferred to germination drums or compartments for 5 to 8 days, where the starch and protein are solubilized; it is then cooled under controlled temperature and moisture conditions, conveyed to drying floors, and slowly dried or "kilned" for 2 to 4 days for the development of optimum moisture, aroma, and flavor characteristics.

In the production of ale, beer, and malt liquor, processed malt and various combinations of cooked corn or rice are added with hot water (150 to 160°F) (65.5 to 71.0°C) to the "mash tun," where the enzymatic changes of the mashing operation take place. Spent grains are removed from the brewing liquid, usually by settling and screening in a "lauter tun." The liquid is decanted, sent to the brewing kettle, boiled, and hops are added. The resulting "wort" is sent to a "hop filter," then to a "hot wort" holding tank, then cooled and sent to the "cold wort" tank for temporary storage. It is then passed through a diatomaceous earth filter to remove suspended solids, mixed with yeast in a fermentation tank, fermented, passed again through diatomaceous earth filters where surplus yeast and proteinaceous materials are removed, cold stored, carbonated, filtered, and packaged in bottles, cans, or kegs.

In the distilled spirits industry, grain is milled into meal, charged to a cooker with about 20 gal of water per bushel (mash), pressure-cooked for about 15 min, and cooled. Milled malt is added to convert the mash starches to sugar; the mixture is further cooled and pumped to fermenters. Yeast and "back slop" are added at this point, and the mix is fermented for 3 or 4 days to convert the sugar to alcohol. The resulting "beer" is dropped into a beer well and pumped to the top of a still; the alcohol is steam stripped and condensed as whiskey or other spirits. Some distillate is cut with deionized water, stored in barrels, and bottled. Bourbon whiskey is steam heated in a kettle "doubler" in which double distillation takes place to remove various impurities.

In the wine industry, grapes are destemmed and crushed into "must." Sulfur dioxide is added to control undesirable microorganism growth, to inhibit browning enzymes, and to serve as an antioxidant. For white wines, skins, seeds, and pomace are separated from the juice before it is sent to primary fermentation. For red wines, the must is first fermented and then pressed to remove the pomace. Pomace from these operations may be further fermented or distilled for the recovery of alcohol.

Fermentation is allowed to proceed to the desired point, and the wine is then separated from the "lees" by racking. Sometimes this is followed by a secondary fermentation, and the wine may be racked several times. It is then "fined" by use of flocculants to remove dispersed elements and colloids, cooled, subjected to ion exchange, chemically stabilized, and bottled. Some wine is fortified with alcohol to produce dessert wines. Some is distilled to produce brandy.

Wastewater Characteristics. Weighted average data from malt house surveys indicate effluent characteristics as shown in Table 13-17.[38] The major portion of the raw wastewater emanates from the steeping operation, with lesser contributions from the germination operation.

TABLE 13-17. Malt House Effluent Characteristics[38]

Process Effluent Parameter	Mean	Standard Deviation
Suspended solids	0.041 lb/bbp[a]	0.019 lb/bbp
	117 mg/l	77 mg/l
BOD_5	0.24 lb/bbp	0.13 lb/bbp
	700 mg/l	500 mg/l
pH	6.74	0.25
Temperature	59.2°F	3.0°F
Volume	42.2 gal/bbp	19.6 gal/bbp
Cooling water effluent temperature	58.9°F	4.4°F

[a]bbp: bushel barley processed.

Similar data for the brewery industry are presented in Table 13-18.[39] The major sources of waste from the brewing processes are liquors pressed from the wet grain, liquors from yeast recovery, wash waters, and cooling waters. Table 13-19[40] contains typical wastewater data from the distillery industry. Wastes originate in the mashing, cooking, fermenting, and distilling operations, and include evaporator condensate, cooling tower overflow, plant and equipment wash waters, and doubler wastes. Table 13-20 contains data from a California winery.[41]

Waste Treatment. The predominant practice in the malt industry is to dispose of wastewaters to municipal sewerage systems, either with or without pretreatment. The usual pretreatment

TABLE 13-18. Brewery Effluent Characteristics[39]

Process Effluent Parameter	Mean
Suspended solids	722 mg/l
	1.24 lb/barrel beer
BOD	1622 mg/l
	3.05 lb/barrel beer
COD	2944 mg/l
	3.37 lb/barrel beer
pH	7.6
Temperature	82°F
Volume	8.3 barrel/barrel beer
Cooling water volume	1.42 barrel/barrel beer

TABLE 13-19. Distillery and Dryhouse Wastes from a Typical Plant—3000 Bushels Capacity[40]

Type of Waste	Volume (gals/1000 bushels)	TOTAL SOLIDS (mg/l)	TOTAL SOLIDS (lbs/1000 bushels)	BOD (mg/l)	BOD (lbs/1000 bushels)	Percent of Total	pH	Temp.°F
Evaporator condensate	15,000	130	16	600	75	42	3.9	125
Washes	7000	1050	61	1000	59	33	6.0	70
Cooling tower overflow	8000	1100	74	550	36	20	8.0	115
Doubler	1000	240	2	1000	8	5	5.0	170
Total	31,000	590	153	690	178	100	5.0	110

consists of equalization and screening. One full-scale treatment plant, including a screenhouse, two trickling filters in series, and a reactor-clarifier with sludge digestion, reported typical reductions of 78 percent of the BOD_5 and 40 percent of the suspended solids.[38] Breweries also practice disposal into municipal systems, and expect to continue this practice with extensive investigation of in-plant procedures to reduce loads. Brewery effluents have no adverse effects on properly sized and operated municipal systems.

A 5.2 mgd treatment facility with 60 percent of its hydraulic load contributed by a brewery has been reported. Its process consists of bar screening, grit removal, neutralization, and anionic polymer addition; primary clarification; activated sludge treatment by a modified Kraus process; secondary clarification; sludge concentration by flotation with cationic polymer addition; vacuum filtration with lime and ferric chloride addition; and landfill. A summary of operating parameters and performance data is presented in Table 13-21.[39]

Distillery wastes are amenable to biological waste treatment, with typical plants including primary settling tanks, aerated lagoons or trickling filters or activated sludge processes, and final settling tanks. These plants have obtained BOD_5 removals of 80-95 percent. One plant has been reported as treating 0.2 mgd of waste with an average BOD_5 load of 920 lb/

TABLE 13-20. Winery Wastewater Characteristics[41]

Characteristic	CRUSHING SEASON Range	CRUSHING SEASON Mean	NONCRUSHING SEASON Range	NONCRUSHING SEASON Mean
pH	3.5–5.5	4.1	–	4.8
Dissolved oxygen (mg/l)	0	0	0	0
BOD_5 (mg/l)	2000–5000	2500	2000–5000	2400
COD (mg/l)	4000–10,000	5000	4000–10,000	4000
Grease (mg/l)	5–30	15	5–50	40
Settleable solids (ml/l)	25–100	80	2–10	2.5
Suspended solids (mg/l)	200–800	500	100–400	400
Volatile suspended solids (mg/l)	150–700	450	80–350	300
Dissolved solids (mg/l)	300–600	800	400–800	700
Nitrogen (mg/l)	5–40	20	10–50	40
Phosphorus (mg/l)	5–10	10	10–25	25
Sodium (mg/l)	100–200	150	100	140
Alkalinity (mg/l as $CaCO_3$)	40–120	115	10–100	50
Chloride (mg/l)	100–250	150	100–250	150
Sulfate (mg/l)	20–75	50	20–75	50
Boron (mg/l)	0–0.2	0.1	0–0.2	0.1
Coliforms (MPN[a])	4000–40,000	10,000	4000–40,000	10,000

[a]Sanitary wastewater.

TABLE 13-21. Summary Description of Brewery Unit Operating Parameters and Performance[39]

Unit Operation	Loading	OPERATING PARAMETERS Flow or Mass Loading	Efficiency
Primary clarification	O. R. = 800 to 900 gpd/ft^2	5.2 mgd 81,335 lb BOD/day	26 percent removal BOD 63 percent removal SS
Activated sludge	F/M = 1.5 200 lb BOD/1000 ft^3	5.14 mgd 60,235 lb BOD/day 8,640 lb SS/day	85–95 percent overall BOD removal
Secondary clarification	O. R. = 440 gpd/ft^2	6.47 mgd 42,164 lb SS/day	80–95 percent removal of SS
Flotation	22 lb/ft^2/day	40,000 lb sludge/day	Concentrated to 4 percent solids
Vacuum filtration	5 lb/ft^2/hr	53,450 lb dry solids/day	Concentrated to 15–25 percent solids

day in a 300 X 80 X 10 ft deep aerated lagoon providing 9 days detention, and equipped with two 15-hp aerators. The loading was 1670 lb BOD_5/acre/day, and BOD removal averaged 60 percent under summer conditions. This treatment was followed by a 300 X 105 X 5 ft deep oxidation pond, providing 6 days detention and loaded at 440 lb BOD_5/acre/day. The oxidation pond averaged 85 percent BOD_5 removal, for an overall system removal of 95 percent.

Winery wastewaters also are readily treated by secondary biological treatment processes. Typical treatments and performance data are shown in Table 13-22.

Pulp and Paper. The U.S. pulp, paper, and paperboard industry is a large and complex one. It is comprised of over 700 establishments which in 1972 produced over 46 X 10^6 tons of pulp and nearly 60 X 10^6 tons of paper and paperboard. The industry includes mills which produce pulp only, paper and/or paperboard only (nonintegrated mills), and both pulp and paper and/or paperboard (integrated mills).

In the pulping process, cellulose fiber, the basic component of paper, is separated from other constituents of a fiber source, which today is primarily wood. Only a very small quantity of rags, cotton linters, and other

TABLE 13-22. Treatment Results on Liquid Winery Wastewaters

Treatment Process		BOD (mg/l) Average	BOD (mg/l) Range	COD (mg/l) Average	COD (mg/l) Range	SUSPENDED SOLIDS (mg/l) Average	SUSPENDED SOLIDS (mg/l) Range	pH Average	pH Range
Anaerobic lagoons	Influent	12,000	6600–17,700	22,000	13,000–29,000				
	Effluent	5600	2800–12,600	16,000					
Anaerobic digestion				94 percent reduction					
Detartration		50 percent reduction							
Lime precipitation	Influent	4200							
	Effluent	2700							
Trickling filters (2-stage)	Influent		340–1000						
	Effluent	37	10–210						
Trickling filters (5-stage)	Influent	3500	3400–3600						
	Effluent	60	35–100						
Aerated lagoons	Influent	2000	131–5000			170	5–448	8.0	4.9–9.5
	Effluent	80	23–127			100	50–346	7.5	6.0–9.5
Activated sludge	Influent	636		994		80		5.7	
	Effluent	42		82				7.0	

fibrous materials are still pulped in the United States.

There are several methods of wood pulping. In some, the wood is cooked in a "digester" with chemicals under controlled conditions of temperature, pressure, and time. The various processes utilize different chemicals or combinations of them. In other methods, wood is reduced to a fibrous state by mechanical means alone or a combination of chemical and mechanical action.

The different processes also utilize different species of softwood and hardwood, and produce a variety of pulps, each of which has special advantages for certain products. For example, sulfite pulp is distinctly superior for fine paper manufacture.

The purely mechanical processes are stone groundwood, in which lengths of logs are forced against rotating grindstones in the presence of water, and refiner groundwood, in which wood chips are ground in disc refiners. In other mechanical processes, the grinding action is preceded by a chemical soak to reduce energy requirements; however, the principal force utilized is still mechanical. These are the chemi-groundwood (logs soaked in dilute sodium sulfite solution and stone ground) and cold soda (chips steeped in a caustic solution and refined) methods.

Semichemical pulping in this country involves three major steps: impregnation of chips with a neutral sodium sulfite solution, cooking at high temperatures, and mechanical fiberizing.

In chemical pulping, wood chips are cooked under pressure with various chemical solutions to the point at which noncellulosic constituents are dissolved, and the fibers are liberated by the force of their ejection from the pressure-cooking vessel. No further mechanical energy is required. The primary chemical pulping processes are sulfite (calcium, ammonia, sodium, or magnesium base); kraft (caustic soda and sodium sulfide); and soda (caustic soda). The kraft process is illustrated in Fig. 13-15.

Waste papers are deinked for reuse in newsprint and fine papers by employing a caustic soda cooking process very similar to the chemical pulping of wood.

A large percentage of pulp is bleached to remove its natural brown color in order to manufacture light-colored or white papers. Bleaching is commonly performed in several stages of chemical application and alkaline extraction. Chemicals used vary with the type of pulp bleached, but may include chlorine, calcium or sodium hypochlorite, and chlorine dioxide for chemical pulps, and hydro-sulfites and peroxides for mechanical pulps. Bleaching agents employed for deinked pulp depend upon the constituents of the waste paper raw material, but generally are comprised of one or more of the above. Semichemical pulps are seldom bleached.

The bleaching process is also used to produce dissolving pulp, the alpha cellulose component of rayon and explosives.

The papermaking process is essentially the same regardless of the pulp used or the end product produced. A layer of fiber is deposited from a dilute water suspension of pulp on a fine screen. This screen permits the water to drain through but retains the fiber layer. This endless mat is then pressed and dried in a series of rollers. The fourdrinier machine, which has a flat sheet-forming surface, is generally used for making paper; the cylinder machine is more commonly associated with the manufacture of paperboard due to its ability to make multi-layered sheets. The type of machine used, however, has little bearing on the raw waste load.

Many familiar paperboard products, such as folding bottle cartons, are made from waste papers which undergo relatively little pretreatment, in contrast to those which are deinked for higher grade products. The waste papers are converted to secondary fiber by adding water, chemicals, and steam to a pulper where the paper is defibered. Waste paper is similarly used in the manufacture of building papers and roofing felts.

Chemical recovery from the spent cooking liquor of chemical pulping is virtually essential in order for these mills to meet federal effluent limitations. In the kraft process, it is also an economic necessity due to the high cost of the chemicals employed and the concentrations required. The number of calcium base sulfite mills is rapidly declining because of the difficulty and cost of recovering this relatively inexpensive

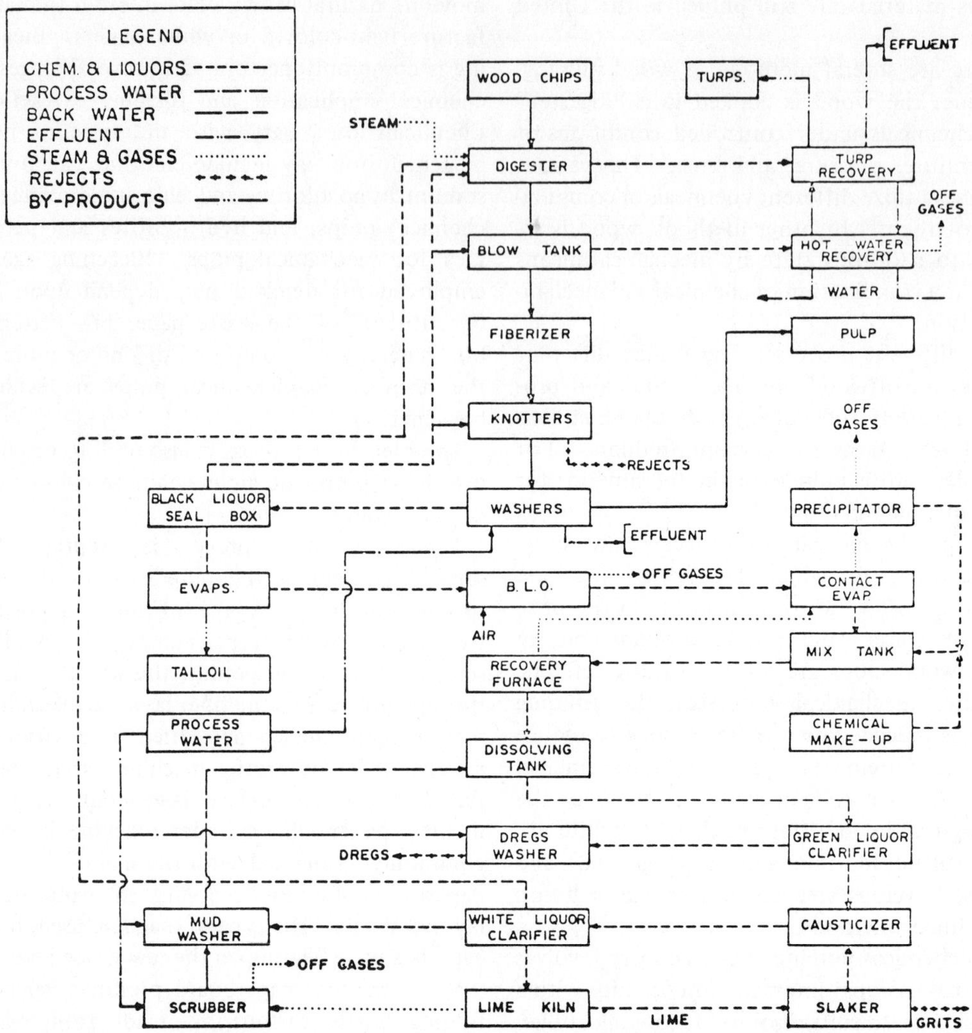

Fig. 13-15. Kraft pulping process diagram (From *State-of-the-Art Review of Pulp and Paper Waste Treatment*, U.S. Environmental Protection Agency, 1973).

base. Recovery is also difficult for sodium base neutral sulfite semichemical (NSSC) mills. The simplest and most economic solution to this difficulty is to recover the NSSC chemicals in an adjacent kraft mill recovery system.

Evaporation and burning of the spent liquor is a viable alternative to recovery in some cases. This procedure is practiced in ammonia base sulfite and NSSC mills.

Wastewater Characteristics. Pulp and paper industry wastewaters contain dissolved organic components of wood, including lignins, tannins, sugars, and cellulose; additives such as adhesives, sizing materials, starches, and resinates; and inorganic cooking chemicals.

Significant quantities of wastewaters result from wood preparation if bark is removed from logs in a wet barking operation. This practice, however, is declining in the United States and the trend is toward the use of delivered chips. The range of effluent flow from hydraulic barking is typically 3000 to 4000 gal/ton of pulp produced; drum barking produces only a fraction of this flow but is much higher in BOD and suspended solids. The typical range of these pollutants in hydraulic barking is from 2-4 lb and 20-30 lb/ton of pulp produced, respectively, and the drum barking range is from 15-20 lb and 80-140 lb/ton, respectively. Log washing also contributes a small flow which is significant only in its suspended solids content of silt.

Effluents from all of the mechanical pulping processes are similar in character and volume. Their pollutional ingredients are suspended solids and dissolved organic matter plus electrolytes in the case of chemi-groundwood and cold soda. Effluent volume can be expected to range between 2000 and 4000 gal/ton of pulp. The suspended solids and BOD_5 ranges for the mechanical pulping processes are as follows:

	Total SS (lbs/ton)	BOD_5 (lbs/ton)
Stone	11-21	8-19
Refiner	30-59	18-32
Chemi-groundwood	15-32	69-81
Cold soda	24-37	73-101

Wastewaters from the NSSC process are low in volume but high in BOD_5 because of the high degree of recycling commonly practiced. Loadings of 57 lb BOD_5/ton at a flow of 1700 gal/ton have been reported as typical. Average values from 13 NSSC corrugating mills with liquor recovery indicated effluent volumes of 11,300 gal, 58 lb of BOD_5, and 32 lb of total suspended solids/ton of pulp.[42]

Modern sulfite mills which practice chemical recovery and/or spent liquor burning discharge between 15,000 and 30,000 gal of effluent/ton of pulp. Overall losses amount to about 800 lb of total solids, 60 lb of suspended solids, and nearly 300 lb of BOD_5/ton of pulp. The pH value is in the range of 2.5 to 3.2. Effluent color values, for which the lignin fraction is largely responsible, range on the chloroplatinate scale from 100 to 750 mg/l; the lower value is typical of magnesium base sulfite and the higher, the ammonia base. Sulfite liquor recovery is not as effective as kraft recovery due to a considerable quantity of acetic acid in the condensates. This material is responsible for a substantial portion of the BOD_5 waste load.

Typical effluent loadings from individual processes in a sulfite pulp mill, including bleaching, are shown in Table 13-23. Bleachery losses from the production of sulfite dissolving pulp rise sharply as follows:

Effluent volume	45,000-60,000 gal/ton
BOD	200-450 lb/ton
Total suspended solids	100-200 lb/ton
Color	>3000 mg/l
pH range	1-3

The volume of kraft pulping effluents in modern mills ranges from 10,000 to 25,000 gal/ton of product. The major waste streams are digester and evaporator condensates, decker seal pit water, and the recovery plant effluent. Pollutional components include suspended solids, dissolved organics, and electrolytes. Also, inorganic ions are attached to organic compounds such as lignins, resins, and fatty acids.

The total suspended solids content of kraft effluents can vary widely, but averages are be-

TABLE 13-23. Typical Effluent Loadings from Individual Processes in a Sulfite Pulp Mill

Process Source	Volume (gals/ton)	Total Solids (lbs/ton)	Suspended Solids (lbs/ton)	BOD_5 (lbs/ton)	COD (lbs/ton)	pH Range
Blow tank	1900	247	1.0	116	–	2.2-2.9
Condensates	1100	47	0.1	66	–	2.3-3.1
Liquor losses	7500	105	21.0	53	–	2.2-2.6
Screening	6000	27	8.0	8	–	5.4-5.7
Washing and thickening	7500	131	8.0	18	–	2.4-3.9
Acid plant wastes	300	10	5.0	Neg.	–	1.2
Boiler blow down	100	22	2.0	Neg.	–	–
Total	24,400	589	45.1	261	618	2.5-3.2
Bleaching	12,000	200	15.0	30	–	5.0-5.8

tween 40 and 60 lb/ton of product for well operated mills. The average BOD_5 load amounts to 25 to 50 lb/ton. The nondegradable organic fraction is largely responsible for the color of kraft effluents. Values for this parameter range from 500 to 1500 mg/l of color units. Both degradable and refractory organics contribute to the COD, which, together with some inorganic sulfides, produce 75-100 lb COD/ton. The electrolyte content normally ranges from 1500 to 3000 mg/l. The pH is on the alkaline side, ranging normally from 9.0 to 10.0. Kraft bleaching wastes are described in Table 13-24.[42]

Wastewaters from waste paperboard mills are generated primarily in the stock preparation area and typically amount to 3300 to 24,000 gal/ton of product. These wastewaters usually contain 8 to 123 lb of total suspended solids and 10 to 75 lb of BOD_5/ton of product.[42]

Wastewater from the manufacture of fine paper without fillers (clay, calcium carbonate, etc.) contains from 150 to 300 ppm of suspended matter, 90 percent of which is organic. The suspended matter is primarily pulp fines. BOD_5 values are in the same range. Filled and coated sheets produce effluents of much higher suspended solids content because of the presence of the inorganic additives, and BOD values are also frequently higher because of the dispersants and adhesives used to retain the filler or coating in or on the paper. The inorganics impart a high turbidity, although the true color is low and the pH is in the neutral range. Flow ranges widely from 7 up to 70 gal/ton.

The volume of water used in the manufacture of tissue and related papers is large and the effluents weak, ranging in total suspended solids content from 72 to 250 mg/l and BOD from 35 to 100 mg/l.

In modern mills, a high degree of recycling of paper machine wastewaters is practiced and "save-alls" are employed to capture fibers lost through the machine wire.

Waste Treatment. In-plant measures to reduce raw waste pollutant loads include recycling, installing save-alls to recover fibers and other suspended material, reducing pump seal water usage, minimizing fresh water usage, and reducing cooling water wastage.

External treatment of suspended solids has employed screens, mechanical clarifiers, sedimentation lagoons, and dissolved air flotation. The most widely used method is the mechanically-cleaned quiescent sedimentation basin, with diameters from 30 to 350 ft, overflow rates from 392 to 2030 gpd/ft^2 (600 to 700 typical design, 300 to 500 conservative design), detention time 4-hr, and solids loading of 200 to 800 ft^2/ton of dry solids/day. Properly designed and operated clarifiers remove 95 percent of the settleable suspended solids and 20 to 30 percent of the BOD_5.

Additional BOD_5 reduction is generally accomplished in a biological system, with large storage oxidation basins and aerated stabilization basins predominating. Activated sludge and trickling filters are employed to a lesser extent. A loading rate for oxidation basins of 50 lb BOD_5/acre/day achieves 85 to 90 percent removal in warmer climates.

Aerated stabilization basins typically employ 2 acres/mgd and 5- to 15-day detention times. They generate 0.1 to 0.2 lb of sludge/lb of BOD_5 removed, require nitrogen addition (and sometimes phosphorus), use turbine aerators, axial flow pump aerators, or diffused aerators (producing from 1.0 to 2.5 lb of oxygen transfer/hp/hr to systems requiring 1.1 to 1.3 lb of

TABLE 13-24. Volume and Characteristics of Kraft Bleachery Wastes

	Effluent Volume (1000 gal/ton product)	BOD (lbs/ton product)	Total Suspended Solids (lbs/ton product)	Color (mg/l)	pH Range
Semi-bleaching	18–25	30–35	15–20	2500–3000	4–5
High-bleaching	25–35	40–60	20–30	4000–6000	3–4
Dissolving pulp (softwood)	50–60	120–150	130–150	> 5000	2–3
Dissolving pulp (hardwood)	55–70	500–700	190–200	> 5000	2–3

oxygen/lb of BOD_5), and operate at 50–200 mg/l MLSS.

Activated sludge plants treating pulp and paper wastewaters have been operated using loadings of 15 to 150 lb of $BOD_5/1000\ ft^3$, detention times of 2.5 to 8.5 hr, and nitrogen and phosphorus additions. An optimum loading of 0.2 to 0.7 lb of BOD_5 of MLSS has been recommended. Sludge generation is about 1 lb/lb of BOD_5 removed. Mechanical aeration is preferred, with oxygen requirements of 1 lb/lb of BOD_5 removed. Sludge dewatering and disposal is a major problem and can cost more than the treatment itself.

Other disposal practices that have been applied to pulp and paper process wastewaters with varying degrees of success are: by-product usage as soil supplements and in the production of bacterial protein; irrigation and land disposal of effluents; and color removal using lime precipitation, activated carbon, reverse osmosis, and alum precipitation.

EPA has recommended, as best practicable control technology currently available, specific internal controls plus suspended solids reduction using screens and (1) an earthen stilling basin, (2) mechanical clarification and sludge removal, or (3) dissolved air flotation. BOD_5 reduction is to be effected using (1) high-rate activated sludge or (2) aerated stabilization basins, with or without roughing filters, and natural oxidation ponds following biological treatment. Recommended secondary solids removal technology includes a secondary mechanical clarification process or a stilling pond. Sludge disposal consists of stilling ponds with dredging, removal, and disposal, or mechanical clarification with thickening by vacuum filtration or centrifugation and dewatering. Final sludge disposal consists of either land disposal or incineration in a sludge incinerator, power boiler, or bark burner.[44,45]

The best available technology economically achievable adds (1) further internal controls, (2) suspended solids removal by coagulation and flocculation followed by settling and filtration, or by reverse osmosis, (3) BOD_5 reduction using two-stage biological treatment, (4) carbon adsorption or chlorination to remove trace organics, (5) lime precipitation or carbon adsorption to reduce color, and (6) ion exchange, biological nitrification-denitrification, or ammonia stripping to reduce nitrogen in specific wastes where this parameter is a problem. (Most pulp and paper wastewater contains little or no nitrogen or phosphorus.)

Textiles. The textile industry uses large volumes of water in its operations, approximately 90 percent of it for process purposes. It is used throughout the industry to wash, scour, dye, print, bleach, and finish the cloth or textile being produced. All of these operations form wastewaters which ultimately need treatment.

The textile industry has three principal categories: wool, cotton, and synthetic fibers. The latter is further divided into cellulosics and noncellulosics. The processing steps (illustrated in Fig. 13-16) used to produce various fabrics are similar for all fibers, but the pollution loadings are diverse. The industry is large; 3.4×10^9 lb of fibers were woven into cloth in 1966.

Wool is received from the shearing operation containing large amounts of dirt, grass, burrs, wool grease, and perspiration or "suint." The wool is first "scoured" or washed with soap or detergent in warm water. The scouring is done countercurrently in a series of bowls. Insolubles collect in hoppers set in the bowls, and spent scour solution containing from 2000 to 4000 mg/l BOD and up to 28,000 mg/l grease is sent to wastewater treatment. After scouring, the wool can be dyed, or spun and woven and then dyed. Spent dye solutions and rinse waters from dyeing are discarded. The principal source of BOD_5 from dyeing is the carrier agent. Color, dissolved solids, and hexavalent chromium are also pollutional factors in these wastewaters.

The wool is next processed through a "fulling" step by working the wool mechanically while in a bath of detergent and water. It is then washed with detergent, and the spent bath is discarded. At some point in the process, the wool is soaked in a weak sulfuric acid solution, squeezed dry, and heated to carbonize vegetable matter. Spent sulfuric acid, discarded at intervals, is the only wastewater from this step. A final step, if needed, is bleaching with sodium hypochlorite or hydrogen peroxide. Spent

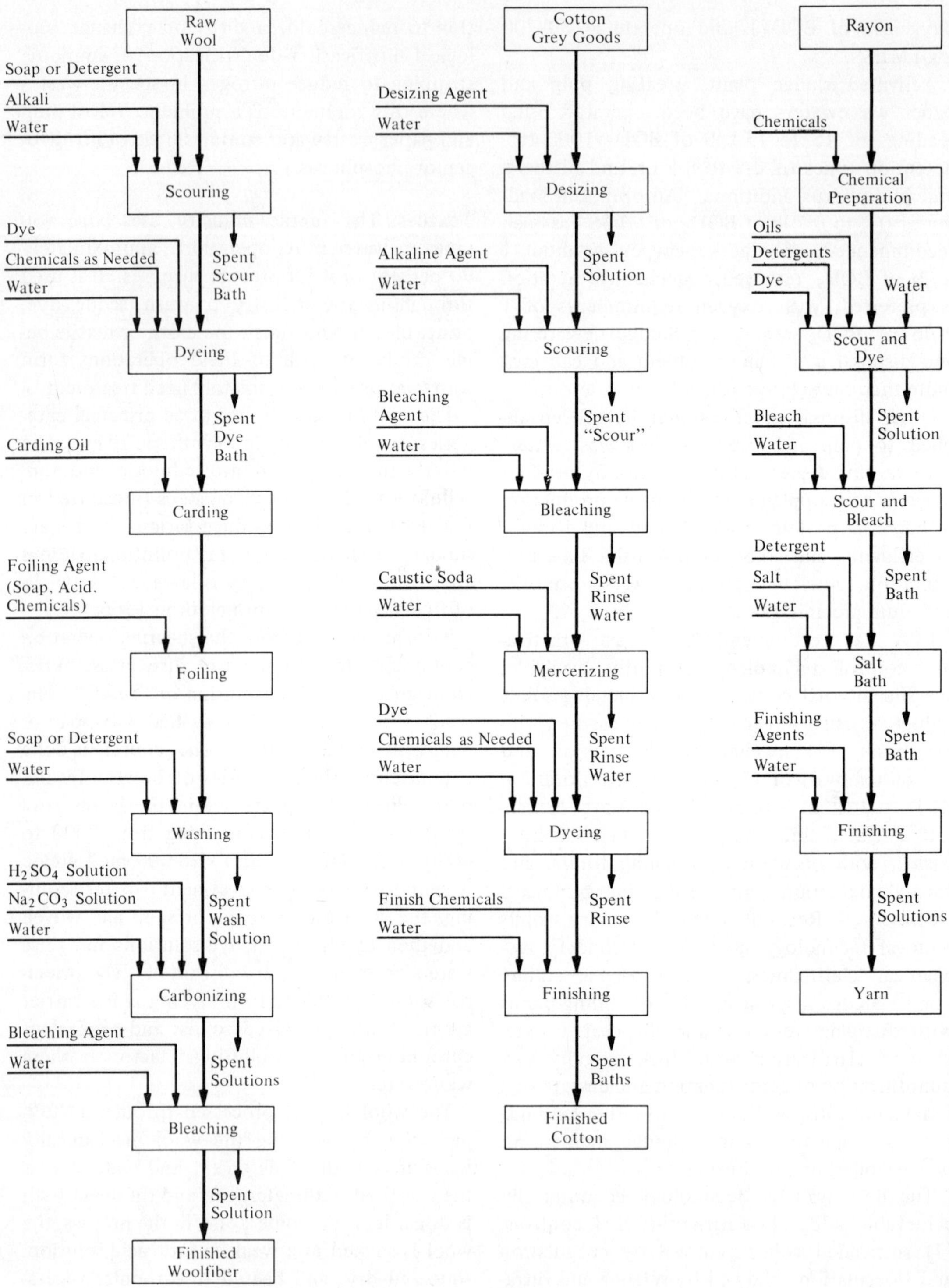

Fig. 13-16. Flow diagrams for typical textile manufacturing.

bleach solutions and rinse waters contribute to the total wastewater.

Cotton, as received at the mill, is usually coated with starch or sizing. The cotton is processed through an acid or chemical bath to remove the starch in a desizing step. The spent solution is discarded. The cotton is then scoured with water and an alkaline reagent to remove natural waxy materials. Spent scour is discarded.

After spinning, the yarn is bleached with sodium hypochlorite or hydrogen peroxide solution; the spent bleach bath is discarded. If needed, the yarn is mercerized by soaking in a strong caustic solution. It is then dyed, rinsed, and coated with a resin solution for finishing. Spent mercerizing, dye, and finish solutions, and various rinse waters contribute to the total waste load.

Rayon is processed through a combination scour and dye bath containing oils, detergents, and dyes. At some point, the yarn is bleached and processed through a bath containing salt, detergent, and water. The yarn is then finished in a procedure similar to that for cotton. Discarded baths from the scour, dye, bleach, salt bath, and finish solution, along with their rinse waters, comprise the effluents from rayon processing operations.

Nylon yarn is scoured with detergent and water, dyed, or bleached with hydrogen peroxide or sodium hypochlorite. The yarn is scoured again, heated to set the yarn, and finished with a chemical. Spent baths from the scouring, dyeing, bleaching, and finishing, with various rinse waters, make up the wastewaters from nylon processing.

Wastewater Characteristics. Wastewaters from the textile fiber industry are atypical. The processing steps for various fibers are similar, but are different in wastewater discharge characteristics. The wastewaters can contain high dissolved solids concentrations, high grease loadings, fibrous materials, inorganic salts, organic chemicals which often are toxic or biologically refractory, and heavy metals. The various individual operations lead to batch discharges that cause wide variations in hydraulic and pollution loadings. Some representative loadings are noted in Table 13-25.[46,47]

Wastewater Treatment. Wool grease recovery is carried out by adding acid to the batch to break the emulsion and free the grease. The aqueous phase is discarded after centrifuging the grease, or by evaporation to recover the grease.

Screening is commonly applied to remove fibrous materials. The screens may be rotary, stationary, or proprietary items such as vibrating or oscillating types. Fixed screens such as the Bauer Hydrasieve may also be used.

Sedimentation devices include batch tanks or settling basins. The addition of polyelectrolytes to improve solids separation is being practiced.

Air flotation units are used to remove suspended solids, often in conjunction with polyelectrolytes or following a coagulation process. The air separation removes solids, oils, grease, and fibers, but does not affect dissolved materials such as dyestuffs.

Centrifuging and filtration are standard operations in use to remove and recover solids from textile wastewaters. A chemical, usually copperas, lime, or alum, is added to the wastewater to coagulate small particles and colloids into larger particles to improve suspended solids separation.

Chemical precipitation is practiced in conjunction with lagooning. Chemicals are added to the wastewater to precipitate dissolved solids, which then settle in the lagoon.

Effluents from the textile industry are frequently alkaline, though some acid streams exist. Neutralization by adding acidic or alkaline materials is necessary to meet regulatory requirements or to protect subsequent treatment methods.

Equalization is necessary to level hydraulic and pollutional loads. It is often achieved by a holding tank or a lagoon.

Carbon adsorption has been practiced to remove color. The wastewater is usually passed through a moving bed in countercurrent operation. Batch treatment is also practiced but is costly for large volume flows.[48]

Biological oxidation is usually the last treatment step. The biological treatment may be

TABLE 13-25. Wastewater Profiles for Typical Textile Industry Subcategories (all values in units/1000 lb finished yarn)

	BOD (lbs)	Suspended Solids (lbs)	Total Dissolved Solids (lbs)	Volume (gals)
Wool waste				
Scouring	330	–	–	3300
Dyeing	19	–	–	3100
Washing	360	–	–	44,300
Carbonizing	2	–	–	16,500
Total	711	–	–	67,200
Cotton waste				
Desizing	119	62	105	5500
Scouring	105	47	130	9500
Bleaching	15	10	75	9500
Mercerizing	10	3	45	19,000
Dyeing	61	25	75	40,000
Printing	15	4	20	4000
Total	325	151	450	87,500
Synthetics				
Scour[a,b,c]	15–90	5–50	12–50	3000–8000
Scour and dye[d,e]	40–70	0–3	25–40	2000–6000
Dye[a,b,c]	2–800	1–40	6–200	2000–4000
Salt bath[d]	0–3	–	20–200	500–1500
Final scour[b,c]	10–25	2–7	4–50	2000–10,000
Special finishing[a,b,c,d,e]	2–80	3–50	3–100	500–7000
Total	20–250	20–160	20–600	3000–29,000

[a]Nylon.
[b]Acrylic.
[c]Polyester.
[d]Rayon.
[e]Acetate.

performed at the mill or, more often, the wastewater is discharged to a municipal sewage treatment plant. Trickling filters, activated sludge, lagoons, and aerated lagoons have all been used for this purpose.

Chemical oxidation with air, chlorine, or hydrogen peroxide is used to remove some dyes and to destroy various biological organisms.

EPA has stated that the textile industry can attain the 1977 guideline requirements by installing the equivalent of primary and secondary municipal sewage treatment. EPA indicates that the 1983 goal is zero discharge. Presumably, this will be achieved by reducing water use through recycling, and employing secondary biological treatment, clarification to remove suspended solids, and reverse osmosis. These treatments would be followed by a three-stage evaporation of the concentrated solution to a condensate suitable for recycling and a solid residue suitable for disposal.[49]

Tanning. Tanning is the process of converting animal skins into leather by treating the collagen fibers with such agents as chromium salts, alum, and tannin. The number of tanneries operating in the United States has steadily decreased, from about 7500 in 1865 to about 200 wet process tanners and 250 essentially dry process finishers in 1972. Since 1960, export of hides has more than doubled; hides from 47 percent of the 36.5×10^6 cattle slaughtered annually in the United States are shipped to foreign tanners.

Cattle hides constitute the bulk of the tanning done in this country. Sheep and lamb are next, pigskin third, and the remainder is accounted for by smaller quantities of goat, kid, hairsheep, and horse hides, and limited quantities of deer, elk, moose, antelope, rabbit, alligator, crocodile, seal, shark, and kangaroo. A cattle hide tanning operation consists of beamhouse, tanhouse, retan, color, fatliquor, and finishing. Sheepskin tanning consists of tanhouse, retan, color, fatliquor, and finishing. Pigskin tanning includes tanhouse, color and fatliquor, and finishing; other skins employ similar processes with modifications for the particular skin or product manufactured.

Beamhouse Process. Green hides received at the tannery are trimmed, graded, and cured by adding salt and storing for 10 to 30 days. Brined hides are prepared at the packinghouse by agitating fresh hides in a saturated brine solution for 2 to 3 days, draining, and bundling. At the tannery, the hides are trimmed (trimmings go to glue or other by-products), "sided" or halved, washed and soaked, fleshed, and unhaired in lime slurries with sharpeners added such as sodium sulfide or sodium sulfhydrate. Mechanical pulling or "pulping," or "burning" (dissolving) is employed for hair removal.

Tanhouse Process. Hides are bated in solutions of ammonium salts and enzymes to de-lime, reduce swelling, peptize fibers, and remove protein degradation products. They are then pickled in brine and acid, tanned in vegetable tannins or proprietary mixtures of basic chromium sulfate or alum, and split to produce a grain-side piece and a flesh side layer, which are processed separately.

Retan, Color, and Fatliquor Processes. Retanning imparts special characteristics to the finished leather through the use of chrome, vegetable, or synthetic tanning agents. The hides can be bleached, colored, and fatliquored (oils are added for pliability).

Finishing Process. Finishing operations consist of drying, wet-in coating, staking or tacking, and plating.

Wastewater Characteristics. Materials which appear in tannery wastewaters include hair, hide scraps, pieces of flesh, blood, manure, dirt, salt, lime, soluble proteins, sulfides, amines, tannin, soda ash, sugars and starches, oils (fats and grease), surface active agents, mineral acids, dyes, and solvents. Parameters used to define wastewaters include BOD_5, COD, total solids, suspended solids, total nitrogen, chromium, oil and grease, sulfide, total alkalinity, and pH. Typical tannery process wastewater characteristics are shown in Table 13-26.[50]

Waste Treatment. In-process methods of reducing tanning wastes include water conservation practices, process solution recovery and rinse, and treatment of specific waste constituents. Tanning wastes are also frequently subjected to pretreatment to increase their suitability for introduction into a municipal or metropolitan

TABLE 13-26. Tanning Wastewater Profile (based on processing 500 hides/day at 60 lb/hide)

Plant process	BOD (lbs/day)	Suspended Solids (lbs/day)	Total Dissolved Solids (lbs/day)	Wastewater Volume (million gal/day)
Washing and soaking	370	1340	3180	0.097
Unhairing	1830	5160	5700	0.112
Bating (paddle or drum)	210	50	420	0.032
Pickling (paddle or drum)	–	–	420	0.016
Tanning (vegetable or chrome)	150	200	900	0.020
Finishing	50	50	100	0.015
Total	2670	6800	10720	0.292

sanitary district sewerage system. Pretreatment operations consist of one of several processes or combinations of them. These include screening; equalization; plain sedimentation; chemical treatment for coagulation; sedimentation using alum, lime, iron salts, and polymers; carbonation; pH adjustment; and sludge disposal.

Fine screening removes hair particles, wool, fleshings, and hide trimmings. Equalization of the flow, strength, and pH of strong wastes requires detention periods usually less than 1 day. Plain sedimentation reduces suspended solids concentrations by approximately 40 to 90 percent, BOD_5 by 30 to 60 percent, total chromium by 53 percent, total alkalinity (as $CaCO_3$) by approximately 27 percent, and grease by 90 percent. Chemical treatment can effect suspended solids removals from 50 to over 95 percent, BOD_5 removals from 50 to 90 percent, sulfides and chromium removals up to 90 percent, and color removals from 93 to 99 percent. Carbonation is used to treat alkaline wastes, and can effect 95 percent suspended solids removal, 84 percent BOD_5 removal, and removal of excess alkalinity. Sludge handling and disposal includes sludge lagoons, landfills, dumps, and spreading on land.

Secondary treatment processes applied to tannery wastewaters include trickling filter systems effecting 80 to 95 percent BOD_5 reduction. Aerobic lagoon systems can remove 92 percent of the BOD_5, 64 percent of the COD, 75 percent of the total Kjeldahl nitrogen, and accomplish complete destruction of sulfide at loadings of 4.5 lb $BOD_5/100$ ft_3/day. Anaerobic-aerobic lagoon systems produced removals of 87 percent of the BOD_5, 74 percent of the suspended solids, 68 percent of the COD, 24 percent of the sulfide, and 33 percent of the total Kjeldahl nitrogen. Full-scale activated sludge plants treating tannery wastewaters are plagued by operating difficulties, and exemplary operation has not yet been achieved on these wastes.

One 1-mgd facility reported the following: two 35-ft diameter clarifiers with surface skimmers loaded at 460 gpd/ft^2, four 1-million gal concrete-lined lagoons with 30-hp aeration capacity per lagoon, return sludge recycling to each lagoon and to primary clarifiers, final settling in two 40-ft diameter clarifiers, chlorination, and discharge to a nearby watercourse. Incomplete operating data indicated removals of 76 percent of the BOD_5 and 89 percent of the suspended solids. Another full-scale facility, operating at only 45 percent of its design capacity, reported removals of 93 percent of the BOD_5 and suspended solids, 88 percent of the COD, total sulfide removal, 34 percent removal of total Kjeldahl nitrogen, 47 percent of organic nitrogen, 9 percent of ammonia nitrogen, and 73 percent of the alkalinity. Possible toxic effects of high ammonia concentrations (162 mg/l) on nitrifying bacteria were indicated.

EPA has recommended, as the best practicable technology currently available, pretreatment using water conservation, recycling of tanning solutions, collection and oxidation of sulfide, fine screening, equalization, primary settling for oil and grease removal and chromium precipitation, pH adjustment, and sludge handling and disposal using thickening, chemical conditioning, dewatering, and landfill for final disposal.[51]

Complete on-site treatment uses all of the above, plus aeration and secondary settling, deep-bed mixed-media filtration, chlorination, and discharge to receiving water. The 1983 goals best available technology economically achievable) add aeration to nitrify organic and ammonia nitrogen, settling, mixing with a carbon source to cause denitrification, aeration flume to assist in nitrogen gas removal, and a final settling tank. One such category, hide curing, which results in very low flows with high concentrations of dissolved solids, requires evaporation techniques resulting in zero discharge.

Steam Power Generation. This section relates only to the portion of the U.S. power generating industry which burns fossil fuels and not to nuclear plants. This industry can be divided into two subcategories, steam generating plants and steam-powered electric power plants. The steam generating plants supply high pressure steam for subsequent power or heat usage, while electric power plants normally supply electricity only, although some of the latter plants supply both steam and electricity. The United States capacity to generate electricity is about 300 × 10^6 kilowatts (3 × 10^5 megawatts). Actual annual production of electricity is 1.5 × 10^{12} kilowatt-hours, 80 percent of which is generated in steam-electric plants. This operation is illustrated in Fig. 13-17.

Wastewater Sources and Characteristics. Total wastewaters from this industry will vary in some respects according to the fuel used—coal, oil, or natural gas. However, wastewaters may originate from:

1. Chemical and physical treatment of boiler feedwaters, including both fresh water and condensate return waters.
2. Boiler blow-downs.
3. Ash handling systems.
4. Chemical cleaning.
5. A number of miscellaneous sources.

Thermal pollution problems arising from the large volumes of cooling water will be covered separately in Chapter 18.

Dissolved solids and gases in intake waters and return condensates would be detrimental to the boiler under the intense heat and pressure in the system. If permitted to remain in the water, the solids would coat the heat transfer surfaces, thus leading to hot spots and blowout holes; they would also cause problems of erosion and corrosion, and might cause blockages. Dissolved gases also become highly corrosive at the high temperatures and pressures in effect in the boilers.

Standard practice is to pretreat the fresh feedwater to remove suspended solids, dissolved minerals, and gases by clarification, filtration, softening, ion exchange, evaporation, and, recently, by reverse osmosis. Each of these operations has a wastewater stream that requires disposal. These streams contain suspended matter and various dissolved solids removed from the water by treatment, as well as water treatment chemicals, flocculants, and backwash solutions.

Condensate waters are a valuable source of feedwater for the boiler system, and their treatment prior to reuse does not have to be as extensive as that for fresh water. It usually consists of passage through an ion exchanger or filtration. The wastewater from these operations is similar to that encountered with fresh water.

Boiler water is usually treated with various chemicals to reduce scaling, to prevent corrosion, and to tie up dissolved gases in the water. Solids precipitate from the water as a byproduct of the evaporation effect while generating steam. The solids are periodically removed by blowing down the boiler; this blow-down water plus casual leakage is the main wastewater source from boiler operation. These waters contain suspended solids and various boiler water treatment chemicals.

Boiler cleaning is done at periodic intervals to remove scale and buildup in the boiler tubes, drums, and auxiliary equipment. The cleaning solution is usually an inhibited acid which dissolves scale material from the equipment. The

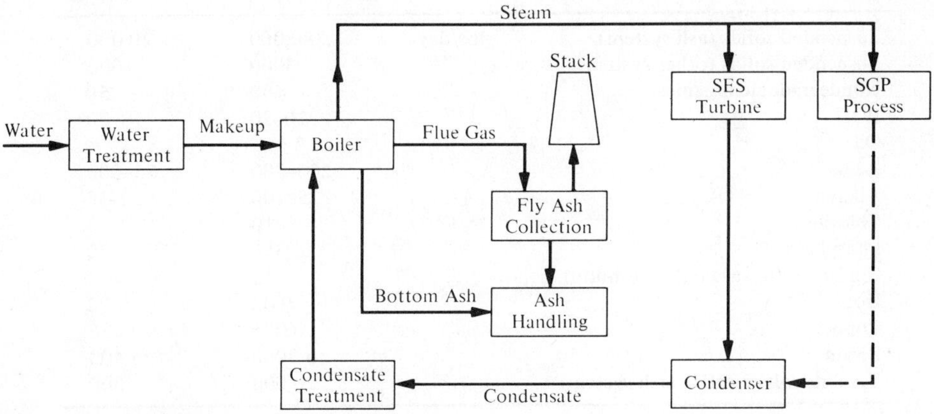

Fig. 13-17. Flow diagram for a typical steam electric power plant (From "Current Developments," Bureau of National Affairs.)

wastewater contains heavy metals as well as some suspended solids and the spent acid. The cleaned boiler is rinsed with water and alkaline solutions to remove all traces of the cleaning solution, and these waters too must be disposed of.

When coal is burned, ash handling develops wastewaters since boiler ash or "bottom ash" is frequently sluiced from the boiler with water. In addition, fly ash, collected mechanically from the flue gases, is processed through wet systems. These waters collectively contain suspended solids and dissolved matter from the ash. There is no ash with natural gas use and very little with oil.

The sulfur dioxide removal systems now being installed on flue gas stacks in the industry sometimes generate sludges that are concentrated prior to disposal. The water from this concentration step is similar to the waters from the ash handling step and is treated similarly. The sulfur dioxide removal systems are not yet in general use and are not required for natural gas.

Miscellaneous wastewaters in the industry include drainage from coal piles, fuel oil tank water decants, sanitary wastewaters, cooling tower blow-downs, compressed air system blowdowns, and the like. Knowledge of the characteristics of these waters is not as advanced as other power plant effluents, and is usually surmised based on data acquired on similar streams in other industries.

Wastewater characteristics of steam boiler wastewaters are described in Table 13-27.[52] The principal parameters of pollution are suspended solids, dissolved solids, heavy metals, acids, alkalies, and nonbiodegradable organics.

Waste Treatment. Wastewater treatment for power plant effluents is the same as for other industries having similar types of waste. Solids reduction is the principal recommended solution to these wastewaters.

Suspended solids are removed by sedimentation, with or without flocculating agents to improve the operation. Air flotation can be used to recover fine solids precipitated in feedwater treatment.

Oil-bearing wastewaters are processed through an oil-water separator of API design equipped with a skimmer device. Separation can be improved by adding coagulants such as alum or copperas, or polyelectrolytes. Very fine oil particles call for treatment by air flotation. Emulsified oils need emulsion breaking by pH adjustment or heating.

Acid and alkaline wastewaters should be mixed, and the mix neutralized. BOD in the

TABLE 13-27. Typical Wastewater Characteristics of a Steam Power Plant

Parameter	Unit	Steam Electric	Steam Generating
Suspended solids (ash system)	lbs/day	1,280,000	20,000
Suspended solids (other systems)	"	3000	1000
Nondegradable organics	"	400	50
BOD	"	10–20	10–20
Oil	"	?	?
Acids	"	200–800	75–300
Alkalies	"	25–100	10–30
Chlorine	"	210	–
Phosphate	"	to 350	10
Chromate (hexavalent chromium)	"	15	–
Iron	"	0.02	0.02
Copper	"	0.035	0.035
Boron	"	to 3000	to 400
Total dissolved solids (ash system)	"	20,000	300

TABLE 13-28. Steam Electric Power Plants Chemical Waste Effluent Limitations Guidelines Based on Best Practicable Control Technology Currently Available

Waste Stream	Parameters	Guidelines and Effluent Limitations	Best Practicable (Level I) Control Technology Currently Available
Condenser cooling system			
Once-through	Cl_2	$Cl_2 < 0.2$ mg/l	Monitor chlorine concentration in the effluent and control chlorine dosage
Recirculating	Cl_2	$Cl_2 < 0.2$ mg/l	Monitor chlorine concentration in the effluent and control chlorine dosage
	Chromates	Below detectable limits	Substitution of other corrosion inhibitors
Water treatment			
Clarification		No discharge	Sludge dewatering and recycling of water
Softening		No discharge	Sludge dewatering and recycling of water
Ion exchange	pH, TSS	$6 < pH < 8.5$; TSS < 15 mg/l	Neutralization and sedimentation
Evaporator	pH, TSS	$6 < pH < 8.5$; TSS < 15 mg/l	Neutralization and sedimentation
Boiler blow-down			
	pH, TSS	$6 < pH < 8.5$; TSS < 15 mg/l	Neutralization and sedimentation
Chemical cleaning			
Boiler tubes	pH; TSS; Cu, Fe, oxidizing agents, oil	$6 < pH < 8.5$; TSS < 15 mg/l Cu < 3 mg/l; Fe < 3 mg/l; oxidizing agents < 5 mg/l equivalent oxygen; oil < 15 mg/l	Neutralization, sedimentation, and oil–water separation
Boiler fireside air preheater, misc. small equipment, stack	pH; TSS; oil	$6 < pH < 8.5$; TSS < 15 mg/l; oil < 15 mg/l	Neutralization, sedimentation, and oil–water separation
Cooling tower basin	pH; TSS	$6 < pH < 8.5$ TSS < 15 mg/l	Neutralization and sedimentation
Ash handling			
Oil-fired plants	pH; TSS; oil	$6 < pH < 8.5$; TSS not to exceed influent; oil < 15 mg/l	Neutralization, sedimentation, and oil–water separation
Coal-fired plants	pH; TSS	$6 < pH < 8.5$; TSS not to exceed influent	Neutralization and sedimentation
Coal and oil-fired plants	pH; TSS; oil	$6 < pH < 8.5$; TSS not to exceed influent; oil < 15 mg/l	Neutralization, sedimentation, and oil–water separation
Drainage			
Coal-pile	pH; TSS	$6 < pH < 8.5$; TSS < 25 mg/l	Collection, neutralization, and sedimentation
Floor and yard drains	pH; TSS; oil	Oil < 15 mg/l; TSS < 25 mg/l; $6 < pH < 8.5$	Collection, oil–water separation, neutralization, and sedimentation

TABLE 13-28. (*Continued*)

Waste Stream	Parameters	Guidelines and Effluent Limitations	Best Practicable (Level I) Control Technology Currently Available
Air pollution control scrubbing devices		No discharge. This guideline is only for non-recovery alkali scrubbing processes. For other types of processes, guidelines will have to be developed in the future	Neutralization, sedimentation, and sludge dewatering
Miscellaneous waste streams			
Sanitary wastes	BOD_5; TSS	90 percent reduction	Packaged sewage treatment plant
Plant laboratory and sampling streams	pH; TSS	$6 < pH < 8.5$; $TSS < 15$ mg/l	Neutralization and sedimentation
Intake screen backwash		No discharge of collected debris, fish, leaves, etc.	Collection of debris and disposal of debris
Closed cooling water	pH; TSS; oil	$6 < pH < 8.5$; $TSS < 15$ mg/l; $oil < 15$ mg/l	Neutralization, sedimentation, and oil–water separation
	Chromates	Below detectable limits	Substitution of other corrosion inhibitors
Low-level radwastes		No guidelines established because low-level radwastes contain radioactive materials which are beyond the scope of this project	
Construction activity	TSS; oil	$TSS < 50$ mg/l $Oil < 50$ mg/l	Collection; sedimentation and oil skimming

Source: Bureau of National Affairs, Inc.

wastewaters can be treated by the equivalent of primary and secondary treatment as practiced by municipalities. Phosphate reduction can be achieved by treating the wastewater with lime slurry or aluminum chloride solution and processing it through a coagulation settler. Heavy metals can be reduced by the same lime treatment used for phosphates. Chromium chemicals must be reduced to the trivalent state prior to treatment for heavy metal removal. This can be done by adding sulfur dioxide or sodium bisulfite to the wastewater or by ion exchange.

EPA requirements for application of the best practical control technology to the chemical wastes of steam electric power plants are shown in Table 13-28. The 1983 guidelines call for zero discharge of chemical pollutants. It is stated that this can be achieved by (1) improving water use management and (2) using water in a "cascade-like" manner with existing wastewater treatment technology.[53]

REFERENCES

1. *Standard Methods for the Examination of Water and Wastewater*, 13th ed., APHA, AWWA, and WPCF, New York (1971).
2. "Industry Profile Study on Blast Furnace and Basic Steel Products," EPA Contract No. 68-01-0006 (draft) (June, 1971).
3. "Summary Report on the Iron and Steel Industry (SIC 3312)," EPA (Jan. 19, 1972).
4. "Effluent Limitation Guidance for the Refuse Act Permit Program–Steel Industry," EPA (Oct., 1972).
5. "Development Document for Proposed Effluent Limitations Guidelines and New Source Performance Standards for the Iron and Steel Industry," EPA Contract No. 68-01-1507 (draft) (June, 1973).
6. "Industrial Waste Survey of the Aluminum Industry," EPA (Aug., 1971).
7. "Development Document for Proposed Effluent Limitations Guidelines and New Source Standards of Performance for the Nonferrous Metals Industry," EPA Contract No. 68-01-1518 (draft) (June, 1973).

8. "Development Document for Proposed Effluent Limitations Guidelines and New Source Performance Standards for the Bauxite Refining Subcategory of the Aluminum Segment of the Nonferrous Metals Manufacturing Point Source Category," EPA 440/1-73/019 (Oct., 1973).
9. "Nonferrous Metals Manufacturing Point Source Category–Proposed Effluent Limitations Guidelines," *Federal Register* 38, 33,170–33,183 (Nov. 30, 1973).
10. "Industrial Waste Study of the Basic Nonferrous Metals Industries, Part III, The Lead and Zinc Industries," EPA Contract No. 68-01-0019 (May, 1972).
11. "Industrial Waste Study of the Basic Nonferrous Metals Industries, Part II, The Copper and Brass Industry," EPA Contract No. 68-01-0019 (May, 1972).
12. "Development Document for Proposed Effluent Limitations Guidelines and New Source Performance Standards for the Copper, Nickel, Chromium, and Zinc Segments of the Electroplating Point Source Category," EPA 440/1-73-003 (Aug, 1973).
13. Federal Water Pollution Control Administration, "The Cost of Clean Water–Vol. III, Industrial Waste Profile No. 2, Motor Vehicles and Parts," FWPCA, U.S. Dept. of the Interior (1967).
14. "Upgrading Metal Finishing Facilities to Reduce Pollution," EPA Technology Transfer Program (series of seminar papers) (1972).
15. Federal Water Pollution Control Administration, "The Cost of Clean Water–Vol. III, Industrial Waste Profile No. 5, Petroleum Refining," FWPCA, U.S. Dept. of the Interior (1967).
16. "Development Document for Proposed Effluent Limitations Guidelines and New Source Performance Standards for the Petroleum Refining Industry," EPA 440/1-73/014 (June, 1973).
17. "Petroleum Refining Point Source Category–Effluent Limitations Guidelines and New Source Performance Standards," *Federal Register* 38, 34,542–34,558 (Dec. 14, 1973).
18. "Development Document for Proposed Effluent Limitations Guidelines and New Source Performance Products for Inorganic Chemicals, Alkali and Chlorine Industry," EPA Contract No. 68-01-1513 (draft) (June, 1973).
19. "Development Document for Proposed Effluent Limitations Guidelines and New Source Performance Standards for the Organic Chemicals Industry," EPA Contract No. 68-01-1509 (draft) (June, 1973).
20. "Development Document for Proposed Effluent Limitations Guidelines and New Source Performance Standards for the Fertilizer and Phosphate Manufacturing Industry," EPA Contract No. 68-01-1508 (draft) (June, 1973).
21. "Development Document for Proposed Effluent Limitations Guidelines and New Source Performance Standards for the Plastic and Synthetics Industry," EPA Contract No. 68-01-1500 (draft) (June, 1973).
22. "Development Document for Proposed Effluent Limitations Guidelines and New Source Performance Standards for the Major Inorganic Products Segment of the Inorganic Chemicals Manufacturing Point Source Category," EPA 440/1-73/007 (Aug., 1973).
23. "Development Document for Proposed Effluent Limitations Guidelines and New Source Performance Standards for the Phosphorus-Derived Chemicals Segment of the Phosphate Manufacturing Point Source Category," EPA 440/1-73/006 (Aug., 1973).
24. "Development Document for Proposed Effluent Limitations Guidelines and New Source Performance Standards for the Synthetic Resins Segment of the Plastics and Synthetic Materials Manufacturing Point Source Category," EPA 440/1-73/010 (Aug., 1973).
25. "Industrial Waste Study of the Meat Products Industry," EPA Contract No. 68-01-0031 (July 8, 1971).
26. "Development Document for Proposed Effluent Limitations Guidelines and New Source Performance Standards for the Canned and Preserved Fish and Seafoods Processing Industry," EPA Contract No. 68-01-1526 (July, 1973).
27. Porges, R., and Struzeski, E. J., Jr., "Waste from the Poultry Industry," U.S. Public Health Service Technical Report W62-3 (1962).
28. "Upgrading Poultry Processing Facilities to Reduce Pollution," EPA Technology Transfer Program (1973).
29. "Upgrading Meat Packing Facilities to Reduce Pollution," EPA Technology Transfer Program (1973).
30. "Development Document for Proposed Effluent Limitations Guidelines and New Source Performance Standards for the Meat Packing Industry," EPA Contract No. 68-01-0593 (draft) (June, 1973).
31. "Meat Products Point Source Category–Proposed Effluent Limitation Guidelines for Existing Sources and Standards of Performance for New Sources," *Federal Register* 38, 29,858–29,868 (Oct. 29, 1973).
32. Federal Water Pollution Control Administration, "The Cost of Clean Water–Vol. III, Industrial Waste Profile No. 9, Dairies," FWPCA, U.S. Dept. of the Interior (1967).
33. "Development Document for Proposed Effluent Limitations Guidelines and New Source Performance Standards for the Dairy Products Processing Point Source Category," EPA 440/1-73/021 (Jan., 1974).
34. "Proposed Effluent Limitations Guidelines for

Existing Sources and Standards for New Sources–Dairy Products Processing Industry," *Federal Register* 38, 34,954–34,967 (Dec. 20, 1973).

35. "Development Document for Proposed Effluent Limitations Guidelines and New Source Performance Standards for the Citrus, Apple and Potato Segment of the Canned and Preserved Fruits and Vegetables Processing Point Source Category," EPA 440/1-73/027 (Nov., 1973).
36. Federal Water Pollution Control Administration, "The Cost of Clean Water–Vol. III, Industrial Waste Profile No. 6, Canned and Frozen Fruits and Vegetables," FWPCA, U.S. Dept. of the Interior (1967).
37. "Proposed Effluent Limitations Guidelines for Existing Sources and Standards for New Sources–Canned and Preserved Fruits and Vegetables Processing Industry," *Federal Register* 88, 31,076–31,084 (Nov. 9, 1973).
38. "Industrial Waste Survey of the Malt Industry," EPA (Aug., 1971).
39. "Industrial Waste Survey of the Malt Liquor Industry," EPA (Aug., 1971).
40. "Industrial Waste Survey of the Distilled Spirits Industry," EPA (Aug., 1971).
41. "Industrial Waste Survey of the Wine Industry," EPA (Aug., 1971).
42. Gehm, H. W., "State-of-the-Art Review of Pulp and Paper Waste Treatment," EPA Projects 12040 GLV and 12040 HAR (April, 1973).
43. Federal Water Pollution Control Administration, "The Cost of Clean Water–Vol. III, Industrial Waste Profile No. 3, Paper Mills Except Building," FWPCA, U.S. Dept. of the Interior (1967).
44. "Development Document for Proposed Effluent Limitations Guidelines and New Source Performance Standards for the Unbleached Kraft and Semichemical Pulp Segment of the Pulp, Paper, and Paperboard Mills Point Source Category," EPA 440/1-74/025 (Jan., 1974).
45. "Development Document for Proposed Effluent Limitations Guidelines and New Source Performance Standards for the Builders Paper and Roofing Felt Segment of the Builders Paper and Board Mills Point Source Category," EPA 440/1-74/-26 (Jan., 1974).
46. Federal Water Pollution Control Administration, "The Cost of Clean Water–Vol. III, Industrial Waste Profile No. 4, Textile Products," FWPCA, U.S. Dept. of the Interior (1967).
47. "Industrial Waste Studies Program–Textile Mill Processes," EPA Rough Draft Report (May 28, 1971).
48. "Upgrading Textile Operations to Reduce Pollution," EPA Technology Transfer Program (1973).
49. "Development Document for Proposed Effluent Limitations Guidelines and New Source Performance Standards for the Textile Mills," EPA 440/1-74/022 (Jan., 1974).
50. Federal Water Pollution Control Administration, "The Cost of Clean Water–Vol. III, Industrial Waste Profile No. 7, Tanning and Finishing," FWPCA, U.S. Dept. of the Interior (1967).
51. "Development Document for Proposed Effluent Limitations Guidelines and New Source Performance Standards for the Leather Tanning and Finishing Industry," EPA Contract No. 68-01-0594 (draft) (June, 1973).
52. "Industrial Waste Studies: Steam Generating Plants," EPA Contract No. 68-01-0032 (draft) (May, 1971).
53. "Development Document for Proposed Effluent Limitations Guidelines and New Source Performance Standards for Steam Electric Power Plants," EPA 440/1-73/029 (June, 1973).

14

Status of Advanced Waste Treatment

I. J. Kugelman
Wastewater Research Division
Municipal Environmental Research Laboratory
U.S. Environmental Protection Agency

INTRODUCTION

As a result of the research activities undertaken in recent years, new wastewater treatment techniques have been developed and the performance of existing processes improved. In this chapter, a general review of advanced waste treatment is presented, with special emphasis on those developments that have been applied or are ready for full-scale engineering application.

WASTEWATER CHARACTERISTICS

An excellent starting point for this discussion is a review of the characteristics of wastewater. Contrary to popular belief, sewage is only slightly contaminated water. In fact, the usual domestic wastewater in this country is 99.9 percent water. Unfortunately, water for most uses must contain much lower levels of impurities. Hence, although waste treatment technology need extract only small quantities of contaminants from wastewater, it must reduce these contaminants to very low levels. This procedure is inherently expensive and, due to the law of diminishing returns, places waste treatment technology at a disadvantage in comparison with other areas of industrial chemical processing.

A second major characteristic of wastewater which must be considered is the great diversity of specific compounds which make up the impurities. In fact, the variety of contaminants is so great and their concentrations so low that only a few substances exist at a measurable level. Much effort is being expended on developing reliable analytical methodology for rigorous analysis of wastewater; however, progress is not as rapid as would be desirable. Consequently, wastewater analysis is based on identifying classes of substances which have similar environmental effects. The six major classes of contaminants are: suspended solids, organics (BOD, COD, TOC), phosphorus compounds, nitrogen compounds, microorganisms, and electrolytes.

The terms BOD, COD, and TOC refer to the three most common tests utilized to measure the organic content of wastewater. BOD measures organics in terms of the oxygen required by microorganisms to oxidize the organics, COD measures the oxygen required to chemically oxidize the organics, and TOC directly measures the amount of organic carbon present. None of these tests gives a complete determination of organics in wastewater. Over the last few years, the TOC test has started to gain predominance over the others because of the rapidity with which it can be conducted.

CONVENTIONAL TREATMENT TECHNOLOGY

At present, the accepted method of wastewater treatment is that illustrated in Fig. 14-1. This system is referred to by a variety of titles such as conventional treatment, biological treatment, and primary plus secondary treatment. This treatment train has been specifically developed to remove suspended solids, biodegradable organics, and microorganisms from wastewater.

Suspended solids produce sludge banks in rivers, biodegradable organics lower the oxygen resources of lakes and rivers, and microorganisms are the source of most waterborne diseases. In the past, it was necessary only to remove most of these three classes of pollutants from wastewater prior to discharge to avoid adverse environmental effects. This is no longer generally true because of expansion in population, expansion of industry, and much greater demand for recreational and aesthetic uses of water resources.

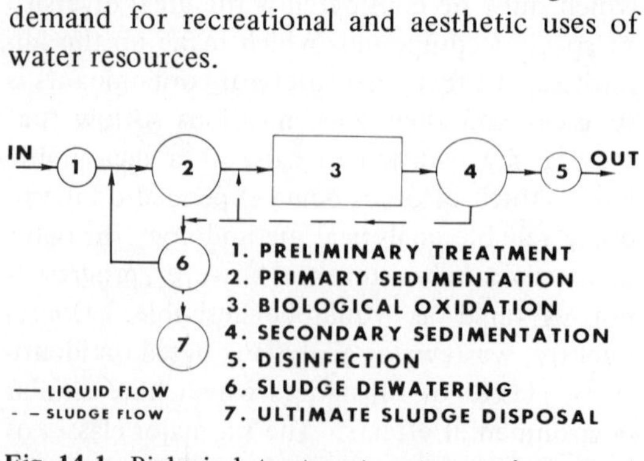

Fig. 14-1 Biological treatment system: primary and secondary.

Although in many locations the conventional system no longer is sufficient of itself, it is the base upon which systems providing a higher degree of treatment are constructed. Therefore, a short description of such a system, and the degree of treatment which it can provide, are both important here.

As illustrated in Fig. 14-1, wastewater is first passed through preliminary treatment of screening and grit removal. Preliminary treatment is utilized to protect pumps and pipes downstream from harmful large articles and abrasives which are often found in sewage. Next, primary sedimentation is provided to remove relatively large organic solids. The following step is biological oxidation, in which a large mass of microorganisms is contacted with the sewage in an aerobic environment. The microorganisms consume the soluble and colloidal organics producing more microbes, carbon dioxide, and water. Two systems which are used for biological oxidation are the trickling filter and activated sludge processes.

The trickling filter process is essentially a fixed-film contacting system. Microorganisms grow on the surface of an inert medium over which the wastewater is sprayed or trickled. Mass transfer of oxygen from the air to the liquid is encouraged by distributing the liquid in thin films. A high active mass of organisms is maintained by employing a medium with a high surface-to-volume ratio. Clogging is prevented by providing large void spaces in the configuration of inert medium packing and utilizing liquid shear to keep the microorganism film thickness down. Older installations use 2-in. to 3-in. stone as the medium. Newer plants employ proprietary corrugated plastic media.

The activated sludge process is a suspended growth system in which a mass culture of organisms is contacted with the wastewater in a mixed flooded reactor tank. Oxygen is introduced into the reactor by diffused air injection or by spraying the liquid through the air with propeller or turbine mixers. In dissolving the oxygen in the liquid, sufficient energy is introduced into the reactor to keep the biological culture in suspension as well.

The microorganisms which are active in biological oxidation are separated from the flow in the secondary sedimentation tank in the form of a slurry. Since they are organic matter, they must not be discharged with the effluent. Some are recycled to the head end of the biological oxidation process in order to maintain an adequate population in this unit, the remainder being sent to sludge handling. After passage through the secondary sedimentation tank, the flow is disinfected, usually by the application of chlorine, and discharged. No attempt is made to sterilize the effluent since all that is required is the elimination of pathogens (disease-producing microbes.) Most pathogens are more susceptible to the effects of chemical disinfectants than non-pathogens. It is important to note that the microorganisms which function in the biological oxidation process are non-pathogens.

The remaining steps in the conventional treatment train are sludge handling and disposal, which can account for up to half its cost. Sludge is a thick suspension of organic solids which is drawn from the bottom of the primary and secondary clarifiers. Disposal usually involves two steps: dewatering and ultimate discharge to the environment. Dewatering is generally necessary because the cost of the second step, "ultimate disposal," is proportional to the volume of the sludge. To facilitate dewatering, sludge must be conditioned to break down its natural gelatinous texture. Procedures applied to condition sludge include treatment with inorganic flocculants and/or organic polymers, biological treatment (anaerobic digestion or aerobic digestion), and heat treatment at atmospheric or elevated pressure. After conditioning with one or more of these techniques, dewatering is accomplished by gravity sedimentation (thickening), gravity draining combined with evaporation (drying beds), or mechanical dewatering (vacuum filtration, centrifugation, or pressure filtration). Gravity thickening often precedes the conditioning techniques.

The liquid removed in the thickening and dewatering steps is termed "supernatant." In the past, it was recycled to the primary clarifier and processed through the treatment plant again. Recent studies have indicated that the pollutant concentration in the supernatant is quite high, especially if the sludge is subjected to anaerobic digestion or heat treatment. The recycling of the pollutant-laden supernatant often produces a shock load on the treatment plant or loads the plant continuously with difficult-to-remove colloidal materials. Consequently, treatment of the supernatant prior to recycling is now recommended as good practice.

Ultimate disposal of fully dewatered or partially dewatered sludge can be accomplished by land spreading, landfill, incineration, wet oxidation, and processing for the recovery of useful by-products.

Some of the advanced waste treatment techniques significantly change the quantity and character of sludges generated in treatment plants. These problems will be addressed where pertinent in later sections of this chapter.

Table 14-1 illustrates the effluent quality which can be expected of a typical well-operated conventional treatment plant. The results have been presented both as percent removal and typical effluent characteristics. In general, the effluent characteristics column is a more fundamental testament to the potential of this process scheme than the percent removal column. To reiterate, its function is to remove suspended solids, organics, and microorganisms; removal of phosphorus, nitrogen, or salts is insignificant.

The cost of conventional treatment, expressed as cents per thousand gallons, is presented in Table 14-2. These costs represent the total waste treatment cost; i.e., capital amortization, operation, and maintenance. All cost data in this paper were prepared by the Treatment Optimization Group of the Advanced Waste Treatment Program of the Environmental Pro-

TABLE 14-1. Typical Performance—Conventional Treatment

Pollutant	Effluent (mg/l)	Percent Removal
Suspended solids	20–30	80–90
BOD	15–25	80–90
COD	30–60	70–80
Ammonia-N	15–25	0–10
Phosphorus	6–10	0–40
Coliform	1 per ml	99.999
Electrolytes	(Depend to a large degree on original water supply)	

TABLE 14-2. Treatment Cost Conventional System (total plant cost—cents/1000 gal. as a function of sludge disposal procedure)

Plant Size (mgd)	Land Disposal	Incineration
1	36	56
10	11.5	17
100	6.5	8

tection Agency (EPA). A uniform amortization schedule was used and costs are referenced to October, 1970. The data in Table 14-2 include costs for a wide size range of plants with land disposal or incineration for sludge disposal. Incineration is the most expensive form of ultimate disposal, and land disposal is the least costly; thus, these costs bracket all usual situations.

BIOLOGICAL–PHYSICAL TREATMENT

The inability of gravity sedimentation in the final clarifiers to remove small, light particles places a limit on the capability of conventional treatment to reduce suspended solids and BOD

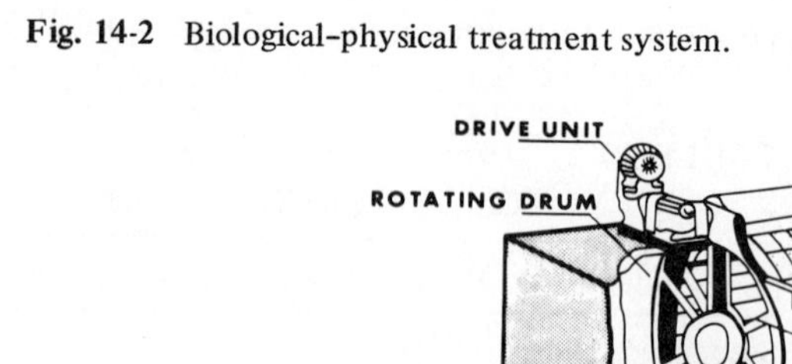

Fig. 14-2 Biological-physical treatment system.

in the treatment of wastewater. The usual range of this limit is indicated by the figures presented for typical effluent characteristics in Table 14-1. Integration of a better liquid-solid separation device into the treatment scheme is thus one method of upgrading conventional treatment.

Figure 14-2 illustrates the simplest type of advanced waste treatment, the biological-physical treatment system. This system is identical to the conventional treatment system except that an additional solids removal step is provided after secondary sedimentation. The three types of systems that have been widely used to upgrade the performance of conventional systems are discussed below.

Microstrainers. The simplest system is the microstrainer, illustrated in Fig. 14-3. Microstrainers are rotating drums on which woven filter fabrics, usually of stainless steel, are mounted. Incoming wastewater flows into the drums along the axis and then passes through the filter cloth which forms the drum surface. Filter rates are 6–10 gpm/ft^2 with a head differential at about 6 in. of water. Control of head loss is obtained by washing the solids from the screen as that portion of the screen rotates to the top of the device. The backflow liquid discharges into a trough in the interior of the drum from which it is recycled to the sedimentation basin. Ultraviolet light mounted above the screen is used to prevent biological growth from blinding the screen.

Since microstraining is fundamentally a screening process, the performance is a function of screen size. Table 14-3 illustrates this

Fig. 14-3 Typical microstrainer unit.

TABLE 14-3. Microstrainer Performance on Secondary Effluent

Location	Screen Size (microns)	PERCENT REMOVAL Suspended Solids	BOD
Luton, England	35	55	30
Bracknell, England	35	66	32
Chicago, Ill.	23	71	74
Brampton, Ontario	23	57	54

effect.[1] It can be seen that substantial improvement in effluent characteristics can be obtained by this technique. BOD removal closely parallels suspended solids removal, as most of the BOD exerted by a conventional plant effluent is due to biological floc which escaped removal in the secondary tank. An extensive study of microstrainers was conducted in Chicago[2]; based on this study, a 15 mgd plant was designed and installed.

Cost data which will be presented later indicate that a microstrainer installation is the most economical of the upgrading techniques. However, microstrainers do not perform well under shock load conditions.

Deep Bed Filtration. A somewhat more costly but more reliable upgrading technique is the use of deep bed filtration. A typical deep bed filter installation is shown in Fig. 14-4. In this technique, the wastewater is forced at 2 to 10 gpm/ft^2 through several feet of a granular material wherein the suspended solids are deposited. Most of the removal takes place on the surfaces of the grains rather than in the bed pores. The accumulation of solids in the bed eventually produces an excessive pressure drop. At this point, the filtration run is terminated and the bed is cleaned. Cleaning always involves an upward flow of water at a rate sufficient to expand the media, thus allowing for removal of the entrapped floc. In addition, an air scour and/or a water jet surface wash may be employed to remove solids which adhere to the grains. The nature of wastewater floc is such that it is almost mandatory for air scour or surface wash to supplement normal water backwash in order to achieve adequate cleaning.

Filter media range in size from 10 to 80 mesh and until recently only a single type of medium was employed. After backwash, however, a single-medium filter is graded with the smallest size particles at the surface. This restricts filtration to the top layers of the bed. To counteract this effect, dual-media or multimedia filter beds are now coming into use.

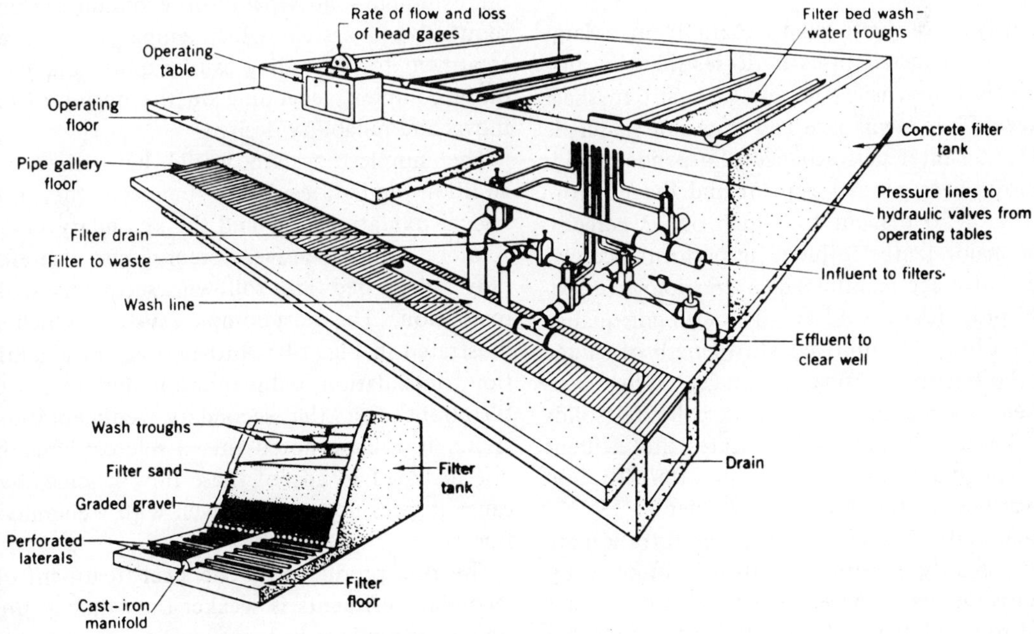

Fig. 14-4 Typical rapid sand filter.

These employ granular materials of such sizes and specific gravities that after backwash the coarser material is always at the top of the filter. Typical combinations are coarse coal and fine sand, or coarse coal, medium sand, and fine garnet. These new combinations allow much longer filter runs at higher flow rates than single-medium systems since the bulk of the solids are retained in the coarse upper layers, thus protecting the fine underlayer from rapid clogging or coating over.

Design guides for deep bed filters treating secondary effluent are as follows:[3]

1. The filter should be designed for a run length of 6 to 8 hr under peak hydraulic and suspended solids load.
2. The filter media size on the influent side should be 1.0 to 1.2 mm to allow reasonable depth of penetration into the filter. This assumes that dual media or multi media are used to approximate a coarse-to-fine gradation.
3. The backwash flow rate must be sufficient to fluidize the coarsest size of each medium used.
4. Air scour at rates of 3 to 5 $ft^3/min/ft^2$ for a 3 to 5 min duration should be provided.

Analysis of performance data from a large number of installations indicates that, because biological floc is extremely resistant to shear forces, filter media size and flow rate had little effect on filter performance. Suspended solids removal averaged 70 percent and ranged from a low of 50 percent to a high of 90 percent. The major factor influencing performance was the sludge age maintained in the biological process preceding the filter. The effluent quality after filtration improved with length of sludge age, probably because at longer sludge ages sludge flocculation is best and colloidal solids are lowest. Table 14-4 indicates the effluent suspended solids to be expected after filtration of various types of secondary effluents.

Removals obtained by deep bed filtration are only slightly superior to those obtained by microstrainers; however, deep bed filters react well to shock loads. Cost data for these mechanisms are presented in a later section.

TABLE 14-4. Expected Effluent Suspended Solids after Filtration of Secondary Effluent

	Effluent SS (mg/l)
High-rate trickling filter	10–20
Two-stage trickling filter	6–15
Contact stabilization	6–15
Conventional activated sludge	3–10
Extended aeration	1–5

Chemical Treatment. Suspended solids and BOD removal across a microstrainer or deep bed filter will rarely exceed 80 percent because these processes cannot remove colloidal matter from wastewater. Chemical coagulation is the only feasible method for removal of colloids. In this process, chemical coagulants added to the wastewater entrap the colloids into a floc which can then be removed by sedimentation or filtration.[1] A large variety of chemical agents have been found useful as coagulants, including salts of iron, salts of aluminum, lime, and organic polymers. Chemical treatment will also coagulate the suspended solids in sewage into large, dense flocs which are likewise amenable to removal by sedimentation.

Removal of these coagulated impurities requires liquid–solid separation techniques. Consequently, a system which employs chemical treatment for upgrading can be quite simple or quite complex depending on the degree of removal and reliability desired.

The simplest system would have chemical addition and flocculation between the biological oxidation unit and the secondary tank. A more comprehensive system would also employ a deep bed filter following secondary sedimentation. The most complex system, which is illustrated in Fig. 14-5, provides chemical addition, flocculation, sedimentation, and deep bed filtration after the secondary sedimentation step. It will be noted that a microstrainer is not included in any of these flow schemes because it does not perform well with a chemical floc.

The floc produced by chemical treatment of secondary effluents is weaker than that of untreated secondary effluents and tends to penetrate granular media filters much more readily.

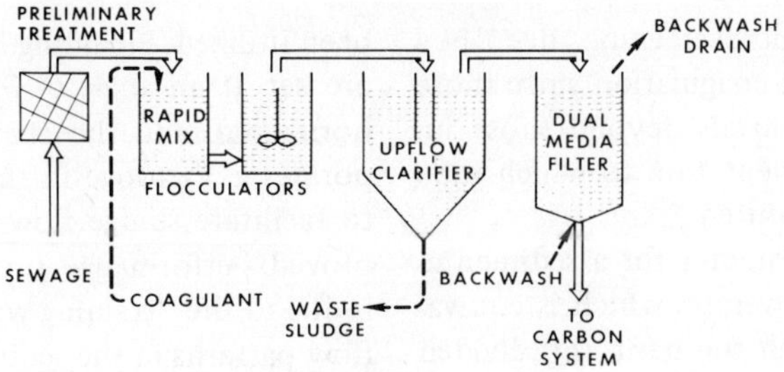

Fig. 14-5 Typical flow diagram of clarification system.

TABLE 14-5. Chemical Treatment of Secondary Effluent

Location	Treatment[a]	Effluent Quality	References
Nassau County, N.Y.	C, (F), S, F	0.4 JTU[b]	7
Lake Tahoe, Cal.	C, (F), S, F	0.3 JTU, 5 BOD	6
Lebanon, Ohio	C, (F), S, F	0.07 to 0.14 JTU	8
Washington, D.C.	C, (F), S, F	6 mg/l SS, 6 BOD	9
Richardson, Tex.	C, S	7 mg/l SS, 5.5 BOD	25

[a]C = coagulation; S = sedimentation; (F) flocculation; F = filtration.
[b]Jackson Turbidity Units.

Filter design must take this into consideration. For this weak floc, flow rates should be kept below 3 gpm/ft^2, and serious consideration should be given to the addition of floc-strengthening agents such as cationic polymers.[3] It is also advantageous to use finer filtration media. It has been found that tri-media filters (which utilize 60 to 80 mesh garnet) perform better than dual-media filters (whose finest grain size is 40 to 60 mesh sand) for filtration of chemically-treated secondary effluent.[5]

Table 14-5 summarizes data from a number of installations in which chemical treatment is practiced as an upgrading technique.

Estimated capital, operation, and maintenance costs for the three types of biological-physical and upgrading techniques are given in Table 14-6. Microstraining, which is the least versatile system, is the cheapest and its cost is almost independent of plant size. Although chemical treatment systems are the most versatile, the most reliable, and give the highest degree of treatment, they are the most expensive. A range of costs is presented for the chemical treatment systems at each plant size because of the variety of flow schemes included in this class. The low cost represents the system where chemical treatment is applied after biological treatment and all liquid–solid separation takes place in the secondary tank. The highest cost presented is for the system which employs chemical addition, flocculation, sedimentation, and filtration after secondary sedimentation. Although the chemical treatment system can be much more expensive than microstrainers or deep bed filters, it must be the system of choice where a high degree of colloidal material exists in the wastewater. This is often the case when the biological system is overloaded.

TABLE 14-6. Treatment Cost for Upgrading Conventional Systems by Suspended Solids Removal

	COST (CENTS/1000 gal)		
	PLANT SIZE (mgd)		
Process	1	10	100
Microstrainer	1.4	1.1	0.9
Deep bed filter	6.8	3.0	1.4
Chemical coagulation–flocculation	6–20	3–10	2–5

Tube Settlers. An additional liquid–solid separation method involves the use of tube settlers to obviate decreased removals during peak flows. A discussion of this concept was not included with that of microscreening, deep bed filtration, and chemical coagulation since those techniques achieve removals beyond those inherent to gravity sedimentation to which tube settler performance is limited.

The major design parameter for a sedimentation tank is the overflow rate, which is equivalent to the flow through the basin (Q) divided by the top area of the basin (A). The term Q/A is a velocity term and is the sedimentation rate that a particle must exceed to be removed by gravity sedimentation in the basin. The flow through most treatment plants follows a diurnal cycle, with peak dry weather flow about double the average flow and peak wet weather flow about triple the average flow. Obviously, as flow increases, the removal obtained in a sedimentation tank decreases unless a mechanism exists for increasing the top area of the tank. In a horizontal flow tank, this can be accomplished by the installation of horizontal trays in the basin. The equivalent area in the Q/A term is then the top area plus the tray area.

A practical method of placing trays in a horizontal sedimentation basin is the installation of tube settler bundles near the outlet weirs. Figure 14-6 illustrates 60°-angle tube settlers installed in a clarifier of an activated sludge system. The steep angle is used to ensure that sludge drops out of the tube settler bundle into the clarifier sludge collector. It has been reported that use of these devices allowed equivalent performance of secondary sedimentation basins at overflow rates up to 3 times normal design.[10] A number of similar techniques have

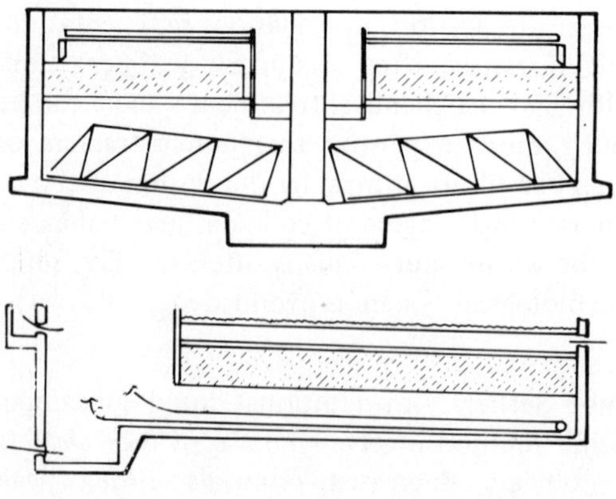

Fig. 14-6 Tube settlers in existing clarifier.

been utilized, including lamella separators which are set at an angle of 32° downward from the horizontal, and the chevron type which incorporates a V-groove in the bottom of each tube to facilitate sludge flow.[11] A portion of the improved performance provided by these devices is due to the "training wall effect" they have on flow patterns in the sedimentation basin.

It is obviously difficult to cost out tube settlers on a cents-per-thousand-gallons basis. Economic justification for use of this concept must be derived from a comparison of the capital cost of the settlers with the savings resulting from the use of smaller sedimentation tanks.

PHOSPHORUS REMOVAL

The major sources of phosphorus contributing to eutrophication are domestic sewage and agricultural runoff. Domestic sewage is the primary source in critical areas, and only this source will be discussed here.

Phosphorus gains entrance to sewage from human body wastes (primarily urine) and through the use of condensed inorganic phosphate compounds as builders in detergent formulations. Each of these sources accounts for about half of the phosphorus in domestic sewage. Thus, while elimination of phosphorus from detergent formulations would be helpful, it would not be the total answer to the eutrophication problem. Treatment of domestic sewage to remove a significant portion of the phosphorus contributed by human wastes and detergent builders would, however, have a significant effect on eutrophication rates.

Phosphorus Removal in Conventional Treatment.

Removal of any pollutant from wastewater requires that it be converted to either an insoluble gas or an insoluble solid. Because none of the chemically stable forms of phosphorus is a gas at normal temperature and pressure, removal from wastewater is dependent on formation of an insoluble solid. Less than 10 percent of the phosphorus discharged into municipal sewerage systems is insoluble, and none of the conventional treatment techniques other than chemical treatment is particularly effective in insolubilizing this nutrient. Thus, phosphorus removal in most conventional treatment plants is limited. Primary treatment can remove only the 10 percent of the phosphorus which is initially insoluble. During secondary treatment, phosphorus removal is achieved by synthesis into the biomass followed by sedimentation and sludge wasting. However, municipal sewage contains an amount of phosphorus considerably in excess of that required for biomass synthesis during complete utilization of the organic carbon present; thus, removals are generally confined to 20–40 percent.[14] Studies indicate that biological systems have the capacity for much higher removals through the mechanism of "luxury uptake."[15] However, attempts to implement this phenomenon in actual plants have not been successful.[16] Indeed, a recent study has illustrated that luxury uptake may actually be a chemical precipitation rather than a biologically mediated phenomenon.[17]

Phosphorus Removal by Chemical Precipitation. Fortunately, phosphorus forms almost completely insoluble precipitates with a number of substances, so that high levels of removal can be obtained when appropriate doses of the proper chemicals are applied. A large variety of chemicals can be utilized for this purpose, but economic factors dictate the use of salts of iron, salts of aluminum, or lime.

For the salts of aluminum and iron, the required dose is set by similar factors. The major control is the stoichiometric reaction between the metal salt and the phosphorus compound, as illustrated in Eq. 1:

$$M^{3+} + PO_4{}^{3-} = MPO_4 \quad (1)$$

This reaction indicates a requirement of 1 mole of the metallic cation for each mole of phosphorus in the wastewater. However, a number of factors act to increase the dose of metallic salt required. As indicated by Eq. 2, the metal salt reacts with the alkalinity in the wastewater to form insoluble hydroxide:

$$M^{3+} + 3HCO_3{}^- = M(OH)_3 + 3CO_2 \quad (2)$$

Although this reaction competes with the phosphorus precipitation, the metal consumed is not wasted since the hydroxide formed is an excellent flocculant of the phosphorus precipitates of Eq. 1.

Another factor which tends to increase metal dosage over that theoretically required is pH. Figure 14-7 illustrates that aluminum phosphate is least soluble at pH 6 and iron phosphate is best precipitated at pH 4.[18] Economics usually dictate use of additional metal salt rather than adjustment of the pH of the sewage from its normal range of 7–8.

Still another factor which influences coagulant dosage is the presence of suspended organic solids in the wastewater. A portion of the metal salt added will be required for coagula-

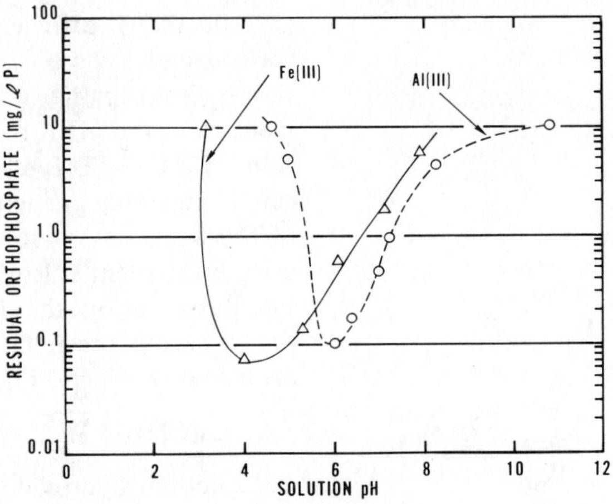

Fig. 14-7 Precipitation of orthophosphate with Al (111) at a 2 : 1 cation-to-orthophosphate molar ratio.[17]

TABLE 14-7. Average Metal Salt Dosage Required for Specified Percent Phosphorus Removal

Percent P Removal	Metal Salt Dosage (mole ratio M^{3+}/P)
75	1.4
85	1.7
95	2.3

tion of these materials, again increasing the dosage requirement.

Thus, the dosage required to achieve any specified degree of phosphorus removal cannot be established here with great precision. It is best to conduct bench-scale jar tests on the wastewater to be treated in order to specify the metal salt dosage requirement. Averages of results from many installations which can serve as a rough guide are given in Table 14-7.[12]

The reaction of lime with wastewater constituents is given in Eqs. 3 and 4:

$$3HPO_4^{2-} + 5Ca^{2+} + 4OH^- = Ca_5(OH)(PO_4)_3\downarrow + 3H_2O \quad (3)$$

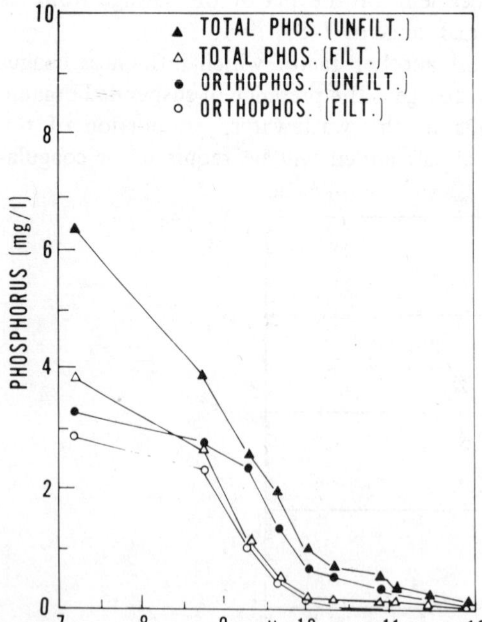

Fig. 14-8 Lime precipitation and phosphorous removal from raw wastewater.

$$Ca(OH)_2 + Ca(HCO_3)_2 = 2CaCO_3\downarrow + 2H_2O \quad (4)$$

Equation 3 represents the formation of the insoluble compound calcium hydroxyapatite. The kinetics of this reaction as well as the stoichiometry are pH-dependent. Elevated pH values, 9.5-11.5, are required to ensure good phosphorus removal by precipitation and liquid solids separation techniques. Figure 14-8 illustrates this pH effect.[19] Thus, while pH was a secondary factor in establishing the dosage requirement with iron or aluminum as the precipitant, it is the primary consideration when calcium is utilized.

In order to reach the pH range which will ensure good phosphorus removals, the alkalinity demand of the wastewater must be met, as illustrated in Eq. 4. Because the bicarbonate concentration in wastewater is much higher than the phosphorus concentration (several hundred milligrams per liter versus about 10 mg/l), most of the lime dosage required to achieve effective phosphorus removal is used to meet the alkalinity demand. Here a situation exists similar to that occurring when metal salts are used for phosphorus removal; i.e., the product of the cation alkalinity reaction serves as a flocculant of the phosphorus precipitate. Thus, the reaction of the lime with the alkalinity does not completely waste the lime.

When Eq. 4 is complete, the pH of the wastewater is in the range of 9.5-10.0. For wastewater with moderate to high levels of alkalinity (> 200 mg/l), sufficient calcium carbonate is formed under these conditions to bring down the hydroxyapatite effectively. After sedimentation, the pH is reduced to 8 by the addition of CO_2. This process is referred to as single-stage lime precipitation. If the alkalinity of the wastewater is lower than 150 mg/l, however, insufficient calcium carbonate is formed. For this situation, the pH must be raised above 11 to ensure precipitation of magnesium hydroxide according to Eq. 5:

$$Ca(OH)_2 + Mg^{2+} = Ca^{2+} + Mg(OH)_2\downarrow \quad (5)$$

Magnesium hydroxide is a gelatinous precipitate which will effectively scavenge hydroxyapatite from solution. Adding excess lime to raise the pH to high levels results in a wastewater containing excess calcium. The calcium is reduced by carbonation with CO_2. As illustrated in Eq. 6, this reduces the pH and results in a second precipitation of calcium carbonate:

$$CO_2 + Ca^{2+} + 2OH^- = CaCO_3\downarrow + H_2O \quad (6)$$

The second precipitation is carried out in a sedimentation tank separate from that in which hydroxyapatite, calcium carbonate, and magnesium hydroxide precipitation takes place. The recarbonation reaction is complete at pH 9.5–10.0, which is the same end point pH as in the single-stage process. A second dose of CO_2 is then applied to reduce the pH below 8. The final pH reduction in the two-stage and single-stage processes is utilized to prevent post-precipitation of $CaCO_3$ in other treatment units. Flow diagrams of the single-stage and two-stage lime precipitation processes are given in Fig. 14-9.

Table 14-8 gives a list of chemicals which are useful for phosphorus precipitation. Small doses of organic polymer may also be required to aid flocculation. In selecting the chemical for use at any particular site, the factors listed in Table 14-9 should be taken into account. It is clear that the cost of the chemical is only one consideration among many which bear on the

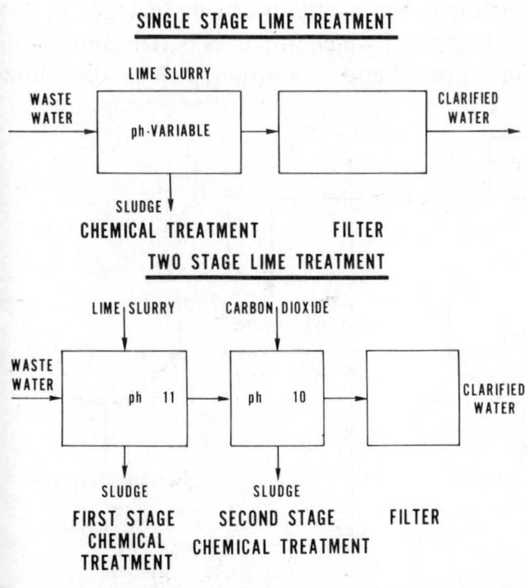

Fig. 14-9 Single-stage and two-stage lime precipitation process.

TABLE 14-8. Chemicals for Phosphorus Removal

Ferric chloride	$FeCl_3$
Ferric sulfate	$Fe_2(SO_4)_3$
Ferrous chloride	$FeCl_2$
Ferrous sulfate	$FeSO_4$
Alum	$Al_2(SO_4)_3$
Sodium aluminate	$NaAlO_2$
Steel mill pickling liquor	$FeCl_2 + FeSO_4$
Lime	$Ca(OH)_2$

ultimate cost of phosphorus removal. Of prime importance is dewatering and ultimate disposal of sludges. Here lime has an advantage because lime sludges are more easily dewatered than iron or aluminum sludges. In addition, lime sludge can be recalcined to reusable lime in an incinerator. However, large capital expenditures such as an incinerator can only be justified at large treatment plants.

The last two factors in Table 14-9 are often, unfortunately, overlooked when the chemical is selected. As will be discussed below, phosphorus removal is integrated with normal operations at conventional plants. It is thus important to make sure that the chemical will not adversely affect normal treatment. For example, the use of acid salts of aluminum or iron may result in a depression of pH below that at which biological treatment units can function. Conversely, too high a pH will occur in biological units if lime is used directly in the plant secondary section.

The use of lime in the primary treatment may result in a calcium carbonate buildup in the anaerobic digestion unit unless adequate

TABLE 14-9. Factors Affecting Choice of Chemical for Phosphorus Removal

1. Influent phosphorus level.
2. Wastewater suspended solids and alkalinity.
3. Chemical cost including transportation.
4. Reliability of chemical supply.
5. Sludge handling facilities.
6. Ultimate disposal methods.
7. Compatibility with other treatment processes in plant.
8. Potential adverse environmental effects of chemical used.

mixing is provided. This may produce digestion problems since the reduction in volume of the digester will reduce detention time. A similar situation may result when iron or aluminum is used in the primary treatment because the resulting sludges are generally of a lower solids content than normal primary sludge. The detention time in the anaerobic digestion tank will therefore be reduced unless the sludge is thickened. Occasionally, toxicity of iron, aluminum, or lime is suggested when a digester treating these sludges fails. These substances are not toxic; indeed, they are not in solution. The failure is the result of a reduction in detention time.

Cations added to sewage to precipitate phosphorus are all accompanied by an anion. This anion does not precipitate except for the hydroxyapatite or carbonate. Thus, the use of iron or aluminum salts increases the total dissolved solids (TDS) in the sewage and may adversely affect its reuse potential. Lime generally softens the sewage and thus lowers the total dissolved solids. The degree of increase of TDS can be calculated easily from the chemical dosage.

A final aspect to consider is that occasionally $Fe(OH)_3$ will slowly precipitate and may not come out of solution until the effluent discharge point. This will result in "red water" downstream of the treatment plant.[21]

In phosphorus removal, precipitation is followed by liquid-solids separation. For the most part, the usual clarification equipment employed in a treatment plant can be utilized. This results in a considerable savings in capital as well as facilitating integration of phosphorus removal with conventional treatment plant operation. In addition, it has been found that the use of chemical precipitants in conventional treatment can markedly upgrade performance of the treatment plant. This results from coagulation of organic, suspended, and colloidal solids by the chemicals added to precipitate phosphorus.

Figure 14-10 illustrates the three general sections in which phosphorus removal can be carried out in a conventional treatment plant. Chemicals can be added either:

1. Just before the primary tank with removal taking place in the primary tank.
2. In the secondary (biological) section of the plant with removal in the secondary sedimentation tank.
3. In a tertiary stage as was discussed in the section on suspended solids removal.

Table 14-10 illustrates typical results obtained with chemical addition in the primary, secondary, and tertiary stages. As can be seen, good removals are obtained in all; however, the lowest levels of remaining phosphorus are achieved by the tertiary addition. One reason for this is that a filter is usually included in the tertiary plant which produces better removal of fine precipitates. Further, when the flow

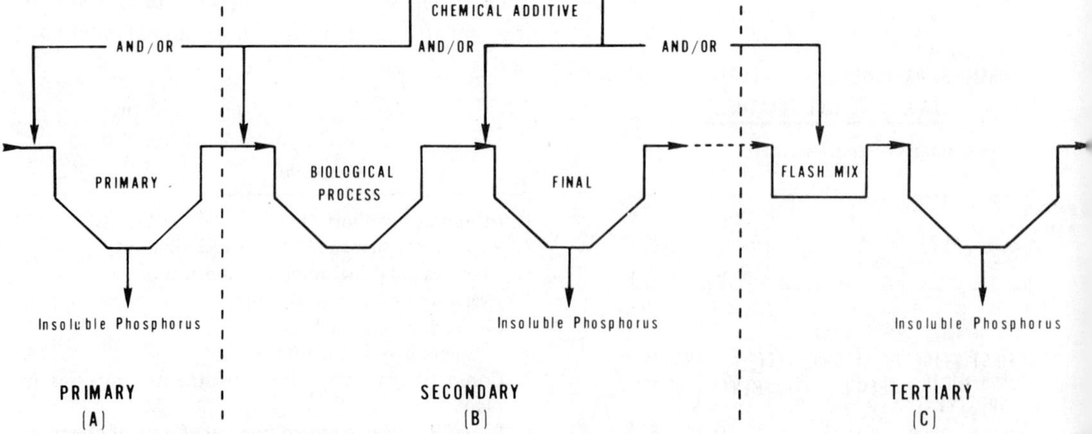

Fig. 14-10 Locations for chemical control of phosphorus.

TABLE 14-10. Phosphorus Removal Experiences

Place	Chemical	P Percent Removal	P Effluent	BOD Percent Removal	SS Percent Removal	Reference
Primary						
Grayling, Mich.	$FeCl_2$	72	4.4	78	58	20
Washington, D.C.	Lime	95	0.45	82	88	21
Lebanon, Ohio	Alum	93	0.7	–	84	22
Secondary						
Pomona, Cal.	Alum	80–93				23
Manassas, Va.	Alum		1.0			24
Richardson, Tex.	Alum		0.5–0.75			25
Tertiary						
Nassau County, N.Y.	Alum		0.08			7
Lake Tahoe, Cal.	Lime		0.14			6
Lebanon, Ohio	Lime		0.1–0.5			26

reaches this section of the plant, all the complex phosphorus forms which are more difficult to precipitate have been hydrolyzed to orthophosphate which is the easiest to precipitate.

In plants where removal in the primary stage is practiced, a major effect is the significant increase in BOD and suspended solids reduction. This may be of important assistance in meeting water quality standards for these pollutants if the treatment plant is overloaded.

In plants where the chemical is added in the secondary section, it has been observed that much more stable operation of the activated sludge is obtained than before chemical addition. The effect of the chemical is to weigh the sludge down, preventing its loss when a filamentous or dispersed growth predominates. Even in plants which have historically exhibited excellent performance, chemical addition has helped to maintain a higher concentration of activated sludge in the aeration tank. This is illustrated by the data in Table 14-11 from parallel operations for 1 year at Penn State University.[27] It has been found that it is best to add the chemical between the biological reactor and the final sedimentation tank rather than at the head end of the secondary tank.

TABLE 14-11. Comparison of Activated Sludge Performance With and Without Alum Addition

Parameter	Influent (mg/l)	EFFLUENT (mg/l) Alum	EFFLUENT (mg/l) Normal
SS	110	22	26
BOD	71	9	13
COD	172	55	68
Soluble P	6.7	0.36	6.7
Total P	10.0	1.4	7.3

Table 14-12 summarizes the advantages and disadvantages of carrying out phosphorus removal in the various sections of a treatment plant.

Split treatment—i.e., a portion of the chemical added in the primary or secondary phase and the remainder added in a tertiary phase—might be expected to be the most effective system. In this situation, conditions could be tailored in each section to achieve optimum performance. Data from a number of plants[24,25,32] have indicated that this supposition is correct. Lower phosphorus residuals at lower overall chemical dosages with less sludge production were achieved with split treatment.

At one site, alum was added ahead of a filter to precipitate phosphorus in the tertiary section of the plant. It was found that control of pH was most important if very low phosphorus residuals—about 0.5 mg/l—were to be obtained. The residual phosphorus versus pH relationship was qualitatively similar to that illustrated in Fig. 14-6, indicating that as the pH increased

TABLE 14-12. Advantages and Disadvantages of Phosphorus Removal in Various Sections of a Treatment Plant

Primary	Secondary	Tertiary
ADVANTAGES		
1. Applicable to all plants. 2. Increased BOD and suspended solids removal. 3. Lowest degree of metal leakage.	1. Lowest capital. 2. Lower chemical dosage than primary. 3. Improved stability of activated sludge. 4. Polymer not required.	1. Lowest phosphorus in effluent. 2. Most efficient metal use. 3. Lime recovery demonstrated.
DISADVANTAGES		
1. Least efficient utilization of metal. 2. Polymer required for flocculation. 3. Sludge more difficult to dewater than primary sludge.	1. Careful pH control to get phosphorus < 1 mg/l. 2. Overdose of metal may cause low-pH toxicity. 3. Cannot use lime because of excessive pH.	1. Highest capital cost. 2. Highest metal leakage.

above 6.5, a phosphorus precipitate was not formed. Measurement of total and soluble phosphorus in the effluent confirmed that the pH effect was an inhibition of phosphorus precipitation rather than an inhibition of filtration performance.[5]

The cost of adding phosphorus removal to existing treatment plants is given in Table 14-13. The lowest cost option is addition in the secondary section because in this mode capital expenditure for chemical feed equipment and extra sludge handling equipment are least. Removals below 1 mg/l phosphorus are difficult, however. The most costly option is the tertiary because of the high capital cost. Nonetheless, an excellent quality effluent is obtained which is very low in phosphorus, BOD, and suspended solids. The tertiary cost is higher here than for the upgrading system due to higher chemical dosages and additional sludge handling difficulties. Table 14-14 lists a number of plants where phosphorus removal is being conducted or is in the planning stage.

TABLE 14-13. Economics of Phosphorus Removal (cost in cents–1000 gal)

Chemical Addition Point	PLANT SIZE (mgd) 1	10	100
Primary	6–8	3–5	3–4
Secondary	6–7	3.5–4.0	3.0–3.5
Tertiary	10–25	6–12	3–7

Sludge Production from Chemical Phosphorus Removal. Chemical treatment for phosphorus removal results in the production of significantly larger quantities of sludge than with conventional treatment. The primary reason is the precipitation of inorganics as illustrated in Eqs. 1 through 6. The approximate quantity of

TABLE 14-14. Installations Practicing or Planning Phosphorus Removal

Location	Capacity (mgd)	Status
Chicago, Ill.	30	Design
Seattle, Wash.	20	Design
Pomona, Cal.	2	In operation
South Lake Tahoe, Cal.	7.5	In operation
Piscataway, Md.	5	Construction
Colorado Springs, Colo.	2	In operation
Richardson, Tex.	1	In operation
Blue Plains, Wash., D.C.	300	In operation
El Lago, Tex.	0.5	Construction
Hatfield Twp., Pa.	5.0	In operation

chemical sludge can be calculated from the stoichiometry of these equations. According to one source,[28] the quantities of sludge with iron or aluminum as the precipitant are about 35 percent higher because of the incorporation of heavy metals and anions into the ligand structure of ferric and aluminum hydroxides. Another source[29] has indicated that the chemical sludge quantity can be expressed as 4 times the aluminum dose in milligrams per liter and 2.5 times the iron dose in milligrams per liter. The sludge quantity is expressed as milligrams of sludge per liter of treated sewage.

In addition to the production of chemical sludge, a change in the quantity of organic sludge will result. Chemical addition in the primary tank substantially increases suspended solids removal. Normally, a primary tank will remove 40 to 60 percent of the suspended solids in sewage; with chemical treatment, 80 to 90 percent removal will be realized. In addition, adsorption of soluble organics of 30 to 50 percent has been reported on chemical floc, increasing the weight of the primary sludge.[30]

Addition of chemicals into the secondary tank may or may not increase the quantity of organic secondary sludge. Although capture of suspended solids in the secondary tank will be improved, these extra solids may not be removed from the system as sludge. Rather, the system may be able to run at a longer "sludge age" which will result in a higher rate of sludge destruction by endogenous respiration. In general, a modest increase in organic sludge will also result.

If the chemical is added in a tertiary phase, a small increase in organic sludge can be expected as a result of improved capture of effluent suspended solids–20 to 30 mg/l. This figure includes some adsorption of soluble organics.

The substantial increases in nonchemical sludge in the primary stage are not as detrimental as it might appear. The extra reduction in BOD obtained in the primary tank reduces the load and consequently the sludge production of the secondary tank. For example, it has been calculated that when sufficient alum is added to the primary tank of a conventional plant to achieve 90 percent removal of phosphorus, primary sludge, including chemical sludge, increased 170 percent but waste activated sludge decreased 65 percent.[28] The total organic sludge increase was only 28 percent; the total sludge increase, including chemical, was 67 percent.

Although the extra weight of sludge produced is significant, the volume, which is not proportional to the weight, is equally as important in assessing sludge disposal costs. Data on sludge volumes are much more difficult to obtain because no theoretical relationships exist such as those for assessment of sludge weight.

In general, iron and aluminum sludges do not compact or thicken as well as primary sludges, nor do they dewater as well in mechanical dewatering devices. As a rule, primary sludges contain 3 to 5 percent solids when drawn from a primary tank. Iron and aluminum sludges usually will not thicken to more then 1 to 2 percent. Lime sludges which are free of $Mg(OH)_2$, however, thicken quite well and are blown down from primary tanks at 10 to 20 percent solids. They also dewater readily. Lime sludges containing $Mg(OH)_2$ are less amenable to thickening and dewatering.

In contrast to the problems created by iron and aluminum in primary sludge, secondary sludges containing these substances are easier to handle than normal secondary sludge.[24] These sludges thicken much more readily in final clarifiers and are easier to dewater with mechanical equipment since the chemical precipitate tends to act as a filter aid.

Few data exist on the characteristics of tertiary sludges except for lime sludge. This material is an almost purely chemical sludge and is easier to handle than a mixed primary-chemical sludge.[6] Some data on alum tertiary sludges indicate great difficulty in thickening and dewatering.[31]

NITROGEN CONTROL

Nitrogen can exist in the aquatic environment in any of four forms: organic-N, ammonia-N, nitrite-N, and nitrate-N. In sewage, it is found primarily in the first two forms. In nature, biologically-mediated reactions convert organic-N to ammonia-N which in turn is biologically oxidized to nitrite-N and nitrate-N. The latter conversions are illustrated in Eqs. 7 and 8:

$$NH_4^+ + 1.5O_2 \xrightarrow{Nitrosomonas\ sp.} NO_2^- + H_2O + 2H^+ \quad (7)$$

$$NO_2^- + 0.5O_2 \xrightarrow{Nitrobacter\ sp.} NO_3^- \quad (8)$$

The major water quality objective of nitrogen control is to prevent excessive depletion of the dissolved oxygen resources of streams by the biological oxidation of ammonia-N to nitrate-N. Table 14-15 shows the significance of the nitrogen oxygen demand (NOD) of a wastewater by comparing the oxygen demand of raw wastewater to that of well-treated secondary effluent. These data illustrate that 90 percent removal of organic oxygen demand (OOD) only results in a 74 percent removal of the total biological oxygen demand (TBOD).[33] Nitrogen control is also important because ammonia-N exerts a chlorine demand which reduces disinfection efficiency, is toxic to fish and other aquatic life, and stimulates corrosion of copper plumbing.

The least expensive method of preventing these adverse actions is to carry out the biological oxidation to nitrate under controlled conditions in the treatment plant. It is possible to achieve control of ammonia-N and provide a significant degree of nitrification by altering the operation of the conventional activated sludge process. The modifications involve increases in the degree of aeration, contact time, and sludge age, as well as maintenance of the pH at 8.0–8.4.[34] The latter is quite important in view of the tendency of the nitrification reaction to lower the pH. These reactions are quite temperature-sensitive, showing significant rate decreases as the temperature drops below 18°C. As the temperature decreases, contact time and sludge age must be increased to compensate for the reduced reaction rates. Of all the factors listed above, sludge age is the most significant as it can be used to control the population of nitrifiers maintained in the system.

When the temperature is low and a significant diurnal flow variation exists, it is difficult to maintain both organic carbon oxidation and nitrification in a single reactor. In such situations, a two-stage system consisting of organic carbon oxidation and nitrification has shown a significant superiority of performance.[35] The stability of such a system is much greater because conditions in each stage can be established to favor the specific types of organisms desired in that stage. Separate sedimentation and sludge recycling is used with each stage.

On the basis of extensive experience, suspended growth nitrification systems have been developed, and the following observations can be made:[35,36]

1. The reactor flow pattern should be plug flow as the reaction is zero order.
2. pH should be kept near 8.0. If the pH is expected to be below 7.6, adjustments in the detention time and mixed liquor suspended solids should be made to compensate for the lower rate of oxidation.
3. The nitrification rate is approximately linear and doubles between 10 and 20°C.
4. At pHs between 7.6 and 8.0 and at a temperature of 20°C, nitrification rates between 0.15 to 0.3 gm ammonia-N per gram of mixed liquor volatile suspended solids (MLVSS) per day have been reported.
5. MLVSS between 1000 and 2500 mg/l

TABLE 14-15. Oxygen Demand in Raw and Treated Sewage

Component	Raw (mg/l)	Treated (mg/l)	Percent Removal
Organic matter	250	25	90
Organic oxygen demand (OOD)	375	37	90
Ammonia-N	25	20	20
Nitrogen oxygen demand (NOD)	112	90	21
Total BOD	487	127	74

can be maintained if the sedimentation tank overflow rate is kept between 400–600 gpd/ft^2 at average flow and if sludge recycling of 50 to 100 percent is provided.

6. The oxygen requirement is approximately 4.6 lb/lb ammonia-N and 1.5 lb/lb of BOD entering the reactor.
7. Little excess nitrifying sludge is produced because the net synthesis of these organisms is low.
8. Alkalinity destruction from nitrification is 7.2 lb/lb of ammonia-N oxidized. A portion or all of this destruction must be compensated for to keep the pH near 8.0.

Although the suspended growth systems have operated well, they suffer a number of disadvantages. The nitrifying sludge is quite light and tends to be lost from the system if the flow suddenly increases or if an excess of carbonaceous BOD enters the nitrification tank. The loss under the latter condition is due to selective sedimentation of the flora which use carbonaceous BOD as their energy source. In addition, the low growth rate of the nitrifying organisms restricts the system's ability to cope with sudden changes in loading, pH, and temperature. Consequently, interest in the use of fixed-film systems has developed. In a fixed-film system, the organisms stay attached to an inert medium. The maintenance of an adequate mass of organisms in the system is thus not dependent on a sedimentation tank. In addition, excess mass tends to remain on the surfaces and comes rapidly into play as the temperatures fall or the pH deviates from the optimum.

A trickling filter filled with "a proprietary plastic" packing has been used successfully.[37] Greater than 90 percent nitrification of secondary effluent was obtained at loading rates of 0.05 gpm/ft^3. Little excess sludge was produced, and it was determined that filtration of the tower effluent was economically feasible without preclarification.

In an evaluation of fixed-film nitrification in a flooded or submerged filter packed with 1 in. to 5 in. quartzite stone, removals of 90 percent were obtained at hydraulic detention times of 60 min based on flow exclusive of recycling and void volume of the filter. BOD loadings of 100 mg/l were successfully handled; this quantity is much higher than that which is usually recommended as input to nitrification stages. Oxygen addition was either external to the unit, with a recycling of treated effluent to maintain aerobic conditions, or by direct introduction of gaseous oxygen into the filter. With the external aeration, better effluents were obtained; with the internal aeration system, less difficulty with clogging was manifest. In these systems, equivalent operation was obtained at all pH values between 6.5 and 8.5.

Nitrification of sewage may not be sufficient for nitrogen control in some instances because an excess of nitrate in water will disqualify it from use as a potable water supply[39] and because nitrogen in any form can serve as an algal nutrient. Although phosphorus is usually the controlling nutrient in eutrophication, nitrogen can be the controlling nutrient under some circumstances. Areas where nitrogen is the controlling nutrient include San Francisco Bay, Lake Tahoe, and the Potomac Estuary.[40] Removal of various forms of nitrogen from wastewater can be achieved by both biological and physical-chemical means.

Biological Denitrification. Under anaerobic conditions, facultative heterotrophs will utilize the nitrate ion as a hydrogen acceptor for the degradation of organic matter. Equation 9 illustrates this reaction with methanol as the organic carbon source:

$$5CH_3OH + 6NO_3^- = 5CO_2 + 3N_2 + 7H_2 + 6OH^- \quad (9)$$

The end product of the nitrogen reduction is nitrogen gas which is essentially insoluble in water. Thus, a combination of nitrification followed by dentrification can achieve nitrogen removal from wastewater. In this combination, pH and temperature have similar significant effects on reaction rates to those of the nitrification reaction. The optimum pH is between 6.5 and 7.5. In the range 6.0 to 8.0, the rate is always 80 percent of the optimum.[35] The effect of temperature is even more pronounced than for nitrification with the rate increasing 4 to 5 times between 10 and 20°C. However, the rate of denitrification at low temperatures is generally higher than that of nitrification.[35] Denitrification is carried out by the same organisms which oxidize organic carbon, and higher rates of organism growth and greater amounts of sludge must therefore be handled.

A three-stage biological system has been developed to provide nitrogen removal.[4] The first stage is a high-rate, short aeration time (approximately 2 hr) biological reactor for organic carbon oxidation and hydrolysis of organic nitrogen to ammonia. The second stage provides approximately 3 hr of detention and achieves essentially complete nitrification. The third stage is for denitrification of nitrate to nitrogen gas. An organic source must be added to the third stage to force the denitrification reaction to take place. A variety of substances have been evaluated for this use and methanol has been found to be far superior to all others. It is relatively inexpensive, reacts rapidly, and provides only a minimum of energy for growth of new organisms. The theoretical requirement is 1.9 mg methanol/mg of nitrate-N. In practice, a dose close to 3 mg methanol/mg nitrate-N is required to compensate for synthesis and the demand exerted by dissolved oxygen remaining in the wastewater after the nitrification stage.

A diagram illustrating this system is given in Fig.14-11. Three alternate contactors are illustrated in the denitrification stage. Contactor 1 is a suspended growth contactor similar to that used in the first two stages. This type was used in the original development by Barth. Solids control is somewhat difficult in this system, and it is often followed by a deep bed filter.

Design information for suspended growth denitrification systems is summarized as follows:[35]

1. Tanks are 12–15 ft deep with stirring mechanisms of the type used for flocculation in water treatment plants. It is important to mix the tanks but not vigorously enough to dissolve much oxygen from the air, as the oxygen will cause an additional methanol demand beyond that required to reduce the nitrate.
2. MLSS in the range of 2000 to 3000 mg/l can be maintained without excessive return sludge rates. Return sludge capacity should be designed for between 50 and 100 percent of sewage flow. Final sedimentation tanks should be designed for overflow rates of 1200 gpd/ft^2 at peak flow.
3. Detention time in the denitrification tank

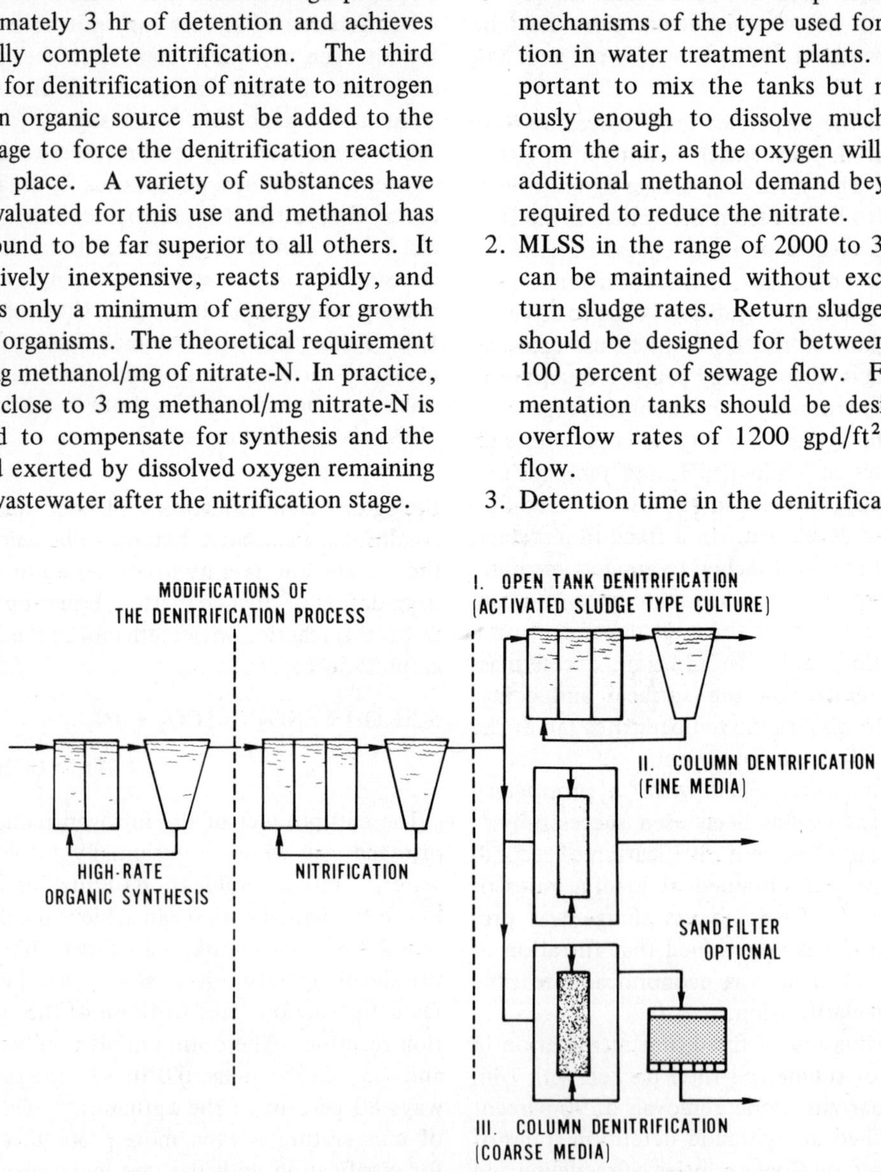

Fig. 14-11 Biological denitrification—three-stage system.

TABLE 14-16. Performance Summary for Three-Stage Biological Treatment Utilizing Mineral Addition and a Suspended Growth Reactor[a]

Constituent	Raw Wastewater (mg/l)	Primary Effluent (mg/l)	High-Rate Effluent (mg/l)	Nitrified Effluent (mg/l)	Denitrified Effluent (mg/l)	Denitrified Effluent (filtered) (mg/l)
COD	320	218	64	43	44	38
Organic-N	10.3	5.9	0.8	0.4	0.4	0.2
Ammonia-N	11.3	13.7	7.7	0.6	0.3	0.4
Nitrite-N	–	–	1.1	0.3	0.3	0.4
Nitrate-N	–	–	4.3	11.5	0.9	0.9
Total phosphorus	12.6	11.9	2.8	2.6	1.5	1.1
SS	157	90	8.8	7.1	7.1	1.7
pH (mode)	(7.2)	(7.3)	(7.5)	(7.5)	(7.7)	(7.7)

[a] All values are averages except pH and are based on 40 determinations.

should be 1 to 2 hr at average flow, followed by 15 min in an aerated contact tank to oxidize surplus methanol.

4. Sludge production will approximate 0.2 lb/lb of methanol fed.

The other two systems supply a contact medium on which a heavy biological growth can develop. These systems require less contact time than that required in the suspended growth reactor, and they can accept higher hydraulic rates of application without fear of a washout of organisms. The coarse medium (1 in.-2 in. gravel) upflow system pioneered by McCarty[42] requires a contact time of 1 hr and may require a supplemental filter. The fine medium downflow system developed by Smith, *et al.*[43] utilizes particles as small as 3 mm in beds 10–20 ft deep. Flow rate is that of a high-rate filter–7 gpm/ft^2–with a bed contact of 10–20 min. Backwash of the fine medium units is required once or twice per day to relieve clogging. In all of the denitrification systems discussed, essentially complete denitrification has been obtained in pilot-scale experiments. An upflow expanded bed of carbon granules has been successfully used for denitrification.[44] High degrees of removal were reported with a contact time of several minutes. This system entails little clogging, but the carbon media tend to float out of the column. As with the nitrification systems, the fixed-film reactors exhibit better stability under varying conditions.

Most experience has been with the three-sludge suspended growth system. Results from a 100,000 gpd pilot plant in Washington, D.C. are presented in Table 14-16. This table includes data on phosphorus removal since addition of alum during the first stage was practiced both to maintain solids control and to achieve phosphorus removal. A number of full-scale installations employing the three-contactor system with mineral addition for phosphorus removal are planned. These are given in Table 14-17. It is anticipated that these systems will reduce nitrogen to 1–2 mg/l.

Breakpoint Chlorination. Chlorine reacts with ammonia-N to yield a variety of amines and inorganic compounds by a variety of paths. Equation 10 below illustrates the preferred reaction for nitrogen removal:

$$3Cl_2 + 2NH_4^+ = N_2 + 8H^+ + 6Cl^- \quad (10)$$

The reaction requires a theoretical dose of 7.6 gm chlorine/gm ammonia-N. The reaction occurs in a stepwise fashion; various chloramines initially form which in turn rapidly react

TABLE 14-17. Planned Installations for Biological Nitrification-Denitrification Systems

Location	Capacity (mgd)
Salt Creek, Ill.	50.0
Tampa, Fla.	90.0
El Lago, Tex.	0.5
Clermont County, Ohio	2.0
Washington, D.C.	300.0

with additional chlorine to yield nitrogen gas. However, it is possible for either of the side reactions (Eqs. 11 and 12) to occur rather than Eq. 10:

$$3Cl_2 + NH_4^+ = NCl_3 + 4H^+ + 3Cl^- \quad (11)$$

$$4Cl_2 + NH_4^+ = NO_3^- + 10H^+ + 8Cl^- \quad (12)$$

Equation 11 is undesirable because it produces the odorous gas nitrogen trichloride. Equation 12 is undesirable because nitrogen is not eliminated and chlorine consumption is significantly higher than when nitrogen gas is the end product.

It has been found that pH is the controlling factor in maintaining Eq. 10 as the predominant reaction.[45] Figure 14-12 illustrates that NCl_3 and NO_3^- formation is kept to a minimum when the pH of the reaction is maintained between 6 and 7. Provision must be made to continuously monitor and adjust pH while this operation is being conducted, as the pH tends to be depressed by Eq. 10.

A system to control pH and keep base addition at minimum levels has been developed at Blue Plains.[47] Gaseous chlorine is added in two stages. In the first stage, chlorine at a dose of 5 parts chlorine/part of ammonia-N is added. The pH will generally stay in the range 6 to 7 due to the buffering action of the bicarbonate. The 5 to 1 dose is just sufficient to form monochloramine, which tends to prevent NCl_3 formation should the pH drop below 6. The CO_2 formed by the bicarbonate destruction is stripped from the solution with air; then base and the remainder of the chlorine dose are added to reach the breakpoint. Sufficient hydroxide is added to balance the acidity produced by the chlorine, maintaining the pH in the 6 to 7 range.

Figure 14-13 illustrates a typical breakpoint curve for nitrogen removal by chlorination of lime-clarified and filtered secondary effluent. The characteristic chlorine residual change with dose is observed, with the breakpoint occurring at a dose of 8 parts of chlorine to 1 part of ammonia-N. This dose is quite close to the theoretical value of 7.6 to 1. In general, a dose close to the theoretical is sufficient. Deviations are caused by the presence of suspended solids and/or organics in wastewater. For example, secondary effluent generally requires a dose of 9 to 1, and primary effluent, a dose close to 10 to 1.[46]

Data from studies at Washington, D.C.[47] and at Owosso, Michigan[48] indicate that the ammonia-N concentration can be reduced to 0.1 mg/l if adequate mixing, dosing, and pH control is achieved. Total removal of nitrogen is not achieved because the organic nitrogen in wastewater is not removed by the reaction with chlorine.

Two potential difficulties in this process are chlorine overdose and mineralization of the water. The potential adverse effect on water quality of an overdose of chlorine can be prevented by the installation of a short contact time (approximately 15 min) activated carbon column. Activated carbon converts residual

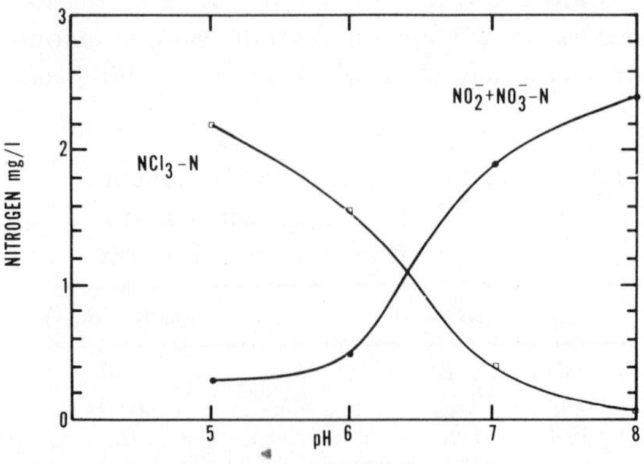

Fig. 14-12 Effect of pH on NO_2^--NO_3 and NCl_3 Formation.

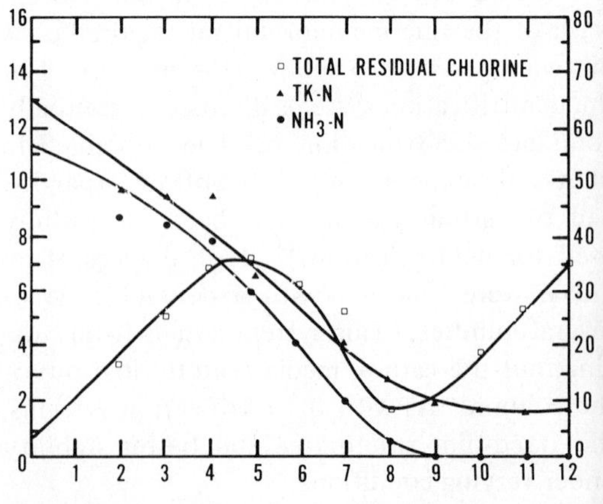

Fig. 14-13 Cl:NH_3 = N wt. Ratio chlorination o lime clarified.

chlorine to the chloride ion. Use of the full dose of 8 to 10 parts of chlorine/part of ammonia-N results in a significant buildup of chloride in the wastewater. Originally it was thought that the buildup would limit the technique to wastewaters which could accept this degree of mineralization. It was also thought that the technique could be used only as a polishing process after another ammonia-N removal technique. However, at Owosso[48] it was found that activated carbon can reduce chloramines to nitrogen gas; thus, doses of approximately 5 parts of chlorine/part of ammonia-N can be used to form chloramines which can then be converted to nitrogen gas and chloride by passing the chlorinated sewage through carbon columns. This reaction does not occur on fresh carbon; rather, the carbon must be aged by contact with chloramines for several days. A published description of a mechanism designed to explain this phenomenon[49] ascribes it to the formation of surface oxides on the carbon. Additional research and demonstration are required before design can be predicated on it; however, the results to date are encouraging.

Ammonia Removal by Air Stripping. At neutral pH, ammonia-N is present in wastewater in the soluble ionic form NH_4^+. As the pH rises, however, the ammonium ion is converted to the un-ionized form NH_3 which is a slightly soluble gas. The solubility is a function of temperature and follows Henry's law. Significant removals of ammonia-N from water are thus achievable by raising the pH of wastewater above 11 and stripping out the ammonia with air. The potential for the integration of air stripping with a lime precipitation process is obvious. Removal of ammonia by stripping is thus a function of three parameters—pH, temperature, and efficiency of contacting the air with the liquid; i.e., efficiency of mass transfer. The higher the pH up to 11.5, the better the removal will be, since the proportion of ammonia-N in the gaseous form will be greater. Similarly, higher temperatures promote better removal since they favor lower solubility of ammonia gas in water. Higher rates of mass transfer are favored by contacting systems which achieve high degrees of gas dispersion into the liquid and which maintain a high ratio of gas to liquid. In the design of a system, less than ideal conditions for one of these parameters can be compensated by keeping the others near optimum. For example, the inability to raise the pH above 11 could be at least partially offset by heating the air or water.

The initial pilot work in this area involved air stripping in a counterflow tower with a standard cooling tower packing.[50] Figure 14-14 illustrates the effect of increasing the ratio of air to liquid flow and increasing the tower depth (both techniques for increasing mass transfer). Stripping efficiency was lowered by 1.5 percent per 1°F (0.56°C) drop in wastewater temperature.

Based on these data, a 3.75 mgd crossflow stripping tower 24 ft high was installed at South Lake Tahoe. Sufficient air capacity (800 ft^3/gal sewage) was provided to obtain 90 percent removal at 0°C. Initially, performance was excellent but two problems soon became evident: freezing and scaling, both of which reduced tower efficiency.

A temperature drop can be compensated by increasing the air-to-liquid ratio; alternatively, the tower depth can be sized to achieve adequate removal under low temperature conditions. When the wet bulb air temperature falls

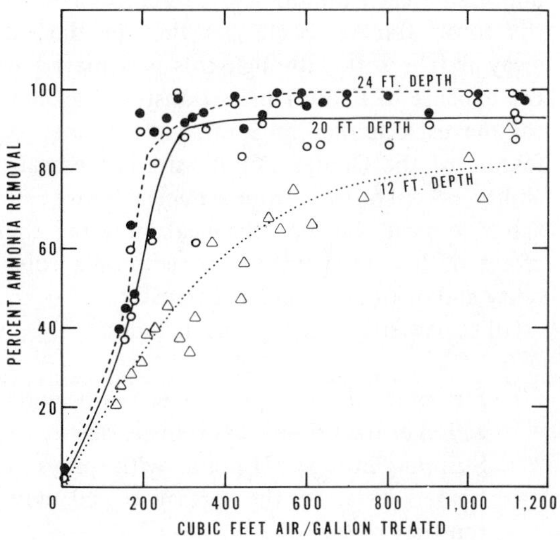

Fig. 14-14 Percent ammonia removal vs. air supply for various depths of packing.

to the freezing point, however, the liquid will start to freeze even if the dry bulb temperature is above freezing. The formation of ice will seriously disrupt mass transfer by interfering with the pattern of droplet splashing and will place an added structural load on the tower. Although this problem can be solved by heating the air, the expense is usually prohibitive.

Scaling results when the hydroxide and calcium in the high pH liquid react with CO_2 in the air and form insoluable $CaCO_3$. This coating can also put an undesirable structural load on the tower and interferes with mass transfer efficiency. If the pH is kept below 10.5, scale formation is inhibited, but this pH is marginal for efficient stripping. To retard scaling, it might be best to keep the pH down and use the stripping technique as a first-stage roughing procedure in advance of a polishing technique.

At Tahoe, a soft scale formed which could have been easily flushed from the tower, but portions of the packing were inaccessible. At Blue Plains,[51] the scale formed was extremely hard and could only be removed by mechanical methods.

Recent studies and designs have attempted to compensate the effects of freezing and scaling. At Orange County, California,[52] the packing is formed of alternating layers of PVC pipe formed in bundles 6 × 6 × 8 ft. The smooth surface tends to discourage scale formation but, more important, each bundle can be removed from the tower frame. Scale can thus be flushed away if it is soft, although this is achieved at the expense of a lower mass transfer efficiency for the packing than at South Lake Tahoe. A feature of the Orange County site is the availability of waste heat from a nuclear power reactor to heat the air; this heat prevents the effect of low temperature on ammonia solubility and of freezing on mass transfer.

Other new stripping systems include:[52,53]

1. Stripping in high pH ponds with wind action as the prime mass transfer agent.
2. Stripping in high pH ponds with sprays of treated water on the surface to aid mass transfer.
3. Aeration of high pH basins with diffused air.
4. Surface stirring of high pH ponds.
5. Use of the spray tower concept rather than packed towers.

All of these techniques have lower mass transfer efficiency than packed towers, but they can function even if scaling and freezing take place.

In general, ammonia stripping is an attractive and economical technique for removal of nitrogen from wastewater. However, until the newer techniques discussed above are proven, the process will be restricted to warmer areas of the country and to special situations where waste heat is available.

One final factor to consider in the study of ammonia stripping is the possibly adverse effect of the plume of ammonia-laden air leaving the stripper. Although the average discharge from a tower is lower than the background level of ammonia from natural causes, localized high concentrations downwind may occur. It would not be advisable to locate a tower upwind of a body of water or a populated area.

Selective Ion Exchange. Ion exchange for removal of ammonium ions from wastewater has been traditionally considered impractical because of the greater affinity of ion-exchange resins for divalent and most other monovalent ions over ammonium. While cation exchange processes removed ammonium ions, they also concurrently removed of all other cations in the wastewater. As ammonium represents at most 10 percent of the total cation equivalents in wastewater, it is clear why ion exchange was not economically feasible. More recently, a naturally-occurring zeolite, clinoptilolite, was found to have ion-exchange properties which favored the exchange of ammonium over all other cations except potassium. Fortunately, potassium is normally present in wastewater at much lower concentrations than ammonium. With this exchange material, ion exchange is economically feasible for ammonium removal. An optimization study of the process by the University of California has indicated the following:[54]

1. Optimum pH is in the range of 4 to 8.
2. Ammonia exchange capacity is a function of the cationic strength of the wastewater

up to 0.01 moles/liter, effluent ammonia concentration, and of the regenerant used.
3. The sodium resin form has twice the capacity of the calcium resin on a breakthrough rather than total capacity basis.
4. Flow up to 15 bed volumes/hr can be used.
5. NaCl is better than $CaCl_2$ as the regenerant. NaCl concentrations are optimum at 0.1 lb NaCl/gal at pH 12.5 and 0.17 lb NaCl/gal at pH 11.5.
6. Regeneration efficiency is better at pH 12.5 than 11.5 or 12.0; however, resin attrition is higher at pH 12.5.
7. Flow rate for regeneration should not exceed 20 bed volumes/hr.
8. Attrition varies from 0.25 to 0.55 percent per cycle.

A 100,000 gpd pilot facility, consisting of two columns of clinoptilolite in series with a third as a standby, was constructed by Battelle Memorial Institute and operated by EPA personnel at the Blue Plains Treatment Plant in Washington, D.C.[46] Operation was at 4 gpm/ft^2 downflow under pressure. The system was operated until ammonium ions started to break through the second column in series (24–30 hr of operation). The lead column was then replaced by a freshly regenerated column which was placed second in line, thus assuring countercurrent contact. Regeneration was achieved by circulating a solution of lime water and sodium chloride through an exhausted column until the ammonia reached approximately 500 mg/l. The ammonia was then air-stripped from the regenerant in a small packed tower. Because the volume of air required for stripping here is much less than for direct stripping in wastewater (since the ammonia concentration is 30–50 times as high), it is economically feasible to heat the stripping air.

A second technique for recovery of the regenerant and final disposal of ammonia is breakpoint chlorination with electrolytically generated chlorine. The advantage of this technique is that only electricity is consumed and nitrogen gas produced. The chloride is alternately oxidized and reduced during the renovation of the regenerant. In a study by Battelle Northwest, the chlorine was generated at 6 V, 1000 amps in cells with lead dioxide anodes. Power requirement varied from 35 to 54 watt-hours/gm ammonia-N. Scaling of the cell electrodes by $Mg(OH)_2$ and $Ca(OH)_2$ took place, requiring periodic cleaning.[55] A third technique, to be tested at Rosemont, Minnesota, is steam stripping with recovery of the ammonia.

The work at Blue Plains[46] and other localities[57,58] has indicated that ammonia-N removal to levels of 0.1 to 0.5 mg/l is technically feasible. Overall performance reported at Blue Plains was not very good because of valve malfunctions in the regeneration system which impaired regeneration efficiency. The capacity of clinoptilolite for ammonium depends on the concentration of other cations in the wastewater, the effluent ammonia desired, and the degree of regeneration obtained. On a practical basis, a capacity in the range 0.1 to 1 lb of ammonia-N/ft^3 of resin should be realized.

Table 14-18 gives the projected cost of various nitrogen control systems. Breakpoint chlorination and ion exchange are the systems which were most recently developed. Hence, cost data for these systems are less precise. In terms of cost effectiveness and treatment efficiency, it might be best to consider combinations of these processes rather than the exclusive use of one. For example, the cost of air stripping consists almost completely of capital expenditure, and the process cannot achieve high degrees of removal at low temperature. Breakpoint chlorination is very low in capital cost but is high in operating costs. The cost of breakpoint is almost directly proportional to ammonia-N levels. The advantages of a com-

TABLE 14-18. Costs of Nitrogen Control (total cost in cents/1000 gal)

	PLANT SIZE		
Process	1	10	100
Nitrification[a]	11.5	3.9	3.1
Nitrification–Denitrification[a]	21	8.5	5.5
Air Stripping[b]	7	4.0	3
Breakpoint Chlorination[b]		7–10	
Ion Exchange[b]		10–15	

[a]Additional cost to add these stages to activated sludge.
[b]Based on ammonia-N concentrations of 20 mg/l.

TABLE 14-19. Treatment Plants Planning Physical-Chemical Nitrogen Removal

Plant Location	Process	Capacity (mgd)
Orange County, Cal.	Stripping tower	15
South Lake Tahoe, Cal.	Spray tower and ponds	7.5
Rosemont, Minn.	Clinoptilolite	0.6
Occoquan, Va.	Clinoptilolite and chlorination	18
Montgomery County, Md.	Clinoptilolite and chlorination	60
Owosso, Mich.	Breakpoint chlorination	6

bination of these two processes is obvious. Table 14-19 gives information on plants planning nitrogen removal by physical-chemical techniques.

ORGANIC CARBON REMOVAL

Organic substances are removed by most of the processes discussed in the previous sections of this chapter. However, some of the organics called refractory organics cannot be removed by coagulation and sedimentation, nor are they amenable to biological oxidation. Two economically feasible methods for removing them have been developed: activated carbon adsorption and ozonation.

Recent studies have shown that these processes are just as applicable to other types of organics present in wastewater; as a result, the procedure will be employed in the future for general organic removal.

Activated Carbon Adsorption. The ability of activated carbon to remove soluble organics from wastewater is a consequence of the similarity in surface chemistry between the activated carbon and the organic molecules. There are many other substances capable of adsorbing organics; the characteristic of activated carbon which makes it unique is that it has a much higher adsorption capacity than other materials, a property which is due to the extensive internal microporous structure formed during the activation process.

A generalized activated carbon treatment process is illustrated in Fig. 14-15. The carbon and wastewater are contacted for a sufficient period of time for adsorption to take place and are then separated. Eventually the activated carbon is exhausted and is removed from the contact vessel to a regeneration step. During regeneration, some of the carbon is consumed or otherwise lost so that make-up must be added.

Commercial grades of activated carbon are granular (sizes 8 × 30 mesh or 12 × 40 mesh) or powdered (N 300 mesh). Both forms have been found essentially equivalent in their ability to remove organics, but each requires a different method of contacting and regeneration.

Granular carbon is contacted with sewage in columns through which the sewage flows. The columns may be either pressure vessels or gravity contactors. The former provide better operational flexibility, but the latter are more economical. Flow may be upward through an expanded bed, upward through a packed bed, or downward through a packed bed. Packed-bed operation provides filtration as well as adsorption but requires frequent backwash. Upflow systems allow for periodic removal of a

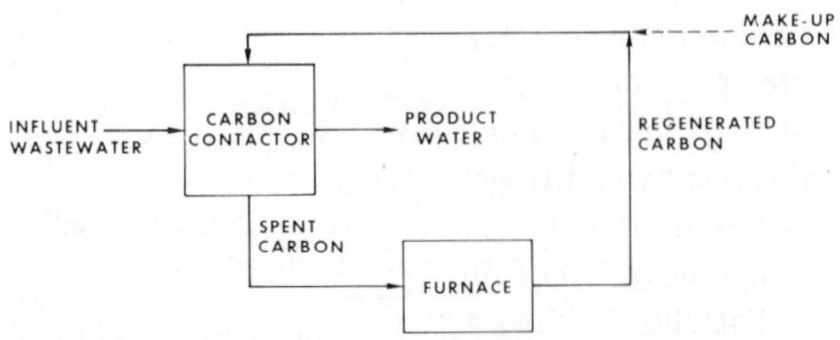

Fig. 14-15 Regeneration and reuse of activated carbon.

portion of the carbon at the base, thereby providing countercurrent contact in a single vessel. However, good flow distribution is more difficult to achieve upflow than downflow. Columns can be connected in series as with the clinoptilolite system to achieve countercurrent contact. Alternatively, a parallel flow arrangement can be used to the same end. With the parallel flow arrangement, the starting time in service is staggered so that whenever one column is near exhaustion, one or more other columns are relatively fresh, and the remainder are at some in between stage. The flow from each column is blended to produce the final effluent.

Regardless of the contacting details, the fundamental design parameter for the system is contact time. The specific contact time to be used is a function of the effluent quality desired and the wastewater characteristics. In general it has been found that as contact time increases, the rate at which organic removal takes place decreases. The point at which there is essentially no additional removal generally is located between 30 and 60 min. Contact time is always rated on an empty bed basis.

Regeneration of granular carbon is conducted in a multiple hearth furnace, a device which has had a long history of successful use for incineration of sewage sludge. Figure 14-16 illustrates a typical granular carbon regeneration scheme. Spent carbon is removed from the contactor by eduction, stored, partially dewatered, and then fed into the furnace. As the carbon moves from hearth to hearth, it is dried, baked, and

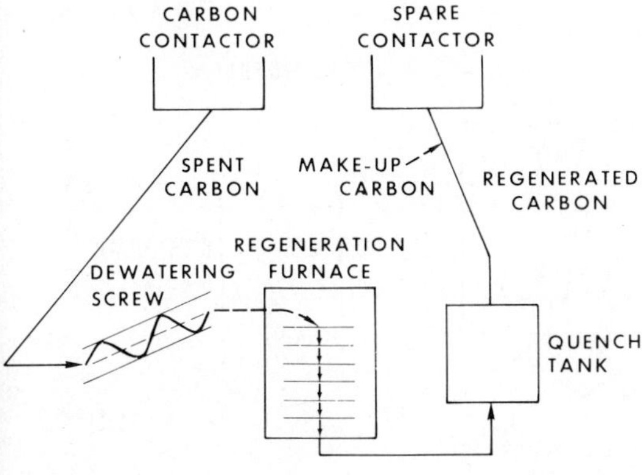

Fig. 14-16 Activated carbon regeneration system.

finally reactivated. During the reactivation stage, the temperature is in the range of 1500–1700°F (815 to 925°C). After regeneration, the carbon is quenched and returned to a contactor. The furnace is rated at 80 to 110 lb carbon/ft^2/day. Air pollution control equipment, such as an afterburner and wet scrubber or cyclone, is usually required to purify the exhaust. Experience with this regeneration system at essentially full scale indicates a carbon regeneration loss of 5 to 10 percent per regeneration cycle. The losses are due to physical attrition in the regeneration loop and burnoff in the furnace.[59]

In systems utilizing powdered carbon, contact is brought about by mixing the carbon and the wastewater in a reactor–clarifier contactor. The apparatus is illustrated in Fig. 14-17. The carbon and wastewater are flash-mixed and flocculated in the center well of the apparatus. The mixture then flows into the outer section where gravity separation of the carbon and wastewater take place. A pool of carbon in a thick slurry form is maintained at the bottom of this contactor.

In this type of contactor, the carbon comes into equilibrium with the organic concentration in the effluent wastewater, whereas in a granular contactor it comes into equilibrium with the organic concentration in the incoming liquid. The organic adsorption capacity of activated carbon is proportional to the concentration of organics with which it comes into equilibrium, making it more necessary to achieve countercurrent contact in powdered carbon systems than in granular systems. Figure 14-18 illustrates schematically a two-stage counterflow contacting system which is utilized with powdered carbon treatment. The carbon is thickened (not shown) between stages to reduce pumping, and the effluent is filtered to remove residual carbon fines.

Figure 14-19 illustrates the most advanced system for powdered carbon regeneration. The slurry containing spent carbon is thickened, dewatered in a centrifuge or on a vacuum filter, and injected into a fluidized bed furnace. The regenerated carbon is reslurried and pumped back into the system. In the furnace, the temperature is maintained at the same level as in

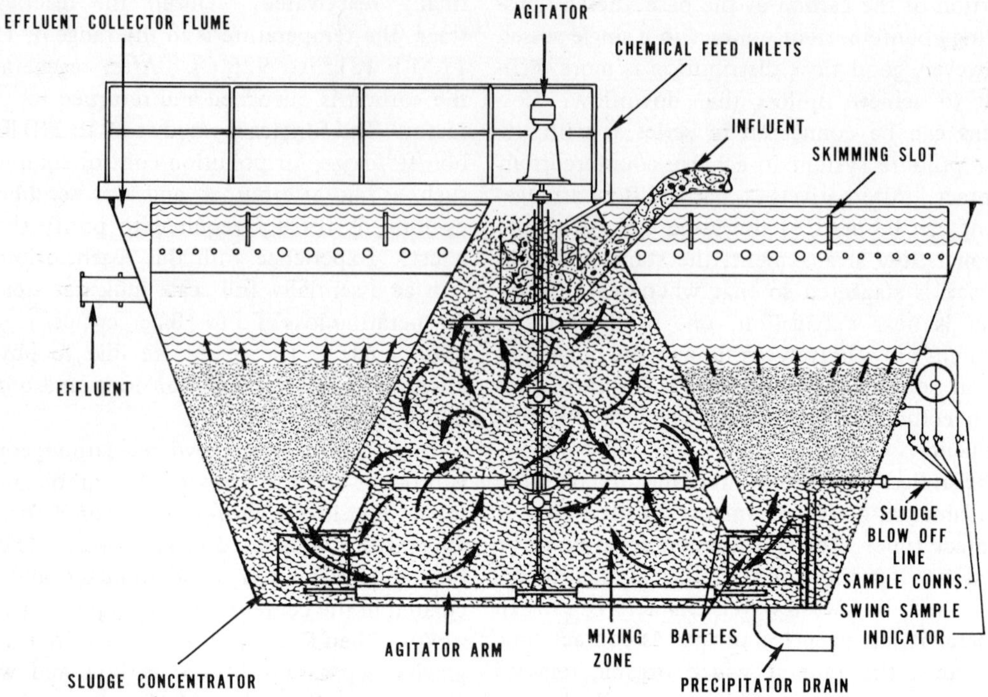

Fig. 14-17 Solids contact clarifier with sludge blanket filtration.

the multiple hearth furnace. Contact time of a carbon particle in the furnace averages several seconds. Shorter time results in only partial regeneration; longer time results in particle incineration. Thus, tight control must be exercised with this regeneration scheme. To date, small pilot studies have indicated its feasibility with losses averaging 15 percent.[60] A larger version of this furnace is now being evaluated at Salt Lake City.[61] Initial difficulty was experienced in keeping losses down because the temperature in the furnace could not be kept in the proper range. An off-gas recycling system was installed, achieving successful temperature control, although the furnace capacity was reduced by more than 50 percent. Successful regeneration of sugar carbon on a large scale has been demonstrated in a transport reactor system.[63]

Both granular carbon and powdered carbon contact are excellent processes for removing

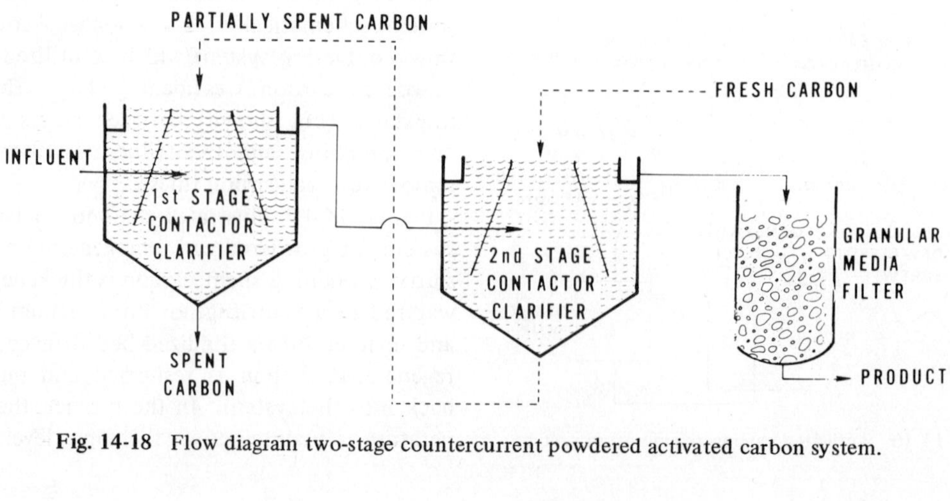

Fig. 14-18 Flow diagram two-stage countercurrent powdered activated carbon system.

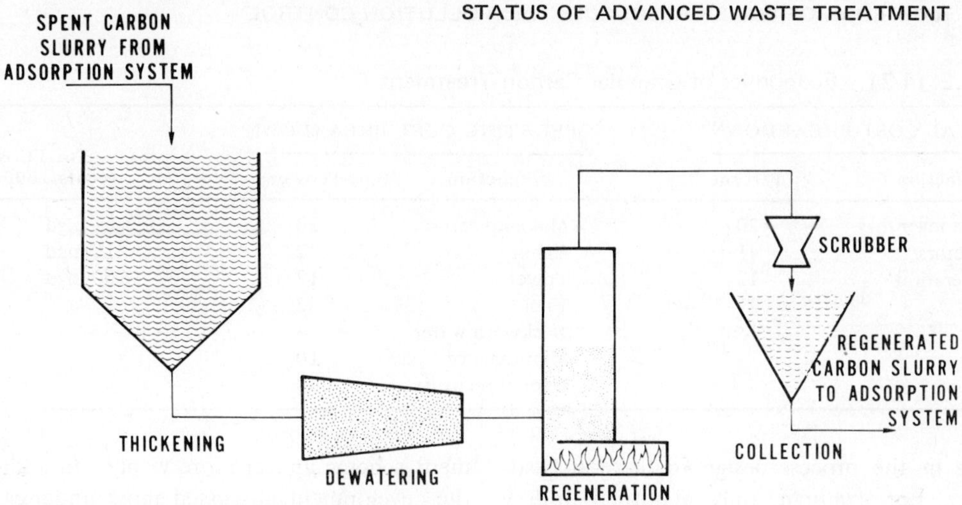

Fig. 14-19 Thermal regeneration powdered carbon flow diagram.

organics from wastewater. The advantages of powdered carbon units are: inexpensiveness ($0.10/lb versus $0.30/lb for granular), and powdered carbon plants require a much smaller carbon inventory and can be dosed to meet the demand. Granular carbon, however, exhibits higher practical adsorption capacity, does not require complex dewatering procedures prior to regeneration, and most important, has been successfully regenerated at full scale in sewage application. Economics demand that regeneration be incorporated into all but the smallest scale carbon adsorption plants. An engineer designing a treatment plant using carbon adsorption would therefore most likely choose granular carbon.

Extensive processing experience has been acquired using activated carbon for treatment of wastewater after biological treatment (tertiary application). Data from Pomona,[64] Lebanon,[65] Colorado Springs,[66] South Lake Tahoe,[6] and Tucson,[67] are summarized in Table 14-20. It can be seen that organic removal in the range of 70 to 80 percent was achieved, which is typical of the capability of activated carbon. Data on adsorption capacity indicates a considerable spread, 0.25 to 0.87 lb COD/lb activated carbon.[59] This wide range occurs because carbon capacity is a function of desired effluent quality and influent organic concentration. Most noteworthy, however, is that in all cases the capacity was significantly higher than that projected by an adsorption isotherm test. This phenomenon was recently confirmed at a number of plants in Pennsylvania.[68] It is attributed to biological activity on the carbon which, it is theorized, accomplishes partial regeneration *in situ*.[69] Naturally, it has had a favorable impact on the economics of carbon adsorption and illustrates that regeneration and make-up carbon account for about 25 percent of the process cost. The economics of granular carbon treatment are presented in Table 14-21.

The biological action has necessitated adjust-

TABLE 14-20. Activated Carbon Treatment of Wastewater—Tertiary Application

Location	Carbon	ORGANICS (mg/l) In	Out	Percent Removal	Organics Measurement
Pomona, Cal.	Granular	43	10	77	COD
Colorado Springs, Colo.	Granular	43	13	70	TOC
South Lake Tahoe, Cal.	Granular	12	3	75	TOC
Lebanon, Ohio	Powdered	20	4	80	TOC
Tucson, Ariz.	Powdered	27	7.5	72.5	COD

TABLE 14-21. Economics of Granular Carbon Treatment

CAPITAL COST BREAKDOWN		OPERATING COST BREAKDOWN			
Function	Percent	Function	Percent	Total Cost (cents/1000 gal)	
Carbon inventory	20	Make-up carbon	20	1 mgd	30
Contactors	41	Labor	12	10 mgd	10
Regeneration	12	Power	12	100 mgd	4
Pumps	27	Fuel	2		
		Backwash water	4		
		Maintenance	10		
		Capital recovery	40		

ments in the process design for carbon treatment. For example, only a normal upflow backwash was provided in early granular carbon contactor design. This allowance was found to be insufficient to ensure cleaning of the carbon column, so that air scour and surface wash are frequently added. In addition, more frequent backwash (at least once per day) must be planned.

Under heavy loads, carbon columns become septic, producing hydrogen sulfide. Aeration of the feed or extrachlorination to oxidize the effluent sulfide may be needed. Expanded bed operation under heavy loads becomes superior to packed beds because clogging is less of a problem and aeration is easier to accomplish.[70] Successful control of sulfides has been achieved at Pomona by addition of nitrate as an oxygen source.[71]

Powdered carbon contact design has also been affected by the realization that biological action may occur. It has been shown that, when the extensive biological action is encouraged, only one carbon contacting stage rather than two is required to achieve high adsorption capacity.[61] Control of sulfides can be achieved by keeping the carbon retention time below 3 days.[61]

Physical-Chemical Treatment. Physical-chemical treatment of raw wastewater—or, more accurately, chemical clarification-carbon adsorption—was first evaluated as a treatment technique in the 1940s and abandoned for being more costly than conventional processes, although it produced a good quality effluent. However, the greater stringency of subsequent water quality standards have made this treatment scheme appear more viable. In addition, the developments discussed above under carbon adsorption techniques, particularly the technology of regeneration, have tended to lower the cost of physical-chemical treatment and make it a practical alternative. Finally, removals achieved by chemical clarification have been greater than anticipated, and this result has also improved the economics of physical-chemical treatment.

Figure 14-20 is a general diagram of the clarification-carbon treatment system. Following the standard type of preliminary treatment, the waste is dosed with chemicals sufficient to achieve the desired level of suspended solids and/or organics and/or phosphorus removal. The same chemicals are used as for phosphorus removal in the primary stage. The significant removals or organics, suspended solids, and phosphorus obtained by chemical clarification are illustrated in Table 14-22.[72] Due to coagulation and possible chemical adsorption of colloidal and soluble organics, the organic removal is higher than would be expected from complete removal of suspended solids. Up to 50 percent of the "soluble" organics in sewage may be removed by chemical coagulation.[73]

After clarification is complete, the wastewater is passed to the carbon adsorption step. Table 14-23 gives data from a number of plants which attest to the very good quality effluents achieved as well as the high carbon adsorption capacity.[47,61,72] As indicated previously, this is the result of biological action on the carbon. In physical-chemical treatment systems, the carbon is subjected to a heavier load than in tertiary plants. Thus, extra care in design must be taken to ensure that the disadvantageous fea-

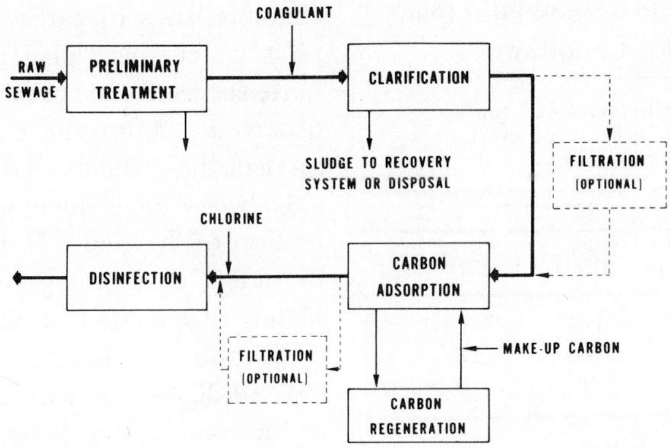

Fig. 14-20 Flow diagram of a physical-chemical treatment system.

tures of biological action on carbon can be mitigated.

Table 14-23 indicates that at Pomona[74] the capacity of the carbon was significantly higher than at any other treatment plant. Indeed, over a 20-month period, the carbon was not regenerated but the effluent was always approximately 20 mg/l (90 percent removal overall, 75 percent removal in the carbon). The major reason for this result was the stimulation of biological activity on the carbon. Sodium nitrate was added to the chemically clarified sewage to prevent sulfide formation. A level of 5 mg/l as nitrogen was sufficient. The nitrate acted as an excellent hydrogen acceptor (oxygen source) and was reduced to nitrogen gas. The biological

TABLE 14-22. Achievements of Chemical Clarification

Plant	Chemical	Organic Removal (percent)	SS Removal (percent)	P Removal (percent)
Ewing-Lawrence	170 mg/l $FeCl_3$	80	95	90
New Rochelle (ZM)	Lime pH 11.5	80	98	98
Westgate, Va.	125 mg/l $FeCl_3$	70	–	–
Salt Lake City	80–100 mg/l $FeCl_3$	75	–	80
Blue Plains	Lime pH 11.5	80	90	95

TABLE 14-23. Activated Carbon Performance in Physical-Chemical Treatment Plants

Plant	Effluent Organics (mg/l)	CARBON CAPACITY lbs TOC/ lb A.C.	CARBON CAPACITY lbs COD/ lb A.C.
Blue Plains (G)	TOC = 6	0.15	0.41
Lebanon (G)	TOC = 6	0.22	0.50
Ewing-Lawrence (G)	TOC = 3–5	0.3	–
New Rochelle (G)	COD = 8	–	0.6
Owosso (G)	COD = 25	–	0.65
Salt Lake City (P)	COD = 22	–	0.36
Pomona (G)	COD = 20	–	3.54

(G)–granular carbon.
(P)–powdered carbon.

TABLE 14-24. Powdered Carbon Pilot Plant Operating Conditions

Flow rate 50 gpm, Chemical 425 mg/l, Lime to pH 10.8, Carbon 150 mg/l + 0.4 mg/l polymer

	RESULTS			
	COD	BOD	SS	P
Raw sewage	222	144	200	7.3
Clarified effluent	65	47	28	1.4
Final effluent	35	13	7	0.4

activity on the carbon was sufficient to maintain the adsorption capacity of the carbon at a high level. The carbon was regenerated in September, 1973 to ascertain the post-regeneration characteristics of carbon from an independent physical-chemical plant. The regeneration was quite successful in that only 2.5 percent loss was observed. After the carbon was put back in service, the effluent COD was 8 mg/l, considerably below the effluent COD level before regeneration of 20 mg/l. This indicates, as would be expected, that biological activity cannot completely regenerate carbon.

Filtration is presented as an optional step in Fig. 14-20, but it would be wise to include it in the treatment scheme. Filtration acts as the safety factor in the solids removal step, and its use can therefore bring a high degree of reliability to the treatment process. In addition, it can be used to prevent an overload of solids on the carbon which could foster excessive biological

TABLE 14-25. Rocky River Waste Treatment Plant Clarification-Carbon Process

	Raw	Polymer Clarification	CARBON CONTACT TIME (minutes) 14	23.4	32.6	Percent Removed
Suspended solids (mg/l)	107	65	13	15	7	93.3
BOD	118	57	21	11	8	93.3
COD	235	177	67	50	44	81.3

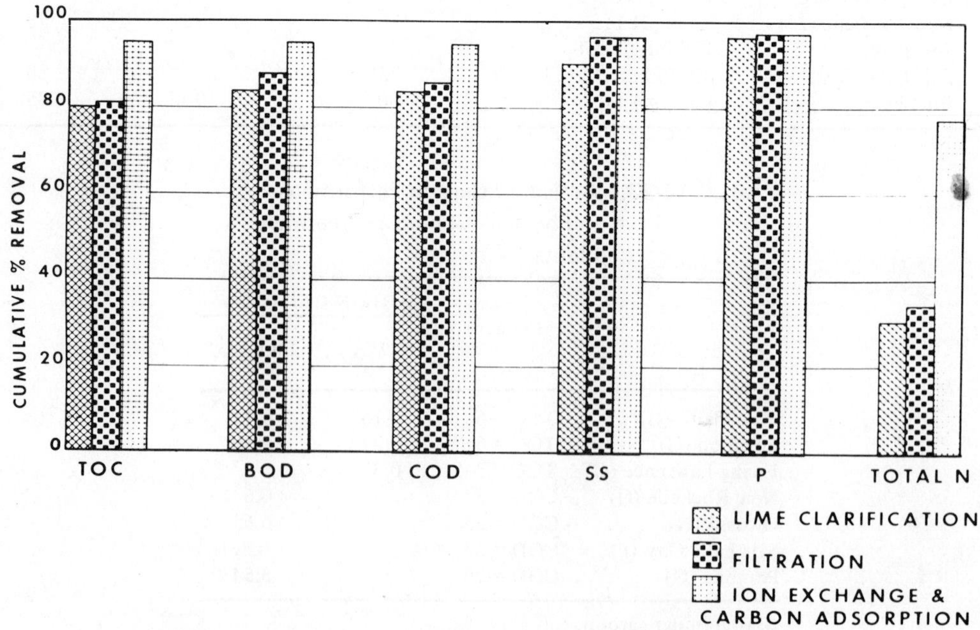

Fig. 14-21 Removal efficiencies of physical treatment–Blue Plains pilot plant.

TABLE 14-26. Preliminary Cost Estimate—Chemical Treatment (total amortization + O and M, cents/1000 gal)

Plant Size (mgd)	5	10	100
Chemical clarification[a]	9.5–13.5	7.3–9.6	4.0–5.3
Carbon adsorption	11.5–18.0	9.1–13.5	4.5–7.8
Filtration	2.9–4.5	2.1–3.3	1.0–1.4
Total	23.9–36.0	18.5–26.4	9.5–14.5

[a]Two-stage lime; recalcination of sludge.

action. The positioning of the filter prior to or after the carbon contact is dictated by the method of carbon contact. Expanded bed and powdered carbon systems usually require filtration after the carbon step. Packed bed systems will benefit from filtration before the carbon. One plant has been designed with filtration before and after carbon contact.[56] Several plants have been designed without a filtration step.[68]

Tables 14-24 and 14-25 and Fig. 14-21[72] provide information on the performance of three physical-chemical treatment pilot plants. These data illustrate removals at various stages of treatment for a variety of pollutants. The Blue Plains data include nitrogen removal by clinoptilolite.

Cost estimates for physical-chemical treatment are presented in Table 14-26. These should be compared to the cost of conventional treatment with phosphorus removal added; i.e., the data in Tables 14-2 and 14-13, since the treatment performance of both systems is similar.

Table 14-27 compares the features of physical-chemical treatment with conventional treatment systems. Because of the advantages, including comparably favorable costs of physical-chemical treatment, a number of communities are planning to establish this type of system; they are listed in Table 14-28.

TABLE 14-27. Advantages of Physical–Chemical Treatment vs. Conventional Primary Plus Secondary Treatment

1. Less area requirement—$\frac{1}{2}$ to $\frac{1}{4}$.
2. Lower sensitivity to diurnal variation.
3. Not affected by toxic substances.
4. Potential for significant heavy metal removal.
5. Superior removal of P compounds.
6. Greater flexibility in design and operation.
7. Superior organic removal.

TABLE 14-28. Physical–Chemical Treatment Plants

Site	Capacity (mgd)	Status
Rocky River, Ohio	10	Under construction
Cleveland, Ohio	10	Bids taken
Fitchburg, Mass.	15	Under construction
Cortland, N.Y.	10	Planned
Niagara Falls, N.Y.	60	Planned
Garland, Tex.	30	Design
Owosso, Mich.	6	Bids taken

Ozone Oxidation. In addition to carbon adsorption, another physical-chemical method of organics removal, ozonation, has been evaluated and found worthy of development. This procedure, however, is several years behind activated carbon in development for lack of necessary information.

Ozone is a powerful oxidant which is generated by passing a properly conditioned air stream between electrodes, which produce a corona discharge. It must be generated on site and used immediately. Ozone is poorly soluble in water so that special contacting techniques must be used to ensure optimum utilization.

At Blue Plains a laboratory-scale study of ozone performance in treating raw sewage and effluents from various treatment stages was conducted.[75] It was found that:

1. Ozone treatment of raw sewage required impractical ozone dosages. Fifty percent COD reduction required 200 mg/l.
2. Doses of 100 mg/l achieved 70 percent COD removal from lime clarified and filtered secondary effluents, lime clarified and filtered raw sewage, and independent physical-chemical effluents. All of these had a COD of 50 mg/l prior to being treated with ozone.
3. Significant additional removal required doses of 200–300 mg/l.
4. For COD reduction up to 70 percent, ozone oxidation efficiency based on 1

TABLE 14-29. Disinfection with Ozone[a]

Feed Type	Ozone Dose	Biological Parameter	In	Out
Filtered activated sludge	5	Total count	2.8×10^6/100 ml	37×10^2/100 ml
Filtered activated sludge	5	Coliform	7.2×10^3/100 ml	5/100 ml
Coagulated filtered secondary	5	Total count	2.6×10^6/100 ml	80×10^2/100 ml
Coagulated filtered secondary	5	f-2 phage	1700 pfu	13 pfu
Secondary	4.8	Fecal coliform	5000/100 ml	1.3/100 ml
Secondary	5.4	Fecal coliform	5000/100 ml	0

[a]These data indicate adequate coliform removals at ozone doses of 5 mg/l.

molecule of active oxygen per molecule of ozone was somewhat greater than 100 percent for secondary effluents and 50 percent for carbon column effluents.

5. Ozone oxidation efficiency for sewage containing ammonia was lower than for ammonia-free wastewater.
6. Ammonia is oxidized to nitrate by ozone with increasing efficiency as the pH increases up to pH 11.
7. The order of ease of oxidation is approximately as follows: readily oxidizable organics (70 percent COD), ammonia, organic nitrogen, and ozone refractory organics. The characteristics of the latter are not well defined.
8. TOC removal was less than COD reduction, indicating that the carbon was oxidized but not removed from solution.

Based on the preceding study, and investigations by others,[76] a 50,000 gpd, six-contacting-stage ozonation pilot plant was installed at Blue Plains.[77] The results were virtually the same as those reported above from the laboratory study. Again, ozone consumed averaged 3 lb/lb of COD destroyed. Total power required was approximately 6 kilowatt-hours/lb of ozone. Only 40 to 60 percent of this power was required for direct generation of ozone, and the remainder was used to dry and pump the gas and to pump sewage.

It is obvious that the cost of ozone treatment is close to a direct function of the dose required. Thus, it is not economical to use high doses of ozone. Although efficiency of ozone use is lower (6 lb ozone/1 lb COD destroyed) for well-treated effluents, economy favors the use of ozone for this purpose because of the lower dose required.

At present, ozone is most widely used for low-dose applications such as disinfection and color removal. Color removal from activated sludge effluents at doses from 5 to 15 mg/l has averaged between 50 and 75 percent. Table 14-29 gives data on disinfection results[78] which indicate adequate coliform removals at ozone doses of 5 mg/l.

The capital cost of a 10 mgd disinfection plant (5 mg/l dose) has been estimated at $515,000.[79] Direct operating costs were about $0.01/1000 gal. Costs of ozonation can be reduced by using a "bone dry" supply of pure oxygen rather than air as the raw material for ozone manufacture.

Pure Oxygen Activated Sludge. The last improvement in procedures for removing organics from wastewater which will be discussed is the application of pure oxygen in the activated sludge process. The use of pure oxygen does not permit activated sludge to go beyond its inherent capability, but allows a closer approach to this limit within an acceptable economic framework.

Early work in the biological treatment field pointed out the advantages of pure oxygen over air as the source of oxygen in activated sludge.[80] However, because the cost of pure oxygen was high and a system of efficient utilization had not yet been found, the concept lay dormant. Over the last decade, the cost of pure oxygen has decreased and systems have been devised which promise high utilization efficiency. Figure 14-22 illustrates the UNOX* system developed by Union Carbide. The key features of this system are: the use of covered reactors to

*Mention of a manufacturer does not constitute endorsement by the Environmental Protection Agency.

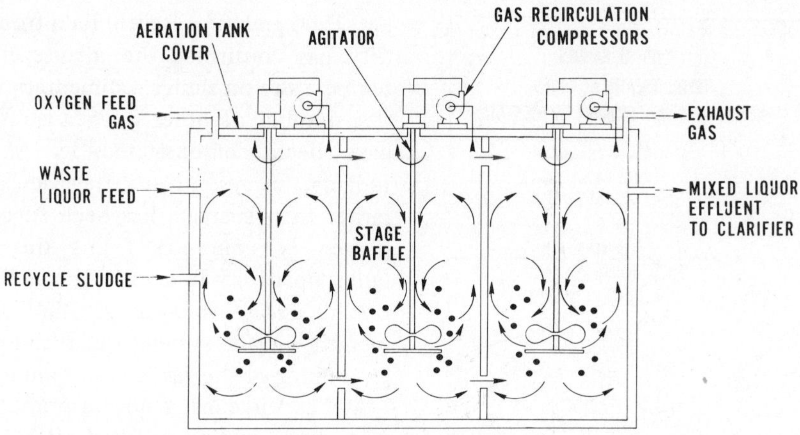

Fig. 14-22 Schematic diagram of multi-stage oxygenation system.

prevent loss of oxygen, a baffled reactor to prevent short circuiting, and recompression and recycling of gas in each chamber. In all other respects, the system is virtually identical to conventional activated sludge treatment.

The performance of this system was evaluated in parallel with that of a conventional activated sludge plant in Batavia, N.Y.[80] The study was conducted in three phases. In Phase I, one-half of the activated sludge tankage of the plant was run with pure oxygen, the other half with air. In Phase II, all flow was routed through the half with pure oxygen feed. In Phase III, one-fourth of the plant was on pure oxygen, one-fourth was on air, and the other half of the plant was out-of-service. Pertinent data from this study are presented in Table 14-30.

These data confirm the superiority of the pure oxygen system over conventional aeration. Under much heavier volumetric loading rates, the oxygen system produced a better effluent than the air system. The key to the improved performance is the development of a sludge which can settle quite well even at high concentration. The pure oxygen system developed higher sludge concentrations in the aeration tank; consequently, the actual load per organism (F/M) was lower than in the air system. In effect, the pure oxygen system could operate at a greater sludge age than the conventional system. According to theory, the greater the sludge age the better the BOD removal in a biological waste treatment system, and the lower the net sludge production.

The economic advantage of a pure oxygen system is attributable to the higher volumetric loadings allowable and the lower excess sludge quantities produced, both in mass and volume.

TABLE 14-30. Batavia, N.Y. Pure Oxygen Activated Sludge Study

	PHASE I		PHASE II	PHASE III	
	Air	Pure O_2	Pure O_2	Air	Pure O_2
Sewage flow (mgd)	1.97	1.91	2.53	1.29	1.40
Aeration time (hrs)	3.5	3.4	1.2	2.6	2.0
BOD, in	159	159	220	262	262
BOD, out	16	11	23	30	14
MLSS	2440	3060	6980	3640	6190
SVI	76	64	36	63	49
Volumetric loading (lbs BOD/1000 ft^3)	60	58	212	129	145
F/M (lbs BOD/lbs MLSS)	0.57	0.41	0.79	0.84	0.55
Excess sludge (lbs VSS/lbs BOD)	0.87	0.48	0.41	0.99	0.13
O_2 Utilization (percent)	–	95.5	92.7	–	91.4

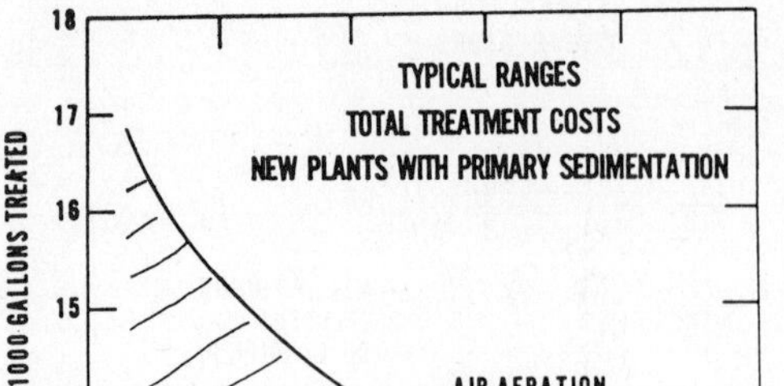

Fig. 14-23 Typical ranges—total treatment costs new plants with primary sedimentation.

As Fig. 14-23 indicates, these are sufficient to more than offset the extra cost of supplying pure oxygen rather than air.

Additional studies of this system have been carried out at Blue Plains.[81] The results of this study are illustrated in Table 14-31. These data support most of the observations from Batavia—i.e., low sludge yield, high sludge age, and excellent BOD removal with a short aeration period (1.95 hr). During the winter, however, sludge sedimentation deteriorated seriously. This problem can be overcome by proper design of the final sedimentation basin. The final clarifier overflow rate should be kept below 900 gpd/ft^2, rather than as originally projected; i.e., 1500 gpd/ft^2. Recent data from New York City has confirmed the adverse effect of low temperature on sludge sedimentation.

TABLE 14-31. Summary of Removal Efficiency Results with Oxygen at Washington, D.C.

Parameter	Primary	Secondary Effluent[a]
BOD (mg/l)	87–132	11–32 (2–3 soluble)
COD (mg/l)	244–284	45–99
TOC (mg/l)	65–106	14–26
SS (mg/l)	–	24–58

[a]1.95 hr detention.

A number of other oxygen aeration systems have been demonstrated.[82] At Speedway, Indiana, a system utilizing surface aerators rather than spargers has been successfully employed. Economics favors this system for tanks up to 15 ft deep. The tanks are partitioned and covered as in the UNOX* submerged turbine system, but recompression and repumping of the gas are not required.

At Las Virgenes, a system primarily aimed at upgrading existing diffused air aeration plants is being demonstrated. Called the "Simplex"* system, it uses conventional air blowers and coarse bubble diffusers with an inflated dome cover to contain the oxygen-rich atmosphere. Loadings in this type of system are lower than in the UNOX* type described above. Extensive nitrification was obtained (70 to 97 percent) in contrast to more highly loaded pure oxygen systems.

An uncovered tank system developed by FMC* Corporation is now being evaluated at Denver, Colorado. Fine bubble diffusers are used in tanks deep enough so that virtually all the oxygen is dissolved in the wastewater during the time of travel to the surface. It has been found that a scum forms on the surface of the tanks which aids in preventing bubble escape.

Table 14-32 presents some data from these plants as well as from the Newtown Creek plant in New York City. At the latter, one 20 mgd module of the 160 mgd modified aeration plant (1.5 hr aerator detention time) was converted to pure oxygen. The modified aeration plant only reduces BOD by 65 to 70 percent.[82]

Table 14-33 provides a list of large-scale treatment plants in which some form of pure oxygen aeration will be or has been installed.

DISSOLVED INORGANICS REMOVAL

An increment of dissolved mineral matter is added to water during domestic or industrial use. The magnitude varies from 100 to 500 mg/l per use depending generally on local con

TABLE 14-32. Performance of Pure Oxygen Plants

Plant	Loading (lbs BOD/ lbs MLVSS)	Sludge (lbs VSS/ lbs BOD)	BOD Removal (percent)	SS Removal (percent)	Loading (lbs BOD/ 1000 ft^3)
Newton Creek	0.65–2.44	0.6–0.94	87–94	84–92	110–379
Las Virgenes	0.07–0.46	0.19–0.27	89–97	86–93	13–63
Denver	0.18	0.33	92	91	55

TABLE 14-33. Future Pure Oxygen Activated Sludge Plants

Location	Size (mgd)	Aeration System
Detroit, Mich.	300	Submerged turbine
New York, N.Y.	20	Submerged turbine
Danville, Va.	24	Surface aerator
New Orleans, La.	122	Surface aerator
Speedway, Ind.	12	Surface aerator
Wyandotte, Mich.	100	Submerged turbine

ditions. Extensive reuse of wastewater even for industrial purposes requires at least partial demineralization. A number of demineralization processes have been evaluated in the EPA advanced waste treatment programs, and only three—electrodialysis, ion exchange, and reverse osmosis—were promising enough for additional studies.

Ion Exchange. Ion exchange is the most advanced demineralization system because the basic technology has been in wide use for over 25 years in water treatment applications. Cognizance must be taken, however, of differences between water and wastewater applications. In waste treatment, the wastewater should be well clarified and treated with carbon before ion exchange. Otherwise, extensive resin fouling will occur. Even with thorough pretreatment, use of strong base anion exchange resins is not recommended.

The resin cycles which can be successfully used in waste treatment applications include:

1. Strong acid cation which exchanges all cations for hydrogen and thus produces a strongly acidic effluent:

$$R{-}H + NaCl = R{-}Na + HCl \quad (13)$$

Regeneration is accomplished by putting a strong acid such as HCl or H_2SO_4 on the resin, thus reversing the reaction.

2. Weak acid cation which exchanges only cations associated with alkalinity:

$$NaCl + NaHCO_3 + R{-}COOH = NaCl + H_2O + CO_2 + R{-}COONa \quad (14)$$

Regeneration is the same as for strong acid exchangers but lower concentrations of acid are required.

3. Weak base anion (free form) which adsorbs strong acids and leaves salts unaffected.

$$HCl + R{-}NH_2 = (R{-}NH_2)HCl \quad (15)$$

Regeneration is with ammonium hydroxide.

4. Weak base anion (bicarbonate form) which exchanges anions for bicarbonate:

$$(R{-}NH)HCO_3 + NaCl = (R{-}NH)Cl + NaHCO_3 \quad (16)$$

$$(R{-}NH)HCO_3 + HCl = (R{-}NH)Cl + H_2O + CO_2 \quad (17)$$

Regeneration occurs in two stages—first with ammonium hydroxide and then with carbon dioxide. This cycle is only feasible where the TDS of the wastewater is greater than 1000 mg/l.

Demineralization can be obtained by various combinations of the above cycles. For example:

1. Strong acid cation followed by weak base anion for high-level demineralization.
2. Weak acid cation followed by weak base anion for partial demineralization.
3. Weak acid cation followed by CO_2 stripping for partial demineralization.
4. Split treatment of high alkalinity water with a portion passed through strong acid cation and the CO_2 stripped from the

blended effluent for partial demineralization.

5. Weak base anion (bicarbonate form) followed by weak acid cation for high-level demineralization.
6. Weak acid cation followed by strong acid cation followed by weak base anion for high-level demineralization with more efficient use of acid regenerant. The same batch of acid can regenerate both acid resins.
7. Two-stage ion exchange: cation followed by anion followed by a second cation and second anion stage. This system will produce a lower TDS than the single-stage cation–anion resin system at the same level of regeneration.

Choice of the system will ultimately be determined by economics. The economic decision is, in turn, influenced by several factors. A primary one is the degree of demineralization required. If only partial demineralization is needed, systems 2, 3, or 4 above may be used instead of one of the more complete TDS removal systems. Of course, split treatment with the complete TDS removal systems can be used followed by blending with the untreated feed to reach the desired product. Other factors to consider are the ratios of hardness to total cations and alkalinity to total anions.

A final major factor is the cost of chemicals for regeneration which constitutes the single largest cost item for ion exchange. The concentration and quantity of regenerant must be higher than the stoichiometric amounts in order to maintain an adequate driving force during regeneration. Thus, some of the regenerant is wasted from the system.

The least costly system is often that which provides for the highest degree of regenerant utilization efficiency. However, the attainment of maximum resin capacity and lowest TDS in effluents requires high regenerant levels.

The most extensive experience with ion exchange has been gained at Pomona,[83,84] where a two-stage, strong cation–weak anion system has been in operation for several years. Feed to the system is carbon-filtered secondary effluent with a TDS of approximately 600 mg/l. Typical results are given in Table 14-34. TDS removal was 70 percent in the first stage and 89 percent overall. Regenerant level used was 1.1 lb H_2SO_4/ft^3 of cation resin and 0.6 lb NH_4OH/ft^3 of anion resin. These levels were chosen to ensure high efficiency of regenerant usage (85 percent for cation resin; 90 percent for anion resin) at an acceptable leakage (5 percent). This level of leakage is normally not acceptable in water treatment systems but is allowable in sewage treatment where very high degrees of demineralization are not required.

TABLE 14-34. Performance of the Two-Stage Ion Exchange System Operating on 610 mg/l TDS Feedwater[84]

Parameter (mg/l)		Carbon Column Effluent	Primary Cation Effluent	Primary Anion Effluent	Secondary Cation Effluent	Secondary Anion Effluent	Total Removal (percent)
Calcium	(Ca)	53	2.0	1.7	1.1	0.60	99
Magnesium	(Mg)	17	0.59	0.56	0.38	0.0	100
Sodium	(Na)	126	61	59	16	15	88
Potassium	(K)	14	7.3	7.1	1.9	1.9	86
Ammonium	(NH_4–N)	20	9.6	9.2	4.0	3.8	81
Sulfate	(SO_4)	72	72	3.6	3.6	1.3	98
Chloride	(Cl)	135	.132	84	83	14	90
Nitrate	(NO_3–N)	2.9	2.8	1.6	1.5	0.35	88
Orthophosphate	(PO_4)	27	27	15	14	0.25	99
pH		7.4	2.7	5.7	2.8	5.8	
Conductivity	(μ mhos/cm)	1040	1390	390	1040	100	90
TDS	(mg/l)	610	298	198	104	70	89
Alkalinity	($CaCO_3$)	218		51		39	
Acidity	($CaCO_3$)		110		110		

Tests with similar waters of TDS levels up to 1600 mg/l yielded similar results to those obtained with the 600 mg/l water.

In order to ascertain resin lifetime, an accelerated single-stage system was set up at Pomona.[84] Performance was similar to that observed in the first stage of the two-stage system; i.e., 70 percent TDS removal. Based on the accelerated life test it was determined that resin replacement would not exceed 10 percent for the cation system and 20 percent for the anion system.

A number of cost estimates were prepared for the Pomona system. These are detailed in Table 14-35 for the conditions described. Pretreatment and brine disposal are not included. Costs were insensitive to plant size above 5 mgd.

Costs are included for a lower level of removal because this level satisfies drinking water standards at this site. These costs are based on achieving 60 to 90 percent removal and blending with untreated water.

It is difficult to generalize on ion exchange costs because they are quite sensitive to such local conditions as resin systems, feed characteristics, effluent requirements, allowable leakage, and regenerant levels. In general, operation costs are about 75 percent of the total system cost.

Electrodialysis. In sewage application, electrodialysis suffers severe limitations. The anion plates foul quite easily, drastically reducing treatment performance unless the feed has received extensive pretreatment.[85] In addition, as the sewage is demineralized, its electrical resistance increases. This limits the degree of demineralization which can be obtained economically. It is difficult to reduce the TDS of water much below 300–400 mg/l with electrodialysis. For these reasons it is no longer considered a candidate process.

Reverse Osmosis. In the reverse osmosis process, water is forced to flow through a semipermeable membrane (permeable to water but not to salts) by application of high pressure. The process is in its infancy since the first applicable membranes were developed only 15 years ago. At present, it is the most expensive of the three demineralization approaches, although its potential is considerably superior to that of the other two. This is especially true since membranes which desalt water remove virtually every other pollutant from wastewater. Reverse osmosis may therefore become an all-purpose pollution control system.

The process requires pressures of several hundred pounds per square inch to achieve fluxes on the order of 10 gpd/ft^2. The degree of demineralization can approach 99+ percent and increases with the pressure applied.

The heart of this process is the membrane. At present, the best membranes are cast from a mix of cellulose acetate, acetone, formamide, and magnesium perchlorate. Special techniques form a membrane with a thin (1 μ) skin which provides the desalting surface and a thick (100 μ) porous sublayer which supplies a structural backing. This membrane is compressible, hydrolyzes at low and high pH, and is easily damaged. Efforts are underway to develop new, tougher membranes.

Even if stronger membrane materials are devised, the engineering problems of the process will remain unchanged. These problems consist of the need for a structural support system to absorb the several hundred pounds per square inch pressure drop across the membrane and for control of fouling on membrane surfaces. The

TABLE 14-35. Pomona Ion Exchange Costs

Conditions	Cost (cents/1000 gal)
Two-stage TDS = 600, 90 percent removal	22.4
Single-stage TDS = 600, 66 percent removal	17.8
Two-stage TDS = 600, 33 percent removal	9.4
Single-stage TDS = 600, 33 percent removal	9.7

support system must not only support the membrane but should provide high membrane area per unit volume for maximum economy.

Three configurations are in use—tubular, spiral-wound, and hollow fine fiber. The tubular system has the membrane positioned along the inner wall of a porous $\frac{1}{2}$-in. diameter tube. The spiral-wound module uses a stack of flat membranes separated by spacers and rolled into a jelly-roll form. The hollow fiber uses microscopic fibers of the membrane which are essentially thick-walled microcylinders. Here, pressure is applied outside the fiber and water permeates into the hollow core. As a rough guide, membrane area per unit volume is 20 ft^2/ft^3 for tubular systems, 250 ft^2/ft^3 for spiral-wound systems, and 2000-5000 ft^2/ft^3 for hollow fine fibers. The systems are in reverse order on the basis of fouling resistance. In general, the tubular system is preferred for dirty-water application and the remaining two systems for clean-water use.

Fouling of reverse osmosis membranes is similar to that which occurs on anion exchange resins and electrodialysis stacks. It is due to an accumulation of organic and inorganic materials at the membrane surface. As the water permeates the membrane, these materials are left behind and build up in the boundary layer next to the membrane surface. Concentrations can become high enough to cause precipitation or to stimulate significant biological activity. Fouling has an adverse effect on both permeation rate and membrane rejection. Indeed, fouling has been implicated not only in functional failure of membranes but in physical failure as well.

Several methods of controlling fouling have been investigated and show promise[86,87]; however, a complete solution is not yet available. An increase of turbulence at the membrane surface will speed diffusion back to the bulk solution being treated, a process that is accomplished by increasing flow velocity and/or use of turbulence promotor inserts. Progress here is achieved at the expense of increased system pressure drop. Reduction of fouling is also brought about by pretreatment to remove solids and organics prior to the reverse osmosis process. The standard advanced waste treatment techniques—coagulation, filtration, carbon adsorption, etc.—have been used with some degree of success. Chlorination of the feed to control biological growth, pH adjustment to 5.5 to retard inorganic precipitation, and slow membrane hydrolysis have also been employed. The benefits of pretreatment, however, are accompanied by increased processing costs.

The most economical and successful method of controlling fouling is *in situ* membrane cleaning. Cleaning solutions of enzymatic detergent solutions, wetting agents, and complexing agents have been successfully used for long periods of time on a once-a-week basis. The system must be shut down and depressurized for cleaning. Often a sponge ball is pumped through tubular systems to remove fouling scales after cleaning by physical means. Despite the success of chemical cleaning, research for the development of a fouling-resistant membrane is continuing.

Results of a number of studies on reverse osmosis systems using membranes with high salt rejection properties indicate that removals of 93 to 95 percent of the TDS, 95 to 99 percent of the phosphate, 80 to 90 percent of the NH_3-N, 60 to 70 percent of the NO_3-N, 99 to 100 percent of the suspended solids, and 90 to 95 percent of the TOC are achievable. Estimated costs of reverse osmosis are rather high: $0.30 to $0.60/1000 gal exclusive of pretreatment and brine disposal. The method, however, provides excellent treatment of a wide variety of pollutants. Membrane lifetime is a major factor in calculating cost. At present, lifetimes of 1 to 2 years are projected, but pilot data only supports the lesser value.[87] One method suggested to reduce costs is development of higher flux membranes with lower salt rejection properties. Here the pressure required would be lower and equipment size would decrease. Although membranes of this type with no salt rejection properties have been successfully employed for organics removal (ultra-filters), satisfactory intermediate level membranes have not been produced. Those which have been evaluated are not structurally sound for long periods and tend to foul very easily.

Demineralization Summary. Although techniques to demineralize wastewater have been developed, the processing cost ranged from \$0.20 to \$0.60/1000 gal, exclusive of brine residue disposal. Brine disposal costs may be higher[88] than the processing costs, except for areas that are arid or adjacent to the ocean. Widespread adoption of demineralization technology cannot be expected until the processes become more economical. It will, however, probably be applied in a few special situations.

TERTIARY TREATMENT

The term "tertiary treatment" creates some confusion due to the variety of its meanings. In the general sense, it represents any stage of treatment or any treatment step applied after secondary treatment. This definition encompasses most of the unit operations and processes discussed in this chapter. In a more restricted sense, it connotes a specific combination of treatment procedures applied after secondary treatment. In this restricted usage, tertiary treatment is the combination of chemical coagulation, flocculation, sedimentation, deep bed filtration, and activated carbon adsorption. Nitrogen procedures are optional in this treatment train. Figure 14-24 illustrates primary, secondary, and tertiary treatment (restricted sense) combined in a system which provides biological, physical, and chemical treatment.

The ultimate goal of the combination of all these treatment procedures is to produce renovated wastewater; i.e., wastewater which can be reused. Existing plants of this type include a 2.0 mgd pilot plant in Contra Costa, California, which has provided feed to test the feasibility for industrial cooling water,[89] a 2 mgd plant at Colorado Springs, Colorado, which provides irrigation flow for public parks and power plant cooling water,[90] and a 7.5 mgd plant at South Lake Tahoe, California, where the effluent is used to provide a recreational lake.[4] The Lake Tahoe plant has been in operation the longest of the three. Table 14-36 illustrates the performance of this plant at various stages of treatment. These data indicate the excellent quality of water which can be produced in such a plant. In most respects, the quality is superior to that required by U.S. Public Health Service standards.[39]. Much more testing and evaluation will be required before this water qualifies as a public drinking water supply, but it has been approved for fishing and water sports.

Table 14-37 provides cost estimates for this type of treatment system. These data do not include demineralization costs. A range of values is presented for each plant size to account for various options in the treatment train and because costs are not precisely known for this system.

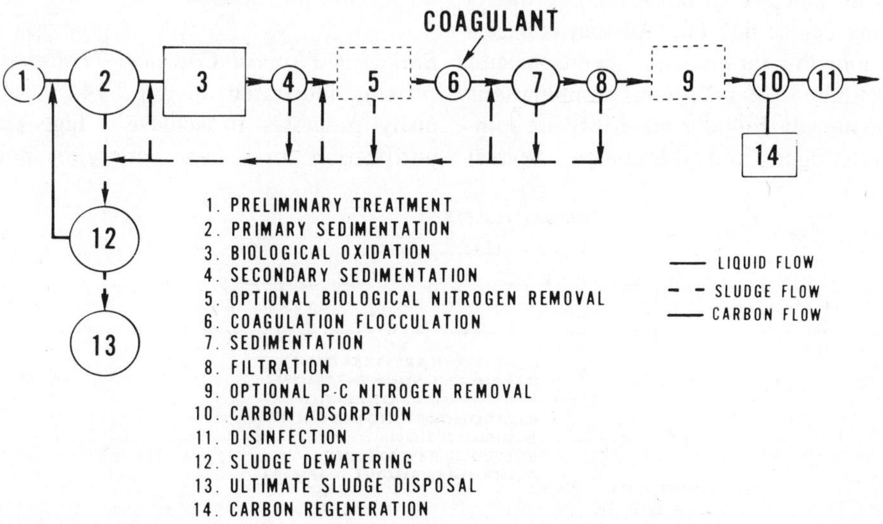

Fig. 14-24 Biological-chemical-physical treatment system—primary, secondary, and tertiary.

TABLE 14-36. Water Quality at Various Stages in Treatment

Parameter	Raw Wastewater	Primary Effluent	Secondary Effluent	Chemical Clarifier Effluent	Filter Effluent	Carbon Effluent	Chlorinated Final Effluent
BOD (mg/l)	300	100	30		3	1	0.7
COD (mg/l)	280	220	40		25	10	10
SS (mg/l)	230	100	26	10	0	0	0
Turbidity (JTU)	250	150	50	10	0.3	0.3	0.3
MBAS (mg/l)	7	6	2.0		0.5	0.10	0.10
Phosphorus (mg/l)	12	9	6	2	0.10	0.10	0.10
Coliform (MPN/100 ml)	50 million	15 million	2.5 million		50	50	<2.0

TABLE 14-37. Treatment Cost: Physical-Chemical-Biological System

Plant size (mgd)	1	10	100
Cost (cents/1000 gal)	92–120	45–60	25–35

SUMMARY

In this chapter, many of the newest developments in wastewater treatment technology have been reviewed. In the interest of brevity, some areas such as microorganism removal and ultimate disposal have not been covered, and the coverage of others has been less than complete. It should be clear, however, that the wastewater treatment engineer now has at his disposal several alternative treatment techniques for meeting most of today's water quality standards. It is well within the realm of technical possibility to conceive of purifying any wastewater to any degree desired. All that is necessary is to join the various unit operations and processes required. A number of combinations which have already found some utility are summarized in the figures and tables below. Several processes and systems such as clinoptilolite, ozone oxidation, reverse osmosis, and single nitrification require further evaluation to exploit their potential to best economic advantage.

Biological-Physical Treatment. This system employs a filter to upgrade the performance of conventional treatment. It is illustrated in Fig. 14-2.

Biological-Chemical Treatment. In this system, illustrated in Fig. 14-25, chemical coagulants can be added at various points for phosphorus removal and improved solids removal.

Physical-Chemical Treatment. In this system, illustrated in Fig. 14-26, chemical clarification and carbon adsorption are combined to achieve wastewater purification.

Biological-Physical-Chemical Treatment. This system, illustrated in Fig. 14-24, combines many processes to achieve a high degree of purification for a large variety of pollutants.

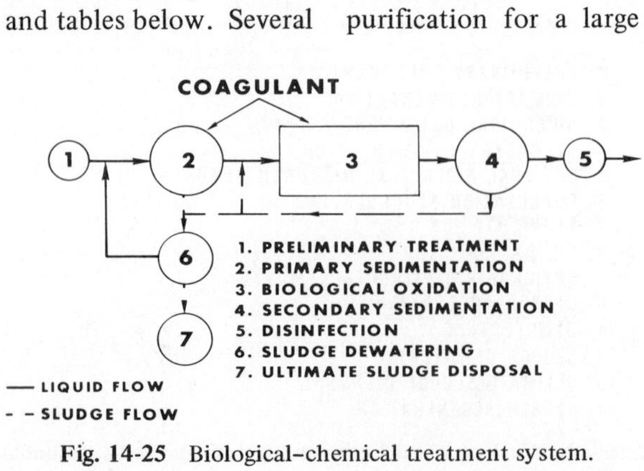

Fig. 14-25 Biological-chemical treatment system.

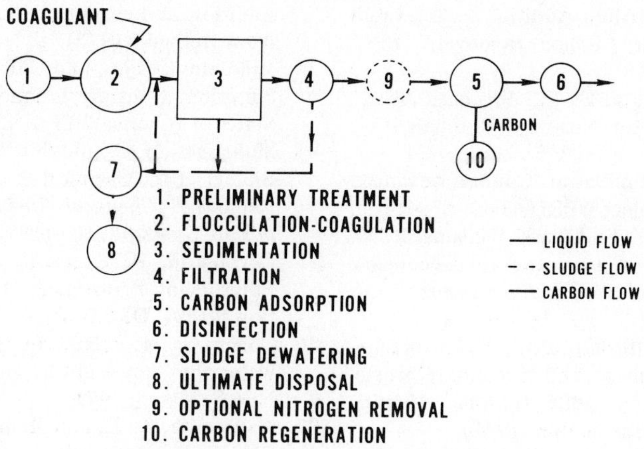

Fig. 14-26 Physical-chemical treatment.

TABLE 14-38. Treatment Performance of Selected Wastewater Treatment Systems (percent removal)

System Parameter	Conventional	Biological-Physical	Biological-Chemical	Physical-Chemical	Biological-Physical-Chemical
Organics	80-90	85-95	90-95	95-99	99+
Suspended solids	85-95	90-95	95-99	95-99	99+
Phosphorus	0-40	0-50	90-95	95-99	99+

Table 14-38 summarizes the performance to be expected from these treatment schemes.

Economic Perspective. National average sewerage costs (i.e., collection and treatment) are given in Table 14-39.[50] It can be seen that the actual cost of treatment is only a minor part of the U.S. wastewater control cost. The cost of treatment is \$0.012/capita/day or \$0.12/1000 gal. Nationwide adoption of the most complete form of advanced waste treatment (i.e., tertiary treatment with nitrogen removal) would increase treatment costs to about \$0.50/1000 gal or about \$18/person/year. The net effect would be only a doubling of the nation's wastewater treatment bill because amortization of the cost of sewage collection accounts for most of the present cost. If a lesser degree of treatment were utilized—and it is probable that in most areas water quality standards could be met by less elaborate forms of advanced wastewater treatment—the economic burden would be lower. In addition, installation of advanced wastewater treatment will yield significant benefits in terms of environmental improvement and water resources conservation by providing at least partial water reuse.

TABLE 14-39. National Average Sewerage Costs

Activity	Cost (\$/person/year)
Collection (sewers)	13.34
Treatment (conventional)	4.38
Overhead	2.08
Total	19.80

REFERENCES

1. "Process Design Manual for Suspended Solids Removal," EPA Technology Transfer Program, (Oct., 1971).
2. Lynam, B., Ettlet, G., and McAloan, T. J., "Tertiary Treatment at Metro Chicago by Means of Rapid Sand Filters and Microstrainers," *Journal WPCF* **41**, 247 (Feb., 1969).
3. "Process Design Manual for Suspended Solids Removal," EPA Technology Transfer Program, pp. 9-18, (Oct., 1971).
4. Kreissl, J. F., "Granular Media Filtration of Wastewater, An Assessment," EPA-NERC (1973).

5. Hais, A. B., *et al.*, "Alum Addition to Activated Sludge with Tertiary Solids Removal," EPA-NERC (1971).
6. Culp, R. L., and Culp, G. L., *Advanced Wastewater Treatment*, Van Nostrand Reinhold Co., New York (1971).
7. Progress Reports, EPA-Nassau County Advanced Waste Treatment Project WPRD 61-01-67 (1967).
8. Berg, E. L., Brunner, C. A., and Williams, R. T., "Single Stage Lime Clarification of Secondary Effluent," *Water and Wastes Engineering* 7, 43 (March, 1970).
9. O'Farrell, T. P., and Bishop, D. F., "Lime Precipitation in Raw, Primary and Secondary Wastewater," presented at 68th National Meeting AIChE, Houston, Texas (March, 1971).
10. Slechta, A. F., and Convley, W. R., "Recent Experiences in Plant Scale Application of the Settling Tube Concept," *Journal WPCF* **43**, 1724 (1971).
11. Tofflemire, T. J., and Heling, L. J., "A Guide to Chemical and Clarifier Selection for Wastewater Treatment," N.Y. State Department of Environmental Conservation, Technical Paper No. 29 (1973).
12. Sawyer, C. N., "ABC's of Cultural Eutrophication, Part I: Cultural Changes," *Water and Sewage Works* **120**, 278 (Sept., 1971).
13. Sawyer, C. N., "ABC's of Cultural Eutrophication, Part II: Wastewaters," *Water and Sewage Works* **120**, 322 (Oct., 1971).
14. "Process Design Manual for Phosphorus Removal," EPA Technology Transfer Program (Oct., 1971).
15. Levin, G. V., and Shapiro, J., "Metabolic Uptake of Phosphorus by Wastewater Organisms," *Journal WPCF* **37**, 800 (1965).
16. Mulbarger, M. C., Shifflett, D. G., Murphy, M. C., and Huffman, D. O., "Phosphorus Removal by Luxury Uptake," *Journal WPCF* **43**, 1617 (1971).
17. Menar, A. B., and Jenkins, D., "Fate of Phosphorus in Waste Treatment Processes; Enhanced Removal of Phosphate by Activated Sludge," *Environmental Science and Technology* **4**, 1115 (1970).
18. Recht, H. L., and Ghassemi, M., "Kinetics and Mechanism of Precipitation and Nature of Precipitate Obtained in Phosphate Removal from Wastewater Using A1 (III) and Fe (III) Salts," EPA Report 17010 EKI (1970).
19. Stamberg, J. B., Bishop, D. F., Warner, H. P., and Griggs, S. H., "Lime Precipitation in Municipal Wastewaters," *Chemical Engineering Progress Symposium Series* **67**, 310, (1970).
20. Kreissl, J. F., "Phosphorus Removal Today," presented at the Sanitary Engineering Institute, University of Wisconsin, Madison (March 9–10, (1971).
21. Bishop, D. F., O'Farrell, T. P., and Stamberg, J. B., "Physical Chemical Treatment of Municipal Wastewater," *Journal WPCF* **44**, 361 (1972).
22. Feige, W. A., and Berg, E. L., "Full-Scale Mineral Addition at Lebanon, Ohio," *Water and Sewage Works* ref. ed. (1973).
23. Middleton, F. M., and Convery, J. J., "Municipal Pollution Control Technology in the United States of America," EPA-NERC (1971).
24. Mulbarger, M. C., and Shifflett, D. G., "Combined Biological and Chemical Treatment for Phosphorus Removal," *Chemical Engineering Progress Symposium Series* **67**, 107 (1970).
25. Derrington, R. E., *et al.*, "Enhancing Trickling Filter Plant Performance by Chemical Precipitation," EPA, OR&D, 670/2-73-060, (Aug., 1973).
26. Cohen, J. M., "Nutrient Removal from Wastewater by Physical-Chemical Processes," EPA-NERC (March, 1971).
27. Tofflemire, T. J., and Heling, L. J., "A Guide to Chemical and Clarifier Selection for Wastewater Treatment," N.Y. State Department of Environmental Conservation, Technical Paper No. 29, pp. 7–13 (1973).
28. Farrell, J., "Increase in Sludge Mass when Aluminum or Iron Salts are Used in Sewage Treatment Processes," Internal Memo AWTRL-EPA-NERC (1973).
29. Isgard, E., "Chemical Methods in Present Sewage Purification Techniques," presented at the 7th Effluent and Water Treatment Exhibition and Convention, London, England (June, 1971).
30. Burns, D. E., and Shell, G. L., "Physical Chemical Treatment of Municipal Wastewater Using Powdered Carbon," EPA, OR&D, R-2-73-264 (Aug., 1973).
31. Bell, R., *et al.*, "Phosphorus Removal Using Chemical Coagulation and a Continuous Countercurrent Filtration Process," EPA, OR&D, 17010 EDO (June, 1970).
32. Bishop, D. F., "Progress Reports," EPA pilot plant, Washington, D.C. (1973).
33. Barth, E. F., "Nitrogen Removal by Biological Suspended Growth Reactors," Advanced Waste Treatment and Reuse Program, vol. 2, Dallas Texas (Jan., 1971).
34. Wild, H. E., Jr., Sawyer, C. N., and McMahon T. C., "Factors Affecting Nitrification Kinetics," *Journal WPCF* **43**, 1845 (1971).
35. Mulbarger, M., "The Three Sludge System for Nitrogen and Phosphorus Removal," EPA-NERC (Apr., 1972).
36. Sawyer, C. N., *et al.*, "Nitrification and Denitrification Facilities," EPA Technology Transfer Program (Aug., 1973).
37. Duddles, G. A., and Richardson, S. E., "Application of Plastic Media Trickling Filters for Biological Nitrification Systems," EPA, OR&D, 17010 SJF (1972).
38. McHarness, D., and McCarty, P. L., "Field Study of Nitrification with the Submerged Filter," EPA OR&D, 17010 EPM (1973).
39. U.S. Public Health Service Drinking Water Standards, USPHS, U.S. Dept. of Health, Education and Welfare (1962).

40. Kreissl, J. F., "Nitrogen Removal Today," presented at Sanitary Engineering Institute, University of Wisconsin, Madison (Mar., 1971).
41. Barth, E. F., *et al.*, "Chemical Biological Control of Nitrogen and Phosphorus in Wastewater Effluents," *Journal WPCF* **40**, 2040 (1968).
42. McCarty, P. L., Beck, L., St. Amant, P. P., "Biological Denitrification of Wastewaters by Addition of Organic Materials, *Proceedings of the 24th Industrial Waste Conference*, Purdue University, Lafayette, Indiana (May, 1969).
43. Smith, J. M., Masse, A. N., Feige, W. A., and Kamphake, L. J., "Nitrogen Removal from Municipal Wastewater by Columnar Denitrification," *Environmental Science and Technology* **6**, 260 (1972).
44. Jeris, J., Dept. of Civil Engineering, Manhattan College, New York, Personal communication (1973).
45. Pressley, T. A., Bishop, D. F., and Roan, S. G., "Nitrogen Removal by Breakpoint Chlorination," EPA-NERC (Sept., 1970).
46. Cassel, A. F., Pressley, T. A., Schuck, W. W., and Bishop, D. F., "Physical-Chemical Nitrogen Removal from Municipal Wastewater," EPA-NERC (March, 1971).
47. Pressley, T. A., *et al.*, "Ammonia Removal in a Physical-Chemical Wastewater Treatment System," EPA-OR&D, 11010EYM (1973).
48. Atkins, P. F., *et al.*, "Ammonia Removal in a Physical-Chemical Wastewater Treatment Plant," EPA, OR&D, R2-72-123 (1972).
49. Bauer, R. D., and Snoeyink, V. L., "Ammonia Removal by Chlorination Followed by Contact with Activated Carbon," *Journal WPCF* (in press), paper presented (1972).
50. Smith, R., and Eilers, R. G., "Cost to the Consumer of Collecting and Treating Wastewater in the United States," EPA-NERC (July, 1970).
51. O'Farrell, T. P., *et al.*, "Nitrogen Removal by Ammonia Stripping," *Journal WPCF* **44**, 1527 (1972).
52. Gonzales, J. G., and Culp, R. L., "New Developments in Ammonia Stripping, I," *Public Works* (May, 1973).
53. S.E.R.L., University of California, "Optimization of Ammonia Removal by Ion Exchange Using Clinoptilolite," EPA, OR&D, 17010 ECZ (1971).
54. S.E.R.L., University of California, "Optimization of Ammonia Removal by Ion Exchange Using Clinoptilolite," EPA, OR&D, 17080 DAK (1971).
55. Battelle Northwest, "Wastewater Ammonia Removal by Ion Exchange," EPA, OR&D, 17010 ECZ (1971).
56. "Regional Agency Starts Up Physical Chemical Treatment Plant," *Environmental Science and Technology* **7**, 804 (Sept., 1973).
57. Final Report "Wastewater Ammonia Removal by Ion Exchange," Project 17010 ECZ, EPA (Feb., 1971).
58. Mercer, B. W., Ames, L. L., Touhill, C. J., Van Slyke, W. J., and Dean, R. B., "Ammonia Removal from Secondary Effluents by Selective Ion Exchange," *Journal WPCF* **42**, R95 (1970).
59. "Process Design Manual for Carbon Adsorption," EPA Technology Transfer Program, (Oct., 1971).
60. Berg, E. L., Villiers, R. V., Masse, A. N., and Winslow, L. A., "Thermal Regeneration of Spent Powdered Carbon Using Fluidized Bed and Transport Reactors," *Chemical Engineering Progress Symposium Series* **67**, 154 (1970).
61. Burns, D. E., and Shell, G. L., "Physical Chemical Treatment of a Municipal Wastewater Using Powdered Activated Carbon," presented at the 44th Annual WPCF Conference, San Francisco, California (Oct., 1971).
62. Progress Reports, "Physical Chemical Treatment of Municipal Waste," Contract 68-01-0183, EPA, Eimco Corp. (1973).
63. Koches, C. F., and Smith, S. B., "Reactivating Powdered Carbon," *Chemical Engineering* **79**, 46 (1972).
64. English, J. N., Masse, A. N., Carry, C. W., Pitkin, J. B., and Haskins, J. E., "Removal of Organics from Wastewater by Activated Carbon," *Chemical Engineering Progress Symposium Series* **67**, 147 (1970).
65. Masse, A. N., "Removal of Organics by Activated Carbon," EPA-NERC (Aug., 1968).
66. Progress Report, EPA–Colorado Springs Project 17080 DJE (Dec., 1971).
67. "Advanced Wastewater Treatment Using Powdered Activated Carbon in Recirculating Slurry Contactor-Clarifiers," Final Report, EPA Project 17020 FKB (July, 1970).
68. Hager, D. G., and Reilly, D. B., "Clarification-Adsorption in the Treatment of Municipal Waste Waters," *Journal WPCF* **42**, 794 (1970).
69. Weber, W. J., *et al.*, "Biologically Extended Physiochemical Treatment," presented at 6th International Conference, IAWPR, Jerusalem (June, 1972).
70. Hopkins, C. B., *et al.*, "Granular Carbon Treatment of Raw Sewage," EPA, OR&D, 17050 DAL (1970).
71. Directo, L., "Interim Report on Independent Physical-Chemical Treatment at Pomona, California," EPA-LACSD (Sept., 1973).
72. Kugelman, I. J., and Cohen, J. M., "Physical-Chemical Processes," presented at EPA Design Seminar for Municipal Wastewater Treatment Facilities, Cleveland, Ohio (Apr., 1971).
73. Plummer, A. H., DiFilippo, Glass, P., "Chemical-Physical Wastewater Treatment for the City of Garland, Texas," presented at EPA Technology Transfer Program, Dallas, Texas (July, 1971).
74. Directo, L., "Monthly Progress Report on Independent Physical-Chemical Treatment, Pomona, California," EPA-LACSD, (Oct., 1973).
75. Roan, S. G., *et al.*, "Laboratory Ozonation of Municipal Wastewater," EPA, OR&D, 11010 EYM (1973).

76. "Ozone Treatment of Secondary Effluents from Wastewater Treatment Plants," TWRC-4 NERC, Cincinnati, Ohio (1969).
77. Progress Report, "Pilot Plant Investigation of the Ozone Treatment of Secondary Effluents from Wastewater Treatment Plants," EPA and Airco, Inc., Contract 14-12-597 (Jan., 1972).
78. Ghan, H., "Monthly Progress Reports on Ozone Treatment of Secondary Effluent, Pomona, California," EPA, OR&D (Aug., 1973).
79. Rosen, H., *et al.*, "Optimization of Ozone Treatment of Water and Wastewater; Disinfection and Quality Improvement," presented at 46th National Meeting, WPCF, Cleveland, Ohio (Oct., 1973).
80. "Investigation of the Use of High Purity Oxygen Aeration in the Conventional Activated Sludge Process," Water Pollution Control Research Series, EPA, 17050 DNW (May, 1970).
81. Stamberg, J. B., Bishop, D. F., and Kumke, G., "Activated Sludge Treatment with Oxygen," presented at the 68th National Meeting of the AIChE, Houston, Texas (March, 1971).
82. Brenner, R., "EPA Experiences in Oxygen Activated Sludge," EPA Technology Transfer Program (1973).
83. Dryden, F. D., "Mineral Removal by Ion Exchange, Reverse Osmosis, and Electrodialysis," Advanced Waste Treatment and Water Reuse Program, Vol. 1, Dallas, Texas (Jan., 1971).
84. Chen, C., "Wastewater Demineralization of the Two-Stage Fixed Bed Ion Exchange Process," EPA-LACSD (1973).
85. Brunner, C. A., "Pilot Plant Experiences in Demineralization of Secondary Effluent Using Electrodialysis," *Journal WPCF* 39, R1 (1967).
86. Smith, J., *et al.*, "Renovation of Municipal Wastewater by Reverse Osmosis," EPA Internal Report (May, 1970).
87. Bridgeford, F., and Johannson, G., "Reverse Osmosis of Treated and Untreated Secondary Effluent," EPA, 17040 DSR (1973).
88. Burns and Roes, Inc., "Disposal of Brines Produced in Renovation of Municipal Wastewater," EPA, 17070 DLY (May, 1970).
89. Horstkotte, G. A., "Pilot Demonstration Project for Industrial Reuse of Renovated Municipal Wastewater," EPA, OR&D, 670/2-73-064 (1974).
90. Progress Reports, Grant WPRD 233-01-66, EPA-City of Colorado Springs, Colorado (1971).

15

Cost Allocation and Recovery for Combined Municipal and Industrial Waste Treatment Facilities

E. Downey Brill, Jr.
Assistant Professor of Civil Engineering and
in the Institute for Environmental Studies
University of Illinois
Urbana, Illinois

John C. Geyer
Professor of Environmental Engineering
The Johns Hopkins University
Baltimore, Maryland

COMBINED MUNICIPAL AND INDUSTRIAL WASTE TREATMENT SYSTEMS

Industrial and municipal wastes are frequently combined for joint treatment to obtain economies of scale. Dilution and chemical equalization of industrial wastes may also be desirable. Although the municipality may own and operate the combined waste treatment facility, the total cost should be borne by the waste contributors or users and other beneficiaries in order to assure economic efficiency and equity. The municipality must determine the cost allocations for the various users and choose a method for recovering these costs.

The total cost of the waste treatment system includes capital costs (interest on first cost plus depreciation) and operation and maintenance costs. Typical cost items are given in Table 15-1. Estimates of total cost can be made using general cost indices, curves, or Eqs. 1, 2, 3, 4, 5, 6, and 7.

System benefits accrue to users such as residences, commercial establishments, industries, and public properties, and owners of undeveloped lands whose value is enhanced by the presence of the system.[8]

Two criteria for evaluating alternative approaches to allocating and recovering costs are economic efficiency and equity. Economic efficiency is measured by the total cost of meeting waste treatment goals. The total cost includes the cost of the combined facility and the costs of waste reduction by individual waste producers. The most efficient program minimizes this total. Equity involves two considerations: whether users of the combined waste treatment system bear its costs and whether users pay in proportion to use.[9]

COST ALLOCATION AMONG USERS

The total cost of combined waste treatment depends, directly or indirectly, on wastewater flow and waste content. Waste contents gen-

TABLE 15-1. Typical Cost Items for Waste Treatment Facility

Pumps	Sludge heater
Plant water supply	Aerator diffusers
Roads and grounds	Final sedimentation tanks
Plumbing and heating	Primary scum pipe
Comminuter	Chlorinator
Grit chamber	Control house–laboratory
Parshall flume meter	Shop
Primary collector	Aerator blower
Digesters	Vacuum filter

erally include biochemical oxygen demand (BOD), suspended solids, and possibly nitrogen, phosphorus, grease, and chlorine demand. Once costs are apportioned among the waste-loading parameters, they can be allocated to individual or groups of users according to waste loads.

The portion of the total cost directly attributable to each waste parameter can be estimated by several methods. One procedure is to assign the unit cost for items such as those listed in Table 15-1 to the parameter that primarily affects the design. For example, pumps and piping costs could be assigned to total flow, sludge disposal costs to suspended solids, and aeration tank costs to BOD. Or, fractions of the cost for any item can be assigned to two or more parameters. As an example, the cost of operating and maintaining a trickling filter could be divided between flow and BOD. In any case, engineering judgment must be used in assigning costs to individual parameters. Examples of this type of calculation are given in Ref. 8 and 10.

Some costs are not directly attributable to flow or waste qualities and are best assigned to flow and loading parameters as fixed percentage "overhead" based on the other costs.[8,9] Examples of such "overhead" costs are given in Table 15-2.

Once the total cost is apportioned to the appropriate loading parameters, the users' shares can be calculated on the basis of the same parameters. For example, the cost assigned to the BOD parameter can be allocated to each user's contribution of BOD.

Costs are generally allocated on an average-cost rather than on an incremental-cost basis.

TABLE 15-2. Examples of Overhead Costs

Capital, operation, and maintenance costs for:
Plant water supply
Plant heating
Capital and maintenance costs for:
Land
Buildings
Roads
Operation costs for:
Plant supervision

The method is illustrated in Example 1. For small waste load contributors, it may be impractical to measure each waste parameter. Such contributors can be grouped and the group cost allocation apportioned according to the estimated combined waste loads.

EXAMPLE 1: Given that:

TC = total cost* of waste treatment facility,

TC_V = total cost assigned to flow,

TC_{BOD} = total cost assigned to BOD,

TC_{SS} = total cost assigned to suspended solids,

$TC = TC_V + TC_{BOD} + TC_{SS}$,

and that for a particular industry:

f_V = fraction of total flow contributed by the industry,

f_{BOD} = fraction of total BOD contributed by the industry, and

f_{SS} = fraction of total suspended solids contributed by the industry,

then the total cost allocated to the industry using the average-cost method is:

$$TC_I = f_V \times TC_V + f_{BOD} \times TC_{BOD} + f_{SS} \times TC_{SS} \quad (1)$$

Inefficiencies can result from using the average-cost method. An individual waste producer might choose to treat his own wastes if that were less expensive than the portion of average costs allocated to him. However, assuming economies of scale, the incremental cost for adding this waste producer to the system would be less than the cost allocation based on the average

*Costs are assumed to be annual.

cost. If the potential contributor's cost were greater than this incremental cost, it would then be more economically efficient for him to join the system.

Allocation of costs on an incremental-cost basis would encourage efficient combinations. Where there is more than one potential contributor, however, two problems may arise. First, inequities are likely since the cost per unit waste decreases for each contributor joining the system because of the economies of scale. Second, inefficiencies may still occur if no potential contributor is willing to bear the cost for the next increment of waste loading, although two or more contributors may be willing to accept the cost for their combined additional loading. Also, calculation of incremental costs requires information that may not be readily available, thus placing an additional burden on the analyst.

ALTERNATIVE METHODS OF RECOVERING COSTS

Methods of allocating costs to users singly or in groups were discussed in the preceding section. Costs thus allocated to individual users call for equivalent user rates. Negotiated contracts based on use are a special form of user rates. If costs are allocated to groups of users, other methods of cost recovery, such as property taxes and general tax revenues, can also be applied. The choice of method involves such considerations as economic efficiency, equity, and practicality.

User Charges. A user charge is based on the use of or benefit from the waste treatment facility. The unit-user charge or rate is expressed as dollars/unit waste. The total charge, here called user charge, for a given contributor is the sum of the unit-user charges multiplied by their corresponding waste loads.

User Charges and Efficiency. User charges tend to be economically efficient since the cost burden is placed on the waste contributors. This efficiency results not only from the economies of scale that can be achieved by a combined system but also from the incentives felt by con-

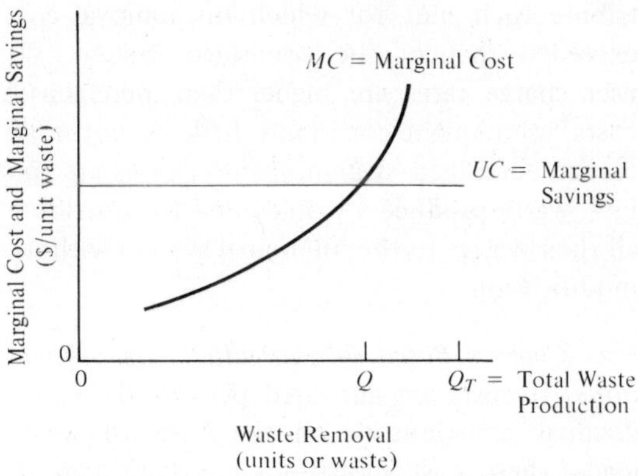

Fig. 15-1 Response of a hypothetical industry to a user charge.

tributors who join the system. Users are encouraged to lessen excessive waste loads that are due to careless industrial waste management and to effect partial waste reduction by changes in production process, raw materials, and product mix, and even by partial waste treatment where economical.

For example, Fig. 15-1 gives the range of possible waste removal levels for a particular type of waste in a hypothetical industry. The unit charge for the waste is represented by UC. By providing an additional unit of waste removal, the industry avoids paying the unit charge. This incremental savings is called the marginal savings. The incremental cost, or marginal cost, of providing waste removal is represented by the curve labeled MC*. The industry would provide Q units of waste removal since up to that point it is less costly to provide each unit of waste removal than it is to pay the unit charge. Beyond the level of Q, it is more costly; therefore, the industry would contribute (Q_T - Q) units of waste.

The most economically efficient levels for waste contributions result from unit charges equivalent to incremental unit costs for the combined system at the operation level. In this case, each user would remove each unit of waste for which his cost was less than the cost of the combined system, and would con-

*The marginal cost function is the derivative of the total cost function with respect to the particular waste variable. The total cost is the cost for waste treatment and changes in raw materials, production process, and product mix.

tribute each unit for which his removal cost exceeded that of the combined system. If user charge rates are higher than incremental costs, users might contribute little or not at all to the combined system. If user rates are too low, waste producers would tend to contribute all their wastes to the municipal system without modification.

User Charges Based on Individual Cost Allocations. If costs are allocated periodically to individual contributors on the basis of waste loads, these cost allocations are equivalent to the individual user charges. In Example 1, the cost allocation (TC_I) for the industry, and therefore the user charge (C_I), are given in Eq. 1 as:

$$C_I = TC_I = f_V \times TC_V + f_{BOD} \times TC_{BOD} + f_{SS} \times TC_{SS} \quad (2)$$

As shown below, each charge component can be interpreted as a unit charge which is equal to the average unit cost multiplied by the corresponding waste load. Equation 2 can be rewritten as:

$$C_I = \frac{V_I}{V} \times TC_V + \frac{BOD_I}{BOD} \times TC_{BOD} + \frac{SS_I}{SS} \times TC_{SS} \quad (3)$$

where

V_I = flow contributed by the industry,
V = total flow,
BOD_I = BOD contributed by the industry,
BOD = total BOD,
SS_I = suspended solids contributed by the industry,
SS = total suspended solids.

Rearranging Eq. 3 yields the following expression for the industrial user charge:

$$C_I = v \times V_I + b \times BOD_I + s \times SS_I \quad (4)$$

where

$v = \frac{TC_V}{V}$, the average unit cost and unit charge for flow,
$b = TC_{BOD}$, the average unit cost and unit charge for BOD, and
$s = \frac{TC_{SS}}{SS}$, the average unit cost and unit charge for suspended solids.

The average unit cost factors are equivalent to the unit charges v, b, and s.

Example 1 was developed using the average-cost method of allocating costs. Since the same average unit costs apply to all users, the same unit charges (v, b, and s) could be applied to all users as well, and the total cost would be recovered. This approach is equitable since the cost burden is placed on the waste contributors. The unit charges are the same for all contributors, and each contributor pays in proportion to use. As discussed previously, however, inefficiencies can result. Potential contributors could be discouraged from joining the system when it would be most economically efficient for them to join. Also, users might be prompted to provide too much in-plant waste reduction since the average unit costs exceed the incremental unit costs at the operating level.

If individual cost allocations are calculated by means of the incremental-cost method as above, the total revenue from the user charges would equal the total costs. For each user, the portion of the cost attributable to a given waste divided by the waste load would give the unit charge for that waste. Due to economies of scale, however, inequities could result as the units costs and corresponding unit charges decrease for each contributor added to the system. The first user to join the system would face the highest unit costs and thus be penalized for all time.

As noted earlier, it may not be practical to calculate incremental unit costs, and inefficiencies could still discourage potential contributors from joining the combined system. In addition, all system users, including the last to join, would face unit charges greater than the unit incremental costs at the operation level. Consequently, the incentives for these users to provide additional in-plant waste reduction are compelling, and additional inefficiencies could result.

User Charges Based on Incremental Unit Costs at the Operation Level. As discussed above, user charges based on individual incremental unit costs for the different waste parameters at the operation level would produce the most efficient combined system. Such charges would also be equitable in the sense that all users would be assigned the same unit charges. It should be noted, however, that this method of calculating user charges does not take cost allocation into account.

The approach presents several difficulties. First, economies of scale would cause the incremental unit costs to be less than the average unit costs; thus, the total cost of the system would not be recovered. In fact, there would be no proof that a combined system is justified unless the deficit were paid by the users. Furthermore, if the plant were already constructed, the capital costs would be "sunk" costs. The incremental cost at the operation level would be the incremental operation and maintenance costs. The capital costs would not be borne by the users and this could be considered inequitable. Second, the operation level would not be known until the contributor's reactions to the user charges were observed; yet the user charges are calculated using the operation level. Because some of the decisions concerning system capacity and operation level are irreversible, it is not possible to iterate through several changes in user charges to arrive at the most efficient operation level and set of user charges. Third, fluctuating loadings and operating procedures as well as difficulties in assigning costs prevent the precise determination of incremental unit costs. Continued updating of these charges, moreover, would burden the program administration.

Summary of User Charges Based on Individual Average- and Incremental-Cost Allocations and on Incremental Unit Costs at the Operation Level. User charges based on average unit costs recover the total cost and are equitable in that the users bear the cost and are assessed the same unit charges. Also, it is relatively simple to make average unit cost estimates. Inefficiencies can occur, however, since the unit charges exceed the incremental unit costs at the operation level.

While user charges based on individual incremental costs are equitable in that users bear the cost, these charges are more difficult to determine and are likely to be inequitable in that users are assigned different unit costs. In addition, incremental costs are more difficult to calculate than average costs, and inefficiencies can still arise since the unit charges for all users exceed the incremental unit costs at the operation level.

User charges based on the incremental unit costs at the operation level are inequitable in that the users do not bear the total cost but equitable in that all users are assessed the same unit charges. In a situation of fluctuating procedures, incremental unit costs would be difficult to calculate accurately and to administer.

User Charges Based on Group Cost Allocations. Costs can be allocated to groups of users when it would not be practical to measure several waste load parameters for each individual. For a group consisting of all households, commercial establishments, and industries contributing wastes similar in character to domestic wastes, the costs could be allocated on the basis of estimates for the total flow, BOD, suspended solids, and other relevant parameters and a uniform rate applied on the basis of volume of waste produced.

Example 2 demonstrates the recovery of a group's cost allocation by a unit charge on flow only, where the unit charge would include the unit cost for flow plus a surcharge for the average BOD and suspended solids concentration costs. Furthermore, for households and industries with no significant water losses and/or alternate supplies, the waste flows can be assumed to equal the water meter readings.

EXAMPLE 2: Assuming costs are allocated using the average-cost method, the cost allocated to a group of users is given as:

$$TC_G = v \times V_G + b \times BOD_G + s \times SS_G \quad (5)$$

where

TC_G = total cost allocated to the group,
V_G = total group flow,
BOD_G = total group BOD,
SS_G = total group SS,
v, b, and s = average unit costs as given in Eq. 4.

The TC_G could be assessed by a unit charge, c, on flow represented by:

$$c = \frac{TC_G}{V_G} = v + \frac{b \times BOD_G + s \times SS_G}{V_G} \qquad (6)$$

This type of unit charge would not necessarily prove economically efficient by generating incentives, since reduction would be encouraged for only those waste parameters that are specifically charged. Nonetheless, such a program would be practical, capable of recovering the appropriate group cost, and equitable in that the users bear the cost. The method is inequitable only to the degree that individual users in the group produce waste of different concentrations. Of course, users with very different waste concentrations can be placed in separate groups.

The formulation and calculation of *user* charges and surcharges is considered further at the end of this chapter.

Negotiated Contracts. In cases where there are a few major industrial contributors to a municipal system, the municipality could negotiate a contract with each, taking into account the nature of the individual wastes. If the contracts are renegotiated periodically or contain provisions for payments which vary with waste loads, they constitute a special case of user charges and provide an incentive for in-plant waste management.

Contracts can be especially useful in the planning of a new facility. Each industrial user may be charged the capital costs for that portion of the design capacity allocated to his use. Also, an industry anticipating expansion has the opportunity to contract for excess capacity.

Negotiated contracts could be used to avoid inefficiencies that would result from average-cost user charges. Since the municipality would not be restricted to offering this set of unit charges, it would be possible to reach an agreement with an industry which chose not to accept these charges. As long as the industry pays at least the incremental cost for being the last contributor added to the system, the inefficiencies are avoided. The total cost, however, is recovered only if other contributors are willing to pay for the more costly preceding increments.

Contracts are equitable to the degree that each contributor pays the cost of his own waste treatment. Inequities may occur if different contributors are unable to negotiate similar contracts for similar waste loads. However, provisions of the contracts for several contributors may vary somewhat and still be equitable since each contract would be negotiated and accepted by the contributor.

Property Taxes. Property taxes generally are an inefficient and inequitable method of paying for the actual utilization of waste treatment facilities. Inefficiencies would result from excessive waste loads contributed to the system, since there would be little incentive for individual waste management. Inequities would occur since non-users also bear a share of the cost. Furthermore, the users of the system would not pay in proportion to use.

Property taxes can provide an equitable means of paying for unused capacity in the system since all property owners conceivably could benefit from regional growth made possible through increased property values.[8,9] The cost, however, of any unused capacity set aside for individual users should be borne by those users.

General Tax Revenues. As a means of paying for utilization of a waste treatment system, general tax revenues would be inefficient and inequitable for the same reasons as property taxes. It would be equitable to use these revenues to pay for excess capacity if the resulting growth appears to benefit the public at large.

FEDERAL REGULATIONS AND GUIDELINES

Municipalities receiving federal grants for the construction of sewage treatment systems are subject to Title 40, Subpart E,[11] of the Code of Federal Regulations (CFR). The section on award limitations contains the following regulations concerning user charges and industrial cost recovery:

35.925—Limitations on award.
"Before approving a grant for any project for any treatment works, the Regional Administrator shall determine: . . .

35.925-11—User charge system.
That, prior to award of any grant after March 1, 1973, for a project which includes the building and erection of a treatment works (Step 3) the applicant (a) has adopted or will adopt a system of charges to assure that each recipient of waste treatment service will pay its proportionate share of the costs of operation and maintenance (including replacement); (b) has received firm written commitments satisfactory to the Regional Administrator for the payment to such applicant by the industrial users for their proportionate share of the Federal share of capital costs of the project allocable to the treatment of such industrial wastes to the extent attributable to the Federal share of the cost of construction; and (c) has legal, institutional, managerial, and financial capability to insure adequate construction, operation, and maintenance of treatment works throughout the applicant's jurisdiction. Grants awarded prior to March 2, 1973, are subject to Sec. 35.835-5 requirements in lieu of the above."

The Environmental Protection Agency has proposed to amend 40 CFR, Part 35, to include regulations for user charges and industrial cost recovery.[12] Proposed modifications of Section 35.925-11 include:

1. In (a), the applicant must have developed "an approvable plan and schedule of implementation for a system of user charges . . . "
2. In (b), signed letters of intent must be received from each "significant industrial user to pay that portion of the grant amount allocable to the treatment of its wastes. Each such letter shall also include a statement of the industrial user's intended period of use of the treatment works. A significant industrial user is one that will contribute greater than 10 percent of the design flow or design pollutant loading of the treatment works. In addition, the applicant must agree to require all industrial users to pay that portion of the grant amount allocable to the treatment of wastes from such users."

The proposed requirements for the industrial cost-recovery system are given in Section 35.928 and include:

1. "(a) Each year during the recovery period, each industrial user of the treatment works shall pay its share of the total grant amount divided by the recovery period."
2. "(d) An industrial user's share shall be based on all factors which significantly influence the cost of the treatment works. Factors such as strength, volume, and delivery flow rate characteristics shall be considered and included to insure a proportional distribution of the grant amount allocable to industrial use to all industrial users of the treatment works."
3. "(e) If there is a substantial change in the strength, volume, or delivery flow rate characteristics introduced into the treatment works by an industrial user, such user's share shall be adjusted accordingly."
4. "(f) If there is an expansion of the treatment works capacity, each existing industrial user's share shall be reduced accordingly."
5. "(g) An industrial user's share shall not include any portion of the grant amount allocable to unused or unreserved capacity."
6. "(h) An industrial user's share shall include any firm commitment to the grantee of increased use by such user."

The proposed regulations also add Section 35.935-13, which places the following condition on federal grants:

"(b) The grantee will obtain the approval of the Regional Administrator of the system of user charges and the system of industrial cost recovery based on the Federal guidelines on user charges for operation and maintenance of publicly owned treatment works and Sec. 35.928 of this subpart."

The guidelines for the user charge for operation and maintenance costs include the following:

1. Costs can be allocated to individual users or to groups of similar users.
2. User charges should be in proportion to

waste contributions; i.e., average-cost user charges.

3. User charges should be reviewed annually and revised periodically to reflect actual costs.
4. Three models of acceptable user charges are:
   a. If waste concentrations of all users are approximately equal, user charges can be determined on volume alone.
   b. User charges can be based on average unit costs for flow, BOD, suspended solids, and/or any other important wastes. (This method is discussed earlier in this chapter.)
   c. Surcharges for "excess" wastes can be added to a base charge for "normal" wastes. (This method has also been analyzed previously.)

Thus, average-cost user charges are to be applied to all users for operation and maintenance costs. Moreover, the industrial users must pay their share of the grant portion of the capital costs. Since these payments would be based on use and would be subject to changing conditions, they represent a form of user charges. Methods of determining user charges are considered further below.

FORMULATION AND CALCULATION OF USER CHARGES

As discussed earlier, user charges based on average costs are both equitable and practical and can recover total costs. Such charges are encouraged in the federal regulations and guidelines. This section considers the formulation and calculation of average-cost user charges and surcharge factors. It is assumed throughout that costs are allocated on the basis of flow, BOD, and suspended solids.

In Eq. 4, the average-unit costs for flow, BOD, and suspended solids are represented as v, b, and s, respectively, where:

$$v = \frac{\mathrm{TC_V}}{\mathrm{V}},$$

$$b = \frac{\mathrm{TC_{BOD}}}{\mathrm{BOD}}, \text{ and}$$

$$s = \frac{\mathrm{TC_{SS}}}{\mathrm{SS}}$$

For individual industries with large waste loads, these unit costs can be applied as individual unit-user charges. For examples of this approach, see Ref. 8 and 10.

As pointed out in the foregoing discussion on user charges based on group cost allocations, it may not be practical to measure each waste parameter for each individual in the case of small waste load contributors, although group estimates can be made. In such instances, user charges can be based on fewer parameters, and surcharges can be added to the other parameters. For example, Eq. 6 illustrates a unit user charge, c, based on flow only:

$$c = v + \frac{b \times \mathrm{BOD_G} + s \times \mathrm{SS_G}}{\mathrm{V_G}} \tag{7}$$

The unit charge, c, would be applied to the flow for each member in the group.

For individually calculated unit charges, the total payment is given in Eq. 4 as:

$$C_i = v \times \mathrm{V}_i + b \times \mathrm{BOD}_i + s \times \mathrm{SS}_i \tag{8}$$

As shown by Maystre and Geyer,[13] this equation can be rewritten in two forms using waste concentrations:

$$C_i = \frac{\mathrm{TC}}{\mathrm{V}} \times \mathrm{V}_i \left[k_\mathrm{V} + k_\mathrm{BOD} \frac{(\mathrm{BOD}_i)}{(\mathrm{BOD})} + k_\mathrm{SS} \frac{(\mathrm{SS}_i)}{(\mathrm{SS})} \right] \tag{9}$$

where

k_V = fraction of total cost, TC, assigned to flow,
k_BOD = fraction of total cost, TC, assigned to BOD,
k_SS = fraction of total cost, TC, assigned to suspended solids,
Q = concentration of the waste Q,

or, since $k_\mathrm{V} + k_\mathrm{BOD} + k_\mathrm{SS} = 1$:

$$C_i = \frac{\mathrm{TC}}{\mathrm{V}} \times \mathrm{V}_i \left[1 + k_\mathrm{BOD} \frac{[(\mathrm{BOD}_i) - (\mathrm{BOD})]}{(\mathrm{BOD})} + k_\mathrm{SS} \frac{[(\mathrm{SS}_i) - (\mathrm{SS})]}{(\mathrm{SS})} \right] \tag{10}$$

Note that the charge can be written as:[13]

$$C_i = F \times W \times V_i \qquad (11)$$

where

$W = \frac{TC}{V}$, the total cost per unit volume, and

F = a charge adjustment factor for BOD and suspended solids.

The charge adjustment factor, F, would be given by the term in brackets in either Eq. 9 or 10. For some individual users, F could be less than 1.0 and for others it could be greater.

Either equation also can be used to calculate a group charge-adjustment factor. In the above equation, the subscript "i" can be replaced with "G." The charge adjustment factor, F, calculated by means of average waste concentrations for the group, G, would then be applied to all users in the group. In calculating individual user charges or group adjustment factors, the charges are identical regardless of the form of Eq. 8 used.

In practice, many municipalities apply a variation of Eq. 10 in calculating a charge-adjustment factor for industries with waste concentrations above a fixed level.[13] Industries with lower waste concentrations are grouped with households, offices, and other users and typically are charged on the basis of flow at a rate for "normal" sewage. The charge-adjustment factor for high concentration industries is generally applied for waste loads in excess of "normal."

In summary, user charges on waste discharges produce incentives for waste reduction. Seagraves,[14,15] Ethridge,[16] and Elliot[17] demonstrated that waste loads contributed to combined systems can be reduced significantly by the application of such charges. Furthermore, the level of water use may also decrease as waterborne wastes decrease.

REFERENCES

1. Patterson, W. L., and Banker, R. F., "Estimating Costs and Manpower Requirements for Conventional Wastewater Treatment Facilities," Office of Research and Monitoring, EPA, Report No. 17090 DAN 10/71 (Oct., 1971).
2. Eckenfelder, W. W., and Adams, C. E., "Design and Economics of Joint Wastewater Treatment," *Proceedings ASCE, Sanitary Engineering Division* **SA1** 98, 153–166 (Feb., 1972).
3. Michel, R. L., "Costs and Manpower for Municipal Wastewater Treatment Plant Operation and Maintenance, 1967–1968," *Journal WPCF* **42** (11) (Nov., 1970).
4. Smith, R., "Preliminary Design and Simulation of Conventional Waste Renovation System Using the Digital Computer," FWPCA, U.S. Dept. of the Interior, Report No. Wp-20-9 (March, 1968).
5. Smith, R., "Cost of Conventional and Advanced Treatment of Wastewater," *Journal WPCF* **40** (9) (Sept., 1968).
6. Di Gregario, D., "Cost of Wastewater Treatment Processes," Robert A. Taft Water Research Center, Cincinnati, Ohio, Report No. TWRC-6 (Dec., 1968).
7. Frankel, R. J., "Water Quality Management: An Engineering Economic Model for Domestic Waste Disposal," Ph.D. thesis, University of California (Jan., 1965).
8. Schroepfer, G. J., "Determination of Fair Sewage Service Charges for Industrial Wastes," *Sewage and Industrial Wastes* **23** (12) (Dec., 1951). Reprinted in Ref. 9.
9. "Federal Guidelines, Equitable Recovery of Industrial Waste Treatment Costs in Municipal Systems," Office of Water Programs, EPA (Oct., 1971).
10. Anderson, N. E., and Sosewitz, B., "Chicago Industrial Waste Surcharge Ordinance," *Journal WPCF* **43** (8) (Aug., 1971).
11. *Federal Register* **38** (39) (Feb. 28, 1973).
12. "Proposed Rules," *Federal Register* **38** (98) (May 22, 1973).
13. Maystre, Y., and Geyer, J. C., "Charges for Treating Industrial Wastewater in Municipal Plants," *Journal WPCF* **42** (7) (July, 1970).
14. Seagraves, J. A., "Sewer Surcharges and Their Effect on Water Use," *Journal AWWA* **64** (8) (Aug., 1972).
15. Seagraves, J. A., "Industrial Waste Charges," *Proceedings ASCE, Environmental Engineering Division* **99**, EE6, 873–881 (Dec., 1973).
16. Ethridge, D., "User Charges as a Means for Pollution Control: the Case of Sewer Surcharges," *The Bell Journal of Economics and Management Science* **3** (1) (Spring, 1972).
17. Elliot, R. D., "The Industrial Demand for Water and Waste Treatment in Selected U.S. Cities Which Are Levying Surcharges," Ph.D. thesis, North Carolina State University at Raleigh (1972).

16

Desalination

Staff of WAPORA, Inc.

POTENTIAL OF DESALINATION

Desalination as a Means of Increasing World Water Supply. Only about 0.5 percent of the earth's water is potable. The 97 percent which is ocean water and the 2.5 percent which is classified as brackish are naturally impotable because of their dissolved salt content. However, if these contaminated or saline waters can be converted to a potable state, they offer a virtually unlimited potential for future water supply.

Current State-of-the-Art. Realization of the potential of the desalination technique will require many future solutions reached in concert by engineers, chemists, physicists, economists, mathematicians, and others. However, the technical approaches to desalination and their governing principles are relatively straightforward. It is possible to trade off technical advantages and disadvantages among the various desalination processes to suit a given desalination context, and to balance the practicality of existing equipment and the promises of conceptualized possibilities.

In all desalination methods currently available, salt and water, or relatively pure water and the more salty water from which it originated, are separated by one means or another into their two distinct phases. Common methods used are: distillation, reverse osmosis, electrodialysis, freezing, and solar evaporation.

The desalting plants in operation in 1970 are listed by type of process in Table 16-1, by geographic location in Table 16-2, and by size in Table 16-3.[1] Pilot plants in which various processes are tested at the Office of Saline Water (OSW) Research and Development Test Station at Wrightsville Beach, North Carolina, are pictured in Fig. 16-1.

Cost Projections. The technical considerations involved in desalination practices require evaluation in terms of their costs in order for the practicality of their application to be assessed. Except for certain local situations, present de-

TABLE 16-1. Desalting Plants by Type of Process

Process	No. of Plants	Plant Capacity (mgd)	Percent Capacity
Total all processes	712	244.6	100
Distillation			
Single-stage flash (F)	49	6.6	2.7
Multi-stage flash (MSF)	196	158.4	64.9
Vertical tube evaporator (VTE)	102	30.5	12.3
Submerged tube (ST)	293	33.5	13.7
Vapor compression (VC)	22	2.9	1.2
Membrane			
Electrodialysis (ED)	44	12.2	5.0
Reverse osmosis (RO)	3	0.2	0.1
Freezing			
Vacuum freezing-vapor compression (VFVC)	3	0.3	0.1

TABLE 16-2. Desalting Plants by Geographic Location

Area	No. of Plants	Plant Capacity (mgd)
United States	301	43.1
United States territories	14	10.3
North America except USA and its territories	12	8.5
Carribbean	31	24.1
South America	23	4.1
England and Ireland	63	15.9
Europe (continental)	97	31.7
Union of Soviet Socialist Republics (USSR)	9	12.6
Middle East	78	69.1
Africa	50	19.7
Asia	27	4.2
Australia	7	1.3
Totals	712	244.6

TABLE 16-3. Desalting Plants by Size

Size Range (thousands gpd)	No. Plants	Plant Capacity (mgd)
25–99	359	18.4
100–499	263	50.7
500–999	32	21.6
1000–4999	51	108.7
5000 and greater	7	45.2
Total	712	244.6

salination costs are still too high to permit desalting techniques to achieve wide practice. Projections to the future offer encouraging possibilities, however. Figures 16-2 and 16-3 prepared by the Office of Saline Water present the future cost possibilities associated with brackish and seawater desalting.

DESALINATION BY DISTILLATION

The process used to the largest extent today to convert saline water to potable water is distillation. It has demonstrated its practicality in producing about 6 mgd of fresh water at a reasonable cost.

In this process, saline water is boiled or evaporated in a suitable vessel to produce vapors of practically pure water and a liquid residue of more highly concentrated brine. The vapors are condensed to potable water and are collected in a separate part of the system.

While distillation is a familiar technique, there are theoretical and practical limitations common to its application to desalination.

Theoretical Considerations. There is an absolute minimum energy requirement (independent of the kind of process) for separating a homogeneous mixture of materials into its component parts. This energy requirement depends upon the physical properties of the mixture at the start of the separation and those of the

Fig. 16-1 OSW Research and Development Test Station, Wrightsville Beach, N.C. (*Courtesy of OSW*)

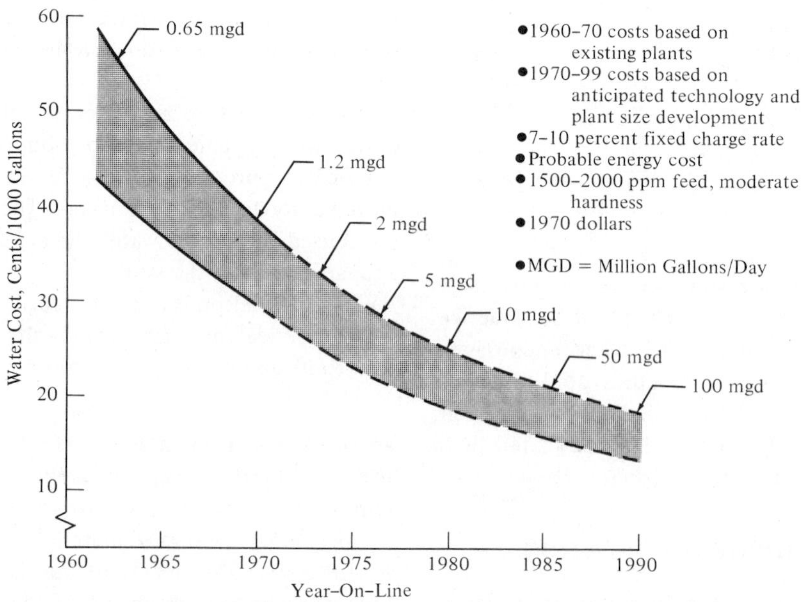

Fig. 16-2 Brackish water desalting cost. (*Courtesy of OSW*)

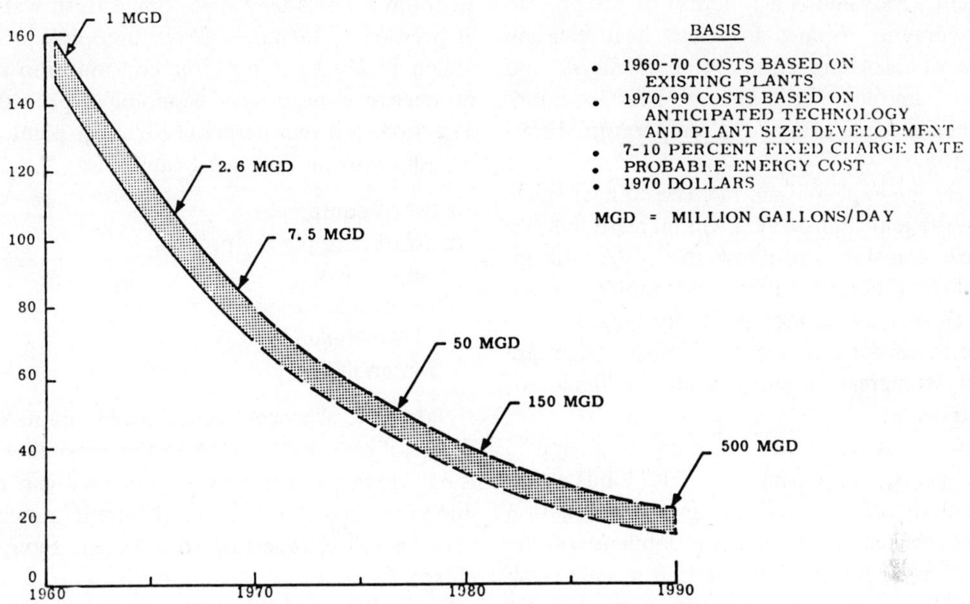

Fig. 16-3 Seawater desalting cost. *(Courtesy of OSW)*

separated component parts at the end of the process, and is commonly known as the free energy of separation.

For example: the initial mixture is an extremely large body of saline water (at 25°C), and the final components are 1000 gal of potable water (at 25°C) and a body of saline water which is only reduced in size by 1000 gal whose salt content is basically unchanged (at 25°C). The minimum amount of energy necessary to effect this separation is about 2.5 kilowatt-hours or about 8500 Btu. Therefore, an engineering process which will effect a separation cannot be constructed to operate more efficiently than these numbers indicate.

If the simple process of boiling saline water and condensing the pure water vapors as product is examined in this light, a severe disparity is apparent. The simple process consumes about 1000 Btu/lb of water produced, while the minimum theoretical requirement is of the order of 1 Btu/lb of water produced. The boiling process, however, can enjoy considerable engineering modification to bring it closer to the theoretical limit.

Perhaps the most direct method of reducing the actual separation energy requirements is to operate a multieffect system. For this process, the pure water vapors generated in the first system of boiling brine are introduced in an appropriate manner into a second system. In the second system, the original water vapor, upon condensation, releases sufficient heat to boil the brine and produce approximately one more pound of vapor. This process can, in principle, be continued indefinitely. In the limit, though, about 1 Btu/lb of water produced will have to be expended.

Practical Considerations. Before establishing engineering designs of distillation, the practical considerations which affect these designs should also be examined. In most cases, it is these practical considerations which limit efficiencies long before the theoretical limit is approached.

The heat transfer process itself is of fundamental concern. Generally speaking, the overall rate of heat transfer (Q, Btu/hr) is directly proportional to the area available for transfer (A, ft^2), and the overall temperature driving force (ΔT, °F):

$$Q = UA\Delta T$$

The proportionality constant U, (Btu/hr/ft^2/°F) is termed the overall heat transfer coefficient. Its value is a function of the prevailing hydraulic regimes for both heat-releasing and heat-absorbing fluids, the geometry and material content of the heat transfer area, and, in general, a function of the temperature difference, too.

As can be expected, an increase of heat transfer coefficient and/or area will increase the rate of heat transfer, and allow the system to approach its theoretical limit more rapidly. However, an increase in area generally means an increase in equipment size; thus, most designs are aimed at increasing the overall coefficient of heat transfer.

Seawater is a complex mixture of dissolved solids, gases, and marine life. The solids play a particularly deleterious role in the distillation process because some of them condense on the heat transfer surfaces and deposit a scale (scaling). Marine life can attach itself to the heat transfer areas and develop an insulating blanket (fouling). Both these phenomena can dramatically reduce the overall heat transfer coefficient, which in turn reduces the heat transfer rate and the process efficiency. Naturally, as the seawater becomes more and more concentrated through the process, these problems become more and more severe.

In order to minimize the effects of these constituents, the seawater is pretreated to either remove the harmful materials or render them inactive. The most common scale-forming salts are calcium sulfate, magnesium hydroxide, and calcium carbonate. Pretreatment of the raw brine through pH control and decarbonation will prevent the formation of magnesium hydroxide and calcium carbonate scale. The limiting temperature and brine concentration will therefore be determined by the concentrations of calcium and sulfate ions in the brine since calcium sulfate becomes *less* soluble as the temperature increases. Although higher temperatures are possible—a more efficient process—by precipitating calcium sulfate in the pretreatment phase, the choice is an economic one between higher efficiencies and the increased cost of pretreatment. Marine life is eliminated by chlorinating the raw feed to kill any existing organisms and to prevent subsequent growth.

Economic Considerations. Since fresh water is at present a cheap natural resource, any process which produces it for large consumption must of necessity be a very economical procedure. The financial requirements for most plants fall broadly into the following categories:

- Cost of equipment
- Cost of land for the process
- Energy costs
- Operational expenses
  - Maintenance
  - Materials

Naturally, a process which could minimize all of the above costs would be the most economical. However, since most processes can only minimize some of them, different processes may be more appealing to different economic climates.

Generally speaking, high efficiency plants in terms of heat exchange are more exotic in design. Thus, while they tend to reduce energy costs, they do so with relatively large investments in equipment.

Process designs which more effectively utilize heat transfer areas will tend to make equipment size and cost smaller. However, these arrangements tend to have fairly complex geometry, and the costs of cleaning and replacement (maintenance) tend to rise.

Obviously, a good design will minimize the overall plant expenses. If energy is a cheap commodity, a less efficient plant would be suggested, whereas high energy costs would dictate more efficient design. Likewise, if land is accessible and cheap, plants of greater areal extent should be considered over more compact units necessary in areas where land is a valuable commodity.

Since there are so many parametric considerations in economically optimizing a desalination plant, this optimization has constantly been a subject of extensive mathematical modeling.

Promising Distillation Processes.

Vertical-Tube Evaporator. A simplified process flow diagram for a multiple-effect vertical-tube evaporator (VTE) is shown in Fig. 16-4.[2,3]

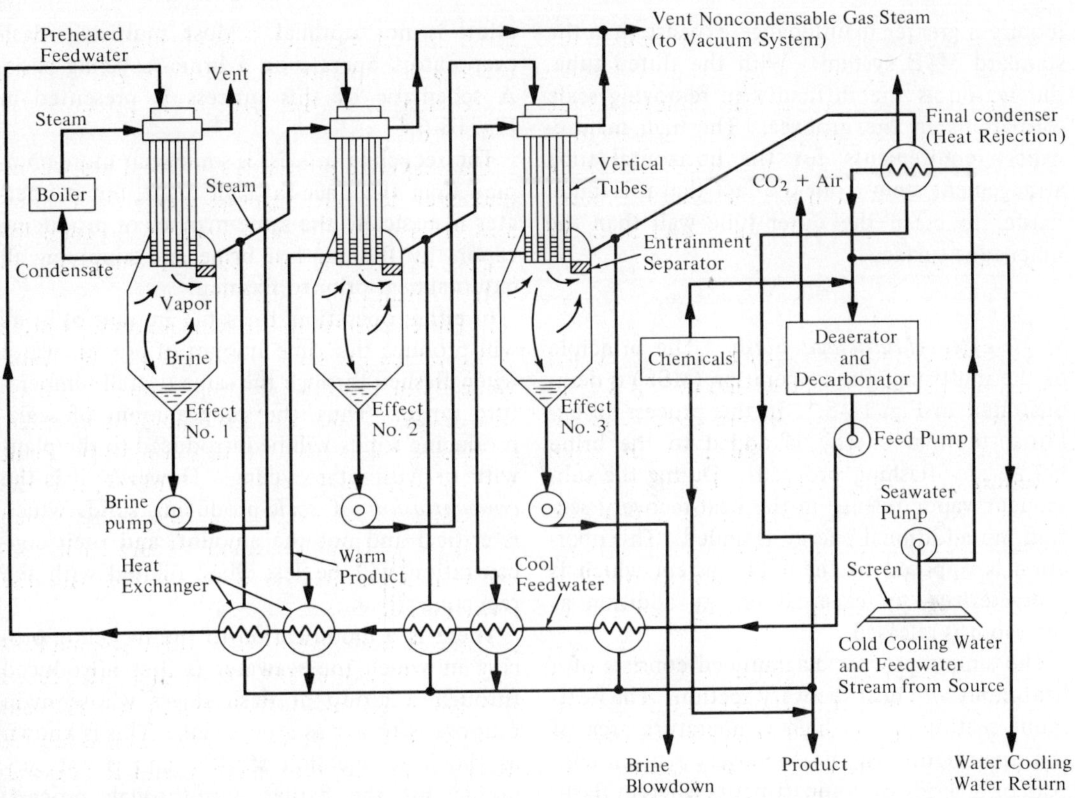

Fig. 16-4 Triple-effect vertical-tube evaporator distillation plant (From *Desalting Handbook for Planners*, 1st Ed., OSW, 1972).

In this process, pretreated seawater is fed to the first of a series of stages, and flows downward (or upward) through a series of tubes as a film coating the inside of the tubes. Externally-supplied steam condenses on the outside of the tubes of the first effect and supplies the necessary thermal energy to vaporize the brine flowing within the tubes. The condensed steam is returned to a boiler to be revaporized. The water vapor produced in the first stage serves as the steam supply for the second stage and the unflashed and more concentrated brine of the first stage serves as the feed for the second stage. The condensed vapors for the second and subsequent stages are withdrawn as product water.

The process is continued for as many stages as is economically practical, with the final water vapor produced being condensed by heat exchange with seawater, and the most concentrated brine discharged as waste.

A recent development in this process has been the use of fluted tubes in place of smooth round tubes.[4] These tubes increase the overall heat transfer performance 2 or 3 times. Thus, they can make either the heat transfer process of a given system more efficient, or allow smaller units to accomplish the same separation as a larger unit with conventional tubing.

Another method of increasing the heat transfer coefficient is to use an array of staggered horizontal tubes (HTME).[5] The brine falls onto the tubes and part of it vaporizes (steam condensing on the tube interior). The more concentrated brine flows to the underside of the tube and drips off. This arrangement yields a bare metal surface on the upper tube surface and a thickened liquid film on the underside. Apparently the increase in heat transfer on the top more than compensates the reduced heat transfer underneath.[6]

Both fluted and horizontal tube arrangements

require a greater maintenance expense than the standard VTE system. With the fluted tube, the reason is the difficulty in removing scale formations in the grooves. The high maintenance requirements for the horizontal tube arrangement stem from the fact that it is much harder to clean the outer tube wall than the inner tube surface.

Single-Effect Multistage Flash. The principle of the multistage-flash operation (MSF) is demonstrated in Fig. 16-5.[4] In this process all external thermal energy is added to the brine before any flashing proceeds. During the subsequent vapor release in the heat-recovery section, no additional energy is added. This operation is opposed to the VTE system which is characterized by external energy addition as the vapor is released.

The simple process diagrammed consists of a heat-input and heat-recovery section. The heat-input section is the high temperature area of the operation. The heat-recovery section consists of a series of compartments wherein flashing of the brine occurs. Each section is operated at a lower pressure than the previous. Since the brine is becoming more concentrated as it passes from one compartment to the next, a lower pressure is needed to induce flashing. The flashed vapor condenses by heat exchange with the feed, and the condensate is removed as product.

Although the system discussed above is adequate for illustrative purposes, such an operation is not optimal. Most multistage-flash evaporators operate on a brine-recycling basis. A schematic of this process is presented in Fig. 16-6.[3]

The recycling process is somewhat more complex than the once-through plant, but it operates in basically the same manner of producing vapors by flashing hot brine and supplying all external heat prior to flashing.

In either operation, the same amount of brine will produce the same amount of potable water when flashed through the same overall temperature range. Thus, the same amount of scale-producing solids will be introduced to the plant with or without recycling. However, it is the *concentration* of scale-producing solids which is critical and not the amount, and their concentration will be less when diluted with the recycling stream.

There is a modification to the recycling process in which the seawater is first introduced through a group of flash stages whose main purpose is to act as a heat sink. This is known as the heat rejection section, and is not economic for the earlier once-through process. The heated brine feed is split, part discharged as effluent, and the make-up is pretreated and returned to the recycling stream.

From an economic viewpoint, the recycling process is superior to the once-through process for high-capacity plants. Feed treatment costs are smaller and more efficient utilization is made of the external heat supplied. A disadvantage is the reduction of brine temperature because of its higher dissolved solids content.

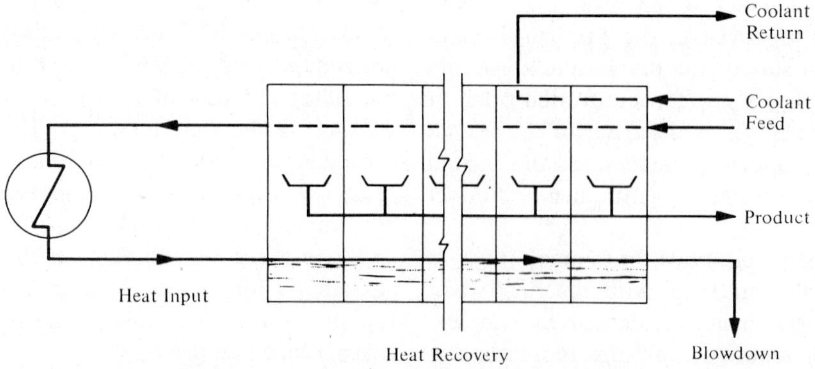

Fig. 16-5 Once-through multistage flash operation. (From *Distillation Plant Data Book*, Houston Research Inc. for OSW, Contract No. 14-01-0001-2099, 1971)

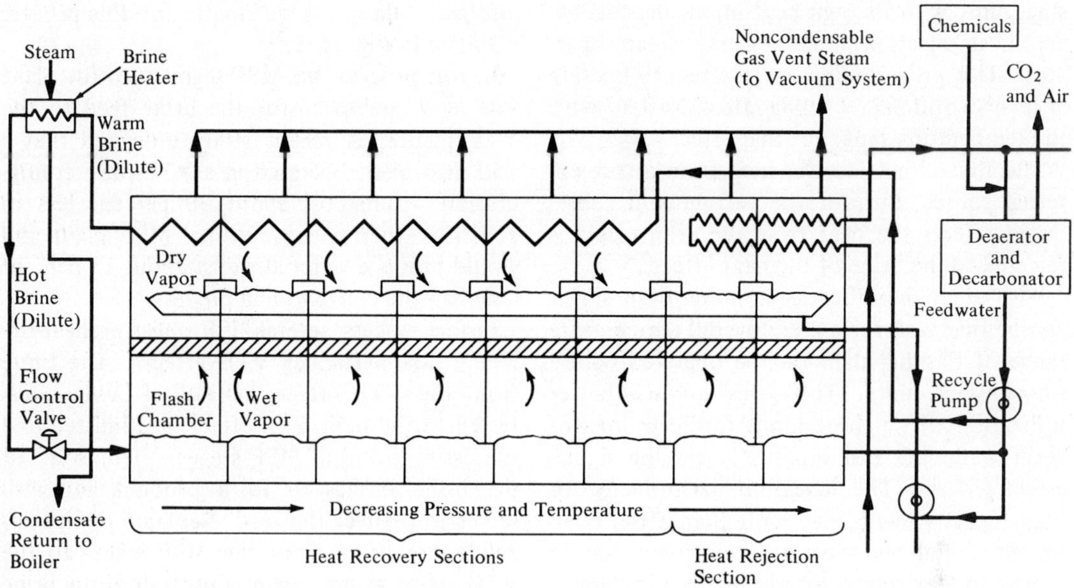

Fig. 16-6 Multistage-flash evaporator with brine recycle. (From *Handbook for Desalting Processes*, Hittman Associates, Inc. for OSW, publication pending)

Multieffect-Multistage. The recycling process discussed above is still considered a single-effect with external heat input at one end and heat rejection at the other. Up to a point, this process becomes more and more efficient for a given overall temperature range as the number of stages increases. The pressure difference needed to move the brine from chamber to chamber at the low temperature, low pressure end of the cycle is the limiting factor in the recycling rate and number of stages to be employed. In order to overcome this limitation, a multistage-multieffect process can be operated as indicated in Fig. 16-7.[2,7]

Operating in this fashion, it is possible to break up the overall temperature flashing range into a number of temperature intervals. Within each interval, an individual single-effect multi-

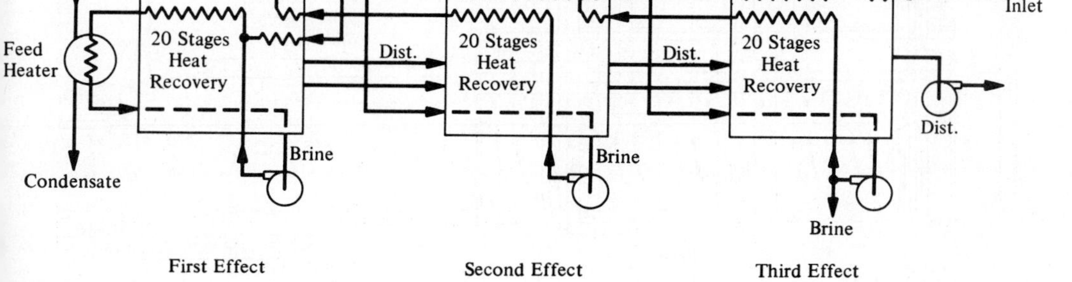

Fig. 16-7 Multieffect-multistage distillation process. (From *Desalting Handbook for Planners*, 1st Ed., OSW, 1972.)

stage unit with its own heat input, heat recovery, heat rejection, and recycling stream operates. This process allows each effect to operate at its own optimal recycling rate consistent with the temperature range for that effect.

The first effect receives heat from a true external source, but for the second and subsequent effects the heat from the prior effect is rejected to the brine of the next effect.

With these modifications, more flash stages can be operated for a given overall temperature range of flashing than can be operated with a single-effect unit. This gives a much better utilization of net heat input (more gallons of fresh water per Btu supplied) over the single-effect process. The increase in performance for high-capacity plants may more than offset additional capital and operating expenses due to more complex staging arrangements and pumps. This would be particularly true in an area where fuel costs are extremely high.

Combined Processes (*VTE/MSF*). One of the most recent innovations in desalting processes is a combination of vertical-tube evaporator and multistage-flash. A schematic for this process is shown in Fig. 16-8.[2,8]

In this process, the MSF segment of the plant acts as a preheater for the brine feed to the VTE plant. A recent study indicated that a 250 mgd plant operated in a VTE/MSF configuration would cost about 30 percent less to construct than a conventional MSF plant and would produce water at a cost about 15 percent less than the conventional process.

In this process, several MSF stages preheat the feed for a particular VTE effect. The brine from the VTE effect, instead of being transferred to the next VTE effect, is admitted to a successive group of MSF stages. Here, some of the brine flashes to form product and also serves to preheat the feed. Separate pumps circulate the brine from the MSF stages to the VTE effect where, again, a portion of the brine is vaporized and the remainder flows by gravity to a cooler, lower pressure series of MSF stages.

All studies indicate that this combined system will be the most prominent distillation process for large-capacity plants of the future. Preliminary operation of a small unit by the Office

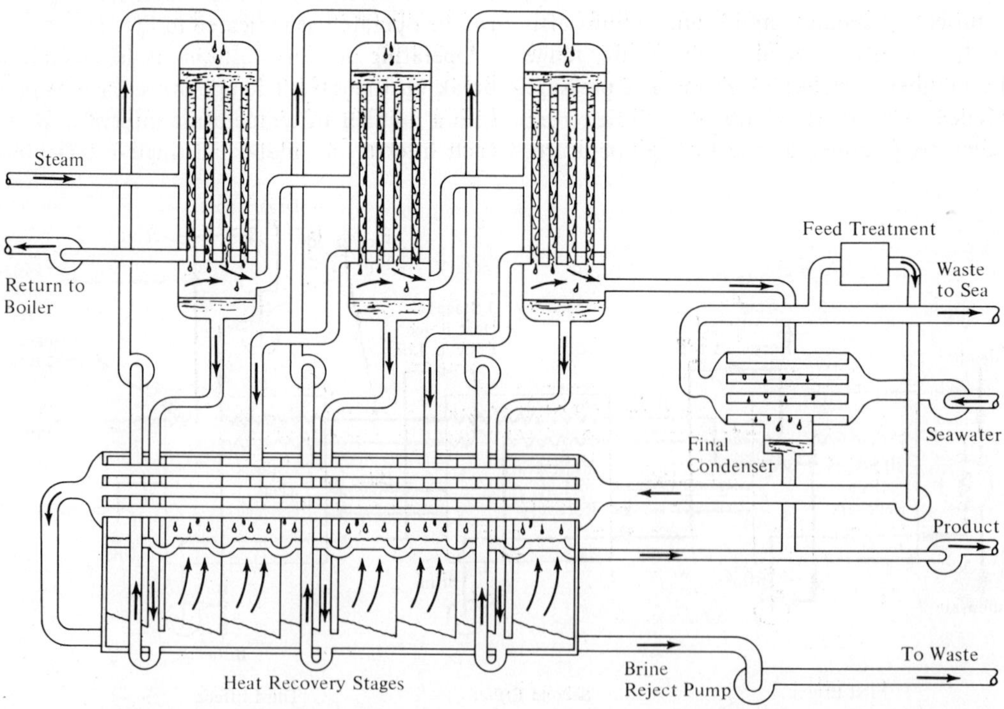

Fig. 16-8 Combination VTE-MSF distillation process (From *Desalting Handbook for Planners*, 1st Ed., OSW, 1972)

of Saline Water at their Freeport, Texas, facility (1 mgd) indicated smooth, reliable performance with a minimum amount of maintenance.[9] As a result, OSW has entered into a joint agreement with the Orange County Water District (Orange County, California) to construct a combined plant to produce about 3 mgd of fresh water.[5]

Summary of Advantages and Disadvantages of Distillation Processes.[6]

Vertical-Tube Evaporator (VTE)

Advantages:
1. Low power consumption.
2. High performance ratio.*
3. Efficient operation at part load.
4. Higher brine discharge concentration (less feed/unit of distillate, therefore, lower pretreatment costs) due to once-through nature.
5. Less land space.
6. Low capital costs.

Disadvantages:
1. Scale deposits induced by metallic heat transfer surfaces in contact with ebulliating brine.
2. Unappealing aesthetics of huge vertical height.

Single-Effect Multistage Flash (SEMS)

Advantages:
1. Substantial background of operating experience (most common unit in operation).
2. Greatest economy for small capacities (<2 mgd).
3. Easiest tube cleaning (no metallic heat transfer surface in contact with evaporating brine).
4. Low maintenance requirements.

Disadvantages:
1. Low performance ratio.
2. Limited operation due to calcium sulfate solubility characteristics.
3. High power consumption requirements.
4. Larger amounts of land required.

*Gallons of water produced/1000 Btu of heat input.

Multieffect-Multistage Flash (MEMS)

Advantages:
1. Higher performance ratio than SEMS.
2. Toleration of high calcium sulfate solubility characteristics.
3. Better performance control for each effect.

Disadvantages:
1. Limited applicability (only one plant in operation).
2. Exceedingly complex operation requiring high-trained operating personnel.
3. Long start-up and shut-down times (low stream factor).

Combined VTE/MSF

Advantages:
1. All of those discussed above in the individual VTE and SEMS processes.
2. Realization of about 30 percent reduction in capital costs and 15 percent reduction in product water costs over conventional MSF operations in a larger capacity plant using MSF as a brine preheater.
3. Smooth, easy operation with no detrimental interactions between the two processes (indicated by limited experience).

Disadvantages:
1. All of those discussed above in the individual VTE and SEMS processes.
2. Limited operating experience, presently confined to pilot or semi-works capacities.

DESALINATION BY REVERSE OSMOSIS

If a suitable membrane is fashioned as a boundary of a vessel containing salt water, it will be found that upon the application of an appropriate pressure to the saline solution, almost pure water will be forced through the membrane. This phenomenon is variously called desalination by reverse osmosis or by hyperfiltration. In actual fact, although the process is considerably advanced in its engineering aspects and in its practical applications, its mechanism is still uncertain.

In ordinary filtrations, comparatively gross particles are separated from their fluid matrices by the sieve action of the filter medium. The mechanism involved is the common one whereby particles of a given size are restrained by their bulk from passing through apertures of smaller dimensions. In the ultrafiltrative reference, particles of a much smaller size, in the micron to fraction-of-a-micron range, are separated from fluids containing them by filters whose sieve action depends upon holes, pores, or passages that are smaller than those that are encountered in the usual filtration experience. The microporous membranes of the Millipore type constitute such filters.

The term hyperfiltration implies the downward extension in size of the usual filtration technique for the removal of particles or entities in solutions by the use of filters. These filters may be polymeric membranes wherein the intersegmental spacings of the constituting polymer serve as the separating sieve. The implication of this term to the desalination phenomenon is that the solute ions (in their hydration states, if appropriate) are "filtered" from their encompassing water molecules by the spatial interference of their radii with the geometry of the membrane's intersegmental interstices. As will be seen later, there is certain evidence that the mechanism of reverse osmosis desalination is not at variance with hyperfiltration and many investigators favor the latter term. Others consider reverse osmosis more apt.[10,11]

If an aqueous salt solution is to be separated from pure water by a membrane that is permeable to water but not to salt, it will be found that water will traverse the membrane into the saline side of the system. Dilution of the salt concentration will occur. Indeed, the extent of that consequence is the measure of the osmotic drive to obliterate the concentration gradient. This is the prime mover in the mechanism. The motivating osmosis can be measured in pressure terms, and is dependent in its magnitude on the concentration of the salt solution. A point of balance is reached in the osmosis phenomenon where the tendency of the salt solution to imbibe water is balanced by the osmotic pressure of the system. The process is reversible. The imposition of pressure on the salt solution will cause water to traverse the membrane (permeable to water but not to salt) in the other direction in an effort to reestablish an equilibrium position in keeping with the new "osmotic pressure," one component of which is the artificially applied pressure. The consequences, then, of the imposed pressure on the salt solution are to cause pure water to be separated through the membrane, and to leave a more concentrated salt solution behind. This experience, the reverse of normal osmosis, exactly describes the desalination situation.[12]

Whether by means of hyperfiltration or by some other mechanism, reverse osmosis eventuates in a separation of pure water from aqueous salt solutions. The nature of the membrane is central to the process.

Membrane Characteristics. The first polymer to indicate suitability as a membrane for desalination by reverse osmosis was cellulose diacetate. It was formulated into a promising membrane, and was eventually developed and engineered into the present standard membrane offering. Research studies on certain other polymeric materials (funded primarily by OSW) showed that cellulose diacetate as a polymer is not unique in its desalination abilities. However, these new polymers were compared at their research level of development with the performance of the optimized and engineered cellulose diacetate standard. The new polymers proved that there are alternatives to cellulose diacetate, but evidenced no advantages compelling enough to warrant development funds for their engineering optimization to the point already achieved for cellulose diacetate. Hence, in effect, cellulose diacetate remained essentially unchallenged.

The exact seat of desalination action in the effective polymers is not understood with scientific preciseness. It is known, however, that the property stems from a balance of influences that carefully governs the degree of water swelling by the polymer. More than this critical amount of water imbibition causes the salt discrimination of the membrane to be reduced, less swelling, and the flux, the volume of water passed by a given area of membrane per unit time, to be diminished.

Cellulose diacetate was optimized by development into an asymmetric membrane of the Loeb type. An entirely different technique, based on the extrusion of hollow fibers, was commercialized by the DuPont organization using an aromatic polyamide polymer, a type of nylon.

Asymmetric Cellulose Diacetate Membrane. Loeb and his associates pioneered the development of the asymmetric or "skinned" cellulose diacetate membrane.[13,14,15,16,17,18,19] It consists of a thin layer of dense cellulose diacetate superimposed upon a thicker "honeycomb" or porous layer of the polymer. The virtue of the membrane lies in the salt-rejection qualities given by the denser layer, a minimum of 98 percent of salt discrimination at the seawater concentration level with a flux of from 1 to 2 cc/hr at 1500 psi.

The overall dimensions can be varied at will, but the skin or dense layer is usually from 25 to 100 atoms, and the overall membrane thickness is from 4 to 10 mils. The thinner the dense layer, the less its inhibitory effect on the flux. Although thin, the layer is continuous and, if unflawed by disruptions, pinholes, etc., provides an adequacy of salt rejection. The less dense the thin layer, the greater the flux, but the lower the amount of salt discrimination. The formulation of the casting composition of polymer, solvent, and various addenda, as well as the casting technique and the post-treatment of the membrane can all be manipulated to give a skin of greater or less density and thickness.

The spongy underside of the membrane serves essentially as a support for the ultrathin, effective denser layer. It is too open to possess salt-rejecting properties or to limit the membrane's flux qualities.

When the asymmetric membrane is used in its flat form, there is a correct "up" and "down" side. The dense side must face the salt solution with the open spongy layer resting upon the membrane-support plate. With the reverse arrangement, the salt concentration builds up in the spongy layer, and polarization effects result. Additionally, the pressure's influence on the intimate contact of thin skin and membrane-support is more likely to damage the continuity of the dense layer, with a concomitant disruption of its salt-barrier properties.

In actual practice, the membrane is seldom used in flat construction since tubular designs offer more effective surface area per unit volume of space. Whether flat or tubular, however, the casting formula and its manipulative techniques are handled so that a dense layer of the minimum thickness sufficient to give an adequacy of salt rejection without exacting an excessive price in the reduction of product-water flux is manufactured.

The attainment of this desirable proportionality is an exercise in practical empiricism. Its achievement is necessary to the success of this method of desalination, since the application of this method is judged, as are others, not only by its technical possibilities, but by what proportion of these is realized for the price paid in terms of dollars, effort, and time.

Preparation of Asymmetric Cellulose Diacetate Membranes. Loeb, his colleagues, and his successors have established numerous formulations from which acceptable asymmetric membranes can be cast. The ingredients commonly contain cellulose diacetate polymer, a solvent or combinations of solvents, and often a "casting salt." Casting, depending upon the formulation, is carried out at near 0°C when more volatile solvents are used. However, the use of higher boiling solvent mixes permits the convenience of casting at room temperature.

After being cast, the wet membrane is exposed to air for a measured time interval before being immersed in water, at different temperatures for different formulations. Following a period of leaching of solvents in the water bath, the membrane is removed and "conditioned" by being exposed for a period of time to heated water. The temperature and duration of the conditioning step fix the final properties of the wet, finished membrane.

The chemistry and physics of the preparative steps are complicated. Many of the interactions still require clarification. Furthermore, the elucidation of this complexity is hardly helped by the vigor of the claims and counterclaims. However, a focus on a useful working hypothesis covering membrane formation techniques and their corresponding influences on membrane structure is available from the companion field of porous membrane preparation.[20]

When the casting "dope" is laid down as a wet sheet, evaporation of the more volatile components commences from its interface with the air. These constituents are invariably "good" solvents for the polymer. As a consequence, a decreasing concentration gradient in the "good" solvent from the center of the membrane outward is created. This provides a driving force for the migration of the "good" or more effective solvents into the top layers of the wet sheet from within the membrane. The evaporation of the "good" solvent leaves a dense membrane skin behind. The remaining gel, becoming progressively richer in nonsolvent, undergoes a phase inversion when it can no longer solublize the polymer. At this point, the wet membrane is finally plunged into water. Its final gel form is set by the massive leaching of whatever solvent remains in the wet gel.

The setting of the top skin of the membrane structure finds its counterpart in a similar finalization of the structure of the bottom portion of the membrane. This area, bereft of "good" solvent by its migration into the top, becomes progressively less dense. This is a consequence of the properties of solutions of polymers in "poor" solvents. In any case, massive dilution with water converts the "poor" solvent into nonsolvent, and fixes the open structure of the honeycombed bottom portion of the membrane.

In the conditioning step, the membrane, unrestricted in its tendency to shrink, undergoes densification as a consequence of heat treatment. As a result, its salt-rejecting qualities are increased, and its flux properties are decreased. Heat treatment also has certain desirable ancillary effects. Casting stresses tend to be relieved, and residual solvent that could serve as plasticizer is more thoroughly removed. This conditioning step finds an exact analogy in porous membrane manufacture where its precise role is appreciated. In that practice, however, the membrane is restrained against shrinkage so that casting stresses and residual solvent are removed without membrane densification taking place.

The above hypothesis is by no means universally endorsed, and it may even offer distortions in not properly weighting the influences of the several factors it considers. Nevertheless, it does explain the membrane formulative and casting process in a useful manner.

Polymer chemists, aware of the cohesive energy density relationships of polymers and solvents, have sought to bring a greater degree of system to the present state of empiricism. However, the complexity of the process has so far proved too great. Likewise, quantitation of relative solvencies and rates of evaporation derived from the lacquer makers' technique have thus far served only as rough first guides. The practice of membrane manufacture remains essentially an empirical art embodying a high degree of technology.

Mechanism of Reverse Osmosis Desalination. Any mechanism that is advanced to interpret the desalination phenomenon by reverse osmosis must explain how water permeates the membrane and why the ions resulting from the dissolved salts do not. As stated earlier, there is no agreement on the mechanism that is, in fact, at work in the process, although several theories have been advanced.

It may well be that some theories differ in semantics, even the exactness of science being limited by the vagueness of new concepts and the imprecisions of verbal expositions. The proponents of the hyperfiltration mechanism do not argue simplistically that the intermolecular spacings within the polymeric structure serve as sieve holes to differentiate among water molecules and hydrated ions. These entities are generally considered too similar in size to lend themselves to such a neat steric differentiation.

It is advocated rather that it is the difference in effects that the force fields of the membrane exert on the transport properties of water molecules and ions that cause their separative filtration. Yet it has been proposed that spatial separation based on geometric considerations are indeed involved.[21]

Cellophane, a regenerated cellulose film, is found to increasingly discriminate against salt ions as the saline concentration diminishes. In this view, the diluter solutions permit the more plentiful hydration of the sodium ions. These, as a consequence of their more extensive hydration, become larger in size, and so are more effectively screened out by the interstitial spacings of the membrane polymer.

Keilin and his associates have elaborated a mechanism involving the transport of water across the membrane based on a hydrogen bonding of the water molecules to the oxygen groupings of the polymer.[22] The function of the applied pressure is to cause the flow of this water through the membrane by motivating the shift of the hydrogen bond allegiances from one polymeric site to another. The failure of the salt ions to traverse the membrane is linked to the critical degree of membrane swelling. It may be helpful to think of this action in terms of "free" and "bound" water, however artificial the concept.

It is known that cellulosics bond water with varying degrees of tightness. Most of the water of a fully swollen cellulosic membrane is rather easily removed by a conventional drying procedure. However, a final portion of water is so strongly held as to defy removal except by the most drastic means. This latter type of water is called, appropriately, bound or primary water.

The bound water is seen as being capable of hydrogen bonding to other water molecules, thus permitting their permeation through the membrane. It is, however, postulated as being unavailable for ligand formation to sodium ions, for instance. Hence, sodium ions that can form hydrates with and traverse areas of free water are incapable of permeating membranes, whether cellulosics or otherwise, that are restricted by their degree of swelling to the exclusion of secondary or free water.

To the extent that the terms "free" and "bound" water imply a step function in their distinctiveness, the concept is misleading. Actually, as the quantity of water diminishes, the remainder is held more and more tightly by the polymer. The point is that in the polymer swollen to the proper degree, the water is available for hydrogen bonding by water molecules, but not to ligand formation with sodium ions. The result is water flux characterized by salt rejection.

Problems of Reverse Osmosis. The requirements for applied pressures in the neighborhood of 100 atm to effect reverse osmosis on salt solutions having the salinity of seawater, and even the lower pressures (50 atm) needed for brackish waters, serve to compact the membrane. Membrane compaction results in a densification that decreases the system's flux.

Research work on this subject has met with some success, particularly efforts to raise the glass transition point of the membrane polymer as by *in situ* cross linking. This approach renders the membrane less responsive to the creep influences of pressure. The commercially available membranes, however, do undergo membrane compaction with its concomitant diminution of flux.

Polarization is the second of the technical problems that is associated with the reverse osmosis desalination operation. As a result of the reversal of the osmotic process, pure water flows through the membrane from the salt solution leaving a more concentrated brine behind. It is found in actual operations that despite the flow of the incoming salt solution feed, there is built up a concentration of high salinity local to the membrane surface. This serves to increase the pressure requirements for the actual reverse osmosis. Put in another way, the constant applied pressure that was sufficient to provide an adequate flux at the original saline level of the feed solution produces a lower flux at the higher local salinities at the membrane caused by salt polarization.

Cell design seeks to minimize polarization effects by arranging for a sweeping of the membrane surface by the incoming feedwater. Turbulent flow is also arranged to stir up and remove the polarization layers. In the operation of commercial units, the problem has been managed as well as can be by the selected equipment design inherent in the commercial device.

Inherent membrane weaknesses post limitations to the reverse osmosis operation. Both cellulose diacetate and the aromatic nylons that are used commercially have hydrolytic stabilities that are pH dependent. This requires regulation of the pH of the feed stocks, or the risk of eventual hydrolytic membrane decrepitude. The state of the membrane art could with advantage be extended to preclude hydrolytic membrane decrepitude and to advance hydrolytically stable reverse osmosis membranes.

The oxidative stabilities of the membranes seem also to have been slighted in developmental efforts on the reverse osmosis process. The justification for this neglect is that, even for chlorinated waters, the limitations of the operations do not in practice find expression in the oxidative compromise of the membrane components. Nevertheless, polymeric oxidative degradation is a reality, and its prevention, if only by the inclusion of antioxidant stabilizers in the formulations, would seem to be well advised.

The possible need to adjust the pH of the feed solution to protect the membranes against hydrolytic degeneration is only one facet of the broader and important problem of raw water pretreatment. The feed stocks may well require management before being introduced into the commercially-available reverse osmosis units. Particulates must be removed lest they clog the hollow fibers or block or scour tubular surfaces. The pH levels may require adjustment and the possible consequences in terms of scale formation must be considered. The hardness of the feedwaters may also merit attention. The point, specifically, is that the optimum operation of commercially-available reverse osmosis units requires a linking to feedwater considerations and to their adequate pretreatments.

Practical Reverse Osmosis Desalination Units. The nub of the successful application of reverse osmosis membranes to desalination is the cost of the installation. The surface area of the unit is, in turn, often the key to its operational cost.

Tubular Design. Certain commercial units are based on tubular membrane designs. This arrangement promotes the turbulent flows that minimize salt-concentration polarization. Tubular dispositions also handle particulate matter better. This is of importance in nonwater contexts, such as with cheese whey. The tubes are also more amenable to clearing devices, such as a sponge ball. The standard Leob-type cellulose diacetate membrane, whether flat or tubular, has a flux of from 15 to 20 gpd/ft^2 at a pressure of about 600 lb/in.^{2} Operating on water having a salt content of from 2000 to 4000 ppm, it will give a product water reflecting a salt rejection of about 90 to 95 percent.

The Abcor Corporation has developed prepackaged reverse osmosis modular units. Each module contains 38 ft^2 of membrane, and is expandable by modular additions to 117 ft^2. Each 38 ft^2 module has a 400 gpd capacity. The enclosure houses a high-pressure diaphragm pump, flow and pressure controls, and electrical circuitry. This and other Abcor models utilize tubular modules. Each module is made up of a bundle of porous fiber-glass reinforced epoxy tubes approximately $\frac{1}{2}$ in. in diameter. Cellulose acetate membranes are curled into the shape of a cylinder with a slight overlap and placed into the tube.

A unique feature of the unit claimed by the manufacturer is the ease of replacing membranes. Because of fouling, degradation, and compression effects, membrane replacement can vary from 6 mo to 2 yr depending on the feed solution.

Spiral-Wound Design. The importance of the surface area of membrane contained within the confines of the given commercial unit is recognized in the spiral-wound design. It enables the utilization of far more membrane surface area than would be offered by a tubular sheet construction within the same size package.

The General Atomic Division of General Dynamics utilizes a spiral-wound cellulose acetate membrane module known as the ROGA. With this design (Fig. 16-9) several hundred feet of membrane can be accommodated in 1 ft^3 of pressure vessel volume.

A typical diagram for their system is shown in

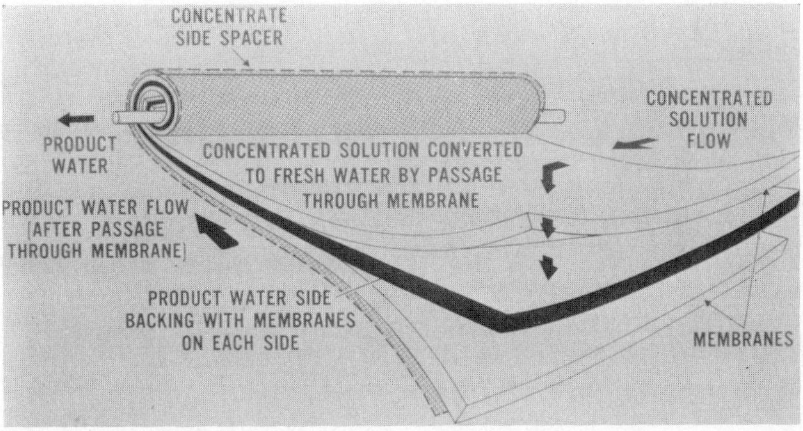

Fig. 16-9 **Spiral-wound cellulose acetate membrane module (ROGA).** *(Courtesy OSW)*

Fig. 16-10. Raw water receives preliminary filtration and acid treatment, flows to the pumps where it is pressurized, and then passes through the reverse osmosis step consisting of membranes, support structure, and piping. Power-recovery turbines may be used in plants having capacities of upwards of 500,000 gpd. The mechanical energy requirement may be suppled by an electric motor-driven pump.

Small units can be used to supply high-purity water for homes, hospitals, hotels, and to meet military requirements in strategic areas. An operating 6000 gpd system of Gulf Environmental Systems Co. is shown in Fig. 16-11.

Hollow-Fiber Design. As artful as the spiral-wound design is, and as useful in application as the commercial units are which incorporate its features, the hollow-fiber technique for desalination is several magnitudes larger in enhancing the key parameter of membrane surface area per unit space.

The hollow-fiber design of commercial utility in the desalination context is the "Permasep" device offered by the Dupont Company. Previously, the Dow Chemical Company had explored hollow-fibers of cellulose acetate for this very purpose. The Dupont devices, however, are based upon their H polymer,

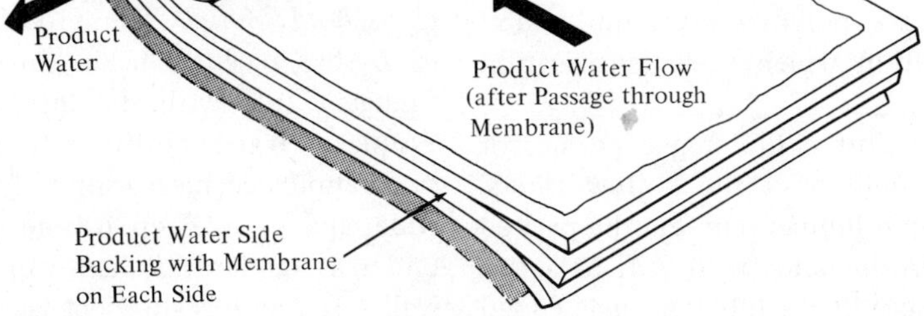

Fig. 16-10 **Cutaway of spiral-wound module.** *(Courtesy of ROGA Systems General Atomic Division, General Dynamics)*

Fig. 16-11 6000 gpd reverse osmosis desalination system with spiral-wound modules. (*Courtesy of Gulf Environmental Systems Co.*)

designed originally as an ablative material for space rocket needs. It can be said, therefore, that the current commercially-available hollow-fiber desalination units are a by-product of space-age research. The hollow fibers have a life of from 2 to 3 years, and in this respect are superior to cellulose acetate.

DuPont married their aromatic polyamide polymer to hollow-fiber spinning techniques to obtain a skinned or asymmetric (Loeb-type) membrane having good salt-rejecting properties. The flux is low, being only from 2 to 4 gpd/ft^2 at 400 psi pressure. The corresponding salt rejection is 90 to 95 percent. However, a single spinnaret produces a hank or rope of some 150 hollow fibers at once, and the enormous extent of their surface within the given unit space more than compensates for the low flux.

Additionally, sophisticated polymer technological applications are used to enhance the flux without needlessly compromising the salt rejection. The polymer is a mixture of *meta* and *para* aromatic polyamides. The blend presumably minimizes polymer crystallinity and the flux diminution which such densification would occasion.

In a Permasep unit containing a permeator unit, about 20 million of the hollow fibers, each smaller than a human hair, can be packed in a shell 1 ft in diameter with 7 ft of active fiber length. The unit resembles a single-ended shell and tube heat exchangers as illustrated in Fig. 16-12.

A 12-in. permeator provides 9000 ft^2 of permeation surface/ft^3 of volume. This is about 50,000 ft^2 of surface in each permeator—more than an acre and larger than the area of a football field. This larger surface is contained, however, in a unit only 5.5 ft^3 in active volume—about the size of a single football stadium seat. Total length of the hollow fibers in a 12-in. permeator is about 28,000 mi, more than enough to circle the earth.

The nominal capacity for 12-in. × 7 ft permeators is 7500 gpd/unit, based on the currently attainable flux (product water rate) of 0.15 gpd/ft^2. A tenfold increase is believed possible with the improved membranes currently being studied. As stated, the current flux is low compared to that of other techniques. However, it is greatly compensated by the strength and anticipated life of the special polymer membranes and the tremendously increased surface area involved through application of the hollow-fiber concept.

Household Model. The process of reverse osmosis has recently been "brought home" by Culligan, Inc. with the introduction of their "Aqua 1 Water Purifier" (Figure 16-13). A small appliance, measuring 19$\frac{3}{4}$ in. high, 10$\frac{1}{2}$ in. wide, and 5$\frac{5}{8}$ in. deep, it is designed to fit on or under a standard kitchen wall cabinet, or on a wall. It requires no electrical connection and operates on water line pressure. A $\frac{3}{8}$-in. or $\frac{1}{4}$-in. copper line supplies water to the unit. A tiny

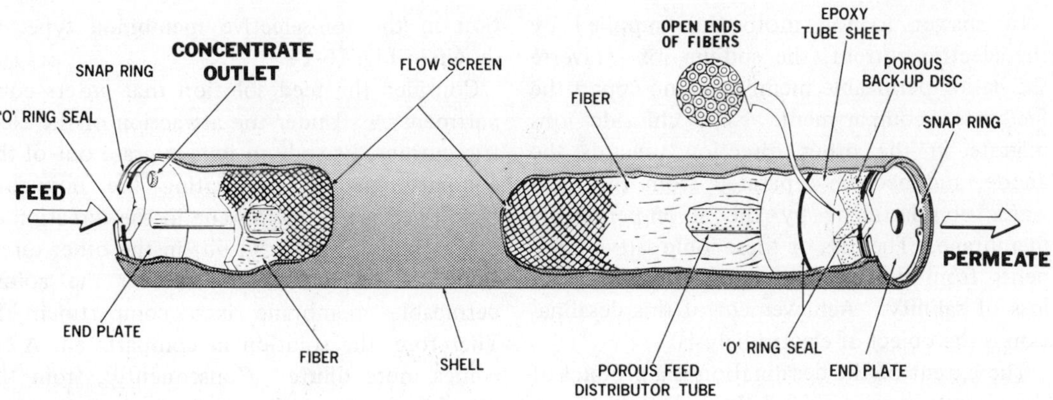

Fig. 16-12 Cut-away drawing of Permasep Permeator. (*Courtesy of E. I. Dupont De Nemours & Co., Inc. "Permasep" Products*)

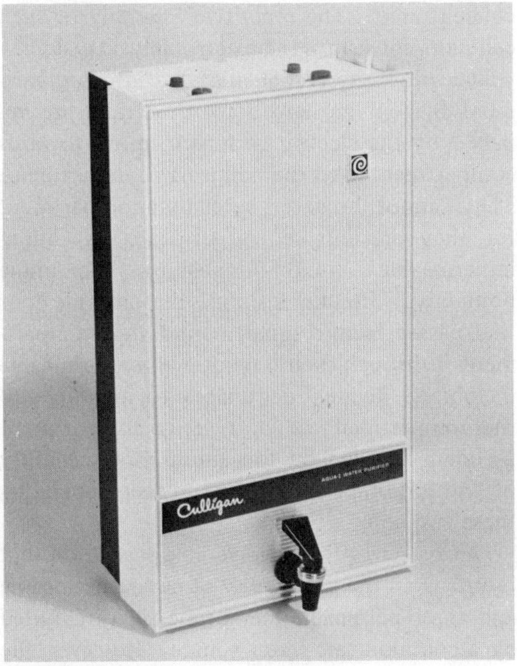

Fig. 16-13 Reverse osmosis household water purifier. (*Courtesy of Culligan USA*)

capillary tube drains away the wastewater containing the contaminants removed. By restricting water flow, a capillary tube helps maintain pressure on the membrane. Water passes through a filter prior to entering the tubular reverse osmosis module which contains 5 ft^2 of membrane. The unit also contains a carbon filter for final removal of taste and odors. Designed to treat water for a wide variety of household uses (for steam irons, humidifiers, aquariums, infant formulas, salt-free diets, etc.) the unit is capable of producing up to 3 gal of purified water a day.

Other Models. Aerojet General Corporation has used flat cellulose acetate desalination membranes in a multiple-plate construction design to produce units having capacities of up to 1000 gpd. Skid-mounted units of 100,000 gpd weigh some 5000 lb and require a 220/440 volt electrical source.

DESALINATION BY ELECTRODIALYSIS

In the electrodialytic desalination process, salt ions are removed from saline feed stocks under the impetus of an electric current.[23] The aqueous salt solution is introduced into a narrow channel, one of whose walls is an anion-permeable membrane, and another which is a cation-permeable membrane. These ion-selective properties are consequences of the functional groupings residing on the molecular backbones constituting the polymers of which the membranes are comprised. Thus, the anion-permeable membrane possesses cationic or positive-charged groupings. It is, therefore, also cation-impermeable, for like-charged bodies repel one another. Almost without exception, the cationic substituents are quaternized amine groups. The anion groups are usually sulfonic acid units (strongly acidic) or carboxylic acid radicals (weakly acidic). They may, however, also be phosphoric acid or other acidic functional entities.

In answer to the motivation supplied by the electric current, the sodium ions traverse the cation-permeable membrane and depart the feedwater compartment. The chloride ions migrate in the other direction towards the anode, and become separated from the feedwater compartment by the anion-permeable membrane. The loss of these ionic salt components from the compartment results in a net loss of salinity. Achievement of this desalination is the object of electrodialysis.

The extent of the desalination is a product of the electric current passed through the cell. A measure of its efficiency is how much of the applied power is translated into the useful transport of the ions, as opposed to its dissipation as heat in overcoming the internal electrical resistivity of the system, or its wastage in other forms of electrochemical expression. Accordingly, one of the requirements made of the membranes is that they easily pass ions in order to minimize internal resistance.

The second requirement, that they possess a low transference for water, is necessary in order to secure as high a yield of desalinated water as possible. The importance of the third requirement, that membranes be able to distinguish between ions of opposite charges, will become evident from the consideration of the hypothetical unit cell or cell pair, batteries of which constitue an electrodialytic desalination apparatus.

Cell Pair. Consider a cell divided into two compartments by an anion-permeable membrane. The boundaries of the two compartments are cation-permeable membranes. As a result, each compartment becomes defined by an alternation in the ion-selective membrane type, as shown in Fig. 16-14.

Consider the feed solution that enters compartment A. Under the attraction of the electric current, its sodium ions migrate out of the compartment by negotiating the boundary cation-permeable membrane in the direction of the cathode. Similarly, but in the other direction, the chloride ions traverse the anion-permeable membrane into compartment B. Therefore, the solution in compartment A becomes more dilute. Consequently, from the cell-pair to the right of the unit being considered, sodium ions migrate into compartment B. Thus, while compartment A grows more dilute in salt, compartment B becomes more concentrated. The repetitive assembly of such cell pairs constitute the electrodialysis system.

The sodium ions that migrated into compartment B from its neighbor on the right are impelled by the electric current to move towards compartment A, the diluting compartment. They cannot, however, enter that compartment for they are unable to permeate its cation-impermeable (anion-permeable) membrane boundary. In like fashion, the chloride ions that passed from compartment A into compartment B through their common anion-permeable membrane divider are unable to migrate into the compartment to the right of compartment B, being repelled by the anion-impermeability of the cation-permeable membrane separating these two compartments.

The battery of successive compartments that results from the periodicity of cation-permeable and anion-permeable membranes is initially fed by a common salt water source. However, the

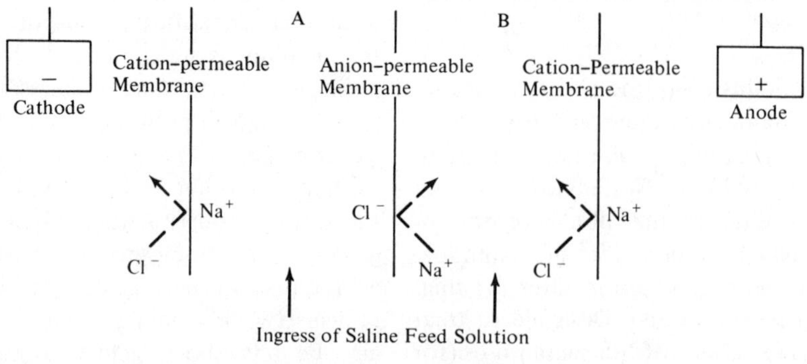

Fig. 16-14 Cell pair.

electrodialysis process causes the solutions in alternating channels to become more dilute or more concentrated in the salt content of their effluent streams. The extent of desalination which occurs in the alternate channels depends upon practical and design considerations.

Efficiency Considerations. Electrodialysis, like all other processes, has its inefficiencies. There are impediments to the passage of the electric current within the stack; the membrane-solution interfaces present such resistances. The membranes themselves are not perfect in their selectivity, and power is wasted in the heating of the membrane and the electrolyte solution. To these power losses must be added those that derive from the polarization effects of salt concentrations. Not surprisingly, then, the theoretical energy levels required for the electrodialytic practice are exceeded in the actual process.

The overall efficiency of the process results from several factors that influence current efficiency, current utilization, voltage requirements, and polarization. As stated, departures from the theoretical and ideal current efficiency level are occasioned by the involvement of hydrogen and hydroxyl ions in the current-carrying process, by the less than perfect selectivity of the membranes, by a large water transfer across the membranes, and by the existence of parallel alternative current paths through the stacks.

An overall current efficiency may be defined in terms of the actual current required to produce a quantity of fresh water, as compared to the amount of current theoretically needed for this same attainment.

Current utilization relates the degree of desalination achieved to the amount of current passed through the electrodialysis stack. It is, then, a measure of the salt removal process. This, in turn, reflects the differences between the transport numbers of the ions in the membranes and in the solutions. It is also, obviously, a function of the number of membranes used to exercise their ion-selectivity in the process.

The voltages involved in an actual electrodialysis mirror numerous operating influences. Some of these, like the concentration potentials across the membranes, merit emphasis in the importance of their influence. Others, such as the electrode overvoltage and the electrode potentials, are significantly minor in their impact to permit their being lumped together. Voltage considerations can be reduced to these classifications: those associated with concentration potentials across membranes and at membrane-solution interfaces, those stemming from the IR drops in the membranes and solutions, and those connected with electrodes and electrode processes.

Polarization. Polarization is the invariant consequence of the passage of the electric current. It is the result of the ionic movements towards the transference number change area under the impetus of the electric current. Polarization effects are controlled in actual contexts by cell designs that minimize the boundary layer. Additionally, polarization is maintained within practical limits of acceptance by regulating the ratio of the current density to the concentration of the diluting stream.

Concern with polarization reflects not only its technical but also its economic consequences. Thus, the amelioration of its effects is measured by the economic gain involved. The use of circulating pumps characterized by high flow rates is relied on to minimize the boundary layer. Additionally, turbulent flow and the mixing effects of its local eddying action are achieved by the use of adroitly-arranged spacers in the stack compartments. Although both of these devices increase the power costs of the circulation process, they reduce the polarization and minimize the overall power requirements of the electrodialysis stack.

Stack Efficiency. Subject to the cost-modifying influences of high-rate pump operations, the supplying of excess feedwater, the power costs of instrumental monitoring and controls, and the like, stack efficiencies are the results of an interplay of some half-dozen parameters. Of these, the membranes employed define the permselectivity and the water transport qualities of the system. The electrode reactions and the manifold leakage are governed by such items of stack design as the manifold configuration and the number of cell pairs. The salinity and current density that characterize the operating conditions define the polarization potential. Such stack design features as cell spacings and membrane thicknesses regulate the total resistance. The overall stack efficiency reflects all these influences.

It is thus seen that for the electrodialysis process, as for the other methods for achieving desalination, opportunities exist for engineering trade-offs and for the selection of the particular aspects of design and operation optimal for specific conditions or objectives. Once again, no single answer suffices for all circumstances. Discretion and judgment demand their cost in critical evaluation for each particular circumstance. In turn, they offer the attainment of an operation optimized in each and every instance.

Electrode Reactions and their Consequences. Within the solution, the electric current is carried by the ions whose membrane-regulated containment eventuates in the separation of fresh product water and its more concentrated saline companion stream. It is the generation and separating of these two aqueous streams, disparate in their salt content, that is the intention of the electrodialytic exercise.

The ionic movement that is at once the electric current and its consequence undergoes termination at the stack electrodes. At the anodic pole, the anions surrender their electrons, those particles of matter that in their aggregate constitute the electric current. At the cathodic electrode, the cations acquire the electrons essential to their transformation into neutral atoms, etc. The traffic in electrons that occurs at the electrodes results in chemical reactions that must be accommodated in the stack design, so that the constancy sought in the electrodialysis operation will not be compromised.

Thus, when nonreactive metals, such as platinum, are used as electrodes in conjunction with seawater or other naturally occuring brines, the chloride ions in company with hydroxyl ions simultaneously surrender their electrons at the anode to produce hypochlorite:

$$Cl^- + 2OH^- - 2e^- = OCl^- + H_2O \qquad (1)$$

In an acidic environment, chlorine gas will be liberated:

$$2Cl^- - 2e^- = Cl_2 \text{ (gaseous)} \qquad (2)$$

That portion of the chlorine escaping the aqueous medium as a gas offers a hazard. The portion that remains dissolved in the water yields an equilibrium mixture of hydrochloric and hypochlorous acids that is an expression of its concentration. The hypochlorous acid in its dissociated form gives rise to the hypochlorite ion. This, like chlorine itself, is an oxidizing agent that is strong enough to alter, and eventually to degrade, the organic components of the stack assembly.

When sulfate rather than chloride is present, as in low-chloride, high-sulfate waters, the electrode reaction will involve the hydroxyl ions or the dissociation of water. Oxygen gas will be evolved along with the generation of hydrogen ions:

$$H_2O - 2e^- = \tfrac{1}{2}O_2 + 2H^+ \qquad (3)$$

The consequences of this occurence in terms of both oxidative degradation and of acidic corrosion are self-evident. In fact, typical brackish waters contain large concentrations of both sulfate and chloride. As a result, both oxygen and chlorine gas may be evolved.

The electrodes used need not be impervious, as is platinum, to the chemical reactions that occur at their surfaces. Thus, when iron or carbon anodes are used, they will gradually be consumed; the iron will be oxidized into iron oxides, the carbon into carbon dioxide.

Reactions at the cathode generally offer less variety because the cations usually present are more resistant to electrolyte involvement than are hydrogen ions or water molecules, being above hydrogen in the electromotive series. Cations, as the hydrogen ions themselves, result in hydrogen gas evolution:

$$2H_2O + 2e^- = H_2 + 2OH^- \qquad (4)$$

A concomitant hydroxyl ion buildup occurs. This increase in alkalinity could cause operational problems. Cations, such as the heavy metal ions, that are below hydrogen in the electromotive force series plate out upon the cathode. This is the result of acquiring at that pole the electrons essential to their charge neutralization.

The electrode reactions require consideration on two counts. First, their incipient threats to the stack components require an antidotal action. Usually, this is accomplished by the hydraulic isolation of the electrode compartments. To this is coupled a copious flushing with the feedwaters. In specific instances, the redress of pH change is sought by appropriate nuetralization. So too, the build up in chlorine concentration is countered by the use of such inexpensive reducing agents as sodium bisulfite.

The second effect of the electrode reactions is on the economics of the process. The electrode reactions consume electrical power. Additionally, their amelioration involves expenses. Given the need for economic viability, these costs must be considered in the overall assessment of desalination by the electrodialysis method.

A list of suppliers of electrodialysis components appears in Table 16-4.

DESALINATION BY FREEZING

The antithesis of liberating pure water from its salt solutions by the application of heat is to be found in the attainment of this same objective by the use of cold, specifically of freezing.[24,25] The freezing of saline solutions results in the formation of essentially pure crystals of ice from the salt solution. The removal of these ice crystals and their subsequent remelting yields salt-free water. As compared to the use of heat, freezing possesses both advantages and disadvantages.

Among the advantages inherent in the technique of desalination by freezing is the smaller thermal effort involved. The latent heat of fusion is only one-seventh the latent heat of vaporization. The cost advantages of this fact are enhanced by the operation of the freezing process at temperatures much closer to the ambient than those of evaporative techniques. Since, however, the cost of Btu in refrigeration is much greater than in heating, the safeguarding of this advantage requires added outlays for improving process efficiency.

Two particularly irksome characteristics of thermal desalination are absent or muted in freezing operations. The deleterious effects of heated saline solutions on materials of construction, such as their corrosive influences on steel and concrete, are significantly reduced at freezing temperatures. This lessens the cost of materials used in the vessels and equipment. Operational down-time is minimized by the absence of scale formation on heat transfer surfaces in the freezing process.

Heat transfer through surfaces is of prime consideration in the various thermal desalination techniques. The attainment of high coefficients of heat transfer is costly. In the freez-

TABLE 16-4. Suppliers of Electrodialysis Components

Company Name	Address	Membranes	Stack Parts	Plants
Permutit Co. Ltd.	Gunnersburg Avenue, London, W.4, England	X		X
Ionics, Inc.	Grove St., Watertown, Mass.	X	X	X
Aqua-Chem.	Waukesha, Wis.		X	X
Amf-Cuno	Meriden, Conn.	X	X	X
Electrodics	1227 Delaware Ave., Phila., Pa.			
DuPont	Yerkes Plant, Buffalo, N.Y.			
Amicon Corp.	Hartwell Ave., Lexington, Mass.	X		
Asahi Chemical Co.	8345 Yakocho Daishagawara, Kawasaki, Japan	X	X	X
Asahi Glass	Chiyoda-Ku, Tokyo, Japan	X		
Tokuyama Soda Co.	Tokuyama City, Japan	X	X	
Japan Orsano Co.	Tokyo, Japan	X		
William Boby and Co.	Rickmansworth, England		X	X
Ionac Chemical Co.	Birmingham, New Jersey	X		
T.N.O.	The Hague, Holland	X	X	X
Polymer Research Corp.	Brooklyn, N.Y.	X		
Universal Water Co.	New York, N.Y.			X

ing process, the primary mechanism of heat transfer is the direct contact of the fluids of different temperatures. As a result of this, the freezing process obtains a high volumetric coefficient of heat transfer. The expenditures for engineering devices designed to enhance heat transfer are thereby diminished.

There are, however, inherent drawbacks to the freezing process. The pure water is obtained in the form of ice, and the separation of pure ice crystals requires a discrete processing step.

The latent heat of crystallization involved in the transformation of liquid water into ice can be used eventually to help melt the ice. This is the counterpart of the heat of condensation in the vapor-compression operation being utilized to furnish the latent heat of vaporization. However, the freezing system requires a large-size installation to be multistaged in order to reuse the latent heat involved in the phase change. This is so because the freezing operation must be carried out at a fixed temperature. Unlike the evaporative processes, pressure changes (within practical limits) do not produce corresponding latitudes in processing temperatures.

One important characteristic in particular detracts from the freezing process. It does not lend itself to the separation of the last of the saline solution phase.

Process improvements which would significantly increase its practicality are: (1) increasing the size of the ice crystals in order to facilitate their wasting and to minimize the surface area available to salt water adsorption, (2) improving the means of handling the ice crystals (generally about 1 mm in size), and (3) optimization of the rate of freezing. Most of these considerations relate to batch process operations, however.

A freezing operation at the OSW Wrightsville Beach, North Carolina Research and Development Test Station, is pictured in Fig. 16-15.

Fig. 16-15 OSW Freezing Plant, Wrightsville Beach, N.C. (*Courtesy of OSW*)

Principles of Desalination by Freezing. The separation of ice crystals from a salt solution subjected to a freezing operation is in response to the characteristics of an equilibrium system comprised only of liquid and solid phases. The general qualities of such a system are given by the phase diagram shown in Fig. 16-16. Material A represents water, material B salt. Many salts, such as sodium sulfate, form eutectics with water. These are generally representative of particular species of hydrates. In the generalized diagram, such a eutectic is typified by point C. For desalination purposes, only the left-hand portion of the phase diagram need be considered, the region between 100 percent pure compound A (water) and the eutectic C representing here some combination of salt and water.

At any composition, such as X, between pure A and C resident at temperature t, the lowering of the temperature to point t^1 will cause the single liquid phase (salt solution) to separate into two phases, namely, pure compound A (ice), plus a solution of salt. By definition, the appearance of pure compound A as the solid phase (ice) represents the phenomenon of freezing. In accordance with the lever rule, the relative quantities of ice and salt solution produced by freezing to temperature t^n are represented inversely by the lengths of the lines running from point x^n to the phase-change line AC, and from x^n to the line representing 100 percent compound A. If the temperature for the salt solution having composition X is low-

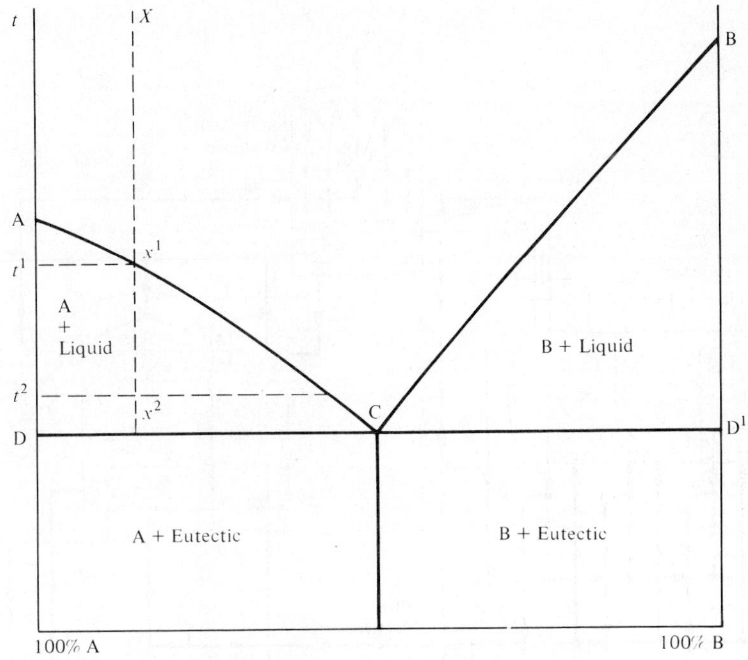

Fig. 16-16 Phase diagram of an equilibrium system.

ered to t^2, then, in accordance with the lever rule, at point x^2 a larger quantity of ice (more 100 percent compound A) will be produced. This illustrates the common experience that lower temperatures produce more freezing.

The Use of Refrigerants. Liquefied hydrocarbons can be used as refrigerants in direct contact with the brine. Their low temperatures serve to freeze the salt solution. An example of such a material is *n*-butane. Its solubility in salt water is slight, and its boiling point at atmospheric pressure is roughly at the freezing point of seawater. Thus, as it vaporizes, it serves further to cool the brine being treated. It has the added advantage of low cost; replacement of the quantities whose losses inevitably occur in the freezing operation, therefore, do not add appreciably to the cost of the product water.

In the direct contact refrigerating system, the refrigerated butane is mixed with the seawater. Heat exchange occurs, freezing the water, and warming the hydrocarbon. The butane is removed as a vapor by lowering the pressure over the mixing vessel. The heat of crystallization released in the formation of the ice serves to help vaporize the butane in its phase change. For process efficiency, in order to maximize the freezing effect of the evaporating butane, prechilled salt water is used. The ice crystals formed are removed from the brine–ice slurry in an ice countercurrent wash separation column, as shown in Fig. 16-17.[26]

The saline solution adsorbed on the ice crystals is removed by freshwater washing. The ice crystals are then removed to a melter-condenser by being formed into a slurry with fresh water. In the melter, butane vapors encounter the melting ice crystals, thereby becoming chilled as the ice adsorbs from them the heat necessary to its melting. This conservation of cooling effects is an essential feature to the economic feasibility of the process. Heat management is also practiced in the return of the brine that is separated from the ice crystals. This is done suitably in a heat exchanger wherein the refrigerated brine prechills the seawater being readied for freezing by liquid butane.

In order to achieve the objective of ice-crystal growth with the avoidance of excessive quantities of seed crystals, the design and operation of the freezer are carefully contrived. The freezer is divided into four compartments by perforated

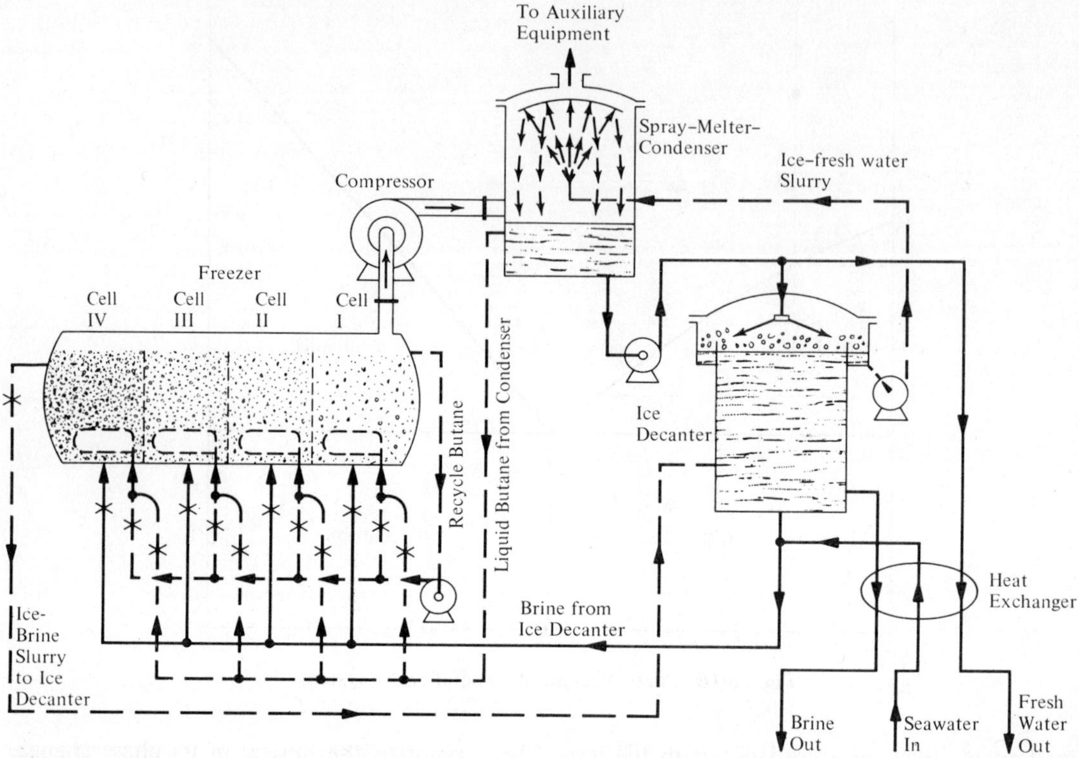

Fig. 16-17 Freezing by direct contact with secondary refrigerant. (*Courtesy of Kirk-Othmer Encyclopedia of Chemical Technology, Vol. 22. Interscience Publishers. Reproduced with permission*)

barriers. These dividers permit a slow flow of ice crystals and brine to take place in the direction of the ice decanter, but serve to minimize backflow and the indiscriminate directional diffusion that would occur in the absence of these barriers. The perforated partitions extend only partway above the liquid level within the freezer. This enables a layer of liquid butane to overlay the four compartments and to flow countercurrent to the brine-ice slurry movement in its progress to the ice decanter.

Ordinarily in its countercurrent flow, incrementally larger quantities of butane would become vaporized. This would bring the remaining refrigerant lower in temperature, for the progressive evaporation would involve a corresponding removal of heat in accordance with the heat of vaporization requirements attendant upon the phase transition. This situation is to be avoided for it would result in the progressive chilling of the brine in the freezer in the direction counter to its flow. If this were to eventuate, large quantities of seed crystals would be produced exactly where the feed enters the freezer, the reverse of the situation desired. What is wanted is a progressive increase in the rate of cooling of the brine as it slowly flows towards the ice decanter. As the crystals grow and the ice quantity increases, the remaining brine becomes progressively richer in its salt content; the cell nearest the decanter becomes the most concentrated, and has the lowest freezing point in consequence.

In order to avoid the circumstances wherein the refrigerant is coldest when it first encounters the brine as it feeds into the freezer, the boiling point of the butane is progressively elevated in its countercurrent flow. This is done by the incremental addition of a naphtha fraction to the butane. As the butane flows from cell to cell (over the perforated cell dividers) it is enriched more and more with naphtha, until at the last cell which the regrigerant encounters (but the first one that the countercurrent brine enters) it has the highest boiling point, being richest in naphtha. This progressive raising of the refrigerant's boiling point serves correspondingly to minimize the cooling of its temperature in that the vaporization process and its accompanying cooling effect through the adsorption of heat for vaporization are successively diminished.

The control of the boiling point of the butane refrigerant through the metered admixture with higher-boiling naphtha is the key to the process. It enables a sufficiency of freezing to take place in the first cell the brine enters in the freezer to produce a minimum crystal surface area in the form of small or seed crystals. As these slowly flow through the perforated partitions into the second cell on their eventual way to the ice decanter, they encounter a blanket of colder refrigerant, a lower-boiling butane solution containing less naphtha. The brine, now more concentrated because of its loss of water through ice formation, has a lower freezing point, and requires the lowered-temperature refrigerant to nuture the growth of its ice crystals. This procedure is successively repeated in the next two cells, with the naphtha quantity and therefore the boiling point of the refrigerant progressively decreasing, and the crystal size, quantity of ice, and brine concentration correspondingly increasing. The exercise in the controlled progressive elevation of the butane boiling point from cell to cell more than balances the successive incremental lowerings of the brine's freezing points (consequent to the salt solution's progressive concentration increases), and results in the desired progressive lowerings in the temperatures of the later cells. The final eventuation is the maximization of the weight of ice of maximum crystal size that is produced per unit freezer volume, the engineering goal sought.

The economic viability of the operation requires, of course, the maximum recovery of the butane and naphtha. The recycling of the butane–naphtha solution is engineered to feed the cells whose freezing temperature requires the ameliorating influences of the naphtha; namely, those cells nearest the ice decanter and those containing the most concentrated brine, and therefore requiring the coldest refrigerant. In each cell, the refrigerant is introduced through spargers at the bottom. This serves to mix the cell contents through its bubbling action, and in so doing supplies the agitation required for good heat transfer effects.

Following the pumped removal of the butane, the product ice is pumped in its brine-ice slurry form to the ice decanter where it is separated from the salt solution. (The brine is recycled and/or discarded, as required.) The ice is then washed free of absorbed brine (the wash water is recycled), and moved in the form of an ice–fresh water slurry to the melter for conversion into product water.

Pressure Freezing Process. An interesting variant on the above process has been designed by two Chinese investigators in Taiwan. It is based on water's unique property of expanding as it freezes. By LeChatelier's principle, therefore, ice melts as it is subjected to pressure. Stated another way, under a pressure increment of 1 atm, the freezing point of water is lowered by 0.01°C. Most substances have a higher freezing (or melting) point at higher pressures. The divergence between the behavior of ice and of frozen hydrocarbons under pressure serves as the basis for a novel approach to desalination by freezing.

A high-boiling hydrocarbon is frozen and mixed with prechilled seawater. A slurry of ice, frozen hydrocarbon, seawater, and liquid hydrocarbon results. The ice is decanted from the mixture after the frozen hydrocarbon is melted, and is washed free of adsorbed salinity in the customary fashion. All the above occurs at atmospheric pressure. The ice, however, instead of being directly melted, is mixed with the liquid hydrocarbon and is subjected to a pressure of 2000 psi. The hydrocarbon freezes in consequence, giving off heat of crystallization as it does so, and raising the temperature to between 3 and 4°C. The product water from the melted ice is decanted, and the solid–liquid hydrocarbon slurry is used to freeze more ice from another batch of prechilled seawater.

Several advantages are gained in this process. First, using a high-boiling hydrocarbon (250°C) minimizes the hazards attendant upon the use of materials such as butane with low flash points. Second, it requires neither a gas compressor to liquefy the hydrocarbon nor a vacuum device to remove it by evaporation. Thus, lower energy requirements would seem to be involved in the operation. Additionally, large ice crystals are formed in the process. This reduces washing costs. Since the rate of ice crystal formation is high, and no vaporization steps are involved, the equipment requires a relatively small volume. The desirable feature of large production per unit size of equipment therefore characterizes this scheme.

It does, however, also offer drawbacks. The irksome and costly steps of ice separation and washing are doubled in this process.

As previously stated, the feasibility of a desalination process is judged on economic rather than on technical grounds alone. The economic viability of this variant of the freezing process for desalination has yet to be assessed in a pilot-plant context. It would seem to offer advantages. Whether or not this is so on balance has not yet been put to an evaluative test.

Refrigeration by Water Vapor. Instead of the use of a direct contact refrigerant, water vapor itself has been utilized to effect a freezing of seawater. This is done in the Zarchin-Colt freezing process is illustrated in Fig. 16-18.[26]

Prechilled seawater is introduced into a freezer, and is reduced in temperature below the freezing point of pure water and almost to the freezing point of the salt solution. At this point, the evaporative influence of a compressor is used to pump water vapor from the freezer. This suction action causes additional vapor formation to restore the vapor–liquid equilibrium, and the heat of vaporization thus drawn from the seawater causes its freezing. The centerpiece of this operation is the inexpensive compressor developed explicitly for this application. It operates on a larger capacity rotor, and is suitable for the pressures and the compression ratio involved. The pressures are quite low, water exerting at the freezing point of seawater a vapor pressure of only 0.05 atm.

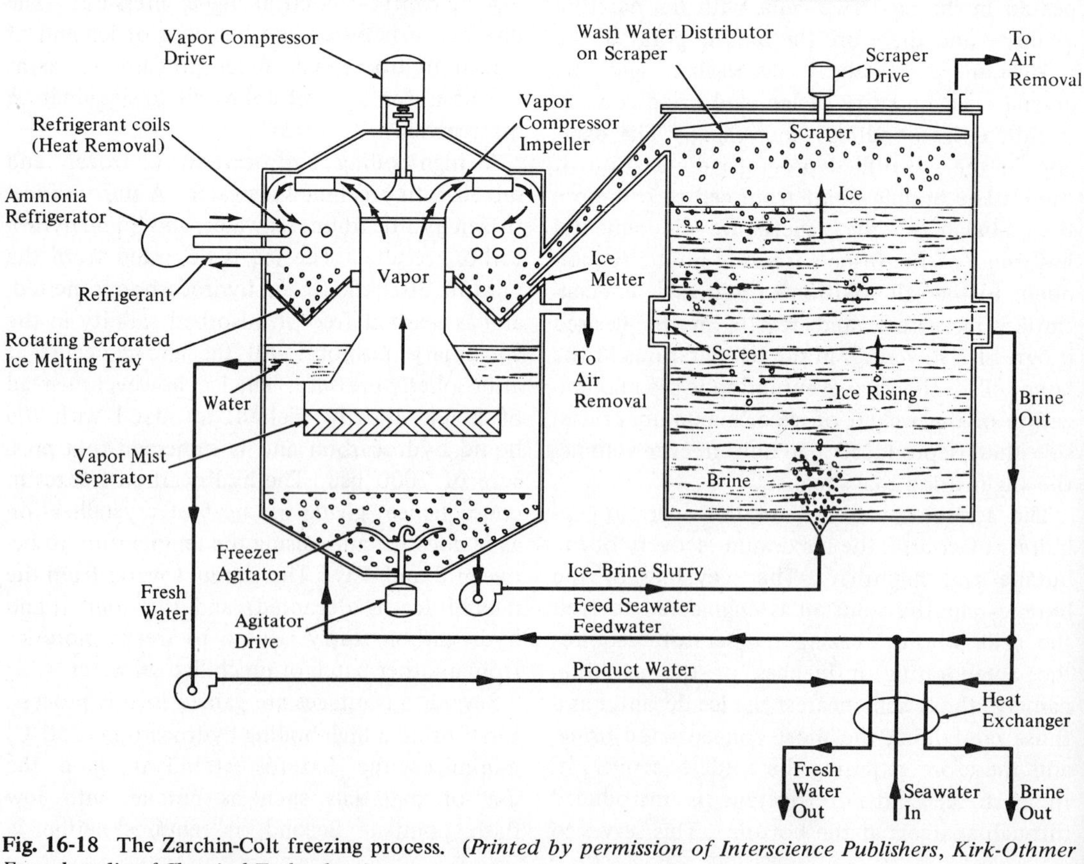

Fig. 16-18 The Zarchin-Colt freezing process. (*Printed by permission of Interscience Publishers, Kirk-Othmer Encyclopedia of Chemical Technology*)

Lyphilization is the term designating chilling by such evaporative effects. In the desalination process, it yields seed crystals that grow to about 0.5 mm on a side dimension in from 20 to 60 min. These are separated on an ice decanter, as is customary, and are then washed, melted, etc. In the melting step, the evaporated water vapor, drawn off through a demister to exclude entrained salt solution, encounters the ice crystals. The ensuing heat exchange serves not only to melt the ice, but also to condense the water vapors. The quantity of product water is thereby enhanced. The prudent conservation of the heat energies involved is also managed in the engineered encounter between the seawater and the cold product water. This serves to prechill the brine prior to its further cooling and lyphilization.

Potential of Freezing Methods. For waters containing more than 3000 ppm of salinity, the freezing process may yet prove to be the method of choice. Unlike desalination by distillation where equipment has long been available to test conceptual designs as they evolve, the freezing method requires the development of compressors and other articles of equipment whose conceptual design is advocated to achieve the process' goals.

Such equipment developments now seem likely. The Office of Saline Water is putting new money into the freezing process to enable a realistic assessment of what its advocates believe it can accomplish. Given its lower energy consumption requirements, the advantages of continuous process designs, the potential for multistaging, and the development of suitable compressors with OSW fund backing, the freezing method of desalination should reveal its potentials and promises.

DESALINATION BY SOLAR EVAPORATION

One of the costly ingredients in any of the distillative desalination processes is the price of the thermal energy involved. This consideration renders the use of solar radiation as a source of heat energy very attractive.[27]

It has been calculated that the amount of the sun's energy that impinges upon the earth at locations where the radiation strikes most directly (that is, at angles normal to the earth's surface) could be as high as 10,000 Btu/day/ft^2. For practical purposes, the value is considerably less. Some of the sun's heat is dissipated by the atmosphere, by clouds, etc., and over much of the earth's extent of latitude the optimal incidence of direct sunlight impingement is not realized.

In any case, the average value of the solar energy striking the earth is calculated to be about 2500 Btu/day/ft^2, or about 100 Btu/hr/ft^2. Whatever the accuracy of this figure, however, the employment of the sun's energy to desalination finds encouraging application mostly where the intensity of the sunlight is high.[28]

In addition to the virtue of utilizing free solar energy, desalination by solar evaporation requires only rather simple equipment. An experimental greenhouse-type still erected at the OSW Solar Research Station at Daytona Beach, Florida, is shown in Fig. 16-19.[29] Under good sunshine conditions, it yields approximately 1 lb of fresh water/day/ft^2 of basin surface.

In essence, the solar energy is adsorbed on blackened bottoms of shallow troughs containing saline water. Evaporated vapors rise to con-

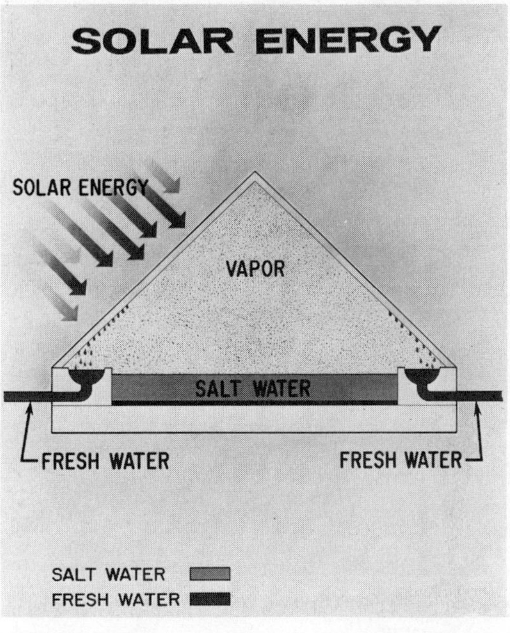

Fig. 16-19 OSW greenhouse-type still, Daytona Beach, Fla. (*Courtesy of OSW*)

dense on the surfaces of glass or plastic which are so inclined as to cause the collected water to flow to a common reservoir. Since the condensing surfaces overlie the evaporation troughs, they must be transparent to permit the passage of the sun's rays into the black-coated boiler. Even so, some sunlight is inevitably absorbed by the condensing surfaces. Their serious elevation in temperature is, however, forestalled by convective atmospheric cooling. The wasting of the incident energy is promoted by dust and dirt particles that collect on the upper portions of the condensing surfaces. Cleanliness and the labor costs it entails must be considered.

Various engineering designs have been utilized to enhance the capture of the incident solar energy. Usually, these are concerned with minimizing the losses due to reflection. Thus, angular saline water surfaces, and even rotating surfaces have been used. Also, the preheating of the salt water and the use of multiple-effect apparatus have been tried.

Under the arrangements used in practical contexts in such warm climates as North Africa, the southern states, and India, it has been calculated that from 0.4 to 0.8 gpd/ft^2 can be realized on an average throughout the entire year. Significantly, the thermal efficiencies assumed in these calculations are only 50 percent of the solar energy reaching the apparatus. Thus far, practical considerations have limited higher values from being reached.

To date, the archetype solar evaporative desalination unit is the one at Patmos, Greece. On a site of 130,000 ft^2, an evaporating area of 93,000 ft^2 is involved. This area consists of double bays 10 × 131 ft in size. Aluminum distillation troughs lined with black butyl rubber are used. The condensing surfaces are of window glass, slightly less than $\frac{1}{8}$ in. in thickness, sloped some 12° facing south. The basins are filled with seawater. Their concentrated residues are periodically removed at a frequency dictated by the rate of evaporation; weekly in the winter, every other day in the summer. Some 27 gal/yr/ft^2 of evaporated water is collected. This equates to about 7000 gpd. To this must be added the rainwater that is col-

Fig. 16-20 4000-gpd solar still (built on Greek Island of Symi by Church World Services with Technical Assistance by OSW).

lected by the equipment. This is a bonus that enhances the value of the process, for the rainwater would otherwise escape collection.

A somewhat smaller plant having a capacity of 4000 gpd is shown in Fig. 16-20.

Since Patmos is in a higher latitude than areas such as Israel, India, South Africa, etc., where solar evaporation has been advocated, these figures are encouraging for potential lower-latitude installations. However, the limitations of the process are also apparent. For the daily production of 1000 gal, as much as 5000 to 15,000 ft^2 or more of evaporating and condensing surfaces may be required. This means a large plant site, indeed, plus significant outlays for accoutrements such as pipes, pumps, glass, plastic, rubber, cement, aluminum, etc. In short, solar evaporation seems best suited to smaller type units of some few gallons per day rather than for thousands of gallons. In such contexts, the high ancillary equipment costs are minimized and the virtue of the free thermal energy asserts itself. This operation would seem especially to commend itself to nonindustrialized regions, where the costly generation of thermal energy would be especially burdensome. It would seem ideal for small family-unit purposes.

REFERENCES

1. Desalting Plants Inventory Report No. 3, Office of Saline Water, U.S.Dept. of the Interior (1971).
2. *Desalting Handbook for Planners,* 1st ed., Bureau of Reclamation and Office of Saline Water, U.S. Dept. of the Interior (1972).
3. *Water Users' Handbook for Desalting Processes,* Hittman Associates for Office of Saline Water (in press).
4. *Distillation Plant Data Book,* Houston Research Inc. for Office of Saline Water, U.S. Dept. of the Interior, Contract No. 14-01-0001-2099 (1971).
5. "Saline Water Conversion Report for 1970–1971," Office of Saline Water, U. S. Dept. of the Interior (1971).
6. Evans, R. H., Acting Chief, Distillation Division, Office of Saline Water, Personal Communication (1972).
7. "The A-B-Seas of Desalting," Office of Saline Water, U.S. Dept. of the Interior (1968).
8. "New Water," Office of Saline Water, U.S. Dept. of the Interior (1970).
9. "Saline Water Conversion Summary Report 1971–1972," Office of Saline Water, U.S. Dept. of the Interior (1972).
10. Spiegler, K. S., ed., *Principles of Desalination,* Academic Press, Inc., New York, Chapter 8 (1966).
11. Glasstone, S., *Textbook of Physical Chemistry,* Van Nostrand Co., New York (1946).
12. Merten, U., ed., *Desalination by Reverse Osmosis,* MIT Press, Cambridge, Mass. (1966).
13. Loeb, S., and Sourirajan, S., UCLA Dept. of Engineering Report 60-60 (1960).
14. Loeb, S., UCLA Dept. of Engineering Report 63-22 (1963).
15. Loeb, S., UCLA Dept. of Engineering Report 62-26 (1962).
16. Loeb, S., and Manjikian, S., UCLA Dept. of Engineering Report 63-22 (1963).
17. *Industrial Engineering Chemistry Products Research and Development* 4, 114 (1965).
18. Loeb, S., UCLA Dept. of Engineering Report 62-41 (1962).
19. *Industrial Engineering Chemistry, Process Design and Development* 4, 207 (1965).
20. Kesting, R. E., *Synthetic Polymeric Membranes,* McGraw-Hill, New York, Chapter 5 (1971).
21. Spiegler, K. S., ed., "Reverse Osmosis," *Principles of Desalination,* Academic Press, Inc. New York (1966).
22. OSW Research and Development Progress Report 84, PB-181571 (1970).
23. "Electrodialysis," *Principles of Desalination,* Academic Press, Inc., New York (1966).
24. *Principles of Desalination*, Academic Press, Inc., New York, Chapter 7 (1966).
25. Ranney, M. W., *Desalination by Freeze Concentration,* John McDermott, Park Ridge, N.J., Noyes Data Corp. (1971).
26. *Kirk-Othmer Encyclopedia of Chemical Technology,* Vol. 22, pp. 48-58, Interscience Publishers (1970).
27. Spiegler, K. S., ed., *Principles of Desalination,* Academic Press, Inc., New York, Chapter 5 (1966).
28. "Solar Desalination," *Kirk-Othmer Encyclopedia of Chemical Technology,* Vol. 22, Interscience Publishers (1970).
29. "Manual on Solar Distillation of Saline Water," OSW Research and Development Progress Report No. 546 (1970).

17

Sludge Handling and Disposal

Harry W. Gehm
WAPORA, Inc.

SOURCES AND TYPES OF SLUDGES

Sources of Sludges. Sludges are produced when naturally suspended or precipitated dissolved solids are separated from potable or spent process waters and sewage. Major sources include water clarification and softening, sanitary sewage treatment, mining process water handling, and treatment of industrial wastes generated by pulp and paper manufacture, food preparation and its chemical processing, chemical production, petroleum processing, and other industrial activities.

Types of Sludges. Sludges produced from water and waste treatment, can be classified under the following general headings: (1) inorganic sludges, (2) organic sludges, (3) mixed sludges, and (4) oily and fatty sludges.

Inorganic Sludges. The most common inorganic sludges include those obtained from clarification and softening of water supplies, mining and ore preparation, inorganic chemical and pigment manufacture, process water reclamation, and acid neutralization residues. While these sludges may contain a small percentage of organic matter, the solids are predominantly inorganic, frequently consisting of one primary substance. The sludges from water softening and causticizing, which consist mainly of calcium carbonate, are good examples.

Organic Sludges. Organic sludges vary greatly in their physical properties, both in the kinds of constituents they contain and the proportions in which these substances are found. Among their components are carbohydrates, proteins, fats, waxes, and fibers.

Mixed Sludges. Mixed sludges are those containing relatively equal proportions of inorganic and organic matter. Sources of this type of sludge include sugar clarification, log washing, certain chemical manufacture, pharmaceutical plants, and some fine paper operations (i.e., deinking and broke reclamation).

Oily and Fatty Sludges. Oily and fatty sludges are produced by petroleum processing, wool scouring, slaughterhouses, metal forming, and edible oil production.

SLUDGE DEWATERING

Separation as well as compaction of the sludge is generally accomplished by sedimentation, flotation, filtration, centrifugation, or pressing.[1,2,3] In order to improve the sludge consistency to make it suitable for disposal or recycling to a manufacturing process, it is often necessary to use two or more of these unit processes sequentially.

Separation of suspended matter can generally be enhanced by the use of coagulants when fines are present. To a degree, the consistency obtained during dewatering can also be raised by the addition of conditioning agents or filter aids, although the sludge consistency also depends on the design and operating capabilities of the separating equipment and the natural water-holding properties of the solids.

Thickening. Although the terms clarification and thickening are frequently applied to the same operation,[4] in wet solids separation clarification refers to the supernatant liquid, and thickening to the slurry fraction or underflow.

The consistency of the underflow is important because of its influence on subsequent dewatering processes.[5] Vacuum filters, centrifuges, and sludge digesters all operate more efficiently and at higher through-put when treating a thick sludge. This is illustrated by the curve presented in Fig. 17-1 for a centrifuge operating on raw sewage sludge[6] which shows that centrifuge loadings increase in a linear fashion with feed concentration. As shown in Fig. 17-2, the same is true for vacuum filter loadings.

It is possible, in treating some wastes, to obtain a high consistency sludge directly from the clarifier if the unit is designed to accomplish this.[7,8] It must provide a sufficient retention

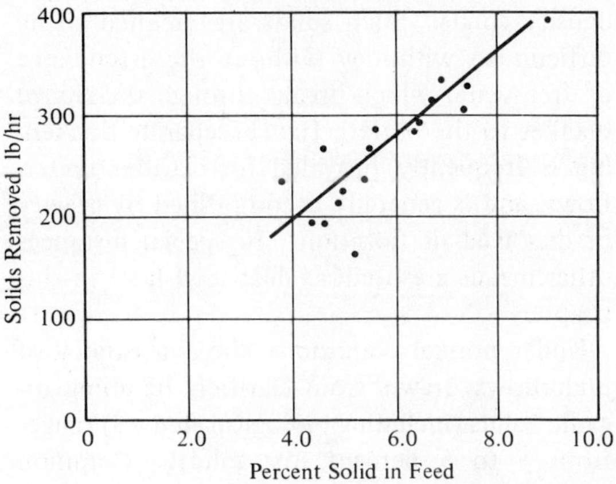

Fig. 17-1 Relationship between feed consistency and solids removal by centrifuges.

capacity to allow the settlings to compress and compact. Circular units are preferably equipped with a collection well center, and rectangular units should be equipped with an enlarged hopper at one end.

However, it is frequently undesirable to attempt both settling and thickening in a single unit because of operating problems, particularly in the case of large units handling low-

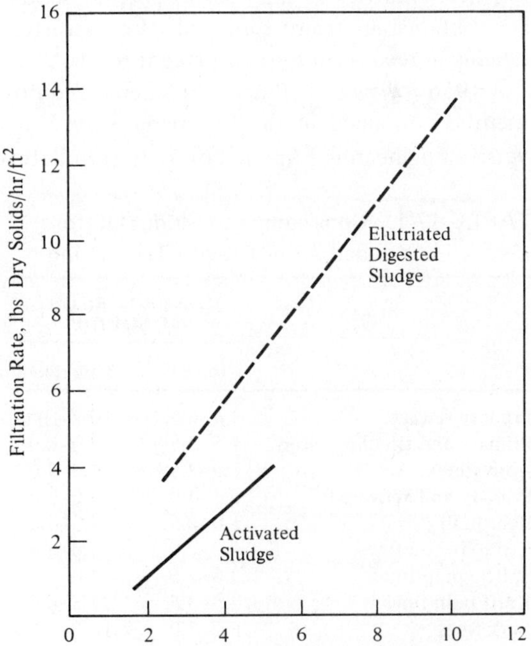

Fig. 17-2 Relationship between feed concentration and filtration rate of sewage sludges.

density solids. Such solids are inclined to be difficult to withdraw without the attendance of free water which breaks through the sludge blanket to the outlet. Hence, separate thickening is frequently provided for clarifier underflows, and is generally accomplished by gravity or dissolved-air flotation. In special instances other means are used, as described later in this chapter.

Under normal conditions, the consistency of underflows drawn from clarifiers handling organic solids (including biological sludges) ranges from 1 to 5 percent dry solids. Common methods of external thickening can raise these values to 4 to 10 percent as indicated by data presented in Table 17-1.[9,10]

Sludges which contain both organic material and a substantial percentage of inorganic solids–such as digested sewage sludge, deinking sludge, and certain chemical manufacturing residues–can be concentrated from underflow consistencies of from 4 to 7 percent to 8 to 12 percent by conventional means.[11,12] Sludges consisting almost wholly of inorganics–i.e., phosphate slimes, red mud, and lime mud–show a wide variation in thickening characteristics. Slurries as high as 60 percent solids are obtainable from some of these wastes ranging in feed from 5 to 15 percent solids.[13]

A 1956 review of patents on thickening equipment is contained in the Louisiana State University Engineering Experiment Station Bulletin No. 54.[14] Thickening practice is discussed in the *WPCF Manual of Practice No. 20*.[3]

TABLE 17-1. Consistency of Sludges Before and After Gravity Thickening

	RANGE PERCENT OF SOLIDS	
	Initial	Thickened
Primary sewage	3.5–6.5	6.2–10.0
Primary and trickling filter	4.5–5.5	7.0–8.0
Activated	1.0–2.5	4.0–5.0
Primary and activated	3.5–4.0	5.0–7.0
Boardmill	1.0–2.5	4.0–4.5
Groundwood fines	1.0–2.0	3.0–3.5
Sulfite pulp fines	2.0–2.5	4.0–5.0
Kraft pulp fines	2.5–3.5	4.5–6.0
Bark fines	5.0–7.0	10.0–12.0
Red mud	3.0–4.0	10.0–25.0
Potash slimes	1.0–5.0	8.0–25.0
Lime mud	8.0–10.0	25.0–50.0

Gravity Thickening. Gravity is the most common method employed for thickening. While it is slow and space-consuming, it requires simple equipment, consumes little if any power, and rarely needs the addition of chemical coagulants. Thus, it costs relatively little to install and operate.[15]

There are generally four basic zones of activity in a gravity thickening vessel. These are:

1. A clear zone near the surface of the liquid.
2. A "settling" zone in which the solids are settling at a constant rate.
3. A so-called "compression" zone in which the settling rate decreases.
4. A "compaction" zone in which water is slowly squeezed from the voids as the particles fall or align themselves closer together.

This sequential action is illustrated in Fig. 17-3. The degree to which thickening occurs in terms

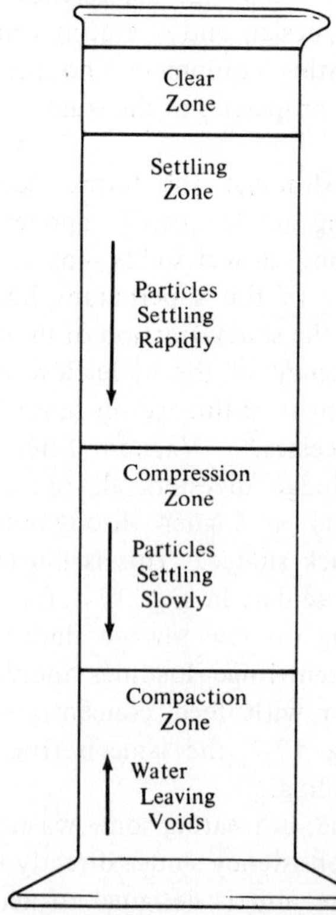

Fig. 17-3 Settling and compaction of sludge.

of the consistency of the slurry drawn from the zone of compaction depends to a large extent on the character of the sludge handled. This, of course, varies from flocculated pulps and slimes to relatively coarse, heavy particulates. The initial solids concentration is the primary factor in thickener performance,[16] and must be taken into account in thickener design.

Temperature, up to a point, is of minor importance in thickening digested sewage and industrial sludges, with significant differences noted only with fresh sewage. It must be pointed out, however, that high temperatures and resultant surface cooling effects can cause a disruption of the hydraulics of gravity devices, rendering them ineffective. Insulation of side walls and the covering or enclosure of these units will serve to extend their effective temperature range in both directions.

There are several laboratory techniques for computing gravity thickener surface area requirements. These are substantially the same

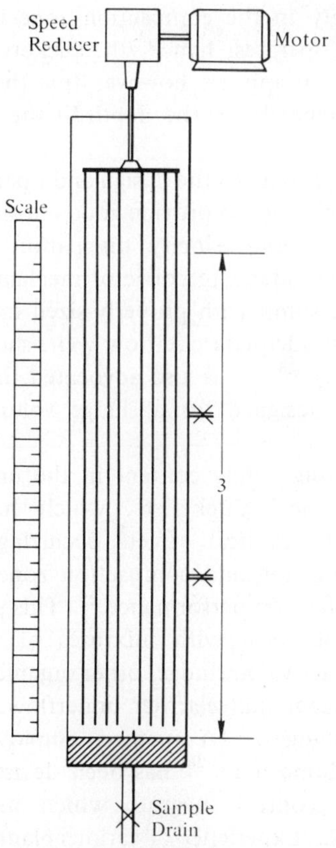

Fig. 17-4 Thickening test apparatus.

except for depth and diameter of the tube employed in the tests. Determination of sizing depends upon two factors: first, that the diameter of the tube be great enough to minimize wall effects, and second, that the depth be sufficient to simulate the slurry blanket depth necessary in a full-scale unit for maximum effectiveness. Most investigators are in agreement that a tube 8 in. in diameter providing a 3-ft slurry depth meets these requirements. These tubes are provided with agitators turning at a speed of about 0.25 rpm. This is shown in Fig. 17-4, a line drawing of the entire apparatus.[17]

To obtain the data necessary for area computation, the cylinder is filled with the slurry to a depth of 3 ft and the agitator is started. The height of the slurry-liquid interface is read at 15-min intervals during the first hour, half-hourly intervals over the next 3 hrs, and hourly intervals for the next 4 hrs. This schedule of readings may be modified to suit the characteristics of the slurry under test, although a sufficient number of readings must be taken to allow accurate computation of the velocity of compaction. When the process is complete, a sample is withdrawn for solids analysis. In some compression studies, sampling ports are located at intervals down the length of the tube to permit sampling during the operation.

It is good practice to conduct all compaction studies on biological sludges in the field where ample quantities of fresh slurry are available for testing. Gasification which occurs during storage of these sludges interferes with thickening and can cause flotation.

The primary technique used today for computing surface area requirements for sewage and industrial sludges was originally developed in 1917 for ore slimes.[18] An example of the modification and application of this procedure is represented in the following formula:[19]

$$A = \frac{Q_0 C_0}{V}\left[\frac{1}{C} - \frac{1}{C_n}\right] \tag{1}$$

where

A = area of thickener in square centimeters,
Q_0 = influent flow rate in milliliters per slurry per minute,
C_0 = influent solids concentration in grams per cubic centimeter,

C = solids concentration of slurry in grams per cubic centimeter,
C_n = final solids concentration in grams per cubic centimeter,
V = solids settling velocity in centimeters per minute.

The velocity (V) is computed from a series of at least 10 batch tests with solids concentrations ranging from C_0 to C_n, the solids interface level being plotted against time readings for each test. Substituting the concentration (C), corresponding interface velocity (V), and the selected values for C_0, Q_0, and C_n into the equation, the required thickener surface area (A) can be computed for each test. The maximum value of A is considered the minimum design figure for the projected unit. Simplifications of this formula have been developed.[20]

Gravity thickeners in general use in the waste treatment field are of the simpler types which do not incorporate washing or other processes integral to their operation. Although these are less efficient or require greater volumetric capacity than the mechanized type, they are effective. They are operated on either a fill-and-draw or a continuous basis.[21,22] The former practice is employed when sludge storage is required or operation is intermittent. Figure 17-5 illustrates the simplest type of thickener, the hopper bottom tank without a collector mechanism.

Experience dictates a hopper slope of no less than 60° from the horizontal so that sludge will not hang up on the sides. If a hopper is to operate on a continuous basis, it should be designed to have a minimum detention period of 4 hrs. Its surface area in relation to feed solids handled is computed in the same manner as for mechanized units.

Mechanically-operated gravity thickeners are generally of the "picket-fence" type similar to that illustrated in Fig. 17-6. The bars suspended vertically to the collector rake assist the thickening process by applying gentle agitation to the slurry under compression and compaction; this action promotes separation of more water from the sludge than would take place naturally. The effectiveness of these units is disputed by some who argue that agitation is useful only in the compaction zone in which the rake operates; hence, the pickets are not necessary. It appears, however, that their effectiveness depends on the depth of the compaction zone.

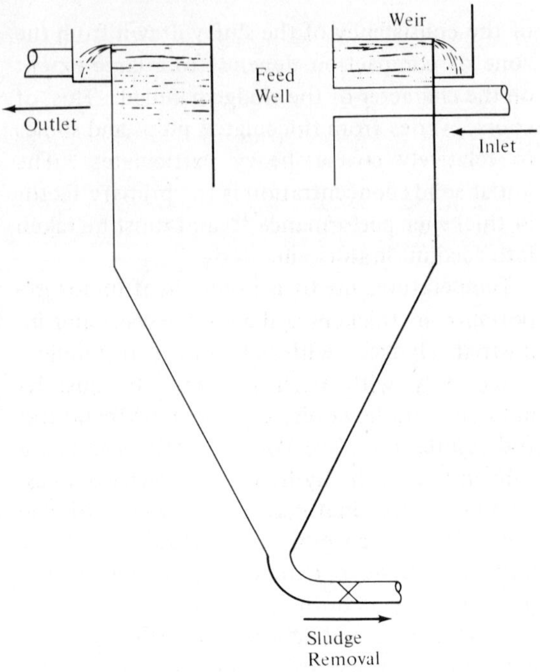

Fig. 17-5 Conical thickening tank.

It is important to the design and operation of mechanical units to provide good entrance flow distribution and velocity dissipation arrangements. An adequate collector mechanism and collection sump with properly-sized underflow piping and adequate overflow weirs should also be provided.[23] It is also advocated that these tanks be designed for a sludge volume ratio of 2.[24]

Fluctuating solids content in the underflow of mechanical thickeners, which has been ascribed to cyclical effects occurring within the compression and compaction zones, has a serious effect on performance.[25] This phenomenon is observed with mixtures of raw and modified activated sludge, other organic sludges such as paper mill clarifier underflow, and inorganic sludges. A method employing the sludge volume index[26] has been devised as an operating control measure which minimizes this effect. Experience at various plants has indicated that deep, circular units equipped with

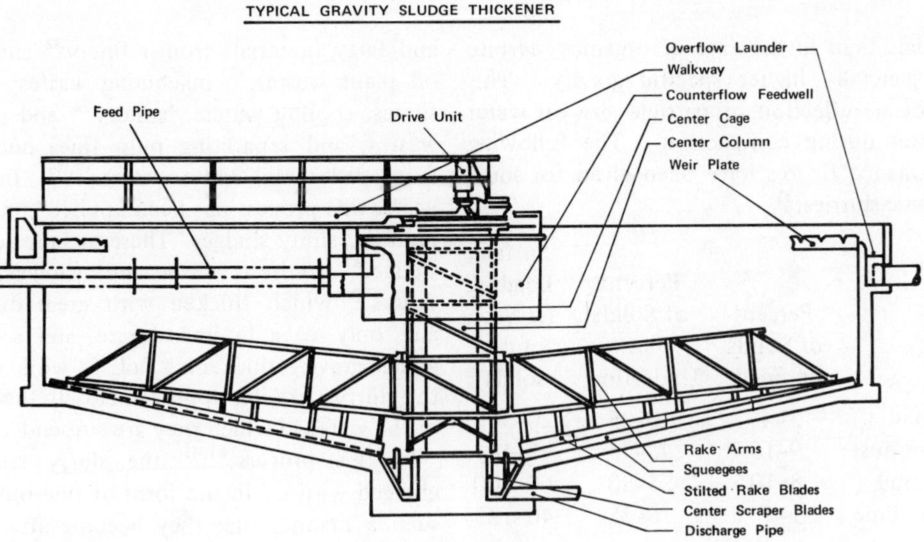

Fig. 17-6 Typical gravity sludge thickener. (*Printed by permission of Dorr-Oliver, Inc.*)

the "picket fence" mechanism are best for thickening very dilute and gelatinous sludge.[27]

Experiments with chlorine, copper sulfate, and calcium hypochlorite for controlling biological activity and subsequent gas evolution (which affects thickening adversely) have indicated that these substances are of doubtful value for this purpose. Experiments with coagulants and weighting agents resulted in the same conclusions. These results were also similar to those reported from experiments with a wide variety of materials, including aluminum and iron salts, lime, limestone, roasted copperas, ground oyster shell, magnesite, and a host of others.[28] Polyelectrolytes have not been found to enhance thickening of sewage or some paper mill sludges to any appreciable degree, but frequently increase overflow clarity.[29,30] The effectiveness of one of these is shown in Fig. 17-7.

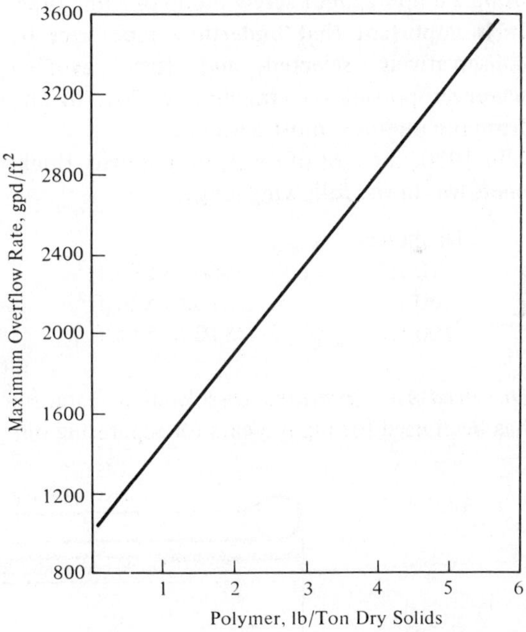

Fig. 17-7 Effect of a polymer on thickening.

Mechanical thickeners for handling organic sludges generally have a range of 4 to 8 hr in detention time with solids loadings ranging from 100 to 800 ft^2 of surface area/ton of dry solids handled/day. Hydraulic loadings range from 400 to 500 gal/ft^2 of surface area, depending upon the compression and compaction characteristics of the particular type of slurry. In practice, the above loading ranges are about doubled for mixed sludges with the exception of the solids rate which may be considerably higher. With inorganic sludges, the capacity is highly variable, ranging from 16 to 225 ft^2 of surface/ton of dry solids/day, so that design factors must be determined by tests of the particular slurry handled over the span of required feed concentration variation.

Peculiarly enough, more variation also appears to occur in the thickening rate of inorganic materials than in the rate of organics, despite their generally higher specific gravity. This may be a reflection of particle size on water rejection during compaction. The following performance figures have been given for some inorganic slurries:[31]

	Percent of Solids in Feed	Percent of Solids in Underflow	Surface Loading (ft^2/ton of dry solids)
Red mud	3-4	10-25	20-30
Cement dust	9-10	45-55	3-18
Lime mud	8-10	25-50	60-100
Potash slime	1-5	6-25	40-125

The underflow from gravity thickeners is generally withdrawn with positive displacement using a diaphragm or screw pump of some type. It is important that underflow pipe sizes be conservatively selected and turns avoided wherever possible—a straight-line flow to and from the pumps is most desirable.

In 1970, the cost of mechanical gravity thickeners was in the following range:

Diameter	
10 ft^2	\$60 to \$80/ft^2
50 ft^2	\$14 to \$20/ft^2
100 ft^2	\$10 to \$14/ft^2

Dissolved-Air Flotation. Dissolved-air flotation has been used for many years for separating oils and fatty materials from refinery[32] and edible oil plant wastes;[33] machining wastes; mineral slimes, cooling waters, laundry[34] and scouring wastes; and separating pulp fines and fillers from paper mill white waters.[35,36] In recent years, this process has been applied to thickening thin, slimy sludges. These include activated sludge,[37] hydrous tissue mill sludges,[38] and others[39] which thicken with great difficulty and only to a limited degree, and some for which gravity thickening fails to work because the slurry particles approach closely the gravity of the water in which they are suspended.

In this process,[40,41] the slurry stream is charged with air in the form of fine bubbles in such a manner that they become attached to sludge particles in sufficient quantity to cause them to float when released into an open chamber. The ratio of air to solids is an important factor since a form of hindered settling in reverse occurs as the particles rise to the surface of the chamber from which the float is skimmed mechanically. A line drawing of a dissolved-air flotation system is presented in Fig. 17-8.

Because of the mechanical equipment involved for introducing the air, this process is more costly to build than gravity-thickening units. The operating cost is also considerably higher because of the air requirement and the chemical coagulants which are generally used. However, its use for specific jobs can be justified at times on the basis of low space require-

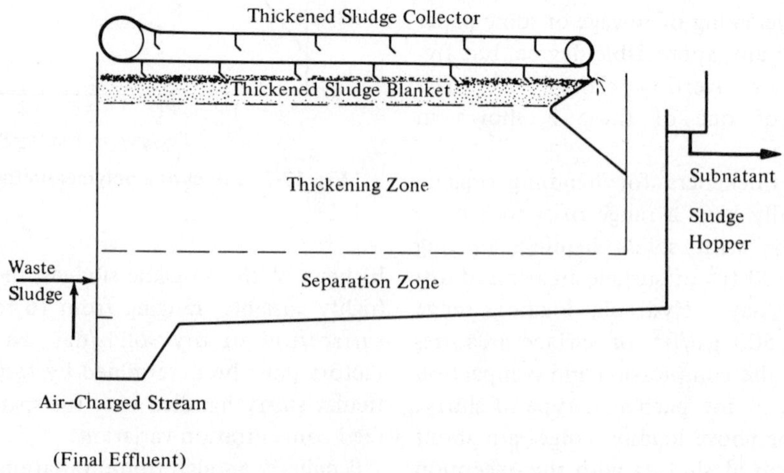

Fig. 17-8 A dissolved-air flotation system.

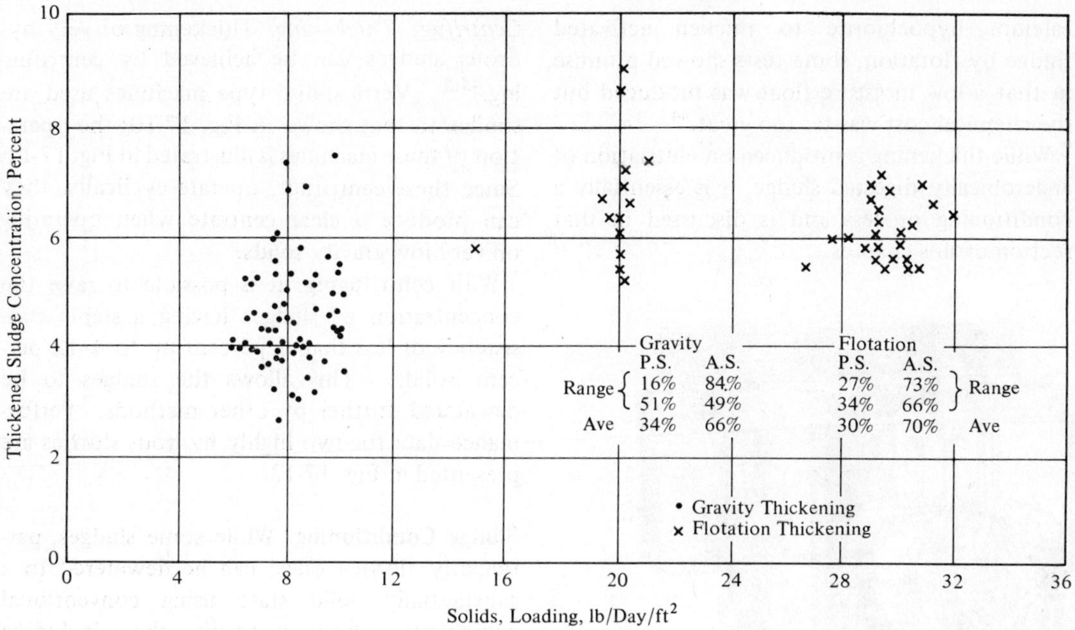

Fig. 17-9 **Comparison of Gravity Thickening and Flotation Thickening of Mixtures of Activated and Primary Sludges. (From Chain-Belt Co.)**

ments and because it can handle some sludges which are not amenable to gravity thickening; with others, it can provide an effluent slurry of higher consistency with greater reliability.

Figure 17-9 compares results of dissolved-air flotation with gravity thickening. It will be noted from this figure that dissolved-air flotation produced a slurry averaging about 50 percent higher concentration at surface loading rates approaching 4 times those of gravity thickening. Data for a number of different sludges are presented in Table 17-2. Results with dissolved-air flotation for thickening activated sludge were improved in Chicago with the addition of polymer. Final sludge concentrations of from 3.1 to 3.5 percent were produced without polymer; when 12 lb of polymer/ton of dry solids were added, final concentrations of 3.5 to 4.0 percent were obtained.[42]

Other Flotation Methods. In a gas flotation process[43] devised for thickening digested sludge prior to spray drying, sludge was pumped to a decantation tank along with an alum solution. This solution served both as a coagulant and acid reactant, liberating carbon dioxide from the bicarbonates present in the sludge. Final concentrations approaching 15 percent solids were obtained, and sludge consistency increased two- to threefold. In laboratory studies using

TABLE 17-2. Flotation Thickening of Organic Sludges

Type of Sludge	CONSISTENCY (PERCENT) Initial	Final
Activated sludge	0.5	4.0
" "	0.8	6.5
" "	0.5	4.6
" "	0.8	3.7
" "	0.8	4.9
Activated and primary	0.6	8.6
" "	2.3	7.1
" "	1.9	7.4
" "	1.3	6.1
Chemical coagulated	0.8	4.1
" "	0.5	4.1
Paper mill sludge	2.1	4.0
" " "	1.9	3.9
Boardmill sludge	2.0	4.0
" "	2.1	3.8
" "	1.7	4.1
Boardmill activated sludge	0.9	4.0
" " "	0.8	3.9

calcium hypochlorite to thicken activated sludge by flotation, some tests showed promise in that a low moisture float was produced but the chemical cost was far too great.[44]

While thickening is produced on elutriation of anaerobically digested sludge, it is essentially a conditioning process and is discussed in that section of this chapter.

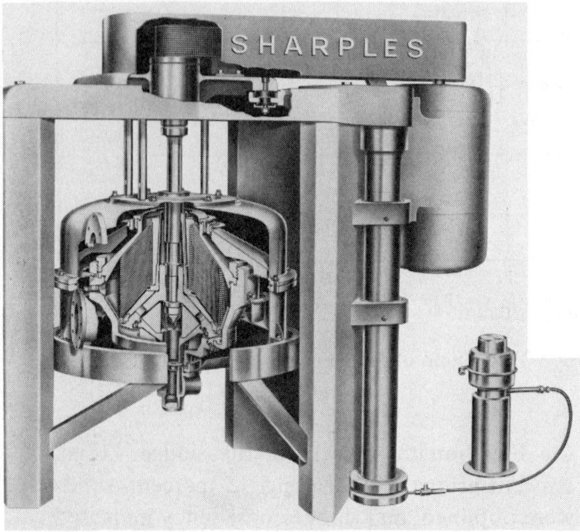

Fig. 17-10 Stacked-disc vertical centrifuge. (*Printed by permission of Sharples Div., Pennwalt Corp.*)

Centrifuge Thickening. Thickening of very hydrous sludges can be achieved by centrifuging.[45,2] Vertical-disc type machines used are similar to that shown in Fig. 17-10; the operation of these machines is illustrated in Fig. 17-11. Since these centrifuges operate cyclically, they can produce a clear centrate when operating on very low gravity solids.

With centrifuging, it is possible to raise the concentration of sludges having a stable consistency of less than 1 percent up to 6–12 percent solids. This allows the sludges to be dewatered further by other methods. Performance data for two highly hydrous slurries are presented in Fig. 17-12.

Sludge Conditioning. While some sludges, particularly fibrous ones, can be dewatered to a substantially solid state using conventional equipment without treatment, others, including most sewage sludges, require conditioning. Conditioning methods employed are either physical or chemical in nature. Most techniques have been in practice for many years, but some new ones are now being applied or are in the advanced stages of development or demonstration.[3,46]

Conditioning is applied to sludges for de-

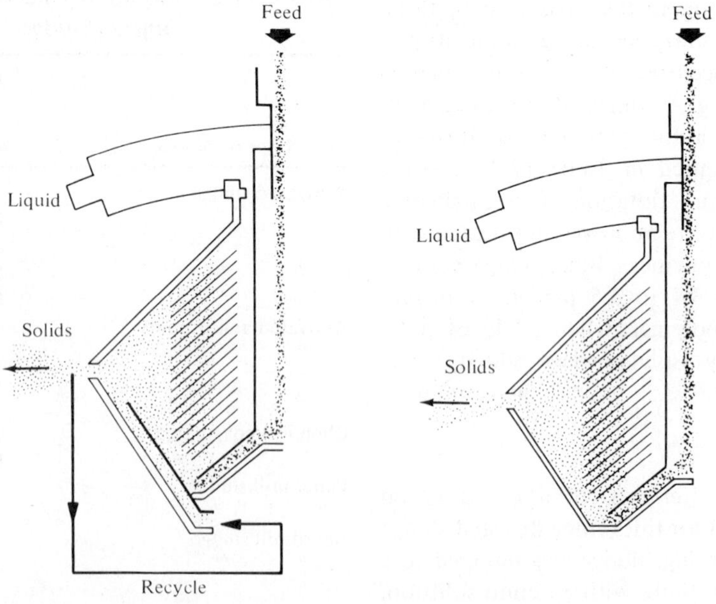

Fig. 17-11 Operation of a disc centrifuge. (From FWPCA Publication No. 20-4-210 [1968]).

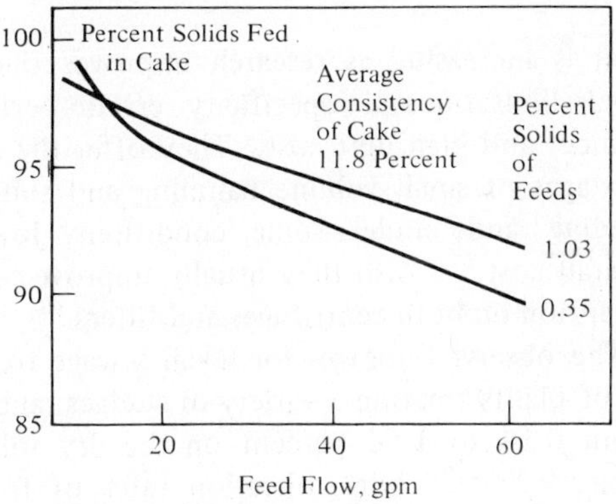

Fig. 17-12 Sludge thickening performance of a vertical-disc centrifuge. (*Technical Section, Canadian Pulp and Paper Assn.*)

watering on sand beds, vacuum filters, pressure filters, and centrifuges, as well as on "wet" machines and some specialized devices employing drainage and pressing. In the sewage treatment field, conditioning gained impetus in the 1920's with the wide application of vacuum filters for dewatering sludge.[47,48,49,50,51]

Chemical Conditioning. Chemical conditioners can be divided into two classes, inorganic and organic. Each of these may be further subdivided into active and inert materials.

Among the chemically-active inorganics, the trivalent cations are the most effective. Because of their low cost, ferric chloride and alum are the most widely used. However, others available as waste materials at some locations are also employed. Aluminum chloride and chlorohydrate, ferric sulfate, and copperas, which is chlorinated to make the iron trivalent, have all been used. Sulfuric acid and sulfur dioxide have been found to be effective for sludges high in fats and greases.

Ferric chloride is most commonly used for vacuum filtration and alum for sand-bed dewatering. Ferric chloride is sold in the anhydrous form in drums and is shipped in solution in rail and truck tankers. Alum is sold in bags and barrels, and in bulk, and is also shipped in solution in tankers.

Small plants use the dry materials which must be dissolved for application either in mixed tanks or in the discharge of dry feeders. Large plants use the solutions since their tank storage facilities make liquid handling and feeding easier and less costly than the use of the dry form.

The next most widely-used active inorganic chemical conditioning agent is lime. It is employed alone as a conditioner for some industrial waste slurries and is used as a supplemental conditioner in the ferric chloride process for raw and digested sewage sludges.

Small plants purchase this as calcium hydrate in bags from which a slurry for feeding is prepared in agitated tanks. Large plants employ calcium oxide which is shipped in bulk form in rail cars or trucks. This is converted to calcium hydroxide in slakers, from which it passes to agitated storage tanks for feeding. Some chemical process plants have these chemicals available on the premises, in which case they are piped or conveyed to the waste treatment facilities for application. Some waste carbide lime is also used where it is available in close proximity. Details on the handling and feeding of lime can be found in two bulletins published by the National Lime Association.[52,53]

The dosage of ferric chloride employed to condition most sewage sludges ranges from 5 to 20 percent on the dry solids basis, depending on the type. The requirement varies with individual sources and supplemental treatment. Alum requirements range from 1 to 6 percent as Al_2O_3 and lime usage runs from 5 to 15 percent of the dry solids as CaO. Actual dosage data for a number of sludges are presented in Table 17-3; chemical usage with specific dewatering equipment is given elsewhere in the text.

Two new processes employing active inorganic chemicals are under development for both sewage and industrial waste sludges. While the chemicals used in them have been used pre-

TABLE 17-3. Chemical Conditioning

Type of Sludge	Percent of Dry Solids		
	Al_2O_3	$FeCl_3$	CaO
Raw sewage	0.7–1.7	5.0–12.0	0–10
Digested	2.1–5.1	8.0–16.0	0–15
Elutriated	0.6–1.8	4.0–8.0	–
Activated	–	10.0–20.0	–

viously for this purpose, the physical treatment accompanying their application is unique.

One process subjects sludges to chlorine under low pressure (30 to 40 psi) for a few minutes, after which it is neutralized with lime and settled.[54,55] To date, this process has been used primarily to treat paperboard mill sludges, which after treatment are practically sterile and light in color and can be returned to process without difficulty. While chemical cost is high, the sludge disposal problem is eliminated and about 3 percent of the mill furnish is reclaimed. In addition, some of the supernatant water can be returned. These factors serve to offset the cost considerably.

The second process[56] is suggested primarily for slimy sludges such as activated sludge or others derived from biological treatment. In this operation, sludge is heated under pressure in the presence of sulfur dioxide at a concentration of 0.5 percent and is partially hydrolyzed. After treatment, the solids are readily filterable, and the filtrate is a raffinate which could be sold on the molasses market.

Methods have been suggested for reclaiming inorganic chemicals, particularly lime and alum, for reuse[57,58,59]; however, none of these have become widely used.

Active organic conditioning agents are actually active in more of a physical than a chemical sense, as they influence the surface chemistry of the sludge system.[3] These materials are for the most part synthesized long chain polymers,[60,61,62] and many are acrylic in nature. However, complex carbohydrates and proteins of large molecular size and chain length such as starches,[63] sugar gum,[64] and glue[65] are also in use. Most of these substances were developed as clarification aids but some have exhibited sufficiently strong coagulating activity, either alone or in conjunction with iron salts or alum, to make them useful as conditioning agents.[66]

The synthetic polymer type falls into three classes: anionic, cationic, and nonionic, according to the predominant charge they carry. Natural materials can generally be classed as nonionic unless associated in their composition with a strongly-charged inorganic ion.

Active organics are not as widely used as the active inorganic conditioners but their application is increasing as research improves them. Their limitations are specificity, erratic performance, and high unit cost. They offer the advantages of small volume handling and simple feeding and, under some conditions, lower overall cost.[67] Also they usually improve cake retention on both centrifuges and filters.[68]

The observed dosages for seven sewage treatment plants treating a variety of sludges ranged from 0.14 to 1.56 percent on the dry solids basis,[69,70,71,72] with filtration rates of from 2.5 to 8.9 $lb/ft^2/hr$. In conditioning pulp and paper mill sludges with these materials,[73] it has been found that, under the most favorable conditions, filter loadings can be increased 3 times, although with some sludges the sole action of the polymer is to improve filtrate quality. Specificity was found to be very high with these slurries. Performance of one of these is compared with ferric chloride for vacuum filtration in Fig. 17-13.

Organic dewatering agents are furnished in dry form or in strong solutions. The former are dissolved in water and the latter are diluted in accordance with manufacturer's instructions. In vacuum filtration, metering pumps are generally used for feeding to a mixing tank; with

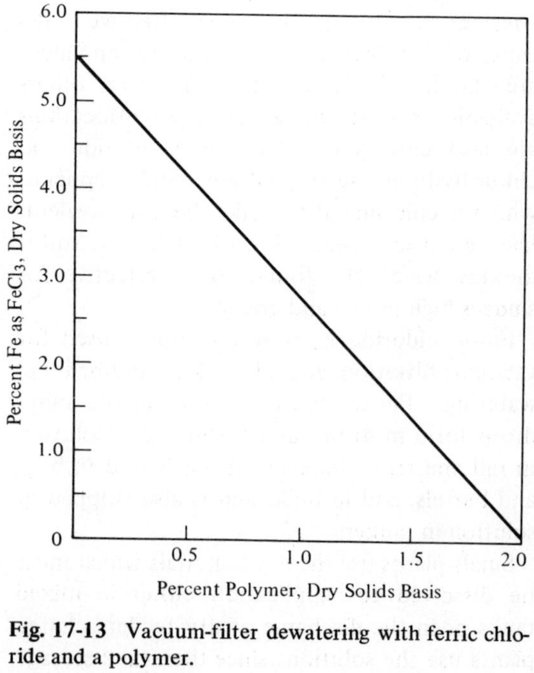

Fig. 17-13 Vacuum-filter dewatering with ferric chloride and a polymer.

centrifuges, feeding is directly to the bowl of the machine through a small-diameter line.

The performance of these chemicals with heterogeneous sludges cannot be reliably predicted. For example, there is often little if any relationship between the reaction of the polymer charge and the sludge particle's predominating charge. Thus, performance is determined on a trial and error basis. Manufacturers are attempting to produce compounds having predictable behavior and uniformly reliable performance with a single sludge.

Many inert materials, both inorganic and organic, have been tested for use in the conditioning of sewage and industrial waste sludges. Some of them are:

Organic	Inorganic	
Straw	Clay	Coke
Sawdust	Bentonite	Coal
Dry sludge	Red mud	Sand
Pulping rejects	Lime mud	Phosphate rock
Peat	Bone ash	Bark ash
		Iron oxide

Only in rare instances have any of the above materials been effective as conditioners, since rather than aiding in the release of water from sludges, they absorb or adsorb it themselves, and thus only serve to swell the cake bulk. Their single area of usefulness appears to be in aiding the release of "sticky" cakes from filters. However, the quantity necessary to achieve this frequently exceeds 100 percent of the dry solids in the sludge.

Of the inerts which have achieved some limited use as conditioners, the most attention has been directed toward fly ash since it is frequently readily available and represents a disposal problem in itself. Although it often causes dusting, this can be allayed by admixture with fibrous and gelatinous sludges. It is mixed best by addition to slurries prior to dewatering.

The most effective use of inorganics has been in their application as precoats to filters. Diatomites are most commonly employed. While these operations are successful, the ratio of precoat to cake solids is high—about 5 to 1—making it a costly operation. In experiments with fly ash as a filter precoat,[74] it was found that the ash from some sources worked satisfactorily on small-scale equipment but that ash from others did not. Paper pulp has been used as a precoat material in pressure filtration of some small-volume industrial waste residues. Interest has been shown recently in the use of bark fines as a conditioning agent with ultimate drying and burning of the cakes.

Elutriation. Of the purely physical processes applied to sludge conditioning, elutriation is the most common, although it is used only on anaerobically digested sewage sludge and is actually an accessory process to chemical conditioning.[75,76,77,78]

In this process, digested sludge is washed with plant effluent or surface or well water to reduce the concentration of bicarbonates since the reaction of bicarbonates with the coagulant hinders its effectiveness. By removing most of them from the sludge, the chemical requirement is substantially reduced, and dewatering is improved. This effect is illustrated in Fig. 17-14.

The three different methods used for washing the sludge[79] are:

1. A single-stage system using one thickener.
2. A two-stage system using two thickeners in which two separate washings are accomplished.
3. A countercurrent system using two thick-

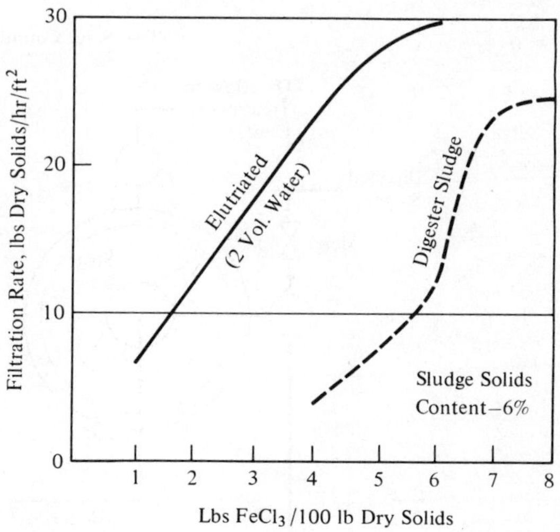

Fig. 17-14 Effect of elutriation on conditioning chemical requirements.

eners in which the wash water is introduced into the influent of the second thickener and the overflow from this unit is used as wash water in the first unit.

The following equations are used for determining the reduction in bicarbonate alkalinity achieved by each of these alternatives:

Single Stage

$$E = \frac{D + RW}{R + 1} \quad (2)$$

Dual Stage

$$E = \frac{D + W(R + 1)^n - 1}{(R + 1)^n} \quad (3)$$

Countercurrent

$$E = \frac{E + (R^2 - R)\,W}{R^2 + R + 1} \quad (4)$$

where

D = sludge liquor alkalinity in milligrams per liter,
E = elutriated sludge alkalinity in milligrams per liter,
W = elutriating water alkalinity in milligrams per liter,
R = ratio of elutriating water to sludge.

The single-stage and two-stage countercurrent processes are illustrated in Fig. 17-15.

Elutriation eliminates the need for lime in conditioning. It also alters ferric salt and alum requirements. These have been determined for many sludges from a number of sources.[80]

Elutriated sludge has been successfully dewa-

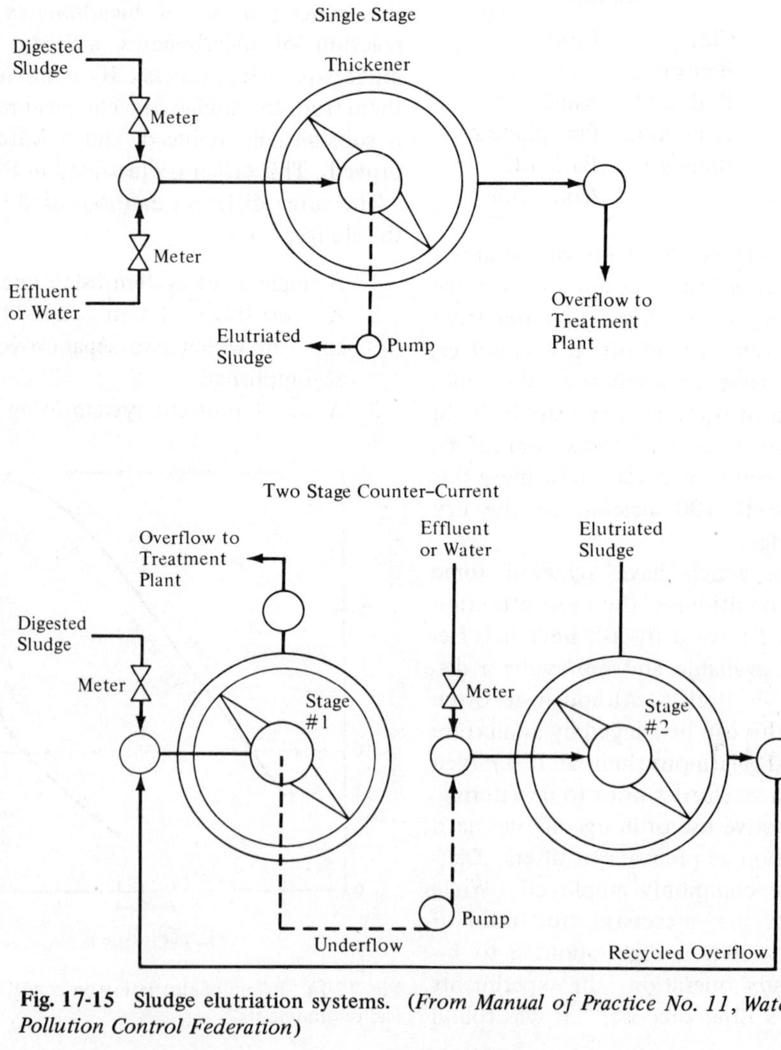

Fig. 17-15 Sludge elutriation systems. (*From Manual of Practice No. 11, Water Pollution Control Federation*)

tered with polymers. The application of certain of these polymers increased filter performance over 200 percent after elutriation.[81,82]

Heat Conditioning. In the first full-scale operation utilizing heat for dewatering,[83] raw sewage sludge was heated to between 290 and 370°F (143 and 188°C) with live steam, held for one-half to three-quarters of an hour, and then evacuated from the reactor through heat exchangers which transferred part of its heat content to incoming sludge. The heated sludge was dewatered on filter presses at a pressure of about 200 psi. Operating with a sludge averaging 5 percent solids, a final filter cake averaging 52 percent solids was produced. This cake was suitable for use as a fuel without supplementation.

Early experiments indicated that chemicals could enhance the performance of the process, particularly sulfuric acid. This is probably because of the high grease content of the sludge treated.

An illustration of the type of reactor unit employed in two plants utilizing thermal conditioning[84] is shown in Fig. 17-16.

Heat treatment processes have also been developed to overcome the difficulty in dewatering activated sludge.[85] One utilizes a low-temperature coking process operating at from 400 to 700°F (200 to 370°C) under pressures of 300 to 3200 psi; after separation of the coke, the liquid phase is subjected to air oxidation.[86] Another is a partial wet-air oxidation process which heats sludges to between 400 and 500°F (200 to 260°C), at which point their solids content becomes granular and readily filterable. A full-scale application of this process is described in Bulletin No. 2130[87] of the Zimpro Corporation, the manufacturer of the equipment. An illustration of this equipment is shown in Fig. 17-17.

A third process heats activated sludge in an exchanger.[88] It is claimed that after conditioning, the sludge settles to one-quarter of its original volume and will centrifuge to a cake containing 25 to 40 percent solids without chemical addition. The relationship between degree of oxidation and the specific resistance of activated sludge is presented in Fig. 17-18.

In a detailed laboratory study of heat syneresis it was found that optimum drainage of raw sew-

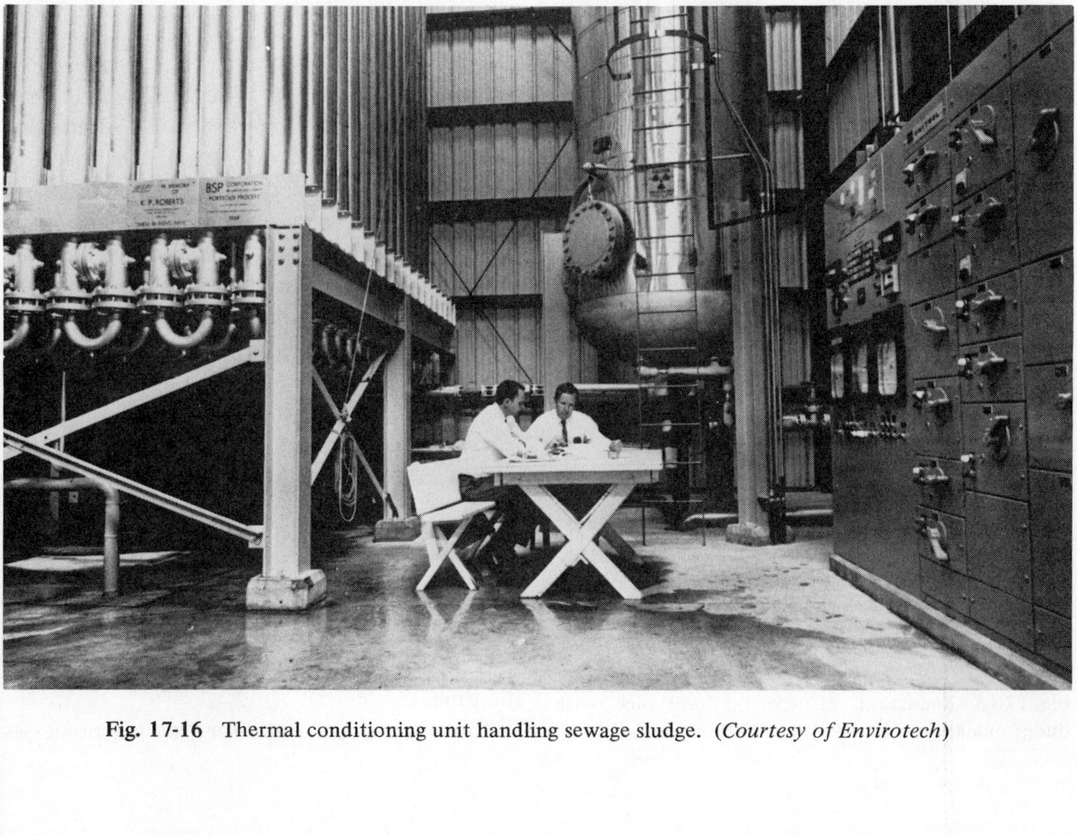

Fig. 17-16 Thermal conditioning unit handling sewage sludge. (*Courtesy of Envirotech*)

Fig. 17-17 Thermal sludge conditioning unit (*Courtesy of ZIMPRO, Inc.*)

age sludge was obtained at around 350°F (176.5°C) and that humus tank sludge did not respond well to heat treatment.[89,90] However, it was concluded that heat treatment of some sludges is attractive as compared to conventional methods such as digestion and bed drying, and can be accomplished with a considerable saving in space.

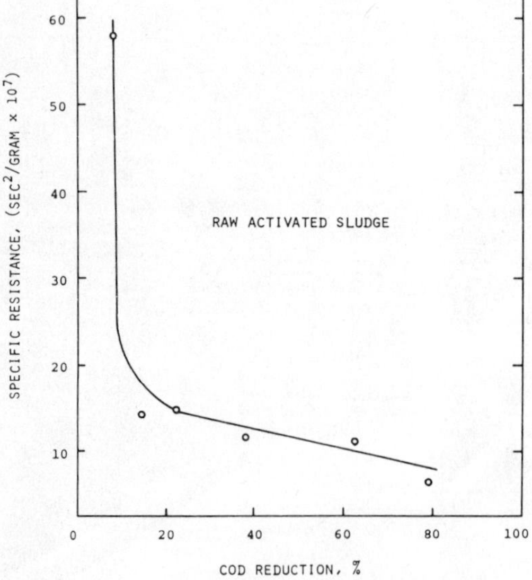

Fig. 17-18 Reduction of specific resistance with sludge oxidation. (*Courtesy of ZIMPRO, Inc.*)

Freezing and Thawing. The first experimental work on a process to apply the dehydrating effect of freezing and thawing on hydrogels to sewage sludge was conducted in the 1950s in England. The most extensive studies[91] were carried out in an apparatus especially designed for the purpose. Both digested and activated sludge were conditioned in this unit with and without the addition of coagulants such as chlorinated copperas, alum, and chlorine. It was demonstrated that slow and complete freezing was necessary to produce satisfactory results but that the rate of thawing was immaterial. Cakes having a solids content of between 23 and 30 percent were produced on vacuum filtration after thawing. All the chemicals applied raised the drainage rate somewhat and improved filtrate clarity. Cost of application of the process was considered excessive at the time.

More recently, patents for freezing processes

have been received in Poland[92] and in England.[93] In this country, a filtration rate of 55 lb/ft^2/hr on a 40/80 mesh screen was obtained after freezing and thawing activated sludge.[94,95] The filtrate produced was relatively clear and the fact that dehydration occurs simultaneously with freezing was established.

Cakes with a solids content of as high as 15 percent have been produced by freezing, thawing, and vacuum filtrating activated sludge from paper mill wastes and pulp fines with an initial solids content below 3 percent.[96] However, the extreme fragility of the floc formed in this manner made it evident that specialized equipment would be required for application of the process. An economic evaluation[97] indicated that, assuming such equipment could be successfully built and operated, capital and operating costs would be high. However, it has demonstrated the ability to perform a difficult job and may find a place in future practice.

Electrolytic Processes. Although electrolytic processes have been subject to intensive study, none have been adopted for general use. One which employed electrodialysis to treat sewage and industrial wastes and condition the resulting sludges in a single unit was successful in that an operating process was developed.[98] However, it did not offer significant economies over established methods. Similar efforts with electro-osmosis[99,100] met with no greater success. Nevertheless interest in these techniques continues.

A great many processes have been proposed in which sacrificial electrodes in electrolytic cells are used to condition sludges.[101] As metals brought into solution by the current coagulate the sludge, gases released on electrolysis float the solids to the surface. The cost of coagulants manufactured directly from the base metal in this manner is considerably higher than the purchase of salts and they are no more effective.

Supersonic waves[102] and magnetic techniques in which paramagnetic particles such as ground magnetite are introduced into slurries have been suggested.[103,104] It is supposed that the magnetic particles will attach themselves to sludge particles and the two will be drawn toward a magnetic source. No application of this principle is in operation at present.

No specific technique has been developed for the application of atomic radiation to sludge conditioning and other waste treatment functions.

Solvent Extraction. A highly-specialized process is employed for handling sludges high in fatty substances.[105] In this process, oil is mixed with the sludge, the mixture is centrifuged, and the bulk of the extraneous water passes off in the centrate. The oil fraction is then evaporated to remove the remaining water. The evaporate then undergoes a separation process through which the oil is returned to the system and the residue is handled in the most suitable manner, generally by incineration.

Summary of Conditioning Practices. It can be stated that active inorganic chemicals are the most effective conditioning agents in use today. Ferric salts, aluminum salts, and lime are widely used to condition both sewage and industrial waste sludges for dewatering. In recent years, some active organic agents of both natural and synthetic origin have become available which can partially or wholly replace the inorganics. These offer the advantage of simpler handling and feeding since only small quantities of them are usually required as compared to the inorganic materials. Unfortunately, the organics are frequently erratic in performance but a strenuous research effort is underway to rectify this shortcoming.

Of the physical processes, elutriation is the most widely practiced at many installations. There is renewed interest in processes employing heat syneresis and partial oxidation, both of which are effective conditioning procedures. Research on freezing and thawing and electrolytic processes continues but none have established themselves to date. One solvent extraction process has been established for fatty sludges, particularly those from which valuables can be extracted since its application is costly.

Filtration. The most common mechanical method employed for dewatering sewage and industrial waste sludges is filtration. This is usually of the vacuum type, although pressure and gravity filtration processes are used under some conditions for some slurries. Gravity filters are so specialized and variable in performance that no discussion of them is included here.

The first vacuum drum-type filters employed were of the blow-back or string-discharge type.

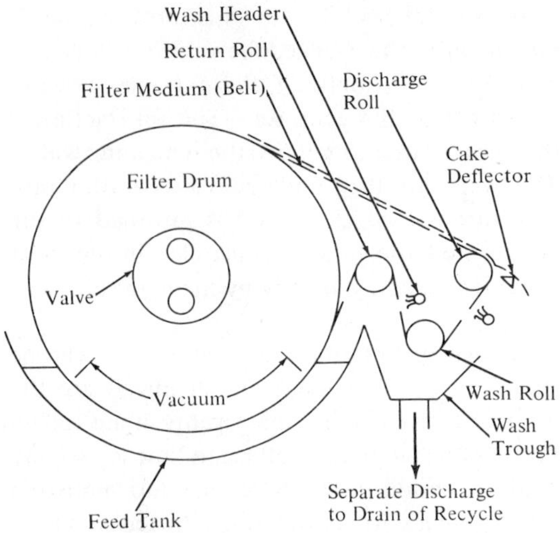

Fig. 17-19 Belt-type filter with facilities for continuous or intermittent washing of belt media.

These are seldom used today, however, and have been replaced by the belt-type filter. The latter, whether it is equipped with a continuous belt or wire, plastic-filter media, or coil springs, affords superior cake removal over the older types. It also provides a constant clean filtering surface since the drum is showered on each revolution. A filter of this type is diagrammed in Fig. 17-19 and pictured in Fig. 17-20.

Disc-type filters are frequently used for handling very readily filterable slurries such as bark fines, and crystalline, granular, or fibrous solids. They provide considerably greater filter area per unit of floor space and are very readily serviced since segments of each disc can easily be removed and replaced. One such filter is shown in Fig. 17-21.

Filter presses in use are of the common plate-and-frame variety, and all except the very small ones are power operated. Wherever possible, aluminum plates and frames are used to facilitate cake removal and cleaning.

A wide variety of filter media are employed. Paper is desirable since it is disposed of with the cake and replaced on each cycle, thus shortening down-time and assuring a clean filter surface.

Vacuum filters may be equipped with a num-

Fig. 17-20 Belt-type vacuum filter. (*Courtesy of Dorr-Oliver, Inc.*)

Fig. 17-21 Disc-type vacuum filter. (*Courtesy of Eimco Corp.*)

ber of auxiliary devices which improve their performance in handling certain slurries with particular properties.[106] The three most common in the waste treatment field are the "cloudy port," compression roll, and precoat devices. A third-valve port is of value when filtrate clarity is an important consideration, since it delivers the first filtrate of the formation cycle, which is inclined to be high in solids, for return to the vat.

Compression rolls are employed to increase moisture removal from cakes which respond to pressing. They may consist of a simple roller bearing on the drum or a more sophisticated device employing a blanket to carry the cake through wringer rolls to squeeze out more water.

Precoat devices deposit a thick layer of diatomite onto a filter drum of the conventional type, rotating under vacuum, by feeding a slurry of it to the vat. When a mat is formed, sludge is introduced into the vat. On each revolution, a thin layer of filter aid is doctored from the drum, removing the deposited solids with it. The doctor blade moves progressively nearer the drum automatically until there is no longer an effective layer of filter aid. The filter is then cleaned and the process repeated. This type of filter is used for oily and other particularly difficult sludges.

Conventional filters handling waste sludges are generally designed to operate in a drum submergence range of 20 to 30 percent. The exceptions are those employed for low-consistency fibrous slurries which have a submergence up to 75 percent and employ a cooch roll for removing the cake.

Vacuum filters are available in a wide variety of materials, and selection depends upon the corrosion resistance requirements. For example, vats are available in wood, mild steel, rubber, fiber glass-lined steel, and stainless steel of several varieties. Filter drums and their decks are made from similar materials. Drainage grids are now commonly fabricated from plastics or rubber compositions, replacing the wood spacers and metal grids of the past, and are available in readily replaceable prefabricated sections. Stainless steel or plastic internal piping is almost universally used. Valves are fabricated from cast iron, nickel-bearing cast iron, and stainless steels. Carrier rolls of stainless steel or rubber-coated steel are available. External filter station appurtenances are supplied in the same materials as the filter units themselves, depending upon specific conditions. A complete filter station appears in Fig. 17-22.

Vat agitators are generally designed for 13.5 strokes/min. However, the necessity for their use and the most desirable speed are deter-

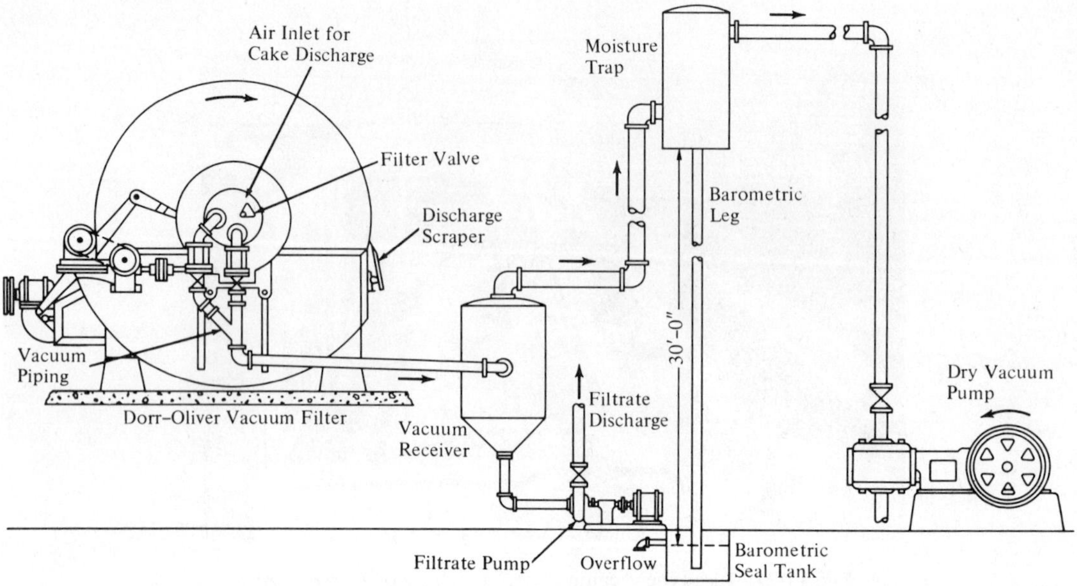

Fig. 17-22 Ancillary equipment typical of a vacuum filtration system. (*Courtesy of Dorr-Oliver, Inc.*)

mined by the nature of the slurry, the filter size, and the operating technique. While they tend to assure an even cake thickness across the drum, they can deteriorate conditioning, particularly when ferric or aluminum chloride is used.

Self-priming filtrate pumps are in common use. Vacuum is supplied by wet-vacuum pumps, which have largely replaced the old dry type, or by using a barometric leg. Vacuum breakers are almost universally used.

Filtration Rates. Equations for computing the filtration rate of slurries have been developed as well as a number of suggested modifications.[107, 108, 109, 110, 111, 112] The model in its basic form is summarized as follows:[113]

$$\frac{dV}{dt} = \frac{PA^2}{u(rcV + RmA)} \tag{5}$$

where

V = volume of filtrate,
t = cake formation time,
P = vacuum in inches of mercury,
A = filtration area,
r = specific resistance,
c = weight of solids per unit volume of filtrate,
Rm = initial resistance of filter media. (Since this is generally negligible, it can be omitted.)

The specific resistance is a measure of sludge filterability and is numerically equal to the pressure difference required to produce a unit rate of filtrate flow of unit viscosity through a unit weight of cake.

Integration and rearrangement of this equation yields this expression:

$$\frac{t}{v} = \frac{urc}{2PA^2} \; \frac{uRm}{PA} \tag{6}$$

From this, a linear relationship results from a plot of t/v versus V from the slope of which the specific resistance can be computed:

$$r = \frac{2bPA^2}{uc} \tag{7}$$

where b is the slope of the plot of t/v versus V. The weight of solids per unit volume of filtrate C is computed from the equation:

$$C = \frac{1}{C_i(100 - C_i) - C_f(100 - C_f)} \tag{8}$$

where

C_i = initial moisture content of the sludge,
C_f = final moisture content of the sludge.

While there is belief that specific resistance is a good index of filtration,[114,115] it is also pointed out that it is of limited value for calculating filtration rates on drum filters.[116] It does appear useful, however, for evaluating the variables involved in vacuum filtration. This observation is in agreement with the thinking of a number of specialists in the field. The specific resistance values for both sewage sludges and some typical chemical sludges are presented in Table 17-4.[116]

Filter Performance Prediction Methods. Three methods are available for predicting the performance of vacuum filters: the filter-leaf test, the Buchner funnel test, and the capillary suction technique, although the latter is not in common use in this country.

While the Buchner funnel test is frequently used to quickly establish probable optimum operating conditions, the filter-leaf test gives a better indication of overall performance[113] and is most widely recommended and used by filter manufacturers.

The broad application of the filter-leaf test is as follows:

1. A sample of about 2 liters of the sludge under test is poured into a pan of a sufficient size to accommodate the test leaf and of a sufficiently small diameter to allow a sludge depth of about 4 in.
2. The sample is then conditioned, if necessary, with the chemicals used in the desired dosage and order.
3. Without delay, and using apparatus similar to that shown in Fig. 17-23, the vacuum is turned on and the leaf is submerged to a depth of approximately an inch in the sludge after the maximum reading is obtained on the gauge.
4. At the end of the established form time for the test, the leaf is removed from the sludge and held face up for a period equivalent to the desired drying time.
5. The vacuum is released at the end of this time and the cake is lifted to a tared watch glass and weighed wet after its thickness is measured.
6. The cake is then dried in an oven at 105°C, cooled in a desiccator, and reweighed.
7. The filtration rate is then computed as follows:

Filtration rate (dry/lbs/ft²/hr)

$$= \frac{\text{Dry wt sludge (gm)} \times \text{cycles/hr}}{453.6 \times \text{leaf area in ft}^2} \qquad (9)$$

Filter Operation. Field data obtained from a number of sewage treatment installations operating vacuum filters have been plotted and the curves that were obtained have been compared, when possible, to ideal ones from carefully controlled experimental data.[117] The variables included cycle time, form vacuum, feed solids

TABLE 17-4. Specific Resistance Values of Sludges

Raw sewage sludge	10–20
Digested sewage sludge	3–30
Ferric hydroxide	1.5
Red mud	15.0
Aluminum hydroxide	2.2
Magnesium hydroxide	0.3
Clay	0.5
Ferric oxide	0.08
Shale fines	0.02
Kieselguhr	0.00017
Vegetable waste	4.6
Lime neutralized mine water	0.3
Tanning waste	0.15
Water treatment sludge	5.3

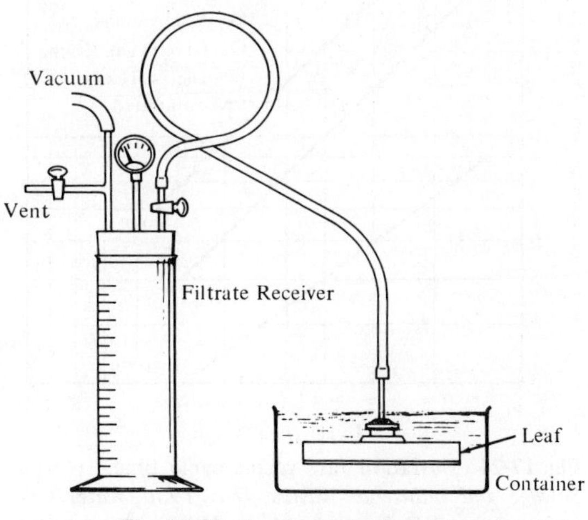

Fig. 17-23 Leaf test equipment.

concentration, and the degree of chemical conditioning employed. Several types of sludge were included and all tests were carried out at 25 percent drum submergence.

Such a comparison is shown in Fig. 17-24 in which cycle time is plotted against filter loading. While the trend of the field data curves parallels that of the ideal one, their slope is -0.5 rather than -0.75. This difference is ascribed to variations in individual sludge characteristics.

It was anticipated from experimental work that the slope of the curves representative of the effect of form vacuuming as related to the filtration rate would range from 0.3 to 0.5. Field data confirmed this prediction as shown by Fig. 17-25 in which plots for three digested primary and activated sludge mixtures are compared with the ideal curve. It was concluded from these expressions that vacuum filters should be operated at the maximum vacuum obtainable throughout the form cycle if maximum production is to be obtained.

Since filter loadings increase in proportion to sludge consistency, prethickening is justified under some conditions. It is obvious from the plots presented in Fig. 17-26 that a linear relationship exists between the two variables for different sludges, although the slope of the curve obtained for a particular sludge can vary considerably from the ideal. This could be anticipated since performance is dependent on the porosity of the cake which varies greatly not only for different sludges, but in different samples of the same sludge collected at different times. This relationship assumes the form of a curve when a wide range of solids concentration is involved as shown in Fig. 17-27.

The curves presented for conditioning of three digested and elutriated sludges are shown in Fig. 17-28. From the loading standpoint, the most efficient operation is obtained at the peak

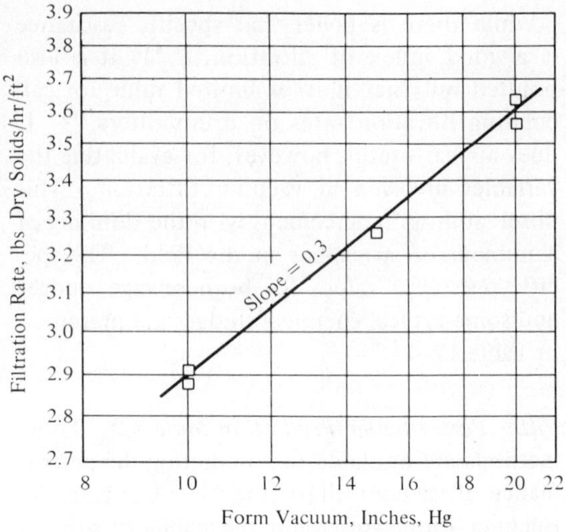

Fig. 17-25 Filtration rate versus form vacuum. (*From Sewage and Industrial Wastes, Dec. 1956, Water Pollution Control Federation*)

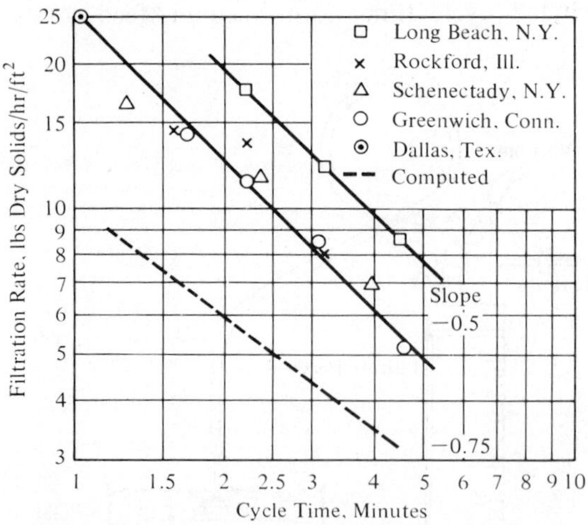

Fig. 17-24 Filtration rate versus cycle time. (*From Sewage and Industrial Wastes, Dec. 1956, Water Pollution Control Federation. Amended by the author.*)

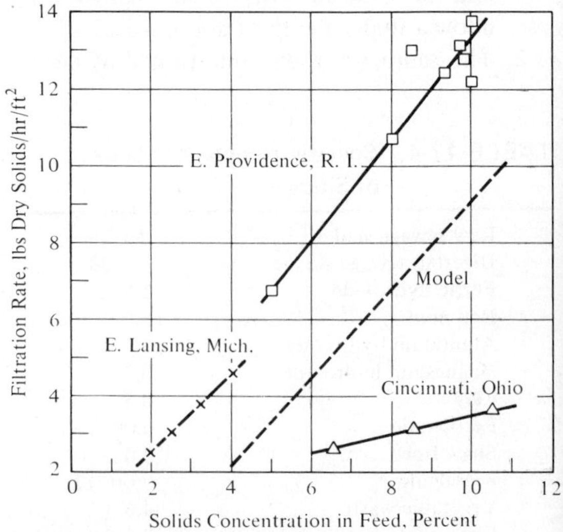

Fig. 17-26 Filtration rate versus feed solids concentration. (*From Sewage and Industrial Wastes, Dec. 1956. Amended by the author.*)

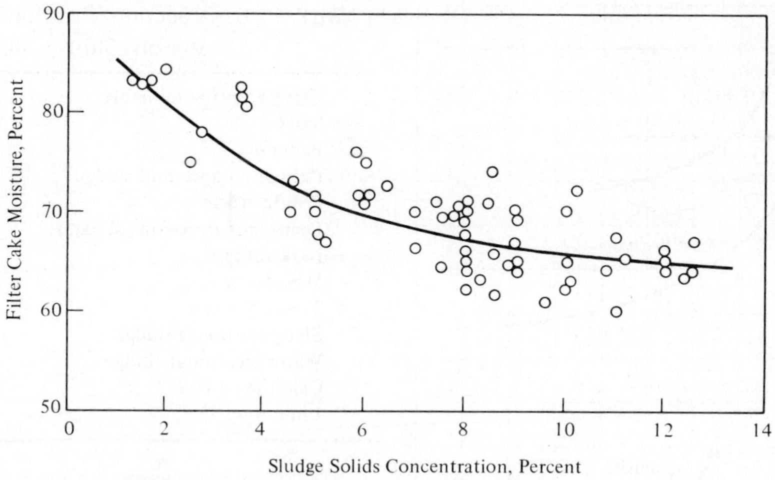

Fig. 17-27 Relationship between average sludge solids concentration and vacuum filter cake moisture content reported by various plants (*Private Communication*)

of the curves obtained on plotting ferric chloride dosage against filter loading. However, it is sometimes more economical to operate at a reduced loading level to save chemicals if adequate operating time is available.

Curves illustrating the effect of the length of the drying cycle on cake moisture for sewage sludges at four plants are plotted in Fig. 17-29. It can be concluded from these curves that cake moisture is a function of drying cycle time, although it also depends on the degree of vacuuming and conditioning. Vacuuming is perhaps the most important factor and should be maintained at a maximum during the drying cycle. The dual-vacuum receiver system is recommended to prevent large volumes of air being drawn through cracks in the cake which are likely to develop under these conditions.

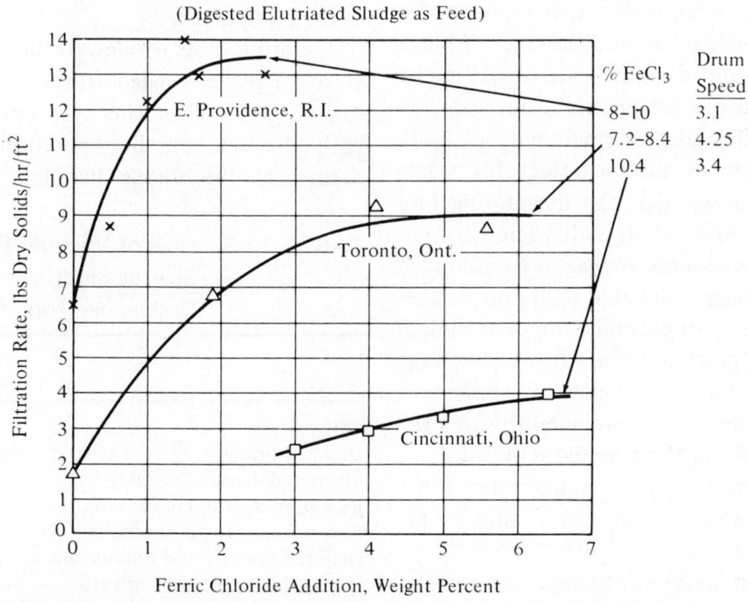

Fig. 17-28 Filtration rate versus ferric chloride addition. (*From Sludge Filtration Fundamentals, Water Pollution Control Federation. Amended by the author.*)

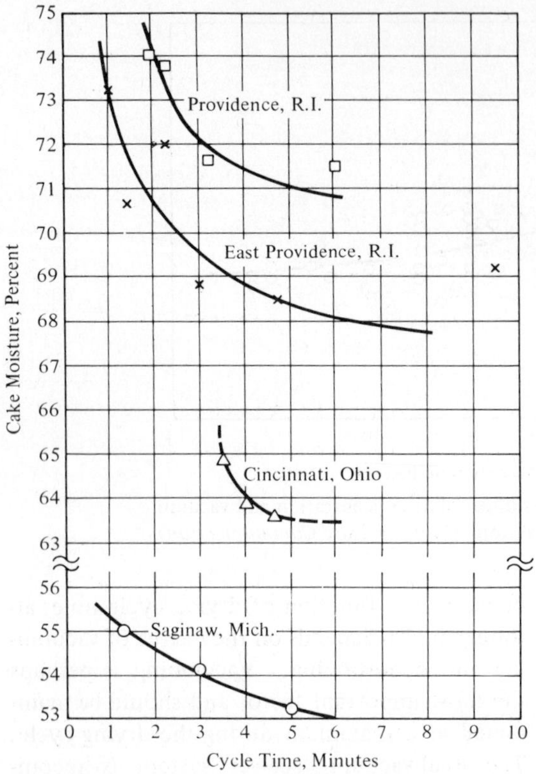

Fig. 17-29 Cake moisture versus drying time. (*From Sewage and Industrial Wastes, Dec. 1956, Water Pollution Control Federation*)

Vacuum filter performance is also affected by the methods of coagulant addition (see "Sludge Conditioning") and entrained air or gas in the sludge which have a detrimental effect on performance. While numerical data are not available to establish an adverse effect for waste sludges, it is known that the dewatering rate of pulp slurries and others is hindered by entrained air and measures are taken to minimize it. Thus, it appears advisable to pump sludges with a minimum of air entrainment. For sludges which contain gases such as digested sewage sludge, means–such as open head boxes or agitated tanks–should be provided through the filter feed system for their partial removal.

Observed vacuum filter loading rates for a number of organic and inorganic sludges are shown in Table 17-5.

Variables such as viscosity and surface tension appear to have little significant effect on vacuum filtration of most waste sludges.

Field and laboratory data indicate that the filtration rate of pulp and paper mill sludges is dependent upon the quantity of fiber present and its lengh.[11,118,119] Conditioning agents such as lime, ferric chloride, or polymers are necessary when activated sludge is incorporated since this has a decidedly adverse effect on dewatering.[120]

TABLE 17-5. Vacuum Filtration Rate of Various Slurries ($lb/ft^2/hr$)

Clays and fine minerals	16–80
Salts	130–400
Pigments	8–20
Pulp and paper mill sludge	5–20
Potato waste	5–6
Deinking–decoating sludges	8–15
Bark sludge	10–30
Myceleum	5–10
Tannery waste	1–3
Slaughterhouse sludge	3–6
Water treatment sludge	2–15
Calcium sulfate	40–70
Lime mud	40–80

Costs of Filtration. The costs of vacuum filtration of different types of sewage sludges at small and large plants are compared in Table 17-6.[15] These figures are somewhat higher than others[121] since they include capitalization of the equipment.

Operating costs for dewatering sewage sludges at seven plants ranged from $5.34 to $30.17/ton of dry solids, the cost naturally varying with the size of the operation, the specific nature of the sludge handled, and the per-

TABLE 17-6. Cost of Vacuum Filtration of Sewage Sludges in 1968 Dollars per Ton of Dry Solids

	Small Plants	Large Plants
Raw primary	17.50	7.50
Digested primary	38.70	13.75
Elutriated digested primary	10.00	8.75
Raw primary and humus tank	25.50	16.30
Raw primary and activated	–	26.20
Digested primary and humus tank	53.80	23.80
Digested primary and activated	32.50	31.25
Raw activated	–	16.30
Elutriated digested primary and activated	–	21.28

centage of the time the filters were operated.[121] The costs were broken down as follows:

	Percentage of Total Direct Cost
Labor and supervision	39
Chemicals and supplies	37
Power	8
Maintenance	16

The cost of dewatering several chemical-type sludges on vacuum filters was as follows:[122]

	Dollars/Ton/ Dry Solids
Lime mud	2.20
Gypsum	1.34
Salts	0.89
Steroids	3.18

Costs for dewatering pulp and paper mill sludges are estimated at between $6 and $10/ton.[123] The cost of conditioning chemicals adds between $1.20 and $4.00/ton of dry solids.

The cost of complete filter stations as of 1966 is illustrated in Fig. 17-30, in which size is plotted against unit cost. These figures include all necessary appurtenances but neither housing nor installation costs.

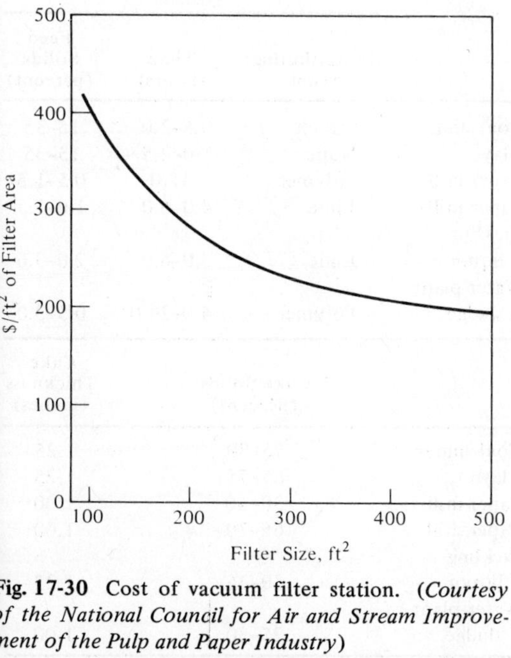

Fig. 17-30 Cost of vacuum filter station. (*Courtesy of the National Council for Air and Stream Improvement of the Pulp and Paper Industry*)

Pressure Dewatering. Pressure dewatering is accomplished by the use of pressure filtration, compression rolls to squeeze additional water from partially dewatered sludges, or direct pressing of either sludge or cake in a screw or disc press of the type commonly used for sugar cane or bark.

Filter Presses. Filter presses are used in handling sludge although to date the application of this method has been limited to dewatering heat-treated primary sewage sludge and some small-volume chemical or metallurgical slurries.[124,125,145] Most of these operations use a pressure in the 200 psi range with the cycle ranging from 2 to 4 hr. From a feed concentration of around 20 percent solids, they produce a cake containing 45 to 55 percent solids about 2 in. thick. A filter press of the type commonly used is shown in Fig. 17-31. This process is not economically practical for dewatering digested and chemically-conditioned sewage sludges due to the long cycle required.

Data on pressure filtration of industrial waste sludges with a pressure of 200 psi and suitable chemical conditioning are presented in Table 17-7.[126]

Compression Rolls. Compression rolls are not used directly on vacuum filters handling sewage sludges since these sludges do not lend themselves to further drying by this method. However, some paper mill sludges have been dewatered and dried further on "wet machines" with traveling belts. The sludges handled in this manner were a type which did not drain well on the normal filter-drying cycle.

One device which uses compression-roll dewatering is the Roto-plug machine.[127] This is a mesh-covered drum using gravity filtration to pick up a cake, which when doctored from the drum in wet form, is dewatered on compression rolls. This machine has been used for sewage sludge, boardmill sludge, and others. Wet

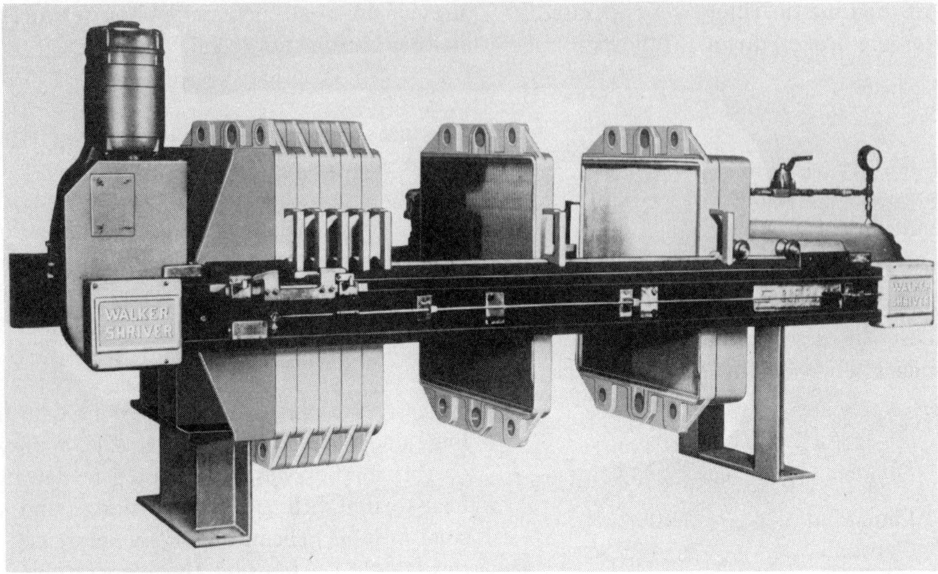

Fig. 17-31 Plate and frame filter press. (*Courtesy of Walker Process Equipment, Div. of Chicago Bridge & Iron Co.*)

cake leaving the drum at 8 to 10 percent solids is pressed to 19 to 26 percent solids through two stages of compression. Recovery of solids over the entire unit ran from 86 to 95 percent.[11]

Presses. Presses of various types have been applied to sludges of a fibrous nature, such as those found in the forest products and food industries, and to filter and centrifuge cakes. They are not in use for handling sewage sludges. Small-scale pilot studies have been conducted[128] on the effectiveness of pressing paper mill sludges of both low and high ash content after their initial dewatering by vacuum filtration. In the samples tested, the final solids concentration obtained by pressing filter cake from boardmill and deinking operations ranged from 28 to 50 percent and 39 to 64 percent, respectively. Optimal pressure-time conditioning was about 300 psi for 5 min. The presence of organic matter in the cake reduced the drainage rate and the ultimate dryness as well. Loss of solids in the filtrates was negligible.

Horizontal and vertical screw presses and disc presses are used for pressing filter or centrifuge cakes. A sectional drawing of one of the former is shown in Fig. 17-32 and a picture of the latter type is shown in Fig. 17-33. A screw press operating on filter cake from a plant treat-

TABLE 17-7. Pressure Filtration of Waste Sludges

	FILTRATION		
	Conditioning Agent	Time (hours)	Feed Solids (percent)
Coal slimes	Starch	0.5–2.0	15–35
Clay	None	1.0–2.5	15–35
Paper mill	Polymer	12.0	0.5–1.5
Paper mill	Lime	4.0–6.0	1.0–1.5
Pickling liquor	Lime	3.0–6.0	2.0–3.0
Water plant sludge	Polymer	4.0–24.0	0.5–3.0

	Cake Solids (percent)	Cake Thickness (inches)
Coal slimes	75–80	1.25
Clay	65–75	1.25
Paper mill	30–40	1.00
Paper mill	65–70	1.00
Pickling liquor	30–35	1.25
Water plant sludge	25–30	1.00

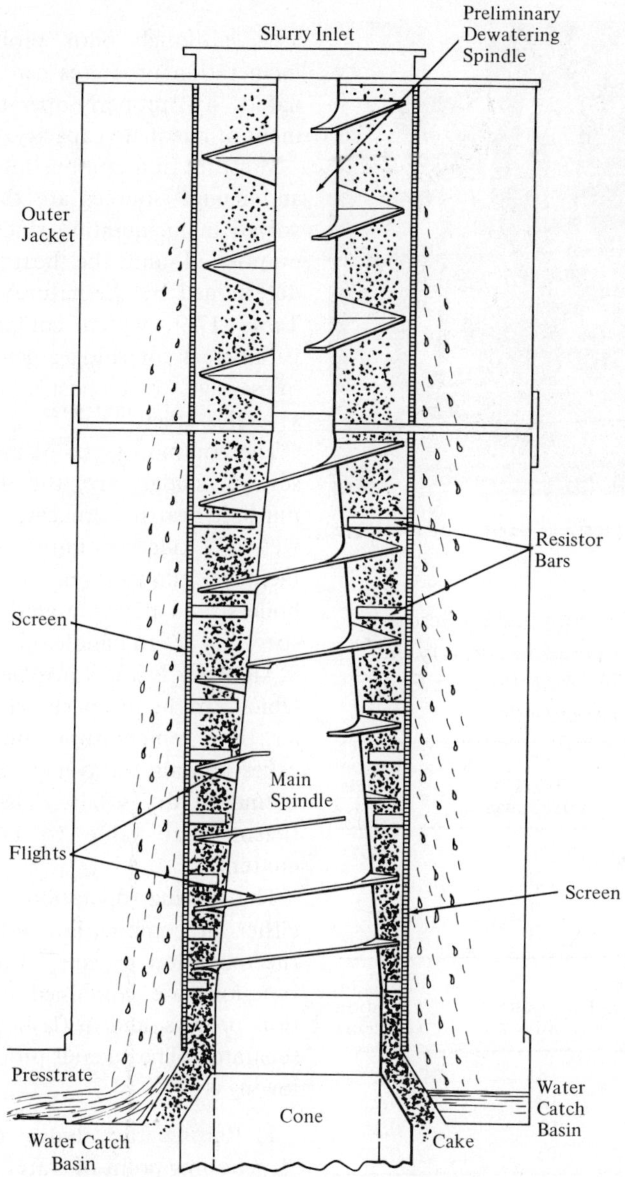

Fig. 17-32 Cross section of a vertical screw press.

ing effluent from a glassine paper mill produced a press cake of about 40 percent solids that could be incinerated by blowing it directly into an oil-fired boiler.[128]

Where possible, it is most desirable to operate screw presses directly on clarifier underflows of low-solids content from kraft pulp and paper-making operations since this method eliminates an expensive intermediate unit process.[129,130] The solids involved must be reasonably free and not too hydrous if this procedure is to succeed. Results obtained at four kraft mills are summarized in Table 17-8.

Press installations cost about $5000 per installed horsepower of capacity for units in the 40 to 60 hp size range. Operating costs can run as low as $3 and as high as $10/ton of dry solids for fibrous solids depending upon the installation size, the character of the sludge handled, and whether or not prethickening or grit separation is necessary. These 1970 figures do not include capitalization.

Fig. 17-33 Filter cake press. (*Courtesy of the Rietz Manufacturing Co.*)

TABLE 17-8. Results Obtained by Screw-Press Operation Directly on Clarifier Underflows at Four Kraft Paper Mills

Mill No.	No. of Presses	Total Horsepower	Percent Solids in Feed
1	1	–	4.0
2	2	120	3.2
3	2	60	3.1
4	1	50	3.5

Mill No.	Percent Solids in Cake	Tons Cake Daily	Solids in Filtration (percent)
1	30	8	–
2	35	30	0.25
3	32	40	0.15
4	32	60	0.20

SLUDGE DISPOSAL

Incineration. Since 1935 the combustion of sewage and industrial waste sludges has become common practice in the United States and has been increasing in popularity as a disposal method.[131,132] This is a result of the decrease in available land areas suitable for disposal, the increased cost of land, and the greater volumes of sludge requiring disposal because of the population increase.

How widely it is adopted in the future may depend to some degree upon air quality standards, although odor problems attendant to incineration processes can be practically eliminated by properly-operated suitable equipment of adequate capacity.

Moisture and combustible content of sludges and sludge sources are the major factors involved in incineration since the water must be evaporated and the heat value of sludges is determined by the nature of the combustibles. Table 17-9, which compares the important parameters for sludges generated on treatment of sewage, was constructed from data from several sources.[133,134,135]

The common types of incinerators in use for sewage sludge are the flash-drying system, multiple-hearth furnaces, fluidized beds, and wet-air oxidation units. Industrial use also includes kiln-type units, traveling-grate bark boilers, and power boilers,[136] as well as some special types to handle oily sludges.[137]

All these systems involve drying and burning which, to be properly effected, require fuel, air, time, temperature, and turbulence. Sludge cakes contain an average of 3 lb of water per pound of dry solids. The latter supplies sufficient heat value to evaporate the water content.

The drying operation can be carried out either in combination with or separate from the oxidation process. For example, the two functions are combined in the wet-air oxidation process, and in flash-drying units they are separate. The overall process includes the following steps:

1. Raising the sludge temperature to the boiling point of water.

TABLE 17-9. Important Parameters for Sludges Generated on Treatment of Sewage

	Percent Combustible	Percent Ash	Btu/Lb Dry Solid
Grease and scum	85–95	5–15	15–17,000
Fresh solids	75–85	15–25	9–11,000
Fine screenings	80–90	10–20	8–10,000
Digested sludge	55–60	40–45	5–6000
Grit	30–40	60–70	3–4000

2. Evaporating off the water content of the sludge.
3. Raising the water vapor content and temperature of the gases.
4. Raising the temperature of the sludge combustibles to the ignition point.

Excess air over and above that required for combustion is needed for complete oxidation, but the quantity must be carefully controlled since it tends to reduce the burning temperature and increase heat loss. Type of equipment used, sludge characteristics, and disposition of off-gases all influence the excess air requirements.

Portions of the heat produced by the combustion process are lost in the stack gases, by radiation from the incinerator, and a small amount which remains in the ash is discharged. The remainder is available for preheating feed sludge and the air supply, and the combustion is frequently self-sustained once the incinerator is brought up to temperature with auxiliary fuel.

The major end products of sludge burning are carbon dioxide, water, sulfur dioxide, and residual noncombustible material or ash. This frequently contains a small percentage of unburned carbon.

Drying costs have been shown to be about double those for incineration, the latter amounting to about \$16 to \$22/ton of dry solids in 1970.

Multiple-Hearth Units. The multiple-hearth incinerator is most commonly used for burning sanitary sewage sludges.[132,137,138,139] This is an adaptation of a type of furnace used for

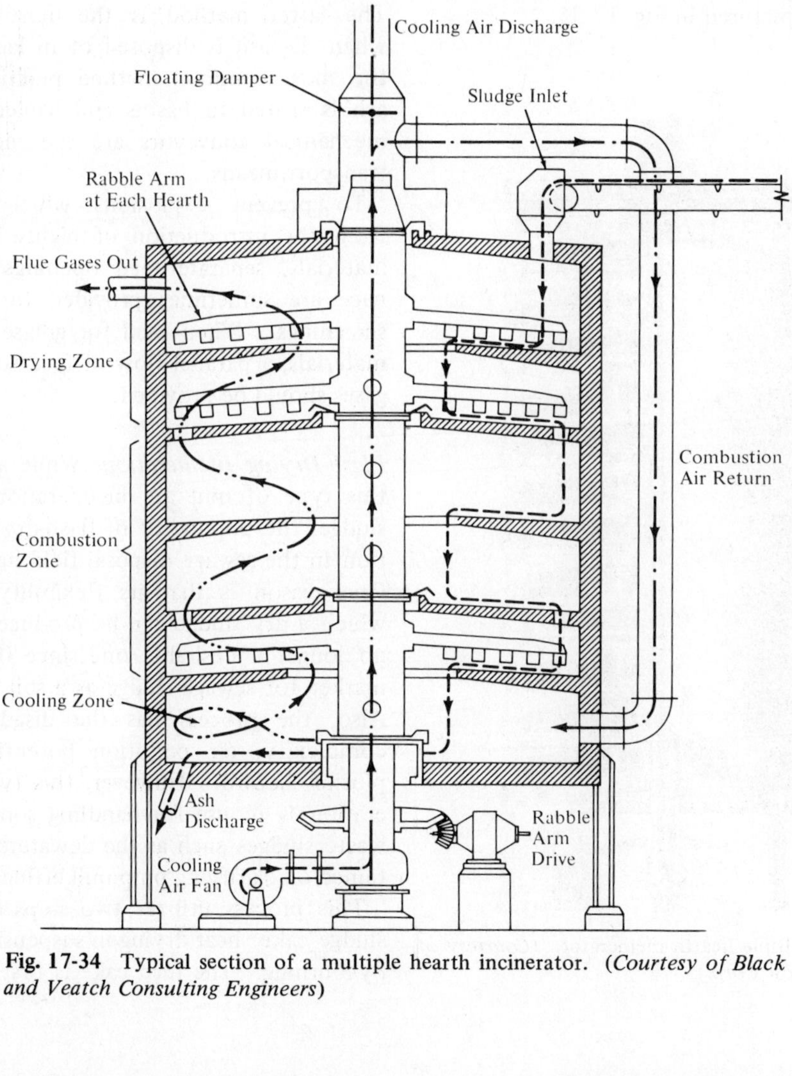

Fig. 17-34 Typical section of a multiple hearth incinerator. (*Courtesy of Black and Veatch Consulting Engineers*)

many years for roasting ores. It has the advantage of containing in one unit means for preheating both the sludge and the combustion air, and drying, burning, and cooling the resultant ash. Also, it can readily be equipped for odor control of the off-gases.

These furnaces consist of a refractory-lined vertical steel shell containing a series of from 4 to 11 refractory-covered trays or hearths surrounding a central rotating shaft. The shaft is equipped with air-cooled rabble arms which rake the sludge across each hearth, break it up, and expose it to the air supply. The annular space between the tube carrying air to the arms and the central shaft serve to preheat the combustion air. Auxiliary fuel burners fed by either oil or gas are located in the shell surrounding the burning hearths. A drawing of this type of furnace appears in Fig. 17-34, and a unit is pictured in Fig. 17-35.

Fig. 17-35 Multiple hearth incinerator. (*Courtesy of Bartlett and Snow Corp.*)

The flow of sludge cake is downward through the unit. Heating and drying occur in the upper hearths, incineration in the middle ones, and ash cooling in the bottommost level. Shaft cooling air is recycled to the bottom hearths to cool the ash. Temperatures in these units generally range from 1000°F (538°C) in the top levels, to 1600 to 1800°F (870 to 982°C) in the burning hearths, to around 600°F (315°C) at the bottom. The off-gases of some units are heated to 1250 to 1500°F (676 to 815°C) to destroy odorous substances. This is sometimes done by passing the gases through a chamber where there is open burning of auxiliary fuel. While effective, this method of odor control is expensive, costing about \$0.30 to \$0.50/ton of dry solids handled.

Ash is moved from the incinerator area mechanically, pneumatically, or hydraulically. The latter method is the most satisfactory when the ash is disposed of in lagoons and is the most common method practiced. When ash is stored in basins and hauled away, dry mechanical conveyors are the most efficient transport means.

To prevent explosions which can result from the introduction of highly combustible materials, separate feed openings in the furnace are sometimes provided for grease and screenings. When used for grease and similar materials, a parallel flow of feed solids and hot gases should be provided.

Flash-Drying Incineration. While a number of this type of unit are in operation on sewage sludge, the popularity of flash-drying incineration in the sewage disposal field has waned.[132] One reason is that its flexibility feature in which a dry sludge can be produced for sale is no longer a desirable one since there is little market for sewage sludge as a soil conditioner. Also, the process has the disadvantages of complexity, air pollution potential, and explosion hazard. However, this type of incineration is in use for handling some industrial waste sludges such as the dewatered slurry obtained on clarifying pulp mill effluents.[140]

This process utilizes two steps for handling sludge cake: heat drying in suspension followed by burning. The feed cake is first mixed with dry sludge to reduce its entrance moisture content, and is then contacted at high velocity with a gas stream generally exceeding 1100°F (593°C) which lowers its moisture content to about 10 percent in a matter of seconds. Burning temperature approaches 1400°F (760°C), and the off-gases are employed in the drying operation. The dry sludge is separated from the gas stream in a cyclone and blown into a furnace in much the same manner as is pulverized coal. An illustration of an incinerator of this type is shown in Fig. 17-36.

Fibrous sludges require the use of a fluffing device to prepare them for introduction into the furnace.

Mechanically thickened sludges (14–19 percent solids) are sometimes applied to these units after admixture with dry sludge and pulverized coal.[141]

More auxiliary fuel than is introduced with the sludge is generally required. This addition amounts to about 6000 Btu/lb of dry sludge solids incinerated.

Fluidized-Bed Incinerators. In the application of this process to sewage sludge,[142,143,132] a bed of hot sand is maintained in suspension above a distribution plate by air blowing from a plenum beneath the plate. The bed is heated to 1200°F (649°C) with oil or gas and the sludge is introduced directly into it by screw or piston-type conveyors. The combustion is then self-supporting.

The sludge is rapidly broken up, dried, and burned within the bed by the intense heat and violent mixing conditions. The transfer of heat is both rapid and efficient because of the large surface area available.[144,149]

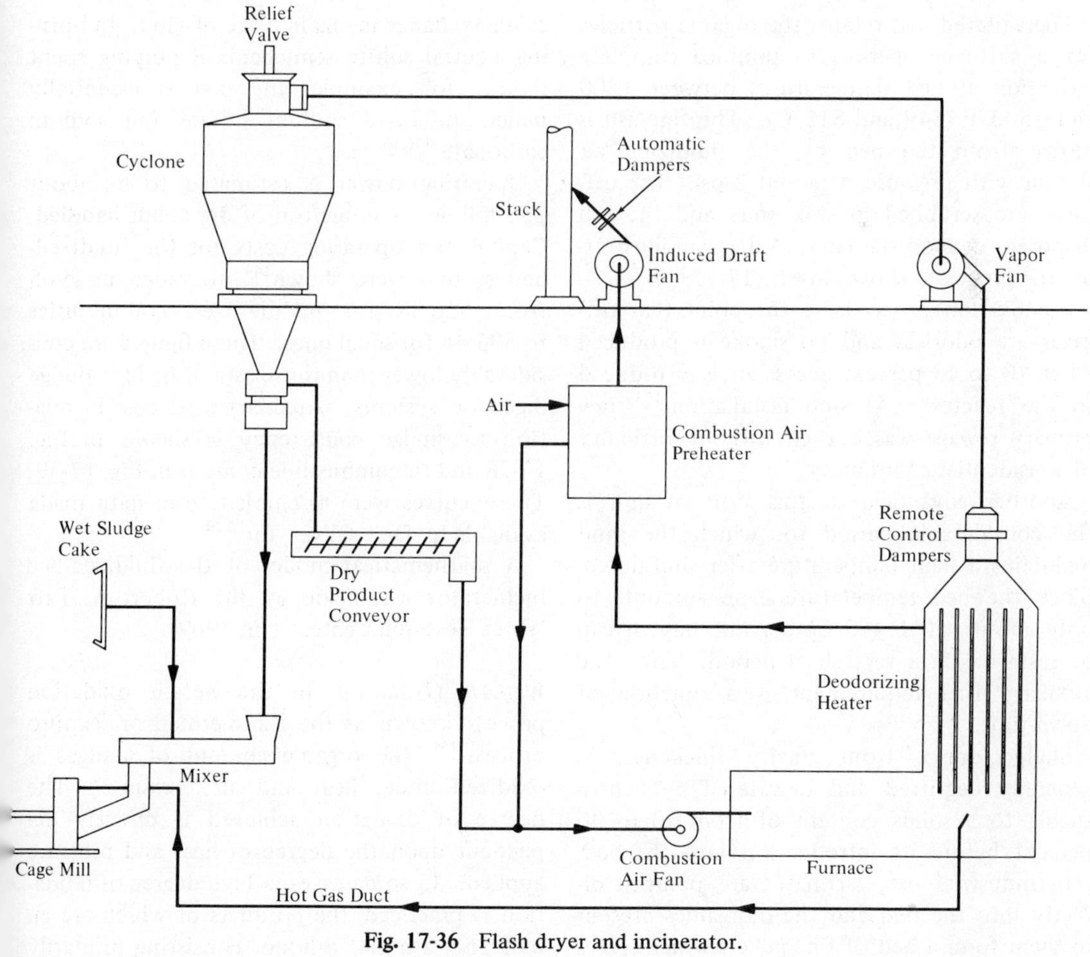

Fig. 17-36 Flash dryer and incinerator.

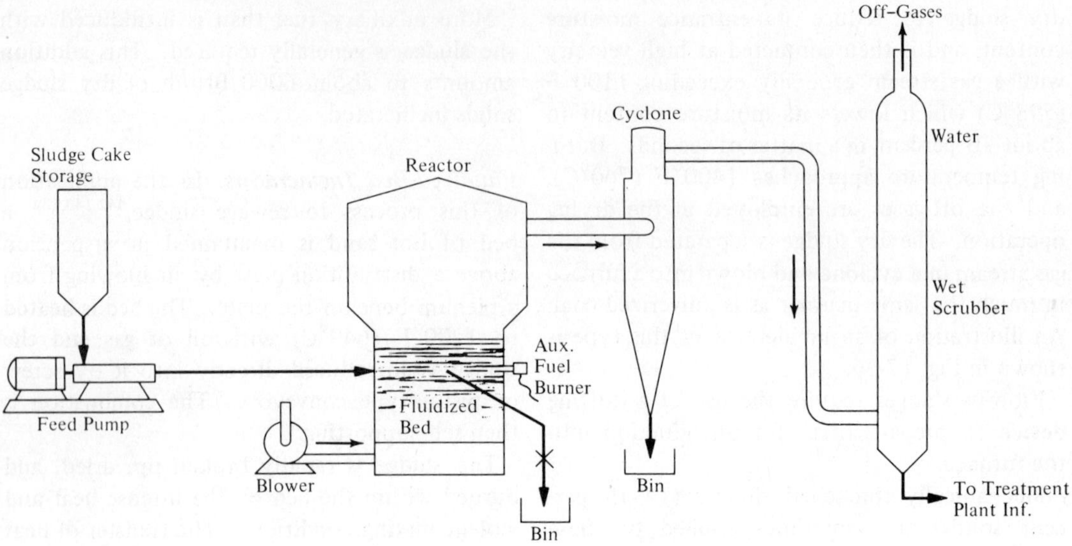

Fig. 17-37 Fluidized-bed reactor.

The agitated bed retains the organic particles for a sufficient period to produce complete oxidation at bed temperatures between 1200 and 1500°F (649 and 815°C). The fine ash is blown from the bed by the fluidizing air stream with pressure of about 2 psi. The off-gases are scrubbed in wet units and the ash disposed of upon the land. A diagram illustrating the process is shown in Fig. 17-37.

Combustion proceeds to the point that off-gases are odorless and no smoke is produced when 10 to 15 percent excess air is introduced to the reactor. At one installation,[143] raw primary sewage was burned without difficulty in a residential community.

Another advantage of this type of unit is the considerable period for which the sand maintains a high temperature after shut-down. Since the bed temperature drop amounts to only about 150°F (65°C) per idle day, it can be restarted in a very short period. Thus, the auxiliary fuel requirement is a function of down-time.

Sludge drawn from gravity thickeners is generally degritted and dewatered by centrifuging to a solids content of from 20 to 30 percent before its introduction into the bed.

In industrial use, raffinates are pumped directly into the bed and the inorganics present in them form a bed of fine pellets which serves as an exchange media in place of sand. In burning neutral sulfite semichemical pulping spent liquor, for example, the bed is essentially pellets or fused sodium sulfate and sodium carbonate.[146,147]

Operating power is estimated to be about 250 killowatt–hours/ton of dry solids handled. Capital and operating costs for the fluidized-bed system were shown[144] to range in 1965 from $26.38 for middle-sized communities to $35.44 for small ones. These figures are considerably lower than those established for sludge digestion systems. Auxiliary fuel cost in relation to sludge consistency is shown in Fig. 17-38 and to combustible content in Fig. 17-39. These curves were assembled from data made available by Dorr-Oliver, Inc.[150]

A mathematical model of the fluidized-bed incinerator was made by the Robert A. Taft Water Research Center[151] in 1969.

Wet-Air Oxidation. In the wet-air oxidation process known as the Zimmerman or Zimpro process,[152] the organic content of sludges is oxidized under heat and air pressure. The degree of oxidation achieved is directly dependent upon the degree of heat and pressure applied. In some cases, a high degree of oxidation is practiced, the products of which are an inorganic ash and a liquor consisting primarily

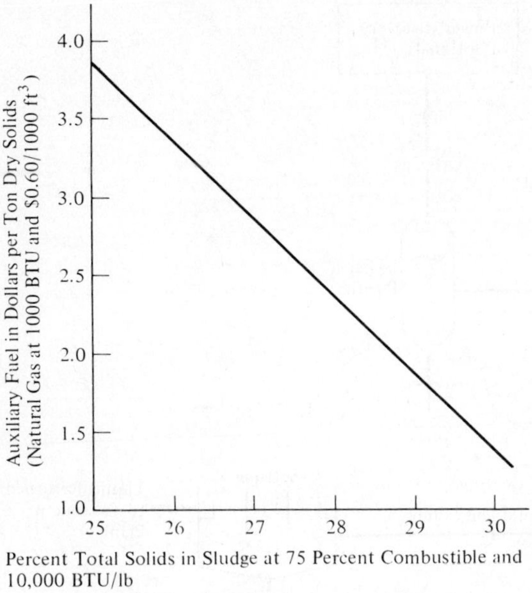

Fig. 17-38 **Percent total solid versus auxiliary fuel.**

of acetates. In others, a partial oxidation is achieved which destroys the putrescible compounds and leaves a carbonaceous, easily dewatered residue which can be disposed of on the land without nuisance.

When a high degree of oxidation is practiced (90 percent), this process is competitive with other incineration systems. Partial

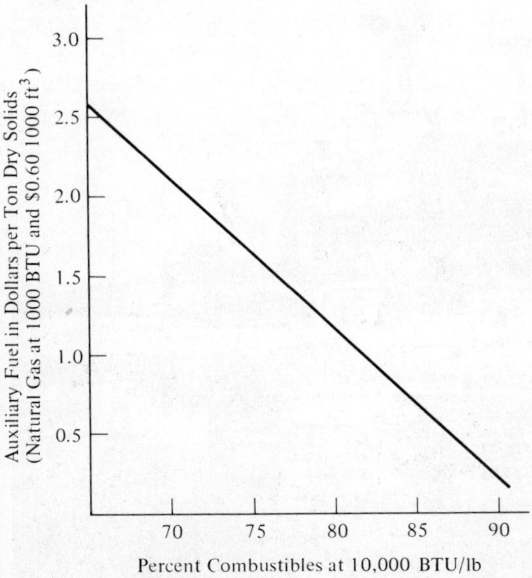

Fig. 17-39 **Percent combustibles in sludge versus auxiliary fuel requirement.**

oxidation can best be compared with sludge digestion and chemical-conditioned dewatering since it produces a reasonable dry, stable material.[153] It is discussed in the conditioning section of this chapter.

Oxidation takes place chemically by direct reaction of dissolved oxygen in the liquid phase with organic compounds in the waste. The theory of this reaction is discussed in an ASCE committee report.[154] A flow sheet of the process is shown in Fig. 17-40 and a picture of a Zimpro plant is shown in Fig. 17-41.

Sludge thickened to 8 to 12 percent solids is required for economical operation of this process. After passage through a grinder, the sludge is pressurized to the operating level and is preheated in exchangers by reactor gases, steam, and water. Air is then injected as it passes to the reactor where the organic matter is oxidized to carbon dioxide, water, and some simple aliphatic compounds such as acetates, which are resistant to oxidation.[153] When blown from the reactor, gases and solids are separated from the liquor which is returned to the treatment process. The solids are disposed of on the land.

It is apparent that COD reduction is related to temperature and that the reactor detention time required to produce a given degree of oxidation decreases as the temperature increases.[155,156,157] Heat value of the sludge and the degree of oxidation required determines the oxygen requirement,[154] and since process economy is a function of air input, it is important that this be determined for the sludge handled. The Btu value and oxygen requirements for some sludges and raffinates are presented in Table 17-10.

Incineration of Fibrous Sludges. All types of common combustion processes can be utilized successfully to incinerate sludges from the manufacture of pulp and paper if they are adapted to do so.[158] The kiln-type unit is recommended for sludges high in ash, and burning in a power boiler is suggested for those sludges low in ash that can be dewatered to substantial degree. Heat values for various types are shown in Table 17-11.

The incineration of primary pulp mill sludges

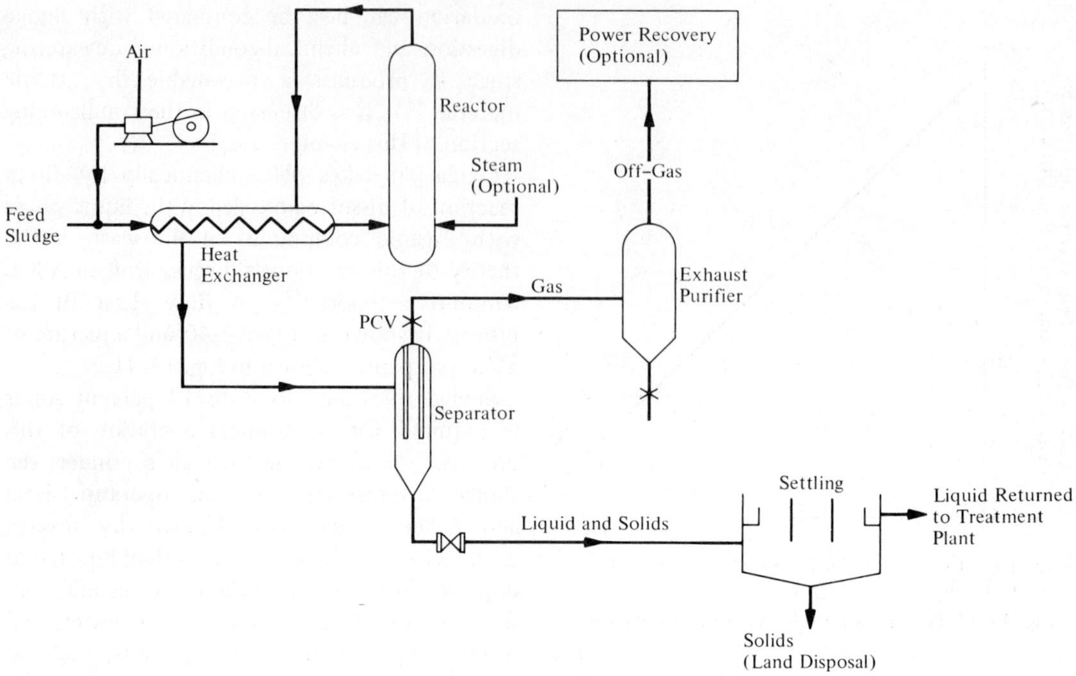

Fig. 17-40 Wet-air oxidation process diagram.

by the flash-drying method at a large kraft pulp mill cost $12.50/ton of dry solids.[159,160] At one mill, a mixture of pulp mill, activated sludge, and bark fines is burned.[161] Two parts of bark fines per pound of sludge solids on the dry solids basis are mixed and dewatered on belt filters. The 24 percent solids cake is pressed to 41 percent, dried, and burned on

Fig. 17-41 Wet-air oxidation plant. (*Courtesy of ZIMPRO, Inc.*)

TABLE 17-10. Heat Value and Oxygen Requirements of some Sludges, Raffinates, and Fuel Oils

Material	Heat Value (Btu/lb)	Oxygen Requirement (lb/lb dry solids)
Primary sludge	7820	1.334
Activated sludge	6540	1.191
NSSC liquor	5812	0.955
Fuel oil	19,376	3.260
Pulp mill sludge	5100	0.715
Casein	10,550	1.750

TABLE 17-11. Heat Value of Paper Mill Sludges

Type of Sludge	Percent Ash	Heat Value (Btu/lb)	Btu/lb Combustibles
Paper mill	37.3	5623	8990
Paper mill	44.9	4377	7960
Boardmill	18.3	6651	8150
Rag pulp mill	42.2	3748	6510
Deinking	65.8	2151	6500

traveling grates. Cost for disposing of 26 tons of sludge daily amounts to about $20/ton. Several pulp mill installations burn pressed sludge with bark and wood wastes[136] and others apply fluidized-bed incineration to similar industrial sludges. In the incineration of tannery waste sludge along with sewage sludge,[162] burning the sticky filter cake obtained from treatment required the use of special arms in the first hearth of the multiple-hearth furnace used.

Incineration of Oily and Greasy Sludges. Sludges containing oil and grease can be incinerated in fluidized-bed units quite readily, although they require special handling when fed to multiple-hearth incinerators. In burning scum from a sewage treatment plant high in grease content with a 45 percent moisture content,[163] water sprays emitting a fine mist were employed to prevent the thermostatically-controlled hearths from attaining too high a temperature. Attempts to burn screenings with the scum were not successful.

Scum-burning operations at Minneapolis-St. Paul use a cascade-type incinerator which is equipped with cast-iron tuyers cooled by the combustion air and a secondary checkered-brick combustion chamber.[164] This unit is started at 1300°F (705°C) by preheating with four oil burners and operates at 1600°F (871°C), burning 1100 lb of dry solids/hr.

A hearth-type incinerator known as the "greaseburn" is employed for small operations (less than 420 lb of dry solids/hr).[137] Skimmings from which free water is decanted are pumped through a rotating pipe to a combustion chamber, which is previously brought up to temperature by auxiliary fuel. When equilibrium is established, combustion becomes self-supporting. Off-gases are discharged through a dilution stack, and the small amount of fine ash formed either drops to the hearth or passes off into the atmosphere. Larger units specifically designed for handling grease and scum are of the fluidized-bed type and are equipped with dust collectors and wet scrubbers.

The oil-bearing sludges of petroleum refineries are sometimes burned.[165,166] In some instances, hearth-type incinerators are used and, in others, the waste is sprayed into the furnace with air and burned like fuel oil. Burn pits were once in common use but these are disappearing due to restrictions on open burning. Fluidized-bed units have also been applied to petroleum sludges.[167]

Some sludges from the petrochemical industry are incinerated by methods similar to those employed for refinery wastes.[168] Some grease-bearing sludges from food processing are treated by the Carter–Greenfield oil extraction process and burned in a furnace especially designed for the purpose.[169]

Sludge Digestion. Sludge digestion by microbial decomposition can be considered a conditioning technique since it renders sludges non-putrescible, substantially inert, and more readily dewaterable. This process generally employs anaerobic decomposition which is substantially a methane fermentation,[170] but aerobic digestion has been practiced for some industrial waste sludges with limited success.

In anaerobic decomposition, fats, proteins, and cellulosic materials are converted largely to methane and carbon dioxide.[171] After the destruction of a large porportion of the organic content of the sludge, it is readily dewaterable on sand beds, upon the soil, or by mechanical means.[172] Soil disposal of this material does not create a nuisance, and to the contrary, conditions the soil.

Raw sludge is pumped to a digestion tank, which in most modern practice is both heated and agitated. Sludge temperature may be brought to desired levels in three ways: preheating it in an exchanger, heating it in the digestion tank, or keeping the entire tank contents up to temperature through circulation through an exchanger. Mixing of the tank contents is accomplished by recirculation, mechanical mixing, or gas diffusion with recycled digester off-gases. Figure 17-42 is a cross section of a digester employing gas diffusion agitation and external heating.

Decomposition results in liquefaction and gasification of the sludge solids and a breakdown of their colloidal structure. This is accomplished by a mixed culture of microorganisms which develop naturally under suitable conditions of anaerobiosis, pH, and temperature. During digestion, intermediate products are formed, such as organic acids, which in turn are further decomposed to methane, carbon dioxide, and water. Sulfates are reduced, to a degree, to hydrogen sulfide and proteins to ammonia and CO_2.

Benefits of this process are:[15]

1. Nuisance prevention by decomposing organic solids to a more acceptable form.
2. Sludge volume reduction by converting organic solids to gases and liquid.
3. Volume reduction by compaction of the residual solids.
4. Sludge storage to accommodate fluctuations in wastewater flows and to permit flexibility in subsequent dewatering operations.
5. Homogenizing sludge solids to facilitate subsequent handling steps.
6. Production of useful by-products such as gas and soil conditioner.
7. Reduction of pathogenic organisms.
8. Production of more easily dewatered solids.

The combustible solids and their nature, digestion time, temperature, effectiveness of mixing, feed concentration, and loading rate

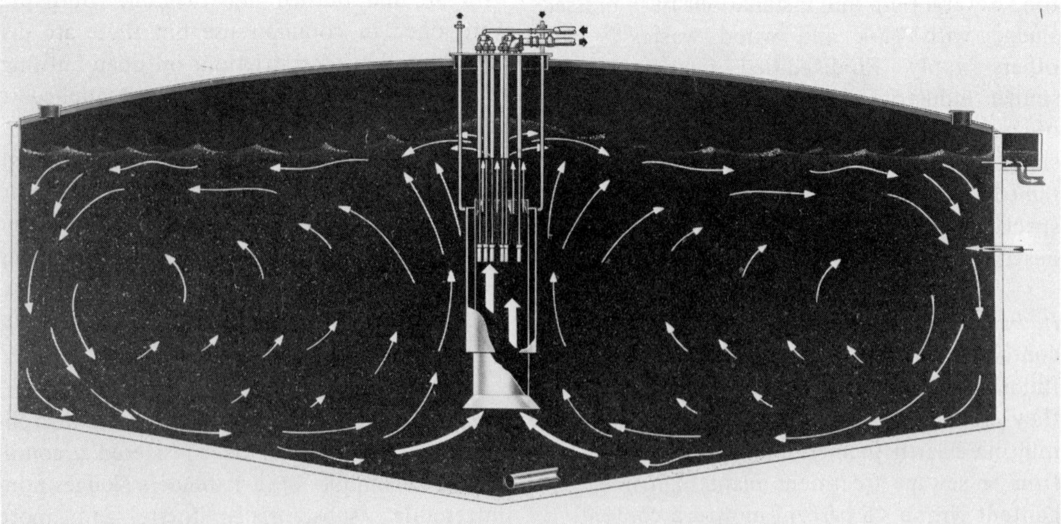

Fig. 17-42 Gas-agitated sludge digestion tank. (*Courtesy of Walker Process Equipment, Div. of Chicago Bridge & Iron Co.*)

all affect digester performance and must be considered in its design. The objective is to provide a system which will produce a sludge with a low-combustible content (preferably under 50 percent) and which will dewater well in the dewatering equipment provided. Design is based on the pounds of combustible solids per cubic foot of capacity applied per day. For normal-rate systems, this is between 0.04 and 0.1 lb/day, and for high-rate, dual-tank systems, between 0.15 and 0.4 lb/day. Detention time in normal-rate systems ranges from 35 to 45 days, and for high-rate systems from 14 to 20 days.

Curves which indicate combustible matter reduction in relation to digestion time and the temperature–time relationship were published in the Water Pollution Control Federation Manual of Practice No. 11,[173] and are presented in Figs. 17-43 and 17-44. These curves were established from data available from numerous investigations.[174,175]

It will be noted in Fig. 17-44 that the temperature ranges up to 140°F (60°C) in high-rate systems. This is because the process can be made to operate in the thermophilic zone

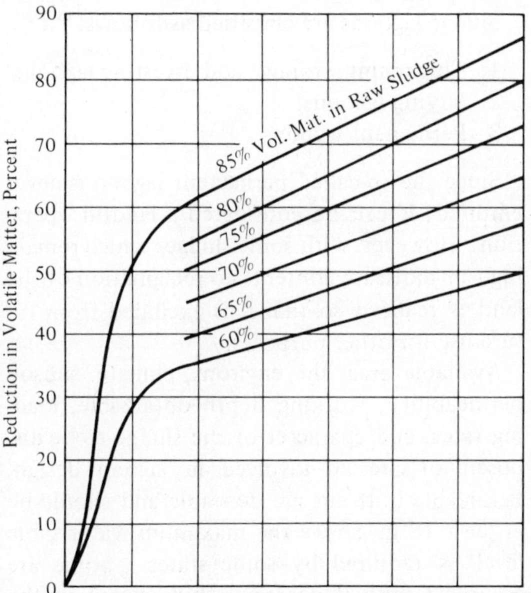

Fig. 17-43 Reduction in volatile matter for various sludges and digestion times. (*From Manual of Practice No. 11, Water Pollution Control Federation*)

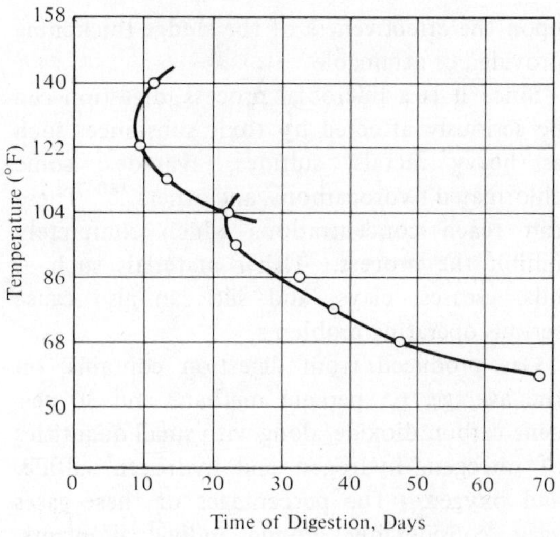

Fig. 17-44 Digestion time-temperature relationship. (*From Manual of Practice No. 11, Water Pollution Control Federation*)

after acclimation and development of organisms operating at high temperature.[176,177] The rate of digestion can be accelerated by temperatures in the 110 to 130°F (43.5 to 55°C) temperature range and it has been demonstrated[178] that the methane content of the gas is higher. However, sludge digested thermophilically does not dewater as well as mesophilically digested sludge, and in practice the separation and withdrawals of superliquor is less satisfactory. Also, the supernatant withdrawn is high in suspended solids.

In order to combat the problem of a less compactable sludge from high-rate digestion, prethickening of the sludge feed or elutriation of the digested sludge is practiced. This difficulty is also overcome to some degree by carrying out the process in two digestion tanks operated in series. Two-thirds of the gasification takes place in the first unit, so supernatant withdrawal from the more placid second unit can be more readily achieved and a higher consistency carried in the system. Heating and mixing facilities are generally confined to the first stage.

It is felt by some[179] that high-rate digestion could be carried out at 25 percent of the capacity generally required at the normal-rate. There are operating data to confirm this. However, this degree of improvement is dependent upon the effectiveness of the sludge thickening provided or attainable.

Since it is a microbial process, digestion can be seriously affected by toxic substances such as heavy metals, sulfides, cyanides, some chlorinated hydrocarbons, and others.[180] These can reach concentrations which completely inhibit the process. Other materials such as oils, greases, clays, and silt can also cause serious operating problems.

Gas produced from digestion contains, on the average, 65 percent methane and 30 percent carbon dioxide along with small quantities of nitrogen, hydrogen and hydrogen sulfide, and oxygen. The percentages of these gases vary considerably among individual plants. This gas is used for heating water to elevate sludge temperatures in exchangers, plant heating, or in operating gas engines, driving pumps, compressors, or generators. In the latter instance, hydrogen sulfide is removed from the gas before use by passing it through a scrubber.

The methane content of the gas is a good index of the digestion operation since it decreases rapidly when the fatty acid content of the digesting sludge reaches a level which begins to inhibit the methane producing microorganisms.

Gas can be stored within the digester if it is equipped with a floating or gas-holder type of cover. It can also be compressed and stored in spherical gas holders independent of the digestion system.

High-rate anaerobic digestion has been tried for the treatment of wastes relatively low in total combustible solids and high in dissolved material, such as yeast waste, distillery waste, potato starch waste, and strawboard waste. Full-scale application of the process to this kind of waste has been successful only for strawboard waste. Anaerobic digestion is also used for treating meat packing waste.[181]

The application of aerobic digestion to sewage and organic industrial waste sludges[182] has not made headway, largely because it does not reduce combustible solids rapidly, nor markedly improve dewaterability. It can, however, reduce putrescence to a degree where the residual sludge can be applied to the land, since the readily oxidizable substances decompose quickly.

Land Disposal. Lagooning, filling, and distribution upon the land are the most common means of disposing of industrial and water treatment plant sludges. Digested sewage sludge is occasionally lagooned and sludge cake is dumped on the land. At one time, trenches were used for disposal of fresh solids, but except in emergencies, this practice is no longer followed because of odor and insect breeding. Landfill, together with other suitable materials, constitutes an excellent disposal method if properly controlled and the disposal area is close enough to the source.

The use of lagoons for sludge digestion is a lengthy process and is a declining practice. However, if sufficient suitable land is available, lagoons can serve as a satisfactory temporary measure while adequate digester capacity or other disposal facilities are built. Some lagoons also act as dewatering devices since, on evaporation and drainage, the contents become solid and can be removed and spread upon the land, thus returning the basin to service.

Sludge lagoons are classified as follows:[25]

1. Thickening, storage, and digesting lagoons.
2. Drying lagoons.
3. Permanent lagoons.

Since the so-called permanent lagoon is never emptied, it can be considered a landfill operation. However, with some sludges which remain high in moisture content, no reclamation of the land is realized so that it is excluded from future use for other purposes.

Available area, the environs, climate, subsoil permeability, working depth obtainable, loading rates, and character of the sludge to be disposed of are all involved in lagoon design. Permeable bottoms are desirable and should be at least 18 in. above the maximum water table level as required by some states. Some are equipped with underdrains, but except in the case of very "free" sludges, these do not accelerate the drying process.

Lagoons should be designed so that the sludge applied will be distributed evenly over the surface. Dike construction and slope depends upon the nature of the local soil and the availability of clay for use as core walls within the embankments. A 1:2 slope for exterior walls and a 1:3 slope for interior walls are suggested if erosion is to be prevented.[172] A ramp to allow access for cleaning equipment should be provided in other than permanent lagoons, and the top width should be sufficient to serve as a road for trucks. Construction of the lagoon wall also depends upon depth. This is commonly from 10 to 12 ft but can be deeper. However, the deeper the basin, the greater must be the slope of the walls and the care taken in construction. Deep basins should in most instances have clay-core walls.

In some units, provision is made for decantation of free water separating from the sludge, while in others, it is pumped out by dropping a suction line from a portable pump into the basin at the depth to which free water has collected.

Sludge removed from lagoons is generally used for landfill or spread upon wasteland. Here it dries further and ultimately becomes similar to some soils. One such land-spreading operation is illustrated in Fig. 17-45.

Increasing attention will be given to more scientific land disposal, particularly for industrial wastes. Soil mechanics will be employed for determining landfill mixtures which can reclaim land most effectively in the shortest period of time and dispose of the most waste in a given area.

Sludge Utilization. A considerable quantity of sewage sludge is ultimately disposed of as a soil conditioner or fertilizer on agricultural land, golf courses, and lawns.[183,184,185] Digested sludge from small plants is frequently handled in this manner and some large plants produce an activated sludge in dry form for marketing. Since both types have been freed of pathogens by processing, they can be used safely on soil producing any type crop. The dried products are granular and readily applied, but digested sludge as it comes from drying beds is difficult to distribute on the soil. Hence, it is frequently ground at the plant in a grinder in order to facilitate its use and make it more acceptable to the public. In some instances, it is applied in a wet state from tank trucks.

Digested sludge can best be classed as a soil conditioner rather than a fertilizer because of the low concentration of nutrients present. It contains about 2.5 percent nitrogen and 1.5 percent available phosphoric acid. The presence of activated or humus tank sludge will raise these values somewhat.[186,187] The content of stabilized organic matter present, which ranges between 45 and 60 percent, is probably the

Fig. 17-45 Lagoon receiving deinking mill clarifier sludge. *(Courtesy of Walker Process Equipment, Div. of Chicago Bridge & Iron Co.)*

greatest value of the material because of its tendency to improve porosity as well as water-holding capacity of soils.

Activated sludge is sufficiently high in nutrients to be considered a fertilizer, with a nitrogen content averaging 6.2 percent and the available phosphoric acid 2.93 percent.[188] Several large sewage treatment plants in the United States have dried and sold activated sludge on a large scale, primarily to golf courses and fertilizer manufacturers. However, unstable markets, rising shipping costs, and advances in the use of the low-cost inorganic fertilizers have made this method of disposal not only unprofitable, but costly. It is possible that this situation may change in the future because of the vast and rapidly increasing amounts of sludge from metropolitan areas, limitations on burning due to air pollution control regulations, and the increasing need to reclaim wasteland in some areas. Subsidies for use of this disposal method could also increase its attractiveness as the cost of incineration increases. Experiments in large-scale, wet-sludge hauling by train are now underway in Chicago.[189]

Lime mud is probably the industrial sludge most frequently recovered in the largest quantity. All kraft and soda pulp mills[190] and some other industries dewater and reburn calcium carbonate to form calcium oxide for reuse in the process as do some water softening plants.

Ocean Disposal. Ocean disposal[191] is rapidly running into disfavor with governmental bodies and environmentalists because of claims that large areas of the ocean are being adversely affected and that coastal fish and shellfish habitats are being destroyed. As the result of studies by the federal government, it has been recommended that the practice be banned. However, no alternative has been forthcoming, and in light of the magnitude of these operations, replacement by an alternate disposal method would prove to be a gargantuan undertaking. Therefore, ocean dumping is likely to be continued, subject to greater regulation as to disposal site and the time span of dumping operations.

Much of the sludge barged to sea is undigested fresh solids and activated sludge. Industrial wastes include acid liquors, copperas, cannery wastes, aluminum waste, shells, and dredging spoil.

Some of the operating problems involved in sludge barging are pumping through long lines, sludge concentration, and barge scheduling.[192,193,194] Since the volume barged is most important cost-wise, the advisability of delivering a sludge of maximum consistency is very important.[195]

Under proper conditions of dilution and dispersion in large bodies of water, digested sludge can be disposed of by pipeline without damaging water quality or creating a nuisance.[196,197] It is mixed with the primary effluent and discharged through a common submarine outfall. However, since the disposal of fresh solids can give rise to floating debris on the ocean's surface which may wash up on beaches, screening is recommended. The success of outfall lines depends upon design, placement, and the diffusion arrangement.[198]

Pipeline disposal of sludge will undoubtedly continue for many years. On the other hand, rapidly rising costs of barging operations and the uncertainty of obtaining a permit in the future make it impossible to evaluate the long-term future of this procedure.

The cost of pipeline transport of digested sludge has been computed at $0.56/1000 gal for 10 mi and $3.90 for 100 mi at 1964 price levels.[199] This amounts to $2.50 to $3.50/ton of dry solids at 5 percent consistency for 10 mi and from $15.00 to $22.00 for 100 mi.

Composting. Composting of sewage sludge with and without the addition of ground garbage or other organic wastes has been the subject of considerable investigation, field trials, and some limited application. This process produces a humus type of material through aerobic thermophilic digestion.

Although the process is workable and sound conservation-wise, the economic uncertainties have severly curtailed its use. Like anaerobically digested sludge, composted sludge is very low in nutrients, the addition of which raises the cost to a point where it is priced out of the market. Widespread use is also inhibited by the high moisture content of sludge and sludge cake and the limited market for this material in this country. Several processes are described in the literature.[159,200]

Composting has been applied to a limited degree to industrial wastes including vegetable and fruit processing wastes[169] and organic residues from the manufacture of pharmaceuticals.[148]

REFERENCES

1. Anable, A., *Perry's Chemical Engineering Handbook*, 4th ed., McGraw-Hill, New York, Sec. 19, p. 1619 (1963).
2. Gehm, H. W., *Pulp and Paper Magazine of Canada*, T-189 (1960).
3. *WPCF Manual of Practice No. 20* (1969).
4. Rich, L. G., *Unit Operations of Sanitary Engineering*, John Wiley and Sons, Inc., New York (1961).
5. McCarty, P. L., *Journal WPCF* 38, 493 (1966).
6. Budd, L. F., *Proceedings, Municipal and Industrial Wastes Conference* **163** (1959).
7. Black, S. A., Ontario Water Research Commission Research Pub. No. 24 (1967).
8. Dick, R. I., *Water and Wastes Engineering* **6**, 47 (1969).
9. Keefer, C. E., *Water and Sewage Works* **RN 107**, R 336 (1960).
10. Sparr, A. E., M.S. thesis, New York University (1942).
11. Follette, R., and Gehm, H. W., NCASI Technical Bulletin No. 190 (1966).
12. AWWA Foundation Report, "Disposal of Wastes from Water Plants," AWWA (1969).
13. Gehm, H. W., and Blosser, R. O., NCASI Technical Bulletin No. 129 (1960).
14. Borhan, W. J., Matherne, J. L., and Keller, A. G., Louisiana State University Engineering Experiment Station Bulletin No. 54 (1956).
15. Burd, R. S., FWPCA Pub. No. 20-4, p. 23 (1968).
16. Vesilind, P. A., *Water and Sewage Works* **115**, 302, 8 418 (1968).
17. Vesilind, P. A., Ph.D. thesis, University of North Carolina, Chapel Hill (1968).
18. Lutkin, P. A., *Proceedings, 1st Northeast Regional Anti-Pollution Conference*, University of Rhode Island (1968).
19. Vesilind, P. A., *Water and Sewage Works,* **115**, 7, 302 (1968).
20. Eckenfelder, W. W., Jr., and Mancini, J. L., *National Symposium on Sanitary Engineering*, Pennsylvania State University (1965).
21. Fitch, B., *Industrial and Engineering Chemistry* **60**, 7, 8 (1968).
22. Shannon, P. T., and Dehass, R. D., *Industrial and Engineering Chemistry Fundamentals* **3**, 250 (1964).
23. Newton, D., University of Michigan Continued Education Pub. No. 113, p. 4 (1963).
24. Brisbin, S. G., *Sewage and Industrial Wastes* **28**, 158 (1956).
25. Torpe, W. N., *Proceedings, ASCE* **433**, 80, 1 (1954).
26. Dick, R. I., and Vesilind, P. A., *Journal WPCF* **41**, 1285 (1969).
27. Sparr, A. E., and Grippi, V., *Journal WPCF* **41**, 1886 (1969).
28. Goudey, R. F., and Bennett, S. M., *Water Works and Sewage* **80**, 179 (1933).
29. Colin, F., *Terres Eaux* (France) **21**, 55, 27 (1968).
30. Burke, J. T., *Proceedings, 21st Industrial Waste Conference*, Purdue University (1966).
31. Irvin, D. F., *Perry's Chemical Engineering Handbook*, 4th ed., McGraw-Hill, New York, Sec. 15, p. 1653 (1963).
32. Rohlich, G. A., *Industrial and Engineering Chemistry* **46** (1954).
33. Sessler, R. E., *Sewage and Industrial Wastes* **37** (1955).
34. Wollner, H. J. *et al.*, *Sewage and Industrial Wastes* **36** (1954).
35. Mayo, E., *Proceedings, 38th Conference WPCF* (1968).
36. Chase, E. S., *Sewage and Industrial Wastes* **30**, 783 (1958).
37. Hurwitz, E., and Katz, W. J., *Wastes Engineering* **730** (1959).
38. Lardieri, N. J., *Environmental Science and Technology* **1**, 380 (1967).
39. Clement, M. *et al.*, *Fortschn Verfahrentech* (Germany) 8, 842 (1968).
40. Katz, W. J., *Proceedings, 12th Industrial Waste Conference*, Purdue University, p. 163 (1957).
41. Katz, W. J., and Genopolis, A., University of Michigan Continued Education Series, Bulletin No. 133, p. 17 (1963).
42. Ettelt, G. A., and Kennedy, T. J., *Journal WPCF* **38**, 248 (1966).
43. Downs, J. R., *Water Wastes and Sewage* **81**, 210 (1934).
44. Rudolfs, W., *Sewage Works Journal* **15**, 642 (1943).
45. Woodruff, P. H. *et al.*, Sanitary Engineering Conference, Vanderbilt University, Nashville (1967).
46. Burd, R. S., FWPCA Pub. WP-20-4-210 (1968).
47. Thompson, W. B., *Wastes Engineering* 297 (1965).
48. Rudolfs, W. *et al.*, *Sewage Works Journal* **1** 398 (1929).
49. Sperry, W. A., *Sewage Works Journal* **13**, 855 (1941).
50. Keefer, C. E., *Sewage Treatment Works*, McGraw-Hill, New York (1940).
51. Conway, R. A., *Water and Sewage Works* **109**, 342 (1962).
52. National Lime Association Bulletin No. 214, Washington, D.C. (1964).
53. National Lime Association Bulletin No. 213, Washington D.C. (1949).
54. Purifax, Inc. decriptive bulletin (1970).
55. Sparr, A. E., *Journal WPCF* **40**, 1438 (1968).

56. *Chemical and Engineering News* **48**, 33, 15 (1970).
57. Davis, N. S., and Faust, O. J., U.S. Patent 3,440, 166 (1969).
58. Marbarger, M. C. *et al.*, *Journal WPCF* **41**, 2070 (1969).
59. Schuster, H. *et al.*, *Chemical Abstracts* **71**, 18, 84378f (1969).
60. Sherbeck, J. M., *Journal WPCF* **37**, 1180 (1965).
61. Goodman, B. L., University of Michigan Continued Education Series No. 113, p. 50 (1964).
62. Hopkins, G. T., and Jackson, R. L., *Proceedings, 43rd Annual Conference WPCF* (1970).
63. Vogh, R. P. *et al.*, *Journal AWWA* **61** 276 (1969).
64. Hahn, D. J. *et al.*, *Water and Sewage Works* **116**, 321 (1969).
65. Rudolfs W., and Gehm, H. W., U.S. Patent 2,393,269 (1946).
66. Mogelnicki, S., and Gatza, E. M., U.S. Patent 3,409,546 (1968).
67. Carr, R. L., *Water and Sewage Works* **114**, R64 (1967).
68. Mogelnicki, S., *10th Sanitary Engineering Conference*, University of Illinois, p. 47 (1968).
69. Burd, R. S., *Proceedings, 2nd Vanderbilt University Sanitary Engineering Conference* (1963).
70. Morris, R. H., *Water Works and Wastes Engineering* **2**, 68 (1965).
71. Sharman, L., *Water and Wastes Engineering* **4**, 50 (1967).
72. Swanwick, J. D., and Davidson, M. F., *Water and Waste Treatment Journal* (British) **8**, 386 (1961).
73. Carpenter, W. L., and Caron, A. L., *Proceedings 19th Industrial Waste Conference*, Purdue University, p. 139 (1964).
74. Carpenter, W. L., NCASI Technical Bulletin No. 158 (1962).
75. Genter, A. L., U.S. Patent 2,259,688 (1941).
76. Genter, A. L., *Sewage Works Journal* **13**, 1164 (1941).
77. Genter, A. L., *Sewage Works Journal* **18**, 580 (1946).
78. Genter, A. L., *Sewage Works Journal* **18**, 27 (1946).
79. Chasick, A. H., and Dewling, R. T., *Journal WPCF* **34**, 390 (1962).
80. Swanwick, J. D., Institute of Sewage Purification, Water Purification Laboratory Paper, Stevenage (1962).
81. Kennedy, R. R., University of Michigan Continued Education Series No. 113, p. 37 (1964).
82. Goodman, B. L., and Wichter, C. P., *Journal WPCF* **37**, 1643 (1955).
83. Lumb, C., *Journal Institute of Sewage Purification*, Part I., p. 5 (1951).
84. Lancond, F., *Bull. Tech. Suisse* (Swiss) **91**, 211 (1965).
85. Cardinal, P., Jr., "Porteous Process Description" (1969).
86. Cole, E. L., U.S. Patent 3,507,788 (1968).
87. ZIMPRO Corp., Bulletin No. 2130, Rothschild, Wisconsin (1969).
88. Evans, R. R., *Paper Trade Journal* **154**, 31 (1970).
89. Harrison, J., M.S. and Ph.D. theses, Virginia Poly-Technic Institute (1964 and 1967).
90. Harrison, J., and Bungay, H. R., *Water and Sewage Works* **115**, 217, 268 (1968).
91. Clements, G. S., *Journal Institute of Sewage Purification* (British) Pt. IV, 318 (1950).
92. Dozanska, W., *Gas Woda Tech. Sanit.* (Poland) **42**, 160 (1968).
93. Moss, G., British Patent 1,097,900 (1968).
94. Benn, D., and Doe, P. W., *Filtration and Separation* (British) **6**, 4, 383 (1969).
95. Katz, W. L., and Mason, D. S., *16th Southern Research and Pollution Control Conference*, p. 119 (1967).
96. NCASI, Private communication (1970).
97. Bishop, F., Personal communication (1969).
98. Slagel, E. A., and Roberts, L. M., *Sewage Works Journal* **14**, 1021 (1942).
99. Cooling, L. F., *et al.*, *Sewage Works Journal* **25**, 242 (1953).
100. Cooling, L. F. *et al.*, *The Water and Sanitary Engineer* (British) (Nov., 1952).
101. Kretchetov, I. V., *Chemical Abstracts* **67**, 7615 (1967).
102. Husmann, W., *Gesundheit Ing.* **73**, 127 (1952).
103. Kellerwessell, H., and Cordes, H., *Fortschr. Verfahrentech* (Germany) **8**, 816 (1968).
104. Peck, D. F., and McBride, T. J., *Water and Sewage Works* **116**, 11, IW-5 (1969).
105. Greenfield, C., *Proceedings, Annual Meeting WPCF* (1966).
106. Dickey, C. D., *Filtration*, Reinhold Publishing Co., New York (1937).
107. Crook, M. D., and Jones, W. D., *British Chemical Engineering* **13**, 94 (1968).
108. Hermia, J., and Lannory, F., *Chim. Ind.* (*Gen. Chim.*) (French) **99**, 511 (1968); *Chemical Abstracts* **41**, 7, 1294 (1968).
109. Brown, J. M., *Water and Sewage Works* **107**, 324 (1960).
110. Baird, R. L., and Perry, M. G., *Filtration and Separations* (British) **4**, 471 (1967).
111. Perry, M. G., and Baird, R. L., *Filtration and Separations* (British) **5**, 546 (1968).
112. Dell, C. C., and Kaynar, M. B., *Filtration and Separations* (British) **5**, 323 (1968).
113. Eckenfelder, W. W., Jr., *Industrial Water Pollution Control*, McGraw-Hill, New York (1966).
114. Swanwick, J. D. *et al.*, *Journal Institute of Sewage Purification* (British) **6**, 487 (1961).
115. Swanwick, J. D., and Davidson, M. F., *Water and Wastes Treatment Journal* (British) **8**, 386 (1961).
116. Gale, R. S., *Filtration and Separations* (British) **5**, 133 (1968).
117. Shepman, B. A., and Cornell, C. F., *Sewage and Industrial Wastes* **28**, 1443 (1956).

118. Linsey, A. M., NCASI Technical Bulletin No. 209, Pt. IV (1967).
119. Bing, K. E., NCASI Technical Bulletin No. 238, Pt. IV (1970).
120. Caron, A. L., NCASI Technical Bulletin No. 223 (1968).
121. Simpson, G. D., and Sutton, S. H., University of Michigan Continued Education Series No. 113, p. 126 (1964).
122. Duffield, G. H., *Filtration and Separations* (British) **5**, 347 (1968).
123. Caron, A. L., NCASI Technical Bulletin No. 238, Pt. XI (1970).
124. Forrest, R. A., *Water and Wastes Trade Journal* (British) **7**, 156 (1968).
125. Triebel, W., *Gas-Wawwerfach* (German) **105**, 35 (1964).
126. Hooton, J. A., and Thomas, C. M., *Filtration and Separations* (British) **5**, 238 (1968).
127. Carpenter, W. L., NCASI Technical Bulletin No. 174 (1964).
128. Slaby, F., paper presented to TAPPI 3rd Annual Air and Water Conference (1966).
129. Stovall, J. H., and Berry, D. A., TAPPI 6th Annual Air and Water Conference, 23-1 (1969).
130. Billings, R. M., *Paper Processing* **6**, 14 (1970).
131. Corey, R. C., *Principles and Practices of Incineration*, Wiley Interscience, New York (1969).
132. Federal Water Pollution Control Administration, "A Study of Sludge Handling and Disposal," Water Pollution Control Series, FWPCA Pub. WP-20-4 (1968).
133. Owens, M. B., *Proceedings ASCE, Sanitary Engineering Division*, Paper No. 1172 (Feb., 1957).
134. Fair, G. M., and Moore, E. W., *Engineering News Record* **114**, 681 (1961).
135. Kempa, B., *Sbornic Nysoke Uceni Tech. Brni*. (Czech); *Chemical Abstracts* **71**, 6, 24577K (1969).
136. Gellman, I., NCASI Technical Bulletin No. 238 (1970).
137. Gordan, C. W., *Proceedings, 10th Sanitary Engineering Conference*, University of Illionois (1968).
138. Bartlett-Snow-Pacific, Inc., Bulletin No. Hl-2M-1-65 (1964).
139. Cardinal, P. J., *Water and Wastes Treatment Journal* (British) **12**, 62 (1968).
140. "Sludge Concentration, Filtration and Incineration," University of Michigan School of Public Health, Continued Education Series No. 113 (1964).
141. Coogan, F. J., and Stovall, J. H., *TAPPI* **48**, 94A (1965).
142. Evans, R. D., *Paper Trade Journal* **154**, 28 (1970).
143. Copeland, G. G., *Water and Wastes Engineering* **5**, 60 (1968).
144. Solir, W. H. *et al.*, *Water Works and Waste Engineering* **2**, 90 (1965).
145. Telelzke, G. H., 16th Annual Sanitary Engineering Conference, Bulletin No. 56, University of Kansas (1965).
146. Laboon, J. F., *Civil Engineering* **24**, 45 (1954).
147. Miles, H., *Paper Trade Journal* **26** (1970).
148. Gabaccia, A. J., *Compost Science* **18**, (Summer, 1960).
149. Albertson, O. E., *Proceedings, 9th Great Plains Sewage Works Design Conference* (1965).
150. Dorr-Oliver, Inc., Bulletin No. 6051, Stamford, Conneticut (1965).
151. Federal Water Pollution Control Administration, "Mathematical Model of Sewage Sludge Fluidized-Bed Incinerator Capacities and Costs," FWPCA, Report No. TWRC-10 (1969).
152. ZIMPRO, "Wet-Air Oxidation Unit," Sterling Drug, Inc. (1964).
153. Telezke, G. H., *Processings, 16th Sanitary Engineering Conference*, University of Kansas (1966).
154. Sanitary Engineering Committee Report, *Proceedings ASCE, Sanitary Engineering Division* **85**, 13 (1959).
155. Hurwitz, E., *Journal WPCF* **32**, 918 (1960).
156. Hurwitz, E., and Dundas, W. A., *Proceedings, 14th Industrial Waste Conference*, Purdue University, p. 211 (1959).
157. Takamatsu, T. *et al.*, 5th Internation Conference on Pollution Research, San Francisco (1970).
158. Blosser, R. O., NCASI Technical Bulletin No. 130 (1960).
159. *Proceedings, National Conference on Solid Wastes*, Chicago (1963).
160. Coogan, F. J., and Stovall, J. H., NCASI Technical Bulletin No. 185, Part III (1965).
161. Biship, F. W., and Drew, A. E., *TAPPI* **54**, 1830 (1971).
162. Hazeltine, T. R., *Sewage and Industrial Wastes* **30**, 65 (1958).
163. Ross, E. E., *Journal WPCF* **42**, 733 (1970).
164. Sager, J. C., *Journal WPCF* **37**, 1243 (1965).
165. Murman, R. G. *et al.*, *Sewage and Industrial Wastes* **33**, 1155 (1961).
166. Rudolfs, W., *ACS Monograph #118*, Reinhold Publishing Co., New York, Chapter 15 (1953).
167. Warshawski, J. *et al.*, "Fluidized-Bed Incineration of Oily Wastes," Tech. NPCEC-68-29, p. 1 (1968).
168. Federal Water Pollution Control Administration, "Petrochemical Effluents–Treatment Practices," Water Pollution Control Series FWPCA Pub. 12020 (1970).
169. Mercer, W. A. *et al.*, *Compost Science* (Autumn, 1962).
170. Buswell, A. M., *Sewage Works Journal* **19**, 28 (1947).
171. Buswell, A. M., and Boruff, C. S., *Sewage Works Journal* **4**, 454 (1932).
172. MacLaren, J. W., *Canadian Municipal Utilities*, p. 23, 51 (1961).
173. *WPCF Manual of Practice No. 11* (1968).
174. Fair, G. M., *Sewage Works Journal* **4**, 589 (1932).

175. Fair, G. M., *Sewage Works Journal* **4**, 755 (1932).
176. Heukelekian, H., *Sewage Works Journal* **2**, 219 (1930).
177. Heukelekian, H., *Sewage Works Journal* **3**, 3 (1931).
178. Garber, W. F., *Sewage and Industrial Wastes* **26**, 1202 (1954).
179. Torpe, W., *Sewage and Industrail Wastes* **26**, 479 (1954).
180. Rudolfs, W., *Sewage Works Journal* **11**, 928 (1937).
181. Steffen, A. J., and Lemen, R. M., *Wastes Engineering* **152** (March, 1960).
182. Lawton, G. W., and Norman, J. D., *Journal WPCF* **36**, 495 (1964).
183. Van Kleek, L. W., *Sewage Works Journal* **15**, 788 (1943).
184. Kraus, L. S., *Sewage Works Journal* **9**, 642 (1937).
185. De Turk, E. E., *Sewage Works Journal* **7**, 597 (1935).
186. Rudolfs, W., and Gehm, H. W., N. J. Agricultural Experiment Station Bulletin No. 699 (1942).
187. Rudolfs, W., and Ingols, R. S., *Sewage Works Journal* **10**, 653 (1938).
188. Hoover, S. R. *et al.*, *Sewage and Industrial Wastes* **24**, 38 (1966).
189. *Compost Science* **13**, 23 (1972).
190. TAPPI Monograph No. 32 (1968).
191. *Bulk Transport of Waste Slurries to Inland and Ocean Disposal Sites*, Three Vol. P. B.-198, 756-7-8-9 (1970).
192. West, L. E., *Sewage Works Journal* **24**, 785 (1952).
193. O'Leary, C., *Civil Engineering* **29**, 52 (1959).
194. Clark, B. D. *et al.*, *The Barged Ocean Disposal of Wastes; A Review of Common Practice and Methods of Evaluation*, EPA Pacific Northwest Laboratory, Corvallis, Oregon (1971).
195. Rudolfs, W., *Sewage Works Journal* **16**, 617 (1944).
196. Miller, D. R., *Sewage and Industrial Wastes* **30**, 1426 (1958).
197. Hume, N. B., and Garber, W. F., *Proceedings, 3rd International Conference on Water Pollution Research* **3**, 243 (1967).
198. Rawn, A. M., and Bauerman, F. R., *Sewage and Industrial Wastes* **26**, 1309 (1954).
199. Koenig, L., USPHS, AWTR-8, Pub. No. 999-WP-10 (1964).
200. Davies, A. G., *Compost Science* **9**, (Autumn, 1960).

18

Thermal Discharges

S. P. Mathur, P.E.

WAPORA, Inc.
Washington, D.C.

DEFINITION OF THERMAL POLLUTION

Thermal pollution can be defined as excessive changes in natural or ambient water temperature caused by addition or removal of heat through man's activities. Heat can be classified as a pollutant if its addition or removal from water alters biochemical reactions, damages aquatic life, changes ecology, or otherwise degrades water quality. Heat can be expressed in terms of temperature, in British thermal units (Btus), or in calories (cals). One Btu is defined[1] as the quantity of heat required to raise the temperature of 1 lb of water 1°F at, or near, its point of maximum density (density at 4°C or 39.2°F).

The major waste heat-producing industries are: steam-electric generating plants (fossil fuel or nuclear), petroleum refineries, steel mills, chemical plants, pulp and paper mills, and distilleries. Industries such as ice plants, milk processing plants, breweries, etc. are minor sources of waste heat which may be of importance locally because of the size of the receiving streams. Stratified reservoirs may also change the natural temperatures of water. However, this chapter will be confined to thermal pollution in receiving waters due to addition of heat of artificial origin, the major source of which is the discharge of condenser cooling waters from steam-electric generating plants.

The primary vehicle for conveying waste heat from the source to the disposal area is water. Therefore, the indices for thermal pollution potential are the volume of water used for cooling and the future demand for steam-electric generation. Table 18-1[2] gives past and future water requirements for various uses. It is important to differentiate between withdrawal and consumption because most of the water withdrawn is returned and only a small percentage is consumed, or lost, to the source. It can be clearly seen from Table 18-1 that steam-electric plants will be the major user of water; however, most

TABLE 18-1. Present and Future Water Use (billion gal/day)

	Used	FUTURE USE			Used	FUTURE USE		
	1965	1980	2000	2020	1965	1980	2000	2020
Type of Use	Withdrawls				Consumption			
Domestic and municipal	26	36	54	78	7	12	19	27
Industrial	47	75	128	310	3	6	11	16
Steam-electric	84	193	470	814	1	2	5	13
Agriculture	113	138	153	166	66	84	93	101
Total	270	442	805	1,368	77	104	128	157

of the water withdrawn will be returned. The consumptive use will be the smallest. Figure 18-1 is the graphic presentation of past and future water use. Future trends in power generation are shown in Figs. 18-2[3] and 18-3.[4]

Available sites for hydroelectric generation in this nation are almost exhausted. Pumped storage sites may be available, but for economic reasons they may be used only for peaking purposes. Thermal plants will generate most of the

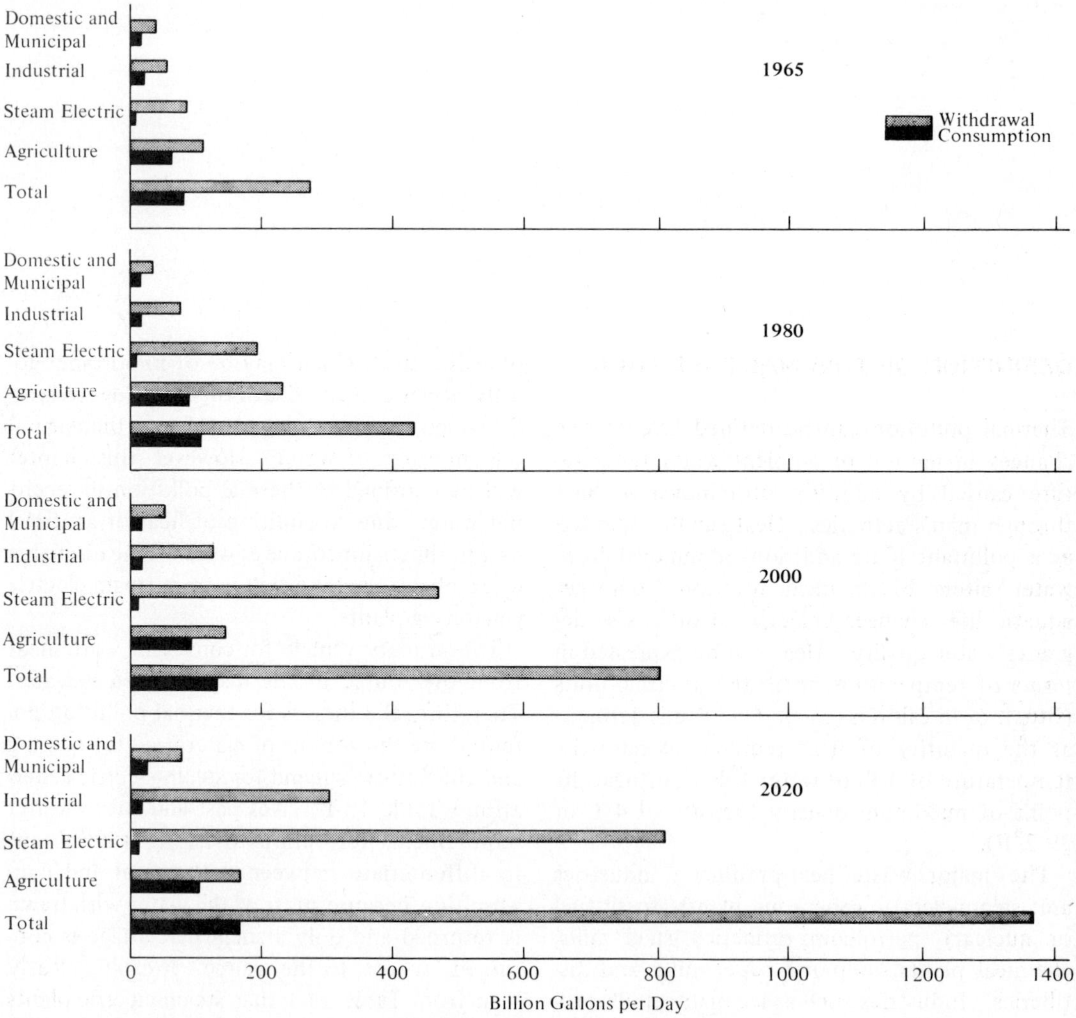

Fig. 18.1 Present and future water use.

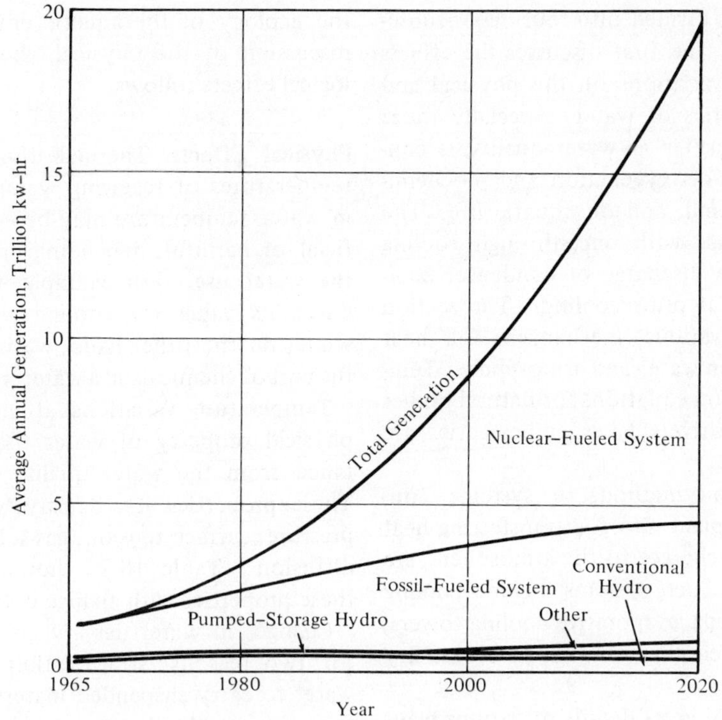

Fig. 18.2 Projected electric generation by types of prime mover (Courtesy of Vanderbilt University.)

electricity in the U.S. in the foreseeable future; it is anticipated that by the year 2000 they will account for 92 percent of the electricity produced; the projected thermal capacity for that year is about 1.8 billion kilowatts. Figure 18-4[4] shows a graphic representation of past and future heat rejection from thermal plants. It can be seen from the figure that in 1965 the heat rejection was approximately 1 trillion Btu/hr, and by the year 2000, the heat rejection is expected to be approximately 8.4 trillion Btu/hr.

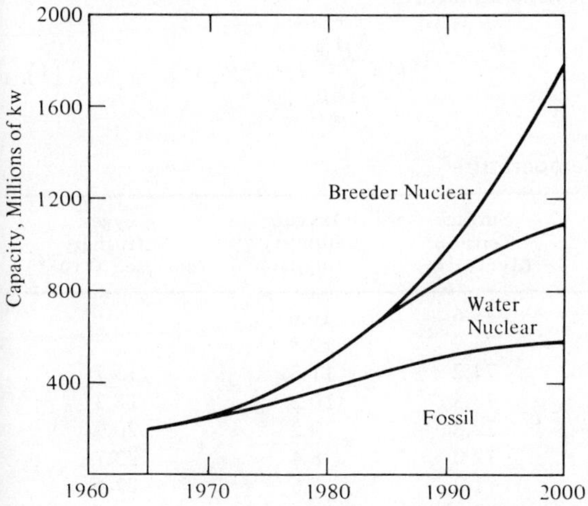

Fig. 18.3 Projected thermal generating capacity.[4]

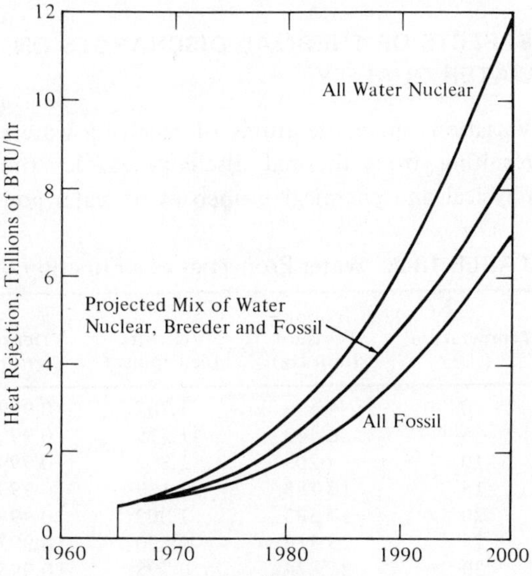

Fig. 18.4 Projected heat rejection from thermal generating plants (U.S. Senate hearings).

This chapter is divided into four major subsequent sections. The first discusses the effects of temperature variations on the physical and chemical properties of water, especially those of importance as far as water quality is concerned, such as deoxygenation and biochemical oxygen demand, and on aquatic life. The next section deals with once-through cooling systems and the discharge of condenser cooling water without prior cooling. The section also describes various mechanisms of heat exchange between water and atmosphere. Temperature prediction equations for natural bodies of water, such as rivers, lakes, and estuaries, are also presented.

Heat dissipation methods or systems, constructed at the plant site for transferring heat from thermal discharges to the atmosphere, are discussed next. Such systems include cooling ponds, spray ponds, evaporative cooling towers, nonevaporative cooling towers, and wet-dry cooling towers.

The last section gives details of various ways in which thermal discharges are and can be utilized beneficially. These applications include: greenhouses, agriculture, aquaculture, space heating, and others. The primary function of these is to improve utilization of energy; heat dissipation or temperature reduction are secondary to this major consideration.

EFFECTS OF THERMAL DISCHARGES ON WATER QUALITY

Variations in temperatures of receiving waters resulting from thermal discharges affect the physical and chemical properties of water and the ecology of the aquatic environs. A brief discussion of the physical, chemical, and biological effects follows.

Physical Effects. Thermal discharges increase temperatures of receiving waters. An increase in water temperature may be considered beneficial or harmful, depending on the nature of the water use. For example, warm water reduces its value for cooling and palatability, while, on the other hand, warm water reduces the cost of chemicals in a water treatment plant.

Temperature variations affect almost every physical property of water that is of importance from the water quality point of view. These properties are: density, viscosity, vapor pressure, surface tension, gas solubility, and gas diffusion. Table 18-2[3] shows the changes in these properties with change in temperature.

Changes in water density are of significance for two reasons: stratification and ability of water to carry suspended materials. Slight differences in density are sufficient to cause thermal stratification, although water movement and depth are also contributing factors. Stratification inhibits vertical mixing and oxygen transfer from epilimnion to hypolimnion. Figure 18-5[5] shows the variation of water density with temperature.

As the temperature increases, the ability of water to carry suspended materials decreases. Stokes' law describes the velocity of settling particles in a nonturbulent medium by the following equation:[3]

$$V = \frac{d^2 g}{18\mu} (P_s - P_w) \qquad (1)$$

TABLE 18-2. Water Properties as a Function of Temperature[3]

Temperature (°C)	Vapor Pressure (mm Hg)	Viscosity (centipoise)	Density (gm/ml)	Surface Tension (dynes/cm)	Oxygen Solubility (mg/l)	Oxygen Diffusivity (cm²/sec × 10⁻⁶)
0	4.579	1.787	0.99984	75.6	14.6	
5	6.543	1.519	0.99997	74.9	12.8	
10	9.209	1.307	0.99970	74.2	11.3	15.7
15	12.788	1.139	0.99910	73.5	10.2	18.3
20	17.535	1.002	0.99820	72.8	9.2	20.9
25	23.756	0.890	0.99704	72.0	8.4	23.7
30	31.824	0.798	0.99565	71.2	7.6	27.4
35	42.175	0.719	0.99406		7.1	
40	55.324	0.653	0.99224	69.6	6.6	

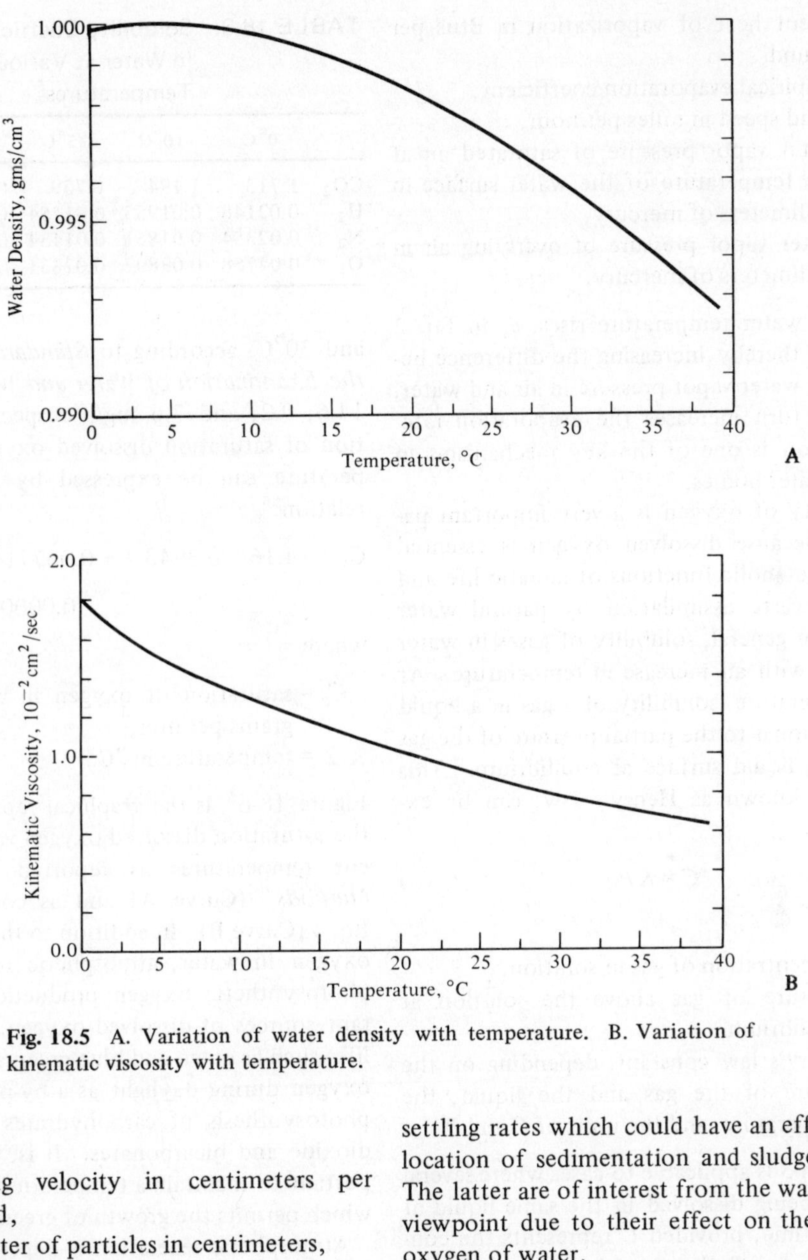

Fig. 18.5 A. Variation of water density with temperature. B. Variation of kinematic viscosity with temperature.

where

V = settling velocity in centimeters per second,

d = diameter of particles in centimeters,

g = gravitational acceleration = 980 cm^2 sec,

P_s = density of settling particles in grams per cubic centimeter,

P_w = density of water in grams per cubic centimeter,

μ = viscosity of liquid, poises.

According to Eq. 1, the settling velocity is directly related to water density and inversely proportional to viscosity as shown in Fig. 18-5A. Both of these properties of water decrease as the temperature increases, resulting in increased settling rates which could have an effect on the location of sedimentation and sludge deposits. The latter are of interest from the water quality viewpoint due to their effect on the dissolved oxygen of water.

The rate of evaporation increases as the water temperature increases because the difference between water vapor in air and water increases. Evaporation flux can be expressed by the following equation:[3]

$$F = \frac{CW}{L}(e_s - e_a) \qquad (2)$$

where

F = evaporation flux in pounds per square foot per day,

L = latent heat of vaporization in Btus per pound,
C = empirical evaporation coefficient,
W = wind speed in miles per hour,
e_s = water vapor pressure of saturated air at the temperature of the water surface in millimeters of mercury,
e_a = water vapor pressure of overlying air in millimeters of mercury.

As the water temperature rises, e_s in Eq. 2 increases, thereby increasing the difference between the water vapor pressure in air and water, which in turn increases the evaporation rate. Evaporation is one of the key mechanisms in cooling water bodies.

Solubility of oxygen is a very important parameter because dissolved oxygen is essential for the metabolic functions of aquatic life and also for waste assimilation by natural water bodies. In general, solubility of gases in water decreases with an increase in temperature. At any temperature, solubility of a gas in a liquid is proportional to the partial pressure of the gas above the liquid surface at equilibrium. This principle, known as Henry's law, can be expressed as:[5]

$$C = KP \tag{3}$$

where

C = concentration of gas in solution,
P = pressure of gas above the solution at equilibrium,
K = Henry's law constant; depending on the nature of the gas and the liquid, the temperature, and the units of C and P.

Equation 3 is applicable to cases where several gases are being dissolved in the same liquid at the same time, provided C represents the concentration and P the partial pressure of each gas; the constant K is different for each gas. Table 18-3[5] shows the solubility coefficients of various gases in water at different temperatures.

The most important of the above gases with respect to stream pollution is oxygen. As the temperature increases, the saturation concentration of dissolved oxygen in water decreases. For example, at standard pressure, and assuming no chlorides present in the water, the saturation dissolved oxygen values for 0°C, 20°C, and 30°C, according to *Standard Methods for the Examination of Water and Wastewater*, are 14.6, 9.2, and 7.6 mg/l, respectively. Variation of saturation dissolved oxygen with temperature can be expressed by the following relation:[5]

TABLE 18-3. Solubility Coefficients of Gases in Water at Various Temperatures[5]

	0°C	10°C	25°C	50°C	100°C
CO_2	1.713	1.194	0.759	0.436	–
H_2	0.02148	0.01955	0.01754	0.01608	0.0160
N_2	0.02354	0.01861	0.01434	0.01088	0.0095
O_2	0.04758	0.03802	0.02831	0.02090	0.0170

$$C_s = 14.16 - 0.3943\,T + 0.007714\,T^2 - 0.0000646\,T^3 \tag{4}$$

where

C_s = saturation of oxygen in water in milligrams per liter,
T = temperature in °C.

Figure 18-6[5] is the graphical representation of the saturation dissolved oxygen values at different temperatures as reported by *Standard Methods*[7] (Curve A) and as computed from Eq. 4 (Curve B). In addition to the solubility of oxygen in water, atmospheric reaeration and photosynthetic oxygen production are important sources of dissolved oxygen. Green plant life, such as green and blue-green algae, produce oxygen during daylight as a by-product of the photosynthesis of carbohydrates from carbon dioxide and bicarbonates. It is, therefore, important to maintain a temperature environment which permits the growth of green plants.

Atmospheric reaeration is dependent on oxygen diffusivity, surface tension, turbulence, and temperature. In general, the reaeration coefficient increases as the temperature increases, primarily because the coefficient of oxygen diffusion or oxygen diffusivity increases with temperature rise, as shown in Table 18-2. The relationship between the reaeration coefficient and temperature can be expressed by the following equation:[5]

$$K_2^1 = K_2\,\theta^{(T_1 - T_2)} \tag{5}$$

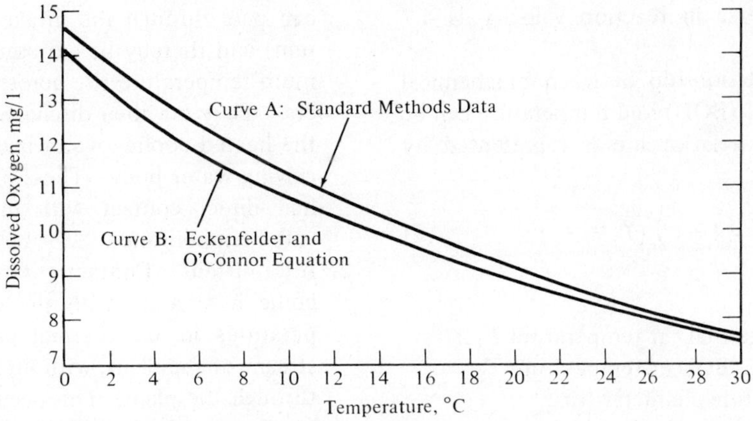

Fig. 18.6 Variation of saturation dissolved oxygen with temperature.

where

K_2^1 = reoxygenation constant at temperature T_1,
K_2 = reoxygenation constant at temperature T_2,
θ = temperature coefficient.

The generally used value for θ, the temperature coefficient, is 1.0241. This value indicates that the rate of aeration should increase at a geometric rate of 2.41 percent/°C. According to Table 18-2, the oxygen diffusivity increases by about 4 percent/°C. Since the reaeration coefficient is dependent on oxygen diffusivity, the corresponding increase in reaeration coefficient is more than 2 percent/°C.

Chemical Effects. The chemical effect of importance from the water quality viewpoint is the variation in the rate of chemical reactions with temperature. Rates of chemical reactions depend on the reacting substances, temperatures, and the concentrations of reactants. An increase in temperature increases the reaction velocity and the rate constant; in general, the rate of reaction increases twofold for every 10°C rise in temperature.

The variation of the rate constant with temperature is expressed by the van't Hoff–Arrhenius law and is expressed by the following equation:[5]

$$\frac{d}{dT}(\ln k) = \frac{E}{RT^2} \tag{6}$$

where

k = reaction rate constant,
T = absolute temperature,
R = gas constant,
E = energy of activation.

Integration of Eq. 6 between temperatures T_1 and T_2 yields:

$$\ln \frac{k}{k'} = \frac{E(T_2 - T_1)}{RT_1 T_2} \tag{7}$$

Activation energy, E, is a measure of the required energy for a molecule to take part in a chemical reaction. For many reactions, E can be determined by plotting $\ln k$ against $1/T$; the slope of the line is $-E/R$.

Modification of Eq. 7 results in the following relationship:

$$\frac{k}{k^1} = \theta^{(T_2 - T_1)} \tag{8}$$

where

k = rate constant at temperature T_2,
k^1 = rate constant at temperature T_1,
θ = temperature constant.

Equation 8 is used for determining the rate of oxygen depletion at various temperatures; k represents the rate of deoxygenation. The generally used value for θ, as far as the deoxygenation relationship is concerned, is 1.047 for a temperature range of 10 to 37°C. This value of θ means that for every 1°C increase in temperature, the increase in reaction velocity is 4.7 percent.

A similar relationship between biochemical oxygen demand (BOD) and temperature can be derived. This relationship is represented by Eq. 9:[5]

$$\frac{L_1}{L_2} = 1 + C_L (T_1 - T_2) \qquad (9)$$

where

L_1 = first-stage BOD at temperature T_1,
L_2 = first-stage BOD at temperature T_2,
C_L = temperature characteristic.

A commonly used value of C_L is 0.02. Equation 9 indicates an increase in BOD by 2 percent for every degree centigrade of temperature difference.

Biological Effects. Numerous studies have been carried out over the past several years to identify the theoretical and real effects of thermal effluents on biological systems. Much of this data has been gathered together in a report submitted to the Congress by the U.S. Environmental Protection Agency (EPA) in November, 1973. The report is entitled "Effects and Methods of Control of Thermal Discharges, Part I, Effects of Thermal Discharges on Water Quality."

All organisms consist of a matrix of many proteins which are heat labile, meaning that temperatures are critical to their proper functioning. An excess of heat will destroy protein and the enzymatic systems involved with the propagation of the organism. Therefore, it can be seen that heat will become a pollutant in terms of its effect on organisms when the temperature is sufficient to interfere with enzymatic systems. This can be manifested in a number of ways, including a change in growth rate, respiration rate, reproductive rate, or death.

The nature of electric generating stations is that cool water enters the plant, becomes heated in the condensers, and then is discharged into the receiving water body. This means that organisms can be affected in one of three basic ways:

1. Entrainment: Organisms can be drawn into the plant (the smaller organisms that can pass through the intake filtering system) and thereby be exposed to the maximum temperature rise across the condensers. They are then discharged along with the heated cooling water back into the receiving water body. These organisms have had direct contact with maximally elevated temperatures.
2. Intermittent Contact: Organisms can come in contact with the elevated temperatures in the thermal plumes, downstream of the plant, without having passed through the plant. This occurs commonly in all water bodies which exhibit a directional flow. It is well known that fish and other mobile organisms can move into and out of the heated plume at will.
3. "Far-field" Effect: Respiration rates, spawning periods, etc. of the organisms can be affected even when they do not enter the thermal plume at the warmest portion near the plant. Such effects can be far-reaching in that the whole water body may be elevated in temperature to a slight degree. These effects are most difficult to measure with field studies because of their low level and the high sensitivity programs required. It is also difficult to separate out these effects from background cyclical biological phenomena.

Duration of exposure to thermally elevated water is also important. It has been found that if the exposure time can be kept to a minimum, particularly for entrained organisms such as plankton, fairly high temperature increases—as great as 20°F (7°C)—can be withstood without any damage. However, the exposure time must be kept rather short—e.g., 1 to 2 min. In general, the larger the organism, the longer it can be exposed without harm because of the laws of thermodynamics which prevent it from heating as quickly.

Whether or not an individual thermal effluent will have an effect on the biota of a receiving stream depends largely upon the location, morphology, and water quality of the particular stream in question, as well as the biota existing within the water body. If few or no organisms are present, there can be little or no effect no matter what heat load is discharged into the stream. Heat discharged into a highly productive stream may have a deleterious effect depending on the assimilative characteristics of the stream itself and the tolerances exhibited by the species found there. It should be emphasized that no definitive statement can be made about thermal discharges for every location in the country since there are too many variables that must be taken into account.

In light of the above discussion, only a summary of the biological effects[8] is provided here:

1. The temperature of stream water, even during the summer, usually is below the optimum–95-100°F (35-37.5°C)–for pollution-associated bacteria. Increasing the water temperature by the addition of excess heat may increase the bacterial multiplication rate when the environment is favorable and the food supply is abundant. Increasing the water temperature beyond the growth range of the bacteria causes a more rapid die-off when the food supply is limited, as it is in nonpolluted waters.
2. Warm water fish can survive temporarily in waters heated artificially to 93°F (34°C) or greater. Some fish populations, such as roach, perch, and carp, are reduced if these high temperatures are maintained for an extended period. In cold weather, stream temperatures should be substantially below 93°F (34°C) to prevent mortalities when fish move from colder water into heated areas. Cold water nonanadromous fish populations such as trout should not be subjected to temperatures exceeding 58°F (14.5°C). In cold weather, stream temperatures should be below 58°F (14.5°C) to prevent mortality among cold water fishes.
3. Changes in temperature, particularly a sudden drop, can be more harmful to some species of fish than continued exposure to a higher temperature.
4. Fish have the ability to adapt to higher temperatures faster than to lower temperatures.
5. The maximum temperature for a given species of fish varies with the fish's rate of heating, size, and physiological condition.
6. Fish may starve at elevated temperatures because they may become moribund and unable to capture food. Respiratory rates and food requirements increase with increased temperatures.
7. Fish seek out a preferred temperature at which they can best survive, which is usually several degrees below their lethal temperature.
8. The toxic effects of certain materials on fish increase with temperature (synergism).
9. Temperature changes are most important to fish in enclosed areas of the marine environment such as estuaries and bays as opposed to open areas. Tolerance to temperature fluctuations, however, is greater in freshwater and estuarine forms than in open water marine species.
10. There are restricted ranges of temperature within which fish can reproduce successfully; larval development especially requires narrow ranges of temperature. A fish population may exist in an excessively heated area only by continued immigration from the outside. Fish may be absent from such areas during warm summer months and present in cold winter months.
11. Increased temperatures may block the migrations of anadromous fish, if the heated zone forms a complete barrier in the stream.
12. Cold is as significant to fish populations as heat because of the inability of fish to acclimate quickly to rapid decreases in temperature. Thus, in some areas fish populations may be limited by decreases as well as increases in temperature. The growth rate of fishes is reduced in waters colder than the optimum temperature range for the species.
13. When water temperatures increase and remain constant over an extended period, the predominant algal species change from diatoms to green algae and finally at high temperatures to blue-green algae.
14. The number and distribution of bottom organisms decrease as water temperatures

increase above 90°F (32°C), which is close to the tolerance limit for a "balanced" population. The adult stage of many species is better able to tolerate higher temperatures than the eggs or young.

15. A benefit of discharging heated effluents is the defouling of intake pipes which is accomplished when the flow of water through the pipes is reversed.
16. Certain benefits–including open water winter fishing in otherwise ice-covered areas, and cold water fisheries downstream from deep reservoirs–can be derived from artificially-induced temperature changes. The benefits to fisherman of fish being attracted to heated water in the winter months may be negligible compared to fish mortalities that may result when the fish return to the cooler water; lethal temperatures may result from heated discharges in the summer months.

Much of the information provided in this section, particularly with respect to physical, chemical, and biological effects, is of a theoretical nature or is the result of laboratory experiments. It is, therefore, important to actually measure and evaluate the effects of thermal discharges on the parameters describing water quality and on aquatic life propagating in waters receiving thermal discharges. As a starting point, readers and researchers in this area are referred to Ref. 12.

MECHANISMS OF HEAT DISSIPATION

Heat transfer from condenser cooling water to the atmosphere is commonly accomplished by two mechanisms: once-through cooling and auxiliary cooling. Once-through cooling, the simplest method for heat disposal, is defined as the direct discharge of cooling water into natural bodies of water, allowing the heat dissipation to occur naturally. Auxiliary cooling involves the provision of cooling devices, such as cooling ponds or cooling towers, for heat dissipation. Auxiliary cooling systems are described in detail in the next section. This section will deal with once-through cooling systems and temperature prediction models.

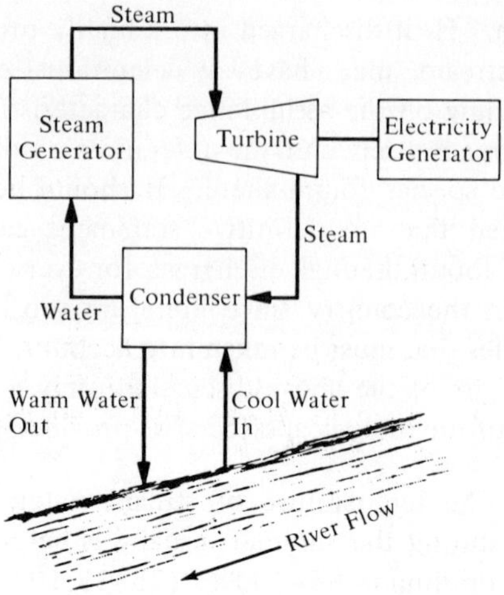

Fig. 18.7 Once-through cooling system.

Once-through cooling systems, Fig. 18-7, are of two basic types: those with surface discharges and those with subsurface discharges. The surface discharge system consists of a pipe through which the thermal discharge flows to the surface of the receiving water, or a weir over which it flows into the water. A subsurface discharge system consists of a pipe and a series of nozzles located below the water surface. This enables the thermal discharge to exit at high velocities, thereby entrapping cooler water in the plume for heat dissipation. In either case, physical characteristics of the receiving waters such as width, length, depth, slope, nature and volume of flow, available dilution, ambient water temperature, and climatic conditions (such as air temperature, humidity, rate of evaporation, and vapor pressure) are important in determining the rate of heat dissipation and in predicting temperatures at various points downstream from the thermal discharge.

As the condenser cooling water flows away from the point of discharge, it dissipates heat to the atmosphere. The rate of heat dissipation depends on the climatic conditions and the characteristics of the receiving waters enumerated above. Several investigators have described the mechanisms involved in heat transfer from water to air. A description of these mechanisms follows below.

Velz and Gannon.[5] The authors state that four major mechanisms are involved in heat transfer and temperature change of a body of water exposed to the atmosphere. These mechanisms are: heat losses due to evaporation, convection, radiation, and heat gain through incoming solar radiation. Losses due to infiltration to and into the enclosing earth may be minor mechanisms of heat transfer but can be neglected in the case of flowing stream or well-sealed bottoms. The heat-balance formulation proposed by them is represented by Eq. 10:

$$H = H_e + H_c + H_r - H_s \tag{10}$$

where

H = net heat loss,
H_e = heat loss by evaporation,
H_c = heat loss by convection,
H_r = heat loss by radiation,
H_s = heat gain through solar radiation.

Definition of terms in Eq. 10:

Evaporation Loss: The authors used Meyer's formula for estimating heat loss due to evaporation. Meyer's formula (Eq. 11) is expressed:

$$I = C(1 + 0.1W)(V_w - V_a) \tag{11}$$

where

I = evaporation in inches per month,
W = mean wind velocity in miles per hour, measured at about 25 ft above the water surface or above the surrounding land area,
V_w = vapor pressure in inches of mercury corresponding to the water surface temperature taken at about 1 ft below the surface,
V_a = the mean absolute vapor pressure prevailing in the overlying atmosphere 25 ft above the water surface or the surrounding land area,
C = a constant, ranging from 10 to 15, depending on the depth and the exposure of the body of water and frequency of meteorologic measurements. For large, deep lakes and reservoirs, C is about 10; for shallow ponds and surface accumulations, C is about 15. Generally, the value of C used for streams of moderate depth and velocity is 14.

Equation 11 can be converted to give loss of heat due to evaporation, as expressed by Eq. 12:

$$H_e = 0.00722 H_v C \cdot (1 + 0.1W)(V_w - V_a) \tag{12}$$

where

H_e = heat loss due to evaporation in Btu/hr/ft^2 of water surface,
H_v = latent heat of vaporization.

Convection Loss: Heat loss by convection depends on the temperature differential between water and air temperatures, and on the wind velocity at the water surface. The general equation for heat loss by convection is:

$$H_c = (a + bW)(T_w - T_a) \tag{13}$$

where

H_c = heat loss due to convection,
T_w = water temperature at the surface,
T_a = mean air temperature,
W = mean wind velocity,
a = coefficient depending on the temperature differential,
b = coefficient depending on wind velocity.

In still air, convection losses from a flat surface range from 0.5 Btu/hr/ft^2/°F for a temperature differential of a few degrees to 1.0 for a temperature differential of 50 to 100°F. The authors decided to use 0.8 for a, based on Lanhaars' suggestion for cooling ponds. The value of the air movement coefficient, b, varies from 0.16 to 0.32 for each mile per hour of wind velocity. The authors selected a value of 0.32 for flowing streams. Further, they recommended a conversion factor of $\frac{1}{2}$ for converting wind velocity measured by a nearby U.S. Weather Bureau Station to wind velocity at the water's surface.

Equation 13 can be rewritten as:

$$H_c = \left(0.8 + 0.32\frac{W}{2}\right)(T_w - T_a) \tag{14}$$

where

H_c = heat loss due to convection in Btus per hour per square foot,
T_w = water temperature at the surface in °F,

T_a = air temperature in °F as reported by the U.S. Weather Bureau,

W = mean wind velocity in miles per hour as reported by the U.S. Weather Bureau.

Radiation Loss: Heat loss by radiation from the water surface depends on the difference between water and air temperatures, and on radiations from the surrounding objects and terrain. The authors, in developing an equation for radiation loss, assumed that the temperature of the surrounding objects and terrain was the same as the air temperature. They further assumed emissivity from the water surface as being 1.0. The equation for radiation loss is as follows:

$$H_r = 1.0(T_w - T_a) \tag{15}$$

where

H_r = heat loss due to radiation in Btus per hour per square foot. The remaining terms have been defined previously.

Solar Radiation Gain: Heat gained by water through solar radiation is dependent on the amount of solar radiation incident on the water surface, and on the coefficient of heat absorption by water. The amount of solar radiation reaching the water surface is dependent on the cloud cover and local effects, such as shade trees along the banks of a river, air pollution, etc. Since the amount of heat gained cannot be computed by meteorological data, the authors recommend that the solar radiation on a horizontal surface, measured by the U.S. Weather Bureau, be used directly, if the exposures are reasonably similar.

Substitution of values for H_e, H_c, and H_r in Eq. 10 gives the heat balance equation, as follows:

$$\begin{aligned} H = {} & 0.00722 H_v C(1 + 0.1W)(V_w - V_a) \\ & + \left(0.8 + 0.32 \frac{W}{2}\right)(T_w - T_a) \\ & + 1.0(T_w - T_a) - H_s \end{aligned} \tag{16}$$

Development of the Temperature Prediction Equation. In the development of the temperature prediction equation, Velz and Gannon used the concept of "equilibrium temperature." They define the equilibrium temperature as a temperature which will balance the heat gained by solar radiation with the heat loss by evaporation, convection, and radiation—i.e., the net heat loss is zero. Equation 10 then becomes:

$$H = 0$$

or

$$H_e + H_c + H_r = H_s \tag{17}$$

At equilibrium, the water temperature $T_w = E$, and the vapor pressure $V_w = V_E$. Substitution and combining Eqs. 10, 12, 14, and 15 gives the following equations:

$$\begin{aligned} H_s = {} & 0.00722 H_v C(1 + 0.1W)(V_E - V_a) \\ & + (0.8 + 0.16W)(E - T_a) \\ & + 1.0(E - T_a) \end{aligned} \tag{18}$$

$$\begin{aligned} = {} & 0.00722 CH_v(1 + 0.1W)(V_E - V_a) \\ & + (1.8 + 0.16W)(E - T_a) \end{aligned} \tag{19}$$

Equation 19 cannot be solved directly for E, but, by successive approximations, a proper value of E can be determined.

Theoretically, an infinite time of exposure and an infinite surface area will be required to cool the warm water introduced into the stream to the equilibrium temperature. A close approximation can be made, however, to the equilibrium temperature. The temperature prediction equation is based on the following relation:

$$\frac{dT_w}{dt} = \frac{H}{62.4b} \tag{20}$$

where

dT_w = an increment of water temperature decline in time dt,

H = heat loss from the corresponding increment of surface area divided by the weight of the underlying water of depth b.

Let A = area in square feet per cubic feet per second. Then integration of Eq. 20 gives:

$$A = -224{,}640 \int_{T_1}^{T_2} \frac{dT_w}{H} \tag{21}$$

where

A is defined as: $t = bA/3600$ in hours.

Substitution of the values of H_e, H_r, H_c, and H_s in Eq. 10 gives the following expression for the total heat loss, H:

$$H = 0.00722\ CH_v(1 + 0.1W)(V_E - V_a) + (1.8 + 0.16W)(T_w - E) \quad (22)$$

Equation 21 can be expressed as follows:

$$A = -224{,}640 \int_{T_1}^{T_2} \frac{dT_w}{p(V_w - V_E) + q(T_w - E)} \quad (23)$$

where

$p = 0.00722\ CH_v(1 + 0.1W)$
$q = (1.8 + 0.16W)$

Equation 23 cannot be integrated directly, but can be solved by dividing the temperature interval $(T_1 - T_2)$ into a number of equal intervals of ΔT_w and performing the integration by successive summations according to the following equation:

$$A = -224{,}640 \sum_{T_1}^{T_2} \frac{\Delta T_w}{p(V_w - V_E) + q(T_w - E)} \quad (24)$$

Equation 24 can be used to predict the temperature profile along the course of the stream for a particular runoff from channel cross-sectioning. The authors found a very good correlation between the actual water temperature and the predicted temperature for the Tittabawassee Basin in Michigan.

Jerome M. Raphael.[5] Raphael states that there are seven different heat transfer phenomena that have to be accounted for in water temperature computations. These are as follows:

1. Direct and diffused short-wave radiation from the sun.
2. Reflected short-wave and long-wave radiation.
3. Long-wave radiation from the water.
4. Long-wave radiation from the atmosphere.
5. Evaporation at the water surface.
6. Heat conducted from air to water.
7. Heat advected into water by rain and other inflows.

The energy budget equation used by Raphael is as follows:

$$Q_t = Q_s - Q_r - Q_b - Q_h - Q_e + Q_v \quad (25)$$

where

Q_t = increase in energy stored in the body of water,
Q_s = the solar radiation incident to the water surface,
Q_r = the reflected solar radiation,
Q_b = the back radiation, or the net energy lost by the body of water through the exchange of long-wave radiation between the body of water and the atmosphere,
Q_h = the energy conducted from the body of water to the atmosphere as sensible heat,
Q_e = the energy used for evaporation,
Q_v = the net energy advected into the body of water by rain and inflows from springs and brooks.

Definition of terms in Eq. 25:

Solar Radiation (Q_s): Radiant energy received on a surface normal to the sun's rays and outside the earth's atmosphere is approximately 430 Btu/hr/ft^2. This is called the "solar constant." As the solar radiation passes through the earth's atmosphere, a portion of the radiant energy is absorbed by the gases, moisture, and solid particles in the air. Part of the direct radiant energy lost in the earth's atmosphere is recovered in diffuse or sky radiation.

The angles of incidence of the sun's rays on the surface of water change every hour of the day, every day of the year, and with the latitude of the water surface. Thus, the angle of incidence modifies the value of direct and diffused radiation by controlling the length of the atmospheric path through which it travels. Also, it controls the relative surface area on which it impinges. The angle of incidence is given by the following equation:

$$\sin a = \sin \phi \sin d + \cos \phi \cos d \cos h \quad (26)$$

where

a = solar altitude in degrees,
ϕ = latitude of the water surface in degrees,
d = hour angle of the sun in degrees,
h = hour angle of the sun (positive before noon and negative in the afternoon) in degrees.

Solar altitudes can be calculated for various times of the year and for various locations with accuracy.

The U.S. Weather Bureau has collected data for the intensity of direct and diffused radiation from the sun incident on a surface perpendicular to the sun's rays. These data are shown in Table 18-4 and Fig. 18-8, based on the computations of Parry Moon.

Reflection of Short-wave Radiation: A portion of the short-wave radiation from the sun that reaches the surface of the body of water is returned unchanged by reflection at the surface and by scattering from bubbles and suspended particles under the water surface. Factors affecting the reflectivity of a water surface are: solar altitude, type and amount of cloud cover, the turbidity in the atmosphere, and the wind velocity. Raphael states that with increasing wind, an increase in total reflected energy can be assumed. He further states that the Lake

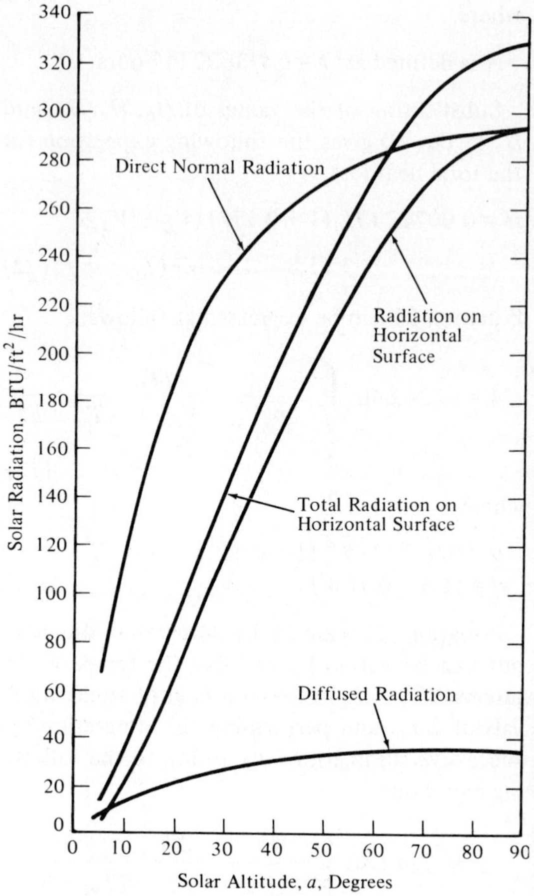

Fig. 18.8 Solar radiation vs. solar altitude.

TABLE 18.4. Solar Radiation on a Horizontal Surface

Solar Altitude *a* (degrees)	Direct Normal Radiation[a, b]	Sin *a*	Radiation on Horizontal Surface[a]	Diffuse Radiation[a, b]	Total Radiation[a]
5	67	0.087	6	7	13
10	123	0.174	21	14	35
15	166	0.259	43	19	62
20	197	0.342	67	23	90
25	218	0.423	92	26	118
30	235	0.500	118	28	146
35	248	0.574	142	30	172
40	258	0.643	166	31	197
45	266	0.707	188	32	220
50	273	0.766	209	33	240
60	283	0.866	245	34	279
70	289	0.940	272	35	307
80	292	0.985	288	35	323
90	294	1.000	294	35	329

[a] Btus per hour per square foot.
[b] Values after Parry Moon.

Hefner studies reported by E. R. Anderson indicate the absence of any wind effect or critical wind speed over the range of wind speeds shown, and the independence of wind speed and reflectivity. Raphael concluded that reflectivity under clear skies was independent of air-mass turbidity, except in extreme cases such as dust-laden or smoke-laden air.

Raphael states that the reflectivity from a water surface is a function of cloud cover and the difference between the reflectivities under high and low cloud cover conditions is very small, as shown in Fig. 18-9. Raphael concluded that for engineering computations, calculations based on total effective incoming radiation ($Q_s - Q_r$) may be more efficient than calculating solar radiation Q_s and the reflected radiation Q_r separately, as shown in Fig. 18-10.

An empirical equation, developed by H. Mosby, for calculations of incoming radiation on a horizontal surface in terms of average solar altitude and average cloudiness is as follows:

$$Q_s = k(1 - 0.071\,C)\,a \qquad (27)$$

where

Q_s = solar insolation,
k = a constant and a function of latitude,
C = average cloud cover in tenths of sky covered,
a = average altitude of the sun in degrees.

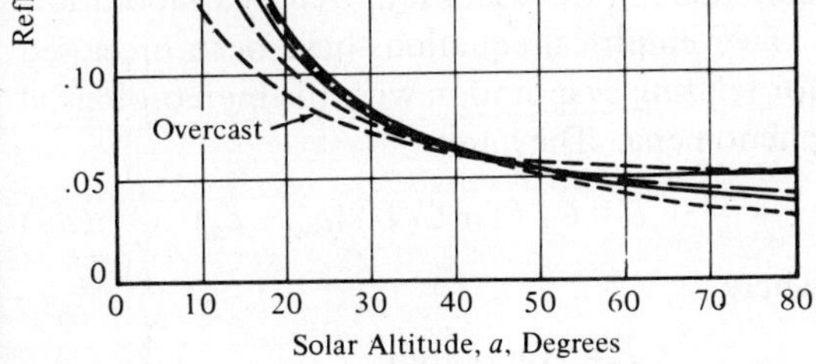

Fig. 18.9 Effect of solar altitude and cloud cover on reflectivity.

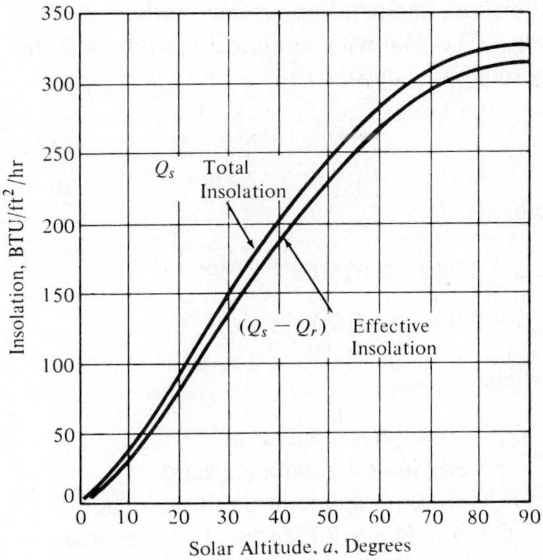

Fig. 18.10 Maximum effective insolation on horizontal water surface.

Equation 27 suggests that the minimum insolation is 29 percent of the maximum, and that the insolation varies linearly with C, the cloud cover. S. Fritz observed, however, that the insolation varied linearly with S, the percentage of sunshine, and not with C. Raphael suggested that for mathematical computations the following equation should be used:

$$Q_i = (1 - 0.0071\,C^2)(Q_s - Q_r) \qquad (28)$$

where

Q_i = the net short-wave insolation, including effect of cloud cover, in Btus per hour per square foot,
C = the cloud cover in tenths of sky covered.

Back Radiation (Q_b): Raphael defines the back radiation as the difference between Q_w, the long-wave radiation leaving the body of water, and Q_a, the long-wave radiation from the atmosphere being absorbed by the body of water. Q_w can be directly calculated by using the Stefan–Boltzmann radiation law, while the calculations of Q_a will need some modifications of the Stefan–Boltzmann law. Equations for the computations of these two factors are developed below.

The atmospheric radiation is dependent on many variables such as the distribution of moisture, temperature, ozone, carbon dioxide, etc. The following empirical relation was suggested by Anderson:

$$\frac{Q_a}{\sigma T_a^4} = a + be_a = \beta \tag{29}$$

where

$$a = 0.740 + 0.025\ Ce^{-0.0584h}$$

$$b = 0.00490 - 0.00054\ Ce^{-0.060h}$$

where

Q_a = atmospheric radiation,
σ = Stefan–Boltzmann constant,
T_a = absolute temperature of air in °K,
C = cloud cover in tenths of sky covered,
e = Naperian base,
h = cloud height in feet and $1600 < h < \infty$,
β = atmospheric radiation factor,
e_a = vapor pressure in millibars.

Figure 18-11 was developed using the original equation and figures of Anderson and the modifications proposed by W. V. Burt. In Fig. 18-11, the atmospheric radiation factor β was plotted against vapor pressure without regard to cloud height. The atmospheric radiation factor β can be calculated once the cloud cover and the vapor pressure of air are known. The Lake Hefner studies showed, however, that the radiations reflected from the water surface had to be taken into account in the calculations

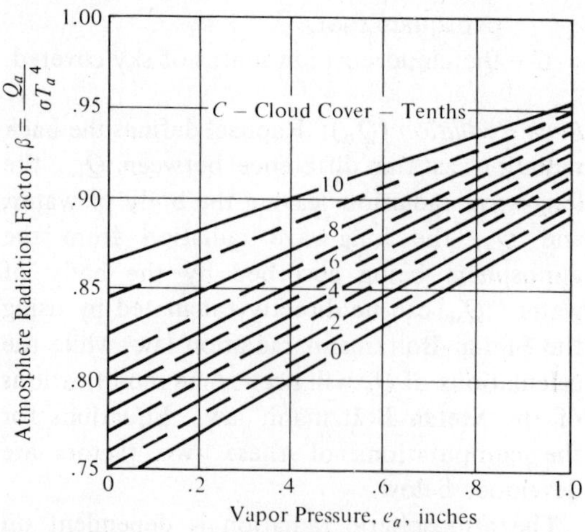

Fig. 18.11 Atmospheric radiation factor, β.

of the atmospheric radiation. Accordingly, Eq. 29 was modified as follows:

$$Q_a = 0.970\ \sigma\beta T_a^4 \tag{30}$$

Equation 30 was used by Raphael in his computations for temperature predictions.

The radiation from the body of water can be calculated from the Stefan–Boltzmann law, provided suitable correction for the emissivity of the surface is applied. The Physical Standards Laboratory of the University of California concluded that the emissivity of water was 0.970 and that it was independent of the water temperature and the dissolved solids. Hence, the equation for the radiation from water is as follows:

$$Q_w = 0.970\ \sigma T_w^4 \tag{31}$$

where

Q_w = radiation from water,
T_w = absolute temperature of water in °K,
σ = Stefan–Boltzmann constant.

According to Raphael's definition of back radiation, Q_b can be expressed by the following relation:

$$Q_b = Q_w - Q_a = 0.970\ \sigma(T_w^4 - \beta T_a^4) \tag{32}$$

where

Q_b = back radiation in Btus per hour per square foot.

Evaporation (Q_e): Water evaporates into the air when the vapor pressure of the air is less than the saturated vapor pressure of air at the temperature of the surface of the water. Removal of heat from the water by evaporation is accomplished in two ways: one is through the energy required to evaporate the water, and the second is by the bodily removal of the heat contained in the water removed by evaporation.

Two empirical equations have been proposed for relating evaporation with the meteorological phenomena. They are:

$$E = C_1\ (1 + C_2 U)\ (e_w - e_a) \tag{33}$$

where

E = evaporation,
C_1 and C_2 = experimental constants,
U = wind velocity in knots,

e_w = vapor pressure of water in saturated air at the temperature of the water,
e_a = vapor pressure of water in the actual air.

and:

$$E = CU(e_w - e_a) \quad (34)$$

where

C = the experimental constant.

According to Eq. (33), the evaporation takes place in the absence of wind. This can occur only by molecular diffusion which is a very slow process. Anderson showed that the evaporation by molecular diffusion was negligible unless the thermal gradient was so great that free convection could be ignored. The results obtained by Eqs. 33 and 34 were almost the same.

Lake Hefner studies showed that Eq. 34 provided the best fit for the experimental data. Equation 34 was modified as follows:

$$E = 0.00177\,U(e_w - e_a) \quad (35)$$

where

E = evaporation in inches per day,
U = wind speed in miles per hour,
e_w = vapor pressure of water in saturated air at the temperature of the water at the surface in millibars,
e_a = vapor pressure of water in the actual air in millibars.

The energy required to evaporate this water can be determined from:

$$Q_e = \lambda \frac{E\,62.5}{12} \quad (36)$$

where

λ = the latent heat of vaporization, averaging about 1060 Btu/lb.

Equation 35 can be rewritten as:

$$Q_e = 12\,U(e_w - e_a) \quad (37)$$

where Q_e is expressed in Btus per hour per square foot, U in knots, and e_w and e_a in inches of mercury.

Conduction (Q_h): Sensible heat is conducted to or from water or from or to air whenever a temperature differential exists between air and water. The amount of heat conducted depends on the temperature difference between air and water and the wind velocity at the interface. I. S. Bowen proposed a ratio relating conducted heat with the energy utilized by evaporation. Bowen's ratio can be expressed as:

$$R = \frac{Q_h}{Q_e} = C\frac{(t_w - t_a)}{(e_w - e_a)} \times \frac{P}{1000} \quad (38)$$

where

P = atmospheric pressure expressed in the same units as e_w and e_a,
C = a constant, varying from 0.58 to 0.66, depending on the state of the atmosphere. Bowen proposed 0.61.

Combination of Eqs. 37 and 38 will yield an equation for conduction as follows:

$$Q_h = 0.00407\,UP(t_a - t_w) \quad (39)$$

where Q_h is expressed in Btus per hour per square foot, U in knots, P in inches of mercury, and t_a and t_w in °F.

Heat will be conducted from air to water if the air temperature is greater than the water temperature and vice versa.

Advected Heat (Q_v): Advected heat is principally from the heat contained in any precipitation that falls directly over the body of water. If no direct data are available on the temperature of the precipitation, then the current wet-bulb temperature of the air can be used. Inflows from springs and underground water are other sources of advected heat.

In summary,

$$Q_t = Q_s - Q_r - Q_b - Q_h - Q_e + Q_v \quad (25)$$

where

$$Q_i = (1 - 0.0071\,C^2)(Q_s - Q_r) \quad (28)$$
$$Q_b = 0.970\,\sigma(T_w^4 - \beta T_a^4) \quad (32)$$
$$Q_e = 12\,U(e_w - e_a) \quad (37)$$
$$Q_h = 0.00407\,UP(t_a - t_w) \quad (39)$$

The equations developed by Raphael for heat transfer are as follows:

$$V_1 \dot{t}_w = \frac{Q_t A \theta}{62.5} + V_i(t_i - t_w) \qquad (40)$$

when the inflow is from one source only; and:

$$V_1 \dot{t}_w = \frac{Q_t A \theta}{62.5} + V_{ia}(t_{ia} - t_w) + V_{ib}(t_{ib} - t_w) \qquad (41)$$

when the inflows are from sources *a* and *b*.

where

V_1 = mean lake volume in acre-feet,
$\dot{t}_w$ = rate of change of temperature of the lake water in degrees per time interval θ,
Q_t = energy budget equation, defined by Eq. 25,
A = total surface area of the lake in acres,
V_i = inflow volume in acre-feet per time interval θ (subscripts *a* and *b* are to identify the two sources of inflows),
t_i = temperature of the inflowing water in °F,
t_w = temperature of the water in the lake in °F.

Equations 40 and 41 were used by Raphael for developing temperature profiles for reservoirs receiving discharges of heated water from power plants.

M. LeBosquet, Jr.[5] LeBosquet presented a mathematical formulation for predicting heat losses from a flowing stream. The derivation of the formula is as follows:

by definition:

$$\text{Btu} = F \times L \qquad (42)$$

where

F = excess temperature in °F of water over air at a distance D miles,
L = pounds of water.

Equation 42 can be rewritten as

$$F = \frac{\text{Btu}}{L} \qquad (43)$$

where

$$\text{Btu} = KA_sFt$$

where

K = heat loss in Btu per square foot per degree Fahrenheit of excess temperature (of water over air) per hour,
A_s = surface area of the stream in square feet,
t = time in hours

Substituting the value of Btu in Eq. 43:

$$F = \frac{KA_sFT}{L} \qquad (44)$$

Differentiating Eq. 44 with respect to time:

$$-dF = \frac{KA_sF}{L}\,dt \qquad (45)$$

(Note: the negative sign is used to indicate decrease in the temperature differential with time.)
Now

$$\frac{L}{A_s} = h(62.5)$$

where

h = average depth of water in feet.

or,

$$\frac{L}{A_s} = \frac{A_x}{W}(62.5)$$

where

A_x = cross-sectional area in square feet,
W = average width of the stream in feet.

or,

$$\frac{L}{A_s} = \frac{Q(62.5)}{WV}$$

where

Q = flow in cubic feet per second,
V = average velocity in feet per second.

Also,

$$t = \frac{\text{Distance in miles}}{\text{Velocity in miles per hour}} = \frac{D}{V\left(\frac{60}{88}\right)} = \frac{88D}{60V}$$

Expressing t in differentials:

$$dt = \frac{88}{60V}\,dD$$

Substituting the values of L/A_s and dt in Eq. 45:

$$-dF = \frac{88KFWV\,dD}{(62.5)(60)\,QV} = \frac{0.02347\,KW\,FdD}{Q}$$

$$= CFdD$$

where

$$C = \frac{0.02347\,KW}{Q} \qquad (46)$$

Integration of Eq. 46 between the limits F_a and F, where F_a is the initial excess temperature in °F, gives:

$$-\int_{F_a}^{F} \frac{dF}{F} = C\int_0^D dD$$

or,

$$F = F_a\,10^{-C'D} \qquad (47)$$

where

$$C' = 0.4343C$$

Substituting the value of C in Eq. (47) and solving it for K, the equation yields:

$$K = \frac{Q \log_{10} F_a/F}{0.0102\,WD} \qquad (48)$$

The graph of Eq. 48, plotted on a semilogarithmic paper with distance D along the abscissa and excess temperature F along the ordinate, is a straight line if W is assumed constant. This graph also assumes that the value of the heat loss coefficient K is known. To solve Eq. 48, the value of K has to be derived from observed river temperature profiles. According to LeBosquet, K varies from 6 Btu for slow, sluggish streams to 18 Btu for rapid, shallow streams. Once K and the river characteristics are known, predictions can be made for other heat loads and runoff conditions.

Edinger and Geyer.[9] Edinger and Geyer have suggested the mechanism for heat transfer as shown in Fig. 18-12. They proposed that the transfer of heat from water to atmosphere is due to radiation, evaporation, and conduction, and that the transfer of heat from atmosphere to water is due to short-wave solar radiation and long-wave atmospheric radiation. The various factors shown in the Fig. 18-12 can be represented by the following equation:

$$\Delta H = (H_s + H_a - H_{sr} - H_{ar}) - (H_{br} \pm H_c + H_e) \qquad (49)$$

where

ΔH = net rate at which heat crosses the water surface,

H_s = short-wave solar radiation,

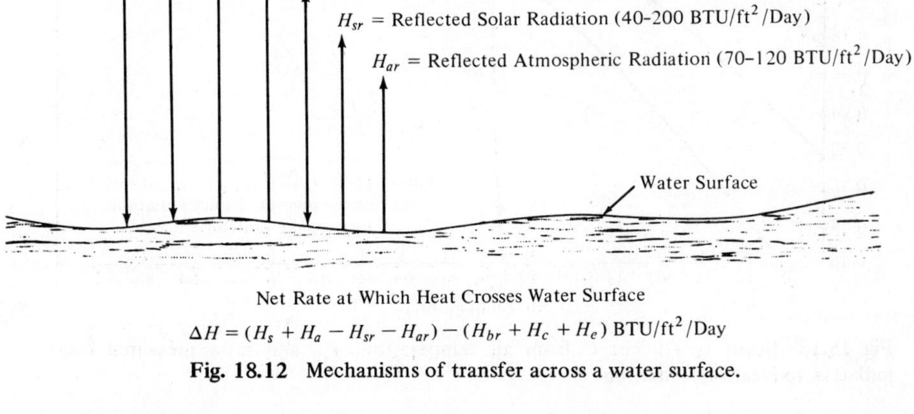

Fig. 18.12 Mechanisms of transfer across a water surface.

H_a = long-wave atmospheric radiation,
H_{sr} = reflected solar radiation,
H_{ar} = reflected atmospheric radiation,
H_{br} = back radiation,
H_c = heat loss due to conduction,
H_e = heat loss due to evaporation.

Definition of terms in Eq. 49:

Short-Wave Radiation (H_s): The short-wave solar radiation is the radiant energy which passes directly from the sun to the earth. Part of the short-wave energy is dissipated in the earth's outer atmosphere due to: (1) absorption by ozone in the upper atmosphere, (2) scattering by dry air, (3) absorption, scattering, and diffuse reflection by suspended particles, and (4) absorption and scattering by water vapor.

The amount of short-wave radiation reaching the Earth's surface is calculated from the "solar constant" which is 10,200 Btu/day/ft^2. The amount of radiation reaching the earth's atmosphere is dependent on the latitude, the time of the day, and the season of the year. This type of radiation can be very easily measured by a pryeheliometer.

Long-Wave Atmospheric Radiation (H_a): Atmospheric radiation is dependent on air temperature and humidity primarily, and increases with the increase in moisture content of the air. Many empirical formulations have been suggested for the estimation of H_a, but the authors feel that Brunt's formula is the most useful one. Brunt's formula is expressed as:

$$H_a = 4.15 \times 10^{-8}\ (T_a + 460)^4 (C + 0.031 \sqrt{e_a}) \qquad (50)$$

where

H_a = long-wave atmospheric radiation in Btus per day per square foot,
T_a = air temperature in °F, measured about 6 ft above the water surface,
e_a = air-vapor pressure in millimeters of mercury, measured about 6 ft above the water surface,
C = Brunt's coefficient, determined from the air temperature and ratio of the measured solar radiation to the clear-sky radiation (Fig. 18-13).

Reflected Solar and Atmospheric Radiation (H_{sr} and H_{ar}): The fractions of the solar and atmospheric radiations from a water surface are computed by using a reflectivity coefficient which is defined as the ratio of the reflected

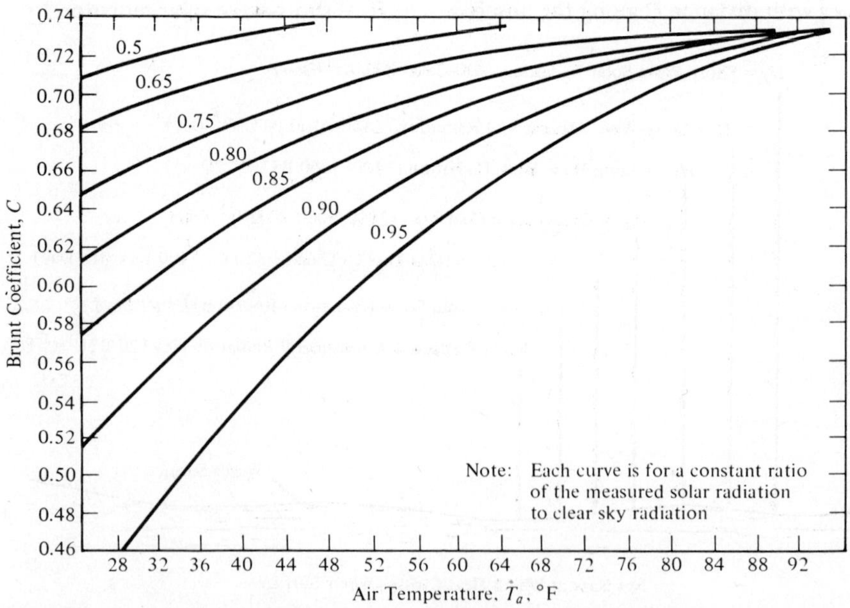

Fig. 18.13 Brunt coefficient C from air temperature, T_c, and ratio measured solar radiation to clear sky radiation.

radiation to the incident radiation. The solar and atmospheric reflectivities are expressed as:

$$R_{sr} = \frac{H_{sr}}{H_s}; R_{ar} = \frac{H_{ar}}{H_a} \qquad (51)$$

The solar reflectivity, R_{sr}, is a function of the sun's altitude and the type and amount of cloud cover. The atmospheric reflectivity, R_{ar}, is relatively constant. Lake Hefner studies showed that solar reflectivity is relatively constant at 0.05 for solar altitudes greater than 40°, and that the atmospheric reflectivity was constant at 0.03.

Absorbed Radiation (H_R): The four terms discussed above can be integrated into one term, the absorbed radiation, H_R, by the following equation:

$$H_R = H_s + H_a - H_{sr} - H_{ar} \qquad (52)$$

From the definition of the individual terms, it will be observed that they are independent of the temperature of water; hence, the absorbed radiation is also independent of the water temperature. The authors suggested that Cummings radiation integrator can be used to measure absorbed radiation. They further suggested that this instrument gave reasonable values for weekly measurements but should not be used for shorter time periods.

Back Radiation (H_{br}): Water sends the energy back to the atmosphere in the form of long-wave radiation. Water can be considered almost like a perfect black body. The rate at which water loses heat can be computed by the Stefan-Boltzmann law, expressed as:

$$H_{br} = r_w \, \sigma (T_s + 460)^4 \qquad (53)$$

where

H_{br} = rate of back radiation in Btu per day per square foot,
r_w = emissivity of water = 0.97.
σ = Stefan-Boltzmann constant = 4.15×10^{-8} Btu/day/ft/°K^4,
T_s = water surface temperature in °F.

Heat Loss Due to Evaporation (H_e): Heat is lost from a body of water to the atmosphere by evaporation of the water. Every pound of water that leaves as vapor carries with it a latent heat of evaporation of 970 Btu. Several empirical formulations have been suggested for estimating the quantity of heat lost by evaporation. The most general equation for heat loss calculations is of the form:

$$H_e = (a + bW)(e_s - e_a) \qquad (54)$$

where

H_e = heat loss due to evaporation in Btus per day per square foot,
a, b = coefficients depending on the evaporation formula used,
e_a = saturation vapor pressure of water determined from the water surface temperature T_s,
e_s = air-vapor pressure in millimeters of mercury,
W = wind speed in miles per hour.

In Eq. 54, the term $(e_s - e_a)$ is the driving force, expressed as the difference in vapor pressures, and is often called the "Dalton difference." In the above expression, if $W = 0$ and $(e_s - e_a) = 1$, then the coefficient a represents the molecular diffusion rate of evaporation. The coefficient b defines the wind effect on the evaporation rate.

A few of the studies that evaluated the coefficients a and b are given below:

Lake Hefner studies:

$$H_e = 11.4W(e_s - e_a) \qquad (55)$$

Lake Colorado studies:

$$H_e = 16.8W(e_s - e_a) \qquad (56)$$

Meyer's equation:

$$H_e = (73 + 7.3W)(e_s - e_a) \qquad (57)$$

In the Lake Hefner studies, the wind speed and the air-vapor pressure measurements were averaged over a period of 3 hr and measured at 24 ft above the water surface. For the Lake Colorado studies, average daily values for wind speed and air-vapor pressure measured at a height of 24 ft were used. No definite heights were specified for Meyer's formula and the monthly mean values for the wind speed were used. Some of the other empirical evaporation formulas are shown in Table 18-5.

TABLE 18-5. Representative Semi-Empirical Evaporation Equations[10]

Reference	Equation	Remarks
Albertson, *et al.*	$E = 6.30 \times 10^{-3} W(e_w - e_a)$	$L = 585$ cal/gm.
Lake Hefner report	$E = 4.50 \times 10^{-3} W_8(e_w - e_8)$	Used by Boyer, Burt, Duttweiler, *et al.*, Raphael.
Gunaji, *et al.*	$E = 6.33 \times 10^{-3} W_2(e_w - e_a)$	
Duttweiler	$E = (0.0207 + 4.14 \times 10^{-3} W_2)(e_w - e_2)$	
Harbeck	$E = 5.70 \times 10^{-3} W_2(e_w - e_a)$	Based on Lake Monona area = 3460 *A*.
	$E = 7.60 \times 10^{-3} W_2(e_w - e_a)$	Based on open water area = 11 *A*.
McAlister	$E = 5.75 \times 10^{-3} W(e_w - e_a)$	Evaporation pan average.
Velz and Gannon	$E = (0.03 + 3.0 \times 10^{-3} W_8)(e_w - e_8)$	Based on climatological data with $C = 12$ in Meyer formula.
Rohwer from Laevastu	$E = (0.026 + 6.89 \times 10^{-3} W_0)(e_w - e_a)$	From Laevastu for sea level
Leavastu	$E = (0.047 + 10.0 \times 10^{-3} W)(e_w - e_a)$	Lake Hefner evaporation pan data.
Penman	$E = 0.0265 + 6.04 \times 10^{-3} W_2)(e_w - e_4)$	Open tanks.
Lake Mead report	$E = (0.013 + 6.16 \times 10^{-3} W_8)(e_w - e_8)$	Lake Hefner field data.
	$E = 5.85 \times 10^{-3} W_2(e_w - e_2)$	

E = evaporation in centimeters per day,
e_w = saturation vapor pressure at the temperature of the water surface in millibars,
a, b = coefficients evaluated empirically,
e_a = ambient vapor pressure in millibars; a number instead of a represents the height above the water surface in meters at which vapor pressure was measured,
W = wind speed in miles per hour.

Heat Conduction (H_c): Heat can enter or leave a body of water by conduction if the air temperature is greater than or less than the water temperature. The rate at which heat is conducted between two media is equal to the product of the heat transfer coefficient and the difference in temperature of the two media. Bowen developed a proportionality between heat conduction and heat lost by evaporation, expressed as follows:

$$B = \frac{H_c}{H_e} \tag{58}$$

where

$$B = \text{Bowen ratio} = C_1 \frac{(T_s - T_a)}{(e_s - e_a)} \times \frac{P}{760} \tag{59}$$

where

T_a = air temperature in °F,
T_s = water temperature in °F,
e_a = air-vapor pressure in millimeters of mercury,
e_s = saturated water-vapor pressure in millimeters of mercury,
P = barometric pressure in millimeters of mercury,
C_1 = coefficient determined from experiments = 0.26.

For standard barometric pressure ($P = 760$ mm of Hg), H_c can be expressed as:

$$H_c = 0.26(a + bW)(T_s - T_a) \tag{60}$$

Studies were made for evaluating the Bowen coefficient, C_1, and the conclusions were that C_1 varied from 0.24 for smooth water surfaces to 0.28 for rough surfaces, and further that $C_1 = 0.26$ is a good average value.

According to the authors, the fundamental heat transfer equation is:

$$\Delta H = H_R - 1801\left(\frac{T_s}{460} + 1\right)^4 - (a + bW) \cdot (e_s - e_a) - 0.26(a + bW)(T_s - T_a) \tag{61}$$

The first term on the right-hand side of Eq. 61 is dependent on the solar radiation, while the rest of the terms depend on the surface temperature of water, T_s, or its related saturation-vapor pressure, e_s. Another point of interest in Eq. 61 is that when the absorbed radiation, H_R, is greater than the heat loss due to conduction, evaporation, and back radiation, the net rate of heat transfer, ΔH, has positive values implying that heat is added to the water body. On the other hand, if H_R is less than the heat lost due to conduction, evaporation, and back radiation, heat is lost from the body of water.

The authors have brought out a very important concept in their report, which they call "equilibrium temperature." They have based the entire theory of temperature prediction on the basis of "equilibrium temperature." They define E, the "equilibrium temperature," as the water surface temperature at which there is no transfer of heat between the atmosphere and the water body. They developed two expressions for E. They are:

1. Based on Newton's law of cooling, the expression for E is:

$$\Delta H = -K(T_s - E) \qquad (62)$$

where

ΔH = net rate of heat transfer in Btus per day per square foot,
K = exchange coefficient in Btus per day per square foot,
T_s = water surface temperature in °F,
E = equilibrium temperature in °F.

2. Equation 61 can be rewritten as follows, when $\Delta H = 0$:

$$H_R = 1801\left(\frac{E}{460} + 1\right)^4 + (a + bW)(e_E - e_a) + 0.26(a + bW)(E - T_a) \qquad (63)$$

where

E = equilibrium temperature in °F,
a, b = constants previously defined,
e_E = saturation vapor pressure at the equilibrium temperature E.

Combining Eqs. 61 and 63 and with some approximations, the net rate of heat transfer can be expressed as:

$$\Delta H = -[15.71(T_s - E) + 0.051(T_s^2 - E^2) + (a + bW)\beta(T_s - E) + 0.26(a + bW)(T_s - E)] \qquad (64)$$

or:

$$\Delta H = \{[15.7 + (0.26 + \beta)(a + bW)] \cdot (T_s - E) + 0.051(T_s^2 - E^2)\} \qquad (65)$$

Neglecting the quadratic terms in temperatures, the exchange coefficient K (in terms of Btus per day per square foot per degree Fahrenheit) can be expressed as:

$$K = 15.7 + (0.26 + \beta)(a + bW) \qquad (66)$$

where β is defined as follows:

$$(e_s - e_E) = \beta(T_s - E) \qquad (67)$$

where

β = a proportionality factor in millimeters of mercury per °F.

Equation (67) can be rewritten in the following form:

$$e_E = \beta E + C(\beta) \qquad (68)$$

The equation for the equilibrium temperature can now be written as:

$$E + \frac{0.051E^2}{K} = \frac{H_R - 1801}{K} + \frac{K - 15.7}{K}\left[\frac{e_a - C(\beta)}{0.26 + \beta} + \frac{0.26T_a}{0.26 + \beta}\right] \qquad (69)$$

or:

$$M = E_1 + F(K)(E_2 + E_3) \qquad (70)$$

where

$$M = E + \frac{0.051E^2}{K}$$

$$E_1 = \frac{H_R - 1801}{K}$$

$$E_2 = \frac{0.26T_a}{0.26 + \beta}$$

$$E_3 = \frac{e_a - C(\beta)}{(0.26 + \beta)}$$

$$F(K) = \frac{K - 15.7}{K}$$

The authors have drawn graphs relating air temperature (T_a), vapor pressure (e_a), wind speed, exchange coefficient, and the factors M,

E_1, E_2, E_3, and $F(K)$. These graphs provide the solution of the equilibrium temperature by trial and error. Once the temperature E is calculated, the temperature of the water surface can then be computed from Eq. 62, where ΔH is obtained from the meteorological conditions. Edinger and Geyer have applied the concept of equilibrium temperature and the exchange coefficient in developing temperature prediction equations for ponds and streams receiving heated discharges, as follows:

Completely Mixed Ponds:

$$\rho c_p Q_p (T_0 - T) = KA(T - E) \quad (71)$$

where

ρ = density of water = 62.4 lb/ft^3,
c_p = specific heat of water = 1.0 Btu/lb/°F,
Q_p = plant pumping rate in cubic feet per day,
T_0 = condenser discharge temperature in °F,
T = pond temperature in °F,
K = exchange coefficient,
A = area of the pond in square feet,
E = equilibrium temperature in °F.

Flow Through Pond:

$$-\rho c_p Q_p \frac{dT}{dA} = K(T - E) \quad (72)$$

where

dT = the temperature drop along the pond per increment of surface area dA. The rest of the symbols have the same meaning as before.

Nonstratified One-Dimensional Temperature Distribution: This is the simplest form of temperature distribution, because in reality the temperature distribution occurs in three dimensions; vertical, along the width of the channel, and along the length of the stream. The equation for temperature distribution for vertically-mixed channels is:

$$\rho c_p Ud \frac{\partial T}{\partial X_1} = -K(T - E) \quad (73)$$

The solution of Eq. 73 is:

$$\frac{T - E}{T_m - E} = \exp\left[-\frac{KX_1}{\rho c_p Ud}\right] \quad (74)$$

where

d = mean depth of the water column in feet,
U = mean stream velocity in feet per day,
$\frac{\partial T}{\partial X_1}$ = longitudinal temperature gradient in °F per foot,
X_1 = distance along the stream in feet.

Nonstratified Two-Dimensional Temperature Distribution: This type of a situation is found in a shallow stream, where a portion of the river flow is withdrawn by the plant, and upon being returned, gradually mixes with the remaining flow. The temperature varies across the river as well as downstream from the discharge. This model takes into account longitudinal direction and lateral mixing. Assuming the river reach under consideration to be in a steady state and the lateral mixing coefficient to be constant, the model for the temperature distribution is:

$$U\frac{\partial T}{\partial X_1} - D_2 \frac{\partial^2 T}{\partial X_2^2} + \frac{k}{\rho c_p d}(T - E) = 0 \quad (75)$$

The solution of Eq. 75 is:

$$\frac{T - E}{T_0 - E} = \exp\left[-\frac{C_2 + \alpha^2}{C_1} X_1\right] \cdot \exp[-\alpha X_2] \frac{\exp[-2\alpha(w - X_2)]}{1 + \exp[-2\alpha w]} \quad (76)$$

where

$\frac{\partial T}{\partial X_1}$ = longitudinal temperature gradient in °F per foot,
T = water temperature in °F,
X_1 = longitudinal distance of the stream in feet,
D_2 = lateral mixing coefficient, assumed constant over the channel, in square feet per day,
X_2 = distance across the stream in feet,
T_0 = temperature of thermal discharge at point $x_1 = x_2 = 0$ in °F,
$C_1 = \frac{U}{D_2}$ per foot,
$C_2 = \frac{k}{c_p d D_2}$

α = coefficient depending on boundary conditions, per foot,
w = stream width in feet,
ρ = density of water in pounds per cubic foot.

The following equations summarize those used by some of the workers in the field of temperature prediction:

Velz and Gannon:

$$A = -224{,}640 \sum_{T_1}^{T_2} \cdot \frac{\Delta T_w}{p(V_w - V_E) + q(T_w - E)} \quad (24)$$

Jerome M. Raphael:

$$V_1 t_w = \frac{Q_t A \theta}{62.5} + V_i (t_i - t_w) \quad (40)$$

when the inflow is from one source only; and,

$$V_1 t_w = \frac{Q_t A \theta}{62.5} + V_{ia} (t_{ia} - t_w) + V_{ib} (t_{ib} - t_w) \quad (41)$$

when the inflows are from sources *a* and *b*.

LeBosquet:

$$K = \frac{Q \log_{10} F_a/F}{0.0102\, WD} \quad (48)$$

Edinger and Geyer:

$$\rho c_p Q_p (T_0 - T) = KA(T - E) \quad (71)$$

for completely mixed ponds;

$$-\rho c_p Q_p \frac{dT}{dA} = K(T - E) \quad (72)$$

for flow-through ponds;

$$\rho c_p U d \frac{\partial T}{\partial X_1} = -K(T - E) \quad (73)$$

for nonstratified one-dimensional temperature distribution; and

$$U \frac{\partial T}{\partial X_1} - D_2 \frac{\partial^2 T}{\partial X_2^2} + \frac{k}{\rho c_p d} (T - E) = 0 \quad (75)$$

for nonstratified two-dimensional temperature distribution.

The temperature prediction models described thus far are for surface discharges of heated waters. Use of these models requires determination of meteorological conditions at the site of thermal discharges; power plant operating data in terms of cooling water inflow and discharge, and temperature rise across the condenser; and receiving water data, such as flow for streams or turn-over time for lakes, critical flow, and ambient temperature conditions. Much of the meteorological data can be obtained from U.S. Weather Bureau stations; however, these stations may not be close to the locations of thermal discharges. Hence, a judgment factor is involved in extrapolating the data. Receiving water data can be obtained from the U.S. Geological Survey, state agencies, and interstate compacts such as ORSANCO.

There are computer models available for defining thermal plumes and for predicting temperatures. The types of data described above are used for predicting temperatures. Such models are available from various sources, such as Vanderbilt University, Massachusetts Institute of Technology, the U.S. Environmental Protection Agency, etc. Care must be taken, however, in selecting models because the models must be verified with actual field measurements (within the limitations of measuring instruments). Computer models can also be used for calculating the heat exchange coefficient, K, and equilibrium temperature, E, from the field data. This type of determination of K and E may be more accurate than the calculations from meteorological data from distant weather stations. These values of K and E may not be applicable to all seasons of the year, however. The computer models can be used for predicting temperatures in rivers, lakes, reservoirs, and estuaries.

During the past few years, many state agencies and the federal government have developed temperature standards, generally applicable at the surface of the receiving water bodies. To meet these standards, subsurface jet-type diffuser systems are being used for thermal discharges. The theory behind subsurface discharges is that due to buoyancy, the heated waters will travel upward toward the surface, and due to the high velocity of the jets (ranging from 10 to 25 ft/sec), the warm water plume will entrain cooler water from the surroundings, thereby dissipating a large portion of heat before the plume emerges at the surface. The discharge system may consist of one port or multiports. The ports may be circular or rectangular. Readers are referred to the following publications for designing subsurface systems.

1. Shirazi, M. A., and Davis, L. R., *Workbook of Thermal Plume Prediction, Volume I, Submerged Discharge*, Pacific Northwest Water Laboratory, National Environmental Research Center, EPA, Corvallis, Oregon (Aug., 1972).
2. Ditmars, J. D., "Submerged Discharges–Single-port and Multiport Diffusers," *Engineering Aspects of Heat Disposal From Power Generation*, Chapter 4, R. M. Parsons Laboratory for Water Resources and Hydrodynamics, Dept. of Civil Engineering, Massachusetts Institute of Technology (June, 1972).
3. Edinger J. E., "Temperature Distribution Analysis," *Effects and Methods of Control of Thermal Dishcarges*, Report to the Congress, EPA (Nov., 1973).
4. Benedict, B. A., Anderson, J. L., and Yandell, E. L., Jr., "Analytical Modeling of Thermal Discharges: A Review of the State of the Art," Argonne National Laboratory Environmental Statement Project, ANL/ES-18, April, 1974.

Computer models for designing subsurface discharges are available also. In addition, there are facilities available in the United States for designing these systems with the use of hydraulic models. These hydraulic models are scaled-down versions of the receiving-water bodies, simulating their geometry and hydraulics. Thermal discharges are introduced into the model and plumes are measured. This procedure is carried out for a variety of discharge systems and the optimum design is selected. Among other advantages, a hydraulic model provides flexibility in that the discharge system, including flow and temperature, can be varied to simulate actual operating conditions. Also, synergistic effects of two or more simultaneous thermal discharges can be easily steadied. However, the accuracy of the results are dependent on the accuracy of instrumentation, and more importantly on the ability to accurately duplicate the receiving waters. Among other places, these facilities are available at Worcester Polytechnic Institute, the U.S. Corps of Engineers at Vicksburg, Mississippi, California Institute of Technology, Massachusetts Institute of Technology, and places in Iowa and Minnesota.

COOLING SYSTEMS

Historically, steam-electric generating plants have used once-through cooling systems. Cooling water, withdrawn from streams, lakes, or estuaries, was passed through a condenser to condense exhaust steam from turbines and discharged to the source. The plants were smaller than many which exist today and were generally located on large bodies of water, so that the environmental impact of thermal discharges was minimal. With the power demand doubling every 10 years or less, and the trend toward constructing larger power plants (500 megawatt and larger), the problems associated with thermal discharges have come to a sharp focus. Removal of waste heat prior to discharge of cooling water to a body of water has become necessary, especially in light of state and federal water quality standards which allow only prescribed temperature increases over the ambient water temperatures. As a result, there are a limited number of sites available to large power plants that could accommodate their thermal discharges on a once-through basis and hold temperature rises within the prescribed standards.

For these reasons, the provision of auxiliary cooling of thermal discharges or the reduction of the temperature of condenser cooling water prior to its discharge to the receiving waters is gaining increased attention. In the United States, the evaporative cooling system is prevalent, as opposed to the situation in Europe where both evaporative and nonevaporative cooling systems are in use. These types of auxiliary cooling systems have inherent difficulties also. They require energy to operate, the amount of which varies with the system used; they require make-up water because of the water losses due to evaporation; cooling system blow-downs require treatment prior to disposal; and the cost of construction, operation, and maintenance is higher than the once-through cooling system.

Further, the installation of auxiliary cooling systems requires modifications in design and operation of power plants. Generally, the cooling system operates on a "closed cycle," which means that the colder water, as received from the cooling system, is pumped back into the condenser. This results in warmer water (compared to that of once-through systems) being passed through the condensers. This causes the condensation temperature and pressure on the turbine to increase, thereby reducing the efficiency of the plants. This loss in energy requires an increased supply of thermal energy to the plant to produce the same electrical output and increase in waste heat.

The two types of auxiliary cooling systems will be described although the evaporative system is far more common, as noted above.

Evaporative Cooling. Condenser cooling water loses its heat by conduction, convection, radiation, and through evaporation, which is the major cause of heat loss. Heat loss through evaporation occurs because the heat required to convert water to vapor (the latent heat of vaporization) is obtained from the heated water. A minor portion of heat required for vaporization is abstracted from the surrounding air.

Several types of evaporative cooling systems are described below. These systems operate on the principle of intimate contact between air and heated water.

Cooling Ponds. Cooling ponds are the simplest and cheapest method of heat dissipation if land is available at a reasonable cost. Earth dike cooling ponds, which are the most economic and easiest to construct, range from 6 to 8 ft deep and can operate for extended periods of time without make-up water requirements. There are three general types of cooling ponds:

1. Completely mixed ponds which have the same temperature throughout the pond.
2. Flow through ponds where the temperature continuously decreases as the water moves from the intake to the outlet. The plant intake may be located at the outlet of the pond, thereby providing a closed-cycle operation, or the pond outlet may discharge the water into a natural water body, thereby providing an open-cycle system (see Figs. 18-14 and 18-15).
3. Internally circulating ponds, where the warm water "rides" on top of the colder water.

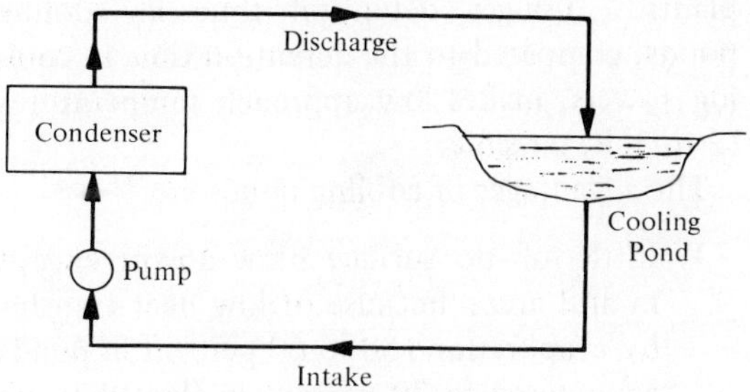

Fig. 18.14 Closed cycle cooling pond.

In cooling pond designs, the most significant parameter is the surface area required to cool the thermal discharge, since the heat transfer predominantly occurs at the surface. Factors necessary for determining the required surface area are: air temperature, relative humidity, wind speed, heat transfer coefficient, and water temperature. Although mathematical models have been developed with which to design the cooling ponds, a widely used rule-of-thumb for the surface area requirement is 1 acre/installed megawatt capacity for fossil fuel plants and $1\frac{1}{2}$ to 2 acres/megawatt for nuclear power

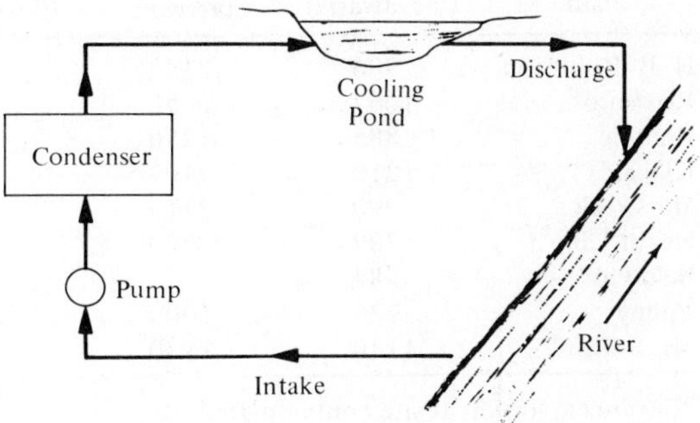

Fig. 18.15 Open cycle cooling pond.

plants. Longer detention time in cooling ponds, compared to the detention time in cooling towers, makes low approach temperatures (2 to 3°F) possible.

The advantages of cooling ponds are:[11]

1. Little or no surface blow-down, except in arid areas, because of low heat transfer by evaporation (30 to 60 percent in ponds as opposed to 70 percent in flow through evaporative cooling towers).
2. Lower capital costs, if land is available at reasonable cost.
3. Ability to operate for extended periods without make-up water.
4. Dual function as settling basins for suspended materials.
5. Availability of ponds for other purposes, such as recreation, fishing, etc.
6. Potential for avoiding chemical treatment of recirculating water in some instances.

The disadvantages are:[11]

1. Large surface areas required for heat dissipation.
2. Requirement for low permeability of sides and bottoms to avoid excessive seepage.
3. Fogging and icing of areas adjacent to the cooling ponds.
4. Debris entering the water (due to the fact that it is open to the atmosphere).
5. Possible maintenance problems due to weed and algae growth.

Examples of the capital costs of cooling ponds and related information are shown in Table 18-6.[11] The capital costs are estimated[11] to be \$8 to \$11/kilowatt or \$2000 to \$2700/million Btu/hr of heat rejected, and the total annual operating cost ranges from 0.15 to 0.28 mills/kilowatt-hour.

Spray Ponds. A spray pond is a modified cooling pond in that it contains a spray system consisting of nozzles in perforated pipes (Fig. 18-16). The spray system breaks the heated water into fine droplets, 0.38 to 0.50 in. in size. These droplets provide maximum water surface to facilitate more efficient heat transfer to the atmosphere than the heat transfer obtained from cooling ponds. In some instances, spray ponds have louvered fences to reduce drift losses.

Heat transfer between the water droplets and air occurs by conduction, convection, and radiation. Design of the spray ponds will depend on the climatic conditions, wind speeds, and land availability. Performance of a spray pond is dependent on the proper design and location of the nozzles for the spray system. The heat transfer rate for spray ponds ranges from 2600 to 5500 Btu/hr/ft^2 according to Russian experience and from 4000 to 10,000 Btu/hr/ft^2 according to American and British experience.[3]

The advantages of spray ponds are:[11]

1. Reduction in land requirements by a fac-

TABLE 18-6. Capital Costs of Selected Cooling Ponds[11]

Plant	1970 Capacity (megawatts)	Surface Area (acres)b	Capital Cost of Cooling Pond (\$)	UNIT COST \$/Megawatt	\$/Acre	Comments
H. B. Robinson	906	2145	4,800,000	5298	2238	1971 capacity
Roxboro[a]	1067	3750	4,831,000	4528	1288	
Braunig	885	1250	4,717,000	5330	3774	
Kincaid	1319	2400	3,819,000	2895	1591	
Mt. Creek	990	2710	4,333,000	4377	1599	
North Lake	709	800	3,555,000	5014	4444	
Baldwin[a]	584		3,000,000	5137		
Valley	725	1000	918,209	1266	918	
Mt. Storm[a]	1140	1120	6,523,000	5722	5824	

[a]Future expansion at site contemplated.
[b]One acre equals 4047m^2.
[c]Covers land and land rights costs only.

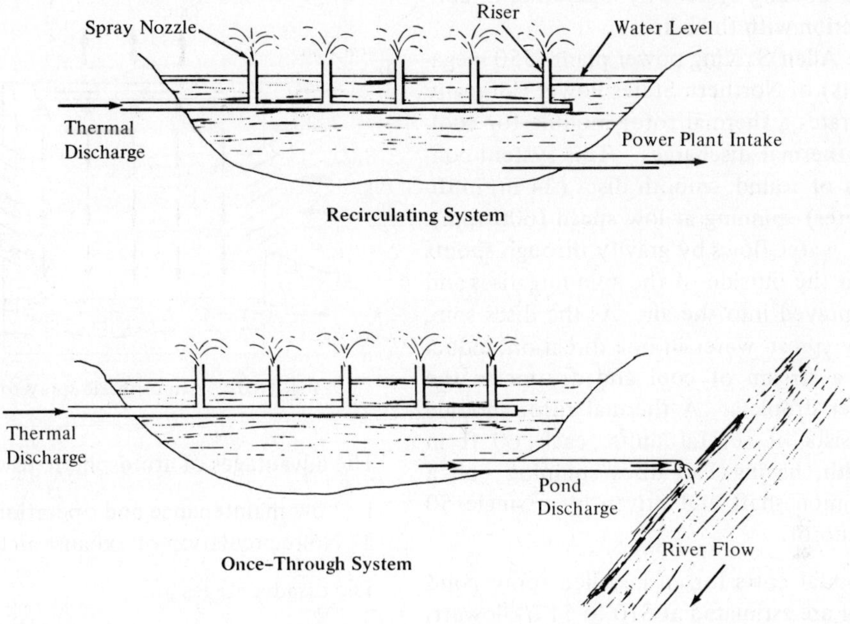

Fig. 18.16 Spray ponds.

tor of 20 compared with cooling ponds (on the average).

2. Relative economics.
3. No requirement for chemical treatment of recirculating water in many cases.

The disadvantages are:[11]

1. Increase in water losses compared to cooling ponds.
2. More severe fogging.
3. Performance directly related to wind speed, wet-bulb temperature, and nozzle design.
4. Maintenance of mechanical equipment.
5. Required blow-down.
6. Increased load requirements compared to cooling towers.

A spray pond may consist of a fixed spray system, in which locations of the pipes and nozzles are fixed, or a modular spray system where units containing several nozzles are used for spraying heated water. The number of units operated at any given time is dependent on the amount of heat to be dissipated. Some examples of spray ponds follow:[12]

1. Camadys Station of the South Carolina Electric and Gas Company uses a 10-acre conventional spray pond to cool condenser cooling water from a 490-megawatt plant. The pond can cool 180,000 gpm of water from 101 to 88°F (38 to 31°C) with an ambient wet-bulb temperature of 78°F (25.5°C). This spray pond was built into the once-through cooling system.
2. Merrimack Station of the Public Service Company of New Hampshire and Chesterfield Station of the Virginia Electric and Power Company have a powered spray module system produced by the Ceramic Cooling Tower Company. This system consists of individual units, each having four spray nozzles. The number of units to be brought on-line depends on the heat load to the pond.
3. Dresden Units No. 2 and 3 of the Commonwealth Edison Company temporarily installed a powered spray module for cooling thermal discharges while a 1375-acre cooling lake was under construction. The spray system will be optimized into the

final cooling system by operating in conjunction with the lake.

4. The Allen S. King power plant (550 megawatts) of Northern States Power Company operates a thermal rotor module for cooling thermal discharges. This system consists of round, smooth discs (24 in. in diameter) spinning at low speed (600 rpm). Hot water flows by gravity through spouts onto the outside of the spinning discs and is sprayed into the air. As the discs spin, they spread water in one direction, inducing a stream of cool and dry air to the water droplets. A thermal rotor module consists of several units, each 60 ft in length, having 150 discs spinning on a common shaft and driven by a single 50 hp motor.

The capital costs for an installed spray pond and canal are estimated at \$10 to \$17/kilowatt, operating and maintenance costs are estimated at 0.10 to 0.12 mills/kilowatt-hour, and total annual costs are estimated at 0.27 to 0.41 mills/kilowatt-hour.[11]

Wet Cooling Towers. A wet cooling tower is a heat exchanger which utilizes evaporation by direct contact of air and water. Heat dissipation is accomplished by maintaining a steady stream of air through the cooling tower, by increasing water surface through the use of sprinklers or splashboards, and by bringing air and water into intimate contact. A description of various types of wet towers follows.

Atmospheric Towers. In an atmospheric tower, the heated water is pumped to the top and is sprayed downward through nozzles. Air enters the tower through louvered sides and flows across in a transverse direction. Warm and moist air rises upward and is discharged through the top. The air movement through the tower is dependent on the wind velocity outside the tower. There are two types of atmospheric towers: spray towers (Fig. 18-17), where nozzles break the water into small droplets, and packed towers (Fig. 18-18), where the tower is packed with baffles which are used as splashboards to increase water surface.

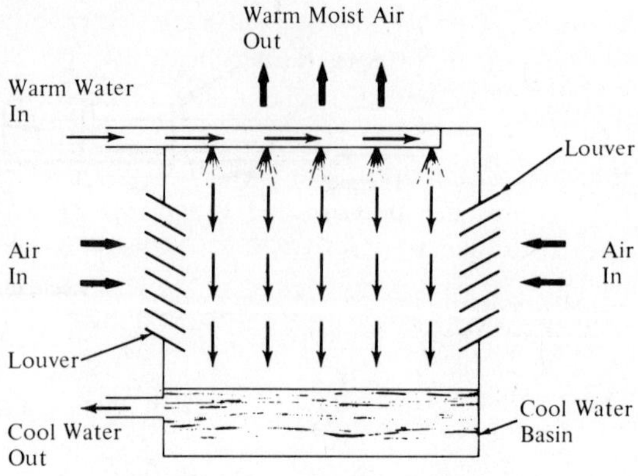

Fig. 18.17 Atmospheric spray tower.

The advantages of atmospheric towers are:

1. Low maintenance and operation costs.
2. No recirculation of exhaust air.

The disadvantages are:

1. Necessary construction in open areas because of the dependence on wind movements.
2. Limited cooling range.
3. Required height to compensate for narrow width.
4. Pumping pressures needed for achieving final spray through the nozzles.

These types of towers are not common today and are used where strict controls on temperatures are not required.

Mechanical Draft Evaporative Towers. A mechanical draft evaporative tower generally consists of a wood/concrete structure packed with

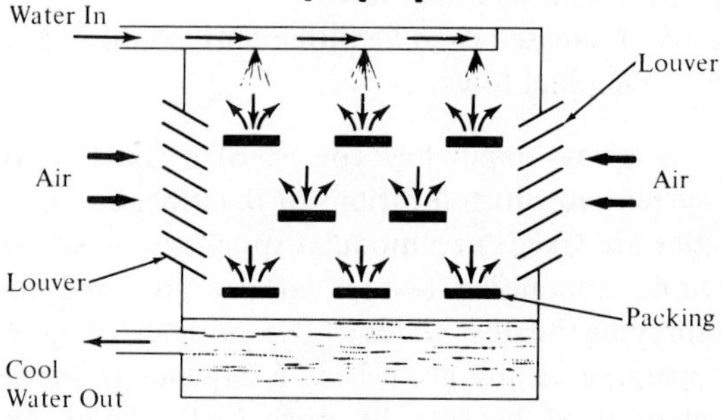

Fig. 18.18 Atmospheric packed tower.

wood or plastic fill over which warm water falls to form small droplets. This type of tower uses fans to provide air movement within the structure. Fans may be located either at the intake, thereby forcing the air through the structure, or the outlet end, thereby inducing flow of air. There are two types of packing available: splash packing, where water splashes down on the media and breaks into small droplets, and film packing, which enables the water to fall in thin, closely-spaced sheets. Splash packing is more prevalent. Drift eliminators are installed above the tower packing to prevent excessive loss of water through the tower and into the air. Drift losses can generally be maintained below 0.008 percent of the recirculating water flow. However, the cooling tower manufacturers guarantee drift losses to be less than 0.002 percent. These drift losses are responsible for causing local fogging and icing problems and care must be taken in selecting the site for cooling towers to minimize the problem.

The advantages of mechanical draft towers are:

1. Close control of cold water temperature.
2. Low land requirements.
3. Generally low pumping head required.
4. Positive control over air movement.
5. More available fill per unit volume of tower.
6. Lower capital cost compared to the cost of natural draft towers.
7. A closer approach and longer cooling range possible.

The disadvantages are:

1. Power required to operate fans, which consume up to 0.8 percent of generating capacity.
2. Mechanical failure potential.
3. High operation and maintenance costs compared to costs for natural draft towers.
4. Possible recirculation of hot and humid exhaust air into the air intakes.
5. Possibility of vapor and stack gases mixing outside the tower, which could be troublesome.
6. Fogging and icing problems due to exhaust air.
7. Potential blow-down disposal.

There are two types of mechanical draft towers: (1) forced draft towers, and (2) induced draft towers.

A forced draft tower has one or more fans located in the air intake, thereby forcing the air through the tower structure. The forced draft tower is illustrated in Fig. 18-19.[8]

The advantages of a forced draft tower are:

1. Low vibration due to solid foundation for equipment.
2. Slightly greater efficiency than induced draft towers.
3. Decreased possibility of moisture condensing on the equipment.

The disadvantages are:

1. Possible recirculation of hot and humid air.
2. Limited fan size, and thus a requirement for more fans.

An induced draft tower has one or more fans located at the outlet end of the tower, thereby inducing flow of air through the tower structure. There are two types of induced draft towers: (1) counterflow towers and (2) crossflow towers.

In a counterflow tower, air is drawn upward by the fan located in the outlet structure and water flows down through the packing. A typical counterflow induced draft tower is shown in Fig. 18-20.[8]

The advantage of a counterflow tower is that the coldest water comes in contact with the direct air and the warmest water comes in con-

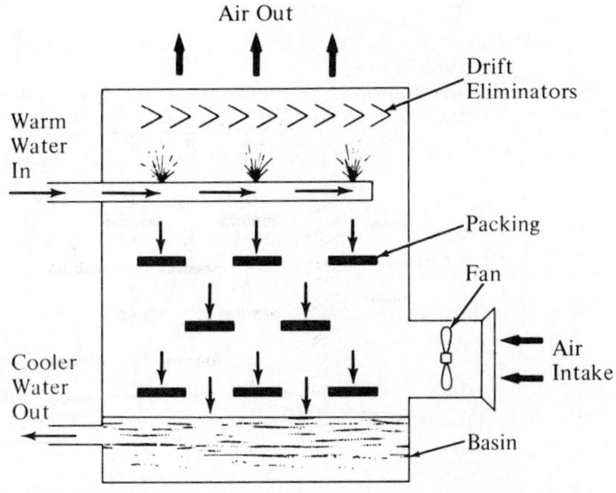

Fig. 18.19 Forced draft tower.

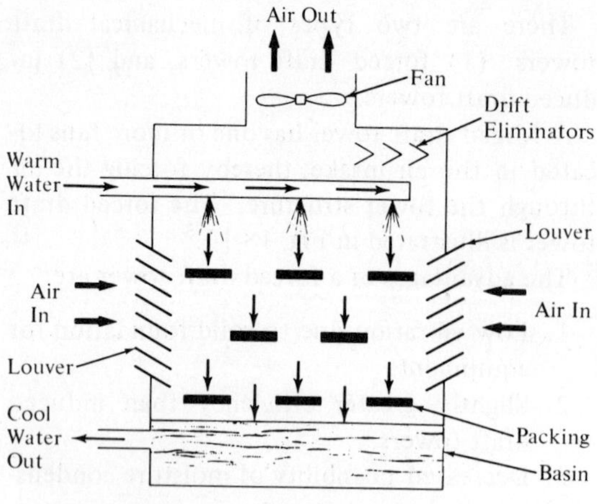

Fig. 18.20 Counterflow induced draft tower.

tact with the most humid air. This provides maximum performance.

The disadvantages are:

1. High energy requirements to enable the fan to overcome resistance to upward flow of air through falling water.
2. Higher static pressure loss due to resistance to upward air flow against the falling water.
3. Uneven distribution of air through the packing, little air movement near the walls.
4. High pumping head necessary due to the height of the tower. The tower is tall because the packing must be placed high to keep the air intake area unobstructed.
5. Limited water load capacity.
6. Hot water system not readily accessible for maintenance.
7. Possibility of airborne debris entering the tower due to high inlet velocity.

In a crossflow induced draft tower, the air, induced by the fans located in the outlet structure, flows horizontally through the tower structure, while the water falls vertically down through the packing. Such a tower is shown in Fig. 18-21.[8]

The advantages of a crossflow induced draft tower are:

1. Low pumping head.
2. Low static pressure drop on the air side.
3. Convenient arrangement of water distribution system for maintenance.
4. Packing height approximately the height of the tower.
5. More air per fan horsepower.
6. Possible higher water loadings for a given height.
7. Ability to provide water at any desired temperature since the water temperatures in the basin vary from the center to the edges.
8. Capacity for larger diameter fans so that fewer cells are required.

The disadvantages are:

1. Insufficient pressure head on the distribution pans to keep the nozzles from clogging.
2. A need for a substantial crossflow correctional factor applied to the driving force

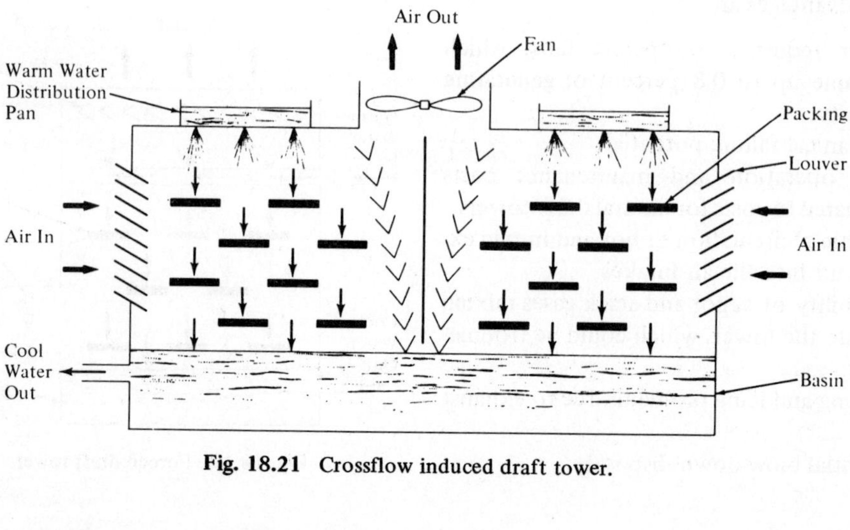

Fig. 18.21 Crossflow induced draft tower.

where long range and close approach performances are required.

Capital costs of mechanical draft evaporative cooling towers are dependent on many factors, such as design, wet-bulb temperature, range, approach, relative humidity, land, construction costs, etc. The cost of the tower, basins, pumps, fans, and make-up water system are estimated at $6.50 to $8.60/kilowatt, or $1600 to $2200/million Btu/hr of heat rejected. The operating costs for power usage, maintenance, labor, and ammortized capital costs range from 0.3 to 0.4 mills/kilowatt-hour.

Natural Draft Evaporative Towers. Although natural draft evaporative towers have been used in Europe for several decades, the first such tower was constructed in the United States in 1962 at the Big Sandy plant of the Kentucky Power Company. Since then, several natural draft towers have been built and several are under construction.

Natural draft towers are hyperbolic in shape for aerodynamic and structural reasons. This type of a tower, like a mechanical draft tower, dissipates heat through air-water contact. Unlike the mechanical draft tower, howerver, a natural draft tower utilizes a high hyperbolic stack or chimney to induce flow of air through the structure instead of fans. Air flows through the tower naturally because of: (1) the difference in density between internal and external air, and (2) the atmospheric pressure differential due to the height of the stack. Air absorbs heat and water vapor as it cools the warm water, which reduces air density and causes it to rise through the tower. As the warm and humid air rises, more air is drawn through the tower, thereby maintaining a continuous flow.

These towers can be as high as 450 ft, most of the height being used as the stack or chimney. The bottom 20-30 ft are used for fill or packing, the water distribution system, and openings for flow of air. There are two types of natural draft towers: counterflow and crossflow, which are shown in Figs. 18-22 and 18-23.[6] These towers can be up to 500 ft in diameter at the base.

In a counterflow tower, the water distribution system and the packing are located within

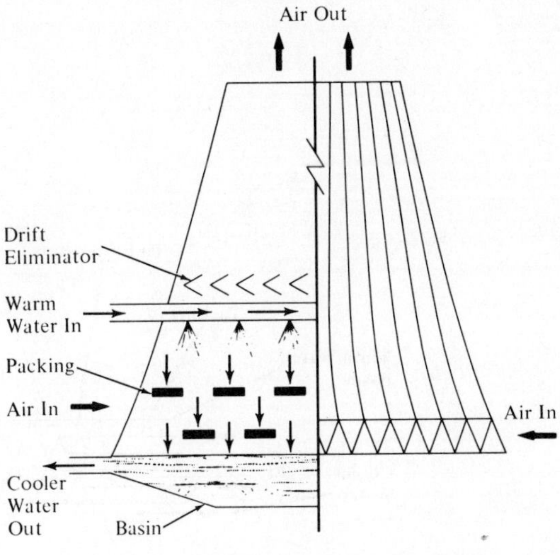

Fig. 18.22 Counterflow natural draft tower.

the tower. As the warm water falls on the packing and flows vertically downwards in thin sheets, it imparts heat to the vertically rising air. On the other hand, in a crossflow tower, the water distribution system and the packing are located along the outside periphery of the shell of the stack. Warm water is broken into droplets as it falls on the packing and flows vertically downwards. The induced air flows horizontally through the falling water and then rises in the stack.

The advantages of a natural draft tower are:

1. The same cooling capability for a given heat load as a mechanical draft tower without the mechanical parts and power required to run them.
2. Low maintenance cost.
3. Independence of external wind velocity.
4. Small land area required.
5. Large water loading capacity.
6. High cooling efficiency in counterflow towers due to an air flow opposite to water flow.

The disadvantages are:

1. The need to minimize resistance to air flow.
2. Great heights required to produce induced flow of air.
3. Necessity for higher inlet hot water temperature than the air dry-bulb temperature.

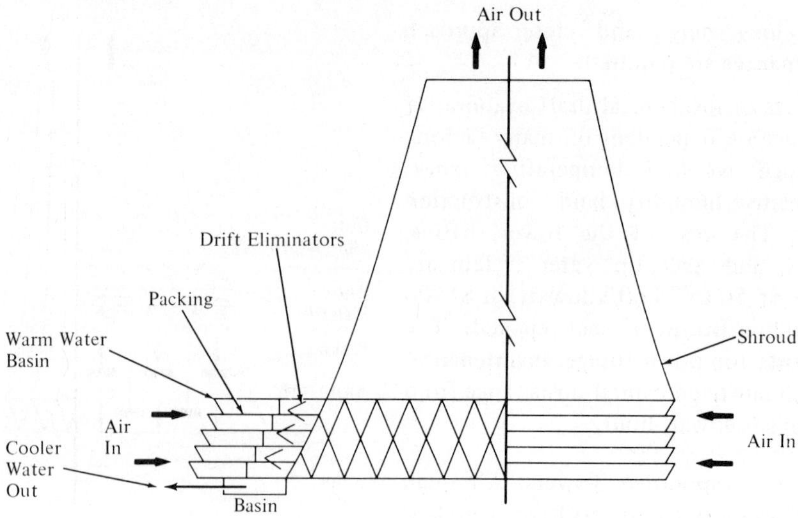

Fig. 18.23 Crossflow natural draft tower.

4. Difficulty in achieving exact control of outlet water temperature.

Nonevaporative Cooling. In a nonevaporative or dry cooling system, heat is transferred to air by conduction and convection as opposed to evaporation. The flow of air maintained through the cooling system does not come in direct contact with the condenser cooling water or steam to be condensed.

This use of air for cooling or condensing purposes has been practiced for the past 50 years. However, until now this application was limited to small industries and small power plants. Table 18-7[13] shows existing power plants using dry cooling towers. In addition, a dry tower for a 340 megawatt power plant is under construction in the Black Hills of South Dakota.

There are two basic types of air-cooled condensing systems: indirect systems and direct systems. Details of the two systems follow.

Indirect System. This system utilizes a direct contact condenser to condense the exhaust steam from the turbine. Water from the condenser is pumped to the dry cooling tower and the cooled water is recirculated to the spray system in the direct contact condenser. This system is often referred to as the Heller system, since this concept was presented by Dr. Laszlo Heller at the World Power Conference in Vienna in 1956.

The major components of the indirect system are: direct contact condenser, circulating water pumps, water recovery turbine, cooling coils, and either a mechanical draft or a natural draft tower to provide the required air movement. Figures 18-24 and 18-25 show the indirect system with mechanical draft and natural draft towers, respectively.

The indirect system operates as follows: cooled water from the cooling coils is sprayed into the direct contact condenser which receives exhaust steam from the turbine. The cooled water mixes directly with the steam and condenses it. This mixture of condensed steam and water from the tower is removed from the condenser by circulating and condensate pumps. Water equal to the amount of exhaust steam from the turbine flows to the boiler for reevaporation. This requires that it be of condensate purity. The remaining volume is directed to the cooling coils for heat dissipation.

A positive pressure of approximately 3 ft is generally maintained in the system to prevent air leakage. This is accomplished either by a throttline valve or a water recovery turbine located in the discharge pipe from the tower.

Direct System. In the direct system, the exhaust steam from the turbine is condensed in

TABLE 18-7. Generating Plants with Dry Cooling Towers[13]

Location	Rating (megawatts)	Type of Dry Towers	Year Commissioned
Rugeley, England	120	Heller	1962
Ibbenburen, Germany	150	Heller	1967
Wolfsburg, Germany	3–50	GEA direct	1961–67
Grootvlei, South Africa	200	MAN/Birwelco (indirect)	1971
Gyongyos, Hungary	2–100	Heller	1969
	2–100	Heller	Under construction
Razdaan, USSR	3–220	Heller	1970–72
Wyodak, Wyoming	22	GEA direct	1969
	3	Direct	1962
Utrillas, Spain	160	GEA direct	1970
Onetta, West Pakistan	7.5	Baldwin-Lima-Hamilton (direct)	1964
Bavaria	40	GEA direct	1960
Windhoek, South Africa	3–30	GEA direct	1971
Switzerland	4.3	GEA direct	1969
Luxembourg	13	GEA direct	1956
Rome, Italy	2–30	GEA direct	1957
Cologne, Germany	28	GEA direct	1958
Sindelfingen, Germany	11 and 15	GEA direct	1960–61
Worms, Germany	5	GEA direct	1962
Chile	3.6	GEA direct	1963
Ludwigshafen, Germany	38	GEA direct	1966
Eilenburg, Germany	5.3	Heller	NA
Dunaújváros, Hungary	16	Heller	1961

the cooling coils rather than in a direct condenser. The main components are: exhaust steam header, cooling coils, fans for air movement, and condensate pumps. Figure 18-26[13] shows a schematic diagram of a direct system tower.

Exhaust steam from the turbine is conveyed to the cooling coils through the exhaust steam header. The cooling coils have fins which provide the necessary surface area for heat transfer. A fan blows the air as required for steam condensation. The finned cooling coils may be

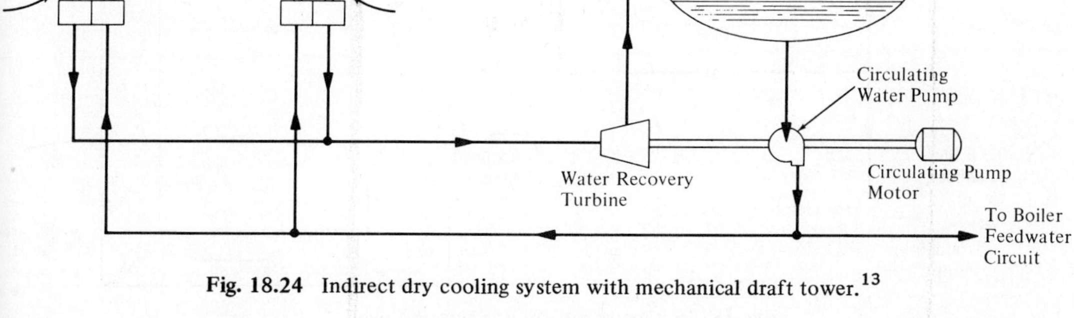

Fig. 18.24 Indirect dry cooling system with mechanical draft tower.[13]

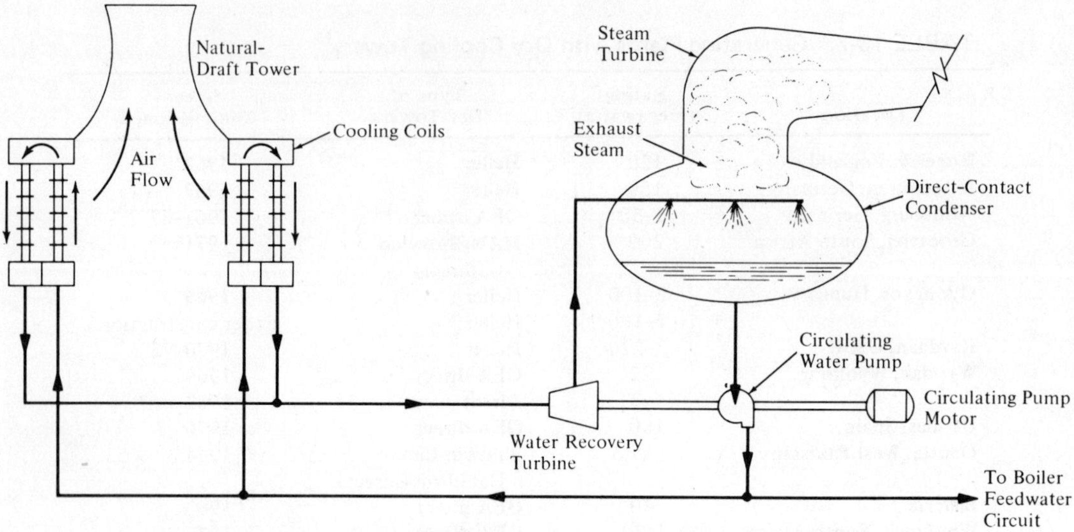

Fig. 18.25 Indirect dry cooling system with natural draft tower.[13]

constructed as shown in Fig. 18-26 or as horizontal tube bundles. The steam and the condensate flow in the same direction, thereby minimizing pressure loss and maximizing heat transfer.

GEA of Bochum, Germany uses a direct system, in which a certain percentage of steam, after passing through the main cooling coils, enters the bottom headers of the aftercooling coils. This results in the steam and the condensate flowing in the opposite direction, which provides better control of condensate temperature during cold weather. The condensed steam from the cooling coils flows by gravity to a reservoir from which it is pumped back to the boiler.

The advantages of a dry cooling tower are:

1. No blow-down problem.
2. No icing or fogging of the surrounding areas.
3. No water losses due to evaporation or drifts.
4. No upper limit to which air can be heated.

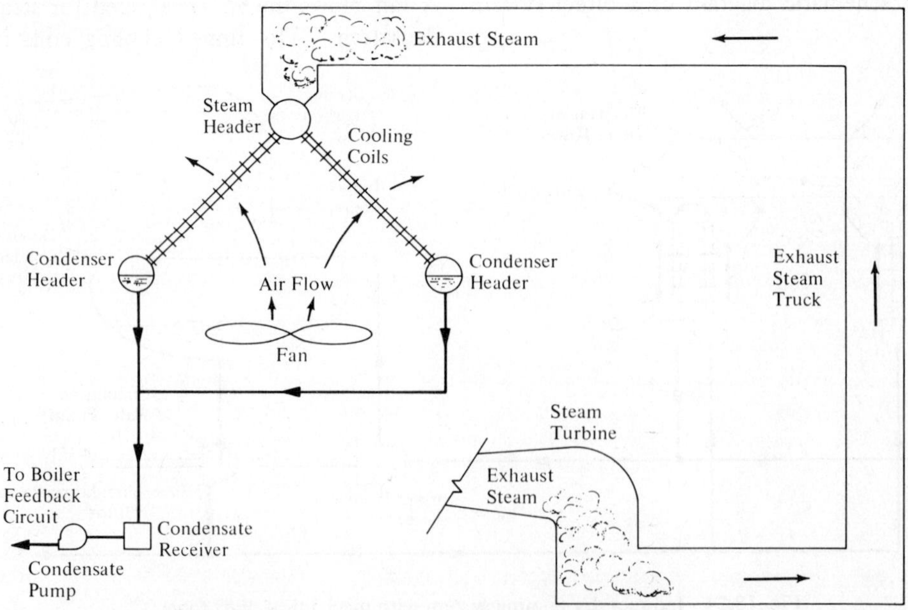

Fig. 18. 26 Direct dry cooling system tower.[13]

The disadvantages are:

1. Higher capital costs compared to evaporaative cooling tower costs.
2. High maintenance costs.
3. Large volumes of air to be circulated, since the specific heat of air is one-fourth that of water.
4. Decreased power plant output at high dry-bulb temperatures.
5. Larger land area requirement than for the evaporative towers.

The estimated capital cost for dry towers is \$20 to \$30/kilowatt, or \$5000 to \$7500/million Btu/hr of heat rejected. The operating and maintenance costs are estimated[11] at \$440 to \$880/yr/million Btu/hr, or 0.20 to 0.40 mills/kilowatt-hour.

New Developments in Cooling Towers. Cooling tower suppliers are working on various combinations of hyperbolic and mechanical draft towers and wet and dry towers to provide a solution that will be best suited to a given location, reduce construction costs, and enhance environmental considerations. Two such developments are described below.

Fan-Assisted Natural Draft Cooling Towers. This hybrid tower will have the appearance of a hyperbolic tower, but will be smaller in size, and will incorporate the advantages of both the hyperbolic and mechanical draft towers. Large motor-driven fans will be located along the circumference of the conventional hyperbolic shell. Since air movement will be created by the fans, tower height requirements will be reduced by half, and the base diameter reduced by one-third of the height and diameter of the natural draft tower required to dissipate the same amount of heat.

The capital cost and land requirements of this hybrid tower are expected to be less than that for the natural draft tower, but the operating costs may be higher because of the power requirements of the fans.

Wet-Dry Cooling Towers. A combination wet-dry cooling system was developed to incorporate desirable features and characteristics of the two types of systems into one. The wet-dry cooling towers can be used to: reduce or eliminate plumes from wet towers; minimize water losses generally associated with wet towers; reduce the condenser pressure; and provide flexibility of operation in terms of ratio of wet-to-dry cooling.

The wet and dry components of the combined tower can operate either in parallel or in series. In the parallel flow design, the air passes through the dry and wet sections simultaneously through separate passages, and the dry, warm air is mixed with the warm saturated air within the tower. This mixture of air is discharged by fans located at the top of the tower. In series flow design, the dry section can be located either behind or ahead of the wet section.

A parallel path wet-dry cooling tower, developed by the Marley Company, is shown in Fig. 18-27. The thermal discharge first passes through the dry section, which consists of finned cooling tubes. Then, the wet section is utilized for further reduction in heat. Air passes through the dry and wet sections in parallel paths. The exhaust air from the two sections mixes within the cooling tower before it is discharged to the atmosphere. The visible plume that normally results from the wet towers is substantially reduced or virtually eliminated, depending on the weather conditions.

This type of tower has been in operation since 1955 at a 30-megawatt plant of the St. Joseph Light and Power Company of Missouri; a wet-dry tower is under construction for a 450-megawatt plant of Northeast Utilities located in Middletown, Connecticut.

Although no wet-dry towers have been built for 1000-megawatt power plants, it is estimated[11] that the capital costs for such units would be \$3000 to \$4500/million Btu/hr of heat rejected, of \$12 to \$18/kilowatt-hour. Operating and maintenance costs could be about \$307 to \$350/million Btu/hr, or 0.14 to 0.16 mills/kilowatt-hour.

Environmental Impact of Cooling Tower Operations. Cooling towers, evaporative and nonevaporative types, are being used for dissipation of heat from cooling water to the atmosphere, thereby reducing one "pollutant," specifically, heat. However, these towers have

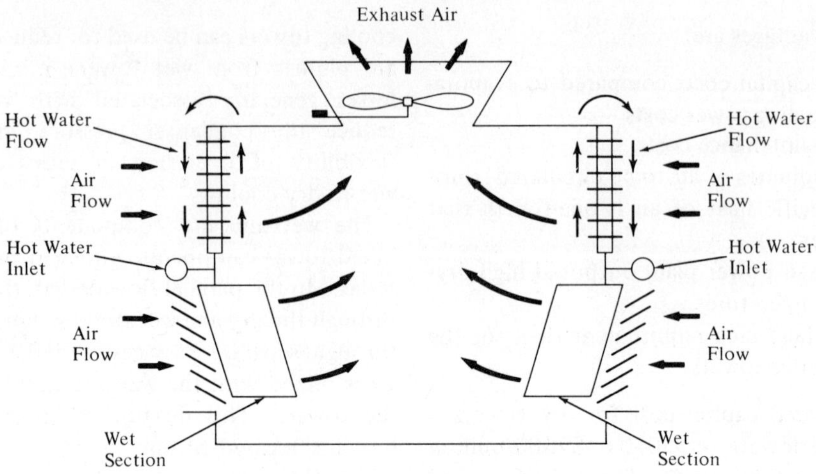

Fig. 18.27 Parallel path wet-dry tower.[13]

their own impact on the environment due to their operating characteristics and mechanisms of heat transfer.

Impact on the Climate. Depending on the cooling system used, heat and/or moisture is transferred to the atmosphere. In the case of a nonevaporative cooling system, heat transfer only takes place while heat and moisture are dissipated from the evaporative cooling system (once-through or wet towers). This addition of heat and/or moisture, over and beyond the natural heat and moisture content of the atmosphere and heat dissipated by other man-made sources, such as cities, could result in increased local temperature and rainfalls. Also, rainfalls containing dissolved and suspended pollutants could result from the mixing of cooling tower plumes with other stack emissions. However, further research is needed to delineate these effects as they relate to cooling tower operations specifically.

Fogging and icing of areas immediately adjacent to evaporative towers have been observed, especially from the mechanical draft evaporative towers. One corrective measure is to build, wherever feasible, natural draft hyperbolic towers which discharge the plume high enough in the atmosphere to prevent these conditions. On the other hand, heated air exhausted from dry towers may help in fog dissipation.

Noise. The primary source of noise associated with cooling towers is the operation of large, high-speed fans used for air movement in the mechanical draft type. This noise is characteristic of the fans and of air movement and very little can be done to abate it. Some alternatives are: location of the cooling tower in remote areas which negates the advantage of size of the tower, operation of fans at lower speeds which may interfere with the heat dissipation capacity of the tower, construction of a cooling tower plenum with sound attenuation material which will increase the cost of the cooling tower, and construction of natural draft towers instead of mechanically induced towers,. which will result in limiting available sites.

Water Pollution. Natural bodies of water contain dissolved and suspended solids and biological materials which are deposited on the various components of cooling systems as the water passes through them. Water pollution problems arising from cooling tower operations derive primarily from efforts to protect wooden and metal parts from damage by these substances in order to maintain an efficient heat transfer from water to air. For example, in a once-through system, chlorine is generally used to prevent biological growths in the condenser tubes.

In a recirculating system, the use of other chemicals is common for the following reasons:

formation of scale on pipe walls and heat exchange surfaces, corrosion of metal parts, biological fouling of heat exchange surfaces, and damage to wooden components. In the normal operation of these systems, the number of times cooling water is recycled depends on the maximum concentration level of dissolved solids which can be tolerated. When this level of concentration is reached, the water is removed and treated prior to its discharge to receiving waters. This procedure is often called "blow-down."

The common types of scales are: calcium carbonate, calcium sulfate, silica, magnesium compounds, iron and manganese compounds, and phosphate compounds. Scale can be prevented by: softening of the make-up water, pH adjustments, and addition of chemicals to retard scale deposition, such as polyphosphates or organics.

Corrosion is the physiochemical destruction of metals caused by the natural air and water environment. Anodic, cathodic, and organic inhibitors have been used to reduce corrosion problems. Typical anodic inhibitors are: sodium and potassium chromates, phosphates, silicates, nitrites, ferrocyanides, and molybdates. Cathodic inhibitors include salts of zinc, nickel, manganese, and trivalent chromium. Organic inhibitors are used in conjunction with polyphosphates and chromates; however, phosphates are nutrients for algae which limits its use as an inhibitor.

Cooling water provides an ideal environment for biological growth. Biological fouling plugs screens, restricts flow, creates growth, and releases oxygen through photosynthesis, which accelerates corrosion unless biocides are used to control it. Chlorine is the most commonly used biocide, although other biocides are frequently used. These include: hypochlorites, chlorophenols, quarternary amines, organosulfur, organo-tin compounds, and organothiocyanate.

Wood components used in cooling towers should be specially treated for structural performance. Various algaecides and fungicides listed above can be used.

Blow-down, as described earlier, is cooling water drained from the closed-cycle evaporative cooling system when dissolved solids have reached prescribed concentrations. The blowdown contains, in addition to the dissolved solids, the various chemicals used in the preservation of timber and metal components and in the prevention of biological fouling. The disposal of blow-down is a serious problem because the chemicals in it are toxic to aquatic life. The disposal methods include: dilution by receiving waters if water quality regulations are not contravened, retention in ponds for subsequent release, evaporation and concentration in ponds, disposal into underground formations, chemical treatment, and desalting the blow-down, thereby permitting water recycling.

BENEFICIAL USES OF THERMAL DISCHARGES

There is general agreement that beneficial use of waste heat is highly desirable. At the same time that efforts are being made to utilize solar energy or to develop geothermal energy sources, a vast amount of energy from power plants is literally being wasted and little effort is made to harness it.

Further, the various methods of heat dissipation to the environment discussed previously are costly, clumsy, inefficient in some cases, and may cause other environmental problems such as fogging, icing, and loss of water through drift losses. In addition, the operation of some of these systems requires substantial amounts of energy expenditure. In essence, most of them are energy intensive and are liabilities to the electric utilities and hence their customers.

There is no question about the quality of heat, low-level or low-temperature, as it is presently being discharged by the power plants. However, if the utilization of this heat were contemplated or planned, the quality of heat could be tailored, within constraints, of course, for the particular use. It is not essential to discard this amount of energy as "unwanted" or as a "pollutant," especially now when most of the world is energy-short.

The potential uses of waste heat may be classified as follows:

1. Greenhouses.
2. Agriculture.

3. Aquaculture.
4. Space heating.
5. Additional applications.

Greenhouses. The practice of growing fruits and vegetables in heated greenhouses can be traced back to Roman days.[14] Commercial greenhouses have since come to be used for flower nurseries throughout the world, but only recently attention has been given to growing vegetables in greenhouses on a commercial scale. Ventures of considerable size have been successful in Sweden, Holland, and Iceland. The largest commercial operation in the United States is in Cleveland, where approximately 700 acres are utilized for growing flowers and vegetables.

A greenhouse can be considered as a horizontal cooling tower, where the heat from thermal discharges can be dissipated by evaporative cooling, nonevaporative cooling, soil heating, or any combination of the three.

Evaporative cooling utilizes the same principle as an evaporative cooling tower, whereby the warm water is allowed to flow down on pads, fill, or packing material while a flow of air is maintained across the falling water. Figure 18-28 shows a typical evaporative system used in greenhouses. As the water loses its heat through evaporation, the temperature and humidity in the greenhouse are increased. The desired level of humidity is maintained by the use of exhaust fans, located either at the side or in the ceiling.

A nonevaporative system entails the passage of thermal discharges through pipes equipped with fins for heat dissipation. This system is similar to the one used in dry cooling towers and is shown in Fig. 18-29. In this type of system, only the temperature within the greenhouse is changed; there is no effect on the humidity. The finned tube can be placed either vertically along the smaller side of the greenhouse or horizontally on the ground along the length of the greenhouse.

A combination of evaporative and nonevaporative systems can be installed within a greenhouse to provide the required temperature and humidity conditions. Sometimes a plastic sheet is used to provide a false ceiling to circulate warm and humid air. Generally, the false ceiling is used to maintain 100 percent relative humidity conditions. This type of an arrangement is shown in Fig. 18-30.[15]

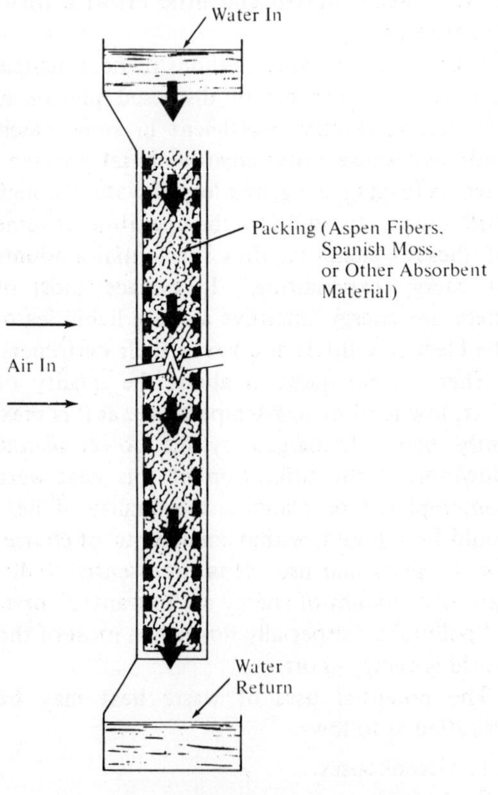

Fig. 18.28 Evaporative cooling system.

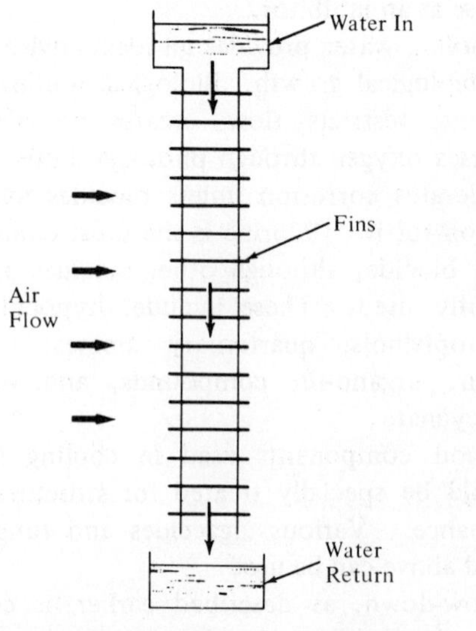

Fig. 18.29 Nonevaporative cooling system.

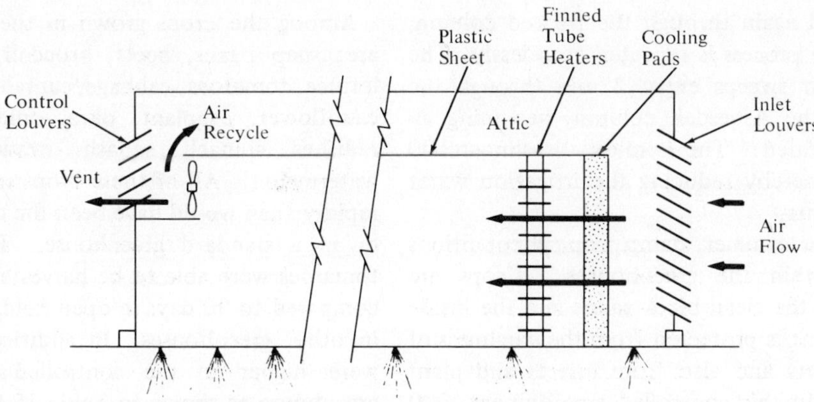

Fig. 18.30 Air flow system in greenhouse.

Soil heating is accomplished by passing warm water through buried pipes which are sometimes perforated to maintain desired moisture level in the soil. Soil heating will be discussed in greater detail under the agricultural uses.

Iceland. Greenhouse heating by natural hot water (primarily geothermal) began in Iceland in the 1920s and has been increasing in importance ever since.[16] By 1960, approximately 115,000 m^2 of glass-covered greenhouses were in use. Approximately 70 percent of the area was used for growing vegetables, such as tomatoes, cucumbers, and lettuce, and the remaining 30 percent of the area was used for flowers, such as roses, carnations, tulips, and chrysanthemums, and various potted plants.

The total use of geothermal energy for greenhouse purposes was 120 Tcal/year (476×10^9 Btu/yr) at an energy price of \$0.43/Gcal (\$0.11/million Btu). The construction cost of greenhouses is approximately \$15/m^2 (\$1.40/ft^2) for timber greenhouses and \$20/m^2 (\$1.86 ft^2) for aluminum frame houses. The annual value of the greenhouse production is estimated to be \$910,000 (1969).

University of Arizona, Tucson. In 1966, the Environmental Research Laboratory of the University of Arizona began a series of experiments with closed environment greenhouses at the University's Puerto Penasco, Sonora, Mexico facility. These greenhouses are part of a power-water-food complex where waste thermal energy is utilized for a desalination plant.[17,18]

There are several greenhouse units at the Puerto Penasco facility, each unit consisting of two air-inflated polyethylene cells. Each cell is 100 ft long and 23 ft wide, connected at the front and back by concrete tunnels. These cells provide a growing area of 4600 ft^2/greenhouse unit. The greenhouses are entered through an air-locked system which keeps the greenhouse air separate from the outside air. The growing medium is sand (beach or desert) instead of the normal soil of open-field farming.

Inside the greenhouse, fans force air up through packed-column heat exchangers–low, honeycomblike stacks of corrugated asbestos (Asbesdek)–through which seawater is sprayed at 80°F (26.5°C). The column is a direct-contact heat exchanger rather than an evaporative cooler because the Asbesdek increases the heat-transfer area. Ordinarily, a flow rate of about 120 gpm is maintained. On cold nights (when the greenhouse radiates heat back to the atmosphere) and on hot days, the flow can be increased to 140 gpm to bring the interior temperature closer to the seawater temperature. If lower night temperatures are desired, then the flow of water can be completely turned off. On extremely cold days, 94°F (35°C) blowdown seawater from the desalting plant can be piped into the greenhouse for necessary heat. From the packed column, the moistened air flows the length of the greenhouse and is deflected across the front tunnel and back down the opposite side. It then goes across the rear tunnel and again through the packed column. This entire process is repeated ceaselessly. The trapped air sweeps every 2 min through the spray of the Asbesdek column, becoming almost saturated. The humidity is almost 100 percent, thereby reducing the irrigation water requirements.

Winter or summer, damp tropical conditions prevail within the greenhouses. Crops are planted in the clean black sands and the inside environment is protected from the harshness of open deserts and also from insects and plant diseases. In this controlled environment, 300 varieties of 20 different vegetables and fruits have been grown.

Among the crops grown in the greenhouses are: snap beans, beets, broccoli, cucumbers, lettuce, tomatoes, cabbage, cantaloupe, carrots, cauliflower, eggplant, okra, onions, peppers, radishes, spinach, squash, strawberries, and watermelon. All of these crops matured more rapidly than would have been the case outdoors or in a standard greenhouse. For example, tomatoes were able to be harvested in 80 days, compared to 90 days in open fields or 100 days in other greenhouses. In addition, the yields were higher in the controlled environment greenhouse as shown in Table 18-8. Cucumber and lettuce production were at least 6 times higher than yields from open-field farming.

TABLE 18-8. Vegetable Crops Grown in Greenhouses in Puerto Penasco, Sonora, Mexico (1969-1970)

PUERTO PENASCO GREENHOUSES		COMPARATIVE DATA FOR U.S. (acres)		
Kind of Vegetable	Marketable Yield/Acre	Approximate Average Yield from Greenhouses	Approximate Average Yield Outdoors/Year	Good Yield Outdoors/Year
Cucumbers (European type)[c]				
Fall crop	6600 bu @ 48 lb/bu	–	185 bu[a]	500 bu[b]
Spring crop	7290 bu[c]			
Eggplant				
Fall crop	4000 bu[c]	–	433 bu[a]	500 bu[b]
Spring crop	4000 bu[c] @ 33 lb			
Lettuce				
Bibb and leaf				
Winter crop	3500 ctn @ 2 dozen	3500 ctn	–	–
Okra				
Winter crop	40 ton	–	–	5 ton[b]
Peppers				
Bell				
Winter crop	1200 bu @ 25 lb	–	372 bu[a]	500 bu[b]
Radishes				
Winter crop	40,000 bnch @ 12/bnch	40,000 bnch	–	20,000 bnch[b]
Squash				
Zucchini				
Spring crop	2000 bu[c] @ 45 lb	–	–	400 bu[b]
Tomatoes				
Fall crop	75 ton[c]	40 ton[c]	6.8 ton[a]	30 ton
Spring crop	65 ton	60 ton[c]	–	–

[a]*Source:* United States Dept. of Agriculture, *Agricultural Statistics* (1969).
[b]*Source:* Knot, J. E., *Handbook for Vegetable Growers*, John Wiley and Sons, Inc. New York (1962).
[c]Based on a harvest period of 90 days.

The increased yields were further enhanced by piping carbon dioxide from the diesel exhaust into the greenhouse.

Estimated capital investment for a 1-acre, framed, open-environment greenhouse for tomato or cucumber production is approximately $87,000, including about $17,500 for heating. Capital costs for a greenhouse using waste heat would probably not exceed those for conventionally-heated installations. The availability of free or low-cost heat energy could save $6000 to $8000 annually for fuel. The estimated annual operating cost per acre is approximately $55,700. The return per acre for harvesting tomatoes is $86,400, based on 12,000 plants/acre, a yield of 24 lb/plant, and a wholesale price of $0.30/lb.[17]

Based on its successful operation of closed-environment greenhouses, the Environmental Research Laboratory has constructed a 5-acre greenhouse complex in Abu Dhabi in the Middle East. This greenhouse complex is a part of a power-water-food system for the nation. Waste heat from power generation and desalination operations is used to create and maintain the required environment for growing vegetables and fruits in desert sand.

TVA Program. TVA has set aside 180 acres of land at Brown's Ferry nuclear plant for testing various ways of using waste heat.[19,20] This nuclear plant site offers many advantages: temperature rise across the condenser of about 27°F, highly productive agricultural land within the area, and proximity to TVA's agricultural and environmental research and development complexes at Muscle Shoals, Alabama.

Several methods of waste heat utilization are under investigation or are planned for investigation by TVA. This includes: (1) heating and cooling greenhouses, (2) environmental control for livestock housing, (3) extending the crop growing season, (4) recycling nutrients from livestock wastes, and (5) fish farming. Heating and cooling greenhouses will be discussed here and the other uses will be discussed under their respective headings elsewhere in this section.

A pilot greenhouse (25 by 100 ft), under construction at Muscle Shoals, will be used to grow high value horticultural crops utilizing thermal discharges for heating and cooling. The warm water will be used as a part of a closed-cycle system in which the cooled water will be returned to the power plant for reuse. Chemical compositions of the water will be monitored for any changes. Thermal discharges, used for heating and cooling the greenhouse, will not be used for irrigating the crops.

The heating and cooling system, developed by Oak Ridge National Laboratory, consists of evaporative pads and finned-tube heat exchangers. In this pilot greenhouse, conventionally-heated water will be used to simulate thermal discharges from the Brown's Ferry plant. Initially, vegetables such as lettuce, tomatoes, cucumbers, and radishes will be grown, followed by crops such as strawberries for the fresh out-of-season markets, flowers, and ornamentals. Biological requirements for these crops and market economics will be determined. Based on Ohio conditions, preliminary calculations indicate a gross sale potential of $40,000 to $60,000/acre/yr is attainable on a 12-month growing basis. The potential gross sale for different crops is shown in Table 18-9.

While the pilot greenhouse is under construction, sections of an existing greenhouse at Muscle Shoals are partitioned off with plastic to simulate high humidity conditions to be anticipated in the pilot greenhouse. Two varieties of tomatoes and two varieties of seedless cucumbers were grown hydroponically. The relative humidity within the plastic chamber was usually 97 to 98 percent at night and 85 to 90 percent during the day, compared to about 60 percent in the rest of the greenhouse. The temperature within the plastic chamber was generally 8°F lower than the temperature in the rest of the greenhouse. Productions of tomatoes and cucumbers in the

TABLE 18-9. Potential Gross Sales from Greenhouse Production

Crops	$/Acre/Yr
Tomatoes–2 crops	44,000
Lettuce–tomatoes–lettuce	52,000
Bibb lettuce–6 crops	56,000
Radishes–7 crops	59,000

plastic chamber was lower, although 68 tons/acre and 53 tons/acre, respectively, were produced.

Calculations show that about 6 percent of the thermal discharges from the Browns Ferry plant can be utilized for 32 acres of greenhouse tomato production and 58 acres of lettuce harvesting. These two crops, thus grown, will meet 1.4 percent of the total requirements in the Southeast. The lettuce production would amount to 30 percent of that now shipped into the combined Atlanta, Nashville, Memphis, and Birmingham markets. The tomatoes will account for about 24 percent of the total.

Conclusions. It can be concluded that a properly designed greenhouse can utilize heat from thermal discharges to warm and cool the inside environment and maintain required temperature and humidity conditions for growing vegetables, fruits, and flowers. The design of the greenhouse will include: heat transfer system, air circulation system, and irrigation system. The operation of the greenhouse will include: determination of temperature and humidity conditions necessary for the crops, nutrient requirements, insect and disease controls, market conditions and program management for optimizing returns from the greenhouse, and an adequate monitoring system.

Agriculture. The discussion of agricultural uses of thermal discharges will be limited to open-field crop farming and animal farming (confined to animal houses). Thermal discharges can be used to heat the crops and animal housing during cold weather and cool them during summer months.

Open-field Crop Farming Applications. Thermal discharges can be used for the following purposes: frost protection, plant cooling, extending growing seasons, and soil heating.

Frost Protection. Frost protection using a sprinkler system is based on the principle that when water freezes, its latent heat of fusion is transferred to the surroundings. As long as the ice, formed on limbs, leaves, or buds, is kept wet by the sprinkling water, an equilibrium between the liquid-solid interfacial temperatures is maintained to hold the ice at its freezing point. Therefore, even when the air temperatures are well below freezing, the temperature of limb, bud, or leaves will not fall below 30.5°F (-0.8°C). The ice covering on buds and limbs protect them from the winter hazards during the cold months. However, care must be taken in designing the sprinker system to prevent limb breakages.

When warm water is sprinkled over an orchard, a thermal blanket, which provides additional protection, is created because of the evaporative cooling which increases humidity and temperature of the air. Also, the thermal discharges lose heat at a faster rate from evaporative cooling. For example, the rate of evaporation, which requires about 970 Btu/lb, is about 7 times greater for water at 70°F (21°C) than for water at 40°F (4.5°C).

Plant Cooling. High air temperatures and low relative humidity conditions cause stress on plants, resulting in a reduction in productivity, both in terms of quality and quantity. Soil moisture is transferred by roots to the leaves, which in turn lose it to the atmosphere. Part of the water absorbed by the plant is used for its growth and production, but a larger percentage of the absorbed water passes out of the stomas and through evaporative cooling helps to maintain the temperature of the plant. As the air temperature rises beyond a critical point, balancing the availability of water from soil against loss through evaporative cooling becomes difficult, and the plant shows stress and a drop in productivity. A properly designed sprinkler system can restore the productivity of the plant by providing moisture to the leaves and replenishing water in the soil. Either cold or warm water can accomplish this; in the case of warm water, the heat will be lost to the atmosphere by the time water reaches the plant from the sprinklers.[21]

Soil Heating. Soil heating with thermal discharges is probably the most beneficial use of such discharges.[22,23] It is accomplished by subsurface application of heat by circulating thermal discharges through a network of buried pipes. The purpose of this is to warm the soil at the root level to a temperature that is optimal for the growth of the crop. Heat moves from the pipes to the soil and ultimately to the atmosphere. The size of the pipes and their spacings are critical in order to provide the amount of heat necessary to the surrounding soil to result in the desired soil temperatures. This type of a system becomes especially attractive where soil temperatures inhibit plant growth. Greenhouses can be constructed over the ground with buried pipes and kept warm by the heat escaping the soil. Figure 18-31 shows the relationship between temperature and percent of possible growth of various crops.

In order to obtain the increases in productivity as shown in Fig. 18-31, it is essential to obtain the corresponding soil temperatures. The soil temperatures are dependent on: the heat transfer from the warm water to the pipe surface to the soil, the spacing of the pipes, the depth at which the pipes are buried, the surface area per unit of length of pipe, the thermal conductivity of the soil, and the number of parallel pipes in the area under consideration.

The following examples are representative of the systems described above. This list is not all-inclusive, however.

1. Thermal Water Horticultural Demonstration Project. The Thermal Water Horticultural Demonstration Project[24] started in the fall of 1968 and is sponsored by the Eugene, Oregon Water and Electric Board and the U.S. Environmental Protection Agency. The project is located just east of Springfield, Oregon on 170

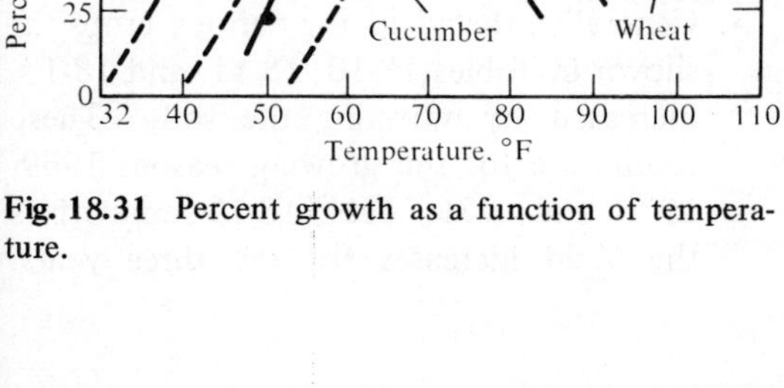

Fig. 18.31 Percent growth as a function of temperature.

acres of farmland. The land is owned and operated by seven farmers who are managing their crops under the direction of the Agricultural Division of Vitro.

Warm water is obtained from the Weyerhaeuser pulp and paper mill located about 2 mi from the project site. The water temperature varies from 90 to 130°F (32 to 55°C); however, 105°F (40°C) occurs most frequently. At present, three pumps with a total capacity of 4000 gpm pump the water to the project site through a 16-in. steel main. Installation of another pump would increase the capacity to 5000 gpm. The 16-in. main is reduced to an 8-in. line as it continues across the project for another $\frac{3}{4}$ mi. At the end of the piping network is a bleed valve that is continuously spraying 125 gpm of the warm water into the air. This continuous flow of water through the pipe network does not permit the water to cool down substantially, thereby providing constant temperature to the project. The water from the bleed valve forms a puddle at the base, and then seeps through approximately 60 ft of soil before entering the McKenzie River at a temperature near or below the river water temperature. The warm water loses between 4 and 7°F in temperature as it flows from the paper mill to the bleed valve, an approximate distance of $2\frac{3}{4}$ mi.

Each farmer controls a submain line which branches off the main and crosses his property. The buried submains are PVC pipes and are equipped with valves and hydrants. The entire project has approximately 21,000 ft of buried main and submain lines.

Approximately 50 acres of orchard crops are protected from early spring frost and high temperatures in summer by sprinklers placed on 6-ft, 10-ft, or 15-ft risers depending on the sizes of the trees to be protected, and spaced on a 40 by 40 ft staggered spacing. The sprinklers are Rain Bird 14 V TNT sprinklers with $\frac{3}{32}$-in. nozzles and are operating at 60 psi. The water application rate is approximately 0.12 in./hr.

A weather station is located in the center of the project. Air temperature, wind speed and direction, humidity, and precipitation are recorded. Soil moisture is monitored throughout the project and irrigation applications are scheduled as the available soil moisture is depleted. A frost-alarm system, located at the weather station, activates an alarm in each farmer's home if the temperature falls to near the critical point.

In addition to frost protection and plant cooling, thermal discharges have also been used for crop irrigation and soil heating at this demonstration project. Irrigation was carried out with sprinkler systems covering areas 30 by 40 ft, 40 by 40 ft, and 40 by 55 ft, depending on the crop and water application rates.

The results of the demonstration project can be summarized as follows:[25]

1. Frost protection. In three full growing seasons, no crops were lost due to frost; adjoining fields, unprotected by thermal water, suffered losses up to 50 percent.
2. Plant cooling and replenishing of soil moisture. No crops were lost due to excessive heat or lack of moisture in the soil.
3. Increased yields. Some crops yielded up to 22 percent more with the use of warm water.
4. Hastened yields. Peach trees, thermally treated, bear fruit in 3 years instead of the 5 years required under natural conditions.
5. Double-cropping. Following a potato harvest from 18 acres of land, turnip seeds were planted and harvested from 14 acres, and rutabagas and bunch beans were planted in the remaining 4 acres during the same year.

2. Water Resources Research Institute Demonstration Project. The Water Resources Research Institute, Oregon State University, through a matching grant from the Office of Water Resources Research, U.S. Department of the Interior, and the Pacific Power and Light Company, conducted a project to study and evaluate the effect on plant growth to be achieved by circulating warm water through an underground pipe network.[23] This project involved laboratory experiments as well as field studies.

The underground network of warm water pipes was simulated by a network of heating cables. These cables were installed at a depth of 92 cm in parallel lines at 183 cm intervals. The cables were rated at 65.5 watts/meter. Thermostats, installed on each cable network, controlled the temperature of the cable. Thermonetic thermistors, with an R_0 of 10 K-ohms at 25°C, were used for temperature measurements. The measurements were made with a digital data acquisition system, consisting of a Hewlett-Packard model 1012B data acquisition system (capable of measuring DC microvolts, DC volts, AC volts, and resistance) and two digital recording devices. The system was capable of scaling 200 guarded channels, with program control for selected range and functions at rates up to 15 channels/sec.

The results of the experimental work can be summarized as follows:

1. Temperature distribution. The soil temperatures of the heated plots are almost always within the optimum range for plant growth. During the cooling portion of the diurnal cycle, the temperatures of the heated plots remained several degrees above those of the unheated plots near the soil surface. The applied heat penetrated down to a considerable depth. The temperature increase near the soil surface is small in relationship to the temperature closer to the piping system; thus, to obtain a larger temperature increase near the soil surface, the heat sources should be placed at a shallower depth and at narrower spacing. The depth and spacing used in the experimental work were adequate to warm the root zone to the optimum soil temperature. Shallower depth and narrower spacing increases the rate of heat dissipation from water to soil to the atmosphere.
2. Increasing root temperature resulted in an increase in the nitrogen, phosphorus, and potassium content in both shoot and root material. The calcium content decreased, however.
3. Generally, the yield for various crops, as shown in Tables 18-10, 18-11, and 18-12, increased by warming the soil. These results are for the growing seasons 1969, 1970, and 1971. Table 18-13 summarizes the yield increases for the three years.

TABLE 18-10. Effects in Yield Due to Soil Warming, Growing Season 1969 (near Corvallis, Oregon)

Crop	YIELD Natural (tons/ hectare)	Heated (tons/ hectare)	Yield Increase (percent)
Corn			
Stover	2.02	2.94	45
Grain	1.17	1.58	34
Tomatoes	11.8	17.7	50
Soybeans	0.83	1.37	66
Bush beans			
Planted May 12	2.36	2.86	21
Planted July 24	1.21	2.09	73
Lima beans	0.81	1.78	14

TABLE 18-11. Increase in Yield Due to Soil Warming, Growing Season 1970 (near Corvallis, Oregon)

Crop	YIELD Natural (tons/ hectare)	Heated (tons/ hectare)	Yield Increase (percent)
Strawberries	2.56	3.45	35
Tomatoes	12.8	16.4	28
Bush beans			
Planted May 30	1.84	28.6	56
Planted July 1	3.37	3.96	-10
Corn			
Stover	1.63	1.69	4
Grain	1.92	2.19	14
Corn (sub-irrigated)			
Stover	1.28	1.59	24
Grain	1.69	1.88	11

The variations in increases are primarily due to variations in the climatic conditions for the three growing seasons.

3. TVA Project. TVA conducted an experiment in which electric cables were used to simulate pipes carrying thermal discharges.[19] The experimental plots were 6 by 116 ft with electric cables buried at a depth of 9 in. and placed 8 in. apart. The soil in the plots were subirrigated with porous plastic pipes (Borg-Wasner Micro Por). These pipes were placed 8 in. deep and 36 in. apart. The results of soil

TABLE 18-12. Increase in Yield Due to Soil Warming, Growing Season 1971 (near Corvallis, Oregon)

Crop	YIELD Natural (tons/ hectare)	Heated (tons/ hectare)	Yield Increase (percent)
Corn			
Stover	1.55	1.89	22
Grain	0.69	1.24	85
Corn (high plant population: total dry matter)	1.24	1.65	34
Tomatoes	9.69	12.66	31
Strawberries	7.42	8.88	20
Bush beans	3.63	4.15	14
Green peppers	1.04	1.46	41
Broccoli	0.37	0.80	115
Ryegrass	1.13	1.33	18
Alta fescue	2.83	3.17	12
Sudan grass	1.97	2.75	40
Sorghum Sudan hybrid	1.95	2.87	47
Crimson clover	0.49	0.81	64
Lima beans	0.84	1.13	41

TABLE 18-13. Summary of Yield Increases, 1969, 1970, and 1971 Growing Seasons (near Corvallis, Oregon)

Crop	YIELD INCREASE 1969 (percent)	1970 (percent)	1971 (percent)
Corn			
Stover	45	14	22
Grain	34	13	85
Tomatoes	50	28	31
Strawberries	–	35	20
Bush beans	21	56	14
Soybeans	66	–	–
Lima beans	14	–	41
Green peppers	–	–	41
Broccoli	–	–	115
Ryegrass	–	–	18
Alta fescue	–	–	12
Sudan grass	–	–	40
Sudan hybrid	–	–	47
Crimson clover	–	–	64

TABLE 18-14. Effects of Soil Heating and Subirrigation, 1971 (Muscle Shoals, Alabama)

	YIELD, (tons/acre)			
	IRRIGATION		NO IRRIGATION	
	Heat	No Heat	Heat	No Heat
String beans	8.1	4.0	6.9	2.7
Sweet corn	9.0	5.0	6.2	3.2
Summer squash	30.6	26.9	20.6	17.6

heating and subirrigation are shown in Table 18-14.

Heating the soil more than doubled the yields of string beans on both irrigated and nonirrigated soils, and the yield of corn almost doubled. An increase in the yield of summer squash was also observed which was considerably higher during the early growth period, however.

Conclusions. Soil heating, frost protection, plant cooling, and subirrigation with thermal discharges are technically feasible. However, the systems have to be chosen on an individual basis, designed to fit the needs of the selected site, and operated in a prescribed manner. In addition, the economic feasibility must be considered, including market conditions for the produce and site selection and evaluation. A properly designed and operated system could result in increased productivity from the soil, improved quality of produce, and a general reduction in agricultural pollution.

Animal Farming. The discussion on the use of thermal discharges for animal farming will be restricted to animals reared and maintained in enclosures. Also, the discussion will be limited to theoretical applications only, since no definite examples of demonstration projects or commercial operations are available at this time.

Agricultural operations in the United States are tending towards increased mechanization to minimize labor-intensive activities. This trend is true for animal husbandry also, especially for egg and broiler chicken production and hog farming. The egg layers and broilers are reared in mechanized enclosures which, among other things, has reduced feed requirements from 5 lb to 2 lb for a 1-lb chicken. However, even though many improvements have been made in animal husbandry, little consideration has been given to environmental factors which could improve yields and quality of food.

Some energy is required to maintain comfort levels in cold and hot environments. Energy is lost to the environment by heat transfer, evaporation, etc., and this amount of energy has to be replenished for normal metabolic functions. If sufficient energy is not available, the metabolic functions are then disturbed and the growing process is affected. Environmental controls of animal enclosures provides an opportunity to use thermal discharges for monitoring comfort levels of the animals, both in winter and summer months.

Temperature is an important environmental factor that affects the normal life of animals, including humans. Effects of temperatures can be readily seen from Fig. 18-32,[15] where percentage of growth at optimum temperature versus temperature are plotted for milk production, chicken (broilers), and swine. The growth rate of these animals is reduced if the temperatures are lower or greater than the optimum temperature. Another important conclusion that can be derived from the graph is that the optimum temperatures (comfort zone of the animals) fall within the range of normal operating condenser cooling water temperatures.

Another consideration for temperature control, beside the comfort level, is the efficiency of feed-to-food conversion. "Feed," in this context, refers to the feed provided to the animals as a part of their daily diet, and "food" refers to the meat on the animals. The efficiency of feed conversion varies with temperature. For example, at 24°C (75.2°F), the conversion efficiency, based on pounds of weight gained per pound of feed consumed, is 39 percent for chicken as compared to the efficiency of 48 percent at 30°C (86°F), and 52 percent efficiency at 33.3°C (92°F). Table 18-15[15] shows the effect of temperature on feed conversion.

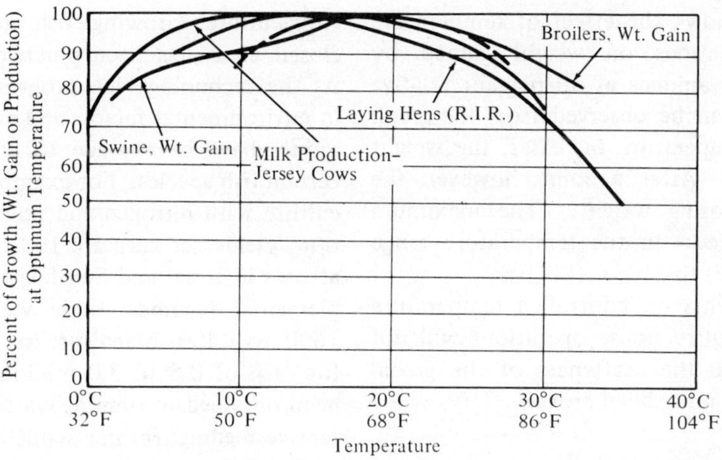

Fig. 18.32 Effect of temperature on growth or production of food animals.

TABLE 18-15. Effect of Temperature on Feed Conversion[15]

Animal	TEMPERATURE °C	TEMPERATURE °F	Pounds Feed/ Pound Weight Gain	Feed Efficiency[a] (percent)
Chickens	24	75.2	2.7	39
	30	86	2.1	48
	34	93.2	1.9	52
Hogs				
Individual hog	3	37.4	5.9	17
Group	3	37.4	4.3	23
Group	8	46.4	3.8	27
Group	15, 23	59, 73.4	3.4	29
Milk Cows (Jersey)	−13	8.6	2.1	48
	10	50	0.65	128
	38	100.4	0.55	180

[a]52 percent efficiency for chickens means 1 lb of growth/1.9 lb of feed.

There are no commercial size operations of this type in use at present. However, experimental enclosures, using thermal discharges for heating and cooling, are being planned at TVA's Browns Ferry plant. The system for cooling and heating feedlots and poultry houses is the same as that used in greenhouses. Research data show that feed conversion and efficiency for livestock and poultry production are greater in facilities with controlled temperatures. Results from Purdue University, presented in Fig. 18-33,[19] indicate that temperature extremes can affect the efficiency of hog production.

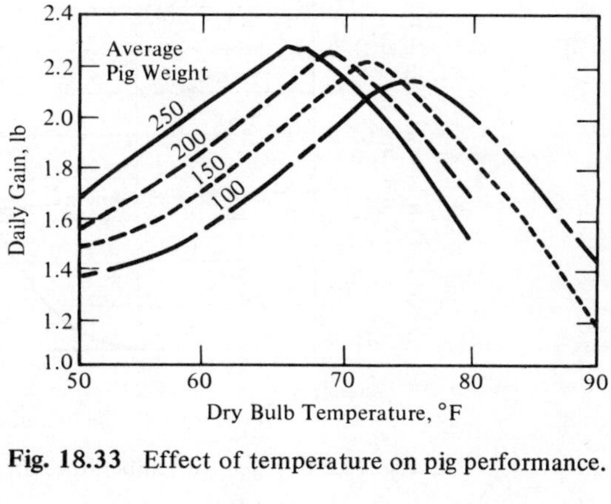

Fig. 18.33 Effect of temperature on pig performance.

The figure shows the effect of temperature (dry-bulb) variations on weight gained by various sizes of animals at 50 percent relative humidity. It can be observed from the graph that as the temperature increases, the weight gain increases. After a point, however, the animals start losing weight. The maximum weight gain occurs in the temperature range of 18.3°C (65°F) to 21.2°C (70°F).

The experiments on controlled temperature feedlot and poultry house operations will not be started until the usefulness of the greenhouse application has been proven.

Conclusions. Thermal discharges can be used to cool animal enclosures in summer and heat them in winter to maintain the optimum temperatures for maximum weight gain and for maximum feed conversion efficiency. However, to make best use of the technology, changes in animal farming practices will have to take place, from open-field farming to farming in enclosures. The economic feasibility and market opportunities must be evaluated prior to these changes. Along with the technological and economic feasibilities, the environmental impact of rearing animals in enclosures must be considered also. Environmental factors such as waste disposal and odors are of significant importance, as is the fact that animal wastes can be used productively to generate methane and produce fertilizer.

Aquaculture. Growing fish in artificially enclosed areas has been practiced for centuries. As the technology has progressed, refinements in environmental factors and management techniques have been made to increase yields of certain fish species. For example:[26] (a) in pond culture with nitrogen and phosphorus fertilization, yields for carp are 100 to 600 lb/acre/yr at sites in Israel and Southeast Asia. With supplemental feedings, these yields increased to 1600 to 2400 lb/acre/yr; (b) in Japan, yields for carp of 0.8 to 3.0 million lb/acre/yr have been obtained in running water culture with intensive feeding; (c) in a semicontrolled environment, catfish culture in ponds may yield 2000 lb/acre/yr, while under intensive culture and high environmental controls in flowing streams, yields as high as 2 million lb/acre/yr for catfish and trout could be achieved.

In the above examples, no attempt was made to maintain temperatures of the culture system in the optimum range for fish growth. The effect of temperature variations on fish growth is shown in Fig. 18-34.[26] It can be seen from the graph that the temperature for optimum growth of shrimp and catfish is about 30°C (86°F). Also, the growth rate increases as the temperature increases till the optimum temperature is attained. If the temperature increases beyond this point, the growth rate starts decreasing. The optimum temperatures, generally, are within the range of thermal discharges.

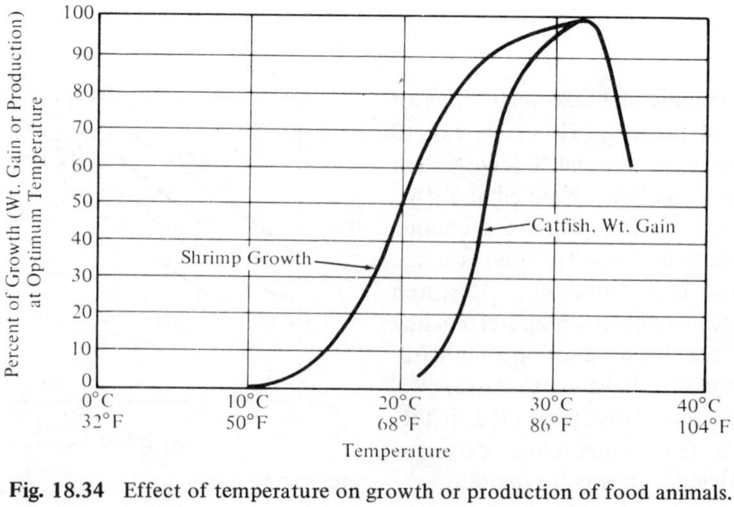

Fig. 18.34 Effect of temperature on growth or production of food animals.

Aquaculture can be carried out in a stationary system such as a pond or a dynamic system such as flowing water. The ponds can be natural ponds or artificially constructed ponds. In the natural pond, the stocking rate depends on the food naturally available. In the artificial ponds, it may be necessary to fertilize the pond with nutrients and food for the fish. It may also be necessary to aerate the water to provide the required amount of oxygen for the metabolic processes of the fish. As the fish density in the pond increases, it will be necessary to remove or change the water in the pond because of fish waste buildup. Fish waters exert biochemical oxygen demand, thereby rapidly depleting the oxygen supply. Moreover, the nutrients in fish waste may accelerate growth of aquatic life such as algae which may interfere with fish propagation.

A dynamic system could consist of natural lakes or rivers, cooling ponds, or cooling channels. Aquaculture can be practiced either by confining the fish in cages or by stocking the fish in multiple channels or raceways. In the cage culture, fish are stocked in cages which are placed in large bodies of water. Flowing water, occurring in streams or fast volume changing lakes, permits higher stocking density. However, fish diseases may be of concern in natural bodies of water. Raceways appear to be a better system, as far as environmental controls are concerned, because raceways are artificially constructed channels through which a desired flow can be maintained. Nutrients and food are added to the water at appropriate rates and concentrations. The danger of fish diseases is reduced in raceways because the stocked fish are not exposed to other life occurring normally in natural water bodies. In a dynamic system, dissolved oxygen is more uniformly distributed and the fish wastes are flushed away. The dynamic system, however, is more expensive than the stationary system but provides a better control over the environment.

Warmer water generally stimulates growth of aquatic life, increases the growth rates, and shortens the time required to produce a marketable product. However, for a successful aquaculture operation using thermal discharges, certain conditions and power plant operating practices must be taken into account. These factors are:[27]

1. Relationships among temperature, food utilization, and growth.
2. Relationship between temperature and carrying capacity of the water body. Carrying capacity of water involves: influences of water exchange rates, oxygen content, waste removal, nuisance growths of algae or fungi, and size, metabolism, and activity of organisms.
3. Differing thermal requirements of various life stages of aquatic organisms.
4. Influences of temperature on diseases and parasites as well as on growth.
5. Biological responses to temperature rises across the condensers and rapid effluent temperature changes due to daily plant operations.
6. Tolerances of organisms to biocides, such as chlorine, and corrosion inhibitors.
7. Tolerances of commercial markets for traces of radionuclides (discharged by nuclear power plants), which may be concentrated by aquatic organisms.
8. Impact of heated water enriched with waste products from the aquaculture facility on the receiving water.

A brief description of a few examples of aquaculture practices follows:

Aquaculture in the United Kingdom. In 1963, the White Fish Authority in the United Kingdom decided to conduct research on improving the quality and quantity of marine fish and the cost to the consumers.[28] The Authority selected species of marine fish for which markets existed, which were already in short supply, and which were likely to be available at a considerably higher cost in the future. To continue the research work of the Fisheries and Food Laboratory of the Ministry of Agriculture, the Authority constructed a hatchery at Port Erin on the Isle of Man, where good quality seawater was available.

The experiments conducted at this hatchery were on the eggs and larvae of plaice and sole during which a controlled environment was provided for the hatching of the eggs. The environmental controls included: control of water temperature, control of light during the gastrulation period through to hatching (to prevent growth of algae which increases mortality), and control of bacteria by means of antibiotics, especially those that attach to eggs. During the period of experimentation which lasted five seasons, an adequate supply of nutritionally balanced food was provided. The technique was perfected and used at the fish farms at the Hunterston Nuclear Power Station.

The next set of experiments was carried out on growing hatched fish in warm water from power plants, both coal-fired and nuclear. These involved observing the effects of temperature variation on plaice and sole. It was found that plaice grew well and continuously at 15-16°C and sole at 18-20°C. Above these temperatures, signs of distress, disease, physical abnormalities, and mortalities occurred. At lower temperatures, the size attained by the fish was less.

In the United Kingdom, power generating stations discharge water between 6 and 10°C above the ambient seawater temperatures. Temperatures can be maintained at desired levels by mixing unheated seawater with the thermal discharges. Therefore, thermal discharges provide a cheap source of warm water for fish propagation.

The next set of experiments was designed to observe the effects of chemicals used in power plant operations on fish. Generally, chlorine and ferrous sulfide are injected into the intake pipe to prevent settlement of mussels in the intake and condenser pipes. For these experiments, a continuous system for injecting chlorine into the intake pipes was developed, and the fish were grown in the thermal discharge of nuclear and coal-fired power plants. No attempt was made to maintain uniform temperature throughout the period of experiments, so the fish were subjected to the diurnal variations in the water temperatures.

However, the temperatures were higher than the ambient seawater temperatures. Both plaice and sole seemed generally to benefit from the higher temperatures, daily feeding, and care they received and did not appear to be affected by low levels of chlorine when they were exposed to free residuals of 0.02 to 0.1 ppm. However, during the summer months when the temperatures were high, the plaice showed signs of distress, but the sole were unaffected.

These experiments started in winter and utilized 9-month old fish, measuring between 5 and 9 cm in length. A year later most of the fish were of marketable size, 23 and 24 cm in length.

These encouraging results led the Authority to build a series of tanks 14.4 by 7.2 by 1.0 m deep for the purpose of carrying out such studies on a larger scale. Since their initial construction, the design of the tanks has been changed and the water inlet and outlet devices have been modified to streamline the operation. In addition, the design changes allow for better servicing of the tanks, for improved tank hygiene, better light controls, better water flow and temperature regulation, and more efficient chlorine and oxygen levels.

In the next set of experiments, 5 acres of the upper arm of Scottish Loch were enclosed to create a seawater pond. The pond was constructed in a manner which allowed interchange between pond water and open seawater. Initially the fish, artificially hatched plaice of 30 mm size, ate only the naturally occurring food. After some experimentation, it was decided that the best method of fish propagation was to use netting enclosures or netting cages and high density stocking, which requires artificial feeding. With these modifications, it was observed that the fish grew to marketable size in 18 months, as compared to 24 months required under the natural environment. The reasons for this acceleration in growth are: supplemental food which is nutritionally balanced, and year-round warm temperatures of the water.

The White Fish Authority has also experimented with other types of fish as well as mollusks and crustaceans. In most of these cases, the experiments began with young animals and not the eggs and larvae. They have had some successes, however, with the latter, although more experiments have to be conducted to perfect the marine fish husbandry techniques.

It can be concluded that with proper aquaculture techniques, fish can be stocked and raised under high density conditions and in controlled environments. The basic requirements are: artificial feeding with nutritionally balanced diets, optimum temperature for the fish species, light control to prevent the growth of undesirable aquatic life, maintenance of proper oxygen levels, removal of fish waste where needed, and proper designs of the hatchery and tanks. The British experience found that the growing time for plaice and sole was reduced to almost half under controlled conditions as opposed to their growth in nature.

Mariculture in Japan. Japan has the world's largest annual fish catch and in 1969 had a total catch of approximately 8.6 million tons. In 1967, the total was 7.8 million tons, valued at nearly $2 billion. This catch included a mariculture production of 470,000 tons, valued at nearly $300 million.

Experiments using thermal discharges have been conducted in Japan since 1964.[29] Some operators of power plants have utilized heated waters in aquaculture while others have been studying the effects of thermal discharges on inshore fisheries. At present, culture experiments are being carried out at five power plants as shown in Fig. 18-35. In addition, experiments are planned at two additional power plants—the Toyama power plant and the Tokai atomic power plant.

Shrimp Culture. Shrimp are cultured to the larvae stage in special larval culture tanks which can produce over a million postlarval shrimp in a 200-ton tank capacity. In normal commercial operations, when the postlarvae are about 20 days old (0.02 gm in weight), they are planted in farming tanks or ponds and are allowed to grow to commercial size (about 16 gm and up). The types of enclosures used in Japan are: fenced-in shore waters, which are cheapest to construct but are susceptible to damage by storms; ponds formed by dikes, where the pond water exchange takes place due to tidal differences (these are expensive to construct and could cover an area of about 20 acres/pond); and large, cement-walled ponds, where the water exchange is effected by expensive continuous pumping. The latter procedure produces the best shrimp color of any method. The usual spawning season of the wild stock of kuruma shrimp is from late April to early Sep-

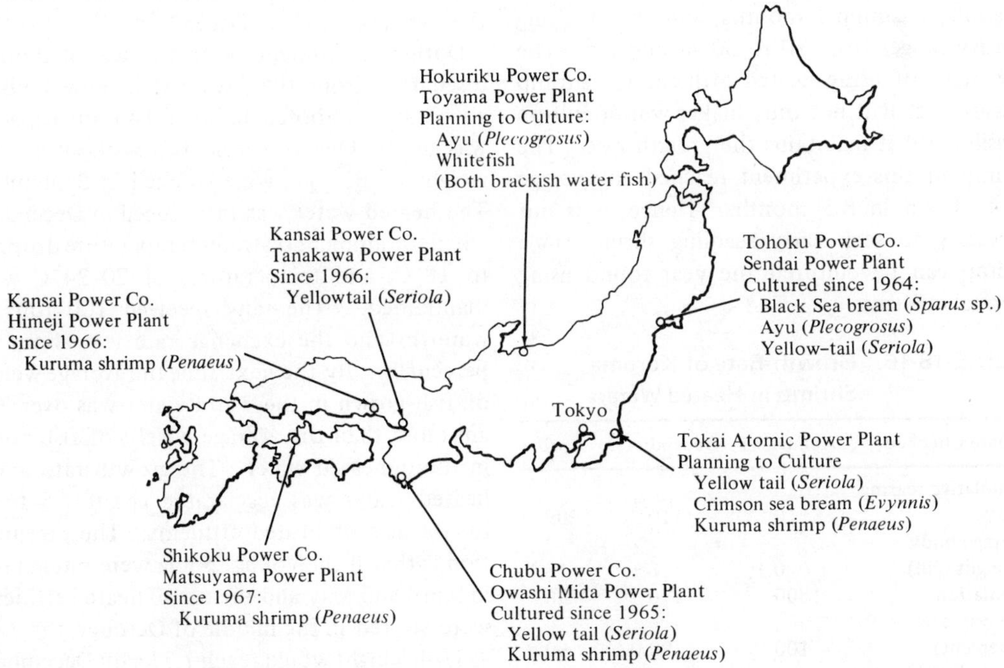

Fig. 18.35 The locations of power plants and species of marine organisms cultures using heated effluent water in Japan.

tember. It requires 6 months for the postlarvae (0.02 gm in weight) to reach marketable size. Farming requires pond water temperatures ranging from 20 to 33°C.

In 1968, experiments with the use of thermal discharges were started. A mixture of warm water and natural seawater was introduced into tanks at the flow rate of 9 tons/hr/tank. The temperature in the tanks was maintained at 20°C. The postlarvae (0.1 gm) shrimp were introduced into the tanks in August, the mixture of thermal discharges and natural seawater was introduced into the tanks in October, and the commerical size shrimp (18.1 gm) were harvested on April 1, 1969. The growth rate of shrimp is shown in Table 18-16.

The low survival rate was attributed to improper handling during monthly checks and may also be due in part to disturbance of the bottom sands. However, the growth rate obtained by using thermal discharges in the winter months was better than that of the ordinary commercially cultured shrimp, despite the high original stocking density of 54.7 shrimp/m^2. In commercial pond culture, without using the thermal discharges, the time required for the postlarvae to reach commercial size (16–17 gm) is about 6 summer months, and the stocking density ranges from 20 to 30 shrimp/m^2. The advantage of using heated effluent for shrimp culture is that it not only makes winter culture possible but it maintains the growth rate. The shrimp in this experiment reached an average of 18.1 gm in 8.5 months. Hence, it is not necessary to seek early seedling shrimp since shrimp can be cultured the year round using thermal effluent.

TABLE 18-16. Growth Rate of Kuruma Shrimp in Heated Waters

Date Checked	August 14	December 3	April 1
Cumulative rearing days	0	110	209
Average body weight (gm)	0.1	8.1	18.1
Population	1300	1170	790
Survival rate (percent)	100	90.0	60.8
Population density (per m^2)	54.7	49.3	33.3

Seriola Culture. Seriola is a migratory fish occurring along the coast of Japan which spawns in February, migrates northward, and returns in fall for spawning. In April and May, schools of young seriola appear along the southern coast of Japan and hide in the seaweeds until they are about 50 gm in weight, after which they leave the seaweed hideout. Fisherman catch the young and either culture them or sell them to the culturists.

There are three types of enclosures used for raising these young seriola to commercial size: narrow-mouthed bays enclosed by dikes, shallow water areas enclosed by net fencing, and cages. The maximum cage size is 20 by 20 by 10 m deep. Seriola cage culture is a good example of hyperintensive culture. The stocking density ranges from 115 fish of less than 25 gm to 7 fish weighing 1000 gm or more each in 1 cubic meter tank capacity. The rate of growth of the fish is higher in the second year than in the first year. Therefore, it is more profitable to hold the fish for sale into the second year. However, the lowest temperature that seriola can tolerate is 12–13°C, and cold water creates problems. As a result, most of the fish culturists prefer to sell their fish after the first year, around December.

During experiments with the use of thermal discharges from the Owase-Mida power plant, an octagonal roofed tank of 143-ton capacity was used. One thousand fish with an average weight of 625 gm were stocked in September. The heated water was introduced in December, when the ambient seawater temperature dropped to 18°C, and temperatures of 20–24°C were maintained. The tank received 100 tons of water/hr and the exchange rate was about 80 percent/hr. By the next June the average weight of fish grown in the warm water was over 700 gm more than the average weight of fish raised in the unheated water. The growth rate in the heated water was accelerated about 1.5 times by the use of heated effluents. The presumption is that if the young seriola were put in tanks in April and May and the use of heated effluents were started in the middle of October, the average fish weight would reach 1.7 kg by December.

It can be concluded that the developments in the utilization of heated effluents from power plants have shown that rapid growth of fish occurs during the usually nonproductive winter months with no loss in commercially desirable characteristics. Usually, the growth rates in the heated waters are many times higher than those in natural waters. In addition, eurythermal organisms may be cultured even in the summer months by blending the thermal effluents with normal seawater.

The importance of these findings is underscored by the fact that there are 68 power plants in Japan using seawater in cooling systems. The effluent from these plants is discharged at 6–8°C higher than the ambient seawater temperature. About 2 million tons of heated effluent are discharged in this country every hour; thus, about 15 billion kcal of energy are wasted every hour.

Aquaculture in the United States. There are several examples of aquaculture practiced in the United States where warm water from power plants is being utilized. Some of them are briefly described below.[26]

Rainbow Trout. The Thousand Springs Trout Company of Buhl, Idaho has a large commercial intensive aquaculture farm that produces several million pounds of trout a year (4–5 million pounds in 1969). The trout is cultured on a year-round basis because of a constant supply of 250,000 gpm of spring water at a constant temperature (60°F; 15.5°C). About a quarter of the flow is diverted and distributed into channels where high density culture is practiced. The stocking density is about 2 lb/ft^3. Nutritionally balanced food is provided to the fish and the food conversion ratio is 1.5 lb of dry feed/lb of wet fish produced. The fish wastes are flushed away due to the flowing water in the channels. All of these factors combined produce high quality fish.

Oyster Farming. Long Island Oyster Farms of Northport, Long Island operate a commercial oyster farm utilizing the thermal effluent from a Long Island Lighting Company power plant. The normal growing period of 4 years has been reduced to 2.5 years by selective breeding, spawning, larvae growth, and seeding oysters in the hatchery. Oysters are cultured in the discharge channel of the power plant for a period of about 4 to 6 months and then the cages are removed to the colder waters on the eastern part of the Long Island Sound.

Catfish Farming. Catfish has been cultured in cages and in raceways by utilizing thermal discharges from power plants. Growth rates equivalent to 100 tons/acre/yr have occurred in cages at the Texas Electric Service Company plant at Lake Colorado City, Texas. A research project is underway by the Trans-Tennessee Industries of Nashville, Tennessee at the TVA plant in Gallatin, Tennessee. Thermal discharges are circulated through nine out of ten channels which are 4 ft wide, 4 ft deep, and 50 ft long. The channels are covered to prevent algae growth. Different stocking densities are being tried. Nutritionally balanced food is provided to the catfish culture. Yields of up to 1000 tons/acre/yr have been extrapolated in several of the raceways. The company is planning a 230-channel facility that would provide 60,000 lb of dressed catfish per week.

Fish Farming in Iceland. Fish farming in Iceland started about 25 years ago and now over 300,000 smolt are being produced annually.[16] The largest fish culture station, constructed in 1961, is the Kollafjord Experimental Fish Farm which is operated by the Ministry of Agriculture. Its main purpose is rearing salmon young to the smolt stage for liberation both inside the farm, from which they can migrate to the sea, and into the inland rivers.

The farm uses 7 liters/sec of 70°C geothermal water for indirect heating in the hatchery, where a 12–14°C range is maintained. These facilities are designed for hatching 3 million salmon eggs, and the yearly production of smolt is about 160,000.

Conclusions. In summary, fish, both freshwater and marine, have been successfully raised in a controlled environment. The factors controlled include: water temperature, light, feed, bacteria, and waste removal. There are definite requirements for the design, operation, and management of fish farm and aquaculture facilities and these should be viewed in the same light as these considerations are for any other industry. There are also problems of waste managements, but they are no more severe than those of other industries or municipalities. And, research and commercial ventures have proven that there are definite benefits to be derived from the use of thermal discharges in these operations.

As is the case with any industry, the market for the product has to be evaluated and established and the technological feasibility determined. Technological considerations include, in addition to the environmental controls enumerated earlier, the location of the fish hatchery, ambient air and water conditions, the species of fish prevalent in the location, the type of fish-growing system to be utilized (tanks, ponds, raceways, or diked bays), the availability of nutritionally balanced food for the fish, the optimum temperatures for the various stages of fish life (from eggs to commercial size), and the waste disposal mechanism that will prevent oxygen depletion in the tanks.

The economic feasibility of aquaculture is determined or affected by the availability of a food supply, the cost and quality of food, and other factors that come into play until the fish is on the consumers' tables. These factors should be evaluated at the planning stage.

In the overall concept of this use of thermal discharges, the overriding factors are the quality and quantity of food it can provide to the world and the maximum utilization of the hundreds of billions of BTUs per hour which otherwise would be discarded as waste.

Space Heating. Space heating in this context refers to heating homes, buildings, and other structures with heated water or steam. This type of heating is practiced in Europe, Iceland, and the United States. The methods used and the source of energy, however, may vary. In the general system, hot water is piped into the structure through a recirculating system–i.e., the hot water is withdrawn from the main pipe, used for heating, and then returned to the main pipe.

There are technical difficulties involved in this type system which include: source of hot water, transmission of hot water without loss in temperature until the intended use is accomplished, and, finally, the disposal of warm water. Initial temperatures required to heat residential buildings range from 180°F (82°C) to almost 212°F (100°C). As a result, the temperature of water to be disposed of will vary with use and may be considerably higher than the temperature of thermal discharges from power plants. The warm water can be disposed of through various means such as greenhouses or agriculture (crop farming or animal husbandry), aquaculture (freshwater fish or marine fish), or a combination of any of these systems, or the warm water may be returned to the heat source to start the cycle.

Two suggested methods of steam extraction from power plants at higher temperatures indicate that it may be possible to meet the requirements of hot water or steam users and, at the same time, reduce the demands on other fuels.

A report prepared by the Oak Ridge National Laboratory[30] points out that presently 40 percent of the energy input to a conventional power plant is converted to electricity and the remaining 60 percent, excluding any losses of heat from the fuel prior to its reaching the turbine, is discarded as thermal discharges at a temperature of 95°F (35°C). One suggestion for extracting more heat from these facilities is to extract a portion of steam from the turbine after it has generated a considerable amount of electricity. In this case, 35 percent of the heat input would be withdrawn as steam at 250°F (121°C), 35 percent of the energy input would generate electricity, and the remaining 30 percent would be discarded to the condensers. The overall conversion of input energy to electricity would decrease from 40 to 35 percent, but the overall energy utilization would increase from 40 to 70 percent. The steam, withdrawn at 250°F, could be utilized for beneficial purposes.

The second scheme for increasing overall energy utilization is to employ a back-pressure turbine and remove all the steam from the turbine after its temperature reaches 250°F. In this case, the efficiency of electricity generation would be reduced to 30 percent, but the total

TABLE 18-17. Temperature Requirements

Energy Component	Supply Temperature °F
Low temperature heat	
Space heat	200
Domestic hot water	200
Absorption air conditioning	250
Water distillation	265
Industry	300
Snow and ice melting	212
Transportation	300
Waste heat	
Secondary sewage treatment plants	95
Agriculture–aquaculture	95

energy use would approach 100 percent. These approaches merit consideration because of their potential for conservation of energy, reduction of air pollution, and more efficient overall operation. Temperature requirements for various uses of energy are shown in Table 18-17.[30]

District Heating in Iceland. Iceland has no deposits of fossil fuels but has an abundance of geothermal energy. The first attempt made to use this energy early in the twentieth century was for heating a single farmhouse. In 1928, the first borehole was drilled in the vicinity of Reykjavik which produced 40 liters/sec of 40°C hot water. This hot water was used to heat 70 homes, a swimming hall, an open-air swimming pool, and a public school by piping it through a 3-km pipeline and feeding it into a district heating system.

The success of this operation led to more boreholes in the Reykir thermal springs area in 1933, about 18 km outside the city of Reykjavik. This area produced about 200 liters/sec of 86°C water which was distributed to the city's district heating system through a network of underground pipes, serving 2300 houses with about 30,000 inhabitants and all the public buildings.

A total of 72 boreholes were drilled which produced a total of 330 liters/sec of 86°C water. These boreholes varied in depth from 300 to 700 m.

By the end of 1969, geothermal heating systems were installed in other areas of Iceland and a total of 81,000 people, or 40 percent of the total population of the country, lived in houses heated by geothermal energy. In addition, 80 swimming pools are heated either as a part of a district heating system or as an independent unit. The details of the Reykjavik District Heating System are as follows:

Climatic Data:

Mean temperature of the year	+ 5.0°C
Mean temperature of the year (warmest month–July)	+11.2°C
Mean temperature of the year (coldest month–January)	- 0.4°C

Available Heat Resources (December, 1969):

Reykir geothermal area– 1000 m^3/hr at 80°C	40 Gcal/hr
Reykjavik geothermal area– 1700 m^3/hr at 119°C	135 Gcal/hr
Own peak power boiler plants (oil-fired)	30 Gcal/hr
National power company boiler plants (oil-fired) (available at electrical off-peak hours only)	20 Gcal/hr
	Total 225 Gcal/hr

Storage Capacity:

Tanks (1943)	8000 m^3
Tanks (1969)	26,000 m^3

Heat Load:

Volume of houses connected (December, 1969)	10.6×10^6 m^3
Number of houses connected (December, 1969)	8700
Heat load at -10°C outside and +20°C inside	190 Gcal/hr
Specific load at -10°C outside and +20°C inside	19 Gcal/hr

System Data:

Installed horsepower in pumping plants	5115 hp
Area served by the distribution system	11.2 km^2

Length of Pipelines:

Collection mains	14.2 km
Supply mains	29.1 km
Street mains	125.2 km
House connections	120.2 km

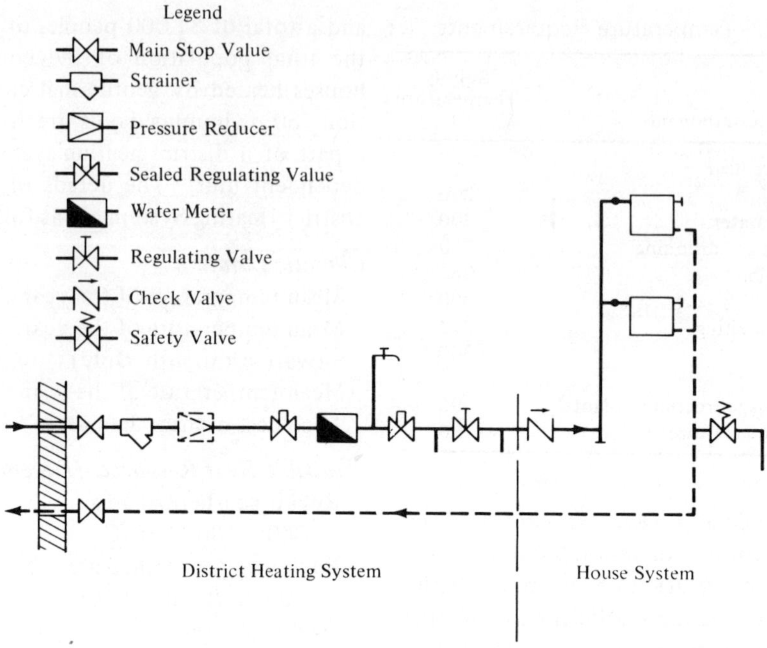

Fig. 18.36 House connection–dual pipe system.

Average density of population	643 inhabitants/km^2
Average load density	17 Gcal/hr/km^2

Yearly Heat Production:

Geothermal energy (1968)	960 Tcal/yr
Peak power station (1968)	80 Tcal/yr

Figures 18-36 and 18-37 show house connections.

The economic feasibility of using geothermal energy for heating or cooling depends on whether it can compete with other available sources of energy. Such comparison, however, should be based on the total cost to the ulti-

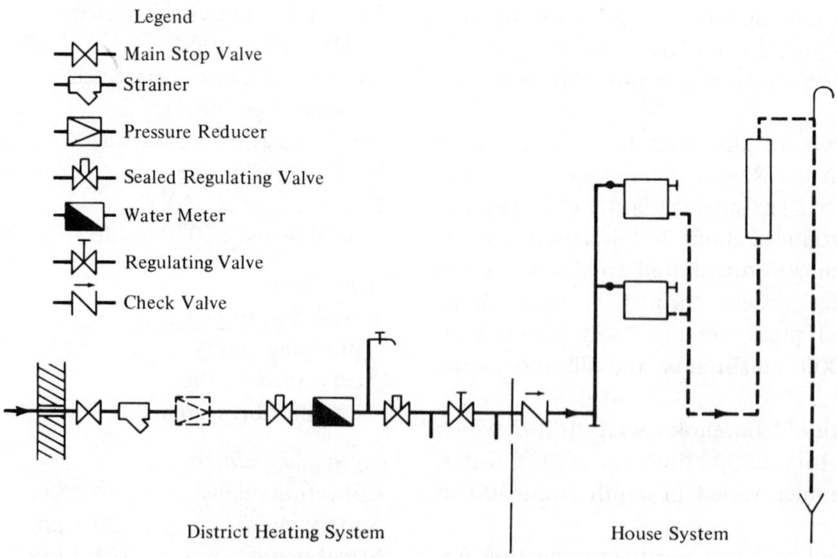

Fig. 18.37 House connection–single pipe system.

mate user per unit of net energy utilized. This comparison should include: costs of the entire system, desired return on capital investment, operating costs, annual load factors, energy utilization efficiency, etc.

The principal cost factors in the use of geothermal energy are:

1. Drilling costs per unit of energy production (\$/Gcal-hr).
2. Temperature of the available geothermal field to the center of gravity of the market.
3. Load density of the market (Gcal/hr/km^2).
4. Annual load factor of the system.
5. Power of the system (Gcal/hr).

The capital investment for the Reykjavik system is:

Heat production	\$29,400/Gcal/hr
Distribution system	\$58,000/Gcal/hr
Total	\$87,400/Gcal/hr

Estimated replacement cost of the system (225 Gcal/hr) is about \$17 million. The energy price paid by the customers is \$0.16/m^3 of 80°C water. Based on average utilization, this corresponds to \$3.45/Gcal. The Reykjavik District Heating System saves about 150,000 tons of oil annually. This oil would have to be imported and would cost consumers about 40 percent more.

Additional Applications. In addition to the uses described so far, warm water could be used for a variety of other purposes. These include: sewage treatment; deicing and defogging of airports, runways, and streets; desalination; and industrial uses. A brief description of these uses follows.

Sewage Treatment. The advantages of higher temperatures in sewage treatment are:[31]

1. Increased rates of chemical and biological reaction.
2. Increased efficiency (greater percentage destruction of volatile solids than at lower temperatures).
3. Improved solids-liquid separation.
4. Increased destruction of pathogens.
5. Nitrification-denitrification.
6. Potential economic advantages in operating a combined cooling and sewage treatment system.

These advantages are primarily applicable to the activated sludge system, where the secondary system is continuously aerated to maintain oxygen levels in the wastewaters. The combined system must be properly designed to handle both the municipal wastewater and the cooling water from power plants. The design should provide for: temperature controls for an optimum environment for the bacteria; a supply of additional oxygen as required to meet the demand of increased bacterial activity; maintenance of at least minimum temperatures for nitrification and denitrification; removal of chemicals used in the power plant operations that may be toxic to the types of bacteria needed for sewage treatment; and routine power plant shut-downs, either for maintenance or refueling of nuclear reactors, and, conversely, shut-down of the waste treatment facilities. These design factors are not insurmountable and, with proper planning, a system could be developed that could be useful to a municipality as well as a power company.

Another suggested application is evaporation of the wastewater in a vertical tube evaporator using exhaust steam from a power plant.[15] It is indicated that a city could reduce the volume of water withdrawn and treated for its potable water supply by judiciously selecting the amount of water to be evaporated and mixed for reuse with the remaining quantity of adequately treated and filtered wastewater.

Airport Defogging. Waste heat in the form of hot water or steam could be piped into heat exchangers equipped with fans. The hot air from the exchangers would then be blown across the approachways and runways. The heated air would rise and heat the fog droplets, causing them to evaporate.

There are two disadvantages to this type of a system: first, for dissipating fog from LaGuardia Airport in New York, approximately 700 megawatts of heat would be required from Consolidated Edison's Astoria plant; the installation of this system would cost approximately \$3 million. Second, New

York averages only 53 foggy days in a year, and as a result this system would be very expensive. However, this type of use is feasible, at least in theory; the details would have to be worked out based on location of the runways and power plants and the number of foggy days experienced at the airports.

Deicing of Runways and Roadways. Theoretically, it is possible to deice airport runways and streets by passing warm water or steam from power plants through pipes buried below the pavement. On the basis of melting 1 in./hr of ice, the thermal discharge from a 1000 megawatt nuclear power plant could deice 2 mi^2 of runway.

Conclusions. There are possibilities of using thermal discharges from power plants for waste treatment, defogging, or deicing. In the practical application of these uses, technological and economical feasibilities will be the governing factors. After the feasibility has been investigated, the overriding factor will be the benefits to society as a whole. For example, the question that may need careful evaluation is: is the safety of passengers on foggy days worth $3 million per runway at LaGuardia Airport?

Summary. In this section, a variety of uses for thermal discharges from power plants are described. It is not intended to suggest that these uses will solve all thermal pollution problems. The intent is to suggest means for improving the overall use of energy resources and thereby enhancing the quality of life. The choice of any particular use or combination of uses will depend on the site, the resources of the site, and the needs. It may be quite practical in some locations to use thermal discharges for district heating, greenhouses, and aquaculture, all as a part of one total system. In any event, the economics and feasibility of these uses should be rigorously examined before dissipating heat to the environment. The environment has a great capacity for waste assimilation, but once the energy resource is discarded, it is lost forever.

REFERENCES

1. *Handbook of Chemistry and Physics*, 42nd ed. (1961–62).
2. Vogely, W. A., and Mack H. B., "A 30-Year Review of Water Supply and Demand," National Conference on Complete WateReuse (April 23–27, 1973).
3. Parker, F. L., and Krenkel, P. A., "Thermal Pollution: Status of the Art," Report No. 3, Vanderbilt University (1969).
4. Kolflat, T., Statement before the Subcommittee on Air and Water Pollution of the Committee on Public Works, U. S. Senate, Thermal Pollution–Part I (1968).
5. Mathur, S. P., "Temperature Prediction of Streams Receiving Heated Discharges," M.S. thesis, University of Cincinnati (1968).
6. Federal Water Pollution Control Administration, "Industrial Waste Guide on Thermal Pollution," FWPCA, U.S. Dept. of the Interior, Pacific Northwest Water Laboratory, Corvallis, Oregon (1968).
7. *Standard Methods for the Examination of Water and Wastewater*, 11th ed., APHA, AWWA, WPCF, New York (1960).
8. Mathur, S. P., "Waste Heat from Steam-Electric Generating Plants Using Fossil Fuels and Its Control," FWPCA, U.S. Dept. of the Interior (1968).
9. Edinger, J. E., and Geyer, J. C., "Heat Exchange in the Environment," Cooling Water Studies for Edison Electric Institute, Johns Hopkins University (June, 1969).
10. Zeller, R. W., "Cooling Water Discharge into Lake Monona," Ph.D. thesis, University of Wisconsin (1967).
11. Boies, D., Levin, J. E., and Baratz, B., "Technical and Economic Evaluations of Cooling System Blow-down Control Techniques," EPA/NERC Pacific Northwest Environmental Reasearch Laboratory, Corvallis, Oregon, EPA Contract No. 68-03-0233 (1973).
12. Hittman Associates, Inc., "Methods of Control of Thermal Discharges," *Effects and Methods of Control of the Discharge*, Report to Congress (Nov., 1973).
13. Rossie, J. P., and Cecil, E. A., "Research on Dry-Type Cooling Towers for Thermal Electric Generation; Part I," R. W. Beck and Associates for Water Quality Office, EPA (Nov., 1970).
14. Beall, S. E., Jr., and Samuels, G., "The Use of Warm Water for Heating and Cooling Plant and Animal Enclosures," Oak Ridge National Laboratory, Oak Ridge, Tennessee, ORNL-TM-3381 (June, 1971).
15. Beall, S. E., Jr., "Agricultural and Urban Uses of Low-Temperature Heat," *Proceedings of the Conference on the Beneficial Uses of Thermal Discharges*, New York State Dept. of Environmental Conservation, Albany, N. Y. (1970).

16. Matthiasson, M., "Beneficial Use of Heat in Iceland, Technical and Economical Aspects and Future Prospects," *Ibid*.
17. Hodges, C. N., *et al*., "Waste Heat Use in Controlled-Environment Greenhouses," *Ibid*.
18. Jensen, M. H., "The Use of Waste Heat in Agriculture," Waste Heat Utilization, *Proceedings of the National Conference*, Gatlinburg, Tennessee (Oct., 1971).
19. Williams, G. G., "TVA Programs–Waste Heat Utilization in Greenhouses and Other Agriculturally Related Projects," *Ibid*.
20. Bond, B. J., *et al*., "TVA Projects–Beneficial Uses of Waste Heat," *Ibid*.
21. "The 1970 Annual Report of the Thermal Water Demonstration Project," EPA and Eugene, Oregon Water and Electric Board (1970).
22. Boersma, L., "Warm Water Utilization," *Proceedings of the Conference on the Beneficial Uses of Thermal Discharges*, New York State Dept. of Environmental Conservation, Albany, N. Y. (1970).
23. Boersma, L., *et al*., "Use of Reactor Cooling Water from Nuclear Power Plants for Irrigation of Agricultural Crops," Water Resources Research Institute, Oregon State University, Corvallis, Oregon (Apr., 1972).
24. Miller, H., Jr., "The Thermal Water Horticultural Demonstration Project at Springfield, Oregon," *Proceedings of the Conference on the Beneficial Uses of Thermal Discharges*, New York State Dept. of Environmental Conservation, Albany, N.Y. (1970).
25. Price, B. L., "Thermal Water Demonstration Project," Waste Heat Utilization, *Proceedings of the National Conference*, Gatlinburg, Tennessee (Oct., 1971).
26. Yee, W. C., "Thermal Aquaculture: Potential and Problems," Waste Heat Utilization, *Ibid*.
27. Contant, C. C., "Biological Limitations on the Use of Waste Heat in Aquaculture," *Proceedings of the Conference on the Beneficial Uses of Thermal Discharges*, N.Y. State Dept. of Environmental Conservation, Albany, N.Y. (Sept. 1970).
28. Richardson, I. D., "Development of Systems of Marine Fish Cultivation In the United Kingdom," *Ibid*.
29. Yang, W. T., "Mariculture in Japan Using Heated Effluent Water," *Ibid*.
30. Miller, A. J., "Urban Use of Thermal Energy From Steam-Electric Plants," Waste Heat Utilization, *Proceedings of the National Conference*, Gatlinburg, Tennessee (Oct., 1971).
31. Boser, L. J., and Grounds, H. C., "A Proposed WateReuse Concept Wherein the Treatment of Raw Municipal Sewage and the Closed Loop Cooling for a Large Power Plant are Combined," National Conference on Complete WateReuse (Apr. 23–27, 1973).
32. Bell, R. A., *et al*., "Combination Urban-Power Systems Utilizing Waste Heat," *Proceedings of the Conference on Beneficial Uses of Thermal Discharges*, N.Y. State Department of Environmental Conservation, Albany, N.Y. (Sept., 1970).

19

The Water Pollution Control Act (As Amended in 1972, P.L. 92-500)

J. I. Bregman
Gene Beeland
WAPORA, Inc.

ENGINEERING SIGNIFICANCE OF THE FEDERAL WATER POLLUTION CONTROL ACT (P. L. 92-500)

It is fitting that a book devoted to water resources problems should include a chapter on water pollution control legislation. Much of the heightened activity in the drive to clean up U.S. waters is due to the antipollution laws enacted by the Congress. It is therefore imperative to understand the laws, their philosophies, the requirements they set forth, and the changes which those requirements are likely to bring about. This knowledge enables the engineer to gauge the magnitude of the effort, as well as the types of technology that will be required to comply.

While there is much legislation that might be considered in this chapter, the one that is by far the most critical is the Federal Water Pollution Control Act, as amended in 1972. A detailed discussion of that act will therefore be presented here on an item-by-item basis. Titles and section headings used in the act will be cited so that the discussion may be readily compared with the act itself.

TITLE I—RESEARCH AND RELATED PROGRAMS

Sec.101—Declaration of Goals and Policy. This section of the act establishes as a national objective the restoration and maintenance of the chemical, physical, and biological integrity of the nation's waters. This is the heart and philosophy of the act.

The achievement of this objective is supported by broad national goals and policies. The goals are:

1. Elimination of pollutant discharge by 1985.
2. Achievement of water quality, where attainable, which provides for the protection

and propagation of fish, shellfish, and wildlife, and which allows body-contact water recreation by July 1, 1983.

The national policies require:

1. Prohibition of toxic pollutant discharge.
2. Provision of federal financial assistance for construction of public waste treatment plants.
3. Development and implementation of area-wide waste treatment plans to assure adequate control of pollution sources in each state.
4. Federal assistance to major research and demonstration efforts to develop technology necessary to eliminate pollutant discharge into the navigable waters, waters of the contiguous zone, and the oceans.
5. Recognition of the primary responsibility and right of the states to prevent water pollution and to plan for the development and use of land and water resources.
6. Provision of federal technical services and financial aid to state and interstate agencies and municipalities.
7. Presidential action to obtain meaningful effort on the part of foreign governments to prevent pollution in their waters and in international waters at least to the same extent that the U.S. pursues its goals.
8. Drastic minimization of paper work and interagency decision procedures and the best use of available manpower and funds so as to prevent needless duplication and unnecessary delays at all levels of government.

The distinction between "goals" and "policies" arose because of controversy in the Congress over "zero pollution" by 1985 and water quality by 1983 to support aquatic life and recreation. These issues were adopted as goals.

It is the intent of Congress that they be achieved. However, the legislative history of this act recognizes that situations may exist where the goals are unattainable in relation to specific water bodies for specific pollutant discharges. In such instances, the Environmental Protection Agency (EPA) has discretionary authority. On the other hand, it is directed to carry out the policies without exception.

Sec. 102–Comprehensive Programs for Water Pollution Control. This section authorizes the administrator of EPA to develop programs designed to eliminate pollution of navigable waters and groundwaters. It also provides for 50 percent matching federal-state grants for river basin planning.

It permits storage for regulation of streamflows from reservoirs, but specifically bans dilution as an alternative to waste treatment. However, the act recognizes streamflow augmentation as a useful means of reducing the environmental impact of runoff from nonpoint sources, as well as for recreational, navigational, and other values. The Corps of Engineers, the Bureau of Reclamation, or other federal agencies are to determine the need and value of storage for streamflow purposes other than water quality control. The latter determination rests with the administrator of EPA, and his views are to be included as part of any report or presentation to Congress proposing authorization or construction of a reservoir.

Sec. 103–Interstate Cooperation and Uniform Laws. This section allows states to enter into interstate compacts, subject to approval by Congress, and establishes a policy for active federal promotion of cooperative efforts among states. Such efforts include programs to promote model legislation and uniform laws.

Sec. 104–Research, Investigation, Training, and Information. The administrator is granted general authority to participate in and encourage research, investigations, training, demonstrations, surveys, and studies relating to the causes, effects, extent, prevention, reduction, and elimination of water pollution. Research proposals and progress are to be evaluated by advisory committees of recognized experts appointed by the administrator.

Broad authority is also extended to the administrator to fund, through grants to and contracts with appropriate public and private agencies, programs as needed to implement the requirements of this section. In addition, there is specific funding authority applicable to various studies in which the participation or cooperation of governmental or private entities is mandated or desired.

The administrator is specifically required to conduct studies on:

1. The removal of oil and hazardous substances from water.
2. The disposal of waste (used) oil.
3. Pollution in lakes, and, specifically, the Great Lakes.
4. Pesticides and the control of their release into the environment.
5. The effects of thermal discharges and methods to minimize their adverse effects and maximize beneficial effects.
6. Pollution problems, including sedimentation, in estuaries and estuarine zones including consideration of demographic trends, exploitation of mineral resources and fossil fuels, land and industrial development, navigation, and erosion control.
7. Pollution from agriculture.
8. Pollution problems from sewage in rural areas where conventional sewage collection systems are impractical and soil conditions preclude the use of septic tank and drainage field systems.
9. The harmful effects of pollution on human health and welfare, and techniques for measuring these effects.
10. The social and economic costs and benefits of activities subject to regulation under the act.

Among the university-oriented financial assistance programs authorized by this section are:

1. Research fellowships.
2. Research on freshwater and aquatic ecosystems.
3. "River study centers" at regionally-located universities to conduct studies on the nature of river systems, including their hydrology, biology, ecology, economics, the relationship between river and land uses, and the effects of developments within river basins on the river systems, as well as the value of water resources and water-related activities.

Experimental facilities will also include six field laboratories at specified locations in the U.S. and the National Marine Water Quality Laboratory authorized by the act.

In connection with waste treatment the administrator is to:

1. Conduct comprehensive research on this problem.
2. Develop and demonstrate practicable means of treating municipal sewage and other waterborne wastes.
3. Investigate devices and methods to reduce sewage flows such as minimizing unnecessary water consumption.
4. Conduct pilot projects on methods of treating sewage and other liquid wastes in combination with treatment and disposal of solid wastes.

In the important area of ensuring an adequate supply of trained manpower in water pollution control programs, the administrator is to:

1. Finance pilot programs to train personnel for operation and maintenance of treatment works.
2. Forecast the supply of and demand for control agency manpower.
3. Make grants for training projects.
4. Establish EPA research fellowships.
5. Train personnel of public agencies and others who are technically qualified.

The administrator is also required to develop a national water quality surveillance system for monitoring the quality of navigable waters and groundwaters, the contiguous zone, and the oceans. To the extent practicable, this surveillance effort is to utilize the resources of the National Aeronautics and Space Administration, the National Oceanic and Atmospheric Administration, the Geological Survey, and the Coast Guard. The Coast Guard is given the additional responsibility for research on the development of shipboard waste treatment units.

The administrator is authorized to develop prototype pollution control devices and methods for evaluating the use of augmented streamflows to control pollution not susceptible to other means of prevention.

The administrator is subject to specified reporting requirements on the research and development projects which variously direct him to provide information to the Congress, to the states, or to the public.

Sec. 105—Grants for Research and Development. This section authorizes the administrator to conduct in-house demonstration projects or make grants to state, municipal, or regional agencies for projects designed to eliminate pollution reaching navigable waters through stormwater runoff or industrial activity. It further authorizes him to undertake a model-river project to demonstrate advanced pollution control and in-stream enhancement techniques.

The administrator may also demonstrate methods of controlling agricultural pollution, preventing and abating discharge from combined sewers, and new or improved methods of joint treatment systems for municipal and industrial wastes. He is to give priority to the study of the impact of specific discharges on receiving water quality.

At least 10 percent of the funds appropriated over 2 years for implementing this section must be allocated for research and demonstration of new or improved methods to control agricultural pollution.

Sec. 106—Grants for Pollution Control Programs. This section authorizes grants to state and interstate agencies for carrying out programs for the prevention, reduction, and elimination of pollution. It specifically includes grants for administering enforcement programs. Allotment of the funds for each fiscal year will be based on the extent of the pollution problems in each of the states.

Payments will be made in accordance with an agency's allotment or the reasonable costs of developing and carrying out its control program during the fiscal year, whichever is lesser. Fiscal 1971 is set as a baseline for the expenditure level by state and interstate agencies of nonfederal funds for the recurrent expenses of their control programs. If an agency reduces its spending below this amount, the administrator will discontinue grants to that state or agency. Further, the act prohibits the administrator from making a grant to any state, beginning with fiscal 1974, if the state:

1. Does not utilize, as part of its program, appropriate devices, methods, systems, and procedures necessary to compile and analyze data on the quality of navigable waters and, to the extent practicable, groundwaters.
2. Does not have authority comparable to that in Section 504 of the act (relating to emergency situations) and adequate contingency plans to implement such authority.

Finally, to be eligible for a program grant, a state or interstate agency must file a summary of its current program and a description of a proposed program with the administrator within 120 days after the enactment of the bill. The administrator may re-allot any funds authorized by this section not expended during a fiscal year.

Sec. 107—Mine Water Pollution Control Demonstrations. The administrator is authorized to conduct, in cooperation with the Appalachian Regional Commission, demonstration projects for control of mine water runoff and related water pollution problems. It calls for a study on the feasibility of using sewage sludge to prevent mine water pollution and to restore affected lands to usefulness for beneficial purposes. Funds authorized are to be available until expended.

Sec. 108—Pollution Control in Great Lakes. The administrator may carry out projects, demonstrate new methods, and develop preliminary plans for the elimination or control of pollution within all of the watersheds of the Great Lakes. The act requires states or other political subdivisions participating in the study to pay at least 35 percent of the actual project costs. Funds for the study will be available until expended.

In addition, the Corps of Engineers is directed to design and develop a $5 million demonstration wastewater management program for Lake Erie.

Sec. 109—Training Grants and Contracts. Grants to colleges and universities are authorized for programs to train students in the design, operation, and maintenance of sewage treatment plants and other water pollution control facilities. It provides for the construction costs of treatment works to carry out the training program and sets the limit for the federal cost for each training program at $250,000.

Sec. 110–Application for Training Grant or Contract; Allocation of Grants or Contracts. This section sets forth procedures for making these applications and prescribes ground rules for allocating the grants and contracts.

Sec. 111–Award of Scholarships. This section provides for the distribution of scholarships for undergraduate study leading to an occupation involving the operation and maintenance of treatment works and outlines the program requirements of participating institutions. Any person awarded such a scholarship will be required by regulation to remain in a relevant occupation for a fixed period of time.

Sec. 112–Definitions and Authorizations. This section defines terms used in Sections 109 through 112 and authorizes funds for administering them.

Sec. 113–Alaska Village Demonstration Projects. Provision is made for programs to demonstrate methods for central community facilities for safe water and elimination or control of pollution in those native villages of Alaska without such facilities.

Sec. 114–Pollution Control in Lake Tahoe. A study is mandated to redefine the optimal relationships between the various governmental bodies involved in the preservation of the ecology of Lake Tahoe.

Sec. 115–In-Place Toxic Pollutants. The administrator is directed to identify and locate in-place pollutants, with emphasis on toxic pollutants in harbors and navigable waterways, and authorize contracts for the removal and appropriate disposal of such materials from critical port and harbor areas.

TITLE II–GRANTS FOR CONSTRUCTION OF TREATMENT WORKS

Sec. 201–Purpose. Plans for treatment works for which grants are requested must provide for the application of the "best practicable" technology, including recycling of water and confined disposal of pollutants. Preference is for treatment on an area-wide basis for the control of all point and nonpoint sources of pollution, including in-place or accumulated pollution sources. Construction of revenue-producing facilities which provide for recycling of sewage pollutants through agricultural practices and for environmentally safe sludge disposal is encouraged.

If possible, facilities for sewage treatment and recycling are to be integrated with facilities for the treatment, disposal, or utilization of other municipal and industrial wastes, including, but not limited to, solid wastes and waste heat. Combining "open space" and recreational considerations with waste treatment management is recommended.

Beginning in fiscal 1975, construction grant applications will be rejected if they result in any discharge of pollutants, unless it has been demonstrated that alternative techniques were considered and the methods proposed are the best practicable treatment. No grant may be approved after July 1, 1973 unless the applicant demonstrates that each sewer collection system discharging into a treatment facility is not subject to excessive infiltration.

Sec. 202–Federal Share. Provision is made for federal grants of 75 percent of the cost of treatment plant construction. This level of funding is retroactive to projects approved by the administrator between January 1 and July 1, 1971, on which construction was not commenced prior to July 1, 1971.

Sec. 203–Plans, Specifications, Estimates, and Payments. The administrator may commit payment of the federal government's portion of the cost of treatment facilities when he approves an applicant's plans, specifications, and estimates. Phased approaches to the construction of treatment plants are encouraged, and payments may be made on the basis of costs incurred to date.

Sec. 204—Limitations and Conditions. This section sets forth a number of grant conditions to assure that treatment facilities are constructed, operated, and maintained to produce the best practicable application of treatment technology. The plants must implement any applicable area-wide plans and must be certified by the state as entitled to priority. The administrator must approve the reserve capacity and the system of charges for services. Industrial users of treatment works must pay a proportionate share of the costs of construction.

The administrator is to issue guidelines concerning payment of waste treatment costs by industrial and nonindustrial users. Portions of the payments from industrial users are to be retained by the municipality against funds necessary for future expansion and reconstruction of the project.

Sec. 205—Allotment. The administrator may allot construction funds on the basis of states' needs, and funds not obligated by a state shall be reallocated on the basis of need. The initial ratio of funds allocated is shown in Table 19.

Sec. 206—Reimbursement and Advanced Construction. Reimbursement is provided for treatment works started between July 1, 1966 and June 30, 1972 which met the then prevailing requirements. Up to 50 percent of the cost may be reimbursed, and the total may be 55 percent if the treatment works were constructed in conformity with a comprehensive metropolitan treatment plan. Those constructed between June 30, 1956 and June 30, 1966 are eligible for a total of 30 percent. Applications for assistance under this section must be filed within 1 year after the enactment of the 1972 Amendments.

Sec. 207—Authorization. This section authorized $5 billion for construction grants for fiscal 1973, $6 billion for fiscal 1974, and $7 billion for fiscal 1975.

Sec. 208—Area-Wide Waste Treatment Management. Section 208 instructs the administrator to publish guidelines for defining areas with substantial water quality control problems. After publication of those guidelines, the governor of each state must identify such areas within his state. Following that identification, the governor is to designate the boundaries of each area and create an organization to develop effective area-wide waste treatment management plans for each area. He may identify additional areas later.

TABLE 19-1. Estimated Construction Cost of Sewage Treatment Facilities for the Period Fiscal Year 1972–74 (in millions of dollars).

Alabama	52.6	New Hampshire	121.0
Alaska	32.8	New Jersey	1121.9
Arizona	19.6	New Mexico	30.7
Arkansas	51.5	New York	1610.3
California	1429.7	North Carolina	134.4
Colorado	46.1	North Dakota	6.8
Connecticut	244.8	Ohio	840.8
Delaware	95.6	Oklahoma	67.1
District of Columbia	103.6	Oregon	123.7
Florida	528.1	Pennsylvania	789.5
Georgia	141.7	Rhode Island	71.2
Hawaii	48.1	South Carolina	94.0
Idaho	31.7	South Dakota	13.8
Illinois	910.0	Tennessee	169.0
Indiana	490.2	Texas	403.3
Iowa	168.3	Utah	20.5
Kansas	54.5	Vermont	32.3
Kentucky	96.1	Virginia	424.4
Louisiana	137.3	Washington	129.7
Maine	140.9	West Virginia	72.8
Maryland	620.1	Wisconsin	253.6
Massachusetts	547.2	Wyoming	3.9
Michigan	1162.3	American Samoa	0.7
Minnesota	295.9	Guam	12.7
Mississippi	57.3	Puerto Rico	128.8
Missouri	241.1	Trust Territory of the Pacific Islands	5.5
Montana	24.2	Virgin Islands	13.0
Nebraska	54.0		
Nevada	41.9	Total	14,562.6

In the case of an interstate area, the governors of the respective states shall follow the same procedures jointly. If a governor in the case of an intrastate area, or governors in the case of an interstate situation, do not act within the designated time deadline, the chief elected officials of local governments within the area may, by agreement, act in their stead.

Not later than 1 year after it is organized, the planning group must have in operation a continuing area-wide waste treatment management planning process. The initial plan is to be certified by the governor and submitted to the administrator not less than 2 years after the planning process is in operation.

The plan, which must be updated and certified by the governor annually, is to include:

1. Treatment works to meet anticipated municipal and industrial needs of the area over a 20-year period, including requirements for land acquisition, necessary waste water collection and urban stormwater runoff systems, and a financial program.
2. Establishment of construction priorities and time schedules.
3. A regulatory program.
4. Identification of implementing agencies.
5. Identification of implementation measures and time frame, costs, and economic, social, and environmental impacts resulting from implementing the plan within such time.
6. Identification of agriculturally and sylviculturally related nonpoint sources of pollution, including run-off from manure disposal areas, and land used for livestock and crop production, and procedures for their control.
7. Identification and control of mine-related pollution.
8. Identification and control of pollution from construction.
9. Identification and control of salt water intrusion.
10. Control of the disposal of residual wastes that could affect water quality.
11. Control of disposal of pollutants in excavations.

The governor of each state is required to select one or more agencies to provide waste treatment management for each designated area, and to submit his selections along with the plan to the administrator. The administrator is required to accept them, unless he finds that the management agency (or agencies) chosen does not have adequate authority to carry out its mission.

The remainder of this section states that after a waste treatment management agency is in operation and the plan for the area has been approved, all grants for construction of publicly-owned treatment works will be made only to that agency and only for works in conformance with the plan. No permits will be issued for any point sources of pollution in conflict with the plan.

The grants shall be for 100 percent of the costs of developing and operating the area-wide waste treatment management planning process for each of the fiscal years ending June 30, 1973, 1974, and 1975, and shall be up to 75 percent thereafter. The administrator and the Corps of Engineers are to provide any needed technical assistance.

Sec. 209–Basin Planning. This section directs the president, acting through the Water Resources Council, to prepare comprehensive regional or river basin plans (Level B under the Water Resources Planning Act) for all areas of the nation by 1980.

Sec. 210–Annual Survey. Section 210 requires the administrator to make an annual survey of treatment works funded under the act to compare their operating and maintenance efficiency with planned efficiency.

Sec. 211–Sewage Collection Systems. Grants are allowed to existing communities for sewage collection systems only if the grant (1) is for replacement or major rehabilitation of an existing system or (2) it is for a new collection system where there is sufficient existing or planned capacity to adequately treat the collected sewage.

Sec. 212–Definitions. The terms "construction," "treatment works," and "replacement" are defined. The administrator is to publish guidelines for evaluating methods of treatment.

TITLE III–STANDARDS AND ENFORCEMENT

Sec. 301–Effluent Limitations. The discharge of any pollutant is made illegal except those in compliance with:

1. Effluent limitations (Sec. 301 and 302).
2. National standards of performance (Sec. 306).

3. Toxic and pretreatment effluent standards (Sec. 307).
4. Aquaculture projects (Sec. 318).
5. Discharge permits (Sec. 402).
6. Permits for discharge of dredged or fill material (Sec. 404).

July 1, 1977, is set as the date for achievement of effluent limitations based upon best practicable control technology for point sources other than publicly-owned treatment works. For the latter, the limitations are based on secondary treatment by that date. Any more stringent limitations necessary to meet any state law or other federal laws or regulations must be met by July 1, 1977.

By July 1, 1983, effluent limits for all point sources other than publicly-owned treatment works require application of the best available technology economically achievable. This is to result in reasonable progress toward the national goal of eliminating the discharge of all pollutants. The administrator shall issue regulations for effluent limits to require the elimination of discharges of all pollutants if he finds that this elimination is technically and economically achievable for a category or class of point sources. The 1983 deadline is also applicable to the introduction of industrial pollutants into publicly-owned treatment works and may require pretreatment or any other method of eliminating pollutants which are incompatible with municipal treatment. At the same time, all publicly-owned treatment works must provide for the application of the best practicable waste treatment technology.

The administrator may waive the requirements of this section with respect to any point source for which a permit application is filed after July 1, 1977 if the owner or operator can prove that the modified requirement will represent the maximum use of the technology within economic capability and will result in reasonable further progress toward elimination of the discharge of pollutants.

Effluent limitations for 1983 must be reviewed at least every 5 years.

It shall be unlawful to discharge any radiological, chemical, or biological warfare agent or high-level radioactive waste into navigable waters.

Sec. 302—Water Quality Related Effluent Limitations. More stringent standards than those required by Section 301 may be set if those effluent limits would interfere with attaining a 1983 interim goal. This interim goal requires water quality assuring protection and propagation of fish, shellfish, and wildlife, and provides for recreation in and on the waters. Before a more restrictive standard can be promulgated, the balance between the economic and social costs and benefits of the new limitation must be determined at a public EPA hearing.

Sec. 303—Water Quality Standards and Implementation Plans. Provision is made to retain existing water quality standards for interstate and intrastate waters. Where these are not in accord with the 1972 Amendments, the administrator is required to so notify the states. The states must then make acceptable changes. If they do not within required time limits, the administrator is empowered to promulgate changes. States which have not yet adopted intrastate standards are to do so in a manner satisfactory to the administrator. Otherwise, the administrator is to set the standards. A time frame is set up for publication of proposed standards by the administrator and their final adoption.

Each state is to review its water quality standards at least once every 3 years. Any new standards will consist of the designated use of the navigable waters involved and water quality criteria based on those uses. The uses are public water supply, propagation of fish and wildlife, recreational purposes, agricultural, industrial, and other purposes as well as navigation. If a state's revisions are not satisfactory, the administrator must notify the state and suggest appropriate changes. If the state does not accept these changes, the administrator is to publish new standards and to adopt them for the state.

In the balance of this section, the following requirements are set forth:

1. The states are to identify waters for which the effluent limitations required in this section are not sufficiently stringent to implement a water quality standard. A priority ranking is to be established for those waters. For each of these waters, the total

maximum daily pollution load that would be suitable to implement water quality standards with seasonal variations and a margin of safety is to be identified.

2. Each state is to identify waters for which controls on thermal discharges are not sufficiently stringent to protect a balanced indigenous population of shellfish, fish, and wildlife. For those waters, the state is to estimate the total maximum daily thermal load required to assure protection and propagation. The estimates are to take into account normal water temperature, flow rates, seasonal variations, existing sources of heat input, and dissipative capacity. In addition, they shall include a calculation of maximum heat input, including a margin of safety.
3. Approval for the above actions must be obtained from the administrator of EPA.
4. Each state is to institute a continuing planning process and to submit a plan to the administrator for approval. The administrator must set up a continuing review process for each state's plan. He is not to approve any state permit program under Title IV for any state not having such an approved planning process. These plans are to include the following:

    a. Effluent limitations and schedules of compliance.
    b. Area-wide waste management plans.
    c. Total maximum daily load for pollutants.
    d. Procedures for revision.
    e. Adequate authority for intergovernmental cooperation.
    f. Adequate implementation, including schedules of compliance, for revised and new water quality standards.
    g. Controls over the disposition of all residual waste from any water treatment processing.
    h. An inventory and ranking, in order of priority, of the need for construction of waste treatment works.

Sec. 304–Information and Guidelines. The administrator must publish criteria for water quality, accurately reflecting the latest scientific knowledge. He must also publish information on the expected effects of pollutants on plankton, fish, shellfish, wildlife, plant life, shorelines, beaches, aesthetics, and recreation; on the concentration and dispersal of pollutants or their by-products through biological, physical, and chemical processes; and on the effects of pollutants on biological community diversity, productivity, and stability, including rates of eutrophication and rates of sedimentation for various waters.

The administrator is also to publish information (1) on the factors necessary to restore and maintain the chemical, physical, and biological integrity of navigable waters, groundwaters, waters of the contiguous zone, and the oceans; (2) on the factors necessary for the protection and propagation of shellfish, fish, and wildlife for all classes of receiving waters and to allow recreational activities in and on them; (3) on the measurement and classification of water quality; and (4) on the identification of pollutants suitable for maximum daily load measurement correlated with the achievement of water quality objectives.

The administrator must issue regulations providing guidelines for effluent limitations and revise them annually. The regulations must:

1. Identify the degree of effluent reduction attainable through the application of the best practicable control technology for point sources except municipal treatment works.
2. Specify factors to be considered such as cost in relation to benefits, equipment age, process effects, nonwater quality environment impact (including energy requirements), and others.
3. Identify the degree of effluent reduction attainable through the application of the best control measures achievable and specify additional factors such as those above.
4. Identify control measures available to eliminate pollutant discharge from categories and classes of point sources taking cost into account.

The administrator is to publish, and from time to time update:

1. Information on technology to achieve pol-

lution elimination or reduction to implement standards of performance (see Sec. 306).
2. Information on the degree of effluent reduction attainable through the application of secondary treatment, as well as on alternative waste treatment management techniques and systems.
3. Guidelines for pretreatment of pollutants that are not susceptible to treatment by municipal works or which might be incompatible with those works.
4. Analytical test procedures for pollutants.
5. Information on techniques for restoration and enhancement of the quality of lakes.

The administrator is to issue guidelines identifying nonpoint sources of pollutants and techniques to control pollution resulting from:

1. Agricultural and silvicultural activities.
2. Mining activities.
3. Construction activities.
4. Deepwell or subsurface disposal.
5. Salt water intrusion.
6. Changes in the movement, flow, or circulation of waters, such as those that may result from construction of dams, levees, etc.

Guidelines must also establish uniform application forms and requirements for the acquisition of information from applicants for permits under Section 402. Minimum elements of state permit programs are to be promulgated including:

1. Monitoring requirements.
2. Reporting requirements.
3. Enforcement provisions.
4. Funding, personnel qualifications, and manpower requirements.

Sec. 305–Water Quality Inventory. This section requires that the administrator report to Congress by the end of 1973 on the specific quality of all U.S. waters during that year. The report should identify and inventory point sources of discharge, and provide an analysis of each discharge. It should identify all navigable waters which presently allow recreational activities and provide protection for fish propagation, those waters that will meet such standards by 1977 or 1983, and those that can be reasonably expected to attain such levels at a later date.

Each state is required to submit by January 1, 1975 and to revise annually a report on the existing water quality of all its navigable waters. The report shall analyze to what extent the waters now provide for swimming and fish protection. An estimate is to be made of the extent to which the requirements of the act will attain its goals for those waters. The report is to include an estimate of the environmental impact, the economic and social costs and benefits, and the date of achievement. In addition, there is to be a description of the nature and extent of nonpoint sources of pollution, recommendations for control programs, and an estimate of costs.

Sec. 306–National Standards of Performance. New sources of pollution in at least twenty-seven specified industries must be constructed to meet effluent standards reflecting the greatest degree of reduction the administrator determines to be achievable through application of the best available demonstrated control means, including, where practicable, no discharge of pollutants. Those industries are as follows:

Pulp and paper mills
Paperboard, builders paper and board mills
Meat product and rendering processing
Dairy product processing
Grain mills
Canned and preserved fruit and vegetable processing
Canned and preserved seafood processing
Sugar processing
Textile mills
Cement manufacturing
Feedlots
Electroplating
Organic chemicals manufacturing
Inorganic chemicals manufacturing
Plastic and synthetic materials manufacturing
Soap and detergent manufacturing
Fertilizer manufacturing
Petroleum refining
Iron and steel manufacturing
Nonferrous metals manufacturing
Phosphate manufacturing

Steam-electric power plants
Ferroalloy manufacturing
Leather tanning and finishing
Glass and asbestos manufacturing
Rubber processing
Timber products processing

"New source" is defined as one on which construction is begun after the publication of proposed regulations prescribing a standard of performance.

Such standards must consider the costs and benefits of attaining the degree of effluent reduction required and may distinguish among classes and sizes of new sources. The requirements also apply to any government owned or operated pollution sources.

If the administrator finds that a state has developed procedures for applying and enforcing standards of performance for new sources at least equal to those described above, he may authorize the state to apply and enforce those standards.

Point sources constructed in accordance with the requirements of this section shall not be subject to more stringent standards of performance for 10 years. However, after the effective date of the performance standards, it will be unlawful for the owner or operator of any new source to operate in violation of them.

Sec 307–Toxic and Pretreatment Effluent Standards. The administrator is directed to publish a list of toxic pollutants and proposed effluent standards for each toxic pollutant enumerated. The standards may include a prohibition of the discharge of the respective toxic pollutants or combinations of pollutants. A category or categories of sources to which the standards of prohibition shall apply must be designated. Disposal of dredged material may be included only after consultation with the secretary of the army.

The administrator is to issue regulations for pretreatment of pollutants going into public treatment systems that would interfere with, pass through, or otherwise be incompatible with the system. A maximum of 3 years will be allowed for compliance.

Pretreatment standards for new sources are to be promulgated simultaneously with the standards of performance.

Sec. 308–Inspections, Monitoring, and Energy. To assist the administrator in developing effluent limitations and standards or in determining violations of prohibitions, the administrator is permitted to require that pollution sources perform the following duties:

1. Keep appropriate records.
2. Monitor effluents (including use of the biological monitoring method).
3. Sample effluents in accordance with prescribed methods.

Sec. 309–Federal Enforcement. This section requires the administrator to notify both the polluter and the state upon discovering a violation of any effluent limitation. He is also directed to issue a compliance order and is authorized to initiate civil action, such as a permanent or temporary injunction. When such an order relates to inspections, monitoring, or entry to a polluter's premises, the owner or operator is entitled to confer with the administrator before the order takes effect.

If the administrator discovers widespread violations of the limitations, he must notify the state. If the state fails to act within 30 days, he is required, after public notice, to assume enforcement powers over all effluent limitation requirements in the state.

Civil penalties upon conviction of noncompliance with provisions of the act may not exceed $10,000 per day of violation. Penalties upon criminal conviction (willful or negligent violation) range from $2500 to $25,000 per day and/or 1 year's imprisonment for a first offense to a maximum $50,000 per day of violation and/or 2 years imprisonment thereafter.

Anyone who knowingly makes a false statement on any application or report, or who tampers with a monitoring device, is liable to a $10,000 fine and 6 months imprisonment.

Sec. 310–International Pollution Abatement. Under Section 310, the secretary of state may request abatement of pollution from a United States source that endangers the health or welfare of persons in a foreign country. The administrator must notify the state where the discharge originates.

When the pollution is in sufficient quantity to warrant such action, and the foreign nation has given the United States reciprocal rights over pollution originating in that nation, the administrator will call a hearing. If the hearing board so recommends, the administrator will initiate abatement action.

Sec. 311–Oil and Hazardous Substance Liability. National policy is established that there should be no discharge of oil or hazardous substances into or upon the navigable waters of the United States, adjoining shorelines, or into or upon the waters of the contiguous zone.

The administrator is to develop regulations designating as hazardous substances, other than oil, those which present an imminent and substantial danger to fish, shellfish, wildlife, shorelines, and beaches. The regulations shall state whether the hazardous substance can actually be removed. If not, the discharger is liable, during the first 2 years after enactment of the act, to a civil penalty per discharge based on toxicity, degradability, and disposal characteristics of the substance in an amount not to exceed $50,000. If the discharge was due to willful negligence or misconduct privy to the owner, a larger civil penalty may be imposed without limit.

The administrator shall establish by regulation, for each hazardous substance, a unit of measurement based on usual trade practice for purposes of establishing penalties after expiration of the 2-year period.

Thereafter, the owner or operator of any vessel, onshore facility, or offshore facility who discharges a nonremovable hazardous substance will be liable to a penalty of from $500 to $5000 or to a penalty derived from the number of units discharged multiplied by an amount to be fixed by the administrator of from $100 to $1000 up to a maximum of $5 million in the case of a discharge from a vessel, and $500,000 in the case of a discharge from an onshore or offshore facility.

The president is to determine the amount of oil and (removable) hazardous substances which if discharged under given conditions will be harmful to the public health and welfare. Any discharge thus deemed of harmful quantity is prohibited except those permitted:

1. Under the International Convention for the Prevention of Pollution of the Sea by Oil (in the contiguous zone where only those discharges which threaten fishing resources of the zone or which may pollute the territory of the United States or its territorial sea may be determined to be harmful).
2. Under conditions as the president may, by regulation, determine not to be harmful.

Notification of a discharge of oil or hazardous substance in an unlawful quantity from a vessel or facility must be made to the appropriate government agency immediately. Failure to do so results, upon conviction, in a fine of not more than $10,000 and/or imprisonment for not more than 1 year. Such notification cannot be used against the person reporting the discharge in any criminal case except a prosecution for perjury.

Discharges of harmful quantities can result in a fine of not more then $5000 per offense, and only after the person charged is given a hearing. Clearance for any vessel whose owner or operator is subject to penalty may be withheld by the secretary of the treasury until a bond or other satisfactory surety is filed.

When a spill occurs, the president is authorized to act to remove the material or arrange for its removal unless he determines that removal will be properly executed by the owner or operator of the vessel or facility from which the spill occurs.

The president is to prepare and publish a National Contingency Plan for the removal of oil and hazardous substances which shall include:

1. Assignment of duties and responsibilities.
2. Identification, procurement, and storage of equipment and supplies.
3. Establishment of a "strike force" including trained and equipped task forces at major ports.
4. A system of surveillance and rapid notice.

5. Establishment of a national center to coordinate and direct clean-up operations.
6. Procedures and techniques to be employed in identifying, containing, dispersing, and removing oil and hazardous substances.
7. Information concerning the useful dispersants and their proper utilization.
8. A system whereby states that act to remove discharges may be reimbursed from the fund created by this section for the reasonable costs incurred in the removal.

Regulations consistent with the plan are to be issued by the president which establish:

1. Methods and procedures for removal of discharged oil or hazardous substances.
2. Criteria for local contingency plans.
3. Requirements for equipment to prevent and contain discharges.
4. Procedures for inspection of vessels carrying cargoes of oil or hazardous substances.

Persons failing to comply with these regulations will be subject to a civil penalty of not more than $5000 per violation.

In the case of a marine disaster which creates a substantial threat of hazardous pollution, the federal government may coordinate and direct all public and private efforts directed at the removal or elimination of the threat, and remove or, if necessary, destroy the vessel by whatever means available. Liability for expenses incurred is described below.

In the case of an imminent or actual discharge from an onshore or offshore facility, the U.S. attorney of the district in which the threat occurs may secure any necessary relief and the district courts shall have jurisdiction to grant relief.

Except where an owner or operator of a vessel can prove that a discharge was caused solely by (a) an act of God, (b) an act of war, (c) negligence on the part of the federal government, or (d) an act of omission of a third party, or a combination of these circumstances, the owner or operator is liable for the cost of removal of the oil or hazardous substance by the government in an amount not to exceed $100 per gross ton of his vessel or $14 million, whichever is lesser. Where the discharge is proven to be the result of willful negligence or misconduct, the owner or operator must pay the entire cost of removal. Such costs constitute a maritime lien on the vessel.

Owners or operators of onshore or offshore facilities which discharge oil or hazardous substances are liable for the actual costs for the removal of the oil or substance by the government, subject to the above exceptions, in an amount not to exceed $8 million. If the discharge is due to willful negligence or misconduct, the full amount of such costs may be recovered. Lower limits of liability may be applied to onshore facilities having a total fixed storage capacity of 1000 barrels or less which do not present a substantial risk for the discharge of oil or a hazardous substance. The liability provisions do not apply to offshore facilities in situations covered by the Outer Continental Shelf Lands Act.

The liability of third parties found at fault for discharges are similar to those described above for owners and operators of vessels and onshore and offshore facilities except when they can prove that the discharge was caused by an act of God or war, negligence on the part of the federal government, or omission of another party.

When the discharge is not legally the fault of the vessel or onshore or offshore facilities from which it derives, and the owner or operator of the source of discharge acts to remove the oil or hazardous substance in accordance with regulations promulgated as part of the National Contingency Plan, then he may recover the cost from the federal government.

These costs will be paid out of a revolving fund established in the Treasury Department to carry out the purposes of this section. The fund is set up with a $35 million appropriation; in the future it will be the depository of all funds received pursuant to this section of the act.

The administration of the president's authority under this section may be delegated to any department or agency he deems appropriate. Anyone designated to enforce its provisions may board and inspect any private vessels, make arrests, and execute warrants. Action may then be taken in the various district courts.

This section does not negate the responsibility of a discharger for payment of damages to property. Nor is any state or political subdivision preempted from imposing any requirement for liability with respect to the discharge.

Any vessel over 300 gross tons, including self-propelled barges that carry oil and hazardous substances, using any port in the United States or its navigable waters, is required to maintain evidence of financial responsibility of $100 per gross ton or $14 million, whichever is the lesser. Where an owner or operator owns or operates more than one such vessel, financial responsibility need only be established to meet the maximum liability which could be assessed against the largest vessel. Evidence of insurance, surety bonds, qualification of a self-insurer, or other evidence of financial responsibility acceptable to the president may be utilized.

Any vessel owner or operator who fails to establish financial responsibility is subject to a fine of not more than $10,000 and his vessel may be refused required clearance. Similarily, entry to U.S. ports or navigable waters may be denied and the ship may be detained in the absence of evidence of financial responsibility.

Sec. 312—Marine Sanitation Devices. This section requires the EPA administrator to promulgate federal standards of performance for marine sanitation devices to prevent the discharge of improperly treated sewage into navigable waters from new and existing vessels, except vessels not equipped with toilet facilities. The secretary of the department in which the Coast Guard is operating shall promulgate regulations under those standards governing the design, construction, installation, and operation of the devices.

Sanitation devices in existence on the date of promulgation of the initial standards and regulations and are in compliance with them will be deemed in compliance until such time as the devices are replaced or are found not to be in compliance. Initial standards and regulations will become effective for new vessels 2 years after promulgation and for existing vessels 5 years after promulgation.

The secretary of the department in which the Coast Guard is operating may waive applicability of standards and regulations for certain vessels. The standards and regulations apply to Defense Department vessels, unless the secretary of defense finds that compliance would not be in the interest of national security. The regulations and certifications for them shall be promulgated and issued by the secretary of defense.

The standards and regulations preempt state regulations governing marine sanitation devices. However, the state may elect to prohibit the discharge of sewage from all vessels, whether treated or not, to specific waters. In this case, the administrator must determine that adequate facilities for the removal and treatment of sewage from vessels operating in those areas are reasonably available.

Before a marine sanitation device may be sold, it must be in all material respects substantially the same as a test device certified under this section. The test device is to be submitted to the Coast Guard for evaluation and comparison with the performance standards. The Coast Guard will then issue the necessary certification. Records and reports to show that the manufacturer is acting in compliance with this section must be kept and made available to the Coast Guard, although all trade secrets will be kept confidential.

After the effective date of standards and regulations, it will be unlawful (1) to sell any vessels not properly equipped, (2) to remove or render inoperative any certified marine sanitation device or element of design of that device installed in the vessel, (3) to refuse to permit access to or copying of records or to make reports, (4) to operate vessels subject to the standards on the navigable waters of the United States if they are not equipped with the proper devices.

Legal action may be brought in the district courts. The sale of noncomplying devices and nonequipped vessels may be penalized by a $5000 fine for each violation. Operation of vessels with inadequate sanitary devices carries a penalty of not more than $2000. The Coast Guard has the authority to assess and compromise penalties and is given the responsibility of enforcing this section. Its personnel may be authorized to board and inspect any nonpublic vessel upon the navigable waters of the United States and execute any required warrants.

Section 313–Federal Facilities Pollution Control. Federal facilities are required to meet the same effluent limitations as other sources of pollution, unless the facility is specifically exempt by the president. The president, however, cannot exempt a federal source from national performance standards or toxic and pretreatment standards.

Section 314–Clean Lakes. This section requires each state to submit an identification and classification according to the eutrophic condition of all publicly-owned freshwater lakes. The state also must develop methods such as land use considerations for preventing pollution in the lakes.

Section 315–National Study Commission. A National Study Commission is established to investigate all the technological aspects, as well as the economic, social, and environmental effects, of the effluent limits and goals set for 1983. The Commission is to have fifteen members–five from the Senate, five from the House, and five from the public. The study commission is authorized to contract with the National Academy of Sciences, the National Academy of Engineering, the National Institute of Ecology, Brookings Institution, and other nongovernmental entities. The report is to be submitted to Congress in 1975.

Section 316–Thermal Discharges. If it can be shown that an effluent limitation proposed for the control of the thermal component of a discharge from any point source requires values more stringent than necessary to assure the protection and propagation of a balanced indigenous population of shellfish, fish, and wildlife, the administrator may impose a more moderate limitation which will still provide the necessary protection. Any standard established with regard to thermal discharge shall require that the location, design, construction, and capacity of cooling water intake structures reflect the best technology available for minimizing adverse environmental impact.

Modifications of point sources having thermal components begun after enactment of the act which meet the requirements of the act will not be subject to any more stringent limitations with regard to the thermal components for 10 years after the modification is completed.

Section 317–Financing Study. The administrator is directed to continue his investigation and study of alternative methods of financing the cost of preventing, controlling, and abating pollution, specifically including financing of programs after fiscal 1976.

Section 318–Aquaculture. The administrator is authorized to permit, after public hearings, discharge of specific pollutants under controlled conditions to an approved aquaculture project subject to procedures and guidelines which he is directed to issue to carry out this provision.

TITLE IV–PERMITS AND LICENSES

Section 401–Certification. Applicants for a federal license or permit to conduct an activity which may result in a discharge into navigable waters are required by this section to provide the licensing or permitting agency with a certification from the responsible state or interstate agency showing that the discharge will comply with the applicable provisions of Section 301, 302, 306, and 307 of the act. For activities where there is not an applicable effluent limitation or standard under those sections, the state shall so certify, but this certification shall not be deemed to satisfy Section 511 of the act. Public notice must be given of all applications for certification by the state and hearings held in connection with specific applications, where appropriate. Where the state has no authority to grant certification, the administrator shall act.

If the state, interstate agency, or administrator, as the case may be, fails or refuses to act on a request for certification within a reasonable period of time (which shall not exceed 1 year), the certification requirements shall be waived with respect to the federal application permit. No federal license or permit is to be granted until the certification required by this section has been obtained or has been waived; no license or permit shall be granted for which certification has been denied.

The administrator must consider the potential interstate impact of license or permit applications. If he determines that the resulting discharge is liable to affect the quality of the water of another state, he is to give that state an opportunity to comment on the proposed discharge. The state may enter an objection to the application and request a public hearing by the licensing or permitting agency at which the administrator's evaluation and recommendations will be heard. The agency is then to condition the license or permit in whatever manner is necessary to insure compliance with applicable water quality requirements. If the imposition of conditions cannot insure such compliance, no license or permit shall be granted.

If any other federal license or permit is needed for operating a facility, the original certification will apply unless the agency to whom an additional application is addressed is notified by the state, interstate agency, or the administrator that the discharge is no longer in conformance with the act. This may occur because of changes in the construction or operation of the facility, the characteristics of the receiving water, water quality criteria applicable to those waters, or applicable effluent limitations or other requirements.

In the case of a plant for which certification was obtained prior to initial operation, but where the plant does not need a federal license or permit for the operation itself, the state or other appropriate authority will still have an opportunity to review the operation and to determine if water quality requirements are likely to be violated. If so, the certificate will be suspended until corrective measures are taken.

Any federal license or permit may be suspended or revoked if there is evidence that the provisions of the act are being violated.

Certification for licenses and permits issued for operation of facilities whose construction was underway prior to April 3, 1970 was not required until April 3, 1973.

The Corps of Engineers may permit utilization of spoil disposal areas under their jurisdiction for disposition of dredged material by federal licensees and permittees, and to charge for their use.

Any certifications provided under this section must spell out effluent limits and monitoring requirements.

Section 402–National Pollutant Discharge Elimination System. This section transfers the 1899 Refuse Act permit program from the Corps of Engineers to the administrator. He is thus authorized to issue permits for the discharge of pollutants into navigable waters, the contiguous zone, or the oceans on condition that the discharge meet the requirements of Sections 301, 302, 306, 307, 308, 316, and 403 of the act. Permits previously issued under the Refuse Act are considered valid and applications pending under that act are to be deemed applications under this title. However, no subsequent permits are to be issued under the Refuse Act.

During the interim period between enactment of the act and publication of guidelines under Section 304, or the date of approval of a state permit program, whichever occurs first, the administrator can authorize a state to issue permits if that state has the necessary capability. The administrator can veto any such interim permit.

Provision is made for a state to administer its own permit program in lieu of the federal program. The administrator is required to approve a proposed state program unless he finds that the state does not have adequate authority to:

1. Issue permits (a) that are in compliance with the act, (b) for fixed terms not to exceed 5 years, (c) that can be terminated or modified for cause, and (d) that control the disposal of pollutants into wells.
2. Inspect, monitor, enter, and require reports, to allow public hearings, and to insure that the issuance of the permit will not impair operation, anchorage, and navigation in navigable waters.
3. Abate permit violations, and insure that changes in publicly-owned treatment works are called to the attention of the permit agency allowing it to reconsider the permit.

The administrator is given 90 days to approve a state program after which time it becomes automatically approved. If he rejects it, he must advise the state as to appropriate corrections and modifications. If the administrator determines that a state is not administering an approved program properly, he must notify the state and, if appropriate corrective action is not taken within 90 days, withdraw approval of the program.

The states are directed to transmit to the administrator copies of pending permit applications and notification of proposed state action. If the administrator objects within 90 days, the permit is not to be issued. However, the administrator may waive these requirements.

Permits for discharge of pollutants into navigable waters from a vessel or other floating craft are subject to regulations promulgated by the Coast Guard relating to safe transportation, handling, carriage, storage, and stowage of pollutants.

If a permit for discharges from a public treatment works is violated, the introduction of any subsequent new pollutants into the treatment plant may be prohibited.

Permit applications are to be made public. Until December 31, 1974, a discharge for which a permit application has been filed, but on which final administrative disposition has not been made, is not in violation of the act, unless the reason for lack of disposition is failure on the part of the applicant to furnish required information. Point sources for which permits had not been requested are given 180 days to file applications.

The administrator may veto any proposed state permit. Any permit must include, where appropriate, a schedule of compliance for achievement of its various requirements.

Section 403–Ocean Discharge Criteria. The administrator is to promulgate guidelines for determining the degradation of the waters of the territorial seas, the contiguous zone, and the ocean. These guidelines are to consider the effects of pollutants on human health or welfare and marine life; the mechanism by which the pollutants affect marine life; the effect of the disposal of pollutants on aesthetic, recreational, and economic values; the persistence and permanence of the effects of the disposal of pollutants; other possible locations and methods of disposal, including recycling and land-based alternatives; and the effects of disposal of pollutants on other uses of the ocean, such as mineral exploitation and scientific study.

Any permits under Section 402 for discharge into the ocean will be subject to the guidelines. Permits may be issued before the guidelines are published if the administrator determines that it is in the public interest. Where insufficient information exists on the proposed discharge to make a reasonable judgment on any of the guidelines, no permit shall be issued.

Section 404–Permits for Dredged or Fill Material. The Corps of Engineers is authorized to issue permits for the discharge of dredged or fill material into navigable waters at specific disposal sites. The sites will be determined through the application of guidelines to be formulated by the administrator and the Corps. In addition to the normal pollution considerations, these guidelines must consider the economic impact of the site on navigation and anchorage. The administrator, after consultation with the Corps, may prohibit the specification of any defined area as a disposal site if he determines that the discharge of pollutants at that point will have an unacceptable effect on municipal water supplies, shellfish beds and fishery areas, wildlife, or recreational areas.

Section 405–Disposal of Sewage Sludge. When the disposal of sewage sludge from the operation of public treatment works would result in pollution of navigable waters, such disposal is prohibited except in accordance with a permit. The administrator is to issue regulations governing those permits applying the same criteria as for Section 402 permits. If a state wishes to administer its own permit program for disposal of sewage sludge, it must obtain approval of its program from the administrator.

TITLE V–GENERAL PROVISIONS

Section 501–Administration. This section authorizes the administrator to prescribe regulations necessary to carry out his functions under the act. He may use officers and employees from other federal agencies to assist him.

Each recipient of financial assistance under this act is to keep detailed records prescribed by the administrator. He, the controller general of the United States, or their representatives shall have access, for the purpose of audit and examination, to any pertinent financial records of recipients.

Provision is made to extend recognition to industrial organizations and political subdivisions of states for outstanding technological achievements in their waste treatment and pollution abatement programs.

Upon request of a state water pollution control agency, EPA personnel may be detailed to them for assistance in connection with administering the act.

Section 502–General Definitions. Section 502 defines the following terms: (1) state water pollution control agency, (2) interstate agency, (3) state, (4) municipality, (5) person, (6) pollutant, (7) navigable waters, (8) territorial seas, (9) contiguous zone, (10) ocean, (11) effluent limitation, (12) discharge of a pollutant, (13) toxic pollutant, (14) point source, (15) biological monitoring, (16) discharge, (17) schedule of compliance, (18) industrial user, (19) pollution.

Section 503–Water Pollution Control Advisory Board. A Water Pollution Control Advisory Board is established within EPA composed of the administrator or his designee as chairman and nine members appointed by the president, none of whom may be federal officers or employees. They are to be selected among representatives of state, interstate, or local governmental agencies, public or private interests contributing to or affected by pollution, and other public or private groups actively interested in the field, as well as other individuals expert in the field.

Each member will be appointed for a 3-year term and receive compensation at a rate not to exceed $100 per diem, including travel time, plus actual travel expenses. The board is to advise, consult with, and make recommendations to the administrator on matters of policy relating to the activities and functions of the administrator under this act. All necessary clerical and technical assistance is to be provided from the personnel of EPA.

Section 504–Emergency Powers. The administrator is authorized to bring suit on behalf of the United States when he determines that a pollution source presents an imminent and substantial danger:

1. To human health.
2. To the livelihood of individuals, such as the inability to market shellfish.

Section 505–Citizen Suits. Under this section, any citizen may commence on his own behalf a civil action against any person or governmental agency alleged to be in violation of an effluent standard or limitation, or an order issued by the administrator or state with respect to such a standard or limitation. Further, suits may be entered against the administrator when there is an alleged failure on his part to perform any act or duty under this law which is not discretionary. The district court shall have the authority to enforce effluent standards or limits, or order the administrator to do so.

Cititzen suits may not be instituted prior to 60 days after notice has been given of the alleged violation to the administrator, the state in which it occurs, and the alleged violator. Citizen suits are also prohibited if the administrator or the state has commenced a civil or criminal action to require compliance. In the latter case, a citizen may intervene in the action as a matter of right. There are exceptions to the 60-day delay in the case of violations of performance and toxic effluent standards.

The action must be brought in the judicial district in which the source is located, and the administrator, if not a party, may intervene. In any final order, the court may award costs of litigation to any party. It also may require the filing of a bond.

Provision is made for the governor of a state to bring civil action against the administrator when an alleged failure on his part to enforce an effluent standard or limitation in another state is resulting in an adverse effect on the public health and welfare or a violation of water quality requirements in the state of the governor filing suit.

Section 506–Appearance. The administrator is to request the attorney general to represent the United States in any civil or criminal action to which he is a party. However, if the attorney general fails to notify the administrator of his intent to appear in a civil action within a reasonable time, EPA attorneys may represent the government in such actions.

Section 507–Employee Protection. This section offers protection to employees who believe they have been fired or discriminated against as a result of the fact that they have testified in a proceeding or brought suit under this act. The employee has recourse to the secretary of labor for a review and a hearing of his case. If the secretary rules in the employee's favor, he may issue an order for the employee to be rehired, or to be otherwise compensated. This section does not apply to an employee who acts without direction from his employer in deliberately violating provisions of the act.

The administrator is required to conduct continuing evaluations of potential loss or shifts in employment and to investigate threatened plant closures or reduction of employment allegedly resulting from any effluent limitation or order under the act. At public hearings, the employer will be required to present information relating to the alleged discharge, layoff, or discrimination. The administrator is then to make findings of facts and recommendations. Nothing in this provision shall be construed as requiring or authorizing the administrator to modify or withdraw an effluent limitation or order.

Section 508–Federal Procurement. The federal government is directed not to patronize or subsidize polluters through its procurement practices and policies. No federal agency may enter into a contract involving any facility convicted under Section 309. The prohibition would continue until the administrator certifies that the violation which led to the conviction no longer exists.

The president may exempt any contract if the exemption is in the paramount interest of the United States. The president is required to submit an annual report to Congress on measures taken in compliance with this section.

Section 509–Administrative Procedure and Judicial Review. Section 509 allows the administrator to issue subpoenas for witnesses and relevant documents. Except for effluent data, the administrator is permitted to consider certain information as trade secrets. Nevertheless, those trade secrets may be disclosed to federal officials responsible for carrying out the act. A failure to appear before the administrator and to produce the required papers, books, and documents may be taken to court and a contempt citation issued.

A review of any of the administrator's actions relative to performance standards, permit programs, effluent limitations, and permit approval or denial may be instituted by any interested person in the circuit court of appeals for the federal judicial district in which that person resides or transacts business. Any such application must be made within 90 days after the administrator's action. If new evidence is introduced subsequent to this hearing, the administrator may, after consideration, modify his findings or make new findings as a result of this new evidence.

Section 510–State Authority. Section 510 provides that states, political subdivisions, and interstate agencies retain the right to set more restrictive standards and limitations than those imposed under this act, except where their authority is expressly preempted.

Section 511–Other Affected Authority. This section states that the act is not to limit the authority or functions of the United States under any other law or regulation that is not inconsistent with the act. Specifically, it is not to affect the authority of the secretary of the army to maintain navigation under the Act of March 3, 1899, except that the permits issued under Section 404 of this act shall be conclusive as to the effect on water quality of any discharge resulting from any activity subject to the 1899 Act. Further, this act shall not be construed as to affect or impair the provisions of any U.S. treaty.

Discharges of pollutants into navigable waters subject to the River and Harbors Act of 1910 and the Supervisory Harbors Act of 1888 shall be regulated pursuant to this act, and are subject to those statutes only as to their effect on navigation and anchorage.

Except for provision of federal financial assistance for constructing municipal sewage treatment plants and the issuance of permits for discharge of pollutants by new sources, no action taken by the administrator under this act shall be considered a major federal action significantly affecting the quality of the human environment within the meaning of the National Environmental Policy Act of 1969. In other words, with those exceptions, no environmental impact statements need be made.

Further, nothing in the National Environmental Policy Act of 1969 shall allow a federal agency issuing licenses or permits to review any effluent limitations or other requirements established under this act or authorize that agency to impose, as a condition precedent to the issuance of any license or permit, any effluent limitation other than the limitation established pursuant to this act.

Section 512–Separability. The invalidation of any provision of this act or the application of any provision to a person or circumstance is not to affect the application of that provision to other persons or circumstances and the remainder of the act. This is a standard clause which allows a portion of an act or its application to be deemed invalid without endangering the entire act.

Section 513–Labor Standards. Application of the Davis-Bacon Act to treatment works for which grants are made under this act is required. This insures that workers employed by contractors and subcontractors are paid wages at rates not less than those prevailing for the same type of work on similar construction in the immediate locality.

Section 514–Public Health Agency Coordination. This section provides that the agency issuing a permit under Section 402 shall assist the permit applicant in coordinating the requirements of this act with the requirements of the appropriate public health agencies.

Section 515–Effluent Standards and Water Quality Information Advisory Committee. An Effluent Standards and Water Quality Information Advisory Committee is established to be composed of a chairman and eight members appointed by the administrator. The members are to be selected from the scientific community and be well qualified technically. They will serve for 4 years and may be reappointed.

The administrator is expected to notify the committee of proposed effluent limitations, standards of performance for new sources, and toxic effluent standards 180 days before he is required to publish them. The committee is to transmit to the administrator such scientific and technical information as is in its possession related to the subject matter, including that presented at public hearings which it may sponsor. This material is to be considered by the administrator in making final determinations. In preparing information for transmittal, the committee is to use the services of any federal agency, including specifically the United States Geological Survey, as well as any national environmental laboratories that may be established.

The committee may appoint employees including a secretary and legal counsel. Members of the committee are to be paid by the government for the time spent on its work. Five members constitute a quorum and the affirmative vote of at least five members are required for any action.

Section 516–Reports to Congress. Within 90 days after each session of Congress convenes, the administrator is to submit a report describing measures toward implementing the objectives of this act. The report is to include the following items:

1. Development of comprehensive plans under Section 102, area-wide plans under Section 208, basin plans under Section 209, and state plans under Section 303(e).
2. Progress in water pollution control research, experiments, studies, and related

matter by the federal government and its grantees or contractors.

3. Progress and problems associated with the development of effluent limits and recommended control techniques.
4. Status of state program plans for development and enforcement of water quality requirements.
5. Enforcement actions pending or completed during the preceding year.
6. Status of the state, interstate, and local pollution control programs established pursuant to and assisted by the act.
7. Results of the annual survey of the efficiency of treatment plants funded under the act.
8. Activities and recommendations pertaining to training grants and scholarships.
9. Reports and recommendations made by the Water Pollution Control Advisory Board.

In cooperation with the states, the administrator is to make (1) a detailed estimate of the cost of carrying out the provisions of the act, (2) a detailed estimate, biennially revised, of the cost of construction of all needed municipal treatment plants, both on a nationwide and individual state basis, (3) a study of the economic impact on affected units of government of the cost of installation of treatment facilities, and (4) a comprehensive analysis of the national requirements for and the cost of treating municipal, industrial, and other effluents to obtain the water quality objectives of the act. This detailed estimate and study of costs is to be submitted to Congress not later than February 10 of each odd-numbered year.

Section 517–General Authorization. Section 517 authorizes funds for fiscal 1973, 1974, and 1975 for implementing those provisions of the act for which there are no specific authorizations.

Section 518–Short Title. This section states that the act may be called the "Federal Water Pollution Control Act."

Section 3–Authorization for Fiscal Year 1972. Funding was provided for fiscal 1972 to carry out provisions of the act as it existed prior to its amendment.

Section 4–Savings Provision. Pending suits, actions, and public proceedings instituted under the act prior to the 1972 amendments are not to abate by reason of the amendments. It is further provided that rules, regulations, determinations, contracts, certifications, authorizations, delegations, or other actions taken under the previous act are to continue in full force and effect until modified by the new amendments.

The previous act was made applicable to all grants for funds authorized for fiscal 1972 and prior fiscal years, including increases in the monetary amount of such grants that might be paid from authorizations for subsequent fiscal years except as otherwise provided in Section 202 and Section 3 of the amended act.

Section 5–Oversight Study. The comptroller general is directed to conduct a study and review of research, pilot, and demonstration programs related to water pollution control being conducted, supported, or assisted by any agency of the federal government, and to assess conflicts between, and the coordination and efficacy of, such programs.

Section 6–International Trade Study. Section 6 requires the secretary of commerce to undertake a study to determine the extent to which pollution abatement and control programs will be imposed on, or voluntarily undertaken by, U.S. manufacturers in the near future and the probable short- and long-range effects of the costs of such programs on production and market costs.

This study is to include an examination of the probable effect to which comparable programs will be implemented in foreign industrial nations and the extent to which production costs of foreign manufacturers will be affected. It is also to assess the probable competitive advantage of foreign-manufactured goods over comparable articles made in the United States in those cases where (1) the foreign nation does not require pollution control programs; (2) requires a lesser degree of control, or (3) reimburses or subsidizes its manufacturers for the costs of these programs.

Means by which the competitive advantages accruing to foreign products can be equalized are to be determined as well as the impact of imposing compensating tariffs or other equalizing measures in encouraging foreign nations to implement pollution control programs. The initial report under this section is to be made within 6 months, and thereafter at least every 12 months.

Section 7—International Agreements. This section requires the president to undertake to enter into international agreements to apply uniform standards of performance for the control of the discharge and emission of pollutants from new sources and of toxic pollutants, and over the discharges of pollutants into the ocean.

Section 8—Loans to Small Business Concerns for Water Pollution Control Facilities. Section 7 of the Small Business Act is amended to authorize small loans to assist small business concerns in making additions to or alterations in equipment, facilities, or other methods of operation to meet requirements established by the act if the administrator determines that the concern is liable to suffer substantial economic injury otherwise. The loan is to be made only if the applicant furnishes to the Small Business Administration a statement in writing from the EPA or the state that the additions or alterations are necessary to comply with the act. The administrator is required to prescribe regulations establishing uniform rules for the issuance of these statements.

Section 9—Environmental Court. The president, acting through the attorney general, is required to investigate and study the feasibility of a separate court having jurisdiction over environmental matters and to report to Congress within a year.

Section 10—National Policies and Goals Study. The president is to investigate all national environmental policies and goals established by law for the purpose of determining the optimum relationship between them, taking into account the nation's resources. The report is required within 2 years.

Section 11—Efficiency Study. The president is required to conduct, utilizing the General Accounting Office, an investigation and study of ways and means of utilizing the resources, facilities, and personnel of the federal government most efficiently in carrying out the objectives of this act.

Section 12—Environmental Financing. A corporation known as the Environmental Financing Authority is created subject to the direction of the secretary of the treasury and the makeup of the body is prescribed. It is authorized until July 1, 1975 to make commitments for and to purchase any obligation or participation therein issued by a state or local public body to finance the nonfederal share of the cost of construction of a waste treatment plant if it is eligible for federal assistance under this act.

No commitment shall be entered into or purchase made unless the administrator has: (1) certified that the public body is unable to obtain on reasonable terms sufficient credit to finance its actual needs, (2) approved the project as eligible under the act, and (3) agreed to guarantee timely payment of principal and interest on the obligation. The secretary of the treasury is authorized to advance funds for initial capital.

The authority is authorized to issue obligations and the secretary of the treasury is authorized to purchase or agree to purchase these obligations. He is also authorized to use as a public debt transaction the proceeds of the sale of any securities issued in the future under the Second Liberty Bond Act, and the purposes of securities issued under that act are extended to include these purposes.

The secretary of the treasury shall make annual payments to the authority in amounts necessary to balance interest expenses accrued because of its obligations with interest income accrued.

The authority is given the usual powers granted corporations. The authority itself, its property, franchise, and funds are exempt from all federal, state, or local taxation; however, real property, any tangible personal property, and obligations issued by the authority are taxable. Annual reports are required, and budget and audit provisions of the Government Corporation Control Act are applicable.

It was not intended that the Environmental Financing Authority compete with private bond underwriters in the municipal bond market. The authority cannot provide financing assistance to a community that can borrow money on the open market at reasonable rates. The authorization provided in this section is to be exercised only as a last resort.

Section 13—Sex Discrimination. Sex discrimination is prohibited under any provision of the act. It is to be enforced through agency provisions similar to those already established under Title VI of the Civil Rights Act. However, these provisions do not prejudice or cut off any other legal remedies available to the person being discriminated against.

FUNDING FOR IMPLEMENTATION OF THE ACT

The total price tag on this legislation is a financial milestone in pollution control. The $18 billion authorized for federal contributions to the construction of sewage treatment plants is an unprecedented sum. The additional amounts authorized for research, demonstration of new or improved control technology, training of manpower, support of state, interstate, and local programs, and other EPA programs are less, but still represent a substantial increase over past authorizations for these functions.

Actual dollar amounts of funds authorized for various purposes under the act have, however, for the most part been omitted in this chapter. They largely apply only to the first 3 years of the life of the legislation and their usefulness in a permanent reference work would be minimal.

Further, actual expenditure is subject to some uncertainty. In this case, the $18 billion authorization for sewage treatment plant construction was not made subject to the usual congressional appropriations procedure—i.e., the basic law authorizing *expenditure* and subsequent legislation *appropriating funds*. Instead, a relatively new "contract authorization" procedure was utilized. Under this approach, EPA has the authority to contract up to $18 billion for construction over 3 years, within specified amounts for each fiscal year. Appropriations will then be made only in like amount to that actually contracted. In other words, there is no appropriations control over increases or decreases in amounts contracted—within the prescribed limitations—and the appropriations after the fact simply validate the contract.

The fact that expenditures may or may not be made up to the full authorization has already been demonstrated. Of the $11 billion authorized to be spent over the first 2 years, the executive branch elected to expend only $5 billion.

The other funds authorized by the act are in essence the operating funds of EPA. They allow on-going programs to continue and new ones required by the act to be instituted. These are handled in the usual agency appropriations bills, and the appropriations do not necessarily match the authorizations. In any case, information on the current status of program expenditures is best obtained from EPA regional offices.

AVAILABILITY OF DOCUMENTS REQUIRED BY THE ACT

The act and subsequent effluent limitations, standards, guidelines, research reports, and cost estimates will provide the basis on which the great bulk of water-related activities in the United States will be undertaken. Those instruments—i.e., regulations, standards, etc.—which have the force of law will become part of the Code of Federal Regulations. Reports to Congress are printed upon presentation as numbered congressional documents. Reports on technical studies and investigations will be printed for sale by the National Technical Information Service or the Government Printing Office.

Together they provide for the implementation and enforcement of the act's provisions and are essential to those in industry and government responsible for almost any aspect of water resources management.

Index